Guide to Tables in Mathematical Statistics

This GUIDE catalogues a large selection of tables belonging to the field of mathematical statistics, and a small selection of mathematical tables lying outside statistics but often used together with statistical tables. The bulk of the tables treated were published between 1900 and 1954; occasional entries relate to works as early as 1799 and as late as 1960. As well as filling an important need for those actively engaged in the computational side of mathematical statistics, this work offers valuable reference to the professional computer faced with a statistical problem, and the statistician called upon to compute. It is to be expected that this work will aid in the rapid selection and location of tables most useful to the purposes of specific programs of analysis and computation, and will also serve to reduce unnecessary and costly duplication of tabulations already in existence but not widely known.

This is the first book exclusively devoted to the tables of mathematical statistics; except for the normal distribution, which has been catalogued in a monograph by the National Bureau of Standards, these tables have been relegated to sections and subsections in general handbooks of mathematical tables.

Among other features, the present guide contains the following:

(1) Sources and characteristics of the principal published tables of random numbers.

(2) A ninety-page list of tables of combinatorial patterns used in the design of experiments.

(3) Annotated tables of contents of sixteen collections of statistical tables.

(4) An exhaustive subject index.

J. ARTHUR GREENWOOD is a Research Associate in Mathematics at Princeton University. H. O. HARTLEY is a Professor of Statistics at Iowa State University of Science and Technology (Ames).

GUIDE TO TABLES IN MATHEMATICAL STATISTICS

GUIDE TO TABLES IN MATHEMATICAL STATISTICS

J. ARTHUR GREENWOOD
RESEARCH ASSOCIATE IN MATHEMATICS
PRINCETON UNIVERSITY

H. O. HARTLEY
PROFESSOR OF STATISTICS
IOWA STATE UNIVERSITY OF SCIENCE AND TECHNOLOGY (AMES)

1962
PRINCETON UNIVERSITY PRESS
PRINCETON, NEW JERSEY

Copyright © 1962 by Princeton University Press
London: Oxford University Press

L.C. Card 62-7040

All rights reserved. No part of
this book may be reproduced in any
manner without written permission from
the Publisher, except for any purpose
of the United States Government.

Printed in the United States of America

TABLE OF CONTENTS

THE COMMITTEE'S FOREWORD	xxxiij	
PREFACE	xxxvij	
INTRODUCTION	xxxxj	
List of abbreviations	lxj	
List of symbols concerning means of interpolation . .	lxij	
DESCRIPTIVE CATALOGUE OF TABLES	1	630

1. THE NORMAL DISTRIBUTION

1.0	Introduction	1	
1.1	Tables with the normal deviate x as argument		
1.11	Area, log area and ordinate as functions of x		
1.111	Central area from $-x$ to $+x$	3	
1.112	Semi-central area from 0 to x	3	
1.1121	$P(x) - \frac{1}{2}$	4	630
1.1122	$(2\pi)^{\frac{1}{2}}[P(x)-\frac{1}{2}]$	5	
1.113	Two-tail area: beyond $-x$ and beyond $+x$	5	
1.1131	$2Q(x)$	5	630
1.1132	$[1- 2Q(x)]/Q(x)$	6	
1.1133	$-\log_{10} 2Q(x)$	6	
1.114	Single-tail area beyond x	6	
1.1141	$Q(x)$	6	630
1.1142	$(2\pi)^{\frac{1}{2}}Q(x)$	7	
1.1143	$-\log_e Q(x)$	7	
1.1144	$\pm\log_{10} Q(x)$	7	
1.115	Single-tail area up to x	7	
1.116	Ordinates		
1.1161	$Z(x) = (2\pi)^{-\frac{1}{2}} e^{-\frac{1}{2}x^2}$	9	
1.1162	$(2\pi)^{\frac{1}{2}} Z(x) = e^{-\frac{1}{2}x^2}$	10	

[v]

1.12	Area and log area as functions of $2^{-\frac{1}{2}}x$ and of x/Probable Error		
1.121	Area and log area as functions of $2^{-\frac{1}{2}}x$	10	
1.1211	$2P(2^{\frac{1}{2}}x) - 1 = 2\pi^{-\frac{1}{2}}\int_0^x e^{-t^2}dt$	11	630
1.1212	$\pi^{\frac{1}{2}}[P(2^{\frac{1}{2}}x)-\frac{1}{2}]$	11	
1.1213	$2Q(2^{\frac{1}{2}}x) = 2\pi^{-\frac{1}{2}}\int_x^\infty e^{-t^2}dt$	11	
1.1214	$\pi^{\frac{1}{2}}Q(2^{\frac{1}{2}}x) = \int_x^\infty e^{-t^2}dt$	12	
1.1215	$2\pi^{\frac{1}{2}}Q(2^{\frac{1}{2}}x) = 2\int_x^\infty e^{-t^2}dt$	12	
1.1216	$\log_{10}\pi^{\frac{1}{2}}Q(2^{\frac{1}{2}}x)$	12	
1.1217	$P(2^{-\frac{1}{2}}x) - \frac{1}{2}$		630
1.122	Area in terms of x/[Probable Error]	12	
1.1221	$P(tx*) - \frac{1}{2}$	12	
1.1222	$2P(tx*) - 1$	12	
1.13	Ratio of area to bounding ordinate and related functions	13	
1.131	$Q(x)/Z(x)$	13	630
1.132	$Z(x)/Q(x)$	13	630
1.133	Miscellaneous ratios	13	631
1.19	Miscellanea	13	
1.2	Tables having the normal area P as argument; tables for probit analysis		
1.21	Deviates as functions of P: percentage points and probits		
1.211	x as a function of P or Q		
1.2111	x(P) for general P	19	631
1.2112	Tail area: x(P) for small Q	20	631
1.2113	x(P) for P = m/n		631
1.2119	Approximations to x(P)	20	
1.212	x as a function of (2P-1) or 2Q		
1.2121	$x(1-\frac{1}{2}p)$ for general p	20	631
1.2122	Double tail area: $x(1-\frac{1}{2}p)$ for small p	20	632
1.213	Percentage points as multiples of the probable error . .	21	
1.22	Ordinates as functions of area P; ratios as functions of P	21	

1.221	$Z(x)$ as a function of $P(x)$	21	
1.222	Ratio of area to bounding ordinate as a function of area	22	632
1.23	Tables for Probit Analysis	22	
1.231	Working probits; range; weighting coefficients; weighted probits		
1.2311	Double-entry tables of working probits	24	
1.2312	Full single-entry tables of working probits	24	632
1.2313	Other tables of working probits	26	632
1.232	Tables of probits: $Y = x(P) + 5$	27	632
1.233	Tables for special problems of probit analysis	28	633
1.3	Moments; integrals; derivatives		
1.31	Incomplete normal moments and related functions . . .	31	
1.32	Derivatives of $Z(x)$; Hermite polynomials; Hermite functions; tetrachoric functions; Gram-Charlier series of Type A		
1.321	Derivatives of $Z(x)$	32	
1.3211	$Z^{(n)}(x) = \dfrac{d^n Z(x)}{dx^n}$	33	634
1.3212	$2^{-\frac{1}{2}n+2} Z^{(n-1)}(2^{\frac{1}{2}}t)$	34	
1.3213	$(-1)^{n-1}(2\pi)^{\frac{1}{2}} Z^{(n-1)}(x)$	34	
1.322	Hermite functions	34	
1.323	Hermite polynomials	34	
1.324	Tetrachoric functions		
1.3241	$\tau_n = (-1)^n (n!)^{-\frac{1}{2}} Z^{(n-1)}(x)$, $\tau_0 = Q(x)$	35	
1.3249	Miscellaneous tetrachorics	35	634
1.325	Gram-Charlier series of Type A	36	634
1.33	Repeated normal integrals; truncated normal distribution		
1.331	Repeated normal integrals	37	
1.332	Truncated normal distribution	39	635
1.4	Rankings of normal samples; normal order statistics; the median; quartiles and other quantiles		
1.40	Definitions and notation	46	
1.41	Median and other quantiles	50	636
1.42	Distribution of straggler and straggler deviate; powers of normal tail area	54	637
1.43	Ordered normal deviates; normal ranking scores; moments of order statistics; order statistic tests	62	638

1.44	Studentized distributions involving order statistics	. .	68	639
1.5	Measures of location other than the arithmetic mean; measures of dispersion other than the standard deviation		70	
1.51	Measures of location		70	641
1.52	The mean deviation from mean and median and related measures	. .	71	643
1.53	Measures of normal dispersion based on variate differences, Gini's mean difference, and Nair's j-statistics			
1.531	Variate differences		75	643
1.532	Gini's mean difference		80	
1.533	Nair's j-statistics		81	
1.54	The range and midrange; Galton differences; measures of dispersion based on selected order statistics			
1.541	The range		82	645
1.542	The midrange		88	647
1.543	Galton differences		90	
1.544	Measures of dispersion based on selected order statistics		92	647
1.549	Miscellaneous functions of the range		95	
1.55	Studentized statistics involving range; analysis of variance based on range; asymptotic distribution of normal range			
1.551	Studentized statistics involving range			
1.5511	The studentized range		95	649
1.5512	Studentized statistics with range in denominator	. . .	98	650
1.552	Analysis of variance based on range		104	651
1.553	Approximate distributions of normal range			
1.5531	Approximations for moderate n		107	652
1.5532	Approximations for large n		108	
1.6	Tests of normality		111	
1.61	$b_1^{\frac{1}{2}}$			
1.611	Moments of $b_1^{\frac{1}{2}}$		111	
1.612	Distribution of $b_1^{\frac{1}{2}}$		112	
1.613	Percentage points of $b_1^{\frac{1}{2}}$		112	
1.62	b_2			
1.621	Moments of b_2		113	
1.622	Distribution of b_2		114	

1.623	Percentage points of b_2	114	
1.63	Geary's ratio	115	
1.631	Moments of Geary's ratio	115	
1.633	Percentage points of Geary's ratio	116	
1.69	Miscellaneous tests of normality	117	652
1.7	The bivariate normal distribution	119	
1.71	Ordinates of the bivariate normal distribution	119	
1.72	Volumes under the bivariate normal distribution . . .	119	
1.721	$p(x,y;\rho)$ for positive ρ	120	
1.722	$p(x,y;\rho)$ for negative ρ	120	
1.723	$V(h,q)$	121	
1.729	Miscellaneous volumes	122	
1.73	Moments	122	
1.78	Miscellaneous tables connected with the bivariate normal distribution	123	652
1.79	The trivariate normal distribution	125	654
1.795	The normal distribution in more than three variates . .		654
1.8	The error integral of complex argument		
1.81	The error integral of pure imaginary argument, alias Poisson's or Dawson's integral		
1.811	$\int_0^x e^{t^2} dt$	126	
1.812	$e^{-x^2} \int_0^x e^{t^2} dt$	126	
1.813	$e^{-y} \int_0^{y^{\frac{1}{2}}} e^{t^2} dt$	126	
1.82	The error integral with semi-imaginary argument $i^{\frac{1}{2}}x$; the Fresnel integrals	127	
1.83	The error integral for general complex argument . . .	127	
1.9	Miscellanea	129	655
2.	THE χ^2 - DISTRIBUTION AND THE INCOMPLETE Γ- FUNCTION; THE PEARSON TYPE III CURVE; THE POISSON DISTRIBUTION; THE COMPLETE Γ- FUNCTION AND THE FACTORIAL; THE χ^2- TEST OF GOODNESS OF FIT		
2.0	Introduction	130	
2.1	The χ^2-integral and the Poisson sum	131	

2.11
2.611

2.11	The χ^2-integral and the incomplete Γ-function		
2.111	The incomplete Γ-function for general exponent	131	
2.112	Areas under the Pearson Type III curve	132	656
2.113	The χ^2-distribution: the incomplete Γ-function for integer and half-integer exponent	132	
2.1131	$p(\chi^2, \nu)$	132	656
2.1132	$p(\chi^2, n'-1)$	132	
2.1133	$p(\nu s^2, \nu)$	133	
2.1134	$x^{-n-1} \int_x^\infty t^n e^{-t} dt$	133	656
2.119	Miscellaneous tables of the χ^2-distribution	133	
2.12	The cumulative Poisson distribution	134	
2.121	$P(m,i)$	134	
2.122	$1 - P(m,i)$	135	656
2.129	Miscellaneous tables of the cumulative Poisson distribution	136	656
2.2	Ordinates of the Pearson Type III frequency curve; individual Poisson frequencies		
2.20	Ordinates of the χ^2-distribution	137	
2.21	Ordinates of the Pearson Type III frequency curve . . .	137	
2.22	Individual Poisson frequencies		
2.221	Direct tabulations of $e^{-m} m^i / i!$	138	
2.229	Miscellaneous Poisson frequencies	138	
2.3	Percentage points of the χ^2- and Poisson distributions; confidence intervals for χ^2 and m		
2.31	Percentage points of χ^2; confidence intervals for σ^2 .	140	
2.311	Tables of percentage points of χ^2	140	
2.3111	Tables with material from both families A and B . . .	140	
2.3112	Tables of family A	141	
2.3113	Tables of family B	141	
2.3114	Short tables	143	658
2.3115	Tables for special ν	143	
2.312	Tables of percentage points of χ^2/ν	143	
2.313	Percentage points of χ	144	658
2.314	Approximations to percentage points of χ^2	144	
2.315	Confidence limits for σ	148	658
2.319	Miscellaneous percentage points of χ^2	150	658

2.32	Percentage points of the Poisson distribution; confidence limits for m		
2.321	Percentage points of the Poisson distribution	151	
2.322	Confidence limits for m	151	659
2.4	Laguerre series; non-central χ^2; power of χ^2 test		
2.41	Laguerre series	154	660
2.42	Non-central χ^2	155	660
2.43	Power of χ^2 test	158	
2.5	The factorial and the complete Γ-function; moments of functions of χ^2		
2.51	The factorial and the complete Γ-function	160	
2.511	Natural values of factorials		
2.5111	$x! = \Gamma(1+x)$	160	661
2.5112	$n! = \Gamma(1+n)$	161	661
2.5113	Double factorials $(2n-1)!!$ and $(2n)!!$	161	
2.512	Reciprocals of factorials		
2.5121	$1/x! = 1/\Gamma(1+x)$	161	662
2.5122	$1/n! = 1/\Gamma(1+n)$	161	
2.513	Logarithms of factorials		
2.5131	$\log_{10} x! = \log_{10}\Gamma(1+x)$	162	662
2.5132	$-\log_{10} x! = -\log_{10}\Gamma(1+x)$	162	
2.5133	$\log_e x! = \log_e \Gamma(1+x)$	162	
2.5134	$\log_{10} n! = \log_{10}\Gamma(1+n)$	163	662
2.5135	$\log_{10}(2n)!!$ and $\log_{10}(2n-1)!!$	163	
2.514	The polygamma functions	163	
2.5141	$\psi(x) = F(x) = \frac{d}{dx}(\log_e x!) = \frac{\Gamma'(1+x)}{\Gamma(1+x)}$	163	
2.5142	$\psi^{(n-1)}(x) = \left(\frac{d}{dx}\right)^n (\log_e x!) = \left(\frac{d}{dx}\right)^{n-1}\frac{\Gamma'(1+x)}{\Gamma(1+x)}$ . . .	164	
2.52	Moments of functions of χ^2	165	
2.521	Moments of χ^2	166	
2.522	Moments of χ	167	
2.529	Moments of other functions of χ^2	172	662
2.53	Moments of the Poisson distribution	173	
2.6	Tables facilitating the χ^2 goodness of fit test . . .	174	
2.61	The 2×2 table	174	
2.611	Exact tests of significance in 2×2 tables	174	663

2.612	Approximate tests of significance in 2×2 tables . . .	176	663
2.613	Power of the χ^2 test in 2×2 tables	177	663
2.62	Other contingency tables		
2.63	Fully specified tables		664
2.69	Miscellaneous goodness-of-fit tests involving χ^2 . . .	179	664

3. THE INCOMPLETE B − FUNCTION; THE BINOMIAL DISTRIBUTION

3.0	Definitions and notation	181	
3.1	Tables of the cumulative binomial sum; tables of the incomplete-beta-function ratio		
3.11	The cumulative binomial sum		
3.111	$Q(p, n, r-\tfrac{1}{2}) = \sum_{i=r}^{n} \binom{n}{i} p^i q^{n-i} = I_p(r, n-r+1)$	183	664
3.112	$P(p, n, r+\tfrac{1}{2}) = \sum_{i=0}^{r} \binom{n}{i} p^i q^{n-i} = I_q(n-r, r+1)$	183	664
3.119	Miscellaneous tables of the cumulative binomial sum . .	183	
3.12	Tables of the incomplete-beta-function ratio	185	665
3.2	Binomial coefficients; individual binomial frequencies		
3.21	Binomial coefficients	186	
3.211	$\binom{n}{i} = \frac{n!}{i!(n-i)!} = {}^{n}C_i$	186	
3.212	$\log_{10} \binom{n}{i}$	186	
3.22	Individual binomial frequencies		
3.221	Tables of $\binom{n}{i} p^i q^{n-i}$	187	
3.229	Miscellaneous binomial frequencies	188	
3.3	Percentage points of the Beta-function and confidence intervals for binomial distribution		
3.30	Introduction	190	
3.31	Percentage points of the beta distribution		
3.311	Tables of $x(P; a, b)$	191	
3.319	Miscellaneous percentage points of the beta distribution	192	665
3.32	Confidence limits for binomial distribution		
3.321	Direct tables of p_L and p_U	193	
3.322	np_L and np_U	195	
3.33	Percentage points of the binomial distribution . . .	195	

3.4	Approximations to the binomial distribution and the Incomplete Beta-function; auxiliary tables for computing $I_x(a,b)$.	197	
3.41	Approximations to percentage points of the beta distribution . . .	197	
3.42	Approximations to the cumulative binomial summation . .	198	667
3.43	Approximations to the incomplete beta-function	199	668
3.44	Auxiliary tables for computing the incomplete beta-function	200	
3.5	The complete B-function; moments of the beta distribution; moments of the binomial distribution	203	
3.51	The complete B-function	204	
3.52	Moments of the beta distribution	204	668
3.53	Moments of the binomial distribution	204	668
3.54	Moments of estimated binomial success rate	205	
3.6	Tables for the binomial sign test	206	669
4.	**THE t-, F-, AND z- DISTRIBUTIONS AND RELATED DISTRIBUTIONS**		
4.1	The t-distribution		
4.11	The probability integral of t	207	
4.111	Tables with argument t		
4.1111	$P(t,\nu)$	208	670
4.1112	$2P(t, n-1) - 1$	209	
4.1113	$2Q(t,\nu)$	209	670
4.1114	$10^6 Q(t,\nu)$	209	
4.112	Tables with argument $t\nu^{-\frac{1}{2}}$	209	671
4.113	Tables with other arguments	210	
4.114	$P(t,\nu)/Q(t,\nu)$	210	
4.115	Approximations to the t integral	210	671
4.12	Ordinates of the t-distribution	210	
4.13	Percentage points of the t-distribution	210	
4.131	Tables of $t(Q,\nu)$	211	
4.1311	Double tail percentage points: $t(Q,\nu)$ as a function of $2Q$	211	671
4.1312	Double tail percentage points: $t(Q,\nu)$ as a function of $1-2Q$ . . .	213	671
4.1313	Single tail percentage points: $t(Q,\nu)$ as a function of Q	213	671
4.1314	Single tail percentage points: $t(Q,\nu)$ as a function of P	213	
4.132	Approximations to percentage points of t	213	672

4.133	Confidence intervals based on percentage points of t . . . 216	
4.139	Miscellaneous significance tests related to t 217	673
4.14	The non-central t-distribution and the power of the t-test; the confluent hypergeometric function; the distribution of t from non-normal parents	
4.141	The non-central t distribution; the power of the t-test . 218	
4.1411	Direct tables of the non-central t-distribution . . . 219	
4.1412	The power of the t-test: the probability integral of non-central t evaluated at the percentage points of central t 220	
4.1413	The power of the t-test: miscellaneous tabulations . . 222	
4.1414	Percentage points of non-central t 224	
4.1419	Miscellaneous applications of non-central t 225	
4.142	The confluent hypergeometric function 226	
4.1421	$M(\alpha,\gamma,x)$ for general α and γ	673
4.1422	$M(\alpha,\gamma,x)$ for integer and half-integer α and γ 227	
4.1423	Functions related to $M(\alpha,\tfrac{1}{2},x)$ 228	
4.143	The distribution of t in samples from non-normal parents 228	
4.15	Two-sample tests for unequal variances 231	
4.151	The Fisher-Behrens test 232	
4.152	Welch's test 237	674
4.153	Scheffe's test 239	
4.16	Moments of t 240	
4.17	Multivariate t-distributions 241	674
4.18	Confidence intervals for a weighted mean	674
4.19	Miscellaneous tabulations connected with t 242	675
4.2	The F-distribution and multivariate variance tests	
4.20	Definition of F and relation to other statistics . . . 245	
4.21	The area of the F-distribution 246	
4.23	Percentage points of F and of $F^{\tfrac{1}{2}}$ 246	
4.231	Tables of percentage points of F 247	
4.2311	Tables with material from both families A and B . . . 248	676
4.2312	Tables of family A 248	676
4.2313	Tables of family B 249	676
4.2314	Short tables 249	
4.232	Tables of percentage points of $F^{\tfrac{1}{2}}$ 250	
4.233	Errors in tables of percentage points of F 252	

			4.133
			5.5
4.239	Miscellaneous combinations of percentage points of F . .	253	677
4.24	Power of the F-test	255	
4.241	Power of the F-test, in terms of the central F-distribution	256	677
4.242	Power of the F-test, in terms of the non-central F-distribution . .	258	678
4.25	Ratios involving ordered mean squares	260	
4.251	The largest variance ratio	261	681
4.252	The smallest variance ratio	261	
4.253	s^2_{max}/s^2_{min}	262	
4.254	The ranking of mean squares		681
4.26	Moments and cumulants of F	262	
4.27	Approximations to the distribution of F	263	
4.28	Effect of non-normality, heteroscedasticity and dependence on the distribution of F	264	681
4.3	The z-distribution		
4.30	Definition of z and relation to other statistics . . .	270	
4.33	Percentage points of z and of $\log_{10} F$		
4.331	Tables of percentage points of z	250	676
4.332	Tables of percentage points of $\log_{10} F = 2z \log_{10} e$ . .	251	
4.333	Errors in tables of percentage points of z	270	
4.339	Miscellaneous significance points of z	272	
4.34	Cumulants of $\log \chi^2$ and of z	272	
4.35	Approximate formulas for percentage points of z . . .	273	
5.	**VARIOUS DISCRETE AND GEOMETRICAL PROBABILITIES**		682
5.1	The hypergeometric distribution	275	
5.2	Generalized Poisson distributions		
5.21	Contagious distributions	278	683
5.22	The bivariate Poisson distribution	282	
5.23	Distribution of the difference between two Poisson variates	282	
5.3	The negative binomial distribution and the logarithmic series	284	
5.31	Tables of $\binom{N+r-1}{r} = {}^{N}H_{r}$	284	683
5.32	The negative binomial distribution	285	683
5.33	The logarithmic series	285	
5.4	Multinomials	287	683
5.5	Some geometrical probabilities	288	

5.6	Probabilities in card games		
5.61	Packs of 52 cards		684
5.62	Other packs of cards		685
5.9	Miscellanea	292	688

6. LIKELIHOOD AND OTHER STATISTICS USED IN TESTING HYPOTHESES AND IN ESTIMATION

6.1	Tests applied to normal populations	295	
6.11	Tests applied to univariate normal populations		
6.111	Neyman and Pearson's one- and two-sample tests	296	
6.112	Neyman and Pearson's k-sample tests L_0 and L_1	297	
6.113	Bartlett's modification of the L_1 test	300	688
6.114	Other tests of homogeneity of variance	302	688
6.115	Sequential t-tests	303	
6.119	Miscellanea	304	688
6.12	Tests applied to bivariate normal populations	305	690
6.13	Tests applied to general multivariate normal populations	306	
6.131	The D^2-statistic	306	
6.132	Other tests of homogeneity	308	692
6.133	Canonical correlations; independence of groups of variates	311	
6.139	Miscellanea	313	
6.2	Tests and estimators applied to binomial populations		
6.21	Tests applied to binomial populations	314	692
6.22	Estimators of parameters in binomial populations . . .	314	
6.3	Tests and estimators applied to Poisson populations		
6.31	Tests applied to Poisson populations	315	693
6.32	Estimators of parameters in Poisson populations . . .	316	
6.4	Tests applied to rectangular populations	318	694

7. CORRELATION, SERIAL CORRELATION AND COVARIANCE

7.1	Distribution of the product-moment correlation coefficient	319	
7.11	Areas and ordinates of the distribution of r in normal samples .	320	
7.12	Percentage points of r and confidence limits for ρ		
7.121	Confidence limits for ρ ; percentage points of r for general ρ .	322	
7.122	Percentage points of r for $\rho = 0$	323	

7.1221	Double tail percentage points: $r(\frac{1}{2}Q, \nu+2)$	323	
7.1222	Double tail percentage points: $r(\frac{1}{2}Q, n)$	323	
7.1223	$(\nu+1)^{\frac{1}{2}} r(\frac{1}{2}Q, \nu+2)$	323	
7.1229	Miscellaneous percentage points of r	324	
7.13	Moments and other distribution constants of r and related statistics	324	
7.131	Moments of r	325	695
7.132	"Probable error" of r	326	
7.133	Mode of r	326	
7.134	Moments of $t = r(1-r^2)^{-\frac{1}{2}}$	327	
7.14	Distribution of r from non-normal parents	328	
7.19	Miscellaneous tables related to r	328	
7.2	The hyperbolic-tangent transformation	330	
7.21	$\tanh x$ and $\tanh^{-1} x$	330	
7.211	$\tanh x = \dfrac{e^x - e^{-x}}{e^x + e^{-x}}$	330	695
7.212	$\tanh^{-1} x = \frac{1}{2} \log_e \dfrac{1+x}{1-x}$	331	695
7.22	Percentage points of z	331	
7.23	Moments of z	331	
7.29	Miscellaneous tables connected with z	332	
7.3	Tables of $1-r^2$, $(1-r^2)^{\frac{1}{2}}$, $(1-r^2)^{-\frac{1}{2}}$ and related functions	334	
7.31	$1-r^2$	334	695
7.32	$(1-r^2)^{\frac{1}{2}}$	334	695
7.33	$(1-r^2)^{-\frac{1}{2}}$	334	696
7.4	Multiple correlation coefficient	335	
7.5	Serial correlation	336	
7.51	Circular serial correlation	336	696
7.52	Non-circular serial correlation	338	
7.53	Variate differences as a measure of serial correlation	339	
7.54	Correlation between time series		696
7.6	Covariance; covariance ratio; regression coefficient	340	697
7.9	Other measures of correlation	343	697
7.91	Biserial correlation	343	697
7.92	Mean square contingency	343	
7.93	Correlation ratio	343	
7.94	Tetrachoric correlation	343	698

7.99	Miscellanea	344	699
8.	**RANK CORRELATION; ORDER STATISTICS**	345	
8.1	Rank correlation	345	
8.10	Definitions and notation	345	
8.11	Spearman's Rho		
8.111	Distribution of S_r: null case		
8.1111	Individual terms	348	
8.1112	Cumulative sums: exact probabilities	349	
8.1113	Cumulative sums: approximate probabilities	349	
8.112	Percentage points of S_r and r_s: null case	349	
8.113	Treatment of ties in the calculation of r_s	351	
8.114	Distribution of r_s under non-null parent normal correlation	351	
8.119	Miscellaneous tables of r_s	351	
8.12	Kendall's Tau		
8.121	Distribution of S_t: null case		
8.1211	Individual terms	352	
8.1212	Cumulative sums	352	
8.122	Percentage points of S_t	352	
8.123	Treatment of ties in the calculation of S_t	353	
8.125	Moments of S_t	353	
8.13	Kendall's concordance coefficient W		
8.131	Distribution of S_w and related statistics: null case		
8.1312	Distribution of S_w: cumulative sums	353	
8.1313	Distribution of $\chi_r^2 = 12 S_w /mn(n-1)$	354	
8.132	Percentage points of W	354	
8.14	Paired comparisons; inconsistent triads	355	699
8.15	Partial and multiple rank correlation	356	
8.19	Other non-parametric measures of correlation	356	
8.2	Order statistics from non-normal parents	360	
8.21	Continuous rectangular parent	360	699
8.22	Discrete rectangular parent	361	
8.23	K. Pearson's frequency curves as parents	362	700
8.24	Iterated exponential parent		701
8.27	Comparison of ranges from various parents		701
8.28	Arbitrary parent		702

[xviij]

8.3	Tables facilitating the asymptotic theory of statistical extreme values	364	
8.31	The extreme value		
8.311	Direct tables		
8.3111	$e^{-e^{-x}}$ and $e^{-[x+e^{-x}]}$	364	
8.3112	$-\log_{10}[1-e^{-e^{-x}}]$	364	
8.312	Inverse tables		
8.3121	$-\log_e(-\log_e x)$	364	
8.3122	$-x.\log_e x$		703
8.3123	$-x.\log_2 x$	365	
8.3129	Miscellaneous combinations of $x \log x$		703
8.319	Miscellaneous tables of the distribution of the extreme value		703
8.32	The mth extreme value		
8.321	Direct tables		703
8.3211	$(1+2e^{-x})e^{-2e^{-x}}$	365	
8.322	Inverse tables	365	
8.33	The range	365	
8.331	Direct tables: $\psi = 2e^{-\frac{1}{2}x}.K_1(2e^{-\frac{1}{2}x})$ and $\psi = 2e^{-x}.K_0(2e^{-\frac{1}{2}x})$	365	
8.332	Inverse tables: x as a function of $\psi = 2e^{-\frac{1}{2}x}.K_1(2e^{-\frac{1}{2}x})$.	365	
9.	**NON-PARAMETRIC TESTS**	366	704
9.1	Tests of location		704
9.11	One-sample tests		
9.111	The Walsh test	366	704
9.112	The signed-rank test	367	704
9.118	Tests of regression		705
9.12	Two-sample tests		
9.121	The Wilcoxon test	368	706
9.122	Other rank-sum tests	369	
9.123	The median test	370	
9.129	Other two-sample tests	370	706
9.13	Three-sample tests	370	707
9.14	k-sample tests	371	707
9.2	Tests of spread	373	708

9.3				
10.043				
9.3	Tests of randomness of occurrence	. . . .	374	
9.31	One-sample tests			
9.311	Runs		374	709
9.312	Tests based on signs of variate differences		376	
9.319	Other tests of randomness			711
9.32	Two-sample tests		377	713
9.33	Tests of association			713
9.4	Tests of goodness of fit of a sample to a population, and analogous tests of the homogeneity of two samples		380	
9.41	The Kolmogorov-Smirnov tests			
9.411	Two-sided tests: asymptotic distribution		380	
9.4111	Tables of $\vartheta_4(0, e^{-2z^2})$		380	714
9.4112	Other tables of $\vartheta_4(0,q)$		381	
9.4113	Inverse tables: z as a function of $\vartheta_4(0, e^{-2z^2})$ . . .		381	714
9.412	Two-sided tests: exact distributions			
9.4121	Distributiom of D_n		381	714
9.4122	Distribution of $D_{m,n}$		382	715
9.413	Power of two-sided tests		384	
9.414	Three-sample tests			715
9.415	One-sided tests			
9.4151	One-sided tests of one sample against a population . .		384	715
9.4152	One-sided tests of slippage between two samples . .		385	715
9.416	The maximum relative discrepancy		385	
9.42	The Cramer-Mises-Smirnov tests			
9.421	The ω^2 test: asymptotic distribution		386	
9.4211	Tables of the asymptotic distribution of $n\omega^2$		386	
9.4212	Tables of the asymptotic percentage points of $n\omega^2$ . . .		386	
9.429	Miscellaneous tests related to ω^2		386	
9.9	Miscellaneous non-parametric tests		387	
10.	**SYSTEMS OF FREQUENCY CURVES; MOMENTS AND OTHER SYMMETRIC FUNCTIONS** .			715
10.0 } 10.1 }	K. Pearson's frequency curves		388	
10.00	Pearson curves in general		388	
10.001	Ordinates and areas of Pearson curves		389	

10.002	Percentage points of Pearson curves	389	715
10.004	Tables and charts for determining the type of a Pearson curve from its moments	389	
10.006	Moments of Pearson curves	391	
10.007	Standard errors of constants of Pearson curves	391	
10.009	Miscellaneous tables of Pearson curves		716
10.01	Pearson curves of Type I (Elderton's 'first main type') .	392	
10.011	Areas of the Type I distribution	392	
10.012	Percentage points of the Type I distribution	392	
10.013	Auxiliary tables for computing ordinates, areas and percentage points of the Type I distribution	392	
10.016	Moments, etc., of the Type I distribution	393	
10.017	Distribution of statistics in samples from a Type I population . . .	393	
10.018	Degenerate cases of Type I curves	393	
10.02	Pearson curves of Type II	393	
10.021	Areas of the Type II distribution	393	
10.022	Percentage points of the Type II distribution	394	
10.023	Auxiliary tables for computing ordinates and areas of the Type II distribution	394	716
10.024	The fitting of Type II curves by the method of moments .	394	
10.028	Degenerate cases of Type II curves	394	
10.03	Pearson curves of Type III	394	
10.031	Ordinates and areas of the Type III distribution . . .	394	
10.032	Percentage points of the Type III distribution	395	
10.033	Auxiliary tables for computing ordinates of the Type III distribution .	395	716
10.035	The fitting of Type III curves by the method of maximum likelihood .	395	716
10.036	Moments, etc., of the Type III distribution	396	
10.037	Distribution of statistics in samples from a Type III population	396	
10.038	Degenerate cases of Type III curves	396	
10.039	Miscellaneous tables of Type III curves	396	717
10.04	Pearson curves of Type IV (Elderton's 'second main type')	397	
10.043	Auxiliary tables for computing ordinates of the Type IV distribution .	397	

10.0431		
10.261		

10.0431	The $G(r,\nu)$ integrals	397	717
10.0432	$\tan^{-1} x$ and $\log_{10} (1+x^2)$	398	
10.048	Degenerate cases of Type IV curves	398	
10.05	Pearson curves of Type V	398	
10.052	Percentage points of the Type V distribution	398	
10.055	The fitting of Type V curves by the method of maximum likelihood . .	398	
10.058	Degenerate cases of Type V curves	398	
10.059	Miscellaneous tables connected with the Type V distribution	399	
10.06	Pearson curves of Type VI (Elderton's 'third main type')	399	
10.061	Areas of the Type VI distribution	399	
10.062	Percentage points of the Type VI distribution	399	
10.068	Degenerate cases of Type VI curves	399	
10.07	Pearson curves of Type VII	399	
10.071	Ordinates and areas of the Type VII distribution . . .	400	
10.072	Percentage points of the Type VII distribution	400	
10.073	Auxiliary tables for computing ordinates and areas of the Type VII distribution	400	717
10.074	The fitting of Type VII curves by methods other than moments . . .	401	
10.078	Degenerate case of Type VII curves	401	
10.08	Pearson curves of Type VIII	401	
10.087	Distribution of statistics in samples from a Type VIII population . . .	401	
10.088	Degenerate case of Type VIII curves	402	
10.09	Pearson curves of Type IX	402	
10.095	The fitting of Type IX curves by methods other than moments		717
10.098	Degenerate cases of Type IX curves	402	
10.10	The Pearson curve of Type X	402	
10.101	Ordinates and areas of the Type X distribution	402	
10.102	Percentage points of the Type X distribution	402	
10.103	Auxiliary tables for computing percentage points of the Type X distribution	403	
10.107	Distribution of statistics in samples from a Type X population . . .	403	
10.109	Generalizations of the Type X distribution		718
10.11	Pearson curves of Type XI	403	

10.117	Distribution of statistics in samples from a Type XI population	403	
10.118	Degenerate case of Type XI curves	403	
10.12	Pearson curves of Type XII	403	
10.123	Auxiliary tables for computing ordinates of the Type XII distribution	403	
10.128	Degenerate case of Type XII curves	404	
10.13	The normal distribution *qua* Pearson curve (Elderton's 'normal curve of error')	404	
10.131	Ordinates and areas of the normal distribution	404	
10.132	Percentage points of the normal distribution	404	
10.135	The fitting of the normal distribution by statistics other than the sample mean and variance	404	
10.136	Moments of the normal distribution	404	
10.137	Distribution of statistics in normal samples	405	
10.18	Generalizations of K. Pearson's frequency curves	405	718
10.19	Bivariate Pearson frequency surfaces	406	
10.2	Other frequency curves	407	
10.21	Gram-Charlier series		
10.211	Gram-Charlier series of Type A	407	
10.212	Gram-Charlier series of Type B	409	
10.213	Gram-Charlier series of Type C	410	
10.218	Generalized Gram-Charlier series	410	
10.219	Bivariate Gram-Charlier series	410	
10.22	Translation systems	410	719
10.221	Polynomial translation	411	719
10.222	Logarithmic and hyperbolic translation	412	719
10.2221	S_L curves: i.e. the log-normal distribution	412	720
10.2222	S_B curves	414	
10.2223	S_U curves	414	
10.229	Bivariate translation systems	414	
10.23	The fourth-degree exponential	415	
10.24	Cumulative distribution functions	416	720
10.25	Bessel-function distributions	417	
10.26	The rectangular distribution and its convolutions		
10.261	The continuous rectangular distribution	418	720

10.262				
12.53				
10.262	The discrete rectangular distribution	. . . .	418	722
10.28	The circular normal distribution	. . . .	418	722
10.29	Miscellaneous frequency curves	. . . .	420	
10.3	Symmetric functions: moments, cumulants and k-statistics			
10.31	Symmetric functions per se	. . . .	421	
10.311	Comprehensive tables of symmetric functions	. . . .	422	722
10.312	Other tables of symmetric functions	. . . .	424	
10.3121	MS tables	. . . .	424	
10.3123	MU tables	. . . .	424	
10.3124	UM tables	. . . .	425	
10.3125	UH and HU tables	. . . .	425	
10.3126	MH tables	. . . .	425	
10.3127	HM tables	. . . .	425	
10.3128	SU and SH tables	. . . .	425	
10.3129	US and HS tables	. . . .	425	
10.32	Moments and cumulants	. . . .	425	
10.33	Moments from grouped distributions			
10.331	Sheppard's corrections	. . . .	426	722
10.332	Corrections for abruptness	. . . .	427	
10.333	Corrections for grouping of discrete distributions	. .	428	
10.34	k-statistics: infinite population	. . . .	429	
10.341	k_r in terms of power sums	. . . .	430	
10.342	Formulas for ℓ-statistics	. . . .	430	
10.343	Sampling cumulants of k-statistics	. . . .	430	723
10.349	Moments of moments	. . . .	430	
10.35	k-statistics: finite population	. . . .	431	
10.36	Multivariate symmetric functions and moments	. . . .	431	723
10.37	Gini's generalized means	. . . .	432	
10.39	Miscellaneous symmetric functions	. . . .	432	
10.4	Cebysev's inequality and related inequalities	. . . .	433	724
10.9	Miscellaneous tables related to frequency curves	. . .	434	
11.	**THE FITTING OF REGRESSION AND OTHER CURVES**			
11.1	The fitting of polynomials			
11.11	Orthogonal polynomials	. . . .	436	
11.110	Formulas	. . . .	436	

11.111	Tables of orthogonal polynomials	437	
11.112	Errors in tables of orthogonal polynomials	438	
11.113	Tables for the summation method	438	
11.119	Miscellaneous tables of orthogonal polynomials	439	724
11.12	Power polynomials	440	725
11.125	The fitting of power polynomials without moments		725
11.13	Factorial polynomials	441	
11.19	Miscellaneous polynomial fitting	441	725
11.2	Harmonic analysis: i.e. the fitting of trigonometric polynomials	442	
11.21	Angles: $2\pi r/s$	442	
11.22	$\cos 2\pi r/s$ and $\sin 2\pi r/s$	442	
11.23	$a \cos 2\pi r/s$ and $a \sin 2\pi r/s$	443	
11.24	Tests of significance in harmonic analysis	443	726
11.28	Miscellaneous tables for harmonic analysis		727
11.29	Periodograms	443	
11.3	The fitting of exponential curves	444	
12.	**VARIATE TRANSFORMATIONS**		
12.0	Introduction	445	
12.1	The angular transformation	445	
12.11	$\sin^{-1} x^{\frac{1}{2}}$: degrees	445	728
12.12	$\sin^{-1}[(x/n)^{\frac{1}{2}}]$: degrees	445	
12.13	$\sin^{-1} x^{\frac{1}{2}}$: radians	445	728
12.14	Working angles	446	
12.15	Special tables of the angular transformation	446	
12.2	The square-root transformation	448	
12.3	The inverse hyperbolic-sine transformation	449	
12.31	$\sinh^{-1} x^{\frac{1}{2}}$	449	
12.32	$\gamma^{-\frac{1}{2}} \sinh^{-1} (\gamma x)^{\frac{1}{2}}$	449	
12.33	Special tables of the $\sinh^{-1}$ transformation	449	
12.4	The probit transformation	449	
12.5	The logit transformation	450	
12.51	$x = \log_e[p/(1-p)]$	450	
12.52	$p = (1 + e^{-x})^{-1}$	450	
12.53	Working logits and weights	450	

12.54	Special tables of the logit transformation	451	728
12.6	The loglog transformation		
12.61	Tables of loglogs	452	
12.62	Working loglogs and weights	452	
12.7	The Legit transformation	453	
12.9	Other transformations	453	

13. TABLES OF RANDOM SAMPLES

13.0	Introduction	454	
13.1	Random digits	456	
13.11	Finding list of tables of random digits	456	729
13.12	Original tables of random digits		
13.121	Original tables of random digits from physical sources		
13.1211	Tables published as obtained	457	729
13.1212	Tables subjected to 'improvement' before publication	457	
13.122	Original tables of random digits produced by rearrangement		
13.1221	Rearrangement of non-random digits	458	
13.1222	Rearrangement of random digits	459	
13.13	Secondary tables of random digits	459	
13.2	Random permutations	460	
13.3	Random normal numbers	462	729
13.4	Random samples from populations other than rectangular and normal	464	
13.5	Random samples from artificial time series	465	
13.6	Generation of random and pseudo-random numbers	468	
13.61	Generation of random numbers	468	
13.62	Generation of pseudo-random numbers	468	
13.63	Testing of random and pseudo-random numbers	468	

14. ACCEPTANCE SAMPLING; CONTROL CHARTS; TOLERANCE LIMITS

14.1	Acceptance sampling	469	
14.11	Lot inspection by attributes	470	
14.111	Comprehensive tables	470	730
14.112	Other tables	473	730
14.12	Lot inspection by variables	477	
14.121	Comprehensive tables	477	730
14.122	Other tables: controlling per cent defective	479	

14.123	Other tables: controlling the mean		731
14.124	Other tables: controlling variance	480	
14.129	Auxiliary tables for the design of variables plans . .	481	731
14.13	Continuous inspection by attributes	481	
14.15	Life testing	483	
14.151	Tables of life-testing plans	483	732
14.152	Auxiliary tables of exponentials	484	
14.159	Other auxiliary tables	485	
14.2	Control charts		
14.21	Control charts for number or per cent defective	487	732
14.22	Control charts for variables		
14.221	Factors for constructing charts for mean and standard deviation, or for mean and range: A, B, C,	488	733
14.222	Factors for constructing charts: other forms	490	733
14.229	Miscellaneous tables of control charts for variables . .	490	733
14.3	Tolerance limits		
14.31	Non-parametric tolerance limits	492	734
14.32	Normal tolerance limits	492	
15.	**TABLES FOR THE DESIGN OF EXPERIMENTS**		
15.0	Introduction	495	
15.1	Factorial patterns: Latin squares, etc.		
15.10	Introduction: blocking and unblocking; effects and estimates	496	
15.11	Finding list of factorial patterns	502	735
15.112	Factorial patterns of the 2^a system	502	735
15.113	Factorial patterns of the 3^a system	516	
15.1132	Factorial patterns of the $3^a.2^b$ system	520	735
15.114	Factorial patterns with one or more factors at 4 levels .	526	735
15.115	Factorial patterns with one or more factors at 5 levels .	533	736
15.116	Factorial patterns with one or more factors at 6 levels .	535	736
15.117	Factorial patterns with one or more factors at 7 levels .	536	736
15.118	Factorial patterns with one or more factors at 8 levels .	537	736
15.119	Factorial patterns with one or more factors at 9 levels .	541	
15.1199	Factorial patterns with one or more factors at 10 or more levels . .	543	736

15.12	Supplementary finding list: factorial patterns derivable from incomplete block patterns	547
15.13	Patterns without control of heterogeneity	
15.131	Saturated patterns: weighing designs	553
15.1311	Saturated fractions of 2^a factorials	553
15.1312	Complete sets of orthogonal squares and cubes	554
15.1313	Orthogonal matrices	554
15.138	Patterns for investigating response surfaces	554
15.14	Patterns with 1-way control of heterogeneity	
15.141	Fractional factorials confounded in blocks	555
15.142	Factorials confounded in blocks	556
15.1421	Some interactions completely confounded	556
15.1422	Balanced patterns: all interactions partially confounded	556
15.143	Lattices	556
15.1431	Balanced lattices	556
15.1432	Unbalanced lattices	557
15.148	Patterns for investigating response surfaces	557
15.15	Patterns with 2-way control of heterogeneity	
15.151	Squares	558
15.1511	Latin squares and cubes	558
15.1512	Enumeration and randomization of Latin squares	558
15.1513	Graeco-Latin squares; unsaturated sets of orthogonal squares	560
15.1514	Orthogonal partitions of Latin squares	561
15.152	Double confounding	561
15.153	Lattice squares	561
15.16	Patterns with factors applied to split plots	562
15.161	Split-plot patterns with 1-way control of heterogeneity	562
15.162	Split-plot patterns with 2-way control of heterogeneity	
15.1621	Latin squares with split plots	562
15.1622	Half-plaid squares	562
15.1623	Plaid squares	562
15.2	Incomplete block patterns	
15.20	Introduction	563
15.21	Finding list of incomplete block patterns	566
15.2102	Patterns in blocks of 2 plots	566

15.2103	Patterns in blocks of 3 plots	570
15.2104	Patterns in blocks of 4 plots	577
15.2105	Patterns in blocks of 5 plots	583
15.2106	Patterns in blocks of 6 plots	587
15.2107	Patterns in blocks of 7 plots	592
15.2108	Patterns in blocks of 8 plots	594
15.2109	Patterns in blocks of 9 plots	598
15.2110	Patterns in blocks of 10 plots	601
15.212	Patterns in blocks of 11 or more plots	604
15.24	Patterns with 1-way control of heterogeneity	
15.241	Balanced incomplete block patterns	
15.2411	Balance on pairs	605
15.2412	Balance on pairs and triples	606
15.2413	Balanced incomplete block patterns in which a second factor is confounded with replications	736
15.243	Square lattices	606
15.244	Rectangular lattices	606
15.245	Cubic lattices	606
15.246	Chain blocks	607
15.248	Partly balanced incomplete blocks	607
15.25	Patterns with 2-way control of heterogeneity	
15.251	Youden squares and generalizations	607
15.2511	Incomplete Latin squares	607
15.2512	Redundant Latin squares	608
15.2513	Mutilated Latin squares	608
15.253	Lattice squares	608
15.256	Generalized chain blocks	609
15.258	Partially balanced incomplete block patterns with 2-way control of heterogeneity	609
15.2581	Single columns are replications	609
15.2582	Sets of columns are replications	609
15.28	Tables for the analysis of incomplete block experiments .	610
15.3	Tables facilitating the choice of sample size	611 736
16.	**SUNDRY MATHEMATICAL TABLES**	
16.00	Introduction	614

16.02			
16.234			
16.02	Powers: positive, negative and fractional	614	
16.0211	Squares: exact values	614	
16.0213	Squares: abridged values	615	
16.0214	Quadratic polynomials	615	
16.022	Cubes	615	
16.023	Higher integral powers	615	
16.0233	kn^4	616	
16.024	Reciprocals	616	
16.0244	Miscellaneous rational fractions	616	737
16.025	n^{-p}	617	
16.0261	Square roots	617	
16.0262	Reciprocal square roots	617	
16.0263	$n^{\frac{1}{2}p}$	618	
16.0264	Square roots of rational functions	618	
16.027	Cube roots	619	
16.03	Factorials; binomial coefficients; partitions	619	
16.0318	Iron numbers		738
16.036	Partitions	619	
16.04	Sums of powers; Bernoulli numbers; differences of zero		
16.041	Bernoulli numbers	619	
16.044	Sums of powers of integers	619	
16.0444	$S_n(x) = \sum_{r=1}^{x} r^n$	620	
16.0445	Sums of powers of odd integers	620	
16.046	Sums of reciprocals	620	
16.0492	Differences and derivatives of zero	620	
16.05	Mathematical constants; roots of equations	620	
16.0564	Solution of cubic equations	620	
16.0568	Golden numbers		738
16.06	Common logarithms and related functions	620	
16.061	$\log_{10} x$	621	
16.0618	$\log_{10} \log_{10} x$	621	
16.064	10^x	621	
16.068	Addition and subtraction logarithms	621	
16.07	Natural trigonometric functions	622	

[xxx]

15.2103	Patterns in blocks of 3 plots	570
15.2104	Patterns in blocks of 4 plots	577
15.2105	Patterns in blocks of 5 plots	583
15.2106	Patterns in blocks of 6 plots	587
15.2107	Patterns in blocks of 7 plots	592
15.2108	Patterns in blocks of 8 plots	594
15.2109	Patterns in blocks of 9 plots	598
15.2110	Patterns in blocks of 10 plots	601
15.212	Patterns in blocks of 11 or more plots	604
15.24	Patterns with 1-way control of heterogeneity	
15.241	Balanced incomplete block patterns	
15.2411	Balance on pairs	605
15.2412	Balance on pairs and triples	606
15.2413	Balanced incomplete block patterns in which a second factor is confounded with replications	736
15.243	Square lattices	606
15.244	Rectangular lattices	606
15.245	Cubic lattices	606
15.246	Chain blocks	607
15.248	Partly balanced incomplete blocks	607
15.25	Patterns with 2-way control of heterogeneity	
15.251	Youden squares and generalizations	607
15.2511	Incomplete Latin squares	607
15.2512	Redundant Latin squares	608
15.2513	Mutilated Latin squares	608
15.253	Lattice squares	608
15.256	Generalized chain blocks	609
15.258	Partially balanced incomplete block patterns with 2-way control of heterogeneity	609
15.2581	Single columns are replications	609
15.2582	Sets of columns are replications	609
15.28	Tables for the analysis of incomplete block experiments .	610
15.3	Tables facilitating the choice of sample size	611 736

16. SUNDRY MATHEMATICAL TABLES

16.00	Introduction	614

16.02	Powers: positive, negative and fractional	. . . . .	614	
16.0211	Squares: exact values	614	
16.0213	Squares: abridged values	615	
16.0214	Quadratic polynomials	615	
16.022	Cubes	615	
16.023	Higher integral powers	615	
16.0233	kn^4	616	
16.024	Reciprocals	616	
16.0244	Miscellaneous rational fractions	616	737
16.025	n^{-p}	617	
16.0261	Square roots	617	
16.0262	Reciprocal square roots	617	
16.0263	$n^{\frac{1}{2}p}$	618	
16.0264	Square roots of rational functions	618	
16.027	Cube roots	619	
16.03	Factorials; binomial coefficients; partitions	619	
16.0318	Iron numbers		738
16.036	Partitions	619	
16.04	Sums of powers; Bernoulli numbers; differences of zero		
16.041	Bernoulli numbers	619	
16.044	Sums of powers of integers	619	
16.0444	$S_n(x) = \sum_{r=1}^{x} r^n$	620	
16.0445	Sums of powers of odd integers	620	
16.046	Sums of reciprocals	620	
16.0492	Differences and derivatives of zero	620	
16.05	Mathematical constants; roots of equations	620	
16.0564	Solution of cubic equations	620	
16.0568	Golden numbers		738
16.06	Common logarithms and related functions	620	
16.061	$\log_{10} x$	621	
16.0618	$\log_{10} \log_{10} x$	621	
16.064	10^x	621	
16.068	Addition and subtraction logarithms	621	
16.07	Natural trigonometric functions	622	

16.0711	$\sin x$ and $\cos x$ (radians)	622
16.0712	$\tan x$ and $\cot x$ (radians)	622
16.0721	$\sin x$ (degrees)	622
16.0722	$\tan x$ (degrees)	622
16.0723	$\sec x$ (degrees)	623
16.073	Tables with centesimal argument	623
16.0735	$\sin[\pi(2x-\tfrac{1}{2})]$	623
16.079	Miscellaneous circular functions	623
16.08	Logarithms of trigonometric functions	623
16.10	Exponential and hyperbolic functions	623
16.101	e^x	623
16.102	e^{-x}	624 738
16.103	$\log_{10} e^x$	624
16.1041	$\sinh x$ and $\cosh x$	624
16.1043	$\coth x$	624
16.1051	$\log \sinh x$ and $\log \cosh x$	624
16.1052	$\log \tanh x$ and $\log \coth x$	625
16.11	Natural logarithms	625
16.111	Natural logarithms of integers	625
16.112	Natural logarithms of decimal numbers	625
16.1149	Miscellaneous natural logarithms	626
16.12	Combinations of circular and hyperbolic functions	
16.1231	Segmental functions	626
16.1232	$e^{-x} \sin y$ and $e^{-x} \cos y$	626
16.17	Bessel functions	626
16.21	Elliptic functions	627
16.219	Functions related to theta functions	627
16.23	Interpolation; numerical integration	627
16.2311	Lagrange interpolation coefficients	627
16.2313	Gregory-Newton interpolation coefficients . . .	628
16.2315	Stirling interpolation coefficients	628
16.2316	Bessel interpolation coefficients	628
16.2317	Everett interpolation coefficients	628
16.2319	Lagrange interpolation coefficients for special unequally spaced pivotal points	628
16.234	Numerical differentiation	629

16.236	Numerical integration	629
16.2369	Lubbock coefficients	629

APPENDICES

1. SUPPLEMENT TO THE DESCRIPTIVE CATALOGUE 630
[For the detailed contents of Appendix 1, see the extreme right column of pp. v-xxxj above.]

2. CONTENTS OF BOOKS OF TABLES

Arkin & Colton 1950		739
Burington & May 1953		740
Czechowski et al. 1957		741
Dixon & Massey	1951	743
	1957	744
Fisher & Yates	1938, 1943, 1948, 1953	746
	1957	749
Glover 1923		752
Graf & Henning 1954		753
Hald 1952		754
Jahnke & Emde	1909, 1933, 1938	755
	1948, 1952, 1960	756
Kelley	1938	756
	1948	757
Kitagawa & Mitome 1953		757
Lindley & Miller 1953		758
Pearson & Hartley 1954		759
K. Pearson	Tables for S & B I	761
	Tables for S & B II	764
Siegel 1956		767
Vianelli 1959		768

3. MATERIAL TREATED BOTH IN THIS <u>GUIDE</u> AND IN FLETCHER ET AL. 1946   786

AUTHOR INDEX   790

SUBJECT INDEX   953

THE COMMITTEE'S FOREWORD

The preparation of this <u>Guide to Tables in Mathematical Statistics</u> has extended over a period of more than twenty years. Its sponsorship originated in the National Academy of Sciences--National Research Council and has remained there throughout the entire period, during which the structure of the National Research Council has undergone considerable evolution. We sketch here the corresponding evolution in the sponsorship of the <u>Guide</u>.

In 1936 the Division of Physical Sciences of the National Research Council established its Committee on Bibliography of Mathematical Tables and Aids to Computation with the responsibility to arrange for the preparation of reports on existing mathematical tables. The Chairman of this Committee, during its first three years, was A. A. Bennett of Brown University. In 1939 the Committee received a $15,000. subvention from the Rockefeller Foundation. At the same time Professor Bennett was succeeded as Chairman of the Committee by R. C. Archibald of Brown University, who proceeded to enlarge the Committee and to set up subcommittees to prepare reports on mathematical tables in various fields.

Professor Archibald also laid plans to start a quarterly journal "to serve as a clearing-house for information concerning mathematical tables and other aids to computation" as it appeared in the current literature. This journal, called <u>Mathematical Tables and Other Aids to Computation</u> (MTAC), was established early in 1943 with Professor Archibald as Editor, with his enlarged Committee as editorial advisory committee, and with D. H. Lehmer of Lehigh University (now at the University of California, Berkeley) as Coeditor. After MTAC had become firmly established, Archibald retired from the editorship of MTAC at the end of 1949 and was succeeded by Professor Lehmer. With Professor Archibald's retirement in 1949 from the Chairmanship of the Committee on Bibliography of Mathematical Tables and Aids to Computation, the said Committee was superseded by a six-member

Committee whose function was to serve as the editorial board for MTAC. This MTAC Editorial Committee was transferred from the Division of Physical Sciences to the Division of Mathematics upon the creation of the latter Division in 1951.

Early in 1940 Professor Archibald established a subcommittee on statistical tables to prepare an index and guide to existing tables in the field of probability and statistics with the following initial membership:

- W. G. Cochran, Iowa State College (now at Harvard)
- A. T. Craig, State University of Iowa
- C. Eisenhart, University of Wisconsin (now at the National Bureau of Standards)
- W. A. Shewhart, Bell Telephone Laboratories (now retired)
- S. S. Wilks, Chairman, Princeton University.

The work of this subcommittee was interrupted by World War II almost before its work was organized and under way. With the resumption of its work after the War, two of its members (Professor Craig and Dr. Shewhart) resigned. W. Feller of Cornell (now at Princeton) and P. G. Hoel of the University of California at Los Angeles were appointed as members of the subcommittee.

During the period 1946-1949 most of the preliminary work of searching the literature and recording on cards information concerning statistical tables then in existence was done by graduate students under the supervision of the members of the subcommittee. In 1949 the subcommittee was extremely fortunate in attracting the interest of Dr. H. O. Hartley of University College, London (now at Iowa State University at Ames) in this project. He came to Princeton for a six-month period to work full time in organizing the material which had been collected and drafting it into manuscript form. Dr. Hartley made considerable progress during this period, establishing the overall structure and coverage of the *Guide*, but much work remained to be done. During the summers from 1950 to 1958 Dr. Hartley continued this work as time permitted with the assistance of graduate students at Princeton and Ames. In 1958 he recommended that, in order to bring this enormous task to completion in a reasonable time, it would be necessary for a competent coauthor to work full-time for at least a year.

In mid-1959 Dr. J. A. Greenwood undertook a full-time effort to draw the available material into final form.

After the termination of the Committee on Bibliography of Mathematical Tables and Aids to Computation, together with its subcommittees in 1949 with the retirement of its Chairman, R. C. Archibald, the Guide was continued under the sponsorship of the Committee on Applied Mathematical Statistics of the Academy-Research Council. This Committee had been established by the Academy-Research Council in 1942 to "provide statistical advice and information on problems concerning statistical personnel and organization, and advice on technical statistical problems in research, development, testing and production, upon request" by government agencies, particularly by the war-time and military agencies. The Committee on Applied Mathematical Statistics had the following initial membership:

    L. P. Eisenhart, Chairman, Princeton University (now retired)
    E. U. Condon, Westinghouse Research Laboratories
                                    (now at Washington University)
    L. J. Reed, Johns Hopkins University (now retired)
    W. A. Shewhart, Bell Telephone Laboratories (now retired)
    H. M. Smallwood, United States Rubber Company (now deceased)
    J. M. Stalnaker, College Entrance Examination Board
                (now with the National Merit Scholarship Corporation)
    S. S. Wilks, Princeton University.

In 1956 this Committee was discharged at its own request. Upon its recommendation, the present Committee on Statistics was established in 1957 in the Division of Mathematics of the Academy-Research Council with functions similar to those assigned to the Committee on Applied Mathematical Statistics.

In the manner outlined above, the present Committee on Statistics of the Division of Mathematics has inherited the sponsorship of the Guide to Tables in Mathematical Statistics. The Committee is pleased to serve as the final sponsoring Committee in the Academy-Research Council for this monumental piece of work, which Dr. Greenwood and Professor Hartley have so effectively prepared. On behalf of its past and present members, and also on behalf of the many members of its two predecessor Committees, this Committee expresses its deep appreciation to the authors for the great service they have thus rendered to the mathematical and statistical community.

The Committee believes that statisticians everywhere will find this <u>Guide</u> indispensable. Since probability and statistical methods have become so widely used in science and technology the Committee also believes that scientists and technologists in many fields will find it a valuable reference to the widely scattered tables of probability and statistics needed in their work.

As a member of the two predecessor sponsoring Committees, the Chairman of the present Committee takes this opportunity to express his appreciation for the valuable advice given by Professor E. S. Pearson of University College, London, and Dr. J. Wishart (deceased) of Cambridge University, concerning the <u>Guide</u> at the formative stage of its development. Professor Pearson was most cooperative in arranging for a six-month leave of absence for Dr. Hartley to begin his work on the <u>Guide</u> in 1949. The Chairman is also pleased to express his warm appreciation to his colleagues at Princeton, F. J. Anscombe and J. W. Tukey, for their continuing interest in and advice on this project.

Finally, the Committee wishes to acknowledge the continuing financial support of the Office of Naval Research to statistical research at Princeton University which has made the preparation of this <u>Guide</u> possible.

Committee on Statistics
Division of Mathematics
National Academy of Sciences—
National Research Council

June 5, 1961

T. W. Anderson
David Blackwell
R. A. Bradley
T. E. Caywood
I. R. Savage
Jacob Wolfowitz
S. S. Wilks, <u>Chairman</u>

PREFACE

The present Guide is a sequel to the guides to mathematical tables produced by and for the Committee on Mathematical Tables and Aids to Computation of the National Academy of Sciences—National Research Council of the United States, viz: number theory (D. H. Lehmer 1941); Bessel functions (Bateman & Archibald 1944); elliptic functions (Fletcher 1948). As Professor Wilks has explained on p. xxxiij of the Foreword, that Committee was discharged in 1949; to our best knowledge, no further surveys of mathematical tables are in progress in the Academy-Research Council.

During the years between 1952, when the first Sections went to the typist, and 1961, when this Guide was delivered to the printer, statistical table-making did not stand still. In addition to the annual accrual of tables in Journals, a number of important volumes containing tables appeared: notably Pearson & Hartley 1954, and the second editions (both 1957) of Cochran & Cox and Dixon & Massey. In an attempt to balance the desiderata of cataloguing these late entries and maintaining a flow of copy to the typist, the following procedure was adopted:

Beginning in September of 1959, we took the 16 major Sections of the book in hand, one at a time. After reviewing the script of a section, and making those additions that seemed desirable, we closed the section and sent it to be typed; any relevant matter coming to our notice after closing was filed for eventual inclusion in Appendix I. The sections were closed in the order 1-3, 5, 13, 4, 6-12, 14, 16, 15.

In March of 1960 Dr. Churchill Eisenhart, of the Statistical Engineering Laboratory, National Bureau of Standards, lent us a file of abstracts, reviews, and catalogue entries concerning statistical tables, which Mrs. Lola S. Deming of that laboratory had compiled from MTAC and Mathematical Reviews. We have incorporated in this Guide items from that file that we had not previously noticed, in three ways: (1) a line or narrative entry, prepared upon examination of the item; (2) a review or description, at length or abridged, from MTAC; (3) for some works reviewed in Mathematical Reviews and not in MTAC, only bibliographical information with a tentative statement of our decimal classification of the tables and a page reference to Mathematical Reviews.

Appendix I was compiled in 1961, in strict numerical order of sections, by combining: entries discovered after closing; relevant tables, not previously covered, from the books treated in Appendix II; certain reviews and descriptions from MTAC, as mentioned above.

In 1958 Silvio Vianelli, of the University of Palermo, solicited permission from many statisticians to reprint tables that they had published; both Hartley and Greenwood were among the authors approached. The tables so reproduced; examples of their use, often abridged from the author's accounts and translated into Italian; and some new tables of Sicilian manufacture; make up a substantial volume (Vianelli 1959). We saw this volume late in 1960, when the body of our text was mostly in typescript. A systematic check of Vianelli discovered several tables that he had reprinted and that our previous search had not found; we have catalogued these tables in Appendix I.

We thank the following authors and proprietors for their generous permission to use published material:

G. E. P. Box and the Biometrika trustees for the quotations from Box on pp. 264-265 and 295-296 below.

Churchill Eisenhart and Lola S. Deming, of the Statistical Engineering Laboratory, U. S. National Bureau of Standards, for the free use we have made of N. B. S. 1952 in preparing sections 1.1 and 1.2 of this Guide, as more fully set forth in the Introduction.

A. Fletcher, J. C. P. Miller, L. Rosenhead, and Scientific Computing Service Ltd., for our extensive borrowing from Fletcher et al. 1946, as set forth in Appendix III.

Professor M. G. Kendall and Charles Griffin & Co. Ltd., for the quotation from Kendall's Rank Correlation Methods (2d edition, 1955) on p. 356 below, and the quotation from Kendall's The Advanced Theory of Statistics (Vol. 1, 1943 edition) on p. 390 below.

The editorial committee of Mathematics of Computation for certain reviews and descriptions of statistical tables from MTAC, reprinted on pp. 633-735 below.

Professor J. W. Tukey and the editor of <u>Operations Research</u> for the quotation from Tukey on p. 458 below.

Mr. Robert White, when a graduate student at Iowa State College, prepared an elaborate draft--amounting almost to a monograph--of the section on Experimental Designs. Messrs. D. M. Brown, B. D. Bucher, J. A. Lechner, C. R. Ohman, D. R. Wallace, E. F. Whittlesey and T. H. Wonnacott, when graduate students at Princeton, contributed to the progress of this <u>Guide</u>. Professor Cochran, when a member of the subcommittee named on p. xxxiv of the Foreword, enjoyed the services of R. L. Anderson; Dr. C. Eisenhart, the services of K. J. Arnold. We acknowledge the efforts of all these workers.

We thank Dr. Churchill Eisenhart and Mrs. Lola S. Deming of the U. S. National Bureau of Standards, for putting the aforementioned file of abstracts etc. at our disposal; our colleagues on the Committee on Mathematical Tables of the Institute of Mathematical Statistics, and on its subcommittees, for voluminous comments on several Sections of this <u>Guide</u>; and the authors of certain tables who have endeavoured to clarify our abstracts. Among these authors we may particularly mention S. S. Gupta, R. B. Murphy, and Milton Sobel, who contributed extended <u>viva voce</u> discussions; and C. I. Bliss, who forwarded for inspection a copy of his scarce table of the angular transformation (Bliss 1937 a).

Professor S. S. Wilks, as Director of the Section of Mathematical Statistics at Princeton University, and director of successive projects under the Office of Naval Research, has expended much time, over many years, in promoting and facilitating work on this <u>Guide</u>. Without his persistent effort it would never have come into being.

The works cited herein were examined at the Library of Congress, the library of Iowa State University of Science and Technology, the New York Public Library, the libraries of the New York Academy of Medicine, the Engineering Societies, Princeton University, Rutgers University (including the New Jersey Agricultural Experiment Station), the Academy of Natural Sciences (Philadelphia), and the University of Pennsylvania. The intelligent and patient cooperation of the staffs of these institutions is gratefully acknowledged.

The research ensuing in this <u>Guide</u> was generally supported, from November 1946 to March 1956, by ONR Contract Nonr 270, Task Order I; and, subsequently, by ONR Contract Nonr-1858(05), Project ONR 042-023; monitored by the U. S. Office of Naval Research. The preparation of the catalogues of combinatorial designs in Section 15 was supported by contract no. DA 36-034-ORD-2297 monitored by the U. S. Army Research Office. The work on Section 15 was also supported by Bell Telephone Laboratories, Incorporated.

During the extended period from 1952 to 1961, a succession of harassed typists worked over this <u>Guide</u>; among them, we may acknowledge Mmes. Verna Bertrand, Maxine Bogue, Shirley Gilbert, Peggy Mott, Lucille Myers and Barbara Nelson. The line drawings on pp. 1-245 were executed by Mr. R. F. Westover. The fragments so produced were assembled into reproducible pages by Mrs. Marilyn Cohan.

The Subject Index leans heavily on the set of subject-entry cards relentlessly compiled by Miss Dorothy T. Angell. The accuracy and completeness of the Author Index and Subject Index have profited from extensive checking by Mmes. Lorie Brillinger, Marilyn Cohan and Jean Doten; the errors and omissions remaining are our own.

Finally we must attempt to indicate our deep indebtedness to John Tukey, who, far beyond his official duties as professor at Princeton, director of the Army Research Office project, and promoter of the Committee on Mathematical Tables of the Institute of Mathematical Statistics, has labored unceasingly to expedite and harmonize our work.

INTRODUCTION

As mathematical statistics is a younger discipline than mathematical analysis generally, so the tables reviewed herein are, on the average, of more recent date than those treated in a guide to general mathematical tables. The oldest mathematical table of current, as distinct from historical, interest is perhaps the 15-decimal canon of natural sines in Briggs & Gellibrand 1633 (now fortunately available in modern dress: N.B.S. 1949 z); the oldest table of statistical relevancy—but not in a form convenient to modern statistical usage—is Kramp 1799; say half as old.

Fletcher et al. 1946, p. 1, stated that "few extensive tables can have been made in the last 25 years without mechanical aid." In repeating this statement we would point out that since 1946 the very meaning of the words <u>computing machines</u> has radically changed. Where a K. Pearson, a Comrie, or a Lowan marshalled relays of human operators at crank-driven machines, the statistical computer today programs—not always wisely—large electronic machines.

The chief published sources of information about statistical tables may be classified as the journal MTAC; compiled volumes of statistical tables; catalogues of general mathematical tables; lists devoted to specific subjects.

The quarterly journal MTAC, founded in 1943 as <u>Mathematical Tables and Aids to Computation</u>, and known from 1944 to 1959 as <u>Mathematical Tables and other Aids to Computation</u>, is now entitled <u>Mathematics of Computation</u>. Numbers 1 to 74 (1943-1961) contain 477 reviews of and catalogue entries for statistical tables.

The text accompanying many of the volumes listed in Appendix II contains lists of sources; we have drawn on the lists in E.S. Pearson & Hartley 1954, K. Pearson 1930 a, 1931 a, Jahnke & Emde 1960, Vianelli 1959 (beware of misprints). Davis 1933, 1935 give copious references to earlier tables of interpolation coefficients, gamma and polygamma functions, and least-squares fitting of polynomials.

Easily the largest single source of information for this Guide was Fletcher et al. 1946, i.e. An Index of mathematical tables by Fletcher, Miller & Rosenhead. We are happy to record their cooperation and that of Scientific Computing Service, their publishers. Many articles in this Guide, treating tables ancillary to statistics, have been severely abridged, and supplemented with a rubric directing the reader to Fletcher for further details. A tabular statement of our further indebtedness will be found in Appendix III. Their coverage of statistical tables proper does not extend much beyond tables of the normal distribution and functions reducible to incomplete Eulerian integrals; they state (Fletcher et al. 1946, pp. 14-15): "We have not indexed . . . the more technical kinds of statistical table."

We have made occasional use of Lebedev & Fëdorova 1956 and its English version Lebedev & Fëdorova 1960. The Russian authors disclaim (Lebedev & Fëdorova 1960, p. iij) "any description of tables on the theory of numbers, mathematical statistics, astronomy and geodesy"; in fact their coverage of statistical tables is roughly coextensive with Fletcher's though less precise. For example, they describe (incomprehensibly) the χ^2 table of Sluckiĭ 1950 on p. 157. Their imperfect classification of Fresnel integrals is noted on p. 127 of this Guide. The supplemental volume (Burunova 1959, 1960) came out too late to be of great use to us; we note with interest that (Burunova, p. 49) the Fresnel muddle continues and that (Burunova, pp. 44, 45, 46, 160; catalogue nos. [35], [36]) the Russians seem to have reprinted N.B.S. 1941, 1942. We believe that an entry in a Russian handbook is good evidence for the existence of the table catalogued, but should not be taken without checking as a definitive description of such table. We are aware of the existence of Davis & Fisher 1949 and Schütte 1955, but have not examined them.

The organization and typographical arrangement of §§1.1-1.2 of our Guide, their use of line drawings to indicate the region of integration, and some individual entries, derive recognizably from N.B.S. 1952 (Guide to tables of the normal probability integral). Section 9.4, dealing with

Smirnov's and related tests, was mostly based on Darling 1957. We call attention to the catalogues of I.R. Savage 1953 on non-parametric tests and of Mendenhall 1958 on life testing; these works do not, however, single out papers containing tables.

Scope

It is not easy to say what the scope of this <u>Guide</u> is. Let us at once warn anyone who was not already deterred by the word <u>mathematical</u> in the title that it will not provide references to tables of statistical data arising from official or unofficial censuses and surveys. Nor have tables relating to statistical mechanics or quantum theory been sought for: those that have been listed are here because the functions tabled are identical with, or closely related to, functions used by statisticians; see Sherman & Ewell 1942, and the papers by Kotani and his colleagues. Again, while we have noticed the Legit transformation, Cavalli's transformation, and segmental functions, the more recondite tables of mathematical genetics are outside our scope.

We do not treat actuarial tables. For compound interest and for annuities certain, we can recommend Kent & Kent 1926 (good typesetting) and Gushee 1942 (all functions for one interest rate at one opening). Finite differences is a specialist subject that we have not attempted to cover fully; we have given some references to graduation methods in §11.1, and to interpolation and numerical integration in §16.23. Among tables which might be included in mathematical statistics, we are aware of two wholesale exclusions: tables relating to queueing theory; and tables of the <u>results</u> of random sampling experiments, as contrasted with tables of random numbers for <u>use</u> in such experiments.

This <u>Guide</u> is clearly not complete. Such questions as what functions to include, and whether a list of formulas ranks as a table, were settled article by article; consistency would have been difficult to achieve at best, and the difficulty was doubled by the method, alluded to in the Preface, of closing off sections one at a time. Our coverage is strongly dependent on date of publication: roughly, up to 1948 we made a comprehensive search; from 1949 to 1954 we think we have covered the leading

American and English statistical journals; thereafter, tables included were either known personally to us, recommended by a colleague, taken from Mrs Deming's catalogue, or discovered while checking through Cochran & Cox 1957, Darling 1957, Dixon & Massey 1957, Fisher & Yates 1957, Mendenhall 1958, Siegel 1956 and Vianelli 1959.

The tables catalogued in §§1-12 give (approximate) values of quantities capable of mathematical definition; this definition, for some tables, would cast little light either on what the quantity tabled was used for or on how to compute it. The random numbers in §13 are <u>ex hypothesi</u> not capable of mathematical definition; we state such facts about the origin of these numbers, and the tests they have passed, as may permit the reader to use them with confidence; but we face the paradox that the pseudo-random numbers of §13.62, which are generated by specifiable mathematical processes, are used with equal or greater confidence for many statistical purposes. The tables of acceptance sampling plans in §14, while based on an underlying distribution (hypergeometric, binomial, Poisson, normal or exponential), frequently embody empirical elements as well as clearly defined parameters. The tables of combinatorial patterns in §15 are lists, complete or incomplete, of solutions to mathematically specifiable problems. Finally, §16 gives an eclectic and even personal selection of mathematical tables; it could be subtitled "Mathematical tables that we think the statistical computer should know about."

As a rule we have ignored graphs, nomographs and abacs. The chief exceptions are the graphs etc. in the books listed in Appendix II. When, as with E.S. Pearson & Hartley 1951, the most usable table of some function <u>is</u> a graph, such graph is duly catalogued. Koller 1943 gives a large collection of abacs for common significance tests; Dunlap & Kurtz 1932 contains nomographs adapted to the statistical methodology of their time, with emphasis on mental measurements.

Most textbooks of statistical methodology contain some tables, often abridged from Fisher & Yates and other standard sources; we have included those that came to hand. The inclusion of a textbook in the Author Index of this <u>Guide</u> says neither that it is better nor worse than another textbook that does not appear; nor does inclusion guarantee that we have catalogued all the tables in a textbook, or even the most interesting tables.

Arrangement

We describe the arrangement of this work under six heads: the Table of Contents; the body of the text and Appendix I; Appendix II; Appendix III; the Author Index; the Subject Index.

The Table of Contents sets forth all the section headings used in the body and Appendix I. These are arranged on the decimal system: the dot in the middle of 13.62 is a decimal point; §9.118 precedes §9.2. The following abridged list of headings of major Sections may serve as an introduction to the Table of Contents:

1. The normal distribution
2. The chi-squared and Poisson distributions
3. The beta and binomial distributions
4. The t-, F- and z-distributions
5. Various discrete distributions
6. Likelihood-test statistics
7. Correlation, mostly product-moment
8. Rank correlation; asymptotic theory of extreme values
9. Non-parametric tests
10. Frequency curves; symmetric functions
11. Regression and other curves
12. Variate transformations
13. Random numbers
14. Quality control
15. Design of experiments, etc.
16. Sundry mathematical tables

Whenever possible we have given section numbers ending with 0 to introductory articles; and with 9, to articles cataloguing miscellaneous or unclassified tables.

The first column of page numbers in the Table of Contents refers to the body of the text; the second column, to Appendix I. If a number appears only in the first column, the article appears only in the body; if only in the second column, only in Appendix I. If no page number appears, then there is no article with that section number and name; the number and name appear, for the purpose of clarifying the hierarchy of subsections, on the first page cited in the first column below the blank or blanks.

The text matter is divided into a body and Appendix I; the method of division is sketched in the Preface, p. xxxvij. We regret the inconvenience incident to this division, and have attempted to mitigate it by copious cross-references; these are to be found at the top of each left-hand page in the body and under each section heading in Appendix I.

Whenever the matter permitted, we have adopted the columnar arrangement discussed by Fletcher et al. 1946, pp. 8-12. In discussing this arrangement we shall follow Fletcher; dwelling on those details where our usage departs from their standard.

Each item normally occupies one line, and gives, in order from left to right, information about:

(1) number of decimals or figures;
(2) the function tabled, if the article is slightly heterogeneous;
(3) interval and range of argument or arguments;
(4) facilities for interpolation, if any;
(5) author, date, etc.

We have ordinarily arranged items in order of number of decimals (dec.)* or figures (fig.)*; the principal exception is the tables of percentage points of χ^2, t, F and z, where variations in how many and what percentage points are tabled are more important to the user than variations in the number of figures given. "0 dec." denotes approximate numbers rounded to the nearest integer. "5 dec. or 2 fig." signifies that 5 decimals are tabled, except that, when the function is less than 10^{-4}, enough additional decimals are given to make 2 significant figures. Some authors commendably give one figure in excess of the nominal accuracy when the leading digit is 1; we have taken no account of this practice in our listings. Values stated to the nearest 1% have been described as to 2 dec.; to the nearest 0.01% as to 4 dec., etc. When function values are described as "exact vulgar fractions", both numerator and denominator are given some-

---

*A number of reviews, reprinted unaltered from MTAC, use the abbreviations D for decimals and S for significant figures.

where on the page cited; perhaps the commonest arrangement is a row or column of numerators followed by a common denominator.

When an upper tail sum or area begins with a string of zeroes after the decimal point, the complementary lower tail sum or area begins with a string of nines. The description of many tables is simplified by treating such initial nines as non-significant. A headnote reminds the reader when we have used this convention; note that in §1.1 we have followed the practice of Fletcher et al. 1946, Section 15, and of N.B.S. 1952, and have counted initial nines as "figures".

Some articles, such as §12.13, group several closely related functions; we then state in each entry the function tabled.

In argument columns, to save space, we have often moved the literal argument into the column heading; when the literal argument of a function of one variable is entirely clear from the context, we have omitted it. We have followed the convention of indicating argument intervals within brackets. Thus "0(.01).99(.001)1" means every hundredth from 0 to .99, thence every thousandth to 1. Many tables of percentage points of F and related statistics facilitate interpolation by taking arguments in harmonic progression, a practice introduced by Fisher & Yates; we describe such a sequence by translating it into an arithmetic progression. Thus if a table has arguments $\nu = 1(1)10(2)20, 24, 30, 40, 60, 120, \infty$, our description will read $\nu = 1(1)10(2)20, [120/\nu = 0(1)6]$. To facilitate comparisons, we use the common numerator 120, even where a smaller numerator would suffice. Arguments not in readily describable progression are listed in extenso and are separated by commas; these commas must not be confused with the semicolons discussed below.

To avoid repetition in the listings, we have used braces and semicolons. For example, the braced entry

$$\left. \begin{array}{lll} 12 \text{ dec.} & 1(.01)10 & \delta^4 \\ 12 \text{ dec.} & 10(.1)100 & \text{No } \Delta \\ 10 \text{ dec.} & 0(.0001)1 & \Delta^2 \end{array} \right\} \text{Davis 1933}$$

in §2.5131 on p. 162 specifies that H.T. Davis, Tables of the higher mathematical functions, Vol. I, contains tables of $\log_{10} x!$ to 10 dec. for $x = 0(.0001)1$ with first and second differences; to 12 dec. for

$x = 1(.01)10$ with second and fourth differences; to 12 dec. for $x = 10(.1)100$ without differences. The use of semicolons (except in §§1.1-1.2, where a different convention, mentioned below, governs) can be explained by taking an entry containing semicolons and translating it into braced notation. For example, the first entry in §7.212 on p. 331 reads

$$\left.\begin{array}{ll} 20 \text{ dec.} & 0(.00001).001 \\ 10; \ 9 \text{ dec.} & .001(.0001).0999; \ .1(.001)1 \end{array}\right\} \quad \text{No } \Delta \quad \text{Hayashi 1926.}$$

If rewritten without semicolons, it would read

$$\left.\begin{array}{ll} 20 \text{ dec.} & 0(.00001).001 \\ 10 \text{ dec.} & .001(.0001).0999 \\ 9 \text{ dec.} & .1(.001)1 \end{array}\right\} \quad \text{No } \Delta \quad \text{Hayashi 1926.}$$

The two forms are synonymous; either specifies that Hayashis Sieben- und mehrstellige Tafeln give $\tanh^{-1} x$ to 20 dec. for $x = 0(10^{-5}).001$; to 10 dec. for $x = .001(.0001).0999$; to 9 dec. for $x = .1(.001)1$; all without differences.

Many tables of the area under the normal distribution increase the number of decimals given as such area approaches its limiting value, i.e. as the abscissa approaches infinity; many tables of the abscissa as a function of the area decrease the number of decimals given as the abscissa approaches infinity. Fletcher et al. 1946, p. 212, introduce a special semicolon convention, which we adopt, in §§1.1-1.2, wherever appropriate, to indicate such changes; for example, our entry for Yule & Kendall on p. 8,

4; 5, 6, 7 dec.   0(.01)3.99; 2.5, 3.1, 3.73   No Δ,

can be expanded to read

$$\left.\begin{array}{ll} 4 \text{ dec.} & 0(.01)2.49 \\ 5 \text{ dec.} & 2.5(.01)3.09 \\ 6 \text{ dec.} & 3.1(.01)3.72 \\ 7 \text{ dec.} & 3.73(.01)3.99 \end{array}\right\} \quad \text{No } \Delta.$$

The column giving facilities for interpolation is only occasionally used in this Guide. Unless specifically stated to the contrary, tables with two or more arguments are not supplied with differences; the most conspicuous exceptions are K. Pearson's and Sluckii's tables of the incomplete Γ-function. When the differences supplied vary between parts of a table, we have used braces or semicolons; cf. the Davis entry on p. 162 discussed above. The symbols concerning differences are listed on the page facing p. 1.

The arrangement of the author column is: initials, if given; surname or surnames; date, and, if used, reference letter; page or table number, if given.

We have often given initials when there is another statistician of the same name as the author cited, even when the homonym has not published tables; but in this column we have listed the Biometrika tables for statisticians of E. S. Pearson & H. O. Hartley simply as "Pearson & Hartley 1954". Three corporate authors have been cited by initials: "A.S.T.M." for American Society for Testing Materials; "B.A." for British Association for the Advancement of Science; "N.B.S." for [U.S.] National Bureau of Standards.

Two joint authors have normally been cited by two surnames with ampersand between; three or more by the senior author et al., except where the senior author was K. Pearson. Throughout this Guide, the ampersand is used only between joint authors and in the names of publishing companies. In order to bring closely related papers together, Fletcher et al. 1946, pp. 219, 222, catalogue two papers by Krause & Conrad as by Conrad, Krause &; we have followed them, but have not used this device elsewhere. In narrow columns the junior author may suffer abbreviation: as, Kaarsemaker & v.W. for Kaarsemaker & van Wijngaarden.

The date and reference letter are the key to locating an item in the Author Index, and will be treated in discussing the Author Index below.

We have given page or table numbers, in brackets, whenever convenient; omission of such numbers signifies that the cited table can readily be found without a page reference, but not conversely. Table numbers are preceded by the letter T.; page numbers are unmarked. Table numbers given in Roman numerals have been translated into Arabic with two chief exceptions. Burington & May use three-part Arabic numerals for tables interspersed in their text, and Roman numerals for tables grouped at the end of the book. Fisher & Yates use Roman numerals for main tables (or tables present in the 1938 edition) and suffixed Arabic numerals for secondary tables (or later additions).

The grouping of several authors after one entry is done to save space, and does not necessarily imply that the author first cited is the source for the others.

Many statistical tables are not readily described in terms of well-known mathematical functions. In cataloguing such tables we find line entries unworkable, and have devoted a short article to each table, or to a group of closely related tables contained in the same publication; the arrangement of articles within a section is normally chronological. Each article begins with the author, date etc. of either the original publication or the most accessible publication. The article may contain a brief discussion of the problem germane to the tables catalogued (when possible, we have put such discussion in articles introductory to a section); it will contain a list of the contents of each table, with relevant table and page numbers.* Reprints of tables, both identical and with minor variations, are listed at the end of articles; substantially altered reprints are accorded separate articles. In cataloguing reprints the following abbreviations are common:

Tables for S & B I, II = Tables for statisticians and biometricians: Part I, Part II; i.e. K. Pearson 1930 a, 1931 a.

Pro., Pri. = _Prontuario_, _Prontuari_; i.e. the 300 major tables in Vianelli 1959.

Appendix II was prompted by a policy statement by Fletcher et al. 1946, p. 13:

"We have indexed practically all tables contained in standard collections such as those of Milne-Thomson & Comrie [1931], Jahnke & Emde [1909 etc.], and the latter part of Dale [1903], because we think that any user of this _Index_ [of mathematical tables] will wish to be reminded if a table which he requires is contained in volumes which he probably has at hand."

Feeling that the statistician may well need a similar reminder, we have given in Appendix II annotated tables of contents of a number of volumes of tables, somewhat widely and arbitrarily chosen. Specifically, Appendix II treats

---

*Reviews from MTAC give a statement of inclusive page numbers in a headnote, but omit page references to individual tables.

Arkin & Colton 1950	Jahnke & Emde 1909 etc.
Burington & May 1953	Kelley 1938, 1948
Czechowski et al. 1957	Kitagawa & Mitome 1953
Dixon & Massey 1951, 1957	Lindley & Miller 1953
Fisher & Yates 1938 etc.	E. S. Pearson & Hartley 1954
Glover 1923 (1930)	K. Pearson 1930 a, 1931 a
Graf & Henning 1954	Siegel 1956
Hald 1952 (1948)	Vianelli 1959

Each line entry in Appendix II comprises the table number, if any; page numbers; title; and references to the text (body and Appendix I) in the form used in the Author Index and discussed below. The contents are complete, and titles of tables (not inclosed in brackets) are the authors' or literal translations thereof, with the following exceptions:

Burington & May 1953. Eclectic coverage of the tables scattered in Chapters 1 to 17; complete coverage of the 23 numbered tables at the back of the book.

Czechowski et al. 1957. Titles of TT. 30, 31 were literally translated from the Polish; other titles reflect our own assessment of their contents.

Glover 1923. Complete coverage of pp. 392-675. The first half of the book comprises actuarial tables. These were compiled for classroom use; and include logarithms of compound interest and annuities certain, and tables, based on combinations of mortality and interest that never prevailed in U. S. practice, of annuities etc.

Jahnke & Emde 1909 etc. Only tables of statistical interest are covered; we call attention to the tables of Bessel functions, but give no details.

Vianelli 1959. Almost complete coverage of the 300 Prontuari; we omit 4 tables from queueing theory and 20 actuarial tables. Eclectic coverage (28 out of 263) of the supplementary tables; the titles for these reflect our own assessment of their contents.

Appendix III lists, in order of our section numbers, material common to this Guide and Fletcher et al. 1946. In general, we have drawn on Fletcher directly in preparing §§1.12, 1.22, 1.3, 2.51, 3.21, 3.5, 4.1422, 5.31, 7.21, 11.2, 11.3, 14.152, 16; we have drawn on Fletcher, both directly and as reflected in N.B.S. 1952, in preparing §§1.11, 1.21.

The Author Index contains an entry for every work cited by author and date anywhere in this Guide. The parts of an entry are:

(1) name of author or authors;
(2) date and reference letter;
(3) title;
(4) facts of publication;
(5) citations to text;
(6) citations to Introduction.

The normal form for an author is surname and initials. We have supplied given names, when ascertained: for women; to avoid confusion between identical or similarly pronounced surnames; and where inaccurate statements of given names are on record. We have preferred citing a personal author to citing a corporate author: thus we cite Birnbaum et al. 1955; Eisenhart, Hastay & Wallis 1947; Freeman, Friedman, Mosteller & Wallis 1948; Meyer 1956; where a librarian could prefer Biometrika 1955; Columbia Univ. 1947, 1948; Florida Univ. 1956. Cross references from the forms not used are provided.

The following particulars may be noted in the alphabetization of authors: All diacritical marks have been disregarded. The prefix Mc is treated as if spelled Mac; the prefixes de, van, and von have been generally alphabetized under d and v (but v. Bortkiewicz is treated as Bortkiewicz). Double surnames, both English and Spanish, have generally been entered under the first element. Joint papers are alphabetized under the name of the senior author (i.e. the author named first on the title page), and follow that name as sole author; among joint authorships with a given senior author, the order is alphabetical by junior author: the exception to this rule is the two papers by Krause & Conrad, discussed above. Russian and Ukrainian names are transliterated by the system of Mathematical Reviews.

Cross references, interspersed with the main author listings, are of two kinds. References from a junior author to the complex of joint authors; from a corporate author to the personal author cited; from variant forms, especially of married women's names, to the form appearing on title pages; from the form or forms of double or prepositional names not used to the form used; from plausible alternative Cyrillic transliterations to the Mathematical Reviews form; from variant spellings of Asiatic names to the form we have arbitrarily adopted; consist of SEE (in small capitals) followed by the author or authors cited, and refer implicitly to all papers by such author or authors. References from the names of actual computers, which we have inserted whenever we have observed the credit line; and references from the names of translators, or other contributors who do not rank as joint authors; consist of SEE followed by author or authors and date (and reference letter if necessary), and refer only to the work designated. Cross references from a name follow citations, if any, to that name as sole author; and precede citations, if any, to that name as senior author.

We have little to add to the admirable account of the difficulties of dating given by Fletcher et al. 1946, pp. 5-8. In general we have dated British Association reports by the spine, other books by the title page, and journals by the wrapper when available. For examples of dating by secondary sources, see the footnotes to Lagrange 1774 and Tallqvist 1908. When a second edition of a textbook containing tables differs materially in contents, especially in pagination, from the first edition, we have often made separate entries for the two editions. An asterisk before a date does not query the date: it signifies that we have not seen the work; so that both the entries in the text, if any, and the entry in the Author Index are from secondary sources. The omission of such asterisk guarantees that the entry in the Author Index was written with the work before us; the entries in the text, if any, may have been taken from Fletcher et al. 1946 or from MTAC.

Reference letters have been attached whenever we cite two or more works of the same date and authorship. The only principle informing our use of letters is that the combination authorship-date-letter must uniquely identify a work: two works of one year may be cited either as 1930, 1930 a or as 1930 a, 1930 b; when an item was cancelled at a late (post-1958) stage of editing, we did not, in general, readjust letters; we sometimes use the letter y or z, for an item added at a late stage, to insure against collision with letters previously assigned.

Titles of journal articles are given in full; occasionally we have been impatient with the length of a book title, and abridgement is then shown either by ellipsis marks or by the abbreviation etc. We have tacitly supplied colons between the title and subtitle of books; all other editorial additions to titles are inclosed in square brackets. Round brackets set off words used in the running head of (right-hand) journal pages but not in the title-and-author block at the beginning of the article.

The normal sequence of facts of publication for a journal article is: abbreviated name of journal, volume, inclusive pagination. We have followed the abbreviatory style of Mathematical Reviews 17, 1423-1435. Librarians are warned that in this style the words Bull(etin), J(ournal), Proc(eedings) etc. normally stand before the name of the issuing agency; the most conspicuous exception is Calcutta Statist. Assoc. Bull. We give fascicule numbers or other designations only when pages are not consecutively numbered through a volume. To aid in the identification of several obscure journals, we add footnotes referring to catalogue listings therefor.

The normal sequence of facts of publication for a book is: city, publisher. We add the state or country, where not obvious, in brackets after the city. For out-of-print books we have followed the title and copyright pages; we do not follow the library practice of bracketing information from the copyright page. For in-print books we amend the imprint (in square brackets) to give the publisher's current (1960) name and address. The word Verlag and analogous Scandinavian forms are omitted; Izdatel'stvo is abbreviated Izdat. Terms indicating the corporate form of the publisher are omitted, except that Mc Graw-Hill Book Co. is so cited to prevent confusion with Mc Graw-Hill (newspaper) Publishing Co. The word University is abbreviated, in the names of university presses, to Univ.; the city is omitted when the publisher is the Cambridge, Oxford or Princeton Univ. Press. For works in the English language, we have listed both a U.K. and a U.S. publisher whenever the information was available. The abbreviation etc. calls attention to further cities or publishers not listed; in particular, "Cambridge Univ. Press, etc." represents a long statement whose 1960 form is "London and New York: Cambridge University Press;

agents for Canada, Brett-Macmillan; agents for India and Pakistan, Macmillan."

Citations to the text are arranged in ascending order of page numbers. Each citation consists of a section number, a solidus (/), a page number, possibly a notation in brackets, and possibly a subscript letter. The following notations are used: an arabic numeral counts multiple line citations; (intro.) directs attention to the introductory article of a section; (ftn.) to a footnote. An author's name indicates that the citation occurs in the text of an article headed with that name; note that the headline may be on the page cited or the preceding page. The subscript letter F identifies citations to descriptions taken from Fletcher et al. 1946; the letter M, from MTAC.

Citations to the introduction consist of the page number, in roman numerals, possibly followed by an arabic numeral in brackets counting multiple citations to one page.

The <u>Subject Index</u> was constructed with two principles in view:

(1) Index entries shall not unnecessarily duplicate the Table of Contents.

(2) Subject to the restriction in (1), no informative permutation of an index entry was consciously excluded.

For example, let the reader be seeking information on tables of percentage points of the non-central t distribution. Plausible subject entries are:

> tables of percentage points of non-central t
> percentage points of non-central t
> non-central t, percentage points of
> t, non-central, percentage points of
> t-distribution, non-central, percentage points of

The Subject Index, entered at <u>tables</u>, yields merely the general injunction
> table (see under the function tabled)

Entered at <u>percentage points</u>, it yields
> percentage points
> . . . .
> of non-central t   224 (§4.1414)

[1v]

; indicating the page and section reference. Entered at <u>non-central</u>, it yields

        non-central t-distribution
            218-225 (§4.141*); 200, 477, 682

The asterisk calls attention to the footnote

        *This section is subdivided;
      see Table of Contents.

The Table of Contents indicates (p. xiv) that §4.1414, Percentage points of non-central t, begins on p. 224. Entered at <u>t</u>, the Subject Index yields

    t (Student's)   415
    . . . .
      non-central
        218-225 (§4.141*); 477, 682

If non-central t were treated only in §4.141, a subdivision of §4.1 (devoted to t), we would omit this entry; since, however, we must give two page references (477, 682) outside §4.141, we take advantage of the entry to give the principal reference as well.* Entered at <u>t-distribution</u>, the Subject Index yields

      t-distribution   207-244, 670-675
        (§4.1*); 320, 344, 389, 708

The asterisk has the purpose mentioned above; the subdivisions of §4.1 occupy some forty lines on pp. xiij-xiv. Further down the reader finds

      non-central
        218-225 (§4.141*); 226, 477, 682

; this entry appears for the same reasons as that under <u>t, non-central</u>.

The parts of an entry in the Subject Index are:

    (1-3) primary, secondary, tertiary entry words;
    (4-5) major page references to body of text, to Appendix I;
    (6)   section references;
    (7)   minor page references.

The hierarchy of entry words is indicated by indention: thus the entries

---

*For an example uncomplicated by minor references outside the section devoted to a function, we note that <u>logarithms of factorials</u> will be found in the Subject Index; but not <u>factorials, logarithms of</u>; which must be sought in the Table of Contents as a subdivision of factorials (§2.51).

> correlation   319-344, 695-699 (§7*)
> . . . .
>   serial   336-339 (§7.5*)
> . . . .
>   non-circular   338 (§7.52); 79

relate to correlation, serial correlation, and non-circular serial correlation. In reading a secondary entry, insert a comma after the primary entry words; in reading a tertiary entry, insert commas after the primary and secondary entry words: a literal reading of the last example is <u>correlation, serial, non-circular</u>. Exceptions to this rule are indicated by colons. A colon after a man's name is a possessive sign:

> Sheppard, W. F.:
> . . . .
>   quadrature formula   629

may be read <u>W. F. Sheppard's quadrature formula</u>. A colon after a secondary entry directs the omission of a comma in reading:

> approximation
> . . . .
>   beta, to distribution:
> . . . .
>   of likelihood test   297

is to be read <u>approximation, beta, to distribution of likelihood test</u>: i.e. beta approximation to the distribution of a likelihood test.

    The alphabetization of entry words is word by word, up to the first punctuation mark; thus

>   clear of cubic effects, estimates of
>   linear and quadratic effects
>   555

precedes

>   clear of cubic effects and blocks,
>   estimates of linear and
>   quadratic effects   557

Articles (when printed) and prepositions are regarded in alphabetizing; the possessive sign 's is disregarded. Hyphenated words are alphabetized by treating the hyphen as a letter preceding <u>a</u> in the alphabet. Substantive nouns printed in the singular are to be read in the plural, and vice versa, when context demands: for example,

> deviate
> . . . .
>   normal, ordered
>   62-68, 638-639 (§1.43); 49

may be read <u>ordered normal deviates</u>; and

    moments 425-426 (§10.32)
    . . . .
    fourth standard, as test of
     normality 113-114 (§1.62)

may be read <u>the fourth standard moment as a test of normality</u>, i.e. b_2. All substantive nouns, whether printed in the singular or in the plural, are alphabetized in the singular:

       normal deviates 2, 14, 63, 144
precedes
       normal deviate test 367

The word <u>distribution</u> is not used as a primary entry word; <u>functions</u>, <u>miscellanea</u>, and <u>tables</u>, are not used as entry words. We have given particularly full lists at the primary entries <u>approximation</u>, <u>moments</u>, <u>percentage points</u>, <u>tests</u>. We have distinguished the following homonyms among primary entries:

    cubes [Latin]; cubes of numbers

    factorial, i.e. combinatorial pattern (for experimental designs);
                 factorial, i.e. Γ-function

    loglogs [$\log_e \log_e x$]; lologs [$\log_{10} \log_{10} x$]

    squares, i.e. combinatorial patterns; squares, of numbers

    t (Student's); $t = r(1-r^2)^{-\frac{1}{2}}$; t [number of treatments]

    $z = \frac{1}{2} \log_e F$; $z = \tanh^{-1} r$; z (Student's)

The order of page and section references can be seen from the example

     balanced incomplete block patterns
      605-606, 736 (§15.241*); 566

; which indicates: that the major treatment of b.i.b. patterns is at pp. 605-606 in the body of the text, and p. 736 in Appendix I, these pages comprising §15.241; and that b.i.b. patterns are also treated on p. 566. The hyphen between page numbers is used only when the pages so joined make up one or more complete sections. The asterisk calls attention to the footnote, quoted on p. lvj above, about subdivision. When the asterisk, or a reference to consecutive sections, such as

     double tail percentage points
    . . . .
     of t 211-213, 671 (§§4.1311-4.1312)

is used, the Table of Contents will occasionally reveal that the consecutive articles cited include some that are not strictly relevant: thus the Subject Index reads

$$\text{correlation} \quad 319\text{-}344, \; 695\text{-}699 \; (\S 7^*)$$

although the Table of Contents shows that §7.6 is given over to covariance, covariance ratio, and regression. When two section numbers <u>not</u> in ascending numerical order, such as

$$\begin{array}{l}\text{exponential functions} \\ \quad 623\text{-}624, \; 738 \; (\S\S 16.101\text{-}16.103); \\ \quad 450 \; (\S 12.52); \; 364, \; 365, \; 402, \; 404\end{array}$$

are cited, we imply that the section cited second is less important than the section cited first.

We close by mentioning the journals that print abstracts of new statistical tables. As of 1961 these are:

    International Journal of Abstracts on Statistical Methods in Industry
    International Journal of Abstracts: Statistical Theory and Method
    Mathematical Reviews
    Mathematics of Computation [MTAC]
    Quality Control and Applied Statistics*
    Referativnyi Zurnal: Matematika
    Zentralblatt fur Mathematik

---

*In addition to abstracts, this monthly loose-leaf service often reprints tables at length.

ABBREVIATIONS

bal.	balanced incomplete blocks. See p. 605.
D, dec.	decimal places
d.f.	degrees of freedom
fig.	(significant) figures
Fig.	figure, i.e. a line drawing
G.P.O.	[U.S.] Government Printing Office
latt	square lattice. See p. 606.
L sq	lattice square (<u>not</u> Latin square). See p. 608.
mtx.	matrix
ortho.	orthogonal. See p. 502.
p., pp.	page, pages
part.	partial confounding. See p. 501.
Pro., Pri	Prontuario, Prontuari. See p. [fifty] above.
rect	rectangular lattice. See p. 606
S	significant figures
sat.	saturated. See p. 500.
S & B	Statisticians and Biometricians. See p. [fifty] above.
sq.	square
s.v.	<u>sub voce</u>, i.e. at the indicated catchword in a glossary etc.
T.	table (Tabelle, tablica, tavola)
TT.	tables (Tabellen, tablicy, tavole)
Y sq	Youden square. See p. 607

SYMBOLS CONCERNING MEANS OF INTERPOLATION

The list below is abridged from Fletcher et al. 1946, pp. 16-17.

No Δ	Plain table, giving values of argument and of function only.
Δ	First differences.
Δ^2	First and second differences.
Δ^n	First, second, ..., n^{th} differences.
$I\Delta$	Indication of first difference.
δ^2	Second differences.
δ^{2n}	Second, fourth, ..., $(2n)^{th}$ differences. If $2n$ is not replaced by a number, as many even differences as needed to interpolate the table.

¶ the subscript $_m$ in δ^2_m, δ^{2n}_m etc., indicates that some or all differences have been modified.

D	First derivatives.
D^n	Derivatives of the orders specified. If not specified, as many derivatives as needed to interpolate the table.
v	First variations, i.e. hD, where h is the tabular interval.
v^n	Variations $h^n D^n/n!$, as many as needed to interpolate the table.
fPPd	Full proportional parts, all tenths of all differences, to one decimal.
PPMD	Tenths of mean differences, rounded to nearest integer.
PP_6MD	Sixths of mean differences, rounded to nearest integer.

[1xij]

SECTION 1

THE NORMAL DISTRIBUTION

1.0 Introduction

As standard form for the normal or Gaussian frequency function, we adopt:

$$Z(x) = (2\pi)^{-\frac{1}{2}} e^{-\frac{1}{2}x^2}$$

which is the frequency function of a normal variate x having zero mean and unit standard deviation. The probability integral of x is then written as the lower tail area

$$P(x) = \int_{-\infty}^{x} Z(u)\, du$$

and as the upper tail area

$$Q(x) = \int_{x}^{+\infty} Z(u)\, du = 1 - P(x).$$

The relation between Z, P, Q and x is shown in fig. 1.1.

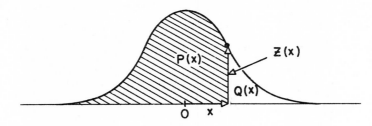

Fig. 1.1

The above notation is advocated by Pearson & Hartley 1954 (p. 1) and has found some acceptance among statisticians. Many mathematical physicists have tabled the 'error function', i.e. the integral of e^{-t^2}; we have listed such tables, which are less convenient for statistical use than direct tables of P(x) or Q(x), in section 1.12.

All functions occurring in sections 1.11-1.13 are expressed in terms of P, Q, Z defined above: for other notations, in particular for no less than five meanings attached to Erf x or erf x, see Fletcher et al. 1946, pp. 212-213.

1.0
1.112
The Normal deviate $x(P)$ is defined as the inverse of $P(x)$, i.e. is the root x of $P = P(x)$ for a given area P. In terms of this function we define:

Double tail percentage points $x(1 - \tfrac{1}{2}p)$ where $100p\%$ is the 'percentage level' (e.g. $p = .05$, $100p\% = 5\%$)

Single tail percentage points $x(1 - p)$. These are illustrated in Figures 1.2 and 1.3 below.

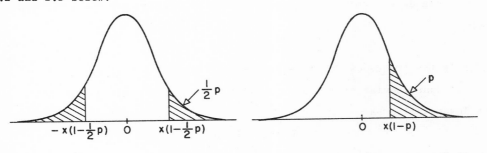

Fig. 1.2   Fig. 1.3

A special form of the normal deviate function is the 'probit' $Y = x(P) + 5$. Tables of the probits Y and other derived quantities used in probit analysis are given in section 1.23 where further definitions will be found.

1.1 Tables with the normal deviate x as argument

1.11 Area, log area and ordinate as functions of x

1.111 Central area from -x to +x

$2[P(x) - \frac{1}{2}] = \alpha_x$

(α_x is Sheppard's notation)

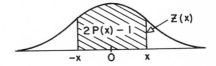

Fig. 1.4

15 dec.	0(.0001)1(.001)8.112	$\frac{1}{2}$D	N.B.S. 1942(2)*
5 dec.	0(.1).4(.05).7(.01) .8(.05)1.1, 1.5(.5)3.5	No Δ	Jones 1924(284)
5 dec.	0(.1)4	$\frac{1}{2}$D	Burington & May 1953(89)
4 dec.	0(.01)4.09	Δ	Dwight 1947(237)
4; 5, 6 dec.	0(.01)3.99; 2.6, 3.6	No Δ	Vianelli 1959(134)
4; 5, 6 etc. dec.	0(.01)3(.1)6; 4, 4.5 etc.	No Δ	Charlier 1906(43)
4 dec.	0(.01).3(.02)1.5(.05) 3.5(.1)3.8	No Δ	Graf & Henning 1953(T.2)
4: 5, 7 dec.	0(.05)3; 4, 5	$\frac{1}{2}$D	Dixon & Massey 1951(305), 1957(381)

1.112 Semi-central area from 0 to x

$P(x) - \frac{1}{2} = \frac{1}{2}\alpha_x$

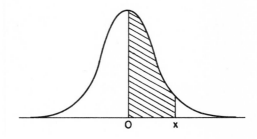

Fig. 1.5

---

* In the 1942 printing <u>only</u>, the arguments .2235 and .3070 are misprinted.

[3]

1.1121
1.1131

1.1121 $P(x) - \frac{1}{2}$ [see also p. 630]

6 dec.	0(.004)4.892	D^5	Harvard Univ. 1952(3)
6 dec.	0(.01)4.99	D^n, n = 1,3,4,5	Rietz 1924(209)
5 dec.	0(.001)2.5(.01)4(.1)5	IΔ, PPd	Davenport & Ekas 1936(166)
5 dec.	0(.01)5	D	Rosenbach, Whitman & Moskovitz 1943(187)
5 dec.	0(.01)4.99	D^n, n = 1, 3(1)9	Glover 1923(392)
5 dec.	0(.01)3.99	D	Hoel 1947(243), Kenney 1939 a (225), Kenney 1939 b (191), Kenney & Keeping 1951 (407), Goedicke 1953(267)
5 dec.	0(.01)3.99	No Δ	Davis & Nelson 1935(398)
5 dec.	0(.01)3.49	No Δ	Mills 1938(699)
5; 7, 10 dec.	0(.01)3(.5)5; 3.5, 5	D	Smith & Duncan 1944(693)
5 dec.	0(.1)2, 2.2, 2.5, 3, 4	No Δ	R.W. Burgess 1927(296)
4 dec.	0(.01)4.99	D^n, n = 1,3,4,5	Hodgman 1957(238)
4; 6, 7 dec.	0(.01)4.4(.05)4.7(.1)5; 3, 3.5	No Δ	Peters & Van Voorhis 1940(485)
4 dec.	0(.01)4.29	D	Richardson 1934(468)
4 dec.	0(.01)4.09	No Δ	Mode 1951(361)
4 dec.	0(.01)3.99	No Δ	Holzinger 1925(70)
4; 5, 6, 7, 8, 9 dec.	0(.01)3.5(.1)6; 3.01, 3.6, 4.8, 5.3, 5.7	No Δ	Morton 1928(T.6)
4 dec.	0(.01)3.87	D^n, n = 1,4,5	Smith & Duncan 1945(469)
4 dec.	0(.01)3.69	D	Snedecor 1946(T 8.6)
4; 7, 10 dec.	0(.01)3.2(.1)4(.5)5; 3.2, 5	No Δ	Smith 1934(494), Croxton & Cowden 1939 (873), Hagood 1941(901), Freeman 1942(166), Clark 1953(258), E.E. Lewis 1953(160)
4; 5, 7 dec.	0(.01)3.2(.1)4(.5)5; 3, 3.2	No Δ	Garrett 1937(110, 467)

4 dec.	0(.01)3.19 ⎫	No Δ	Arkin & Colton 1950(114)
5 dec.	3(.1)4 ⎭		
4 dec.	0(.01)3.19	No Δ	Weatherburn 1946(56)
4 dec.	0(.01)3.09	No Δ	Neyman 1950(345)
4 dec.	0(.01)3(.1)4	No Δ	Shewhart 1931(90)
4 dec.	0(.01)3	D	Walker 1943(338)
4; 5, 6 dec.	0(.01)2(.02)3(.2)4(.5)5; 3, 3.6	No Δ	Uspensky 1937(407)*
4; 5, 6 dec.	0(.01)2(.02)3(.2)4, 4.5; 3, 4.5	No Δ	Bowley 1937(271)
4; 5 dec.	0(.02)1.5(.05)2(.1)3(.2)4; 3	No Δ	Aitken 1939(144)
3; 4, 5 dec.	0(.002)2(.01)4; 2.88, 3.29	graphical	Koller 1943(73)

1.1122

| 5 dec. | 0(.1)4.5 | D^5 | Thiele 1903(18) |

1.113 Two-tail area: beyond $-x$ and beyond $+x$

$2Q(x) = 1 - \alpha_x$

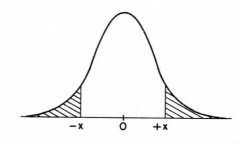

Fig. 1.6

1.1131 $2Q(x)$

7 fig.	6(.01)10	$\tfrac{1}{2}$D	N.B.S. 1942(340)
5 dec.	0(.1)4.5	Δ^2	Yule & Kendall 1937(533)
5 dec.	0(.2)4	No Δ	Cramér 1946(538), Burington & May 1953(88)
4 dec.	0(.01)3(.1)4	No Δ	Fry 1928(453)
4 dec.	0(.05)3	No Δ	Dixon & Massey 1957(381)

---

* Error: for $x = 5.0$ read .4999997 instead of .499997.

1.1131
1.115   [see also p. 630]

4 dec.	0(.1)4.1	No Δ	Kelley 1938(116, n = 1), 1948(202, n = 1)
3 fig.	0(.01)3(.1)5	No Δ	Davies 1949(266)
2-4 fig.	.67449, .7(.1)4(1)7	No Δ	Pearl 1940(473), Arkin & Colton 1950(141), Hodgman 1957(236)

1.1132 $[1 - 2Q(x)]/Q(x)$

2-4 fig.	.67449, .7(.1)4(1)7	No Δ	Pearl 1940(473), Arkin & Colton 1950(141), Hodgman 1957(236)

1.1133 $-\log_{10} 2Q(x)$

5 dec.	0(.1)5	δ^2	Seal 1941(46)

1.114   Single-tail area beyond x

$$Q(x) = 1 - P(x) = \tfrac{1}{2}(1 - \alpha_x)$$

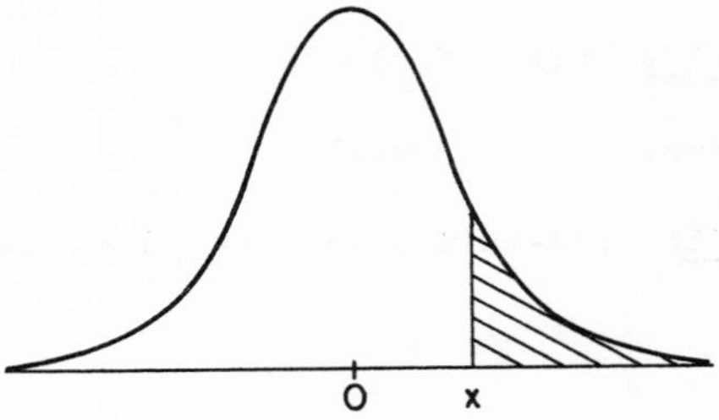

Fig. 1.7

1.1141 $Q(x)$

7 dec.	0(.01)4; called τ_0	No Δ	Lee 1925(351), K. Pearson 1931a (74)
7; 10 dec.	3(.5)5; 4	No Δ	Rider 1939(199)
5 fig.	6(1)10		
5 fig.	0(.01)4.99	No Δ	Fisher & Yates 1953 (T. VIII 4), 1957(T. II 1)
5 dec.	0(.05)3(.25)4	No Δ	Mc Nemar 1949(346)
4 fig.	-4.99(.01)+4.99	No Δ	Czechowski et al. 1957(42)
4 fig.	0(.01)4.99	No Δ	Hald 1952(T.2)
4 dec.	0(.01)3.99	D	Rider 1939(194)
4 dec.	0(.01)3.09	No Δ	Eisenhart et al. 1947(31)
4 dec.	-3.25(.05)+3.25	No Δ	Dixon & Massey 1957(382)

4 dec.	0(.1)3.9	No Δ	E.S. Pearson 1935 c (120)
about 4 dec.	−4(.1)+4	No Δ	Quenouille 1950(223)
3 fig.	0(.01)3(.1)5	No Δ	Davies 1949(264)
2-3 fig.	0(.2)3	No Δ	Dudding & Jennett 1944(58)

<u>1.1142</u> $(2\pi)^{\frac{1}{2}} Q(x)$

10 dec.	−7(.1)+6.6	D^7	B.A. 1, 1931(63)

<u>1.1143</u> $-\log_e Q(x)$

24 dec.	0(1)10	No Δ	Sheppard 1939(23)
16 dec.	0(.1)10	∇^n	Sheppard 1939(24)

<u>1.1144</u> $\pm \log_{10} Q(x)$

12 dec.	0(.1)10	∇^n	Sheppard 1939(28)
8 dec.	0(.01)10	δ^2	Sheppard 1939(30)
5 dec.	5(1)50(10)100(50)500	No Δ	K. Pearson 1930 a (11), Pearson & Hartley 1954(111)

<u>1.115</u> Single-tail area up to x

$$P(x) = \tfrac{1}{2}(1 + \alpha_x)$$

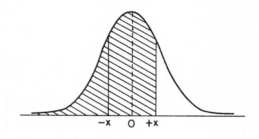

Fig. 1.8

7; 10 dec.	0(.01)6; 4.5	Δ^2 (to 4.5)	Sheppard 1903(182), K. Pearson 1930 a (T.2), E.S. Pearson & Hartley 1954(104)
7 dec.	0(.01)4	D^7	Jørgensen 1916(178)
7 dec.	0(.01)4	D^n, n = 1, 3(1)7	Charlier 1947(103)
7 dec.	0(.1)5.4	No Δ	Gosset 1925(114)
6 dec.	−4.89(.01)+4.89	No Δ	Salvosa 1930(α_3 = 0)

1.115
1.1161

6 dec.	0(.1)4.5	D	Feller 1950(132)
5 dec.	0(.001)2.499	No Δ	Berkson 1955(T.5), Vianelli 1959(391)
5 dec.	−4(.01)+4	No Δ	Baten 1938(282)
5 dec.	0(.01)3.99	No Δ	Vianelli 1959(136)
5 dec.	0(.01)3(.1)4	Δ	Lindley & Miller 1953(T.1)
5 dec.	0(.02)3	$\tfrac{1}{2}\Delta$	Jahnke & Emde 1960(33)
5 dec.	0(.1)4.8	D	N.L. Johnson & Tetley 1949(280)
5 dec.	0(.1)4.4	Δ^2	Yule & Kendall 1937(532), M.G. Kendall 1943(439)
5 dec.	0(.1)4	D^7	Fry 1928(456), Cramér 1946(557)
5 dec.	0(.1)4	D	Burington & May 1953(89)
5 dec.	0(.1)4	No Δ	Arley & Buch 1950(216)
5 dec.	0(.1)4(.05).7(.01).8(.05)1.1, 1.5(.5)3.5	No Δ	Jones 1924(284)
4 fig. or less	0(.01)5.99	PPMD	Camp 1931(380)
4 fig.	−4.99(.01)+4.99	No Δ	Czechowski et al. 1957(42)
4; 5, 6, 7, 8, 9, 10 dec.	0(.01)4.99; 1.29, 2.33, 3.10, 3.72, 4.27, 4.76	No Δ	Hald 1952(T.2)
4; 5, 6, 7 dec.	0(.01)3.99; 2.5, 3.1, 3.73	No Δ	Yule & Kendall 1950(664)
4 dec. 5 dec.	−2.99(.01)+2.99 −3.59(.01)−3, +3(.01)+3.59	No Δ	Grant 1946(534), 1952(510)
4 dec.	0(.01)3.49	D	Mood 1950(423), A.J. Duncan 1952(610)
4 dec.	0(.05)3.95	D^n, n = 1, 4(1)7	Arne Fisher 1922(280)
4 dec.	0(.05)3.7	No Δ	P.O. Johnson 1949(359)
4 dec.	−3(.1)+3	No Δ	Wilks 1948(145), Dixon & Massey 1951(306), Burington & May 1953(82)

1.116 Ordinates

1.1161 $Z(x) = (2\pi)^{-\frac{1}{2}} e^{-\frac{1}{2}x^2}$

15 dec.	0(.0001)1(.001)8.285	No Δ	N.B.S. 1942(2)
7; 10 dec.	0(.001)2(.01)6; 4.51	No Δ	Vianelli 1959(121)
7; 10 dec.	0(.01)6; 4.5	Δ^2 (to 4.5)	Sheppard 1903(182), K. Pearson 1930a (T.2), E.S. Pearson & Hartley 1954(104)
7 dec.	0(.01)4	D^6	Jørgensen 1916(178)
7 dec.	0(.01)4	D^n, n = 2(1)6	Charlier 1947(103)
7 dec.	0(.01)4; called τ_1	No Δ	Lee 1925(351), K. Pearson 1931a (74)
7 fig.	6(.01)10	No Δ	N.B.S. 1942(340)
6 dec.	0(.004)5.216	D^4	Harvard Univ. 1952(3)
6 dec.	0(.1)4.5	No Δ	Feller 1950(132)
5 dec.	0(.01)4.99	D^n, n = 2(1)8	Glover 1923(392)
5 dec.	0(.01)4.99	D^n, n = 2(1)4	Rietz 1924(209)
5 dec.	0(.01)4.49	D^n, n = 2(1)8	Vianelli 1959(989)
5 dec.	0(.01)3.99	D^n, n = 2(1)4	Nemčinov 1946(118)
5 dec.	0(.01)3.99	No Δ	Kenney 1939a (225), Kenney 1939b (191), Hoel 1947(243), Kenney & Keeping 1951(407), Goedicke 1953(266)
5 dec.	0(.02)3	$\frac{1}{2}\Delta$	Jahnke & Emde 1960(33)
5 dec.	0(.1)4.8	Δ^2	Yule & Kendall 1937(531), M.G. Kendall 1943(438), Yule & Kendall 1950(663), M.G. Kendall & Stuart 1958(398)
5 dec.	0(.1)4.8	No Δ	N.L. Johnson & Tetley 1949(280)
5 dec.	0(.1)4	D^6	Fry 1928(456), Cramér 1946(557)
5 dec.	0(.1)4	No Δ	Arley & Buch 1950, Burington & May 1953(89)
4 fig.	0(.01)4.99	No Δ	Hald 1952(T.1), Czechowski et al. 1957(41)
4 dec.	0(.01)4.99	D^4	Hodgman 1957(238)

1.1161
1.1213   [see also p. 630]

4 dec.	0(.01)4.99	D^3, D^4 only	Smith & Duncan 1945(469)
4; 6, 7 dec.	0(.01)4.4(.05)4.7(.1)5; 3, 3.5	No Δ	Peters & Van Voorhis 1940(485)
4 dec.	0(.01)4.29	PPMD	Camp 1931(384)
4 dec.	0(.01)4.09	No Δ	Mode 1951(360)
4 dec.	0(.01)4(.05)5	D^6	Burington & May 1953(267)
4 dec.	0(.01)3.99	No Δ	Rider 1939(194), Weatherburn 1946(53), Mood 1950(422), A. J. Duncan 1952(610)
4 dec.	0(.01)3(.1)4.9	No Δ	Snedecor 1946(T.8.5)
4 dec.	0(.01)3(.1)3.9	PPMD	Fisher & Yates 1938-57(T.II)
4 dec.	0(.01)3	No Δ	Walker 1943(338)
4 dec.	0(.01)2.99	No Δ	Royo López & Ferrer Martín 1954(199), 1955(26)
4 dec.	0(.01)2.75(.05)4	D^3	Scarborough & Wagner 1948(130)
4; 5 dec.	0(.02)3; 1.68	No Δ	Graf & Henning 1953(T.1)
2 fig.	3.5(.5)5, 6, 8, 10, 20		
4 dec.	0(.1)4	No Δ	Lindley & Miller 1953(T.1)
3; 4 dec.	0(.02)3; 1.68	graphical	Koller 1943(69)
3; 4, 6 dec.	0(.05)3(1)5; 1.5, 5	No Δ	Dixon & Massey 1957(381)

1.1162 $(2\pi)^{\frac{1}{2}} Z(x) = e^{-\frac{1}{2}x^2}$

10 dec.	-7(.1)+6.6	D^6	B.A.1, 1931(60)
5 dec.	0(.01)3(.1)5	No Δ	Smith 1934(493), Davenport & Ekas 1936(164), Garrett 1937(127, 469)
5 dec.	0(.01)3, 4	No Δ	Arkin & Colton 1950(115)
5 dec.	0(.05)4(.1)5	Δ	Sheppard 1898(153)
4; 5, 6, 7, 8 dec.	0(.01)3.5(.1)6; 3.01, 3.6, 5, 5.8	No Δ	Morton 1928(T.6)

1.12   Area and log area as functions of $2^{-\frac{1}{2}}x$ and of x/Probable Error

1.121 Area and log area as functions of $2^{-\frac{1}{2}}x$

For tables of the ordinate $2\pi^{-\frac{1}{2}}e^{-x^2}$ see Fletcher et al. 1946, articles 15.11 to 15.132, pp. 213-214.

1.1211 $\quad 2P(2^{\frac{1}{2}}x) - 1 = 2\pi^{-\frac{1}{2}} \int_0^x e^{-t^2} dt$

19-22 dec.	1, 1.25, 1.4, 1.5, 1.6, 1.8, 2, 2.5, 3	No Δ	J. Burgess 1898(273)
15 dec.	0(.0001)1(.001)6	D	N.B.S. 1941(2)
15 dec.	1(.001)1.5(.002)3	Δ^4	J. Burgess 1898(296)[*]
15 dec.	3(.1)5(.5)6	No Δ	J. Burgess 1898(321)[‡]
9 dec.	0(.001)1.249	Δ^2	J. Burgess 1898(283)[‖]
7 dec.	0(.01)3	Δ^2	De Morgan 1838(App.34)
7 dec.	0(.01)2.97, 3	Δ^2	Galloway 1839(212)
7 dec.	0(.01)2	Δ^2	De Morgan 1845(483), Kelvin 1890(434)
5 dec.	0(.001)1.5(.002)2.5(.01)3.09	No Δ	Peirce 1929(116)
5 dec.	0(.001)1.5(.01)3.09	IΔ	Dwight 1941(210)
5; 6 etc. dec.	0(.01)2.6(.05)3.6(.1)5(.5)6; 2.85 etc.	No Δ	Brunt 1917(214), 1931(234)
5 dec.	0(.02)3	No Δ	Levy & Roth 1936(197)
4 dec.	0(.001)1.51(.01)2.87	No Δ	Jahnke & Emde 1909(35)
4; 5, 6 dec.	0(.01)2.99; 2.0, 2.8	IΔ	Jahnke & Emde 1933(98), 1938(24), 1948(24)
4; 5 dec.	0(.1)3; 2.7	No Δ	Burnside 1928(103)
4 dec.	0(.1)2.5	No Δ	Whittaker & Robinson 1924(181)

1.1212 $\quad \pi^{\frac{1}{2}}[P(2^{\frac{1}{2}}x) - \frac{1}{2}]$

10 dec.	0(.01)4.52	No Δ	Oppolzer 1880(587)

1.1213 $\quad 2Q(2^{\frac{1}{2}}x) = 2\pi^{-\frac{1}{2}} \int_x^\infty e^{-t^2} dt$

8 fig.	4(.01)10	D	N.B.S. 1941(296)

---

[*] N.B.S. 1942 reports major errors at 1.308, 1.564, 1.798, 1.966, 1.998.
[‡] N.B.S. 1942 reports major errors at 3.5, 4.1, 4.7, 6.
[‖] N.B.S. 1942 reports major errors at .015, .055, .291, .367.

1.1214
1.19   [see also pp. 630-631]

__1.1214__ $\pi^{\frac{1}{2}}Q(2^{\frac{1}{2}}x) = \int_x^\infty e^{-t^2} dt$

15; 21 dec.   3(.1)5(.5)6; 5.5   No Δ J. Burgess 1898(321)

11 dec.   0(.001)3(.01)4.8 Δ^n Markoff 1888(1)

11; 13, 14 dec. 3(.01)4.5; 3.51, 4.01   No Δ Glaisher 1871(436)

__1.1215__ $2\pi^{\frac{1}{2}}Q(2^{\frac{1}{2}}x) = 2\int_x^\infty e^{-t^2} dt$

10 dec. $\begin{cases} x = 0(.01)5 \\ e^{-x^2} = 0(.01).8 \end{cases}$ Δ^n Legendre 1826(520)

__1.1216__ $\log_{10} \pi^{\frac{1}{2}}Q(2^{\frac{1}{2}}x)$

7 dec.   0(.01)3 Δ^2 Kramp 1799(203),
                                                        De Morgan 1845(477)

1.122 Area in terms of x/[Probable Error]

In the early stages of statistical methodology, much use was made of the 'probable error' of a normal variate. The probable error is defined as the deviate x^*, say, that is exceeded in absolute value with frequency 50%. Thus $2Q(x^*) = 0.5$ and $x^* = 0.67448\ 97501\ 96081\ \ldots$. For an alternative definition see Fletcher et al. 1946, p. 212.

Our listing of tables based on the probable error is selective: for further references see Conrad, Krause &, 1937b.

The argument in the tables below is t.

__1.1221__ $P(tx^*) - \frac{1}{2}$

8; 9(1)13 dec.   0(.01)9(.1)10; 3.81, Δ Conrad, Krause &,
                 7.43, 8.08, 8.68, 9   1938a (400)

4; 5(1)13 dec.   0(.01)5(.1)10;   No Δ Conrad & Krause 1938b (492)
                 4.57, 5.5 etc.

4; 5 etc. dec.   0(.05)5.95; 5.1 etc.   No Δ Garrett 1937(111, 468)

__1.1222__ $2P(tx^*) - 1$

7 dec.   0(.01)7   No Δ Conrad, Krause &,
                                                           1937b (56)

5 dec.   0(.01)3.5(.1)5 Δ Chauvenet 1874(594)

5 dec.   0(.01)3(.1)5.9   No Δ Fowle 1933(57)

4 dec.	0(.01)3.5(.1)5.9		IΔ	Merriman 1910(221)
4 dec.	0(.01)3.5(.1)5.9		No Δ	Mellor 1931(622)

1.13 Ratio of area to bounding ordinate and related functions

For specialized tables of these ratios, used in probit analysis, see section 1.23. For further references, particularly to tables with argument $2^{-\frac{1}{2}}x$, see Fletcher et al. 1946, pp. 220-221.

1.131 $Q(x)/Z(x)$

24 dec.	0(.1)10	v^n	Sheppard 1939(16)
12 dec.	0(.01)10	v^n	Sheppard 1939(2)
5 dec.	0(.01)4(.05)5(.1)10	Δ	J.P. Mills 1926(396), K. Pearson 1931a(11)

1.132 $Z(x)/Q(x)$

15 dec.	0(.1)10	v^{n-1}/n	Sheppard 1939(24, L')
5 dec.	-2(.1)+2.5	δ^2	Stevens 1937a(815)

1.133 $\log_{10}[Q(x)/Z(x)]$

4 dec.	0(.01)4(.1)9.9	PPMD	Camp 1931(404)

1.139 Miscellaneous ratios

Pearson, K. 1930a

<u>Tables</u> <u>for</u> <u>S</u> & <u>B</u> I, T.21, pp. xlij, 33

Diagram to facilitate computation of

$$\sigma_r = \frac{1}{\sqrt{N}} \sqrt{\frac{P_1 Q_1}{Z_1^2}} \sqrt{\frac{P_2 Q_2}{Z_2^2}}$$

where $Q_i = 1 - P_i$ and $Z_i = Z_i(P_i)$

Accuracy: 2 figures + .

The quantities with which the chart must be entered are N, Q_1 and Q_2.
The range being $100 \leq N \leq 10000$, $.001 \leq \frac{Q_1}{Q_2} \leq .999$.

1.19 Miscellanea

In this section we have assembled abstracts of tables, not elsewhere classified, with the normal deviate as argument.

1.19

Ferris, C.D., Grubbs, F.E. and Weaver, C.L. 1946

Annals of Math. Stat. 17, 178-197

Power function for normal deviate test for mean.

The hypotheses considered are:

$H_0: \xi = a$

$H_1: \xi = a + \lambda\sigma \qquad \lambda > 0$

Figure 6 gives the Type II error, $(2\pi)^{-\frac{1}{2}} \int_{\lambda n - 1.96}^{\lambda n + 1.96} e^{-\frac{1}{2}x^2} dx$,

for $n = 1(1)8, 10(5)20(10)50, 75, 100$ plotted against λ for $0 \leq \lambda \leq 5$.

Jeffreys, H., 1939

The Theory of Probability

Table I (p. 361) gives roots x^2 of $K = 2\sqrt{n}\, Z(x)$ to 1 dec. for

$K = 10^{-\frac{1}{2}i}$

$i = 0(1)4$

$n = 5(1)20(10)100, 200, 500, 1000, 2000, 5000, 10\,000, 20\,000, 50\,000, 100\,000$.

Table II (p. 362) gives roots x^2 of $K = (nx^2 - n)^{\frac{1}{2}} \exp(-\frac{1}{2}x^2)$ to 1 dec. for K, i, n as above, omitting $n = 5, 6$, and with some other omissions.

Livermore, T. R. and Neely, W., 1933

Am. Sc. Agron, Jn. 25, 573

The table on p. 575 gives for the normal deviate D, percentage points in terms of the probable error, i.e., roots $\frac{D}{P.E._D}$ of $\frac{1}{m+1} = 2Q\left(\frac{D}{P.E._D}\right)$ to 2 dec. for m 12(2)70.

From these values of $\frac{D}{P.E._D}$ have been computed ratios

$F = \left(\frac{D}{P.E._D}\right) \cdot \frac{\sqrt{2}}{D}$ to 4 dec. for

$$D = 5.0(2.5)20.0(5)40$$

and

$$m = 12(2)70.$$

Example of use of table:

Given a normal population with mean M and probable error PE_s expressed as percent of M, let D denote the difference between the means of two independent samples of n expressed as percent of M; then D will be judged "significant" at the odds $m:1$ if $n \geq (F \cdot PE_s)^2$ where F is the value given in the table for m and D.

Hastings, C., Jr., Hayward, Jeanne T. and Wong, J.P., Jr. 1955
Approximations for digital computers

Pages 151-153 offer three approximations to $Z(x)$, with maximum error ranging from .003 down to .0003; pp. 167-169, three approximations to $2Q(2^{\frac{1}{2}}x)$, with maximum error ranging from .00002 down to .00000015 (these approximations require the computation of $8^{\frac{1}{2}}Z(2^{\frac{1}{2}}x)$); pp. 185-187, three approximations to $2Q(2^{\frac{1}{2}}x)$, with maximum error ranging from .0005 down to .0000003. The merit of these last three approximations, and alternatives to them, are discussed on pp. 92-93. An error curve is provided for each approximation.

Sheppard, W. F., 1898
Royal So London, Phil. Trans. A, 192, 101

The equations $Z = \exp\left(-\tfrac{1}{2}x^2 \sec^2 2\pi\theta\right)$

$$\alpha(x) = \frac{1}{\sqrt{2\pi}} \int_{-x}^{x} e^{-\tfrac{1}{2}t^2} dt$$

yield Z as a function of α and θ.

Table III (p. 156) gives Z to 5 dec. for

$$\theta = .00(.01).25$$
$$\alpha = .0(.1).9.$$

1.19

Table IV (pp. 157-158) gives θ to 5 dec. for $Z = 0(.01)1$ and $\alpha = 0(.1).9$.

Soper, H.E. 1915

<u>Biometrika</u> 10, 384-390

    also in

Pearson, K. 1931a

Tables for S & B II, T. XVI, XVI<u>bis</u>

Tables concerning the biserial correlation coefficient. The quantities tabled are:

$$\lambda_1 = \tfrac{1}{4} + \tfrac{1}{2} \tfrac{PQ}{Z^2} - (1 - \tfrac{xP}{Z})(1 \tfrac{xQ}{Z})$$

$$\lambda_2^2 = \tfrac{PQ}{Z^2}$$

$$\lambda_3 = \tfrac{3}{2} + (1 - \tfrac{xP}{Z})(1 + \tfrac{xQ}{Z})$$

to 4 dec. all as functions of q for $q = .010(.005).095$, $.10(.01).50$

where $P = \dfrac{1}{\sqrt{2\pi}} \displaystyle\int_{-\infty}^{x} e^{-\tfrac{1}{2}t^2} dt$, $Q = 1 - P$, $Z = \dfrac{1}{\sqrt{2\pi}} e^{-\tfrac{1}{2}x^2}$

No Δ.

In table XVI <u>bis</u> two decimal values of $\sqrt{\lambda_2^2 - \lambda_3 r^2 + r^4}$ are given for

    $Q = .500, .309, .159, .067, .023, .006$

and

    $r = .00(.25)1.00$

The chosen levels of Q correspond to $Q = Q(\tfrac{1}{2}i)$, $i = 0(1)5$.

On p. 390 (of <u>Biometrika</u> 10) are given 1-2 decimal values of $\lambda_2 - r^2$ for $r = 0(.25)1.00$, $Q = .500, .309, .159, .067, .023, .006$.

Notation:	K. Pearson	Ours	Soper
	λ_1	λ_1	ϕ_a
	λ_2^2	λ_2^2	χ_a^2
	λ_3	λ_3	ψ_a
	$\tfrac{1}{2}(1-\alpha)$	Q	$\tfrac{1}{2}(1-\alpha)$

Wald, A. and Wolfowitz, J. 1946

Annals of Math. Stat. 17, 208-215

Tolerance limits for the normal distribution.

The table on p. 213 gives 3 decimal values of

$$\lambda^*(\beta,\gamma) = \sqrt{\frac{\nu}{\chi^2(\beta,\nu)}} \; r$$

where $\chi^2(\beta,\nu)$ is the $100\beta\%$ point of χ^2 for ν d.f. and r is the root of

$$\frac{1}{\sqrt{2\pi}} \int_{(\nu+1)^{-\frac{1}{2}}-r}^{(\nu+1)^{-\frac{1}{2}}+r} e^{-\frac{1}{2}x^2} dx = \gamma$$

for the following values

$\nu+1$	γ	β
2	.95	.95
9	.95	.99
25	.95	.95
25	.95	.99

The tolerance limits are then given by $\bar{x} \pm \lambda^* s$, s = sample variance.

Bowker, A.H. 1946

Annals of Math. Stat. 17, 238-240

Tolerance limits for the normal distribution (for notation, see the preceding abstract).

Bowker derives an approximate formula for $\lambda^*(\beta,\gamma)$, viz:

$$\lambda^*(\beta,\gamma) \sim x(\frac{1+\gamma}{2})\left(1 - \frac{x(1-\beta)}{2(\nu+1)} + \frac{5x^2(1-\beta)+10}{12(\nu+1)}\right)$$

Table 1 (p. 239) gives the exact $\lambda^*(\beta,\gamma)$, the approximation and the difference to 5 dec. for $N = \nu+1 = 50, 100, 160, 500, 800, 1000$; $\beta = .75, .95, .99$; $\gamma = .75, .95, .99$.

1.19
1.2111 [see also p. 631]

Eisenhart, C., Hastay, M.W. and Wallis, W.A. 1947

Techniques of statistical analysis

Tolerance factors for normal distributions

Table 2.1 (pp. 102-107) gives tolerance factors to 3 dec. for covering $P = .75, .90, .95, .99$ and $.999$ of the population with confidence $\gamma = .75, .90, .95$ and $.99$ for sample sizes $= 2(1)102(2)180(5)300(10)400(25)750(50)1000, \infty$.

The interval $\bar{x} \pm sk$, where k is the appropriate entry in the table, will cover 100P% of the population with probability γ.

The table is reprinted as T. 16 in Dixon & Massey 1951, 1957.

Williams, J.D. 1946

Annals of Math. Stat. 17, 363-365

"An approximation to the [normal] probability integral"

Williams shows that
$$(2\pi)^{-\frac{1}{2}} \int_{-x}^{+x} e^{-\frac{1}{2}t^2} dt \leq [1 - e^{-2x^2/\pi}]^{\frac{1}{2}} \qquad (*)$$
and that the equality is never in error by as much as three-fourths of one percent.

Table 1 gives, for $x = 0(.1)2$, values of
$$\alpha(x), \quad p'(x), \quad \frac{p'(x)}{\alpha(x)} - 1, \quad Z(x), \quad \tfrac{1}{2} dp'(x)/dx, \quad \frac{2Z(x)}{dp'(x)/dx} - 1,$$
where $p'(x)$ denotes the right-hand side of $(*)$.

1.2 Tables having the normal area P as argument; tables for probit analysis

1.21 Deviates as functions of P: percentage points and probits

1.211 x as a function of P or Q

1.2111 x(P) for general P

Fisher & Yates give PPMD; other authors give no differences.

10 dec.	.5(.001).999	Kondo & Elderton 1931(368), K. Pearson 1931a(2)
8 dec.	.5(.0001).9999	Kelley 1938(14), 1948(38)
6 dec.	.5(.001).999	Kelley 1923(373), 1947(639)
5 dec.	Q = .001(.001).5	Everitt 1910(442, called h), K. Pearson 1930a(42, called h)
5 dec.	Q = .001(.001).01(.01).49	Eisenhart, Hastay & Wallis 1947(18)
4 dec.	.001(.001).98(.0001).9999	Fisher & Yates 1938 etc. (T.IX, x+5 tabulated)
4 dec.	.98(.0001).9999	E.S. Pearson & Hartley 1954(111)
4 dec.	.5(.001).999	Sheppard 1907(405), K. Pearson 1930a(1), E.S. Pearson & Hartley 1954(112)
4 dec.	Q = .001(.001).5	Peters & Van Voorhis 1940(481)
4 dec.	Q = .005(.005).5	Rider 1939(198)
4 dec.	.01(.01).99	Charlier 1920(122), 1947(102)
4 fig.	.5(.001).999	Neyman 1950(346)
3 dec.	.5(.001).999	Holzinger 1925(72), 1928(212)
3 dec.	.0001(.0001).025(.001) .975(.0001).9999	Hald 1952(T.3, x+5 tabulated)
3 dec.	.00001, .0001, .001, .005, .01, .02, .025, .03(.01).1(.05).9(.01).97, .975, .98, .99, .995, .999, .9999, .99999	Dixon & Massey 1957(383)

1.2112
1.221
                                           [see also pp. 631-632]

1.2112 Tail area: x(P) for small Q

No differences are given with these tables.

8-5 dec.	$Q = 10^{-i}, 5 \cdot 10^{-i}, i = 5(1)9$	Kelley 1938(114), 1948(36)
6 dec.	$P = .95(.0001).9999$	
6; 5, 4 dec.	$Q = r \cdot 10^{-i},$ $r = 9(1)1, i = 5(1)10;$ $9 \cdot 10^{-6}, 9 \cdot 10^{-9}$	Conrad & Krause 1937a (279)
5 dec.	$Q = 5 \cdot 10^{-i}, i = 3(1)10$	Rider 1939(199)
3 dec.	$P = .9, .95, .975, .99, .995,$ $.999, .9995, .99995, .999995$	Mood 1950(423)

1.2119 Approximations to x(P)

Hastings, C., Jr., Hayward, Jeanne T. and Wong, J.P., Jr. 1955
<u>Approximations for digital computers</u>

Pages 191-192 offer two approximations to x(P), with maximum error .003 and .0005. The rationale of these approximations is briefly indicated on page 113. An error curve is given for each approximation.

1.212 x as a function of (2P - 1) or 2Q

1.2121 $x(1 - \frac{1}{2}p)$ for general p

Sheppard 1903 gives Δ^3; other authors give no differences.

10 dec.	$1 - p = .1(.1).9$	Sheppard 1898(156)
7 dec.	$1 - p = 0(.01).8$	Sheppard 1903(189), K. Pearson 1930a (9)
6 dec.	$p = .01(.01)1$	Fisher & Yates 1938 etc. (T.I)
5 dec.	$1 - p = 0(.01).99$	Sheppard 1898(167)
4 dec.	$p = .01(.01)1$	Tippett 1952(386)
4 dec.	$p = .0001, .001, .01, .05(.05)1$	Cramér 1946(558)
4 dec.	$p = .05(.05)1, 10^{-i}, i = 2(1)7$	Burington & May 1953(88)

1.2122 Double tail area: $x(1 - \frac{1}{2}p)$ for small p

No differences are given with these tables.

5 dec.	$p = 10^{-i}, i = 3(1)9$	Fisher & Yates 1938 etc.(T.I), Arley & Buch 1950(216)
3 dec.	$p = 10^{-i}, i = 1(1)6$	Fry 1928(455)

1.213 Percentage points as multiples of the probable error

No differences are given with these tables of $x(P)/x^*$, where x^*, the probable error, is defined in section 1.122.

7 dec.	$.5(.001).95$	⎫
6 dec.	$.95(.0001).9999$	⎬ Conrad, Krause &,
6; 5, 4 dec.	$Q = r \cdot 10^{-i}$, $r = 9(1)1$, $i = 5(1)11;$ $9 \cdot 10^{-6}$, $9 \cdot 10^{-10}$	⎭ 1938 a (416)
2 dec.	$.5(.001).95(.0001).9999$	⎫ Conrad & Krause 1938 b (496)
2 dec.	$Q = r \cdot 10^{-i}$, $r = 9(1)1$, $i = 5(1)11$	⎭

1.22 Ordinates as functions of Area P; ratios as functions of P.

Let $x(P)$ denote, as usual, the normal deviate corresponding to a lower tail area P, i.e. $x(P)$ is the inverse of $P = \frac{1}{\sqrt{2\pi}} \int_{-\infty}^{x} \exp(-\tfrac{1}{2}t^2)dt$. We may then tabulate the ordinate $Z(x) = \frac{1}{\sqrt{2\pi}} \exp(-\tfrac{1}{2}x^2)$ as a function of P.

1.221 Z(x) as a function of P(x)

Sheppard 1903 gives Δ^3; other authors give no differences.

10 dec.	$P = .5(.001)1$	Kondo & Elderton 1931(368), K. Pearson 1931 a (2)
8 dec.	$P = .5(.0001)1$	Kelley 1938(14), 1948(38)
7 dec.	$2P - 1 = 0(.01).8$	Sheppard 1903(189), K. Pearson 1930 a (9)
6 dec.	$P - \tfrac{1}{2} = 0(.001).499$	Kelley 1923(373), 1947(639)
6-8 fig.	$Q = 10^{-i}$, $5 \cdot 10^{-i}$, $i = 5(1)9$	Kelley 1938(114), 1948(36)
5 dec.	$P = 0(.001)1$	K. Pearson & Soper 1921(428), K. Pearson 1931 a (1), E.S. Pearson & Hartley 1954(113)
5 dec.	$Q = 0(.001).5$, Z called τ_1	Everitt 1910(442), K. Pearson 1930 a (42)
4 dec.	$Q = 0(.001).5$	Peters & Van Voorhis 1940(481)
4 dec.	$2P - 1 = 0(.1).9, .95, .99,$ $.999, .9999$	Dixon & Massey 1957(381)
3 dec.	$P - \tfrac{1}{2} = 0(.01).5$	Holzinger 1928(212), Garrett 1937(370, 473)

[see also p. 632]

1.222 Ratio of area to bounding ordinate as a function of area

Kondo, T. and Elderton, Ethel M.

Biometrika 22, 368-376

The ratios Q/Z, P/Z, Z/Q and Z/P are tabled to 10 dec., without differences, for P = .5(.001)1. Reprinted in Tables for S & B II, T.2.

1.23 Tables for Probit Analysis

For a complete account of the theory and applications of probit analysis, see Bliss 1935 or Finney 1952a. Here we confine ourselves to illustrating the basic assumptions and definitions.

Consider the estimation of a dosage mortality curve for an insecticide. It is first assumed that each individual insect has attached to it a 'critical dose' or 'tolerance dose' which is the minimum dose (i.e. concentration) of the insecticide required to kill that particular insect under specified conditions of exposure. It is then assumed that for a specified population of insects the values of x = log (critical dose) are normally distributed with mean μ and standard deviation σ which have to be estimated from experimental data. The estimate of mean μ represents a measure of the potency of the insecticide.

If n insects are exposed at log (dose) x* and, of these, d insects are killed, then the expected value of the proportion killed d/n is given by the expected proportion of insects whose x values are below x* i.e. by $P(\frac{x^* - \mu}{\sigma})$ while the distribution of d is a binomial with binomial fraction given by $P(\frac{x^* - \mu}{\sigma})$.

If k experiments are made (numbered t = 1, 2, ..., k) and if in the tth experiment n_t insects are exposed at log (dose) x_t giving rise to respective 'observed proportions killed' $p_t = d_t/n_t$ and 'observed proportions surviving' $q_t = 1 - p_t$ the unknown parameters μ and σ can be estimated by the method of maximum likelihood. To this end we introduce the following notations and auxiliary quantities:

The normal equivalent deviate $(x_t - \mu)/\sigma$

The corresponding normal ordinate $Z_t = Z[(x_t - \mu)/\sigma]$

The corresponding normal ordinate $\quad Z_t = Z(\frac{x_t-\mu}{\sigma})$

The expected proportion killed at log (dose) $x_t \quad P_t = P(\frac{x_t-\mu}{\sigma})$

" " " of survivors " " " " $\quad Q_t = 1-P_t$

The <u>weighting coefficient</u> " " " " $\quad w_t = Z_t^2/P_t Q_t$

The <u>probit</u> " " " " $\quad Y_t = 5 + (x_t-\mu)/\sigma$

The <u>working probit</u> at log (dose) $x_t \quad y_t = Y_t + \dfrac{p_t - P_t}{Z_t}$

$$= Y_t + \frac{(Q_t - q_t)}{Z_t}$$

The **maximum likelihood** solutions are then given by the equations

$$\left. \begin{array}{l} b = \text{estimate of } \dfrac{1}{\sigma} = \dfrac{[nw(x-\bar{x})(y-\bar{y})]}{[nw(x-\bar{x})^2]} = \dfrac{[nw][nwxy] - [nwx][nwy]}{[nw][nwx^2] - [nwx]^2} \\[2ex] a = \text{estimate of } (5 - \mu/\sigma) = \bar{y} - b\bar{x} \end{array} \right\} \quad \text{(A)}$$

where: the subscript t has been suppressed, square brackets denote summation over t, and $\bar{x}$ and $\bar{y}$ are the weighted means

$$\bar{x} = [nwx]/[nw], \quad \bar{y} = [nwy]/[nw].$$

When we substitute for y_t and w_t their expressions (in terms of Y_t) in equations (A) two simultaneous equations for the unknown μ and σ arise which must be solved. In practice this is done by an iterative process:- Initial values of probits Y_t (initial) $= x(p_t) + 5$ are computed from the observed proportions killed and plotted against x_t to estimate trial values of $\dfrac{1}{\sigma}$ and $\dfrac{\mu}{\sigma}$ from which, in turn, trial values of Y_t, w_t and y_t can be computed. These are then substituted in (A) to revise the estimates of $\dfrac{1}{\sigma}$ and $\dfrac{\mu}{\sigma}$, the process continuing until stable values are reached. In order to facilitate these calculations, tables for probit analysis have been published, comprising:

Tables of probits $\quad Y = x(P) + 5$

These are, of course, tables of normal deviates $x(P)$ increased by 5 and are only required for the initial stage of the above calculations.

Tables of Maximum working probit $\quad y_{max} = Y + Q/Z$

Minimum working probit $\quad y_{min} = Y - P/Z$

Range $\quad 1/Z$

1.23
1.2312
[see also p 632]

The weighting coefficient $w = Z^2/PQ$

All the above quantities are functions of the normal equivalent deviate $(x_t - \mu)/\sigma$ and are tabulated with the probit $Y_t = (x_t - \mu)/\sigma + 5$ as argument. The first 3 quantities are required for the calculation of working probits y_t from $y_t = y_{min} + p_t \frac{1}{Z}$ or $y_t = y_{max} - q_t \frac{1}{Z}$ using the Y_t based on trial μ, σ and the observed p_t, q_t. Certain tables provide these working probits y_t directly in the form of double entry tables with arguments Y_t and p_t.

Fisher & Yates 1953, pp. 14-17, 70-75, give tables of working values and weights for applying the method of equations (A) to the logit, the angular, and the complementary loglog transformations; these and similar tables are discussed in section 12.

1.231 Working probits; range; weighting coefficients; weighted probits

1.2311 Double-entry tables of working probits

Finney, D. J. 1947
Probit analysis

Table 4, pp. 239-248, gives working probits

$$y = Y + (p-P)/Z$$

to 3 dec. for provisional probits $Y = 2(.1)7.9$ and observed percentage kill $100p = 0(1)100$. Values of y outside the range 0 to 10 are omitted.

Reprinted in Finney 1952a (297), 1952b (626).

1.2312 Full single-entry tables of working probits

Finney, D. J. and Stevens, W. L. 1948
Biometrika 35, 191

Minimum and maximum working probits, range, and weighting coefficients, all to 4 dec., for $Y = 1(.01)9$. Minimum and maximum working probits are given only if they are between 0 and 10.

Reprinted in E.S. Pearson & Hartley 1954(114).

1.23
1.2312

Finney, D.J. 1947

<u>Probit</u> <u>analysis</u>

    Table 3, p. 238, gives:

Y + Q/Z	4 dec.	Y = 3.5(.1)8.9
1/Z	4-5 fig.	1.1(.1)8.9
Y - P/Z	4 dec.	1.1(.1)6.5
Z^2/PQ	5 dec.	1.1(.1)8.9

    Reprinted in Finney 1952a (295), 1952b (625).

Fisher & Yates 1938, 1943(T.XI)

Y + Q/Z	4 dec.	Y = 5(.1)8.9
1/Z	4; 3, 2, 1, 0 dec.	5(.1)8.9; 6.5, 7.5, 8, 8.5
Z^2/PQ	5 dec.	5(.1)8.9

Fisher & Yates 1948, 1953(T.XI), 1957(T. IX 2)

Y - P/Z	4; 3, 2, 1, 0 dec.	Y = 1.1(.1)8.9; 6.5, 7.5, 8, 8.5
1/Z	4-5 fig.	1.1(.1)8.9
Y + Q/Z	0; 1, 2, 3, 4 dec.	1.1(.1)8.9; 1.6, 2.1, 2.6, 3.6
Z^2/PQ	5 dec.	5(.1)8.9

Bliss, C.I. 1935

<u>Annals</u> <u>of</u> <u>Applied</u> <u>Biology</u> 22, 134

    The table on p. 149 gives:

Y + Q/Z	4 dec.	Δ	Y = 1.5(.1)8.9
P/Z	4 dec.	No Δ	1.5(.1)8.9

    The table on pp. 151-152 gives:

The expected percentage kill   100 P(x)   to 3 dec.

Weighting coefficients Z^2/PQ   to 2-3 sig. fig., usually 3 dec.

'Relative number of individuals required for equal weight'

$$= 10 \frac{Z^2(o)}{P(o)\,Q(o)} \bigg/ \frac{Z^2(x)}{P(x)\,Q(x)} \quad \text{to nearest integer}$$

$$Z^2/PQ \quad \text{(repeated) to 5 dec.}$$

all for Y = 1.5(.1)8.9.

Δ provided for the weighting coefficients.

1.2312 [see also p. 632]
1.232

Black, A.N. 1950

<u>Biometrika</u> 37, 158-167

The table on pp. 164-167 gives:

The expected proportion killed	P
The weighting coefficient	$W = Z^2/PQ$
The minimum working probit x weight	$w^1_{min} = (Y - P/Z)\dfrac{Z^2}{PQ}$
The maximum working probit x weight	$w^1_{max} = (Y + Q/Z)\dfrac{Z^2}{PQ}$

all to 3 dec. for $Y = 1.00(.02)8.00$. No Δ.

This table is used in conjunction with a machine method of computing working probits and weighting coefficients by

setting	and	multiplying by
W and w^1_{min}		$n-d$
W and w^1_{max}	to obtain	d
nW and nWy		

<u>1.2313</u> Other tables of working probits

Kelley, T. L., 1947

<u>Fundamentals</u> <u>of</u> <u>Statistics</u>

The Table on pp. 639-52 gives the ratios

Z/Q to 4 dec. ⎫
Z/P to 5 dec. ⎬ as functions of the expected proportion killed, P, for
also PQ to 6 dec. ⎭ $P - \tfrac{1}{2} = 0(.001).499$

Pearson, K. 1913 a

<u>Biometrika</u> 9, 22-27

$(PQ)^{\frac{1}{2}}/Z$ to 4 dec. for $P = .5(.01).98(.005).99(.001).999$.

Reprinted in Tables for S & B I, T.24.

Cornfield, J. and Mantel, N., 1950

Am. Stat. Ass., **Jn.** 45, 181

The table on pp. 185-8 gives

$$-\Delta_{min} = Z/Q$$
$$\Delta_{max} = Z/P$$
$$w^1_{min} = (Z^2/Q^2) - (ZY/Q)$$
$$w^1_{max} = (Z^2/P^2) - (ZY/P)$$

all to 5 decimals for $Y = 5.00(.01)10.00$ and therefore, by symmetry, also for $Y = .00(.01)5.00$

The table is based on a modification of the standard iteration process (first suggested by Garwood 1941) in which the coefficients of the maximum likelihood equations are computed at the observed percentage kills, and not at the expected percentage kills as in the standard method. The modified method requires modified weighting coefficients computed from

$$w^1 = \frac{1}{n}\left[(n-d)\,w^1_{min} + d w^1_{max}\,\zeta\,\right]$$

which is equivalent to

$$w^1 = q\left(\frac{Z^2}{Q^2} - \frac{ZY}{Q}\right) + p\left(\frac{Z^2}{P^2} + \frac{ZY}{P}\right)$$

and reduces to the standard weighting coefficient $W = Z^2/PQ$ when q and p are replaced respectively by Q and P. The tabulated Δ are required for computing the standard right hand sides of the maximum likelihood equations in the simplified form of weighted deviations of working probits, y, from the provisional probits, Y, from

$$nW(y - Y) = (n - d)\,\Delta^1_{min} + d\,\Delta^1_{max}.$$

See also sections 1.131, 1.222.

<u>1.232</u>   Tables of probits: $Y = x(P) + 5$

4 dec.	.001(.001).98(.0001).9999	PPMD	Bliss 1935(138), Fisher & Yates 1938 etc.(T.IX), Finney 1947(T.1)
4 dec.	.001(.001).98(.0001).9999	No Δ	Finney 1952 a (264)
2 dec.	.01(.01).97(.001).999	No Δ	Finney 1952 b (624)
2 dec.	.01(.01).99(.001).999	No Δ	Finney 1947(22)

See also section 1.2111.

1.233   [see also pp. 633-634]

1.233 Tables for special problems of probit analysis

Finney, D.J. 1947

Probit analysis

Table 2, pp. 226-237, gives weighting coefficients in the presence of natural mortality. If in a toxicity test a fraction C of the test subjects would die without any poison, the standard weighting coefficients must be replaced by

$$w^\dagger = Z^2/Q(P - \frac{C}{1-C})$$

The table gives $w^\dagger$ to 5 decimals for provisional probits
Y = 1.1(.1)9.0 and natural mortality rates 100C = 0(1)40 ,
The values of $w^\dagger$ in the column 100C = 0 are identical with the
standard weighting coefficients W = Z^2/PQ.
Also shown are the ratios

Q/Z (to 4 dec.) for Y = 1.1(.1)9.0

which are required in the maximum likelihood equations for the simultaneous estimation of the unknown parameters C, μ and σ .

Table 5, p. 249, gives Z and Z^2 to 5 dec. for Y = 1(.1)9 .

Finney, D.J. 1949 a

Biometrika 36, 239-256

Tables of the weighting coefficient Z^2/Q. If in dosage mortality work s insects have survived from an unknown number, n, of insects exposed to dose x, it is sometimes assumed that n follows a Poisson distribution with mean N (say). The probability p(s) of observing s survivors in a sample subjected to a treatment with kill probability P is

$$p(s) = \sum_{n=s}^{\infty} e^{-N}\left(\frac{N^n}{n!}\right)\binom{n}{s}P^{n-s}Q^s$$

If P follows the usual normal tolerance distribution

$$P = \int_{-\infty}^{(\mu-x)/\sigma} Z(v)\,dv \; ; \; Z(v) = (2\pi)^{-\frac{1}{2}}e^{-\frac{1}{2}v^2} ,$$

the simultaneous estimation of μ, σ and N by maximum likelihood requires the weighting coefficient $Z^2/(1-P)$, given on p. 252 to 5 dec. or 1 fig. for $Y = 5 + v = 1.1(.1)\overset{\circ}{9}$.

Reprinted in Finney 1952 a (309).

Finney, D.J. 1952 a

Probit analysis

Table 2, pp. 268-294 gives $w^\dagger$ (defined on the preceding page) to 5 dec. for $Y = 1.1(.1)9$ and $100 C = 0(1)90$.

Table 5, p. 306, gives:

P	5; 6, 7, 8, 9 dec.	$Y = 5(.1)9$; 6.3, 7.4, 8.1, 8.8
Z	5 fig.	$5(.1)9$
Z^2	5 dec.	$5(.1)8.3$

If sampling at log dose x_t is continued until a sample of n_t is reached, containing a specified number, s_t, of survivors, maximum likelihood estimation requires the working probits $y_t = Y_t - Q_t/Z_t + n_t Q_t^2 / s_t Z_t$ and the weights $w_t = Z_t^2 / P_t Q_t^2$. Table 9, p. 310, gives:

$\left.\begin{array}{ll} Y - Q/Z & \text{4 dec.} \\ Q^2/Z & \text{5 fig.} \\ Z^2/PQ^2 & \text{4 dec.} \end{array}\right\} \quad Y = 5(.1)9$

Healy, M. J. R. 1949

Biometrics 5, 330

Nomogram for determining the relative potency of two preparations in a 6 point assay with 3 readings at each dose.

If S_0, S_1 and S_2 are the totals of n readings for the 'standard' at doses 0, 1 and 2, U_0, U_1 and U_2 the corresponding totals of n readings for the preparation of unknown strength, and I the dilution factor, and, if further

$$D = U_2 + U_1 + U_0 - S_2 - S_1 - S_0$$

$$B = U_2 - U_0 + S_2 - S_0,$$

then the relative potency is given by $(4D \log_e I)/3B$ and, letting

$$x = (2D^2/3B^2)^{\frac{1}{2}},$$

[29]

1.233
1.31

fiducial limits for x are the roots of the quadratic

$$D^2(2^{\frac{1}{2}} - 3^{\frac{1}{2}}x)^2 = \text{const.} \, nu^2 \bar{R}^2 (1+x^2)$$

where $\bar{R}$ is the mean range of n readings and u is a 'range substitute for t' (Lord 1947). The nomogram on p. 332 gives 95% fiducial limits, together with the relative potency, for I = 1.5 and n = 3.

Fisher & Yates 1948, 1953(T. XI 1), 1957(T. IX 3)

w (defined on p. 28) is tabled to 3 dec. for Y = 1.1(.1)8.9 and C = 0(.01).02(.02).1(.05).4.

Wilcoxon, F. and McCallan, S.E.A. 1939
Contributions from Boyce Thompson Institute 10, 329

The calculation of 95% confidence limits for lethal dose 50 (LD 50) and LD 95 can be facilitated by tables of $x^{0.146}$ and $x^{0.26}$ respectively. These powers are tabled on p. 333 for x = 0(.1)5.9.

A line fitted to all dosages with mortality other than 0 and 100% will provide an 'expected percentage' $\bar{P}$ at levels when 0% or 100% are observed. These are then converted to 'corrected' percentages P' by finding the probit Y corresponding to $\bar{P}$, the 'maximum working probit' Y + Q/Z, and finally the percentage P' corresponding to Y + Q/Z as probit. In our notation P' and $\bar{P}$ are therefore related by

$$P' = 100P\left(x + \frac{Q(x)}{Z(x)}\right) \text{ where } x = x\left(\frac{\bar{P}}{100}\right).$$

Table II (p. 334) gives P' to 1 dec. for $\bar{P}$ = 50(1)99
to 2 dec. for $\bar{P}$ = 99.50(.01)99.99

A nomogram for computing contributions to x^2 from dosage mortality curves is given on page 336. Three scales are drawn, viz:

100 p = expected percentage,

Δ = difference between observed and expected,

x^2 = contribution to x^2 ;

these are aligned by the solution $x^2 = \Delta^2/p(1-p)$.

1.3 Moments; integrals; derivatives

1.31 Incomplete normal moments and related functions

Following K. Pearson & Lee 1908, we define the incomplete normal moment

$$m_n(x) = [(n-1)!!]^{-1}(2\pi)^{-\frac{1}{2}}\int_0^x t^n e^{-\frac{1}{2}t^2} dt = [(n-1)!!]^{-1}\mu_n(x)$$

where $(n-1)!! = (n-1)(n-3)\ldots \begin{cases} 2 & \text{for } n \text{ odd} \\ 1 & \text{for } n \text{ even} \end{cases}$

We note the following relations to the χ^2 integral and the incomplete Γ function:-

If $\Gamma_x(a) = \int_0^x t^{a-1} e^{-t} dt$ and if we denote by $P(\chi^2, \nu)$ the probability that the χ^2 statistic based on ν degrees of freedom exceeds χ^2, then

$$m_n(x) \begin{cases} \frac{1}{2}(1-P(\chi^2,\nu)) & \text{for } n = \text{even}; \quad \nu = a+1 \text{ odd} \\ \frac{1}{\sqrt{2\pi}}(1-P(\chi^2,\nu)) & \text{for } n = \text{odd}; \quad \nu = n+1 \text{ even} \end{cases}$$

and $x = \frac{1}{2}\chi^2$

$$m_n(x) \begin{cases} \frac{1}{2}\Gamma_x(\frac{1}{2}(n+1))/\Gamma(\frac{1}{2}(n+1)) & n = \text{even} \\ \frac{1}{\sqrt{2\pi}}\Gamma_x(\frac{1}{2}(n+1))/\Gamma(\frac{1}{2}(n+1)) & n = \text{odd}. \end{cases}$$

Tables of the χ^2 integral and the incomplete Γ function are indexed in section 2.1.

Pearson, K. and Lee, Alice   1908

<u>Biometrika</u> 6, 59

$m_n(x)$ 7 dec. $n = 1(1)10$ $x = 0(.1)5, \infty$

Reprinted in Tables for S & B I, T. 9, pp. 22-23; Vianelli 1959, Pro. 214, pp. 699-700.

1.31
1.3211   [see also p. 634]

Pearson, K., 1931 a

Tables for Statisticians and Biometricians Part II (T.XIII)

$m_n(x)$ 7 dec. $n = 11, 12$ δ^2

$x = 0(.1)5, \infty$

Pearson, K. 1930 a

Tables for S & B I, T. 18-20, 22

Table 18 gives $-\log_{10} P(r, {}_o\sigma_r)$ where

$$P(r, {}_o\sigma_r) = \frac{2}{{}_o\sigma_r \sqrt{2m}} \left(\Delta_0 - \frac{1}{8m} \Delta_4 - \frac{1}{24m^2} \Delta_6 + \frac{1}{128m^2} \Delta_8 \right)$$

$$\Delta_i = \mu_i(\sqrt{2m}) - \mu_i(\sqrt{2m}\, r)$$

μ_i is an incomplete normal moment function (m_i in our notation) and

$$m = \tfrac{1}{2}\left(\frac{1}{{}_o\sigma_r^2} - 3\right).$$

The table gives $-\log_{10} P(r, {}_o\nu_r)$ to 3 dec.

for $r = .05, .075, .1, .15, .2(.1).9, .95$ and

${}_o\sigma_r$ $.01(.01).08$.

Table 19 gives percentage points of x^2 for 3 d.f. and for probability levels $p = P(r, {}_o\sigma_r)$. The table was prepared on the assumption that there were 3 degrees of freedom in a 2x2 contingency table. Table 20 gives the logarithms of the x^2 values from table 19.

1.32   <u>Derivatives of Z(x) ; Hermite polynomials; Hermite functions;</u>
       <u>tetrachoric functions; Gram - Charlier series of Type A</u>

1.321   Derivatives of Z(x)

The derivatives of the normal ordinate $Z(x)$ are used in expanding distribution functions in Gram - Charlier series of Type A; for references see Harvard Comp. Lab. 1952, pp. xxij-xxvij. A number of tables of the derivatives of e^{-t^2}, where $t = 2^{-\frac{1}{2}} x$, are listed below; these tables were computed for use in the classical expansion in the complete series of Hermite functions. The exact relation between Gram - Charlier and Hermite series has been the subject of some confusion; a correct statement is found in Milne 1949, pp. 309-313. We adopt as standard form for derivatives

$$Z^{(n)}(x) = \frac{d^n Z(x)}{dx^n}$$

and proceed to define several related functions in terms of $Z^{(n)}$.

Tetrachoric function $\quad \tau_n(x) = (-1)^{n-1} Z^{(n-1)}(x)/\sqrt{n!}$

Hh functions $\quad Hh_{-n}(x) = (-1)^{n-1} \sqrt{2\pi}\, Z^{(n-1)}(x)$

Hermite polynomials $\quad h_n(x) = (-1)^n e^{\frac{1}{2}x^2} \sqrt{2\pi}\, Z^{(n)}(x)$

Hermite Polynomials $\quad H_n(t) = (-1)^n e^{t^2} \dfrac{d^n}{dt^n} e^{-t^2}$

$\qquad\qquad\qquad\qquad\quad = \sqrt{2\pi}\, (-1)^n e^{t^2} 2^{\frac{1}{2}n} Z^{(n)}(\sqrt{2}t)$

Hermite functions $\quad e^{-\frac{1}{2}t^2} H_n(t) = \sqrt{2\pi}\, (-1)^n e^{\frac{1}{2}t^2} 2^{\frac{1}{2}n} Z^{(n)}(\sqrt{2}t)$

1.3211 $\quad Z^{(n)}(x) = \dfrac{d^n Z(x)}{dx^n}$

	n	x	
7 dec.	1(1)6	0(.01)4	Jørgensen 1916(178)

An extensive list of errors in Semon 1951b.

	n	x	
6 dec.	1(1)4	0(.004)6.468	Harvard Comp. Lab. 1952(3)
6 dec.	5(1)10	0(.004)8.236	Harvard Comp. Lab. 1952(35)
7 fig.	11(1)15	0(.002)6.198	Harvard Comp. Lab. 1952(77)
6 dec.	11(1)15	6.2(.002)9.61	Harvard Comp. Lab. 1952(139)
7 fig.	16(1)20	0(.002)8.398	Harvard Comp. Lab. 1952(171)
6 dec.	16(1)20	8.4(.002)10.902	Harvard Comp. Lab. 1952(255)
7 dec.	2, 3, 5	0(.01)4	Charlier 1947(103)
4 dec.	4, 6		
5 dec.	2(1)8	0(.01)4.99	Glover 1923(392)

7 errors noted in Semon 1951a.

	n	x	
5 dec.	2(1)4	0(.01)4.99	Rietz 1924(209)
5 dec.	1(1)6	0(.1)4	Fry 1928(456), Cramér 1946(577)
4 dec.	2,3	0(.01)4.99	PPMD ⎫ Camp 1931(386-391)
	4	0(.1)3.9	No Δ ⎭

(estimates of error involved in using PPMD given)

1.3211
1.3249   [see also p. 634]

4;5 dec.	3	0(.01)3(.1)6;5.2	} I △	Charlier 1906(45,46)
4 dec.	4	0(.01)3(.1)6		
4 dec.	3,4	0(.01)3(.1)4.9	No △	Charlier 1920(124)
4 dec.	2(1)4	0(.01)4.99		Hodgman 1957(238)
4 dec.	3, 4	0(.01)4.99		Smith & Duncan 1945(469)
4 dec.	3(1)6	0(.05)3.95		A. Fisher 1922(280)
3 dec.	2(1)4	0(.01)2.75(.05)4		Scarborough & Wagner 1948(130)

1.3212 $2^{-\frac{1}{2}n+2} Z^{(n-1)}(2^{\frac{1}{2}}t)$

4 dec.	n = 1(1)6	t = 0(.01)4	Bruns 1906(A.6),
			Czuber 1914(444)
4 dec.	n = 1(1)6	t = 0(.01)3	Jahnke-Emde 1909(37),
			1933(99), 1938(26)

1.3213 $(-1)^{n-1}(2\pi)^{\frac{1}{2}} Z^{(n-1)}(x)$

10 dec.	1(1)3	} -7(.1)+6.6	B.A.1, 1931(60)
(13 - n) dec.	4(1)7		
(5 - n) dec.	1(1)5	0(.1)4.5	Thiele 1903(18)

1.322 Hermite functions

Russell 1933(292) tables $e^{-\frac{1}{2}t^2} H_n(t) = (2\pi)^{\frac{1}{2}}(-1)^n 2^{\frac{1}{2}n} e^{\frac{1}{2}t^2} Z^{(n)}(2^{\frac{1}{2}}t)$ to 5 fig. for n = 0(1)11, t = 0(.04)1(.1)4(.2)7(.5)8. Jahnke & Emde 1948(28), 1960(107), table the parabolic cylinder function

$$\psi_n(t) = (-1)^n \pi^{-1/4} 2^{-\frac{1}{2}n}(n!)^{-1} e^{\frac{1}{2}t^2} \frac{d^n}{dt^n} e^{-t^2}$$

to 5 fig. for the same n and t; their table was derived (by S. Kerridge) from Russell 1933, and its accuracy is restricted accordingly.

1.323 Hermite polynomials

Jørgensen 1916(196) gives $(-1)^n h_n(x) = (2\pi)^{\frac{1}{2}} e^{\frac{1}{2}x^2} Z^{(n)}(x)$ exact, for n = 2(1)6 and x = 0(.01)4. Vianelli 1959(1466) gives $h_n(x)$ exact, for n = 3(1)5 and x = 0(.01)4. See also the abstract of Peters & Van Voorhis 1940 on p. 35 below.

1.324 Tetrachoric functions

1.3241 $\tau_n = (-1)^{n-1}(n!)^{-\frac{1}{2}} Z^{(n-1)}(x)$, $\qquad \tau_0 = Q(x)$

7 dec. $\quad n = 0(1)19 \quad x = 0(.1)4 \qquad$ Lee 1925(351), K. Pearson 1931a (74)

5 dec. $\quad n = 0(1)6 \quad \tau_0 = 0(.001).5 \quad$ Everitt 1910(442),
$\qquad\qquad\qquad\qquad\qquad\qquad\qquad$ K. Pearson 1930a (42),
$\qquad\qquad\qquad\qquad\qquad\qquad\qquad$ Vianelli 1959(702)

1.3249 Miscellaneous tetrachorics

Peters & Van Voorhis 1940(501) table τ_n/τ_1 for $n = 2(1)25$ and $\tau_0 = .05(.01).4$.

Lee, A. 1925

<u>Biometrika</u> 17, 343-354

Differences of tetrachorics from higher tetrachorics.

The finite difference formulae of converting differentials to differences result in the approximate equations

$$\delta^2 \tau_n = \sqrt{\frac{(n+1)(n+2)}{100}} \tau_{n+2} + \sqrt{\frac{(n+1)(n+2)(n+3)(n+4)}{120,000}} \tau_{n+4}$$

$$\delta^4 \tau_n = \sqrt{\frac{(n+1)(n+2)(n+3)(n+4)}{10,000}} \tau_{n+4} + \sqrt{\frac{(n+1)(n+2)\ldots(n+6)}{6,000,000}} \tau_{n+6}$$

The 4 coefficients (called $_1X_n \ldots _4X_n$) are tabulated to 8 dec. for $n = 0(1)15$ on p. 345.

Reprinted in Tables for S & B II, T.6.

Coefficients in the recurrence formula for tetrachoric functions

$$\tau_n(x) = x \frac{1}{\sqrt{n}} \tau_{n-1}(x) - \frac{n-2}{\sqrt{n(n-1)}} \tau_{n-2}(x)$$

This is a transformation of the well-known recurrence formula for Hermite polynomials. The table on p. 346 gives the coefficients

$p_n = 1/\sqrt{n}$
$q_n = \dfrac{n-2}{\sqrt{n(n-1)}} \qquad$ 10 dec. $\quad n = 2(1)25$

1.3249
1.331   [see also p. 634]

   Reprinted in Tables for S & B II, T.5. A 7 dec. table of these coefficients is given in K. Pearson & E.S. Pearson 1922.

<u>1.325</u> Gram-Charlier series of Type A
Edgeworth, F. Y. and Bowley, A. L. 1902
Royal Stat. So. <u>Jn</u>., 65, 325

The tables on pp. 333-334 give

$$\tfrac{1}{2}\alpha(\sqrt{2}x) = \pi^{-\tfrac{1}{2}} \int_0^x e^{-t^2} dt \text{ to 3 dec. for } x = .00(.01)1.50(.02)2.04$$
$$\text{to 4 dec. for } x \quad 2.06(.02)2.10(.05)3.00$$

$$f(x) = \frac{\sqrt{2}}{3}(1-2x^2)Z(\sqrt{2}x) \text{ to 3 dec. for } x = .00(.01)1.50(.02)2.10$$
$$(.05)2.70$$
$$\text{to 4 dec. for } x = 2.75(.05)3.00$$

$$Y(x) = \tfrac{1}{2}\alpha(\sqrt{2}x) \mp j/3\sqrt{\pi} \pm jf(x) \text{ to 3 dec. for } x = -2.2(.2)-1.2(.1)$$
(upper or lower signs taken according as $1.2(.2)2.2$
 $x > 0$ or $x < 0$). $j = 0(.02).30$

These probability integrals are used for approximately normal data which are, however, somewhat skew. If y'_x is the number of observations (out of N) between the 'mean' and the deviation cx where

$$c = \sqrt{2(\mu_2 - \tfrac{1}{12}h^2)},$$

then the following are approximations to y'_x

1. $Y'_x = \dfrac{N}{\sqrt{\pi}} \int_0^x e^{-t^2} dt = N \cdot \tfrac{1}{2}\alpha(\sqrt{2}x)$ assuming normality

2. $Y'_x = N\left(\tfrac{1}{2}\alpha(\sqrt{2}x) \mp \dfrac{j}{3\sqrt{\pi}} \pm j \cdot f(x)\right)$ where $j = \dfrac{\mu_3}{c^3}$

3. $Y'_x = N\left[\tfrac{1}{2}\alpha(\sqrt{2}x) \mp \dfrac{j}{3\sqrt{\pi}} \pm jf(x) + \dfrac{e^{-x^2}}{\sqrt{\pi}}\left(i(\dfrac{x}{2} - \dfrac{x^3}{3}) \right.\right.$
$$\left.\left. + j^2(\dfrac{-5x}{3} + \dfrac{20x^3}{9} - \dfrac{4x^5}{3}))\right]$$

where $i = c^{-4}m_4 - 3/4$,

m_4 being the 4th moment corrected for grouping.

[36]

Henderson, J., 1922

__Biometrika__ 14, 157

Gram Charlier A or Edgeworth expansion of certain incomplete Γ - and B - integrals. The expansions are taken either at the mean or at the mode of the distribution function:

Expansion of	at	in Table	page	terms up to order	to decimals
I(29.4,49)	mean 49	I	176	30	7
I(42,49)	mean 49	II	178	30	7
I(29.4,49)	mode 48	III	180	30	7
I(42,49)	mode 48	IV	181	30	7
$I_{.5}(15,5)$	mean 175	V	181	8	7
$I_{.5}(4,1.5)$	mean 8/11	VI	182	9	7
$I_{.1}(4,1.5)$	mean 8/11	VII	182	9	7
I(49,49)	mean 49	VIII	184	15 odd only	7

1.33 Repeated normal integrals; truncated normal distributions

1.331 Repeated normal integrals

Repeated integrals of the normal tail area

$$Q(x) = \frac{1}{\sqrt{2\pi}} \int_x^\infty e^{-\frac{1}{2}u^2} du .$$

are required in a number of applications, notably the fitting of a normal distribution to truncated data or the fitting of a truncated normal distribution. The repeated integrals are also related to the incomplete normal moments and the χ^2 integral.

Following B.A.1 , 1931, we write:

$$Hh_0(x) = \sqrt{2\pi}\, Q(x) = \int_x^\infty e^{-\frac{1}{2}u^2} du$$

and define recurrently the repeated integrals $Hh_n(x) = \int_x^\infty Hh_{n-1}(u)\, du$ so that

1.331
1.332   [see also pp. 635-636]

$$Hh_n(x) = \frac{1}{n!} \int_x^\infty (u-x)^n e^{-\frac{1}{2}u^2} du$$

$$= \frac{\sqrt{2\pi}}{n!} \sum_{i=0}^n \binom{n}{i} (-1)^{n-i} x^{n-i} (\mu_i(\infty) - \mu_i(x))$$

where $\mu_i(x) = \frac{1}{\sqrt{2\pi}} \int_0^x u^i e^{-\frac{1}{2}u^2} du$

is the incomplete normal moment discussed in Section 1.31. The above representation of $Hh_n(x)$ also shows that it is related to the nth differential of the ratio $Q(x)/Z(x)$ mentioned in Sections 1.13 and 1.23: Writing $F(x) = Q(x)/Z(x)$ it can be verified by induction that

$$\frac{d^n F(x)}{dx^n} = (-1)^n n! \, Hh_n(x)/Z(x) \sqrt{2\pi}$$

Thus the tables of 'reduced derivatives' of $F(x)$ in Sheppard 1939 are relevant: see Fletcher et al. 1946, p. 224, for details.

The standard table of Hh functions is B.A.1, 1931, T.15, giving 10 dec. for n and x as follows:

n	x	n	x	n	x
0	-7(.1)+6.6	7	-7(.1)+4.5	14	-5(.1)+2.3
1	-7(.1)+6.3	8	-7(.1)+4.2	15	-5(.1)+2
2	-7(.1)+6	9	-7(.1)+3.9	16	-5(.1)+1.6
3	-7(.1)+5.7	10	-7(.1)+3.6	17	-5(.1)+1.3
4	-7(.1)+5.4	11	-7(.1)+3.3	18	-2.5(.1)+1
5	-7(.1)+5	12	-5(.1)+2.5	19	-2.5(.1)+.7
6	-7(.1)+4.8	13	-5(.1)+2.5	20	-2.5(.1)+.4
				21	-2.5(.1)0

All derivatives necessary for interpolation are given. A preliminary version of the table, with inferior typography and without facilities for interpolation, is given in B.A. 1928, p. 325. The 1931 printing (but not reprints) of B.A. 1 contains an extensive introduction by R. A. Fisher on the properties and use of Hh functions.

For tables of repeated integrals of e^{-t^2}, see Fletcher et al. 1946, p. 225.

1.332 Truncated normal distribution

A random sample of $n + a$ observations has been drawn from a normal population. The n smallest observations $x_1 \ldots x_n$ are available and $\leq x$, while the a largest $x_{n+1}, \ldots x_{n+a}$ are not available and are $\geq x$. Required to estimate the mean and variance of the normal population.

Two main cases arise:

A: The scale-point of truncation, x, is predetermined and known.

B: The scale point, x, is not predetermined, has arisen in the experimental process and must also be estimated.

Case A has been completely treated and may be divided into two sub-cases: A_1, $(n+a)$ unknown; A_2, $(n+a)$ known. The epithet <u>truncated</u> is attached to A_1 and the epithet <u>censored</u> to A_2.

Notation:

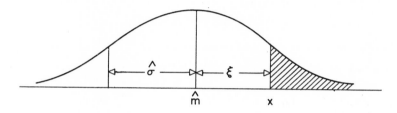

Fig. 1.9

$\hat{\xi}$ = maximum likelihood estimate of distance of x from population mean, m. Location parameter estimate.

$\hat{\sigma}$ = maximum likelihood estimate of σ.

$$\hat{\phi} = \hat{\xi} / \hat{\sigma} \qquad (1.33.1)$$

1.332 [see also pp. 635-636]

A solution for sub-case A_1 is given by R.A. Fisher 1931 and Hald 1949. Compute, from the trunk data $x_1, \ldots, x_n$ the statistics

$$S_1 = \sum x_i, \quad S_2 = \sum x_i^2, \quad y = nS_2 \Big/ 2S_1^2 ; \tag{1.33.2}$$

then $\hat{\phi}$ is the root of
$$y = \frac{Hh_o(\phi) \, Hh_2(\phi)}{Hh_1^2(\phi)} \tag{1.33.3}$$

Tables of y as a function of ϕ are given in B.A. 1, 1931(T.16), K. Pearson 1930 a(T.11), 1931 a(T.12); tables of ϕ as a function of y, in Hald 1949 (130), 1952(T.9), Vianelli 1959(215).

With the help of these functions we obtain $\hat{\phi}$, the estimate of ϕ, and hence $\hat{\sigma}$ from

$$\hat{\sigma} = \frac{1}{n} |S_1| \frac{Hh_o(\hat{\phi})}{Hh_1(\hat{\phi})} = \frac{1}{n} |S_1| \, g(\hat{\phi}) \quad \text{(say)} \tag{1.33.4}$$

where $g(\hat{\phi}) = Hh_o(\hat{\phi})/Hh_1(\hat{\phi})$ is tabled in Hald 1949(131), 1952(T.9), K. Pearson 1930 a(T.11), 1931 a(T.12), Vianelli 1959(216).

The maximum likelihood variances and covariances of $\hat{\phi}$ and $\hat{\sigma}$ are the elements of the inverse of the information matrix A given by

$$(A) = \begin{pmatrix} 2\dfrac{Hh_2}{Hh_o} - \left(\dfrac{Hh_1}{Hh_o}\right)^2 & , & -\dfrac{Hh_1}{Hh_o} \\[2ex] -\dfrac{Hh_1}{Hh_o} & , & 2\dfrac{Hh_2}{Hh_o} + 1 \end{pmatrix} \tag{1.33.5}$$

The values given by Hald as $\mu_{11} \, \mu_{12} \, \mu_{22}$ are, however, the variances and covariances of $\hat{\xi} = -\hat{\phi}\hat{\sigma}$ and $\hat{\sigma}$ and are therefore the elements of the inverse of the transformed information matrix B given by

$$(B) = \begin{pmatrix} 2\dfrac{Hh_2}{Hh_o} - \left(\dfrac{Hh_1}{Hh_o}\right)^2 & , & \phi\left(\dfrac{2Hh_2}{Hh_o} - \left(\dfrac{Hh_1}{Hh_o}\right)^2\right) + \dfrac{Hh_1}{Hh_o} \\[2ex] \phi\left(2\dfrac{Hh_2}{Hh_o} - \left(\dfrac{Hh_1}{Hh_o}\right)^2\right) + \dfrac{Hh_1}{Hh_o} & , & \phi^2\left(2\dfrac{Hh_2}{Hh_o} - \left(\dfrac{Hh_1}{Hh_o}\right)^2\right) + \phi\dfrac{Hh_1}{Hh_o} + 2 \end{pmatrix}$$

A solution for sub-case A_2 is given by Stevens 1937a and Hald 1949. Compute the statistics of (1.33.2). The maximum likelihood estimate of is then given by the root $\hat{\phi}$ of

$$2y = \frac{\phi}{\left(\phi - \frac{h}{1-h} \frac{Hh_{-1}(\phi)}{\sqrt{2\pi} - Hh_0(\phi)}\right)} + \frac{1}{\left(\phi - \frac{h}{1-h} \frac{Hh_{-1}(\phi)}{\sqrt{2\pi} - Hh_0(\phi)}\right)^2} \quad (1.33.6)$$

where $h = a/(n+a)$.

Hald 1949(132), 1952(T.10), Vianelli 1959(217) table[*] $\phi = f(y, h)$ to 3 dec. for $y = .5(.005).6(.01).8(.05)1.5$ and $h = .05(.05).5$. The maximum likelihood estimate of σ is then given by

$$\hat{\sigma} = \frac{S_1}{n-a} g(h, \hat{\phi}) \quad (1.33.7)$$

where

$$g(h, \phi) = \left[\frac{h\psi'(\phi)}{1-h} - \phi\right]^{-1} \quad (1.33.8)$$

and the ratio

$$\psi'(\phi) = Hh_{-1}(\phi) \Big/ [(2\pi)^{\frac{1}{2}} - Hh_0(\phi)] \quad (1.33.9)$$

is tabled to 5 fig. for $\phi = -3(.1)+2$ by Hald 1949(134), 1952(T.10), Vianelli 1959(218). The variance-covariance matrix of $\hat{\phi}$ and $\hat{\sigma}$ is $(2\pi)^{\frac{1}{2}}\sigma^2/n$ times the inverse of the matrix

$$\begin{bmatrix} Hh_0 + J & -Hh_{-1} + J \\ -Hh_{-1} + J & 2Hh_0 - Hh_{-1} + \phi^2 J \end{bmatrix} \quad (1.33.10)$$

where $J = Hh_{-2} + Hh_{-1}^2 \Big/ [(2\pi)^{\frac{1}{2}} - Hh_0]$ and the argument of all the Hh functions is $-\phi$. This matrix is equal to $(2\pi)^{\frac{1}{2}}$ times

---

[*] The functions f and g of Hald 1946 are entirely different from the functions f and g discussed above.

1.332   [see also pp. 635-636]

$$\left\{ \begin{array}{ll} \mu_2(\phi) + \tfrac{1}{2} + \dfrac{z^2}{Q}, & \mu_3(\phi) - \mu_1(\phi) - \dfrac{1}{\sqrt{2\pi}} + \phi\dfrac{z^2}{Q} \\ \mu_3(\phi) - \mu_1(\phi) - \dfrac{1}{\sqrt{2\pi}} + \dfrac{\phi z^2}{Q}, & \mu_4(\phi) - 2\mu_2(\phi) + \mu_0(\phi) + 1 + \dfrac{\phi^2 z^2}{Q} \end{array} \right\}$$

where Q is the upper tail area and the $\mu_i(\phi)$ are incomplete normal moments:

$$\mu_i(\phi) = \dfrac{1}{\sqrt{2\pi}} \int_0^\phi x^i \, e^{-\tfrac{1}{2}x^2} \, dx$$

Hald 1949(134), 1952(T.10) tables $\mu_{11}(\phi) = n\sigma^{-2}\text{Var}\,\hat{\phi}$, $\mu_{12}(\phi) = n\sigma^{-2}\text{Cov}(\hat{\phi},\hat{\sigma})$ and $\mu_{22}(\phi) = n\sigma^{-2}\text{Var}\,\hat{\sigma}$ to 4 fig.; and $\rho(\phi)$, the correlation between $\hat{\phi}$ and $\hat{\sigma}$, to 3 dec.; all for $\phi = -3(.1)+2$.

Reprinted in Vianelli 1959, <u>Prontuario</u> 72, p. 218.

Stevens, W.L. 1937a

<u>Annals</u> <u>of</u> <u>Applied</u> <u>Biology</u> 24, 844-847

The elements of the inverse of (1.33.10) are tabled to 5 dec. with δ^2 and δ^4 for $\phi = -2(.1)+2.5$.

Lee, A. 1914

<u>Biometrika</u> 10, 208-214

$$\left. \begin{array}{l} \psi_1 = 2\,\dfrac{Hh_0(x)\,Hh_2(x)}{Hh_1^2(x)} - 1 \\[2ex] \psi_2 = \dfrac{Hh_0(x)}{Hh_1(x)} \\[2ex] \psi_3 = \dfrac{\sqrt{2\pi}}{Hh_0(x)} \end{array} \right\} \quad \begin{array}{l} 3 \text{ dec. } -x = .00(.01).10(.1)3.0 \\[2ex] \Delta \end{array}$$

Reprinted in Tables for S & B II(T.12).

Keyfitz, N. 1938

<u>Annals</u> <u>of</u> <u>Mathematical</u> <u>Statistics</u> 9, 66-67

ψ_1, ψ_2 $\begin{cases} 4 \text{ dec.} \quad x = 0(.1)3 \quad \Delta \\ 3\text{-}7 \text{ fig.} \quad x = 3(.5)5 \quad \text{No } \Delta \end{cases}$

Pearson, K. 1930 a

<u>Tables</u> <u>for</u> <u>S</u> & <u>B</u> I, T.11, pp. xxviij - xxxj, 35

ψ_1, ψ_2, ψ_3 3 dec. $x = 0(.01).1(.1)3, 3.5$

Pearson, K. and Lee, Alice 1908

<u>Biometrika</u> 6, 59-68

Table 2 gives:

ψ_1, ψ_2 3 dec. $x = .1(.1)3$

B.A. 1, 1931, T. 16, p. 72

$Hh_0(x) Hh_2(x) / [Hh_1(x)]^2$ 9; 8-3 dec. $-7(.1)+5$; 0, +1, 2, 3, 4, 4.5

Gupta, A. K., 1952

Biometrika 39, 260

Maximum likelihood estimates of μ and σ in censored samples of Type 2.

Let $x_1 \ldots x_k$ denote the k smallest observations in a sample of n from a normal population with mean μ and standard deviation σ while $x_{k+1} \ldots x_n$ are not known. The maximum likelihood estimates of μ and σ require the root Z of the simultaneous equations in x and z:

$$z = x + (\frac{1}{p} - 1) \frac{Z(x)}{Q(x)} \; ; \quad \psi = \frac{1 + xz - z^2}{1 + xz}$$

where $p = k/n$ and ψ is computed from the $x_1 \ldots x_k$; in fact

$$\psi = s^2/(s^2 + d^2) \quad \text{where} \quad d = x_k - \frac{1}{k}\sum_{i=1}^{k} x_i \quad \text{and} \quad s^2 = \frac{1}{k}\sum_{i=1}^{k}(x_i - \bar{x})^2$$

1.332   [see also pp. 635-636]

Table 1 (p. 262) gives z to 4 dec. for $\psi = .05(.05).95$, $p = 0.1(.1)1.0$. Having found z from this table the **maximum likelihood estimates** are given by

$$\hat{\sigma} = d/z \; ; \quad \hat{\mu} = \bar{x} + (\hat{\sigma}^2 - s^2)/d$$

while their large sample variances and covariances v_{ij} are given in Table 2 (p. 263) to 5 dec. for $p = .05(.05).95(.01).99$.
These are computed from the formulas:-

$$v_{11} = p + Z(\tfrac{Z}{Q}) - \hat{x})$$

$$v_{12} = -Z + xZ(\tfrac{Z}{Q} - \hat{x})$$

$$v_{22} = 2p - \hat{x}Z + \hat{x}^2 Z(\tfrac{Z}{Q} - \hat{x})$$

where $\hat{x} = (x_k - \hat{\mu})/\hat{\sigma}$ and Z and Q are taken at **argument** $\hat{x}$. These formulas can be transformed to Hald's **maximum likelihood equations** for μ and σ.

Thompson, H. R. 1951
Biometrika 38, 414

Moment fit of truncated log normal distribution.

Let r be the observed variate and let $x = \log_e r$ be normally distributed with mean μ and standard deviation σ which are to be estimated. Let a be the point of truncation on the x-scale, and let $t = (a - \mu)/\sigma$. The estimation of μ and σ from the first two moments of x about a is facilitated by tables of

$$\Phi_1(t) = Z(t) - tQ(t),$$

$$\Phi_2(t) = [\Phi_1(t)]^2 / [(t^2+1)Q(t) - tZ(t)].$$

Table 1 (p. 417) gives these two functions with Δ^2 to 4 dec. for $t = -4(.1)+2$.
If data are grouped in categories of integer width in r, the group centers for x are given by $x = \tfrac{1}{2}[\log_e r + \log_e(r+1)]$. Table 3 (p. 417) gives x and x^2 to 3 dec. for $r = 1(1)100$ to facilitate the computation of grouped moments.

Table 2 (p. 418, 419) gives adjustments for grouping for μ and σ based on empirical calculations.

Kimball, B.F. 1944

Annals of Mathematical Statistics 15, 423-427

Asymptotic distribution of the sum of absolute values of n observations from a normal parent with mean zero.

Let $F_n(u) = \dfrac{d}{du}\left[(2\pi)^{-\frac{1}{2}n}\int_{R'} \exp\left(-\frac{1}{2}\sum_{i=1}^{n} x_i^2\right) dx_1 \ldots dx_n\right]$, where R' is the region bounded by $x_i \geq 0$ and by $\sum x_i \leq u$. Kimball proves that $\lim\limits_{u \to \infty} n^{\frac{1}{2}} F_n(u)/Z(n^{-\frac{1}{2}}u) = 1$. Defining u' as the least value of u such that $\left| n^{\frac{1}{2}} F_n(u)/Z(n^{-\frac{1}{2}}u) - 1 \right| \leq \delta$ for $u \geq u'$, Kimball tables $n^{-\frac{1}{2}}u'$ on p. 426 to 2 dec. for $\delta = .5, .25, .1, .01, .001$ and $n = 2, 3, 4, 5$.

Gjeddebaek, N. F. 1949

Skandinavisk Aktuarietidskrift 32, 135-159

"Contribution to the study of grouped observations. Application of the method of maximum likelihood in case of normally distributed observations."

Table 1, pp. 151-154, gives $z_1(x,y) = \dfrac{Z(x) - Z(x+y)}{P(x+y) - P(x)}$ to 4 dec. for $y = 0(.1)4$ and $x = -.1[5y+1](.1)4$.

Table 2, pp. 155-158, gives $z_2(x,y) = \dfrac{Z'(x+y) - Z'(x)}{P(x+y) - P(x)}$ to 4 dec. for the above x and y.

Table 3, p. 159, gives $z_1(x,\infty) = Z(x)/Q(x)$, and $z_2(x,\infty) = -xZ(x)/Q(x)$, to 4 dec. for $x = -4.5(.1)+4.5$.

1.4
1.40

[see also pp. 635-636]

1.4 Rankings of normal samples; normal order statistics; the median; quartiles and other quantiles

1.40 Definitions and notation

Let $x_1, x_2, \ldots, x_n$ denote a random sample drawn from a normal population with mean 0 and $\sigma = 1 (N(0, 1))$ and arranged in order of magnitude so that x_1 is the smallest and x_n is the largest. In general we shall call the ith value x_i the ith order statistic in a sample of n and when necessary write this more fully as $x_{i|n}$. (See Fig. 1.10).

We shall abbreviate $x_{i|n}$ to x_i when no confusion can arise; and similarly we may omit the suffix $|n$ from the notations introduced below.

We further use the following notation:

The ith ranking score = expected value of $x_{i|n} = E(x_{i|n}) = \xi_{i|n}$

Its standard deviation $\left[E\left((x_{i|n} - \xi_{i|n})^2 \right) \right]^{\frac{1}{2}} = \sigma_{i|n}$

The correlation between i^{th} and j^{th} order statistics

$$\sigma_i^{-1} \sigma_j^{-1} E\left[(x_i - \xi_i)(x_j - \xi_j) \right] = \rho_{ij|n}$$

In particular we have

'Upper straggler' (or largest observation) = $x_{n|n}$

'Lower straggler' (or smallest observation) = $x_{1|n}$

For a symmetrical distribution like the normal, it is usually permissible to speak about the distribution of the 'straggler' which is that of $x_{n|n}$ or $x_{1|n}$.

Further special cases are:

Median $x = x_{\frac{1}{2}(n+1)|n}$ for n odd

$= \frac{1}{2}\left[x_{\frac{1}{2}n|n} + x_{\frac{1}{2}n+1|n} \right]$ for n even

1.4
1.40

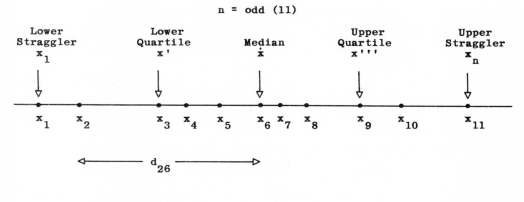

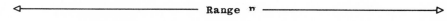

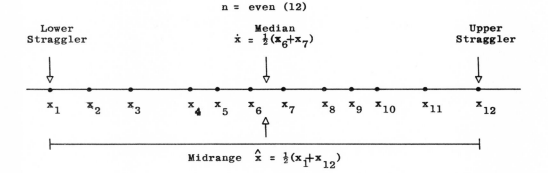

Fig. 1.10

[47]

1.40 [see also pp. 635-636]

Obviously $E(\dot{x}) = 0$ and $\sigma_{\dot{x}} = [E(\dot{x}^2)]^{\frac{1}{2}}$.

The quartiles[*]:

Lower quartile $\quad x' = x_{m+1} \quad$ for $\quad n = 4m+3$
$\hspace{11em}$ or $\quad n = 4m+2$

$\hspace{6em} = \frac{1}{2}(x_m + x_{m+1}) \quad$ for $\quad n = 4m+1$
$\hspace{11em}$ or $\quad n = 4m$

Upper quartile $\quad x''' = x_{3m+3} \quad$ for $\quad n = 4m+3$

$\hspace{6em} = x_{3m+2} \quad$ for $\quad n = 4m+2$

$\hspace{6em} = \frac{1}{2}(x_{3m+1} + x_{3m+2}) \quad$ for $\quad n = 4m+1$

$\hspace{6em} = \frac{1}{2}(x_{3m} + x_{3m+1}) \quad$ for $\quad n = 4m$

The rank intervals $d_{ij|n} = x_{j|n} - x_{i|n}$ have mean $E(d_{ij|n}) = \xi_{i|n} - \xi_{j|n}$ and **standard deviation** $\quad \sigma_{ij|n} = [\sigma_i^2 + \sigma_j^2 - 2\rho_{ij}\sigma_i\sigma_j]^{\frac{1}{2}}$.

Other special combinations of order statistics are the range $R_n = d_{1n|n}$, the ith gap $d_{i,i+1|n}$, and the midrange $\hat{x} = \frac{1}{2}[x_{1|n} + x_{n|n}]$.

The subscript, in our notation for moments of order statistics, denotes the degree of the moment. Thus:

$$\mu'_r(i|n) = E[x^r_{i|n}],$$

in particular $\quad \mu'_1(i|n) = \xi_{i|n}$;

$$\mu_r(i|n) = E[(x_{i|n} - \xi_{i|n})^r],$$

in particular $\quad \mu_2(i|n) = \sigma^2_{i|n}$;

$$\mu'_r(ij|n) = E[d^r_{ij|n}],$$

in particular $\quad \mu'_1(ij|n) = \xi_{j|n} - \xi_{i|n}$;

$$\mu_r(ij|n) = E[(d_{ij|n} - \xi_{j|n} + \xi_{i|n})^r],$$

in particular $\quad \mu_2(ij|n) = \sigma^2_{ij|n}$.

---

[*] These somewhat unsymmetrical definitions of quartiles are due to Hojo 1931, p. 332. Quartiles for $n \neq 4m+3$ are not often used now; accordingly, they are shown in Fig. 1.10 only for $n = 4m+3$.

We define the ordered deviate

$$x'_{i|n} = x_{i|n} - \bar{x} \quad \text{(or short } x'_i\text{)}$$

so that

$$E(x'_{i|n}) = \xi_{i|n}.$$

In particular x'_1 or x'_n is the **straggler deviate**.

Let s_ν^2 denote a sample variance distributed as χ^2/ν with ν d.f. and independent of the sample from which the order statistics were computed. We then define the studentized order statistics

$$t(i|n\nu) = x_{i|n}/s_\nu$$

$$t(ij|n\nu) = (x_{j|n} - x_{i|n})/s_\nu$$

the studentized range

$$q(n,\nu) = t(i,n|n,\nu) = (x_{n|n} - x_{1|n})/s_\nu = R_n/s_\nu$$

and the studentized ordered deviates

$$t'(i|n\nu) = x'_{i|n}/s_\nu = (x_{i|n} - \bar{x})/s_\nu \quad \text{(or short } t'_i\text{)}.$$

In particular t'_1 or t'_n is the **studentized straggler deviate**.

The following formulas for quantiles[*] in large samples are based on K. Pearson 1920. Define the sample quantile x_p as the largest observation such that the number of $x_i \geq x_p$ equals or exceeds np. (If $P = 100p$, we speak of the Pth percentile) Then $E(x_p)$ is approximately $x(p)$, and σ_p is approximately

$$\frac{[p(1-p)]^{\frac{1}{2}}}{n^{\frac{1}{2}} Z[x(p)]}.$$

---

[*] The definition of quantile given is appropriate only in large samples, and in modern practice quantiles are used only in large samples. For attempts to distinguish <u>quantile</u> from <u>tantile</u>, and <u>centile</u> from <u>percentile</u>, and for the term <u>fractile</u>, see Kendall & Buckland 1957.

1.40
1.41   [see also pp. 635-636]

Define the sample quantile range $x_{p_1 p_2} = x_{p_1} - x_{p_2}$; then $E(x_{p_1 p_2})$ is approximately $x(p_1) - x(p_2)$, and $\sigma_{p_1 p_2}$ is approximately

$$n^{-\frac{1}{2}} \left[\frac{p_1(1-p_1)}{z_1^2} + \frac{p_2(1-p_2)}{z_2^2} - \frac{2p_1(1-p_2)}{z_1 z_2} \right]^{\frac{1}{2}}, \text{ where } Z_i = Z[x(p_i)]. \ddagger$$

The correlation between quantiles, $\rho_{p_1 p_2}$, is approximately

$[p_1(1-p_2)]^{\frac{1}{2}}[p_2(1-p_1)]^{-\frac{1}{2}}$. Defining $x_{pq} = x_p - x_{1-p}$, $\hat{x}_p = \frac{1}{2}(x_p + x_{1-p})$, we find that approximately

$$\sigma_{pq} = [2p(1-2p)/n]^{\frac{1}{2}}/Z[x(p)], \quad \sigma_{\hat{p}} = (p/2n)^{\frac{1}{2}}/Z[x(p)].$$

1.41 Median and other quantiles

Again, let $x_1, x_2, \ldots, x_n$ denote a random sample drawn from $N(0,1)$ and arranged in order of magnitude. When the sample size is odd, i.e., $n = 2k - 1$, the median is $\dot{x} = x_k$, and the probability that $\dot{x} \leq x^*$ is given by $I_{P(x^*)}(k,k)$ where I denotes the incomplete Beta function ratio (see Section 3) and $P(x)$ the normal probability integral. The density function of $\dot{x}$ is given by $[B(k,k)]^{-1}[P(1-P)]^{k-1} Z(\dot{x})$. Since $E(\dot{x}) = 0$, its central moments reduce to ordinary moments:

$$\mu_i(\dot{x}) = [B(k,k)]^{-1} \int_0^1 [x(P)]^i [P(1-P)]^{k-1} dP.$$

For $i = 2$ this yields $\sigma_{\dot{x}}^2$. No closed form exists for these moments; various quadrature formulas have been used.

Fisher & Yates 1938, 1943, 1948, 1953 (T.X), 1957 (T. IX 1), give $x(1-Q)$ to 4 dec. where Q is a vulgar fraction, in lowest terms, not exceeding $\frac{1}{2}$, and with denominator not exceeding 30.

---

‡These formulas demand that p_1 be less than p_2; the next set of formulas demand that p be less than q.

Pearson, K. 1920

Biometrika 13, 113-132

Large-sample standard errors of quantiles and interquantile ranges

Table 1, p. 118, gives

$$\sqrt{n}\ \sigma_p = \sqrt{p(1-p)}\, Z^{-1}(p)\ ^*$$
$$\sqrt{n}\ \sigma_p^* = \sqrt{1 + \tfrac{1}{2}x^2(p)}$$
$$\sigma_p / \sigma_p^*$$

all to 4 dec. for $p = .5, .6, .7, .75, .8, .9$.

Table 2, p. 119, gives

Range $2x(p)$

$$\sqrt{2n}\ \sigma_{pq} = 2Z^{-1}(p)\sqrt{p(1-2p)}$$
$$\sqrt{2n}\ \sigma = \sqrt{2n}\ \sigma_{pq}/2|x(p)|$$

to 4 dec. for $p = \tfrac{1}{40}, \tfrac{1}{30}, \tfrac{1}{20}, \tfrac{1}{15}, \tfrac{1}{14}, \tfrac{1}{13}, \tfrac{1}{12}, \tfrac{1}{10}, \tfrac{2}{10}, \tfrac{1}{4}, \tfrac{3}{10}, \tfrac{4}{10}$

Table 3, p. 120, gives

Range $(x(1-p_1) - x(1-p_2))$

$$\sqrt{2n}\ \sigma_{p_1 p_2}$$
$$\sqrt{2n}\ \sigma_{p_1 p_2}/(x(1-p_1) - x(1-p_2))$$

to 4 dec. for the following pairs p_1, p_2

$p_1 =$	.1	.2	.25	.3	.4	.3	.1	$\tfrac{1}{14}$	$\tfrac{1}{20}$
$p_2 =$	.5	.5	.5	.5	.5	.9	.8	.9	.9

Hojo, T. 1931

Biometrika 23, 315-360

"Distribution of the mean, quartiles and interquartile distance"

Tables II^a and II^b (p. 327) give

$\sigma_{\dot{x}}$ to 5 dec.
$\sigma_{\dot{x}}^2$ to 7 dec.   for $n = 1(1)12$
$\sigma_{\dot{x}} \sqrt{n}$ to 4 dec.

$\mu_4(\dot{x})$ to 6 dec.
$\beta_2(\dot{x})$ to 4 dec.   for $n = 1(1)8$

These values are all obtained from the exact formulas (see pp. 317-324) by numerical quadrature.

---

* $Z^{-1}(p)$ is an abbreviation for $1/Z[x(p)]$.

1.41 [see also pp. 635-636]

In table II^c (pp. 328) the exact values of $\sigma_{\dot{x}}\sqrt{n}$ are compared with those obtained from an empirical formula of the form

$$a_0 + a_1/n + a_2/n^2 + a_3/n^3 \tag{1}$$

fitted by moments to the values for $n \leq 12$ and $n = \infty$ as well as with K. Pearson's approximate formula (equation (iv) of appendix) obtained by a Taylor expansion of $x(p)$ in powers of p.

All values to 4 dec. for $n = 1(2)19,25,31,35,41,45,55,81,101,161,\infty$ with some ommissions.

Table II^d (p. 329) gives a comparison of exact values of $\sigma_{\dot{x}}\sqrt{n}$ with (1) to 4 dec. for $n = 2(2)20,30(10)100,\infty$. Tables III^a, III^b (p. 329-330) give a comparison of the theoretical values of $\sigma_{\dot{x}}\sqrt{n}$ and $\beta_2(\dot{x})$ to 4 dec. with those obtained from a sampling experiment using 1000 samples of $n = 3,4,7,10,15,20$ and 500 samples of $n = 30,40$.

Values for $n > 12$ are obtained from various approximations.

Table IV^a (p. 341) gives 6 decimal values of $E(x''')$, $\sigma_{x'''}$, $\mu_2(x''')$, $\mu_3(x''')$, $\mu_4(x''')$ for $n = 1(1)13$. (μ_3 and μ_4 only for $n = 1(1)10$). These have been computed for the exact formulas (pp. 333-340) by numerical quadrature.

Table IV^b (p. 344) gives $n^{\frac{1}{2}}\sigma_{x'''}$ to 4 dec. for $n = 1(1)13$.

Table IV^c (p. 344) gives $\beta_1(x''')$, $\beta_2(x''')$ to 4 dec. for $n = 1(1)10$.

Table V gives the results of a sampling experiment with 1000 samples of $n = 4,10,20$ and 500 samples of $n = 30,40$ giving $E(x''')$, $\mu_i(x''')$ $i = 2,3,4$ β_1, β_2, $\sigma_{x'''}$, $\sigma_{x'''}\sqrt{n}$ as well as the same results for x' and for x' and x''' combined.

Pearson, K. and Pearson, Margaret V. 1931
Biometrika 23, 364-397

Table 6, p. 382, gives $\sigma_{\dot{x}}$ and $n^{\frac{1}{2}}\sigma_{\dot{x}}$ to 6 dec.: from K. Pearson's approximate formula (xviij), p. 381, for $n = 1(2)25, 31, 33, 51, 53, 71, 73, 99, 101$; from the exact formula of Hojo 1931, p. 317, for $n = 1(2)11$;

and from the asymptotic formula of section 1.40, above, for n = 71, 73, 99, 101. Hojo's and Pearson's formulas are compared graphically on pp. 383-384.

See also section 1.43 .

Hojo, T. 1933
Biometrika 25, 79-90

Define x^* to equal $\frac{1}{2}(x' + x''')$. Table 1, p. 80, gives $n^{\frac{1}{2}}\sigma_{x'}$ and $\rho_{x'x'''}$ to 4 dec. and $n^{\frac{1}{2}}\sigma_{\dot{x}}$ and $n^{\frac{1}{2}}\sigma_{x^*}$ to 3 dec. for n = 4, 7, 10, 12, 22, ∞ . Table 2, p. 84, gives $\sigma_{\dot{x}}$, $\sigma_{x'}$, and cov($\dot{x}$, x') to 5 dec. and $\rho_{\dot{x}x'}$ to 4 dec. for n = 1(1)12 . An attempt to link these exact results with the large-sample formulas of K. Pearson 1931z leads to four empirical formulas, for n = 4m, 4m+1, 4m+2, 4m+3, for $\sigma_{x'}$, and to four formulas for $\rho_{\dot{x}x'}$; from these formulas Hojo computed Table 3, p. 87, giving

$$\sigma_{\dot{x}}\sqrt{n} \qquad \sigma_{x'}\sqrt{n} \qquad \rho_{\dot{x}x'} \qquad \sigma_{x'-\dot{x}}\sqrt{2n}$$

to 4 dec. for n = 1(1)20(5)60,99,100;101,102,500,∞ , also values for n = 49,50,52,99,100,101,102,500 obtained by different methods.

The discontinuity in formula as n goes through a cycle of 4 caused Pearson to propose alternative formulas for sample quartiles: K. & M.V. Pearson 1931, p. 377.

Table 4, p. 88, gives the standard errors of three competitive estimates of the population standard deviation, viz:

$$\frac{s}{E(s)} = \frac{\sqrt{\sum(x_i-\bar{x})^2}\,\Gamma(\frac{n-1}{2})}{\sqrt{2}\,\Gamma(\frac{1}{2}n)} , \quad \frac{x'''-x'}{E(x'''-x')} \quad \text{and} \quad \frac{x'''-\dot{x}}{E(x''')}$$

The standard deviations of these estimates, multiplied by $(2n)^{\frac{1}{2}}$, are tabled to 3 dec. for n = 2(1)12, ∞ .

Gumbel, E. J. 1943
Annals of Mathematical Statistics 14, 163-179

Table 1, p. 177, gives $(\frac{1}{2}n)^{\frac{1}{2}}\sigma_p = [\frac{1}{2}p(1-p)]^{\frac{1}{2}}/Z(x)$ for $2^{-\frac{1}{2}}x = 0(.2)1$.

1.42 [see also pp. 637-638]

1.42 **Distribution of straggler and straggler deviate; powers of normal tail area**

Let x_n (or more precisely $x_{n|n}$) denote the straggler, i.e. the largest observation in a sample of n observations from $N(0,1)$. The chance for $x_n \leq x^*$ (say) is given by $P^n(x^*)$ so that the distribution of x_n is given by $nP^{n-1}(x_n) Z(x_n)$ and its moments by

$$\mu'_i(x_n) = n \int_{-\infty}^{+\infty} x_n^i P^{n-1}(x_n) Z(x_n) dx_n$$

$$= n \int_0^1 x^i(P) P^{n-1} dP .$$

Sheppard, W.F. 1899

Proceedings of the London Mathematical Society 31, 70-99

Rejection of univariate, bivariate, and trivariate normal stragglers

(a) Single variate

Let $x_1, x_2, \ldots, x_n$ denote an ordered sample of size n drawn from a normal parent $N(0,1)$. Then the probability that $\max\left(|x_1|, |x_n|\right) \leq x$ is given by

$$(2P(x) - 1)^n.$$

Table I (p. 95) gives the median of this distribution (viz. roots x of $(2P(x) - 1)^n = \frac{1}{2}$ to 5 dec. for $\log_{10} n = 1.0(.1)6.0$ Δ^2. Also shown are 5 dec. values of

$$(2P(x) - 1)/(2P(x) - 1 - 2xZ(x)) = P \text{ (say)}$$

and 5 dec. values of

$10 + \log_{10}(P - 1)$, the latter with Δ^2.

These are required for converting the mean square computed from the retained (n or $n-1$) x_i into a 'most probable' mean square for the original parent (under H_o).

(b) Two correlated normal variates

The critical regions are ellipses of equal frequency which by linear transformation are transformed into circles of radius r, so that the chance for all n points obtained by linear transformation from the x_{ij} to lie within the circle of radius r is given by $(1 - e^{-\frac{1}{2}r^2})^n$.

Table II (pp. 96-97) gives the median of this distribution (viz. roots r of $\left|1 - e^{-\frac{1}{2}r^2}\right|^n = \frac{1}{2}$) to 5 dec. with Δ^2 for $\log_{10} n = 1(.1)6$; also shown are r^2 to 5 dec. with Δ^2, $P = [1 - e^{-\frac{1}{2}r^2}]/[1 - (1+\frac{1}{2}r^2)e^{-\frac{1}{2}r^2}]$ to 5 dec. and $10 + \log_{10}(P - 1)$ to 5 dec. with Δ^2.

These are required to correct the variances and covariances computed from the retained n or $n - 1$ pairs (x_i, y_i).

(c) Three correlated normal variates.

A similar argument leads to the tabulation in Table III (pp. 98-99) of

$$R = \text{root of } \left[1 - \sqrt{\frac{2}{\pi}} \int_R^\infty e^{-\frac{1}{2}x^2} x^2 \, dx \right]^n = \frac{1}{2}$$

to 5 dec. for $\log_{10} n = 1.0(.1)6.0, \Delta^2$; R^2 to 5 dec., Δ^2.

$$P = 3 \int_0^R e^{-\frac{1}{2}x^2} x^2 \, dx \bigg/ \int_0^R e^{-\frac{1}{2}x^2} x^4 \, dx$$

to 5 dec. and $10 + \log_{10}(P-1)$ to 5 dec., Δ^2.

1.42 [see also pp. 637-638]

Tippett, L.H.C. 1925

<u>Biometrika</u> 17, 364-387

"On the extreme individuals and the range of samples taken from a normal population"

Table I (p. 366) gives

$\xi_{n|n}$ to 5 dec.

$\sigma_{n|n}$ to 4 dec.   for $n = 2, 5, 10, 20, 60, 100, 200, 500, 1000$

$\beta_1(x_{n|n})$
$\beta_2(x_{n|n})$ to 3 dec.

all obtained by numerical quadrature.

Table II (p. 367) gives upper percentage points of the straggler, i.e. roots x of $[P(x)]^n = 1 - p$, to 4 dec. for $p = .05, .01$ and $n = 10, 20, 50, 100, 200, 500, 1000$.

10,000 observations were drawn from a grouped normal distribution (h = 0.1) and formed into 1000 samples of 10. For each sample the extremes x_{10} and x_1 were noted, and in Table VII (p. 381) the joint frequency distribution of x_{10} and x_1 is given. (Each sample pair has been counted twice, once as x_{10}, x_1 and once as $-x_1, -x_{10}$.) A similar table (Table VIII (p. 382)) is set out for samples of 20.

Table IX (pp. 384-385) gives:

Powers of normal area $P^n(x) = [\tfrac{1}{2}(1 + \alpha(x))]^n$

7 dec. $n = 3, 5, 10$; $x = -2.6(0.2)5.8$

$n = 20, 30, 50, 100(100)1000$; $x = -0.1(0.1)6.5$

No Δ.

Error: For $n = 20$, $x = 2.7$

For $P^n(x) = .9328910$ read $P^n(x) = .9328975$.

Reprinted in Tables for S & B II, T.21.

Pearson, E. S., 1926

Biometrika 18, 173

On p. 178 the following integrals are given to 7 dec.:

$$\sqrt{2\pi}\, I_i = \int_{-\infty}^{+\infty} P^i(x)\, \sqrt{2\pi}\, Z(\sqrt{2}x)\, dx \qquad i = 2(1)4$$

$$\sqrt{2\pi}\, T_i = \int_{-\infty}^{+\infty} P^i(x)\, \sqrt{2\pi}\, Z(\sqrt{3}x)\, dx \qquad i = 2,3$$

$$\sqrt{2\pi}\, K_2 = \int_{-\infty}^{+\infty} P^2(x)\, \sqrt{2\pi}\, Z(2x)\, dx \;,\; \int_{-\infty}^{+\infty} P^2(x)\, \sqrt{2\pi}\, Z(\tfrac{1}{\sqrt{2}}x)\, dx \;,$$

$$\int_{-\infty}^{+\infty} P(x)\, P(\sqrt{2}x)\, \pi\sqrt{2}\, Z(\sqrt{3}x)\, dx \;,$$

$$\int_{-\infty}^{+\infty} P(x)\, P(\sqrt{3}x)\, \frac{2\pi}{\sqrt{3}}\, Z(\sqrt{2}x)\, dx$$

Hojo, T., 1931

Biometrika 23, 315

Table I (p. 325-6) gives the following integrals all to 8 dec. accuracy:

$$T_i = \int_{-\infty}^{+\infty} P^i(x)\, Z(\sqrt{3}x)\, dx \qquad i = 2(1)10$$

$$I_i = \int_{-\infty}^{+\infty} P^i(x)\, Z(\sqrt{2}x)\, dx \qquad i = 2(1)11$$

$$R_i = \int_{-\infty}^{+\infty} P^i(x)\, P(\sqrt{2}x)\, Z(\sqrt{2}x)\, dx \qquad i = 1(1)8$$

1.42 [see also pp. 637-638]

$$_iI_j = \int_{-\infty}^{+\infty} P^i(x)\, Z(\sqrt{2}x) \int_{-\infty}^{x} P^j(u)\, Z(\sqrt{2}u)\, du\, dx$$

$j = 1$, $i = 2(1)8$
$j = 2$, $i = 3(1)6$
$j = 3$, $i = 4,5$

$$S_i = \int_{-\infty}^{+\infty} P^i(x)\, Z(\sqrt{5}x)\, dx \qquad i = 2(1)5$$

$$K_i = \int_{-\infty}^{+\infty} P^i(x)\, Z(2x)\, dx \qquad i = 2(1)6$$

$$L_{ij} = \int_{-\infty}^{+\infty} P^i(x)\, P(\sqrt{2}x)\, Z(\sqrt{j}x)\, dx \qquad i = 2,\ j = 2,4$$
$$\int_{-\infty}^{+\infty} P(x)\, P(\sqrt{3}x)\, Z(\sqrt{3}x)\, dx \qquad\qquad i = 3,\ j = 3,4$$
$\qquad\qquad\qquad\qquad\qquad\qquad\qquad\qquad\qquad i = 4,\ j = 3$

Pearson, K. 1931a

Tables for S & B II, T.XXI bis , pp. cx-cxix, 164

Upper percentage points of straggler in normal samples of size n.

Roots x of $P^n(x) = 1 - p$.

3 dec. $p = 0.005, 0.01, 0.05, 0.10$

$n = 1(1)25, 30(10)150, 200(100)1000$

partly Δ **partly** δ^2.

De Finetti, B. 1934

Metron 9, No. 3, 127-138

Table 1 (pp. 135-136) gives the density function of the maximum of n normal deviates, viz: $nZ(x)[P(x)]^{n-1}$ to 3 dec. for $n = 1(1)10$ and $x = -3.5(.1)+4$.

Table 2 (p. 137) gives the median of the maximum of n normal deviates, viz: the root of $[P(x)]^n = \tfrac{1}{2}$,

(a) to 4 dec. for $n = 1(1)50(10)500(100)1000$
(b) to 2 dec. for $n = n' \cdot 10^i$ with
$\qquad\qquad i = 3(1)8$
$\qquad\qquad n' = 1, 1.2, 1.5, 2, 2.5, 3(1)9$.

Daniels, H.E. 1941

Biometrika 32, 194-195

Using values of the mean $\xi_{n|n}$ and the standard deviation $\sigma_{n|n}$ of the straggler in normal samples of size n, given by Tippett 1925, the relation

$$\xi_{n|n} = 2 \cot\left(\tfrac{1}{2} \pi \, \sigma_{n|n}\right)$$

was found to hold empirically.

Table 1 (p. 195) gives $\xi_{n|n}$, $\sigma_{n|n}$, $2 \cot\left(\tfrac{1}{2} \pi \, \sigma_{n|n}\right)$ and $\xi_{n|n} - 2 \cot\left(\tfrac{1}{2} \pi \, \sigma_{1|n}\right)$ to 4 dec. for n = 1,2,5,10,20,60,100,200,500,1000 and Table 2 (p. 195) gives the same for n = 7228, 639.10^3, 264.10^9.

Nair, K.R. 1948 b

Biometrika 35, 118-144

Let $x_1 \ldots x_n$ represent an ordered sample from a normal parent with $\sigma = 1$. The probability $P_n(u)$ for $x_n - \bar{x} \leq u$ is given by

$$P_n(u) = \sqrt{n}(2\pi)^{-\frac{n-1}{2}} G_{n-1}(n\,u)$$

where the G functions are defined by the recurrence

$$G_o(x) \equiv 1; \quad G_r(x) = \int_o^x \exp\left[-\frac{t^2}{2r(r+1)}\right] G_{r-1}(t)\, dt.$$

Table 1 (pp. 131-140) gives $P_n(u)$ to 6 dec. for n = 3(1)9
u = 0.00(.01)4.70

Table 2(p. 140) gives roots of
$$P_n(u) = p \tag{1}$$
to 2 dec. for p = .001, .005, .01, .025, .05, .1, .9, .95, .975, .99, .995, .999, and n = 3(1)9. Reproduced in Pearson & Hartley 1954, T.25, p. 172.

Table 3 (p. 122) gives a comparison of exact upper percentage points with approximate percentage points computed from

$$P_n(u) \sim 1 - n + nP\left[\left(\frac{nu^2}{n-1}\right)^{\frac{1}{2}}\right] \tag{2}$$

The roots of (1) and (2) are tabled to 4 dec. for p = .95, .99 and n = 3(1)9.

1.42   [see also pp. 637-638]

Table 4(p. 123) gives approximations to $P_n(u)$ for small u, viz:

$$P_n(u) \sim \frac{n^{5/2}}{2} \left(\frac{u}{\sqrt{2\pi}}\right)^{n-1}, \quad n > 2$$

$$P_n(u) \sim \frac{\sqrt{n}}{(n-1)!} \left(\frac{nu}{\sqrt{2\pi}}\right)^{n-1}, \quad n \geq 2$$

$$P_n(u) \sim \frac{\sqrt{n}}{(n-1)!} \left(\frac{nu}{\sqrt{2\pi}}\right)^{n-1} \left[1 - \frac{n(n-1)u^2}{2(n+1)}\right], n \geq 2.$$

These three approximations and the exact $P_n(u)$ are tabled to 6 dec. for u = .05, .1, .2, and n = 3, 4, 5.

Godwin, H. J., 1949(a)

Biometrika 36, 92

Fundamental Integrals for evaluation of normal order statistics.

Table 1 (p. 97) gives

$$\psi(i) = \int_{-\infty}^{+\infty} P^i(x) \, Q^i(x) \, dx$$

$$\psi(i,j) = \int_{-\infty}^{+\infty} P^i(x) \left[\int_{x}^{+\infty} Q(y)^j \, dy\right] dx$$

to 10 decimals for i = 1(1)5; $i \leq j \leq 10-i$.

Howell, J.M. 1949

Annals of Mathematical Statistics 20, 305-309

"Control chart for largest and smallest values"

Let d_n denote the mean range in samples of size n from a normal population with unit standard deviation; and let $\sigma_{1|n}$ denote the standard deviation of the straggler. Let L and S denote the largest and smallest values, respectively, in a sample of n test records assumed to follow a normal distribution; and let $\bar{L}$ and $\bar{S}$ denote their averages over k pilot samples. To set up control charts for future samples, compute 3σ limits for L and S from $\frac{1}{2}(\bar{L}+\bar{S}) + A_3(\bar{L}-\bar{S})$ and $\frac{1}{2}(\bar{L}+\bar{S}) - A_3(\bar{L}-\bar{S})$ respectively.

Table 1 (p. 306) gives the following quantities for $n = 2(1)10$;

d_n (called d_2) to 3 dec.

$\sigma_{1|n}$ (called d_4) to 3 dec.

$A_2 = 3n^{-\frac{1}{2}}/d_n$ to 3 dec.

$A_3 = \frac{1}{2} + 3\sigma_{1|n}/d_n$ to 2 dec.

$A_4 = \frac{1}{2}d_n + 3\sigma_{1|n}$ to 2 dec.

Table 2 (p. 308) gives information on the power of the L & S control chart, compared to the customary $\bar{X}$ & R chart, to detect a slippage of the mean or a percentage increase of the standard deviation.

Certain errors in Table 1 are corrected in Howell 1950, p. 616.

Grubbs, F.E. 1950

<u>Annals of Mathematical Statistics</u> 21, 27-58

Distribution of extreme deviate from mean $x'_n \equiv x_n - \bar{x}$, percentage points and moment constants.

Let $x_1, \ldots, x_n$ denote an ordered sample from a normal parent with $\sigma = 1$. The probability $P_n(u)$ for $x'_n \equiv x_n - \bar{x} \leq u$ is given by

$$P_n(u) = \sqrt{n}\,(2u)^{-(\frac{n-1}{2})} G_{n-1}(nu)$$

where the G functions are defined by the recurrence

$$G_0(x) \equiv 1, \quad G_r(x) = \int_0^x \exp\left[-\frac{t^2}{2r(r+1)}\right] G_{r-1}(t)\, dt.$$

Cf. Mc Kay 1935 and Nair 1948b.

Table II (pp. 31-37) gives $P_n(u)$ to 5 dec. for $n = 2(1)25$, $u = 0(.05)4.9$.

Table III (p. 45) gives the upper $100\alpha\%$ points of $x'_n = x_n - \bar{x}$, to 3 dec., for $\alpha = .005, .01, .05, .1$; $n = 2(1)25$.

Table IV (p.46) gives the mean, standard deviation, β_1 and β_2 of x'_n to 4 dec. for $n = 2(1)15$, computed by quadrature from Table II; and the same quantities to 3 dec. for $n = 20, 60, 100, 200, 500, 1000$, computed by the formula of Mc Kay 1935 from the moments of x_n given by Tippett 1925.

1.42
1.43   [see also pp. 637-639]

Pearson, E. S. and Hartley, H. O. 1954

Table 24, p. 172, computed by Jean H. Thompson and Joyce M. May, gives upper and lower 0.1%, 0.5%, 1%, 2.5%, 5% and 10% points of $x_{n|n}$, the extreme deviate from the mean of a normal population, to 3 dec. for $n = 1(1)30$.

1.43 <u>Ordered normal deviates; normal ranking scores; moments of normal order statistics; order statistic tests</u>

Let $x_1, x_2, \ldots, x_n$ denote a random sample of observations from $N(0,1)$ and arranged in ascending order of magnitude. The following are formulas for x_i (or more precisely $x_{i|n}$) the ith order statistic:-
The probability integral of x_i i.e., the chance that $x_i \leq x^*$ (say)

$$P_r[x_i \leq x^*] = \frac{n!}{(n-i)!(i-1)!} \int_0^{P(x^*)} P^{i-1}(1-P)^{n-i} dP = I_{P(x^*)}(i, n-i+1)$$

where $I_p(a,b)$ is the incomplete Beta Function Ratio (see Section 3). The density function of x_i:-

$$\frac{n!}{(n-i)!(i-1)!} P^{i-1}(x_i) Q^{n-i}(x_i) Z(x_i)$$

The rth moment about the origin

$$\mu_r'(i|n) = \frac{n!}{(n-i)!(i-1)!} \int_0^1 P^{i-1}(1-P)^{n-i} x^r(P) dP$$

and in particular the ith ranking score

$$\xi_{i|n} = \mu_1'(i|n)$$

$$\sigma_{i|n}^2 = \mu_2'(i|n) - [\mu_1'(i|n)]^2$$

We also note the distribution-free recurrence formula

$$\frac{n+1-i}{n+1} \mu_r'(i|n+1) + \frac{i}{n+1} \mu_r'(i+1|n+1) = \mu_r'(i|n)$$

which has been extensively used for calculating normal ranking scores $(r = 1)$.

Pearson, K. and Pearson, M. V. 1931

Biometrika 23, 364-397

By reverting the ascending power series for P(x), Pearson obtains a power series for x(P), with the help of whose coefficients he expresses

$$\xi_{i|n} = \frac{n!}{(n-i)!(i-1)!} \int_0^1 P^{i-1}(1-P)^{n-i} x(P) \, dP$$

in a series of B-functions. Hence he computes $E(x''')$, which is given to 6 dec. in Table 1, p. 372, for n = 1(1)25, 30(1)33, 50(1)53, 70(1)73, 98(1)101.

Table 2, p. 374, gives $\xi_{3m|4m}$, $\xi_{3m+1|4m}$ and $\frac{1}{2}[\xi_{3m|4m} + \xi_{3m+1|4m}]$; and also $\xi_{3m+1|4m+1}$, $\xi_{3m+2|4m+1}$ and $\frac{1}{4}[3\xi_{3m+1|4m+1} + \xi_{3m+2|4m+1}]$; to 6 dec. for m = 1(1)6, 8(5)18, 25. Table 3, p. 375, gives $\xi_{3m-1|4m-1}$, $\xi_{3m|4m-1}$ and $\frac{1}{4}[\xi_{3m-1|4m-1} + 3\xi_{3m|4m-1}]$ to 6 dec. for the same m. A further attempt to define upper quartiles is given on p. 377.

By squaring the series for x(P), and substituting in

$$\mu_2'(i|n) = \frac{n!}{(n-i)!(i-1)!} \int_0^1 P^{i-1}(1-P)^{n-i} x^2(P) \, dP$$

Pearson obtains a formula for $\sigma^2_{i|n}$. Table 5, p. 380, gives $\sigma_{x'''}$ and $n^{\frac{1}{2}}\sigma_{x'''}$ for n = 2(4)22, 30(20)70, 98.

Camp, B. H. 1931

The mathematical part of elementary statistics

Camp tabulates the mean abscissa of the normal deviate of rank i, i.e. the average value of the normal deviate when integrated between $x\left(\frac{i-1}{n}\right)$ and $x(i/n)$. This is equal to $n[Z(i/n - 1/n) - Z(i/n)]$, and is given in Table 6, pp. 396-403, to 3 dec. for n = 1(1)50 and i = 1(1)n.

McKay, A. T. 1935

Biometrika 27, 466-470

Let $x_1, x_2, \ldots, x_n$ denote an ordered normal sample and let $X = \sum_1^r a_j x_j$. Let $u = X - \bar{x}$. Then all cumulants of u are equal to the corresponding cumulants of X, except the second; and $\text{var } u = \text{var } X + n^{-1}\left(1 - 2\Sigma a_j\right)$.

[63]

1.43   [see also pp. 638-639]

On p. 468 Mc Kay tables $\sigma(x_1 - \bar{x}) = \left(\sigma_1^2 - \frac{1}{n}\right)^{\frac{1}{2}}$ to 4 dec. for n = 2, 5, 10, 20, 60, 100, 200, 500, 1000; and the correlation between x_1 and $\bar{x}$ to 3 dec. for the same n.

Fisher, R. A. and Yates, F. 1938, 1943, 1948, 1953, 1957

Table XX gives expectations of normal order statistics:

$\xi_{n-i|n}$ 2 dec.

  n = 2(1)50

  i = 1(1)$\left[\frac{n}{2}\right]$.

Table XXI gives the sum of squares of these expectations:

$\sum_{i=1}^{n} \xi^2_{i|n}$ to 4 dec. for n = 1(1)50.

The quantity tabled is the 4 dec. sum of the 4 dec. squares of the 2 dec. expectations from Table XX.

Sandon, F. 1946

Annals **of** **Eugenics** 13, 118-121

The nomogram on p. 120 permits $\xi_{i|n}$ to be read to about $1\frac{1}{2}$ dec., and $50 + 15\xi_{i|n}$ to about $\frac{1}{2}$ dec., for n = 1(1)250 and i = 1(1)n.

Walsh, J. E. 1946

Annals **of** **Mathematical** **Statistics** 17, 44-52

Substitute for t-test using order statistics.

Let $y_1, \ldots, y_m$ be a normal sample from parent (η, σ)

  x be a normal observation from parent (ξ, σ).

The criteria for testing $\eta < \xi$ are of the form

  $x > (j+1)\bar{y} - jy_{i|m}$, $x > (j-1)\bar{y} + jy_{m+1-i|m}$, where $j = (m+1)^{\frac{1}{2}}$.

The criteria for testing $\eta > \xi$ are similar.

The significance level, as a function of i and m, is given explicitly on p. 46 for i = 1(1)4; and implicitly in formula (12), p. 48, for general i. Table 1, p. 49, gives the power of the test in four simple and useful cases, compared with the classical t-test; Table 2, p. 50, gives percentage efficiencies derived from the powers in Table 1.

Hastings, C., Jr., Mosteller, F., Tukey, J. W. and Winsor, C. P. 1947
Annals of Mathematical Statistics 18, 413-426
"Low moments for small samples: A comparative study of order statistics"

Means, variances and covariances of order statistics were computed for three parent distributions: the normal distribution N, the rectangular distribution U, and a special long-tailed distribution S. S may be defined as the distribution of $x = (1-u)^{-1/10} - u^{-1/10}$, where u is uniformly distributed on $[0, 1]$. All distributions were adjusted to have zero mean and unit variance.

Table I (pp. 415-417) gives

$$\left. \begin{array}{l} \xi_{n-i|n} = E(x_{n-i|n}) \\ \sigma_{n-i|n} \end{array} \right\} \text{to 5 dec. for} \quad \begin{array}{l} n = 1(1)10 \\ i = 1(1)\left[\frac{n+1}{2}\right] \end{array}$$

Also shown are 4 dec. values of approximations (see pp. 423-4) for parents N and S.

Tables II to V give values of $E\left[(x_{n-i|n} - \xi_{n-i|n})(x_{n-j|n} - \xi_{n-j|n})\right]$
for $n = 1(1)10$, $i = 1(1)\left[\frac{n+1}{2}\right]$, $j = 1(1)n$ as follows:

Table	Page	Decimal Accuracy	Parent
II	418	2	N and asymptotic formula
III	419	5	U
IV	420	5	S
V	421	5.	S using asymptotic formula.

Table VI (p. 422) gives $100 \, \rho_{n-i,n-j|n}$ as integers for
$n = 1(1)10$, $i = 1(1)\left[\frac{n}{2}\right]$, $j = i+1(1)n-i+1$ and parents U, N, S.

1.43 [see also pp. 638-639]

The corrections listed below, all for the normal distribution, were made by comparison with Godwin 1949b.

Table I: corrections of 1 in the fifth place are omitted.

Means	Standard Deviations
(3/10) = .65606	(4/10) = .41299
(4/10) = .37576	(5/10) = .40756
(5/10) = .12267	(3/10) = .41833
	(4/10) = .39742
	(5/10) = .38866

Corrections to Tables III and VI are all 1 in the second place.

Table III: (i,j,n)

(1,3,3) = .16	(1,5,8) = .07	(3,6,9) = .09
(1,2,4) = .25	(1,7,8) = .05	(4,6,9) = .11
(1,4,4) = .10	(2,3,8) = .16	(2,3,10) = .15
(1,3,6) = .14	(2,7,8) = .06	(3,7,10) = .07
(1,4,6) = .10	(3,6,8) = .10	(4,5,10) = .13
(1,5,6) = .08	(2,3,9) = .15	
(2,5,6) = .11	(2,4,9) = .12	

Table VI: (i,j,n)

(1,5,5) = .17	(4,6,10) = .68
(1,6,6) = .14	
(2,6,10) = .41	
(3,8,10) = .36	
(4,5,10) = .83	

Godwin, H. J. 1949a

<u>Biometrika</u> 36, 92-100

Table 2, p. 98, gives the following first and second moments of the difference between two consecutive normal order statistics:

$E[x_{i+1|n} - x_{i|n}] = \xi_{i+1|n} - \xi_{i|n}$, $E[(x_{i+1|n} - x_{i|n})(x_{j+1|n} - x_{j|n})]$, to $11 - [2n/3]$ dec. for $n = 2(1)10$, $i = 1(1)\frac{1}{2}[n-1]$, $j = i(1)n-i$.

Godwin, H. J. 1949 b

Annals of Mathematical Statistics 20, 279-285

"Some low moments of order statistics"

Table 1 (p. 281) gives $\xi_{i|n}$ to 7 dec. for $n = 2(1)10$, $i = 1(1)\left[\frac{n+1}{2}\right]$.

Table 3 (p. 283) gives correlations between order statistics $\rho_{ij|n}$ to 4 dec. for $n = 2(1)10$, $i = 1(1)\left[\frac{n-1}{2}\right]$, $j = i+1(1)n-i+1$.

Table 2 (p. 282) gives the corresponding values of variances and covariances $\sigma^2_{i|n}$ and $\rho_{ij|n}\sigma_{i|n}\sigma_{j|n}$ to 5 dec. for $n = 2(1)10$, $i = 1(1)\left[\frac{n+1}{2}\right]$, $j = i(1)n-i+1$.

Table 4 (p. 284) gives formulas for those of the above parameters which can be expressed in closed form. The formulas involve six constants which are tabled to 10 dec. on p. 285.

Chandler, K. N. 1952

Journal of the Royal Statistical Society, Series B 14, 220

Let x_n, $n = 1, 2, 3, \ldots$, denote a time series of independent unit normal variates. Define the 'i^{th} lowest record value' $X_i = x_{u(i)}$ and its serial number $u(i)$: $X_1 = x_1$, $u(1) = 1$. Having found X_{i-1} and $u(i-1)$, take $u(i)$ as the smallest serial number such that

$$X_i = x_{u(i)} > x_{u(i-1)} = X_{i-1}.$$

Table 1, p. 222, gives the upper and lower $\frac{1}{2}\%$, 1%, 10% and 50% points of X_i to 3 dec. for $i = 2(1)9$.

Pearson, E. S. and Hartley, H. O. 1954

Table 28, p. 175, gives:

$-\xi_{i|n}$ 3; 2 dec. $n = 2(1)20$; $21(1)26(2)50$ $i = 1(1)[n/2]$

Vianelli, S. 1959

Tavola 77, p. 1216, gives:

$-\xi_{i|n}$ 2 dec. $n = 21(1)26(2)50$ $i = 1(1)[n/2]$

1.43
1.44 [see also pp. 638-641]

Dixon, W. J. and Massey, F. J. 1957

Table 8 b 5, p. 407, gives:

$\xi_{i|n}$ 3 dec.   n = 2(1)10; 11(1)20   i = 1(1)n; 1(1)10

1.44 Studentized distributions involving order statistics

Thompson, W. R. 1935

<u>Annals of Mathematical Statistics</u> 6, 214-219

Let $x_1, x_2, \ldots, x_n$ denote an ordered normal sample and

$$s'^2 = n^{-1}\left(\sum_{i=1}^{n} (x_i - \bar{x})^2\right).$$

As a criterion for rejection of the straggler x_n the ratio

$$\tau = (x_n - \bar{x})/s'$$

is proposed. Thompson does not consider the ordering of the observations, and so erroneously concludes that $t = (n-2)^{\frac{1}{2}}\tau/(n-1-\tau^2)^{\frac{1}{2}}$ has the t-distribution with n-2 degrees of freedom. Two tables are devoted to alleged percentage points of τ.

Pearson, E. S. and Chandra Sekhar, C. 1936

<u>Biometrika</u> 28, 308-319

"a criterion for the rejection of outlying observations"

Let $x_1, x_2, \ldots, x_n$ denote an ordered random sample from a normal parent and let

$$\tau^{(i)} = \sqrt{n}\,(x_i - \bar{x})\Big/\sqrt{\sum(x_j - \bar{x})^2} \qquad i = 1, 2, \ldots, n \qquad (1)$$

Scott 1936 has shown that it follows from (1) that only $\tau^{(1)}$ and $\tau^{(n)}$ can exceed $[\frac{1}{2}(n-2)]^{\frac{1}{2}}$ in absolute value; so that for moderate n the method of Thompson 1935 (see the previous abstract) yields correct upper percentage points of $\tau^{(n)}$.

The table on p. 318 gives $\tau^{(n)}(p)$ to 4 dec. for p = .1, .05, .025, .01, and n = 3, 4; and to 3 dec. for:

p = .1   .05   .025   .01
n = 5(1)11   5(1)14   5(1)16   5(1)19

Nair, K. R. 1948 b

Biometrika 35, 118-144

"The distribution of the extreme deviate from the sample mean and its studentized form"

Let $x_1, \ldots, x_n$ be an ordered sample from a normal parent, and let s^2 be an estimate of variance based on ν degrees of freedom and independent of the ordered sample. Let $_\nu P_n(T)$ denote the probability that

$$t'_{n|n} = (x_n - \bar{x})/s \leq T;$$

and let $P_n(x)$ denote the probability that

$$\frac{x_n - \bar{x}}{\sigma} \leq x.$$

Then we have the studentized expansion

$$_\nu P_n(T) = a_o + a_1/\nu + a_2/\nu^2 + \ldots$$

where

$$a_o = P_n(T)$$

$$a_1 = \frac{1}{4}\left[T^2 P_n''(T) - T P_n'(T)\right]$$

$$a_2 = \frac{1}{96}\left[3T^4 P_n^{iv}(T) - 2T^3 P_n'''(T)\right] - \frac{1}{8} a_1$$

Table 5, pp. 141-142, gives these coefficients:

a_0 6 dec.   6 dec.   6 dec.
a_1 5 dec.   4 dec.   4 dec.
a_2 3 dec.   3 dec.   3 dec.
n = 3   4(1)6   7(1)9
T = 0(.2)4 0(.2)4   0(.2)4.6

Table 6 A, p. 143, gives lower percentage points of t' to 2 dec. for $p = .01, .05$; $n = 3(1)9$; $\nu = 10, 15, 30, \infty$.

Table 6 B, p. 143, gives upper percentage points of t' to 2 dec. for $p = .95, .99$; $n = 3(1)9$; $\begin{cases} \nu = 10(1)20 \\ 120/\nu = 0(1)6 \end{cases}$.

Tables 6 A and 6 B are reproduced in Pearson & Hartley 1954, T. 26, p. 173.

[69]

1.5
1.52

1.5 Measures of location other than the arithmetic mean; measures of dispersion other than the standard deviation

The mean of a normal distribution is itself normally distributed; for the distribution of the standard deviation see section 2. For notation and definitions used in this section, see section 1.40.

1.51 Measures of location

For the median see also section 1.41; for the mid-range see also section 1.54.

Pearson, K. 1920
<u>Biometrika</u> 13, 113-132

Pearson attempts to estimate the mean of a normal distribution by means of the midpercentile $\hat{x}_p$. Table IV, p. 121, gives $n^{\frac{1}{2}}\hat{\sigma}_p$ to 4 dec. for

$$p = \frac{1}{14}, \frac{1}{10}, \frac{2}{10}, .25, .270268, \frac{3}{10}, \frac{4}{10}, \frac{5}{10}.$$

As further improved estimates the optimum weighted averages of any two mid-percentiles are considered:

$$m_{p_1 p_2} = \tfrac{1}{2}\lambda_1(x_{p_1} + x_{q_1}) + \tfrac{1}{2}\lambda_2(x_{p_2} + x_{q_2}), \quad \lambda_1 + \lambda_2 = 1.$$

The weights λ_1, λ_2 are chosen in such a way that the variance of $m_{p_1 p_2}$ is at a minimum. The λ_1, λ_2 depend, therefore, on the pair p_1, p_2 chosen. To obtain $\sigma(m_{p_1 p_2})$ it is necessary first to table

$$\tfrac{1}{4}E\left[\left((x_{p_1} + x_{q_1}) - (x(p_1) + x(q_1))\right)\left((x_{p_2} + x_{q_2}) - (x(p_2) + x(q_2))\right)\right]$$

which is given in Table V (p. 123) to 6 dec. for

p_1, p_2 = 1/14, .1, .2, .25, .270268, .3, .4, .5.

Table VI (p. 123) gives for the same range of p_1, p_2:—

The optimum λ_1 λ_2 to 5 dec. and

$\sqrt{n}\ \sigma(m_{p_1 p_2})$ to 4 dec.

Mosteller, F. 1946

Annals of Mathematical Statistics 17, 377-406

"On some useful 'inefficient' statistics"

Tables I and II (p. 389) are concerned with unweighted averages of k selected order statistics for estimation of mean. The methods of selection are:

(i) Spacing uniformly in probability. (Choose the $x_{i|n}$ such that $P(\xi_{i|n})$ is nearest to $i/(k+1)$)

(ii) Choose the $x_{i|n}$ whose $\xi_{i|n}$ are nearest to $\xi_{i|k}$.

(iii) Choose the $x_{i|n}$ such that $P(\xi_{i|n})$ is nearest to $(i - \tfrac{1}{2})/k$.

(iv) Minimum variance.

Table I gives the large sample expectations of the order statistics selected in accordance with (i) ... (iv) as well as $\tfrac{1}{2}(1 + \alpha(\xi_i))$ to 4 dec. for $k = 1,2,3$.

Table II gives the efficiency compared to using sample mean to 3 dec. for (i) ... (iii), $k = 1(1)10$ and (iv), $k = 1,2,3$.

The formulas for method (iii) are given explicitly for $n = 1(1)5, 10$, together with efficiencies rounded to 2 dec. from Mosteller's Table II, in Dixon & Massey 1951 (T.8 a 1, p. 314), 1957 (T. 8 a 1, p. 404).

Table III, p. 390, facilitates method (iii) by giving $P(\xi_{i|n})$ to 4 dec. for $n = 1(1)10$, $i = 1(1)n$.

1.52 The mean deviation from mean and median and related measures

Let x_i denote a random sample from a normal population; then the above measures of dispersion are defined as follows:-

The mean deviation from the mean

$$m = \frac{1}{n} \sum_{i=1}^{n} |x_i - \bar{x}|$$

The mean deviation from the median

$$M = \frac{1}{n} \sum_{i=1}^{n} |x_i - \dot{x}|$$

1.52 [see also p. 643]

If the x_i are drawn from $N(0,1)$ the moment constants of m are as follows:—

$$E(m) = \left[2(n-1)/\pi n\right]^{\frac{1}{2}}$$

$$\sigma^2(m) = \frac{2(n-1)}{n^2 \pi}\left[\tfrac{1}{2}\pi + \sqrt{n(n-2)} - n + \sin^{-1}\frac{1}{n-1}\right]$$

$$\mu_3(m) = \left(\frac{n-1}{n}\right)^{\frac{3}{2}}\left[\sqrt{\frac{2}{\pi}}\left(\frac{3}{n^2(n-1)} - \frac{1}{n^2}\right)\right.$$

$$+ \left(\frac{2}{\pi}\right)^{\frac{3}{2}}\left(\frac{n-2}{n}\left(\frac{n-3}{n-1}\right)^{\frac{1}{2}} + \frac{3(n-2)^2}{n^2(n-1)}\sin^{-1}\frac{1}{n-2} + 2\right.$$

$$\left.\left. - \frac{3(n-1)}{n}\left(1 - \frac{1}{(n-1)^2}\right)^{\frac{1}{2}} - \frac{3}{n}\sin^{-1}\frac{1}{n-1}\right)\right]$$

For $n = 3$

$$\mu_4(m) = \frac{9}{16}\left[\frac{7}{32} - \frac{11}{3\pi} + \frac{1}{2\pi}\sin^{-1}\frac{1}{3} + \frac{13\sqrt{2}}{9\pi} + \frac{12\sqrt{2}}{\pi^2} + \frac{6}{\pi^2}\sin^{-1}\frac{1}{3} - \frac{12}{\pi^2}\right]$$

For $n = 4$

$$\mu_4'(m) = \left[\frac{9\pi(5\pi - 7\alpha)}{512} + \frac{165\pi}{384\sqrt{2}}\right]\frac{4}{\pi^2}$$

where $\alpha = \cos^{-1}\frac{1}{3}$

The last three results are due to A. R. Kamat. R.A. Fisher 1929 b gave the exact distribution of m for n = 4, but his formula for the fourth moment of m is in error. Geary 1936 a gives approximate expansions of $\beta_1(m)$ and $\beta_2(m)$; Godwin 1948 makes some corrections in Geary's formulas.

The mean deviation from the median is a special case of what K.R. Nair 1950 has called 'j statistics'. The first and second moments of M are clearly reducible to the first and second moments of order statistics, because

$$nM = \begin{cases} \sum_{i=\frac{1}{2}(n+3)}^{n} x_i - \sum_{i=1}^{\frac{1}{2}(n-1)} x_i & \text{for } n \text{ odd} \\ \sum_{i=\frac{1}{2}n+1}^{n} x_i - \sum_{i=1}^{\frac{1}{2}n} x_i & \text{for } n \text{ even} \end{cases}$$

Davies, O. L. and Pearson, E. S., 1934

Royal Stat. So., Jn., Suppl. 1, 76

Estimation of σ from mean deviation $m = n^{-1} \sum_{i\ 1}^{n} |x_i - \bar{x}|$.

Consider k samples of size n:- x_{ti} $t = 1, 2, \ldots, k$; $i = 1, 2, \ldots, n$

and the extimator of σ (=1),

$$s_4 = \frac{\sum m_t}{kE(m)} = \frac{\sum m_t}{k\left(2(n-1)/\pi n\right)^{\frac{1}{2}}}$$

having a standard deviation

$$\sigma(s_4) = \frac{1}{\sqrt{nk}} \left(\frac{\pi}{2} + \sqrt{n(n-2)} - n + \sin^{-1} \frac{1}{n-1} \right)^{\frac{1}{2}}$$

Table I (p. 90) gives $\sigma(s_4)$ to 3 dec. for

k = 1(2)6,10,20,50

n = 2(2)6,10,15,20,50

Table II (p. 91) gives $\frac{1}{f_n} = E(m)$ to 3 dec. and $1/E(m)$ to 4 dec.

Pearson, E. S., 1935(c)

The Application of Statistical Methods to Industrial Standardization and Quality Control

British Standards Institution, No. 600

Table 23 (p. 122) gives the ratio of the standard deviation to the expected value of the mean deviation taken from normal samples of size n = 2(1)10,15,20,50,∞ . The ratios are given to 3 dec.

Vianelli 1959, Prontuario 76, p. 221, reprints inter al. T. 20 of Pearson & Hartley 1954 (abstracted on p. 75 below) but credits it to Hald 1952.

1.52
1.531

[see also pp. 643-645]

Pearson, E. S., Godwin, H. J. and Hartley, H. O.   1945

Biometrika 33, 252-265

"The probability integral of the mean deviation"

Godwin's formula for the ordinate $f_n(m)$ of the distribution of the mean deviation m in samples of n is

$$f_n(m) = \tfrac{1}{2} n^{3/2} (2\pi)^{-\frac{1}{2}(n-1)} \left[\sum_{k=1}^{n-1} \binom{n}{k} \exp\left(-\frac{m^2 n^3}{8k(n-k)}\right) G_{k-1}\left(\tfrac{nm}{2}\right) G_{n-k-1}\left(\tfrac{nm}{2}\right) \right]$$

where the G-functions are defined by the recurrence

$$G_r(x) = \int_0^x \exp\left(-\frac{t^2}{2r(r+1)}\right) G_{r-1}(t)\, dt, \quad G_0(x) \equiv 1$$

Table 1 (pp. 260-4) gives the probability integral $P_n(m) = \int_0^m f_n(m)\, dm$

to 5 dec. for n = 2(1)10, m = .00(.01) as far as required.

Reprinted in Vianelli 1959, Prontuario 68, pp. 210-214.

Table 2 (p. 265) gives percentage points, i.e., roots m of

$P_n(m)$ = 1-p to 3 dec. for

p = .999,.995,.99,.975,.95,.9;.1,.05,.025,.01,.005,.001

n = 2(1)10

For n = 10 the normal approximation to m is shown also.

Reprinted in E.S. Pearson & Hartley 1954, T. 21, p. 165.

Nair, K. R. 1949

Biometrika 36, 234-235

Table 1, p. 235, gives the coefficients of variation of m and M to 2 dec. for n = 2(1)10.

Nair, K. R. 1950

Biometrika 37, 182-183

Table 1, p. 182, gives inter al. the percent efficiency of M against the sample standard deviation s, viz:

$$100\ \sigma^2(s)\ E(M)^2 \Big/ \sigma^2(M)\ E(s)^2$$

to 2 dec. for n = 2(1)8 and to 1 dec. for n = 9,10.

Johnson, N. L. and Tetley, H., 1950

<u>Statistics</u>, Vol. II

The table on page 103 gives $E(m)$ and σ_m, where m is the mean deviation in normal samples of size n, to 3 dec. for $n = 5(1)8$. Also given are upper and lower .5%, 1% and 2.5% points of the distribution of m to 3 dec. for the above n.

Dixon, W. J. and Massey, F. J. 1951

Table 8b2, p. 315, gives $1/E(M)$ to 4 dec. and
$$\frac{\sigma^2(\frac{M}{E(M)})}{\sigma^2(s_1)}$$
to 2 dec. for $n = 2(1)10$,

where $M = \sum |x_i - \dot{x}|$ and $\dot{x}$ = median.

Reprinted in Dixon & Massey 1957, T. 8b2, p. 405.

Pearson, E. S. and Hartley, H. O. 1954

Table 20, p. 164, computed by Joyce M. May, gives $E(m)$ to 6 dec., $\sigma(m)$ to 4 dec., $\sigma^2(m)$ to 5 dec., $\beta_1(m)$ to 3 fig., $\beta_2(m) - 3$ to 3 fig., for $m = 2(1)20, 30, 60$.

1.53 <u>Measures of normal dispersion based on variate differences, Gini's mean difference, and Nair's j-statistics</u>

1.531 Variate differences

This section is concerned with estimators of the standard deviation based on the kth variate differences of a sequence of observations: i.e. on

$$\Delta^k x_i = \sum_{j=0}^{k} \binom{k}{j} (-)^j x_{i+j} .$$

In addition to using variate differences of different order, k, various powers of their numerical values, $|\Delta^k x_i|^p$, are employed and the resulting estimators of σ are usually based on averages of the form

$$\frac{1}{n-k} \sum_{i=1}^{n-k} |\Delta^k x_i|^p$$

1.531 [see also pp. 643-645]

For $p = 2$ a comprehensive study of the means and variances of these estimators is given by Morse & Grubbs 1947; for $p = 1, 2$ by Guest 1951; for $p = 1(1)4$ by Kamat 1953 b.

Von Neumann, Kent, Bellinson & Hart 1941, von Neumann 1941, 1942, Hart & von Neumann 1942, Hart 1942 have treated the mean square successive difference

$$\delta^2 = \frac{1}{n-1} \sum (x_{i+1} - x_i)^2 ;$$

Kamat 1953 a has treated the mean successive difference

$$d = \frac{1}{n-1} \sum |x_{i+1} - x_i| .$$

Young, L. C. 1941
Annals of Mathematical Statistics 12, 293-300

Young defines $C = 1 - \delta^2/2s^2$, and expresses the moments of C in terms of the moments of the parent distribution. For a normal parent he fits a Pearson Type II distribution to the second and fourth moments of C, and hence derives approximate upper (and lower) 5% and 1% points of C and of C/σ_C, given to 4 dec. for $n = 8(1)25$ in Table 1, p. 297.

The exact results of Hart & von Neumann 1942 confirm Young's approximations.

Hart, B. I. and von Neumann, J. 1942
Annals of Mathematical Statistics 13, 207-214
"Tabulation of probabilities for the ratio of the mean square successive difference to the variance"

Following von Neumann 1941, define

$$s'^2 = n^{-1} \sum_{i=1}^{n} (x_i - \bar{x})^2$$

and write

$$\delta^2 / s'^2 = 2n(n-1)^{-1}(1-\epsilon)$$

so that ϵ is identical with C, used by Young 1941. The distribution of ϵ was found by von Neumann 1941; Hart & von Neumann give explicit formulas for $n = 3(1)7$ on pp. 207-209.

Table II (p. 212) gives $\Pr[\delta^2 \leq ks'^2]$ to 5 dec. for $n = 7$ and $k = .25(.05)1.2$, computed both from the exact distribution and from an approximation due to R. H. Kent and given on pp. 210-211.

Table III (p. 213) gives the above probability, computed exactly, to 5 dec. for $\begin{cases} n = 4 & 5 & 6 & 7 \\ k = .8(.05)1.15 & .5(.05)1.15 & .35(.05)1.2 & .25(.05)1.2 \end{cases}$; and Kent's approximation to 5 dec. for $\begin{cases} n = 8 & 9, 10 \\ k = .25(.05)1.2 & .25(.05)1.25 \end{cases}$ $\begin{cases} 11, 12 & 15 & 20 & 25 & 30 & 40 \\ .3(.05)1.3 & .35(.05)1.4 & .5(.05)1.45 & .6(.05)1.5 & .7(.05)1.55 & .8(.05)1.6 \end{cases}$ $\begin{cases} 50 & 60 \\ .9(.05)1.65 & 1(.05)1.7 \end{cases}$. Table III also gives the lower limit of the distribution of δ^2/s'^2 for $n = 4(1)12, 15(5)30(10)60$.

Table I (p. 212) gives the eighth and tenth moments of the exact distribution, and of Kent's approximation, to 5 dec. for $n = 7(1)9$.

Hart, B. I. 1942
Annals of Mathematical Statistics 13, 445-447
"Significance levels for the ratio of the mean square successive difference to the variance"

Upper and lower 0.1%, 1% and 5% points for $\delta^{2'}/s'^2$ to 4 dec. for $n = 4(1)60$.

Morse, A. P. and Grubbs, F. E. 1947
Annals of Mathematical Statistics 18, 194-214
"The estimation of dispersion from differences"

Let $[x_i]$ be a sequence of observations uniformly ordered in time with $x_i = \eta_i + \epsilon$ where $E(x_i) = \eta_i = f(x_i)$ and the $[\epsilon_i]$ are normally and independently distributed with mean 0 and variance σ^2. The mean square k^{th} backward difference

$$\delta^2_{n,k} = \binom{2k}{k}^{-1} (n-k)^{-1} \sum_{i=k+1}^{n} (\Delta^k \epsilon_i)^2$$

is an unbiased estimate of σ^2. Its efficiency $W(n,k)$ (relative to s^2, when $f(x_i) \equiv 0$) is given by equations (12), (13), (14) and is tabulated in Table I (pp. 196-197) to 5 dec. for $k = 1(1)10$, $n = k+1(1)40(2)58(4)82(8)122(16)170(32)266(64)394, 522, 778, 1290, 2314, \infty$.

1.531 [see also pp 643-645]

An upper bound for

$$\nu^2_{n,k} = \binom{2k}{k}^{-1} (n-k)^{-1} \sum_{i=k+1}^{n} (\Delta^k \eta_i)^2$$

in the case of $f(t) = \sin t$ is given by

$$\nu^2_{n,k} \leq \frac{\pi}{\binom{2k}{k}(n-k)} \left(\frac{2\pi}{n-1}\right)^{2k-1}$$

which is tabulated in Table II (p. 202) to 3 dec. for $n = 5(1)10$, $k = 1(1)5$.

Table IV (pp. 210-211) gives the probable error of

$$d_{n,k} = \left[\binom{2k}{k}^{-1}(n-k)^{-1}\sum_{i=k+1}^{n}(\Delta^k x_i)^2\right]^{\frac{1}{2}}$$

from the approximate formula

$$\text{P.E.}(d_{n,k}) = .6745\, \sigma\left[2(n-1)W(n,k)\right]^{-\frac{1}{2}}$$

to 4 dec. for the same k and n as Table I: the tabular values must be multiplied by σ.

Durbin, J. and Watson, G. S. 1951
<u>Biometrika</u> 38, 159-178

Consider a regression model of the form $y = \beta_0 + \beta_1 x_1 + \ldots + \beta_k x_k + \epsilon$ where the x_i are fixed and the ϵ are error variates with zero mean and constant variance. Let z denote the residual of y from the estimated regression computed from n sets of observations $[y, x_1, \ldots x_k]$; and let Δz denote the first differences of z. To test for serial correlation in the ϵ Durbin & Watson propose the criterion $d = \Sigma \Delta z^2 / \Sigma z^2$.

Durbin & Watson 1950 showed that d belongs to a class of statistics whose distributions are not unique; and gave formulas for the upper and lower bounds of these distributions. From these formulas upper and lower bounds, d_U and d_L, for percentage points have now been calculated.

Tables 4-6, pp. 173-175, give 1%, $2\frac{1}{2}$% and 5% points, both d_U and d_L, to 2 dec. for $k = 1(1)5$ and $n = 15(1)40(5)100$. These points were computed from an approximate expansion in Jacobi polynomials.

Table 7, p. 176, is concerned with a modified criterion d' that yields exact formulas (cf. Watson & Durbin 1951) for the bounds on its distribution. 5% points d_U' and d_L' are given to 2 dec. for $k = 1(1)5$ and $n = 13(2)23$.

Watson, G. S. and Durbin, J. 1951
Annals of Mathematical Statistics 22, 446-451
"Exact tests of serial correlation using noncircular statistics"

Given a sample of size $2m$, Watson & Durbin define a modified serial correlation coefficient at lag 1:

$$c_1 = [\Sigma x_i x_{i+1} - x_m x_{m+1}]/\Sigma x_i^2,$$

which differs from the usual coefficient by the omission of a term in the numerator. This modification makes it possible to obtain the exact distribution of c_1. A similar treatment yields the exact distribution of a number of other ratios, including a modification of von Neumann's ratio of mean square successive difference to variance:

$$d_3 = \left[\sum(x_i - x_{i+1})^2 - (x_m - x_{m+1})^2\right]/\sum(x_i - \bar{x})^2.$$

The tables on p. 451 give 5% points of c_1 to 3 dec. for $n = 10(2)22$ and 5% points of d_3 to 3 fig. for $n = 12(2)30$.

Kamat, A. R. 1953 a
Biometrika 40, 116-127
"On the mean successive difference and its ratio to the root mean square"

Table 1, p. 120, gives the standard deviation and β_1 and β_2 of d/σ to 4 dec. for $n = 3(1)10(5)30(10)50$; the last figure in β_1 and β_2 is not reliable.

Table 2, p. 120, gives upper and lower $\frac{1}{2}$%, 1%, $2\frac{1}{2}$% and 5% points of d/σ to 2 dec. for the above n. The percentage points for $n = 3$ are exact; for $n = 4(1)8$ a Pearson Type I curve was fitted to four moments; for $n \geq 8$ a Pearson Type III curve was fitted to three moments.

1.531
1.533

[see also pp. 643-645]

To form a studentized d, Kamat adopts von Neumann's definition of s' (see abstract of Hart & von Neumann 1942). Table 4, p. 126, gives the mean, standard deviation, β_1 and β_2 of d/s' to 4, 4, 3 and 2 dec. for n = 5(5)30(10)50. Table 5, p. 120, gives upper and lower $\frac{1}{2}$%, 1%, 2$\frac{1}{2}$% and 5% points of d/s' to 2 dec. for n = 10(5)30(10)50, computed from Pearson-curve approximations.

1.532 Gini's mean difference

Nair, U. S. 1936

Biometrika 28, 428-435

"The standard error of Gini's mean difference"

Let $x_1, x_2, \ldots x_n$ denote an ordered sample from some distribution. Then Gini's mean difference is defined as*

$$\Delta = \frac{2}{n(n-1)} \sum_{\substack{i > j \\ i,j=1}}^{n} (x_i - x_j)$$

Formulas for $E(\Delta) = \overline{\Delta}$ and σ_Δ are derived for samples from normal, exponential, and rectangular distributions. For the normal we have

$$\overline{\Delta} = 2\sigma/\sqrt{\pi}$$

$$\sigma_\Delta = \frac{2\sigma}{\sqrt{n(n-1)}} \left[\frac{n+1}{3} + \frac{2(n-2)\sqrt{3}}{\pi} - \frac{2(2n-3)}{\pi} \right]^{\frac{1}{2}}$$

If an observed $\frac{\Delta \sqrt{\pi}}{2}$ is used as an estimate of σ its standard deviation is given by $\sigma_\Delta \frac{\sqrt{\pi}}{2}$ and the ratio of this to σ, viz: $\frac{\sigma_\Delta}{\sigma} \frac{\sqrt{\pi}}{2}$ is tabulated as E_5 in Table II (p. 435).

---

*This definition is substantially Nair's. It will be seen to be equivalent to Gini's definition

$$\Delta = [n(n-1)]^{-1} \Sigma_i \Sigma_j |x_i - x_j| .$$

1.533 Nair's j-statistics

The notation $j_{(r)}$ was adopted by K. R. Nair in memory of Dr A. E. Jones († 7 May 1948). Jones himself wrote $S(n,r)$.

Jones, A. E. 1946

Biometrika 33, 274-282

Let $x_1, \ldots x_n$ denote an ordered sample and let

$$j(r) = x_n + x_{n-1} + \ldots + x_{n-r+1} - x_r - x_{r-1} - \ldots - x_1$$

be the difference between the sum of the r largest observations and the sum of the r smallest observations. For a normal parent population, table 1, p. 280, gives $E\, j(r)$ to 1 dec. and $\mathrm{Var}\, j(r)$ to 2 dec. for

$$\begin{cases} n = 100 & 200 & 400(200)1000 \\ r = 5, 7 & 5, 7, 10, 12 & 5, 7, 10, 12(4)20 \end{cases}.$$

Table 2 gives the coefficient of variation of $j(r)$ for the same n and r, and the coefficient of variation of the standard deviation for $n = 100, 200(200)1000$.

Table 3 gives the auxiliary quantity

$$\exp\left(-\sum_{k=r+1}^{n} \frac{1}{k}\right)$$

to 5 dec. for $\begin{cases} n = 100 & 200 & 400(200)1000 \\ r = 5(1)8 & 5(1)10(2)14 & 5(1)10(2)20 \end{cases}$; and table 4 gives

$$\sum_{k=r+1}^{n} \frac{1}{k^2}$$

to 4 dec. for the same n and r.

Nair, K. R. 1950

Biometrika 37, 182

"Efficiencies of certain linear systematic[*] statistics for estimating dispersion from normal samples"

The percentage efficiency of $j(r)$, relative to the sample variance s^2, is given by

$$100\left[\tfrac{1}{2}(n-1)\left(\frac{\Gamma(\tfrac{1}{2}(n-1))}{\Gamma(\tfrac{1}{2}n)}\right)^2 - 1\right] \Big/ \left[\frac{E(j^2(r))}{E(j(r))^2} - 1\right]$$

---

[*] The epithet <u>systematic</u>, applied by Mosteller to linear combinations of order statistics, is deprecated by Kendall & Buckland 1957, p. 289.

1.533
1.541   [see also pp. 645-647]

Table 1, p. 182, gives this percentage efficiency to 2 dec. for $n = 2(1)10$, $r = 1(1)[\frac{1}{2}n]$; Table 3, p. 183, gives it to 1 dec. for

$$\begin{cases} n = 100 & 200 & 400(200)1000 \\ r = 5.\ 7 & 5,\ 7,\ 10,\ 12 & 5,\ 7,\ 10,\ 12(4)20 \end{cases}$$

1.54 The range and midrange; Galton differences; measures of dispersion based on selected order statistics

1.541 The range

Let $x_1, \ldots, x_n$ denote a random sample of n drawn from a normal population and arranged in ascending order of magnitude and denote the difference between the largest (x_n) and smallest (x_1) observation by the 'range' $R = R_n = x_n - x_1$.

If $\sigma = 1$ the probability integral of range $P_n(R) = \Pr\left[R_n \leq R\right]$ is given by

$$P_n(R) = n \int_{-\infty}^{+\infty} Z(x)(P(x+R) - P(x))^{n-1} dx$$

its distribution $f_n(R) = \dfrac{d}{dR} P_n(R)$ by

$$f_n(R) = n(n-1) \int_{-\infty}^{+\infty} Z(x)\, Z(x+R)(P(x+R) - P(x))^{n-2} dx$$

its mean

$$d_n = \int_0^\infty f_n(R)\, R\, dR = \int_0^\infty (1 - P_n(R))\, dR$$

$$= \xi_{n|n} - \xi_{1|n}$$

and its variance by

$$\sigma^2_{R|n} = \int_0^\infty 2(1 - P_n(R))\, R\, dR - d_n^2$$

                                                                    1.533
                                                                    1.541
or alternatively, in terms of moments of order statistics by

$$\sigma^2_{R|n} = \sigma^2_{1|n}(1 - 2\rho_{1n|n})$$

The percentage points of range, $R(p,n)$ are defined as the roots of $P_n(R) = 1 - p$.

For short tables of mean range, intended for use in constructing control charts, see section 14.221.

Tippett, L. H. C. 1925

<u>Biometrika</u> 17, 364-387

Table 10, pp. 386-387, gives d_n to 5 dec. for $n = 2(1)1000$. Reprinted in Tables for S & B II, T. 22, pp. 165-166. Abridged in Pearson & Hartley 1954, T. 27, p. 174: $n = 2(1)500(10)1000$. Abridged in Vianelli 1959, Pro. 77, p. 222: $n = 2(1)499$.

Table 4, p. 376, gives $\sigma_{R|n}$ to 3 dec. for $n = 2, 10, 20, 60, 100, 200, 500, 1000$. Values of $\beta_1(R)$ and $\beta_2(R)$ in Table 4 have been superseded by a later computation by E. S. Pearson 1926.

Pearson, E. S. 1926

<u>Biometrika</u> 18, 173-194

Table 8, p. 192, gives:

d_n 5 dec.   5 dec. (from Tippett 1925)

$\sigma_{R|n}$, $\beta_1(R)$ 4 dec.   3 dec.

$\beta_2(R)$ 4 dec.   2 dec.

$n =$ 2(1)6   10, 20, 60, 100, 200, 500, 1000.

Shewhart, W. A. 1928 b

<u>Jn. of Forestry</u> 26, 899

On page 901 are given:

<u>Curve</u> II:

Graph showing mean range d_n for samples of $n = 2$ to $n = 1000$. Graph can be read to .05 units.

1.541

Dotted curve:

Graph of the proportion of the area of a normal curve within the region about the mean bounded by $\pm \frac{1}{2}$ expected range for same values of n. Graph can be read to .005 units.

Pearson, K. and Pearson, Margaret V. 1931
Biometrika 23, 364-397

The table on p. 367 (by Tippett) gives d_n to 6 dec. for n = 2(1)31.

Pearson, E. S. 1932
Biometrika 24, 404-417

Using the moments of R from E.S. Pearson 1926, Pearson curves were fitted to the distribution of R; hence approximate percentage points were computed. Table A, p. 416, gives the $\frac{1}{2}\%$, 1%, 5% and 10% points of R, both upper and lower, to 2 dec.; d_n to 5 dec.; and $\sigma_{R|n}$ to 3-4 dec.; all for n = 2(1)30(5)100.

McKay, A. T. and Pearson, E. S., 1933
Biometrika 25, 415

Exact distribution of normal range R_n for n = 3 and approximations for large n. The distribution of range R_3 is derived as

$$f_3(R) = \frac{6}{\sqrt{2}} \left(2P\left[\frac{R}{\sqrt{6}}\right] - 1 \right) Z\left(\frac{R}{\sqrt{2}}\right) \qquad (1)$$

and is given in Table I (p. 418) as 100 $f_3(R)$ to 1 dec. for R=0.1(0.1)4.8 along with corresponding values obtained from a Type I fit to the exact moments.

Table II (p. 419) gives percentage points, viz: roots R of

$$1 - P_3(R) = \int_R^\infty f_3(R) \, dR = p \qquad (2)$$

for p = .005, .01, .05, .1, .9, .95, .99, .995 to 2 dec. computed from (A) (equation (2) above), (B) Pearson Type I fit, (C) fitted normal.

1.541

For large n the limiting approximation

$$1 - P_n(R) = \binom{n}{2} 2 \, Q\left(\frac{R}{\sqrt{2}}\right)$$

is derived; the table on p. 420 gives the 5%, 1% and ½% points to 2 dec. for n = 50, 100, computed from the limiting approximation above and from the Pearson curve approximation of E.S. Pearson 1932.

Davies, O. L. and Pearson, E. S. 1934
Supplement to the Journal of the Royal Statistical Society 1, 76-93

Let $\bar{R}(k,n)$ denote the mean range over k samples each of size n. The population standard deviation may be estimated by $S_3 = \bar{R}(k,n)/d_n$; and $\sigma(S_3) = k^{-\frac{1}{2}} d_n^{-1} \sigma_{R|n}$.

Table 1, p. 90, gives $\sigma(S_3)$ to 3 dec. for n = 2(1)6, 10(5)20, 50 ; k = 1(1)6, 10, 20, 50. Table 2, p. 91, gives d_n to 3 dec. and d_n^{-1} to 4 dec. for n = 2(1)15. Table 5, p. 93, gives $\beta_1(S_3)$ and $\beta_2(S_3)$ to 2 dec. for n = 2(1)6, 20 ; k = 1(1)4.

Freeman, H. A. 1942
Industrial statistics

Table 6, p. 168, gives d_n to 5 dec. for n = 2(1)40(5)110(10)200(50)500(100)1000.

Pearson, E. S. and Hartley, H. O. 1942
Biometrika 32, 301-310

Table 1, pp. 302-307, gives $P_n(R)$ to 4 dec. for n = 2(1)20 and R = 0(.05)7.25. Reprinted in Pearson & Hartley 1954, T. 23, pp. 166-171; Vianelli 1959, Pro. 119, pp. 321-326.

Table 2, p. 308, gives upper and lower 0.1%, ½%, 1%, 2½%, 5% and 10% points of R to 2 dec.; and d_n^{-1} to 4 dec.; for n = 2(1)12.

1.541 [see also pp 645-647]

Smith, J. G. and Duncan, A. J., 1945
Sampling Statistics and Applications

Table XIII (p. 482) gives the following:

d_n, $1/d_n$, $\sigma_{R|n}$ to 3 dec.
for n = 2(1)20

R(p,n) to 2 dec.
for p = .001,.005,.01,.025,.1,.999,.995,.99,.975,.9
n = 2(1)20

Grant, E. L. 1946
Statistical quality control

Table B, p. 536, gives d_n to 3 dec. for n = 2(1)25(5)100. Reprinted in Grant 1952, p. 512.

Snedecor, G. W., 1946
Statistical Methods

Mean range and upper 5% and 1% points of range in normal samples of size n.

Table 5.5 (p. 98) gives

mean range d_n { to 2 dec. for n = 2(1)10,15,20,30,50,75,100
and R(p,n)   { to 1 dec. for n = 150,200,300,500,700

p = .05, .01

Ferber, R. 1949
Statistical techniques in market research

Table 7, p. 488, based on E.S. Pearson 1932 and E.S. Pearson & Hartley 1942, gives reciprocals of the upper and lower 1%, $2\frac{1}{2}$% and 5% points of range to 3 dec., and d_n^{-1} to 4 dec., for n = 2(1)20.

Johnson, N. L. and Tetley, H., 1950
Statistics Vol II (p. 96)

d_n, $\sigma_{R|n}$ and $\dfrac{1}{d_n}$ to 3 dec. for n = 2(1)12.

[86]

Mosteller, F. 1946

Annals of Mathematical Statistics 17, 377-406

"On some useful 'inefficient' statistics"

Table IV, p. 391, gives the coefficient of variation of the range, the coefficient of variation of the standard deviation, and the efficiency of the range; all to 3 fig. for n = 2(1)10.

Dixon, W. J. and Massey, F. J. 1951

Table 8 b 1, p. 315, gives d_n^{-1} to 3 dec., and the efficiency of the range to 2 dec., for n = 2(1)10.

Duncan, A. J. 1952

Quality control and industrial statistics

Table D, p. 614, gives upper and lower 0.1%, $\frac{1}{2}$%, 1%, 2$\frac{1}{2}$% and 5% points of R to 2 dec.; d_n to 3 dec,; and $\sigma_{R|n}$ to 3-4 dec.; for n = 2(1)12. Reprinted in Duncan 1959, T. D, p. 872.

Hald, A. 1952

Table VIII gives d_n, $\sigma_{R|n}$ and $d_n^{-1}\sigma_{R|n}$ to 3 dec. for n = 2(1)20; and R(p,n) to 2 dec. for p = .0005, .001, .005, .01, .025, .05, .1(.1).9, .95, .975, .99, .995, .999, .9995, and n = 2(1)20.

Hartley, H. O. and Pearson, E. S. 1951
Biometrika 38, 463

For the distribution of range R_n in normal samples the following moment constants are given.

For n = 2(1)20:

d_n = E(R_n) (5 dec.) $\sigma_{R|n}$ (4 dec.) $\mu_2 = \sigma^2_{R|n}$ (5 dec.)

β_1 (4 dec.) β_2 (3 dec.) μ_3 (4 dec.) μ_4 (3 dec.)

For n = 2(1)12:

μ_5 (2 dec.) μ_6 (1 dec.) k_4 (3 dec.) k_5, k_6 (2 dec.)

Lindley & Miller 1953, T. 6, p. 7, based on Tippett 1925, gives d_n^{-1} to 4 dec. for n = 2(1)13.

1.541
1.542   [see also pp. 645-647]

Tippett, L. H. C. 1952

The methods of statistics

   Table A, p. 385, gives d_n and $\sigma_{R|n}$ to 3 dec. for n = 2(1)10, 20, 50, 100.

Burington, R. S. and May, D. C. 1953

   Table 13.86.1, p. 167, gives:

d_n 3; 2; 1 dec.   n = 2(1)20; 30, 50(25)100; 150, 200

$\sigma_{R|n}$ 3 dec.   n = 2(1)20

R(p,n)   2 dec. $\begin{cases} p = .001, .005, .01, .025, .05, .1; 1-p \text{ the same} \\ n = 2(1)20 \end{cases}$

Pearson, E. S. and Hartley, H. O. 1954

   Table 20, p. 164, gives d_n to 5 dec., $\sigma_{R|n}$ to 4 dec., $\sigma^2_{R|n}$ to 5 dec., $\beta_1(R)$ to 4 dec., $\beta_2(R)$ to 3 dec., $d_n/\sigma^2_{R|n}$ to 2 dec., and $d^2_n/\sigma^2_{R|n}$ to 3 fig. for n = 2(1)20.

   Table 22, p. 165, gives d_n^{-1} to 4 dec.; and the upper and lower 0.1%, $\frac{1}{2}$%, 1%, $2\frac{1}{2}$%, 5% and 10% points of R to 2 dec.; for n = 2(1)20.

Dixon, W. J. and Massey, F. J. 1957

   Table 8 b 1, pp. 404-405, gives d_n^{-1} to 3 dec., $\sigma^2_{R|n}$ to 3 fig., efficiency of range to 3 dec., and upper and lower 0.1%, $\frac{1}{2}$%, 1%, $2\frac{1}{2}$%, 5% and 10% points of R to 2 dec.; all for n = 2(1)20.

1.542   The midrange

   See also Howell 1949; for the range-midrange test see Section 1.5512.

Pearson, E. S. 1926

Biometrika 18, 173-194

   Table I (p. 179) gives

$$\xi_{n|n}, \sigma_{n|n}, \mu_r(n|n), r = 2(1)4 \text{ to 6 dec. and}$$

$$\beta_1(x_{n|n}), \beta_2(x_{n|n}) \text{ to 4 dec.}$$

for n = 2(1)6.

1.541
1.542

The same values apply to $x_{1|n}$ (apart from sign, see p. 173). Also given are $\rho_{1n|n}$ and

$$E\left[(x_{n|n} - \xi_{n|n})^i (x_{1|n} - \xi_{1|n})^j\right] \text{ for } \begin{array}{l} i = 1\ 1\ 1\ 2 \\ j = 1\ 2\ 3\ 2 \end{array}$$

all to 6 dec. for above n, and all obtained by numerical quadrature.

Table III (p. 181) gives for the regression of x_n on x_1

$$\left.\begin{array}{l} E(x_n|x_1) \text{ to 3 dec.} \\ \sigma^2(x_n|x_1) \text{ to 4 dec.} \\ \sigma(x_n|x_1) \text{ to 3 dec.} \end{array}\right\} \begin{array}{l} \text{for } n = 6 \\ x = -3.0(.5)1.0 \end{array}$$

and certain approximations to the regression data.
These tables can readily be transformed to yield the joint distribution of the range and midrange.

Pillai, K. C. S. 1950

<u>Annals</u> <u>of</u> <u>Mathematical</u> <u>Statistics</u> 21, 100-105

"On the distributions of midrange and semi-range in samples from a normal population"

If a sample of size n is drawn from a normal population, the midrange and semi-range are given by $M = \frac{1}{2}(x_{1|n} + x_{n|n})$ and $W = \frac{1}{2}(x_{n|n} - x_{1|n})$ respectively. The density function of M can be written

$$p(M) = n(n-1)\pi^{-1} e^{-\frac{1}{2}nM^2} \sum_{i=0}^{\infty} B_i M^{2i}$$

where the B_i are linear functions of

$$I(s,q,m) = \sqrt{2/\pi} \int [\alpha_x]^s x^q e^{-\frac{mx^2}{2}} dx .$$

Table 1, p. 102, gives $I(s,0,m)$:

m =	2	2	2	2	4	4	6	8	
s =	1	2	3	4(1)8	1, 2	3(1)6	1(1)4	1, 2	
	8	7	6	5	8	7	8	8	dec.

[89]

1.542
1.543   [see also pp. 645-647]

    Table 2, p. 102, gives B_i for $i = 0(1)4$:

$n = 3 \quad 4, 5 \quad 6(1)10$
    $8 \quad 6 \quad\;\; 5 \quad\quad\;\;$ dec.

    Table 4, p. 104, gives coefficients used in computing the distribution of W.

Dixon, W. J. and Massey, F. J. 1957

    Table 8 b 4, p. 406, gives, inter al., the variance of the midrange and its efficiency as an estimate of the mean to 3 dec. for $n = 2(1)20$.

1.543 Galton differences

The term 'Galton differences' has been applied to the rank intervals d_{ij} of p. 49, and in particular to d_{12}, d_{13} and d_{23}.

Pearson, K. 1902
<u>Biometrika</u> 1, 390-399

    The table on p. 397 gives

$$\left. \begin{array}{l} d_{12|n}, d_{23|n} \quad \text{to 4 dec.} \\ d_{13|n}/(d_{13|n} + d_{23|n}) \\ d_{23|n}/(d_{13|n} + d_{23|n}) \end{array} \right\} \text{to 3 dec.} \left.\vphantom{\begin{array}{l}a\\b\\c\end{array}}\right\} \begin{array}{l} \text{for} \\ n = 3, 10, 50, 100, 1000 \end{array}$$

as well as quantities required in their derivation.

Irwin, J. O. 1925 a
<u>Biometrika</u> 17, 100-128

    The tables on p. 107 give

$$E(d_{12}) = \xi_{2|n} - \xi_{1|n} \quad \text{and} \quad E(d_{23}) = \xi_{3|n} - \xi_{2|n} \quad \text{to 4 dec.}$$

$$\left. \begin{array}{ll} \mu'_r(1,2|n) & \mu'_r(2,3|n) \\ \mu_r(1,2|n) & \mu_r(2,3|n) \end{array} \right\} r = 2, 3, 4 \quad \text{to 4 dec.}$$

$\beta_1(1,2|n)$ $\beta_2(1,2|n)$; $\beta_1(2,3|n)$, $\beta_2(2,3|n)$ to 2 dec.

σ_{12} σ_{23} to 4 dec.

100 $\sigma_{12}/(\xi_2 - \xi_1)$,   100 $\sigma_{23}/(\xi_3 - \xi_2)$ to 2 dec.

All for n = 3, 10, 50, 100, 500, 1000.

The approximate distribution

$$y = y_o \exp\left(-\tfrac{1}{2} \frac{(x+h)^2 - h^2}{\Sigma^2}\right)$$

is fitted to mean and standard deviation of both d_{12} and d_{23}; and the fitted h, Σ, y_o for d_{12} are tabled on p. 118 to 3 dec. for n = 2, 3, 10, 50, 100, 500, 1000; and for d_{23}, for n = 3, 10, 50, 100, 500, 1000, 1200. Page 119 contains tabular comparisons of the exact and approximate distributions.

Irwin, J. O. 1925 b

Biometrika 17, 238-250

"On a criterion for the rejection of outlying observations"

(Notation as in the preceding abstract)

Table 1, p. 239, gives h to 2 dec. and Σ to 3 dec. for n = 3, 10(10)100(100)1000.

Table 2, p. 241, gives $P_1(\lambda) = \Pr[d_{12} \le \lambda]$ to 3 dec. for λ = .1(.1)5 and n = 2, 3, 10(10)100(100)1000.

Table 3, p. 242, gives $P_2(\lambda) = \Pr[d_{23} \le \lambda]$ to 3 dec. for λ = .1(.1)4 and n = 3, 10(10)100(100)1000.

Reprinted in Tables for S & B II, TT. 18-20, pp. 159-161.

1.544 [see also pp. 647-649]

1.544 Measures of dispersion based on selected order statistics

Pearson, K. 1920
<u>Biometrika</u> 13, 113-132

The standard deviation of a normal distribution may be estimated as the weighted sum of two interpercentile ranges:

$$s_{p_1 p_2} = \lambda_1 \frac{x_{p_1} - x_{q_1}}{x(p_1) - x(q_1)} + \lambda_2 \frac{x_{p_2} - x_{q_2}}{x(p_2) - x(q_2)}$$

where λ_1 and λ_2 are chosen such that $\sigma(s_{p_1 p_2})$ is a minimum. To this end Table VII gives 6 dec. values of

$$E\left[\left((x_{p_1} - x_{q_1}) - (x(p_1) - x(q_1))\right)\left((x_{p_2} - x_{q_2}) - (x(p_2) - x(q_2))\right)\right] \Big/ \sigma_{p_1 q_1} \sigma_{p_2 q_2}$$

for $p_1, p_2 = \frac{1}{40}, \frac{1}{30}, \frac{1}{20}, \frac{1}{14}, \frac{1}{10}, \frac{2}{10}, \frac{1}{4}, \frac{3}{10}, \frac{4}{10}$

and Table VIII (p. 126) gives 5 dec. values of the optimum λ_1 and λ_2 and $\sigma(s_{p_1 p_2})$ to 4 dec. for the above combinations of p_1 and p_2.

Hojo, T. 1931
<u>Biometrika</u> 23, 315

Distribution constants of interquartile range in normal samples

Table VI^a (p. 354) gives

$E(x''' - x')$ to 6 dec.
$\sigma_{x''' - x'}$ to 5 dec.
$\sigma^2_{x''' - x'}$ to 7 dec. ⎱ for n = 2(1)12

$\rho_{x''' x'} \cdot \sigma^2_{x'''}$ to 6 dec.

μ_3, μ_4 of $x''' - x'$ to 6 dec.
β_1, β_2 of $x''' - x'$ to 4 dec. ⎱ for n = 2(1)6

$E\left[(x''' - E(x'''))^i (x' - E(x'))^j\right]$ to 6 dec. for n = 2(1)6

$i = 1, j = 2, 3$
$i = 2, j = 2;$

the latter were required for the computation of β_1 and β_2.

1.544

Table VI[b] (p. 357) gives $\sigma_{x''' - x'}\sqrt{2n}$ and $\rho_{x'''x'}$ to 4 dec. for $n = 2(1)12, 15, 20, 30, 40$ partly theoretical, partly values from experimental sampling, partly both. The experimental sampling results are based on 1000 samples of $n = 4, 7, 10, 15, 20$ and 500 samples of $n = 30, 40$, and the distribution constants of these experimental samples are summarized in Table VII (p. 358).

Table VIII (p. 359) gives theoretical comparison of standard error of estimating σ from the interquartile distance and from the sample standard deviation s, i.e.

$$\frac{\sigma_{x''' - x'}\sqrt{2n}}{E(x''' - x')} \text{ and } \left[\frac{n-1}{2}\frac{\left(\Gamma(\frac{n-1}{2})\right)^2}{\left(\Gamma(\frac{n}{2})\right)^2} - 1\right]^{\frac{1}{2}} \sqrt{2n}$$

to 4 dec. for $n = 2(1)12, \infty$; also their ratios to 4 dec.

Godwin, H. J. 1949 a

<u>Biometrika</u> 36, 92-100

"On the estimation of dispersion by linear systematic statistics"

Godwin constructs linear estimators of σ of the form

$$d = \sum_{i=1}^{n-1} a_i \, d_{i+1, i \mid n} \qquad (1)$$

where $d_{i+1, i \mid n}$ is the difference $x_{i+1 \mid n} - x_{i \mid n}$ of the $(i+1)^{st}$ and i^{th} order statistic and the a_i are constant multipliers such that d is unbiased, i.e., such that

$$\sum_{i=1}^{n-1} a_i (\xi_{i+1 \mid n} - \xi_{i \mid n}) = \sigma \qquad (2).$$

Table 3 (p. 99) gives the 'best choice' of the a_i, i.e., the solution of var $(d) =$ minimum subject to (2) for $n = 2(1)10$, $i = 1(1)\left[\frac{n}{2}\right]$.

1.544
1.5511

[See also 647-649]

Clearly $a_{n-i} = a_i$; the number of decimals range from 5 to 10. Table 4 (p. 99) gives the 'efficiency' of the best linear estimate of d compared with the unbiased estimate based on the sum of squares of deviations,

$$\frac{\Gamma(\tfrac{1}{2}(n-1)) \left[\sum (x-\bar{x})^2 \right]^{\tfrac{1}{2}}}{\sqrt{2}\ \Gamma(\tfrac{1}{2}n)} ;$$

so that the efficiency is given by

$$\left[\tfrac{1}{2}(n-1) \left(\frac{\Gamma(\tfrac{1}{2}(n-1))}{\Gamma(\tfrac{1}{2}n)} \right)^2 - 1 \right] \Big/ \mathrm{var}\ (d).$$

Corresponding values of efficiency are shown for:

The mean deviation, m, from the mean

The mean deviation, M, from the median

The range R

Also for $x_{n-1} - x_2$, $x_{n-2} - x_3$, $x_{n-3} - x_4$, $x_{n-4} - x_5$.

Values of efficiency are given as percentages with 1 or 2 dec. for the above D

Dixon, W. J. and Massey, F. J. 1951

Table 8 a 2, p. 314, gives formulas for estimating the standard deviation from j selected quantiles, $j = 2(1)10$, with their efficiencies to 2 dec. Table 8 a 3, p. 314, gives analogous formulas for simultaneous estimation of the mean and standard deviation. Reprinted in Dixon & Massey 1957, p. 404.

Table 8 b 3, p. 315, gives formulas for estimating the standard deviation from j selected order statistics: $j = 2; 4: n = 2(1)5; 6(1)10$. Their efficiency, which is never less than .96, is given to 2 dec.

Dixon, W. J. and Massey, F. J. 1957

Table 8 b 3, p. 406, gives formulas for estimating the standard deviation from j selected order statistics:

$j = 2 \quad 4 \quad 6 \quad 8$
$n = 2(1)5 \quad 6(1)10 \quad 11(1)16 \quad 17(1)20$. } The variance of these estimators is

given to 3 fig. and their efficiency to 3 dec. The lowest efficiency shown is .955 for n = 5.

Table 8b6, p. 407, gives formulas for the 'best linear estimate' of the standard deviation, i.e. a weighted sum of all the order statistics for n = 2(2)10; of all the order statistics except the median for n = 3(2)9. The efficiency of the 'best linear estimate', which is never less than .988, is given to 3 dec.

1.549 Miscellaneous functions of the range

Hartley, H. O. 1942

<u>Biometrika</u> 32, 334-348

Consider an equidistant grid of group end points $\xi_i = \xi + ih$, i = 0, ±1, ±2,... and a random sample of size n, $x_1, x_2, \ldots x_n$ drawn from N(0,1) and arranged in order of magnitude. Let ξ_1^* and ξ_n^* denote, respectively, the <u>centers</u> of the group intervals containing x_1 and x_n and define the 'grouped range' as $R(h, \xi; n) = \xi_n^* - \xi_1^*$

Hartley shows on p. 338 that

$$E[R(h,\xi;n)] = h \sum_{i=-\infty}^{+\infty} \left[1 - [P(\xi_i)]^n - [Q(\xi_i)]^n \right];$$

this is tabled on p. 339 to 5 dec. for n = 5, 10, 20; h = .2(.4)2.2; $\xi = 0(.2)h$.

1.55 <u>Studentized statistics involving range; analysis of variance based on range; asymptotic distribution of normal range</u>

1.551 Studentized statistics involving range

1.5511 The studentized range

Let $x_1, \ldots, x_n$ denote a sample from $N(\mu, \sigma)$ arranged in order of magnitude. Using the probability integral of range $P_n(R)$ in samples from $N(\mu, 1)$ we have

$$\Pr\left[x_n - x_1 \leq R\sigma \right] = P_n(R). \quad \text{(see section 1.54)}$$

1.5511 [see also p. 649]

Denote now by s_ν a statistic independent of the above sample and distributed as $\sqrt{\chi_\nu^2 \sigma^2 / \nu}$ and introduce the studentized range

$$q(n,\nu) = (x_n - x_1)/s_\nu ;$$

then the probability integral of $q(n,\nu)$ is given by

$$\Pr[q(n,\nu) \leq Q] = {}_\nu P_n(Q) = \int_0^\infty P_n(Qs)\, s^{-1}\left[4s\, Z(s)\right]^\nu ds$$

where $Z(s) = \dfrac{1}{\sqrt{2\pi}} e^{-\frac{1}{2}s^2}$ so that $s^{-1}\left[4s\, Z(s)\right]^\nu$ will be recognized as the distribution ordinate of $s = \sqrt{\chi^2/\nu}$.
The above integral can be evaluated by numerical quadrature;

Newman, D. 1939
Biometrika 31, 20-30

Using the approximation to the distribution of range given by E. S. Pearson 1932, Newman writes the distribution of the studentized range as

$$\Pr[q(n,\nu) \geq Q] = 1 - {}_\nu P_n(Q) = \int_0^\infty P_n'(R) \int_0^{R/Q} f_\nu(s)\, ds\, dR$$

where $f_\nu(s)$ is the distribution of the independent sample standard deviation s_ν based on ν d.f.

Table of roots Q of ${}_\nu P_n(Q) = p$ to 2 dec. (Table II, p. 24)

        for $p = .95, .99$

        $n = 2, 3, 4, 6, 10, 20$

        $\nu = 5, 10, 20, 30, \infty$

to 1 dec. (Table III, IV, p. 25) for $p = .95, .99$

        $n = 2(1)12, 20$

        $\nu = 5(1)20, 24, 30, 40, 60, \infty$

2 dec. for $n = 2$ and $\nu = \infty$

The second decimal place is not reliable for small ν.

Pearson, E. S. and Hartley, H. O. 1943

Biometrika 33, 89-99

"Tables of the probability integral of the studentized range"

Let $x_1, \ldots, x_n$ denote an ordered normal sample of size n; and s^2, an independent mean square based on ν degrees of freedom. Let $P_n(R)$ denote the probability integral of range, i.e., the chance of $x_n - x_1 \leq R\sigma$. Then the probability integral of the studentized range, i.e., the chance $_\nu P_n(Q)$ for $q(n,\nu) = \dfrac{x_n - x_1}{s\nu} \leq Q$ is approximately given by

$$_\nu P_n(Q) = P_n(Q) + \frac{a_n(Q)}{\nu} + \frac{b_n(Q)}{\nu^2}$$

where

$$a_n(Q) = \tfrac{1}{4}\left[Q^2 P_n''(Q) - Q P_n'(Q)\right]$$

$$b_n(Q) = \frac{1}{16}\left[\frac{Q^4}{2} P_n^{IV}(Q) - \frac{Q^3}{3} P_n'''(Q)\right] - \tfrac{1}{8} a_n(Q)$$

the superscripts denoting differentiation.

Table 1, pp. 95-97, gives P_n; a_n; b_n to 4; 2; 1 dec. for n = 3(1)20 and Q = 0(.25)6.5 . Reprinted in Vianelli 1959, Prontuario 78, pp. 223-225.

Table 2, pp. 98-99, gives upper and lower 1% and 5% points of Q for n = 2(1)20

to 2 dec. for   p = .01, .05 ν = 10(1)20,24,30,40,60,120,∞

          and   p = .95, .99 ν =   24, 30, 40, 60, 120,∞

to 1 dec.[*] for   p = .95, .99 ν = 10(1)19

Note misprint on p. 93 in formula for $b_n(Q)$. Read $\dfrac{Q}{2}$ in place of Q. The above expansions of $_\nu P_n(Q)$ in powers of $1/\nu$ break down for small ν when Q is large. The tabulated upper percentage points are therefore in error for $\nu \leq 20$; the upper 1% points are seriously in error for $\nu \leq 15$.

---

[*] 2 dec. for n = 2.

1.5511
1.5512

[see also pp. 649-651]

Hartley, H. O. 1950

Biometrika 37, 271-280

Table 6, p. 278, reproduces the 95% and 99% points of Q from E. S. Pearson & Hartley 1943, p. 99, for the same ν and n but with improved typography.

Dixon, W. J. and Massey, F. J. 1951

Table 18a, p. 342, gives upper 5% points of Q:

ν = 5(1)19   5(1)19   120/ν = 0(1)6
n = 2   3(1)20   2(1)20
    2 dec.   1 dec.   2 dec.

Table 18b, p. 343, gives upper 1% points of Q:

ν = 10(1)19   10(1)19   120/ν = 0(1)6
n = 2   3(1)20   2(1)20
    2 dec.   1 dec.   2 dec.

These tables are credited to Newman 1939, E.S. Pearson & Hartley 1943, and Biometrika tables not specified.

May, Joyce M. 1952

Biometrika 39, 192-193

"Extended and corrected tables of the upper percentage points of the 'Studentized' range"

Upper 1% and 5% points of Q for n = 2(1)20; $\begin{cases} \nu = 1(1)20 \\ 120/\nu = 0(1)6 \end{cases}$ 3 fig.

Extensive errata in Hartley 1953.

Pillai, K. C. S. 1952

Biometrika 39, 194-195

Using a series expansion suitable for small ν and n, Pillai computed upper and lower 5% and 1% points of q to 2 dec. for n = 2(1)8 and ν = 1(1)9 The lower percentage points are printed; the upper percentage points were incorporated into May's table (see the preceding abstract).

Hartley, H. O. 1953

Biometrika 40, 236

Corrected percentage points to May 1952 for ν = 5, 7.

Pearson, E. S. and Hartley, H. O. 1954

Table 29, pp. 176-177, compiled from E.S. Pearson & Hartley 1943, May 1952, and Hartley 1953, gives upper and lower 5% and 1% points of q for n = 2(1)20 : the lower points to 2 dec. for = 10(1)20 and $120/\nu = 0(1)6$; the upper points to 3 fig. for $\nu = 1(1)20$ and $120/\nu = 0(1)6$.

Reprinted in Vianelli 1959, <u>Prontuario</u> 79, pp. 226-227. The same information, but with quite different typography, is given in Dixon & Massey 1957, Table 18, pp. 440-442.

1.5512 Studentized statistics with range in denominator

Daly, J. F. 1946

<u>Annals</u> <u>of</u> <u>Mathematical</u> <u>Statistics</u> 17, 71-74

"On the use of the sample range in an analogue of Student's t-test"

Daly proposes the criterion

$$\frac{\bar{x} - \xi}{R_n} = u(1,n)/d_n = G \quad \text{(Daly's notation)}.$$

Table 1, p. 73, gives 5% points of G, computed by numerical quadrature from the distribution of range in E.S. Pearson & Hartley 1942, to 2 dec. for n = 3, 5, 7, 10 .

Table 3, p. 74, gives 5% points of G, computed by an empirical formula confirmed by the above quadratures, to 3 dec. for n = 3(1)10 .

Lord, E. 1947

<u>Biometrika</u> 34, 41-67

"The use of range in place of standard deviation in the t-test"

Let $\bar{R}(m,n)$ denote the mean of m sample ranges each from a sample of size n, x a normal deviate, and

$$u(m,n) = \frac{x \cdot d_n}{\bar{R}(m,n)} .$$

1.5512   [see also pp. 649-651]

The ratio $u(m,n)$ serves as a substitute for t in testing the significance of x. Since x and $\bar{R}(m,n)$ are independent the probability integral $p_{m,n}(u)$ of $u(m,n)$ is obtained by numerical quadrature from

$$p_{m,n}(u) = \int_0^\infty f_{m,n}(\bar{R}) \; P\left(\frac{\bar{R}u}{d_n}\right) d\bar{R}$$

where $P(x)$ is the normal integral and $f_{m,n}(R)$ is the density function of $\bar{R}(m,n)$.

Table 1 (p. 47) gives two tail percentage points for $m = 1$, viz. roots u of

$$p_{1,n}(u) = 1 - \tfrac{1}{2}p$$

to 2-4 dec. for $p = .1, .05, .02, .01, .002, .001$
$n = 2, 3, 4, 6, 10, 16, 20, \infty$.

Table 2 (p. 51) gives the same for $m = 2$, i.e. gives roots u of $p_{2,n}(u) = 1 - \tfrac{1}{2}p$ (∞ omitted).

Tables 3-6 (pp. 63-4) give the roots u of

$$p_{m,n}(u) = 1 - \tfrac{1}{2}p$$

to 2 dec. for $p = .1, .05, .02, .01$
$m = 1(1)8, 10, 15, 20, 30, 60$
$n = 2(1)10(2)20$

Tables 7-8 (p. 65) give the roots u of

$$p_{m,n}(u) = 1 - \tfrac{1}{2}p$$

to 1 dec. for $p = .002, .001$ and $m = 1(1)6(2)10, 15, 20, 30$
$n = 2(1)10, 12, 15, 20.$

Table 9 (p. 66) gives the roots v of $p_{1,n}(n^{\frac{1}{2}}vd_n) = 1 - \tfrac{1}{2}p$ for $n = 2(1)20$; to 3 dec. for $p = .1, .05, .02, .01$; to 2 dec. for $p = .002, .001$. This table is for testing the significance of $v = (\bar{x}_n - \mu)/\bar{R}(1,n)$, i.e. the significance of the ratio of the mean to the range of the same sample of n observations.

Walsh, J. E. 1949 a

Annals of Mathematical Statistics 20, 257-267

"the range-midrange test"

Following E.S. Pearson & Adyanthāya 1929, pp. 280-286, Walsh proposes to test the hypothesis that a normal distribution have zero mean by using the statistic $D = (x_n + x_1)/2(x_n - x_1)$.

Table 3, p. 261, gives some values of the power of D, compared to t. Table 4, p. 261, gives the efficiency of D relative to t to the nearest $\frac{1}{2}\%$ for p = .05, .01 and n = 2(1)10 .

Table 5, p. 261, gives $\frac{1}{2}\%$, 1%, $2\frac{1}{2}\%$ and 5% points of D to 2 dec. for n = 2(1)10 .

Link, R. F. 1950

Annals of Mathematical Statistics 21, 112-116

"The sampling distribution of the ratio of two ranges from independent samples"

Let R_{n_1} and R_{n_2} denote the ranges in two independent samples, of sizes n_1 and n_2, from the same normal population. Table 2, pp. 114-115, gives lower percentage points of $Q^* = R_{n_1}/R_{n_2}$ to 2 fig. for p = .005, .01, .025, .05, .1 ; n_1, n_2 = 2(1)10 .

Reprinted in Vianelli 1959, Prontuario 118, p. 320.

Lord, E. 1950

Biometrika 37, 64-77

"Power of the modified t-test (u-test) based on range"

Denote by $\bar{R}(m,n)$ the mean of m independent ranges, each in a sample of n normal deviates and by $d_n = E(R_n)$ the population mean of R_n and by $f_{m,n}(\bar{R})$ the density function of $\bar{R}(m,n)$. Then
$$u(m,n) = x \, d_n/\bar{R}(m,n)$$
is a substitute for the t statistic (based on range) provided the normal variate x is independent of $\bar{R}(m,n)$.
If x comes from a normal parent with mean ρ and standard deviation 1, the power of u (1,n) is given by

1.5512 [see also pp. 649-651]

$$\beta(\rho) = \beta'(\rho) + \beta''(\rho) \quad \text{with}$$

$$[1]\cdots \beta'(\rho) = \int_0^\infty Z(x-\rho) \left[\int_0^{cx} f_{1,n}(\bar{R}) \, d\bar{R} \right] dx \quad \text{where}$$

$c = d_n/u_{\frac{1}{2}\alpha}$ and $u_{\frac{1}{2}\alpha}$ is the $100(\frac{1}{2}\alpha)$ % point of $u(1,n)$. $\beta''(\rho)$ is obtained from [1] by replacing $-\rho$ by $+\rho$

Table 3 (p. 70) gives $\beta'(\rho)$ and $\beta''(\rho)$ to 5 dec. for $\rho = 1(1)10$; $m = 1$ $n = 2(1)20, \infty$; $\alpha = .05$.

Table 4 (p. 71) gives the 'standardized error' of the u-test, viz. the roots ρ_α of $\beta(\rho) = 1 - \alpha$ to 3 dec. for $\alpha = .05$ and to 2 dec. for $\alpha = .01$ and the above $n(m=1)$.

Table 5 (p. 72) gives corresponding 'standardized errors' for $u(m,n)$ for $m = 1(1)10, 15, 20, 30, 60, 120$; $n = 2(1)20$ and $\alpha = .05$ all to 2 dec.

Table 6 (p. 76) gives the ratio of the 'standardized errors' of the $u(1,n)$ test and the corresponding t-test to 3 dec. for $\alpha = .05, .01$ and $n = 2(1)20, \infty$.

Dixon, W. J. and Massey, F. J. 1951

Table 8 c , p. 316, gives 95%, 97½%, 99% and 99½% points for three substitute t-ratios: all for sample size $n = 2(1)10$.

Table 8 c 1 **gives the percentage points of the t-ratio with the standard deviation replaced by the range,** $\tau_1 = \dfrac{\bar{x}-\mu}{R}$. 3 fig.

Table 8 c 2 **gives the same thing for two samples, the test being difference of means divided by sum of ranges.**

Table 8 c 3 **gives a similar test using midrange divided by range,**

$$\tau_2 = \frac{\frac{1}{2}(x_{1|n} + x_{n|n}) - \mu}{R}$$

Tables 8 c 1 and 8 c 2 are taken from Lord 1947; table 8 c 3 from Walsh 1949 a .

Table 8 d , pp. 317-318, gives 95%, 97½%, 99% and 99½% points for the ratio of two ranges from independent samples, for numerator and denominator sample sizes $2(1)10$. Dixon & Massey state in a footnote on p. 317 that the table is reproduced "from Link, R. F., 'On the ratio of two ranges,' Annals of Mathematical Statistics, vol. 21(1950), p. 112." The idea of this tabulation is apparently due to Link; the figures tabled are certainly not Link's. The values for sample size 2 in numerator were taken from a table of percentage points of studentized range; the other values may have been formed by taking reciprocals of Link's.

Table 8e, p. 319, gives 5% and 1% points of range/range statistics for testing suspected outliers in samples of size 3(1)30.

Pillai, K. C. S. 1951a
Sankhyā 11, 23-28

Let $x_1, \ldots, x_n$ denote an ordered sample from $N(0,1)$. Table 1, p. 25, gives upper 1% and 5% points of $(x_n + x_1)/(x_n - x_1)$ to 3; 2 dec. for $n = 3(1)8; 9, 10$.

If two such samples, of sizes n_1 and n_2, are drawn, denote their ranges by R_1 and R_2. Table 3 gives the upper 1% and 5% points of R_1/R_2 to 2 dec. for $n_1, n_2 = 2(1)8$. Spot checks indicate agreement with the lower percentage points of Link 1950.

Table 4 compares the power of the R_1/R_2 test with the power of the F-test for alternatives comprising two normal distributions with variance σ_1^2 and σ_2^2. The powers are given to 3 dec. for $\sigma_1/\sigma_2 = 1(.5)2.5$ and
$\begin{cases} n_1 = 3 & 4 & 8 & 5 & 8 \\ n_2 = 7 & 4 & 2 & 2 & 3 \end{cases}$

Pillai, K. C. S. 1951b
Annals of Mathematical Statistics 22, 469-472

Let $\bar{x}$ and R denote the mean and range of a random sample from $N(\mu, \sigma^2)$. Following Lord 1947, Pillai uses $u(1,n) = (\bar{x} - \mu)d_n/R$ to compute confidence intervals for μ.

Table 1 gives the width of the confidence interval based on u, the width of the confidence interval based on t, and the efficiency, i.e. the ratio of widths, for various values of the confidence coefficient β:

β = .9, .95   .99   .999
n = 3(1)8(2)14(3)20 3(1)5   3

1.5512
1.552

[see also pp. 650-652]

Terpstra, T. J. 1952

Applied Scientific Research, Series A 3, 297-307

"A confidence interval for the probability that a normally distributed variable exceeds a given value, based on the mean and the mean range of a number of samples"

Given k samples of size n from $N(\mu, \sigma^2)$, 90% central confidence intervals for the probability of the title are constructed as functions of $|x^* - \bar{x}|/\bar{w}$, where $\bar{x}$ is the grand mean over all samples, x^* is the 'given value', and $\bar{w}$ is the mean of the k sample ranges. The intervals, constructed with the help of Patnaik's χ-approximation to the distribution of $\bar{w}$, are graphed for $k = 4(2)10$ and $n = 1(1)5$; the graphs can be read to about 2 dec.

Dixon, W. J. and Massey, F. J. 1957

(notation as in abstract of Dixon & Massey 1951, above)

Table 8c1, p. 408, gives upper (and lower) 5%, 2½%, 1% and ½% points of r_1 to 3 dec., and 0.1% and 0.05% points to 2 dec.; all for sample size $n = 2(1)20$. Table 8c2, p. 409, does the same thing for two samples. These tables are taken from Lord 1947.

Table 8c3, p. 409, is the same as in Dixon & Massey 1951.

Table 8d, pp. 410-411, gives upper and lower 5%, 2½%, 1% and ½% points for the ratio of two ranges from independent samples, for numerator and denominator sample sizes $2(1)10$. The lower percentage points are taken from Link 1950, the upper points from Dixon & Massey 1951.

Table 8e, p. 412, gives 30%, 20%, 10%, 5%, 2%, 1% and ½% points, all to 3 dec., of range/range statistics for testing suspected outliers in samples of size $3(1)25$.

1.552 Analysis of variance by range

Hartley, H. O. 1950

Biometrika 37, 271-280

Let x_{ti} denote the experimental result obtained from the tth treatment ($t = 1, 2, \ldots, k$) on the ith block ($i = 1, 2, \ldots, n$), and assume that

$$x_{ti} = a_t + b_i + z_{ti}$$

where the z_{ti} are random variates from $N(0,\sigma)$.

Form the deviates from treatment means $\bar{x}_{t.}$, viz: form $x_{ti} - \bar{x}_{t.}$; then, for any fixed block (i) form the range of the deviates

$$R_i = \text{Range of } (x_{ti} - \bar{x}_{t.}) = \text{Range of } (z_{ti} - \bar{z}_{t.})$$

and $\bar{R} = \dfrac{1}{n} \sum_{i=1}^{n} R_i$ as an estimator of σ.

We have

$$E(\bar{R}) = \sqrt{(1 - 1/n)}\ d_k\ \sigma$$

$$\text{Var}(\bar{R}) = n^{-1} \sigma^2_{R|k} (1 - 1/n)(1 + (n-1)\rho_w)$$

where ρ_w is the correlation coefficient between any two of the ranges. Table 4 (p. 276) gives ρ_w to 2 significant figures for $k = 2(1)9$ and $1/(n-1) = 0(.1).5, 1$; end figures are not reliable. The distribution of $\bar{R}$ is then approximated by that of $c\chi/\sqrt{\nu}$ fitted by equating means and variances. The fractional degrees of freedom ν of χ^2 and the scale factor c are given respectively to 1 and 2 dec. in Table 5 (p. 276).

David, H. A. 1951
<u>Biometrika</u> 38, 393-409

In extending the range analysis of data in a double classification (Hartley 1950) to the above designs the following tables are revised and/or freshly computed, each dealing with an approximation to the distribution of mean range $\bar{R}$ in residuals in the form $c\chi/\nu$ where ν are the (fractional) degrees of freedom of the χ^2 approximation:

Table I (p. 408) gives ν (1 dec.) and c (2 dec.) for a single classification in k groups of size n for $k = 1(1)5, 10, \infty$; $n = 2(1)10$. See p. 406 for the use of the ν value for $k = \infty$.

Table II (p. 408) gives ν (1 dec.) and c (2 dec.) for a double classification in n 'blocks' and k 'treatments' for $k = 2(1)9$; $n = 2(1)10, 20, \infty$.

1.552
1.5531

[see also pp. 651-652]

Table III (p. 409) is a similar table for the split-plot design with l main treatments, $m = 2(1)10$ blocks and $n = 2(1)9$ sub-treatments.

Table IV (p. 409) caters to the analysis of single classification with unequal cell frequencies, viz: observations x_{ti} classified in k groups ($t = 1, \ldots, k$) with n_t observations in the tth group. Assuming $x_{ti} = \alpha_t + z_{ti}$, where the z_{ti} are independent random variables $N(0, \sigma)$, group ranges $R_t = \max_i x_{ti} - \min_i x_{ti}$ are computed; σ is then estimated from

$$s_R = \frac{\sum_t \left[d_{n_t} / \sigma_{R|n_t}^2 \right]}{\frac{1}{2}k + \sum_t \left[d_{n_t}^2 / \sigma_{R|n_t}^2 \right]}$$

for which, approximately, $E s_R^2 = \sigma^2$. To facilitate the computation of s_R Table IV gives d_n and $d_n / \sigma_{R|n}^2$ to 2 dec.; and $d_n^2 / \sigma_{R|n}^2$ to 1 or 2 dec.; all for $n = 2(1)20$.

Tables I and II are reproduced as tables 30 A and 30 B in Pearson & Hartley 1954, p. 178; and as <u>Prontuario</u> LXXX in Vianelli 1959, p. 228.

Moshman, J. 1952

<u>Annals</u> <u>of</u> <u>Mathematical</u> <u>Statistics</u> 23, 126-132

"Testing a straggler mean in a two way classification using the range"

Let x_{ij} ($i = 1, \ldots, r$; $j = 1, \ldots, k$) denote the observation from the ith row and j^{th} column in a two way classification. To test whether the largest column mean is a straggler, Moshman uses the statistic

$$g = W^{-1} \max_j \left| k \sum_i x_{ij} - \sum \sum_{ij} x_{ij} \right|$$

where W is the sum of the column ranges or of the row ranges, whichever involves fewer summands.

Tables 1 and 2, pp. 129-130, give 5% and 1% points of g to 2 dec. for r, k = 3(1)9. These are based on the approximate χ-distribution of W given by Hartley 1950 and Patnaik 1950.

1.553 Approximate distributions of normal range

1.5531 Approximations for moderate n

Cox, D. R. 1949

Journal of the Royal Statistical Society, Series B 11, 101-114

An approximate χ^2-distribution of the range, R_n, is computed by using the correct start $R_n = 0 \sim \chi^2 = 0$ and equating means and variances, so that R_n is approximately represented as $\tfrac{1}{2} a \chi^2$ for ν degrees of freedom with

$$a = \sigma^2_{R|n}/d_n \ ; \ \nu = 2d_n^2/\sigma^2_{R|n}.$$

Table 1 (p. 102) gives:

$$\begin{cases} \theta_n = \sigma^{-1}_{R|n} d_n & \theta_n^2 d_n = \sigma^{-2}_{R|n} d_n^3 & b_n = \tfrac{1}{2}\nu & \nu & \lambda_n = n^{\tfrac{1}{2}} \sigma^{-1}_{R|n} d_n \\ \text{3 dec.} & \text{3 dec.} & \text{4 fig.} & \text{2 dec.} & \text{3 dec.} \end{cases}$$

$$\begin{cases} \lambda'_n = n(n+1)^{-\tfrac{1}{2}} \sigma^{-1}_{R|n} d_n \\ \text{3 dec.} \end{cases} \text{ for } n = 2(1)10.$$

Table 2 (p. 103) gives comparative values of β_1 (3 dec.) β_2 (2 dec.) for R_n and χ^2 for n = 2(1)10, 15, 20. Graphs illustrate the precision of the fit which is good for $n \geq 8$ but rather poor for smaller n.

Patnaik, P. B. 1950

Biometrika 37, 78-87

Denote by $\bar{R}(m,n)$ the mean of m independent ranges each in a sample of n normal deviates. The distribution of $\bar{R}(m,n)$ is approximated by that of $\dfrac{c\chi}{\sqrt{\nu}}$ where $\chi = \sqrt{\chi^2}$ is based on ν degrees of freedom which, together with the scale factor c are determined by equating the first two moments of $\bar{R}(m,n)$ and $\dfrac{c\chi}{\sqrt{\nu}}$.

Table 1 (p. 80) gives ν (3 dec.) and c(4 dec.) for m = 1(1)5 and n = 3(1)10, but for small m and n the higher decimals are not reliable. Note the difference between this table and the table of Hartley 1950, p. 276, which is based on <u>correlated</u> ranges.

1.5531
1.5532   [see also p. 652]

Pearson, E. S. 1952

Biometrika 39, 130-136

"Comparison of two approximations to the distribution of the range in small samples from normal populations"

The two approximations considered are:

(a) Patnaik's (1950) χ- approximation, i.e. $R \sim c\, \chi_\nu / \sqrt{\nu}$

(b) Cox's (1949) χ^2- approximation, i.e. $R \sim a \tfrac{1}{2} \chi^2_{\nu_2}$

Table 1 (p. 131) gives the exact $P_n(R)$ to 4 dec., n = 4,6,10 and 15 and selected values of R. Alongside are shown the results from the above two approximations and their differences against $P_n(R)$, all to 4 dec. Broadly speaking, the conclusion is that for $n \leq 10$ (a) is better, while for $n > 10$ (b) is better.

1.5532 Approximations for large n

Elfving, G. 1947

Biometrika 34, 111-119

"The asymptotical distribution of range in samples from a normal population"

Let $X = 2n[P(x_1)Q(x_n)]^{\frac{1}{2}}$, $X^* = 2n\, Q(\tfrac{1}{2} R_n)$. Then Elfving shows that, in the limit as $n \to \infty$, $\Pr[X \leq \Xi] = \Pr[X^* \leq \Xi] = 1 - \Xi K_1(\Xi)$, where K_1 is a Hankel function.

The table on p. 116 gives $1 - \Xi K_1(\Xi)$, and $-\frac{d}{d\Xi}\Xi K_1(\Xi) = \Xi K_0(\Xi)$, to 4 dec. for $\Xi = 0(.1)1(.5)3(1)10$.

X is found to approach its asymptotic distribution faster than X^*.

Gumbel, E. J. 1947

Annals of Mathematical Statistics 18, 384-412

Let $x_1, \ldots, x_n$ be an ordered sample drawn from a parent $\phi(x)$ with

$$\Phi(x) = \int_{-\infty}^{x} \phi(\tau)\, d\tau.$$

Let u be the root of
$$\Phi(u) = 1 - \frac{1}{n} \qquad (1)$$
then the 'reduced range' R' is defined by
$$R' = (x_n - x_1 - u)\alpha = (R - u)\alpha \qquad (2)$$
where $\alpha = \dfrac{\phi(u)}{1 - \Phi(u)}$.

The asymptotic distribution $\psi(R')$ and probability integral $\Psi(R')$ of R' are given by certain Hankel functions: specifically,
$$\psi(R') = 2e^{-R'} K_0\left(2e^{-\frac{1}{2}R'}\right) \; ; \; \Psi(R') = 2e^{-\frac{1}{2}R'} K_1\left(2e^{-\frac{1}{2}R'}\right)$$

In Table III (p. 406) α and u are determined for the normal parent $\phi(x)$ by the moment fit
$$\frac{1}{\alpha_n} = \frac{\sqrt{3}}{\pi} \sigma_{R|n} \qquad\qquad 2u_n = d_n - \frac{2\gamma}{\alpha_n}$$
where d_n and $\sigma_{R|n}$ are the known normal mean range and σ of range. Table III gives d_n (3 dec.), $\sigma_{R|n}$ (3-4 dec.), $\dfrac{1}{\alpha_n}$ (3-4 dec.), $2u_n$ (3 dec.), $2\alpha_n u_n$ (2 dec.) for $n = 3, 4, 5, 10, 20, 50, 100$. Also shown are u the roots of (1) and $\tilde{x}$ the roots of
$$\tilde{x} \Phi(\tilde{x}) = (n-1)\phi(\tilde{x})$$
both to 3 dec. for above n.

Table IV (p. 406) gives (with the above values of α_n and u_n) the values of R' to 2 dec. from (2) corresponding to $R' = -3(1)7$ for $n = 5, 10, 20, 50, 100$. Also $\Psi(R')$ to 3 dec.

Figures 4, 5, 6 give comparisons of exact results for normal range with reduced range results.

[We have seen no entirely satisfactory account of the relation between Elfving's and Gumbel's results. It should be pointed out that the natural cologarithm of Elfving's X (or X*) has the same asymptotic distribution as Gumbel's $\frac{1}{2}R'$. The requirement of a normal parent distribution can be relaxed somewhat for Elfving's result, and considerably for Gumbel's; the extent to which the asymptotic distribution of the range is distribution-free is discussed in some detail by Kendall & Stuart 1958, pp. 340-343.]

1.5532
1.611

Cox, D. R. 1948

Biometrika 35, 310

Asymptotic distribution of range.

Let $x_1 \ldots x_n$ be an ordered sample drawn from a parent $\phi(x)$ and let $\Phi(x) = \int_{-\infty}^{x} \phi(\xi)\, d\xi$. Further let us denote by x_n^* and x_1^* respectively the roots of $\Phi(x_n^*) = 1 - \frac{1}{n}$; $\Phi(x_1^*) = \frac{1}{n}$. Then the 'reduced range' can be defined as

$$R' = n\left[(x_n - x_n^*)\phi(x_n^*) - (x_1 - x_1^*)\phi(x_1^*)\right]$$

and has a limiting distribution given by

$$2e^{-R'} K_0\left(2e^{-\frac{1}{2}R'}\right)$$

where K_0 is a modified Bessel function of the second kind. This can be made the basis of an asymptotic distribution of the ordinary range $R = x_n - x_1$ since for symmetrical $\phi(x)$

$$R = 2x_n^* + \frac{R'}{n\phi(x_n^*)} \, . \tag{A}$$

For $\phi(x)$ = normal parent, a comparison with the exact mean, S.D., β_1 and β_2 of R_n has been made in Table 1 (p. 313) for n = 10, 20, 50, 100, 500, 1000. For normal $\phi(x)$, a closer approximation is obtained in the form

$$c_n \cdot \frac{\left[\phi(\tfrac{1}{2}w_n)\right]^2 \left[\Phi(\tfrac{1}{2}w_n) - \Phi(-\tfrac{1}{2}w_n)\right]^{n-3/2}}{\left[\phi'(-\tfrac{1}{2}w_n)\right]^{\tfrac{1}{2}}}$$

(c_n = constant) and for this the above parameters are tabulated for n = 20, 50 and compared with the Bessel-function approximation and with the approximation of Elfving 1947.

The Bessel-function approximation of this paper is formally identical with that of Gumbel 1947 but is numerically different; Cox computes R' by means of equation (A), whereas Gumbel equates the first two moments of the exact and asymptotic distributions.

1.6 Tests of Normality

Given a sample $x_1, x_2, \ldots, x_n$ which is believed to have been drawn from $N(\mu, \sigma)$, the question sometimes arises whether the sample provides any evidence that it has, in fact, been generated by a parent population which is not normal.

In devising tests to detect non-normality the usual procedure is:

(a) To decide upon a suitable statistic (the criterion) to be computed from the sample.

(b) To refer the observed criterion to tables of its random sampling distribution which it would follow if the parent were normal.

Section 1.61 is devoted to the moment ratio $b_1^{\frac{1}{2}} = m_3 m_2^{-3/2}$, where $m_j = n^{-1} \sum (x_i - \bar{x})^j$; this is designed to detect departures from symmetry.

Sections 1.62 and 1.63 cover two criteria based on absolute moments: section 1.62 treats the moment ratio $b_2 = m_4 m_2^{-2}$, and section 1.63 treats the ratio of the mean deviation from the mean to the standard deviation. These two criteria are special cases of Geary's general ratio criterion

$$a(c) = \frac{1}{n} \sum_{i=1}^{n} |x_i - \bar{x}|^c \left[\frac{1}{n} \sum_{i=1}^{n} (x_i - \bar{x})^2 \right]^{-\frac{1}{2}c}$$

Geary 1947 b gives a valuable review of the subject of testing for normality; he stresses that non-normality may vitiate normal-theory tests.

1.61 $b_1^{\frac{1}{2}}$

1.611 Moments of $b_1^{\frac{1}{2}}$

The following formulas for the even moments of $b_1^{\frac{1}{2}}$ may be noted; the odd moments all vanish.

$$\mu_2(\sqrt{b_1}) = 6(n-2)/(n+1)(n+3)$$

$$\mu_4(\sqrt{b_1}) = 108(n-2)(n^2 + 27n - 70)/(n+1)(n+3)(n+5)(n+7)(n+9)$$

$$\mu_6(\sqrt{b_1}) = \frac{3240(n-2)(n^4 + 84n^3 + 2695n^2 - 15168n + 20020)}{(n+1)(n+3)(n+5)\ldots(n+15)}$$

$$\mu_8(\sqrt{b_1}) = 7.5.3^5 .2^4 (n-2)(n^6 + 171n^5 + 13893n^4 + 580401n^3 - 5131014n^2 \\ + 14132268n - 12932920) / (n+1)(n+3)\ldots(n+21)$$

1.611
1.621

The second, fourth and sixth moments are due to R.A. Fisher 1930; the eighth moment, to Pepper 1932.

Pearson, E. S. 1936

<u>Biometrika</u> 28, 306-307

"Note on probability levels for $\sqrt{b_1}$"

$\sigma(b_1)$ to 4 dec. for n = 25(5)50(10)100(25)200(50)1000(200)2000 (500)5000. Reprinted in Geary & Pearson 1938, T. 4, p. 7; Pearson & Hartley 1954, T. 34 B, p. 183.

<u>1.612</u> Distribution of $b_1^{\frac{1}{2}}$

Mc Kay, A. T. 1933 a

<u>Biometrika</u> 25, 204-210

"The distribution of $\sqrt{\beta_1}$ in samples of four from a normal universe"

Mc Kay expresses the frequency curve of $b_1^{\frac{1}{2}}$ in terms of the hypergeometric function F as

$$\phi(\sqrt{b_1}) = \frac{1}{2\sqrt{3}} F(\frac{1}{3}, \frac{2}{3}, 1; 1 - \frac{3}{4}(\sqrt{b_1})^2);$$

it is symmetrical, has an infinite cusp at $\sqrt{b_1} = 0$ and is equal to $\frac{1}{2\sqrt{3}}$ at the terminals $\sqrt{b_1} = \pm \frac{2}{\sqrt{3}}$; $\sigma^2_{\sqrt{b_1}} = \frac{12}{35}$.

In the table (p. 209) grouped frequencies

$$2 \cdot 200 \int_{ih}^{i+1\ h} \phi(\sqrt{b_1})\, d\sqrt{b_1}$$

are given to 2 dec. for h = 0.1, i = 0(1)10 as well as the tail to the terminal $\sqrt{b_1} = \frac{2}{\sqrt{3}}$.

<u>1.613</u> Percentage points of $b_1^{\frac{1}{2}}$

Pearson, E. S. 1930 a

<u>Biometrika</u> 22, 239-249

Table 4, p. 248, gives upper 5% and 1% points of $b_1^{\frac{1}{2}}$, approximated by fitting Pearson Type VII curves to its second and fourth moments; and the squares of these percentage points, which are upper 10% and 2% points for b_1; all to 3 dec. for $n = 50(25)200(50)1000(200)2000(500)5000$. Reprinted in Tables for S & B II, T. XXXVII bis, p. 224.

Williams, P. 1935

<u>Biometrika</u> 27, 269-271

Table 2, p. 271, gives upper 5% and 1% points of $b_1^{\frac{1}{2}}$ to 3 dec. for $n = 25(5)50(10)100$. The values for $n \geq 30$ were approximated from Pearson Type VII curves; for $n = 25$, a three-parameter curve due to Hansmann was used. The 1% column is incorrectly headed 10%.

Reprinted, including the incorrect heading, in Smith & Duncan 1945, T. 10, p. 480.

Pearson, E. S. 1936

<u>Biometrika</u> 28, 306-307

Upper 5% and 1% points of $b_1^{\frac{1}{2}}$ to 3 dec. for $n = 25(5)50(10)100(25)200(50)1000(200)2000(500)5000$. Reprinted in Geary & Pearson 1938, T. 4, p. 7; Pearson & Hartley 1954, T. 34B, p. 183; Freeman 1942, T. 1, p. 164; Vianelli 1959, Pro. 215, p. 701.

Duncan, A. J. 1952

<u>Quality control and industrial statistics</u>

Table F, p. 616, gives upper 5% and 1% points of $b_1^{\frac{1}{2}}$ for $n = 25(5)50(10)100(25)200(50)300(100)500(250)1000$. Reprinted in Duncan 1959, T. F, p. 876.

1.62 b_2

1.621   Moments of b_2

Pearson, E. S. 1930a

<u>Biometrika</u> 22, 239-249

Expansions of $\sigma(b_2)$, $\beta_1(b_2)$, $\beta_2(b_2)$ in terms of n^{-i}, and exact $E(b_2)$ (equations (20) - (23)). Table II (p. 244) gives 4 terms of the expansions to 6 dec. and their sum to 4 dec. for $n = 50, 75, 100(50)250, 500, 1000$.

1.621
1.631
Geary, R. C. 1935

Biometrika 27, 310-332

Table A, p. 312, gives $\beta_1^{\frac{1}{2}}$ and β_2 of b_2 to 2 dec. for n = 6, 11, 26, 36, 51, 101; computed from the formulas of E.S. Pearson 1931 z .

1.622 Distribution of b_2

Mc Kay 1933 b derives formulas, in terms of hypergeometric functions, for the distribution (i.e. the density function) of b_2 in normal samples of 4 ; but gives no numerical results. It is noteworthy that b_2 cannot exceed 7/3 in samples of 4 .

1.623 Percentage points of b_2

Pearson, E. S. 1930 a

Biometrika 22, 239-249

Table 4, p. 248, gives upper and lower 5% and 1% points of b_2, approximated by fitting Pearson Type IV curves from its first four moments, to 2 dec. for n = 100(25)200(50)1000(200)2000(500)5000 . Reprinted in Tables for S & B II, T. 37 bis, p. 224.

Geary, R. C. and Pearson, E. S. 1938
Tests of normality

Table 5, p. 8, is the table of E.S. Pearson 1930 a , abstracted above, with values for n = 100(25)175 omitted. Reprinted in Freeman 1942, T. 2, p. 164; Pearson & Hartley 1954, T. 34 C , p. 184; Vianelli 1959, p. 701.

Smith, J. G. and Duncan, A. J. 1945
Sampling statistics and applications

Table 11, p. 480, abridged from E.S. Pearson 1930 a , gives upper and lower 5% and 1% points of b_2 to 2 dec. for n = 100(25)200(50)300(100)500, 800, 1000, 2000, 5000 .

Duncan, A. J. 1952
Quality control and industrial statistics

Table G, p. 616, abridged from Geary & Pearson 1938, gives upper and lower 5% and 1% points of $\gamma_2 = b_2 - 3$ to 2 dec. for n = 200, 250, 300, 500, 1000, 2000, 5000 .

1.63 Geary's ratio

In section 1.63 we use N for the sample size and $n = N-1$ for the number of degrees of freedom in the standard deviation.

1.631 Moments of Geary's ratio

Geary, R. C. 1936 a

Biometrika 28, 295-305

"Moments of the ratio of the mean deviation to the standard deviation for normal samples"

Let

$$a(1) = \sum_{i=1}^{N} |x_i - \bar{x}| \Big/ \sqrt{n \sum_{i=1}^{N} (x_i - \bar{x})^2}$$

The first four moments and semi-invariants of $a(1)$ are obtained (see equations 25, 26) and used to show that the distribution of $a(1)$ is approximately given by

$$\frac{1}{\sigma\sqrt{2\pi}} \left[1 - \frac{\sqrt{B_1}}{b} \left(\frac{3(a - \lambda'_1)}{\sigma} - \frac{(a - \lambda'_1)^3}{\sigma^3} \right) \right] e^{-\frac{(a - \lambda'_1)^2}{2\sigma^2}} \quad (1)$$

where

$$\lambda'_1 \sim .798 + \frac{.2}{n}$$

$$\sigma^2 \sim \frac{.045}{n}$$

$$-\sqrt{B_1} \sim \frac{1.76}{\sqrt{n}} \left[1 - \frac{2.4}{n} \right]$$

For higher accuracy see equations (25), (26), (27).

Tables 1 and 2, p. 302, give constants of the distribution for $N = 6, 11, 26, 51, 101$. Table 3, p. 303, gives λ'_1 and σ to 5 dec. for $N = 11(5)51(10)101(100)1001$.

Reprinted in Geary & Pearson 1938, T. 3, p. 7; Pearson & Hartley 1954, T. 34 A, p. 183.

1.631
1.69   [see also p. 652]

Geary, R. C.   1947 b

Biometrika 34, 209-242

Tests of normality using

$$a(c) = \frac{1}{N} \sum_{i=1}^{N} |x_i - \bar{x}|^c \bigg/ \left[\frac{1}{N} \sum_{i=1}^{N} (x_i - \bar{x})^2\right]^{\frac{1}{2}c}$$

with auxiliary statistic

$$a_1(c) = \frac{1}{N} \sum_{i=1}^{N} |x_i|^c \bigg/ \left[\frac{1}{N} \sum_{i=1}^{N} x_i^2\right]^{\frac{1}{2}c}$$

Table 4, p. 225, gives moments of $a(2.4)$ in samples of 25 and $a_1(2.4)$ in samples of 24. Formulas for the moments of $a(c)$ from a normal parent are given on pp. 222-224; for the moments of $a_1(c)$ from an arbitrary parent, on pp. 235-236.

Peters, C. C. and Van Voorhis, W. R.   1940

Statistical procedures and their mathematical bases

Table 26, p. 309, gives the mean and standard deviation of $a(1)$ to 4 dec. for n = 5(5)50(25)100, 500, 1000.

1.633   Percentage points of Geary's ratio

Geary, R. C.   1936 a

Biometrika 28, 295-305

Table 3, p. 303, gives upper and lower 10%, 5% and 1% points of $a(1)$ to 4 dec. for N = 11(5)51(10)101(100)1001. Reprinted in Geary & Pearson 1938, T. 3, p. 7; Pearson & Hartley 1954, T. 34A, p. 183; Vianelli 1959, Pro. 215, p. 701.

Freeman, H. A.   1942

Industrial Statistics

Table III (p. 165) gives the upper and lower 10%, 5% and 1% points for $a(1)$ to 4 dec. and also the mean of $a(1)$ to 5 dec. for N = sample size = 11(5)51(10)101(100)1001.

Smith, J. G. and Duncan, A. J. 1945

Sampling Statistics and Applications

Table XII (p. 481) gives the upper and lower 10%, 5% and 1% points of a(1) to 4 dec. for n = sample size - 1 = 10(5)50(10)100(100)300. The upper 10% and 1% points are incorrectly labelled 1% and 10%.

Peters, C. C. and Van Voorhis, W. R. 1940

Statistical procedures and their mathematical bases

Table 26, p. 309, gives upper and lower 5% and 1% points of a(1) to 3 dec. for n = 5(5)50(25)100, 500, 1000.

Duncan, A. J. 1952

Quality Control and Industrial Statistics

Table H (p. 617) gives upper and lower 10%, 5% and 1% points of a(1) to 4 dec. for n = sample size - 1 = 10(5)30(10)80,100(100)500,700,1000.

Reprinted in Duncan 1959, T. H, p. 876.

1.69 Miscellaneous tests of normality

In this section, n again denotes the sample size.

Geary, R. C. 1935

Biometrika 27, 310-332

"The ratio of the mean deviation to the standard deviation as a (new) test of normality"

As a substitute for a(1) with a more readily computable distribution, Geary introduces the statistic w_{n-1}, which depends on an artificial randomization of the sample $x_1, \ldots, x_n$:

$$w_{n-1} = \frac{1}{\sqrt{n-1}} \frac{\sum_{i=1}^{n-1} \left| \frac{x_1 + \cdots + x_i - i x_{i+1}}{\sqrt{i(i+1)}} \right|}{\sqrt{\sum_{i=1}^{n} (x_i - \bar{x})^2}}.$$

1.69
1.72
[see also p. 652]

Closed formulas for the moments of w_{n-1}, and recursion formulas for the semi-invariants, are given on pp. 313-315. Table B, p. 315, gives λ_1, λ_2, λ_3 and λ_4 to 6-9 dec.; and $\beta_1^{\frac{1}{2}} = \lambda_3 / \lambda_2^{3/2}$ and $\beta_2 = \left[\lambda_4 / \lambda_2^2\right] + 3$ to 2 dec.; for n = 5, 10, 25, 50, 100.

$f_n(w)$, the density function of w_{n-1}, is given explicitly on p. 325 for n = 2, 3, 4, 5.

Table C (p. 327) gives $f_n(a)$ to 2-4 dec. (mainly 3 sig. figures) for n = 2(1)10; a = 0.44(.03).98,1.00; Table D (p. 328) the corresponding probability integrals

$$\int_a^{1.00} f_n(w) \, dw$$

to 4 dec. for the same range of n and a and the figure on p. 329 illustrates the normal (and other) approximation to $f_n(w)$ for n = 10 with the 1% and 5% comparison set out to 3 dec. accuracy in Table E (p. 330)

Table F (p. 330) gives upper and lower 1% and 5% points of $f_n(w)$ to 3 dec., the lower end of the limited variate range min w_n, $E(w_n)$ and σ_{w_n} to 4 dec. for n = 5(5)50, 75, 100, 500, 1000.

Comparison with the (theoretically preferable) criterion

$$a(1) = \frac{\sum_{i=1}^{n+1} |x_i - \bar{x}|}{\sqrt{n+1} \sqrt{\sum_{i=1}^{n+1} (x_i - \bar{x})^2}} \quad \text{(Geary's } w_n')$$

is illustrated by comparison of λ_1 and $\sqrt{\lambda_2}$ (4 dec.), λ_3 (8 dec.) $\lambda_3 / \lambda_2^{3/2}$ (2 dec.) for a(1) and w_{n-1} for n = 50 in the table on p. 318. Close agreement is found.

1.7 The bivariate normal distribution

1.71 Ordinates of the bivariate normal distribution

The ordinate of the bivariate normal frequency surface may be written in standardized form, so that both x and y have zero mean and unit variance:

$$Z(x,y;\rho) = \frac{1}{2\pi\sqrt{1-\rho^2}} \exp\left[-\frac{1}{2(1-\rho^2)}\left[x^2 + y^2 - 2\rho xy\right]\right]$$

where ρ is the correlation coefficient between x and y. This function has not been extensively tabulated. It can be calculated from tables of the univariate normal ordinate

$$Z(x) = (2\pi)^{-\frac{1}{2}} \exp\left[-\tfrac{1}{2}x^2\right];$$

$$Z(x,y;\rho) = (1-\rho^2)^{-\frac{1}{2}} Z(x) Z\left(\frac{y-\rho x}{\sqrt{1-\rho^2}}\right).$$

Irwin, J. O. 1923
Tracts for Computers, No. 10
Tables I and II (p. 29 and facing p. 29) give

$$Z(x,y;0) = \frac{1}{2\pi} e^{-\frac{1}{2}(x^2 + y^2)}$$

to 7 dec. for x,y = 0(.25)5.00
9 dec. for x,y = 0(1)5

also Z(x) to 7 dec. for x = 0(.25)5.00
9 dec. for x = 0(1)5

These tables are used as examples for cubature.

1.72 Volumes under the bivariate normal distribution

The most extensively tabulated form of bivariate normal volume is

$$p(x,y;\rho) = \int_x^\infty \int_y^\infty Z(\xi, \eta; \rho)\, d\xi\, d\eta, \quad x,y \geqq 0.$$

If the integral of $Z(x,y;\rho)$ is required for a section of the x,y plane bounded by two lines, a suitable linear transformation will reduce this to $p(x,y;\rho')$ with x, y and ρ' depending on the equations of the lines. By a combination of such integrals the region of integration of $Z(x,y;\rho)$ can be extended to any region bounded by lines.

1.72
1.723

According to Gertrude Blanch (N.B.S. 1959, p. v), the function

$$V(h,q) = \frac{1}{2\pi} \int_0^h e^{-\frac{1}{2}x^2} \left[\int_0^{qx/h} e^{-\frac{1}{2}y^2} dy \right] dx$$

was known to Sheppard 1900. Its tabulation is due to Nicholson 1943 and to H.H. Germond et al. (N.B.S. 1959).

The integral of $Z(x,y;\rho)$ over a general polygon can be reduced to linear combinations of $V(h,q)$ and the univariate normal integrals of section 1.111 - 1.115. The technique is due to Cadwell 1951, and is given in detail, with examples, by D.B. Owen in N.B.S. 1959, pp. xxviij-xxxij. In particular (N.B.S. 1959, p. vij, equations 13 and 14)

$$p(x,y;\rho) = V(x, \tau y - \upsilon x) + V(y, \tau x - \upsilon y) - 1/4 + \tfrac{1}{2}[Q(x) + Q(y)] + (2\pi)^{-1}\sin^{-1}\rho$$

where $\tau = (1-\rho^2)^{-\frac{1}{2}}$, $\upsilon = \rho\tau$, and Q is the univariate normal integral of section 1.1141.

No differences are given with the tables listed below.

1.721 $p(x,y;\rho)$ for positive ρ

6 dec.	$\rho = 0(.05).95(.01)1$	$x,y = 0(.1)4$	N.B.S. 1959(2)
6 dec.	$0(.05)1$	$0(.1)2.6$	Lee 1927, K. Pearson 1931a (T.8)[*]
4 dec.	$.8(.05)1$	$0(.1)2.6$	Everitt 1912, K. Pearson 1930a (T.30), Vianelli 1959(712)

1.722 $p(x,y;\rho)$ for negative ρ

7 dec.	$-\rho = 0(.05).95(.01).99$	$0(.1)4$	N.B.S. 1959(128)
7 dec.	$0(.05).6$		
8 dec.	$.65(.05).75$	$0(.1)2.6$	K. Pearson 1931a (T.9)[*]
7 dec.	$.8(.05).95$		
7 dec.	$0(.05)6$	$0(.1)2.6$	K. Pearson 1930b[*]
8 dec.	$.65(.05).75$		
7 dec.	$.8(.05).95$	$0(.1)2.6$	Lee 1917[*]

[*] List of errors in N.B.S. 1959, p. xlv.

1.723 V(h,q)

Nicholson, C. 1943

<u>Biometrika</u> 33, 59-72

"The probability integral for two variables"

The table on pp. 69-70 gives V(h,q) to 6 dec. for h = .1(.1)3; q = .1(.1)3, ∞ . The linear transformation required to reduce the general bivariate normal to the uncorrelated case is given on pp. 61-62, and is implemented by Table 7, p. 71, containing:

$$\sqrt{1-r^2} = \cos \lambda, \quad 1/\sqrt{1-r^2} = \sec \lambda, \quad r/\sqrt{1-r^2} = \tan \lambda \quad \text{to 5 dec.}$$

and $\frac{1}{2} - \frac{\cos^{-1} r}{\pi} = \frac{\lambda}{\pi}$ to 6 dec., at argument r = sin λ = 0.00(.01)1.00

Owen, D. B. 1956

<u>Annals</u> <u>of</u> <u>Mathematical</u> <u>Statistics</u> 27, 1075-1090

"Tables for computing bivariate normal probabilities"

Define
$$T(h,\lambda) = (2\pi)^{-1} \int_0^\lambda (1+x^2)^{-1} e^{-\frac{1}{2}h^2(1+x^2)} \, dx \;;$$

then $T(h,\lambda) = (2\pi)^{-1} \tan^{-1} \lambda - V(h, \lambda h)$. Owen gives formulas for expressing the bivariate normal integral over polygons, including in particular $p(x,y;\rho)$, in terms of $T(h,\lambda)$.

T(h,λ) is tabled to 6 dec. as follows:

Table A, pp. 1080-1083   h = 0(.01)2(.02)3   λ = 0(.25)1
      B   1084-1086   0(.25)3   0(.01)1, ∞
      C   1087   3(.05)3.5(.1)4.7, 4.76   .1, .2(.05).5(.1)
                                                          .8, 1, ∞

Reprinted in Vianelli 1959, <u>Prontuario</u> 210, pp. 654-660.

N. B. S. 1959

"Tables of the bivariate normal distribution function and related functions"

Table 3, pp. 216-233, gives V(h, λh) to 7 dec., with errors of 2 in the last place, for h = 0(.01)4(.02)4.6(.1)5.6, ∞ ; λ = .1(.1)1. Table 4, pp. 236-255, gives V(λh, h) to 7 dec., with errors of 2 in the last place, for h = 0(.01)4(.02)5.6, ∞ ; λ = .1(.1)1.

1.723
1.78 [see also pp. 652-654]

The values of $\sin^{-1}\rho$ needed to express $p(x,y;\rho)$ in terms of $V(h,q)$ are given in Table 5, p. 258:

$(2\pi)^{-1}\sin^{-1}\rho$ 8 dec. $\begin{cases} \rho = 0(.01).9 & \delta_m^2 \\ \rho = .91(.01)1 & \text{No } \Delta \end{cases}$

1.729 Miscellaneous volumes

Sheppard, W. F. 1900

<u>Transactions of the Cambridge Philosophical Society</u> 19, 23-68

"On the calculation of the double-integral expressing normal correlation"

Table I (p. 67) gives

$$V = \frac{1}{2\pi \sin D} \int_x^\infty \int_x^\infty e^{-\frac{1}{2}(x'^2 - 2x'y'\cos D + y'^2)\csc^2 D} \, dx' \, dy'$$

to 7 dec. for $\frac{D}{\pi} = .00(.01).80$ and x = probable error = .6744897502; Δ^4. The table was originally calculated to 10 dec. and rounded off to 7. The differences in the table are taken from the 10-figure table.

For $\frac{D}{\pi}$ above .75, the differences of V become unwieldy. A supplementary table II (p. 68) gives

$$V \, e^{\frac{1}{2}x^2 \sec^2 \frac{1}{2}D}$$

to 7 dec. for $\frac{D}{\pi} = .75(.01).90$, same x, Δ^3.

1.73 Moments

Pearson, K. and Young, A. W. 1918

<u>Biometrika</u> 12, 86-92

Define the product-moment

$$p_{ij} = \int_{-\infty}^\infty \int_{-\infty}^\infty Z(x,y;\rho) \, x^i y^j \, dx \, dy$$

If $i+j$ is odd, $p_{ij} = 0$. On pp. 87-88 are given formulas for p_{ij}, $0 \leq i \leq j \leq 10$. On pp. 90-91 these p_{ij} are tabled, all for $\rho = 0(.05)1$:

$$\begin{cases} i = 1 & 2 & 3 & 4 & 5 & 6 & 7 & 7 & 8 \\ j = 1(2)9 & 2(2)10 & 3(2)9 & 4(2)10 & 5(2)9 & 6(2)10 & 7 & 9 & 8 \\ \text{exact} & \text{exact} & \text{exact} & \text{exact} & \text{exact} & \text{exact} & \text{5 dec.} & \text{4 dec.} & \text{4 dec.} \end{cases}$$

$$\begin{cases} 8 & 9 & 10 \\ 10 & 9 & 10 \\ \text{3 dec.} & \text{3 dec.} & \text{2 dec.} \end{cases}$$

Error in column heading on p. 90: for $p'_{1,1}$ read $p_{1,1}$.

Nabeya, S. 1951

<u>Annals of the Institute of Statistical Mathematics</u> (Tokyo) 3, 2-6

"Absolute moments in 2-dimensional normal distribution"

Nabeya gives formulas (but no numerical tables) for the absolute moments $E|x^i y^j|$ for $i+j \leq 12$. The absolute moments are polynomials in ρ unless i and j are both odd, when they involve $(1-\rho^2)^{\frac{1}{2}}$ and $\sin^{-1}\rho$.

1.78 Miscellaneous tables connected with the bivariate normal distribution

Pearson, K. 1931a

<u>Tables for S & B</u> II, T. 27

Let (x_i, y_i) be a sample from a bivariate normal population with correlation ρ. The correlation, $\rho_{s_1 s_2}$, between the two sample standard deviations

$$s_1 = \frac{1}{\sqrt{n-1}} \sqrt{\sum (x_i - \bar{x})^2} \quad \text{and} \quad s_2 = \frac{1}{\sqrt{n-1}} \sqrt{\sum (y_i - \bar{y})^2}$$

is given by

$$\rho_{s_1 s_2} = \frac{2 \lambda_n^2 (H(\rho^2) - 1)}{n - 1 - 2 \lambda_n^2}$$

where $\lambda_n = \Gamma(\tfrac{1}{2}n) / \Gamma(\tfrac{1}{2}(n-1))$ and

$H(\rho^2)$ is the Gaussian hypergeometric function $F(-\tfrac{1}{2}, -\tfrac{1}{2}; n-1; \rho^2)$.

$\rho_{s_1 s_2}$ is tabled to 4 dec. for $\rho = 0(.1)1$; $n = 2(1)25, 50, 100, 400$.

1.78
1.79

A previous attempt (K. Pearson 1925 a) to table this correlation was vitiated by an error in the formula for $H(\rho^2)$.

Craig, C. C. 1936

<u>Annals</u> <u>of</u> <u>Mathematical</u> <u>Statistics</u> 7, 1-15

"On the frequency function of xy"

Let x and y be jointly normally distributed with parameters μ_x, μ_y; σ_x, σ_y; ρ. Write $v_1 = \mu_x/\sigma_x$, $v_2 = \mu_y/\sigma_y$. Craig derives a rather cumbersome series, involving Hankel functions and confluent hypergeometric functions, for the frequency function of $z = xy/\sigma_x\sigma_y$. When $v_1 = v_2 = \rho = 0$ all terms of the series, except the first, vanish; and the frequency function of z is simply $\pi^{-1}K_0(z)$.

The table on pp. 12-14 gives the frequency function F(z) to 5 dec., $t = (z - \bar{z})/\sigma_z$ to 2 dec., and $F(z)\sigma_z$ to 5 dec., for:

$v_1 = 0 \quad v_2 = 0 \quad \rho = 0 \quad z = .1(.1)1(.2)2(.4)4(.8)8(1)12$

$\quad\quad 1 \quad\quad\quad 0 \quad\quad\quad 0 \quad\quad\quad\quad .1(.1)1(.2)2(.4)4(.8)8(1)10$

$\quad\quad \tfrac{1}{2} \quad\quad\quad \tfrac{1}{2} \quad\quad\quad 0 \quad\quad \begin{cases} z = .1, .2, .4(.4)4(.8)9.6 \\ -z = .1, .2, .4(.4)4(.8)12 \end{cases}$

For z = 0 the frequency function has a logarithmic singularity.

Aroian, L. A. 1947

<u>Annals</u> <u>of</u> <u>Mathematical</u> <u>Statistics</u> 18, 265-271

(Notation as in the preceding abstract)

Aroian derives formulas for the distribution of z. When $\rho = 0$, he finds that $\quad E(z) = v_1v_2, \quad \sigma_z = \sqrt{v_1^2 + v_2^2 + 1}$

$\lambda_3 = 6v_1v_2(v_1^2 + v_2^2 + 1)^{-3/2}, \quad \lambda_4 = 6\left[2(v_1^2 + v_2^2) + 1\right]\left[v_1^2 + v_2^2 + 1\right]^{-2}$

Table II (p. 269) gives the above (E(z) as an integer, the others to 6 dec.) for v_1, v_2 = 0(2)10. Also shown is the Pearson Type curve criterion of Craig 1936 z

$$\delta = (2\lambda_4 - 3\lambda_3^2) / (6 + \lambda_4) \quad\quad \text{to 4 dec.}$$

Table I gives the exact distribution of z (obtained by quadrature from the formulas (3.1), (3.2)) for the special case $v_1 = 0$, $v_2 = 10$ and compared with normal and Gram-Charlier A approximation in terms of t_z, where $t_z = [z - E(z)] \sigma_z^{-1}$.

Burington, R. S. and May, D. C. 1953
Probability and statistics

Table 11.11.1, pp. 102-105, gives $(2\pi)^{-1} \int\int e^{-\frac{1}{2}(x^2+y^2)} dxdy$ over a circle of radius R with center at distance d from the origin, to 3-4 dec. for R = .1(.1)1(.2)3 and d = 0(.1)3(.2)[R+3].

1.79 The trivariate normal distribution

Steck, G. P. 1958
Annals of Mathematical Statistics 29, 780-800
"A table for computing trivariate normal probabilities"

Let X, Y and Z be random variables with the trivariate normal distribution such that $EX = EY = EZ = 0$, $EX^2 = EY^2 = EZ^2 = 1$, $EXY = \rho_{12}$, $EXZ = \rho_{13}$, $EYZ = \rho_{23}$: denote $\Pr[X \geq x, Y \geq y, Z \geq z]$ by $D(x,y,z,\rho_{12},\rho_{13},\rho_{23})$. Steck gives formulas for reducing this probability to a linear combination of: three univariate normal integrals, six bivariate T-functions (see abstract of Owen 1956), and six trivariate S-functions

$$S(h,a,b) = \int_0^1 b(2\pi)^{-1}(1+b^2t^2)^{-1}(1+a^2+a^2b^2t^2)^{-\frac{1}{2}} P[h(1+a^2+a^2b^2t^2)^{\frac{1}{2}}] dt,$$

where P is the univariate normal integral of section 1.115.

The table on pp. 790-799 gives $S(h,a,b)$ to 7 dec. for b = .1(.1)1 and

$\begin{cases} a = 0(.1)1.2 \quad 1.3, 1.4 \quad 1.5, 1.6 \quad 1.7, 1.8 \quad 1.9, 2 \\ h = 0(.1)1.5, \infty \quad 0(.1)1.4, \infty \quad 0(.1)1.3, \infty \quad 0(.1)1.2, \infty \quad 0(.1)1.1, \infty \end{cases}$

$\begin{cases} 2.2 \quad\quad 2.4, 2.6 \quad 2.8, 3 \quad\quad 3.2, 3.4 \quad 3.6, 3.8 \quad 4(.2)4.6 \\ 0(.1)1, \infty \quad 0(.1).9, \infty \quad 0(.1).8, \infty \quad 0(.1).7, \infty \quad 0(.1).6, \infty \quad 0(.1).5, \infty \end{cases}$

$\begin{cases} 4.8, 5, 5.5 \quad 6(.5)8 \\ 0(.1).4, \infty \quad 0(.1).3, \infty \end{cases}$

Reprinted in Vianelli 1959, Pro. 297, pp. 1063-1072.

[see also pp. 654-655]

1.8
1.83
1.8 **The error integral of complex argument**

1.81 **The error integral of pure imaginary argument, alias Poisson's or Dawson's integral**

1.811 $\int_0^x e^{t^2} dt$

6 dec.	0(.01)2 ⎫	No Δ	Terrill & Sweeny 1944b(220),
7-9 fig.	2(.01)4 ⎭		1944a(496)
6 dec.	0(.01)2	No Δ	Dawson 1898(521)
4; 3, 2 dec.	0(.01)1.99; 1, 1.8	IΔ	Jahnke & Emde 1933(106), 1938(32),
			1948(26), 1952(26)
3-4 fig.	0(.02)2	½Δ	Jahnke & Emde 1960(32)

1.812 $e^{-x^2} \int_0^x e^{t^2} dt$

10 dec.	-.2(.05)+4(.1)6.5(.5)12.5	No Δ	Rosser 1948(190)
6; 8, 9 dec.	0(.01)4(.05)7.5(.1) 10(.2)12; 2, 5	No Δ	W.L. Miller & Gordon 1931(2878)
6 dec.	0(.1)1, 4, 5, 10	D^4	Terazawa 1917(172)[*]
4; 5 dec.	0(.2)12; 5.2	No Δ	Mitchell & Zemansky 1934(322)

1.813 $e^{-y} \int_0^{y^{\frac{1}{2}}} e^{t^2} dt$

6 dec.	1.2(.4)12	[D^4]	Terazawa 1917(173)[*]

The four derivatives are with respect to $x = y^{\frac{1}{2}}$.

3-4 dec.	3(1)10(2)20, 21(3)30	No Δ	Nottingham 1936(85)

---

[*]Fletcher et al. 1946, p. 220, impute "a number of errors" to Terazawa, but do not specify them.

1.82 The error integral with semi-imaginary argument $i^{\frac{1}{2}}x$: the Fresnel integrals

If $z = \frac{1}{2}\sqrt{\pi}(1+i)u = \sqrt{\frac{1}{2}\pi i}\, u$ then

$$\int_0^z e^{-t^2}\,dt = \sqrt{\tfrac{1}{2}\pi i}\int_0^u e^{-\frac{1}{2}\pi i t^2}\,dt$$

$$= \sqrt{\tfrac{1}{2}\pi i}\,\bigl[\,C(u) - i\,S(u)\,\bigr]$$

where
$$C(u) = \int_0^u \cos(\tfrac{1}{2}\pi t^2)\,dt$$

$$S(u) = \int_0^u \sin(\tfrac{1}{2}\pi t^2)\,dt$$

are Fresnel integrals.

For early tables of these integrals see Fletcher et al. 1946, pp. 296-300. Lebedev & Fedorova 1956, pp. 199-202, indicates additional tables: these Russians, in attempting to update Fletcher, ignored the difference between $C(u)$ and $C\!\left(\sqrt{\tfrac{2x}{\pi}}\right)$, so that their statement of the contents of tables of Fresnel integrals is almost totally misleading.

Fresnel integrals can be obtained to 11 dec. from the table of Rosser 1948, pp. 187-189, which gives

$$Rr(u) = [\tfrac{1}{2}-S(u)]\cos\tfrac{1}{2}\pi u^2 - [\tfrac{1}{2}-C(u)]\sin\tfrac{1}{2}\pi u^2,$$
$$Ri(u) = [\tfrac{1}{2}-C(u)]\cos\tfrac{1}{2}\pi u^2 + [\tfrac{1}{2}-S(u)]\sin\tfrac{1}{2}\pi u^2,$$

to 12 dec. for $u = -.06(.02)+3.5$ and $3(.05)5.15$.

1.83 The error integral for general complex argument

Faddeeva, V. N. and Terent'ev, N. M. 1954

Let $u + iv = e^{-z^2}\!\left(1 + 2i\pi^{-\frac{1}{2}}\!\int_0^z e^{t^2}\,dt\right)$, where $z = x + iy$.

Table 1, pp. 21-247, gives u and v to 6 dec. for $y = 0(.02)3$, $x = 0(.02)3$.
Table 2, pp. 250-268, gives them for $\begin{cases} y = 0(.1)2.9 & 3(.1)5 \\ x = 3(.1)5 & 0(.1)5 \end{cases}$. Second differences of u and v are given in the x-direction; by virtue of the Cauchy-Riemann differential equations, these differences also serve for interpolation in the y-direction.

1.83
1.9

Karpov, K. A. 1954 [see also p. 655]

Let $u + iv = e^{-z^2} \int_0^z e^{t^2} dt$, where $z = re^{is}$, and the angle s is expressed in degrees. The tables give u and v to 5 dec. for

$\begin{cases} s = 0 & 2°.5(5°)17°.5 & 5°(5°)15° & 20°, 25° \\ r = 0(.001)2(.01)10 & 3.5(.01)5 & 0(.001)2(.01)5 & 0(.001)2.5(.01)5 \end{cases}$

$\begin{cases} 22°.5 & 27°.5 & 30° & 31°.25 \\ 2(.001)2.5(.01)5 & 1(.001)3(.01)5 & 0(.001)3(.01)5 & 3(.001)4(.01)5 \end{cases}$

$\begin{cases} 32°.5 & 33°.75 & 35° & 35°.625(1°.25)38°.125 \\ 1(.001)4(.01)5 & 2.5(.001)4(.01)5 & 0(.001)4(.01)5 & 3.5(.001)5 \end{cases}$

$\begin{cases} 36°.25 & 37°.5, 42°.5 \quad 38°.75 & 39°.375 \quad 40° & 40°.625, 41°.875 \\ 2(.001)5 & ..5(.001)5 \quad 1.5(.001)5 & 3(.001)5 \quad 0(.001)5 & 2.5(.001)5 \end{cases}$

$\begin{cases} 41°.25 & 43°.125, 44°.375 \quad 43°.75 & 45°,, 90° \\ 1(.001)5 & 2(.001)3(.0002)5 \quad 1(.001)3(.0002)5 & 0(.001)3(.0002)5 \end{cases}$.

Four- and five-point Lagrangian interpolation is recommended.

Karpov, K. A. 1958

Let $u + iv = \int_0^z e^{t^2} dt$, where $z = re^{is}$, and the angle s is expressed in degrees. The tables give u and v to 5 dec. for

$\begin{cases} s = 45°, 50° & 45°.3125, 45°.9375 & 45°.625(1°.25)54°.375 & 46°.25(2°.5)53°.75 \\ r = 0(.001)5 & 3.5(.001)5 & 2(.001)5 & 1(.001)5 \end{cases}$

$\begin{cases} 47°.1875(0°.625)48°.4375 & 47°.5, 52°.5 \quad 55° & 56°.25 \\ 4(.001)5 & .5(.001)5 \quad 0(.001)3(.01)5 & 1.5(.001)3(.01)5 \end{cases}$

$\begin{cases} 57°.5 & 58°.75 & 60° \\ .5(.001)3(.01)4.8(.1)5 & 1.5(.001)3(.01)4.4(.1)5 & 0(.001)3(.01)4.4(.1)5 \end{cases}$

$\begin{cases} 61°.25, 63°.75 & 62°.5 & 65° \\ 1.5(.001)3(.01)4 & .5(.001)3(.01)4 & 0(.001)2(.01)3.9, 4 \end{cases}$

$\begin{cases} 67°.5 & 70° & 72°.5 \\ .5(.001)2(.01)3.6(.1)4 & 0(.001)2(.01)3.6(.1)4 & .5(.001)2(.01)3.5(.1)4 \end{cases}$

$\begin{cases} 75° & 77°.5(5°)87°.5 & 80°(5°)90° \\ 0(.001)2(.01)3.4(.1)4 & 1(.001)2(.01)3 & 0(.001)2(.01)3 \end{cases}$.

Four- and five-point Lagrangian interpolation is recommended.

1.9 Miscellanea

Tippett, L. H. C. 1944

<u>Biometrika</u> 33, 163-172

"The control of industrial processes subject to trends in quality"

Define $S(h,k)$ by the formula

$$\log S(h,k) = \sum_{i=1}^{\infty} \log Q\left[\frac{h-i}{k}\right].$$

Table 1, p. 165, gives $S(h,k)$ to 4 dec. for:

$\begin{cases} k^{-1} = 1 & .5 & .2 & .1 & .05 \\ h = -3.4(.2)+2(.1)3.2 & -6.8(.4)+2(.2)4.4 & -18.5(.5)+6 & -38(1)+6 & -78(2)0 \end{cases}$

$\begin{cases} .03 & .02 \\ -129.5(3)-12.5 & -196.5(4)-28.5 \end{cases}$.

Table 2, p. 166, gives the means and 5% and 1% points of the distributions of Table 1 to 2 dec.

Reprinted in Vianelli 1959, <u>Prontuario</u> CCIL, p. 851, and <u>Tavola</u> CCXXVIII, p. 1408.

Goodwin, E. T. and Staton, J. 1948

<u>Quarterly</u> <u>Journal</u> <u>of</u> <u>Mechanics</u> <u>and</u> <u>Applied</u> <u>Mathematics</u> 1, 319-326

"Table of $\int_0^\infty \frac{e^{-u^2}}{u+x}\,du$"

Denote the integral of the title by $f(x)$, and let $g(x) = f(x) + \log_e x$. The table on p. 325 gives $g(x)$ to 4 dec. for $x = 0(.01)1$; the table on p. 326 gives $f(x)$ to 4 dec. for $x = .02(.02)2(.05)3(.1)10$.

2.0
2.111

SECTION 2

THE χ^2 - DISTRIBUTION AND THE INCOMPLETE Γ - FUNCTION:
THE PEARSON TYPE III CURVE; THE POISSON DISTRIBUTION;
THE COMPLETE Γ - FUNCTION AND THE FACTORIAL; THE χ^2 - TEST OF GOODNESS OF FIT

2.0 Introduction

The χ^2-distribution, the incomplete Γ-function and the Poisson distribution are three different statistical applications of the same mathematical function. The chance that the sum of the squares of ν independent unit normal deviates exceed χ^2 is given by

$$p(\chi^2,\nu) = 2^{-\frac{1}{2}\nu}[\Gamma(\tfrac{1}{2}\nu)]^{-1}\int_{\chi^2}^{\infty} e^{-\frac{1}{2}y^2} y^{\frac{1}{2}\nu-1}\, dy . \qquad (2.1.1)$$

Introducing the variable of integration $t = \tfrac{1}{2}y^2$, we obtain the incomplete Γ-function

$$\Gamma_x(n) = \int_0^x e^{-t} t^{n-1}\, dt = \Gamma(\tfrac{1}{2}\nu)[1 - p(\chi^2,\nu)] \qquad (2.1.2)$$

with

$$x = \tfrac{1}{2}\chi^2, \quad n = \tfrac{1}{2}\nu .$$

When n is an integer, i.e. when ν is an even integer, we can transform the incomplete Γ-function into a Poisson series by repeated integration by parts:

$$p(\chi^2,\nu) = \sum_{j=0}^{i-1} e^{-m} m^j / j! = P(m, i-1) \qquad (2.1.3)$$

where $m = \tfrac{1}{2}\chi^2$ and $i = \tfrac{1}{2}\nu$.

Another form of the incomplete Γ-function is the integral of the Pearson Type III curve. Following Salvosa 1930, set $\nu = 8\alpha_3^{-2}$, $\chi^2 = 8\alpha_3^{-2} + 4\alpha_3^{-1}x$; then

$$\Pr[t \le x] = y_0 \int_{-2\alpha_3^{-1}}^{x} (1+\tfrac{1}{2}\alpha_3 t)^{4\alpha_3^{-2}-1} e^{-2\alpha_3^{-1}t}\, dt \qquad (2.1.4)$$

The relation of incomplete normal moments to the χ^2-distribution has been treated in section 1.31.

2.1 The χ^2-integral and the Poisson sum

The division of tables hereunder rests on the authors' intent (and, we hope, the readers' convenience) rather than on a logical necessity. We have listed tables that were published with χ^2 or the incomplete Γ-function as their primary object in section 2.11; and tables that were published with the Poisson sum as their primary object in section 2.12. It will be seen from equation (2.1.3) above that either tables may be used for the other purpose: in particular the table of Hartley & Pearson 1950, which by an artificial arrangement of double argument columns and two kinds of type serves both as a χ^2- and as a Poisson table, is listed in both sections.

2.11 The χ^2-integral and the incomplete Γ-function

2.111 The incomplete Γ-function for general exponent

Pearson, K. 1922

Tables of the incomplete Γ-function

Define[*] $I(u,\Pi) = \Gamma_x(\Pi+1)/\Gamma(\Pi+1) = 1 - p(\chi^2, 2\Pi+2)$, where $u = x/(\Pi+1)^{\frac{1}{2}}$.

Table 1, pp. 2-115, gives $I(u,\Pi)$ to 7 dec. for $\Pi = 0(.1)5(.2)50$. u varies by increments of .1 over all values for which both $I(u,\Pi)$ and $1 - I(u,\Pi)$ exceed $5 \cdot 10^{-8}$. δ^4 in both arguments.

Table 2, pp. 118-138, does the same for negative Π: $\Pi = -1(.05)0$.

Table 5, pp. 163-164, gives $I(u,\Pi)$ to 5 dec. for $\Pi = -1(.01)-.75$; $u = 0(.1)6$. No Δ.

Table 3, pp. 140-151, gives $\log I'(u,\Pi) = \log I(u,\Pi) - (\Pi+1)\log u$ to 8 dec. for $\Pi = -1(.05)0(.1)10$; $u = 0(.1)1.5$. δ^4 in both arguments.

'Table 5 adjusted', p. xiv, gives

$$\log I''(\xi,\Pi) = \log \Gamma_\xi(\Pi+1) - (\Pi+1)\log \xi$$

to 7 dec. for $\Pi = -1(.01)-.9$; $\xi = 0(.01).3$. No Δ.

Sluckiĭ, E. E. 1950

Define $p(\chi^2,\nu)$ by formula (2.1.1) for arbitrary (fractional) ν. Table 1, pp. 16-29, gives $(\frac{1}{2}\chi^2)^{-\frac{1}{2}\nu}[1- p(\chi^2,\nu)]$ for $\begin{cases} \nu = 0(.05).2(.1)6 \\ \chi^2 = 0(.05).2(.1)1 \end{cases}$. Table 2,

---

[*]We adopt the notation Π instead of Pearson's p, because p and P are used in this section for other purposes.

2.111
2.119   [see also p. 656]

pp. 16-39, gives $p(x^2, \nu)$ for $\begin{cases} \nu = 0(.05).2(.1)6 \quad 0(.1)4 \\ x^2 = 0(.1)3.2 \qquad\qquad 3.2(.2)7(.5)10(1)17 \end{cases}$

$\begin{cases} .6(.2)6 \\ 3.2(.2)7(.5)10(1)34 \end{cases}$. Table 3, pp. 42-55, gives $p\left(\frac{1}{2}[t + (2\nu)^{\frac{1}{2}}]^2, \nu\right)$ for

$\begin{cases} \nu = 6(.5)11 \qquad 12(1)32 \\ t = -3.5(.1)+4.8 \quad -3.9(.1)+4.7 \end{cases}$. Table 4, pp. 58-64, gives

$p\left(\frac{1}{2}[t + 2y^{-1}]^2, 2y^{-2}\right)$ for $\begin{cases} y = 0(.02).2 \qquad .22(.01).25 \\ t = -4.4(.1)+4.6 \quad -4(.1)+4.6 \end{cases}$.

δ^2 and unnecessary δ^4 are given for the arguments x^2 and t; δ^2 for ν and y. The arrangements for interpolation where the tabular interval changes are often inconvenient. The 5 dec. values printed were interpolated in K. Pearson 1922 and are good to no more than 0.52 in the fifth place.

<u>2.112</u> Areas under the Pearson Type III curve

Salvosa, L. R. 1930

<u>Annals of Mathematical Statistics</u> 1, 191-198 + *1-187*

(For notation, see equation (2.1.4) of the introduction)

Table 1, pp. *2-61*, gives $\Pr[t \leq x]$ to 6 dec. for $\alpha_3 = 0(.1)1.1$ and $x = -4.89(.01)+9.51$.

<u>2.113</u> The χ^2-distribution: the incomplete Γ-function for integer and half-integer exponent

For the incomplete Γ-function for negative exponent, see Fletcher et al. 1946, pp. 206-207. No differences are given with the tables below, unless specifically mentioned.

<u>2.1131</u> $p(\chi^2, \nu)$

Hartley & Pearson 1950 table the integral of the title to 5 dec. for
$\begin{cases} \nu = 1(1)30(2)70 \\ \chi^2 = .001(.001).01(.01).1(.1)2(.2)10(.5)20(1)40(2)134 \end{cases}$.

Reprinted in Pearson & Hartley 1954, T. 7, pp. 124-129.

<u>2.1132</u> $p(\chi^2, n'-1)$

	$n' =$	$\chi^2 =$	
10 dec.	3(1)30	1(1)30(10)70	Davis & Nelson 1935(399)
6 dec.	3(1)30	1(1)30(10)70	W.P. Elderton 1902, K. Pearson 1930 a (26),
			Peters & V.V. 1940(498), Vianelli 1959(193)

	n' =	x^2 =	
5 dec.	7(1)15	4(1)15	Jones 1924
5 dec.	1	0(.01)1(.1)10	Yule & Kendall 1937
3 dec.	$\begin{cases}3(1)12 \\ 10(1)20\end{cases}$	$\begin{cases}1(1)4(2)10(5)20 \\ 8(2)20(5)30\end{cases}$	Hodgman 1958(366)

2.1133 $p(\nu s^2, \nu)$

Kelley 1938, p. 116, tables the integral of the title to **4 dec.** for
$\begin{cases}\nu = 1(1)10, 12, 15, 19, 24 \\ s = 0(.1)4.1\end{cases}$

Reprinted in Kelley 1948, p. 202.

2.1134 $x^{-n-1}\int_x^\infty t^n e^{-t} dt$

	n =	x =	
11 fig.	0(1)15	.25(.25)8(.5)13(1)16(2)24, 21	Kotani et al. 1938(26)

List of alleged errors in Kotani & Amemiya 1940, p. 17.

| 6 fig. | $\begin{cases}0(1)8 \\ 0(1)12\end{cases}$ | $\begin{cases}2.75(.5)3.75 \\ 4.25\end{cases}$ | Rosen 1931a (267) |

Rosen 1931a contains other values, superseded by those in Rosen 1931b.

| 6 fig. | $\begin{cases}0(1)10 \\ 11(1)20\end{cases}$ | $\begin{cases}.5(.5)5(1)10(2)14 \\ 1(1)10(2)14\end{cases}$ | Rosen 1931b (2109) |
| 4+ fig. | $\begin{cases}0(1)6 \\ 7(1)11\end{cases}$ | $\begin{cases}1.5(.5)5(1)10 \\ 3(1)10\end{cases}$ | J.H. Bartlett 1931(520) |

2.119 Miscellaneous tables of the χ^2-distribution

Elderton, W. P. 1902

Biometrika 1, 155-163

Table 2, p. 162, gives $\log_{10}(2x^2/\pi)^{\frac{1}{2}} e^{-\frac{1}{2}x^2}$ and $\log_{10} e^{-\frac{1}{2}x^2}$ to 8 dec. for $x^2 = 1(1)100$.

Table 3, p. 163, gives $-\log_{10} \nu!!$ to 8 dec. for $\nu = 1(1)100$, where $(2m)!! = 2^m m!$ and $(2m+1)!! = (2m+1)!/(2m)!!$.

Table 4, p. 163, gives $1-\alpha(x)$ to 7 dec. for $x^2 = 1(1)30$.

Reprinted in Tables for S & B I, TT. 13-15, pp. 29-30.

2.119
2.122   [see also p. 656]

Pearson, K. 1930 a

<u>Tables for S & B I</u>

    Table 17, p. 31, gives $-\log_{10} p(x^2, 3)$ to 3 dec. for
$x^2 = 1(1)50(10)100(50)2100(100)3000(500)25000$.

David, F. N. 1934

<u>Biometrika</u> 26, 1-11

    Table 4, pp. 7-11, gives $1 - p(y \log_e 100, 2n)$ to 4 dec. for
$$\begin{cases} n = 2(1)6 & 7(1)30 \\ y = 0(.125)10.375 & 0(.25)26.25 \end{cases}$$

Fertig, J. W. and Proehl, Elizabeth A. 1937

<u>Annals of Mathematical Statistics</u> 8, 193-205

"A test of a sample variance based on both tail ends of the distribution"

    Let x_1^2 and x_2^2 be the lesser and the greater root of
$$x^2/\nu - \log_e(x^2/\nu) = k \log_e 10.$$
Table 1, pp. 198-200, gives $1 - p(x_1^2, \nu) + p(x_2^2, \nu)$ to 4 dec. for $\nu = 1(1)50$, $k = .435(.005).5(.01).7(.05)1.5(.3)3$.

2.12 The cumulative Poisson distribution

We denote the cumulative sum of Poisson frequencies by

$$P(m,i) = \sum_{j=0}^{i} e^{-m} m^j / j!$$ which is related to the

χ^2 tail area $p(\chi^2, \nu)$ by $p(\chi^2, \nu) = P(m, i-1)$

with $m = \tfrac{1}{2}\chi^2$ and $i = \tfrac{1}{2}\nu$.

    The entry 'all' comprehends all integer values of i that make P different from 0 and 1 to the number of places tabled; the upper limit ∞ is used with an analogous meaning.

<u>2.121</u> P(m,i)

	m =	i =	
6 dec.	.1, 1(1)15	all	Mothes 1952(492)

5 dec.	$\begin{cases} .0005(.0005).005(.005) \\ .05(.05)1(.1)5(.25) \\ 10(.5)15.5 \\ 16(.5)20(1)67 \end{cases}$	$\begin{cases} \text{all} \\ \\ \\ 2(1)34 \end{cases}$	Hartley & Pearson 1950 (318), Pearson & Hartley 1954(122)
5 dec.	1(1)30	0(1)m-1	Whitaker 1914(45), K. Pearson 1930a(122)
3 dec.	.02(.02).1(.05)1(.1)2(.2) 8(.5)15(1)25	all	E. L. Grant 1946(542), 1952(522)
3 dec.	.1(.1)1(.2)2(.5)5	all	Dixon & Massey 1951(333), 1957(435)
3 dec.	$\begin{cases} .1, .5 \\ 1(1)9 \\ 10(1)25 \end{cases}$	$\begin{cases} 0 \\ 0(1)m \\ 0(1)10 \end{cases}$	Przyborowski & Wileński 1935(282)

2.122   1 - P(m,i)

7 dec.	.001(.001).01(.01).3, .4	all	Molina 1942(T.2), Czechowski et al. 1957(64)
6 dec.	.5(.1)15(1)100		
6 dec.	.1(.1)14	all	Vianelli 1959(570)
5 fig.	$\begin{cases} .1(.1)1 \\ 2(1)20 \end{cases}$	$\begin{cases} 0(1)[3+5m] \\ 0(1)[8+9m/5] \end{cases}$	Fry 1928(463)
5 dec.	.001(.001).01(.01).1(.1) 2(1)10	all	E.S. Smith 1953(66)
5 dec.	1(1)30	m(1)∞	Whitaker 1914(45), K. Pearson 1930a(122)
4 dec.	.1(.1)10(1)20	all	Burington & May 1953(263)
3 dec.	.1(.1)15	all	Greenshields, Schapiro & Ericksen 1947(142), Greenshields & Weida 1952(227)
3 dec.	.05, .1, .2, .25, .5, 1, 2, 2.5, 5, 10, 20, 25, 50, 100	all	Przyborowski & Wileński 1935(283)

2.129 [see also pp. 656-657]
2.21

2.129 Miscellaneous tables of the cumulative Poisson distribution

Smith, E. S. 1953

<u>Binomial</u>, <u>normal</u> and <u>Poisson</u> <u>probabilities</u>

Table C 7, p. 66, gives, in addition to material listed in section 2.122, $1 - P(m,i)$ to 5 dec. for

$\begin{cases} m = 20 & 30 & 40 & 50 & 60 & 70 & 80 \\ i = 3(2)35 & 8(3)50 & 11(4)63 & 19(5)74 & 23(6)89 & 34(7)97 & 39(8)111 \end{cases}$

$\begin{cases} 90 & 100 \\ 44(9)125 & 49(10)139 \end{cases}$.

2.2 Ordinates of the Pearson Type III frequency curve; individual Poisson frequencies

2.20 Ordinates of the χ^2-distribution

The ordinate, $(\tfrac{1}{2}\chi^2)^{\tfrac{1}{2}\nu-1} e^{-\tfrac{1}{2}\chi^2}$, can be computed from exponential and logarithmic tables; or, alternatively, from a table of the χ^2-integral by means of the formula

$$\Gamma^{-1}(\tfrac{1}{2}\nu)\, e^{-\tfrac{1}{2}\chi^2} (\tfrac{1}{2}\chi^2)^{\tfrac{1}{2}\nu-1} = -\frac{dp(\chi^2,\nu)}{d(\tfrac{1}{2}\chi^2)} = p(\chi^2,\nu) - p(\chi^2,\nu-2). \qquad (2.21.1)$$

If a tabulation of the ordinates is needed as a function of $s = \sqrt{\chi^2/\nu}$, it is useful to note that

$$e^{-\tfrac{1}{2}\chi^2}(\tfrac{1}{2}\chi^2)^{\tfrac{1}{2}\nu-1} = \text{const } s^{-2}(s\, Z(s))^\nu. \qquad (2.21.2)$$

2.21 Ordinates of the Pearson Type III frequency curve

Burgess, R. W. 1927

Introduction to the mathematics of statistics

The table on p. 296 gives

$$y = \frac{e^{-n}\, n^{n-\tfrac{1}{2}}}{(n-1)!}\left(1 + \frac{x}{\sqrt{n}}\right)^{n-1} e^{-x\sqrt{n}}$$

to 4 dec. for $n = 4, 9, 16$; $x = -4(\text{var.})-2(.1)+2(\text{var.})5$.

Salvosa, L. R., 1930

Annals of Math. Stat. 1, 191

Let the ordinate of the Pearson Type III curve be written in the form

$$y(t) = y_0 (1 + \tfrac{1}{2}\alpha_3 t)^{(2/\alpha_3)^2 - 1} e^{-2t/\alpha_3}.$$

Table 2, pp. 64-125, gives $y(t)$ to 6 dec. for $t = -5.21(.01)+9.81$; $\alpha_3 = 0(.1)1.1$.

2.22
2.229

2.22 Individual Poisson frequencies

2.221 Direct tabulations of $e^{-m}m^i/i!$

No differences are given with these tables. Except Fry 1928, the tables give all i for which the frequency exceeds $\frac{1}{2}$ in the last decimal place tabled.

	m =	
8 dec.	.001(.001)1(.01)5	Kitagawa 1952
7 dec.	5.01(.01)10	
7 dec.	.001(.001).01(.01).3, .4	Molina 1942, Czechowski et al. 1957(59)
6 dec.	.5(.1)15(1)100	
7 dec.	.001, .01, .1	Arkin & Colton 1950(124)
6 dec.	1(1)25(5)50(10)100	
6 dec.	.1(.1)15	Soper 1914, K. Pearson 1930 a (T.51), E.S. Pearson & Hartley 1954(T.39)
6 dec.	.1(.1)14	Vianelli 1959(564)
6 dec.	.1(.1)1(1)11	Darmois 1928(308)
6 dec.	.1,1(1)15	Mothes 1952(492)
5 fig.	.1(.1)1(1)20	Fry 1928

All i are given for which the frequency exceeds 10^{-6}.

| 4 dec. | .1(.1)10(1)20 | Burington & May 1953(259) |
| 4 dec. | .1(.1)10 | Bortkiewicz 1898, Charlier 1906(47), 1947(111) |

2.229 Miscellaneous Poisson frequencies

Bliss, C. I. 1948

Connecticut Agricultural Station, **Bulletin** 513

A sample of N observations from a Poisson distribution, of which only those falling into classes $j = 0, 1, 2, \ldots, i$ have been counted, gives rise to a truncated frequency distribution $f_0, f_1, \ldots, f_i$.

Compute further the truncated frequency

$$f_{oi} = \sum_{j=0}^{i} f_j .$$

The estimate of the Poisson parameter m here used is the root $m = \hat{m}$ of

$$m\left[f_{oi} - \eta(m,i)(N - f_{oi})\right] = \sum_{j=0}^{i} jf_j$$

where

$$\eta(m,i) = \frac{e^{-m} m^i}{i!} \bigg/ \sum_{j=i+1}^{\infty} \frac{e^{-m} m^j}{j!}$$

is the ratio of bounding frequency to the tail sum of the Poisson distribution.

To assist in the computation of $\hat{m}$, tables 1 and 2 (pp. 3-4) give

$$\sum_{j=0}^{i} e^{-m} \frac{m^j}{j!} \quad \text{to 5 dec.} \qquad \text{for} \quad m = .1(.1)6.0$$
$$i = 0(1)3$$

$\eta(m,i)$ to 6 dec.

According to Tippett 1932, the variance of $\hat{m}$ is $N^{-1} \sigma^2_{\hat{m}}$ where

$$\sigma^2_{\hat{m}} = \hat{m} e^{\hat{m}} \left(\sum_{j=0}^{i} \frac{\hat{m}^j}{j!} - \frac{\hat{m}^i}{(i-1)!} + \frac{\hat{m}^i}{i!} \left[(1 + \eta(\hat{m},i))\hat{m} - 1 \right] \right)^{-1}$$

Table 5 (p. 8) gives $\sigma^2_{\hat{m}}$ to 1-5 dec. for $\hat{m} = .1(.1)6$, $i = 0(1)3$.

2.3
2.3113

2.3 Percentage points of the χ^2- and Poisson distributions; confidence intervals for σ^2 and m

2.31 Percentage points of χ^2; confidence intervals for σ^2

The upper $100p\,\%$ point of χ^2 is denoted by $\chi^2(p,\nu)$ and is defined as the root of $p = p(\chi^2,\nu)$ for a given upper tail area p.

2.311 Tables of percentage points of χ^2

Most of the tables of $\chi^2(p,\nu)$ are for selected values of p, mainly in the neighborhood of 0 (upper percentage points) and 1 (lower percentage points).

We classify the tables below according to what percentage points they give: designating as **family A** those tables containing (with minor variations) $\frac{1}{2}\%$, 1%, $2\frac{1}{2}\%$, 5%, 10% and 25% points; and as **family B** those containing 1%, 2%, 5%, 10% and 20% points. The leading representatives of these families are found in Pearson & Hartley (T.7) and Fisher & Yates (T.IV); many short tables in textbooks derive from one or both of these. Within each family the order is by number of percentage points tabled.

A table substantially to a fixed number of decimals may add places for very small χ^2 or drop one place for $\chi^2 > 100$; a table substantially to a fixed number of significant figures may drop figures for very small χ^2: we have not attempted to index all these minor variations.

2.3111 Tables with material from both families A and B

	$\nu =$	$p =$ [1-p =]		
3 dec.	1(1)100	.0005, .001, .005, .01, .025, .05, .1(.1).5 ; [ditto]		Hald & Sinkbæk 1950
3 fig.	1(1)100	.0005, .001, .005, .01, .025, .05, .1(.1).5 ; [ditto]		Hald 1952(T.V)
3 fig.	1(1)30	.001, .005, .01, .025, .05, .1, .2, .3, .5 ; [ditto]		Duncan 1952(613)

2.3112 Tables of family A

	$\nu =$	$p =$ [$1-p =$]	
3 dec.	1(1)30(10)100	.001	E.S. Pearson & Hartley 1954(130)
6 fig.	1(1)30(10)100	.005, .01, .025, .05, .1, .25, .5 ; [ditto]	
3 fig.	1(1)30	.001, .005, .01, .025, .05, .1, .25, .5 [.005, .01, .025, .05, .1, .25]	Davies 1949(268)
6 fig.	1(1)30(10)100	.005, .01, .025, .05, .1, .25, .5 ; [ditto]	C.M. Thompson 1941b
3 fig.	1(1)30(10)60	.005, .01, .025, .05, .1, .25, .5 ; [ditto]	Anderson & Bancroft 1952(383)
3 fig.	1(1)30	.005, .01, .025, .05, .1, .25, .5 ; [ditto]	Mood 1950(424)
2 dec.	1(1)30(10)100	.001, .005, .01, .025, .05, .1 [.005, .01, .025, .05]	Lindley & Miller 1953(T.5)
2 dec.	$\begin{cases} 1(1)16(2)20 \\ 120/\nu = 1(1)6 \end{cases}$	.005, .01, .025, .05, .1 ; [ditto]	Dixon & Massey 1951(308), 1957(385)
4 dec.	1(1)30(10)100	.005, .01, .025, .05 [ditto]	Johnson & Tetley 1950(270)
3 fig.	1(1)30	.001, .005, .01, .025, .05, .1, .25, .5	Brookes & Dick 1951(284)
2 dec.	1(1)20(2)30	.001, .005, .01, .025, .05, .1, .25, .5	Quenouille 1950(229), 1952(392), 1953(339)

2.3113 Tables of family B

	$\nu =$	$p =$ [$1-p =$]	
3 dec.	1(1)30(2)70	.001, .01, .02, .05, .1, .2, .3, .5 [.01, .02, .05, .1, .2, .3]	Fisher & Yates 1957(T.IV)

2.3113
2.312   [see also p. 658]

3 dec.   1(1)30   .001, .01, .02, .05, .1(.1).3   Fisher & Yates
                   [.01, .02, .05, .1(.1).3, .5]   1948*(T.IV), 1953(T.IV);
                                                    Cramér 1946(559),
    P.O. Johnson 1949(361), Arkin & Colton 1950(121), Vianelli 1959(192),
    Burington & May 1953(286), Czechowski et al. 1957(75)

2 dec.   1(1)30   .001, .01, .02, .05, .1(.1).3   Siegel 1956(249)
                   [.01, .02, .05, .1(.1).3, .5]

3 dec.   1(1)30   .01, .02, .05, .1(.1).3, .5   Fisher & Yates 1938, 1943
                   [ditto]   (T.IV); Fry 1928(468),
                                       Rider 1939(202), Freeman 1942(169),
    Smith & Duncan 1945(475), Snedecor 1946(190), Hoel 1947(246),
    Kenney & Keeping 1951(414), C.E. Clark 1953(264), E.E. Lewis 1953(283)

3 dec.   1(1)30   .001, .01, .02, .05, .1, .2   Juran 1951(755)
                   [.01, .02, .05, .1, .2]

2 dec.   1(1)20   .001, .01, .02, .05, .1(.1).3   Mather 1943(240)
         (2)30   [.1(.1).3, .5]

2 dec.   1(1)30   .001, .01, .02, .05, .1(.1).3   Goulden 1952(444)
                   [.01, .05, .5]

2+ dec. 1(1)30   .001, .01, .02, .05, .1, .5   K.A. Brownlee 1947(145)
                   [.01, .02, .05, .1]

1 dec.   1(1)20   .001, .01, .02, .05, .1(.2).5   Finney 1952 a (307)
         (2)30   [.1, .3]

4 dec.   1(1)30   .01, .02, .05, .1, .5; [ditto]   Mode 1951(364)

2 dec.   1(1)30   .001, .01, .02, .05, .1, .2,   Goedicke 1953(268)
                   .3, .5

3 dec.   1(1)30   .01, .05, .1, .2, .3, .5   Snedecor 1937(163)
                   [.01, .05]

2 dec.   1(1)30   .01, .05, .1, .2, .3, .5   Goulden 1939(268),
                   [.01, .05]   Weatherburn 1946(171)

2 dec.   1(1)30   .01, .02, .05, .1, .2, .3, .5   Kenney 1939 b (198)

---

*Norton 1952 reports three errors in the 1948 edition:

Page	p	v	For	Read
33	·001	3	16·268	16·266
		4	18·465	18·467
		5	20·517	20·515

[142]

2.3114 Short tables

The tables below give neither the $2\frac{1}{2}\%$ point nor the 20% point.

	$\nu =$	$p =$	
1 dec.	1(1)20(2)30	.001, .01, .05, .1	Finney 1952 b (623)
3 dec.	1(1)30	.01, .05, .1	Mainland 1948(158)
3 dec.	1(1)30	.001, .01, .05	Graf & Henning 1953(T.5)
3 dec.	1(1)30	.01, .05	Tippett 1952(387)
8 dec.	1(1)30(10)100, 120	.999	T. Lewis 1953(421)

2.3115 Tables for special ν

The tables below do not include a consecutive range of ν from 1 up.

	$\nu =$	$p =$ [1-p =]	
4 dec.	2	.001(.001).999	M.H. Gordon et al. 1952
3 dec.	1	.001, .01, .02, .05, .1, .2, .3, .5 [.01, .05, .1, .3]	Mainland 1948(158)
2 dec.	1, 5(5)170	.001, .01, .05, .5 [ditto]	H.L. Seal 1941(45)
8 dec.	40(10)100, 120	.001	T. Lewis 1953(421)

2.312 Tables of percentage points of χ^2/ν

	$\nu =$	$p =$ [1-p =]	
3 fig.	1(1)20(2)30(5) 50(10)100(20) 200(50)500, 750, 1000, 5000	.0005, .001, .005, .01, .025, .05, .1(.1).5 ; [ditto]	Dixon & Massey 1957(386)
4 dec.	1(1)100(5) 200(10)300(50) 1000(1000)5000, 10000	.0005, .001, .005, .01, .025, .05 [ditto]	Hald 1952(T.6), Vianelli 1959(203)
3 fig.	$\begin{cases} 1(1)16(2)20 \\ 120/\nu = 0(1)6 \end{cases}$	.005, .01, .025, .05, .1 ; [ditto]	Dixon & Massey 1951(309)
3 dec.	1(1)30	.01, .05	Tippett 1952(387)

[see also p. 658]

2.313
2.314

2.313 Percentage points of χ

Pearson, K. 1930a

Tables for S & B, Part I

Table X

Table of median χ (called table of generalized 'Probable Errors')

$\sqrt{\chi^2(\tfrac{1}{2},\nu)}$ 7 dec. $\nu = 1(1)11$

charted for $0 \leq \nu \leq 15$

Grubbs, F. E. 1944

Annals of Math. Stat. 15, 75

Distribution of radial standard deviation.

Let x_i, y_i $i = 1, \ldots, n$ be two independent samples drawn from normals with 0 means and S. D's σ_x and σ_y respectively. If they are the two coordinates of a target hitting point x,y, the 'radial standard deviation' is defined by

$$z = n^{-\tfrac{1}{2}} \sqrt{\sum (x_i - \bar{x})^2 + \sum (y_i - \bar{y})^2}$$

In the general case the distribution of z^2 is obtained as a series of χ^2 integrals (equation (20)).

In the special case $\sigma_x = \sigma_y$, $\dfrac{nz^2}{\sigma^2} = \chi^2$ for $2n-2$ d.f. Table I (p. 80) gives the mean, standard deviation and the 100p% points of $k = z/\sqrt{2}\sigma = \chi_{2n-2}/\sqrt{2n}$ all to 4 dec. for $n = 2(1)15$, $p = .005, .05, .95, .995$.

2.314 Approximations to percentage points of χ^2

The two best known approximate formulas for $\chi^2(p,\nu)$ are:

(a) Fisher's normal approximation for large ν

$$\chi^2(p,\nu) \doteq \tfrac{1}{2}\left[x(1-p) + \sqrt{2\nu - 1} \right]^2 \qquad (2.3.1)$$

where $x(P)$ is the normal deviate.

(b) The Wilson and Hilferty approximation

$$\chi^2(p,\nu) \doteq \nu\left[1 - \frac{2}{9\nu} + x(1-p)\sqrt{\frac{2}{9\nu}}\right]^3 \quad (2.3.2)$$

which is a good approximation even for small ν.

Haldane, J. B. S. 1938
<u>Biometrika</u> 29, 392

Approximation to % points of χ^2 using
    A. The Wilson-Hilferty formula

$$\chi^2(p,\nu) = \nu\left[1 - \frac{2}{9\nu} + x(1-p)\sqrt{\frac{2}{9\nu}}\right]^3$$

    B. A generalization of A, viz: $\chi^2(p,\nu) =$ root of

$$x(1-p) = \left[\left(\frac{13\chi^2 - \nu}{12\nu}\right)^{5/13} + \frac{5}{18\nu}\left(1 + \frac{7}{48\nu}\right) - 1\right]\frac{6\sqrt{2\nu}}{5}\left(1 - \frac{1}{18\nu}\right)$$

Table I (p. 401) gives 3 dec. values of χ^2 from A and B as well as the exact value for $\nu = 10$ and $p = .01, .05, .2, .5, .8$.

For $\nu = 150, 200$, the approximations (a) and (b) are tabled without the exact value (c).

Merrington, Maxine 1941
<u>Biometrika</u> 32, 200

Comparison of approximations to % points of χ^2.
Table of

(a) $\chi^2(p,\nu) \sim \frac{1}{2}\left[x(1-p) + \sqrt{2\nu - 1}\right]^2$ to 3 dec. (Fisher's formula)

(b) $\chi^2(p,\nu) \sim \nu\left[1 - \frac{2}{9\nu} + x(1-p)\sqrt{\frac{2}{9\nu}}\right]^3$ to 3 dec. (Wilson-Hilferty formula)

2.314

(c) $\chi^2(p,\nu)$ exact to 4 dec.

     for   p = .005, .01, .05, .1, .25, .5

               .995, .99, .95, .9, .75

    and ν = 30, 40, 50, 75, 100

Peiser, A. M. 1943

Annals of Mathematical Statistics 14, 56-62

By inverting a Gram-Charlier series. Peiser finds the formula

$$\chi^2(p,\nu) = \nu + x(1-p)\sqrt{2\nu} + \frac{2}{3}\left([x(1-p)]^2 - 1\right) + \frac{[x(1-p)]^3 - 7x(p)}{9\sqrt{2\nu}} \tag{1}$$

which agrees with the Cornish-Fisher formula to order $\nu^{-\frac{1}{2}}$.

Table 1, p. 61, gives the results of (1) and the corresponding exact percentage points to 3 dec. for ν = 10, 30, 50, 100 and p = .01, .05, .1, .5, .9.

Aroian, L. A. 1943

Annals of Mathematical Statistics 14, 93-95

"A new approximation to the levels of significance of the chi-square distribution"

Define the standardized Type III curve by equation (2.1.4) for positive α_3, and by reflection in the point x = 0 for negative α_3. Let $t(p, \alpha_3)$ be the abscissa such that the area from t to $+\infty$ is p. Aroian determines the coefficients b(p) and c(p) in the approximation

$$t(p, \alpha_3) = x(1-p) + b(p)\alpha_3 + c(p)\alpha_3^2 \tag{1}$$

by least squares; and tables them on p. 95 to 4-5 dec. for p = .0001, .001, .005, .01, .025, .1, .2, .25, .3, .4, .5 .

To obtain upper-tail percentage points of χ^2 from (1), set $t = (2\nu)^{-\frac{1}{2}}(\chi^2 - \nu)$, $\alpha_3 = (8/\nu)^{\frac{1}{2}}$; for lower-tail percentage points, set $t = (2\nu)^{-\frac{1}{2}}(\nu - \chi^2)$, $\alpha_3 = -(8/\nu)^{\frac{1}{2}}$.

Table I (p. 94) gives (i) 3 dec. values from the Wilson-Hilferty formula, (ii) 4 dec. values from (1), (iii) 4 dec. of the exact values for p = .005, .01, .05, .1, .25, .5, .995, .99, .95, .9, .75; ν = 30, 40, 50, 75, 100.

Table II gives results for (ii) and (iii) only for ν = 30 and p = .0001, .001, .025, .9999, .999, .975, .2, .3, .4, .8, .7, .6.

Goldberg, H. and Levine, Harriet 1946
<u>Annals</u> <u>of</u> <u>Mathematical</u> <u>Statistics</u> 17, 216-225
"Approximate formulas for the percentage points and normalization of ... χ^2"

By inverting Edgeworth's asymptotic series, Cornish & Fisher 1937 have obtained formulas for the percentage points of a distribution in terms of its moments. Goldberg & Levine here apply these formulas to the χ^2- and t-distributions.

Table 4 (pp. 221-2) gives $\chi^2(p,\nu)$ to 4 dec. for ν = 1, 2, 10, 20(20)100; 1-p = .005, .01, .05, .25, .995, .99, .95, .75; also shown are 4 dec. values from

(a) The 5-term Cornish-Fisher expansion

(b) The 4-term Cornish-Fisher expansion which had previously been given by Peiser 1943

(c) The Wilson-Hilferty formula

(d) The Fisher formula.

Table 3 gives the Cornish-Fisher polynomials required for (a) and (b).

Thomson, D. Halton 1947
<u>Biometrika</u> 34, 368-392

Most of the error in the Wilson-Hilferty approximation to percentage points can be corrected by a term of the form $\nu^{-\frac{1}{2}}C(p)$. Halton Thomson fixes this term by requiring that the residual error vanish for ν = 30. The table on p. 372 gives C(p) to 3 dec. for

$\begin{cases} p = .005, .01, .025, .05, .1, .25, .5 \\ 1-p = \text{ditto} \end{cases}$

2.315

2.315 Confidence limits for σ [see also p. 658]

Pearson, E. S. 1935 c

<u>The Application of Statistical Methods to Industrial Standardization and Quality Control</u>

Confidence interval for population standard deviation σ in a normal parent based on χ^2 distribution:-

$$\sqrt{n}\ s'/\chi(p,\nu) \leq \sigma \leq \sqrt{n}\ s'/\chi(1-p,\nu)$$

where $s' = n^{-\frac{1}{2}} \sqrt{\sum (x_i - \bar{x})^2}$, $\nu = n-1$,

2p is the confidence coefficient.

Table 11 (facing p. 68) gives

$\sqrt{n}/\chi(p,\nu)$ and $\sqrt{n}\ \chi(1-p,\nu)$ (called b_1, b_2) to 3 dec. for 2p = 0.1, .02 and n = ν + 1 = 5(1)30.

Reprinted in Burington & May 1953, T. 13.58.1, p. 155; the account of the table given on p. 154 is erroneous.

Table 13b (p. 86) gives factors for computing control limits for the sample standard deviation using the population standard deviation. Limits are for $s' = s\sqrt{(n-1)/n}$.

Given are

$$B_{1-p} = \frac{\chi(p,\nu)}{\sqrt{n}} \quad \text{to 3 dec. for}$$

1 - p = .001 and .999 (outer limits)
1 - p = .025 and .975 (inner limits)
n = 2(1)30.   ($\nu = n - 1$)

Also given is the mean value of s', b_n, to 4 dec. for n = 2(1)30 and $1/b_n$ to 3 dec. for n = 2(1)30.

2.315

Pearson, E. S. and Haines, J. 1935

Royal Stat. So. Jn. Suppl. 2, 83

Control limits for standard deviation.

From a small sample of n measurements $x_1, \ldots, x_n$

compute the standard deviation $s' = n^{-\frac{1}{2}} \sqrt{\sum (x_i - \bar{x})^2}$

Table II(a) (p. 86) gives:

Factor for mean level $b = E(s') = \dfrac{\Gamma\left(\frac{n}{2}\right)\sqrt{2}}{\Gamma\left(\frac{n-1}{2}\right)\sqrt{n}}$ to 4 dec.

Factor for estimating $\dfrac{1}{b}$ to 3 dec.
σ from s'

Upper and lower control limits $B_{1-p} = \sqrt{\dfrac{\chi^2(p, n-1)}{n}}$ to 3 dec.

for $1-p = .005, .01, .05, .10$ (lower limits)
$1-p = .995, .99, .95, .90$ (upper limits)

Eisenhart, C. 1949

Industrial Quality Control 6, No. 1, 24-26

"Probability center lines for standard deviation and range charts"

If, following usual control chart practice, 'central lines' are drawn at a scale point corresponding to the average σ' (or $\bar{s}'$) more points will, in the long run, fall below this line than above it. Table 2 (p. 25) gives the proportion below this line, i.e. $P\left[s' \leq E(s')\right]$ to 3 dec. for sample sizes $n = 2(1)10, 12, 15, 20, 24, 30$.

If it is desired to draw a 'central line' such that an equal number of points will, in the long run, lie above it and below it, this line must be drawn at scale point $h\bar{s}'$, where $h = $ median s'/mean s' is given in Table I (p. 24) to 4 dec. for the above n.

2.315
2.322   [see also pp. 658-659]

Peach, P. and Littauer, S. B., 1946

<u>Annals</u> <u>of</u> <u>Math</u>. <u>Stat</u>. 17, 81

Single sample inspection plan' percentage points of χ^2 as approximations to percentage points of incomplete B.

Let n = sample size

  c = acceptance number

  p_1 = acceptable fraction defective

  p_2 = objectionable fraction defective

  α = risk of rejecting a lot if $p = p_1$

  β = risk of accepting a lot if $p = p_2$.

Table of

$$p_1/p_2 = \chi^2(\beta, 2c+2) / \chi^2(1-\alpha, 2c+2) \quad \text{to 2-4 sig. fig.}$$

$$np_1 = \tfrac{1}{2} \chi^2(1-\alpha, 2c+2) \quad \text{to 2-3 dec.}$$

for c = 0(1)22,25,30,37,47,63,129,215.

2.319 Miscellaneous percentage points of χ^2

Jeffreys, H., 1948

<u>Theory</u> <u>of</u> <u>Probability</u>

Table I (p. 400) gives χ^2 as the root of

$$K = \left(\frac{2n}{\pi}\right)^{\frac{1}{2}} e^{-\frac{1}{2}\chi^2} \quad \text{to 1 dec. for}$$

  $K = 10^x$, $x = -2(\tfrac{1}{2}) 0$, and n = 5(1)20(10)100, 200, 500, 1000, 2000, 5000, 10000, 20000, 50000, 100000.

Table II (p. 401) gives χ^2 as the root of

$$K = \tfrac{1}{2}\pi n^{\frac{1}{2}} \chi e^{-\frac{1}{2}\chi^2} \quad \text{to 1 dec. for}$$

$K = 10^x$, $x = -2(\tfrac{1}{2})\,0$ and $\nu = 7(1)20(10)100, 200, 500, 1000,$
2000, 5000, 10000, 20000, 50000, 100000.

($\nu = 7(1)19$ are omitted for $x = -\tfrac{3}{2}, -2$)

($\nu = 7(1)11$ are omitted for $x = -1$).

2.32 Percentage points of the Poisson distribution; confidence intervals for m

2.321 Percentage points of the Poisson distribution

Dudding, B. P. and Jennett, W. J., 1944

Quality Control Chart Technique when Manufacturing to a Specification

The General Electric Co. Ltd. of England.

Control Chart limits for number defective C when expected number of defectives is m. Table 7 (p. 61) gives upper 5% and upper 0.5% points for the Poisson distribution, i.e. 'roots' c of

$$\left.\begin{array}{r}.05\\ .005\end{array}\right\} = e^{-m} \sum_{i=c}^{\infty} m^i/i!$$

The 'roots' are given to 1 decimal place for $m = 0.6(0.2)2.0$ and must have been obtained by an unstated process of inverse interpolation.

2.322 Confidence intervals for m

Campbell, G. A., 1923

Bell System Technical Journal, 2(1), 95

Table II (pp. 108-111) gives upper and lower confidence limits for the Poisson mean m based on percentage points of χ^2. If

$$P(m,c) = e^{-m}(1 + \frac{m}{1!} + \frac{m}{2!} + \ldots + \frac{m^c}{c!}),$$

the table gives values of m to 2-4 dec. corresponding to $P(m,c) = $ .000001, .0001, .01, .1, .25, .5, .75, .9, .99, .9999, .999999 and $c = 1(1)101$.

2.322   [see also p. 659]

Przyborowski, J. and Wileński, H. 1935

Biometrika 27, 273-292

If
$$P(m,x) = e^{-m}\left(1 + \frac{m}{1!} + \frac{m^2}{2!} + \ldots + \frac{m^x}{x!}\right)$$

upper confidence limits of m corresponding to an observed x and a confidence level p are calculated from

$$P(m,x) = p(\chi^2, \nu) = p \quad \text{for} \quad \chi^2 = 2m \quad \nu = 2(x+1)$$

and are therefore given by

$$m = \tfrac{1}{2}\chi^2(p, 2x+2)$$

Table V (p. 288) gives these upper confidence limits, m, to 1 dec. for x (denoted by a) = $\tfrac{1}{2}\nu - 1 = 0(1)50$

$$p = 1 - \alpha = .001, .005, .01, .02, .05, .1.$$

Table VI (p. 289) gives 2 dec. values of w given by

$$w = m/m' = \frac{1}{2m'}\chi^2(p, 2x+2)$$

for $p = 1 - \alpha = .001, .005, .01, .02, .05, .1$

$$\begin{cases} x = 0(1)3, 50 \\ m' = 5, 10, 20, 40 \end{cases} \qquad \begin{cases} x = 0 \\ m' = 0.1, 1 \end{cases}$$

Garwood, F. 1936

Biometrika 28, 437-442

"Fiducial limits for the Poisson distribution"

If
$$P(m,x) = e^{-m}\left(1 + \frac{m}{1!} + \frac{m^2}{2!} + \ldots + \frac{m^x}{x!}\right),$$

fiducial limits of x are calculated from the equations

$$P(m, x-1) = 1 - p$$
$$P(m, x) = p$$

and are therefore given by the χ^2 percentage points

$$\tfrac{1}{2}\chi^2(1-p, 2x) \quad \text{and} \quad \tfrac{1}{2}\chi^2(p, 2x+2).$$

2.322

These are tabulated in Table I (p. 439) to 3-4 sig. fig. for
$$x = 0(1)20(5)50 \text{ and } p = .01, .05.$$

Table II (p. 442) gives approximate fiducial limits for $x = 20(10)50$, calculated from the Fisher and Wilson-Hilferty approximations to the percentage points of χ^2.

Grubbs, F. E. 1949
<u>Annals of Mathematical Statistics</u> 20, 242-256

Table 3, p. 256, gives roots

$$m_1 \text{ of } \sum_{i=0}^{c} e^{-m_1} m_1^i/i! = 0.95$$

$$m_2 \text{ of } \sum_{i=0}^{c} e^{-m_2} m_2^i/i! = 0.10$$

to 4 fig. for $c = 0(1)15$. Reprinted in Vianelli 1959, T. 219, p. 1392.

Kitagawa, T. 1952
<u>Tables of Poisson distribution</u>

Let $\bar{x}$ be the mean of a sample of size n from a Poisson distribution with mean m. Wilks 1938 has given the asymptotic formula for the 100 p % confidence limits for m:

$$\left.\begin{array}{r}\bar{m}\\ \underline{m}\end{array}\right\} = \bar{x} + \frac{\left[x(\tfrac{1}{2}P)\right]^2}{2n} \pm \sqrt{\frac{\left[x(\tfrac{1}{2}P)\right]^2}{n} \bar{x} + \frac{\left[x(\tfrac{1}{2}P)\right]^4}{4n^2}}$$

where $x(\tfrac{1}{2}P)$ is the upper $100 \tfrac{1}{2}(1-P)\%$ point of a unit normal distribution.

Table 4, pp. 149-156, gives $\bar{m}$ and $\underline{m}$ to 6 dec. for
$$\begin{cases} P = .95, .99 \\ \bar{x} = .001, .005, .01, .02(.02).1, .2(.2)1(1)10 \\ n/1000 = .05(.01).1(.1).5, 1, 2(2)10, 20, 50, 100 \end{cases}$$

E.S. Pearson & Hartley 1954, T. 40, p. 203, give fiducial limits for a Poisson mean, defined in the abstract of Garwood 1936 on p. 152 above, to 2 dec. or 3 fig. for $p = .001, .005, .01, .025, .05$; $x = 0(1)30(5)50$.

[153]

2.4
2.42

2.4 Laguerre series; non-central χ^2; power of χ^2 test [see also p. 660]

2.41 Laguerre series

It is sometimes convenient to expand the probability integral $P(x)$ of a positive variate x in terms of a linear aggregate of χ^2 distributions. Of these an expansion of the type

$$1 - P(x) = \sum_{i=0}^{\infty} c_i p(\alpha x, \nu + 2i) \quad \text{with} \quad \sum c_i = 1 \qquad (2.4.1)$$

is the most important, is equivalent to an expansion of the distribution into a Laguerre series with leading term $e^{-\frac{1}{2}\alpha x} x^{\frac{1}{2}\nu}$, and has been frequently employed in distribution theory (K. Pearson 1906; Robbins & Pitman 1949). Durand & Greenwood 1957 offer a modified Laguerre series, and pretend that it enjoys the same advantages over the unmodified series that the Edgeworth series enjoys over the Gram-Charlier series. For the use of Laguerre series in fitting curves to empirical data see Romanovsky 1924.

The two papers abstracted below are more or less closely related to Laguerre series. Pearson & Pearson 1935 makes disguised use of Gram-Charlier series with coefficients *not* determined by the usual orthogonal relations; cf. Arne Fisher 1922 a .

Salvosa prefaced his tables of derivatives with the following advertisement (Salvosa 1930, p. 192):

> Furthermore, there will be furnished tables of derivatives of this more effective type of frequency function which will permit its utilization as a generating function in an expansion corresponding to the Gram-Charlier series. In general, the more closely the generating function approaches the function to be graduated, the more rapid the convergence of the derivative series.

The series so obtained involves the Type III ordinate multiplied by a series of polynomials in $(1 + \alpha_3 t)^{-1}$. Such series must be used with great caution because the nth derivative of the Type III ordinate, for $n+1 \geq 4\alpha_3^{-2}$, does not vanish at the left end of its range.

Salvosa, L. R., 1930

__Annals__ __of__ __Math.__ __Stat.__ 1, 191

Higher derivatives of Pearson Type III curves.

Let the ordinate of the Pearson Type III be written in the form

$$y(t) = y_0 (1 + \tfrac{1}{2}\alpha_3 t)^{(4/\alpha_3^2)-1} e^{-2t/\alpha_3}$$

Table 3, pp. *128-187*, gives $\dfrac{d^i y(t)}{dt^i}$ to 6 dec. for $\alpha_3 = 0(.1)1.1$ and

$\begin{cases} i = 1 & 2 & 3 & 4 \\ t = -5.5(.1)+10.1 & -5.8(.1)+10.4 & -6(.1)+10.7 & -6(.1)+11 \end{cases}$

$\begin{cases} 5 & 6 \\ -6(.1)+11.3 & -6(.1)+11.5 \end{cases}$.

Pearson, K. and Pearson, Margaret V. 1935

__Biometrika__ 27, 409-423

Pearson expands $I'(x,a) = [\Gamma(a)]^{-1} \int_0^x e^{-t} t^{n-1} dt$, for x near the mode $a - 1$, in the form

$$I'(x_0, a) = I'(a-1, a) + \frac{\sqrt{2\pi} \, e^{-(a-1)} (a-1)^{a-\tfrac{1}{2}}}{2 \, \Gamma(a)} \alpha(\rho_0) + \sum_{i=2}^{\infty} a_{\tfrac{1}{2}i} \, I'(z_0, \tfrac{1}{2}i+1)$$

where $\rho_0 = \dfrac{x_0 - (a-1)}{\sqrt{a-1}}$, $z_0 = \dfrac{\tfrac{1}{2}(x_0 - (a-1))^2}{(a-1)}$.

2.42   __Non-central__ x^2

Fisher, R. A. 1928

__Proceedings__ __of__ __the__ __Royal__ __Society__ __of__ __London__, Series A 121, 654-673

"The general sampling distribution of the multiple correlation coefficient"

Let R denote the multiple correlation coefficient between y and n independent variables $x_1, \ldots, x_{n_1}$, computed from a sample of $n_1 + n_2 + 1$ observations each consisting of $n_1 + 1$ variates $y, x_1, \ldots x_{n_1}$.

2.42 [see also p. 660]

Let ρ be the population value of R and $B^2 = n_2 R^2$, $\beta^2 = n_2 \rho^2$. Then the limiting form of the distribution of B for large n_2 is given by equation (B), p. 663:

$$df = \frac{(\tfrac{1}{2}B^2)^{\tfrac{1}{2}(n_1-2)}}{\Gamma\left(\tfrac{n_1}{2}\right)} e^{-\tfrac{1}{2}B^2 - \tfrac{1}{2}\beta^2} \left[1 + \frac{1}{n_1} \frac{\beta^2 B^2}{2} + \frac{1}{n_1(n_1+2)} \left(\frac{\beta^2 B^2}{2}\right)^2 \frac{1}{2!} + \ldots \right] d(\tfrac{1}{2}B^2)$$

The upper 5% points of B (i.e. the roots $B_{5\%}$ of

$$\int_0^{\tfrac{1}{2}B^2_{5\%}} df = .95)$$ are given in Table I (p. 665)

to 4 dec. for $\beta = 0(.2)5.0$, $n_1 = 1(1)7$.

To translate these results into the notation of non-central χ^2 (see the next abstract), set

$$B^2 = {\chi'}^2, \quad \beta^2 = \lambda \quad n_1 = n .$$

For further applications, see R. A. Fisher 1938.

Fix, E., 1949

University of California Publications in Statistics, 1, No. 2, 15

'Tables of Noncentral χ^2'

Let x_i i = 1, 2, ..., n denote a random sample from a normal parent with 0 mean and $\sigma = 1$.
Then the non-central χ^2 is defined as

$${\chi'}^2 = \sum_{i=1}^n (x_i + a_i)^2 \quad \text{with non centrality}$$

$$\lambda = \sum_{i=1}^n a_i^2$$

The exact distribution is given by the series

$$f(\chi'^2) = e^{-\frac{1}{2}\chi'^2 - \frac{1}{2}\lambda} \sum_{j=0}^{\infty} \frac{(\chi'^2)^{\frac{1}{2}n+j-1} \lambda^j}{\Gamma(\frac{1}{2}n+j) \, 2^{2j+\frac{1}{2}n} \, j!} \qquad (1)$$

The table gives non-centralities λ satisfying

$$\int_{\chi^2(\alpha,n)}^{\infty} f(\chi'^2) d\chi'^2 = \beta$$

to 3 dec. for $\alpha = .05, .01$

$n = 1(1)20(2)40(5)60(10)100$

$\beta = .1(.1).9$

Here $\chi^2(\alpha,n)$ is the upper 100 % point of χ^2 for n degrees of freedom.

Patnaik, P. B. 1949

Biometrika 36, 202-232

(For notation, see the previous abstract.)

By fitting a Pearson Type III curve, (χ'^2/ρ) is found to be approximately distributed as χ^2_ν where $\rho = (n+2\lambda)/(n+\lambda)$; $\nu = (n+\lambda)^2/(n+2\lambda)$. Table I (p. 207) gives a comparison of this approximation with the exact values computed from the series (1) to 4 decimals for $n = 4, 7, 12, 10, 24$, and selected values of λ and χ'^2. Table 2 (p. 208) gives the comparison in terms of upper and lower 5% points of χ'^2 to 2 decimals for $n = 2, 4, 7$; $\lambda = 1, 4, 16, 25$. Table 3 (p. 210) gives a comparison of exact probability, and those computed from the χ^2 approximation as well as from the normal approximation to 4 decimals for a few selected n, λ, χ'^2 combinations. Table 4 (p. 213) gives closer approximations to the χ'^2 integral with the help of correction terms (see 3.3 of paper) and Table 5 (p. 213) gives the same for % points.
Using the χ^2 approximation with correction terms (second method of 3.3), Table 6 (p. 214) gives 3 decimal values of $\beta(n,\lambda) = 1 - P(\chi'^2 \geq \chi^2(.05,n))$ for $n = 2(1)10(2)20$; $\lambda = 2(2)20$.

2.43 Power of χ^2 test

Ferris, C. D.; Grubbs, F. E.; Weaver, C. L. 1946
Annals of Math. Stat. 17, 178

Power chart for χ^2-test. (central χ^2, change in σ^2 is alternative).
The hypotheses considered are H_0: $\sigma_1 = \sigma$; H_1: $\sigma_1 = \lambda\sigma$
Figure 1 gives the error of the 2nd kind

$$1 - p(\lambda^{-2} \chi^2(.05, \nu), \nu)$$

for $n = \nu + 1 = 2(1)8, 10, 15, 20(10)50, 75, 100$ plotted against λ
for $0 \leq \lambda \leq 4$ in testing $\sigma_1 > \sigma$.
Figure 2 gives the error of the 2nd kind

$$p(\lambda^{-2} \chi^2(.95, \nu), \nu)$$

for $n = \nu + 1 = 2(1)8, 10, 15, 20(10)50, 75, 100$ plotted against λ
for $0 \leq \lambda \leq 2$ in testing $\sigma_1 < \sigma$.

Walsh, J. E. 1947 a
Annals of Math. Stat. 18, 88

Effect of sample correlation on χ^2.
Let $x_1, \ldots, x_n$ represent a single n-variate observation from a multinormal parent in which all means are μ, all variances $\sigma^2 = 1$ and all correlations ρ. The confidence interval for σ^2 is then

$$0 \leq \sigma^2 \leq s^2 \frac{1}{(1-\rho) \chi^2 (1-\epsilon, n-1)}$$

where $s^2 = \sum (x_i - \bar{x})^2$, ϵ is the confidence coefficient and $\chi^2(1-\epsilon, n-1)$ is the $100(1-\epsilon)\%$ point of χ^2 for n-1 d.f.
If for nominal confidence coefficient ϵ, the ordinary confidence

interval $0 \leq \sigma^2 \leq S^2/\chi^2$ (1-ϵ, n-1) is used, define an actual confidence coefficient $\alpha(\rho,n)$ by

$$\chi^2(1-\alpha(\rho,n),n-1) = \chi^2(1-\epsilon, n-1)/(1-\rho)$$

Table IV (p. 92) gives $\alpha(\rho,n)$ to 3 dec. for $\rho = 0(.1).5$, n = 4, 16, 32.

2.5
2.5122

[see also pp. 661-662]

2.5 The factorial and the complete Γ-function; moments of functions of x^2

2.51 The factorial and the complete Γ-function

The complete Γ-function

$$\Gamma(n) \equiv \Gamma_\infty(n) = \int_0^\infty e^{-t} t^{n-1} \, dt$$

is the limit for $x \to \infty$ of the incomplete Γ-function $\Gamma_x(n)$. Following Fletcher et al. 1946, p. 198, we define

$$n! = \Gamma(1+n) \, ;$$

for n an integer, this coincides with the ordinary combinatorial definition of the factorial.

Sections 2.511 to 2.514, so far as they concern tables published before 1944, are severely abridged from Fletcher et al. 1946, pp. 34-37, 198-204. We use the following notational conventions in these sections:

The letter x denotes a general argument. The letter n denotes a positive integer argument; tables so indexed are not supplied with differences. Fletcher et al. 1946, p. 198, point out that "whatever choice of notation is made, much alteration of printed arguments by unity is necessary if the various tables are to be indexed on a uniform basis."

2.511 Natural values of factorials

2.5111 $x! = \Gamma(1+x)$

18 dec.	0(.01)1	No Δ	B.A. 1, 1931 (1946 printing, p. xj)
12 dec.	0(.01)1	δ_m^4	B.A. 1, 1931(T.9)
10 dec.	0(.0001)1	No Δ	Davis 1933 (T.1)
10 fig.	-11(.01)0	No Δ	Davis 1933 (T.5)
8 fig.	.5(1)999.5	No Δ	Salzer 1951
5 dec.	0(.01)1	No Δ	Hodgman 1957(307)
4 dec.	0(.01)2	IΔ	Jahnke & Emde 1933, 1938, 1948
4-5 fig.	2(.01)4		
4 dec.	0(.05)1	No Δ	Jahnke & Emde 1909

2.5112 $n! = \Gamma(1+n)$

Exact	1(1)200	Uhler 1944
Exact	1(1)60	Peters & Stein 1922
Exact	1(1)50	F. Robbins 1917
Exact	1(1)20	Hodgman 1957(193), Burington & May 1953(292)
Exact	1(1)15	Arkin & Colton 1950(108)
16 fig.	1(1)1000	Salzer 1951
8 fig.	1(1)200	Fry 1928
8 fig.	1(1)100	Barlow 1930, 1941
6 fig.	1(1)300	Fisher & Yates 1938 etc. (T.XXX)
6 fig.	1(1)250	E.S. Pearson & Hartley 1954(228)
5 fig.	1(1)100	Dixon & Massey 1957(467), Hodgman 1957(233), Burington & May 1953(290)

2.5113   Double factorials $(2n-1)!!$ and $(2n)!!$

$(2n-1)!!$ stands for $1.3.5\ldots(2n-1)$, and $(2n)!!$ for $2.4.6\ldots 2n$, the differences between consecutive factors being 2, instead of 1 as in ordinary factorials. Double factorials occur in the computation of the gamma function for half-integer argument, and in the computation of complete normal moments.

The table listed below gives both functions; the range shown is for n.

Exact	1(1)25	Potin 1925(840)

2.512   Reciprocals of factorials

2.5121 $\dfrac{1}{x!} = \dfrac{1}{\Gamma(1+x)}$

12 dec.	-2(.1)0	No Δ	Davis 1933(270)
5 fig.	-3(.01)+3	IΔ	Jahnke & Emde 1938(16), 1948(16)

2.5122 $\dfrac{1}{n!} = \dfrac{1}{\Gamma(1+n)}$

475 dec.	1(1)214	⎫
70 fig.	215(1)369	⎬ Uhler 1937
815 dec.	370(1)410	⎭
108 dec.	1(1)74	Van Orstrand 1921

2.5122
2.5141   [see also p. 662]

Recurring dec.   1(1)13 ⎫
                         ⎬   F. Robbins 1917(T.5)
85 dec.   14(1)61 ⎭

54 dec.   1(1)43   Peters & Stein 1922

Recurring dec.   2(1)12 ⎫
                         ⎬   Glaisher 1883
28 fig.   13(1)50 ⎭

12 fig.   1(1)50   K. Pearson 1931a (256)

5 fig.   1(1)20   Burington & May 1953(292), Hodgman 1957(193)

2.513   Logarithms of factorials

2.5131 $\log_{10} x! = \log_{10} \Gamma(1+x)$

20 dec.   0(.01)1   No Δ Gauss 1813, Davis 1933(T.12)

15 dec.   -11(.01)0   No Δ Davis 1933

12 dec.   0(.001)1 Δ^3 Legendre 1826(490),
                                              Legendre & Pearson 1921, Davis 1933

12 dec.   1(.01)10 δ^4 ⎫
                                          ⎬
12 dec.   10(.1)100   No Δ ⎬ Davis 1933
                                          ⎬
10 dec.   0(.0001)1 Δ^2 ⎭

10 dec.   1(.1)4(.2)69 δ^{2n} E. S. Pearson 1922

7 dec.   0(.01)49.99 δ^2 J. Brownlee 1923

7 dec.   0(.001)1 Δ Duffell 1909, K. Pearson 1930a (58)

4 dec.   0(.01)1   No Δ Jahnke & Emde 1909

2.5132 $-\log_{10} x! = -\log_{10} \Gamma(1+x)$

7 dec.   -5(.01)+15 Δ Jørgensen 1916(T.1)

The sixth figure is not everywhere accurate, and the table should properly be used only to five figures: Jørgensen's headnote.

2.5133 $\log_e x! = \log_e \Gamma(1+x)$

10 dec.   0(.005)1   No Δ B.A. 1916

8 dec.   9(.1)10.1 Δ^3 Bellavitis 1874

__2.5134__ $\log_{10} n! = \log_{10} \Gamma(1+n)$

33 dec.	1(1)3000	Duarte 1927
33 dec.	3050(50)10^4(10^4)10^5, 10^6	Duarte 1933
18 dec.	1(1)1200	Peters & Stein 1922
18 dec.	1(1)120	F. Robbins 1917
10 dec.	1(1)1199	E. S. Pearson 1922
7 dec.	1(1)1000	Glover 1923(482), K. Pearson 1930 a(98), Arkin & Colton 1950(109), E. S. Pearson & Hartley 1954(T.51)
7 dec.	1(1)300	Fisher & Yates 1938 etc. (T.XXX)
5 dec.	1(1)100	Burington & May 1953(290), Hodgman 1957(233)
4 dec.	1(1)1000	Hald 1952(T.XIII)
4 dec.	1(1)300	Lindley & Miller 1953(34)

__2.5135__ $\log_{10}(2n)!!$ and $\log_{10}(2n-1)!!$

For definitions see section 2.5113. The tables listed below give both functions; the range shown is for n.

18 dec.	1(1)60	F. Robbins 1917
8 dec.	1(1)50; cologs tabled	Elderton 1902, K. Pearson 1930 a(30)

__2.514__ The polygamma functions

The digamma and trigamma functions occur in the likelihood estimation of the exponent of a Type III curve; polygamma functions of all orders occur in the computation of the cumulants of the distribution of $\log \chi^2$

__2.5141__ $\psi(x) = F(x) = \dfrac{d}{dx}(\log_e x!) = \dfrac{\Gamma'(1+x)}{\Gamma(1+x)}$

18 dec.	0(.01)1	No Δ	Gauss 1813, Davis 1933(T.12)
12 dec.	0(.01)1 and 10(.1)60	δ_m^{2n}	B. A. 1, 1931(T.11)
10 dec.	0(.0001).1	Δ^2	Davis 1933(T.7)
10 dec.	0(.001)1	Δ^3	Davis 1933(T.8)
12 dec.	1(.02)19	δ^{2n}	Davis 1933(T.9)
16 dec.	-.5(.5)+99	No Δ	Davis 1933(T.10)
10 dec.	100(1)449		

2.5141
2.52

15 dec.	$-11(.01)0$	No Δ	Davis 1933(T.11)
8 dec.	$0(.02)20$	δ^{2n}	Pairman 1919
8 dec.	$0(.02)20$	No Δ	Vianelli 1959(1018)
4 fig.	$0(.01)1$	No Δ	Jahnke & Emde 1909
4 dec.	$0(1)10$		
4 dec.	$0(.01)1$	IΔ	Jahnke & Emde 1933, 1938, 1948
4 dec.	$0(.02)1$	$\tfrac{1}{2}\Delta$	Jahnke & Emde 1960(15)

[see also p. 662]

2.5142 $\psi^{(n-1)}(x) = \left(\dfrac{d}{dx}\right)^n (\log_e x!) = \left(\dfrac{d}{dx}\right)^{n-1} \dfrac{\Gamma'(1+x)}{\Gamma(1+x)}$

	$n =$	$x =$		
12 dec.	2	$0(.01)1, 10(.1)60$	δ^{2n}_m	B. A. 1, 1931(T.12)
12 dec.	3	$0(.01)1, 10(.1)60$	δ^{2n}_m	B. A. 1, 1931(T.13)
10 dec.	4	$0(.01)1$	δ^{2n}_m	B. A. 1, 1931(T.14)
12 dec.	4	$10(.1)60$		
10-12 dec.	2(1)5	$-11(.1)-1(.01)0$	No Δ	Davis 1935
10-19 dec.	2(1)5	$0(.01)3(.02)19(.1)99$	δ^{2n}	
8 dec.	2	$0(.02)20$	δ^{2n}	Pairman 1919
8 dec.	2	$0(.02)20$	No Δ	Vianelli 1959(1018)
4 dec.	2	$0(.01)1$	IΔ	Jahnke & Emde 1933, 1938, 1948
4 dec.	2	$0(.02)1$	$\tfrac{1}{2}\Delta$	Jahnke & Emde 1960(15)
3 dec.	2	$0(.05)1$	No Δ	Jahnke & Emde 1909

2.515 Gamma function of complex argument

For some older tables, see Fletcher et al. 1946, pp. 204-205.

2.5151 Real and imaginary parts of $\dfrac{1}{(x+iy)!} = \dfrac{1}{\Gamma(1+x+iy)}$

	$x =$	$y =$		
6 dec.	$-1.5(.01)-0.5$	$0(.01)4$	No Δ	Stanley & Wilkes 1950

2.5152 Real and imaginary parts of $\log_e(x+iy)! = \log_e \Gamma(1+x+iy)$

	$x =$	$y =$		
12 dec.	$-1(.1)+9$	$0(.1)10$	No Δ	N. B. S. 1954a
6 dec.	$0(.01)1$	$0(.01)4$	δ^2	Abramov 1954

2.52 Moments of functions of χ^2

The νth moment (about the origin) of either χ^2 or χ is clearly expressible as a ratio of Γ-functions. We have

$$\mu_r(\chi^2) = \frac{1}{\Gamma(\tfrac{1}{2}\nu)} \int_0^\infty (\chi^2)^r (\tfrac{1}{2}\chi^2)^{\tfrac{1}{2}\nu - 1} e^{-\tfrac{1}{2}\chi^2} d(\tfrac{1}{2}\chi^2)$$

$$= 2^r \, \Gamma(r + \tfrac{1}{2}\nu) / \Gamma(\tfrac{1}{2}\nu)$$

and hence

$$\mu'_r(\chi) = \mu'_{\tfrac{1}{2}r}(\chi^2) = 2^{\tfrac{1}{2}r} \, \Gamma(\tfrac{1}{2}r + \tfrac{1}{2}\nu) / \Gamma(\tfrac{1}{2}\nu)$$

An important corollary of these moments is the mean and variance of the sample variance and the sample standard deviation. Let n, the sample size, equal $\nu+1$; define

$$s^2 = \nu^{-1}\chi^2 = (n-1)^{-1} \Sigma(x_i - \bar{x})^2, \quad s'^2 = n^{-1}\chi^2 = \nu s^2/(\nu+1).$$

Then

$$E(s^2) = 1, \quad \mathrm{Var}(s^2) = 2/\nu;$$
$$E(s) = 2^{\tfrac{1}{2}} \Gamma(\tfrac{1}{2}\nu + \tfrac{1}{2}) / \nu^{\tfrac{1}{2}} \Gamma(\tfrac{1}{2}\nu), \quad \mathrm{Var}(s) = 1 - [E(s)]^2;$$

$$E(s'^2)=\nu/(\nu+1), \quad Var(s'^2)=2\nu/(\nu+1)^2;$$
$$E(s')=2^{\frac{1}{2}}\Gamma(\tfrac{1}{2}\nu+\tfrac{1}{2})/(\nu+1)^{\frac{1}{2}}\Gamma(\tfrac{1}{2}\nu), \quad Var(s')=[\nu/(\nu+1)]-[E(s')]^2.$$

An unbiased estimate of σ^2 is furnished by s^2; both s and s' are biased estimators of σ. There is, moreover, no agreed notation for s/E(s), i.e. for the unbiased quadratic estimate of the standard deviation of the population.

<u>2.521</u> Moments of χ^2

LeRoux, J. M. 1931
<u>Biometrika</u> 23, 134

β_1 and β_2 of the distribution of the sample variance s^2 for Pearson type parents.

For such parents the moments of s^2 are uniquely determined by the β_1 and β_2 of the parent (see equations (1) to (4), p. 136) and the sample size n. Tables I and II (pp. 150-1) give 5 decimal values of $\beta_1(s^2)$ and $\beta_2(s^2)$ for parental $\beta_1 = 0(.1)1.5$

$\beta_2 = 1.7 + 10 \beta_1(.1)5.0$ for $\beta_1(s^2)$

$\beta_2 = 1.7 + 10 \beta_1(.1)\min(4.4 + 2.0 \beta_1, 5.0)$ for $\beta_2(s^2)$

and n = 10.

Bancroft, T. A., 1944
<u>Annals</u> <u>of</u> <u>Math</u>. Stat. 15, 190

Expectation (and Bias) of estimates of σ^2 from estimates based on a preliminary test of significance.

Let s_1^2 and s_2^2 be two sample variances based on ν_1 and ν_2 degrees of freedom and drawn from normal parents with variances σ_1^2 and σ_2^2. If we assume $\sigma_1^2 = \sigma_2^2$, then the best estimate of σ_1^2 is

$$(\nu_1 s_1^2 + \nu_2 s_2^2)/(\nu_1 + \nu_2)$$

If in doubt, we may use the following procedure:

To estimate σ_1^2 use

$$c^* = \begin{cases} (\nu_1 s_1^2 + \nu_2 s_2^2)/(\nu_1 + \nu_2) & \text{if} \quad s_1^2/s_2^2 < F(p; \nu_1, \nu_2) \\ s_1^2 & \text{if} \quad s_1^2/s_2^2 > F(p; \nu_1, \nu_2) \end{cases}$$

Table I (p. 195) gives 3 decimal values of $E(c^*)/\sigma_1^2$ for $\nu_1 = 4$, $\nu_2 = 20$ and $\nu_1 = 12$, $\nu_2 = 10$; $\sigma_2^2/\sigma_1^2 = 0.1(0.1)1.0$; $p = 0, .01, .95, .20, 1$ and for $\lambda = F(p, \nu_1, \nu_2) = 1$.

Table II (p. 197) gives 3 decimal values of the variance about its true mean:- $(V + \text{Bias}^2)/\sigma_1^4$ for the same ν_1, ν_2, p and σ_2^2/σ_1^2.

2.522 Moments of χ

Pearson, K. 1915 b

<u>Biometrika</u> 10, 522-529

The following constants of the distribution of s' are tabled to 4 dec. for $n = 2(1)20(5)100$:

$$\widetilde{s'} = \text{Mode} = \sqrt{\frac{n-2}{n}}$$

$$E(s') = \text{Mean} = \frac{\Gamma\left(\frac{n}{2}\right)}{\Gamma\left(\frac{n-1}{2}\right)} \frac{\sqrt{2}}{\sqrt{n}}$$

$$\sigma_{s'} = \text{Standard deviation} = \sqrt{\frac{n-1}{n} - \frac{2}{n} \frac{\Gamma\left(\frac{n}{2}\right)^2}{\Gamma\left(\frac{n-1}{2}\right)^2}}$$

$$\sqrt{2n} \cdot \text{Standard Deviation} = \sqrt{2(n-1) - \left[2 \frac{\Gamma\left(\frac{n}{2}\right)}{\Gamma\left(\frac{n-1}{2}\right)}\right]^2}$$

$$\text{Skewness} = \frac{E(s') - \widetilde{s'}}{\sigma_{s'}}$$

2.522

$$\beta_1 = \frac{E(s')^2}{n^2} \frac{(1-2n\sigma_{s'}^2)^2}{\sigma_s^6}$$

$$\beta_2 = \frac{1}{(2n\sigma_{s'}^2)^2}\left[5 - 2n(4 - \frac{3}{n})(1 - 2n\sigma_{s'}^2) - 3(1 - 2n\sigma_{s'}^2)^2\right]$$

The accuracy of the fourth place is not guaranteed.

Reprinted in Tables for S & B II, T. 17.

Pearson, K. 1925 a

<u>Biometrika</u> 17, 176-200

The tables on pp. 197-198 give

$$\frac{\bar{\sigma}_{a_1}}{\sigma_1\sqrt{1-\rho^2}} = \sqrt{\frac{2}{n}} \; \frac{\Gamma(\tfrac{1}{2}(n-1))}{\Gamma(\tfrac{1}{2}(n-2))}$$

$$\frac{\sigma_{\sigma_{a_1}}}{\sigma_1\sqrt{1-\rho^2}/\sqrt{2n}} = \sqrt{\frac{n-2}{n} - \frac{2}{n}\frac{\Gamma^2(\tfrac{1}{2}(n-1))}{\Gamma^2(\tfrac{1}{2}(n-2))}}$$

4 dec.

n = 2(1)25, 50, 100, 200, 400

Reprinted in Tables for S & B II, TT. 29-30.

Kondo, T. 1930

<u>Biometrika</u> 22, 36-64

Kondo gives the following formulas for the moments of s':

Accurate formulas

$$E(s') = \frac{\Gamma(\tfrac{1}{2}n)}{\Gamma\left(\frac{n-1}{2}\right)} \frac{\sqrt{2}}{\sqrt{n}}$$

$$E(s'^2) = 1 - \frac{1}{n}$$

$$E(s'^3) = E(s')$$

$$E(s'^4) = (1 - \frac{1}{n^2})$$

Approximate expansions in terms of $\frac{1}{n}$ for moment about mean equations (43), (44)

$$E(s') \cong (1 - \frac{3}{4n} - \frac{7}{32n^2} - \frac{9}{128n^3} - \frac{59}{2048n^4})$$

$$\sigma_{s'}^2 \cong 1 - \frac{1}{n} - E(s')^2$$

Tables I and II (p. 57) give 7 decimal values of $E(s')$ and $\sigma_{s'}^2 = E(s'^2) - E(s')^2$ from accurate formula and from approximate formula to order n^{-2}, n^{-3}, n^{-4} for $n = 5, 7, 9, 10(5)30, 50, 75, 100, 150, 200$.

Table III (p. 38) gives σ_s, exact and from approximate formula to order $n^{-3/2}$, $n^{-5/2}$, $n^{-7/2}$ to 4 dec. for above n.

Tables IV-VII (pp. 59-61) give respectively

$\mu_3(s')$ to 7 dec. ⎫
$\mu_4(s')$ to 6 dec. ⎬ from exact formulas as well as from approximate formulas to various n^{-i}
$\beta_1(s')$ to 4 dec. ⎪
$\beta_2(s')$ to 4 dec. ⎭

Pearson, K. 1931a

<u>Tables</u> <u>for</u> S & B II, T. 27

See abstract on p. 123.

Davies, O. L. and Pearson, E. S. 1934

Royal Stat. So. <u>Jn.</u> <u>Suppl.</u>, 1, 76

Comparison of estimates of σ from k normal samples of size n.

Consider k independent samples of size n $\quad x_{ti} \quad t = 1,2,\ldots,k$
$\quad\quad i = 1,2,\ldots,n$

Let $s_t'^2 = n^{-1} \sum_{i=1}^{n} (x_{ti} - \bar{x}_t)^2$

$s_t^2 = (n-1)^{-1} \sum_{i=1}^{n} (x_{ti} - \bar{x}_t)^2$

and form the following estimates of σ

$$s_{(1)}^2 = \frac{1}{k} \sum_{t=1}^{k} s_t^2 \quad \text{for which} \quad E(s_{(1)}^2) = \sigma^2 = 1$$

$$s_{(2)} = \frac{1}{b_n} \frac{1}{k} \sum_{t=1}^{k} s_t' \quad \text{for which} \quad E(s_{(2)}) = \sigma = 1$$

$$\text{where} \quad b_n = \frac{\Gamma\left(\frac{n}{2}\right)}{\Gamma\left(\frac{n-1}{2}\right)} \frac{\sqrt{2}}{\sqrt{n}}.$$

2.522

$$\sigma(s_{(1)}) = \sqrt{1 - \frac{2}{(n-1)k} \cdot \frac{\Gamma^2\left(\frac{k(n-1)}{2}+1\right)}{\Gamma^2\left(\frac{k(n-1)}{2}\right)}}$$

$$\sigma(s_{(2)}) = \frac{1}{b_n\sqrt{k}} \sqrt{\frac{n-1}{n} - \frac{2}{n} \frac{\Gamma^2\left(\frac{n}{2}\right)}{\Gamma^2\left(\frac{n-1}{2}\right)}}$$

Table I gives the above standard errors to 3 dec. for

　　n = 2(1)6, 10, 15, 20, 50

　　k = 1(1)6, 10, 20, 50

together with those from two other estimates of σ (range and mean deviation).

Table II gives b_n to 4 dec. and $1/b_n$ to 3 dec. for n = 2(1)15.

Table V gives β_1 and β_2 coefficients to 2 dec. for both $s_{(1)}$ and $s_{(2)}$ for n = 2(1), 6, 20, k = 1(1)4.

Bliss, C. I. 1937a

Annals of Applied Biology 24, 815

Table of the variance estimate of the sample standard deviation

$$\text{var } s = \sigma^2 \left[1 - \frac{2}{\nu} \left(\frac{\Gamma\left(\frac{\nu+1}{2}\right)}{\Gamma\left(\frac{\nu}{2}\right)} \right)^2 \right]$$

The bracket [ ] is tabulated on p. 833 to 6 dec. for $\nu = (N-1) = 3(1)49$.

For $\nu > 49$, Bliss recommends the approximation var $s = \sigma^2/(2\nu + \frac{1}{2})$.

　　Graf & Henning 1953, T. 6, p. 66, gives $E(s/\sigma)$ to 3 dec. for N = 2(1)25, 30(10)50(25)100.

Jennett, W. J. and Welch, B. L. 1939

<u>Supplement to the Journal of the Royal Statistical Society</u> 6, 80

Express the mean and variance of s in the form

$$E(s) = a\sigma \quad \text{where} \quad a = \frac{\Gamma\left(\frac{n}{2}\right)}{\Gamma\left(\frac{n-1}{2}\right)} \frac{\sqrt{2}}{\sqrt{n-1}},$$

$$\sigma(s) = \frac{b\sigma}{\sqrt{2(n-1)}} \quad \text{where} \quad b = \sqrt{2(n-1) - \left[\frac{2\,\Gamma\left(\frac{n}{2}\right)}{\Gamma\left(\frac{n-1}{2}\right)}\right]^2}$$

Table I (p. 85) gives

a, b, a^2 and b^2 to 3 dec. for $n = 5(1)15(5)30(10)50, 75, 100$

End figure error: For $n = 5$, $b^2 = 8(1-a^2)$ is not satisfied.

Johnson, N. L. and Welch, B. L. 1939

<u>Biometrika</u> 31, 216-218

"On the calculation of the cumulants of the χ-distribution"

Let χ_ν denote the χ-distribution with ν degrees of freedom. The table on pp. 217-218 gives

$k_i(\chi_\nu)$ to 6 dec. for $i = 1(1)6$

$\nu = 1(1)8$

$\nu^{\frac{1}{2}i} k_i(\chi_\nu)$ to 6, 5, 4, dec. for $i = 2, 4, 6$

$24/\nu = 0(1)4$

$\nu^{\frac{i}{2}-1} k_i(\chi_\nu)$ to 6, 6, 5 dec. for $i = 1, 3, 5$

$24/\nu = 0(1)4$

E.S. Pearson & Hartley 1954, T. 35, p. 184, gives inter al. $E(s)$ to 7 dec.; $\sigma(s)$ to 5 dec.; $(2\nu)^{\frac{1}{2}}\sigma(s)$ to 4 dec.; $\beta_1(s)$ and $\beta_2(s)$ to 4-5 dec.; all for $\nu = 1(1)20(5)50(10)100$. An identical table appears in Vianelli 1959, T. IL, p. 1182.

2.529
2.53

2.529 Moments of other functions of χ^2 [see also pp. 662-663]

Feldman, H. M. 1932

<u>Annals</u> <u>of</u> <u>Mathematical</u> <u>Statistics</u> 3, 20-31

"The distributions of the precision constant and its square in samples of from a normal population"

Define $u = 1/2s'^2$ and $h = u^{\frac{1}{2}}$; denote h as the 'precision constant'. The corresponding population parameters are

$$U = 1/2\sigma^2 \quad \text{and} \quad H = 1/\sqrt{2}\,\sigma.$$

Table I (p. 30) gives the modes of u and h in terms of U and H with coefficients to 3 dec. for n = 4, 10, 25, 100. Also shown are the maximum likelihood estimates $\hat{U}$ and $\hat{H}$ in the form

$$u_{obs.} = \frac{n}{n-1}\hat{U}, \quad h_{obs.} = \sqrt{\frac{n}{n-1}}\hat{H}$$

with $\frac{n}{n-1}$ and $\sqrt{\frac{n}{n-1}}$ given to 4 dec.

Table II (p. 30) gives $\mu'_1(u)/U$, $\mu_2(u)/U^2$, $\mu_3(u)/U^3$, $\mu_4(u)/U^4$ to 3 to 4 significant figures and the same for h, for n = 4, 10, 25, 100.

Table III gives $\beta_1(u)$, $\beta_2(u)$, $\beta_1(h)$, $\beta_2(h)$ to 4 dec. for n = 10, 25, 100, ∞.

Table IV gives Median (u)/U and Quartiles (u)/U to 3 dec. and the same for h for n = 4, 10, 25.

Errors (supplied by author)

The following corrections should be made to formulas and Tables:

$$\mu_4(u) = \frac{12n^4(n+9)}{(n-3)^4(n-5)(n-7)(n-9)} U^4$$

$$\beta_2(u) = \frac{3(n-5)(n+9)}{(n-7)(n-9)}$$

Tables II and III: The following corrections should be made.

n	μ_4 (u)	β_2 (u)
10	63.307	95.0000
25	1.182	7.0833
100	0.00186	3.6707
∞		3.0000

2.53 Moments of the Poisson distribution

Wald, A. and Wolfowitz, J. 1945

See abstract on p. 279 below.

Grab, E. L. and Savage, I. R. 1954

Journal of the American Statistical Association 49, 169-177

The random variable X is said to have a positive Poisson distribution if $Pr[X = x] = e^{-m}(1-e^{-m})^{-1}m^x/x!$ for positive integer x.

Table 2, p. 177, gives $E(1/X \mid m) = e^{-m}(1-e^{-m})^{-1} \sum_{x=1}^{\infty} m^x/x \cdot x!$ to 5 dec. for m = .01, .05(.05)1(.1)2(.2)5(.5)7(1)10(2)20.

Douglas, J. B. 1955

Biometrics 11, 149-173

Define $\mu_n' = e^{-m} \sum_{x=0}^{\infty} x^n m^x/x!$, $p_n = \mu_{n+1}'/\mu_n'$, $q_n = p_n(p_{n+1} - p_n)$. Tables 1-3, pp. 159-173, computed by Misses B. Laby, J.H. Roynane and J.A. Murray, give q_n; p_n to 3-4 fig. for n = 0(1)18; 19 and m = 0(.001).03(.01).3(.1)3.

Reprinted (slightly abridged) in Vianelli 1959, Prontuario 196, pp. 588-593; in Vianelli m = 0(.002).03(.01).3(.1)3.

See also Okamoto 1955.

2.6 Tables related to the χ^2 test of goodness of fit

[see also p. 663]

K. Pearson 1900 invented the χ^2 distribution to provide a significance test for the discrepancy of observed from expected frequencies. In contingency tables with a small number of cells, in particular 2 × 2 tables, the gamma distribution (even with the correct number of degrees of freedom) is only a rough approximation to the actual distribution of χ^2: section 2.61 is largely devoted to exact tables based on the hypergeometric distribution. The relevant theory is given by Yates 1934.

2.61 The 2 × 2 table

Both to simplify the exposition and to condense the numerical tables, it is customary to put the 2 × 2 contingency table into the standard form

a	c	m
b	d	n
r	s	N

where without loss of generality $m \geq n$, $an \geq bm$. The probability of a given configuration, given fixed marginal totals is

$$P(b \mid m,n,r) = \frac{m!\ n!\ r!\ s!}{N!} \cdot \frac{1}{a!\ b!\ c!\ d!}.$$

For the moments of χ^2 computed from a 2 × 2 table, see Haldane 1940, 1945a.

2.611 Exact tests of significance in 2 × 2 tables

Swaroop, S. 1938

Sankhyā 4, 73-84

Section A of Swaroop's tables gives $P(b \mid m,n,r)$ to 5 dec. for
$n = 5(5)20, 30, 50;\ a = 0(1)n;\ b = a(1)n.$

Section B gives the cumulative double tail sums

$$2 \sum_{i=0}^{b} P(i \mid n,n,r-i)$$

also to 5 dec. for the above n, a, b.

Finney, D. J. 1948
Biometrika 35, 145-156

Finney tables the probabilities

$$P^*_{b^*} = \sum_{b=k}^{b^*} P(b \mid m,n,r)$$

where in the summation m, n and r are kept fixed, i.e., $a = r-b$
$k = \max(0, r-m)$

b^* = largest integer such that
(i) $P^*_{b^*} \leq .05$
(ii) $P^*_{b^*} \leq .025$
(iii) $P^*_{b^*} \leq .01$
(iv) $P^*_{b^*} \leq .005$

The table (pp. 149-154) gives $P^*_{b^*}$ to 3 dec. and the integer b^* for $m = 3(1)15$, $n = 3(1)m$ but <u>not</u> as a function of r. The table is entered with a and the value of b^* shown is the highest for which the appropriate inequality is satisfied; the test of significance thus relates to Problem I of E.S. Pearson 1947.

Reprinted in Pearson & Hartley 1954, T. 38, pp. 188-193.

Mainland, D. and Murray, I. M. 1952
Science 116, 591-594
"Tables for use in fourfold contingency tests"

A sample of size 2N is split into two samples, each of size N, and the numbers a_1 and a_2 having characteristic A are noted for the two samples.

Tables 1 and 2, pp. 592-593, based on the exact hypergeometric distribution, give respectively the 5% and 1% points for the difference $a_2 - a_1$ by tabulating all significant pairs a_2, a_1 for $N = 4(1)20(10)100(50)200(100)500$. For large N, pairs a_2, a_1 are omitted if they can be obtained by interpolation at sight.

The typography leaves much to be desired.

Latscha, R. 1953
Biometrika 40, 74-86

Extension to $m = 20$ of the table of Finney 1948.

2.611 [see also pp. 663-664]
2.613
Armsen, P. 1955

Biometrika 42, 494-511

This table extends the tabulations of Finney 1948 and Latscha 1953. For arguments $i = d$, $x = b$, and $y = m - r$, the least integer z is tabulated such that, for all n,

$$\sum_{\Omega} \frac{(n+y+z)!\, n!\, (n+z)!\, (n+y)!}{(2n+y+z)!} \cdot \frac{1}{(z+d)!\,(n-d)!\,(n+y+d)!\,d!} \leq \alpha,$$

for $\alpha = .01$ and $.05$, and for three definitions of Ω:

Ω_1 $d \geq i$;

Ω_2 $d \geq i$ and $d \leq j$, where j is the least integer such that $(z+j)!\,(n-j)!\,(n+y-j)!\,j! \geq (z+i)!\,(n-i)!\,(n+y-i)!\,i!$

Ω_3 $d \geq i$ and $d \leq j$, where j is the least integer such that
$$\sum_{d \leq j} [(z+d)!\,(n-d)!\,(n+y-d)!\,d!]^{-1} \leq \sum_{d \geq i} [(z+d)!\,(n-d)!\,(n+y-d)!\,d!]^{-1}.$$

The table permits all contingency tables to be tested for which $N \leq 50$; but it is not adapted to writing down rapidly, for fixed m, n, r and s, the values of d (say) that reject the hypothesis of independence.

2.612 Approximate tests of significance in 2×2 tables

Fisher, R. A. and Yates, F. 1938, 1943, 1948, 1953, 1957
Table VIII

(Our abstract relates to the improved version of T. VIII, contained in the 1948 and later editions.)

Given a 2×2 table, compute χ^2 with benefit of Yates' correction for continuity, and denote the square root of the result by χ_c. Let m denote the smallest expectation in the table; let p denote the ratio of the smallest expectation to the smallest marginal total.

For given m and p the percentage points of χ_c vary between two (fairly close) limits corresponding to a 'binomial' and 'limiting contingency' distribution. The 'upper' and 'lower' 2.5% and 0.5% points of these two limiting distributions are given to 2 dec. for

$$\begin{cases} m = 1(1)6,\ 8,\ 12,\ 24,\ 48,\ 96 \\ p = 0,\ 0.25,\ 0.5 \end{cases}$$

The omission of values indicates that for such m and p the exact test should be used.

<u>2.613</u> Power of the χ^2 test in 2 × 2 tables

Patnaik, P. B. 1948

<u>Biometrika</u> 35, 157-175

"The power function of the test for the difference between two proportions in a 2 × 2 table"

Let a 2 × 2 table have arisen as specified in Problem II of E. S. Pearson 1947. Construct critical regions in the (a,b) plane from the approximate equation

$$\left(a - \frac{mr}{N}\right) \bigg/ \sqrt{\frac{mnrs}{N^2(N-1)}} = \pm x(1-p) \qquad (1)$$

where $x(p)$ is the single tail normal percentage point corresponding to 100 p% and is taken as $p = \alpha$ or $p = \tfrac{1}{2}\alpha$ according to whether single or double tail test is required. An observed point a,b inside the critical region rejects the hypothesis that $p_1 = p_2$ for the two samples. The power function for a given alternative pair p_1, p_2 is then given by

$$\sum \binom{m}{a} p_1^a (1-p_1)^c \binom{b}{n} p_2^b (1-p_2)^d$$

where the summation is taken over all points a,b inside the region. This power for two sided test is tabulated to 3 dec. for $\alpha = 2p = .10, .02$ in Table   page

2a 2b	160	m = 18, n = 12	p_1, p_2 = 0.1(0.1)0.9								
3a	164	m = n = 15	and	p_1 = .3	.6	.1	.2	.05	.1	.2	.1
				p_2 = .4	.8	.3	.7	.5	.6	.8	.7
3b	165	m = n = 30	and	p_1 = .05	.1	.3	.2	.1	.2		
				p_2 = .3	.4	.7	.6	.5	.7		

In Table 3 only the pairs $p_1\ p_2$ in the same column are tabulated. The exact values are compared with certain approximations.

Tables 5 and 4, described overleaf, are reprinted in Vianelli 1959, <u>Prontuari</u> 91, 90, pp. 256-257, 255.

2.613
2.69

The effect of the approximation on the estimates of required sample size is shown in Table 7 (p. 170). In the case of $m = n$ the power can be computed from the approximation

$$\text{Power} = Q[x(1-p) - h] + Q[x(1-p) + h],$$

where Q is the normal tail area of section 1.1141 and

$$\text{Power}^* = 1 - \tfrac{1}{2}\alpha\bigl[x(1-p) - h\bigr] + \tfrac{1}{2}\alpha\bigl[-x(1-p) - h\bigr]$$

$$h = \frac{\sqrt{2n}}{4}\sqrt{(p_1 + p_2)(2 - p_1 - p_2)}\;\left|\log_e \frac{p_1(1 - p_2)}{p_2(1 - p_1)}\right|$$

Table 5 (p. 175) gives the power to 4 dec. for $h = 0.1(0.1)3.0(0.2)5.0$, $p = .05, .025, .01, .005$ ($\alpha = .1, .05, .02, .01$) and $h/\sqrt{n}$ to 3 dec. for the above h and $n = 10(5)50(10)100, 150$.

Table 4 (p. 174) gives 3 decimal values of $h/\sqrt{n}$ for $p_1, p_2 = .05(.05).95$

Pearson, E. S. and Merrington, Maxine 1948

Biometrika 35, 331-345

"2 x 2 tables; the power function of the test on a randomized experiment"

Let a 2 x 2 table have arisen as specified in Problem I of E. S. Pearson 1947. Construct critical regions from formula (1) of the previous abstract. The alternate hypothesis here considered is

$a = x_1$	$c = y_1 + z_1$	m
$b = x_2 + y_2$	$d = z_2$	n
$X + y_2$	$Z + y_1$	N

with $X, Y = y_1 + y_2 > 0$ and Z assumed fixed and randomly subdivided into their two components, so that the probability of a particular pattern is

$$p(x_1, y_1) = \frac{m!\; n!\; X!\; Y!\; Z!}{x_1!\; x_2!\; y_1!\; y_2!\; z_1!\; z_2!\; N!}$$

and the power for given $X, Y > 0$ is given by summation of $p(x, y)$ over all points in the critical region.

This power, and a normal approximation thereto, is given in Table 4 for $n = m = 15$, $x = 0(5)20$, $\alpha = .05, .15$.

2.62 Other contingency tables

For the moments of χ^2 from a $2 \times k$ table see Haldane 1940 b, 1945 b; for an $n \times k$ table see Haldane 1940 a.

Welch, B. L., 1938(b)

Biometrika 30, 149

Distribution of 'χ^2' in a $2 \times k$ table. For a particular case (2×20 table with total sample size $N = 40$ and all ($k = 20$) marginal totals = 2) the exact distribution of $E^2 = \chi^2/n$ (correlation ratio) has been worked out by considering all possible 3^{20} patterns and computing the % frequencies of patterns giving rise to values of $E^2 \geq E_o^2$ for $E_o^2 = 0, .25(.05).75, 1.00$. These are given to 2 dec. in Table III (p. 157) and compared with the χ^2 distribution with 19 degrees of freedom and with an incomplete-beta-function approximation.

2.69 Miscellaneous goodness-of-fit tests involving χ^2

Neyman, J. and Pearson, E. S. 1931 a

Biometrika 22, 298-305

"Further notes on the χ^2-distribution"

Two sample χ^2 test.

Comparison of minimum χ^2 with likelihood criterion.

Formulas (equations (13), (14)).

Table III p. 303 gives numerical illustrations.

2.69
3.0   [see also p. 664]

Hoel, P. G. 1943

<u>Annals</u> <u>of</u> <u>Mathematical</u> <u>Statistics</u> 14, 155-162

"On indices of dispersion"

Following R. A. Fisher 1929 b, define the binomial dispersion

$$z = \sum_{1}^{N} (x - \bar{x})^2 / \left[\bar{x}\left(1 - \frac{\bar{x}}{n}\right)\right].$$

The exact moments $\mu_i(z)$ are derived for a binomial and Poisson parent and the ratios of these to the corresponding values of $\mu_i(\chi^2)$ are tabulated on p. 161 to 2 dec. for Poisson parents with $m = 1, 2, 5, 25$ and binomial parents with $p = 1/3$; $n = 3, 6, 15, 75$ and various sample sizes N.

David, F. N., 1947

<u>Biometrika</u> 34, 299

A χ^2 smooth test for goodness of fit.

The test makes use of the number T of the sets of equal signs in the r ordered residuals (obs. - exp.). With <u>given</u> r_1 '+signs' and $r_2 = r - r_1$ '- signs' the frequency of T sets is given by

$$\frac{2(r_1-1)!\,(r_2-1)!}{(t-1)!\,(t-1)!\,(r_1-t)!\,(r_2-t)!} \quad \text{for } T = 2t$$

and by

$$\frac{(r_1-1)!\,(r_2-1)!\,(r_1+r_2-2t)}{t!\,(t-1)!\,(r_1-t)!\,(r_2-t)!} \quad \text{for } T = 2t+1$$

Table 1 (p. 302) gives above as exact integers for $r = 2(1)14$, all partitions $r_1+r_2 = r (r_1 \geq r_2)$, $T = 2(1)r$ with frequency totals $\binom{r}{r_1}$ shown so that the chance $P(T \leq T_0)$ can be computed by division.

Table 5 (p. 308) gives 5% and 1% points of distribution of

$$Y = P(T \leq T_0)\, p(\chi_0^2, r-1),$$

the proposed criterion, to 4 dec. for $r = 5(1)14$ and above partitions from formulas on p. 307.

SECTION 3

THE INCOMPLETE B - FUNCTION; THE BINOMIAL DISTRIBUTION

3.0 Definition and Notation

The incomplete Beta-function is defined by

$$B_x(a,b) = \int_0^x u^{a-1}(1-u)^{b-1} du \qquad (3.0.1)$$

where a and b are positive parameters. For $x = 1$ we obtain the complete B-function

$$B(a,b) = B_1(a,b) = \int_0^1 u^{a-1}(1-u)^{b-1} du = \Gamma(a)\Gamma(b)/\Gamma(a+b).$$

In order that the function $B_x(a,b)$ may represent a cumulative distribution of a variate x over the range $0 \leq x \leq 1$ we must 'standardize' it by dividing by $B(a,b)$; we thus obtain the incomplete-beta-function ratio*

$$I_x(a,b) = \frac{B_x(a,b)}{B(a,b)} = \frac{\Gamma(a+b)}{\Gamma(a)\Gamma(b)} \int_0^x u^{a-1}(1-u)^{b-1} du.$$

Clearly

$$I_x(a,b) = 1 - I_{1-x}(b,a). \qquad (3.0.2)$$

The statistical importance of the incomplete-beta-function ratio arises from its relations to the binomial distribution and to the F-distribution. We adopt the standard form

$$Q(p, n, r-\tfrac{1}{2}) = \sum_{i=r}^{n} \binom{n}{i} p^i q^{n-i} = I_p(r, n-r+1), \qquad (3.0.3)$$

where $q = 1 - p$, for the cumulative binomial sum; the complementary sum, with limits 0 and $r-1$, will be written $P(p, n, r-\tfrac{1}{2})$.

---

*Because of the importance of the application to the binomial, where the argument x must be equated to a binomial probability p or q, Pearson & Hartley 1954 (p. 32) altered the traditional notation from $I_x(p,q)$ to $I_x(a,b)$ to avoid confusion; we have followed their example.

3.0
3.119

[see also p. 664]

If s_1^2 and s_2^2 denote two independent mean squares based respectively on ν_1 and ν_2 degrees of freedom (i.e. if $s_1^2 = \chi_{\nu_1}^2/\nu_1$ and $s_2^2 = \chi_{\nu_2}^2/\nu_2$) and if we define the variance ratio $F = s_1^2/s_2^2$, then the tail area of the F-distribution is related to $I_x(a,b)$ by

$$\Pr[F \geq F_o] = I_{x_o}(\tfrac{1}{2}\nu_2, \tfrac{1}{2}\nu_1) \qquad (3.0.4)$$

where $x_o = \nu_2/(\nu_2 + \nu_1 F_o)$.

Tables catering directly to the F-distribution are indexed in Section 4.2.

The χ^2-distribution can be obtained as the limit of the F-distribution. If (say) $\nu_1 \to \infty$, we find that $F \to 1/s_2^2 = \nu_2/\chi_{\nu_2}^2$ where $\chi_{\nu_2}^2$ is based on ν_2 degrees of freedom, whence from (3.0.4) we have

$$\lim_{\substack{\nu_1 \to \infty \\ x_o \nu_1 \to \nu_2/F_o}} I_{x_o}(\tfrac{1}{2}\nu_2, \tfrac{1}{2}\nu_1) = \Pr[\chi_{\nu_2}^2 \leq \nu_2/F_o] = P(\chi^2, \nu_2) \qquad (3.0.5)$$

with $\chi^2 = \nu_2/F_o$.

The Poisson distribution can be obtained as the limit of the binomial distribution:

$$\lim_{\substack{n \to \infty \\ np \to m}} \sum_{i=a}^{n} \binom{n}{i} p^i (1-p)^{n-i} \to \sum_{i=a}^{\infty} e^{-m} m^i/i!$$

Poisson- and χ^2-tables are indexed in Section 2.

Although the individual terms of the binomial series

$$\binom{n}{i} p^i q^{n-i} \qquad (3.0.6)$$

can be obtained by differencing the cumulative sum $Q(p,n,r)$, they have been specially tabulated; such tables are indexed in Section 3.22. The binomial coefficients

$$\binom{n}{i} = {}^nC_i = \frac{n!}{i!(n-i)!} = \frac{\Gamma(n+1)}{\Gamma(i+1)\Gamma(n-i+1)} = \frac{1}{(n+1)B(i+1, n-i+1)}$$

are indexed in Section 3.21. Applications of the binomial distribution to quality control are discussed in Section 15.

3.1 Tables of the cumulative binomial sum; tables of the incomplete-beta-function ratio

3.11 The cumulative binomial sum

3.111
$$Q(p, n, r-\tfrac{1}{2}) = \sum_{i=r}^{n} \binom{n}{i} p^i q^{n-i} = I_p(r, n-r+1)$$

The tables indexed in sections 3.111 and 3.112 give all r for which the function tabled differs from 0 or 1 to the number of decimals given.

	p =	n =	
7 dec.	.01(.01).5	1(1)150	Army Ordnance Corps 1952
7 dec.	.01(.01).5	2(1)49	N.B.S. 1950(T.2)*
5 dec.	.01(.01).5 ; 1/12(1/12)5/12 ; 1/16(1/16)7/16	1(1)50(2)100(10) 200(20)500(50)1000	Harvard Comp. Lab. 1955
5 dec.	.01(.01).1(.05).5	1(1)20	E.S. Smith 1953(59)
4 dec.	.05(.05).5	1(1)20	Burington & May 1953(251)

3.112
$$P(p, n, r+\tfrac{1}{2}) = \sum_{i=0}^{r} \binom{n}{i} p^i q^{n-i} = I_q(n-r, r+1)$$

	p =	n =	
5 dec.	.01(.01).5	50(5)100	Romig 1953

The printed sixth place contains errors up to 6.

4 dec.	.025(.025).1(.05) .9(.025).975	5(5)20, 30	Durham et al. 1929(40), Burn 1937a (277), Burn 1937b (216), Burn et al. 1950(426)

3.119 Miscellaneous tables of the cumulative binomial sum

See also the distribution of the sign test in Dixon & Massey 1957, abstracted in Section 3.6.

---

*List of errors:

Page	n =	r =	p =	for	read
200	6	3	.47	.6984534	.5984534
244	24	22	.45	.0000001	.0000021
327	40	4	.07	.3163132	.3063132

3.119
3.12

[see also pp. 664-665]

Curtiss, J. H. 1946

Annals of Mathematical Statistics 17, 62-70

If, in a single sampling plan, we introduce the notations

P_r = Probability of rejecting lot

c = Acceptance number = largest number of defects permitting acceptance

c + 1 = Rejection number = smallest number of defects requiring rejection of lot

n = size of sample

p = quality of product = fraction defective in lot;

then $P_r = Q(p, n, c+\frac{1}{2})$.

Figures 1, 3, and 5 (pp. 63-69) give respectively charts for a rejection number of c + 1 = 1, 2, and 3. Each figure gives curves of P_r as a function of 100p for n = 1(1)12, 14, 16, 20, 25*, the ranges being $0 \le 100p \le 80$, $0 \le P_r \le 1.0$. Figure 1A (p. 64) gives the section $0 \le 100p \le 4$; $0 \le P_r \le .10$ for c + 1 = 1 to an enlarged scale. Further, we have average outgoing quality

$$AOQ = p[1 - P_r] = p P(p, n, c+\tfrac{1}{2}) .$$

Figures 2, 4, and 6 (pp. 66-70) give respectively charts for a rejection number of c + 1 = 1, 2, and 3. Each figure gives curves of AOQ as a function of 100p for n = 1(1)12, 14, 16, 20, 25*.

van Wijngaarden, A. 1950

Koninklijke Nederlandse Akademie van Wetenschappen, Proceedings 53, 857-868

$$1 - 2P(\tfrac{1}{2}, n, r+\tfrac{1}{2}) = 2^{-n} \sum_{i=r+1}^{n-r} \binom{n}{i}$$ is tabled to 5 dec. for n = 1(1)200

and $r = 0(1)\left[\tfrac{n-1}{2}\right]$.

Hartley, H. O. and Fitch, E. R. 1951

Biometrika 38, 423

See abstract in Section 3.12 .

---

*Curves for n = 25 are not given for c + 1 = 1 .

3.12 Tables of the incomplete-beta-function ratio

Pearson, K. and Stoessiger, Brenda 1931

Biometrika 22, 253-283

"Tables of the probability integrals of symmetrical frequency curves in the case of low powers such as arise in the theory of small samples"

Table 1, pp. 274-282, gives $\frac{1}{2}[1 + I_x(\frac{1}{2},b)]$ to 7 dec. for $\begin{cases} x = 0(.01)1 \\ b = \frac{1}{2}(\frac{1}{2})15 \end{cases}$.

Reprinted in Tables for S & B II, T. 25.

Table 2, p. 283, gives $x^{-\frac{1}{2}} I_x(\frac{1}{2},b)$ to 7 dec. for $\begin{cases} x = 0(.01).1 \\ b = \frac{1}{2}(\frac{1}{2})15 \end{cases}$.

Reprinted in Tables for S & B II, T. 25 bis.

Pearson, K. 1934

Tables of the incomplete beta-function

$I_x(a,b)$ is tabled to 7 dec. for $\begin{cases} x = 0(.01)1 & 0(.01)1 \\ a = b(\frac{1}{2})11(1)50 & b(1)50 \\ b = \frac{1}{2}(\frac{1}{2})10\frac{1}{2} & 11(1)50 \end{cases}$.

Hartley, H. O. and Fitch, E. R.

Biometrika 38, 423-426

"A chart for the incomplete Beta-function and the cumulative binomial distribution"

The chart gives $I_x(a,b)$ to about $1\frac{1}{2}$ fig. in the smaller of I_x and $1-I_x$ for $1 \leq b \leq 60$; $b \leq a \leq 200$; $0 \leq x \leq 1$. The upper part of the chart contains a family of curves for I_x = .0005, .001, .005, .01, .02, .05, .1(.1).5 ; $1 - I_x$ = ditto. The lower part contains a family of curves for a = 2(1)10(2.5)25(5)50(10)100(25)150, 200 . The abscissa scale at the top of the chart covers b = 1(.5)15(5)30(10)60 . The scale of x is provided on a separate ruler. With the origin of the x-scale pivoted at the required point (a,b) in the lower part of the chart, corresponding values of x and I_x are in alignment in the upper chart.

The operation of the chart can easily be reversed to yield x as a function of I_x .

Reprinted in Pearson & Hartley 1954, T. 17, p. 156.

3.2
3.221   [see also p. 665]

3.2 Binomial coefficients; individual binomial frequencies

3.21 Binomial coefficients

The tables in sections 3.211 and 3.212 effectively cover the range $i = 0(1)n$. Some tables save space by omitting coefficients for i greater than $\tfrac{1}{2}n$; missing values are to be supplied from the formula $\binom{n}{i} = \binom{n}{n-i}$. The lower limit stated for n need not be taken very seriously; cf. Fletcher et al. 1946, p. 10.

3.211 $\binom{n}{i} = \dfrac{n!}{i!(n-i)!} = {}^nC_i$

For tables of $\binom{n}{i}$ for small i, see section 5.31.

	$n =$	
Exact	2(1)200	J. C. P. Miller 1954
Exact	2(1)60	Peters & Stein 1922
Exact	2(1)50	Glaisher 1917, Potin 1925
Exact	1(1)35	Royo López & Ferrer Martín 1954(240), 1955(44)
Exact	2(1)25	Dwight 1941
Exact	1(1)20	Burington & May 1953(257), Czechowski et al. 1957(124), Dixon & Massey 1957(467), Hodgman 1957(359)
Exact	1(1)16	Woolhouse 1864
Exact	1(1)12	Barlow 1930, 1941
Exact	1(1)10	Mode 1951(362)
8 fig.	1(1)100	Fry 1928(439)
5 fig.	1(1)50	E. S. Smith 1953(54)

3.212 $\log_{10}\binom{n}{i}$

7 dec.	5(1)64	Glover 1923(459)
5 dec.	1(1)50	E. S. Smith 1953(54)
4 dec.	2(1)100	Hald 1952(T.14), Vianelli 1959(557), Czechowski et al. 1957(124)

3.22 Individual binomial frequencies

3.221 Tables of $\binom{n}{i} p^i q^{n-i}$

Unless otherwise stated, all values of i are tabled for which the frequency differs from 0 to the number of decimals given.

	p =	n =	
7 dec.	.01(.01).5	2(1)49	N.B.S. 1950(T.1)[*]
7 dec.	.1(.1).5	2(1)50	Czechowski et al. 1957(44)
6 dec.	.01(.01).5	50(1)100	Romig 1953[‡]
5 dec.	.01, .02(.02).1(.1).5	5(5)30	E.S. Pearson & Hartley 1954(186), Vianelli 1959(549)
4 dec.	.05(.05).5	1(1)20	Burington & May 1953(247)
4 dec.	$\frac{1}{256}, \frac{1}{64}, \frac{1}{16}, \frac{1}{8}, \frac{3}{16}, \frac{1}{4}, \frac{1}{3}, \frac{7}{16}, \frac{1}{2}$	1(1)50	Warwick 1932(9)
4 dec.	.01, .05(.05).5, 1/3	2(1)10	Dixon & Massey 1957(466)
3 dec.	.5 only	1(1)23	S. Brown 1873(345)

---

[*] Values for $i = n$ are omitted, but can be read from Table 2. List of errors:

Page	n =	i =	p =	for	read
18	14	2	.18	.2624913	.2724913
116	37	0	.09	.0395163	.0305163
134	40	3	.07	.2211640	.2311640
134	40	4	.07	.1709448	.1609448
140	41	5	.06	.0668162	.0628162
140	41	6	.06	.0200573	.0240573
192	49	17	.35	.0289183	.1189183
192	49	18	.35	.2038364	.1138364

[‡] The frequencies have not been forced to sum to 1.

3.229 Miscellaneous binomial frequencies

Whitaker, Lucy 1914

Biometrika 10, 36-71

Table 1, p. 41, gives the following binomial probabilities, and the normal and Poisson approximations thereto:

$\left(\frac{1}{100} + \frac{99}{100}\right)^{1000}$ terms i = 0 to i = 23 to 5 dec.

$\left(\frac{3}{100} + \frac{97}{100}\right)^{1000}$ terms i = 19 to i = 42 to 5 dec.

Shook, B. L. 1930

Annals of Math. Stat. 1, 224

Section of binomial distribution for n = 100,000.

Table XXIII (pp. 250-1) gives $P_i = \binom{100,000}{i}(.008)^i(.992)^{100,000-i}$ to 4 dec. for i = 690(1)911 as well as

$$\sum_{i=0}^{x} P_i$$

to 4 dec. for x = 690(1)911.

Seal, H. L. 1941

Inst. of Actuaries, Jn. 71, 5

Table 3(a) (p. 43-44) gives the binomial terms $\binom{n}{i} p^i q^{n-i}$ to 4 dec. for pairs of n,p with np = 10 and p = .0025, .005, .01, .03, .05, .1.

Alongside are shown, also to 4 dec.:

  (a) The normal approximation

$$\tfrac{1}{2}\left[P([i-np+\tfrac{1}{2}]/\sqrt{npq}) - P([i-np-\tfrac{1}{2}]/\sqrt{npq})\right];$$

(b) The Type III approximation

$$\left(2\left(\frac{4}{\beta_1}\right)^{\frac{4}{\beta_1}-1} \beta_1^{-\frac{1}{2}} e^{-\frac{4}{\beta_1}} \left[\Gamma\left(\frac{4}{\beta_1}\right)\right]^{-1} \int_{i_1}^{i_2} (1+\tfrac{1}{2}\sqrt{\beta_1}\, t)^{\frac{4}{\beta_1}-1-\frac{2t}{\sqrt{\beta_1}}} e^{\frac{4}{\beta_1}-1-\frac{2t}{\sqrt{\beta_1}}}\, dt\right)$$

where

$\beta_1 = (1-4pq)/npq$; $i_1 = (i-np-\tfrac{1}{2})/\sqrt{npq}$; $i_2 = (i-np+\tfrac{1}{2})/\sqrt{npq}$;

(c) The Poisson approximation $\dfrac{e^{-np}(np)^i}{i!}$ which is the same for all p.

The normal approximation produces sensible frequencies for i as small as -2.

Seal's table is reproduced, with different notation and without algebraic exposition of the Type III approximation, in N. L. Johnson & Tetley 1949, pp. 215-217.

Notation:	Seal	Johnson & Tetley	Ours
	E	n	n
	p	p	q
	q	q	p
	θ	r	i

3.3
3.311
3.3 Percentage points of the Beta-function and confidence intervals for binomial distribution.

3.30 Introduction

The percentage points of the Beta-function, $x(P;a,b)$, are defined as the roots x of $I_x(a,b) = P$, so that, once again, P denotes the probability integral or lower tail area; here of the Beta distribution. These percentage points can be related directly to the upper (p_U) and lower (p_L) end points of the confidence interval for the population proportion, p, of the binomial distribution:- If in a random sample of n we observe that c items have a certain characteristic and we wish to compute the confidence interval for p with a confidence coefficient of at least $1 - 2\alpha$, then the end points p_U and p_L of this interval are defined as the roots of

$$\sum_{i=c}^{n} \binom{n}{i} p_L^i (1-p_L)^{n-i} = \alpha \; ; \quad \sum_{i=0}^{c} \binom{n}{i} p_U^i (1-p_U)^{n-i} = \alpha \; ;$$

the situation being illustrated in Fig. 3.1 below.

Fig. 3.1

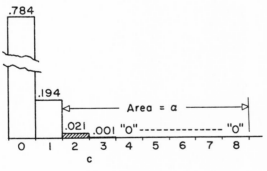

Binomial distribution corresponding to $p_L = .03$
$n = 8; \quad \alpha = .025$

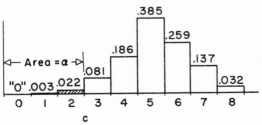

Binomial distribution corresponding to $p_U = .65$
$n = 8 \quad \alpha = .025$

the situation being illustrated in Fig. 3.1.

It follows from (3.0.3) that p_L and p_U are directly given by the $100\alpha\%$ points of the Beta-function, viz by

$$p_L = x(\alpha;c,n-c+1); \quad p_U = 1 - x(\alpha;n-c,c+1);$$

so that these confidence points are mathematically equivalent to $x(P;a,b)$.

3.31 Percentage points of the beta distribution

3.311 Tables of $x(P;a,b)$

Thompson, Catherine M. 1941a
Biometrika 32, 151-181

$x(P;a,b)$ is tabled to 5 fig. for $P = .005, .01, .025, .05, .1, .25, .5$;
$\begin{cases} a = \frac{1}{2}(\frac{1}{2})15 \\ 60/a = 0(1)4 \end{cases}$; $\begin{cases} b = \frac{1}{2}(\frac{1}{2})5 \\ 60/b = 0(1)6(2)12 \end{cases}$.

Reprinted in E.S. Pearson & Hartley 1954, T. 16, pp. 142-155; Vianelli 1959, Pro. 211, pp. 661-674.

Grubbs, F. E., 1949
Annals of Math. Stat. 20, 242

Tables I and II (pp. 247-56).
Percentage points of the Incomplete Beta-function suitable for designing single sampling inspection plans.
Table I gives $1 - x(.95, n-c, c+1) = p_1$
Table II gives $1 - x(.10, n-c, c+1) = p_2$
to 3 fig. for $c = 0(1)9$ and $n = [c+1](1)150$.

Reprinted in Vianelli 1959, **Prontuari** 235-236, pp. 821-826.

Hartley, H. O. and Fitch, E. R. 1951
See abstract in Section 3.12.

Burington, R. S. and May, D. C. 1953
Table III, pp. 255-257, gives $x(P; a,b)$ to 4 dec. for
$P = .005, .01, .025, .05, .1$; $\begin{cases} a = 1(1)15 \\ 60/a = 0(1)4 \end{cases}$; $\begin{cases} b = 1(1)6 \\ 60/b = 1(1)4, 6 \end{cases}$.

3.319
3.321
[see also pp. 665-666]

3.319 Miscellaneous percentage points of the beta distribution

Pearson, E. S. 1925

<u>Biometrika</u> 17, 388-442

Let x be an abbreviation for $x(P;a,b)$, $P = .005, .05, .95, .995$.
Define
$$d = (x - m)/\sigma$$
where $m = a/(a + b)$
$$\sigma^2 = ab/(a+b)^2(a+b+1).$$

The table on pp. 439-442 gives d to 2 dec. for those combinations of a and b such that

$$\beta_1 = \frac{4(a-b)^2(a+b+1)}{ab(a+b+2)^2}$$

$$\beta_2 = \frac{3(a+b+1)[2(a+b)^2 + ab(a+b-6)]}{ab(a+b+2)(a+b+3)}$$

are points of the grid
$$\beta_1 = .00,.01,.03,.05(.05).20(.10)1.00(.20)3.00$$
$$\beta_2 = 2.4(0.2)4.4$$

for which $1 + \beta_1 \leq \beta_2 \leq \tfrac{1}{2}(6 + 3\beta_1)$.

Reprinted in Tables for S & B II, T. 50.

Woo, T. L. 1929

<u>Biometrika</u> 21, 1-66

Define $n = 2a + 2$, $N = 2a + 2b + 2$:

the mean $\bar{x}(a,b) = \dfrac{a}{a + b} = \dfrac{n - 1}{N - 1}$

and the standard deviation

$$\sigma_x(a,b) = \frac{\sqrt{ab}}{(a+b)\sqrt{a+b+1}} = \frac{\sqrt{2\bar{x}(1-\bar{x})}}{\sqrt{N+1}}$$

are tabulated to 6 dec. for

$n = 3(1)20$ or $a = 1(.5)9.5$
$N = 51(1)1000$ $b + a = 25(.5)499.5$

Also tabulated are

$$P_1 = I_{x_1(a,b)}(a,b); \quad P_2 = I_{x_2(a,b)}(a,b)$$

to 3 dec. for the above range of a and b.

Here $\dfrac{x_1(a,b) - \bar{x}(a,b)}{\sigma_x(a,b)} = \lambda_1(a,b); \quad \dfrac{x_2(a,b) - \bar{x}(a,b)}{\sigma_x(a,b)} = \lambda_2(a,b)$

where the two λ's are tabulated at the head of each page column to 2 dec. and are chosen such that $I_{x_1} \sim .01$, $I_{x_2} \sim .02$. The λ's do not change much with b.

Reprinted in Tables for S & B II, T. 4, pp. 16-72.

3.32 Confidence limits for binomial distributions

3.321 Direct tables of p_L and p_U

For definitions of p_L and p_U, see Section 3.30.

Clopper, C. J. and Pearson, E. S. 1934
<u>Biometrika</u> 26, 404-413
"The use of confidence or fiducial limits illustrated in the case of the binomial"

The charts on pp. 410-411 plot p_L and p_U (to about $2\frac{1}{2}$ dec.) against c/n (to about 3 dec.) for $n = 10, 15, 20, 30, 50, 100, 250, 1000$; $\alpha = .005, .025$.

The chart for $\alpha = .025$ is reproduced in Wilks 1948, p. 201. Both charts are reproduced in Deming 1950, and reprinted as Table 41, pp. 204-205, in Pearson & Hartley 1954. This last printing, being quarto, is more legible; and, being stereotyped, it is more accurate.

Snedecor, G. W. 1946
<u>Statistical Methods</u>

Table 1.1, pp. 4-5, gives p_L and p_U to 2 dec. for $\alpha = .005, .025$;
$\begin{cases} n = 10, 15, 20, 30, 50, 100 \\ c = 0(1)n \end{cases}$ $\begin{cases} n = 250, 1000 \\ c/n = 0(.01).5 \end{cases}$.

3.321
3.33

Eisenhart, C., Hastay, M. W. and Wallis, W. A. 1947

Techniques of statistical analysis

Charts of p_L and p_U against c/n for

$\begin{cases} n = 5,\ 25 & 10,\ 15,\ 50,\ 100,\ 250,\ 1000 & 20,\ 30 \\ \alpha = .05,\ .1 & .005,\ .025,\ .05,\ .1 & .005,\ .025 \end{cases}$

The charts for $\alpha = .005,\ .025$, are reproduced from Pearson & Clopper 1934.

Mainland, D. 1948

Canadian Journal of Research 26, Section E, 1-166 + 6 plates

Table IA (p. 104) gives p_L and p_U to 2 fig. or more for

$\alpha = .005,\ .025,\ .10$ and

$n = 1(1)50(5)100(10)200,\ 220,\ 250,\ 300,\ 400,\ 500,\ 700,\ 1000$

when

$c = 0$.

Table IB (p. 105-124) gives similar limits for $c = 1(1)20$.

Table II (p. 125-134) gives limits when $c = 20$ for

$c/n = .001,\ .003,\ .005,\ .007,\ .01,\ .02,\ .03,\ .05(.025).1(.05).5$

and for appropriate values of n ranging to 100,000.

Dixon, W. J. and Massey, F. J. 1951

Table 9, pp. 320-323, gives charts of p_L and p_U against c/n for

$\begin{cases} n = 5 & 10,\ 15,\ 20,\ 30,\ 50,\ 100,\ 250,\ 1000 \\ \alpha = .05,\ .1 & .005,\ .025,\ .05,\ .1 \end{cases}$

The charts for $\alpha = .005,\ .025$, are 'reproduced from' (scil. redrawn after) Pearson & Clopper 1934.

Reprinted in Dixon & Massey 1957, T. 9, pp. 413-416.

Hald, A. 1952

Table XI gives p_L and p_U to 3 dec. for $\alpha = .005,\ .025$;

$\begin{cases} c = 0(1)20(2)30(5)50,\ 60,\ 80,\ 100,\ 200,\ 500 \\ n - c = 1(1)20(2)30(5)50,\ 60,\ 80,\ 100,\ 200,\ 500 \end{cases}$

Burington, R. S. and May, D. C. 1953

Tables 14.57.1 and 14.57.2, pp. 192-195, give p_L and p_U to 2 dec. for $\alpha = .005,\ .025$; $n = 10,\ 15,\ 20,\ 30,\ 50,\ 100$; $c = 0(1)n$.

3.322 np_L and np_U

Fisher, R. A. and Yates, F. 1943

Table VIII 1 gives np_L and np_U to 2 dec. or more for
α = .005, .025, .1 and

$$\begin{cases} c = 0 & 1 & 2 & 3 & 4(1)14 \\ n = & 2(1)5 & 4(1)8 & 6(1)10 \\ c/n = & 0(.05).2 & 0(.05).25 & 0(.1).3 & 0(.1).5 \\ 20/n = 0(1)4 \end{cases}$$

Reprinted in Davies 1949, T. F, pp. 278-279.

Fisher, R. A. and Yates, F. 1948, 1953, 1957

Table VIII 1 gives np_L and np_U to 2 dec. or more for
α = .005, .025, .1 and

$$\begin{cases} c = 0 & 1 & 2 & 3 & 4(1)10 \\ n = & 2(1)5 & 4(1)8 & 6(1)10 \\ c/n = & 0(.05).2 & 0(.05).25 & 0(.1).3 & 0(.1).5 \\ 20/n = 0(1)4 \end{cases}$$

To cover the range $c > 10$, the table gives $np_L - c - [c(1 - c/n)]^{\frac{1}{2}} x(\alpha)$ and $np_U - c + [c(1 - c/n)]^{\frac{1}{2}} x(\alpha)$, where x is the normal deviate defined on page 2, to 2 dec. for c = 9, 16, 36, 144, ∞ ; c/n = 0(.1).5.

3.33 Percentage points of the binomial distribution

Warwick, B. L. 1932

<u>Probability tables for mendelian ratios with small numbers</u>

Table 5, p. 28, gives $\frac{1}{2}$% and $99\frac{1}{2}$% points of the cumulative binomial summation for

$$p = 1/8, \; n = 50(1)75$$
$$p = 1/16, \; n = 50(1)153$$
$$p = 1/64, \; n = 50(1)629$$
$$p = 1/256, \; n = 50(1)2531$$

3.33
3.41
Cochran, W. G. 1937

Journal of the Royal Statistical Society 100, 69-73

"The efficiencies of the binomial series tests of significance of a mean and of a correlation coefficient"

For each integral value of n up to 50, this table gives the smallest integral value of r for which

$$2\left[\frac{1}{2^n}\sum_{i=r}^{n}\binom{n}{i}\right] \leq .05.$$

3.4 Approximations to the binomial distribution and the Incomplete Beta-function; auxiliary tables for computing $I_x(a,b)$.

There are numerous approximations to the incomplete Beta-function. These fall, roughly, into two categories:

(a) Those giving approximations to the percentage points $x(P,a,b)$ directly and (b) those yielding approximate values of the Beta integral $I_x(a,b)$ or the cumulative binomial series. In this latter class falls the well known 'normal' approximation to the binomial distribution in which

$$X = (i + \tfrac{1}{2} - np)/\sqrt{npq}$$

is approximately regarded as a normal deviate.

3.41 Approximations to percentage points of the beta distribution

Thomson, D. Halton 1947

Biometrika 34, 368

"Approximate formulae for the percentage points of the incomplete beta function and of the χ^2 distribution"

The following approximations to $x(P;a,b)$ are tabled:

(A) $\quad x(P;a,b) \doteq \exp\left[-\chi^2(p,2b)/(2a+b-1) \right]$

(B) $\quad x(P;a,b) \doteq \left(\dfrac{2a-1}{2a+2b-1} \right)^k$

with $k = \chi^2(p,2b)/2b$

(noting that the upper tail area p of χ^2 corresponds to the lower tail area P of x.)

(C) $\quad x(P;a,b) = e^{-u} - r(1 + \tfrac{7}{6}r + \tfrac{1}{6n})u + \tfrac{11}{12}ru^2$

with $u = \chi^2(p,2b)/2(a+b-1)$, $n = (2a+2b-1)/2b$,

and $r = (b-1)/2(a+b-1)$

3.41
3.43   [see also pp. 667-668]

(D) $x(P;a,b) = a/(a+be^{2z})$

where z is the Fisher-Cochran approximation to percentage points of the z-distribution.

Table 1 (p. 371) gives the four approximations, together with the exact value of $x(P;a,b)$, to about 5 fig. for

   2a = 120   2b = 2, 4, 10, 20, 30, 40, 60, 120.   P = .995

The discrepancies are also shown.

Table 2 (p.371) gives discrepancies expressed as a percentage of min(x, 1-x) for   2a = 120, 2b = 12, 20, 40, 60, 120;

   2a = 30, 2b = 3, 5, 10, 15, 30;

   P = .005, .5, .995 .

Wise, M. E. 1950

<u>Biometrika</u> 37, 208-218

"The incomplete beta function as a contour integral and a rapidly converging series for its inverse"

Wise develops the formula

$$x(P;a,b) \doteq e^{-y/N}$$

where $N = a + \frac{1}{2}b - \frac{1}{2}$ and

$$y = \frac{1}{2}\chi^2 \left[1 + \frac{(b-1)(b+1+\frac{1}{2}\chi^2)}{24N^2} \right]$$

and $\chi^2 = \chi^2(P;2b)$ is the upper 100P % point of χ^2 for 2b degrees of freedom.

Tables 1 and 2 (pp. 213-4) give, for selected values of a = b and P, a comparison of exact values and approximation.

3.42   <u>Approximations to the cumulative binomial summation</u>

By virtue of equation (3.0.3), the approximations of sections 3.42 and 3.43 are in principle interchangeable; in practice, conversion of a beta-approximation into a binomial approximation may be awkward.

McCallan, S. E. A. and Wilcoxon, F., 1932

Contributions from the Boyce Thompson Institute 4, 233

Let p_1 and p_2 denote two percentages based on n observations each and $\bar{p} = \frac{1}{2}(p_1 + p_2)$.

Using the normal approximation

$$x = |p_1 - p_2| \bigg/ \sqrt{\frac{2\bar{p}(100-\bar{p})}{n}}$$

for a test of significance, the table on p. 240 gives significant differences

$$|p_1 - p_2| = x(P) \sqrt{\frac{2\bar{p}(100-\bar{p})}{n}}$$

where $x(P) = 2.339$ corresponding to the double tail area $P = \frac{1}{51}$ (odds of 50 to 1). $|p_1 - p_2|$ is given to 1 dec. for

$\bar{p} = 1, 3, 5, 10(5)30(10)70(5)95, 97, 99$

$n = 10, 12, 15(5)30(10)60, 75, 100, 120, 150, 200, 250, 300(100)600, 750, 1000(500)3000(1000)6000, 7500, 10,000.$

Haldane 1938 gives a formula, without tables, for approximately normalizing the binomial distribution.

E. S. Smith 1953 gives copious examples, keyed to the tables in his book, of Gram-Charlier Type A and Type B approximations to the binomial distribution.

3.43 Approximations to the incomplete beta-function

Pearson, K. and Pearson, Margaret V. 1935

Biometrika 27, 409-423

"On the numerical evaluation of high order incomplete Eulerian integrals"

Let $I'(x,n) = \Gamma_x(n)/\Gamma(n)$, defined on page 130. For $b > a$ the following expansion is valid:

$$I_x(a+1, b+1) = \frac{\Gamma(a+b+2)}{\Gamma(b+1)} \left[\frac{I'(z_0, a+1)}{b^{a+1}} + c_2 \frac{I'(z_0, a+3)}{b^{a+3}} (a+2)(a+1) \right.$$

$$\left. + \cdots + c_i \frac{I'(z_0, a+i+1)}{b^{a+i+1}} \frac{(a+i)!}{a!} \right]$$

where $z_0 = bx$ and the c_j, rational functions of b, are given on p. 416.

3.43
3.44   [see also p. 668]

Hartley, H. O. 1944

Biometrika 33, 173-180

"Studentization, or the elimination of the standard deviation of the parent population from the random sample-distribution of statistics"

Hartley's equation (30) is an expansion through b^{-2}, of which the terms through b^{-1} are

$$I_x(a,b) = I'(\omega,a) + \frac{1}{2b} e^{-\omega} \omega^a \Gamma(a)^{-1}(a - 1 - \omega) + \ldots$$

where $\omega = bx/(1-x)$. The table on p. 179 gives $I_x(a,b)$, and its approximation from equation (30), to 5-7 dec. for

$$\begin{cases} a = 1 \\ b = 5(1)10(2.5)25(5)50 \\ \omega = 5.555 \end{cases}$$

3.44 <u>Auxiliary tables for computing the incomplete beta-function</u>

Soper, H. E. 1921

The <u>numerical evaluation of the incomplete</u> B-<u>function</u>

The table on p. 15 gives $\log_{10}\left[\left(\frac{x}{\xi}\right)^\xi \left|\frac{1-x}{1-\xi}\right|^{1-\xi}\right]$ to 5 dec. for

$$\begin{cases} x = .1(.1).8 \quad .1(.1).9 \\ \xi = [x+.1](.1).9 \quad 1.1(.1)1.9 \end{cases}; \quad \log_{10}(\xi|1-\xi|/2\pi)^{\frac{1}{2}}(\xi-x)^{-1}$$ to 5 dec. for

$$\begin{cases} x = .1(.1).9 \\ \xi = .1(.1).9, \; 1.1(.1)1.9 \end{cases}; \text{ and } \log_{10}(2\pi\xi|1-\xi|)^{-\frac{1}{2}}$$ to 5 dec. for

$\xi = .1(.1).9, \; 1.1(.1)1.9$.

Wishart, J. 1925a

Biometrika 17, 68-78

Wishart expands the trigonometric form of the incomplete beta-function

$$I_\Theta(n+1) = \frac{1}{2} \int_0^\Theta \cos^{n+1}\Theta \, d\Theta \bigg/ \int_0^{\pi/2} \cos^{n+1}\Theta \, d\Theta$$

$$= c_0 m_0(x) - c_4 m_4(x) - c_6 m_6(x) + c_8 m_8(x) + c_{10} m_{10}(x) -$$
$$- c_{12} m_{12}(x) - \ldots$$

where $x = \sqrt{n} \sin \Theta$ and

$$m_i(x) = \frac{1}{\sqrt{2\pi}} \int_0^x x^i e^{-\frac{1}{2}x^2} dx \bigg/ (r-1)(r-3) \ldots \frac{1}{\text{or}} \; 2$$

He tables c_i to 7 dec. for $i = 0, 4(2)12$; $n = 100(2)400$.

Wishart, J. 1925 b

Biometrika 17, 469-472

Wishart expands $I_\theta(n+1)$, defined above, in terms of aggregates of incomplete normal moments:

$$\left(\int_0^{\frac{\pi}{2}} \cos^{n+1}\tau \, d\tau\right) I_\theta(n+1) = \frac{\sqrt{2\pi}}{\sqrt{n+1}}\left[\phi_0(x) - \frac{1}{n+1}\phi_1(x) + \frac{1}{(n+1)^2}\phi_2(x) - + \ldots\right]$$

where $x = 2\sqrt{n+1} \tan \frac{1}{2}\theta$;

$\phi_0(x) = m_0(x)$

$\phi_1(x) = \frac{1}{4}m_2(x)$

$\phi_2(x) = .1875\, m_4(x) - .15625\, m_6(x)$

$\phi_3(x) = .234375\, m_6(x) - .2734375\, m_8(x)$

$\phi_4(x) = .41015625\, m_8(x) - .984375\, m_{10}(x) + .56396484\, m_{12}(x)$;

and the $m_i(x)$ are defined in the previous abstract. He tables the $\phi_i(x)$ for $x = 0(.1)5, \infty$ and $\begin{Bmatrix} i = 0 & 1(1)4 \\ 8 \text{ dec.} & 7 \text{ dec.} \\ \delta^4 & \delta^2 \end{Bmatrix}$. This is equivalent to a Gram-Charlier Type A expansion of the distribution of t.

3.44
3.5
Wishart, J. 1927

Biometrika 19, 1-39

Wishart derives the expansion

$$I_x(a+1, b+1) = 1 - \left(1 + \sqrt{\frac{b}{a(a+b)}} u\right)^{a+1} \left(1 - \sqrt{\frac{a}{b(a+b)}} u\right)^{b+1}$$

$$\left(\frac{1}{u}\left[B_0 - \frac{B_1}{(u^2+r)} + \frac{B_2}{(u^2+r)(u^2+2r)} - + \ldots\right]\right.$$

$$\left. - \left[\frac{C_1}{(u^2+r)(u^2+2r)} - \frac{C_2}{(u^2+r)(u^2+2r)(u^2+3r)} + \right]\right)$$

where

$$u = \sqrt{\frac{(a+b)^3}{ab}} \left(x - \frac{b}{a+b}\right), \quad r = \frac{a}{b} + \frac{b}{a},$$

$$B_i = \sum_{j=1}^{\infty} {}_jB_i (a+b)^{1-j}, \quad C_i = \sum_{j=1}^{\infty} {}_jC_i (a+b)^{\frac{1}{2}-j}. \quad \text{He tables:}$$

i	j	$_jB_i$	$_jC_i$		i	j	$_jB_i$	$_jC_i$
0	2	7 dec.			3	1	5 dec.	4 dec.
	3	6				2	4	2
1	1	8	6 dec.			3	2	0
	2	6	5		4	1	4	2
	3	5	4			2	2	0
2	1	6	5			3	0	
	2	5	4		5	1	2	
	3	4	2					

Pearson, K. 1931 a

Tables for S & B II

The tables of Wishart 1925 a, 1925 b, 1927 are reprinted as Tables 45, 46, 47.

3.5 The complete B-function; moments of the beta distribution; moments of the binomial distribution

From the moments of the beta distribution

$$\mu_r' = B^{-1}(a,b) \int_0^1 x^{r+a-1}(1-x)^{b-1}dx$$

$$= B(a+r,b)/B(a,b)$$

$$= \Gamma(a+r)\Gamma(a+b)/\Gamma(a)\Gamma(a+r+b)$$

we derive the moment-constants

$$\bar{\mu} = \mu_1' = a/(a+b)$$

$$\sigma = \sqrt{ab}/[(a+b)\sqrt{a+b+1}]$$

$$\beta_1 = \frac{4(a-b)^2(a+b+1)}{ab(a+b+2)^2}$$

$$\beta_2 = \frac{3(a+b+1)[2(a+b)^2 + ab(a+b-6)]}{ab(a+b+2)(a+b+3)} \, ;$$

from the moments and factorial moments of the binomial distribution

$$\mu_r' = \sum_{i=0}^{n} \binom{n}{i} p^i q^{n-i} i^r,$$

$$\mu_{[r]} = \sum_{i=0}^{n} \binom{n}{i} p^i q^{n-i} i(i-1)\ldots(i-r+1)$$

$$= \binom{n}{r}p^r,$$

we derive the moment-constants

$$\bar{\mu} = np$$

$$\sigma = \sqrt{npq}$$

$$\beta_1 = \frac{(q-p)^2}{npq}$$

$$\beta_2 = \frac{3npq + (1 - 6pq)}{n^2 p^2 q^2} \, .$$

[203]

3.51
3.54
[see also pp. 668-669]

3.51 The complete B-function

For definition, see Section 3.0.

Pearson, K. 1931a

Tables for S & B II

Table 26 gives $q_n = I_n = \frac{1}{2}B(\frac{1}{2}, \frac{1}{2}n)$ to 10 dec. for $n = 1(1)105$.

Pearson, K. 1934

Tables of the incomplete beta-function

$B(a,b)$ is given to 8 fig. for $\begin{cases} b = \frac{1}{2}(\frac{1}{2})10\frac{1}{2} & 11(1)50 \\ a = b(\frac{1}{2})11(1)50 & b(1)50 \end{cases}$.

Each value of $B(a,b)$ is given at the head of a column of $I_x(a,b)$.

3.52 Moments of the beta distribution

Pearson, K. 1934

Tables of the incomplete beta-function

Table 2, pp. 433-494, gives the moment-constants $\bar{\mu}$, σ, β_1 and β_2 and the mode $x = (a-1)/(a+b-2)$, to 8 dec. for $a, b = \frac{1}{2}(\frac{1}{2})11(1)50$.

3.53 Moments of the binomial distribution

Larguier, E. H. 1936

Annals of Mathematical Statistics 7, 191-195

"On a method for evaluating the moments of a Bernoulli distribution"

The moments

$$\mu_r = \sum_{i=0}^{n} (x - np)^r \binom{n}{i} p^i q^{n-i}$$

are written in the form

$$\mu_r = \sum_{t=1}^{r} F_{r,t}(n) p^t$$

A recurrence formula for the F (equation (5)) is derived and from this the $F_{rt}(n)$ are tabulated in Table I (p. 193) as polynomials in n with exact integer coefficients for $r = 1(1)8$, $t = 1(1)r$.

Burington, R. S. and May, D. C. 1953

Probability and statistics

Table V, p. 258, gives $[np(1-p)]^{\frac{1}{2}}$ to 4 dec. for n = 1(1)20 and p = .05(.05).5 .

Vianelli, S. 1959

Prontuari per calcoli statistici

Prontuario 36, pp. 158-163, gives the first three cumulants of the binomial distribution as exact decimal fractions for n = 6(1)30(2)60(4)100 and p = .04(.04).96 .

3.54 Moments of estimated binomial success rate

Gevorkiantz, S. R. 1928

Journal of Forestry 26, 1043-1046

The table on p. 1045 gives $[p(1-p)/n]^{\frac{1}{2}}$ to 5 dec. for p = .5(.05).95 and n = 100(100)600(200)1000 .

Schrock, E. M. 1957

Quality control and statistical methods

Table 11.7 a, pp. 85-87, gives $3[p(1-p)/n]^{\frac{1}{2}}$ to 4 dec. for n = 10(10)100(100)4000(200)5000(500)10000 and p = .001, .0025, .005(.005).02(.01).1(.02).2(.05).5 .

Vianelli, S. 1959

Prontuari per calcoli statistici

Prontuario 181, pp. 535-537 , gives $3[p(1-p)/n]^{\frac{1}{2}}$ to 4 dec. for p = .005, .01(.01).2(.02).5 and n = 1(1)10(2)22, 25(5)40(10)100(25)200(50)300, 400(200)1200, 1500(500)3000, 4000(2000)12000, 15000, 20000 .

3.6
4.11

3.6 Tables for the binomial sign test

[The name 'sign test' is also given to a non-parametric test of runs in time series, whose criterion does not follow the binomial distribution: see Section 9.1.]

Dixon, W. J. and Mood, A. M. 1946
Journal of the American Statistical Association 41, 557-566

Let r denote the number of + or − signs, whichever is smaller. Table 1, p. 560, gives 100 P% points of r for P = .01, .05, .1, .25; n = 1(1)100. Formulas for n > 100 are also given. In the following reprints, n = 1(1)90: Dixon & Massey 1951, T. 10, p. 324; Dixon & Massey 1957, T. 10 a, p. 417; Vianelli 1959, T. 88, p. 1231.

Table 2, p. 562, gives minimum values of n necessary to detect significant differences 95% of the time when q = .55(.05).95, P = .01, .05, .1, .25; and corresponding values of r.

Walker, Helen M. and Lev, J. 1953
Statistical inference

Table IV B, p. 458, gives $P(\frac{1}{2}, n, r+\frac{1}{2})$ to 3 dec. for
$\begin{cases} n = 5(1)15 & 16(1)25 \\ r = 0(1)n & 0(1)15 \end{cases}$. Reprinted in Siegel 1956, T. D, p. 250; Vianelli 1959, T. 86, p. 1229.

Dixon, W. J. and Massey, F. J. 1957

Table 10 b, pp. 418-420, gives $P(\frac{1}{2}, n, r+\frac{1}{2})$ for n = 3(1)100 and r = s(1)t, where $P(\frac{1}{2}, n, s+\frac{1}{2}) \leq .005 < P(\frac{1}{2}, n, s+1\frac{1}{2})$; $P(\frac{1}{2}, n, t-\frac{1}{2}) < .125 \leq P(\frac{1}{2}, n, t+\frac{1}{2})$; to 3 dec.

SECTION 4

THE t-, F-, AND z-DISTRIBUTIONS AND RELATED DISTRIBUTIONS

4.1 The t-distribution

4.11 The probability integral of t

Let x denote a normally distributed variable with mean 0 and standard deviation σ and let s_x^2 stand for an independent estimate of σ_x^2 such that $\nu s_x^2/\sigma_x^2$ follows the χ^2 distribution for ν degrees of freedom. Then the t-ratio $t = x/s_x$ follows a distribution law whose frequency ordinate, $f(t,\nu)$, probability integral $P(t,\nu)$ and tail area $Q(t,\nu)$ are respectively given by (see Fig. 4.1):

$$f(t,\nu) = \Gamma\left[\tfrac{1}{2}(\nu+1)\right] \Gamma\left[\tfrac{1}{2}\nu\right]^{-1} (\nu\pi)^{-\tfrac{1}{2}} \left(1+\tfrac{t^2}{\nu}\right)^{-\tfrac{1}{2}(\nu+1)} \quad\text{-----(4.11.1)}$$

$$P(t,\nu) = \int_{-\infty}^{t} f(u,\nu)\,du = \Gamma\left[\tfrac{1}{2}(\nu+1)\right] \Gamma\left(\tfrac{1}{2}\nu\right)^{-1} (\nu\pi)^{-\tfrac{1}{2}} \int_{-\infty}^{t}\left(1+\tfrac{u^2}{\nu}\right)^{-\tfrac{1}{2}(\nu+1)} du$$
$$\text{-----(4.11.2)}$$

and $Q(t,\nu) = \int_{t}^{\infty} f(u,\nu)\,du = 1 - P(t,\nu)$ --------------------(4.11.3)

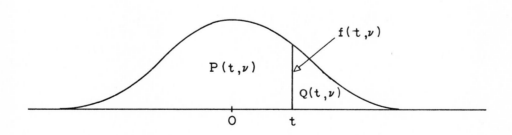

Fig. 4.1

4.11
4.112

The integral $P(t,\nu)$ is related to the Incomplete Beta function ratio:
Writing

$$t^2 = \nu(1-x_o)/x_o \quad \text{or} \quad x_o = \nu/(\nu + t^2) \qquad (4.11.4)$$

we find that

$$1 - P(t,\nu) = \tfrac{1}{2} I_{x_o}(\tfrac{1}{2}\nu, \tfrac{1}{2}) . \qquad (4.11.5)$$

Behrens 1929, T. 2, p. 815, gives algebraic expressions for
$\dfrac{\pi^{\frac{1}{2}}\Gamma(\tfrac{1}{2}n-\tfrac{1}{2})}{\Gamma(\tfrac{1}{2}n)}\left[P[x(n-1)^{\frac{1}{2}}, n-1] - \tfrac{1}{2}\right]$ for $n = 2(1)10$. Error: in $n = 10$, for $\dfrac{105}{385}$ arc tg x read $\dfrac{105}{384}$ arc tg x.

In sections 4.111, 4.112 and 4.113, the upper limit X signifies the greatest value of the argument that makes the function different from 1 to the number of decimals tabled.

4.111 Tables with argument t

4.1111 $P(t,\nu)$

	$\nu =$	$t =$	
6 dec.	3(1)10	See section 4.1114	
5 dec.	1(1)10; 11, 12	0(.1)4(.2)8; 7	Hartley & Pearson 1950
	13(1)19	0(.1)4(.2)X	Pearson & Hartley
	20	0(.05)2(.1)4(.2)5.8	1954(132)
	21(1)24	0(.05)2(.1)4, 5	
	120/ν = 3(1)5		
	120/ν = 0(1)2	0(.05)2(.1)4	
4 dec.	1(1)11; 12(1)20	0(.1)6; 0(.1)X	Gosset 1925(114),
7 dec.	∞	0(.1)5.4	Gosset 1942(118), Peters & Van Voorhis 1940(488)
3 dec.	1(1)6; 7(1)20	0(.1)6; 0(.1)X	Yule & Kendall 1937(536), M.G. Kendall 1943(440)
3 dec.	1(1)4(2)8; 10(5)20	0(.2)5; 0(.2)X	Burington & May 1953(284)

4.1112 $2P(t, n-1) - 1$

Behrens 1929, T. 7, p. 818, gives the central area to 3 dec. for $n = 2(1)10, \infty$; $t = 1(1)6(2)10$.

4.1113 $2Q(t,\nu)$

Treloar 1942, pp. 89-90, tables the double tail area to 3 dec. for

$\begin{cases} \nu = 1, 2 & 3 \\ t = .6(.1)6(.5)9(1)12(2)16(4)28 & .6(.1)6(.5)9(1)12(2)16 \end{cases}$

$\begin{cases} 4 & 5, 6 & 7(1)30, \infty \\ .6(.1)6(.5)9, 10 & .6(.1)6(.5)X & .6(.1)X \end{cases}$

4.1114 $10^6 Q(t,\nu)$

The following short table is arranged from Gosset 1942, p. 120.

t	$\nu=3$	4	5	6	7	8	9	10	t	$\nu=3$	4	5
6·0	4636	1941	923	482	271	162	101	66	11·0	804		
6·5	3697	1445	643	316	167	94		34	12·0	623	138	35
7·0	2993	1096	458	212	106	56	32		14·0	395	76	
7·5	2456	845				35			16·0	265	45	
8·0	2038	662	246	102	46				20·0	137		
8·5	1710								24·0	79		
9·0	1448	422	141	53					28·0	50		
10·0	1064	281	85	29								

4.112 Tables with argument $t\nu^{-\frac{1}{2}}$

$P(z\nu^{\frac{1}{2}},\nu)$

	$\nu =$	$z =$	
4 dec.	1	.1(.1)3.5(.5)10(5)50(10)100(20)140(10)160(20)200(50)500(100)700, 1000, 1500, 2000, 3000	Gosset 1917, 1942(62)
	2, 3	.1(.1)3.5(.5)10(5)X	
	4, 5	.1(.1)3.5(.5)X	
	6(1)29	.1(.1)X	
4 dec.	3(1)7	.1(.1)3	Gosset 1908a (19), 1942(29)
5 dec.	8, 9		

4.113   [see also p. 671]
4.1311

4.113   Tables with other arguments

Behrens, W. U.   1929

<u>Landwirtschaftliche Jahrbücher</u> 68, 807-837

"Ein Beitrag zur Fehlerberechnung bei wenigen Beobachtungen"

Table 3, p. 815, gives $2P\left(x\left(\frac{n}{n+1}\right)^{\frac{1}{2}}, n-1\right) - 1$ to 3 dec. for $n = 2(1)10, \infty$; $x = 1(1)6(\dot{2})10$. Table 5, p. 817, gives $2P(2^{-\frac{1}{2}}x, n-1) - 1$ to 3 dec. for the same n and x.

4.114 $P(t,\nu)/Q(t,\nu)$

The ratio of the left tail to the right tail has been tabled by Love 1924, Miles 1934 and Livermore 1934. These tables were computed by treating Student's 4-decimal values of P as <u>exact numbers</u>; accordingly, the last figures as printed are not necessarily accurate.

4.115   Approximations to the t integral

Gosset, W. S.   1925

<u>Metron</u> 5, No. 3, 113-120

R.A. Fisher 1925 z gives the coefficients $C_1, \ldots, C_5$ of the expansion $P(t,\nu) = P(t,\infty) + C_i \nu^{-i}$ as polynomials in t.

Table 3, pp. 119-120, gives C_i to (8-i) dec. for $i = 1(1)4$, $t = 0(.1)6$. Reprinted in Peters & Van Voorhis 1940, T. 46, p. 493; Gosset 1942, T. 3, p. 120.

4.12   <u>Ordinates of the t-distribution</u>

Sukhatme 1938, pp. 45-48, gives $f(t,\nu)$, defined by (4.11.1), to 7 dec. for $\begin{cases}\nu = 1(1)10 \\ t = .05(.1)7.25\end{cases}$ $\begin{cases}120/\nu = 4(1)6(2)12 & 2 & 0 \\ t = .05(.1)7.25 & .05(.1)6.35 & .05(.1)5.55\end{cases}$.

4.13   Percentage points of the t-distribution

We adopt as standard form the single upper tail $100Q\%$ point $t(Q,\nu)$; i.e. the inverse function to $Q(t,\nu)$. The function most generally tabled is the double tail perentage point, i.e. $t(Q,\nu)$ with argument $2Q$. Percentage points of t can be computed from percentage points of the beta distribution: $t(Q,\nu) = \nu^{\frac{1}{2}}(x^{-1}-1)^{\frac{1}{2}}$, where $x = x(\Pi; \frac{1}{2}\nu; \frac{1}{2})$ and $\Pi = 2Q$.

4.131 Tables of $t(Q,\nu)$

The arrangement in sections 4.1311 and 4.1313 is by number of percentage points tabled.

4.1311 Double tail percentage points: $t(Q,\nu)$ as a function of $2Q$

	$\nu = ;\ [\ 120/\nu = \ ]$	$2Q =$	
3 dec.	1(1)30; [0(1)4]	.001, .01, .02, .05, .1(.1).9	Fisher & Yates 1938 etc.[*] (T.III), Cramér 1946(560),

P.O. Johnson 1949(326), Mood 1950(425), Duncan 1952(612)

3; 5 dec.	1(1)30; ∞	.01, .02, .05, .1(.1).9	R.A. Fisher 1925, Rider 1939(200), Peters & Van Voorhis 1940(173),

Freeman 1942(167), Smith & Duncan 1945(474), Hoel 1947(248), Peach 1947(233), Kenney & Keeping 1951(416), Clark 1953(259)

2 dec.	1(1)20(2)30	.001, .01, .02, .05, .1(.1).3(.2).7(.2).9	Mather 1943(239)
3 dec.	1(1)30; [0(1)4]	.001, .002, .005, .01, .02, .05, .1, .2, .5, .8	Pearson & Hartley 1954(138)
3 dec.	1(1)30(10)60(20) 100, 200, 500, ∞	.001, .002, .01, .02, .05, .1, .2(.2).8	Hald 1952(T.IV)
2 dec.	1(1)10(2)30; [0(1)4]	.001, .01, .02, .05, .1(.2).9	Finney 1952 a (308)
3 dec.	1(1)30; [0(1)4]	.001, .01, .02, .05, .1, .2, .4, .8	N.L. Johnson & Tetley 1950(273), Juran 1951(754)

---

[*] The following errors in the 1948 edition, reported by Norton 1952, have been corrected in the 1953 and 1957 editions:

p	ν	For	Read
·001	3	12·941	12·924
·001	5	6·859	6·869
·001	7	5·405	5·408

4.1311
4.1312   [see also pp. 671-673]

3; 5 dec.	1(1)30; ∞	.01, .02, .05, .1(.1).5	Snedecor 1946(65)
3 dec.	35(5)50(10)100(25)150, 200(100)500, 1000	.01, .05	
3 fig.	1(1)30; [0(1)4]	.001, .005, .01, .025, .05,' .1, .25, .5	Davies 1949(270), Brookes & Dick 1951(283)
2 dec.	1(1)20(2)30(10) 60, ∞	.001, .005, .01, .025, .05, .1, .25, .5	Quenouille 1950(228), 1952(234), 1953(333)
5 fig.	1(1)30; [0(1)4]	.005, .01, .025, .05, .1, .25, .5	Merrington 1942
3 dec.	1(1)30(10)60; [0(1)4]	.01, .05, .1(.1).5	Youden 1951(119)
2 dec.	1(1)30; [0(1)4]	.001, .01, .02, .05, .1	K.A. Brownlee 1947(144)
3 fig.	1(1)30(5)50(10) 100(25)150, 200(100)500, 1000, ∞	.01, .02, .05, .1, .5	Goulden 1939(267), Arkin & Colton 1950(116)
3 fig.	1(1)30(5)50, 60, ∞	.01, .02, .05, .1, .5	Weatherburn 1946(188)
3 fig.	1(1)30	.01, .02, .05, .1, .5	E.E. Lewis 1953(214)
3 dec.	1(1)30(5)50(10) 100; [0(1)4]	.001, .01, .05, .1	Arley & Buch 1950(217)
2 dec.	1(1)10(2)30; [0(1)4]	.001, .01, .05, .1	Finney 1952 b (619)
3 dec.	1(1)30, ∞	.01, .05, .1	Wilks 1948(208)
3 dec.	1(1)30(5)50(10) 100(25)150, 200(100)500, 1000, ∞	.01, .05	Snedecor 1934(88)
3 dec.	1(1)30(2)100	.01, .05	Baldwin 1946
3 dec.	1(1)20(5)30, ∞	.01, .05	Tippett 1952(387)

4.1312 Double tail percentage points: $t(Q,\nu)$ as a function of $1-2Q$

Behrens, W. U. 1929

Landwirtschaftliche Jahrbücher 68, 807-837

"Ein Beitrag zur Fehlerberechnung bei wenigen Beobachtungen"

Table 4, p. 816, gives $(1+1/n)^{\frac{1}{2}}t(Q, n-1)$ to 3 fig. for $n = 2(1)10, \infty$; $1-2Q = .5, .75, .9, .95, .98, .99$. Table 6, p. 817, gives $2^{\frac{1}{2}}t(Q, n-1)$ to 3 fig. for the same n and Q. Table 8, p. 819, gives $t(Q, n-1)$ to 3 fig. for the same n and Q.

4.1313 Single tail percentage points: $t(Q,\nu)$ as a function of Q

	$\nu =$; [$120/\nu =$]	Q =	
3 dec.	1(1)30; [0(1)4]	.005, .01, .025, .05 (.05).2, .35(.05).45	Mode 1951
3 dec.	1(1)30; [0(1)4]	.005, .01, .025, .05, .1(.1).4	Dixon & Massey 1957(384)
2 dec.	1(1)30; [0(1)4]	.0025, .005, .0125, .025, .05	Dixon & Massey 1951(307)
2 dec.	1(1)10	$10^{-3}, 10^{-4}, 10^{-5}, 5 \cdot 10^{-6}$	E.S. Pearson & Hartley 1954(133)

4.1314 Single tail percentage points: $t(Q,\nu)$ as a function of P

	$\nu =$	P =	
3 dec.	1(1)30(10)60(20) 100, 200, 500, ∞	.6(.1).9, .95, .975, .99, .995, .999, .9995	Hald 1952(T.IV)

4.132 Approximations to percentage points of t

Hendricks, W. A. 1936

Annals of Math. Stat. 7, 210

Approximate normalization of t.

The variate transformation

$$x_h(t) = \sqrt{2(\nu+1)}\, c_\nu\, t\, \left[t^2 + 2\nu \right]^{-\frac{1}{2}} \quad \ldots\ldots (1)$$

with

$$c_\nu = \left(\frac{2}{\nu+1} \right)^{\frac{1}{2}} \frac{\Gamma(\frac{1}{2}(\nu+1))}{\Gamma(\frac{1}{2}\nu)}$$

is suggested for normalization of t.

4.132 [see also p. 673]

Table 5(p. 219) gives c_ν to 5 decs for $n = \nu+1 = 100(50)500$ $(100)1000(1000)5000,10000(10000)50000, 100000$. The approximation is tested and compared with normalizations suggested by

Student: $x_s(t) = \left[(\nu-2)/\nu\right]^{\frac{1}{2}} t$ ......(2)

and

Deming and Birge: $x_{db}(t) = \left[(\nu-\frac{1}{2})/\nu\right]^{\frac{1}{2}} t$ ......(3)

Tables 2 and 3 (p. 218) give $t(p,\nu)$ to 3 decs for $\nu = 9$, 29 and $2p = .01, .02, .05, .1(.1).9$ as well as 3 decimal values of t obtained from (1), (2) and (3) by substituting $x(1-p)$ for x and solving for t,

Table 4 (p. 219) gives the ordinates of the distribution of $x_h(t)$, $x_s(t)$ and $x_{db}(t)$ for $\nu = 9$ as well as the normal ordinate $z(x)$ all to 4 decimals for $x = \pm 0(.5)3.0$ for comparison. Table 1 (p. 217) gives $1 - p(t,\nu)$ to 4 decs for $z = t\nu^{-\frac{1}{2}} = \pm 0(.2)2.0$. Alongside are shown the corresponding probability integrals from (1), (2) and (3) by regarding x as a normal deviate.

Hotelling, H. and Frankel, L. R. 1938
<u>Annals</u> <u>of</u> <u>Mathematical</u> <u>Statistics</u> 9, 87-96
"The transformation of statistics to simplify their distribution"

Hotelling and Frankel find the asymptotic expansion

$$x = t\left[1 - \frac{t^2+1}{4\nu} + \frac{13t^4+3t^2+3}{96\nu^2} - \frac{35t^6+19t^4+t^2-15}{384\nu^3} + \frac{6271t^8+3224t^6-102t^4-1680t^2-945}{92,160\nu^4} - \ldots\right]$$

where $t = t(Q,\nu)$ and $x = t(Q,\infty)$. The table on p. 95 gives $t(Q,)$, $t(Q,\infty)$, and the above expansion through ν^{-1}, ν^{-2} and ν^{-3}, for $\nu = 10, 30$ and $Q = .005, .025, .05$. The table on p. 96 gives the same, and the expansion through ν^{-4}, for $\begin{cases} Q = .00005 \quad .0005 \\ \nu = 10, 30, 100 \quad 10, 30 \end{cases}$.

Arley, N. 1940

Det Kgl. Danske Videnskabernes Selskab, <u>Mathematisk-fysiske</u> <u>Meddelelser</u> 18, No. 3

"On the distribution of relative errors, etc."

Let a sample of n observations be drawn from a normal population. If $\bar{x}$ is the sample mean, x a randomly selected item of the sample $x_1 \ldots x_n$, and s the sample standard deviation, then $r = \dfrac{x - \bar{x}}{\sqrt{\dfrac{n-1}{n}}\, s}$ has a distribution given by

$$\phi(r)\, dr = \frac{1}{\sqrt{\pi}} \frac{1}{\sqrt{\nu+1}} \frac{\left(\frac{\nu-1}{2}\right)!}{\left(\frac{\nu+2}{2}\right)!} \left(1 - \frac{r^2}{\nu+1}\right)^{\frac{\nu-2}{2}} dr$$

$$|r| \leq \sqrt{\nu+1},\ \nu = n - 2.$$

The distribution of r is similar to that of Student's t and indeed the statistic $t' = \dfrac{\sqrt{n-2}\ r}{\sqrt{n-1-r^2}}$ is distributed as t with n - 2 degrees of freedom.

The table on p. 61 gives $r(Q,\nu)$:

	2Q =	ν =	120/ν =
4 dec.	0	} 1(1)30(5)60(10)100 {	
3 dec.	{ .01, .02, .05, .1 2(.1).9 }	1(1)30	0(1)4

Peiser, A. M. 1943

<u>Annals</u> <u>of</u> <u>Mathematical</u> <u>Statistics</u> 14, 56-62

Peiser obtains the Cornish-Fisher expansion of percentage points of t through the term in ν^{-1}. Table 2, p. 62, gives exact and approximate values of $t(Q,\nu)$ to 3 dec. for $\begin{cases} Q = .0125,\ .025,\ .05,\ .125,\ .25 \\ \nu = 10,\ 30,\ 60,\ 120 \end{cases}$

4.132 [see also p. 673]
4.139

Goldberg, H. and Levine, Harriet   1946

Annals of Mathematical Statistics 17, 216-225

Cornish & Fisher 1937 have inverted the Edgeworth series to give asymptotic formulas for percentage points in terms of moments.

Table 2, p. 219, gives $t(Q,\nu)$, and Cornish-Fisher approximations through ν^{-1} and ν^{-2}, to 4 dec. for $\begin{cases} p = .0025, .005, .025, .05, .25 \\ \nu = 1, 2, 10, 20, 40, 60, 120 \end{cases}$.

Table 1, p. 218, gives the necessary values of Cornish-Fisher polynomials.

Table 5, p. 2 , gives $P(t,\nu)$, and the approximation thereto obtained from the transformation of Hotelling & Frankel 1938 through ν^{-2}, to 4 dec. for $\begin{cases} t = .1, 1, 3, 5, 6 \\ \nu = 1, 2, 10, 20 \end{cases}$.

Arley, N. and Buch, K. R.   1950

Probability and statistics

(for definitions, see abstract of Arley 1940, above)

Table 4, p. 217, gives $r(Q,\nu)$ to 3 dec. for $\begin{cases} 2Q = .01, .02, .05, 1 \\ \nu = 1(1)30(5)50(10)100 \\ 120/\nu = 0(1)4 \end{cases}$.

4.133   Confidence intervals based on percentage points of t

Pearson, E. S. and Sukhatme, A. V.   1935

Sankhyā 2, 13-32

"An illustration of the use of fiducial limits in determining the characteristics of a sampled batch"

Draw a batch of N observations with batch mean $\overline{X}$ from a normal population; draw a sample of n observations x_i, with sample mean $\bar{x}$, from the batch. Set

$$s^2 = \sum (x_i - \bar{x})^2 / n-1;$$ the statement

$$\bar{x} - \frac{t(Q,n-1)}{\sqrt{n}} s \sqrt{(1 - \frac{n}{N})} \leq \overline{X} \leq \bar{x} + \frac{t(Q,n-1)}{\sqrt{n}} s \sqrt{(1 - \frac{n}{N})} \quad (1)$$

will be correct in $100(1 - 2Q)\%$ of batches.

To facilitate the computation of the confidence limits for $\overline{X}$ given by (1), Table 1, p. 17, gives $n^{-\frac{1}{2}} t(Q, n-1)$ to 4 dec. for $\begin{cases} 2Q = .02, .1 \\ n = 2(1)31 \end{cases}$.

Pearson, E. S. 1935 c

The Application of Statistical Methods to Industrial Standardization and Quality Control

British Standards Institution, No. 600.

Confidence interval for population mean ξ in a normal parent based on t-distribution:

$$\bar{x} - t(p,\nu)s'/\sqrt{\nu} \leq \xi \leq \bar{x} + t(p,\nu)s'/\sqrt{\nu}$$

where $\bar{x}$ is the mean of a sample of size n, $\nu = n-1$, $1-2p$ the confidence coefficient, $t(p,\nu)$ the % point of t (double tail area $2p$ and ν d.f.), and $s' = n^{-1}\sqrt{\sum(x-\bar{x})^2}$. Table 11 p. 69, gives 3 decimal values of $t(p,\nu)/\sqrt{\nu}$ for $2p = 0.1, .02$ and $n = \nu + 1 = 5(1)30$.

4.139 Miscellaneous significance tests related to t

Treloar, A. E. and Wilder, Marian A. 1934

Annals of Mathematical Statistics 5, 324-341

Define

$$\Delta f = \pi^{-\frac{1}{2}} n^{\frac{1}{2}} [\Gamma(\tfrac{1}{2}n - \tfrac{1}{2})]^{-1} \int_B^\infty e^{-\frac{1}{2}nx^2} \int_0^{\frac{1}{2}nx^2 b^{-2}} v^{\frac{1}{2}(n-3)} e^{-v} dv\, dx .$$

Table I (p. 334) gives Δf to 4 decimals (col. (4)) for $B = 1.96/\sqrt{n}$ and $b = (n-1)^{-\frac{1}{2}}t(.025,n-1)$ for $n = 3(2)21, 25, 99$. Also shown are B to 4 decs (col. (2)), b to 3 decs (col. (3)), $.025 - \Delta f$ to 4 decs (col. (5)), $100\Delta f/.025$ to 1 dec. (col. (6)) and $(2.5 - 100\Delta f)/.025$ to 1 dec. (col. (7)). The object is to investigate in relation to the size of sample the proportion of samples for which the normal deviate test is significant but the t test is not, using .05 as the significance level in both cases.

Vianelli, S. 1959

Prontuari per calcoli statistici

Prontuario IC, p. 268, gives $[1+n^{-1}]^{\frac{1}{2}}t(Q, n-1)$ to 3 dec. for $1 - 2Q = .5(.05).95, .98, .99, .999$; $n = 2(1)30$; $120/n = 0(1)4$.

4.139
4.1411

Jeffreys, H. 1948

Theory of Probability

Table III (p. 402) gives t^2 as the root of

$$K = \left(\frac{\pi \nu}{2}\right)^{\frac{1}{2}} \left(1 + \frac{t^2}{\nu}\right)^{-\frac{1}{2}\nu + \frac{1}{2}} \quad \text{to 1 dec.}$$

for $K = 10^x$, $x = -2(\frac{1}{2})0$ and

ν (for most x) = 5(1)20, 50, 100, 200, 500, 1000, 2000, 5000, 10000, 20000, 50000, 100000.

Table IV (p. 403) gives t^2 as the root of

$$K = \tfrac{1}{2}\nu^{\frac{1}{2}} \pi t \left(1 + \frac{t^2}{\nu}\right)^{-\frac{1}{2}\nu} \quad \text{to 1 dec.}$$

for most of the above values of K and ν.

The formulas used for K above are large-sample approximations; Tables IIIA and IVA give t^2, computed from exact small-sample formulas, for $\nu = 1(1)9$ and $1(1)8$ respectively.

4.14 <u>The non-central t-distribution and the power of the t-test;</u>
<u>the confluent hypergeometric function;</u>
<u>the distribution of t from non-normal parents</u>

4.141 The non-central t-distribution; the power of the t-test

Let x be a normal variate with mean μ and variance σ_x^2; denote by S_x^2 an independent estimate of σ_x^2 such that $\nu S_x^2/\sigma_x^2$ follows the χ^2 distribution for ν degrees of freedom. Then the 'non-central t-ratio' $t' = x/S_x$ has the non-central t-distribution depending only on the degrees of freedom and the 'non-centrality' $\delta = \mu/\sigma_x$. The probability density function of t' is given by

$$f(t'; \nu, \delta) = c \exp\left(-\tfrac{1}{2} \frac{\nu \delta^2}{\nu + t'^2}\right) \left(\frac{\nu}{\nu + t'^2}\right)^{\frac{1}{2}(\nu+1)} Hh_\nu\left(-\frac{t'\delta}{\sqrt{\nu + t'^2}}\right)$$

where $c = \nu! \Big/ 2^{\frac{1}{2}(\nu-1)} \Gamma(\tfrac{1}{2}\nu) \sqrt{\pi \nu}$

and the Hh functions were defined in section 1.33.

Clearly when $\mu = 0 = \delta$ the distribution reduces to the t-distribution defined by (4.11.1): $f(t;\nu,0) = f(t,\nu)$. For even degrees of freedom the non-central t-integral can be expressed as a finite sum:

$$\int_0^{t_o} f(t';\nu,\delta)\, dt' =$$

$$(2\pi)^{-\frac{1}{2}} \exp\left[-\tfrac{1}{2}\frac{\delta^2}{1+t_o^2/\nu}\right] \sum_{i=0}^{\frac{1}{2}(\nu-2)} \frac{(2i)!}{(1+t_o^2/\nu)^i 2^i i!}\, Hh_{2i}\left(\frac{-\delta t_o}{\sqrt{\nu+t_o^2}}\right);$$

whence it follows that

$$\Pr\left[\,|t| \le t_o\,|\,\nu,\delta\right] = \int_{-t_o}^{+t_o} f(t';\nu,\delta)\, dt' =$$

$$\sum_{i=0}^{\frac{1}{2}\nu-1} \frac{\Gamma(\tfrac{1}{2}\nu+\tfrac{1}{2})\,\exp[-\tfrac{1}{2}\delta^2]}{\Gamma(\tfrac{1}{2}\nu - i)\,\Gamma(\tfrac{3}{2}+i)}\, R^{i+\tfrac{1}{2}}(1-R)^{\tfrac{1}{2}\nu-i-1}\, M(\tfrac{1}{2}\nu+\tfrac{1}{2},\tfrac{3}{2}+i,\tfrac{1}{2}R\delta^2),$$

where $R = t_o^2/(\nu+t_o^2)$ and M denotes the confluent hypergeometric function. Johnson & Welch 1940, p. 387, report dissatisfaction with these series for numerical work.

4.1411 Direct tables of the non-central t-distribution

Resnikoff, G. J. and Lieberman, G. J. 1957

<u>Tables of the non-central t-distribution</u>

Table [1], pp. 33-175, gives $f(z^{\frac{1}{2}}; \nu; K[\nu+1]^{\frac{1}{2}})$ to 4 dec. for $\nu = 2(1)24(5)49$ and K = 0.674490, 1.036433, 1.281552, 1.514102, 1.750686, 1.959964, 2.326348, 2.652070, 2.807034, 3.090232. z takes all multiples of .05 that make f greater than .00005.

Table [2], pp. 179-380, gives $\int_{-\infty}^{z\nu^{\frac{1}{2}}} f(t'; \nu; K[\nu+1]^{\frac{1}{2}})\, dt'$ to 4 dec. for the above ν and K. For $\nu = 2$, z takes all multiples of .05 such that $.00045 < f < .99955$; for $\nu > 2$, z takes all multiples of .05 such that $.00005 < f < .99995$.

4.1412 The power of the t-test: the probability integral of non-central t evaluated at percentage points of central t

Neyman, J., Iwaszkiewicz, K. and Kołodziejczyk, St. 1935

Supplement to the Journal of the Royal Statistical Society 2, 107-180

Let $\bar{x}(A)$ and $\bar{x}(B)$ denote two means to be tested by $t = [\bar{x}(A) - \bar{x}(B)]/s_d$ at the one-sided 100α % for t, where s_d is an estimate of the standard deviation of the difference σ_d based on ν degrees of freedom. The power of this test to detect a difference Δ between the population means $\xi(A)$ and $\xi(B)$ (hypothesis H_1) is given by the complement to the second kind error

$$P(t \leq t(\alpha,\nu) | \nu, \delta) = 2^{-\frac{1}{2}\nu+1} \Gamma^{-1}(\frac{\nu}{2}) z_0^{-\nu} \int_0^\infty u^{\nu-1} e^{-\frac{1}{2}\frac{u^2}{z_0^2}} (P(u-\delta)) du$$

where $\delta = \Delta/\sigma_d$.

$$P(x) = (2\pi)^{-\frac{1}{2}} \int_{-\infty}^x e^{-\frac{1}{2}t^2} dt.$$

$$z_0 = t(\alpha,\nu)/\sqrt{\nu}.$$

Tables III and IV (pp. 131-132) give $P_{II}(\nu)$ to 3 decimals for ν = 1(1)30, ∞, δ = 0(1)10; α = .05., .01; also the roots of $P_{II}(\delta) = \alpha$ to 2 decimals.

Craig, C. C. 1941

Annals of Mathematical Statistics 12, 224-228

Craig derives the expansion

$$\Pr[t^2 < t_0^2 | \nu, \delta] = e^{-\frac{1}{2}\delta^2} \sum_{i=0}^\infty (i!)^{-1} (\tfrac{1}{2}\delta^2)^i I_{\frac{t_0^2}{\nu+t_0^2}}(i+\tfrac{1}{2}, \tfrac{1}{2}).$$

The table on p. 228 gives $\Pr[t^2 < t_0^2(\tfrac{1}{2}\alpha,\nu) | \nu, \delta]$ to 4 dec. for ν = 18, 38; α = .05, .01; δ = .1, .2, .5.

Lord, E. 1950

Biometrika 37, 64

Power of the t-test

Consider the two alternative hypotheses

$$H_0: x \text{ is drawn from } N(0, \sigma)$$

$$H_1: x \text{ is drawn from } N(\delta\sigma, \sigma)$$

and the usual t criterion

$$t = x/s$$

where s is a mean square estimate of σ based on ν degrees of freedom. Then the power of the t test is given by $\beta(\delta) = \beta'(\delta) + \beta''(\delta)$ where

$$\beta'(\delta) = 1 - \underline{p}(t \leq t(\tfrac{1}{2}\alpha, \nu) | \nu, \delta) = \int_0^\infty Z(x-\delta)\left[1 - p\left(\frac{\nu x^2}{t(\tfrac{1}{2}\alpha, \nu)}, \nu\right)\right] dx$$

and where $p(x^2, \nu)$ is the tail area of χ^2 for ν degrees of freedom, $t(\tfrac{1}{2}\alpha, \nu)$ is the double tail $100\alpha\%$ point of t for ν degrees of freedom and β'' is formed by replacing $-\delta$ by δ in $\beta'(\delta)$.

Table 1 (p.67) gives $\beta'(\delta)$ and $\beta''(\delta)$ to 5 decimals for $\delta = 1(1)10$; $\nu = 1(1)20, \infty$.

Table 2 (p.68) gives the "standardized errors" of the t-test, viz. the roots δ_α of

$$\beta(\delta) = 1 - \alpha \text{ to 3 decimals for } \alpha = .05 \text{ and to 2 decimals}$$

for $\alpha = .01$ for $\nu = 1(1) 20, 30, 60, 120, \infty$.

Pearson, E. S. and Hartley, H. O. 1954

Table 10, p. 135, reprinted from E.S. Pearson & Hartley 1951, gives

$$\Pr\left[|t| \leq t(p, \nu) | \nu, \delta\right] \text{ plotted}$$

against non-centrality $\phi = \delta/\sqrt{2}$

for degrees of freedom $\nu = 6(1) 10, 12, 15, 20, 30, 60, \infty$

and

$$2p = \alpha = .05 \text{ for } 1 \leq \phi \leq 3.5$$

$$2p = \alpha = .01 \text{ for } 2 \leq \phi \leq 5.$$

4.1413 The power of the t-test: miscellaneous tabulations

Neyman, J. and Tokarska, B. 1936

<u>Journal of the American Statistical Association</u> 31 318-326

"Errors of the second kind in testing 'Student's' hypothesis"

Tables 1 and 2, p. 322, give δ as the root of

$$\beta = \Pr\left[\, t' \leq t(Q, \nu) \,|\, \nu, \delta \,\right]$$

for prescribed errors of the 2nd kind β, degrees of freedom, ν, of the t statistic and levels of significance, Q, used for the single tail t-test. δ is given to 2 decimals for $\beta = .01, .05, 1(.1).9$; $\nu = 1(1)\,30, \infty$ and Q = .01, .05.

Harris, Marilyn, Horwitz, D. G. and Mood, A. M. 1948

<u>Journal of the American Statistical Association</u> 43, 391-402

"On the determination of sample sizes in designing experiments"

Suppose that there is available from a preliminary sample an estimate s_1^2, based on m_1 degrees of freedom, of the population variance. It is desired to test the hypothesis that the mean $\mu = \mu_0$ against the alternative $\mu = \mu_0 + a$. Tables 1 and 2, pp. 394-395, determine the degrees of freedom, m_2, required in a second sample to attain prescribed levels of significance and power. For a one-tailed test, calculate a/s_1 and read m_2 from the table, which gives a/s_1 to 3 fig. for
$$\begin{cases} m_1 = 1(1)6,\ 8(4)16(8)32,\ \infty \\ m_2 = 1(1)10(2)20(5)30(10)60(20)100 \\ \beta = .05 \\ \alpha = .8,\ .95 \end{cases}$$

m_2, which must be sought by inverse interpolation, is roughly proportional to $(s_1/a)^2$.

Reprinted in Dixon & Massey 1951, p. 344.

Cochran, W. G. and Cox, Gertrude M. 1950

<u>Experimental Designs</u>

Given an estimate S of the true population σ based on ν d.f., a true difference δ between two population means and given that this difference is to be detected by a two sample t-test at the $100\alpha\%$ level of significance with a probability of β of reaching a significant result, then the

4.1413

minimum sample size ν (for each of the two means) is taken as the smallest integer larger than $2(\frac{\sigma}{\delta})^2 \left[t(\frac{\alpha}{2},\nu) + t((1-\beta),\nu) \right]^2$ (1) and is tabulated (on p. 21) for $\alpha = .05$, $\beta = .8$, .9 and $\alpha = .01$, $\beta = .95$, $\delta = 5(5)30$, $\sigma = 2(1)12(2)20$; $\nu = 3(r-1)$, the formula for ν being that for the error d.f. in a randomized block experiment of 4 treatments in r blocks. Formula (1) is based on the approximate assumption that the experiment happens to yield $S = \sigma$. Corresponding values for the single tail t-test are given on p. 20 (i.e. $t(\alpha/2,\nu)$ in (1) is replaced by $t(\alpha,\nu)$.

The sample size, r, required for obtaining a confidence interval of prescribed length, 2L, for the difference of two population means, is given in T. 2.2, p. 24, for $\begin{cases} L = 3(1)6(2)10 \\ \sigma = 2(1)12(2)20 \\ \alpha = .05, .1, .2 \end{cases}$ according to the formula

$$r \geq 2(1 - \frac{1}{4\nu})^2 \frac{\sigma^2}{L^2} t^2(\frac{\alpha}{2},\nu) \quad \text{(where } \nu = 3(r-1)\text{)} \quad {}^{\textit{f}}$$

Walsh, J. E. 1952

Journal of the American Statistical Association 47, 191-201

See abstract on p. **688**.

Dixon, W. J. and Massey, F. J. 1957

Introduction to statistical analysis

Table 12 b, p. 421, gives values of $d = [\nu+1]^{\frac{1}{2}}\delta$ for a single sample such that a one-sided test at level of significance α has power $1-\beta$, to 2 dec. for $\alpha = .005, .0125, .025, .05$; $1-\beta = .1(.1).9, .95, .975, .99, .995$; $\nu = 4(1)10, 12, 16, 24, 36, \infty$.

Table 12 c, p. 425, gives sample sizes needed to attain power $1-\beta$ in one-sided one- and two-sample tests for $\alpha = .05, .025, .0125, .005$; $1-\beta = .5(.1).9, .95, .99$; $\delta = .1, .2(.2)2, 3$.

${}^{\textit{f}}$ The t-test is assumed to be based on an error mean square from a randomized block experiment with 4 treatments in r blocks.

4.1414
4.1419

4.1414 Percentage points of non-central t

Johnson, N. L. and Welch, B. L.   1940

Biometrika 31, 362

Non-central t-distribution, percentage points and non-centrality as function of probability. For given δ, ν, Q the $100\,Q\%$ points of non-central t' are the roots $t(\nu, \delta, Q)$ of

$$\Pr\left[t' \geq t(\nu, \delta, Q) \mid \nu, \delta\right] = Q$$

and can be written in the form

$$t(\nu, \delta, Q) = \frac{\delta + \lambda\,(1 + \frac{\delta^2}{2\nu} - \frac{\lambda_Q^2}{2\nu})^{\frac{1}{2}}}{(1 - \lambda_Q^2/2\nu)} \qquad (1)$$

where $\lambda_Q \sim x(1-Q)$ but is really a function of ν, δ, Q defined by making (1) exact.

Similarly for given t_o, ν, Q, the root $\delta(\nu, t_o, Q)$ of

$$\Pr[t \geq t_o \mid \nu, \delta(\nu, t_o, Q)] = Q$$

can be written in the form

$$\delta(\nu, t_o, Q) = t_o - \lambda(\nu, t_o, Q)\,(1 + \frac{t_o^2}{2\nu})^{\frac{1}{2}} \qquad (2)$$

where again $\lambda(\nu, t_o, Q) \sim x(1-Q)$.

Table IV (pp. 377-385) gives 3 decimal values of $\lambda(\nu, t_o, Q)$ for $Q = .005, .01, .025, .05, .1(.1).5$; $\nu = 4(1)9, 16, 36, 144, \infty$ and $t_o = \pm(\sqrt{2\nu}\ \sqrt{1-y^2})/y$ with $y' = 0(.1).8$ or $\pm\sqrt{2\nu}\ y'/\sqrt{1-y'^2}$, $y' = 0(.1).6$. For iteration method to obtain the % point $t(\nu, \delta, Q)$ see pp. 372-3, but Table V gives the $\lambda_{.05} = \lambda(\nu, \delta, .05)$ of (1) straight (to 4 dec.) for $\nu = 4(1)9, 16, 36, 144, \infty$ and $n' = -1.0\,(.1)\,1.0$ where

$$n' = \frac{\delta}{\sqrt{2\nu}}\,(1 + \frac{\delta^2}{2\nu})^{-\frac{1}{2}}.$$

Table IV is reprinted in Vianelli 1959, Prontuario XCVIII, pp. 263-267.

Patnaik, P. B. 1955

Sankhyā 15, 343-372

Review by Ingram Olkin (MTAC 12, 67):

The non-central t' distribution is given by

$$p(t', \nu, \delta) = C(\nu)\left(\frac{\nu}{\nu + t'^2}\right)^{\nu+1/2} Hh_\nu\left(-\frac{t'\delta}{(\nu + t'^2)^{1/2}}\right)$$

where $Hh_\nu(x) = \int_0^\infty \frac{u^\nu}{\nu!} \exp \frac{1}{2}(u + x)^2 du$. The solution of $P(t' \leq x) = .05$, $P(t' \geq x) = .05$ is tabulated to 2D for $\nu = 4, 5$, $\delta = 0(.5)4$, and to 3D for $\nu = 6(1)15, 20$, $\delta = 0(.5)4$. The tabulation is based on the tables by Johnson and Welch [1]. An approximation for the upper and lower percentage points is given in terms of the central t-distribution. A number of statistical hypotheses for which the t' statistic can be used are discussed.

Resnikoff, G. J. and Lieberman, G. J.

Tables of the non-central t-distribution

Table [3], pp. 383-389, gives $\nu^{-\frac{1}{2}} t(\nu, K[\nu+1]^{\frac{1}{2}}, Q)$ to 3 dec. for the K stated in the abstract in section 4.1411 and for

$$\begin{cases} Q = .005, .01 & .05, .1, .25(.25).75, .9, .95, .99, .995 \\ \nu = 5(1)24(5)49 & 2(1)24(5)49 \end{cases}$$

4.1419 Miscellaneous applications of non-central t

Pearson, E. S. 1935 c

The application of statistical methods to industrial standardization and quality control

Confidence interval for the coefficient of variation σ/ξ in a normal parent based on the non-central t-distribution. The confidence interval is put in the form

$$\left(c_1\left[1 + \left(\frac{\bar{x}}{s'}\right)^2\right] - 1\right)^{-\frac{1}{2}} \leq \sigma/\xi \leq \left(c_2\left[1 + \left(\frac{\bar{x}}{s'}\right)^2\right] - 1\right)^{-\frac{1}{2}}$$

---

[1] Johnson & Welch 1940

4.1419
4.1422

where $\bar{x}$ is the mean of a sample of size n,

$$s' = n^{-\frac{1}{2}} \sqrt{\sum (x-\bar{x})^2}$$, and c_1 and c_2 depend on the sample size n and are tabulated in Table II (p. 69) to 3 decimals for confidence coefficients .90, .98 and n = 5(1)30 .

Harley, B. I. 1957
<u>Biometrika</u> 44, 219-224

Harley shows, by comparison of the moments of $r(1-r^2)^{-\frac{1}{2}}$ with the moments of non-central t , that $v = (n-2)^{\frac{1}{2}}(2-2\rho^2)^{\frac{1}{2}}(2-\rho^2)^{-\frac{1}{2}}r(1-r^2)^{-\frac{1}{2}}$ is approximately distributed as non-central t with n-2 d.f. and non-centrality $\delta = (2n-3)^{\frac{1}{2}}\rho(2-\rho^2)^{-\frac{1}{2}}$; and hence proposes that the tables of F. N. David 1938 be used to yield approximate values of non-central t .

Table 1, p. 222, compares the first 4 moments of v and t to 3-4 dec. for n = 15, 27 ; ρ = .2, .6 . Table 2, p. 223, gives the upper 40%, 20% and 5% points of t from the tables of N. L. Johnson & Welch 1940 to 3 dec., and the upper 1% points to 2 dec., together with approximate values obtained by transforming David's r s to v s : all for

$\begin{cases} \nu = 7 & 16 \\ \delta = .553, 1.815, 2.657, 3.195 & .821, 2.691, 3.941, 4.739 . \end{cases}$

Table 2 is reprinted in Vianelli 1959, <u>Prontuario</u> 97, p. 263.

See also Marakathavalli 1954.

4.142   The confluent hypergeometric function

The function

$$M(\alpha, \gamma, x) = \sum_{j=0}^{\infty} \frac{\Gamma(\gamma)\Gamma(\alpha+j) x^j}{\Gamma(\alpha)\Gamma(\gamma+j) j!}$$

occurs frequently in statistical contexts. It yields as special cases the Hh functions, the polynomials of Hermite and Laguerre, and the incomplete gamma-function. The confluent hypergeometric function occurs in the moments of non-central χ, the distribution of non-central t for even degrees of freedom (see section 4.141), the construction of sequential t-tests, and the characteristic functions of Pearson frequency curves.

4.1419
4.1422

The tables in sections 4.1421 and 4.1422 give the function M directly, i.e. not disguised as an Hermite polynomial or the like. No differences are provided: interpolation in x is laborious, and interpolation in α and γ seems to have contemplated only by Slater.

<u>4.1422</u> $M(\alpha,\gamma,x)$ for integer and half-integer α and γ

The tables in this section are arranged in ascending order of γ.
The B.A. tables give 5 dec.; N.B.S., Nath and Rushton & Lang give 7 fig.

$\gamma =$	$\alpha =$	$x =$	
$-1\frac{1}{2}$	$-4(\frac{1}{2})+4$	$0(.1)1(.2)3(.5)8$	B.A. 1926(279)
		$0(.02).08, .15(.1).95, 1.1(.2)1.9$	B.A. 1927(221)
$-\frac{1}{2}$	$-4(\frac{1}{2})+4$	$0(.1)1(.2)3(.5)8$	B.A. 1926(279)
		$0(.02).08, .15(.1).95, 1.1(.2)1.9$	B.A. 1927(221)
$+\frac{1}{2}$	$1\frac{1}{2}(1)100\frac{1}{2}$	$0(.01)1$	N.B.S. 1949
	$\frac{1}{2}(1)24\frac{1}{2}$	$.02(.02).1(.1)1(1)10(10)50, 100, 200$	Rushton & Lang 1954
	$1(1)25$	$.02(.02).1(.1).5, 2, 5, 8, 10(10)50$	
	$25\frac{1}{2}(1)44\frac{1}{2}$	$10(10)50, 100, 200$	
	$-4(\frac{1}{2})+4$	$0(.1)1(.2)3(.5)8$	B.A. 1926(279)
		$0(.02).08, .15(.1).95, 1.1(.2)1.9$	B.A. 1927(221)
1	$\frac{1}{2}(1)39\frac{1}{2}$	$.02(.02).1(.1)1$	Rushton & Lang 1954
	$1(1)40$	$.02(.02).1(.1)1(1)10(10)50, 100, 200$	
	$41(1)50$	$.02(.02).08$	
	$-4(\frac{1}{2})+4$	$0(.02).1(.05)1(.1)2(.2)3(.5)8$	B.A. 1927(229)
$1\frac{1}{2}$	$\frac{1}{2}(1)39\frac{1}{2}$	$.02(.02).1(.1)1$	Rushton & Lang 1954
	$40\frac{1}{2}(1)44\frac{1}{2}$	$.02(.02).1(.1)1(1)10(10)50, 100, 200$	
	$-4(\frac{1}{2})+4$	$0(.1)1(.2)3(.5)8$	B.A. 1926(279)
		$0(.02).08, .15(.1).95, 1.1(.2)1.9$	B.A. 1927(221)
2	$\frac{1}{2}(1)39\frac{1}{2}$	$.02(.02).1(.1)1$	Rushton & Lang 1954
	$1(1)40$	$.02(.02).1(.1)1(1)10(10)50, 100, 200$	
	$-4(\frac{1}{2})+4$	$0(.02).1(.05)1(.1)2(.2)3(.5)8$	B.A. 1927(229)
$2\frac{1}{2}$	$\frac{1}{2}(1)44\frac{1}{2}$	$.02(.02).1(.1)1(1)10(10)50, 100, 200$	Rushton & Lang 1954
3	$\frac{1}{2}(1)39\frac{1}{2}$	$.02(.02).08$	Rushton & Lang 1954
	$1(1)40$	$.02(.02).1(.1)1(1)10(10)50, 100, 200$	Nath 1951
	$-4(\frac{1}{2})+4$	$0(.02).1(.05)1(.1)2(.2)3(.5)8$	B.A. 1927(229)

4.1422
4.143

$3\frac{1}{2}$	$\frac{1}{2}(1)49\frac{1}{2}$	.02(.02).1(.1)1(1)10(10)50, 100, 200	Rushton & Lang 1954
4	$\begin{cases}1(1)40 \\ -4(\frac{1}{2})+4\end{cases}$	.02(.02).1(.1)1(1)10(10)50, 100, 200 0(.02).1(.05)1(.1)2(.2)3(.5)8	Nath 1951 B.A. 1927(229)
$4\frac{1}{2}$	$\frac{1}{2}(1)49\frac{1}{2}$	.02(.02).1(.1)1(1)10(10)50, 100, 200	Rushton & Lang 1954

<u>4.1423</u> Functions related to $M(\alpha,\frac{1}{2},x)$

N.B.S. 1949 tables $(4\alpha x)^{-\frac{1}{2}}\log_e M(\alpha,\frac{1}{2},x)$ to 6 dec. for

$\begin{cases}\alpha = 1\frac{1}{2}(1)10\frac{1}{2} \qquad\qquad\qquad\qquad 11\frac{1}{2}(1)100\frac{1}{2} \\ x = .1(.01).6(.05)2(.2)7(1)45(5)100 \quad .1(.01).6(.1)2(.2)7(1)45(5)100\end{cases}$, and

$M(\alpha,\frac{1}{2},x)/\cosh[(4\alpha-1)x]^{\frac{1}{2}}$ to 6 dec. for $\begin{cases}\alpha = 21\frac{1}{2}(1)100\frac{1}{2} \\ x = 0(.01).1 \quad.\end{cases}$

<u>4.143</u> The distribution of t in samples from non-normal parents

Rider, P. R. 1929

<u>Biometrika</u> 21, 124-143

"On the distribution of the ratio of mean to standard deviation in small samples from certain non-normal universes"

Rider gives short tables of the distribution of $z = (n-1)^{\frac{1}{2}}t$ to 2-5 dec.:

Parent	Sample sizes	Table	page
Grouped rectangular (5 groups)	4	I	127
" " "	3	II	128
" " "	2	III	129
" " (10 groups)	4 distr.	IV	130
" " "	4 prob.int.	V V extended	131 142-3
" normal (5 groups)	4	V	131
" " "	4	VIII	134
Binomial $(\frac{1}{2}+\frac{1}{2})^4$	"	4	VIII	134
Grouped rectangular "	4	VIII	134
Binomial $(\frac{3}{5}+\frac{2}{5})^4$	"	4	IX	134

Table VI (p. 132) gives the regression of the variance on means, viz $E(s^2|\bar{x})$ for samples of 4 in a grouped rectangular population of 5 groups for $x = \bar{x} - \mu = \pm\ 0(0.25)\ 2.00$.

Rider, P. R., 1931

Annals of Math. Stat. 2, 48

Distribution of t in samples of four from non-normal parents.

The parents considered are discontinuous distributions

Rectangular	R	$f(x) = 1/10$	for $x = 0(1)9$
Triangular	T	$f(x) = x/55$	for $x = 1(1)10$

U-shaped U

x	0	1	2	. . .	7	8	9
36 f(x)	10	5	1	. . .	1	5	10

For samples of four from these populations exact distribution of the statistics

$$z = 2(x - \mu_1')/\sqrt{\sum(x-\bar{x})^2} \quad \text{and} \quad \bar{x} - \mu_1',$$

were computed. Table II (p. 55) gives the grouped probabilities of z to 4 dec for R, T and U, and a normal parent, for z-group end points $-4.25(.5)+4.25$. Table III (p. 56) gives $\Pr[z \leq z_0]$ to 5 decs for $z_0 = -4.0$, $-3.5, -3.0(.1)+3.0, 3.5, 4.0$ for parents T and U.

Table IV (p. 57) gives the double tail $\Pr[|z| \geq z_0]$ to 4 dec for $z_0 = 0(.1)3.0$ for parents T and the normal.

Table V (p. 58) gives the exact probabilities of $\bar{x} - \mu_1'$ from parent T to 5 decimals for $\bar{x} - \mu_1' = -5.25(.25)3.00$, and Table VI (p. 59) the same for parent U to 4 decimals for $x - \mu_1' = -4.50(.25)4.50$ (also shown are the exact frequencies). Each sample from R may be classified according to its value of

$$p = \int_{\bar{x}-3s'}^{\bar{x}+3s'} f(x)\,dx$$

where $s' = n^{-\frac{1}{2}}\sqrt{\sum(x-x)^2}$. Table VII (p. 60) gives this classification for the 10,000 possible samples from R for $p = 0.1(0.1)1.0$. Similarly Table VIII (p. 61) gives this classification of the 9,150,625 possible samples from T for $p = i/55$; $i = 1(1)55$. Also given in Table VIII are the probabilities and cumulative probabilities for these classifications.

Perlo 1933 gives explicit formulas, and a short table, for the distribution of t in samples of 3 from a rectangular parent.

4.143
4.15

Geary, R. C. 1936

Royal Stat. So. Jn. Suppl. 3, 178

Effect on t-distribution of small deviations from normality in parent.

As an approximate distribution of t the expression

$$P(t \leq t_0) = 1 - p(t_0, \nu) + \lambda_3 \int_{-\infty}^{t_0} \emptyset(t) dt$$

is derived where

$$\emptyset(t) = [3\nu t - (2\nu+1)t^3]/6\nu(\nu+1)^{\frac{1}{2}}(2\pi)^{\frac{1}{2}}(1+t^2/\nu)^{2+\frac{1}{2}\nu},$$

$$\lambda_3 = k_3 (k_2)^{-3/2}$$

The table on p. 183 gives $\int_{-\infty}^{t_0} \emptyset(t) dt$ to 4 dec. for $\nu = 1, 4, 9, 19, 29$

$t_0 = 0(.5)4.0$

Geary, R. C. 1947 b

See abstract in section 4.27..

Gayen, A. K. 1949

Biometrika 36,353

The t-distribution for parental distribution represented by the Edgeworth series.

Consider a sample $x_1, \ldots, x_n$ drawn from a parental distribution of the form

$$f(x) = Z(x) \left[1 + \frac{\lambda_3}{3!} H_3(x) + \frac{\lambda_4}{4!} H_4(x) + \frac{10}{6!} \lambda_3^2 H_6(x) \right]$$

where the $H_\nu(x)$ are Hermitian polynomials. The probability integral of $t = \bar{x}\sqrt{n}/s$ is derived in the form

$$\Pr[t \geq t_0] = P_0(t_0) + \lambda_3 P_{\lambda_3}(t_0) - \lambda_4 P_{\lambda_4}(t_0) \lambda_3^2 P_{\lambda_3^2}(t_0) \quad (1)$$

where the $P_0(t_0), \ldots, P_{\lambda_3^2}(t_0)$ involve incomplete Beta ratios.

Table 1 (p. 361) gives these functions to 4 dec. for $t_o = 0(.5)4$; $n' = n-1 = 1(1)6$, $120/n' = 0(5)20$. Reprinted in Vianelli 1959, Prontuario 63, p. 201.

Table 3 (p. 366) gives 4 dec. values of $\Pr[|t| \geq t_o]$ computed from a two-tailed analogue of (1), with the term $-\lambda_3 P_{\lambda_3}(t_o)$ added: for $\lambda_3^2 = 0, .2, .25, .5(.5)2, 4$; $\lambda_4 = -2(.5)+2(1)6$; $t_o = t(.02, 4)$. Table 4 (p. 366) does the same for $n = 6$; $t_o = t(.015, 5)$.

Uranisi, H. 1950
Bull. Math. Stat. 4,1
Distribution of t in samples from a Gram-Charlier A distribution.
The author obtains the initial terms in the expansion of the distribution of t for a sample $x_1 \ldots x_n$ drawn from a Gram-Charlier A distribution. The coefficients in this expansion for t are given to 6 dec. for $n = 5, 10, 15, 21$ and for $u = (1+t^2/\nu)^{-1} = .05, .1(.1) .9, .95, 1$. With the help of these tables it is possible to compute t-tail errors for given moment constants β_1 β_2 β_3 β_4 of the parental distribution.

See also Gayen 1950 z.

4.15 Two-sample tests for unequal variances

This section deals with tables concerning a problem which has become known as the 'Two sample problem': Given two means $\bar{x}_1$ and $\bar{x}_2$ of two random samples x_{1i} and x_{2j} of size n_1 and n_2 drawn, respectively, from two normal populations with means μ_1 and μ_2 and variances σ_1^2 and σ_2^2. Required a criterion for testing whether $\mu_1 = \mu_2$ or $\mu_1 \neq \mu_2$. When it is known that $\sigma_1^2 = \sigma_2^2$ the familiar t-test is the appropriate criterion but when no such assumption can be made various solutions to the problem have been suggested and have given rise to heated controversy. The three main solutions which

4.15
4.151

have resulted in the preparation of tables are

1. The Fisher–Behrens solution based on a fiducial probability argument.

2. The series-solution by Welch involving the ratio s_1^2/s_2^2 of the two sample variances as an ancilliary statistic.

   This work supersedes earlier work by Welch more concerned with examining the distribution theory of criteria depending on σ_1^2/σ_2^2.

3. Scheffé's solution in which, with $n_1 \leq n_2$, the variance of $\bar{x}_1 - \bar{x}_2$ is directly estimated from the sample variance of the quantities
   $w_i = x_{1i} - (\frac{n_1}{n_2})^{\frac{1}{2}} x_{2i}$ ($i = 1, 2, \ldots, n_1$) and $n_2 - n_1$ values of the x_{2j} are sacrificed in variance estimation.

4.151 The Fisher-Behrens test

We begin this section with an abstract of R.A. Fisher 1941, including a definition of the distribution (in the sense of ordinary, not fiducial, distribution theory) underlying the tables of the Fisher-Behrens statistic d; abstracts of tables of d follow in chronological order.

Fisher, R. A. 1941

<u>Annals of Eugenics</u> 11, 141-172

"The asymptotic approach to Behrens' integral, with further tables for the test of significance"

Given two independent t-variates, t_1 and t_2, with $\nu_1 (= n_1 - 1)$ and $\nu_2 (= n_2 - 1)$ degrees of freedom, and three constants s_1, s_2, and d, required to find the total frequency $Q_{12}(d)$ for which

$$s_1 t_1 - s_2 t_2 \geq d\sqrt{s_1^2 + s_2^2} . \qquad (1)$$

Expanding the t-distribution, $f(t, \nu)$ say, in terms of Hermite functions,

$$f(t, \nu) = (2\pi)^{-\frac{1}{2}} e^{-\frac{1}{2}t^2} \sum_0^\infty \Pi_r(t) \nu^{-r}, \qquad (2)$$

where the Π_r are polynomials in t, given on p. 150. Hence Fisher obtains the distribution of $s_1 t_1 - s_2 t_2$:

$$Q_{12}(d) = Q(d) + dZ(d) \left\{ \sum_{i,j=0}^\infty S_{ij}(d, s, o) \nu_1^{-i} \nu_2^{-j} \right\} \qquad (3)$$

where $c = s_2/\sqrt{s_1^2 + s_2^2}$, $s = s_1/\sqrt{s_1^2 + s_2^2}$, S_{ij} (d, s, c) is an even polynomial in d, s and c, and Q and Z are the normal tail area and ordinate defined on p. 1 of this Guide. Table 1, pp. 153-154, gives the coefficients of S_{ij}, for $i + j \leq 4$, as exact vulgar fractions.

Following the methods of Cornish & Fisher 1937, Fisher reverts the series (3), yielding the percentage points $d(Q_{12})$:

$$d(Q_{12}) = x\left[1 + \sum_{i,j=0}^{\infty} T_{ij}(x,s,c) \nu_1^{-i} \nu_2^{-j}\right] \quad (4)$$

where $Q(x) = Q_{12}$. Table 2, pp. 155-156, gives the coefficients of the (even) polynomials T_{ij}, for $i + j \leq 4$, as exact vulgar fractions.

Fisher converts (4) into a series of cosines of multiples of $\theta = \tan^{-1}(s_1/s_2)$, with coefficients depending on powers of

$$\sigma = \frac{1}{\nu_1} + \frac{1}{\nu_2} \quad \text{and} \quad \delta = \frac{1}{\nu_2} - \frac{1}{\nu_1} \;.$$

Table 3, pp. 162-163, gives the coefficients of $\sigma^i \delta^j \cos k\theta$ to 7 dec. for $i + j \leq 4$ and $2Q_{12} = .002, .005, .01, .02, .05, .1$. Table 4, pp. 164-165, which is intelligible only by reference to T. 3, evaluates the cosines and so gives the coefficients of $\sigma^i \delta^j$ to 5 dec. for $\theta = 0(15°)90°$, and for i, j and $2Q_{12}$ as in T. 3.

For $\nu_1 = \infty$, Fisher writes (4) in the form

$$d(Q_{12}) = x\left[1 + \sum_{j=1}^{\infty} \nu_2^{-j} C_j(\theta, Q_{12})\right]: \quad (5)$$

Table 5, pp. 168-169, gives C_j to 4-6 dec. for $j = 1(1)5$; $\theta = 0(5°)85°$; $2Q_{12}$ as in T. 3. Table 6, pp. 170-171, gives $d(Q_{12})$ to 3 dec. for

$$\begin{cases} \theta = 0(10°)90° \\ 120/\nu_2 = 0(2)12 \\ 2Q_{12} = .002, .005, .01, .02, .05, .1 \;. \end{cases}$$

Table 6 is reprinted in Fisher & Yates 1943, 1948, 1953, T. V2; 1957, T. VI 2.

4.151

Behrens, W. U. 1929

<u>Landwirtschaftliche Jahrbücher</u> 68, 807-837

"Ein Beitrag zur Fehlerberechnung bei wenigen Beobachtungen"

Table 9, p. 822, gives $1 - Q_{12}(d)$ to 3 dec. for

$$\begin{cases} n_1 = n_2 = 2 & 2 & 3 & 3 & 4(1)6, \infty \\ s_1/s_2 = 0, 1 & \tfrac{1}{2} & 0, 1 & \tfrac{1}{2} & 0, 1 \\ d = 1(1)5, 10, 50 & 1(1)4 & 1(1)5, 10 & 1(1)3, 10 & 1(1)5 \end{cases}$$

$$\begin{cases} 4 & 7(1)10 \\ \tfrac{1}{2} & 0 \\ 1, 2, 4, 5 & 1(1)5 \end{cases}.$$

Table 16, p. 835, gives $1 - Q_{12}(d)$ to 2-3 dec. for $s_1 = s_2$ and

$$\begin{cases} d = 0(.2)3 & 4 & 5 \\ n_1 = n_2 = 3(1)6(2)10, \infty & 3(1)6(2)10 & 3, 4 \end{cases}.$$

Sukhatme, P. V. 1938

<u>Sankhyā</u> 4, 39-48

$d(Q_{12})$ is tabled to 3 dec. for $2Q_{12} = .05$

$$\left.\begin{matrix}\nu_1 \\ \nu_2\end{matrix}\right\} = 6, 8, 12, 24, \infty$$

$$\theta = 0° \ (15°) \ 90°$$

Errors in the third place were discovered by Fisher 1941, and definitely assessed by Sukhatme et al. 1951; accordingly, the table in Fisher & Yates 1948 supersedes this table.

Fisher, R. A. and Yates, F. 1943, 1948, 1953

Table V 1 gives $d(Q_{12})$ to 3 dec. for $2Q_{12} = .01, .05$

$$\left.\begin{matrix}\nu_1 \\ \nu_2\end{matrix}\right\} = 6, 8, 12, 24, \infty$$

$$\theta = 0° \ (15°) \ 90°$$

Reprinted as Table VI in Fisher & Yates 1957 and as <u>Prontuario</u> LXIV in Vianelli 1959.

4.151

Chapman, D. G. 1950

<u>Annals</u> <u>of</u> <u>Mathematical</u> <u>Statistics</u> 21, 601-606

Table 1, p. 605, gives $1 - Q_{12}(d)$ to 4 dec. for $\theta = 45°$ and

$\begin{cases} \nu_1 = \nu_2 = 1 & 3 & 5 & 7 \\ d = 0(.5)8(2)12(9)30, 50, 100 & 0(.5)8(2)12(9)30 & 0(.5)7 & 0(.5)6 \end{cases}$

$\begin{cases} 9 & 11 \\ 0(.5)5.5 & 0(.5)4.5 \end{cases}$, Table 2, p. 605, gives corresponding two-tailed 1%
and 5% points of d to 2 dec. For $\nu_1 = \nu_2 = 11$ a normal approximation is also given.

Sukhatme, P. V.; Thawani, V. D.; Pendharkar, V. G. ; Natu, N. P. 1951

<u>Indian</u> <u>So</u>. <u>Agr</u>. <u>Stat</u>. <u>Jn</u>. , 3,9

Revision of end figures of Fisher-Behrens criterion d.

The original tables of 1% and 5% values of d computed by Sukhatme and published in the 1943 edition of the Fisher-Yates tables were compared by Fisher with those evaluated from his asymptotic series expansion (<u>Annals</u> <u>of</u> <u>Eugenics</u> 11, 141) and certain systematic discrepancies found. For the 5% points Sukhatme was able to trace the discrepancies as having been caused by his linear interpolation in the t-tables in his original work and his corrected values were used in the 1948 edition of the Fisher-Yates tables , whilst the 1% points in this edition were adjusted by Fisher on the basis of his expansion. The present investigation suggests that for the 1% points certain of the discrepancies are due to the inadequacy of the expansion used by Fisher. Tables III and IV (p. 20) present finally corrected values to 3 decimals which agree with the 5% values of the Fisher-Yates tables. For the 1% values the following discrepancies are noted:

ν_1	ν_2	θ	Sukhatme	Fisher-Yates
24	24	15°	2.784	2.785
6	6	30°	3.558	3.557
8	8	30°	3.241	3.239

4.151
4.152
[see also p. 674]

[The following account, quoted from Fisher & Yates 1957, is not entirely accordant with the above:

"[In] the third edition [1948] . . . a small positive bias was removed from the 1 per cent. values of Table VI for the significance of the difference between two means." (p. vj)

"Sukhatme's values for 1 per cent. have been revised to remove the very small positive bias, about ·001 in d, detected by comparison with the asymptotic calculation used in constructing Table VI2." (p. 3)]

Finney, D. J. 1952 b

Statistical method in biological assay

Table 3, p. 622, gives $d(Q_{12})$ to   dec. for
$$\begin{cases} \nu_1 = 6, 8, 12, 24, \infty \\ \nu_2 = 6, 8, 12, 24, \infty \\ 2Q_{12} = .01, .05 \\ \theta = 0°, 15°, 30°, 45°. \end{cases}$$

Fisher, R. A. and Healy, M. J. R. 1956

Journal of the Royal Statistical Society, Series B 18, 212-216

"New tables of Behrens' test of significance"

When ν_1 and ν_2 are both uneven, $Q_{12}(d)$ can be written as a trigonometric polynomial in α, where $d = (\nu_1^{\frac{1}{2}} \sin\theta + \nu_2^{\frac{1}{2}} \cos\theta) \tan\alpha$. Table 1, p. 215, exhibits the polynomials for
$$\begin{cases} \nu_1 = 1(2)7 \\ \nu_2 = \nu_1(2)[14-\nu_1] \end{cases}.$$

Fisher, R. A. and Yates, F. 1957

Based on the formulas of Fisher & Healy 1956, T. VI1 gives $d(Q_{12})$

to 5 dec. for
$$\begin{cases} 2Q_{12} = .01, .02, .05, .1 \\ \nu_1 = 1(2)7 \\ \nu_2 = \nu_1(2)7 \\ \theta = 0(15°)90° \end{cases}$$

[Fisher & Yates 1957, p. vij, preface this table with the following remarks:

"In view of the logical and mathematical interest of Behrens' test for the difference between two means, we have added a further table (VI1), based on exact formulæ, which gives values of d for a number of significance levels when n_1 and n_2 are both small and odd. We believe the later test published by Pearson and Hartley (Biometrika Tables for Statisticians, Table 11) to be erroneous."]

4.152 Welch's test

Aspin, Alice A. 1948

Biometrika 35, 88-96

Let $\bar{x}_i$, $i = 1, 2$, be two independent means from normal parents with variances σ_i^2; let s_i^2 be estimates of the σ_i^2, independent of each other and of the means, and distributed as $\sigma_i^2 \chi^2 / \nu_i$. Welch 1947 seeks to find a function $h(s_1^2, s_2^2; P)$ such that the direct probability $\Pr[\bar{x}_1 - \bar{x}_2 \leq h] = P$. He derives the functional equation for h

$$\theta\left(\tfrac{1}{2}\left[1 + \alpha\left(\frac{h}{\sqrt{\tfrac{1}{n_1}\sigma_1^2 + \tfrac{1}{n_2}\sigma_2^2}}\right)\right]\right) = P$$

where θ is the operator

$$\exp[-\Sigma \sigma_i^2 \partial_i - \tfrac{1}{2}\Sigma \nu_i \log(1 - 2\sigma_i^2 \partial_i / \nu_i)],$$

∂_i is the partial differential operator $\partial / \partial(\sigma_i^2)$, and σ_i^2 is replaced by s_i^2 after expanding the operator. Aspin develops h in the series

$$h = h_0 + h_1 + h_2 + \ldots$$

where $h_0 = x(P)\sqrt{\tfrac{1}{n_1}s_1^2 + \tfrac{1}{n_2}s_2^2}$

and h_j is of the order $\dfrac{1}{\nu_1^j}$, $\dfrac{1}{\nu_2^j}$ ...

Table 1, p. 93, gives $h_0^{-1}\sum_{i=0}^{j} h_i$ for

$$\nu_1 = \nu_2 = 6, 12, 18 \qquad P = .95$$

$$j = 1(1)4$$

The solutions depend on the quantity

$$G = \tfrac{1}{n_1}s_1^2 \Big/ (\tfrac{1}{n_1}s_1^2 + \tfrac{1}{n_2}s_2^2) \qquad \text{(when } n_i = \nu_i + 1\text{)}.$$

Aspin investigates the range $G = .5(.1)1$; in the case, here treated, where $n_1 = n_2$, the results for $G = 0(.1).4$ follow from symmetry. Table 2, p. 94, gives $\bar{h} = h_0 + h_1 + h_2 + h_3 + h_4$ in the form

$$\bar{h} \Big/ \sqrt{\tfrac{1}{n_1}s_1^2 + \tfrac{1}{n_2}s_2^2}$$

to 2-4 dec. for above ranges of $\nu_1 = \nu_2$ and of G.

[237]

4.152
4.153
Aspin, Alice A. 1949 [see also p. 674]

Biometrika 36, 290-293

Based on the theory outlined above, Aspin tables percentage points of
$$\nu = h/h_0 = \bar{x}_1 - \bar{x}_2 / (\lambda_1 s_1^2 + \lambda_2 s_2^2)$$
(when λ_1 and λ_2 are here to be interpreted as $1/n_1$ and $1/n_2$)

to 2 dec. for
$$\begin{cases} P = .95 & .99 \\ G = 0(.1)1 & 0(.1)1 \\ \nu_1, \nu_2 = 6(2)10 \\ 120/\nu_1, 120/\nu_2 = 0, 6, 8 & 0, 4(2)12 \end{cases}.$$

Reprinted in Pearson & Hartley 1954, T. 11, pp. 136-137.

[In their introduction to this table, Pearson & Hartley 1954, p. 26, state:

"R.A. Fisher (following Behrens) has proposed a test for the equality of the population means . . . and tables have been computed by P. V. Sukhatme (see Fisher & Yates, 1953, Tables V 1 and V 2). This work, however, involves the use of an argument which is not universally accepted by statisticians. One of the consequences of the use of the Fisher-Behrens test is that, if the null hypothesis is true, the probability (in the usual direct sense) of rejecting it will not equal the figure specified as the level of significance. Fisher, indeed, does not consider this to be a drawback. It should be noted, however, that the table given in the present volume is not related to the Fisher-Behrens test in any way. It is the result of an attempt to produce a test which *does* satisfy the condition that the probability of rejection of the hypothesis tested . . . will be equal to a specified figure."]

Gronow, D. C. G. 1951

Biometrika 35, 252-256

Let $\bar{x}_1$ denote the mean of n_1 observations drawn from a normal parent with μ_1, σ_1 and $\sum_1 = \sum_1^{n_1} (x - \bar{x}_1)^2$ and similarly for $\bar{x}_2$ etc. Then the criteria

$$u = \frac{\bar{x}_1 - \bar{x}_2}{\sqrt{\frac{\sum_1 + \sum_2}{n_1 + n_2 - 2} \left(\frac{1}{n_1} + \frac{1}{n_2}\right)}}$$

$$v = \frac{\bar{x}_1 - \bar{x}_2}{\sqrt{\frac{\sum_1}{n_1(n_1 - 1)} + \frac{\sum_2}{n_2(n_2 - 1)}}},$$

introduced by Welch 1938 a, are examined by Gronow. Several short tables give the exact size of the tests obtained by referring u and v to the 5% and 1% points of t with (n_1+n_2-2) degrees of freedom, and the power of such tests.

Gronow concludes (p. 255):

"(i) The significance levels of 'Student's' t may be very inaccurate; if the variances are nearly equal, the bias is less for u, but when they differ widely v is affected less than u.

"(ii) Whether u or v is employed, for a given difference in means of detectable size, the chance of establishing significance is considerably larger . . . if the larger sample is taken from the population with the larger variance."

4.153 Scheffe's test

Scheffé, H. 1943

Annals of Mathematical Statistics 14, 35-44

Scheffé advocates the confidence interval for the mean

$$|(\bar{x}_1-\bar{x}_2) - (\mu_1-\mu_2)| \leq \Sigma^{\frac{1}{2}}[n_1(n_1-1)]^{-\frac{1}{2}} t(\tfrac{1}{2}\alpha, n_1-1) \qquad (1)$$

where $1 - \alpha$ is the confidence coefficient; $t(\frac{\alpha}{2}, n_1-1)$ is the double-tail 100α % point of t; $\Sigma = \sum(u_i - \bar{u})^2$ with $u_i = x_{1i} - (m_1/m_2)^{\frac{1}{2}} x_{2i}$,

$i = 1, 2, \ldots, n_1;$

$$\bar{x} = n_1^{-1} \sum_{i=1}^{n_1} x_{1i} \ ; \ \bar{x}_2 = n_2^{-1} \sum_{j=1}^{n_2} x_{2j}$$

Note that $x_{2,n+1}, \ldots, x_{2,n_2}$ only occur in the formation of $\bar{x}_2$ and that the criterion is dependent on the (random) order of the x_{2j}.

In the case of known σ_1/σ_2 the most efficient (shortest unbiased) confidence interval can be obtained from a t-test. Let R be the ratio of the expected length of the interval (1) to the expected length of such shortest interval. Tables 1 and 2, p. 43, give R to 2 dec. for

$\alpha = .01, .05$ and $\begin{cases} n_1 = 6 & 11 & 21 \\ n_2 = 6, 11, 21, 41, \infty & 11, 21, 41, \infty & 21, 41, \infty \end{cases}$

$\begin{cases} 41 & \infty \\ 41, \infty & \infty \end{cases}$.

4.153
4.17

Walsh, J. E.   1949 b

Annals of Mathematical Statistics 20, 616-618

Walsh finds an approximate formula for the efficiency, r, of the two-sample test of Scheffe 1943:
$$r = (2N)^{-1}\left[2 + AN + \tfrac{1}{2}x^2 + [(2+AN+\tfrac{1}{2}x^2)^2 - 8AN]^{\tfrac{1}{2}}\right]$$
where $N = n_1 + n_2$, $A = 1 - \tfrac{1}{2}x^2/(n_1-1)$, and x is the normal deviate $x(1-\alpha)$.
Table 1, p. 617, gives r to 3 dec. for
$\begin{cases} n_1, n_2 = 4 \quad 8 \quad 6, 10, 15, 20, 30, 50, 100, \infty \\ \quad \alpha = .05 \quad .01 \quad .01, .05 \end{cases}$

4.16 Moments of t

It is readily verified that the $(2i)^{th}$ absolute moment of t exists only for $2i < \nu$, and is then given by $E\left|t^{2i}\right| = \nu^i \Gamma(\tfrac{1}{2}+i)\Gamma(\tfrac{1}{2}\nu-i)/\Gamma(\tfrac{1}{2})\Gamma(\tfrac{1}{2}\nu)$. For integer i, the above formula gives the moments of even order; the moments of odd order, so far as they exist, vanish.

For $\nu = 1$, the t-distribution is a Cauchy distribution, and possesses no moments of integer order--not even a mean.

Romanowsky, V., 1928

Metron 7, part 3,3

Skewness and Kurtosis of t^2.

The table on p. 31 gives $\beta_1(t^2)$ and $\beta_2(t^2)-3$ to 3 decimals for
$$\nu + 2 = 15, 20 \ (10) \ 60, 80, 100, 150, 200 \ (100) \ 500$$

Pearson, K. 1931 (d)

Biometrika 23, 1

Standard deviation of t under various hypotheses (see paper) leading to the following auxiliary tables.

Table (p.3) of $\dfrac{|\bar{x}-m|}{\sigma_{\bar{x}}}\sqrt{1 - \dfrac{\Gamma^2(\tfrac{1}{2}(n-2))}{\Gamma(\tfrac{1}{2}(n-1))\,\Gamma(\tfrac{1}{2}(n-3))}}$

to 5 dec. for n = 5, 10, 25, 50, 100

$$\dfrac{\bar{x}-m}{\sigma_{\bar{x}}} = 0.5(0.5)3.5$$

also $1/\sqrt{n-3}$ to 5 dec. for above n.

Table (p.7) of $\sqrt{\frac{n-3}{n}} \sum \frac{1}{\sigma}$ to 4 dec. for n = 5, 10, 25, 50, 100, 500, ∞; $\frac{\sigma}{\sum}$ = 0.25, 0.5(0.5)3.5

4.17 Multivariate t-distributions

Dunnett, C. W. and Sobel, M. 1954

Biometrika 41, 153-169

Let x, y be a pair of observations from a bivariate normal distribution with zero means, unit variances and correlation coefficient ρ; let s^2 be a quantity distributed, independently of x and y, as χ^2_ν/ν. Let $P(t,\rho) = \Pr[x < ts, y < ts]$. Then

$$2\pi(1-\rho^2)^{\frac{1}{2}} P(t,\rho) = \int_{-\infty}^{t} \int_{-\infty}^{t} \left[1 + \frac{u^2 - 2\rho uv + v^2}{\nu(1-\rho^2)}\right]^{-\frac{1}{2}(\nu+2)} du\,dv.$$

Table 1, pp. 162-163, gives $P(t,\rho)$ to 5 dec. for $\rho = +0.5$ and

$\begin{cases} \nu = 1 & 2 & 3 & 4 \\ t = 0(.25)2.5(.5)10 & 0(.25)2.5(.5)9 & 0(.25)2.5(.5)5.5 & 0(.25)2.5(.5)4.5 \end{cases}$

$\begin{cases} 5, 6 & 7(1)12 & 13(1)30(3)60(15)120, 150, 300, 600, \infty \\ 0(.25)2.5(.5)4 & 0(.25)2.5(.5)3.5 & 0(.25)2.5, 3 \end{cases}$.

Table 2, pp. 164-165, gives $P(t,\rho)$ to 5 dec. for $\rho = -0.5$ and

$\begin{cases} \nu = 1, 2 & 3 & 4 & 5 \\ t = 0(.25)2.5(.5)10 & 0(.25)2.5(.5)6 & 0(.25)2.5(.5)5 & 0(.25)2.5(.5)4.5 \end{cases}$

$\begin{cases} 6 & 7(1)13 & 14(1)30(3)60(15)120, 150, 300, 600, \infty \\ 0(.25)2.5(.5)4 & 0(.25)2.5(.5)3.5 & 0(.25)2.5, 3 \end{cases}$.

Tables 3 and 4, pp. 166-167, give lower-tail 100 P % points of t for

$\begin{cases} \nu = 1(1)30(3)60(15)120, 150, 300, 600 & \infty \\ \rho = +0.5, -0.5 & +0.5, -0.5 \\ P = .5, .75, .9, .95, .99 & .5, .75, .9, .95, .99 \\ 3 \text{ dec.} & 5 \text{ dec.} \end{cases}$

Dunnett, C. W. and Sobel, M. 1955

Biometrika 42, 258

The two correlated normal deviates of Dunnett & Sobel 1954 here become three or nine deviates; all correlations are taken equal to $+\frac{1}{2}$. Table 1, p. 260, gives lower-tail 100 P % points of 3-variate and 9-variate t for $\nu = 5, \infty$; P = .5, .75, .95, .99; and four approximations to the same.

4.17
4.19 [see also pp. 674-675]

Gupta, S. S. and Sobel, M. 1957

Annals of Mathematical Statistics 28, 957-967

Let z_i be (k-1) normal variates with zero means, unit variances, and all correlations $+\frac{1}{2}$. Let Z denote $\max_i z_i$. Let s^2 be a quantity distributed, independently of the z_i, as χ^2_ν/ν. Let q be the root of

$$\Pr[Z/s < 2^{-\frac{1}{2}}q] = P^*.$$

Table 1, pp. 962-964, computed by Cornish-Fisher expansions, gives

q to 2 dec. for
$\begin{cases} \nu = 15(1)20(10)40,\ 24(12)60(20)120,\ 360,\ \infty \\ k = 2,\ 5,\ 10(1)16(2)20(5)40,\ 50 \\ P^* = .75,\ .9,\ .95,\ .975,\ .99 \end{cases}$

4.19 Miscellaneous tabulations connected with t

Pearson, K., 1931 (c)

Biometrika 22, 405

Correlation between t-values in correlated samples.

Let t_x and t_y denote t-values computed from a sample of n pairs x_i, y_i drawn from a binormal parent with correlation ρ, and let $z_x = t_x \sqrt{n-1}$, $z_y = t_y \sqrt{n-1}$. The formula for the correlation between t_x and t_y is then given by:-

$$\rho_{z_x z_y} = \rho_{t_x t_y} = \Gamma^2(\tfrac{n-2}{2})\ \Gamma^{-1}(\tfrac{n-1}{2})\ \Gamma^{-1}(\tfrac{n-3}{2})\ \rho\ F(\tfrac{1}{2},\tfrac{1}{2},\tfrac{n-1}{2},\rho^2)$$

(equation (v)) and $\rho_{t_x t_y}$ is tabulated (Table I, p.412) to 4 decs for

n = 4(1)30, 50, 52, 100, 102, 400, 402, 1000, ∞ and $\rho = 0.0(0.1)1.0$

The array means of z_x given z_y are given by.

$$\hat{z}_x = \frac{\rho z_y}{(1 + (1-\rho^2)z_y^2)^{\frac{1}{2}}} \left(1 + \frac{1}{n-2}\ \frac{(1-\rho^2)(1+z_y^2)}{1 + (1-\rho^2)z_y^2} \right) \quad (1)$$

which may be approximated by the tangential line

$$\hat{z}_x \sim \frac{n - 1 - \rho^2}{n - 2}\ \rho z_y \quad (2)$$

Table II (p. 417) gives 6 decimals of z_x computed from (1), (2) and

$z_x \sim \rho z_y$ for $z_y = i/\sqrt{n-3}$ i = 1,2,3

$\rho = .2(.2).8$

n = 10, 50, 100, 500.

Table III gives corresponding values of the array standard deviation $\sigma_{z_x | z_y}$ to 6 dec. from the exact formula (equations (xvii) (xviii)) and 3 approximations for the above ρ and n.

Bancroft, T. A. 1944
Annals of Mathematical Statistics 15, 190-204
"On biases in estimation due to the use of preliminary tests of significance"

Given a sample of n triplets $y; x_1, x_2$ and consider the Analysis of Variance Table

			Mean Square
Total	$\sum (y)^2$	n-1	
Reduction due to x_1	$b' \sum (yx_1)$	1	s_1^2
Reduction due to x_2 after fitting x_1	$\sum (yx_1)(b_1 - b') + b_2 \sum (yx_2)$	1	s_2^2
Residual		n-3	s_3^2

where $b' = \sum yx_1 / \sum x_1^2$, b_1 b_2 are the partial regression coefficients and $\sum (yx_1)$ stands for $\sum (y - \bar{y})(x_1 - \bar{x}_1)$.

To estimate the real regression β_1 use

b_1 if $s_2^2 / s_3^2 \leq F$

b_1 if $s_2^2 / s_3^2 > F$.

Table III (p.203) gives 3 decimal values of the bias in estimating β_1 by the above rule for n = 5, 11, 21; F = 1 and F(.05; 1, n-3); β_2 = 0.1, 0.4, 1, 2, 4 and ρ = .2(.2).8 where ρ is the correlation between x_1 and x_2.

Walsh, J. E. 1947 a
Annals of Math. Stat. 18, 88
Effect of intraclass correlation on confidence intervals of t.

If $x_1, \ldots, x_n$ represents a single n-variate observation from a multinormal parent with all means = μ, unit variances and all

4.19
4.20   [see also p. 675]

correlations $= \rho$, the confidence interval for μ from it will be

$$\bar{x} \pm t(\frac{1-\epsilon}{2}, n-1) \; S \sqrt{\frac{1 + (n-1)\rho}{n(n-1)(1-\rho)}} \quad \text{where} \quad S^2 = \sum (x - \bar{x})^2 \; ;$$

ϵ = actual confidence coefficient; $t(\frac{1-\epsilon}{2}, n-1)$ = double tailed $100(1-\epsilon)\%$ point of t for $n-1$ d.f.

If, for nominal confidence coefficient τ, the confidence interval $\bar{x} \pm \dfrac{t(\frac{1-\epsilon}{2}, n-1)S}{\sqrt{n(n-1)}}$ is used, then define an actual confidence coefficient $a(\rho, n)$ by

$$t(\frac{1-a(\rho, n)}{2}, n-1) = \sqrt{\frac{1-\rho}{1 + (n-1)\rho}} \; t(\frac{1-\epsilon}{2}, n-1)$$

Table III (p. 91) gives $a(\rho, n)$ to 2-3 decs for $n = 4, 8, 16, 32, 64, 128$, $\rho = 0, .05, .1(.1).5$ with some omissions for $\epsilon = .95, .99$.

Generalizations to two sample t in Theorem V (p.96).

4.2 The F-distribution and multivariate variance tests

4.20 Definition of F and relation to other statistics

Consider two independent χ^2-values, χ_1^2 and χ_2^2, based respectively on ν_1 and ν_2 degrees of freedom and giving rise to two 'sample variances' or 'mean squares' $S_1^2 = \chi_1^2 / \nu_1$ and $S_2^2 = \chi_2^2 / \nu_2$.

The 'variance ratio F' is then defined by

$$F = S_1^2/S_2^2 = \frac{\chi_1^2}{\nu_1} \bigg/ \frac{\chi_2^2}{\nu_2} \qquad (4.20.1)$$

The frequency ordinate function $f(F; \nu_1, \nu_2)$, the probability integral $P(F; \nu_1, \nu_2)$ and the tail area $Q(F; \nu_1, \nu_2)$ of this statistic are respectively given by

$$f(F; \nu_1, \nu_2) = \Gamma(a+b)[\Gamma(a)\Gamma(b)]^{-1} a^a b^b F^{b-1} (a+bF)^{-a-b} \qquad (4.20.2)$$

where $a = \tfrac{1}{2}\nu_2$, $b = \tfrac{1}{2}\nu_1$;

$$P(F; \nu_1, \nu_2) = \int_0^F f(\phi; \nu_1, \nu_2) \, d\phi \, ;$$

and $\qquad (4.20.3)$

$$Q(F; \nu_1, \nu_2) = \int_F^\infty f(\phi; \nu_1, \nu_2) \, d\phi = 1 - P(F; \nu_1, \nu_2). \qquad (4.20.4)$$

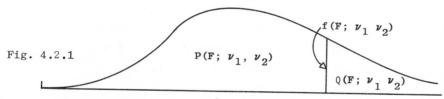

Fig. 4.2.1

Since for $\nu_1 = 1$ the mean square S_1^2 is the square of a unit normal variate we have for this special case that $F = t^2$; and the relation to the tail-area of t is given by $2Q(t,\nu) = Q(F;1,\nu)$ for $F = t^2$. At the other end of the scale we have as $\nu_1 \to \infty$, $F \to \nu_2 / \chi_2^2$ or as $\nu_2 \to \infty$, $F \to \chi_1^2 / \nu_1$ so that

$$p(\chi^2, \nu) = \lim_{\nu_2 \to \infty} Q(F; \nu, \nu_2) \quad \text{for } F = \chi^2/\nu$$

and

$$1 - p(\chi^2, \nu) = \lim_{\nu_1 \to \infty} Q(F; \nu_1, \nu) \quad \text{for } F = \nu/\chi^2$$

4.20
4.231

For general ν_1, ν_2 we have the important relation to the incomplete Beta function

$$Q(F; \nu_1, \nu_2) = I_x(\tfrac{1}{2}\nu_2, \tfrac{1}{2}\nu_1) \qquad (4.20.5)$$

where $x = \nu_2/(\nu_1 F + \nu_2)$.

4.21 The area of the F-distribution

With the help of the above relation (4.20.5) the 'tail area', and hence the probability integral of F, can be evaluated with the help of tables of the Incomplete Beta function.

Pearson, E. S. 1929
<u>Biometrika</u> 21, 337-360

Pearson discusses various methods of obtaining approximations to the F-integral. He illustrates these approximations in TT. 2-3, pp. 354-355.

4.23 Percentage points of F and of $F^{\frac{1}{2}}$

The 100 Q % point of F is defined as the root F of

$$Q = Q(F; \nu_1, \nu_2)$$

for a given percent level 100 Q and a given pair of degrees of freedom ν_1, ν_2. It is denoted by $F(Q; \nu_1, \nu_2)$. More precisely $F(Q; \nu_1, \nu_2)$ should be referred to as the upper 100Q % point of F.

Because of their use in Analysis of Variance, tables of $F(Q; \nu_1, \nu_2)$ are given in numerous text books. The relations to the percentage points of t, χ^2 and the Incomplete Beta function follow immediately from the formulas of 4.20:

$$t(Q,\nu) = F(2Q; 1, \nu)^{\frac{1}{2}}$$

$$\chi^2(p,\nu) = \nu F(p; \nu, \infty)$$

$$x(P;a,b) = a/(b\ F(P; 2b, 2a) + a).$$

4.231   Tables of percentage points of F

It is customary to table only upper percentage points of F ; lower percentage points may be obtained from the relation

$$F(1-Q\,;\,2b\,,\,2a) = 1/F(Q\,;\,2a\,,\,2b)$$

We classify the tables below according to what percentage points they give: designating as family A those tables containing $\tfrac{1}{2}\%$, 1%, $2\tfrac{1}{2}\%$, 5%, 10% and 25% points; and as family B those containing 0.1%, 1%, 5% and 20% points. The leading representatives of these families are found in Merrington & Thompson 1943 and Fisher & Yates (T.V); abridgements of these are common in textbooks. Within each family the order is by number of percentage points tabled.

4.2311 Tables with material from both families A and B

Q =	$v_1 = [120/v_1 =]$	$v_2 = [120/v_2 =]$	
3 fig.	.0005, .001, .005, .01, .025, .05, .1, .25, .5; P = ditto	1(1)12, 50, 100, 200, 500 [0(1)6(2)10]	1(1)12 [0(1)6(2)10]	Dixon & Massey 1957(390)
3 fig.	.0005, .001, .01, .3, .5	1(1)10(5)20, 30, 50, 100, 200, 500, ∞	1(1)20(2)30(10)60(20)100, 200, 500, ∞	} Hald 1952(T.VII)
3 fig.	.005, .01, .025, .05	1(1)20(2)30(5)50, 60(20)100, 200, 500, ∞	1(1)30(2)50(5)70(10)100(25) 150, 200, 300, 500, 1000, ∞	
2 dec.	.001, .005, .01, .025, .05, .1, .25	1(1)10 [0(1)6(2)12]	1(1)30 [0(1)4]	E.S. Pearson & Hartley 1954(157)
3 fig.	.001, .005, .01, .025, .05, .1, .5	1(1)10 [0(1)6(2)12]	1(1)10 [0(1)6(2)12]	Duncan 1952(T.J)
2 dec.	.001, .01, .05, .1, .2	1(1)6 [0(5)20]	1(1)30 [0(1)4]	Fisher & Yates 1948[x], 1953, 1957(T.V)

4.2312 Tables of family A

Q =	$v_1 = [120/v_1 =]$	$v_2 = [120/v_2 =]$	
5 fig.	.005, .01, .025, .05, .1, .25, .5	1(1)10 [0(1)4]	1(1)30 [0(1)4]	Merrington & Thompson 1943[x]
3 fig.	.005, .01, .025, .05, .1	1(1)10 [0(1)6(2)12]	1(1)10 [0(1)6(2)12]	Mood 1950(T.V)
3 fig.	.005, .01, .025, .05	1(1)10 [0(1)6(2)12]	1(1)25 [0(1)5]	Dixon & Massey 1951(310)
2 dec.	.01, .05, .1	1(1)10 [0(1)6(2)12]	1(1)30 [0(1)4]	Davies 1947(T.D)
2 dec.	.01, .05, .1	1(1)6 [0(5)20]	1(1)30, 40	Kempthorne 1952(T.I)

[x] Errors in this table are reported in Section 4.233, below.

4.2213 Tables of family B

	Q =	ν_1 =	[$120/\nu_1$ =]	ν_2 =	[$120/\nu_2$ =]	
2 dec.	.001, .01, .05, .2	1(1)6	[0(5)20]			Fisher & Yates 1938, 1943(T.V)
1 dec.	.001, .01, .05, .2	1(1)6	[0, 5, 10, 20]			Mather 1943(T.IV)
1 dec.	.001, .01, .05, .2	1(1)6	[0, 5, 10, 20]	1(1)20(2)30	[0, 1, 2, 4]	K.A. Brownlee 1947(T.III)
2 dec.	.001, .01, .05	1(1)6	[0(5)20]	1(1)20(2)30	[0, 2, 3, 4]	Yule & Kendall 1950(T.5), Graf & Henning 1953(T.4)
3 dec.	.001 only	1(1)10	[0(1)6(2)12]	1(1)30	[0(1)4]	Banerjee 1936[x]

4.2314 Short tables

	Q =	ν_1 =	[$120/\nu_1$ =]	ν_2 =	[$120/\nu_2$ =]	
2 dec.	.01, .05	1(1)12(2)16(4)24, 30(10)50(25)100, 200, 500, ∞		1(1)30(2)50(5)70, 80, 100(25)150, 200, 400, 1000, ∞		Snedecor 1946(T.10.7), Goulden 1939(T.96), Kenney 1939 b (T.2)

Freeman 1942(T.VIII), Treloar 1942(Appendix II), Smith & Duncan 1945(T.IX), Hoel 1947(T.V),
P.O. Johnson 1949(T.IV), Mann 1949, Arkin & Colton 1950(117), Juran 1951(T.E), Kenney & Keeping 1951(T.2),
Mode 1951(T.L), Burington & May 1953(276), Clark 1953(T.V), E.E. Lewis 1953(T.24), Ferber 1949(512)

2 dec.	.01, .05	1(1)10, 12(4)24, 30(10)60, 75, 100, ∞		1(1)30(10)80, 100, 150, 200, ∞		Quenouille 1950(TT.3,4), 1952(TT.7,8), 1953(TT.2,3)
4 fig.	.01, .05	1(1)10	[0(1)6(2)12]	1(1)30	[0(1)4]	N.L. Johnson & Tetley(T.A.9)
3 fig.	.01, .05	1(1)10	[0(1)6(2)12]	1(1)25	[0(1)5]	Dixon & Massey 1957(388)
2 dec.	.01, .05	1(1)10, 12(4)24, 30, 50, 100, ∞		1(1)20(2)28(4)40, 60, 100, ∞		Anderson & Bancroft 1952(T.IV)
2 dec.	.01, .05	1(1)10, 12		1(1)30(10)60	[0(1)4]	Youden 1951(TT.II,III)

---

[x] Errors in this table are reported in Section 4.233, below.

4.2314 [4.33]
4.232 [4.332]

2 dec.	.01, .05	$\Big\{$ 1(1)6 [5(5)20] ∞ [0]	1(1)30(5)50(10)100(25)150, 200(100)500, 1000, ∞	Snedecor 1934(88)
3 dec.	.01, .05	1(1)6 [0(5)20]	1(1)30(5)50(10)70	Mahalanobis 1932(686)
2 dec.	.01, .05	1(1)6 [0(5)20]	1(1)30 [0, 2, 4]	Weatherburn 1946(T.6)
1 dec.	.01, .05	1(1)4 [0(10)30]	2(1)10(2)20(5)30, 40, 60, 80	Finney 1952 b (T.II)
2 dec.	.05 only	1(1)8(2)16(4)24, 30(10)50(25)100	1(1)10 [0(2)8, 12]	Peach 1947(T.VI)
2 dec.	.05 only	1(1)6 [0(5)20]	1(1)16(2)20, 24, 30(10)80, 100, 200, ∞	
			1(1)30 [0(1)4]	Arley & Buch 1950(T.V)

4.232 Tables of percentage points of $F^{\frac{1}{2}}$

		$\nu_1 =$ [$120/\nu_1 =$]	$\nu_2 =$ [$120/\nu_2 =$]	
	Q =			
4 dec.	.01, .05	1(1)6 [0(5)20]	1(1)30 [0(2)4]	Mahalanobis 1932(692)

4.33 Percentage points of z and of $\log_{10} F$

For definitions see Section 4.30, below.

4.331 Tables of percentage points of z

		$\nu_1 =$ [$120/\nu_1 =$]	$\nu_2 =$ [$120/\nu_2 =$]	
	Q =			
4 dec.	.001, .01, .05, .1, .2	1(1)6 [0(5)20]	1(1)30 [0(1)4]	Fisher & Yates 1948[x], 1953, 1957(T.V)
4 dec.	.001, .01, .05, .2	1(1)6 [0(5)20]	1(1)30 [0(1)4]	Fisher & Yates 1938, 1943(T.V)

[x] Errors in this table are reported in Section 4.333, below.

4.2314 [4.33]
4.232 [4.332]

4 dec.	$\begin{cases}.001 \\ .01, .05\end{cases}$	1(1)6 [0(5)20]	1(1)30 $\begin{cases}[0, 2, 3, 4] \\ [0(2)4]\end{cases}$	Yule & Kendall 1950(671)
4 dec.	.001, .01, .05	1(1)6 [0(5)20]	1(1)30 [0(2)4]	Burington & May 1953(280)
4 dec.	.01, .05	1(1)6 [0(5)20]	1(1)30 [0(2)4]	Rider 1939(204), M.G. Kendall 1943(442), M.G. Kendall & Stuart 1958(404)
4 dec.	.001 only	1(1)10 [0(1)6(2)12]	1(1)30 [0(1)4]	Colcord & Deming 1936(423)[x]
4 dec.	.01 only	1(1)6 [0(5)20]	1(1)15(5)30 [0(1)4]	W.E. Deming 1950(591)

4.332 Tables of percentage points of $\log_{10} F = 2z \log_{10} e$

$\nu_1 = [120/\nu_1 =]$ $\nu_2 = [120/\nu_2 =]$

4 dec.	.01, .05	1(1)6 [0(5)20]	1(1)30 [0(2)4]	Mahalanobis 1932(679)

---

[x]Errors in this table are reported in Section 4.333, below.

4.233
4.239   [see also p. 677]

4.233 Errors in tables of percentage points of F

The order in this section is alphabetical.

Banerjee 1936

Since this table is based on the table of $z = \frac{1}{2}\log_e F$ by Colcord & Deming 1936, giving 4 dec. in z, only 4 fig. in $F = e^{2z}$ are reliable. In addition, the following errors have come to our notice:

ν_1	ν_2	in place of Banerjee's value of	e^{2z} of Colcord and Deming's z - value	$F = e^{2z}$ from Fisher-Yates Table and/or unpublished results.
1	1	406491	405373	405284
3	1	536657	536793	540379
4	16	7.948	7.944	7.94(3)
4	20	7.102	7.097	7.09(7)
1	120	11.262	11.380	11.37(9)

Fisher & Yates 1948

Norton 1952 examined this table thoroughly, and hopes that the following list is complete except for end-figure errors:

Q	ν_1	ν_2	for	read	Q	ν_1	ν_2	for	read
·2	4	1	13·73	13·64	·001	1	7	29·22	29·25
·2	8	1	14·59	14·58	·001	2	5	36·61	37·12
·01	1	2	98·49	98·50	·001	2	120	7·31	7·32
·01	8	2	99·36	99·37	·001	4	6	21·90	21·92
·001	1	3	167·5	167·0	·001	5	19	6·61	6·62
·001	1	5	47·04	47·18	·001	∞	120	1·56	1·54

These errors have been corrected in the 1953 and 1957 editions.

Merrington & Thompson 1943

A casual examination by Norton 1952 found the following errors:

Q	ν_1	ν_2	for	Read
.01	3	120	3.9493	3.9491
.01	8	1	5981.6	5981.1
.01	∞	2	99.501	99.499

4.239 Miscellaneous combinations of percentage points of F

For percentage points of log F, see sections 4.331 and 4.332 above.

Pearson, E. S. and Sukhatme, A. V. 1935
Sankhyā 2, 13-32
"An illustration of the use of fiducial limits in determining the characteristics of a sampled batch"

A batch of N observations $X_1, \ldots, X_N$ with batch mean $\bar{X}$ has been drawn from a normal population and a sample of n observations $x_1, \ldots, x_n$ with sample mean $\bar{x}$ has been drawn from the batch. Let

$$S^2 = \sum (X_i - \bar{X})^2 / N-1; \qquad s^2 = \sum (x_i - \bar{x})^2 / n-1$$

Then the statement $B_1 s \leq S \leq B_2 s$ will be correct in $100(1-2Q)\%$ of batches if

$$B_1 = \left[(n-1) + (N-n)/F(Q,n-1,N-n) \right]^{\frac{1}{2}} / \sqrt{(N-1)}$$

$$B_2 = \left[(n-1) + (N-n)F(Q,N-n,n-1) \right]^{\frac{1}{2}} / \sqrt{(N-1)}$$

Tables 2 and 3 (pp. 20-21) give B_1 and B_2 to 3 decimals for
$$\begin{cases} Q = .01, .05 & .01, .05 \\ n = 5, 7, 9 & \\ 60/n = & 1(1)6 \\ N/n = 2(1)6 & 1.5, 2(1)6 \\ n/N = 0(.025).125 & 0(.025).125 \end{cases}$$

Peters, C. C. and Van Voorhis, W. R. 1940
<u>Statistical Procedures and their Mathematical Bases</u>

Consider a two way classification into n rows and k columns, the columns representing different levels of the variate x and the rows, of y. If σ_y^2 is the overall variance of the y's and σ_c^2 the 'within column' variance, assuming it to be the same for all columns, then the coefficient of curvilinear correlation of y on x is given by

$$\eta_{yx} = \sqrt{1 - \frac{\sigma_c^2}{\sigma_y^2}}$$

4.239
4.24   [see also p. 677]

and an unbiased estimate of η^2 is given by

$$\epsilon^2 = 1 - \frac{s_c^2}{s_y^2}$$

where s_y^2 is the total mean square and s_c^2 is the within column mean square. If the true η_{yx}^2 is zero, then the ratio of the mean squares has an F distribution; the distribution of ϵ^2 is that of

$$\epsilon^2 = \frac{(k-1)\, F - (k-1)}{(k-1)\, F - k(n-1)}$$

Table XLVII (pp. 494-497) gives upper 5% and 1% points for ϵ^2, when the true $\eta^2 = 0$, to 3 dec. for

$\begin{cases} k(n-1) = 1(1)12 & 13 & 14 & 15 & 16(1)19 \\ k - 1 = 1(1)[k(n-1)] & 1(1)12 & 1(1)12,\ 14 & 1(1)12(2)16 & 1(1)12(2)16,\ 20 \end{cases}$

$\begin{cases} 20(1)24 & 25(1)30 & 32,\ 34 \\ 1(1)12(2)16(4)24 & 1(1)12(2)16(4)24,\ 30 & 1(1)12(2)16(4)24,\ 30,\ 40 \end{cases}$

$\begin{cases} 36(2)50(5)70,\ 80,\ 100,\ 125,\ 150,\ 200,\ 400,\ 1000 \\ 1(1)12(2)16(4)24,\ 30,\ 40,\ 50 \end{cases}$

Lower percentage points of ϵ^2, which are negative, are not tabled. Peters & Van Voorhis (p. 336, footnote) pretend that the "ϵ technique tells all that analysis of variance tells and more."

Snedecor, G. W. 1946

Statistical methods

Table 13.6, p. 351, gives upper 5% and 1% points of the multiple correlation coefficient R:

$$R = \left[\frac{1}{1 + \dfrac{\nu_2}{\nu_1 F(p,\ \nu_1,\ \nu_2)}} \right]^{\frac{1}{2}}$$

where ν_1 = number of independent variables fitted, ν_2 = degrees of freedom for residual (usually number of observations $- \nu_1 - 1$) and $F(p, \nu_1, \nu_2)$ is the upper $100p\%$ point of F. Table of the above to 3 decimals for $p = .05, .01$; $\nu_1 = 1(1)4$; $\nu_2 = 1(1)30(5)50(10)100(25)150, 200(100)500, 1000$.

Note: Snedecor's number of variables $= \nu_1 + 1$ as it includes the dependent variable. Reprinted in Arkin & Colton 1950, T. 21, p. 140.

4.24 Power of the F-test

The F tables of the preceeding sections all dealt with the distribution of the variance ratio $F = s_1^2/s_2^2$ under the Null hypothesis, i.e. under the assumption that s_1^2 and s_2^2 are respectively distributed as χ_1^2/ν_1 and χ_2^2/ν_2. The same distribution for F results if s_i^2 is distributed as $\sigma^2 \chi_i^2/\nu_i$ (i = 1,2), provided the __same__ σ^2 applies to __both__ s_1^2 and s_2^2.

In the present section we are concerned with 'alternatives' to the above 'Null hypothesis' and the applications to the analysis of variance suggest the following two alternatives as being of particular importance.

a. The component-of-variance model:

The s_i^2 are independently distributed as $\sigma_i^2 \chi_i^2/\nu_i$ with $\sigma_1^2 > \sigma_2^2$.

For example, in an analysis of variance of data x_{ti} in k groups (t = 1,2 ... k) containing n readings each (i = 1,2 ... n) the model $x_{ti} = a_t + Z_{ti}$ with $Z_{ti} = N(0, \sigma_2^2)$, $a_t = N(\mu, \sigma_a^2)$ would give rise to the above situation with $\sigma_1^2 = n\sigma_a^2 + \sigma_2^2$; $\nu_1 = k-1$, $\nu_2 = k(n-1)$. In this case the variance ratio or mean square ratio s_1^2/s_2^2 is distributed as $(\sigma_1^2/\sigma_2^2)F$ where F is the usual, central, F statistic based on ν_1, ν_2 degrees of freedom. All problems regarding power, sample size, error of the 2nd kind can therefore be answered, in this situation, by reference to the ordinary F distribution, although this may require interpolation in the Tables of the Incomplete Beta function.

b. The linear-hypothesis model:

s_2^2 is distributed as $\sigma_2^2 \chi_2^2/\nu_2$ but

s_1^2 is distributed as $\sigma_2^2 \chi_1'^2/\nu_1$ where $\chi_1'^2$ is a non-central χ^2 based on ν_1 degrees of freedom and non-centrality λ. s_1^2 and s_2^2 are again independent.

For example, if in the analysis of variance situations mentioned under (a) the model is $x_{ti} = \alpha_t + Z_{ti}$ with fixed group means α_t, then the situation in (b) arises with

$$\lambda = \tfrac{1}{2} \sum_{t=1}^{k} (\alpha_t - \bar{\alpha})^2 / \sigma_2^2 \qquad (4.24.1)$$

and $\nu_1 = k-1$, $\nu_2 = k(n-1)$.

4.24
4.241

Here s_1^2/s_2^2 is essentially the ratio of a non-central χ^2 to a central χ^2. This ratio, often called a non-central F-ratio, will be denoted by F'. Using the series expansion for the distribution of non-central χ^2, we find the distribution of F':

$$\beta = P_r[F' \leq F^*] = e^{-\lambda} \sum_{j=0}^{\infty} \lambda^j P(F^*; \nu_1 + 2j; \nu_2)/j! \quad (4.24.2)$$

$$= e^{-\lambda} \sum_{j=0}^{\infty} \lambda^j I_x(\tfrac{1}{2}\nu_1 + j, \tfrac{1}{2}\nu_2)/j!$$

where the non-centrality λ is defined by (4.24.1) for the special case considered above,* and

$$x = \nu_1 F^*/(\nu_1 F^* + \nu_2) \quad (4.24.3).$$

In order to assess the power $(1-\beta)$ of the F-test we must substitute for F^* the percentage point of F corresponding to a given level of significance or error of the first kind α; i.e., we must substitute

$$F^* = F(\alpha; \nu_1, \nu_2) \quad (4.24.4)$$

in (4.24.2) to obtain the relation between the error of the first kind α and the error of the second kind β:

$$\beta = e^{-\lambda} \sum_{j=0}^{\infty} \lambda^j I_x(\tfrac{1}{2}\nu_1 + j, \tfrac{1}{2}\nu_2)$$

(4.24.5)

with

$$x = \nu_1 F(\alpha; \nu_1, \nu_2)/(\nu_1 F(\alpha; \nu_1, \nu_2) + \nu_2)$$

4.241 Power of the F-test, in terms of the central F-distribution

Brown, G. W. 1939

Annals of Mathematical Statistics 10, 119-128

"On the power of the L_1 test for equality of several variances"

The L_1 test for equality of two variances is equivalent to the F-test.

---

*A general definition of λ is given by Tang 1938 and Lehmer 1944.

Tables 1 and 2, p. 126, give the power to 2 dec. for $\alpha = .05$ and

$$\begin{cases} n_1 = 5 \\ n_2 = 5 \end{cases}, \begin{cases} 10 \\ 10 \end{cases} \quad 20 \quad 12$$
$$\phantom{\begin{cases} n_1 = 5 \end{cases}} \qquad \qquad 20 \quad 8$$
$$\sigma_2^2/\sigma_1^2 = 1(1)5, 10, 20 \quad 1(1)5 \quad .6(.2)1.2, 1.5, 2(1)6, 10$$
$$\sigma_1^2/\sigma_2^2 = .75 \qquad \qquad .75 \quad 1(1)5, 10$$

$$\begin{cases} 15 \\ 10 \\ .6, 1(.5)2(1)5, 10 \\ 1(1)5, 10 \end{cases}$$

Scheffé, H. 1942

<u>Annals</u> <u>of</u> <u>Mathematical</u> <u>Statistics</u> 13, 371-388

Let $\nu_1 s_1^2/\sigma_1^2$ and $\nu_2 s_2^2/\sigma_2^2$ denote two independent χ^2 values based on ν_1 and ν_2 degrees of freedom and $\theta = \sigma_1^2/\sigma_2^2$, $F = s_1^2 \sigma_2^2 / s_2^2 \sigma_1^2$ and $T = s_1^2/s_2^2$. Let the null hypothesis H_0 consist of $\theta = \theta_0$. Then various criteria for testing H_0 at level of significance α are suggested all consisting of rejecting H_0 if $T \leq A \theta_0$ or $T \geq B \theta_0$, where A and B depend on ν_1, ν_2 and α.
Four of the six criteria are

    I. $A = F(1-\alpha; \nu_1, \nu_2)$ $B = \infty$

    II. $A = 0$ $B = F(\alpha; \nu_1, \nu_2)$

    III. A and B are roots of

$$B f(B; \nu_1, \nu_2) = A f(A; \nu_1, \nu_2)$$
$$\int_A^B f(F; \nu_1, \nu_2) \, dF = 1 - \alpha$$

    VI. $A = F(1 - \tfrac{1}{2}\alpha; \nu_1, \nu_2)$ $B = F(\tfrac{1}{2}\alpha; \nu_1, \nu_2)$.

The criteria determine confidence intervals for θ as well as tests of H_0.

4.241
4.242   [see also pp. 677-680]

Criterion III is shown to have the logarithmically shortest confidence interval. Table 1, p. 381, gives the ratios θ/θ_0 at which Power VI is a minimum to 3 dec., and corresponding values of Power VI to 3 fig., for $\alpha = .01, .05$; $\nu_1, \nu_2 = 1, 2, 3, 5, 10, 20, 40, \infty$; $\nu_1 \neq \nu_2$.

Eisenhart, C., Hastay, M. W. and Wallis, W. A.   1947
Selected techniques of statistical analysis

If the two 'sample variances' s_i^2 are distributed as $\sigma_i^2 \chi_i^2 / \nu_i$ (i = 1,2) the chance of returning the variance ratio s_1^2/s_2^2 as significant at the $100\alpha\%$ level of significance is given by

$$\Pr\left[s_1^2/s_2^2 \geq F(\alpha; \nu_1, \nu_2)\right] = Q\left(\sigma_2^2 F(\alpha; \nu_1, \nu_2)/\sigma_1^2 \,;\, \nu_1, \nu_2\right)$$

If this chance, i.e. the power of the F test, is fixed at the level $1-\beta$ we must have

$$F(\alpha; \nu_1, \nu_2)/(\sigma_1^2/\sigma_2^2) = F(1-\beta; \nu_1, \nu_2)$$

or

$$\frac{\sigma_1^2}{\sigma_2^2} = F(\alpha; \nu_1, \nu_2)/F(1-\beta; \nu_1, \nu_2) \equiv F(\alpha; \nu_1, \nu_2)F(\beta; \nu_1, \nu_2).$$

Tables 8.3-8.4 (pp. 284-95; 298-309) give σ_1^2/σ_2^2 to 4 significant figures for $\nu_1, \nu_2 = 1(1)10$; $120/\nu_1, 120/\nu_2 = 0(1)6(2)12$; and

$$\begin{cases} \alpha = .01 \quad .05 \quad .01, .05 \\ \beta = \qquad\qquad\qquad .005, .01, .025, .05, .1, .25 \\ 1-\beta = .05 \quad .01 \quad .005, .025, .1, .25, .5 \end{cases}$$

4.242   Power of the F-test, in terms of the non-central F-distribution

Tang, P. C.   1938
Statistical Research Memoirs 2, 126-149 + 8 pages of tables

Tables I and II (following p. 149) give β (of equation 4.24.5) to 3 dec. for $\alpha = .01, .05$

$\nu_1 = 1(1)8$; $\nu_2 = 2(2)6(1)30, 60, \infty$ and at arguments

$$\emptyset = \sqrt{2\lambda/(\nu_1 + 1)} = 1(.5)3(1)8.$$

Values of β less than .001 are omitted.

Reprinted in Mann 1949, pp. 188-195; Kempthorne 1952, pp. 613-628.

4.241
4.242

Lehmer, Emma   1944

Annals of Mathematical Statistics 15, 388-398

If λ is the root of (4.24.5) for prescribed levels of α, β; ν_1, ν_2 the tables on pp. 390-3 give

$$\phi = \sqrt{2\lambda/(\nu_1 + 1)}$$

to 3 dec for $\alpha = .01, .05$; $\beta = .2, .3$ and

$$\nu_1 = 1(1)10, 12, 15, 20, 24, 30, 40, 60, 120, \infty$$

$$\nu_2 = 2(2)20, 24, 30, 40, 60, 80, 120, 240, \infty$$

Certain approximate formulas are tested, viz:

equation

(21) $\quad \phi^2 \sim \dfrac{1 - \left(1 - 2[\chi^2(\alpha, \nu_1) - \chi^2(\beta, \nu_1)]/\nu_1\right)^{\frac{1}{2}}}{\chi^2(\alpha, \nu_1)/\nu_1}$

(22) $\quad \sqrt{\tfrac{1}{2}\nu_1}\, \phi^2 \sim x(1-\alpha) + \dfrac{x^2(\beta)}{\sqrt{\tfrac{1}{2}\nu_1}} - x(\beta)\left(1 + \dfrac{2x(1-\alpha)}{\sqrt{\tfrac{1}{2}\nu_1}} + \dfrac{x^2(\beta)}{\tfrac{1}{2}\nu_1}\right)^{\frac{1}{2}}$

(20) $\quad \lim\limits_{\nu_1 \to \infty} \sqrt{\tfrac{1}{2}\nu_1}\, \phi^2 \sim x(1-\alpha) - x(\beta)$.

Three decimal values from the above three formulas are given in the table on p. 396, together with the exact values for $\nu_1 = 120$ and $(\alpha, \beta) = (.01, .2); (.01, .3); (.05, .2); (.05, .3)$.

Notation:

Lehmer	Ours
$1 - \beta$	β
α	α
f_1	ν_1
f_2	ν_2
λ	λ
φ	ϕ
y(p) (as defined)	x(p)
y(p) (as used)	x(1-p)

4.242
4.252 [see also pp. 678-681]

Dixon, W. J. and Massey, F. J. 1951

Introduction to statistical analysis

Table 13, pp. 328-331, gives ϕ^2, i.e. the square of Lehmer's φ, to 2 dec. for exactly the arguments given by Lehmer 1944 (see the preceding abstract).

Pearson, E. S. and Hartley, H. O. 1951

Biometrika 38, 112-130

"Charts of the power function for analysis-of-variance tests, derived from the non-central F-distribution"

The charts on pp. 115-22 give the power $1-\beta$ (with β given by equation (4.24.5)) plotted against the non-centrality $\emptyset = \sqrt{2\lambda/(\nu_1 + 1)}$ for $\nu_1 = 1(1)8$ $\nu_2 = 6(1)10, 12, 15, 20, 30, 60, \infty$ $\alpha = .01, .05$. β can be read from the charts to about $1\frac{1}{2}$ decimal places.

Eight charts, each covering a value of ν_1, each give two nests of curves, corresponding to $\alpha = .01, .05$, in which $1-\beta$ is plotted against $\emptyset$ with ν_2 as curve-parameter. The range of $\emptyset$ varies and only sections of the power curves for moderately large values of the power are shown. The chart for $\nu_1 = 1$ (t-test) is also given in Pearson, E. S. and Hartley, H. O., 1954, T. 10.

Reproduced in reduced size in Dixon & Massey 1957, T. 13, pp. 426-433.

Notation:

Pearson & Hartley	Ours
λ	2λ
$\emptyset$	$\emptyset$
ν_1	ν_1
ν_2	ν_2
β	$1-\beta$

4.25 Ratios involving ordered mean squares

Let s_i^2, $i = 1, \ldots, k$, denote a set of mean squares, each based on ν degrees of freedom and arranged in ascending order of magnitude. This section is concerned with criteria depending on one, several or all of the <u>ordered</u> s_i^2 and can therefore be regarded as a section on

'χ^2 order statistics'. Some of the approximations employed do, indeed, depend on the reduction to the corresponding theory for 'normal' order statistics by employing the fact that $\log_e s^2$ is approximately normal with mean $\log \sigma^2$ and variance $2/\nu - 1$.

4.251 The largest variance ratio

Finney, D. J. 1941

Annals of Eugenics 11, 136

Let the s_i^2 each have 2 d.f.; let s^2 be an independent mean square with ν d.f. Table 1, p. 139, gives the upper 5% points of s_k^2 / s^2 to 2 dec. for $k = 1(1)3$; $\nu = 1(1)10, 20, \infty$. Table 2, p. 139, compares these with the upper 5% points computed (following Hartley 1939) on the assumption that the s_i^2 / s^2 are independent F-statistics.

Nair, K. R. 1948 a

Biometrika 35, 16-31

Let the s_i^2 each have 1 d.f.; let s^2 be an independent mean square with ν d.f. Table 3, p. 26, gives upper 5% and 1% points of s_k^2 / s^2 to 2 dec. for $k = 1(1)10$; $120/\nu = 0(2)12$. Reprinted in Pearson & Hartley 1954, T. 19, p. 164.

4.252 The smallest variance ratio

Hartley, H. O. 1938

Supplement to the Journal of the Royal Statistical Society 5, 80

Let the s_i^2 each have 1 d.f.; let s^2 be an independent mean square with ν d.f. The table on p. 86 gives the lower 5% points of s_1^2 / s^2 to 7 dec. for $k = 1(1)10(10)30$ and $\nu = \infty$. For finite ν, the tail area exceeds .05 by approximately $\delta(1/4\nu + 1/360\nu^3)$; the table gives δ to 4 dec. for $k = 2(1)10(10)30$.

Nair, K. R. 1948 a

Biometrika 35, 16-31

Let the s_i^2 each have 1 d.f.; let s^2 be an independent mean square with ν d.f. Expand the distribution of s_1^2 / s^2 in the form

$$\Pr\left[s_1^2 / s^2 \leq \Delta^2\right] = a_0 + a_1(1/\nu + 1/8\nu^2) .$$

4.252
4.27

Table 5, p. 29, gives a_0 to 6 dec. and a_1 to 5 dec. for $k = 2(1)10$; $\Delta = 0(.01).1$.

4.253 s^2_{max}/s^2_{min}

Hartley, H. O. 1950 a

<u>Biometrika</u> 37, 308-312

"The maximum F-ratio as a short-cut test for heterogeneity of variance"

Let the s^2_i each have ν d.f. Then $\log_e s^2_k/s^2_1 = \log_e s^2_k - \log_e s^2_1$ is the range of the $\log_e s^2_i$; since these are approximately normally distributed with mean $\log_e \sigma^2$ and variance $2/(\nu-1)$, approximate percentage points of s^2_k/s^2_1 can be computed from percentage points of range in normal samples (see section 1.51).

Table 1, p. 309, gives exact and approximate 5% points of s^2_k/s^2_1 to 3 fig. for $k = 2$ and $\nu = 2(1)10$; $120/\nu = 6(2)12$. Table 2, p. 309, gives exact and approximate 5% points and exact 1% points to 3 fig. for $\nu = 2$ and $k = 2(1)12$. Table 3, p. 310, gives approximate 5% points to 3 fig. for $k = 2(1)12$ and $\nu = 2(1)10$; $120/\nu = 0(2)12$. These approximations, adjusted to agree with the exact values of TT. 1, 2, were found not accurate enough and are superseded by the tables of H.A. David 1952, abstracted below.

David, H. A. 1952

<u>Biometrika</u> 39, 422-424

"Upper 5 and 1% points of the maximum F-ratio"

Table (a), p. 424, gives 5% points of s^2_k/s^2_1 to 3 fig. for $k = 2(1)12$ and $\nu = 2(1)10$; $120/\nu = 0(2)12$. Table (b), p. 424, gives 1% points of

s^2_k/s^2_1 for $\begin{cases} k = 2 & 3(1)12 & 3(1)12 \\ \nu = 2(1)10 & 2 & 3(1)10 \\ 120/\nu = 0(2)12 & & 0(2)12 \\ \text{3 fig.} & \text{0 dec.} & \text{2 fig.} \end{cases}$

Reprinted in Pearson & Hartley 1954, T. 31, p. 179.

4.26 <u>Moments and Cumulants of F</u>

The ith moment of F about the origin, μ'_i, is given by

$$\mu'_i = \int_0^\infty F^i f(F; \nu_1; \nu_2) \, dF = \frac{\Gamma(\tfrac{1}{2}\nu_1 + i)\, \Gamma(\tfrac{1}{2}\nu_2 - i)}{\Gamma(\tfrac{1}{2}\nu_1)\, \Gamma(\tfrac{1}{2}\nu_2)} \left(\frac{\nu_1}{\nu_2}\right)^{-i}$$

which exists for all $i < \tfrac{1}{2}\nu_2$. In particular we have

Mean $(F) \equiv \mu'_1 \equiv \nu_2/(\nu_2-2)$ for $\nu_2 > 2$;

Var $(F) \equiv \mu'_2 - (\mu'_1)^2 = 2\nu_2^2 (\nu_1+\nu_2-2)/(\nu_2-2)^2 \nu_1 (\nu_2-4)$

$\qquad\qquad\qquad\qquad\qquad\qquad\qquad\qquad$ for $\nu_2 > 4$.

Romanowsky, V., 1928

Metron 7, part 3, 3

Skewness and Kurtosis of F

The table on pp. 26 and 37 give $\beta_1(F)$ and $\beta_2(F)$ to 2 decimals for selected degrees of freedom ν_1, ν_2.

Notation:

	Romanowsky	Ours
	m	$\nu_1 + 1$
	n	$\nu_2 + 1$

4.27 Approximations to the distribution of F

Paulson, E. 1942

Annals of Mathematical Statistics 13, 233-235

Applying the result of Geary 1930, as quoted by Fieller 1932, to the approximate normalization of χ^2 of Wilson & Hilferty 1931, Paulson finds that

$$U = \left[\left(1-\frac{2}{9\nu_2}\right)F^{1/3}-\left(1-\frac{2}{9\nu_1}\right)\right]\left[\frac{2}{9\nu_2}F^{2/3}+\frac{2}{9\nu_1}\right]^{-\tfrac{1}{2}} \qquad (1)$$

is approximately a unit normal variate; the approximation is satisfactory for $F > 1$ and $\nu_2 \geq 3$. Two short tables illustrate the approximation.

Kelley, T. L. 1948

The Kelley statistical tables

Kelley, with the assistance of K.J. Arnold, offers the approximate normal variate

$$d = \left[\theta_2 F^{1/3} - \theta_1\right]\left[F^{2/3}/\nu_2 + 1/\nu_1\right]^{-\tfrac{1}{2}} \qquad (61)$$

where $\theta_i = (9/2)^{\tfrac{1}{2}} - (2/9\nu_i)^{\tfrac{1}{2}}$; this is a simple modification of (1) above, and indeed $d \equiv U$. When $\nu_2 = 1$ or 2, Kelley and Arnold find that

$$x = d + .08d^5/\nu_2^3 \qquad (63)$$

is a better approximation than (61).

4.27
4.28

Table F, p. 32, gives $Q^*[F(Q; \nu_1, \nu_2); \nu_1, \nu_2]$, where Q^* is an approximation to Q computed by (61) or (63), for

$$\begin{cases} Q = .0001 & .01, .05, .1, .25, .5 \\ \nu_1, \nu_2 = 1, 2, 10, \infty & 1, 2, 10, \infty \\ \quad 4 \text{ dec.} & 3 \text{ dec.} \end{cases}$$

Table 8, p. 222, gives θ_i to 7 dec. for

$\nu_i = 1(1)100(10)200(20)300(50)500(100)1000, 10000, \infty$.

Kelley, T. L. 1947

Fundamentals of statistics

Table IX E, p. 326, gives θ_i, defined in the abstract above, to 4; 6 dec. for $\nu_i = 1(1)50(10)100(100)1000; \infty$.

4.28 Effect of non-normality, heteroscedasticity and dependence on the distribution of F

Let G be a statistic that, in sampling from a normal population, has the tabular F-distribution. Then the behaviour of G in sampling from a non-normal population is exceedingly different, according as G is:

(a) an F-test applied to the comparison of two independent mean squares computed from non-normal data. This problem arises when the F-criterion is used as a test for heterogeneity of variance.

(b) the analysis-of-variance F-test for the comparison of group means. If such a test is applied to data from a non-normal parent, G is simultaneously affected by two departures from normal conditions:

(j) the mean squares no longer follow the tabular χ^2-distribution;

(ij) the mean squares are no longer independent.

The effects of (j) and (ij) on the distribution of G largely compensate each other.

Box 1953, p. 333, warns:

"It has frequently been suggested that a test of homogeneity of variances should be applied before making an analysis of variance test for homogeneity of means in which homogeneity of variance is assumed. The present research suggests that when, as is usual, little is known of the parent distribution, this practice may well lead to more wrong conclusions than if the preliminary test was omitted. It has been shown . . . that in the commonly occurring case in which the group sizes are equal or not very different, the analysis of variance test is affected surprisingly little by variance inequalities. Since this test is also known to be very insensitive to non-normality it would be best to accept the fact that it can be used safely under most practical conditions. To make the preliminary test on variances is rather like putting to sea in a rowing boat to find out whether conditions are sufficiently calm for an ocean liner to leave port!"

Morgan, W. A. 1939

Biometrika 31, 13

Test for difference between correlated variances.

Let x_i, y_i $i = 1, 2, \ldots, n$ denote a sample of n pairs drawn from a binormal parent with $\sigma_1, \sigma_2, \rho_{12}$, then the sample correlation between the pairs $X_i = \frac{1}{2}(x_i + y_i)$; $Y_i = \frac{1}{2}(x_i - y_i)$ is given by

$$r_{xy} = \frac{s_1^2 - s_2^2}{\left[(s_1^2 + s_2^2)^2 - 4 r_{12}^2 s_1^2 s_2^2 \right]^{\frac{1}{2}}}$$

where s_1 and s_2 are sample variances of the x_i and y_i and r_{12} their sample correlation. If r_{xy} is used as a test criterion for H_o:- $\sigma_1/\sigma_2 = 1$ (i.e. $\rho_{xy} = 0$) its percentage points can be obtained from

$$t = \frac{r_{xy} \sqrt{n-2}}{\sqrt{1 - r_{xy}^2}}$$

independent of ρ_{12}, but its power (equa. (23)) will depend on $\gamma = \sigma_1/\sigma_2$ as well as ρ_{12}. This power is tabulated on p. 18 as a function of n, ρ_{12} and

$$\rho_{xy} = \frac{\gamma - \gamma^{-1}}{\left[(\gamma - \gamma^{-1})^2 + 4(1 - \rho_{12}^2) \right]^{\frac{1}{2}}}$$

to 3 dec. for $\begin{cases} n = 12, 25 & 12, 25 & 100 \\ \rho_{12} = 0, .8 & .5 & 0, .5, .8 \\ \rho_{xy} = 0(.1).4(.2).8 & 0(.1).4, .6 & 0(.1).4 \end{cases}$

The power of r_{xy} is compared with the power of s_1/s_2, which is tabled to 3 dec. for some of the above arguments.

4.28 [see also pp. 681-682]

Walsh, J. E. 1947 a

Annals of Mathematical Statistics 18, 88

Let $x_1, \ldots, x_n; y_1, \ldots, y_m$ be an observation from a multivariate normal distribution in which all means are 0, all x variances σ^2 and all y variances σ'^2 and the correlations between any two $x_i = \rho$, any two $y_i = \rho'$, and y and any $x = \rho''$. The confidence interval for σ^2/σ'^2 is then given by

$$0 \le \sigma^2/\sigma'^2 \le \frac{S^2}{S'^2} \frac{m-1}{n-1} \frac{(1-\rho')}{(1-\rho)} / F_\epsilon$$

where $S^2 = \sum(x_i - \bar{x})^2$, $S'^2 = \sum(y_i - \bar{y})^2$, ϵ = confidence coefficient and F_ϵ is the $100(1-\epsilon)$% point of F for n-1, m-1 degrees of freedom, i.e. $F(\epsilon; n-1, m-1)$.

If the usual confidence interval $0 \le \sigma^2/\sigma'^2 \le \frac{S^2}{S'^2} \frac{m-1}{n-1} \frac{1}{F_\epsilon}$ with nominal confidence coefficient ϵ is used, define the actual confidence coefficient $\alpha = \alpha\left(\frac{1-\rho'}{1-\rho}, n, m\right)$ by

$$F(\alpha; n-1, m-1) = [(1-\rho')/(1-\rho)] F(\epsilon; n-1, m-1)$$

Table V (p. 93) gives $\alpha\left(\frac{1-\rho'}{1-\rho}, n, m\right)$ to 2-3 dec. for $(1-\rho')/(1-\rho) = 1$, 1.25, 1.5, 2.0 and the following pairs of n and m

| n | 4 | 16 | 32 | 4 | 16 | 32 | 4 | 32 |
| m | 4 | 4 | 4 | 16 | 16 | 16 | 32 | 32 |

Error: On page 92, 3rd formula from bottom of page should read

$$0 \le \frac{\sigma^2}{\sigma'^2} \le \frac{S}{S'^2} \frac{(m-1)}{(n-1)} \frac{(1-\rho')}{(1-\rho)} \bigg/ F_\epsilon \ ; \ \text{the } \frac{m-1}{n-1} \text{ is}$$

omitted throughout the paper. The tables are not affected.

Geary, R. C. 1947 b

Biometrika 34, 209-242

Table 1, p. 211, gives approximate values of

$$P = \Pr\left[S_1^2/S_2^2 \ge F(.05; \nu_1, \nu_2)\right]$$ for two <u>independent</u> mean squares S_1^2 and S_2^2 computed from two 'large' samples of size $\nu_1 + 1$ and $\nu_2 + 1$

4.28

drawn, independently, from the same non-normal distribution with Kurtosis β_2 (for $\beta_2 = 1.5(.5)6$). Because of the independence of S_1^2 and S_2^2 (which would not, of course, prevail for Analysis-of-Variance mean squares) P may be considerably different from its normal value of .05. Table 2 (p. 217) gives similar values for a t-test based on two samples of size 10 drawn from a non-normal universe with selected values of β_1, β_2. Here $x_1 - x_2$ is dependent on the Within sample S^2 and the effect of non-normality less pronounced.

David, F. N. and Johnson, N. L.   1951 a
Biometrika 38, 43-57
"The effect of non-normality on the power function of the F-test in the analysis of variance"

The Analysis of Variance considered is that of a single classification into K groups $t = 1, 2, \ldots, k$ of varying size n_t involving $F(k-1, \nu; .05)$ in the familiar 5% F test of the between group mean square S_1^2 based on k-1 degrees of freedom against the within group mean square S_2^2 based on $\nu = \sum n_t - k$ degrees of freedom. Deriving the exact first 4 moments of $u = S_1^2 - F(k-1, \nu; .05) S_2^2$ and fitting a Gram-Charlier series to them the author obtains an approximation to $P_r [u \geq 0]$ i.e. the power of the F test for any set of different non-normal within group distributions characterized by their cumulants, and for the linear hypothesis model (cf. section 4.242). The power of the 1% F test is treated by the same means.

David, F. N. and Johnson, N. L.   1951 b
Trabajos de Estadística 2, 179-188
"The sensitivity of analysis of variance tests with respect to random between groups variation"

By the same methods as in David & Johnson 1951 a, the authors investigate the effect of non-normality and heteroscedasticity on the power of the F-test for the component-of-variance model (cf. section 4.241).

4.28 [see also pp. 681-682]

Box, G. E. P. 1954

Annals of Mathematical Statistics 25, 290-302

"Some theorems on quadratic forms applied in the study of analysis of variance problems, I. Effect of inequality of variance in the one-way classification"

Both exact and approximate results are first derived for the distribution of the ratio of two quadratic forms of independent $N(0,\sigma^2)$ variates $Y = \theta_1/\theta_2$ using the following devices:—

(a) When the canonical form of θ only involves χ^2 statistics with even degrees of freedom θ is exactly distributed as a finite χ^2 series;

(b) When θ is positive definite it can be expressed in terms of an infinite χ^2 series developed by Robbins & Pitman 1949.

(c) When θ_1 and θ_2 are independent and (a) and/or (b) apply to θ_1 and θ_2 the ratio Y has a distribution expressible as a finite or infinite F-series.

(d) When θ_1 and θ_2 are dependent we note that
$$P_r\left[\theta_1/\theta_2 \geq \phi\right] \equiv P_r\left[\theta_1 - \phi\,\theta_2 \geq 0\right]$$
and apply (a) to the form $\theta_1 - \phi\,\theta_2$;

(e) If $\theta_1 \geq 0$, $\theta_2 \geq 0$ and θ_1 and θ_2 are independent we can approximate to the probability integral of Y by the F integral $P(Y/b;\ \nu_1,\nu_2)$ with b, ν_1, ν_2 given on p. 297.

The above results are applied to the single classification analysis of variance with θ_1 between group mean square for k groups, θ_2 within group mean square for $N = \sum n_t$ observations to evaluate $P = P_r\left[\theta_1/\theta_2 \geq F(.05;\ k-1,\ N-k)\right]$ in the Null case. Table 4 (p. 299) gives selected results by both exact methods (a) to (d) and approximate method (e). Since agreement between them is good method (e) permits some wider conclusions summarized in Table 5 (p. 300) for the case of equal n_t.

Box, G. E. P. 1954 a

4.28

<u>Annals</u> <u>of</u> <u>Mathematical</u> <u>Statistics</u> 25, 484-498

"Some theorems on quadratic forms applied in the study of analysis of variance problems, II. Effects of inequality of variance and of correlation between errors in the two-way classification"

Using the results of Box 1954, Box derives formulas for the effects named in the title. Table 2, p. 493, illustrates the effect of differences between column variance on the between-rows test; T. 6, p. 497, illustrates the effect of a first-order within-row serial correlation on both the between-rows test and the between-columns test.

4.3
4.333

4.3 The z-distribution

4.30 Definition of z and relation to other statistics

Fisher originally proposed to compare two mean squares by means of the statistic $z = \frac{1}{2}\log_e\left(s_1^2/s_2^2\right) = \frac{1}{2}\left(\log_e s_1^2 - \log_e s_2^2\right)$.

The transformation of the variance ratio F into $z = \frac{1}{2}\log_e F$ has certain theoretical advantages and for large ν_1, ν_2 the distribution of z is approximately $N(0, \frac{1}{2}(\frac{1}{\nu_1} + \frac{1}{\nu_2}))$. Springing from this approximate normality there result a convenience in harmonic interpolation for large ν_1, ν_2 and a number of approximations to the percentage points of z discussed in Section 4.35. With the publication of extensive tables of F, however, these advantages of z disappear and in practical analysis of variance work the use of F is more direct (as was first recognized by Snedecor).

The distribution ordinate of z is obtained from that of $F = e^{2z}$ and is

$$f(z; \nu_1, \nu_2) = \frac{2\,\Gamma(\frac{1}{2}\nu_1 + \frac{1}{2}\nu_2)}{\Gamma(\frac{1}{2}\nu_1)\,\Gamma(\frac{1}{2}\nu_2)}\, \nu_1^{\frac{1}{2}\nu_1}\, \nu_2^{\frac{1}{2}\nu_2}\, e^{z\nu_1}\,(\nu_2 + \nu_1 e^{2z})^{-\frac{1}{2}(\nu_1+\nu_2)}$$

The probability integral of z is obtained from that of F by evaluating it at argument $F = e^{2z}$; the percentage points of z are given by

$$z(Q; \nu_1, \nu_2) = \frac{1}{2}\log_e F(Q; \nu_1, \nu_2).$$

4.33 Percentage points of z and of $\log_{10} F$

4.331 Tables of percentage points of z
4.332 Tables of percentage points of $\log_{10} F = 2z \log_{10} e$

These sections will be found on pp. 250-251.

4.333 Errors in tables of percentage points of z

Norton 1952 thoroughly examined the z-tables of Colcord & Deming 1936 and Fisher & Yates 1948 and found the following errors. We show Colcord & Deming's value only when it differs from Fisher & Yates's.

p	ν_1	ν_2	Colcord & Deming	Fisher & Yates	Norton
·2	4	1		1·3097	1·3067
·2	8	1		1·3400	1·3397
·2	24	2		·7452	·7453
·2	24	21		·1829	·1831
·1	∞	120		·0081	·0881
·001	1	3		2·5604	2·5591
·001	1	5		1·9255	1·9270
·001	1	7		1·6874	1·6879
·001	1	40		1·2674	1·2672
·001	1	120	1·2159	1·2158	1·215937
·001	2	5		1·8002	1·8071
·001	2	29		1·0903	1·0901
·001	2	60		1·0248	1·0250
·001	2	120	·9948	·9952	·995381
·001	3	1		6·5966	6·6000
·001	3	40		·9435	·9431
·001	3	120	·8783	·8773	·877320
·001	4	6		1·5433	1·5438
·001	4	7		1·4211	1·4224
·001	4	8		1·3332	1·3333
·001	5	6		1·5177	1·5175
·001	5	19		·9422	·9452
·001	5	120	·7425	·7426	·742580
·001	6	120	·6983	·6986	·698586
·001	8	5		1·6596	1·6598
·001	8	29		·7679	·7669
·001	8	120	·6329	·6338*	·633739
·001	12	21		·7735	·7734
·001	24	5		1·6123	1·6121
·001	24	6		1·4134	1·4136
·001	24	8		1·1662	1·1659
·001	24	19		·7277	·7279
·001	24	21		·6964	·6965
·001	24	120	·4369	·4380	·438128
·001	∞	3		2·4081	2·4080
·001	∞	60		·3198	·3184
·001	∞	120	·2224	·2199	·216964

With the exception marked *, these errors have been corrected in the 1953 and 1957 editions of Fisher & Yates.

4.339
4.35

4.339 Miscellaneous significance points of z   [see also pp. 675-676]

Jeffreys, H., 1948

Theory of Probability

Table V (p. 404) gives roots z to 2 dec. from

$$K = \left(\frac{\pi n}{2}\right)^{\frac{1}{2}} \frac{\cosh 2z}{\cosh 2} e^{nz} \exp\left[\tfrac{1}{2} n (1 - e^{2z})\right], \qquad (11)$$

and from

$$\frac{1}{K} = \frac{\sqrt{2}}{\pi} \int_0^\infty \frac{u^2 + 1}{u^4 + 1} u^n \exp\left[\tfrac{1}{2} nb^2 (1 - u^2)\right] du, \qquad (14)$$

where $b = e^z$, $K = 10^x$, for $x = -2(\tfrac{1}{2})0$; $= n-1 = 1(1)10(2)20, 50$.
Table V also gives $K = (\pi n/2)^{\frac{1}{2}}$, corresponding to $z = 0$ in equation (11), for the same n.

4.34 Cumulants of $\log \chi^2$ and of z

It is readily verified that the characteristic function of $\log_e[\tfrac{1}{2}\chi^2]$ is $\Gamma(\tfrac{1}{2}\nu + it)/\Gamma(\tfrac{1}{2}\nu)$. Hence

$$\varkappa_1(\log_e \chi^2) = \log_e 2 + \Psi(\tfrac{1}{2}\nu),$$

$$\varkappa_i(\log_e \chi^2) = \Psi^{(i-1)}(\tfrac{1}{2}\nu), \quad i > 1;$$

$$\varkappa_1(\tfrac{1}{2}\log_e \chi^2/\nu) = \tfrac{1}{2}[\Psi(\tfrac{1}{2}\nu) - \log_e(\tfrac{1}{2}\nu)],$$

$$\varkappa_i(\tfrac{1}{2}\log_e \chi^2/\nu) = 2^{-i}\Psi^{(i-1)}(\tfrac{1}{2}\nu), \quad i > 1;$$

the Ψ function and its differentials are defined on pp. 163-164, and tables thereof are indexed there. Since $z = \tfrac{1}{2}\log_e \chi_1^2/\nu_1 - \tfrac{1}{2}\log_e \chi_2^2/\nu_2$, it follows that

$$\varkappa_1(z) = \tfrac{1}{2}[\Psi(\tfrac{1}{2}\nu_1) - \log_e \nu_1 - \Psi(\tfrac{1}{2}\nu_2) + \log_e \nu_2],$$

$$\varkappa_i(z) = 2^{-i}[\Psi^{(i-1)}(\tfrac{1}{2}\nu_1) + (-1)^i \Psi^{(i-1)}(\tfrac{1}{2}\nu_2)], \quad i > 1.$$

These formulas are multifariously transformed in Wishart 1947.

The Cornish-Fisher approximation to z depends essentially on expanding $\varkappa_i(z)$ in asymptotic series in negative powers of ν_1 and ν_2.

Bartlett, M. S. and Kendall, D. G. 1946

<u>Supplement to the Journal of the Royal Statistical Society</u> 8, 128-138

Denote $\varkappa_i(\log_e \chi^2/\nu)$ by $k_i(\nu)$. Table 1, p. 130, gives, all for $\nu = 1(1)20$, $k_1(\nu)$ to 5 dec., together with five-decimal values computed from its approximate formula $-\left(\dfrac{1}{\nu} + \dfrac{1}{3\nu^2}\right)$; $1/k_2(\nu)$ to 5 significant figures; the ratio $2/(\nu-1)k_2$ to 4 decimals (illustrating that $k_2(\nu) \doteq 2/(\nu-1)$; and $\gamma_1 = k_3(\nu)/k_2^{3/2}(\nu)$ and $\gamma_2 = k_4(\nu)/k_2^2(\nu)$ both to 3 decimals (these are indicative of the slow convergence to normality). Finally the 'efficiency ratio'
$E = 100(2/\nu\, k_2(\nu))$ is given to 2 decimals.

If s_1^2 and s_2^2 are two independent mean squares both based on ν degrees of freedom, then, for large ν, the distribution of $2z^2/k_2(\nu)$ is approximately the χ^2-distribution with 1 d.f. Table 2, p. 131, gives $2[z(\tfrac{1}{2}\alpha;\nu,\nu)]^2/k_2(\nu)$ to 2 dec. for $\nu = 1, 2, 3, 6, 12, \infty$ and $\alpha = .01, .02, .05, .1$. The approximation is seen to be poor for small ν and shown to be worse than the χ^2 approximations to Bartlett's test criterion which are given alongside for the above ν and α.

4.35 Approximate formulas for percentage points of z

R.A. Fisher 1925 prescribed that, in large samples, z be treated as a normal variate with mean 0 and variance $\tfrac{1}{2}(1/\nu_1 + 1/\nu_2)$. For smaller samples, Fisher gave numerical formulas for specific percentage points, which were shown by Cochran 1940 to arise from the general formula

$$z(Q;\nu_1,\nu_2) = x(h-1)^{-\tfrac{1}{2}} + \left(\tfrac{1}{6} - \lambda\right)\left(\dfrac{1}{\nu_1} - \dfrac{1}{\nu_2}\right) \qquad (4.35.1)$$

where x is the normal deviate corresponding to upper tail area Q, $\lambda = (x^2+3)/6$, and $2/h = 1/\nu_1 + 1/\nu_2$. The abstracts below are concerned with refinements of (4.35.1).

Approximations to z can readily be derived from the approximations to F of section 4.27, from the approximations to t of sections 4.115 and 4.332, and from the approximations to the incomplete beta-function of sections 3.41 and 3.43.

4.35
5.1
Cochran, W. G. 1940

Annals of Mathematical Statistics 11, 93-95

By considering the Cornish-Fisher reversion of the Gram-Charlier expansion of z, taken through terms in $\nu^{-3/2}$, Cochran obtains the formula

$$z(Q;\nu_1,\nu_2) = x(h-\lambda)^{-\frac{1}{2}} + \left(\frac{1}{6}-\lambda\right)\left(\frac{1}{\nu_1}-\frac{1}{\nu_2}\right) \qquad (4.35.2)$$

where notation is as in (4.35.1) above. The first table on p. 95 gives λ to 2 dec. for Q = .001, .01, .05, .1(.1).4 . The second table gives $z(Q;24,60)$ to 4 dec.: exact from Fisher & Yates, and approximated by (4.35.1) and (4.35.2); all for Q = .001, .01, .2 .

The footnotes to T. V of Fisher & Yates give λ to 1 dec.; λ can be readily obtained to 4 dec. by adding .1667 to the coefficient of $(1/\nu_1 - 1/\nu_2)$ in those footnotes.

Carter, A. H. 1947

Biometrika 34, 352-358

Carter proposes the formula

$$z = \frac{x\sqrt{h'+\lambda'}}{h'} - \frac{d'}{6}\left(x^2 + 2 - 2s'\right)$$

where $s' = \dfrac{1}{\nu_1-1} + \dfrac{1}{\nu_2-1}$, $d' = \dfrac{1}{\nu_1-1} - \dfrac{1}{\nu_2-1}$

$$\lambda' = \frac{x^2-3}{6}, \quad h' = \frac{2}{s'},$$

and x is the normal deviate corresponding to tail area Q.

Table 1, p. 353, gives z to 4 dec.: exact (from Fisher & Yates), by (4.35.2) and by Carter's formula; all for p = .001, .01, .05, .2 and for

$\begin{cases} \nu_1 = 6 & 12 \quad 20 \quad\quad 24 \quad\quad 36 \quad\quad\quad 60 \quad\quad\quad 100 \\ \nu_2 = 12, 60 & 6 \quad 36, 100 \quad 24, 60 \quad 20, 36, 60 \quad 6, 24, 36 \quad 20 \end{cases}$

Table 1a, p. 353, gives z to 4 dec. from the Cornish-Fisher expansion for p = .001, .01, .05, .2 ; $\begin{cases} \nu_1 = 6 & 12 \quad 24 \quad 60 \\ \nu_2 = 12, 60 & 6 \quad 60 \quad 6, 24 \end{cases}$

Table 2, p. 355, gives x, λ and λ' for p = .001, .01, .05, .2 .

SECTION 5

VARIOUS DISCRETE AND GEOMETRICAL PROBABILITIES

5.1 The hypergeometric distribution

The hypergeometric distribution may be written in the form

$$\Pr[x = r] = \binom{n}{r}\frac{\Gamma(Np+1)\,\Gamma(Nq+1)\,\Gamma(N-n+1)}{\Gamma(Np-r+1)\,\Gamma(Nq-n+r+1)\,\Gamma(N+1)}, \quad r = 0(1)n, \quad p+q = 1. \quad (5.1.1)$$

When the distribution arises in sampling from a finite population, Np and Nq are integers.

Most extant tables of the hypergeometric distribution have been compiled for the purpose of testing independence in 2×2 contingency tables: see section 2.6, particularly Finney 1948, Latscha 1953, Armsen 1955.

Hypergeometric _functions_, and confluent hypergeometric functions, are ubiquitous in statistical computations: see Erdélyi et al. 1953, p. 87, for the representation of the incomplete beta-function as a hypergeometric function, and Erdélyi et al. 1953, pp. 266-267, for the representations of the incomplete gamma-function and the Hermite polynomials as confluent hypergeometric functions. For the representation of the non-central F-distribution in terms of confluent hypergeometric functions, see Tang 1938, Lehmer 1944, N.B.S. 1949, 1951. For a list of tables of confluent hypergeometric functions, see section 4.142.

The epithet 'hypergeometric' is applied to the distribution (5.1.1) because its probability generating function (Kendall & Stuart 1958, p. 133) and its factorial moment generating function (Aitken 1939, p. 57) are hypergeometric functions. This fact, of some importance in statistical combinatorics, is not helpful to the table user, because (Fletcher et al. 1946, p. 335) the "tabulation of the general hypergeometric function $F(\alpha,\beta,\gamma,x)$ would call for tables with 4 arguments, and [they] have not encountered any direct tabulation." Fletcher et.al.

5.1

go on to point out that K. Pearson & Elderton 1923 and K. Pearson 1931b tabulate certain hypergeometric functions.

For the definition of the bivariate hypergeometric distribution see Aitken 1939, p. 84; and for an example, see K. Pearson 1924. For the hypergeometric distribution with negative n see K. Pearson 1928.

Chung, J. H., and DeLury, D. B. 1950
Confidence limits for the hypergeometric distribution

Given a finite population of size N containing, say, X defectives. If a random sample of size n is drawn without replacement, the probability of drawing x defectives is given by the hypergeometric distribution. The cumulative distribution of x given N, X and n is then

$$\sum_{i=0}^{x} \frac{\binom{X}{i}\binom{N-X}{n-i}}{\binom{N}{n}} .$$

A series of charts is given showing two-sided $100p\%$ confidence limits for the proportion defective X/N for $\begin{cases} N = 500, 2500, 10000 \\ p = .9, .95, .99 \\ n/N = .05, .1(.1).9 \end{cases}$. For each combination of p and n/N, three charts are given: in the first, x/n ranges from 0 to .5; in the second, from 0 to .1. In the third chart, confidence limits for X are given corresponding to stated values of x; this chart gives N = 2500 only for n = 125.

According to a tipped-in errata slip, when reading _lower_ confidence limits, observed numbers defective should be diminished by 1, and observed proportions defective by 1/n.

Greenwood, M. 1913
Biometrika 9, 69-90
Posterior binomial probability distribution

Given that in a pilot sample of size n there were i successes, the chance of j successes in a subsequent sample of size m is computed from Bayes' theorem as

$$P(j,m|i,n) = (n+1)\binom{m}{j}\binom{n}{i} / (m+n+1)\binom{m+n}{i+j} .$$

See Condorcet 1785, p. 188.

This is tabled to 6 dec. for $j = 0(1)m$ and

$\begin{cases} n = 5(1)9 & 10,\ 15 & 20,\ 25 & 50 & 100 & 100 \\ m = 5(1)n & 5(5)n & 5(5)n & 5(5)25,\ 50 & 5(10)25,\ 20,\ 50 & 10 \\ i = 0(1)[\tfrac{1}{2}n] & 0(1)[\tfrac{1}{2}n] & 0(1)5 & 0(1)5 & 0(1)10 & 0(1)10(5)50 \end{cases}$

Reprinted in Tables for S & B I, T. 48, pp. 89-97.

Pearson, K. 1928

<u>Biometrika</u> 20 A, 149-174

A sample of n containing r marked members and $s = n-r$ unmarked members has been drawn from a population of N containing R marked members and $S = N-R$ unmarked members. The posterior distribution of R is given by $\binom{R}{r}\binom{N-R}{s} / \binom{N\ 1}{n+1}$.

The table on p. 156 gives $\binom{R}{20}\binom{1000-R}{80} / \binom{1001}{101}$ to 5 dec. for $R = 80(1)363$.

See also Peck & Hazelwood 1958.

5.2 Generalized Poisson distributions

5.21 Contagious distributions

Like the Poisson distribution, discussed in sections 2.12 and 2.22, the distributions of this section are discrete distributions over the non-negative integers; in addition to the mean, they contain one or more additional parameters, which operate to make the variance greater than the mean. A useful introduction to contagious distributions is furnished by pp. 358-366 of Anscombe 1950.

The most common contagious distribution, the negative binomial, is treated in section 5.3.

Wagner, K. W. 1913
Elektrotechnische Zeitschrift 34, 1279-1281

Wagner tables $f(x,y) = \int_{-\infty}^{+\infty} \exp[-y^2 z^2 - z - xe^{-z}]dz$ to about 3 fig. for $x = 0, 0.001, 0.003, 0.008, 0.02, 0.05, 0.1, 0.2, 0.4, 0.7, 1.0, 1.5, 3, 5, 10, 25$; $y = 0(.05)1$.

Reprinted in Jahnke & Emde 1933, p. 113; 1938, 1948, p. 39.

Grundy 1951, p. 428, states that if $x = a$, $y = (2\sigma^2)^{-\frac{1}{2}}$, then $f(x,y) = \sigma\phi_1(2\pi)^{\frac{1}{2}}/a$, where the notation is explained in the abstract of Grundy 1951, below.

Garwood, F. 1940
Supplement to the Journal of the Royal Statistical Society 7, 65-77
"An application of the theory of probability to the operation of vehicular-controlled traffic signals"

The traffic at a simple cross-roads, consisting of a main stream and a subsidiary stream crossing it, is supposed to be controlled by vehicle-actuated signals that keep a vehicle in the subsidiary stream waiting until a time interval I occurs between successive main-stream vehicles, subject to a maximum waiting period M. The main-stream traffic is assumed to follow the Poisson distribution and to arrive at the average rate of N per unit time.

The probability of a waiting period not less than T

$$P(T) = e^{-NT} \sum_{n=1}^{\infty} (NT)^n q_n(I/T),$$

where

$$q_n(a) = \sum_{i=0}^{r} (-)^i \binom{n+1}{i} (1-ia)^n$$

$ra \leq 1 < (r+1)a$.

The mean waiting time

$$\overline{T} = N^{-1} \sum_{i=0}^{r} (-)^i (NM)^i (1-ia)^i [NM(1-ia) + i + 1] e^{-iaNM}/(i+1)!, \text{ where } a = I/M.$$

Table 1, p. 68, gives $P(M)$ to 3 dec., and $\overline{T}$ to 0 dec., for $I = 5(1)9$, $M = 25(5)35$, and 36 $N = 1(1)6$.

Wald, A. and Wolfowitz, J. 1945

Annals of Mathematical Statistics 16, 30-49

The table on pp. 46-47 gives

$$T = \sum_{i=0}^{M-1} e^{-x}(M-i)x^i/i!$$

to 2 dec. for $M = 2(1)16$ and $x = 1(1)25$.

Shenton, L. R. 1949

Biometrika 36, 450-454

"On the efficiency of the method of moments and Neyman's type A distribution"

Neyman's type A generalization of the Poisson distribution is given by

$$P_0 = e^{\lambda' - m_1}, \quad P_x = e^{-m_1} \frac{m_2^x}{x!} \sum_{j=1}^{\infty} \frac{\lambda'^j j^x}{j!}, \quad x > 0, \text{ where } \lambda' = m_1 e^{-m_2}.$$

Table 1, p. 453, gives the percentage efficiency of jointly estimating m_1 and m_2 by the method of moments, using the first two sample moments, for $m_1 = .1, .5, 1, 2, 3, 4, 6, 10$ and $m_2 = .1, .2, .5, 1, 2, 3, 5$.

Thomas, Marjorie 1949

Biometrika 36, 18-25

"A generalization of Poisson's binomial limit for use in ecology"

The 'generalization' considered is the law

$$P_0 = e^{-m}; \quad P_k = e^{-m} \sum_{r=1}^{k} e^{-r} m^r (r\lambda)^{k-r}/r!(k-r)!, \quad k > 0;$$

in which the number of clusters of plants' per quadrat follows a Poisson distribution with

5.21

mean m and the number of plants in excess of 1 per cluster follows a Poisson distribution with mean λ. In a sample of N, the cell frequencies n_0 and n_1 provide maximum likelihood estimates of m and λ; the estimating equations are $\hat{m} = \log_e(N/n_0)$, $\hat{\lambda} = \log_e(n_0 m/n_1)$. The variances of these estimates are $N\hat{\sigma}^2_{\hat{m}} = e^m - 1$, $N\hat{\sigma}^2_{\hat{\lambda}} = m^{-2}[me^{m+\lambda} - 1 + (1-m)^2 e^m]$; these variances can be estimated by substituting $\hat{m}$ and $\hat{\lambda}$ for m and λ. The estimated variances are denoted $s^2_{\hat{m}}$ and $s^2_{\hat{\lambda}}$.

$\hat{m}$ and $\hat{\lambda}$ are tabled to 3 dec., $N^{\frac{1}{2}}s_{\hat{m}}$ and $N^{\frac{1}{2}}s_{\hat{\lambda}}$ to 2 dec., for $n_0/N = .05, .1(.1).9$; $n_1/N = .05, .1(.1).3$.

Bateman, G. I. 1950

<u>Biometrika</u> 37, 59-63

"The power of the χ^2 index of dispersion test when Neyman's contagious distribution is the alternate hypothesis"

Neyman's contagious distribution [of type A] has been defined in the abstract of Shenton 1949, above. Let $x_1, \ldots, x_N$ be a sample from this distribution; compute the index of dispersion $z = \bar{x}^{-1}\sum(x_i - \bar{x})^2$. For a Poisson parent, i.e. for $m_2 = 0$, z is approximately distributed as χ^2 with N-1 degrees of freedom.

For $m_2 > 0$, Bateman obtains approximate expansions for the first four moments of z, and denotes the ratios of these to the corresponding moments of $(1 + m_2)\chi^2$ by $R_1, \ldots, R_4$. Table 1, p. 62, gives these ratios, to somewhat more figures than are significant, for $m_1 m_2 = 5, 10, 20$ and

$\begin{cases} m_2 = 0 & .5 & 1, 2 \\ N = 5, 10, 20, \infty & 5, 10, 20, 50, \infty & 5, 10, 20, 50, 100, \infty. \end{cases}$

Raff, M. S. 1951

<u>Journal</u> <u>of</u> <u>the</u> <u>American</u> <u>Statistical</u> <u>Association</u> 46, 114-123

"Distribution of blocks in an uncongested stream of automobile traffic"

The number, K, of cars passing a fixed location during an interval t is assumed to follow the Poisson distribution with mean Nt.

The waiting time, T, of a car arriving via a side road and trying to enter the stream is measured from the arrival of the car to the occurrence of a gap in the stream: i.e. an interval L in which no cars pass the junction. The cumulative distribution $F(T)$ of the waiting time is found to be

$$F(T) = e^{-NL} \sum_{r=0}^{h} (-e^{-NL})^r [(NT-rNL)^r/r! + (NT-rNL)^{r+1}/(r+1)!]$$

where h is the greatest integer not exceeding T/L.

Table 1, p. 118, gives $F(T)$ to 3 dec. for $NL = 0(.25)2(.5)6$ and $T/L = 0(.5)3(1)7(2)11, 17, 27, 42, 64, 94, 136, 197, 297$. Most values of $F(T)$ that exceed .99 are omitted.

Thomas, Marjorie 1951
Biometrika 38, 102-111

It is assumed that the number of plants, k, counted in a 'quadrat' follows the generalized Poisson distribution of Thomas 1949, abstracted above. The first seven cumulants of this distribution are given on p. 107.

Three criteria are examined for testing the hypothesis $\lambda = 0$, i.e. no clustering:

(j) A general criterion based on n_0, the frequency of quadrats with no plants. This criterion was introduced by Stevens 1937b in another context.

(ij) The index of dispersion z, defined in the abstract of Bateman 1950, above.

(iij) The maximum likelihood estimate of λ based on n_0 and n_1, the frequency of quadrats with 0 and 1 plant, viz the estimate $\hat{\lambda}$, defined in the abstract of Thomas 1949, above.

Tables 1-3, pp. 105-109, give the power of these three criteria.

Grundy, P. M. 1951
Biometrika 38, 427-434

"The expected frequencies in a sample of an animal population in which the abundances of species are log-normally distributed"

The probability of obtaining r individuals of a species with abundance m is assumed to follow the Poisson distribution with mean m. If it is further assumed that in a super-population of species the abundances follow the log-normal distribution

$$f(m) = (2\pi)^{-\frac{1}{2}} (m\sigma)^{-1} e^{-(\log m - \log a)^2/2\sigma^2},$$

then the expected proportion of species represented by r individuals is

$$\phi_r = (r!)^{-1} \int_0^\infty m^r e^{-m} f(m)\, dm.$$

5.21
5.23

Table 1, p. 429, gives $1 - \phi_0$, the expected proportion of species represented in the sample; T. 2, p. 430, gives ϕ_1, the expected proportion of species represented by one individual each; both to 4 dec. for $\log_{10} a = -2(.25)+3$; $\sigma^2 = 2(1)16$.

5.22 The bivariate Poisson distribution

Maritz, J. S. 1950

Psychological Bulletin 47, 434-443

Aitken 1939, pp. 94-95, defines the bivariate Poisson distribution as

$$p(x,y) = \exp(-m_1-m_2+m_3)(m_1-m_3)^x(m_2-m_3)^y(x!y!)^{-1}\sum_{i=1}^{z} i! \binom{x}{i}\binom{y}{i}(m_1-m_3)^{-i}(m_2-m_3)^{-i},$$

where $z = \min(x,y)$. For this distribution $E(x) = \text{Var}(x) = m_1$; $E(y) = \text{Var}(y) = m_2$; $\text{Cov}(x,y) = m_3$; and the marginal distributions of x and y are Poisson.

Table 1, p. 435, illustrates the distribution by giving $500\,p(x,y)$ to 1 dec. for $m_1 = m_2 = 0.5$; $m_3 = 0.4$.

5.23 Distribution of the difference between two Poisson variates

Jørgensen, N. R. 1916

Undersøgelser over Frequensflader og Korrelation

Jørgensen introduces the function $f_r(t) = t^{-r}I_r(t) = i_r(t)$, where the Bessel functions I and i are defined in B.A. 10, 1952, pp. xxxij, xxxv, in equation (30) on p. 17. Table 2, pp. 153-156, gives $\log_{10} f_r(t)$ to 7 dec. for $r = 0(1)11$; $t = 0(.1)6$; with first differences in t.

For application see the following abstract.

Theil, H. 1952

Revue de l'Institut International de Statistique 20, 105-120

Let i and j be independent Poisson variates with means m and μ. The distribution of $d = i - j$ is given (Charlier 1908, p. 2; Jørgensen 1915, p. 6; Jørgensen 1916, p. 7; Skellam 1946) by the formula

$$P = \Pr[i - j = d] = e^{-m-\mu}(m/\mu)^{\frac{1}{2}d}I_{|d|}[2(m\mu)^{\frac{1}{2}}],$$

where the Bessel function I is defined in B.A. 10, 1952, p. xxxij.

5.21
5.23

Table 3, p. 120, gives P to 3 dec. for

$\begin{cases} m = \tfrac{1}{2} & 1 & 2 & 4 \\ \mu = \tfrac{1}{2},\ 1,\ 2,\ 4 & 1,\ 2,\ 4 & 2,\ 4 & 4 \end{cases}.$

5.3
5.33

5.3 The negative binomial distribution and the logarithmic series

[see also p. 683]

The negative binomial distribution may be defined by the probability generating function $(1 + d - dt)^{-h/d}$, so that

$$\Pr[x = r] = \frac{(1+d)^{-h/d}}{\Gamma(h/d)} \frac{\Gamma(r + h/d)}{r!} \left(\frac{d}{1+d}\right)^r . \qquad (5.3.1)$$

$E(x) = h$, $\text{Var}(x) = h(1+d)$.

The cumulative sums of the negative binomial may be expressed in terms of incomplete beta functions:

$$\Pr[x \le r] = I_y(h/d, r+1) = 1 - I_{1-y}(r+1, h/d) ;$$

$$\Pr[x \ge r] = I_{1-y}(r, h/d) = 1 - I_y(h/d, r) ;$$

where $y = (1+d)^{-1}$ and other notation is as in (3.0.2). When h/d is an integer (say $h/d = N$), both the cumulative sums and the individual terms of the negative binomial may be expressed in terms of positive binomials:

$$\Pr[x \ge r] = \sum_{i=r}^{r+N-1} \binom{r+N-1}{i}(1-y)^i y^{r+N-1-i} = Q(1-y, r+N-1, r-\tfrac{1}{2}) = P(y, r+N-1, N-\tfrac{1}{2})$$

$$\Pr[x = r] = (1+d)^{-1}\binom{r+N-1}{r}(1-y)^r y^{N-1} .$$

The logarithmic distribution is obtained by excluding the class $x = 0$ from (5.3.1):

$$\Pr[x = r \mid x > 0] = \frac{1}{\Gamma(h/d)[(1+d)^{h/d}-1]} \frac{\Gamma(r + h/d)}{r!} \left(\frac{d}{1+d}\right)^r ,$$

and then taking the limit as $h \to 0$:

$$\Pr[x = r] = \frac{1}{\log d}\frac{1}{r}\left(\frac{d}{1+d}\right)^r .$$

5.31 Tables of $\binom{N+r-1}{r} = {}^N H_r$

All tables below give exact values.

$r =$	$N =$	
2, 3	1(1)5001 - r	
4, 5	1(1)2001 - r	J. C. P. Miller 1954
6(1)11	1(1)1001 - r	
12 ·	1(1)489	
2(1)12	3(1)133 - r	Sasuly 1934
2(1)10	1(1)111 - r	Jordan 1932
2(1)10	1(1)101 - r	Vianelli 1959(556)
2(1)10	1(1)51 - r	B. A. 9, 1940

5.32 The negative binomial distribution

Sichel, H. S. 1951

Psychometrika 16, 107-127

R. A. Fisher 1941 a has shown that: if the negative binomial distribution be written in the form

$$f(r) = \frac{\Gamma(r+p)}{\Gamma(p)\,\Gamma(r+1)}\left[\frac{p}{p+m}\right]^p\left[\frac{m}{p+m}\right]^r,$$

the optimum estimates of p and m are uncorrelated; the moment estimates of m and p are $\tilde{m} = \bar{r}$, $\tilde{p} = \bar{r}^2/(s_r^2 - \bar{r})$; m is fully efficient; the efficiency of $\tilde{p}$ is equal to

$$\left[2\sum_{r=2}^{\infty} r^{-1} m^{r-2}(p+m)^{2-r}\Gamma(r)\,\Gamma(p+2)/\Gamma(p+r)\right]^{-1}.$$

The maximum likelihood estimate of p, calculated from the sample $x_1, \ldots x_n$, is the root of

$$n^{-1}\sum_{i=1}^{n}\Psi(x_i+p-1) = \Psi(p-1) + \log_e(1 + \bar{x}/p).$$

Table 1, pp. 118-120, gives $\Psi(x+p-1) - \Psi(p-1)$ to 5 dec. for $x = 0(1)35$ and $p = .1(.1)3$; table 2, p. 121, gives $\Psi(p-1)$ to 5 dec. for $p = .1(.1)3$.

L. Katz 1952 a alleges that there are some errors in T. 1, but does not specify them.

5.33 The logarithmic series

Fisher, R. A., Corbet, A. S. and Williams, C. B. 1943

Journal of Animal Ecology 12, 42-58

"The relation between the number of species and the number of individuals in a random sample of an animal population"

Let the expected number of species with n individuals be given by the term $\alpha n^{-1} x^n$ of the logarithmic series. The expected total number of species is then $S = -\alpha \log_e(1-x)$; and the expected total number of individuals is $N = \alpha x/(1-x)$. Table 7, p. 53, gives $S = \alpha\log_e(1 + N/\alpha)$ mostly to 2 dec. for
$\begin{cases}\alpha = 1(1)10(2)20(5)50(10)100,\ 200\\ N = 10,\ 20,\ 50,\ 100,\ 200,\ 500,\ 1000,\ 2000,\ 5000,\ 10000\end{cases}$
$\begin{cases}150\\ 500,\ 1000,\ 2000,\ 5000,\ 10000\end{cases}$ $\begin{matrix}5,\ 10,\ 20,\ 50,\ 100\\ 100000\end{matrix}$. Table 8, p. 53, gives the standard error of α,

$$\alpha^{3/2}\left[(N+\alpha)^2\log_e\frac{2N+\alpha}{N+\alpha} - \alpha N\right]^{\frac{1}{2}}(SN+S\alpha-N\alpha)^{-1},$$

to 3 fig. for $\alpha = 1, 5, 10, 20, 50, 100$; $\log_{10} N = 1(1)5$.

5.33
5.4

[see also pp. 683-684]

Table 9, p. 55, gives $A = \log_{10} N/\alpha$ as a function of $B = \log_{10} N/S$, whence $10^{-B} = 10^{-A}\log_e(1 + 10^A)$, to 5 dec. for $B = 0.40(.01)3.59$.

Table 10 is entitled "The amount of information respecting k, supposed small, according to the numbers of individuals (N) and species (S) observed"; k being the exponent of the negative binomial distribution considered as an alternative to the logarithmic series. According to R.A. Fisher 1950, pp. 43.53 a-b, the function tabled is

$$t^2/12 + \pi^2/6 - 2t^{-1}s_3 + \sum_{r=1}^{\infty} r^{-2} e^{-rt}(1 + 2/rt) - t^2(1-e^{-t})/4(t-1+e^{-t})$$

$$= t^2/12 - t^3/144 + t^4/1080 + t^5/32400 - t^6/27216 - 341\, t^7/228614400 + O(t^8),$$

where s_3 is the sum of the inverse third powers of the natural numbers. Table 10 gives this function to 4 dec. with argument

$$\log_{10} \frac{e^t - 1}{t} = .4(.1)3.5.$$

5.4 Multinomials

Neyman, J. and Pearson, E. S. 1931a
Biometrika 22, 298-305

Table 2, p. 302, gives 3-7 dec. values of

$$P_c = \Sigma \, p(i,j,k) = \Sigma \, 10!(.3)^i(.5)^j(.2)^k / i! \, j! \, k!$$

where the $p(i,j,k)$ are arranged in order of magnitude and progressively summed, starting from the smallest, to form the P_c. These are compared with $p(\chi^2, 2)$, given to 3-8 dec., where $\chi^2 = (i-3)^2/3 + (j-5)^2/5 + (k-2)^2/2$; and with 3-7 dec. values of P_λ, the chance of obtaining a sample with a value of $\lambda = (3/i)^i (5/j)^j (2/k)^k$ smaller than the observed value. P_λ is computed exactly, i.e. <u>not</u> from a χ^2 approximation.

The comparison of P_c with χ^2 is given in greater detail, and slightly different notation, by El Shanawany 1936, p. 181.

El Shanawany, M. R. 1936
Biometrika 28, 179-187

Table 1, p. 181, gives (in the notation of the previous abstract) $p(i,j,k)$ to 5 dec., P_c to 4 dec., χ^2 to 4 dec., $p(\chi^2, 2)$ to 4 dec., and the percentage error of the χ^2 approximation. Table 3, pp. 185-186, gives the same information for the multinomial $20!(.3)^i(.5)^j(.2)^k / i! \, j! \, k!$.

El Shanawany opines that Yates' correction would not improve the approximation.

Kullback, S. 1937
Annals of Mathematical Statistics 8, 127-144

Consider N trials in which n events e_i may occur with frequencies arrayed by the multinomial $(\Sigma p_i t_i)^N$. Count in each set of N trials the number x of events that do not occur and the number y of events that occur exactly once. Kullback derives formulas for the distribution of x and y on pp. 133-136. The table on p. 136 gives the distribution of x, and the table on p. 138 gives the distribution of y, for the parent multinomial distribution

$$(.1 + .1 + .1 + .1 + .1 + .1 + .1 + .1 + .1 + .1)^{10}.$$

5.5 Some geometrical probabilities

For the integral of the multivariate normal distribution over regions bounded by planes, see pp. 119-125; for the integral over spherical regions, see pp. 155-157.

Wicksell's two papers are addressed to the following problem:

"In an opaque body there are suspended a large number of corpuscles of different sizes, the density of the corpuscles and their size-distribution being the same in all parts of the body. To express the distribution of the sizes of the corpuscles in terms of the distribution of the sizes of the contours of corpuscles found in a plane section by which the body is split in two."

Wicksell, S. D. 1925

Biometrika 17, 84-99

The corpuscle problem, defined on p. 86, leads Wicksell to the system of equations

$$f_0 = P_0 a_{00} + \sum_{j=1}^{15} P_j a_{0j}, \quad f_i = P_i a_{i0} + \sum_{j=1}^{15-i} P_{i+j} a_{ij};$$

where $a_{00} = 1/4$, $a_{0j} = j - [j^2 - 1/4]^{\frac{1}{2}}$, $a_{i0} = [i^2 - (i-\frac{1}{2})^2]^{\frac{1}{2}}$,

$a_{ij} = [(i+j)^2 - (i-\frac{1}{2})^2]^{\frac{1}{2}} - [(i+j)^2 - (i+\frac{1}{2})^2]^{\frac{1}{2}}$. The table on p. 95 gives the matrix of this system, and its inverse, to 4 dec.

Wicksell, S. D. 1926

Biometrika 18, 151

Wicksell (eq. 11 bis, p. 157) establishes the law, independent of the exact distribution of corpuscle diameters, $h_0 M'(r^s) = \lambda_s M(\rho^{s+1})$, where ρ is the geometric mean of the principal diameters of an ellipsoidal corpuscle, r is the geometric mean of the principal diameters of a central elliptic section, h_0 is the mean height of the corpuscle, and M and M' denote mean values. He gives (eq. 13, p. 158) general formulas for λ_s in terms of the eccentricities e_2 and e_3 for $s = -1(1)+4$.

Table 1, p. 160, gives 3 dec. values of λ_s for $s = -1(1)+4$ and $e_2 = e_3 = 0(.1).9, .95, .99$. Table 3, p. 162, gives 3 dec. values of $\lambda_{-1}/\lambda_{s-1}$ for the same arguments.

Table 2, p. 161, gives 3 dec. values of λ_s for $e_3 = 0$ and s and e_2 as in Table 1.

Table 4, p. 163, gives 3 dec. values of λ_{-1}/λ_s for the same arguments. The λ-quotients are used in the equation

$$M(\rho^{s+1}) = \frac{k_{-1}\lambda_{-1}M'(r^s)}{k_s\lambda_s M'(r^{-1})}$$

where $k_s = \frac{1}{2}B(1+\frac{1}{2}s, \frac{1}{2})$.

Garwood, F. 1947

Biometrika 34, 1-17

"The variance of the overlap of geometrical figures, with reference to a bombing problem"

A number k of circles are placed at random on a plane so that each circle has some or all of its area inside a fixed square. What are the mean and variance of the area of the part of the square covered by circles? Following Robbins 1944, 1945, Garwood gives (pp. 5-9) means and variances, depending on k and the dimensions of the square and circles, for the above problem and for similar configurations, viz:

(j) circles falling on square
(ij) circles falling on rectangle 2 × 1
(iij) circles falling on rectangle 4 × 1
(iv) circles falling on circle
(v) squares falling on square (sides parallel).

Assuming that k follows a Poisson distribution, he obtains variances which depend on c, the ratio of the falling area to the fixed area, and on m, the mean fraction of area not covered. Table 1, p. 10, gives these variances for the five configurations for c = .2(.8)1.8 and m = .25(.25).75 ..

For fixed k and configuration (j) Garwood finds an empirical formula (64) for the variance. Table 5, p. 15, gives the variance to 4 dec. for k = 1(1)6 and c = .2(.2)1.8, together with approximate values from the empirical formula. Table 6, p. 16, gives the corresponding comparison with k following a Poisson distribution and c and m as in T. 1.

5.5

Oberg, E. N. 1947

Annals of Mathematical Statistics 18, 442-447

"Approximate formulas for the radii of circles which include a specified fraction of a normal bivariate distribution"

Let
$$\phi(x,y) = (2\pi\sigma_x\sigma_y)^{-1} \exp\left[-\frac{x^2}{2\sigma_x^2} - \frac{y^2}{2\sigma_y^2}\right]$$

Required to find approximate formulas for the radius R of circle c with center at origin such that

$$\int_c \phi(x,y)\, dx\, dy = p \tag{1}$$

Three approximate formulas are derived, viz:

$$R_1 = \sqrt{2\sigma_x\sigma_y \log_e 1/(1-p)}$$

$$R_2 = \sqrt{(\sigma_x^2 + \sigma_y^2) \log_e 1/(1-p)}$$

$$R_3 = (\sigma_x + \sigma_y)\sqrt{\tfrac{1}{2} \log_e 1/(1-p)}$$

Tables I, II, III give 4 dec. values of

$$p_j = \iint_{c_j} \phi(x,y)\, dx\, dy$$

where c_j is the circle with radius R_j for $j = 1, 2, 3$; $p = .1, .2, .25, .3(.1).7, .75, .8, .9$; $\sigma_x/\sigma_y = .5(.1)1.0$.

Krishna Iyer, P. V. 1950

Annals of Mathematical Statistics 21, 198-217

"The theory of probability distributions of points on a lattice"

A two-dimensional lattice consists of a rectangular array of m × n points. Each point is of one of k different colors; it is known that n_i points are of the ith color, where $\Sigma n_i = mn$. Let s_{ij} denote the number of times a point of the ith color adjoins a point of the jth color.

For $k = 2$, TT. 1-4, pp. 204-206, give the coefficients of $p^{n_1} q^{n_2}$ in the joint distribution of n_1 and s_{12}, in the lattices: $\begin{Bmatrix} m = 3 & 4 \\ n = 2(1)4 & 2 \end{Bmatrix}$. Tables 7-8, pp. 212-213, gives the same for the three-dimensional lattices $2 \times 2 \times 2$ and $2 \times 3 \times 3$. Table 9, p. 214, gives the distribution of the number of joins of different colors in the lattices $2 \times 2 \times 2$, $2 \times 2 \times 3$, $2 \times 3 \times 3$ and $3 \times 3 \times 3$, when one point is black, one red and the others white; T. 10, p. 215, gives the same when one point is black, one red, one green and the others white.

Daniels, H. E. 1952

Biometrika 39, 137-143

"The covering circle of a sample from a circular normal distribution"

Consider a sample of n points drawn from a bivariate normal population with center at the origin, zero correlation, equal x- and y-variances (and therefore circular symmetry). Define the covering circle as the smallest circle in the x-y plane containing the n points in its interior or boundary. Let r denote its radius; let d denote the distance of its center from the origin. Daniels derives both the joint distribution of r and d and the marginal distribution of r.

Table 1, p. 140, gives the mean and upper and lower 5% and 1% of r/σ to 3 dec., and its variance to 4 dec., for $n = 2(1)15, 20(10)50, 100$.

Gardner, A. 1952

Journal of the Royal Statistical Society, Series B 14, 135-139

M. Greenwood 1946 posed the following problem: A fixed interval of length L is divided into $N+1$ sub-intervals by placing N dividing points at random. Required to find the distribution of the variance of the $N+1$ sub-intervals.

Irwin and Isserlis obtained results, reported by Greenwood, for $N = 2$.

For $N = 3$, Gardner obtains exact formulas for the distribution of S, the sum of squares of the intervals, and the corresponding density. Table 1, p. 139, gives illustrative values to 3 dec. or more.

5.9

[see also p. 688]

5.9 <u>Miscellanea</u>

Waite, H. 1910

<u>Biometrika</u> 7, 421-436

"Mosquitoes and malaria. A study of the relation between the number of mosquitoes in a locality and the malaria rate"

Let

p = average population
m = number of infected persons
a = number of anophelines
n = total number of persons bitten by mosquitoes during a month
r = recovery rate per month

Waite finds, for a stable population, the empirical formulas $r = 1/m - 1/p$, $\log_{10}(1-r) = -.100343/n$, $n = am/192\,p$, which define a as an implicit function of m given p. The tables in Waite's paper are all computed for $p = 1000$.

Table A, p. 426, gives a to the nearest integer for $m = 5, 8, 10(5)25(25)200(50)950$. As approximations to a, T. B, p. 430, gives $44.36\,p^2/(p-m)$ for $m = 0, 10, 50(50)300$; the table on p. 431 gives $44.36[p^2/(p-m) - p/2m - (p-m)/12m^2]$ for $m = 10, 50, 100, 500$.

Tables C to H, pp. 432-434, give the monthly malaria cases m_s from the formula

$$m_s = m_{s-1} R_s^{n_s} + \frac{1 - R_s^{n_s}}{1 - R_s}$$

where $n_s = am_{s-1}/192\,p$, $R_s = 1 - r_s - 1/p$,

$$r_s = \begin{cases} 1 - 10^{-.100343/n_s}, & n_s \geq 1 \\ .2063, & n_s \leq 1 \end{cases};$$

for $\begin{cases} m_0 = 50 & 50 & 500 & 500 & 500 & 500 \\ a = 64000 & 24000 & 96000 & 48000 & 24000 & 0 \\ s = 1(1)70 & 1(1)48 & 1(1)12 & 1(1)60 & 1(1)38 & 1(1)25 \end{cases}$.

Fieller, E. C. 1931

<u>Biometrika</u> 22, 377-404

"The duration of play"

A and B play a game, their respective chances of winning being p and q. A starts with a counters, B with b counters. At each game the loser pays one counter to the winner. When $a = \infty$, B must be ruined eventually. The probability, $_nP_b$, that B is ruined in n games is expressed in terms of incomplete beta-functions: by equation (4) for infinite a and by equation (10.2) for finite a.

The table on p. 399 gives $\hat{n}$, the (fractional) root of $_nP_b = \frac{1}{2}$, to 3 dec. for $a = \infty$; $p = \frac{1}{2}$; $b = 1(1)10(10)100$. n is calculated by two approximate formulas (equations (24) and (28)). For each b, a convenient integer n near $\hat{n}$ is chosen; and $_nP_b$ is evaluated from the exact formula (4) and three approximations:

$$\begin{cases} \text{formula (4)} & (13) & (16) & (19) \\ b = 3(1)6 & 7(1)10(10)100 & 7(1)10(10)100 & 3(1)10(10)100 \\ 8 \text{ dec.} & 8 \text{ dec.} & 5\text{-}6 \text{ dec.} & 8 \text{ dec.} \end{cases}$$

Fisher, R. A. and Yates, F.   1938, 1943, 1948, 1953, 1957
<u>Statistical Tables</u>

$\Delta^r 0^s / r! = \sum_{i=1}^{r} (-)^{r-i} i^s / i!(r-i)$ is the number of ways of dividing s objects into r groups of all possible sizes. Table XXII gives this number as an exact integer for $r \leq s \leq 25$.

For a related table see Tables for S & B II, p. 246. For a guide to related tables see Fletcher et al. 1946, pp. 69-80, who list some 35 varieties. For statistical applications see W.L. Stevens 1937b, 1937z; F.N. David 1950. David's paper is abstracted below.

Dodd, E. L.   1942
<u>Econometrica</u> 10, 249-257

Define the range, in any k-digit number, as the difference between the greatest and the least digit. The probability that the range in a k-digit decimal number shall equal r, $\Pr[R = r] = (10 - r)\phi(r,k)10^{-k}$, where $\phi(r,k) = (r+1)^k - 2r^k + (|r-\frac{1}{2}|-\frac{1}{2})^k$. This is tabled exactly for $r = 0(1)9$ and $k = 2(1)6$.

David, F. N.   1950
<u>Biometrika</u> 37, 97-110

Given N sample units which may fall with equal probability into any of N cells, the chance that exactly z cells are vacant is $N^{-N}\binom{N}{z}\Delta^{N-z}0^N$ and is given in T. 1a, p. 109, to 4 dec. for $N = 3(1)20$ and $z = 0(1)N-1$.

5.9
6.1
[see also p. 688]

Table 1b, p. 110, gives convenient upper and lower significance points of z with corresponding tail sums for $N = 5(1)20$.

Given N sample units which may fall with equal probability into any of 2N cells, the chance that exactly Z cells, out of N specified cells, are vacant is $(2N)^{-N} \binom{N}{Z} \Delta^{N-Z} N^N$ and is given in T. 4a, p. 110, to 4 dec. for $N = 1(1)10$ and $Z = 0(1)N$; in particular, the chance that no cells are vacant is $(2N)^{-N} N!$.

Table 4b gives convenient significance points of Z with corresponding tail sums for $N = 3(1)10$.

Other tables compare the distributions of z and Z with normal approximations. The use of z and Z as criteria for testing whether a sample has come from a completely specified population is discussed.

Katz, L. 1952
<u>Annals</u> <u>of</u> <u>Mathematical</u> <u>Statistics</u> 23, 271-276
"The distribution of the number of isolates in a social group"

Each of a group of N individuals chooses d of the remaining N-1 individuals as associates. An individual not chosen by any of the others as an associate is termed an isolate. The probability, under random choice, of exactly k isolates,

$$P(k;N,d) = \sum_{j=k}^{N-d-1} (-)^{j+k} \binom{j}{k} S_j ,$$

where $S_0 = 1$ and

$$\frac{S_{k+1}}{S_k} = \frac{N-k+d}{k+1} \left[\frac{N-k-d}{N-k}\right]^{k-1} \left[\frac{N-k-d-1}{N-k-1}\right]^{N-k-1} .$$

Table 1, p. 275, gives S_k to 7-9 dec. and $P(k;N,d)$ to 6 dec. for $N = 26$ and $d = 3$, together with Poisson and modified Poisson approximations.

See also Hammersley 1953.

SECTION 6
LIKELIHOOD AND OTHER STATISTICS USED IN TESTING HYPOTHESES AND IN ESTIMATION

6.1 Tests applied to normal populations

Useful surveys of these tests are given by Plackett 1946 and Box 1949. We quote Box's summary (Biometrika 36, 345-346):

For a particular class of likelihood criteria, whose moments appear as the product of Γ-functions, a general method is described for obtaining probability levels when the null hypothesis is true. A number of statistics whose moments appear in this form are referred to, and a general method developed to obtain:

(a) A series which is in close agreement with the exact distribution.
(b) An approximate solution, using a single χ^2 distribution, which is sufficiently accurate for moderate or large samples.
(c) A rather better approximation, using a single F distribution, giving close agreement even when the samples are rather small.

The method is illustrated for the following two general statistics:

(1) Tests for constancy of variance and covariance

(a) Univariate case. The F approximation is of the same order of accuracy as Hartley's (1940) series solution although it requires very much less calculation, and significance may be judged by consulting tables of the significance points of the variance ratio F alone.

(b) Multivariate case. The series solution shows remarkably close agreement with the exact distribution when this is known, and is used in other cases to compare approximations. The χ^2 approximation does not correspond with that found by Bishop (1939), but is, in fact, simpler and more accurate.

The series confirms the accuracy of significance points found by fitting a type I curve to the first two moments of l_1. The calculation of the moments involved in this method renders it too laborious for routine use, and Bishop suggested two working approximations; the F approximation developed here is more accurate than these approximations, whilst it involves no more labour and can be used when the sample sizes are unequal.

(2) Wilks's [1935] test for independence of k groups of variates

The asymptotic series, and χ^2 and F approximations are derived for this case, and the relation of the results with those of Wald & Brookner [1941], Bartlett [1938], and Rao [1948] is discussed. The exact

6.1
6.112

distribution is used to assess the accuracy of the proposed methods in a number of cases. The probabilities given by the series are found to be in excellent agreement with the true values, even for fairly small samples. Providing the sample sizes are not too small, the x^2 and F approximations will be sufficiently accurate, the latter providing the better approximation, and allowing the sample size to be rather smaller than is possible with the x^2 approximation. When the number of variates in each group is one, we have a test criterion for the hypothesis that k variates are mutually independent, and the same procedure provides the series solution and simple approximations for tests of significance.

6.11 Tests applied to univariate normal populations

6.111 Neyman and Pearson's one- and two-sample tests

Neyman, J. and Pearson, E. S. 1928 a

<u>Biometrika</u> 20A, 175-240

To test the hypothesis that a sample of n has been drawn from a normal population with mean μ and variance σ^2, Neyman & Pearson compute the statistics $\bar{x}$, $s^2 = (n-1)^{-1}\sum(x_i-x)^2$, $M = (\bar{x}-\mu)/\sigma$, $S^2 = (n-1)s^2/n\sigma^2$, $k = (M^2+S^2)\log_{10} e - \log_{10} S^2$, $\lambda = (10^k e^{-1})^{-\frac{1}{2}n} = S^n e^{-\frac{1}{2}n(M^2+S^2-1)}$; they tabulate the integral $P_\lambda = 2(\frac{1}{2}n)^{\frac{1}{2}n}\; ^{-\frac{1}{2}}[\Gamma(\frac{1}{2}n-\frac{1}{2})]^{-1}\int\!\!\int_\omega S^{n-2} e^{-\frac{1}{2}n(M^2+S^2)} dM\, dS$, where ω is the exterior of the curve k = constant.

Table 12, p. 235, gives P_λ/λ for $n > 10$ to 4; 3 dec. for k = .435(.005).65; .7(.05)1. Reprinted in Tables for S & B II, T. 36, p. 223.

The table on pp. 238-240 gives P_λ to 4 dec. for $n = 3(1)50$, k = .44(.01).65(.05)1.5(.3)3. Reprinted in Tables for S & B II, T. 35, pp. 221-223.

Figs. 13-14, pp. 236-237, give contours in the (M,S) plane, which can be read to about 2 dec., for k = .44(.01).46(.02).6(.05).8(.1)1.5, 1.8. Reprinted in Tables for S & B II, pp. clxxxij-clxxxiij.

Pearson, E. S. and Neyman, J. 1930

Bulletin de l'Académie Polonaise des Sciences, Série A 1930, 73-96

To test the hypothesis that two samples of sizes n_1 and n_2 have been drawn from the same normal population, Neyman and Pearson compute the statistics $s_1'^2 = \sum(x_1-\bar{x}_1)^2/n_1$, $s_2'^2 = \sum(x_2-\bar{x}_2)^2/n_2$, $\bar{x} = (n_1\bar{x}_1 + n_2\bar{x}_2)/(n_1+n_2)$, $s_0'^2 = \left[\sum(x_1-\bar{x})^2 + \sum(x_2-\bar{x})^2\right]/(n_1+n_2)$, $\lambda = s_1'^{n_1} s_2'^{n_2} s_0'^{-n_1-n_2}$. Thus λ must lie between 0 and 1.

The moments of λ are Γ-function ratios and a beta distribution has been fitted to the mean (m_1) and variance (m_2), given in T. 1, p. 90, by setting $a = m_1[m_1(1-m_1) - m_2]/m_2$, $b = (1-m_1)[m_1(1-m_1) - m_2]/m_2$. With these values of a and b, TT. 2, 3, p. 92, give lower 5% and 1% points of λ to 4 dec. for n_1, n_2 = 5, 10, 20, 50, ∞.

Reprinted in Tables for S & B II, TT. 37[a], 37[b], pp. 223-224.

Sukhatme, P. V. 1935

Proceedings of the Indian Academy of Sciences, Section A 2, 584-604

Set $L = \lambda^{2/N}$, where $N = n_1+n_2$ and λ is defined in the previous abstract. Table 1, pp. 596-597, gives 5% and 1% points of L to 3 dec. for n_1, n_2 = 5(1)10; $120/n_1$, $120/n_2$ = 0(1)6(2)12. The exact distribution of L under the null hypothesis can be reduced to an integral involving incomplete beta functions as integrands: T. 1 was computed from fitted Type I approximations and compared with the exact results in special cases.

6.112 Neyman and Pearson's k-sample tests L_0 and L_1

Neyman & Pearson 1931b, pp. 461-463, contemplate k independent normal samples x_{ti} ($t = 1, \ldots, k$; $i = 1, \ldots, n_t$). They distinguish three hypotheses:

H. The k population means are equal, the k population variances are equal: against an alternative that gives arbitrary means and variances.

H_1. The k population means are arbitrary, but the k population variances are equal: against an alternative that gives arbitrary means and variances.

H_2. The k population means are equal, the k population variances are

6.112

equal: against an alternative that gives arbitrary means and equal variances. H_2 will be recognized as the analysis-of-variance test for a one-way classification.

Neyman & Pearson denoted the statistics arising in likelihood-ratio testing of these hypotheses λ_H, λ_{H_1}, λ_{H_2}. Mahalanobis 1933a, p. 110, introduced the statistics $L_0 = \lambda_H^{2/N}$, $L_1 = \lambda_{H_1}^{2/N}$, $L_2 = \lambda_{H_2}^{2/N}$; later writers have adopted these notations.

In the general case of unequal n_t,

$$N \log L_0 = \sum n_t \log \left[(n_t-1)s_t^2/n_t\right] - N \log \left[(N-1)s_0^2/N\right],$$

$$N \log L_1 = \sum n_t \log \left[(n_t-1)s_t^2/n_t\right] - N \log \left[(N-k)s_a^2/N\right],$$

where $N = \sum n_t$, $s_t^2 = (n_t-1)^{-1}\sum(x_{ti}-\bar{x}_t)^2$, $s_a^2 = (N-k)^{-1}\sum(n_t-1)s_t^2$, $s_0^2 = (N-1)^{-1}\sum\sum(x_{ti}-\bar{x})^2$, and $\bar{x}$ is the grand mean of all the x_{ti}; in the special case of equal n_t (all equal to n, say),

$$k \log L_0 = \sum \log s_t^2 - k \log \left[(N-1)ns_0^2/N(n-1)\right],$$

$$k \log L_1 = \sum \log s_t^2 - k \log s_a^2.$$

Mahalanobis, P. C. 1933a

Sankhyā 1, 109-122

"Tables for L-tests"

Let all the n_t equal n. Tables 1 and 2, pp. 113-114, give means and standard deviations of L_0 and L_1 to 4 dec. for k = 2(1)5, 10, 20, 25, 50 and n = 2(1)5(5)20(10)50. These are computed from the gamma-function ratio formulas of Neyman & Pearson 1931b, p. 476. Tables 3 to 6, pp. 121-122, give lower 5% and 1% points of L_0 and L_1 to 4 dec. for the above k and n. These were obtained by fitting a Type I to the moments of L_0 (or L_1) and then interpolating in a table of percentage points of Fisher's z.

The tables of L_0 for k = 2 are found to agree with those of Pearson & Neyman 1930, p. 92, except for the 5% point for n = 5 and the 1% point for n = 10.

Tables 3 to 6 are reprinted in Freeman 1942, TT. 9-10, pp. 174-175.

Nayer, P. P. N. 1936

Statistical Research Memoirs 1, 38-51

Again let all $n_t = n$. The first two moments of L_1 are given by gamma-function ratios and lead to the approximation of the distribution of L_1 by a Pearson Type I curve.

Table 1, p. 41, gives μ_1' and $\mu_2^{\frac{1}{2}}$ to 4 dec. for $k = n = 10$; $k = 30$, $n = 20$; together with the corresponding values for $L_1' = L_1^n$. Table 2, p. 42, gives, for the same n and k, 5% and 1% points obtained by four Type I approximations, viz: 2- and 3-moment fit to L_1, 2- and 4-moment fit to L_1'.

Table 3, p. 44, gives exact 5% and 1% points for $k = 2$ (from the F-distribution) together with approximate percentage points obtained by a 2-moment Type I fit to L_1, to 4 dec. for $n = 2(1)4, 10, 20, 50$.

Tables 4 and 5, p. 51, give lower 5% and 1% points of L_1 to 3 dec. for $k = 2(1)10(2)30$; $n = 3(1)10, 120/n = 0(2)12$. The values for $k = 2$ were computed from the F-distribution; the others, by a 2-moment Type I fit to L_1. Table 6, p. 45, gives lower 5% and 1% points of L_1 to 3 dec. for $n = 2$ and $k = 10(2)30, 50$. Tables 4, 5, 6 are reprinted in P.O. Johnson 1949, T. 5, pp. 366-367; Vianelli 1959, Pro. 92, p. 258.

Table 7, p. 47, gives 4 dec. values of the lower 5% and 1% points of L_1, computed from the approximation $\chi^2_{k-1} = -N \log_e L_1$, for $n = 30, 60$; $k = 2, 3, 5, 10(10)30$; together with 3 dec. values from TT. 4, 5.

An investigation of unequal n_t on pp. 47-48 leads Nayer to conclude that "if none of the values of n_t are less than 20, the tables may be entered with k and $\bar{n}$", where $\bar{n} = N/k$.

Mood, A. M. 1939

Annals of Mathematical Statistics 10, 187-190

Let λ_i denote the ith cumulant of $-N \log_e L_1$. The table on p. 190 gives $\lambda_i / [(i-1)! 2^{i-1}(k-1)]$ to 3 dec. for $i = 1(1)4$; $n = 20, 100, \infty$; $k = 10, \infty$.

Mood concludes "that the degree of approximation of $[-n \log_e L_1]$ to the χ^2 law with $k-1$ degrees of freedom is mainly dependent on n, and is for all practical purposes independent of k when n is moderately large."

6.113 Bartlett's modification of the L_1 test

Bartlett 1937 proposed a statistic analogous to λ_{H_1}, but using degrees of freedom instead of sample sizes: in the case where the appropriate $\nu_t = n_t - 1$, Bartlett's statistic μ may be defined by the equation

$$-2 \log \mu = (N-k) \log s_a^2 - \sum (n_t - 1) \log s_t^2$$

where notation is as in section 6.112. Pearson and Hartley (Thompson & Merrington 1946, p. 297) set $M = -2 \log_e \mu$; we shall describe the several tables in terms of M. The following notations are on record:

Bartlett 1937	$-2 \log \mu$
Bishop & Nair 1939	$-2 \log \mu = -F \log L_1'$
Pitman 1939	$2L$
Hartley 1940	$-2 \log \mu = -F \log L_1' = x$
Thompson & Merrington 1946	
Pearson & Hartley 1954	M .

For the history of Bartlett's test see two notes by E.S. Pearson: Pitman 1939, p. 215; Pearson & Hartley 1954, p. 57.

Bartlett 1937, p. 274, introduces the correction factor

$$C = 1 + \frac{1}{3(k-1)} \left[\sum \frac{1}{n_t - 1} - \frac{1}{N-k} \right]$$

and shows that M/C is approximately distributed as χ^2 with $(k-1)$ d.f.; for criticism of this approximation, see Hartley 1940, p. 254.

Bishop, D. J. and Nair, U. S. 1939
Supplement to the Journal of the Royal Statistical Society 6, 89-99

Let $F = N-k$. Table 1a, p. 93, gives lower 5% and 1% points of $L_1' = e^{-M/F}$ to 4 dec. for $\nu_1 = \ldots = \nu_k = 1$, computed from:

(1) the exact expansion of U.S. Nair 1939

(2) Type I fit to L_1'

(2A) Type I fit to $L_1'^{1/5}$

(3) Bartlett's M/C approximation.

For methods (2A) and (3), $k = 2(1)10$; for methods (1) and (2), $k = 2(1)5, 10$.

6.113

Table 1b, p. 94, gives the same to 3 dec. for $\nu_1 = \ldots = \nu_k = \nu$ (say):

$$\begin{cases} k = & 2 & 3, 5, 10 & 3, 5, 10 & 20, 50 \\ \nu = & 2(1)4, 9 & 2, 4 & 3, 9 & 2(1)4, 9 \\ \text{method} & 1, 2, 3 & 1, 2, 3 & 2, 3 & 2, 3 \end{cases}$$

Table 2, p. 95, tests the accuracy of the percentage points computed by method 3, by using method 1 to compute the associated probability level, for $k = 2(1)5, 10$ and $\nu_1 = \ldots = \nu_k = 1$.

Table 3, p. 96, gives lower 5% and 1% points of L_1', computed by method 1, to 3 dec., and corresponding values of M to 1 dec., for $\nu_1 = \ldots = \nu_k = 1$ and $k = 2(1)10(5)30$.

Table 4, p. 97, gives lower 5% and 1% points of L_1' for various combinations of unequal ν_t, viz:

$$\begin{cases} \nu_1 = 5 & 6 & 7 & 10 & 14 & 16 & 17 & 5 & 6 & 7 & 7 & 10 & 14 & 16 & 17 & 14 & 16 & 17 \\ \nu_2 = 4 & 4 & 4 & 9 & 9 & 9 & 9 & 5 & 6 & 6 & 7 & 9 & 9 & 9 & 9 & 14 & 16 & 17 \\ \nu_3 = 3 & 2 & 1 & 8 & 4 & 2 & 1 & 4 & 4 & 4 & 4 & 9 & 9 & 9 & 9 & 9 & 9 & 9 \\ \nu_4 = 0 & 0 & 0 & 0 & 0 & 0 & 0 & 3 & 2 & 2 & 1 & 9 & 9 & 9 & 9 & 4 & 2 & 1 \\ \nu_5 = 0 & 0 & 0 & 0 & 0 & 0 & 0 & 3 & 2 & 1 & 1 & 8 & 4 & 2 & 1 & 4 & 2 & 1 \end{cases}.$$

Pitman, E. J. G. 1939

<u>Biometrika</u> 31, 200-215

Table 1, pp. 209-210, gives the exact distribution of M and Bartlett's M/C approximation to 4 dec. for $k = 2$ and

$$\begin{cases} \nu_1 = 1 & 2 & & 10 \\ \nu_2 = 1 & 2, 4, \infty & & 10 \end{cases} ; \text{ and for } k = 3, \nu_1 = \nu_2 = 2, \nu_3 = 4.$$

Notation: $\begin{cases} \text{Pitman's} & 2L & 2m_i & 2M \\ \text{ours} & M & \nu_i & N-k \end{cases}$

Thompson, Catherine M. and Merrington, Maxine 1946

<u>Biometrika</u> 33, 296-304

Hartley 1940 obtained the distribution of M (under homoscedasticity) in the form of a χ^2-expansion depending on the parameters

$$k, \quad c_1 = \sum \nu_t^{-1} - (N-k)^{-1}, \quad c_3 = \sum \nu_t^{-3} - (N-k)^{-3}.$$

[301]

6.114
6.115
[see also p. 688]

Tables 1 and 2, pp. 302-303, based on this expansion, give the 5% and 1% points of M to 2 dec. for $k = 3(1)15$; $c_1 = 0(.5)5(1)10(2)14$, $c_1 \leq k$; $c_3 = k^{-2}c_1^3$ and $c_3 = c_1$. These are approximately the limiting values of c_3 given k and c_1 . Table 3, p. 304, gives $k^{-2}c_1^3$ and $c_1 - k^{-2}c_1^3$ for k and c_1 as in TT. 1 and 2, to facilitate interpolation.

Reprinted in Pearson & Hartley 1954, TT. 32-33, pp. 180-182.

Box, G. E. P. 1949
Biometrika 36, 317-346

See summary on p. 295. Table 1, p. 327 of Box; TT. 3, 4, p. 333; T. 5, p. 334; contain illustrative values for Bartlett's M.

6.114 Other tests of homogeneity of variance

Neyman, J. 1941
Annals of Mathematical Statistics 12, 46-76

Let $s_1^2, \ldots, s_N^2$ be N independent mean squares, each with ν degrees of freedom, so that s_i^2 is distributed as $\chi_\nu^2 \sigma_i^2 / \nu$. To test the hypothesis that all the σ_i^2 are equal, against the alternative that they follow a Pearson Type V distribution, Neyman introduces the statistic $\zeta = \sum s_i^4 / \left[\sum s_i^2\right]^2$. He obtains, under both the null hypothesis and the alternative, the first 4 moments of ζ ; and approximates the distribution of ζ by a bivariate normal integral extended over the interior of a certain parabola. This approximation is suitable when $N > 100$ and $\nu < 5$.

Table 3, p. 75, gives $\Pr[\zeta > z]$ to 4-5 dec. for $\nu = 2$, $N = 100$, and $z_1 = zN\nu/(\nu+2) = .8(.1)1.7$; and 5% and 1% points of z_1 . Table 4, p. 75, gives 5% and 1% points of z and illustrative values of the power of the ζ-test, for the same ν and N.

Rushton, S. 1952
Sankhyā 12, 63-78

"On sequential tests of the equality of variances of two normal populations with known means"

Rushton's test criteria are based on:

(a) cumulative sums of squares about the (assumed known) population means.

(b) cumulative sums of squares about the respective sample means.

(c) cumulative sums of ranges in subsamples of 4 and 8.

For (a) and (b), let U denote the ratio of the two cumulative sums of squares. The hypothesis $\sigma_1 = \sigma_2$ is accepted as soon as $U \leq U'$; the hypothesis $\sigma_1 = \delta\sigma_2$ is accepted as soon as $U \geq U''$. Table 1, pp. 72-74, gives U' and U'' for $= 1(1)10(2)20(5)30$; $\delta = 1.5, 2, 3$; $\alpha = .01, .05, .1$; $\beta = .01, .05, .1$. For (a), $\nu = n$; for (b), $\nu = n-1$.

Table 2, pp. 75-77, gives corresponding critical values R' and R'' for (c): for $\alpha = .01, .05$; $\beta = .01, .05$;
$\begin{cases} \delta = 1.5 & 2, 3 \\ n = 1(1)10(2)20 & 1(1)10(2)30 \end{cases}$; where n denotes the number of subsamples.

For Cochran's test of homogeneity of variance see section 11.24.

6.115 Sequential t-tests

N.B.S. 1951

The principal table, pp. 1-82, gives values of $z(L,n,\delta)$, i.e. the root of $L = \log_e F(\tfrac{1}{2}n, \tfrac{1}{2}, \tfrac{1}{2}\delta^2 z) - \tfrac{1}{2}\delta^2 n$, to 2-3 dec. for certain values of L, n, δ in the ranges $\pm L = 2, \log_e 19, 3, 4, \log_e 99, 5, 6, 7$; $n = 1(1)200$; $\delta = .1(.1)1(.2)2, 2.5$; for which $n^{-1} \leq z \leq n$.

The introduction gives nine tables (pp. xiv-xvij) to assist in the application of the main table to Wald's sequential t-test.

Kitagawa, T., Kitahara, T., Nomachi, Y. and Watanabe, N. 1953
Bulletin of **Mathematical Statistics**, Fukuoka 5, No. 3~4, 35-45
"On the determination of sample size from the two sample theoretical formulation"

The problem of designing a two-sample procedure for a confidence interval of fixed length, d, for the mean of a normal distribution, leads to the function

$$I(n_2;n_1 \mid d^2\sigma^{-2}, \alpha, \beta) = \int_{b(n_2-n_1)}^{b(n_2)} \varphi_{n_1}(s_1;\sigma)\,ds_1 \int_0^{c(n_2)} \varphi_{n_2}(s_2;\sigma)\,ds_2$$

where $c(n) = dn^{\tfrac{1}{2}}/t(\tfrac{1}{2}\alpha, n-1)$, $b(n) = c(n)[F(\beta; n-1, n-1)]^{-\tfrac{1}{2}}$, and $\varphi_n(s;\sigma)$ is the ordinate of the distribution of s for (n-1) degrees of freedom.

6.115
6.12   [see also pp. 688-691]

The table on pp. 38-45 gives this function to 4 dec. for

$$\begin{cases} \alpha = .01 & .01 & .01 & .01 \\ \beta = .01 & .01 & .05 & .05 \\ d^2\sigma^{-2} = .75 & .5 & .75 & .5 \\ n_1 = 21, 25, 31 & 25, 31 & 10, 15, 21, 25, 31 & 21, 25, 31 \\ n_2 = 11(1)60 & 15(1)60 & 7(1)60 & 12(1)60 \end{cases}$$

$$\begin{cases} .05 & .05 & .05 & .05 & .05 \\ .01 & .01 & .01 & .05 & .05 \\ .75 & .5 & .25 & .5 & .25 \\ 15, 21, 25, 31 & 21, 25, 31 & 31 & 10, 15, 21 & 10, 15, 21 \\ 6(1)51 & 9(1)54 & 16(1)61 & 5(1)41 & 9(1)60 \end{cases}$$

6.119 Miscellanea

Albert, G. E. and Johnson, R. B. 1951

Annals of Mathematical Statistics 22, 596-599

Let $\bar{x}$ and s^2 be the mean and variance of a sample of size n from a unit normal distribution. S.S. Wilks 1941 has shown that $E[P(\bar{x}+\lambda s) - P(\bar{x}-\lambda s)] = 1 - 2Q$, where P is the normal integral, $\lambda = (n+1)^{\frac{1}{2}} n^{-\frac{1}{2}} t(Q, n-1)$ and $t(Q, \nu)$ is the single-tail $100 Q \%$ point of t with ν degrees of freedom.

Given Π, α, d_1, d_2, Albert & Johnson propose to find the smallest sample size n for which $Pr[1-\Pi-d_1 \leq P(\bar{x}+\lambda s) - P(\bar{x}-\lambda s) \leq 1-\Pi+d_2] \geq \alpha$. They give a table of such minimum n for

$d_2 =$	$d_1 =$	$\Pi =$	$\alpha =$
.01	.01, .02, .025, .05	$\begin{cases} .01 \\ .05, .25, .5 \end{cases}$	.95, .99 .8, .95, .99
.015 .025	$\left.\begin{array}{l}.035 \\ .025\end{array}\right\}$	.05, .25, .5	.8, .95, .99
.05	.05, .075	$\begin{cases} .05 \\ .25, .5 \end{cases}$	.95, .99 .8, .95, .99 .

6.12 Tests applied to bivariate normal populations

Hsu, C. T. 1940

Annals of Mathematical Statistics 11, 410-426

Consider a sample x_{1i}, x_{2i}, $i = 1, \ldots, n$, from a bivariate normal distribution with parameters ξ_1, ξ_2; σ_1, σ_2; ρ. Denote the sample means, variances and correlation by $\bar{x}_1, \bar{x}_2$; s_1^2, s_2^2; r; compute the statistics

$$R_1 = \frac{2rs_1s_2}{s_1^2 + s_2^2}, \quad R_2 = \frac{2rs_1s_2 - \frac{1}{2}(\bar{x}_1 - \bar{x}_2)^2}{s_1^2 + s_2^2 + \frac{1}{2}(\bar{x}_1 - \bar{x}_2)^2}.$$

Hsu enounces seven hypotheses concerning the parameters of the distribution. He provides tables for the four hypotheses

H_2 $\rho = \rho_0$ given $\sigma_1 = \sigma_2$,

H_4 $\sigma_1 = \sigma_2$ and $\rho = \rho_0$,

H_5 $\sigma_1 = \sigma_2$ and $\xi_1 = \xi_2$,

H_6 $\rho = \rho_0$ given $\sigma_1 = \sigma_2$ and $\xi_1 = \xi_2$;

for which the appropriate test statistics are

$$L_2 = (1-R_1^2)(1-\rho_0 R_1)^{-2},$$

$$L_4 = s_1^2 s_2^2 (1-r^2)[(s_1^2+s_2^2)(1-\rho_0 R_1)]^{-2},$$

$$L_5 = s_1^2 s_2^2 (1-r^2)(1-R_2^2)^{-1}[s_1^2 + s_2^2 + \frac{1}{2}(\bar{x}_1-\bar{x}_2)^2]^{-2},$$

$$L_6 = (1-R_2^2)(1-\rho_0 R_2)^{-2}.$$

The distributions of L_4 and L_5 are found to be proportional to $L^{\frac{1}{2}(n-4)}$; T. 1, p. 415, gives the lower 1% and 5% points of L_4 (or L_5) to 4 dec. for $n = 5(1)10$; $120/n = 0(1)6(2)12$.

Since the statistics L_2 and L_6 are monotone functions of R_1 and R_2, in practice the latter would be used. Table 2, p. 421, gives the power of R_1 as a test of H_2 and the power of R_2 as a test of H_6 for upper- and lower-tail $2\frac{1}{2}\%$ tests: to 4 dec. for $n = 10$, $\rho_0 = .6$ and alternatives $\rho_t = -.8(.2)+.2(.1).9, .95, .975$. Table 3, p. 423, gives the power of R_1, R_2 and R as tests of H_6 for two-tailed 5% tests: to 4 dec. for $n = 10, 20$; $\rho_0 = .6, .8$; $\rho_t = -.6(.2)+.4(.1).9, .95, .975, .99$; with some omissions.

[305]

6.12
6.131   [see also pp. 690-691]

Table 4, p. 424, gives the bias in R_2 as a test of H_2 to 6 dec. for $n = 10$; $\alpha = .05$; $\rho_0 = .6$; $\rho_t = .5, .59(.001).6$.

Box, G. E. P. 1949

<u>Biometrika</u> 36, 317-346

See summary on p. 295. Table 1, p. 327 of Box, contains illustrative values for the bivariate case of the p-variate extension of Bartlett's M test, due to Bishop 1939.

Nanda, D. N. 1951

<u>Journal</u> <u>of</u> <u>the</u> <u>Indian</u> <u>Society</u> <u>of</u> <u>Agricultural</u> <u>Statistics</u> 3, 175-177

"Probability tables of the [larger] root of a determinantal equation with two roots"

The root has the distribution

$$P(x) = I_x(2m+2, 2n+2) - \tfrac{1}{2}x^{m+1}(1-x)^{n+1}B_x(m+1, n+1)/B(2m+2, 2n+2).$$

Upper 5% and 1% points of x are tabled to 2 dec. for $m = 0(\tfrac{1}{2})2$, $n = \tfrac{1}{2}(\tfrac{1}{2})10$.

Pillai, K. C. S. 1956

<u>Biometrika</u> 43, 122-127

Table 1, p. 125, gives upper 5% and 1% points of the distribution defined in the previous abstract for $m = 0(1)4$ and $n = 5(5)30, 40(20)100(30)160, 200, 300, 500, 1000$.

6.13 Tests applied to general multivariate normal populations

Throughout this section, A denotes the matrix with elements a_{ij}; $|A|$ denotes the determinant of A.

6.131 The D^2-statistic

Bose, P. K. 1947

<u>Sankhyā</u> 8, 235-248

The distribution of D^2, found by R.C. Bose 1936, can be put in the form $P(L) = \int_0^L x^{q+1} \lambda^{-q} e^{-\tfrac{1}{2}(x^2+\lambda^2)} I_q(x\lambda) dx$, where $q = \tfrac{1}{2}p - 1$, $L^2 = \tfrac{1}{2}\bar{n}pD^2 + p$, $\lambda^2 = \tfrac{1}{2}\bar{n}p\Delta^2$, and Δ^2 and D^2 are the population and estimated squared distances of two p-variate samples whose harmonic mean size is $\bar{n}$.

Tables 1-4, pp. 245-246, give upper and lower 5% and 1% points of L for $p = 1(1)10$ and $\lambda = 0(.5)3(1)6, 8, 12, 18$; $432/\lambda = 1, 2(2)8, 12, 18$. Table 5, pp. 247-248, gives certain Bessel functions $I_m(x)$, allegedly used in computing TT. 1-4. Errata in P.K. Bose 1951.

Bose, P. K.   1949

Calcutta Statistical Association, <u>Bulletin</u> 2, 131-137

"Incomplete probability integral tables connected with studentised D^2-statistic"

Consider two p-variate normal populations with vectors of means μ_1 and μ_2 and a common covariance matrix Σ. The population squared distance is defined by $p\Delta^2 = (\mu_1-\mu_2)^T \Sigma^{-1}(\mu_1-\mu_2)$.

Draw a sample of size n_i from the ith population, and denote its vector of means by m_i; set $N = n_1+n_2$. Let S be the sample estimate, based on N-2 d.f., of the covariance matrix Σ. Then the sample squared distance is defined by $(1 - 2/N)pD^2 = (m_1-m_2)^T S^{-1}(m_1-m_2)$.

In addition to the notation already introduced, the tables use $c^2 = n_1 n_2 p/(n_1+n_2)$. Table 1, pp. 133-134, gives the upper 5% points of $c^2 D^2/(N + c^2 D^2)$, in the null case $\Delta^2 = 0$, to 3 dec. for $p = 1(1)6$ and $\nu = N-2 = p(1)50(10)90$.

Tables 2-5, pp. 134-137, give the upper 5% points of $c^2 D^2/(N + c^2 D^2)$ in the non-null case to 3 dec. for $\beta = \frac{1}{2}c^2\Delta^2 = 5, 20, 50, 100$; $p = 2, 4, 6$; $\nu = [p+1](2)49$. The tables were computed by recurrence formulas given in P.K. Bose 1947.

$F' = c^2 D^2(N-1-p)/Np$ has the non-central F-distribution with p and N-1-p d.f. and non-centrality $\lambda = c^2\Delta^2 = 2\beta$; thus Bose's tables can be used to extend the non-central F-tables of Tang 1938 and Emma Lehmer 1944.

Rao, C. R.   1949

<u>Sankhyā</u> 9, 343-366

"On some problems arising out of discrimination with multiple characters"

Set $D_p^2 = (1 - 2/N)pD^2$, $\Delta_p^2 = p\Delta^2$, where Δ^2 and D^2 are defined in the abstract of P.K. Bose 1949, above. Table 2, p. 353, gives the power of D_p^2 to 2 dec. for $p = 1(1)8$; $N = 16(4)28$; $\varphi^2 = 2\beta/(p+1) = 1, 2.25, 4$.

6.131
6.132   [see also p. 692]

The values have been computed from Tang's tables of non-central F; comparison of Rao's values with Tang's reveals that Rao's table gives the power of D_p^2 for tests at the 5% significance level.

6.132 **Other tests of homogeneity**

Fertig, J. W. and Leary, Margaret V. 1936

Annals of Mathematical Statistics 7, 113-121

"On a method of testing the hypothesis that an observed sample of n variables and of size N has been drawn from a specified population of the same number of variables"

Let x_{it}, $i = 1, \ldots, k$; $t = 1, \ldots, n$; be a sample of size n from a k-variate normal distribution with density proportional to $e^{-\varphi}$, where φ is a quadratic form. Let

$$s'_{ij} = n^{-1} \sum_{t=1}^{n} (x_{it} - \bar{x}_i)(x_{jt} - \bar{x}_j).$$

Then the likelihood criterion λ for testing whether the sample has come from that distribution is proportional to $|S'|^{\frac{1}{2}n} \cdot \exp[-\sum_t \varphi(x_{1t}, \ldots, x_{kt})]$. The distribution of $x = \lambda^\pi$ is approximated by $I_x(a,b)$; where π is chosen to give an approximately correct third moment, and a and b are given in terms of the first two moments of λ^π on p. 118. The necessary fractional moments of λ are given on p. 117.

The table on p. 119 gives lower 5% and 1% points of λ, based on this approximation, to 2-3 fig. for $\begin{cases} k = 1(1)3 & 4 \\ n = 5(5)20(10)50, \infty & 10(5)20(10)50, \infty \end{cases}$.

Lengyel, B. A. 1939

Annals of Mathematical Statistics 10, 365-375

A sample of size n is drawn from each of two k-variate normal populations. Let S_1 be the covariance matrix of the first sample; S_2 of the second sample; S_0 the covariance matrix with $(2n-1)$ d.f. around the grand mean. If λ is the likelihood criterion for testing the hypothesis that the two populations are identical, then

$$\Lambda = \lambda^{1/n} = (n-1)^n (n-\tfrac{1}{2})^{-n} [|S_1||S_2|]^{\frac{1}{2}} |S_0|^{-1}.$$

The moments of Λ are given by $E(\Lambda^j) = \mu(n)\mu(n+1)\ldots\mu(n+j-1)$, where $\mu(n) = [\Gamma(\tfrac{1}{2}n)/\Gamma(\tfrac{1}{2}n-\tfrac{1}{2}k)]^2 [4/(2n-1)][4/(2n-2)]\ldots[4/(2n-k)]$.

The distribution of Λ is approximated by $I_\Lambda(a,b)$, where

$$a = \mu(n)[1 - \mu(n+1)]/[\mu(n+1) - \mu(n)], \quad b = a[1/\mu(n) - 1].$$

The table on p. 374, based on this approximation, gives lower 1% and 5% points of Λ to 3 dec. for $k = 2(1)4$ and $n = 10(1)20(2)50$.

For $k = 2$ a closed form is found for the distribution of Λ, but this is not tabled.

Notation: $\begin{cases} \text{Lengyel} & N \quad n \quad p \quad q \\ \text{Ours} & n \quad k \quad a \quad b \end{cases}$

Bishop, D. J. 1939
<u>Biometrika</u> 31, 31-55

Given k samples each containing n q-variate observations, let $V_t = [v_{ijt}]$ be the covariance matrix of the tth sample, and $kv_{ij} = \sum_t v_{ijt}$. The likelihood criterion L_1 for testing the hypothesis that the k population covariance matrices are equal is defined by

$$\log L_1 = (2k)^{-1} \sum_{t=1}^{k} \log|V_t| - \tfrac{1}{2}\log|V|.$$

A Pearson Type I curve has been fitted to the moments μ_i of L_1. Table 2, pp. 40-42, gives:

(i) μ_1' and μ_2 of L_1 to 8-10 dec.

(ii) μ_3 and μ_4, both of L_1 and of the fitted curve, to 8-10 dec.

(iii) β_1 of L_1 to 2 dec. and β_2 of L_1 to 1 dec.;

all for $\begin{cases} k = 5 & 2, 5, 10 & 2 \\ q = 1(1)6 & 1(1)6 & 1, 2 \\ n = 10(10)50 & 30 & 10(20)50. \end{cases}$

Table 3, p. 43, compares μ_4 of L_1 to 2-3 fig. and the 5% and 1% points of L_1 to 4 dec., obtained from the Type I fit to 2 moments, with the same quantities obtained from a Type I fit to 3 moments; for $n = 30$, $k = 5$, $q = 1(1)6$. The 3-moment fit is found not to be worth while.

Table 4, p. 45, gives the 5% and 1% points of L_1, computed from the Type I fit, to 4 dec.; and the corresponding true probabilities to 5 dec.; for $k = 2$; $q = 1, 2$; $n = 10, 30, 50$.

6.132
6.133
[see also p. 692]

For large n, $-2n \log_e L_1$ is asymptotically distributed as χ^2 with $\frac{1}{2}q(q+1)(k-1)$ d.f. Table 5, p. 46, gives the 5% and 1% points of L_1 to 4 dec. from the Type I and the χ^2 formulas, as well as the ratio of the discrepancy to σ_{L_1} to 2 dec., for $\begin{cases} k = 2 & 5 & 10 \\ q = 1(2)5 & 1(1)6 & 1, 3 \\ n = 30 & 30(10)50 & 30 \end{cases}$.

Table 6, p. 48, gives the exponents in the Type I approximation to 3 dec. for $\begin{cases} k = 2 & 2 & 5 & 10 & 25 \\ n = 10, 50 & 30 & 10(10)50 & 30 & 30 \\ q = 1, 2 & 1(1)6 & 1(1)6 & 1(1)6 & 1, 2 \end{cases}$. Tables 7-10 are addressed to the merits of various empirical formulas for these exponents.

Wilks, S. S. 1946

<u>Annals of Mathematical Statistics</u> 17, 257-281

"Sample criteria for testing equality of means, equality of variances, and equality of covariances in a normal multivariate distribution"

Let x_{it}, $i = 1, \ldots, k$, $t = 1, \ldots, n$, be a sample of n from a k-variate normal distribution with means ξ_i, variances σ_i^2 and correlations ρ_{ij}. Denote the sample covariances by s_{ij}, and define

$$s^2 = k^{-1}\sum s_{ii}, \quad r = s^{-2}[k(k-1)]^{-1}\sum_{i \neq j} s_{ij}, \quad t = s^2(1-r)(1-n^{-1}).$$

Wilks considers the criterion for equality of variances and covariances $L_{vc} = |S|s^{-2k}[1-r]^{1-k}[1+(k-1)r]^{-1}$, the criterion for equality of means, given equal variances and covariances

$L_m = t[t + (k-1)^{-1}\sum(\bar{x}_i - \bar{x})^2]^{-1}$, and the criterion for equality of means, variances and covariances $L_{mvc} = L_{vc}L_m^{k-1}$.

L_m is distributed as $I_x(a,b)$ with $a = \frac{1}{2}(n-1)(k-1)$ and $b = \frac{1}{2}(k-1)$; i.e. L_m is essentially an F-test. Table 2, p. 264, gives lower 5% and 1% points of L_m to 4 dec. for $\begin{cases} k = 2 \\ n = 2(1)31, 41, 61, 121, \infty \end{cases}$

$\begin{cases} 3 & 4 & 5 \\ 2(1)16, 21, 31, 61, \infty & 2(1)11, 21, 41, \infty & 2(1)8, 11, 16, 31, \infty \end{cases}$.

L_{vc} and L_{mvc} for $k = 2$, and $L_{vc}^{\frac{1}{2}}$ and $L_{mvc}^{\frac{1}{2}}$ for $k = 3$, are also beta-variates; T. 1, p. 263, gives lower 5% and 1% points of L_{vc} and L_{mvc} to 4 dec. for $\begin{cases} k = 2 & 3 \\ n = 3(1)32, 42, 62, 122, \infty & 4(1)18, 23, 33, 63, \infty \end{cases}$.

Tables 1 and 2 are based on the tables of Catherine M. Thompson 1941a.

The criteria $-n \log_e L_{mvc}$, $-n \log_e L_{vc}$, $-n(k-1) \log_e L_m$ are asymptotically distributed as χ^2 with $\frac{1}{2}k(k+3) - 3$, $\frac{1}{2}k(k+1) - 2$, $k-1$ d.f. respectively. Table 3, p. 266, gives upper 5% and 1% points of these three cologarithms for $k = 2(1)6$. Table 4 compares the resulting approximations to L_{mvc}, L_{vc}, L_m with the exact values for selected k and n.

Box, G. E. P. 1949
Biometrika 36, 317-346

See summary on p. 295. Table 2, p. 329 of Box; TT. 6, 7, p. 335; T. 8, p. 336; contain illustrative values for the statistic of Bishop 1939.

6.133 Canonical correlations; independence of groups of variates

Bartlett, M. S. 1938
Proceedings of the Cambridge Philosophical Society 34, 33-40

Let S_1 be the matrix of n observations on an m-variate normal distribution, and S_2 the matrix of n observations on a k-variate distribution. Write the regression of the '2' variates on the '1' variates in the form $S_2 = B_{21} S_1 + S_{2.1}$, where $B_{21} = C_{21} C_{11}^{-1}$ and $C_{ij} = S_i S_j^T$; so that the residual covariance matrix of the '2' variates after fitting the '1' variates is $C_{22.1} = C_{22} - C_{21} C_{11}^{-1} C_{12}$. The criterion for testing association between the '1' and '2' variates is
$\Lambda = |C_{22.1}|/|C_{22}|$.

The distribution of $\chi_1^2 = \frac{1}{2}(k+m-2n+1) \log_e \Lambda$ is approximated by a χ^2-distribution with km d.f. Table 1, p. 39, gives the upper 5% and 1% points of χ_1^2 to 2-3 dec. for $m = 3$; $k = 1, 2$; $n = 10(10)30, \infty$.

Wald, A. and Brookner, R. J. 1941
Annals of Mathematical Statistics 12, 137-152
"On the distribution of Wilks' statistic for testing the independence of several groups of variates"

Let x_{it}, $i = 1, \ldots m$; $t = 1, \ldots, N$ be a sample of size N from an m-variate normal parent. Let the m variates be arranged in k groups of sizes $m_1, m_2 - m_1, \ldots, m_k - m_{k-1}$, so that $m_k = m$; and define $m_0 = 0$.

6.133
6.139

The likelihood criterion for testing the hypothesis that $\rho_{ij} = 0$ if the ith and j^{th} variates are in different groups, found by Wilks 1935, is $\lambda = |R|/[|R_1| \cdot \ldots \cdot |R_k|]$, where R is the matrix of all sample correlations r_{ij} and R_h is the sub-matrix corresponding to the hth group.

By inverting the moment generating function of λ, Wald & Brookner expand the distribution of $v = \log_e \lambda$ in incomplete gamma functions:

$$\Pr[v \geq v_0] = (2/n)^r c_n \sum_{i=0}^{\infty} n^{-i} \beta_i \left[1 - I(\tfrac{1}{2} n v_0 [r+i]^{-\tfrac{1}{2}}, r+i-1)\right] ,$$

where $n = N-1$; the $I(u, \Pi)$ notation is K. Pearson's and is defined on p. 131, above; r and the β_i are functions of the partition of m into groups and do not depend on N. The table on p. 149 gives r, β_1, β_2, β_3, β_4 exact or to 5 dec. for all the partitions of m ($m \leq 6$); the table on pp. 150-152 gives $(2/n)^r c_n$ for the said partitions and for $n = N-1 = 10(1)20(2)30(5)70(10)100$.

Bartlett, M. S. 1941
<u>Biometrika</u> 32, 29-37

See abstract in section 6.4.

Box, G. E. P. 1949
<u>Biometrika</u> 36, 317-346

See summary on p. 295. Table 9, p. 341 of Box; T. 10, p. 344; contain illustrative values of the statistic of Wilks 1935.

Marriott, F. H. C. 1952
<u>Biometrika</u> 39, 58-64

Let $x_1, \ldots, x_p$; $y_1, \ldots y_q$ ($p \leq q$) denote two sets of variates on which n+1 joint observations have been made. The canonical correlations $r_1^2 \geq \ldots \geq r_p^2$ are the p roots of the determinantal equation $|Q - r^2 T| = 0$ where T is the observed covariance matrix of the y s and Q is the covariance matrix due to regression of the y s on the x s. Marriott derives the exact distribution of r_1^2 for $q = 4$ and $p = 2, 3$; and an approximate test for $q = 5$ and $p = 3, 4$. For larger p, q he suggests that $\tfrac{1}{2}(p+q+1-2n)\log_e(1-r_1^2)$ be tested as χ^2 with $p + q - 1 + \tfrac{1}{2}[(p-1)(q-1)]^{2/3}$ d.f.

The tables on pp. 62-63 give asymptotic (for large n) 1% and 5% points of $\tfrac{1}{2}nr_1^2$ to 2 dec. for $\begin{cases} p = 2 & 3 & 4 \\ q = 2(1)12,\ 21 & 3(1)12,\ 21 & 5 \end{cases}$.

6.139 Miscellanea

Mauchly, J. W. 1940

Annals of Mathematical Statistics 11, 204-209

"Signficance test for sphericity of a normal n-variate distribution"

Let x_{it} be a sample of n from a k-variate normal distribuion; and let other notation be as in the abstract of Wilks 1946 on p. 310. The sphericity test is the likelihood criterion for the equality of the population variances and the vanishing of the population covariances, and may be written $L_{nk}^2 = |S|s^{-2k}$. Mauchly derives the distribution of L_{n2} on p. 206.

L_{n3}^2 is approximately distributed as $I_x(a,b)$ with

$a = (9n+5)(n-2)(n-3)/2(9n^2-8n-15)$ and $b = 2(9n-13)(9n+5)/9(9n^2-8n-15)$.

Table 1, p. 208, gives a and b to 4 dec. for $n = 8(2)50(10)100$, and lower 0.1%, 1% and 5% points of L_{n3}^2 for $n = 8(2)50(5)100$.

[see also pp. 692-694]

6.2 Tests and estimators applied to binomial populations

6.21 Tests applied to binomial populations

Noether, G. 1956

<u>Journal of the American Statistical Association</u> 51, 440-450

Table 1, p. 443, computed by D. R. Morrison, gives
$h = \log_e 19 \Big/ 2 \tanh^{-1} 2\epsilon$ to 4 dec.; $s = -\log_e(1-2\epsilon) \Big/ 2 \tanh^{-1} 2\epsilon$ to 4 dec.;
$n_0 = -9 \log_e 19 \Big/ 5 \log_e (1-4\epsilon^2)$ to 2 dec.;
$n_1 = 9 \log_e 19 \Big/ [5 \log_e (1-4\epsilon^2) + 20\epsilon \tanh^{-1} 2\epsilon]$ to 2 dec.; all for
$\epsilon = .01(.01).49$. They are respectively the intercept, the slope, and the expected number of observations under p_0 and p_1, in the Wald sequential test of the binomial $p_0 = \frac{1}{2}$ against $p_1 = \frac{1}{2}+\epsilon$.

6.22 Estimators of parameters in binomial populations

Finney, D. J. 1949

<u>Annals of Eugenics</u> 14, 319-328

"The truncated binomial distribution"

Table 1, p. 321, gives $1/pq$ to 4 dec. for $p = .01(.01).05(.05).5$.

The fitting of the binomial distribution with the 0 class truncated is facilitated by tables of $B = n^2 pq^{n-2}(1-q^n)^{-2}$ and
$W = [n(1-q^n) - n^2 pq^{n-1}]/pq(1-q^n)^2$. These are given in T. 2, pp. 321-324, to 3 dec. for $p = .01(.01).05(.05).95$ and $n = 2(1)20$.

The fitting of the binomial distribution with the 0 and n classes truncated is facilitated by tables of
$B = [n^2 pq(p^{n-2}+q^{n-2}-p^{n-2}q^{n-2}) - np^{n-1}(1-p^n-q^n)]/q(1-p^n-q^n)^2$ and
$W = [n(1-p^n-q^n) - n^2 pq(p^{n-2}+q^{n-2}-p^{n-2}q^{n-2})]/pq(1-p^n-q^n)^2$. These are given in T. 5, pp. 327-328, to 3 dec. for $p = .01(.01).05(.05).5$ and $n = 3(1)20$.

See also Siraždinov 1956.

6.3 Tests and estimators applied to Poisson populations

6.31 Tests applied to Poisson populations

Przyborowski, J. and Wilenski, H. 1940
Biometrika 31, 313-323

Let x_1, x_2 be drawn from Poisson parents with means m_1, m_2. The joint distribution of x_1, x_2 is $\frac{\mu^n e^{-\mu}}{n!}\binom{n}{x_1}\rho^{x_1}(1-\rho)^{n-x_1}$, where $\mu = m_1+m_2$, $n = x_1+x_2$, $\rho = m_1/(m_1+m_2)$. Under the null hypothesis $(m_1 = m_2)$ $\rho = \frac{1}{2}$, and thus the appropriate test is a binomial sign test conditional on n; the convention adopted is that for each n the two-tailed binomial sum shall not exceed the nominal significance level.

Table 1, p. 320, gives the critical region for $n = 1(1)80$ and nominal $\alpha = .01, .05, .1$. Table 2, p. 321, gives the power to 3 dec. for nominal $\alpha = .05, .1$; $\mu = 2(1)15(5)50$; $\rho = 0(.1).5$. Table 3, p. 322, gives the true α for $\mu = 5(5)50$: to 4 dec. for nominal $\alpha = .01$ and 3 dec. for nominal $\alpha = .2$.

Fisher, R. A. 1950a
Biometrics 6, 17-24
"The significance of deviations from expectation in Poisson series"

For testing the significance of the deviations of an observed distribution of n counts, x, with mean count y, from a fitted Poisson distribution, the set of all possible samples of size n with total count $ny = B$ is considered. Under the null (Poisson) hypothesis the conditional distribution within this set is independent of the unknown Poisson parameter, and generates exact conditional distributions of criteria depending on the observed counts: effectively on the a_x, defined as the number of times the count x was obtained. Fisher treats three criteria:

a) the index-of-dispersion χ^2,

b) the Pearson goodness-of-fit χ^2,

c) a likelihood criterion: $\sum_x a_x(\log_e a_x - \log_e m_x)$, where $m_x = ne^{-y}y^x/x!$.

6.31
6.32
[see also pp. 693-694]

Table 1, pp. 22-23, illustrates these criteria in the case n = 140, B = 22, by giving the 45 partitions of the conditional distribution with greatest probability, together with their probabilities to 6-8 dec. and the corresponding values of the 3 criteria.

Pearson, E. S. and Hartley, H. O. 1954

Table 36 A, p. 185, gives the critical region of Przyborowski & Wilenski's test for the equality of Poisson means (abstracted on p. 315, above) for n = 1(1)80 and nominal $\frac{1}{2}\alpha$ = .1, .05, .025, .01, .005. Table 36 B, p. 185, gives the true $\frac{1}{2}\alpha$ for m = $\frac{1}{2}\mu$ = 2.5(2.5)25 and the above nominal $\frac{1}{2}\alpha$ to 2 fig. Reprinted in Vianelli 1959, __Prontuario__ 194, p. 579.

Rao, C. R. and Chakravarti, I. M. 1956
__Biometrics__ 12, 264-282

Let $x_1, \ldots, x_f$ be a sample from an allegedly Poisson distribution, and define $T = \sum x_i$. When T is small, the distribution on the null (Poisson) hypothesis, conditional on f and T, of a statistic intended to detect departures from the Poisson distribution, is very lumpy. Rao & Chakravarti give significance points (nominally 5% points but often more stringent) of 4 such statistics for T = 3(1)10 and f = 3(1)10(10)100:

Table 1, p. 279 $\sum x_i^2$ [effectively the index-of-dispersion χ^2]
Table 2, p. 280 $\sum [x_i \log_e x_i]$
Table 3, p. 281 $\sum [f_r \log_e (f_r!)]$ where f_r is the frequency of the value r.
Table 4, p. 282 f_0.

Each significance point is accompanied by its probability level to 1 or 2 fig.

6.32 Estimators of parameters in Poisson populations

Fisher, R. A. and Yates, F. 1943, 1948, 1953, 1957
__Statistical Tables__

Table VIII 2. Densities of organisms estimated by the dilution method.

In the dilution method a series of s suspensions is prepared, at concentrations 1, a^{-1}, a^{-2}, ..., a^{1-s}: sn samples of equal volume are drawn (n from each level) and each sample is used to inoculate one tube.

6.31
6.32

The number of organisms per tube at the highest concentration, λ, is then estimated from the total number of tubes infected, Y, by solving the equation

$$Y/n = \sum_{j=0}^{s-1} \exp[-\lambda a^{-j}] . \qquad (1)$$

Fisher 1922, p. 366, has shown that this estimator has 88% efficiency.

To solve (1), set $y = Y/n$, $x = s-y$, $\log_{10} \lambda = x \log_{10} a - K$. The tables give K to 3 dec. with arguments a, s, and the lesser of x and y:

$$\begin{Bmatrix} a = 2 & 4 \\ s = 4(1)11 & 4(1)6 \\ x, y = .4(.2)2(.5)\tfrac{1}{2}s & .4(.2)1(.5)\tfrac{1}{2}s \end{Bmatrix}. \text{ For } a = 2 \text{ and } s \geqq 12, \text{ or for } a = 4$$

and $s \geqq 7$, K is constant to 3 dec. An auxiliary table permits the calculation of K for $a = 10$ and $s \geqq 3$.

Mc Crady, M. H. 1915
Journal of Infectious Diseases 17, 183-212

The table on p. 195 gives the maximum likelihood estimates of the number of bacteria per 100 milliliters in terms of the number of tubes infected at each dilution, for two sets of dilutions:

(a) 2 tubes containing 10 ml. each and 10 tubes containing 1 ml. each.

(b) 2 tubes containing 10 ml. each, 10 tubes containing 1 ml. each and 10 tubes containing 0.1 ml. each.

The estimates are rounded to integer counts when under 100 and to the nearest multiple of 5 counts when over 100.

6.4
7.1

[see also pp. 694-695]

6.4 Tests applied to rectangular populations

Bartlett, M. S. 1941

Biometrika 32, 29-37

Let $x_1 \geq x_2$ be two rectangular variates, one with range 0 to 1, the other with range m to 1+m; if $x_1 < 1$, it is not known which variate corresponds to which range. Bartlett investigates two criteria for testing $m > 0$, viz:

(i) $p(x_1)$, (ii) $p(x_1|x_2)$,

with power functions

(i) $P = \alpha + m(1-\alpha)^{\frac{1}{2}}$, (ii) $P' = m + \alpha(1-m+\frac{1}{2}m^2)$.

The most powerful (randomized) test of size α against a specified m has power

$$P'' = \begin{cases} \alpha + m & \alpha \leq (1-m)^2 \\ 1 - m + m^2 + [\alpha-(1-m)^2][1-m]/[2-m], & \alpha \geq (1-m)^2 \end{cases}.$$

Table 1, p. 33, gives P, P', P" to 4 dec. for $\alpha = .05, .1$; $m = 0(.1).9$.

SECTION 7

CORRELATION, SERIAL CORRELATION AND COVARIANCE

7.1 Distribution of the product-moment correlation coefficient

If x_i, y_i $(i = 1, 2, \ldots, n)$ denote a random sample of n paired observations drawn from a bivariate distribution $f(x,y)$ then the sample product-moment correlation coefficient is defined by

$$r = \sum_i (x_i - \bar{x})(y_i - \bar{y}) / \left[\sum_i (x_i - \bar{x})^2 \sum_i (y_i - \bar{y})^2 \right]^{\frac{1}{2}} \qquad (7.1.1)$$

where $\bar{x}$ and $\bar{y}$ are the sample means.

Most of the tables in this section (Sections 7.11-7.13) are concerned with sampled pairs x_i, y_i drawn from the bivariate normal parent

$$f(x',y') = \left[2\pi \sqrt{(1-\rho^2)} \right]^{-1} \exp\left[-\tfrac{1}{2}(1-\rho^2)^{-1} (x'^2 - 2\rho x'y' + y'^2) \right]$$

where $x' = (x - \mu_x)/\sigma_x$, $y' = (y - \mu_y)/\sigma_y$ $\qquad (7.1.2)$

Since r computed from the $x_i y_i$ is identical with that computed from the $x'_i y'_i$ its distribution only depends on the parameters n and ρ the population correlation coefficient. Its frequency ordinate function was first derived by R. A. Fisher 1915:

$$f(r; \rho, n) = \frac{(1-\rho^2)^{\frac{1}{2}(n-1)}}{\pi (n-3)!} (1-r^2)^{\frac{1}{2}(n-4)} \frac{d^{n-2}}{d(r\rho)^{n-2}} \left(\frac{\cos^{-1}(-r\rho)}{\sqrt{(1-r^2\rho^2)}} \right)$$

$$(7.1.3)$$

For the evaluation of $f(r; \rho, n)$ the recurrence formula

$$f(r; \rho, n+2) = \frac{2n-1}{n-1} k_1 f(r; \rho, n+1) + \frac{n-1}{n-2} k_2 f(r; \rho, n) \qquad (7.1.4)$$

7.1
7.11
with

$$k_1 = \rho r \left[(1-\rho^2)(1-r^2) \right]^{\frac{1}{2}} / (1-\rho^2 r^2)$$

$$k_2 = (1-\rho^2)(1-r^2)/(1-\rho^2 r^2)$$

(7.1.5)

may be used; for large n, the asymptotic formula

$$f(r;\rho,n) \doteq \frac{n-2}{\sqrt{n-1}} (1-\rho^2)^{3/2} \chi(\rho,r) \left[1 + \frac{\phi_1}{n-1} + \ldots + \frac{\phi_4}{(n-1)^4} + \ldots \right]$$

(7.1.6)

is given by Soper et al. 1917.

The most extensive tables of ordinates are given by F.N. David 1938; David also tables the probability integral

$$P(r;\rho,n) = \int_{-1}^{r} f(u;\rho,n) \, du$$

(7.1.7)

When $\rho = 0$, (7.1.3) reduces to

$$f(r;0,n) = \frac{\Gamma(\frac{1}{2}n-\frac{1}{2})}{\sqrt{\pi}\,\Gamma(\frac{1}{2}n-1)} (1-r^2)^{\frac{1}{2}(n-4)}$$

(7.1.8)

which, through the relation

$$r^2 = t^2/(\nu + t^2)$$

(7.1.9)

can be reduced to the t-distribution. Thus

$$P(r;0,n) \equiv 1 - Q(t,\nu)$$

(7.1.10)

where $\nu = n-2$ and $t^2 = \nu r^2/(1-r^2)$.

7.11 Areas and ordinates of the distribution of r in normal samples

Fisher, R. A. 1915

Biometrika 10, 507-521

Fisher derives formula (7.1.3) above. The table on p. 512 gives the ordinates $f(r; .66, 4)$ from (7.1.3) and from an approximate formula due to Soper 1913.

Soper, H. E., Young, A. W., Cave, B. M., Lee, Alice and Pearson, K. 1917
Biometrika 11, 328-413

Table A, pp. 379-404, gives $f(r;\rho,n)$ for $\begin{cases} n = 3(1)25 & 50, 100 \\ \rho = 0(.1).9 & 0(.1).8 \\ r = -1(.05)+1 & -1(.05)+1 \\ 5 \text{ dec.} & 5 \text{ dec.} \end{cases}$

$\begin{cases} 50 & 100 & 400 & 400 \\ .9 & .9 & 0(.1).7 & .7(.1).9 \\ .67(.01).99 & .75(.01)1 & -.4(.05)+1 & .54(.01)1 \\ 4 \text{ dec.} & 4 \text{ dec.} & 5 \text{ dec.} & 4 \text{ dec.} \end{cases}$. For $\begin{cases} n = 100 & 400 \\ \rho = .9 & .7(.1).9 \end{cases}$

normal approximations are also tabled. Reprinted in Tables for S & B II, T. 32, pp. 185-211; p. 210 of the reprint is not in the original and gives

$f(r;\rho, 200)$ for $\begin{cases} \rho = 0(.1).7 & .8, .9 \\ r = -.35(.05)+.95 & .63(.01).95 \\ 5 \text{ dec.} & 4 \text{ dec.} \end{cases}$

These ordinates underlie the numerical quadratures of F.N. David 1938.

Table B, pp. 405-411, gives

$$\log_{10} \chi_1 = \log_{10} \frac{1 - \rho r}{\left[(1-\rho^2)(1-r^2) \right]^{\frac{1}{2}}}$$

$$\log_{10} \chi_2 = \log_{10} \frac{\sqrt{2\pi}\left[(1-\rho^2)(1-r^2) \right]^{3/2}}{(1-\rho r)^{\frac{1}{2}}}$$

$$\phi_1 = \frac{1}{8}(\rho r + 2)$$

$$\phi_2 = \frac{1}{128}(3\rho r + 2)^2$$

$$\phi_3 = \frac{5}{1024}\left[15(\rho r)^3 + 18(\rho r)^2 - 4(\rho r) - 8 \right]$$

$$\phi_4 = \frac{21}{32768}\left[175(\rho r)^4 + 200(\rho r)^3 - 120(\rho r)^2 - 160(\rho r) - 16 \right]$$

to 7 dec. for $\rho = 0(.1).9$ and $r = -.95(.05)+.95$; and $\log_{10}[(n-2)(n-1)^{-\frac{1}{2}}]$ to 7 dec. for $n = 3(1)100, 400$. Reprinted in Tables for S & B II, T. 33, pp. 212-218.

7.1
7.1223
Pearson, K. 1933

Biometrika 25, 379-410

    Table 7, p. 396, gives

1000 f(r; ρ ,n) to 2 dec.   for n = 20

and **P(r; ρ ,n)** to 5 dec.   ρ = .277

                                  r = -1.00(.05)1.00

δ^2, δ^4 of P(r; ρ,n).

David, F. N. 1938

Tables of the ordinates and probability integral of the correlation coefficient in small samples

    The tables on pp. 2-55 give P(r;ρ,n) to 5 dec. and f(r;ρ,n) to 4-5 dec.

for $\begin{cases} n = 3(1)25 & 3(1)25 & & 3(1)25 & & 50 \\ \rho = 0(.1).4 & .5(.1).9 & & .9 & & 0(.1).4 \\ r = -1(.05)+1 & -1(.05)+.6(.025)1 & & .8(.01).95(.005)1 & & -.75(.05)+1 \end{cases}$

$\begin{cases} 50 & 50 & 50 & 100 & 100 & 100 \\ .5(.1).7 & .8 & .9 & 0(.1).4 & .5(.1).7 & .8 \\ -.25(.05)+1 & .22(.02)1 & .61(.01)1 & -.5(.05)+1 & 0(.02)1 & .5(.01)1 \end{cases}$

$\begin{cases} 100 & 200 & 200 & 200 & 200 \\ .9 & 0(.1).4 & .5 & .6(.1).8 & .9 \\ .74(.005)1 & -.4(.02)+.7 & .15(.01).79 & .3(.01).94 & .8025(.0025).9625 \end{cases}$

$\begin{cases} 400 & 400 & 400 & 400 & 400 \\ 0(.1).4 & .5, .6 & .7 & .8 & .9 \\ -.3(.02)+.7 & .26(.01).8 & .53(.005).8 & .63(.005).9 & .8325(.0025).96 \end{cases}$.

7.12 Percentage points of r and confidence intervals for ρ

7.121 Confidence intervals for ρ ; percentage points of r for general ρ

David, F. N. 1938

Tables of the ordinates and probability integral of the correlation coefficient in small samples

    Charts 1-4, at the end of the book, give the roots ρ_A and ρ_B ($> \rho_A$) of $\alpha = P(r;\rho_B,n)$ and $1-\alpha = P(r;\rho_A,n)$ plotted against r for $\alpha = .05, .025, .01, .005$; n = 3(1)8(2)12, 15(5)25, 50, 100, 200, 400. Thus the charts yield 90%, 95%, 98% and 99% confidence limits for ρ , given r ; and upper and lower 5%, $2\tfrac{1}{2}$%, 1% and $\tfrac{1}{2}$% points for r , given ρ .

[322]

The charts for 95% and 99% confidence limits are reprinted on a reduced scale in Pearson & Hartley 1954, T. 15, pp. 140-141. The chart for 95% confidence limits is reprinted on a drastically reduced scale in Dixon & Massey 1951, T. 12, p. 327; Dixon & Massey 1957, T. 27, p. 464.

7.122 Percentage points of r for $\rho = 0$

Define $r(Q,n)$ as the root of the equation $P(r;0,n) = 1-Q$. Equation (7.1.9) permits us to calculate this from percentage points of t:
$r(Q,n) = x(\nu+x^2)^{-\frac{1}{2}}$, where $x = t(Q,n-2)$, and $\nu = n-2$.

7.1221 Double tail percentage points: $r(\frac{1}{2}Q, \nu+2)$

The order is by number of percentage points tabled.

	$\nu =$	$Q =$	
3 fig.	1(1)20(5)50(10)100	.001, .005, .01, .02, .05, .1	E.S. Pearson & Hartley 1954(138)
4 fig.	1(1)20(5)50(10)100	.001, .01, .02, .05, .1	Czechowski et al. 1957 (88), Vianelli 1959(198)
4+ dec.	1(1)20(5)50(10)100	.001, .01, .02, .05, .1	Fisher & Yates 1938-1953 (T.VI), 1957(T.VII)
3 dec.	1(1)20(5)50(10)100	.001, .01, .02, .05, .1	K.A. Brownlee 1947(150), Davies 1949(276)
3 dec.	1(1)20(5)50(10)100	.01, .02, .05, .1	Mode 1951(370)
3 dec.	1(1)30(5)50(10)100(25)150, 200(100)500, 1000	.01, .05	Snedecor 1946(351), Arkin & Colton 1950(140)
3 dec.	4(1)20(5)50(10)100	.05	Weatherburn 1946(193)

7.1222 Double tail percentage points: $r(\frac{1}{2}Q, n)$

	$n =$	$Q =$	
3+ dec.	3(1)20(2)30(5)40(10)100, 120, 150, 200(100)400	.01, .05	Quenouille 1952(235)

7.1223 $(\nu+1)^{\frac{1}{2}}r(\frac{1}{2}Q, \nu+2)$

	$\nu = $ [$120/\nu =$]	$Q =$	
3 dec.	1(1)30(5)60(10)100 [0(1)4]	.01, .02, .05, .1	Arley 1940
	1(1)30 [0(1)4]	.2(.1).9	

7.1223
7.131   [see also p. 695]

3 dec.   1(1)30(5)60(10)100   .01, .02, .05, .1   Arley & Buch 1950(217)
         [0(1)4]

3 dec.   1(1)30(5)60(10)100   .001, .01, .05, .1   Burington & May 1953(162)
         [0(1)4]

<u>7.1229</u>   Miscellaneous percentage points of r

Bose, S. S.   1934

<u>Sankhyā</u> 1, 277-288

"Tables for testing the significance (of coefficient) of linear regression in the case of time-series and other single-valued samples"

The tabulated percentage points of b are not computed from the sampling distribution of the sample regression coefficient b; they are merely $\Delta^{\frac{1}{2}}$ times the percentage points of r, where $\Delta = \sigma_y^2 / \sigma_x^2$. Tables 1 and 2, pp. 285-288, give 5% and 1% points so computed for $\Delta = .05(.05)1$ and $n = 3(1)30$.

7.13   <u>Moments and other distribution constants of r and related statistics</u>

In the Introduction to 7.1 we gave (see equation 7.1.3) R. A. Fisher's form of the distribution of the sample correlation coefficient r computed from n pairs x_i, y_i (i = 1,2 ... n) drawn from a bivariate normal distribution with correlation coefficient ρ. Only approximate values can, in general, be worked out for the moments of r. Cook 1951 gives, to order n^{-2},

$$E(r) = \rho \left[1 - \frac{1}{2n} - \frac{3}{8n^2} + \rho^2 \left(\frac{1}{2n} - \frac{3}{4n^2}\right) + \rho^4 \frac{9}{8n^2} \right] + \cdots \quad (7.13.1)$$

$$\sigma^2(r) = \frac{1}{n}(1-\rho^2)^2 \left(1 + \frac{1}{n} + \frac{11\rho^2}{2n}\right) + \cdots \quad (7.13.2)$$

$$\mu_3(r) = \frac{-6}{n^2}\rho(1-\rho^2)^3 + \cdots \quad (7.13.3)$$

$$\mu_4(r) = \frac{3}{n^2}(1-\rho^2)^4 + \cdots \quad (7.13.4)$$

confirming in part earlier results by Soper 1913 and Soper et al. 1917. For $n = 4$, R.A. Fisher 1915 gives the exact result

$$E(r) = \frac{2}{\pi}(\alpha + \cot \alpha - \alpha \cot^2 \alpha) \quad (7.13.5)$$

where $\tan \alpha = \rho(1-\rho^2)^{-\frac{1}{2}}$

7.131 Moments of r

Soper, H. E. 1913

__Biometrika__ 9, 91

Soper derives the approximate formulas

$$E(r) = \rho\left[1 - \frac{1-\rho^2}{2n}\left(1 + \frac{3}{4n}(1 + 3\rho^2)\right)\right]$$

which agrees with (7.13.1);

$$E(r) = \rho\left[1 - \frac{1-\rho^2}{2(n-1)}\left(1 - \frac{1}{4(n-1)}(1 - 9\rho^2)\right)\right];$$

$$\sigma(r) = \frac{1-\rho^2}{\sqrt{n-1}}\left[1 + \frac{11\rho^2}{4(n-1)}\right].$$

The table on p. 109 gives $E(r)^*$, $\sigma(r)$ to 4 decimals for $\rho = 0$, $n = 4, 8$; $\rho = .66$, $n = 4, 8, 30$.

Fisher, R. A. 1915

__Biometrika__ 10, 507-521

The table on p. 517 gives for $n = 4$, $\rho = 0.1(0.1)0.9, 0.95$ the exact $E(r)$ from equation (7.13.5) to 4 decs as well as 4 decimal values from two of Soper's approximate formulas:

(I) $$E(r) = \rho\left[1 - \frac{1-\rho^2}{8}(1 + \frac{3}{16}(1 + 3\rho^2))\right],$$

(II) $$E(r) = \rho\left[1 - \frac{1-\rho^2}{6}(1 - \frac{1}{12}(1 - 9\rho^2))\right].$$

Johnson, P. O. 1949

__Statistical Methods in Research__

Table 17 (p. 52) gives constants β_1, β_2 and $\sigma(r)$ to 3 dec. for the distribution of the sample correlation coefficient r in samples of size $N = 20$ from bivariate normal distributions having correlation coefficient $\rho = 0(.2).8, .9$.

---

*It appears from R.A. Fisher 1915, p. 517, that Soper used (7.13.1).

[325]

7.132
7.134

7.132 "Probable error" of r

The tables in this section permit rapid calculation of $E = .67449 \, n^{-\frac{1}{2}} (1-\rho^2)$. For the factor .67449 cf. p. 12. Kendall & Stuart 1958, p. 236, warn that the "use of the standard error to test the significance of the correlation coefficient is not, however, to be recommended, since the sampling distribution of r tends to normality very slowly."

Heron, D. 1910
<u>Biometrika</u> 7, 410-411

The abac on p. 410 gives lines for $\rho = 0(.1).4(.05).6(.02).98$, and covers the ranges $10 \le n \le 1000$, $.005 \le E \le .22$.

Holzinger, K. 1925 a
<u>Tables of the probable error of the coefficient of correlation as found by the product moment method</u>

Table A, pp. 5-19, gives E to 4 dec. for $n = 20(1)100$ and $\rho = 0(.02)1$ with fPPd in ρ. Table B, pp. 20-35, gives E to 4 dec. for $n = 100(10)1000$ and $\rho = 0(.02)1$ with fPPd in both n and ρ.

7.133 Mode of r

Soper, H. E., Young, A. W., Cave, B. M., Lee, Alice and Pearson, K. 1917
<u>Biometrika</u> 11, 328-413

If

$$I_n = \int_0^\infty (\cosh z - \rho_0^2)^{-n} \, dz$$

and $E_n = I_n/I_{n-1}$

the latter ratios are of help in determining the mode $\breve{r}$ (see pp. 342-4) and are tabulated to 7 dec. for $\rho_0 = 0.6$, $n = 2(1)26$.

Hence Soper et al. expand $\breve{r}$ (pp. 349-350) in the series

$$\breve{r} = \rho + \frac{\nu_1(\rho)}{(n-1)} + \cdots + \frac{\nu_4(\rho)}{(n-1)^4} + \cdots$$

where the $\nu_i(\rho)$ are given by

$$\nu_1(\rho) = \frac{5}{2}\rho(1-\rho^2)$$

$$\nu_2(\rho) = \frac{\rho(61-101\rho^2)(1-\rho^2)}{8}$$

$$\nu_3(\rho) = \frac{\rho(367-1480\rho^2+1273\rho^4)(1-\rho^2)}{16}$$

$$\nu_4(\rho) = \frac{\rho(17606-125727\rho^2+246783\rho^4-143782\rho^6)(1-\rho^2)}{256}$$

They examine the formula for the mode given by Soper 1913, p. 108, and find it invalid. The table on p. 373 gives $\nu_i(\rho)$ to 10 dec. for $i = 1(1)4$ and $\rho = 0(.05)1$.

Reprinted in Tables for S & B II, T. 51a, p. 252.

7.134 Moments of $t = r(1-r^2)^{-\frac{1}{2}}$

Fisher, R. A. 1915

Biometrika 10, 507-521

Let τ be the same function of ρ that t is of r. Fisher obtains the formulas (p. 518):

$$E(t) = \frac{n-2}{n-3}\tau$$

$$\sigma^2(t) = (n-4)^{-1}\left[1 + \tau^2 + \frac{n-2}{(n-3)^2}\tau^2\right]$$

$$\sqrt{\beta_1(t)}\,\sigma^3(t) = \frac{(n-2)\tau}{(n-3)(n-4)(n-5)}\left[3(1+\tau^2) + \frac{2\tau^2(n-1)}{(n-3)^2}\right]$$

$$\beta_2(t)\,\sigma^4(t) = \frac{3}{(n-4)(n-6)}\left[(1+\tau^2)^2 + \frac{6(n-2)\tau^2(1+\tau^2)}{(n-3)(n-5)} + \frac{6(n-2)(3n^2-11n+12)\tau^4}{(n-3)^4(n-5)}\right]$$

The tables on p. 519 give $\sigma^2(t)$ and $\beta_1(t)$ to 4 fig. and $\beta_2(t)$ to 4 dec. for $n = 8(5)23(10)53$ and $\tau^2 = .01, .03, .1, .3, 1, 3, 10, 30, 100, \infty$. Fisher advertises that the tables "are appended for inspection rather than for reference."

See also B. I. Harley 1957.

7.14
7.19

7.14 Distribution of r from non-normal parents

Gayen, A. K. 1951

Biometrika 38, 219-247

Representing the non-normal bivariate parent by means of a truncated bivariate Edgeworth series, Gayen derives the distribution of r as an expansion whose leading term is $f(r; \rho, n)$ (given by (7.1.3) of the Introduction) and whose further terms depend on the derivatives

$$(n-1) \; f^{(i)}(r; \rho, n) \; / \; 8n(n-1) \qquad \text{for } i = 1,2$$

$$(n-2) \; f^{(i)}(r; \rho, n) \; / \; 12n(n-1)(n-3) \qquad \text{for } i = 1,2,3$$

where the superscript (i) denotes ith differentiation with regard to ρ.

The above are given to 5 dec. for

$n = 11$, $\begin{cases} \rho = 0 & r = -1(.05)1 \\ \rho = 0.8 & r = -.75(.05).6(.025)1 \end{cases}$

$n = 21$, $\quad \rho = 0.8 \quad r = -.25(.05).6(.025)1$

Cook, M. B. 1951

Biometrika 38, 368-376

Cook considers a parent population with $\varkappa_{02} = \varkappa_{20} = 1$, $\varkappa_{11} = \rho$, specified values for $\varkappa_{03}$ and $\varkappa_{30}$, and all other cumulants zero. [This is not a realizable population, because it has negative frequencies in the tails.]

Table 1, p. 373, gives $\beta_1(r)$ to 4 dec. for $n = 20, 25, 50$; $\rho = 0$; $\varkappa_{03}^2 = \varkappa_{30}^2 = 0(.2).6$.

7.19 Miscellaneous tables related to r

Soper, H. E., Young, A. W., Cave, B. M., Lee, Alice and Pearson, K. 1917

Biometrika 11, 328-413

Maximum likelihood values $\hat{\rho}$ of the population ρ computed from the distribution of the sample correlation r and called the "most likely" value $\hat{\rho}$.

Formula:

$$\hat{\rho} = r - \frac{\lambda_1(r)}{n-1} - \frac{\lambda_2(r)}{(n-1)^2} - \frac{\lambda_3(r)}{(n-1)^3} - \cdots$$

where $\lambda_1(r) = \frac{1}{2} r(1 - r^2)$

$\lambda_2(r) = \frac{1}{8} r(5r^2 - 1)(1 - r^2)$

$\lambda_3(r) = \frac{1}{16} r(17r^4 - 8r^2 - 1)(1 - r^2)$

The table on p. 375 gives $\lambda_i(r)$ to 10 dec. for $i = 1(1)3$ and $r = 0(.05)1$.

Reprinted in Tables for S & B II, T. 51b, p. 253.

Brandner, F. A. 1933

See abstract in section 7.29.

[see also p. 695]

7.2 The hyperbolic-tangent transformation

R.A. Fisher 1921 proposed that $z = \tanh^{-1} r$ could be treated as a normal variate with mean $\zeta = \tanh^{-1} \rho$ and variance $1/(n-3)$. This transformation has two important properties:

(a) The distribution of z is nearly normal.

(b) The variance of z is virtually independent of the population ρ.

These properties enable us to subject the z-transforms of a set of correlation coefficients to standard analysis-of-variance techniques.

The efficacy of z as a variance-stabilizing transformation is examined by Hotelling 1953, who finds that

$$z - (3z+r)/4n \quad \text{and} \quad z - (3z+r)/4n - (23z+33r-5r^3)/96n^2$$

have variances constant to order n^{-2} and n^{-3} respectively.

There is an historical connexion between the z of this section and $z = \tfrac{1}{2}\log_e F$ of section 4.3: the latter arose from applying the hyperbolic-tangent transformation to the intra-class correlation coefficient. For this archaic form of the analysis of variance, see M.G. Kendall 1943, pp. 358-362.

7.21 tanh x and $\tanh^{-1} x$

The lists below are extremely selective: we confine ourselves to the tables in statistical works, and to a few tables with very many decimals. See section 10.42 of Fletcher et al. 1946 for additional tables of tanh x; sections 11.63 and 11.69 for $\tanh^{-1} x$; sections 10.52, 10.55, 10.64, 12.43, 12.53, 12.63 and 12.68 for related functions.

7.211 $\tanh x = \dfrac{e^x - e^{-x}}{e^x + e^{-x}}$

20 dec.	0(.00001).001		
10 dec.	.001(.0001).0999, 3(.01)10(.1)12.3	No Δ	Hayashi 1926
8 dec.	.1(.001)3		
10 dec.	0(.1)10		
8 fig.	0(.0001)2	No Δ	N.B.S. 1943

7 dec.	0(.01)3	δ^2	Milne-Thomson 1932 (m = 1.0), 1950(114, m = 1.0)
4 dec.	0(.01)1.79	IΔ	Fisher & Yates 1938,
5 dec.	1.8(.01)2.99; 3(.1)4.9	IΔ; No Δ	1943, 1948, 1953 (T.VII), 1957(T.VII 1)
5 dec.	0(.01)3(.1)5	No Δ	Smith & Duncan 1945(483)
4 dec.	0(.01)3(1)5	No Δ	Arkin & Colton 1950(123)
3 dec.	0(.02)1.98	No Δ	Weatherburn 1946(201)
3 dec.	0(.05)1(.1)1.6(.2)3	No Δ	Quenouille 1952(236)

7.212 $\tanh^{-1} x = \frac{1}{2} \log_e \frac{1+x}{1-x}$

20 dec.	0(.00001).001	No Δ	Hayashi 1926
10; 9 dec.	.001(.0001).0999; .1(.001)1		
9 dec.	0(.001).5(.0005).75(.0002) .9(.0001).95(.00005) .975(.00002).99(.00001).99999	Δ^2	Harvard Univ. 1949(T.1)
7 dec.	0(.01).9(.001)1	δ^4	R.A. Fisher 1921
4; 3 dec.	0(.002).748; .75(.002).998	PPMD	Pearson & Hartley 1954(139)
4 dec.	0(.01).9(.001).999	No Δ	Treloar 1942(91)
3 dec.	0(.02).8(.01).94(.001)1	Δ	Lindley & Miller 1953(6)

7.22 Percentage points of z

Allan, F. E. 1930 b

Proceedings of the <u>Cambridge Philosophical Society</u> 26, 536-537

The table on p. 537 gives the root z' of $\Pr(z \geq z' | \zeta) = .95$ to 4 decimals for $\zeta = -3.0(.1)3.0$. By changing the signs of ζ and z, this table also gives the values of z which would be exceeded in 5% of random trials.

7.23 Moments of z

Setting $N = n-3$ in the formulas originally given by R.A. Fisher 1921, and corrected by Gayen 1951, we find, after rearranging terms:

$$E(z) = \zeta + (\rho/2N)[1 + (\rho^2-3)/4N + 3(1-\rho^2)^2/8N^2] + O(N^{-4}) ;$$

7.23
7.29

$$\sigma^2(z) = N^{-1}[1 - \rho^2/2N - (2-6\rho^2+\rho^4)/6N^2] + O(N^{-4}) ;$$

$$\beta_1(z) = N^{-3}\rho^6 + O(N^{-4}) ;$$

$$\beta_2(z) = 3 + 2/N + \rho^2(2-3\rho^2)/N^2 + O(N^{-3}) .$$

Johnson, P. O. 1949

Statistical methods in research

Table 18, p. 52, gives $E(z-\zeta)$, $\sigma(z)$ and $\beta_2(z)$ to 4 fig. and $\beta_1(z)$ to 4 dec. for $n = 20$ and $\rho = 0, .2, .6, .9$.

7.29 Miscellaneous tables connected with z

Brandner, F. A. 1933

Biometrika 25, 102-109

Given two samples of paired observations x,y of respective size n_1 and n_2 and let $s_1^2, s_1'^2, r_1; s_2^2, s_2'^2, r_2$ denote the respective sample variances and correlations; required to test the hypothesis that the samples have been drawn from binormal populations having the same value of ρ. The likelihood criterion is

$$\lambda = \left[\frac{(1-\rho_M^2)^{\frac{1}{2}}(1-r_1^2)^{\frac{1}{2}}}{(1-\rho_M r_1)}\right]^{n_1} \left[\frac{(1-\rho_M^2)^{\frac{1}{2}}(1-r_2^2)^{\frac{1}{2}}}{(1-\rho_M r_2)}\right]^{n_2}$$

where

$$\rho_M = \frac{(n_1+n_2)(1+r_1 r_2) - \left[(n_1+n_2)^2(1-r_1 r_2)^2 - 4n_1 n_2(r_1-r_2)^2\right]^{\frac{1}{2}}}{2(n_1 r_2 + n_2 r_1)}$$

For $n_1 = n_2 = n$ the relation to Fisher's z-transforms is

$$\lambda^{1/2n} = \text{sech } \tfrac{1}{2}(z_1 - z_2)$$

where $z_i = \tanh^{-1} r_i$

For $n_1 \neq n_2$ there is a slight difference between the two criteria, but if ρ_M above is approximated by ρ_a, the root of

$$\tanh^{-1}\rho_a = \frac{n_1 z_1 + n_2 z_2}{n_1 + n_2} ,$$

then λ is again a function of $z_1 - z_2$. Tables I and II (pp. 107-8) give a numerical comparison of ρ_M and ρ_a to 3 decimals for $n_1 = 2n_2, n_1 = \tfrac{1}{2}n_2, n_1 = 6n_2, n_1 = \tfrac{1}{6}n_2$ and various pairs of r_1, r_2.

[332]

Alongside is shown Fisher's 'optimum estimate' of ρ, viz

$$\rho' = \tanh\left[\frac{(n_1-3)z_1 + (n_2-3)z_2}{n_1 + n_2 - 6}\right]$$

which depends on the ratio $\frac{n_1 - 3}{n_2 - 3}$ (and not $\frac{n_1}{n_2}$) and is tabulated for $\min(n_1 n_2) = 10$, the above $\frac{n_1}{n_2}$ ratios and the same pairs of r_1, r_2.

Quenouille, M. H. 1948

<u>Biometrika</u> 35, 261-267

Quenouille applies the hyperbolic-tangent transformation to the distribution, found by Madow 1945, of the serial correlation from a first-order Markov process. He finds z approximately normal with

mean $\quad \zeta - \dfrac{\rho}{n(1-\rho^2)} + \dfrac{\rho(1+\rho^2)}{n^2(1-\rho^2)^2} \quad$ and

variance $\quad \dfrac{1}{n(1-\rho^2)} - \dfrac{2\rho^2}{n(1-\rho^2)^2}$.

See also B. I. Harley 1954.

7.3 Tables of $1-r^2, \sqrt{1-r^2}, 1/\sqrt{1-r^2}$ and related functions

[see also pp. 695-696]

The use of the above tables in correlation work is obvious and we confine ourselves to quoting a few selected usages. $1-r^2$ gives directly the factor required to reduce a population variance to a residual variance when the population correlation is r and $\sqrt{1-r^2}$ gives the corresponding factor for the standard derivation. $(1-r^2)^{-\frac{1}{2}}$ is useful when evaluating partial correlation coefficients by continued multiplication on a calculating machine.

The lists below are not exhaustive; see Fletcher et al. 1946, p. 30, for additional entries and (p. 31) for related functions.

No differences are given with the tables below.

7.31 $1-r^2$

6 dec.	0(.001).999	K. Pearson 1930 a (20)
4 dec.	0(.01)1	Pearson & Hartley 1954(234)

7.32 $(1-r^2)^{\frac{1}{2}}$

A table giving $\sin x$ and $\cos x$ in adjacent columns may be considered to provide selected pairs of r and $(1-r^2)^{\frac{1}{2}}$; in particular we notice N.B.S. 1949 z, which gives 4500 such pairs to 15 dec.

10 dec.	0(.01)1 (called $P_{1,1}(x)$)	Tallqvist 1908(12)
8 dec.	0(.0001)1	Kelley 1938(14), 1948(38)
5 dec.	0(.01).91(.002).99(.001)1	Pearson & Hartley 1954(234)
4 dec.	0(.001)1	Davenport & Ekas 1936(187)
4 dec.	0(.01)1	Garrett 1937(475)

7.33 $(1-r^2)^{-\frac{1}{2}}$

10; 9 dec.	0(.005).1; .1(.001).199	
8 dec.	.2(.001).9	N.B.S. 1941 a (31)
7 dec.	.9(.0005).96(.0002).99(.0001).995(var.).99999 99990	
5 dec.	0(.01).91(.002).99(.001)1	Pearson & Hartley 1954(234)

7.4 Multiple correlation coefficient

Fisher, R. A. 1928

Proceedings of the Royal Society of London, Series A 121, 654-673

    See abstract on p. 155.

Pearson, K. 1931 b

Biometrika 22, 362-367

    According to Wishart 1931, the mean and variance of the multiple correlation coefficient are

$$\text{Mean }(R^2) = \overline{R^2} = 1 - \frac{n-k}{n-2}\, \gamma_1$$

$$\text{Variance }(R^2) = \sigma^2_{R^2} = \frac{(n-k)(n-k+2)}{(n-2)n}\, \gamma_2 - (1-\overline{R^2})^2$$

where

  n = sample size, k = number of variates (including dependent)

$$\gamma_1 = (1-\rho^2)\, F(1, 1, \tfrac{1}{2}(n+1), \rho^2) \cdot (\tfrac{n-2}{n-1})$$

$$\gamma_2 = (1-\rho^2)^2\, F(2, 2, \tfrac{1}{2}(n+3), \rho^2) \cdot (\tfrac{n-2}{n-1}) \cdot (\tfrac{n}{n+1})$$

and ρ^2 is the population value of the multiple correlation.

γ_1 and γ_2 are tabled to 7 dec. for n = 3(1)25, 50, 100, 200, 400; ρ = 0(.1)1. Reprinted in Tables for S & B II, TT. 52, 52 bis.

    For large n and moderate ρ the formulas

$$\overline{R^2} = \rho^2 + 1(1-\rho^2)\frac{k-1-2\rho^2}{m}$$

$$\sigma^2_{R^2} = \frac{4\rho^2(1-\rho^2)^2}{n}$$

of Hall 1927 a may be used.

Snedecor, G. W. 1946

Statistical methods

    See abstract on p. 254.

    See also the abstract of Woo 1929 on p. 192.

7.5
7.51

[see also p. 696]

7.5 Serial correlation

Let $x_1, x_2, \cdots, x_n$ denote a sample from a stationary time series, then the 'sample serial correlation lag ℓ' is defined as the ordinary product moment correlation between x_i and $x_{i-\ell}$ i.e. by

$$_\ell r_n = \sum_{i=1}^{n-\ell} (x_i - \bar{x}_1)(x_{i-\ell} - \bar{x}_2) \left[\sum_{i=1}^{n-\ell} (x_i - \bar{x}_1)^2 \sum_{i=1}^{n-\ell} (x_{i-\ell} - \bar{x}_2)^2 \right]^{-\frac{1}{2}}$$

where

$$\bar{x}_1 = (n-\ell)^{-1} \sum_{i=1}^{n-\ell} x_i \quad ; \quad \bar{x}_2 = (n-\ell)^{-1} \sum_{i=1}^{n-\ell} x_{i+\ell}$$

When $E(x) = 0$, it is lawful to consider

$$_\ell r'_n = \sum_{i=1}^{n-\ell} x_i x_{i+\ell} \left[\sum_{i=1}^{n-\ell} x_i^2 \sum_{i=1}^{n-\ell} x_{i+\ell}^2 \right]^{-\frac{1}{2}}$$

Only asymptotic results are available for the distributions of $_\ell r_n$ and $_\ell r'_n$, notably those by T. Koopmans 1942. When the x_i are independent normal deviates, exact distributions are known for two modified serial correlations:

(a) The circular serial correlation coefficient $_\ell R_n$ first suggested by Hotelling for which the exact distribution theory was developed by R. L. Anderson (see abstract below). With this coefficient the sample series is extended modulo n by defining $x_i = x_{i-n}$ for $i > n$.

(b) Watson & Durbin (see abstract below) in which the sample series is split into two equal size sections which have no x_i in common. This device sacrifices a fraction of the information but does not introduce an assumption as in (a).

7.51 Circular serial correlation

Anderson, R. L. 1942

Annals of Mathematical Statistics 13, 1-13

"Distribution of the serial correlation coefficient"

Let $x_1, x_2, \cdots, x_n$ denote a series of independent variates from the same normal parent, define $x_{n+j} = x_j$ and the circular serial correlation lag ℓ by

$$_\ell R_n = n \sum_{i=1}^{n} x_i x_{i+\ell} - \left(\sum_{i=1}^{n} x_i \right)^2 / nV_n$$

[336]

where $V_n = \sum_{i=1}^{n} (x_i - \bar{x})^2$. For the case $\ell = 1$ general formulas for the joint distribution of $_1R_n$ and V_n are given (pp. 3-5).

Table I (p. 8) is based on these formulas and gives upper and lower 5% and 1% points for $_1R_n$ to 3 decimals for $n = 5(1)15(5)75$. Some of these values were obtained by 'graphical interpolation' in the computed values.

The table on p. 6 gives comparison with the large-sample normal approximation for $n = 45, 75$.

Table 1 (p. 11) gives 3 decimal values of the upper and lower 5% and 1% points of $_\ell R_n$ for $n = 2\ell$ and $n = 3\ell$ with $\ell = 2(1)10(2)20(5)30(10)50$ and computed from the B-function distributions derived (pp. 9-10) in these special cases, Table III (p. 11) the same for $n = 4$, $\ell = 2(1)7$ and Table IV for $n = 10$, $\ell = 2$; $n = 14$, $\ell = 2$. These answers are compared with the corresponding significance levels for $_1R_n$ (from Table I) which serve as an approximation to those of $_\ell R_n$ for large n.

Rubin, H. 1945

Annals of Mathematical Statistics 16, 211-215

"On the distribution of the serial correlation coefficient"

The table on p. 214 gives 3 decimal values of the upper 5% and 1% points of

$$r_1 = \sum_{t=1}^{n} x_t x_{t+1} \bigg/ \sum_{t=1}^{n} x_t^2 \qquad x_{n+1} = x_1$$

to 3 dec. for $n = 3(1)11, 15(5)45$ from

(a) Anderson's exact distribution

(b) Dixon and Koopmans' approximate distribution

$$f(r,n) \sim (1 - r_1^2)^{\frac{1}{2}(n-1)}$$

(c) The normal curve with correct mean and standard deviation.

The table on p. 215 gives the upper 1%, 2%, 3%, 4% and 5% points of r_1 from (b) and (c) to 3 dec. for $n = 15, 20, 25$.

7.51 [see also p. 696]
7.53

Anderson, R. L. and Anderson, T. W. 1950

Annals of Mathematical Statistics 21, 59-81

"Distribution of the circular serial correlation for residuals from a fitted Fourier series"

Given mn equally spaced values of a time series, assume the model $x_{mt+j} = \alpha_j + \epsilon_{mt+j}$. On the null hypothesis the ϵ are normally and independently distributed with constant variances; on the alternative, ϵ_i and ϵ_{i+1} are correlated.

Compute $a_j = n^{-1} \sum_{m=0}^{n-1} x_{mt+j}$ and $e_{mt+j} = x_{mt+j} - a_j$; let R be the circular serial correlation at lag 1 computed from the es. Table 1, p. 65, gives upper and lower 5% and 1% points of R, calculated on the null hypothesis, to 3 dec. for
$\begin{cases} m = 2 & 4, 6, 12 \\ n = 3(1)30 & 2(1)25 \end{cases}$.

Table 2, pp. 66-67, gives upper and lower 5% and 1% points of R, calculated on the null hypothesis for the model

$$x_{mt+j} = \beta_0 + \beta_1 \cos \frac{2\pi j}{m} + \beta_2 \sin \frac{2\pi j}{m} + \epsilon_{mt+j},$$

to 3 dec. for $\begin{cases} m = 3 & 4 & 6, 12 \\ n = 2(2)50 & 2(1)25, 27(3)36 & 2(1)25 \end{cases}$.

7.52 Non-circular serial correlation

Quenouille, M. H. 1948

Biometrika 35, 261-267

See abstract on p. 333.

Watson, G. S. and Durbin, J. 1951

Annals of Mathematical Statistics 22, 446-451

See abstract on p. 79.

Durbin, J. and Watson, G. S. 1951

Biometrika 38, 159-178

See abstract on p. 78.

7.53 **Variate differences as a measure of serial correlation**

For variate differences as a measure of variance, see section 1.531, pp. 75-80.

Pearson, K. and Elderton, Ethel M. 1923
Biometrika 14, 281-310

James Henderson's table, on p. 310, gives two quantities related to the polynomial

$$\phi(n,\epsilon) = 1 - \frac{n}{n+1}\epsilon + \frac{n(n-1)}{(n+1)(n+2)}\epsilon^2 - \cdots (-1)^n \frac{n!}{(n+1)\cdots 2n}\epsilon^n$$

They are

$$\left. \begin{array}{c} 2\phi(n,\epsilon) - 1 \\ \text{and} \\ \left(4 - \frac{2}{n+1}\right)\left[\frac{2\phi(n+1,\epsilon) - 1}{2\phi(n,\epsilon) - 1}\right] \end{array} \right\} \begin{array}{l} \text{to 6 dec} \\ \text{for} \\ n = 0(1)10 \\ \epsilon = -1.0(0.1)1.0 \end{array}$$

Reprinted in Tables for S & B II, T. 44, p. 235. The use of the polynomial is abundantly illustrated on Oskar Anderson's data (Tables for S & B II, pp. ccxj-ccxvij) but the illustrations show no extravagant success for the method.

Quenouille, M. H. 1951
Revue de l'Institut International de Statistique 19, 121-129

Let V_i denote the variance of 'the ith set of variate differences' as defined by the author. An expression is derived for the variance of the mth variate differences of this set in the form

$$\text{Var}\left[\Delta^m V_i\right] = \alpha_{mi} \sigma^4/n$$

where n is the number of observations and σ^2 is the true variance of the random element. α_{mi} is tabled for $i = 1(1)10$: to 4 dec. for $m = 0, 1$ and to 6 dec. for $m = 2(1)5$.

7.6 Covariance; covariance ratio; regression coefficient

Pearson, K., Jeffery, G. B. and Elderton, Ethel M. 1929
Biometrika 21, 164-201

Let k_{11} be the sample covariance in a bivariate normal sample of size n. Define $v = (n-1)k_{11}/\sigma_1\sigma_2(1-\rho^2)$, where σ_1^2, σ_2^2 and ρ are the population variances and correlation. Then the distribution of v is given by $(1-\rho^2)^{m+\frac{1}{2}}e^{\rho v}|T_m(v)|dv$; where $m = \frac{1}{2}n - 1$,

$$T_m(v) = (\tfrac{1}{2}v)^m K_m(v) / \sqrt{\pi}\,\Gamma(m + \tfrac{1}{2}),$$

and $K_m(v)$ is the Bessel function of the 3d kind for imaginary argument. When m is half an odd integer these T functions are elementary; e.g.

$$T_{\frac{1}{2}}(v) = \tfrac{1}{2}e^{-v}, \quad T_{3/2}(v) = \tfrac{1}{4}(v+1)e^{-v}, \quad T_{5/2}(v) = \tfrac{1}{16}(v^2 + 3v + 3)e^{-v}.$$

The table by Elderton on pp. 195-201 gives $T_m(v)$ to 6 fig. for

$$\begin{cases} m = 0(\tfrac{1}{2})5\tfrac{1}{2}, & v = 0(.1)4(.5)16(2)40(5)120 \, ; \\ \\ m = 6\tfrac{1}{2}(\tfrac{1}{2})11\tfrac{1}{2}, & v = 0(.5)16(2)40(5)120 \, . \end{cases}$$

Also for same m and v,

$\log_{10}T_m(v)$ is tabulated to 7 dec.; and except for $m = 0$ and $\tfrac{1}{2}$,

$$\rho = \frac{vT_{m-1}(v)}{(2m-1)T_m(v)} = \frac{K_{m-1}(v)}{K_m(v)} \quad \text{to 5 dec.}$$

Reprinted in Tables for S & B II, T. 10, pp. 138-144.

Certain related functions are tabled in K. Pearson, Stouffer & David 1932.

Hirschfeld, H. O. 1937
Biometrika 29, 65-77

"The distribution of ratio of covariance estimates in two samples drawn from bivariate normal populations"

Let x_i, y_i; X_j, Y_j; be two random samples, of size n and N, drawn from the same bivariate normal parent with correlation ρ; let

$$c = \frac{\sum(x_i-\bar{x})(y_i-\bar{y})}{n-1} \bigg/ \frac{\sum(X_j-\bar{X})(Y_j-\bar{Y})}{N-1}$$

The distribution of c is derived for odd n, N as a finite weighted sum of $I_x(a,b)$ values (equation (1) and (2) p. 67) with weights depending on ρ and mainly dependent on

$$(1-\rho)^{-(a+b)} + (1+\rho)^{-(a+b)}$$

which is given in Table II (p. 72) to 4 fig. for o = .2, .4, .6, .8 and a+b = 2(1)14. Table I (p. 68) gives for n = N = 15,

$$P(c \geq \frac{74}{26} \mid \rho)$$

to 3 dec. for $\rho = -1(.2)+1$.

Quenouille, M. H. 1949
Annals **of** **Math**. **Stat**. 20, 355
Relative efficiency of systematic-, stratified-, and random sampling in two dimensions.

Given a sample of $n_1 n_2$ values of x_{ij} drawn from a two dimensional population of elements x_{ij} (i=1, ..., $n_1 k_1$, j=1, ..., $n_2 k_2$), the variance of the mean $\sum_{i\,j} x_{ij} / n_1 n_2$ is considered under the following sampling schemes:

    (r,r)   Random for both i - coordinate and j - coordinate

    (st,st) Stratified for both i - coordinate and j - coordinate

    (sy,sy) Systematic for both i - coordinate and j - coordinate

    (r,sy)   Random for the i - coordinate, systematic for the j - coordinate

etc. for other combinations as explained by Fig. 2 (p. 360). A two dimensional correlogram function ρ_{uv} is defined (p. 361) and formulas for $\sigma_{\bar{x}}^2 = \sigma^2 / n_1 n_2$ in terms of ρ_{uv}. The particular case $\rho_{uv} = \rho_u \rho_v$ is then considered (an artificial assumption depending on the directions of the coordinate axes) and Table 3 (p. 368) gives results for the double exponential correlogram $\rho_{uv} = \rho_1^{|u|} \rho_2^{|v|}$:- Relative efficiencies σ^2(st st) / σ^2(r r), σ^2(sy sy) / σ^2(r r), σ^2(st st) / σ^2(sy sy) all for independent

7.6
7.94

[see also p. 697]

samples and the ratio $\dfrac{\sigma^2(\text{sy sy}) \text{ for aligned samples}}{\sigma^2(\text{r r}) \text{ for independent samples}}$ are tabled to about 3 fig. for $\rho_1 = 0(.1)1$ and $\rho_2 = 0(.1)\rho_1$.

Cook, M. B. 1951

<u>Biometrika</u> 38, 368-376

Let b be the sample regression of y on x in samples from the bivariate population described in the abstract on p. 328 above. The tables on p. 375 of Cook give $\beta_1(b)$ and $\beta_2(b)$ to 4 dec. for n = 20, 40, 60 as follows:

Table	2	3(a)	3(b)	3(c)
$\rho =$	0	.2	.2	.4
$\varkappa_{30}/\varkappa_{03} =$	1	+1	-1	+1
$\varkappa_{30}^2 = \varkappa_{03}^2 =$	0(.2).4	0(.2).6	0(.2).6	0(.2).6

Pearson, K., Jeffery, G. B. and Elderton, Ethel M. 1929

<u>Biometrika</u> 21, 164-201

[see also the review on p. 340 above]

Tables 1, 2, pp. 186, 187, give the mean, the standard deviation, β_1 and β_2 of the sample correlation and the sample covariance in samples of 5(5)25, 50, 100 for $\rho = 0(.3).9$. Reprinted in Tables for S & B II, Tables 11b, 11a, p. 145.

7.9 Other measures of correlation

For rank correlation see section 8.1.

7.91 Biserial correlation

Soper, H. E. 1915

Biometrika 10, 384-390

See abstract on p. 16.

7.92 Mean square contingency

See also pp. 174-179.

Sheppard, W. F. 1900

Transactions of the Cambridge Philosophical Society 19, 23-68

See abstract on p. 122.

7.93 Correlation ratio

Peters, C. C. and Van Voorhis, W. R. 1940

Statistical procedures and their mathematical bases

See abstract on p. 253.

7.94 Tetrachoric correlation

The basic tool in determining a correlation coefficient from a fourfold table is the tables of the bivariate normal distribution of sections 1.721 and 1.722. The tetrachoric functions of section 1.324 were originally computed as an aid to calculating the bivariate normal: see Tables for S & B I, page fifty of Introduction.

Pearson, K. 1913 a

Biometrika 9, 22-27

See abstract on p. 26.

Guilford, J. P. and Lyons, T. C. 1942

Psychometrika 7, 243-249

Table 1, p. 247, gives $(PQ)^{\frac{1}{2}}/Z$ to 4 dec. for $P = .5(.01).95$.

Table 2, p. 247, gives $[(1-\rho^2)(1-\gamma^2)]^{\frac{1}{2}}$, where $\pi\gamma = 2\sin^{-1}\rho$, to 4 dec. for $\rho = 0(.01).99$.

7.94
8.10
[see also pp. 698-699]

Tables 3 and 4, pp. 248-249, give double tail 'upper 5% and 1% points' of tetrachoric r to 3 dec. for p, p' = .5(.1).9 and n = 100(50)400(100)600(200)1000(500)3000, 5000, 10 000. These have been computed from Student's distribution by an unspecified process.

7.99 Miscellanea

Pitman, E. J. G. 1937b
Supplement to the Journal of the Royal Statistical Society 4, 225-232

Let x_i, y_i, $i = 1, \ldots, n$, denote a sample of n random pairs. Compute the usual correlation coefficient

$$r = \sum (x-\bar{x})(y-\bar{y}) \Big/ \sqrt{\sum (x-\bar{x})^2 \sum (y-\bar{y})^2}$$

and refer to a distribution obtained by forming correlation coefficients for all possible permutations of the observed y_i with regard to the x_i. In the table on p. 226 the example

$$x_i = 0 \quad 1 \quad 2 \quad 4 \quad 8$$
$$y_i = 7 \quad 6 \quad 9 \quad 3 \quad 0$$

is considered and the 9 permutations yielding the largest r^2 are given (the above, observed, permutation yields

$$\left| \sum (x-\bar{x})(y-\bar{y}) \right| = 39 \quad)$$

and the sixth largest r^2 so that it is significant at the 6/5! = 1/20 level of significance.
In the table on p. 228 the exact distribution is compared to 3 decimals with the approximation $I_{r^2}(\frac{1}{2}, \frac{n}{2} - 1)$ which is suggested by the exact moments. In the table on p. 230 a similar randomization distribution is obtained for Spearman's rank correlation (no ties) for n = 6 which is compared with the approximation $I_{\frac{1}{2}(1+r)}(2,2)$ to 3 decimals. The latter had a correction for continuity applied.

SECTION 8

RANK CORRELATION; ORDER STATISTICS

The bibliography of I.R. Savage 1953 should be consulted for additional references bearing on Sections 8 and 9.

8.1 Rank correlation

For a full discussion of rank correlation, with further references, see M.G. Kendall 1948, 1955.

8.10 Definitions and notation

A number of items are said to be ranked if to the items we attach the integers $1, \ldots, n$ (usually in accordance with a judged order) resulting in a permutation $x_1, \ldots, x_i, \ldots, x_n$ where x_i is the rank attached to the ith item. If $y_1, \ldots, y_i, \ldots, y_n$ is a second ranking of the items (say by a different judge) then two measures of the correlation between the two rankings are currently discussed. (For a third measure, Spearman's footrule, see Spearman 1906; Kendall 1948, pp. 23-24; Kendall 1955, pp. 32-33.)

(a) Spearman's rank correlation Rho, here denoted r_s.
This is based on 'Spearman's Score'

$$S_r = \sum_{i=1}^{n} d_i^2 \quad \text{where} \quad d_i = x_i - y_i \quad (8.10.1)$$

ranging between the values of 0 (complete agreement of ranks) and $(n^3-n)/3$ when the ranks show perfect disagreement:

$$x_i \equiv 1, 2, \ldots, n \quad \text{and} \quad y_i \equiv n, n-1, \ldots, 1 \; .$$

Spearman's Rho is defined as

$$r_s = 1 - 6 S_r / (n^3 - n) \quad (8.10.2)$$

and has therefore the range of a correlation coefficient: $-1 \leq r_s \leq 1$. In fact r_s can be seen to be the ordinary product-moment correlation coefficient of the x_i and y_i.

8·10

In repeated pairs of random rankings

$$\left.\begin{aligned} E(r_s) &= 0 \\ \mathrm{Var}(r_s) &= (n-1)^{-1} \\ \mu_3(r_s) &= 0 \\ \mu_4(r_s) &= 3(25n^3-38n^2-35n+72)/25n(n+1)(n-1)^3 \end{aligned}\right\} \quad (8·10·3)$$

and the variate transform

$$t = \sqrt{[(n-2)r_s^2/(1-r_s^2)]} \quad (8·10·4)$$

is approximately distributed as Student's t with $n-2$ d.f. An equivalent statement is that r_s is approximately distributed as a product-moment r so that it may be tested for significance by the tables of §7·122.

If x_i and y_i arise from ranking a sample from a bivariate normal distribution with population correlation R †, then

$$E(r_s) = 6[\sin^{-1}R + (n-2)\sin^{-1}\tfrac{1}{2}R]/\pi(n+1), \quad (8·10·5)$$

$$\mathrm{Var}(r_s) = (n-1)^{-1} + 36\pi^{-2}(n^2-n)^{-1}(n+1)^{-2} \times$$
$$\times [n^3(-0·42863279R^2 + 0·08354697R^4 + 0·04257246R^6 +$$
$$+ 0·01687474R^8 + 0·00664071R^{10} + 0·00270655R^{12}) +$$
$$+ n^2(0·15513010R^2 - 0·05736229R^4 - 0·18443407R^6 -$$
$$- 0·02271732R^8 + 0·00757524R^{10} + 0·01329883R^{12}) +$$
$$+ n(0·36837259R^2 + 0·44738882R^4 - 0·08427574R^6 -$$
$$- 0·27929901R^8 - 0·19943375R^{10} - 0·13861060R^{12}) +$$
$$+ 0·07179677R^2 + 0·06467162R^4 + 0·21015257R^6 +$$
$$+ 0·28589798R^8 + 0·31704425R^{10} + 0·07923733R^{12}]. \quad (8·10·6)$$

(b) Kendall's rank correlation Tau, here denoted t_k.

We consider one of the $\binom{n}{2}$ pairs (i,j) of the n items $(i, j = 1, 2, \ldots, n)$. Then, for each pair, we introduce the score

$$S_{ij} = \mathrm{signum}(x_i-x_j)\mathrm{signum}(y_i-y_j), \quad (8·10·7)$$

viz +1 if the two differences in ranks have the same sign, and −1 if they have opposite signs.

---

†Because the letter ρ is often used to denote r_s, the letter R is used throughout §8·1 for a parent product-moment correlation.

Kendall's score S_t is then defined as the total of these $\tfrac{1}{2} n(n-1)$ individual scores, i.e. by

$$S_t = \sum_{i,j} S_{ij} \qquad (8.10.8)$$

and has the range $\pm \tfrac{1}{2} n(n-1)$. Kendall's Tau is defined as

$$t_k = S_t / \tfrac{1}{2} n(n-1) \qquad (8.10.9)$$

and has the range $-1 \leq t_k \leq +1$.

In repeated pairs of random rankings

$$\left. \begin{array}{l} E(t_k) = 0 \\ \text{Var}(t_k) = 2(2n+5)/9n(n-1) \\ \mu_3(t_k) = 0 \\ \mu_4(t_k) = \dfrac{4(100n^4 + 328n^3 - 127n^2 - 997n - 372)}{675 \, n^3(n-1)^3} \end{array} \right\} \qquad (8.10.10)$$

Expressions for the generating function of t_k for random rankings are given by Moran 1950 and Silverstone 1950; for the moments of t_k in general, by Sundrum 1953.

If x_i and y_i arise from ranking a sample from a bivariate normal distribution with population correlation R, then

$$E(t_k) = 2\pi^{-1} \sin^{-1} R, \qquad (8.10.11)$$

$$\text{Var}(t_k) = 2[9n(n-1)]^{-1}[2n-5 - (6\pi^{-1}\sin^{-1}R)^2 - 2(n-2)(6\pi^{-1}\sin^{-1}\tfrac{1}{2}R)^2]. \qquad (8.10.12)$$

(c) The concordance coefficient W

If n items are ranked by m judges, and x_{ij} denotes the rank number given to the ith item by the jth judge, then Kendall's concordance score

$$S_w = \sum_{i=1}^{n} (x_{i.} - \tfrac{1}{2} m(n+1))^2 \qquad (8.10.13)$$

where $x_{i.} = \sum_j x_{ij}$ is the total of the m ranks given to the ith item and $\tfrac{1}{2}m(n+1)$ is the mean of the $x_{i.}$. The maximum value of S_w is $m^2(n^3-n)/12$, representing perfect agreement between the judges. The minimum value of S_w is 0 when m is even or n is odd; when m is odd and n is even, the minimum value of S_w is $n/4$. S_w can attain only values that are congruent mod 2 to its minimum and maximum values. Accordingly, S_w

runs over odd integers when m is odd and $n = 8k+4$; over numbers $2s+\frac{1}{2}$ when m is odd and $n = 8k+2$; over numbers $2s-\frac{1}{2}$ when m is odd and $n = 8k+6$; over even integers in all other cases. For $m = 2$, all values of S_w permitted by this congruence occur; for $m > 2$, S_w attains no values between $[m^2(n^3-n) - 24(m-1)]/12$ and $m^2(n^3-n)/12$, and there may be other gaps.

Kendall's coefficient of concordance

$$W = 12\,S_w/m^2(n^3-n) \qquad (8\cdot10\cdot14)$$

ranges between 0 and 1: the maximum value of W is 1; the minimum value of W is 0 when m is even or n is odd; when m is odd and n is even, the minimum value of W is $3/(m^2n^2-m^2)$.

If r_s is computed for each of the $\binom{m}{2}$ pairs of judges, then the average value of r_s can be shown to equal $(mW-1)/(m-1)$. For random rankings

$$E(W) = 1/m, \quad \mathrm{Var}(W) = 2(m-1)/m^3(n-1). \qquad (8\cdot10\cdot15)$$

$m(n-1)W$ is approximately distributed as χ^2 with $n-1$ d.f.; more accurately, $\Pr[S_w \leq T] \approx I_x(a,b)$, where $x = 12(T-1)/[m^2(n^3-n)+24]$, $a = (mn-m-2)/2m$, $b = (m-1)a$: see Kendall & Babington Smith 1939a and Friedman 1940.

8·11 Spearman's Rho

8·111 Distribution of S_r: null case

8·1111 Individual terms

The following tables give $n!\Pr[S_r = x]$, which is an integer. 'All x' means all x for which this probability is not zero; 'half x' means that only one tail of the (symmetrical) distribution of S_r is given, together with a conversion formula for computing the other tail.

8.10
8.112

n =

2(1)5; 6, 7	all x; half x		Olds 1938(139)
2(1)6; 7, 8	all x; half x		M. & S. Kendall 1939(255), M. Kendall 1943(396)
9, 10	half x		S.T. David et al. 1951(132)

<u>8.1112</u> Cumulative sums: exact probabilities

The following tables give $\Pr[S_r \geq x]$. 'All x' and 'half x' are used as above. Initial nines are counted as non-significant figures.

n =

4 dec.	2(1)7	all x	Olds 1938(145)
3 dec. or 2 fig.	4(1)8	all x	M.G. Kendall 1943(397), 1948(142)
3 fig.	9, 10	half x	S.T. David et al. 1951(133)
3 dec. or 2 fig.	4; 5 6; 7 8; 9 10	12(2)20; 22(2)40 50(2)70; 74(4)110 108(6)162; 156(8)228 208(10)308	E.S. Pearson & Hartley 1954(T.44)
3 dec. or 2 fig.	4(1)10	half x	M.G. Kendall 1955(172)
3 dec.	4; 5 6; 7 8; 9 10	18, 20; 32(2)40 56(2)70; 82(4)110 120(6)162; 172(8)228 228(10)298	Dixon & Massey 1957(469)

<u>8.1113</u> Cumulative sums: approximate probabilities

Calculated from a Type II fit to the second and fourth moments of S_r:

4 dec.	8; 9, 10	all x; half x	Olds 1938(145)

<u>8.112</u> Percentage points of S_r and r_s : null case

Olds, E. G. 1938

<u>Annals</u> <u>of</u> <u>Mathematical</u> <u>Statistics</u> 9, 133-148

Table 5, p. 148, gives $A(n) \pm B(n)x(p)$, where $A(n) = (n^3-n)/6$, $B(n) = (n-1)^{\frac{1}{2}}n(n+1)/3$, and $x(p)$ is the standard normal deviate of Section 1.0, to 1 dec. for n = 11(1)30 and p = .005, .01, .02, .05, .1. These

[349]

8.112
8.119

numbers are intended for use as lower and upper 100 p % points of S_r. Note errors:

n = 12   add 3 to all entries
n = 14   p = .005   for 125.5 read 129.9
n = 29   p = .01   for 5744.9 read 5844.9

Thornton, G. R. 1943

Psychometrika 8, 211-222

Table 1, p. 212, gives the upper 1%, 2%, 4%, 5%, 10% and 20% points of r_s to 3 dec. for n = 5(1)8; 9, 10; 11(1)30 : calculated from the exact distribution of S_r; Olds' Type II approximation; Olds' normal approximation. Tables 2 and 3, pp. 217-218, attempt to assess the merits of several approximations, all empirical and some dimensionally wrong, to the 5% and 1% points.

Snedecor, G. W. 1946

Statistical methods

Table 7.10, p. 166, gives the upper $2\frac{1}{2}$% and $\frac{1}{2}$% points of r_s to 3 dec. for n = 5(1)8. Reprinted in Ferber 1949, T. 16, p. 523.

Olds, E. G. 1949

Annals of Mathematical Statistics 20, 117

Table 5, p. 118, gives A(n) $\pm$ B(n)x(p) , defined in the abstract of Olds 1938 above, to 1 dec. for n = 11(1)30 and p = .025 .

Dixon, W. J. and Massey, F. J. 1951

Introduction to statistical analysis

Table 17.6 , p. 261, gives upper 5% and 1% points of r_s to 3 dec. for n = 4(1)10(2)30 .

Siegel, S. 1956

Nonparametric statistics

Table P, p. 284, derived from Olds 1938, 1949, gives upper 5% and 1% points of r_s to 3 dec. for r = 5(1)10(2)30 . An identical table appears in Vianelli 1959, T. 43, p. 1175.

8.113 Treatment of ties in the calculation of r_s

Woodbury, M. A. 1940

<u>Annals of Mathematical Statistics</u> 11, 358-362

Let $x_1, \ldots, x_n$
$y_1, \ldots, y_n$

denote two rankings of the integers $1, 2, \ldots, n$ (no ties). Then Spearman's rank correlation is defined by (8.10.2).

In the case of k of the (say) x_i tying, it will be known that a group of k of the x_i may equally claim any of a consecutive sequence of ranks $j, j+1, \ldots, j+k-1$. By allotting these ranks in all possible k! ways, computing r_s from (8.10.2), and taking the average, a measure of rank correlation $\bar{\rho}$ is obtained. To facilitate the evaluation of $\bar{\rho}$ the 'midrank method' is considered in which the 'rank' of all k x_i in the tie is taken as $(2j+k-1)/2$. Let the r_s resulting from this be ρ_M; then $\delta_{n,k} = \rho_M - \bar{\rho}$ only depends on k and is given by $k(k^2-1)/2n(n^2-1)$. The table on p. 361 gives $\delta_{n,k}$ to 4 dec. for $k = 2(1)13$, $n = 3(1)30(5)50(10)100$, $n > k$. If several x_i and/or y_i tie the $\delta_{n,k}$ correction to the midrank ρ_M are additive. Also shown are exact values of $\Delta k = \dfrac{k(k^2-1)}{12}$.

8.114 Distribution of r_s under non-null parent normal correlation

Moran, P. A. P. 1948

<u>Biometrika</u> 35, 203-206

The table on p. 204 gives $E(r_s)$ from (8.10.5) and $E(r)$ from (7.13.1) to 4 dec. for $R = .1(.1).9$ and $n = 5, 10, 20, \infty$.

8.119 Miscellaneous tables of r_s

Lyerly, S. B. 1952

<u>Psychometrika</u> 17, 421-428

N judges rank n items and each judge's ranking is compared against a 'standard rank' (criterion) and the Spearman correlation r_s (from equation (8.10.2)) computed. Averaging these we have $\bar{r} = \sum r_s / N$.

8.119
8.1312

In random rankings $E(\bar{r}) = 0$, $\text{Var}(\bar{r}) = 1/N(n-1)$ and

$$\beta_2 = 3 + \frac{24}{100N} \left(\frac{36 - 5n - 19n^2}{n^3 - n} \right).$$

Table 1 (p. 426) gives β_2 to 2 dec. for $N = 1(1)6$ and $n = 2(1)7$.

Other tables examine the normal approximation to $\bar{r}$ and find it good.

8.12 Kendall's Tau

8.121 Distribution of S_t : null case

8.1211 Individual terms

M.G. Kendall 1938, p. 88; 1943, p. 404; tables $n! \Pr[S_t = x]$, which is an integer, for $n = 1(1)10$. The table covers all non-negative values that S_t can assume; probabilities for negative S_t may be supplied by symmetry.

8.1212 Cumulative sums

The following tables give $\Pr[S_t \geq x]$. The tables are for non-negative x; values for negative x may be supplied by symmetry. 'All x' means all non-negative x for which $\Pr[S_t = x] > 0$.

	$n =$		
3 dec. or 2 fig.	4(1)10	all x	M.G. Kendall 1943(405), 1948(141), 1955(171); Siegel 1956(285), Vianelli 1959(199)
3 dec.	4(1)40	all* x	Kaarsemaker & v.W. 1953(44,49)
3-4 dec.	$\begin{cases} 4(1)7; \ 8 \\ 9; \ 10 \end{cases}$	$\begin{matrix} \text{all x; } 0(2)26 \\ 0(2)30; \ 1(2)35 \end{matrix}$	Pearson & Hartley 1954(T.45)

8.122 Percentage points of S_t

Kaarsemaker & van Wijngaarden 1953, T. 3, p. 53, give the upper $\frac{1}{2}\%$, 1%, $2\frac{1}{2}\%$, 5% and 10% points of S_t for $n = 4(1)40$.

---

*all x for which $\Pr[S_t \geq x] > .0005$. Error: $n = 8$, $S = 22$, for .002 read .003.

8.119
8.1312

8.123   Treatment of ties in the calculation of S_t

Sillitto, G. P.   1947

Biometrika 34, 36-40 + folding plate

In a series of ranked integers each member is compared with all members to its right. In each such comparison it is scored:

$+1$ if it is smaller than
-1 if it is larger than ⎫ the member
0 if it is equal (tie) to ⎭ on its right

and the scores added for all members to form S_t. When no ties occur the maximum value of S_t is $\tfrac{1}{2}n(n-1)$, but when p_2 pairs of ties, p_3 triplets of ties, etc. occur the maximum score S_t as well as its distribution depends on $p_2, p_3, \ldots$.

Table 1 (p. 38) gives the frequencies of occurrences of S_t in samples of $n = 3(1)7$ for all possible p_2 (assuming $p_3 = p_4 = \ldots = 0$).

Table 2 (on folding plate) gives the cumulative sum of this distribution to 2 fig. for $n = 3(1)10$ and for all possible combinations of p_2 and p_3 (assuming $p_4 = p_5 = \ldots = 0$).

8.125   Moments of S_t

Kaarsemaker & van Wijngaarden 1953, T. 4, p. 54, give $(\text{Var } S_t)^{\frac{1}{2}} = [n(n-1)(2n+5)/18]^{\frac{1}{2}}$ to 5 dec. for $n = 40(1)100$.

8.13   **Kendall's concordance coefficient W**

8.131   Distribution of S_w and related statistics: null case

8.1312   Distribution of S_w: cumulative sums

Kendall, M. G. and Babington Smith, B.   1939 a

Annals of Mathematical Statistics 10, 275-287

Tables 1 to 4, pp. 279-282, give $\Pr[S_w \geq x]$ to 3 dec. or 2 fig. for
$\begin{cases} n = 3 & 4 & 5 \\ m = 2(1)10 & 2(1)6 & 3 \end{cases}$; for all x such that $\Pr[S_w = x] > 0$, and a few additional x to secure typographical uniformity.

Reprinted in M.G. Kendall 1943, TT. 16.5-16.8, pp. 412-415; 1948, T. 5, pp. 146-149; 1955, T. 5, pp. 182-185; Vianelli 1959, Pro. 62, pp. 199-200.

8.1312
8.14

[see also p. 699]

Pearson & Hartley 1954, T. 46, p. 211, print an abridgement of Kendall & Babington Smith's table, giving 3-4 dec. for
$\begin{cases} n = 3 & 3 \\ m = 3 & 4 \\ x = 6(\text{all})18 & 8(\text{all})32 \end{cases}$

$\begin{cases} 3 & 3 & 3 & 3 \\ 5 & 6 & 7 & 8 \\ 14(\text{all})50 & 18(\text{all})72 & 24(\text{all})72, 78, 96 & 26(\text{all})50, 56, 72, 78, 86, 98 \end{cases}$

$\begin{cases} 3 & 3 \\ 9 & 10 \\ 32(\text{all})50(6)62, 78, 86, 98(\text{all})114 & 32, 42, 50(6)62(12)86, 96, 104, 122, 126 \end{cases}$

$\begin{cases} 4 & 4 & 4 \\ 3 & 4 & 5 \\ 19(\text{all})45 & 32(4)40, 46(4)54, 62(4)74 & 41, 43, 51, 57, 61, 67, 81, 85, 93, 101, 105 \end{cases}$

$\begin{cases} 4 & 5 \\ 6 & 3 \\ 46, 52, 62(6)80, 100, 108(10)128 & 46, 50, 56, 60, 62, 66, 74(4)86 \end{cases}$.

8.1313 Distribution of $\chi_r^2 = 12 S_w /mn(n+1)$

Friedman, Milton 1937

Journal of the American Statistical Association 32, 675-701

Tables 5-6, pp. 688-689, give $\Pr[\chi_r^2 \geq x]$ to 3 dec. or 2 fig. for $n = 3$, $m = 2(1)9$; $n = 4$, $m = 2(1)4$; and all x for which $\Pr[\chi_r^2 = x] > 0$. Reprinted (more legibly) in Siegel 1956, T. N, pp. 280-281.

8.132 Percentage points of W

Friedman, Milton 1940

Annals of Mathematical Statistics 11, 86-92

Kendall & Babington Smith 1939a recommended that S_w be transformed into an approximate beta-variate by the formula on line 15 of page 348, and that the beta distribution be evaluated by using tables of Fisher's z; Friedman 1937 suggested that χ_r^2, defined in section 8.1313, be referred to the χ^2-distribution with $n-1$ d.f.

Table 1, p. 88, gives the upper 5% and 1% points of χ_r^2: by linear inverse interpolation in Kendall & Babington Smith; by the chi-squared approximation; and by the z-approximation with and without continuity correction; all to 2-3 dec. for $\begin{cases} n = 3 & 4 & 5 \\ m = 8(1)10, \infty & 4(1)6, \infty & 3, \infty \end{cases}$.

Tables 2 and 3 are based on the z-approximation. Table 2, p. 89, gives the upper 5% and 1% points of χ_r^2 to 2-3 dec. for n = 3(1)7 and m = 3(1)6(2)10(5)20, 100, ∞. Table 3, p. 90, gives the upper 5% and 1% points of S_w to 1 dec. for

$$\begin{cases} n = 3 & 4 & 5(1)7 \\ m = 8(1)10(2)14(1)16(2)20 & 4(1)6(2)10(5)20 & 3(1)6(2)10(5)20 \end{cases}.$$

Table 3 is reprinted in M.G. Kendall 1948, T. 6, p. 150; 1955, T. 6, p. 186; Siegel 1956, T. R, p. 286.

8.14 Paired comparisons; inconsistent triads

Kendall, M. G. and Babington Smith, B. 1940
Biometrika 31, 324-345

Consider n objects A, B, etc. A judge compares a pair of objects (say A, B) and puts them in order of preference (say $A > B$), this judgment being carried out for all $\binom{n}{2}$ possible choices of pairs. A triad is defined as an inconsistency in a set of three judgments of the kind $A > B$, $B > C$, $C > A$. Let d be the total number of triads occurring in the set of all $2^{\binom{n}{2}}$ possible judgments. Table II (p. 333) gives the frequencies f (integer) of d for n = 2(1)7 as well as their cumulative relative frequencies (3 decs).

Let there be m such judges and let $\gamma(A,B)$ etc. denote the number of judges who judged $A > B$, etc. Let

$$\Sigma = \sum \binom{\gamma(A,B)}{2}$$

the summation going over all possible n(n-1) pairs (i.e. A,B as well as B,A) so that

$$\max \Sigma = \binom{m}{2}\binom{n}{2}$$

when all judges agree. Tables III to V (pp. 336-7) give the cumulative relative frequencies (3 dec. or 2 fig.) of Σ for

$$\left.\begin{array}{ll} m = 3 & n = 2(1)8 \\ m = 4 & n = 2(1)6 \\ m = 5 & n = 2(1)5 \\ m = 6 & n = 2(1)4 \end{array}\right\} \text{ and all possible } \Sigma.$$

8.14
8.19 [see also p. 699]

The coefficient of agreement u is defined as

$$u = -1 + 8\Sigma/[m(m-1)n(n-1)].$$

Tables II to V are reprinted in Kendall 1943, TT. 16.10-16.14, pp. 428-432; 1948, TT. 9-10, pp. 154-158; 1955, TT. 9-10, pp. 190-194.

8.15 Partial and multiple rank correlation

Following M.G. Kendall 1942, define

$$\tau_{QR.P} = [\tau_{QR} - \tau_{PR}\tau_{QP}][(1-\tau_{PR}^2)(1-\tau_{QP}^2)]^{-\frac{1}{2}}.$$

Certain tables for the distribution of partial Tau are given by Moran 1951. According to M.G. Kendall 1955, p. 122:*

"Moran (1951) has considered partial τ without reaching any clear conclusions other than that the distributional problem is a very complex one.

"Moran also considers a coefficient of multiple correlation defined, for example, in the case of three variates by a formula analagous to the ordinary [product-moment] theory:

$$1 - R_{1(23)}^2 = (1-\tau_{13}^2)(1-\tau_{12.3}^2)$$

For small n the distributional theory is again difficult. Moran suggests a test based on the variance-ratio."

8.19 Other non-parametric measures of correlation

For mean square contingency see sections 2.6 and 7.92.

Hoeffding, W. 1948
<u>Annals</u> <u>of</u> <u>Mathematical</u> <u>Statistics</u> 19, 546-557

Let $x_1, y_1; \ldots; x_n, y_n$ be a random sample of paired observations from a bivariate parent $f(x,y)$. Define the criterion of independence (for $n \geq 5$) by

$$D_n = (n-5)! \, n!^{-1} \sum{}'' \phi(x_{\alpha_1}, y_{\alpha_1}; \ldots; x_{\alpha_5}, y_{\alpha_5})$$

where $\sum{}''$ denotes summation over the $n!/(n-5)!$ choices of α_i out of $1,\ldots,n$, with $\alpha_i \neq \alpha_j$ for $i \neq j$ and where

$$4\phi(x_1, y_1; x_2, y_2; x_3, y_3; x_4, y_4; x_5, y_5)$$
$$= \psi(x_1, x_2, x_3)\psi(x_1, x_4, x_5)\psi(y_1, y_2, y_3)\psi(y_1, y_4, y_5)$$

---

*Reproduced from Kendall's ∈<u>Rank Correlation Methods</u> (2d edition, 1955) by permission of the author and the publisher, Charles Griffin and Co. Ltd.

8.14
8.19

with $\psi(x_1, x_2, x_3) = C(x_1-x_2) - C(x_1-x_3)$

and with $C(u) = \begin{cases} 1 & \text{if } u \geq 0 \\ 0 & \text{if } u < 0 \end{cases}$

Table 1 (p. 554) gives for the case of independence, i.e. $f(x,y) = f(x)g(y)$:

$P[60D_5 \geq x]$ for $x = -1, 0, 2$

$P[180D_6 \geq x]$ for $x = -2(1)3, 6$

$P[1260D_7 \geq x]$ for $x = -11, -8(1)0, 2(1)4(2)8, 9, 12, 14, 18, 24, 30, 42$

all to 4 dec.

Individual probabilities for exact equalities of D_n are also shown alongside.

Whitfield, J. W. 1949
<u>Biometrika</u> 36, 463-467

Suppose 2m objects to have been ranked with regard to one quality and arranged in m pairs with regard to a second. To measure the intraclass rank correlation, i.e. the extent to which objects placed in the same pair have similar ranks, Whitfield arranges the m pairs in order of the higher rank in each pair. For each object, he counts the number of elements with lower rank to its right (not including its mate; from the total of these counts (T, say) he computes the score $S_p = 2T - 3m(m-1)$ and Whitfield's Tau, defined as $S_p/[m(m-1)]$.

The range of S_p is $\pm m(m-1)$; in the absence of correlation, the distribution is symmetrical. The table on p. 466 gives $\Pr[S_p \geq x]$ to 5 dec. or 1 fig. for $m = 3(1)10$ and $x = 2(2)[m(m-1)]$.

The value .50000 shown for $x = 0$ rests on the unusual convention of splitting $S_p = 0$ between positive and negative correlations.

Blomqvist, N. 1950
<u>Annals</u> <u>of</u> <u>Mathematical</u> <u>Statistics</u> 21, 593-600

Let x_i, y_i ($i = 1, \ldots, n$) denote an even number n of paired observations drawn from a bivariate universe and denote by $\dot{x}, \dot{y}$ their respective sample medians. By taking the point $\dot{x}, \dot{y}$ as the origin of

8.19

4 quadrants and contrasting the number of points x_i y_i in the 1st and 3rd quadrants against the number in the 2nd and 4th a measure of dependence is obtained as follows:

Define n_1 = number of x_i y_i for which $(x_i - \dot{x})(y_i - \dot{y}) > 0$

n_2 = "   "   "   "   " $(x_i - \dot{x})(y_i - \dot{y}) < 0$

If it is assumed that all x_i and all y_i differ from one another then for an even sample $x_i \neq \dot{x}$, $y_i \neq \dot{y}$ and $n_1 + n_2 = n$ the criterion for testing dependence, after Mosteller 1946, is $q' = -1 + 2n_1/n$, which for fixed n is a function of n_1.

The table on p. 598 gives $\Pr\left[\left|n_1 - \tfrac{1}{2}n\right| \geq \nu\right]$ to 3-4 dec. for $n = 4(4)48$, $\nu = 0(2)14$ and $n = 6(4)50$, $\nu = 1(2)15$

By a convention the tables are also applicable to odd sample sizes $n + 1$.

Moran, P. A. P. 1950 *at*

Royal Stat. So. <u>Jn</u>. B, 12, 292

Curvilinear Ranking Test.

Of the n! possible rankings of n items the two completely monotonic ones i.e. 1, 2,···,n and n, n-1,···,1 .as well as all those in which one increasing sequence of ranks is followed by one decreasing sequence (e.g. 1234) are called 'perfectly curvilinear'. For all other rankings 'the curvilinear deficit' D is defined as the least number of nearest neighbour interchanges required to change the ranking to one of the 'perfectly curvilinear' class for which, of course, D = 0.

Table 1 (pp. 293-4) gives the frequency distribution of D for all possible n! rankings for $n = 2(1)8$; for $n = 9(1)14$ the table is curtailed, and covers only $D = 0(1)12$. Table 2 (p. 294) gives the cumulative distribution of D to 4 dec.

Blomqvist, N. 1951

<u>Annals</u> <u>of</u> <u>Mathematical</u> <u>Statistics</u> 22, 362-371

"Some tests based on dichotomization"

Blomqvist considers a random sample of n units each having m variates and denotes the jth variate of the ith unit by z_{ij}; he seeks to test whether there is independence among the z_{ij} when their distribution is unknown.

Assume that the range of the z is dichotomized into 'high' and 'low'.

Let y_i be the count of high variates in the ith unit, and compute

$$S = \sum_{i=1}^{n}(y_i-\bar{y})^2 .$$

Table 1, pp. 368-369, gives $Pr[S \geq x]$ to 3-4 dec. for all x and for

$\begin{cases} m = 3 & 4 & 5 & 6(1)8 \\ n = 4(2)16 & 4(2)10 & 4, 6 & 4 \end{cases}$.

See also Olmstead & Tukey 1947.

8.2
8.22   [see also pp. 699-702]

8.2 Order statistics from non-normal parents

This section contains tables analogous to those for normal parents abstracted in sections 1.4 and 1.5. The notation of pp. 46-49 and 82-83 will be used where appropriate.

For some inequalities satisfied by order statistics and ranges, see section 8.28 (in appendix); for tables of the asymptotic distribution of extreme values and ranges, see section 8.3.

8.21 Continuous rectangular parent

Neyman, J. and Pearson, E. S. 1928 a

<u>Biometrika</u> 20 A , 175-240

Table V (p. 218) gives theoretical values of d_n, σ_R, $\beta_1(R)$, $\beta_2(R)$ all to 3 dec. for samples of $n = 3, 4, 6, 10, 20$ drawn from a rectangular parent as well as for samples from a normal parent.

Hastings, C., Jr., Mosteller, F., Tukey, J. W. and Winsor, C. P. 1947

<u>Annals</u> <u>of</u> <u>Mathematical</u> <u>Statistics</u> 18, 413-426

See abstract on p. 65.

Rider, P. R. 1951 a

<u>Journal</u> <u>of</u> <u>the</u> <u>American</u> <u>Statistical</u> <u>Association</u> 46, 502-507

"The distribution of the quotient of ranges in samples from a rectangular population"

Denote by R_1 the range in a sample of m variates drawn from a rectangular distribution, and by R_2 the range in a second sample of n variates from the same distribution and put $u = R_1 / R_2$. By elementary integration the distribution ordinate function of u is found to be

$$f(u) = \begin{cases} c\left[(m+n)\, u^{m-2} - (m+n-2)\, u^{m+1}\right] & \text{for } 0 \le u \le 1 \\ c\left[(m+n)\, u^{-n} - (m+n-2)\, u^{m-1}\right] & \text{for } 1 \le u \le \infty \end{cases}$$

with $c = m(m-1)\,n(n-1)\,/\,(m+n)(m+n-1)(m+n-2)$

Tables 2-4, pp. 506-507, give the upper 10%, 5% and 1% points of u to 3-4 fig: for $m, n = 2(1)10$.

8.22 Discrete rectangular parent

Rider, P. R. 1929

Biometrika 21, 124-143

Rider considers a parent distribution of 10 integer variate values 0 1 2 3 4 5 6 7 8 9 each with probability 1/10. Table 11, p. 136, gives the distribution of the median $\dot{x}$ in samples of 3 to 3 dec., i.e. exact, for $\dot{x} = 0(1)9$. Table 12, p. 136, gives the distribution of the range R_4 in samples of 4 to 4 dec., i.e. exact, for $R_4 = 0(1)10$; and compares it with the distribution of R_4 from a continuous rectangular parent. Table 14, p. 139, gives the joint distribution of the stragglers x_1 and x_4 in samples of 4.

Table 13, p. 138, considers a parent distribution of 5 integers 0 1 2 3 4 each with probability 1/5, and gives the distribution of midrange $\hat{x}$ in samples of 4 to 4 dec., i.e. exact, for $\hat{x} = 0(.5)4$.

Rider concludes (p. 139) that there "appears to be little difference in the distribution of these variates when the universe is continuous and when it is discrete."

Rider, P. R. 1951

Journal of the American Statistical Association 46, 375-378

"The distribution of the range in samples from a discrete rectangular population"

Assume that a variate x attains the N discrete values $x_i = a + ih$ $i = 0, 1, \cdots, N - 1$, with equal probability $\frac{1}{N}$. If a random sample of n is drawn from this 'discrete rectangular' population the chance for the sample range to be precisely R is given by

$$p(R) = N\left[\left(\frac{R+1}{N}\right)^n - 2\left(\frac{R}{N}\right)^n + \left(\frac{R-1}{N}\right)^n\right]\left(1 - \frac{R}{N}\right), R \neq 0 \; ; \quad p(0) = N^{1-n}.$$

Table 1 (p. 378) gives p (R) to 4 decimals for N = 10 (only) n = 2(1)10, R = 0(1)9.

Table 2 (p. 378) gives corresponding probabilities for samples of n from the continuous rectangular population, P (x) = x/10, $0 \leq x \leq 10$.

8.23

8.23 K. Pearson's frequency curves as parents

Pearson, K. and Pearson, Margaret V. 1932

Biometrika 24, 203-279

Pearson gives exact formulas for

$$\xi_{i|n} = \int_0^1 (1-p)^{i-1} p^{n-i} x \, dp$$

$$\mu_r(i|n) = \int_0^1 (1-p)^{i-1} p^{n-i} x^r \, dp$$

where $p = e^{-x/\sigma}$

and gives eleven illustrative tables on pp. 220-224, to 5-6 dec., for $n = 3, 11, 23, 51, 75, 99, 999, \infty$.

Pearson derives general formulas for the expectation, variance and covariance of order statistics from the parent $f(x) = y_0(ab+x)^m$. He tables some twenty illustrative special cases, all for samples of $n = 11$, on pp. 256-267.

Mc Intyre, G. A. 1952

Australian Journal of Agricultural Research 3, 385-390

"A method for unbiased selective sampling, using ranked sets"

Draw n samples, each containing n observations, from a population with cumulative distribution $F(x)$; denote by x_{ij} the ith order statistic in the jth sample. Mc Intyre considers the problem of estimating the moments μ_1', μ_2, μ_3, μ_4 from the x_{ii} only. The form of the estimators is the same as from a random sample of size n; the sampling variances of the x_{ii}-estimators are smaller. The resulting efficiency gains are given as percentages in T. 2, p. 388, for the four moments, $n = 2(1)5$, and the distributions $F(x) = I_x(1,1), I_x(4,4), I_x(4,7), I_x(4,13)$ and $1 - e^{-x}$; the gains are substantial only for the mean.

In samples of 5 from the above populations, T. 1, p. 387, gives the means and standard deviations of the five order statistics to 3 dec.

Jones, H. L. 1953

<u>Journal</u> <u>of</u> <u>the</u> <u>American</u> <u>Statistical</u> <u>Association</u> 48, 113-130

"Approximating the mode from weighted sample values"

Given a t-distribution with unknown mode, Jones considers an estimate of the mode by means of a weighted sum of the n order statistics in a sample of n, with weights so determined that the estimate approximates (in a sense discussed on pp. 116-118) a maximum likelihood estimate.

Table 1, p. 115, gives the weights to 2 dec. for n = 3(1)10 and for t-distributions with kurtosis $\mu_4/\mu_2^2 = 3(.5)5, 6, 9, \infty$; i.e. with ∞, 16, 10, 8, 7, 6, 5, 4 degrees of freedom.

8.3
8.332

[see also p. 703]

8.3 Tables facilitating the asymptotic theory of statistical extreme values

The standard reference for most of these tables is N.B.S. 1953. Although we group the tables under the extreme, the mth extreme, and the range, we describe their contents strictly as mathematical functions. For the relevancy of the functions tabled to the statistics of extreme values, see the papers cited; the introduction to N.B.S. 1953; Kendall & Stuart 1958, pp. 330-343; Gumbel 1954, 1958. Gumbel finds that, notwithstanding the extremely slow convergence of normal extremes to their limiting distribution (Fisher & Tippett 1928), the limiting distribution fits well except in the tails; for an example, see Gumbel 1954, p. 26, fig. 3.4.

8.31 The extreme value

8.311 Direct tables

8.3111 $e^{-e^{-x}}$ and $e^{-[x+e^{-x}]}$

7 dec.	$-3(.1)-2.4(.05)0(.1)+4(.2)8(.5)17$	δ_m^2	N.B.S. 1953(17)
5 dec.	$-2(.25)+6$; $e^{-e^{-x}}$ only	No Δ	Gumbel 1941(173)
5 dec.	$-2.5(.5)+10$	Δ	Gumbel 1935

8.3112 $-\log_{10}[1 - e^{-e^{-x}}]$

3 dec.	$-2(.25)+4$	No Δ	Gumbel 1941(173)

8.312 Inverse tables

8.3121 $-\log_e(-\log_e x)$

5 dec.	$0(.0001).005(.001).05$ $.05(.001).89$ $.89(.001).988(10^{-4}).9994(10^{-5})1$	δ_m^2 No Δ δ_m^2	N.B.S. 1953(19)
5 dec.	$0(10^{-4}).005(.001).988(10^{-4})$ $.9994(10^{-5})1$	No Δ	Vianelli 1959(1052)
4 dec.	$.002(.002).998$	No Δ	Fisher & Yates 1957(T.XII)

According to N.B.S. 1953, p. 10, the 4 dec. table by Mather 1949 is extremely inaccurate.

8.3123 $-x \cdot \log_2 x$

4 dec.   0(.001).999   No Δ   Newman 1951, Senders 1958(532)

8.32 The mth extreme value

8.321 Direct tables

8.3211 $(1 + 2e^{-x})e^{-2e^{-x}}$

5 dec.   -1.95(.05)+5.25(var.)6.45 δ^2 Gumbel & Greenwood 1951

8.322 Inverse tables

Let $y_m(\Phi) = \log_e 2m - \log_e \chi^2(\Phi, 2m)$. N.B.S. 1953, T. 4, p. 28, gives $y_m(\Phi)$ for $m = 1(1)15(5)50$; $\Phi = .005, .01, .025, .05, .1, .25, .5$; $1-\Phi$ = ditto. The values are nominally given to 5 dec.; their accuracy is limited by their source, namely Catherine M. Thompson 1941 b.

8.33 The range

Tables of Bessel functions closely related to those of this section are abstracted on pp. 108-110, above.

8.331 Direct tables: $\Psi = 2e^{-\frac{1}{2}x}K_1(2e^{-\frac{1}{2}x})$ and $\psi = 2e^{-x}K_0(2e^{-\frac{1}{2}x})$

7 dec.   -4.6(.1)-3.3(.05)+11(.5)20 δ_m^2 N.B.S. 1953(29),
                                                      N.B.S. 1953(29),
                                                      Vianelli 1959(1060)

5 fig.   -3.2(.1)+10.6   No Δ   Gumbel 1949(145)

5 dec.   -3(.5)+10.5   Δ   Gumbel 1947(395)

8.332 Inverse tables: x as a function of $\Psi = 2e^{-\frac{1}{2}x}K_1(2e^{-\frac{1}{2}x})$

No differences are given with these tables.

4 dec. Ψ = .0002(.0001).001(.001).01(.01).95(.001).998⎫
3 dec. Ψ = .0001, 999(.0001).9999   ⎬ N.B.S. 1953(32)
                                                             ⎭
2 dec. Ψ = .0002(.0001).001(.001).01(.01).99(.0001).9997   Gumbel 1947(397)

3 dec. Ψ = .0005, .001, .002, .0025, .005, .01, .02,   Gumbel 1949(146)
         .025, .05, .1, .2, .25; $1-\Psi$ = ditto

9.1
9.112
[see also p. 704]

SECTION 9

NON-PARAMETRIC TESTS

For an extended index of non-parametric statistics, see I.R. Savage 1953.

For the χ^2 test of goodness of fit see Section 2.6; for the binomial sign test, 3.6; for rank correlation, 8.1; for tables for the (partly distribution-free) asymptotic theory of extreme values, 8.3.

9.1 Tests of location

9.11 One-sample tests

9.111 The Walsh test

Walsh, J. E. 1949 c

Annals of Mathematical Statistics 20, 64-81

Let $y_1, \ldots, y_n$ be n independent observations each from a symmetrical population (the same or different) with median 0; let $x_1 < \ldots < x_n$ be the order statistics obtained by ordering the ys; let $z_k = \frac{1}{2}(x_i + x_j)$, $k = \frac{1}{2}i(i-1) + j$, $i \leq j$; let f_k be an arbitrary function of k that is 0, +1 or -1 for every integer k, $1 \leq k \leq \frac{1}{2}n(n+1)$. Then Walsh shows that the joint probability $\Pr[f_1 z_1 \geq 0, f_2 z_2 \geq 0, \ldots, f_{\frac{1}{2}n(n+1)} z_{\frac{1}{2}n(n+1)} \geq 0]$ is independent of the n populations from which the ys were drawn; can be evaluated, therefore, by letting the xs be order statistics from the rectangular distribution on $[-\frac{1}{2}, +\frac{1}{2}]$; and is equal to $r.2^{-n}$ for some r. The same three properties hold for any probability obtained by summing $\Pr[\ ]$ over several functions f.

Table 1, pp. 66-67, gives convenient tests constructed by this method for $n = 4(1)15$. We quote an example:

$n = 12$. Denote the population median by ϕ. To test $\phi = \phi_0$ against $\phi < \phi_0$, reject the null hypothesis if $x_4 + x_{12} < 2\phi_0$ and $x_5 + x_{11} < 2\phi_0$. Significance level: 4.7%.

For some of the tests in T. 1, the power-efficiency (for normal populations) relative to the t-test is given. The lowest efficiency shown is $87\frac{1}{2}\%$.

Reprinted in Walsh 1949 d , T. 1, p. 343; reprinted (without column showing efficiency) in Siegel 1956, T. H, p. 255.

Walsh, J. E.   1950

Journal of the American Statistical Association 45, 225-237

Table 1, p. 228, gives tests for $n = 10(1)15$, constructed by the methods of Walsh 1949 a , with their efficiencies relative to the normal deviate test (σ known). Table 2, p. 231, spells out in detail that $[\frac{1}{2}(x_1+x_{n-1}), \frac{1}{2}(x_2+x_n)]$ and $[\frac{1}{2}(x_1+x_{n-2}), \frac{1}{2}(x_3+x_n)]$ are respectively 50% and 75% confidence intervals for the median of a symmetrical distribution. Table 3, p. 234, gives illustrative values of power underlying the efficiencies in T. 1.

Dixon, W. J. and Massey, F. J.   1951

Introduction to statistical analysis

Table 26, p. 361, gives Walsh's tests for $n = 4(1)10$. Efficiencies are not given. Reprinted in Dixon & Massey 1957, T. 26, p. 463.

9.112   The signed-rank test

A convenient test of Walsh's family[*] uses as its statistic S, the excess of the number of positive over the number of negative z_k (defined in the abstract of Walsh 1949 c , p. 366, above). The characteristic function of S is $\prod_{j=1}^{n} \cos jt$. The distribution of S in samples of n, $Q(x,n) = \Pr[S > x]$ may be computed, starting with $Q(-2, 1) = 1$, $Q(0,1) = \frac{1}{2}$, $Q(+2, 1) = 0$, from the recursion $Q(x,n) = \frac{1}{2}[Q(x+n, n-1) + Q(x-n, n-1)]$.

In practice the statistic $T = \frac{1}{2}[\frac{1}{2}n(n+1) - |S|]$ is used. T is most rapidly computed by the method of Wilcoxon 1945, p. 80, rubric <u>Paired Comparisons</u>:

---

[*] The signed rank test seems to have been introduced by Wilcoxon. Walsh's family includes tests that cannot be reduced to the statistic S : see Walsh 1959.

9.112
9.122 [see also pp. 704-705]

Rank the observations y in ascending order of absolute value. T is the sum of the ranks corresponding to positive or to negative ys, whichever is smaller.

Wilcoxon 1945, T. 2, p. 82, gives approximate 1%, 2% and 5% points of T, together with their actual tail sums to 2 fig., for n = 7(1)16.

Wilcoxon 1947, T. 2, p. 121, gives approximate 1%, 2% and 5% points of T for n = 6(1)20.

Wilcoxon 1949, T. I, p. 13, gives approximate 1%, 2% and 5% points of T for n = 6(1)25. Reprinted in Siegel 1956, T. G, p. 254; Senders 1958, T. Q, p. 556; Vianelli 1959, Pro. 127, p. 344.

9.12 Two-sample tests

9.121 The Wilcoxon test

[The eponym <u>Wilcoxon</u> is customarily and conveniently attached to the two-sample rank-sum test. For a discussion of historical priority see Kruskal & Wallis 1952, pp. 602-605; Kruskal 1957 z.]

Given two samples of sizes m, n, $m \leq n$, to test the hypothesis that the two samples are from populations with the same median. Rank all the observations in ascending order of magnitude: let T be the sum of ranks assigned to the sample of size m. Define $U = mn + \tfrac{1}{2}m(m+1) - T$, $d = |2T - m(m+n+1)|/2m = |U - \tfrac{1}{2}mn|/m$.

Wilcoxon 1945, T. 1, p. 81, gives approximate 1%, 2% and 5% points of T, together with their actual tail sums to 2 fig., for m = n = 5(1)10.

Festinger 1946, TT. 1-2, pp. 103-104, gives approximate 1% points (when possible) and 5% points of d to 2 dec. for

$\begin{cases} m = 2 & 3 & 4(1)12 & 13(1)15 \\ n = 8(1)38 & 5(1)37 & m(1)[40-m] & m(1)[30-m] \end{cases}$.

Mann & Whitney 1947 give many useful formulas for the distribution, moments and approximate normalization of U. Their T. 1, pp. 52-54, gives the lower half of the (symmetrical) distribution of U to 3 dec. for n = 3(1)8; m = 1(1)n. Reprinted in Siegel 1956, T. J, pp. 271-273; Vianelli 1959, Pro. 128, pp. 345-346.

Wilcoxon 1947, T. 1, p. 121, gives approximate 1%, 2% and 5% points of T for $m = n = 5(1)20$.

White 1952, pp. 37-39, gives approximate 5% points, and 1% and 0.1% points (when possible), of T for $\begin{cases} m = 2 & 3 & 4(1)15 \\ n = 8(1)28 & 5(1)27 & m(1)[30-m] \end{cases}$. The 5% and 1% points are reprinted in Senders 1958, T. I, pp. 541-542.

Van der Reyden 1952, pp. 102-104, gives approximate 5% points, and 2% and 1% points (when possible), of T for $\begin{cases} m = 2(1)5 & 6(1)12 \\ m+n = 10(1)30 & 2m(1)30 \end{cases}$. Minor errors in White and Van der Reyden are discussed in Kruskal & Wallis 1953, pp. 907-908.

Siegel 1956, T. K, pp. 274-277, based on Auble 1953, gives approximate 0.1%, 1%, $2\frac{1}{2}$% points (when possible) and 5% points of U for $\begin{cases} n = 9(1)18 & 19, 20 \\ m = 2(1)n & 1(1)n \end{cases}$. Reprinted in Vianelli 1959, Pro. 129, pp. 347-348.

9.122 Other rank-sum tests

Terry, M. E. 1952

<u>Annals</u> <u>of</u> <u>Mathematical</u> <u>Statistics</u> 23, 346-366

Using the method of Hoeffding 1951, Terry constructs a criterion $c_1(R)$: the units in two samples of sizes m, n, $m \leq n$, are pooled and ranked in ascending order of magnitude; $\xi_{i|m+n}$, the ith order statistic in a normal sample of size m+n, is attached to the rank i in the pooled sample. Terry points out that Fisher & Yates, T. XX, is a convenient source for the $\xi_{i|m+n}$. $c_1(R)$ is the sum of the $\xi_{i|m+n}$ over the sample of size m. A symmetrical beta-approximation to $c_1(R)$ is compared with the exact distribution on p. 357 for $m+n = 6(1)10$.

Table 1, pp. 358-361, gives the non-negative values of $c_1(R)$ to 2 dec. for $\begin{cases} m = 1(1)5 \\ n = m(1)[10-m] \end{cases}$ and half the permutations of the two samples; together with the corresponding values of the U-statistic of Section 9.121. Table 2, p. 362, gives $\Pr[c_1(R) \geq x]$ to 3 dec. for $\begin{cases} m = 2 & 3 & 4 & 5 \\ n = 5(1)8 & 3(1)7 & 4(1)6 & 5 \end{cases}$ and values of x covering the range of Pr from .01 to .1.

9.123
9.14 [see also pp. 706-708]

9.123 The median test

Pool two samples and compute the median of the pooled sample. Dichotomize each single sample at the pooled median, and analyse the resulting 2 × 2 table as a contingency table. For tabular aids, see Section 2.61; for discussion of the test, see Mathisen 1943; Westenberg 1947, 1948, 1950, 1952; Siegel 1956, pp. 111-116, and works there cited.

9.129 Other two-sample tests

Dixon, W. J. 1940

Annals of Mathematical Statistics 11, 199-204

See abstract in Section 9.31.

9.13 Three-sample tests

Whitney, D. R. 1951

Annals of Mathematical Statistics 22, 274-287

Let $x_1, \ldots, x_l$; $y_1, \ldots, y_m$; $z_1, \ldots z_n$ be independent variables from the same population. Let U be the number of times a y precedes an x when the pooled samples are ranked in ascending order of magnitude; let V be the number of times a z precedes an x. Then the marginal distributions of U and V are the U-distribution of Section 9.121.

For the special case $l = 6$, $m = n = 3$, T. 1, p. 280, gives the frequency of x, y, z sequences having all possible combinations of U, V values; T. 2, pp. 280-281, gives the proportional frequencies cumulated from $U = 0$, $V = 0$, to 3 dec. The corresponding values from a bivariate normal surface are also shown.

Kruskal, W. H. and Wallis, W. A. 1952

Journal of the American Statistical Association 47, 583-621

Three samples of sizes n_1, n_2, n_3 are pooled and ranked in ascending order of magnitude; numbers from 1 up to $N = n_1+n_2+n_3$ are attached to the ranks. Let $\bar{r}_i$ denote the mean rank assigned to the sample of size n_i.

To test whether the three samples have come from the same population, Kruskal & Wallis propose the criterion (p. 587):

$$H = 12(N-1)\sum n_i (\bar{r}_i - \tfrac{1}{2}[N+1])^2 \Big/ N(N^2-1) . \tag{1.5}$$

For transformations of (1.5) and for the treatment of tied ranks, see pp. 586-587. Table 6.1, pp. 614-617, gives exact values of $\Pr[H \geq x]$, and various approximations thereto, to 3 dec. for $n_1 = 2(1)5$, $n_2 = 1(1)n_1$, $n_3 = 1(1)n_2$, and attainable values of x that yield tail sums near .1, .05 and .01.

Errata and additional references in Kruskal & Wallis 1953.

The true probabilities (but not the approximations) are reprinted, with benefit of the corrections, in Siegel 1956, T. O, pp. 282-283.

9.14 k-sample tests

Friedman, Milton 1937

<u>Journal of the American Statistical Association</u> 32, 675-701

See abstract on p. 354.

Mosteller, F. 1948

<u>Annals of Mathematical Statistics</u> 19, 58-65

"A k-sample slippage test for an extreme population"

Given k populations $f(x-a_i)$, $i = 1, 2, \ldots, k$, identical with the possible exception of their location parameters a_i. In order to test whether one of the a_i is larger than the remaining k-1 a_i, a sample of n observations is drawn from each population. Assuming that all the a_i are equal, consider the pooled sample of kn observations. Then the chance that at least the r largest observations belong to the same population is

$P(r) = k\binom{n}{r}\bigg/\binom{kn}{r}$. Table 1, p. 62, gives P(r) for $n = 3(2)7, 10(5)25, \infty$

and $\begin{cases} k = 2 & 3 & 3 & 4 & 4 & 5, 6 & 5, 6 \\ r = 2(1)6 & 2(1)5 & 6 & 2(1)4 & 5, 6 & 2, 3 & 4, 5 \\ 3 \text{ dec.} & 3 \text{ dec.} & 4 \text{ dec.} & 3 \text{ dec.} & 4 \text{ dec.} & 3 \text{ dec.} & 4 \text{ dec.} \end{cases}$

When the k samples are of unequal sizes n_i, and $N = \sum n_i$, then

$P(r) = \sum \binom{n_i}{r} \bigg/ \binom{N}{r}$. For approximate formulas and illustrative tables see Mosteller & Tukey 1950.

9.14
9.2   [see also p. 707-709]

Kruskal, W. H. and Wallis, W. A. 1952

<u>Journal</u> <u>of</u> <u>the</u> <u>American</u> <u>Statistical</u> <u>Association</u> 47, 583-621

The theory presented on pp. 597, 606, 608-610, relates to k samples; the numerical tables are confined to three samples and are abstracted on p. 370 above.

Rijkoort, P. J. 1952

K. Nederlandse Akademie van Wetenschappen, <u>Proceedings</u>, Series A 55, 394-404
"A generalisation of Wilcoxon's test"

k samples of sizes $n_1, n_2, \ldots, n_k$ are pooled and ranked in ascending order of magnitude; numbers from 1 up to $N = \sum n_i$ are attached to the ranks. Let T_i denote the sum of the ranks in the sample of size n_i.

To test whether the k samples have come from the same population, Rijkoort proposes the criterion $S = \sum [T_i - \frac{1}{2} n_i (N+1)]^2$. Kruskal & Wallis 1953, p. 908, point out that Rijkoort's S is equivalent to their H only when all the n_i are equal; then $S = N^2(N+1)H/12k$, where H is defined on p. 370 above.

Table 4.1, pp. 400-402, gives the distribution of S to 3-6 dec. for all n_i equal (to m, say): complete for $\begin{cases} k = 2 & 3(1)5 \\ m = 2(1)4 & 2 \end{cases}$; the upper 5% tail for $\begin{cases} k = 3 & 4 \\ m = 5 & 3 \end{cases}$.

Table 4.2, p. 402, gives approximate 5% points of S, based on χ^2 and F approximations, for $k = 3(1)10$ and $m = 2(1)10$; together with exact 5% points, when available from T. 4.1.

Errata in Kruskal & Wallis 1953, p. 908; Rijkoort 1953.

9.2 Tests of spread

Moses, L. E. 1952

Psychometrika 17, 239-247

Two samples of sizes m and n, drawn respectively from populations A and B, are pooled and ranked in ascending order of magnitude. To test A = B against the alternative that the two populations have different dispersions, Moses uses criteria s_h^* defined as the difference in the ranks attached to the hth largest and h^{th} smallest observation in the sample of size m, increased by unity.

The tables on pp. 244-245 give the tail of the distribution of s_h^* to 3 dec. for h = 3, 4, 5, 8; m = 4h; n = 2h, 3h, 4h, 6h, 8h.

Barton, D. E. and David, F. N. 1958

Annals of Human Genetics 22, 250-257

"A test for birth-order effect"

Given two samples of sizes m, n; to test the hypothesis that the samples are from identical populations, against the alternative that the sample of size m is from a population with greater spread. Pool the two samples and rank in ascending order of magnitude. If m+n = 2p, dichotomize the pooled sample into p+p; if m+n = 2p + 1, into p + [p+1]. Assign scores p, p-1, ..., 2, 1 | 1, 2, ..., p-1, p, (p+1 if necessary) to the ranked items. The criterion S is the sum of scores attached to the sample of size m.

Barton & David give a generating function and moment formulas for S. Tables 1-7, pp. 253-256, give the individual terms of the distribution of S as exact vulgar fractions for m = 2(1)8; n = m(1)[16-m]. For m+n > 16 a normal approximation is recommended.

9.3
9.311   [see also pp. 709-711]

9.3 Tests of randomness of occurrence

See also the papers listed in Section 13.63, which in general contain methodology but no tables.

9.31 One-sample tests

9.311 Runs

For a comprehensive survey of the theory of runs, without tables, see Mood 1940 z . We have not verified the references to Shewhart given by Mosteller 1941, p. 229.

Mosteller, F. 1941
Annals of Mathematical Statistics 12, 228-232

Let $x_1, x_2, \ldots, x_{2n}$ denote a sample ranked in ascending order of magnitude; attach the mark *a* to $x_1, \ldots, x_n$ and the mark *b* to $x_{n+1}, \ldots, x_{2n}$. Now arrange the x_i in the order in which they were drawn: let s_a be the maximum number of consecutive *a*s; s_b, of consecutive *b* s.

The second table on p. 232, due to P. S. Olmstead, gives $\Pr[s_a \geq s]$, $\Pr[\max(s_a, s_b) \geq s]$ and $\Pr[\min(s_a, s_b) \geq s]$ to 3 dec. for 2n = 10, 20, 40 and s = 1(1)13 . The first table on p. 232 gives upper 5% and 1% points of s_a and $\max(s_a, s_b)$ for 2n = 10(10)50 .

Olmstead, P. S. 1946
Annals of Mathematical Statistics 17, 24-33
"Distribution of sample arrangements for runs up and down"

Let $a_1, a_2, \ldots, a_n$ be any set of unequal numbers and let $S = (h_1, h_2, \ldots, h_n)$ be a random permutation of them; each permutation having a chance 1/n! of occurring. Let R be the sequence of + and - signs of the first differences $h_{i+1} - h_i$. A run of length k is a sequence of k consecutive + or - signs.

Table 1, p. 27, gives the number of permutations S resulting in at least one run of length at least k , for n = 2(1)14 and k = 1(1)13 ; T. 2, p. 27, gives the probability of at least one such run to 8 dec. Table 3, p. 29, gives a Poisson approximation to the probability of at least one such run:
$$P_n(r'_k) = 1 - e^{-E(r'_k)} = 1 - e^{-2[(n-k)(k+1)+1]/(k+2)!}$$

to 4 dec. for n = 2(1)15, 20(20)100, 200, 500, 1000, 5000; k = 1(1)10.

The agreement is not very good, and so T. 5, p. 32, gives $P_n(r_k')$ to 4 dec. for n = 14, 15, 20(20)100, 200, 500, 100, 5000; k = 2(1)6; based on an extrapolation formula whose coefficients are given in T. 4, p. 30.

Table 6, p. 32, gives roots n of $P_n(r_k') = p$, computed from the Poisson approximation, for $\begin{cases} p = .01 & .05, .1 & .9, .95, .99 \\ k = 1(1)8 & 1(1)7 & 1(1)6 \end{cases}$.

Table 7, p. 32, gives the same, computed from T. 2 or from the extrapolation formula, for p = .01, .05, .1; 1-p = ditto; k = 2(1)6.

Levene, H. 1952

<u>Annals</u> of <u>Mathematical</u> <u>Statistics</u> 23, 34-56

"On the power function of tests of randomness based on runs up and down"

In the notation of the previous abstract, let a run up denote a sequence of consecutive + signs and a run down a sequence of - signs. Let S be the number of runs up in R and define $E'(s) = \lim n^{-1} E(s)$, $\sigma'^2(S) = \lim n^{-1} \sigma^2(S)$; let k be the number of + signs in R and define $E'(k) = \lim n^{-1} E(k)$, $\sigma'^2(k) = \lim n^{-1} \sigma^2(k)$; where all limits are taken as $n \to \infty$.

Levene considers the null hypothesis H_0 that the h_i are drawn from N(0,1), so that the tables of Olmstead 1946 apply; and the alternative H_1 that the h_i are drawn from $N(i\theta, 1)$. Table 1, p. 45, gives θ to 6 dec., $1 - E'(k)$ and $\sigma'^2(k)$ to 5 dec., and $\sigma'(k)$ to 3 dec. for $2^{-\frac{1}{2}}\theta = 0(.1)4$; and $E'(s)$ to 6 dec. for $2^{-\frac{1}{2}}\theta = 0(.1)2.6$.

Grant, Alison M. 1952

<u>Biometrika</u> 39, 198-204

"Some properties of runs in smoothed random series"

Let x_i be a random time series of independent rectangular variates on [0,1], and construct a moving average of length a:

$$z_r = a^{-1}[x_r + x_{r+1} + \ldots + x_{r+a-1}].$$

Table 1, p. 200, gives relative frequencies of runs up (or down) in the z_r series: to 4 dec. for a = 1, 3, 5 and length of runs n = 0(1)7. It also gives the frequencies for $a \geq 9$; these are 2^{-n-1} to 4 dec.

9.311
9.32 [see also pp. 709-711]

[A test, conditional on m and n, that m zeroes and n ones have been obtained from a binomial process, against an alternative specifying positive or negative autocorrelation, may be constructed by treating the zeroes and ones as two samples and applying one of the tests of Section 9.32: see Swed & Eisenhart 1943.]

9.312 Tests based on signs of variate differences

The tests in this section, closely related to runs up and down (Olmstead 1946), were intended by their authors for time-series analysis.

Wallis, W. A. and Moore, G. H. 1941
A significance test for time series and other ordered observations

The expected number of runs-up-and-down of length 1, 2 and longer than 2 are found to be $5(N-3)/12$, $11(N-4)/60$, $(4N-21)/60$. Let χ_p^2 be the goodness-of-fit χ^2 calculated from the observed numbers of runs-up-and-down and these expectations. Table 3, p. 21, gives the exact distribution of χ_p^2, which does not tend to the Pearson χ^2 distribution as a limit, to 4 dec. for $N = 6(1)12$ and all attainable χ_p^2. When $N > 12$, Wallis & Moore recommend that if $\chi_p^2 < 6.3$, $6\chi_p^2/7$ be referred to the χ^2 distribution with 2 d.f.; and that if $\chi_p^2 > 6.3$, χ_p^2 be referred to the χ^2 distribution with $2\frac{1}{2}$ d.f. Table 5, p. 30, gives $p(\chi^2, 2\frac{1}{2})$ to 4 dec. for $\chi^2 = 5.5(.25)15$; and $\chi^2(p, 2\frac{1}{2})$ to 3 dec. for $p = .001, .005, .01(.01).1$.

Wallis, W. A. and Moore, G. H. 1941a
Journal of the American Statistical Association 36, 401-409
"A significance test for time series analysis"

Table 1, p. 404, reproduces TT. 3 and 5 of Wallis & Moore 1941, abstracted above.

Moore, G. H. and Wallis, W. A. 1943
Journal of the American Statistical Association 38, 153-164
"Time series significance tests based on signs of differences"

The expected number of negative differences in a random time series of N items y_i (assumed to be all different) is $\frac{1}{2}(N-1)$. The observed number of negative differences, m, is here proposed as a criterion of departure from randomness. If $\phi_N(m)$ is the number of permutations of N different y_i that produce exactly m negative differences $y_{i+1} - y_i$, then

[376]

$$\phi_N(m) = (N-m)\phi_{N-1}(m-1) + (m+1)\phi_{N-1}(m).$$

Table 1, p. 156, based on this recursion, gives $2\Pr[m \leq x]$ for $x \leq \frac{1}{2}(N-1)$; and $2\Pr[m \geq x]$ for $x \geq \frac{1}{2}(N-1)$; all to 6 dec. for $N = 2(1)12$ and $x = 0(1)[N-1]$. For $N \geq 12$, Moore & Wallis suggest that $[3/(N+1)]^{\frac{1}{2}}(|2m-N-1| - 1)$ be treated as a unit normal deviate; they illustrate the normal approximation for $N = 12$.

For illustrative values of the power of the test, and an extension to two simultaneous time series, see Stuart 1952.

9.32 Two-sample tests

Dixon, W. J. 1940

Annals of Mathematical Statistics 11, 199-204

Let $x_1, x_2, \ldots, x_n; y_1, y_2, \ldots, y_m$; denote two ordered samples from the same continuous distribution: then the x_i will split the y_j into $n+1$ groups. Set $x_0 = -\infty$, $x_{n+1} = +\infty$; and define m_i as the number of j such that $x_{i-1} \leq y_j < x_i$. For testing the hypothesis that the y_j come from the same population as the x_i, Dixon proposes the criterion $c^2 = m^{-2}\sum m_i^2 - (n+1)^{-1}$, with mean $n(m+n+1)/m(n+1)(n+2)$ and variance $4n(m-1)(m+n+1)(m+n+2)/[m^3(n+2)^2(n+3)(n+4)]$. For small m and n the distribution of c^2 can be evaluated exactly by enumeration.

Table 1, p. 201, gives upper 10% points, and 5% and 1% points when possible, of c^2 to 3 dec. for $\begin{cases} n = 2 & 3 & 4(1)10 \\ m = 7(1)10 & 5(1)10 & n(1)10 \end{cases}$. For larger m and n a χ^2 approximation is proposed; but it is not good at the 1% point.

Swed, F. S. and Eisenhart, C. 1943

Annals of Mathematical Statistics 14, 66-88

Let m markers *a* and n markers *b* be arranged in random order along a line. Of the $\binom{m+n}{n}$ possible arrangements, the number in which *a* and *b* are neighbors u-1 times is $f_u = \binom{m-1}{[\frac{1}{2}u]-1}\binom{n-1}{u-[\frac{1}{2}u]-1} + \binom{m-1}{u-[\frac{1}{2}u]-1}\binom{n-1}{[\frac{1}{2}u]-1}$. The number of arrangements having exactly u groups is, therefore, f_u; so that

$$\Pr[u \geq u'] = \binom{m+n}{n}^{-1}\sum_{u=2}^{u'} f_u.$$

9.32

Table 1, pp. 70-82, gives this probability to 7 dec. for $m = 2(1)20$; $n = m(1)20$; $u' = 2(1)2m$. Table 2, pp. 83-86, gives lower and upper percentage points of u, viz: the largest integer u' such that $Pr[u \leq u'] \leq p$ and the smallest integer u' such that $Pr[u \leq u'] \geq 1-p$, for $p = .005, .01, .025, .05$; $m = 2(1)20$; $n = m(1)20$.

For large m and n, u is approximately normal with mean $1 + 2mn/(m+n)$ and variance $2mn(2mn-m-n)/[(m+n)^2(m+n-1)]$. Table 3, p. 87, gives lower and upper percentage points of u, computed from the normal approximation, for $m = n = 10(1)100$. The five cases where the approximate percentage points differ from those of T. 2 are indicated by underscoring; add an underscore under $u_{.005}(18,18) = 10$, the exact value being 11.

Bateman, G. 1948

Biometrika 35, 97-112

"On the power function of the longest run as a test for randomness in a sequence of alternatives"

Let m markers *a* and n markers *b* be arranged in order along a line. Let g be the length of the longest string of *a*s or *b*s, whichever is longer; let T be the number of such strings, viz the u of Swed & Eisenhart 1943.

Table 1, pp. 98-99, gives individual terms of the distribution of g as exact vulgar fractions for $\begin{cases} m+n = 10, 11 & 12, 13 & 14, 15 & 20 \\ n = 1(1)5 & 1(1)6 & 1(1)7 & 1(1)10 \end{cases}$.

Table 2, p. 101, gives individual terms of the joint distribution of g and T as exact vulgar fractions for $m = 14$, $n = 6$.

Wilks, S. S. 1948

Elementary Statistical Analysis

Table 12.3, p. 227, abridged from Swed & Eisenhart 1943, gives lower and upper 5% and 1% points of u for $m = n = 5(1)30$.

Dixon, W. J. and Massey, F. J. 1951

Introduction to Statistical Analysis

Table 11, pp. 325-326, abridged from Swed & Eisenhart 1943, gives lower and upper $2\frac{1}{2}$% points of u for $\begin{cases} n = 2(1)20 & 21(1)30(2)50(5)100 \\ m = n(1)20 & n \text{ only} \end{cases}$

9.32

Siegel, S. 1956

Nonparametric statistics

Table F, pp. 252-253, abridged from Swed & Eisenhart 1943, gives lower and upper $2\frac{1}{2}\%$ points of u for m, n = 2(1)20. The upper $2\frac{1}{2}\%$ points have been increased by 1 over those in Swed's table (without notice): so that when using Siegel, reject the null hypothesis if $u \leq u_{.025}$ or $u \geq u_{.975}$. Reprinted in Vianelli 1959, Prontuario 137, p. 368.

Czechowski, T. et al. 1957

Tablice statystyczne

Tablica 21, pp. 91-92, abridged from Swed & Eisenhart 1943, gives upper and lower $2\frac{1}{2}\%$ and 5% points of u for m = 2(1)n, n = 2(1)20; and for m = n = 10(1)100.

Dixon, W. J. and Massey, F. J. 1957

Introduction to Statistical Analysis

Table 11, pp. 421-422, abridged from Swed & Eisenhart 1943, gives $\Pr[u \leq u']$ to 3 dec. for m = 2(1)10, n = m(1)10; and the lower and upper $\frac{1}{2}\%$, 1%, $2\frac{1}{2}\%$ and 5% points of u, and the variance and standard deviation of u to 2 dec., for m = n = 11(1)20(5)100.

Senders, V. R. 1958

Measurement and Statistics

Table H, p. 540, abridged from Swed & Eisenhart 1943, gives lower $2\frac{1}{2}\%$ points of u for m, n = 2(1)20.

Vianelli, S. 1959

Prontuari per calcoli statistici

Prontuario 221, pp. 725-732, is abridged from Swed & Eisenhart 1943, T. 1, by omitting u' greater than 33. Prontuario 222, pp. 733-735, reproduces Swed & Eisenhart 1943, TT. 2-3.

9.4
9.4121

9.4 Tests of goodness of fit of a sample to a population, and analogous tests of the homogeneity of two samples

[see also p. 714]

The classical test of goodness of fit is the χ^2 test: K. Pearson 1900. For relevant tables see pp. 132-133, 140-143, 174-180 above. For a general discussion of the application of the test see Cochran 1952; for a remarkable theorem, derived from the work of Mann & Wald 1942, on the large sample inefficiency of χ^2 relative to the tests exposed hereunder, see Kac et al. 1955, especially pp. 190 and 207. See also C.A. Williams 1950.

Our chief source of references for Sections 9.41 and 9.42 is Darling 1957. We have not verified references 11 and 75 given by Darling.

9.41 The Kolmogorov-Smirnov tests

9.411 Two-sided tests: asymptotic distribution

A sample of size n has been drawn from a population with cumulative distribution function $F(x)$. Define the empirical distribution function $F_n^*(x)$ to be the step function $F_n^*(x) = k/n$, where k is the number of observations not greater than x; define $D_n = \max|F_n^*(x) - F(x)|$.

Draw a second sample of size m, construct its empirical distribution function $G_m^*(x)$, and define $D_{n,m} = \max|F_n^*(x) - G_m^*(x)|$. Then (Kolmogorov 1933, p. 84)

$$\lim_{n\to\infty} \Pr[n^{\frac{1}{2}} D_n \leq z] = \vartheta_4(0, e^{-2z^2}) = 1 + 2\sum_{j=1}^{\infty}(-1)^j e^{-2j^2 z^2} \qquad (9.41.1)$$

where ϑ_4 is an elliptic theta function (Fletcher 1948, p. 235); and (Smirnov 1939, p. 4)

$$\lim \Pr\left[\left(\frac{mn}{m+n}\right)^{\frac{1}{2}} D_{m,n} \leq z\right] = \vartheta_4(0, e^{-2z^2}) \qquad (9.41.2)$$

where m and n go to infinity so that $0 < a < m/n < b < \infty$.

9.4111 Tables of $\vartheta_4(0, e^{-2z^2})$

7 dec.	.25(.001)2.299	△	Kunisawa et al. 1951
6; 7 dec.	.28(.01)2.49; 2.5(.05)2.8	No △	Smirnov 1939(15), 1948(280)
8 dec.	2.85(.05)3		
4 dec.; 2 fig.	0(.2)1.8; 2(.2)2.8	No △	Kolmogorov 1933(84)

9.4112 Other tables of $\vartheta_4(0,q)$

Schuler & Gebelein 1955 a, T. 4, p. 100, column $z = -1$, give $\vartheta_4(0,q)$ to 5 dec. for $q = 0(.01).5$, with Δ^2. Schuler & Gebelein 1955 b, T. 4, pp. 131-132, column $z = -1$, give $q^{-1}[\vartheta_4(0,q) - 1]$ to 9 dec. for $q^3 = 0(.002).176$, with Δ.

In the tables of $\vartheta_4(0,q)$ cited by Fletcher 1948, p. 236, q is not a tabular argument.

9.4113 Inverse tables: z as a function of $\vartheta_4(0, e^{-2z^2})$

Kolmogorov 1941 gives z to 2 dec. for $p = \vartheta_4(0, e^{-2z^2}) = .95, .98, .99, .995, .998, .999$. For $p = .95$, z should be **1·36**.

9.412 Two-sided tests: exact distributions

9.4121 Distribution of D_n

Massey, F. J. 1950

<u>Annals of Mathematical Statistics</u> 21, 116-119

Massey shows that $\Pr[D_n < k/n] = n! U_k(n)/n^n$, where the path function $U_k(n)$ satisfies the linear difference equation

$$\sum_{h=1}^{2k-1} (h-2k)^h U_k(2k-1-h+m)/h! = 0.$$

Table 1, p. 118, gives $U_k(n)$ to 4 dec. for

$\begin{cases} n = 5 & 10 & 20 & 25(5)80 \\ k = 1(.5)4 & 1(.5)4(1)6 & 2(1)9 & 5(1)9 \end{cases}$, the values for fractional k

having been interpolated. Table 2, p. 119, gives the resulting $\Pr[n^{\frac{1}{2}}D_n < \lambda]$ to 2 dec. for $\begin{cases} n = 10(10)40 & 50(10)70 & 80 \\ \lambda = .9(.1)1.4 & .9(.1)1.1 & .9, 1 \end{cases}$. To these are added 3 dec. values for $n = \infty$ from (9.41.1).

Massey, F. J. 1951

<u>Journal of the American Statistical Association</u> 46, 68-78

Based on the definitions, formulas and methods of Massey 1950, Table 1, p. 70, gives upper 100α % points $D_n(\alpha)$ for $\alpha = .01, .05(.05).2$; to 3 dec. for $n = 1(1)20$ and 2 dec. for $n = 25(5)35$.

Reprinted in Siegel 1956, T. E, p. 251.

9.4121
9.4122   [see also pp. 714-715]

Dixon, W. J. and Massey, F. J.   1951

Introduction to Statistical Analysis

Table 21, p. 348, gives $D_n(\alpha)$ to 2 dec. for $1-\alpha$ = .8(.05).95, .99 ; n = 5, 10, 20(5)50 .

Reprinted, together with some formulas for the two-sample test of Section 9.4122, in Dixon & Massey 1957, T. 21, p. 450. Reprinted in Senders 1958, T. G, p. 539.

Birnbaum, Z. W.   1952

Journal of the American Statistical Association 47, 425-437

Table 1, pp. 428-430, gives $\Pr[D_n \leq c/n]$ to 5 dec. for n = 1(1)100 and c = 1(1)15 . Table 2, p. 431, gives upper 5% and 1% points of D_n to 4 dec. for n = 2(1)5(5)30(10)100 . These percentage points are compared with the results of the asymptotic formulas $1.3581\, n^{-\frac{1}{2}}$ and $1.6276\, n^{-\frac{1}{2}}$; the asymptotic values are systematically too high.

Maniya, G. M.   1953

Define $D_n(\theta_1, \theta_2) = \max|F_n^*(x) - F(x)|$, $\theta_1 \leq F(x) \leq \theta_2$;

$\Phi(\theta_1, \theta_2; \lambda) = \lim_{n \to \infty} \Pr[D_n(\theta_1, \theta_2) \leq n^{-\frac{1}{2}}\lambda]$.

The table on p. 523 gives $(\theta, 1-\theta; \lambda)$, computed from formulas given on p. 522, to 4 dec. for λ = 1(.05)2 and θ = .1(.05).5 .

L.H. Miller 1956, pp. 119-120, offers evidence that his table of upper p% points of the one-sided Kolmogorov statistic D_n^+, abstracted on p. 384, may be used as an approximate table of upper 2p% points of D_n .

9.4122   Distribution of $D_{m,n}$

Korolyuk, V. S. and Yaroševs'kiĭ, B. I.   1951

The table on p. 246 gives $\Pr[(25/2)^{\frac{1}{2}}D_{25,25} < z]$ to 5 dec. for $2^{-\frac{1}{2}}z$ = .2(.1)1.6 , and the asymptotic distribution of (9.41.2) to the same accuracy for the same range.

Massey, F. J. 1951 a

Annals of Mathematical Statistics 22, 125-128

Table 1, pp. 126-127, gives $\Pr[D_{n,n} \leq k/n]$ to 4-12 dec. (mostly 6 dec.) for $n = 1(1)40$ and $k = 1(1)12$.

Based on this table, Siegel 1956, T. L, p. 278, gives inter al. 5% points of $nD_{n,n}$ for $n = 4(1)30(5)40$, and 1% points for $n = 5(1)30$.

Massey, F. J. 1952

Annals of Mathematical Statistics 23, 435-441

Table 1, pp. 436-439, gives all possible values of d_α and α with $\Pr[D_{m,n} \leq d_\alpha] = \alpha$ for: m and n not greater than 10; $d_\alpha = h/k$, where k is the least common multiple of m and n and h is an integer. The table gives α to 5 dec. Table 2, pp. 440-441, gives additional values of h and α for

$$\begin{cases} m = 3 & 4 & 5 & 6 & 7 & 8 & 9 \\ n = 12 & 12, 16 & 15, 20 & 12(6)24 & 14, 28 & 12, 16, 32 & 12(3)18, 36 \end{cases}$$

$$\begin{cases} 10 & 12 & 15 & 16 \\ 15, 20, 40 & 15, 16(2)20 & 20 & 20 \end{cases}.$$

Drion, E. F. 1952

Annals of Mathematical Statistics 23, 563-574

Table 1, p. 572, gives $P = \Pr[D_{n,n} \geq d]$ to 4 dec. for

$$\begin{cases} n = 20 & 50 & 100 \\ h = nd = 5, 8, 9, 10 & 8, 12, 14, 16 & 12, 17, 19, 23 \end{cases}.$$ Approximate values of P from formula (9.41.2) are shown for comparison.

Tsao, C. K. 1954

Annals of Mathematical Statistics 25, 587-592

Let x_r, y_r be the rth order statistics of two samples of sizes n, m from the same population, and F_n^*, G_m^* the respective sample cumulatives. Define $d_r = \max\limits_{z \leq x_r} |F_n^*(z) - G_m^*(z)|$, $d_r' = \max\limits_{z \leq \max(x_r, y_r)} |F_n^*(z) - G_m^*(z)|$.

Table 1, pp. 589-590, gives $\Pr[d_r \leq c/m]$ to 5 dec. for

$$\begin{cases} m = n = 3(1)10 & 15, 20(10)40 \\ r = 2(1)m & 2(1)10 \\ c = 1(1)m & 1(1)12 \end{cases}$$

Table 2, pp. 591-592, gives $\Pr[d_r' \leq c/m]$ to 5 dec. for the same combinations of m, n, r, c.

9.4122 [see also p. 715]
9.416

Gnedenko, B. V. 1954

<u>Mathematische Nachrichten</u> 12, 29-66

See abstract on p. 385.

9.413 Power of two-sided tests

Dixon, W. J. 1954

<u>Annals of Mathematical Statistics</u> 25, 610-614

Dixon considers the power of the two-sample maximum deviation test against the alternative of normal slippage: $\delta = |\mu_1 - \mu_2|/\sigma$.

Table 1, p. 611, gives, inter al., the large-sample power and power-efficiency of the test for $\alpha = 10/126$ and $\delta = 0(.5)4.5$. Table 2, p. 612, gives, inter al., the power and power-efficiency of the test, randomized to level of significance $\alpha = .025$, for $m = n = 5$ and $\delta = 0(.5)4.5$.

9.415 One-sided tests

9.4151 One-sided tests of one sample against a population

Birnbaum, Z. W. and Tingey, F. H. 1951

<u>Annals of Mathematical Statistics</u> 22, 592-596

Define the upper confidence contour $F_{n,\epsilon}^+(x) = \min[F_n^*(x)+\epsilon, 1]$. Birnbaum & Tingey derive an explicit expression for

$$P_n(\epsilon) = \Pr[\max F_{n,\epsilon}^+(x) - F(x) \geq 0].$$

Table 1, p. 595, gives the roots $\epsilon_{n,\alpha}$ of $P_n(\epsilon_{n,\alpha}) = 1 - \alpha$ to 4-5 dec. for $\alpha = .001, .01, .05, .1$; $n = 5, 8, 10, 20, 40, 50$. Table 2, p. 595, gives Kolmogorov's asymptotic approximations to these percentage points, $(2n \log_e \alpha)^{\frac{1}{2}}$, to 4 dec. for the same α and n.

Miller, L. H. 1956

<u>Journal of the American Statistical Association</u> 51, 111-121

Table 1, pp. 113-115, gives upper 10%, 5%, $2\frac{1}{2}$%, 1% and $\frac{1}{2}$% points of D_n^+ to 5 dec. for $n = 1(1)100$.

9.4152 One-sided tests of slippage between two samples

Gnedenko, B. V. 1954

<u>Mathematische Nachrichten</u> 12, 29-66

Define $D_{m,n} = \max|F_m^*(x) - G_n^*(x)|$, $D_{m,n}^+ = \max[F_m^*(x) - G_n^*(x)]$, $D_{m,n}^- = \max[G_n^*(x) - F_m^*(x)]$, $R_{m,n} = D_{m,n}^+ + D_{m,n}^-$; let ν_n^+ be the number of intervals in which $F_m^*(x) - G_n^*(x) = D_{m,n}^+$.

The table on p. 32 gives upper 5% points of $nD_{n,n}$ for $n = 9, 10, 12, 15(5)35, 50(25)100$. The table on p. 53 gives individual terms of the joint distribution of $nD_{n,n}^+$ and ν_n^+ as exact vulgar fractions for $n = 2(1)4$. The table on p. 55 gives $\tfrac{1}{2}n \operatorname{var} D_{n,n}^+$, $\tfrac{1}{2}n \operatorname{var} D_{n,n}$, $\tfrac{1}{2}n \operatorname{var} R_{n,n}$, and the correlation between $D_{n,n}^+$ and $D_{n,n}^-$, all to 4 dec. for $n = 2(1)6, 10(5)25, 50, \infty$. This last table is also in Gnedenko & Studnev 1952.

Goodman, L. A. 1954

<u>Psychological Bulletin</u> 51, 160-168

"Kolmogorov-Smirnov tests for psychological research"

Table 3, based on the exact formula of Gnedenko & Korolyuk 1951, gives 0.1%, 1%, 5%, 10% points (when possible) and 20% points of $nD_{n,n}^+$ for one-sided two-sample tests, for $n = 2(1)30(5)50$. The 1% and 5% points for $n = 3(1)30(5)40$ are reprinted in Siegel 1956, T. L, p. 278.

9.416 The maximum relative discrepancy

Renyi, A. 1953

<u>Acta Mathematica Academiae Scientiarum Hungaricae</u> 4, 191-231

Theorem 6, p. 207, asserts that

$$\lim_{n \to \infty} \Pr[n^{\tfrac{1}{2}} \max_{F(x) \geq a} |F_n^*(x) - F(x)| < y F(x)] = 4\pi^{-1} \sum_{k=0}^{\infty} (-1)^k (2k+1)^{-1} \exp[-(2k+1)^2 \pi^2 (1-a)/8ay^2].$$

This probability is tabled on pp. 228-229 to 4 dec. for $a = .01(.01).1(.1).5$ and $y = .5(.5)25(1)30(5)40, 43$.

9.42
9.9

9.42 The Cramér-Mises-Smirnov tests

9.421 The ω^2 test: asymptotic distribution

Let $F_n^*(x)$ denote the sample cumulative distribution function for a sample of size n. The Cramér-Mises criterion for testing whether the sample has been drawn from a population with a specified cumulative distribution $F(x)$ is

$$\omega^2 = \int_{-\infty}^{+\infty} [F(x) - F_n^*(x)]^2 dF(x).$$

9.4211 Tables of the asymptotic distribution of $n\omega^2$

Watanabe, Y. 1952

Journal of the Gakugei College, Tokushima University (Natural Science) 2, 21-30

The table on p. 30, computed by Y. Ueda, gives $\lim_{n \to \infty} \Pr[n\omega^2 \leq z]$ to 4 dec. for $z = 0(.01)1(.1)1.8$.

9.4212 Tables of asymptotic percentage points of $n\omega^2$

Anderson, T. W. and Darling, D. A. 1952

Annals of Mathematical Statistics 23, 193-212

The table on p. 203 gives the values of z for which $\lim_{n \to \infty} \Pr[n\omega^2 \leq z] = \alpha$ to 5 dec. for $\alpha = .01(.01).99, .999$.

9.429 Miscellaneous tests related to ω^2

Kac, M., Kiefer, J. and Wolfowitz, J. 1955

Annals of Mathematical Statistics 26, 189-211

Define $w_n = \int [G_n^*(y) - N(y|\bar{x}, s^2)]^2 d_y N(y|\bar{x}, s^2)$; i.e. as the squared distance between a sample and the normal distribution fitted to its first two moments. Then nw_n has a limiting distribution which is not Smirnov's limiting distribution of $n\omega^2$.

Table 2, p. 210, gives, inter al., $\lim_{n \to \infty} \Pr[nw_n \leq x]$ to 3 dec. for $x = .01(.01).16$.

See also Kondo 1954.

9.9 Miscellaneous non-parametric tests

David, F. N. 1950

Biometrika 37, 97-110

    See abstract on p. 293.

Page, E. S. 1955a

Biometrika 42, 523-527

Review by C. A. Bennett [MTAC 11, 34]:

The statistic $m = \max\limits_{0 \leq t \leq n} \{S_r - \min\limits_{0 \leq t < r} S_t\}$, where $S_r = \sum\limits_{j=1}^{r} (x_j - \theta)$, $S_0 = 0$, is suggested to test the hypothesis that the observations x_i, $i = 1, \cdots, k$ (in order of observation) have mean value θ and the observations x_i, $i = k+1, \cdots, n$ have mean value $\theta' > \theta$. For the special case $y_i = \text{sgn}(x_i - \theta)$ and x_i symmetrically distributed, 5% and 1% points are derived for values of n ranging from 21 to 185 (Table 1). The power of this test is compared to that of the usual sign test (values to 3D) for $k = 0$ and $p = \Pr[x_i - \theta > 0 | E(x_i) = \theta' > \theta] = 0.5(.05).8$, and for $p = .75$ and $k = 0(10)50$ (Tables 2 and 3).

10.0
10.004   [see also p. 715]

SECTION 10

SYSTEMS OF FREQUENCY CURVES; MOMENTS AND OTHER SYMMETRIC FUNCTIONS

10.0 and 10.1 K. Pearson's frequency curves

Section 10.00 treats tables relating to the entire Pearson system of frequency curves; sections 10.01 through 10.12 treat Pearson's types I through XII; section 10.13 treats the normal distribution <u>qua</u> Pearson curve, i.e. as the (unnumbered*) thirteenth type of Pearson's system; sections 10.18 and 10.19 treat certain generalizations of Pearson's system.

10.00 Pearson curves in general

The differential equation
$$dy/dx = y(a+x) \Big/ (b_0 + b_1 x + b_2 x^2) \quad (10.0.1)$$
was introduced by K. Pearson 1895, p. 381 (1948, p. 79); its solutions are known as Pearson curves. These curves have been extensively used to graduate empirical data; their largest recent use, however, is to represent probability distributions whose moments are computable but whose functional form is unknown. This use seems to have been introduced by E.S. Pearson 1932.

The parameters a, b_0, b_1, b_2 determine and are determined by the first four moments of a Pearson curve; thus the <u>shape</u> of such a curve is specified completely by $\beta_1^{\frac{1}{2}}$, β_2, and the information that it <u>is</u> a Pearson curve. This fact underlies the classification of Pearson curves into types, and the finding diagrams listed in section 10.004.

---

*Early papers differ in the numeration of the normal curve. It appears as Type V in K. Pearson 1895, p. 360 (1948, p. 58); as Type VII, without specific citation to Pearson but presumably with his approval, in Elderton 1906, pp. 45, 87; Elderton's notation was widely copied.

Type VII of the nomenclature now current was introduced by Rhind 1909, p. 129, on the authority of K. Pearson 1905, p. 74; it is rather casually discussed by K. Pearson 1916, p. 450 (1948, p. 550), who adds: "The formulæ have been given for many years in lecture-notes, and the curves have been frequently used."

For the technique of fitting Pearson curves to empirical data, see Elderton 1938; the 1954 edition corrects some (not all) errors. For Pearson curves as approximate theoretical distributions, see Box 1949; E.S. Pearson & Hartley 1954, pp. 80, 82, 83.

10.001 Ordinates and areas of Pearson curves

With one notable exception (Type IV) the probability integrals of Pearson curves can be reduced to certain known integrals, such as the F-, t- and χ^2-integrals and the Incomplete Gamma- and Beta-functions. K. Pearson 1931a, p. iij, footnote, regrets this exception:

"The series [of tables edited by Pearson] ought to include Tables of the Incomplete G- Function, i.e. the Probability-Integral of the Type IV curve, but the age [74] of the present Editor is likely to preclude his superintending any task, which even exceeds in the magnitude of its calculations that of the Incomplete B - Function."

10.002 Percentage points of Pearson curves

Pearson, E. S. and Merrington, Maxine 1951

<u>Biometrika</u> 38, 4-10

The basic framework of TT. 1-4, pp. 6-7, is the grid comprising all combinations of $\beta_1 = 0, .01(.02).05(.05).2(.1)1$; $\beta_2 = 1.8(.2)5$. For the points of this grid that yield unimodal curves, the tables give the upper and lower 5% and $\frac{1}{2}$% points of $(x-\mu)/\sigma$ to 2 dec.

Pearson, E. S. and Hartley, H. O. 1954

Table 42, pp. 206, 209, reproduces the tables of Pearson & Merrington 1951, abstracted above. Pages 207 and 208 give upper and lower $2\frac{1}{2}$% and 1% points for the same β_1 and β_2.

10.004 Tables and charts for determining the type of a Pearson curve from its moments

Rhind, A. 1909

<u>Biometrika</u> 7, 127-147

Diagram A, p. 131, covers the range $0 \leq \beta_1 \leq 1\cdot 8$ and $1 \leq \beta_2 \leq 8$. In the body of the diagram are marked the regions or lines in the $\beta_1 - \beta_2$ plane that correspond to Types I through VII of Pearson curves, and the boundary

10.004
10.007

of the J-curve region. Reprinted in Tables for S & B I, T. 35, p. 66; redrawn, with major changes in notation, two new lines, and an unfortunate choice of β_2-scale, in von Mises 1931, Abb. 62, p. 273; redrawn, with minor changes in notation and one new line, in E.S. Pearson & Hartley 1954, T. 43, p. 210.

Diagram B, p. 132, gives similar information for the range $0 \leq \beta_1 \leq 60$ and $0 \leq \beta_2 \leq 240$. Reprinted in Tables for S & B I, T. 36, p. 67.

Pearson, K. 1916

Philosophical Transactions of the Royal Society of London, Series A 216, 429-457

Plate 1, following p. 456[?], covers the range $0 \leq \beta_1 \leq 10$ and $0 \leq \beta_2 \leq 24$. The lines on the diagram give the loci of Pearson's Types II, III, V, VII, VIII, IX, XI, XII; the inequality $\beta_2 \geq 1+\beta_1$, satisfied by all frequency curves (not merely Pearson's); the Poisson distribution $\beta_2 = 3+\beta_1$ (Pearson 1916, p. 453; 1948, p. 553); the condition $8\beta_2 < 15\beta_1 + 36$ that a Pearson curve have a finite eighth moment. Concerning this last condition M.G. Kendall 1943, p. 145, remarks: "Curves for which higher moments do not exist were called by Pearson heterotypic; but there is nothing sinister about them except that they do not fall within the Pearsonian system." For a darker view of heterotypy, and for the portrayal of negative moments in the β_1-β_2 diagram, see Kelley 1923, pp. 127, 138-145.

Reprinted in K. Pearson 1931a, frontispice; 1948, plate 1, following p. 557. The 1931 print is hinged to lie flat.

Craig, C. C. 1936z

Annals of Mathematical Statistics 7, 16-28

Define $\delta = (2\beta_2 - 3\beta_1 - 6)/(\beta_2 + 3)$. The diagram on p. 28 plots the Pearson Types as curves and areas in the range $0 \leq \beta_1 \leq 13$ and $-1 \leq \delta \leq +0.4$.

Elderton, W. P. 1938

Frequency curves and correlation

Table 6, facing p. 51, gives a summary of the formulas of the 13 Pearson types: position of origin, transformed equation with origin at mean, criteria for determining the type from β_1 and β_2, and a description of the shape of each curve.

10.006 Moments of Pearson curves

Rhind, A. 1909

<u>Biometrika</u> 7, 127-147

In a Pearson curve the first four moments determine all higher moments: specifically,

$$\mu_{n+2}\mu_2[4\beta_2 - 3\beta_1 - n(2\beta_2-3\beta_1-6)] = (n+1)[\mu_n\mu_2^2(4\beta_2-3\beta_1) + \mu_{n+1}\mu_3(\beta_2+3)],$$

where $1/3 \geq 1/n > (2\beta_2-3\beta_1-6)/(4\beta_2-3\beta_1)$.

Table 6, pp. 146-147, gives $\beta_3 = \mu_5\mu_3/\mu_2^4$, $\beta_4 = \mu_6/\mu_2^3$, $\beta_5 = \mu_7\mu_3/\mu_2^5$, $\beta_6 = \mu_8/\mu_2^4$, to 5-7 fig. for $\beta_1 = 0(\cdot 1)1\cdot 5$; $\beta_2 = 2(\cdot 5)7$; $8\beta_2 < 15\beta_1+36$. Reprinted in Tables for S & B I, T. 42, pp. 78-79; Vianelli 1959, Pro. 204.

Yasukawa, K. 1926

<u>Biometrika</u> 18, 263-292 + 2 plates

Table 1, pp. 268-275, gives β_j, defined in the previous abstract, to 5 dec. or 8 fig. for $j = 3(1)6$; $\beta_1 = 0(\cdot 1)1\cdot 5$; $\beta_2 = 1\cdot 8(\cdot 1)7$; $8\beta_2 < 15\beta_1+36$. Reprinted, allegedly with corrections, in Tables for S & B II, T. 14, pp. 148-155.

10.007 Standard errors of constants of Pearson curves

The tables in this section, according to E.S. Pearson & Hartley 1954 a, are no longer in current use; we confine ourselves to a general indication of their contents. The curious reader will find the titles of K. Pearson 1930 a, TT. 37-41, 43-47; 1931 a, T. 15; listed in Appendix II. All the named quantities in this section are computed from β_1 and β_2 by formulas derived from (10.0.1), and are appropriate to Pearson curves only.

Rhind 1909, TT. 1-5, pp. 136-145, gives the standard errors of β_1 and β_2 and the correlation between them; and the standard errors of the distance between mean and mode and the Pearson measure of skewness. Reprinted in Tables for S & B I, TT. 37-41, pp. 68-77.

Rhind 1910, diagram C and TT. 7-10, pp. 389-397, gives the semi-axes minor and major and the inclination of the correlation ellipse of β_1 and β_2; and the standard error of the <u>criterion</u> for discriminating the types of Pearson curves. Reprinted in Tables for S & B I, TT. 43-47, pp. 80-88.

10.007
10.021

Yasukawa 1926, T. 2, pp. 276-277, gives the standard error of the mode. Reprinted in Tables for S & B II, T. 15, pp. 156-157.

10.01 <u>Pearson curves of Type I (Elderton's 'first main type')</u>

Formula (K. Pearson 1895, p. 367; 1948, p. 65; Elderton 1938, p. 58):
$$y = y_0(1 + x/a_1)^{m_1}(1 - x/a_2)^{m_2} \; ; \quad m_1/a_1 = m_2/a_2 \; .$$
Origin at mode if m_1 and m_2 both positive; at antimode if both negative; outside the range if one negative and one positive (J-shaped curve).

10.011 Areas of the Type I distribution

See section 3.12.

10.012 Percentage points of the Type I distribution

Pearson, E. S. 1925
<u>Biometrika</u> 17, 388-442

See abstract on p. 192. Reprinted in Tables for S & B II, T. 50, pp. 248-251.

Thompson, Catherine M. 1941a
<u>Biometrika</u> 32, 151-181

See abstract on p. 191. Reprinted in E.S. Pearson & Hartley 1954, T. 16, pp. 142-155.

[For other tables of percentage points, see section 3.31]

10.013 Auxiliary tables for computing ordinates, areas
and percentage points of the Type I distribution

Wishart, J. 1927
<u>Biometrika</u> 19, 1-38

See abstract on p. 202. Reprinted in Tables for S & B II, TT. 47-48, pp. 240-245.

Elderton 1938, p. 63, alludes to the usefulness of Gauss-logarithms in computing ordinates: see Fletcher et al. 1946, pp. 117-119. For this particular computation, the arrangement of Wittstein 1866 is advantageous.

[For other auxiliary tables, see section 3.44]

10.007
10.021

10.016 Moments, etc., of the Type I distribution

See section 3.52.

10.017 Distribution of statistics in samples from a Type I population

Irwin, J. O. 1930

<u>Metron</u> 8, no. 4, 51-105

Irwin writes the distribution of means $\bar{x}$, in random samples of n drawn from a beta-distribution $I_x(a,b)$, in the form $k\sum \lambda_{j,s}(n\bar{x}+j)^s/s!$; k and the λ are integers.

Tables 1 to 24, pp. 70-88, give the λ for all required j, s ; $n = 2(1)4$; $a, b = 1(1)4$; omitting $a = b = 1$, i.e. the rectangular distribution, for which we are referred to Hall 1927b and Irwin 1927.

10.018 Degenerate cases of Type I curves

Set $m_1 = m_2$: Type II. Let m_2 go to ∞ : Type III. Set $m_2 = 0$: for negative m_1, Type VIII; for positive m_1, Type IX. Set $m_1 = -m_2$: Type XII.

10.02 <u>Pearson curves of Type II</u>

Formula (K. Pearson 1895, p. 372; 1948, p. 70; Elderton 1938, p. 86):
$y = y_0(1 - x^2/a^2)^m$. Origin at mean.

10.021 Areas of the Type II distribution

Pearson, K. and Stoessiger, Brenda 1931

<u>Biometrika</u> 22, 253-283

See abstract on p. 185. Reprinted in Tables for S & B II, TT. 25, 25 bis, pp. 169-178.

Carlson, J. L. 1932

<u>Annals</u> <u>of</u> <u>Mathematical</u> <u>Statistics</u> 3, 86-107

Let $y(x) = 15(1-x^2)^2/16$. The table on pp. 92-93 gives
$\int_{-x}^{+x} y(t)\, dt = (15x - 10x^3 + 3x^5)/8$ to 9 dec. for $x = 0(.01)1$.

10.022
10.035   [see also pp. 716-717]

10.022   Percentage points of the Type II distribution

See E. S. Pearson & Hartley 1954, T. 13, p. 138. For description of this table and other tables, see section 7.122.

10.023   Auxiliary tables for computing ordinates and areas of the Type II distribution

Wishart, J.   1925 a

Biometrika 17, 68-78

See abstract on p. 200. Reprinted in Tables for S & B II, T. 45, pp. 236-238.

Wishart, J.   1925 b

Biometrika 17, 469-472

See abstract on p. 201. Reprinted in Tables for S & B II, T. 46, p. 239.

10.024   The fitting of Type II curves by the method of moments

Fisher, R. A.   1922

Philosophical Transactions of the Royal Society of London, Series A 222, 309-368

The table on p. 355 gives, inter al., the efficiency of the method of moments in estimating m to 4 dec. for $m = 0(\tfrac{1}{2})6$.

10.028   Degenerate cases of Type II curves

Set $m = 0$: the rectangular distribution. Let m go to ∞ ; the normal distribution.

10.03   Pearson curves of Type III

Formula (K. Pearson 1895, p. 373; 1948, p. 71; Elderton 1938, p. 92):

$$y = y_0(1 + x/a)^{\gamma a} e^{-\gamma x}.$$

Origin at mode if γa positive; outside the range if negative (J-shaped curve).

10.031   Ordinates and areas of the Type III distribution

Salvosa, L. R.   1930

Annals of Mathematical Statistics 1, 191-198 + 187 pages of tables

See abstracts on pp. 132 and 137.

Olshen, A. C. 1938

Annals of Mathematical Statistics 9, 176-200

    See abstract in section 10.039.

    [For other tables of ordinates see sections 2.20-2.21; of areas, 2.11]

10.032 Percentage points of the Type III distribution

Thompson, Catherine M. 1941 b

Biometrika 32, 187-191

    See description in section 2.3112, p. 141. Reprinted in E.S. Pearson & Hartley 1954, T. 8, pp. 130-131.

    [For other tables of percentage points see section 2.31]

10.033 Auxiliary tables for computing ordinates of the Type III distribution

Elderton, W. P. 1903 a

Biometrika 2, 260-272

    Table 1, p. 270, gives $-\log_{10}[(1+x)e^{-x}]$ to 6 dec. for $x = -.95(.05)+2.5$. Reprinted in Tables for S & B I, T. 26, p. 37.

10.035 The fitting of Type III curves by the method of maximum likelihood

    The necessary formulas are given by R.A. Fisher 1922, pp. 332-335. The estimators involve digamma functions; and their large-sample variances, trigamma functions; see sections 2.5141 and 2.5142. The following account of special tables is based on Greenwood & Durand 1960, which discusses their application in some detail.

    Define $\eta(\rho) = \log_e \rho - \psi(\rho-1)$, and let $\phi(\eta)$ be the inverse function. Then $\eta' = \rho^{-1} - \psi'(\rho-1)$.

    Masuyama & Kuroiwa 1951, pp. 21-23, give η, $\eta \log_{10} e$ and $\rho\eta \log_{10} e$ to 7 dec.; e^{η} to 8 fig.; $-1/\eta'$ and $\rho^{-1} - \rho^{-2}/\eta'$ to 8 dec.; for $\rho = .1(.05)3(.1)5(.5)10(5)20(10)50$.

    Chapman 1956, T. 1, p. 500, gives $\phi(\eta)$ to 2; 3 dec. for $\eta = .01(.001).1(.005).5(.01).57; .58(.01)1$.

    Greenwood & Durand 1960, TT. 1A, 1B, pp. 57-58, give $\eta\phi(\eta)$ to 8; 7 dec. for $\eta = 0(.01)1.4; 1.4(.2)18$ with δ^2 or δ_m^2.

[see also p. 717]

10.036 Moments, etc., of the Type III distribution

See section 2.52.

10.037 Distribution of statistics in samples from a Type III population

Pearson, E. S.   1929 b

Biometrika 21, 294-302

Table 1, p. 295, gives the first four terms of the expansions of $E(s')$, $\sigma(s')$ and $\mu_3(s')$ given by Craig 1929, to 6 dec. for n = 5, 10, 20, 50, 100, 250 and population $\beta_1 = 0, .2, .5(.5)1.5$. Table 2, p. 296, gives the first two or three terms of $\beta_1^{-\frac{1}{2}} E(b_1^{\frac{1}{2}})$ and $\sigma^2(b_1^{\frac{1}{2}})$ to 6 dec. for n = 5, 10, 20, 50, 100, 250, 500, 1000 and population $\beta_1 = 0, .2, .5(.5)1.5$. Table 3, p. 297, gives, for n and β_1 as in T. 1, $E(s')$ to 4 dec.; and exact normal-theory values for $\beta_1 = 0$. Table 4, p. 298, gives, for n and β_1 as in T. 1, $\sigma(s')$ to 4 dec.; and 4 dec. values from two approximate formulas. Table 5, p. 299, gives $\beta_1(s')$ and $\beta_2(s')$ to 3-4 dec. for n and population β_1 as in T. 1. Table 6, p. 301, gives $\beta_1^{-\frac{1}{2}} E(b_1^{\frac{1}{2}})$, $\sigma(b_1^{\frac{1}{2}})$ and $\beta_1(b_1^{\frac{1}{2}})$ to 2-4 dec. for n and population β_1 as in T. 2.

Doubtful results are indicated by (?) or (??); unsatisfactory results are omitted.

[For order statistics from Type III populations, see section 4.25]

10.038 Degenerate cases of Type III curves

Set a = 0 : Type X, i.e. the exponential distribution. Let γa go to ∞ : the normal distribution.

10.039 Miscellaneous tables of Type III curves

Salvosa, L. R.   1930

Annals of Mathematical Statistics 1, 191-193 + 187 pages of tables

See abstract on p. 155.

Olshen, A. C.   1938

Annals of Mathematical Statistics 9, 176-200

Table 2, p. 185, gives the Type III ordinate
$\eta(t) = \eta_0 (1 + 3t/10)^{-91/9} e^{-10t/3}$, the variate transform $t' = t^3$ and the transformed ordinate $\eta(t)/3t^2$: all to 6 dec. for t = -3(.5)-2(.25)-1, -.5,

−.25, −.1, −.05, −.02(.02)+.02, .05(.05).15, .25(.25)1(.5)4(1)6 .

Table 3, p. 186, gives the same for $\eta(t) = \eta_0(1+t/2)^3 e^{-2t}$, $t' = t^{1/3}$ and t = −2(.25)−1, −.9, −.75, −.5, −.27, −.08, 0, +.08, .27, .5, .64, .74, .9, 1(.5)5.5 .

Olshen gives on pp. 192-195 a method of normalizing the Type III distribution by means of a polynomial transformation. For example, define t = 9.9(−.0101010 + .1005038 u + .0033670 u^2 + .000082 u^3) . Table 5, p. 197, gives $\eta(t) = \eta_0(1+t/10)^{99} e^{-10 t}$ and $\eta(t)\frac{dt}{du}$ to 6 dec. for u = −2(.2)+2 ; the normal curve that this transformation approximates is also shown to 6 dec.

10.04 Pearson curves of Type IV (Elderton's 'second main type')

Formula (K. Pearson 1895, p. 376; 1948, p. 74; Elderton 1938, p. 66):
$$y = y_0 a^{2m}(x^2+a^2)^{-m} \exp\left(-\nu \tan^{-1} \frac{x}{a}\right) .$$
Origin is $\nu a/(2m-2)$ after mean.

10.043 Auxiliary tables for computing ordinates of the Type IV distribution

10.0431 The $G(r,\nu)$ integrals

Setting x = a tan θ , r = 2m−2 , we find that the Type IV distribution can be made to depend on the integral $\int_{-\frac{1}{2}\pi}^{\xi} \cos^r\theta \, e^{-\nu\theta} d\theta = F_\xi(r,\nu)$ say. This function, we have noted on p. 389 above, K. Pearson did not succeed in tabulating; nor is the project likely to commend itself to a contemporary statistician. An extensive table of the complete integral,
$$F(r,\nu) = \int_{-\frac{1}{2}\pi}^{+\frac{1}{2}\pi} \cos^r\theta \, e^{-\nu\theta} d\theta ,$$
is described below.

B. A. 1899

Define $\phi = \tan^{-1} \nu/r$; $H(r,\nu) = (r-1)^{\frac{1}{2}} e^{-\nu\phi}(\cos\phi)^{-r-1} F(r,\nu)$. The tables on pp. 71-120 were computed by Alice Lee. The table on p. 71 gives $\log_{10} F(r,\nu)$ to 7 dec. for ϕ = 0(1°)45° with Δ^2 . The table on pp. 72-120 gives $\log_{10} F(r,\nu)$ to 7 dec. without differences; and $\log_{10} H(r,\nu)$ to 7 dec. with Δ^2 in ϕ; all for r = 2(1)50 and ϕ = 0(1°)45° . Reprinted in Tables for S & B I, T. 54, pp. 126-142.

10.0431
10.07

Jahnke, P. R. E. and Emde, F. 1909

Funktionentafeln mit Formeln und Kurven

The table on pp. 44-45 gives $\log_{10} F(r,\nu)$ to 4 dec. without differences for $r = 1(1)50$ and $\phi = 0(5°)45°$. All the figures 1, appearing as characteristics for $\phi = 0, 5°, 10°$, should bear a − superscript: $\bar{1}$. This error appears in both the 1909 and the 1923 printing, and vitiates Fig. 12, p. 44.

[For the preliminary table by Lee in B.A. 1896, see Fletcher et al. 1946, p. 341]

10.0432 $\tan^{-1} x$ and $\log_{10}(1+x^2)$

Comrie 1938 gives $\tan^{-1} x$ to 7 dec. for $x = 0(.01)5(.1)20(1)164$; and $\log_{10}(1+x^2)$ to 7 dec. for $x = 0(.01)5(.1)15$; both with δ^2. Reprinted in Vianelli 1959, Pro. 202, pp. 613-619; the columns headed Δ^2 contain δ^2.

10.048 Degenerate cases of Type IV curves

Let ν go to ∞ : Type V. Set $\nu = 0$: Type VII.

10.05 Pearson curves of Type V*

Formula (K. Pearson 1901, p. 448; 1948, p. 382; Elderton 1938, p. 96):
$y = y_0 x^{-p} e^{-\gamma/x}$. Origin at start of curve.

10.052 Percentage points of the Type V distribution

See section 2.315.

10.055 The fitting of Type V curves by the method of maximum likelihood

Masuyama & Kuroiwa 1951, p. 19, point out that the tables abstracted in section 10.035 apply, mutatis mutandis, to the Type V distribution; Greenwood & Durand 1960, p. 61, give a few relevant formulas.

10.058 Degenerate case of Type V curves

Let p go to ∞ : the normal distribution.

---

*See footnote on p. 388.

10.059 Miscellaneous tables connected with the Type V distribution

Neyman, J. 1941

Annals of Mathematical Statistics 12, 46-76

See abstract on p. 302.

10.06 Pearson curves of Type VI (Elderton's 'third main type')

Formula (K. Pearson 1901, p. 450; 1948, p. 364; Elderton 1938, p. 76):
$y = y_0(x-a)^{q_2} x^{-q_1}$. Range from a to ∞.

Kendall & Stuart 1958, p. 151, allege that "since the simple transformation [$z = a/x$] reduces it to the Type I form a separate study [of the Type VI] is unnecessary." The study of the shape of the curves in transformed form affords no difficulty (cf. section 4.20 for the corresponding transformation of the F-distribution): but symmetry would then require that the Type VI be fitted by the method of negative moments (Kelley 1923, pp. 140-144); in practice, the method of positive moments (Elderton, loc.cit.) is always used.

10.061 Areas of the Type VI distribution

See section 4.21.

10.062 Percentage points of the Type VI distribution

See E.S. Pearson & Hartley 1954, T. 18, pp. 157-163. For description of this table and other tables, see section 4.231.

10.068 Degenerate cases of Type VI curves

Let q_1 go to ∞ : Type III. Let q_2 go to ∞ : Type V. Set $q_2 = 0$: Type XI.

10.07 Pearson curves of Type VII[*]

Formula (K. Pearson 1916, p. 450; 1948, p. 550; Elderton 1938, p. 90):
$y = y_0(1 + x^2/a^2)^{-m}$. Origin at mean[‡].

---

[*] See footnote on p. 388.

[‡] If $\frac{1}{2} < m \leq 1$, origin at mode: but in Pearson's system $m > 4\frac{1}{2}$.

10.071
10.087
[see also p. 717]

10.071 Ordinates and areas of the Type VII distribution

Pearson, K. and Stoessiger, Brenda   1931

<u>Biometrika</u> 22, 253-283

See abstract on p. 185. Reprinted in Tables for S & B II, TT. 25, 25 bis, pp. 169-178.

Hartley, H. O. and Pearson, E. S.   1950 a

<u>Biometrika</u> 37, 168-172

See description in section 4.1111, p. 208. Reprinted in E.S. Pearson & Hartley 1954, T. 9, pp. 132-134.

[For tables of ordinates see section 4.12; for other tables of areas see section 4.11]

10.072 Percentage points of the Type VII distribution

See E.S. Pearson & Hartley 1954, T. 9, p. 133; T. 12, p. 138. For description of these and other tables, see section 4.131.

10.073 Auxiliary tables for computing ordinates and areas of the Type VII distribution

Wishart, J.   1925 a

<u>Biometrika</u> 17, 68-78

See abstract on p. 200. Reprinted in Tables for S & B II, T. 45, pp. 236-238.

Wishart, J.   1925 b

<u>Biometrika</u> 17, 469-472

See abstract on p. 201. Reprinted in Tables for S & B II, T. 46, p. 239.

Comrie, L. J.   1938

<u>Tables</u> <u>of</u> $\tan^{-1} \underline{x}$ and $\log(1+\underline{x}^2)$

See abstract on p. 398.

10.074 The fitting of Type VII curves by methods other than moments

Sichel, H. S.   1949

Biometrika 36, 404-425

"The method of frequency moments and its application to the Type VII population"

Write the Type VII curve in the form $y = y_0(1 + x^2/c^2)^{-m}$. Let $p_i = f_i/N$ denote the observed fraction of N observations falling into the ith class. The method of frequency moments employs the statistics $o_n = \sum p_i^n$ as estimators of population parameters; since the o_n are invariant under shift, other methods must be used to estimate the location parameter.

Table 2, p. 415, gives the efficiency of estimating c, given m, from o_2, from $o_{3/2}$, and from the second sample moment; to 3 dec. for $m = .5$, .8, 1(1)10, ∞.

When m and c must be estimated jointly, Sichel uses the statistics o_2 and $a_{3/2} = o_2 / o_{3/2}^2$. Table 3, p. 416, gives the efficiency of estimating m from $a_{3/2}$ and from b_2; to 3 dec. for $m = .5, .7, 1(1)10, \infty$.

Jones, H. L.   1953

Journal of the American Statistical Association 48, 113-130

See abstract on p. 363.

10.078   Degenerate case of Type VII curves

Let m go to ∞: the normal distribution.

10.08   Pearson curves of Type VIII

Formula (K. Pearson 1916, p. 438; 1948, p. 538; Elderton 1938, p. 102): $y = y_0(1 + x/a)^{-m}$. Range from $x = 0$ to $x = -a$.

10.087   Distribution of statistics in samples from a Type VIII population

For some work on order statistics from Type VIII populations, including illustrative tables, see K. & M.V. Pearson 1932, pp. 233-239, 234-260.

10.088
10.123
                                 [see also p. 718]

10.088 Degenerate case of Type VIII curves

Set $m = 0$: the rectangular distribution.

10.09 Pearson curves of Type IX

Formula (K. Pearson 1916, p. 441; 1948, p. 541; Elderton 1938, p. 105):
$y = y_0(1 + x/a)^m$. Range from $x = 0$ to $x = -a$.

10.097 Distribution of statistics in samples from a Type IX population

For some work on order statistics from Type IX populations, including illustrative tables, see K. & M.V. Pearson 1932, pp. 234, 240-241, 263-271.

10.098 Degenerate cases of Type IX curves

Let m go to ∞ : Type X. Set $m = 0$: the rectangular distribution.

10.10 The Pearson curve of Type X[*]

Formula (K. Pearson 1916, p. 443; 1948, p. 543; Elderton 1938, p. 108):
$y = N\sigma^{-1}e^{-x/\sigma}$. Range from 0 to ∞ .

10.101 Ordinates and areas of the Type X distribution

Tables of the descending exponential will furnish such ordinates and areas: see Fletcher et al. 1946, pp. 167-168. For the latest printing of the book there cited as New York W.P.A. 1939 b , see our author index at N.B.S. 1951 z. For the tail of the distribution, see N.B.S. 1955 z , which tables e^{-x} to 20 dec. for $x = 2.5(.001)10$, without differences.

10.102 Percentage points of the Type X distribution

Tables of natural logarithms of decimal fractions will furnish such percentage points. Fisher & Yates 1948, 1953, 1957, T. XXVI, gives, inter al., $-\log_e x$ to 5 dec. for $x = .1(.001).999$, with PPMD. See also Fletcher et al. 1946, pp. 177-178; for the latest printing of the book there cited as New York W.P.A. 1941 c , see our author index at N.B.S. 1953 z.

---

[*] The designation Type X is sometimes loosely applied to the two-tailed exponential distribution of Laplace, $y = \frac{1}{2}e^{-|x|}$; but this distribution does not satisfy Pearson's differential equation (10.0.1).

10.103 Auxiliary tables for computing percentage points of the Type X distribution

If the product $\delta \cdot \log_e 10$ can be formed once for a series of percentage points, a table of decimal cologarithms permits a uniform treatment of all decades: but such tables are neither numerous nor extensive; see Fletcher et al. 1946, p. 113.

10.107 Distribution of statistics in samples from a Type X population

For some work on order statistics from a Type X population, including many illustrative tables, see K. & M.V. Pearson 1932, pp. 210-233, 271-273.

10.11 Pearson curves of Type XI

Formula (K. Pearson 1916, p. 444; 1948, p. 544; Elderton 1938, p. 111):
$y = y_0 x^{-m}$. Range from $x = b > 0$ to $x = \infty$.

10.117 Distribution of statistics in samples from a Type XI population

For some work on order statistics from Type XI populations, including illustrative tables, see K. & M.V. Pearson 1932, pp. 234, 241-243, 273-277.

10.118 Degenerate case of Type XI curves

Let m go to ∞ : Type X.

10.12 Pearson curves of Type XII

If, in J-shaped curves of Type I, $5\beta_2 = 6\beta_1 + 9$, an attempt to follow the standard solution yields the contradiction $a_1 = -a_2$, $a_1 + a_2 =$ [a positive constant]. In fact, the 'mode' of this curve has receded to ∞, and it is customary to write the curve with origin at mean (K. Pearson 1916, p. 446; 1948, p. 546; Elderton 1938, p. 111):

$$y = N(2\pi\delta\gamma)^{-1} \sin(\pi\gamma/\delta)[[\delta(\gamma+\delta)+x]/[\delta(\gamma-\delta)-x]]^{\gamma/\delta},$$

where $\gamma = \beta_1^{\frac{1}{2}}$, $\delta = (3+\beta_1)^{\frac{1}{2}}$.

10.123 Auxiliary tables for computing ordinates of the Type XII distribution

If the distribution is transformed to the range $[-1,+1]$, Weidenbach's logarithms can be used: Fletcher et al. 1946, p. 119.

10.128   [see also pp. 718-719]
10.18

10.128   Degenerate case of Type XII curves

Set $m = 0$: the rectangular distribution.

10.13   The normal distribution *qua* Pearson curve[*]
        (Elderton's 'normal curve of error')

Formula (K. Pearson 1895, p. 356; 1948, p. 54; Elderton 1938, p. 80):
$y = y_0 e^{-x^2/2\sigma^2}$. Origin at mean.

To catalogue here all tables based on the normal distribution would be to rehearse the greater part of sections 1, 2, 3, 4, 6 and 7. E.S. Pearson & Hartley 1954, p. xj, point out that of "the fifty-four tables in [that] volume, the main purpose of perhaps thirty lies in the analysis of data involving normally distributed variables." In particular, Tables for S & B I, TT. 1-6, 9, 11-16, 25, 29; II, TT. 1-7, 12, 13, 17-22, 35-37; Pearson & Hartley 1954, TT. 1-12, 16-35; rest on the normal distribution. Descriptions of these tables can be traced through the author index.

The cross-references below are suggestive only; details must be sought in the table of contents.

10.131   Ordinates and areas of the normal distribution

For ordinates, see section 1.116; for areas, 1.111-1.115.

10.132   Percentage points of the normal distribution

See sections 1.21, 1.232.

10.135   The fitting of the normal distribution
            by statistics other than the sample mean and variance

See sections 1.332, 1.41, 1.51-1.54.

10.136   Moments of the normal distribution

See section 1.31.

[*]See footnote on p. 388.

10.137 Distribution of statistics in normal samples

David, F. N. 1949

Biometrika 36, 383-393

Let $\bar{x}$ and s be the mean and variance of a normal sample of n; let the corresponding population parameters be μ and σ; define $v = s/\bar{x}$.

Writing $v = \left\{1 + \left[\dfrac{s^2-\sigma^2}{\sigma^2}\right]^{\frac{1}{2}}\left[1 + \dfrac{\bar{x}-\mu}{\mu}\right]^{-1}\right\}\dfrac{\sigma}{\mu}$, and expanding the brackets by the binomial theorem, David expands the moments of v in 'moments of moments' and thence in powers of n^{-1}.

Table 2, p. 388, based on these expansions, gives the mean of $v\sigma/\mu$ to 4 dec.; its standard deviation to 3 dec.; its β_1 and β_2 to 2 dec.; all for n = 6, 11(10)41 and μ/σ = 3, 5, 10.

[See also sections 1.4-1.6, 2.113, 2.31, 3.12, 3.31, 4.11, 4.13, 4.141, 4.15, 4.20-4.26, 4.33, 6.1]

10.18 Generalizations of K. Pearson's frequency curves

Romanovsky, V. 1924

Biometrika 16, 106-116

Romanovsky proposes to expand a distribution into a frequency function of Type I (or II) multiplied by a series of Jacobi polynomials; or a frequency function of Type III multiplied by a series of Laguerre (Sonin) polynomials. K. Pearson 1924 z discountenances the proposal; Wishart 1926 numerically illustrates the unaptness of the expansion.

We have seen no tables of Laguerre polynomials useful in computing Romanovsky's series. The Jacobi tables of Karmazina 1954 go only through polynomials of degree 5 : see review in appendix.

Hansmann, G. H. 1934

Biometrika 26, 129-195

Hansmann generates a system of symmetrical frequency curves y(x) from the differential equation $dy/dx = -xy/(c_0 + c_2 x^2 + c_4 x^4)$, and fits the c_i to the observed values of $\beta_2 = \mu_4/\mu_2^2$, $\beta_4 = \mu_6/\mu_2^3$, $\beta_6 = \mu_8/\mu_2^4$.

[Note that the method of moments, following Pearson, would fit the c_i to μ_2, β_2 and β_4]

[see also pp. 718-719]

He classifies the curves (TT. 1, 2, pp. 134, 139) into six main types of the form $y = y_0(q \pm x^2)^k (p \pm x^2)^{-k}$ and fourteen transition types. Tables 4, 5, 6, pp. 145-150, give formulas for y_0, p, q, k etc. in terms of the β_{2i}. Tables 9-21, pp. 157-162, give parametric equations for the surfaces in β_2, β_4, β_6 space representing transition types.

Davies, O. L. 1934
Biometrika 26, 59-107

Davies fits curves of the form $e^{\rho x} y$, where y is the ordinate of a suitable Pearson curve, to the successive terms of the Gaussian hypergeometric series $F(a, b ; c ; x)$. He gives histograms and tables for $F(-30, -50 ; 100 ; \frac{1}{2})$; $F(30, 30 ; 111 ; 3/4)$; $F(-30, 60 ; -81 ; 1/3)$; $F(-20, 20 ; -200 ; \frac{1}{2})$; $F(1, -60 ; -65 ; 9/10)$; $F(-100, -100 ; 1 ; \frac{1}{2})$; and for the curves fitted thereto.

10.19 Bivariate Pearson frequency surfaces

The attempts of Steffensen 1922; 1923, pp. 116-132; Rhodes 1923; Pearson, Filon and Isserlis (summarized by K. Pearson 1923 y); and K. Pearson 1923 z to construct such surfaces are reviewed by Pretorius 1930, pp. 122, 132-137. The only recent investigations we have seen are those of van Uven 1947 a, b; 1948 a, b; van Uven constructs 6 main types and 26 sub-types, much in the manner of K. Pearson 1895, 1901, 1916, and sketches a method of fitting the hopefully more useful of these types.

None of the above systems has found sufficient favor to require the construction of specific tables. We have not investigated how far the tables catalogued in sections 10.00-10.13 may be of use in bivariate work.

10.2 Other frequency curves

Of the curves treated in this section, only the Gram-Charlier curves (section 10.21) and Johnson's curves (section 10.222) are sufficiently general to furnish systems at all comparable with K. Pearson's curves. All distributions treated herein are continuous, except for the Gram-Charlier Type B (p. 409) and the discrete rectangular distribution (p. 418): we have discussed a number of discrete distributions in section 5. The generalized hypergeometric distributions of Kemp & Kemp 1956 are the closest approach we have seen to an extensive system of discrete distributions: for an analogy between generalized hypergeometric distributions and Pearson curves, and a suggestion for the use of truncated generalized hypergeometric distributions, see Carver 1924, pp. 111-113.

10.21 Gram-Charlier series

10.211 Gram-Charlier series of Type A

The series

$$\psi(t) = -\tfrac{1}{2}t^2 + \sum_{s=3}^{\infty} \zeta_s (it)^s / s! \qquad (10.21.1)$$

formally represents the cumulant-generating function, and so the infinite product

$$\varphi(t) = e^{-\tfrac{1}{2}t^2} \cdot \exp\left[\sum_{s=3}^{\infty} \zeta_s (it)^s / s!\right] \qquad (10.21.2)$$

formally represents the moment-generating function, of a distribution having $\varkappa_1 = 0$; $\varkappa_2 = 1$; $\varkappa_s = \zeta_s$, $s \geq 3$. Formal inversion of (10.21.2) yields the density function

$$f(x) = (2\pi)^{-\tfrac{1}{2}} \exp\left[\sum_{s=3}^{\infty} \zeta_s (-D)^s / s!\right] e^{-\tfrac{1}{2}x^2}. \qquad (10.21.3)$$

For an arbitrary specification of the ζ_s, even for a specification corresponding to an existing distribution, none of these series need converge: for practical computing, moreover, we require a finite expansion corresponding to the number of ζ_s specified. Consider the following three truncation rules:

10.211
10.212

(1) Let the upper limit of summation in (10.21.1) be T. Then we have a formal representation of a distribution with $\kappa_1 = 0$; $\kappa_2 = 1$; $\kappa_s = \zeta_s$, $3 \leq s \leq T$; $\kappa_s = 0$, $s > T$. Unfortunately, no such distribution exists, except the trivial case $T = 3$, $\zeta_3 = 0$: Marcinkiewicz 1938.

(2) Expand the second factor in (10.21.1) as a power series in t, and truncate the series to a polynomial of degree T. The result is a Gram-Charlier series of Type A: Gram 1883, pp. 71-73; Charlier 1905 a, p. 8.

(3) Assume that ζ_s is $O(n^{1-\frac{1}{2}s})$, and expand the bracket in (10.21.3) in powers of the ζ_s, keeping all terms of order $n^{1-\frac{1}{2}S}$, $3 \leq S \leq T$. The result is an Edgeworth series: Edgeworth 1905.

We illustrate (2) and (3) by writing explicit forms of Gram-Charlier and Edgeworth series for $T = 6$. The Hermite polynomials h_s are defined in section 1.323.

Gram-Charlier:
$$f(x) = (2\pi)^{-\frac{1}{2}} e^{-\frac{1}{2}x^2} [1 + \zeta_3 h_3/6 + \zeta_4 h_4/24 + \zeta_5 h_5/120 + (\zeta_6 + 10\zeta_3^2) h_6/720] \quad (10.21.4)$$

Edgeworth:
$$f(x) = (2\pi)^{-\frac{1}{2}} e^{-\frac{1}{2}x^2} [1 + \zeta_3 h_3/6 + (\zeta_4 h_4/24 + \zeta_3^2 h_6/72) +$$
$$+ (\zeta_5 h_5/120 + \zeta_3 \zeta_4 h_7/144 + \zeta_3^3 h_9/1296) +$$
$$+ (\zeta_6 h_6/720 + \{8\zeta_3 \zeta_5 + 5\zeta_4^2\} h_8/5760 +$$
$$+ \zeta_3^2 \zeta_4 h_{10}/1728 + \zeta_3^4 h_{12}/31104)] \quad (10.21.5)$$

Only if $T = 3$; or if $T = 4$ and $\zeta_3 = 0$; are the two series identical.

The essential tool for numerical evaluation of these series is a table of derivatives of the normal distribution. Harvard Univ. 1952 is probably the most useful table; for description thereof and for other tables, see section 1.321.

See section 1.325 for some tables of special Gram-Charlier series.

Numerical coefficients for expressing (10.21.3) and (10.21.4), with $T = 4$, in terms of tetrachoric functions (section 1.324) are given by Elderton 1938, pp. 130, 133.

Shenton, L. R. 1951

<u>Biometrika</u> 38, 58-73

Table 1, p. 59, and fig. 1, p. 60, indicate the region in which (10.21.4) with T = 4 yields a non-negative frequency function. (Cf. Charlier 1947, pp. 60, 116.) Tables 2-9, pp. 62-70, give the efficiency of the method of moments in estimating various combinations of parameters, and compares the Gram-Charlier Type A with Pearson's Types IV and VII: the efficiency is almost always low.

Barton, D. E. and Dennis, K. E. 1952

<u>Biometrika</u> 39, 425-427

The diagram on p. 426 indicates the regions in the β_1, β_2 plane in which (10.21.4) and (10.21.5), with T = 4, yield non-negative and unimodal frequency functions. In view of the statement (Elderton 1938, p. 133, footnote) that Edgeworth's series avoids the negative frequencies that may arise with Type A, it is noteworthy that (p. 427) the "positivity region for the Edgeworth curves lies almost entirely within that of the Gram-Charlier and is less than it."

<u>10.212</u> Gram-Charlier series of Type B

The necessary formulas are given, in rather summary fashion, by Elderton 1938, pp. 131-132. The work of Steffensen 1916 has shown that Type B must be taken as a discrete distribution: in the continuous form, involving $\frac{\sin \pi x}{\pi x}$, the moments are without exception <u>divergent</u>. The ideal for use in computing Type B series would be a table of Poisson sums, individual terms, <u>and differences</u>: we have seen no such table. For tables of Poisson sums, see section 2.12 ; for individual Poisson terms, 2.22 .

Jørgensen 1915 introduces, apparently with Charlier's approval, a variant of the Type B series that proceeds by central differences of the distribution of the difference of two Poisson variates: cf. section 5.23 . The continuous form of this distribution (Jørgensen 1915, p. 5, eq. 10) is likewise invalidated by Steffensen 1916; extreme caution, therefore, is necessary in studying the numerical examples in Jørgensen 1916.

10.213
10.221 [see also pp. 719-720]

10.213 Gram-Charlier series of Type C

Charlier 1928 proposed to expand the logarithm of a frequency function in a series of Hermite polynomials. This expansion requires that the frequency function be positive on the range $[-\infty, +\infty]$ and that it vanish at ∞ more slowly than $\exp[-e^{x^2/4}]$.

According to Charlier 1928, p. 20, the "practical advantages of the frequency function do not put it in other respects above other forms of these functions." Subsequent statisticians seem to have shared this opinion of Type C: the only exposition of the series we have seen is a summary of formulas in Charlier 1931, pp. 57-58.

10.218 Generalized Gram-Charlier series

For the work of Romanovsky 1924 see p. 405; for the work of Salvosa 1930, p. 155. The tables of Salvosa 1930 are reprinted in Salvosa 1930 z, which contains formulas and numerical illustrations of their use.

Roa 1923 proposes to expand frequency functions in series of derivatives of $y_0 e^{-(x/\gamma)^q}$ and of $y_0 \text{sech}^m \gamma x$. The details of his numerical processes are obscure. Pages 93-108 of Roa contain tables of various combinations of gamma and polygamma functions, and of the gudermannian: the form of tabulation makes them of little use except for the frequency functions that Roa advocates.

10.219 Bivariate Gram-Charlier series

See Pretorius 1930 and works there cited; also Kendall 1949, pp. 177-178.

10.22 Translation systems

Given a random variable x, we may seek a function ξ of x such that ξ is nearly normally distributed; conversely, given a normal variate ξ, we may take a function x of ξ, usually containing one or more disposable parameters, and study the shape, moments etc. of the distribution of x. For early essays of the method, and in particular for the difficulties arising from non-monotone translation functions, see Pretorius 1930, pp. 115-121, and works there cited.

10.221 Polynomial translation

Edgeworth, F. Y. 1924

Journal of the Royal Statistical Society 87, 571-594

Edgeworth 1899 c, p. 538, proposes to set $x = a(\xi + \kappa\xi^2 + \lambda\xi^3)$, where ξ is normally distributed, and x and ξ are measured from their medians. Edgeworth 1913, p. 430; 1914 a, p. 307; proposes to determine a, κ, λ by the method of moments. This process leads to rather formidable equations: the present paper attempts to render these equations practically computable.

The table on p. 582, computed by Frederick Brown, gives $\beta = \beta_1/8$ and $\epsilon = (\beta_2 - 3)/12$ to 4 dec. for

$$\begin{cases} x = \kappa^2 = 0, .01 & .02(.01).04 & .05, .06 \\ \lambda = -.06(.01)+.1 & [-.07+](.01)+.1 & -.03(.01)+.06 \end{cases}$$

Bowley, A. L. 1928

F. Y. Edgeworth's contributions to mathematical statistics

The table on pp. 123-128, computed by Miss F. J. Nicholas, gives λ and χ to 3 dec. for $\epsilon = -.05(.005)0(.01)+.64$ and $\beta = 0(.01).46$, so far as these λ and χ lie within a somewhat complicated region described on pp. 76-77.

Cornish, E. A. and Fisher, R. A. 1937

Revue de l'Institut International de Statistique 5, 307-320

Generalize (10.21.5) by including terms in ζ_1 and ζ_2, where ζ_1 is $O(n^{-\frac{1}{2}})$ and ζ_2 is $O(n^{-1})$: so that $\kappa_s = \zeta_s$, $s = 1, 3, 4, 5, 6$; $\kappa_2 = 1 + \zeta_2$. The explicit form of $\int_{-\infty}^{x} f(t)dt$ is given on p. 315 (Fisher 1950, p. *30.*9). ξ, a normal deviate, is expanded in the asymptotic series

$$\xi = x + A_{-\frac{1}{2}} + A_{-1} + A_{-1\frac{1}{2}} + A_{-2},$$

where A_p is $O(n^p)$: $A_{-\frac{1}{2}} = \zeta_1 + \zeta_3(x^2-1)/6$,

$A_{-1} = (3\zeta_2 - 2\zeta_1\zeta_3)x/6 + \zeta_4(x^3-3x)/24 - \zeta_3^2(4x^3-7x)/36$. Formulas through A_{-2} on p. 316 (Fisher 1950, p. *30.*10). This series is then reverted:

$x = \xi + B_{-\frac{1}{2}} + B_{-1} + B_{-1\frac{1}{2}} + B_{-2}$: formulas on p. 317 (Fisher 1950, p. *30.*11).

Table 1, p. 318 (Fisher 1950, p. *30.*12) gives the polynomials in ξ,

10.221
10.2221

[see also pp. 719-720]

entering into the B_p, to 5 dec. for ξ equal to the upper 25%, 10%, 5%, 2½%, 1%, ½%, .25%, .1% and .05% points of the standard normal deviate.

[The most readable account of Cornish & Fisher's method is given in Kendall & Stuart 1958, pp. 163-167: our ζ_s is Kendall's I_s. We follow Edgeworth, Bowley, and Kendall & Stuart in writing x for a skew variate and ξ for a normalized variate: Cornish & Fisher interchange the letters]

10.222   Logarithmic and hyperbolic translation

Johnson, N. L.   1949 a

Biometrika 36, 149-176

Johnson considers the translation $\xi = \gamma + \delta g[(x-\mu)/\lambda]$, where g is a monotone function, containing no disposable parameters, and g and g^{-1} are analytically and tabularly pleasant. He specializes on three functions g:

S_L curves. $g(y) = \log_e y$

S_B curves. $g(y) = \log_e[y/(1-y)] = 2 \tanh^{-1}(2y-1)$

S_U curves. $g(y) = \log_e[y + (y^2+1)^{\frac{1}{2}}] = \sinh^{-1} y$.

Figure 2, p. 157, sketches the loci of these curves in the β_1, β_2 plane. In roughly the region in which Pearson curves are U-shaped, S_B curves are bimodal. Figures 3-11, pp. 160-161, give typical forms of the curves.

10.2221 S_L curves: i.e. the log-normal distribution

We follow Aitchison & Brown 1957 and adopt the standard form

$$(2\pi\sigma^2)^{-\frac{1}{2}}(x-\tau)^{-1} e^{-[\log_e(x-\tau) - \mu]^2 / 2\sigma^2} dx \qquad (10.22.1)$$

for the differential of probability of the log-normal distribution. With this notation,

$$\mu_1' = \tau + e^{\mu + \frac{1}{2}\sigma^2} ; \qquad (10.22.2)$$

$$\mu_2 = \left(e^{\mu + \frac{1}{2}\sigma^2}\right)^2 \left(e^{\sigma^2} - 1\right) = \alpha^2 \eta^2 , \text{ say}; \qquad (10.22.3)$$

$$\beta_1^{\frac{1}{2}} = \eta^3 + 3\eta ; \qquad (10.22.4)$$

$$\beta_2 = \eta^8 + 6\eta^6 + 15\eta^4 + 16\eta^2 + 3 . \qquad (10.22.5)$$

Jenkins, T. N. 1932

Annals of Mathematical Statistics 3, 45-55

To facilitate the fitting of the log-normal distribution, T. 1, p. 50, gives $\log_{10}(2n-1)$, $[\log_{10}(2n-1)]^2$ and $\log_{10}\tfrac{1}{2}n$ to 10 dec. for $n = 1(1)50$.
Errors: $n = 1$, insert minus sign in k; $n = 9$, $k = .65321\,25137$.

Yuan, P. T. 1933

Annals of Mathematical Statistics 4, 30-74

It follows from (10.22.4) that $\omega = 1 + \tau^2 = e^{\sigma^2}$ is the real root of the cubic $\omega^3 + 3\omega^2 = 4 + \beta_1$.

Table 4, p. 48, gives $\beta_1^{\frac{1}{2}} = (\omega^3 + 3\omega^2 - 4)^{\frac{1}{2}}$ and $\sigma = (\log_e \omega)^{\frac{1}{2}}$ to 4 dec. for $\omega = 1(.01)1.5$.

Thompson, H. R. 1951

Biometrika 38, 414-422

See abstract on p. 44.

Moshman, J. 1953

Journal of the American Statistical Association 48, 600-609

Table 1, pp. 606-607, gives the upper and lower $\tfrac{1}{2}\%$, 1%, $2\tfrac{1}{2}\%$, 5% and 10% points of $(x-\mu_1')\mu_2'^{-\frac{1}{2}}$, where x is log-normally distributed, to 3 dec. for $\beta_1^{\frac{1}{2}} = 0(.05)3$.

Aitchison, J. and Brown, J. A. C. 1957

The lognormal distribution

Table A 1, pp. 154-155, gives $\eta = \left(e^{\sigma^2}-1\right)^{\frac{1}{2}}$, $\eta^3 + 3\eta$, $\eta^8 + 6\eta^6 + 15\eta^4 + 16\eta^2$, $e^{\frac{1}{2}\sigma^2}$, $e^{3\sigma^2/2}$, $P(\tfrac{1}{2}\sigma)$ and $2P(2^{-\frac{1}{2}}\sigma) - 1$: all to 4 fig. for $\sigma = 0(.05)1(.1)4$.

Table A 2, pp. 156-158, gives the function $\psi_n(t) = 2^m m! i_m[(n-1)(2t/n)^{\frac{1}{2}}]$, where $m = \tfrac{1}{2}(n-3)$, $i_m(x) = x^{-m}I_m(x)$, and I_m is the Bessel function of the first kind of purely imaginary argument, to 4 dec. for $t = .05(.05)2$ and $n = 10(10)200(100)500(250)1000, 5000, \infty$. Table A 3, pp. 159-160, gives $x_n(t) = \psi_n(2t) - \psi_n[(n-2)t/(n-1)]$ to 4 dec. for $t = .05(.05)1$ and n as in T. A 2. These functions are used in the method, introduced by Finney 1941a.

10.2221
10.23

of obtaining efficient estimates of the mean and variance of a log-normal distribution: see pp. 45-46 for the procedure.

Table A 4, p. 161, gives u and $s^2 = \log_e(1+u^2)$, where $u^3 + 3u = k$, to 4 dec. for $k = 0(.2)10(1)24$. If $k = b_1^{\frac{1}{2}}$, u estimates η and s^2 estimates σ^2 in (10.22.4).

[See also the 36 references given by Yuan 1933 and the 217 references given by Aitchison & Brown 1957]

10.2222 S_B curves

We have seen no tables. The method of fitting is discussed by Johnson 1949 a, pp. 158-162. Aitchison & Brown 1957, p. 65, summarize the problem, saying that every "method except the method of [equating selected population and sample] quantiles may be dismissed immediately as mathematically intractable, and even for this method the determination is by no means easy."

10.2223 S_U curves

Johnson, N. L. 1949 a
Biometrika 36, 149-176

If μ_s is the sth moment of y about its mean, then
$2\mu_2 = (\omega-1)[\omega \cosh 2\Omega + 1]$,
$4\mu_3 = -\omega^{\frac{1}{2}}(\omega-1)^2[\omega(\omega+2)\sinh 3\Omega + 3 \sinh \Omega]$,
$8\mu_4 = (\omega-1)^2[\omega^2(\omega^4+2\omega^3+3\omega^2-3)\cosh 4\Omega + 4\omega^2(\omega+2)\cosh 2\Omega + 3(2\omega+1)]$,
where $\Omega = \gamma/\delta$ and $\omega = e^{\delta^{-2}}$.

Figure 12, p. 164, is an abac from which δ and Ω can be read for given values of β_1 and β_2.

10.229 Bivariate translation systems

Both polynomial and logarithmic translation have been applied to bivariate distributions with some success. See Pretorius 1930, pp. 129-132, 167, 169, 175-177, 193, 208-218, and works there cited; Yuan 1933, pp. 53-72 (Yuan translates one marginal distribution logarithmically, and leaves the other margin normal); N.L. Johnson 1949 b ; Yasukawa 1925.

10.23 The fourth-degree exponential

In the course of an excursus on the inefficiency of the method of moments in fitting Pearson curves, R.A. Fisher 1922, p. 356, remarked:

"The system of curves for which the method of moments is the best method of fitting may easily be deduced, for if the frequency in the range dx be

$$y(x, \theta_1, \theta_2, \theta_3, \theta_4) dx,$$

then

$$\frac{\partial}{\partial \theta} \log y$$

must involve x only as polynomials up to the fourth degree; consequently

$$y = e^{-a^2(x^4 + p_1 x^3 + p_2 x^2 + p_3 x + p_4)}, \qquad [10.23.1]$$

the convergence of the probability integral requiring that the coefficient of x^4 should be negative, and the five quantities a, p_1, p_2, p_3, p_4 being connected by a single relation, representing the fact that the total probability is unity.

"Typically these curves are bimodal, and except in the neighbourhood of the normal point are of a very different character from the Pearsonian curves."

Wolfenden 1942, p. 77, considered this bimodality of some importance for the actuary:

"Since Pearson's curves cannot produce a double hump, and it is doubtful whether Edgeworth's or the Gram-Charlier Type A can do so conveniently (cf. [Elderton 1938, pp.] 139 and 140), it is interesting to note that the classes of curves which arise from [10.23.1] are typically bi-modal."

It will be seen that (10.23.1) can be obtained from Charlier's Type C (p. 410) by truncating at H_4. Shenton 1951, p. 72, attributes to Jeffreys 1939, p. 242, the discovery that if we attempt to approximate a Pearson Type IV distribution by a fourth-degree exponential, the coefficient of x^4 is positive and so the integral diverges. Aroian 1948 attempts to meet this objection by fitting (10.23.1) on a finite range: his proposed method of fitting is erroneous in that $x_1^s y(x_1) - x_2^s y(x_2)$ (Aroian 1948, p. 590, eq. 5) cannot, in general, be made to vanish for $s = 0(1)4$ simultaneously. Aroian's statement on p. 591 that the methods of moments and maximum likelihood are not the same for the fourth-degree exponential rests on a tacit reinterpretation of 'method of moments': in the classical papers (K. Pearson 1902 z, pp. 267-270; R.A. Fisher 1922, p. 321) it is clear that

[415]

10.23
10.25
[see also p. 720]

as many moments, <u>and no more</u>, are to be used as there are disposable parameters in the curve to be fitted; Aroian 1948, p. 590, has fitted <u>four</u> parameters to <u>six</u> moments.

O'Toole, A. L. 1933

<u>Annals of Mathematical Statistics</u> 4, 1-29

The table on p. 27 gives $y = 2[\Gamma(1/4)]^{-1} e^{-x^4}$, $x^2 y$ and $x^4 y$ to 7 dec. for $x = 0(.1)2$. The table on p. 29 gives $y = \exp[\tfrac{1}{2}x^2 - x^4]/2.187099$, $x^2 y$ and $x^4 y$ to 7 dec. for $x = 0(.1)2$.

The table on p. 11 gives $I = \int_{-\infty}^{+\infty} e^{2bx^2 - x^4} dx$ to 6 dec. for $b = 0(.1)1$. For the expression of this integral in terms of Bessel functions see p. 8.

O'Toole 1933 is devoted to the descriptive and analytical properties of the curves (10.23.1): for a method of fitting, see O'Toole 1933a. O'Toole's method is a 'method of moments' only in the sense deprecated in our discussion of Aroian 1948, above; in fact O'Toole fits <u>four</u> parameters to <u>seven</u> moments. His numerical example (O'Toole 1933a, pp. 90-91) is not particularly happy: the curve reproduces the two modes of the data, but conspicuously misses both the range and the abscissas of the modes.

10.24 <u>Cumulative distribution functions</u>

Burr, I. W. 1942

<u>Annals of Mathematical Statistics</u> 13, 215-232

Burr considers the cumulative distribution

$$\Pr[t \leq x] = F(x) = 1 - (1+x^c)^{-k}, \quad c, k > 0.$$

Tables 2, 3, pp. 225-226, give the mean and standard deviation of this distribution to 5 dec., $\beta_1^{\frac{1}{2}}$ to 3 dec., and β_2 to 4 fig. for $c = 1(1)10$, $k = 1(1)11$; these quantities do not exist for ck less than 1, 2, 3, 4.

Kendall & Stuart 1958, p. 174, fail to notice that Burr gives formulas for reducing his <u>cumulative moments</u> to ordinary moments.

Hatke, Mary Agnes

<u>Annals of Mathematical Statistics</u> 20, 461-463

To the diagram of Craig 1936z, p. 28, Hatke adds a stippled area indicating the range covered by Burr's $F(x)$ with $1\tfrac{1}{2} \leq c \leq 20$ and $1 \leq k \leq 20$.

Curve-fitters interested in pursuing Burr's system can purchase from the Brown University Library photocopies of a table by Hatke (which we have not seen) giving moments of Burr's $F(x)$, as in the previous abstract, for $c^{-1} = .05(.025).675$ and $k^{-1} = .05(.025)1$.

10.25 Bessel-function distributions

For other statistical distributions involving Bessel functions, see the subject index at <u>Bessel functions</u>. The multiple occurrence of Bessel functions in statistics was amusingly noted by Pearson (K. Pearson, Jeffery & Elderton 1929, p. 193):

"It is interesting to note how, as statistical theory advances, even on the basis of simple normal distributions, we call into requisition more and more functions familiar in physics."

Mc Kay, A. T. 1932

<u>Biometrika</u> 24, 39-44

Mc Kay considers the distributions $y = y_0 e^{-cx/b} x^m I_m(x/b)$, $c > 1$, $0 \leq x < \infty$; and $y = y_0 e^{-cx/b} |x|^m K_m(|x/b|)$, $|c| < 1$, $-\infty < x < +\infty$; in both distributions $b > 0$ and $m > -\frac{1}{2}$. The determination of c, b and m from $\bar{x}$, σ^2 and $\delta = (\beta_2 - 3)/\beta_1$ leads to a cubic equation for c^2, viz:

$$2c^2(c^2+3)^2 \delta = 3(c^2+1)(c^4+6c^2+1).$$

Table 1, p. 43, gives the three positive roots of this equation to 5 fig. for $\delta = 1.5(.002)1.576, 1.57735$; T. 2, p. 44, gives the single positive root to 5 fig. for $\delta = 1.58(.02)3$.

Pearson, K., Stouffer, S. A. and David, F. N. 1932

<u>Biometrika</u> 24, 293-350

The (symmetrical) $T_m(x)$ distribution (K. Pearson, Jeffery & Elderton 1929) is defined by $T_m(x) = \pi^{-\frac{1}{2}} 2^{-m} |x|^m K_m(|x|) / (m-\frac{1}{2})!$ where K_m is the Bessel function of the third kind of pure imaginary argument.

Table 1, pp. 344-346, gives $S_m(x) = \int_0^x T_m(\xi) \, d\xi$ to 6 dec. for

$$\begin{cases} m = 0(\tfrac{1}{2})5\tfrac{1}{2} & 6(\tfrac{1}{2})11\tfrac{1}{2} \\ x = 0(.1)4(.5)18 & 0(.5)18 \end{cases}.$$

10.26
10.28   [see also pp. 720-722]

10.26   The rectangular distribution and its convolutions

10.261   The continuous rectangular distribution

The distribution of the mean of samples from a continuous rectangular population is due to Lagrange 1774, pp. 225-226 (1868, pp. 227-228). For the use of this distribution as the leading term in series expansions, see Mulholland 1951. For certain statistics in samples from a rectangular population, see sections 6.4 and 8.21.

Bhate, D. H.   1951
Calcutta Statistical Association, Bulletin 3, 172-173

Bhate gives upper 5%, 2½% and 1% points of the mean in samples of n from a continuous rectangular population with range [0,1] to 4 dec. for $n = 1(1)16$. For $n > 16$ the normal approximation is stated to yield 4 dec.

10.262   The discrete rectangular distribution

The distribution of the mean of samples from a discrete rectangular population is likewise due to Lagrange 1774, pp. 210-213 (1868, pp. 212-216); Lagrange showed, moreover, that results for the continuous rectangular distribution can be obtained from those for the discrete by a limiting process.

Rider 1929 gives a number of tables concerned with the distribution of Student's t in samples of 3 and 4 from rectangular populations with 5 and 10 classes.

Whitfield, J. W.   1953
British Journal of Statistical Psychology 6, 35-40

Let x_i, $i = 1, 2, \ldots, m$, denote a random sample from the uniform distribution on the integers 1 to n. Whitfield tables, on pp. 38-40, the lower half of the (symmetrical) cumulative distribution of $T = x_1 + \ldots + x_m$ to 5 dec. or 1 fig. for $m = 2(1)8$ and $n = 3(1)8$.

10.28   The circular normal distribution

The tables below relate to the density function (Kendall & Buckland 1957, p. 42)

$$f(\theta) = e^{k\cos(\theta-\theta_0)} / I_0(k), \quad 0 \leq \theta \leq 2\pi, \qquad (10.28.1)$$

[418]

where I_0 is a Bessel function of the first kind of imaginary argument; to its integral; to the estimation of k; and to testing the hypothesis of uniform circular density $f_0(\theta) = 1/2\pi$, when (10.28.1) is an express or at least a plausible alternative. For detailed methods see the papers cited.

von Mises, R. 1918

<u>Physikalische Zeitschrift</u> 19, 490-500

Table 3, p. 494, gives $1/2\ I_0(k)$ to 4 fig., and $I_1(k)/I_0(k)$ and and $kI_1(k)/I_0(k)$ to 4 dec., for $k = 1(1)6$. Table 7, p. 499, computed by S. Weich, gives $2\int_0^x f(\theta)\,d\theta$ to 4 dec. for $k = 0(.5)3$ and $x = 0(15°)180°$.

Gumbel, E. J., Greenwood, J. A. and Durand, D. 1953

<u>Journal</u> <u>of</u> <u>the</u> <u>American</u> <u>Statistical</u> <u>Association</u> 48, 131-152

Table 2, p. 140, gives k as the root of $I_1(k) = aI_0(k)$ to 5 dec. for $a = 0(.01).87$ with δ_m^2. Table 3, p. 146, gives $[I_0(k)]^{-\frac{1}{2}}e^{\frac{1}{2}k\cos x}$ to 3 dec. for $k = 0(.1)4$ and $x = 0(10°)180°$. Table 4, pp. 147-150, gives $2\int_0^x f(\theta)\,d\theta$ to 5 dec. for $k = 0(.2)4$ and $x = 0(5°)180°$, with δ^2 in both arguments.

Gumbel, E. J. 1954a

<u>Journal</u> <u>of</u> <u>the</u> <u>American</u> <u>Statistical</u> <u>Association</u> 49, 267-295

Table 1, p. 271, gives $\int_{x-\pi/12}^{x+\pi/12} f(\theta)\,d\theta$ to 5 dec. for $k = 0(.1)4.2$ and $x = 0(30°)180°$.

Greenwood, J. A. and Durand, D. 1955

<u>Annals</u> <u>of</u> <u>Mathematical</u> <u>Statistics</u> 26, 233-246

Table 1, p. 235, gives $P(r,n) = r\int_0^\infty [J_0(x)]^n J_1(rx)\,dx$ to 5 dec. for $\begin{cases} n = 6(1)12 \quad 13(1)24 \\ r = 0(.5)n \quad 0(.5)12(1)n \end{cases}$. Table 2, p. 236, gives 95% and 99% points of r to 3 dec. for $n = 6(1)24$.

10.29 Miscellaneous frequency curves

Soper, H. E., Young, A. W., Cave, B. M., Lee, A. and Pearson, K. 1917
Biometrika 11, 328-413

The tables of mean, mode, standard deviation, β_1 and β_2 of the distribution of the correlation coefficient (T. 1, p. 361; TT. 5-7, p. 372; and footings to T. A, pp. 379-401) permit these distributions, which lie in the Type I and VI areas of the β_1, β_2 plane but are all of finite range, to be used as graduation curves. Soper et al. state:

"We thus found the moment coefficients and from them the values of β_1 and β_2 for the ten values of ρ from 0 to ·9, and for the values of n, 2 to 25, 50, 100 and 400. Diagram I[*] shows that our 270[‡] frequency curves are fairly well distributed over the most frequently occurring portion of the β_1, β_2 plane. Now our view is that the constants β_1, β_2 describe adequately for statistical purposes the bulk of the usual [frequency] distributions. But we have provided tables of the values of the ordinates for the above 270 curves. Hence <u>by interpolation</u> it will now be possible to determine rapidly ordinates[||] which will graduate with reasonable accuracy any frequency distribution whatever quite apart from the idea of sampling normally correlated variates." (p. 341)

Reprinted in Tables for S & B II, TT. 31-32, pp. 182-211: values for n = 200 have been added.

To facilitate such graduations, T. C, pp. 412-413, gives $\mu_1'(r)/\sigma(r)$ and $.05/\sigma(r)$ to 7 dec. for $\rho = 0(.1).9$ and $n = 3(1)25, 50, 100, 400$. Reprinted in Tables for S & B II, T. 34, pp. 219-220.

Reprinted in Tables for S & B II, TT. 31-32, pp. 182-211: values for n = 200 have been added.

Galvani, L. 1931
Metron 9, no. 1, 3-45

The table on pp. 26-27 gives various characteristics of the linear distribution $\Phi(x) = mx - \frac{1}{2}m + 1$, $0 \leq x \leq 1$, for $m = -2(.2)+2$.

---

[*] Diagram I was unfortunately not published.

[‡] Ten of these are degenerate (two-point) distributions: see Soper et al. 1917, pp. 360-361.

[||] The computer interested in pursuing this suggestion will supplement these ordinates by the tables of areas in F.N. David 1938.

10.3 Symmetric functions: moments, cumulants and k-statistics

10.31 Symmetric functions per se

A symmetric function of the n arguments $x_1, \ldots, x_n$ is a function that is unchanged by any permutation of the x_i. The present article is devoted to rational integral symmetric functions, i.e. to polynomials in the x_i. If, upon multiplying all the x_i by b, the symmetric function is multiplied by b^w, we speak of a function of <u>weight</u> w.

Following F.N. David & Kendall 1949, p. 431, we define:

M*. The monomial symmetric functions

$$(p_1^{\pi_1} p_2^{\pi_2} \cdots p_s^{\pi_s}) = \sum (z_1 \cdots z_{\pi_1})^{p_1} (z_{\pi_1+1} \cdots z_{\pi_1+\pi_2})^{p_2} (z_{\pi_1+\cdots+\pi_{s-1}+1} \cdots z_{\pi_1+\cdots+\pi_s})^{p_s} \qquad (10.31.1)$$

where the z_i are a permutation of $\Pi = \pi_1 + \ldots + \pi_s$ of the x_i, subject to the constraint that the subscripts of the x_i in each bracket run in ascending order, and the summation is over the $n!/\pi_1! \ldots \pi_s!(n-\Pi)!$ such permutations.

A. The augmented monomial symmetric functions

$$[p_1^{\pi_1} \cdots p_s^{\pi_s}] = \pi_1! \cdots \pi_s! (p_1^{\pi_1} \cdots p_s^{\pi_s}) = \sum y_1^{p_1} \cdots y_{\pi_1}^{p_1} y_{\pi_1+1}^{p_2} \cdots y_{\pi_1+\pi_2}^{p_2} \cdots y_{\pi_1+\cdots+\pi_{s-1}+1}^{p_s} \cdots y_{\pi_1+\cdots+\pi_s}^{p_s} \qquad (10.31.2)$$

where the y_i are an unrestricted permutation of Π of the x_i, and the summation is over the $n!/(n-\Pi)!$ such permutations.

U. The unitary symmetric functions

$$a_r = (1^r) = \sum (z_1 \cdots z_r), \qquad (10.31.3)$$

there being $\binom{n}{r}$ terms.

S. The power sums

$$s_r = (r) = x_1^r + \ldots + x_n^r. \qquad (10.31.4)$$

H. The homogeneous product sums

$$h_r = \sum (p_1^{\pi_1} p_2^{\pi_2} \cdots p_s^{\pi_s}) \qquad (10.31.5)$$

where the summation is overaall partitions of the number r: i.e. over all monomial symmetric functions such that $p_1 \pi_1 + \ldots + p_s \pi_s = r$. Alternatively, the h_r may be defined by the generating identity

---

*The letters M, U, S, H are introduced by David & Kendall 1951, p. 435. The letter A is consistent with their other abbreviations, but they do not use it.

10.31
10.311

$$\left[1 + \sum_{r=1}^{n} a_r (-t)^r\right]^{-1} = 1 + \sum_{r=1}^{\infty} h_r t^r .$$

Any rational integral symmetric function of weight w can be written as a linear combination of $\left[p_1^{\pi_1} \ldots p_s^{\pi_s}\right]$ or $\left(p_1^{\pi_1} \ldots p_s^{\pi_s}\right)$, where $p_1 \pi_1 + \ldots + p_s \pi_s = w$; and as a linear combination of $b_{q_1}^{x_1} \ldots b_{q_s}^{x_s}$ where $q_1 x_1 + \ldots + q_s x_s = w$, and b_q represents a_q, s_q or h_q throughout.

We illustrate these definitions by writing out a few symmetric functions of weight 4 on 5 letters.

$(2^2) = x_1^2 x_2^2 + x_1^2 x_3^2 + x_1^2 x_4^2 + x_1^2 x_5^2 + x_2^2 x_3^2 + x_2^2 x_4^2 + x_2^2 x_5^2 + x_3^2 x_4^2 + x_3^2 x_5^2 + x_4^2 x_5^2$.

$[2^2] = 2.(2^2)$.

$a_2^2 = (x_1 x_2 + x_1 x_3 + x_1 x_4 + x_1 x_5 + x_2 x_3 + x_2 x_4 + x_2 x_5 + x_3 x_4 + x_3 x_5 + x_4 x_5)^2$.

$(2)^2 = s_2^2 = (x_1^2 + x_2^2 + x_3^2 + x_4^2 + x_5^2)^2$.

$h_2^2 = (x_1^2 + x_1 x_2 + x_1 x_3 + x_1 x_4 + x_1 x_5 + x_2^2 + x_2 x_3 + x_2 x_4 + x_2 x_5 + x_3^2 + x_3 x_4 + x_3 x_5 + x_4^2 + x_4 x_5 + x_5^2)^2$.

10.311 Comprehensive tables of symmetric functions

The tables in this section permit the conversion of any one of M, U, S, H into any other.

David, F. N. and Kendall, M. G. 1949; 1951; 1953; 1955
<u>Biometrika</u> 36, 431-449; 38, 435-462; 40, 427-446; 42, 223-242

David & Kendall introduce (38, 435) the notation BC for a table expressing symmetric functions of type B as linear combinations of symmetric functions of type C. They offer the following tables: each is divided into 11 sub-tables, covering weights w = 2(1)12.

Table 1 (36, 435-449). AS and SA. By inserting and removing factorials, this also yields MS and SM.

Table 2 (38, 439-454). MU and UM. By inserting and removing factorials, this also yields AU and UA.

Table 3 (38, 455-462). HU and UH.

Table 4 (40, 428-446). MH and HM. By inserting and removing factorials, this also yields AH and HA.

Table 5 (42, 224-242). US and SU. By ignoring all printed minus signs, the US table yields HS ; by replacing $(q_1)^{x_1} \ldots (q_s)^{x_s}$ by $(-)^{x_1 + \ldots + x_s + s} (q_1)^{x_1} \ldots (q_s)^{x_s}$, the SU table yields SH: MacMahon 1915, pp. 5-7.

We illustrate each conversion by a simple example of weight 4.

Table 1.4 (36, 435):

AS. $[1^4] = -6(4) + 8(3)(1) + 3(2)^2 - 6(2)(1)^2 + (1)^4$.

SA. $(1)^4 = [4] + 4[31] + 3[2^2] + 6[21^2] + [1^4]$.

MS. $24.(1^4) = -6(4) + 8(3)(1) + 3(2)^2 - 6(2)(1)^2 + (1)^4$.

SM. $(1)^4 = (4) + 4(31) + 6(2^2) + 12(21^2) + 24(1^4)$.

Table 2.4 (38, 439):

MU. $(4) = -4a_4 + 4a_3 a_1 + 2a_2^2 - 4a_2 a_1^2 + a_1^4$.

UM. $a_2 a_1^2 = 12(1^4) + 5(21^2) + 2(2^2) + (31)$.

AU. $[2^2] = 4a_4 - 4a_3 a_1 + 2a_2^2$.   UA. $2a_2 a_1^2 = [1^4] + 5[21^2] + 2[2^2] + 2[31]$.

Table 3.4 (38, 455):

HU. $h_4 = -a_4 + 2a_3 a_1 + a_2^2 - 3a_2 a_1^2 + a_1^4$.   UH. $a_4 = -h_4 + 2h_3 h_1 + h_2^2 - 3h_2 h_1^2 + h_1^4$.

Table 4.4 (40, 428):

MH. $(2^2) = h_1^4 - 4h_2 h_1^2 + 3h_2^2 + 2h_3 h_1 - 2h_4$.

HM. $h_2^2 = 6(1^4) + 4(21^2) + 3(2^2) + 2(31) + (4)$.

AH. $[2^2] = 2h_1^4 - 8h_2 h_1^2 + 6h_2^2 + 4h_3 h_1 - 4h_4$.

HA. $4h_2^2 = [1^4] + 8[21^2] + 6[2^2] + 8[31] + 4[4]$.

Table 5.4 (42, 224):

SU. $(2)^2 = a_1^4 - 4a_2 a_1^2 + 4a_2^2 + a_4$.   US. $24a^2 = 6(1)^4 - 12(2)(1)^2 + 6(2)^2 + 3(4)$.

SH. $-(4) = h_1^4 - 4h_2 h_1^2 + 2h_2^2 + 4h_3 h_1 - 4h_4$.

HS. $24h_2^2 = 6(1)^4 + 12(2)(1)^2 + 6(2)^2 + 3(4)$.

Roe, Josephine R. 1931

<u>Interfunctional</u> <u>expressibility</u> <u>problems</u> <u>of</u> <u>symmetric</u> <u>functions</u>

Roe's tables cover weights 1(1)10 ; they are grouped on 20 atlas plates. MU: pl. 4, 5. UM: pl. 4, 6. MH: pl. 7, 8. HM: pl. 9, 10, 11. HU and UH: pl. 12, 13. MS: pl. 14, 15, 16. SM: pl. 17, 18. SU and SH: pl. 19, 20. US and HS: pl. 21, 22, 23.

10.311 [see also p. 722]
10.32
Sukhatme, P. V. 1938a

Philosophical Transactions of the Royal Society of London,
Series A 237, 375-409

Sukhatme's tables cover weights 1(1)8; they are scattered through his text. SU and SH: p. 380. US and HS: p. 383. SM: p. 385. UM: p. 389. UH and HU: p. 392. HM: p. 396. MS: p. 399. MU: p. 404. MH: p. 408.

10.312 Other tables of symmetric functions

The lists below are mostly based on F.N. David & Kendall 1951, p. 438.

<u>10.3121</u> MS tables

$w = 12$	Kerawala & Hanafi 1948*	$w = 7(1)9$	Zia-ud-Din 1940‖
$w = 11$	Kerawala & Hanafi 1942	$w = 1(1)8$	Dwyer 1937(30)
$w = 10$	Kerawala & Hanafi 1941‡	$w = 1(1)6$	O'Toole 1931(125)
$w = 9$	Kerawala 1941		

<u>10.3123</u> MU tables

$w = 15$	Decker 1910◊
$w = 14$	Durfee 1887⁰
$w = 13$	MacMahon 1884
$w = 12$	Durfee 1882a (54)
$w = 12$(again!)	Durfee 1882b
$w = 1(1)11$	Faa di Bruno 1876(TT. 1-3, after p. 320), 1881(311)
$w = 1(1)10$	Hirsch 1809, 1827(plates at end of book); Cayley 1858(494), 1889(423); Salmon 1876(307)
$w = 1(1)10$(again!)	Cayley 1885, 1896(264)
$w = 1(1)6$	Salmon 1877(445)
$w = 1(1)6$	MacMahon 1915 (290)

---

*F.N. David & Kendall 1951, p. 435, report 8 errors but do not specify them.

‡For list of 5 errors, see Kerawala & Hanafi 1942, p. 81.

‖Kerawala 1941, p. 51, reports 26 errors but does not specify them.

◊F.N. David & Kendall 1951, p. 438, seem to have confused this work with Shaw 1907.

⁰F.N. David & Kendall 1951, p. 438, cite this table as unpublished.

10.3124 UM tables

w = 12   Durfee 1882 a (46)
w = 12(again!)   Durfee 1882 b
w = 1(1)10   Cayley 1858(494),
                         1889(423)
w = 1(1)6   Mac Mahon 1915 (289)

10.3125 UH and HU tables

w = 1(1)5   E.D. Roe 1904 b (T.3)

10.3126 MH tables

w = 1(1)5   E.D. Roe 1904 b (T.1)

10.3127 HM tables

w = 1(1)5   E.D. Roe 1904 b (T.2)

10.3128 SU and SH tables

w = 1(1)5   E.D. Roe 1904 b (T.5)

10.3129 US and HS tables

w = 1(1)5   E.D. Roe 1904 b (T.6)

10.32 Moments and cumulants

Consider a variate x drawn from a population with cumulative distribution $F(x)$. Postulating the necessary behaviour of $F(x)$ at infinity, we define:

Moments about the origin

$$\mu'_r = \int x^r \, dF(x) = E(x^r) ; \qquad (10.32.1)$$

Moments about the mean

$$\mu_r = \int (x-\mu'_1)^r \, dF(x) = E(x-\mu'_1)^r ; \qquad (10.32.2)$$

Population cumulants

$$\varkappa_r = \text{coefficient of } (it)^r/r! \text{ in } \log_e E(e^{itx}) . \qquad (10.32.3)$$

We note that the expectation of a monomial symmetric function in a sample of n is readily expressed in terms of moments:

$$E[p_1^{\pi_1} \cdots p_s^{\pi_s}] = n! \, \mu'_{p_1}{}^{\pi_1} \cdots \mu'_{p_s}{}^{\pi_s} / (n-\pi)! \qquad (10.32.4)$$

Moments computed about one origin are readily translated to another origin:

$$E(x-b)^r = \sum_{i=0}^{r} \binom{r}{i} (a-b)^{r-i} E(x-a)^i ; \qquad (10.32.5)$$

the most useful cases of this formula are $a = \mu'_1$ and $b = \mu'_1$.

The coefficient of $\varkappa_{p_1}^{\pi_1} \cdots \varkappa_{p_s}^{\pi_s}$ in the expansion of μ'_w equals the coefficient of $[p_1^{\pi_1} \cdots p_s^{\pi_s}]$ in the expansion of $(1)^w$, and can be read immediately from the bottom line of each table in F.N. David & Kendall 1949.

10.32
10.332   [see also p. 722]

The coefficient of $\mu'^{\pi_1}_{p_1}\cdots\mu'^{\pi_s}_{p_s}$ in the expansion of x_w equals the product of the coefficient of $\left[p_1^{\pi_1}\cdots p_s^{\pi_s}\right]$ in the expansion of $(1)^w$ by the coefficient of (w) in the expansion of $\left[p_1^{\pi_1}\cdots p_s^{\pi_s}\right]$, and can be readily obtained from the top and bottom lines of the said tables. Explicit formulas for these two conversions are given (up to w = 10) by M.G. Kendall 1940c, pp. 394-395; 1943, pp. 62-64; M.G. Kendall & Stuart 1958, pp. 69-71.

When moments about the mean are given or required, the formulas are materially simplified: replace μ'_r by μ_r throughout, and cancel all terms containing μ'_1 or x_1 as a factor. These simpler formulas are given, for w = 2(1)10, by Kendall 1943, pp. 63-64; Kendall & Stuart 1958, pp. 70-71. In particular

$$x_2 = \mu_2, \quad x_3 = \mu_3, \quad x_4 = \mu_4 - 3\mu_2^2. \tag{10.32.6}$$

10.33 Moments from grouped distributions

10.331 Sheppard's corrections

Let a distribution be defined by its density function f(x), so that 10.32.1 becomes

$$\mu'_r = \int x^r f(x)\,dx. \tag{10.33.1}$$

Group the distribution into intervals of length h:

$$f_t = \int_{-\frac{1}{2}h}^{+\frac{1}{2}h} f(x_0 + th + \xi)\,d\xi \;;$$

and compute the grouped moments

$$\bar{\mu}'_r = \sum_t (x_0 + th)^r f_t. \tag{10.33.2}$$

Define $\bar{x}_w$ to be the same function of the $\bar{\mu}$'s that x_w is of the μ's. Then

$$\mu'_r = \bar{\mu}'_r + A + B, \tag{10.33.3}$$

where A is a function of moments of order less than r, and B is a periodic function of x_0 with period h. R.A. Fisher 1922, pp. 317-321, expands B in trigonometric series and shows that, for the normal distribution and ordinary sample sizes[*], B is negligible. The explicit form of A is given for r = 1(1)5 by Sheppard 1898a, p. 368; 1907a, p. 453; for r = 2(1)8 by Carver 1924, p. 94.

---

[*] The word <u>billion</u>, which Fisher uses four times on p. 320, denotes 10^{12}; cf. R.A. Fisher 1950, p. 10.320.

[426]

Langdon & Ore 1930 found the Sheppard corrections to cumulants to be simply given by

$$\varkappa_{2s+1} = \bar{\varkappa}_{2s+1} \; ; \quad \varkappa_{2s} = \bar{\varkappa}_{2s} + (-)^s h^{2s} B_s / 2s \; ; \quad (10.33.4)$$

where the Bernoulli numbers are defined in Fletcher et al. 1946, pp. 41-42. In particular, $\varkappa_2 = \bar{\varkappa}_2 - h^2/12$, $\varkappa_4 = \bar{\varkappa}_4 + h^4/120$, $\varkappa_6 = \bar{\varkappa}_6 - h^6/252$, $\varkappa_8 = \bar{\varkappa}_8 + h^8/240$, $\varkappa_{10} = \bar{\varkappa}_{10} - h^{10}/132$, $\varkappa_{12} = \bar{\varkappa}_{12} + 691 h^{12}/32760$.

For the conditions under which Sheppard's corrections are valid see M.G. Kendall 1938a and works there cited.

10.332 Corrections for abruptness

Pairman, Eleanor and Pearson, K. 1919
<u>Biometrika</u> 12, 231-258

Pairman & Pearson derive corrections, as linear functions of the first and last five subrange frequencies, to be applied, in addition to Sheppard's corrections, to reduce $\bar{\mu}'_r$ to μ'_r. Formulas on pp. 233-239, applied successively, effect the corrections for $r = 1(1)6$.

Similar formulas (pp. 251-253) for infinite terminal ordinate have been superseded by the tables of Pearse 1928, abstracted below. See also Elderton 1938, pp. 233-236.

Pearse, Gertrude E.
<u>Biometrika</u> 20A, 314-355

Let $y(x)$ be a frequency function on the range $(0, x_p)$, assumed to have $y(0) = +\infty$, and to be of such form that

$$-\int_0^x y(u)\, du = x^q (A + Bx + Cx^2 + Dx^3 + Ex^4), \quad 0 < q < 1, \quad 0 \le x \le 6 \; ;$$

let $n'_i = n_i/N$, $i = 1(1)6$, denote the first six observed frequencies in the subranges $i-1 \le x < i$. The tables facilitate the evaluation of the complete moments

$$\int_0^{x_p} (x-1)^r y(x)\, dx$$

by estimating q from the n'_i and then expressing Euler-Maclaurin correction terms as functions of the estimated q and linear combinations of the n'_i.

Table 1, p. 349, gives the coefficients a_i in

$$n'_6 = \sum_{i=1}^{5} a_i(q) n'_i$$

10.332
10.34

to 5 dec. for q = 0(.01).1(.1)1 . Table 2, pp. 349-351, gives the coefficients for expressing $\int_0^1 x^r y(x)\, dx$, r = 1(1)4 , as linear combinations of the n_i' ; T. 3, pp. 351-353, gives the abruptness coefficients $a_1 \ldots a_5$ (Pairman & Pearson 1919, p. 252; Pearse 1928, pp. 317-318) as linear combinations of the n_i' ; all to 9 dec. for q as in T. 1. Table 4, pp. 354-355, gives K_r , defined by $N\int_0^x y(x)(x-1)^r\, dx = (N-n_1)(\bar{\mu}_r' + A_r) + K_r$, where $\bar{\mu}_r'$ is the grouped moment about x = 1 of the distribution truncated at x = 1 and A_r is the Sheppard's correction applicable to such moment, in the form $K_r = \sum_{i=1}^{5} {}_r\beta_i(q) n_i$; the ${}_r\beta_i(q)$ are tabled to 9 dec. for r = 1(1)4 and q as in T. 1.

Reprinted in Tables for S & B II, TT. 38-41, pp. 225-231.

Martin, E. S. 1934

Biometrika 26, 12-58

Martin offers corrections to moments when there is a finite ordinate at the terminal and the frequency n_1 is believed to stand on a base of width $\Lambda < 1$, the remaining frequencies standing on bases of width 1 . Λ must be found by trial: either by equating n_6 (cf. abstract of Pearse 1928, above) or by tentatively fitting an exponential curve to the n_i , i = 1(1)s , s = 3(1)7 . The computations are then similar to those of Pearse 1928. Martin gives six tables facilitating the method.

[The usefulness of corrections for abruptness has been obscured by considerable controversy and polemic. A correct estimate can perhaps be obtained by reading the papers cited by Kendall 1946a, p. 45, "The inefficiency of moments", in chronological order, and reworking the numerical examples.]

10.333   Corrections for grouping of discrete distributions

Carver, H. C. 1933

Annals of Mathematical Statistics 4, 229-239

When each k consecutive ordinates of a discrete frequency distribution are grouped into one cell for computing moments, Sheppard's corrections, which are based on the Euler-Maclaurin summation formula, must be replaced by corrections based on Woolhouse's summation formula. The correction to the variance is $\mu_2 = \bar{\mu}_2 - (1-k^{-2})h^2/12$. In correcting δ and $\beta_1^{\frac{1}{2}}$ for grouping, $\omega = (\mu_2/\bar{\mu}_2)^{\frac{1}{2}}$ and ω^{-3} are required.

Tables 3 and 4, pp. 236-239, give ω to 4 dec. and ω^{-3} to 3 dec. for $h/\delta = .01(.01)1$ and $k = 2(1)10, \infty$.

10.34 k-statistics: infinite population

Given a sample of n from an infinite population, the easiest sample symmetric functions to compute are the power sums and moments about the origin

$$s_r = \sum x_i^r, \quad m_r' = s_r/n. \qquad (10.34.1)$$

Moments about the sample mean $\bar{x}$ may then be computed by (10.32.5); but they are <u>biased</u> estimates of the corresponding population moments.

Repeated use of (10.32.4), together with the tables of F.N. David & Kendall 1949, will permit us to write down an unbiased estimate of any polynomial function of population moments, the sampling moments or cumulants of such estimate, and an unbiased estimate of such moments or cumulants; so long as the weight of the expressions does not exceed 12. The algebra involved is rather heavy, and can be materially lightened (for those who wish to make these transformations routinely) by the combinatorial devices introduced by R.A. Fisher 1929 b.

Fisher 1929 b, p. 203, defines k_r to be the symmetric function whose expectation is $\varkappa_r$, viz the function obtained by expanding $\varkappa_r$ in power products of μ_s' and replacing these by monomial symmetric functions in accordance with (10.32.4). Tukey 1950 introduces k_{rst}, the symmetric function whose expectation is $\varkappa_r \varkappa_s \varkappa_t$; to avoid confusion with multivariate k-statistics we follow Kendall & Stuart 1958, p. 303, and denote this symmetric function by l_{rst}.

Fisher's methods lead to simple expressions for $\varkappa(p^\pi)$, viz the πth cumulant of the sampling distribution of k_p, and for $\varkappa(\rho_1^{\pi_1}...\rho_s^{\pi_s})$, viz $\varkappa_{\pi_1...\pi_s}$ of the joint sampling distribution of $k_{p_1},...,k_{p_s}$; for the definitions of multivariate cumulants see section 10.36 below.

10.34
10.36
[see also p. 723]

The combinatorial rules necessary to derive these cumulants and co-cumulants were given by R.A. Fisher 1929 b, pp. 219-226; Kendall & Stuart 1958, pp. 282-318, give an elaborate exposition based on M.G. Kendall 1940 b.

10.341 k_r in terms of power sums

r = 9, 10   Zia-ud-Din 1954

r = 1(1)8   M.G. Kendall 1943(259), Kendall & Stuart 1958(280)

r = 1(1)6   R.A. Fisher 1929 b (203)

10.342 Formulas for *l*-statistics

Wishart 1952, pp. 5-7, expresses power products of *k*- and *l*-statistics in terms of single *k*- and *l*-statistics for weights 2(1)6. Wishart 1952, pp. 10-12, does the same for weights 7 and 8.

Wishart 1952, pp. 2-3, expresses *k*- and *l*-statistics in terms of augmented monomial symmetric functions, and vice versa, for weights 1(1)6. Abdel-Aty 1954 does the same for weight 12, together with rules for writing down formulas of lower weight.

10.343 Sampling cumulants of *k*-statistics

For an arbitrary population the most extensive table is that given by R.A. Fisher 1929 b, pp. 210-214, which contains all formulas of weight 4(1)10, and $\varkappa(2^6)$, $\varkappa(3^4)$, $\varkappa(4^3)$, $\varkappa(6^2)$. Reprinted with corrections in M.G. Kendall 1943, pp. 268-273; R.A. Fisher 1950, pp. *20*.210-*20*.214; Kendall & Stuart 1958, pp. 290-295.

For a normal population, Wishart 1930, pp. 232-235, gives general formulas for $\varkappa(p^q 2^r)$ and $\varkappa(2^r)$, and specific formulas of weight up to 20 obtained by applying these to Fisher's $\varkappa(p^q)$; also $\varkappa(3^6)$, $\varkappa(4^4)$, $\varkappa(4^4 2)$, $\varkappa(4^4 2^2)$, and an approximate value for $\varkappa(4^5)$. Reprinted in M.G. Kendall 1943, p. 274; Kendall & Stuart 1958, pp. 296-297.

10.349 Moments of moments

Dwyer 1937, 1938 b gives a number of formulas up to weight 8. See also works cited by Dwyer 1937, p. 65; 1938 a, pp. 46-47; 1938 b, pp. 131-132.

10.35 *k*-statistics: finite population

See Irwin & Kendall 1944; Tukey 1950; Wishart 1952; Kendall 1952. The principal mathematical tools are the finite analogue of (10.32.4),

$$\binom{N}{\Pi} E\left[p_1^{\pi_1}\cdots p_s^{\pi_s}\right] = \binom{n}{\Pi}\left[p_1^{\pi_1}\cdots p_s^{\pi_s}\right] \qquad (10.35.1)$$

where the augmented symmetric function on the left is computed over permutations of Π out of n letters, and on the right over Π out of N; and the *l*-statistics defined on p. 429.

10.36 Multivariate symmetric functions and moments

In essence, all the concepts of sections 10.31–10.35 are susceptible of multivariate generalizations; we confine ourselves to generalizations that have led to some published work. Given an s-variate distribution, we generalize (10.32.1)–(10.32.3) by defining

$$\mu'_{r_1\cdots r_s} = E\left[x_1^{r_1}\cdots x_s^{r_s}\right], \qquad (10.36.1)$$

$$\mu_{r_1\cdots r_s} = E\left[(x_1-\bar{x}_1)^{r_1}\cdots(x_s-\bar{x}_s)^{r_s}\right], \qquad (10.36.2)$$

$$\kappa_{r_1\cdots r_s} = \text{coefficient of } (it_1)^{r_1}\cdots(it_s)^{r_s}/r_1!\cdots r_s!$$
$$\text{in } \log_e E[\exp(it_1 x_1 + \cdots + it_s x_s)]. \qquad (10.36.3)$$

Cook 1951, pp. 181–183, expands bivariate μ' in terms of κs and vice versa, and bivariate κ in terms of μs and vice versa, all for $r_1 + r_2 \leq 6$; the μ–κ and κ–μ formulas are reproduced in Kendall & Stuart 1958, p. 84.

For bivariate Sheppard corrections see Wold 1934, pp. 252–253, who gives $\mu'_{rs} - \mu'_{rs}$ for $r + s = 0(1)6$, $r \geq s$. Wold shows (pp. 254–255) that Sheppard corrections in any number of variables may be obtained by applying (10.33.4) to the marginal cumulants; the corrections to all co-cumulants vanish.

$k_{r_1\cdots r_s}$ is defined as the s-variate symmetric function whose expectation is $\kappa_{r_1\cdots r_s}$; for specific bivariate formulas up to weight 4, see Kendall & Stuart 1958, p. 308. Kendall & Stuart 1958, pp. 311, 318–322, give a summary, based on Kendall 1940c, Cook 1951 and

10.36
10.4   [see also p. 724]

Kaplan 1952, of results on sampling co-cumulants of multivariate k-statistics. See also Pepper 1929.

10.37 Gini's generalized means*

Gini 1938 introduces ten kinds of means, which can all be derived by specializing c, d, p, q in the media biplana combinatoria potenziata (biplane combinatorial power mean)

$$B^{cp}_{dq} = \left[\frac{c!\,(n-c)!\,(p^c)}{d!\,(n-d)!\,(q^d)} \right]^{1/(pc-qd)},$$

where (p^c) and (q^d) are monomial symmetric functions if p and q are integers, and are defined by (10.31.1) even if p and q are not integers. This mean is shown to contain practically all well-known mean values, in particular the arithmetic mean, geometric mean, and root mean square, as special cases.

Short tables, illustrating the effect of changes in c, d, p, q for fixed xs, are given by Gini 1938, pp. 13-22; Zappa 1940, pp. 51-52; Gini 1949, pp. 22-28.

10.39 Miscellaneous symmetric functions

Dressel, P. L.   1940
Annals of Mathematical Statistics 11, 33-57

By using the general definition of algebraic semi-invariants (p. 34) and certain general principles of constructing semi-invariants, Dressel finds new statistical semi-invariants.

Table 1, p. 40, gives a system of linearly independent semi-invariants of weight up to 8 in terms of moments about the mean. Table 2, p. 45, expresses unbiased estimates of population cumulants and their power products up to weight 8 in terms of power products of raw sample moments. Table 3, p. 51, expresses the same estimates in terms of the semi-invariants of T. 1.

---

*For Gini's mean difference, see p. 80.

10.4 Čebyšev's inequality and related inequalities

The original statements of the Bienaymé-Čebyšev inequality are Bienaymé 1853, pp. 320-321 (1867, pp. 171-172); Čebyšev 1867a, pp. 1-6 (1947, pp. 431-434); 1867b, pp. 177-182 (1899, pp. 687-692).

Pearson, K. 1919
<u>Biometrika</u> 12, 284-296

Pearson derives the inequality

$$P = \int_{\mu_1' - \delta\sigma}^{\mu_1' + \delta\sigma} a(x)\, dx > 1 - (\delta\sigma)^{-2i} \mu_{2i},$$

and applies it to the normal distribution, yielding: $P > 1 - \delta^{-2i}(2i-1)!!$. Table 1, p. 285, gives the exact P and the above lower bound to 4-5 dec. for $\begin{cases} \delta = 1.5, & 2 & 3 & 4 \\ i = 1(1)3 & & 1(1)6 & 1(1)10 \end{cases}$.

Tables on pp. 289 and 295 give analogous but fragmentary results for the bivariate normal distribution, using both circular and elliptical contours.

Berge, P. O. 1932
<u>Skandinavisk Aktuarietidskrift</u> 15, 65-77

Given the first four moments of a distribution, the cumulative distribution can be bounded above and below by (rather widely separated) curves. The table on pp. 75-77 gives the lower and upper bounds to 3 dec. for $x = -3(.2)+3$; $\mu_1' = 0$, $\mu_2 = 1$; and $\begin{cases} \mu_3 = 0 & -.5 \\ \mu_4 = 2.2, 3, 4.6 & 3 \end{cases}$.

Camp, B. H. 1948
<u>Annals of Mathematical Statistics</u> 19, 568-574

Let T denote the point $(t_1, \ldots, t_n)$ in n-space. If Q_λ is the set of points for which $f(T) > \lambda$, and $x = x_\lambda$ is the measure of Q_λ, Camp defines the rth contour moment $\hat{\mu}_r = \int_0^{x_0} x^r f\, dx$, and derives the inequality

$1 - P_\delta \leq \hat{\mu}_{2r}[2r/(2r+1)\delta\hat{\sigma}]^{2r} = J$, say: where $P_\delta = \int_Q f\, dT$, $\hat{\sigma} = \hat{\mu}_2^{\frac{1}{2}}$, and $x_\lambda = \delta\hat{\sigma}$.

10.4
10.9
[see also p. 724]

When the t_i are independent normal variables,

$$\hat{\sigma}^{-2r}\hat{\mu}_{2r} = \Gamma(rn+\tfrac{1}{2}n)[\Gamma(\tfrac{1}{2}n)]^{r-1}[\Gamma(3n/2)]^{-r};$$ this is given in T. 1, p. 573, for $r = 1(1)3$ and $n = 1(1)4$. Table 2, p. 573, gives J to 3 dec. for

$$\begin{cases} \delta = 1 & 2 & 3 & 3 \\ n = 1, 2 & 1 & 1 & 2, 3 \\ r = 1, 2 & 1(1)3 & 1(1)5 & 1(1)3 \end{cases}.$$

Moriguti, S. 1952

<u>Annals</u> <u>of</u> <u>Mathematical</u> <u>Statistics</u> 23, 286-289

Let $f(x)$ be a probability density function with mean 0 and finite variance σ^2. Moriguti shows that

$$\sigma^n \int_{-\infty}^{+\infty} [f(x)]^{n+1}\,dx \geq (2n+2)(3n+2)^{-\tfrac{1}{2}n-1} n^{\tfrac{1}{2}n}[B(1+n^{-1},\tfrac{1}{2})]^{-n}.$$

Table 1, p. 287, gives this lower bound to 5 dec. or 4 fig. for $n = 1(1)9$, together with the corresponding exact values of the integral when $f(x)$ is the normal and the rectangular distribution. An asymptotic expression for the lower bound is also tabled, and happens to be nearly equal to the values for the rectangular distribution.

[See also Savage 1953, p. 899; Godwin 1955 and works there cited; Mallows 1956]

10.9 <u>Miscellaneous tables related to frequency curves</u>

Neyman, J. 1926

<u>Biometrika</u> 18, 401-413

Using the approximate parabolic regression of s^2 on $\bar{x}$ derived by K. Pearson 1921a, Neyman obtains the following formulas.

Coefficient of correlation between s^2 and $\bar{x}$:

$$H_1 = |r_{\bar{x},s^2}| = (n-1)^{\tfrac{1}{2}} |\beta_1^{\tfrac{1}{2}}| [(n-1)\beta_2 - n + 3]^{-\tfrac{1}{2}}.$$

Correlation ratio of s^2 on $\bar{x}$: H_2, where

$$[(n-1)\beta_2 - n + 3][2n + \beta_2 - \beta_1 - 3][1 - H_2^2] = 2n^2[\beta_2 - \beta_1 - 1].$$

Table 1, pp. 411-412, gives H_1 to 4 dec. for $n = 10, 50, 100, 1000$; $\beta_1 = 0, .1, .5, 1$; $\beta_2 = 1.5(.5)4.5$. Table 2, p. 412, gives H_2 to 4 dec. for the same arguments. Table 3, p. 413, gives $(H_2^2 - H_1^2)/H_1^2$ to 3 dec. for the same arguments.

Winsten, C. B. 1946

Biometrika 33, 283-295

Winsten derives inequalities for gauging tail areas of any continuous distribution in terms of mean range. These inequalities are expressed in terms of $R_n(y) = 1 - y^n - z^n$, where $z = 1-y$.

Table 1, p. 284, gives $1/R_n(y)$ to 3 dec. for $n = 2(1)10$ and $y = .05(.025).25(.05).45$. Table 2, p. 286, gives
$$2[1 + y^{n+1} - z^{n+1} - (n+1)y(y^n + z^n)]/[(n+1)R_n(y)]$$
to 3 dec. for the same arguments.

Moriguti, S. 1951

Annals of Mathematical Statistics 22, 523-536

Let x_n denote the extreme observation in a random sample of size n from any symmetrical parent with variance σ^2. Then
$$E(x_n)/\sigma \leq n(4n-2)^{-\frac{1}{2}}\left[1 - 1\bigg/\binom{2n-2}{n-1}\right]^{\frac{1}{2}}.$$
Table 1, p. 525, gives the above bound to 4 dec. for $n = 2(1)20$, together with the corresponding values for a normal and a rectangular parent.

$E(x_n)\big/(\operatorname{Var} x_n)^{\frac{1}{2}} \geq 1 - M_n^{-1}$, where
$$M_{m+1} = \int_{\frac{1}{2}}^{1} (y^m - z^m)^2 (y^m + z^m)^{-1} dy$$
and $z = 1-y$. Table 2, p. 528, gives this bound to 4 dec. for $n = 2(1)6$, together with the corresponding values for a normal and a rectangular parent.

SECTION 11

THE FITTING OF REGRESSION AND OTHER CURVES

11.1 The fitting of polynomials

11.11 Orthogonal polynomials*

11.110 Formulas

If $T_r(x)$ is the rth degree polynomial in an orthogonal set, then for any pair T_r, T_s the orthogonality condition

$$\sum_{t=1}^{n} T_r({}_tx) T_s({}_tx) = 0, \quad r \neq s, \tag{11.11.1}$$

must be satisfied. The polynomial ${}_tY$ to be fitted is then written in the form

$$_tY = \sum_{r=0}^{k} A_r T_r({}_tx) \tag{11.11.2}$$

and the least-squares solutions A_r are given by

$$A_r = \sum_{t=1}^{n} {}_tY\, T_r({}_tx) \bigg/ \sum_{t=1}^{n} [T_r({}_tx)]^2. \tag{11.11.3}$$

Explicit forms for the orthogonal polynomials can be written conveniently only when the ${}_tx$ are equally spaced. The orthogonal polynomial of degree r, referred to the n points $z = 0, 1, \ldots, n-1$, can be written in terms of binomial coefficients:

$$T_r = \sum_{s=0}^{r} (-)^s \binom{r}{s} \binom{z+s}{r} \binom{n-1+r-z-s}{r}. \tag{11.11.4}$$

This form is equivalent to formula (12) of Aitken 1933, p. 59: the

---

*The orthogonal polynomials arising in the least-squares fitting of polynomials to discrete points are known as Čebyšev polynomials (Čebyšev 1855; 1858) or Gram polynomials (Gram 1883, p. 44). Čebyšev was a prolific inventor of polynomials: besides the above and the Čebyšev polynomials proper (Čebyšev 1873; 1874 b; Fletcher et al. 1946, p. 341) we may credit him with the Čebyšev-Hermite polynomials (i.e. the Hermite polynomials of section 1.323: Čebyšev 1860; Hermite 1864 a, b) and the polynomials (Čebyšev 1874 a; Whittaker & Robinson 1924, p. 158) whose zeroes are the abscissas of Čebyšev's quadrature formula.

coefficient of z^r in T_r is $(2r)!(r!)^{-3}$. To obtain a symmetrical form of T_r, set $x = z - \frac{1}{2}(n-1)$.

The form of orthogonal polynomial usual in non-numerical work is $\xi_r = T_r(r!)^3/(2r)!$; the usual form in numerical work is ξ'_r, obtained by removing the greatest common factor from the values of T_r for each r and n.

We have seen the following lists of formulas for such polynomials:

r = 0(1)10   Sasuly 1934(271), Birge & Weinberg 1947(307)

    1(1)10   Allan 1930 a (319)

    0(1)9   Van Der Reyden 1943(369)

    0(1)7   Davis 1933 a (161), 1935(319)

    0(1)6   E.S. Pearson & Hartley 1954(212)

    0(1)5   Eisenhart 1944(148)

    1(1)5   Čebyšev 1859(2), 1899(474), 1947(315)

    1(1)4   Aitken 1933(67)

<u>11.111</u> Tables of orthogonal polynomials

The following tables give ξ'_r for the ranges of n and r shown. When $r \geq n$, $\xi'_r = 0$ for tabular x and is therefore not tabled. As a rule, symmetry is used to eliminate printing of almost half the arguments x. The tables also give $\sum \xi'^2_r$ and ξ'_r/ξ_r (often called λ_r).

n = 3(1)104   r = 1(1)5   R.L. Anderson & Houseman 1942

    3(1)45   1(1)5 ⎫
                                ⎬ Fisher & Yates 1948, 1953, 1957(T.XXIII)
    46(1)75   2(1)5 ⎭

    5(1)52   1(1)9   Van Der Reyden 1943(384)

    3(1)52   1(1)6   E.S. Pearson & Hartley 1954(212)

    3(1)52   1(1)5   Fisher & Yates 1938, 1943(T.XXIII)

    3(1)50   1(1)6   Vianelli 1959(624)

    3(1)30   1(1)5   Birge & Shea 1927(111), Birge & Weinberg 1947(341)

    3(1)26   1(1)25   De Lury 1950(18)

    2(1)20   1(1)5   Jordan 1921(322)

    3(1)12   1(1)3   Quenouille 1952(237)

    2(1)6   1(1)5   Snedecor 1946(410)

n = 8 ; 9 ; 10, 11 ; 12   r = 0(1)6 ; 0(1)7 ; 0(1)8 ; 0(1)10   Sasuly 1934(372)

11.112
11.119   [see also p. 724]

11.112 Errors in tables of orthogonal polynomials

Fisher & Yates 1938

$n = 39$, $r = 2$. For 496.388 read 4.496.388. This error is also in Anderson & Houseman 1942; it is corrected in Fisher & Yates 1943 and later editions.

Anderson & Houseman 1942

Sherman 1951 reports the following allegedly complete list of errors:

Page	Line	n	For	Read
615	14	31	broken type	-585
618	22	39	496388	4496388
618	14	40	-3583	-2583
625	2	57	+42481	+32481
629	17	61	-2648	+2648
642	38	74	13505	135050
649	14	81	+701925	+701935
653	40	85	-88686	+88686
663	44	95	+2107	-2107
665	45	97	-110308	+110308
669	24	101	-26593	-26592

The values given for $n = 72$ on page 640, line 36, are really for $n = 71$. The correct values are: 1 24392, 966 52584, 15878 63880, 39 89066 92520, 3436 29622 27080.

11.113 Tables for the summation method

The tables below essentially give the coefficients for expressing T_r or ξ_r as a weighted sum of $\binom{x}{s}$, $s = 0(1)r$. The utility of this representation rests on the fact that $\sum \binom{x}{s} y$ is readily computable, without a multiplying machine, by successive summation: Kendall & Stuart 1958, pp. 63-66. We list the tables by the range of n and r covered.

$n = 7(1)133$ $r = 0(1)10$ Sasuly 1934(274, 360)

   4(1)25   0(1)5   Aitken 1933(61, 69)

   5(1)25   0(1)5   Goulden 1939(242)

Since both multiplying machines and the tables of section 11.111 are widely available, we cannot recommend the summation method for general use. It shows to best advantage in the following circumstances:

(1) Many polynomial fits are required as a matter of routine analysis.

(2) A printing adding machine, or better yet a multi-register machine (Burroughs; National) is available.

(3) Throwing the fitting problem onto an electronic computer is not a practicable alternative.

The statistician interested in using the summation method should read Aitken 1939, pp. 114-120; and then Aitken 1933. Jordan 1932 gives an exceedingly diffuse account of the expansion of orthogonal polynomials in series of binomial coefficients: but, in view of his statement (p. 282)

"We will give . . . formulae leading directly to Newton's expansion, so that the computer will have nothing to do with the orthogonal polynomials themselves. He will only have to compute the binomial moments . . . , and then deduce the mean orthogonal moments . . . , which will give, with appropriate coefficients, the solution in Newton's series."
it seems more appropriate to catalogue Jordan's tables in section 11.13, below.

11.119 Miscellaneous tables of orthogonal polynomials

Hald, A. 1948

The decomposition of a series of observations

Define $x_{\nu i} = (\nu-1)k + i$, $i = 1(1)k$, $\nu = 1(1)n$. Hald introduces a set of orthogonal polynomials of degree r, $\xi_r(x_{\nu i}) = \xi_{r\nu i}$, with the following properties: $\sum\sum(\xi_{r\nu i} - \xi_{r.i})(\xi_{s\nu i} - \xi_{s.i}) = 0$, $r \neq s$, $\xi_{r.i} = n^{-1}\sum_\nu \xi_{r\nu i}$; $\sum\sum \xi_{r\nu i} = 0$.

The table on pp. 112-134 gives the $\xi_{r\nu i}$, normalized so that the coefficient of x^r is unity, exact or to 6 fig. for r = 1(1)5 and
$$\begin{cases} k = 3 & 4 & 5 & 12 \\ n = 5, 7 & 25 & 5, 7(1)11 & 10 \end{cases}.$$
Also tabled are $s_{rr} = \sum\sum(\xi_{r\nu i} - \xi_{r.i})^2$ and $s'_{rs} = \sum\sum \xi_{r\nu i}\xi_{s\nu i}$ exact or to 6 fig.

De Lury, D. B. 1950

Values and integrals of the orthogonal polynomials

Let ξ'_r be the orthogonal polynomial of degree r for the n points $-\frac{1}{2}[n-1](1)+\frac{1}{2}[n-1]$. Define
$$I_r = \int_{-\frac{1}{2}(n-1)}^{+\frac{1}{2}(n-1)} \xi'_r(x)\, dx, \quad J_r = \int_{-\frac{1}{2}n}^{+\frac{1}{2}n} \xi'_r(x)\, dx.$$

11.119
11.19
[see also pp. 724-726]

If r is odd, $I_r = J_r = 0$. De Lury tables I_r and J_r exact or to 5 dec.

for $\begin{cases} n = 3(1)16 & 17(1)26 \\ r = 0(1)[n-1] & 0(1)14 \end{cases}$.

11.12 Power polynomials

The tables below facilitate the fitting of a polynomial
$$\sum_{s=0}^{r} a_s x^s$$
to a series y when the moments of y through the rth order have been computed. The details of the tables are of little interest: the tables are for practical purposes superseded by the orthogonal polynomials of section 11.11. We confine ourselves to cataloguing the range of n and r covered.

$n =$	$r =$	
2(1)201(2)301	1	Davis 1935(326)
2(1)240	1	Davis 1935(370)
2(1)201	1, 2	Davis 1933a(172)
3(1)201	2	Davis 1935(332)
4(1)151	3	Davis 1935(338)
3(1)120	2	Davis 1935(376)
4(1)101	3	Davis 1933a(180)
14(7)70	6	Campbell 1930(756)*
4(1)60	3	Davis 1935(382)
12(6)60	5	Campbell 1930(756)*
5(1)51	4(1)7	Davis 1933a(182), 1935(342)
8(4)40; 10(5)50	3; 4	Campbell 1930(756)*
5(1)31	1, 2	Roeser 1920
1(1)30	1(1)5	Kerawala 1941a(255)
6(3)30	2	Campbell 1930(756)*
5(1)26; 9(1)26	2; 3	Hale 1882(346; 347)
2(1)10; 4(2)20	0; 1	Campbell 1930(756)*

---

*Campbell's tables relate not to the method of least squares but to the method of zero sum (Mayer 1750, pp. 151-154).

11.13 Factorial polynomials

The tables below facilitate the fitting of a factorial polynomial $\sum_{s=0}^{r} a_s \binom{x}{s}$ to a series y when the factorial moments $\sum \binom{x}{s} y$ up to the $(2r)^{th}$ order have been computed. The operations of interpolation and finite summation are more readily performed on factorial polynomials than on power polynomials. Again we confine ourselves to cataloguing ranges.

$n = 7(1)133 \quad r = 0(1)10 \quad$ Sasuly 1934(248, 360)

$ 3(1)100 0(1)7 \quad$ Jordan 1932(336)

11.19 Miscellaneous polynomial fitting

Greenleaf 1932 seeks to minimize

$$\sum_{t=0}^{n} \binom{n}{t} [y(t) - \Pi_r(t)]^2 ,$$

where Π_r is a polynomial of degree r. He expands Π_r in a series of orthogonal polynomials obtained by the Gram-Schmidt process; these polynomials, suitably normalized, tend to Hermite polynomials as $n \to \infty$. He gives tables facilitating the fit for $n = 3(2)41$ and $r = 1(1)6$.

[For the relation of the method of <u>moving averages</u> to the method of least squares, and for tables of coefficients in moving-average graduation formulas, see Whittaker & Robinson 1924, pp. 291-300; Sasuly 1934, pp. 38-58; Kendall 1946a, pp. 372-395; and works there cited. For many additional references, mostly to the work of Erastus L. De Forest, see Wolfenden 1942, pp. 132-137.]

11.2 [see also p. 726]
11.29

11.2 Harmonic analysis: i.e. the fitting of trigonometric polynomials

For an account of the theory and practice of harmonic analysis consult Stumpff 1937; for other references see Fletcher et al. 1946, p. 367. The tables in sections 11.21-11.23 are concerned with facilitating the evaluation of sums of products

$$a_0 + \sum(a_n \cos nx + b_n \sin nx) \qquad (11.2.1)$$

where $x = 2\pi r/s$.

For <u>schedules</u>, i.e. tabular arrangements intended to simplify the work for a specific s, see Fletcher et al. 1946, pp. 369-371: they list schedules for $s = 3(1)40, 42, 45, 48, 52, 60, 72, 73, 121$. Pollak & Egan 1949 a gives schedules for $s = 3(1)24(2)30, 34(2)38, 42(2)46, 52(8)92$.

11.21 Angles: $2\pi r/s$

These tables, especially those expressing $2\pi r/s$ in degrees, minutes and seconds, must be considered obsolete: a desk calculating machine will yield 8 decimals of a degree immediately, and 17 dec. with a little manipulation.

Pollak & Heilfron 1947 gives $2\pi r/s$, reduced to the first quadrant, for $s = 3(1)100$ and all r, to 6 dec. of a degree and 2 dec. of a second; Pollak & Egan 1949 b, which is not self-explanatory but must be used in conjunction with Pollak & Heilfron 1947, gives the same angles to 12 dec. of a degree and 10 dec. of a second. For earlier tables see Fletcher et al. 1946, p. 368.

11.22 $\cos 2\pi r/s$ and $\sin 2\pi r/s$

The effective range of r in the tables in sections 11.22 and 11.23 is $r = 1(1)[s-1]$.

8 dec.	$s = 3(1)100$	Pollak & Egan 1949 b
7 dec.	$3(1)40$	Pollak 1929
5 dec.	$3(1)100$	Pollak & Heilfron 1947
5 dec.	$3(1)24$	Pollak & Egan 1949 a

See also Fletcher et al. 1946, p. 368.

11.23 $a \cos 2\pi r/s$ and $a \sin 2\pi r/s$

These tables are principally useful to the statistician without a calculating machine.

3 dec. $a = 1(1)1000$ $s = 3(1)24$ Pollak 1926

See also Fletcher et al. 1946, p. 369.

11.24 Tests of significance in harmonic analysis

Let $u_1, \ldots, u_n$ be n independent χ^2 variates, each with r degrees of freedom. Cochran 1941, p. 49, obtains the distribution of $g = \max u_i / \sum u_i$, and tables (p. 50) the upper 5% point of g to 4 dec. for $r = 1(1)6(2)10$ and $n = 3(1)10$.

When $r = 2$, each u_i can be taken as proportional to an estimated harmonic amplitude $a_i^2 + b_i^2$: yielding a test of the null hypothesis that the largest such amplitude represents random fluctuation only. Several special tables have been published:

R.A. Fisher 1929 a, p. 59, gives the upper 5% point of g to 5 dec. for $n = 5(5)50$.

R.A. Fisher 1940 a, p. 17, gives the upper 5% and 1% points of g and of g_2 (the second largest ratio) to 5 dec. for $n = 3(1)10(5)50$. The table is due to W. L. Stevens.

R.A. Fisher 1950, p. *16*.59, gives the upper 5% and 1% points of g to 5 dec. for $n = 5(1)50$. Reprinted in Fisher & Yates 1953, 1957, T. VIII 3.

11.29 Periodograms

Brooks, C. E. P. 1924
<u>Proceedings</u> <u>of</u> <u>the</u> <u>Royal</u> <u>Society</u> <u>of</u> <u>London</u>, Series A 105, 346-359

Brooks gives two tables intended to facilitate the fitting of periodograms. Table 1, p. 359, gives $2C/(C+2)$ and $2C/(C-2)$ to 3 dec. for $C = 3(1)30$. Table 2, p. 359, gives $\tfrac{1}{2}x/\sin^3 x$ to 2; 1 dec. for $x/\pi = .2(.01).7; .71(.01).89$.

Bartlett, M. S. 1950
<u>Biometrika</u> 37, 1-16

See abstract in section 13.5, below.

11.3
12.13   [see also p. 728]

11.3 The fitting of exponential curves

Glover 1923, pp. 469-481, tables $n/(1-r^{-n}) - 1/(1-r^{-1})$ to 4 dec. for $n = 2(1)40$ and $r = 1.001(.001)1.1$. The table is intended for use in fitting $y = ar^x$ by the method of moments.

Stevens, W. L. 1951

Biometrics 7, 247-267

Let a, b, r denote the least-squares estimators of the parameters of the curve $y = \alpha + \beta \rho^x$. The six elements of their variance-covariance matrix can be reduced to functions of r alone. For the case of n equidistant x_i, these functions are tabled on pp. 254-256 to 5 dec. for

$$\begin{cases} n = 5, 6 & 7 \\ r = .25(.01).65(.005).7 & .3(.01).7(.005).75 \end{cases}.$$

Keeping, E. S. 1951

Annals of Mathematical Statistics 22, 180-198

Following a method due to Hotelling 1939, Keeping derives a significance test for the fit of the exponential curves

(1) $Y = be^{px}$ (2) $Y = a + be^{px}$

to a sequence of n values of y_i observed for equidistant x_i: a, b, p are estimated by least squares. The criterion for testing the null hypothesis $b = 0$ in (1) is $R' = y_i Y_i \left(y_i^2 Y_i^2 \right)^{-\frac{1}{2}}$, where $Y_i = \hat{b} e^{\hat{p} x_i}$; the criterion in (2) is the product-moment correlation R between y_i and $Y_i = \hat{a} + \hat{b} e^{\hat{p} x_i}$. Keeping obtains approximate expressions for the upper tails of the null distributions of R' and R.

Table 3, p. 194, gives the upper $100 \alpha \%$ points of R' to 3 dec. or 1 fig. for $\begin{cases} \alpha = .05 & .01 & .001 & .0001 \\ n = 3(1)5 & 3(1)7 & 4(1)10 & 4(1)10, 12, 15 \end{cases}$. Table 4, p. 195, gives the same for R for $\begin{cases} \alpha = .05 & .01 & .001 & .0001 \\ n = 4(1)6 & 4(1)8 & 4(1)10 & 4(1)12 \end{cases}$.

SECTION 12

VARIATE TRANSFORMATIONS

12.0 Introduction

If x denotes a statistical variate and g a mathematical function of one variable, then the statistical variate $y = g(x)$ is called a variate transformation of x. We could deal in this section with the whole family of tabled mathematical functions g; we confine ourselves, however, to transformations that have been seriously proposed and tabled for use in statistical methodology.

12.1 The angular transformation

In n binomial trials with probability p, let x be the number of successes. Then the large-sample variance of $\sin^{-1}[(x/n)^{\frac{1}{2}}]$ is approximately independent of p, and is $1/4n$ (if the angle is expressed in radians) or $820.7/n$ (if the angle is expressed in degrees).

12.11 $\sin^{-1} x^{\frac{1}{2}}$: degrees

$0.^{\circ}01$	0(.0001).01(.001).99(.0001)1	No Δ	Bliss 1937 (70), Snedecor 1946 (449), Ferber 1949(516)
$0.^{\circ}1$	0(.01).97(.001).999	No Δ	Finney 1952 b (640)
$0.^{\circ}1$	0(.01).99	No Δ	Fisher & Yates 1938-53(T.XII), 1957(T. X)

12.12 $\sin^{-1}[(x/n)^{\frac{1}{2}}]$: degrees

$0.^{\circ}1$	n.= 2(1)30 x = 1(1)[$\frac{1}{2}$n]		Fisher & Yates 1938-53(T.XIII), 1957(T.X1)

12.13 $\sin^{-1} x^{\frac{1}{2}}$: radians

4 dec.	$2 \sin^{-1} x^{\frac{1}{2}}$	0(.001)1	No Δ	Hald 1952(T.12)
4 dec.	$2 \sin^{-1} x^{\frac{1}{2}} - \frac{1}{2}\pi$	0(.01)1	No Δ	Kelley 1947(596)
4 dec.	$\sin^{-1} x^{\frac{1}{2}}$	0(.01).5	No Δ	Bartlett 1937 a (132)
3 dec.	$\sin^{-1} x^{\frac{1}{2}}$	0(.01)1	Δ	Lindley & Miller 1953(14)

12.14
12.15

12.14 Working angles

The **weight** used with the method of working angles is a constant: in other respects the computations parallel those for working probits, pp. 22-24 above.

Finney, D. J. 1952 b

Statistical method in biological assay

Table 12, pp. 641-642, gives the minimum working angle $\frac{180}{\pi}(y - \frac{1}{2}\tan y)$; the range $\frac{180}{\pi \sin 2y}$; the maximum working angle $\frac{180}{\pi}(y + \frac{1}{2}\cot y)$; to 1 dec. for $\frac{180\,y}{\pi} = 1(1)65$; $1(1)89$; $25(1)89$. Table 13, pp. 643-644, gives the working angle $\frac{180}{\pi}\left(y - \frac{1}{2}\tan y + \frac{s}{\sin 2y}\right)$ to 1 dec. for $\frac{180\,y}{\pi} = 1(1)10(2)80(1)89$ and $s = 0(.05)1$, omitting values outside the range 0 to 90.

Fisher, R. A. and Yates, F. 1938, 1943, 1948, 1953

Statistical tables

Table XIV gives the maximum working angle and the range to 1 dec. for $\frac{180\,y}{\pi} = 45(1)89$.

Fisher, R. A. and Yates, F. 1957

Statistical tables

Table X2 gives the minimum and maximum working angles and the range to 3 dec. for $\frac{180\,y}{\pi} = 1(1)89$.

12.15 Special tables of the angular transformation

Anscombe, F. J. 1948

Biometrika 35, 246-254

If the variate x has the binomial distribution with parameters n and p, and $y = \sin^{-1}[(8x+3)/(8n+6)]^{\frac{1}{2}}$, then T. 1, p. 253, gives $(n+3/4)\sin^2 E(y) - 3/8$ to 3 dec., $(4n+2)\mathrm{var}\,y$ to 3 dec., $[\beta_1(y)]^{\frac{1}{2}}$ and $\beta_2(y) - 3$ to 2 dec., $[r(x,y)]^2$ to 3 dec.; all for $\begin{cases} n = 5 & 10 \\ p = .2,\ .4 & .1(.1).4 \end{cases}$.

12.14
12.15

Finney, D. J. 1952 b

Statistical method in biological assay

Table 14, p. 645, gives $5 + x(\sin^2 y)$, where $x(P)$ is the standard normal deviate of section 1.21, to 2 dec. for $y = 1°(1°)89°$. Table 15, p. 645, gives $5 + \log_e \tan y$ to 2 dec. for $y = 1°(1°)89°$. These tables are intended to convert angles into probits and logits.

Stevens, W. L. 1953

Biometrika 40, 70-73

Stevens writes the angular transformation in the form
$\theta = 50 - \pi(250)^{\frac{1}{2}} + (4000)^{\frac{1}{2}} \sin^{-1} p^{\frac{1}{2}}$: $\theta(0) = 0.327$, $\theta(\frac{1}{2}) = 50$, $\theta(1) = 99.673$, $\theta(p) + \theta(1-p) = 100$. In binomial trials, $\theta(x/n)$ has asymptotic variance $1000/n$.

Tables 1-2, pp. 71-72, give θ to 3 dec. for $p = 0(.0001).02(.001).499$ with PPMD. Table 3, p. 72, gives $\theta(x/n)$ to 3 dec. for $n = 2(1)30$ and $x = 1(1)[\frac{1}{2}n]$.

12.2
12.4

12.2 The square-root transformation

If x is a Poisson variate, and $E(x)$ is not too small, then $2x^{\frac{1}{2}}$ has approximately unit variance. For small expectations the approximation is not good: $(4x+2)^{\frac{1}{2}}$, $(4x+1\frac{1}{2})^{\frac{1}{2}}$ and $x^{\frac{1}{2}} + (x+1)^{\frac{1}{2}}$ have been proposed as improvements.

Bartlett, M. S. 1936

Supplement to the Journal of the Royal Statistical Society 3, 68-78

"The square root transformation in analysis of variance"

Table 1, p. 69, gives the variances of $x^{\frac{1}{2}}$ and $(x+\frac{1}{2})^{\frac{1}{2}}$, where x is a Poisson variate with mean m; and the variance of $y^{\frac{1}{2}}$, where $2y$ is distributed as χ^2 with $2m$ d.f.; to 3 dec. for $m = 0(.5)1(1)4, 6(3)15$.

Irwin, J. O. 1943

Journal of the Royal Statistical Society 106, 143-144

The table on p. 144 gives var $x^{\frac{1}{2}}$, where x is a Poisson variate with mean m, to 4 dec. for $m = 0(.1)15, 20, 50, 100$.

Anscombe, F. J. 1948

Biometrika 35, 246-254

If x is a Poisson variate with mean m, and $y = (x + 3/8)^{\frac{1}{2}}$, then T. 1, p. 253, gives $[E(y)]^2 - 3/8$ to 3 dec., 4 var y to 3 dec., $[\beta_1(y)]^{\frac{1}{2}}$ and $\beta_2(y) - 3$ to 2 dec., $[r(x,y)]^2$ to 3 dec.; all for $m = 1(1)4, 6, 10, 20, \infty$.

Vianelli, S. 1959

Prontuari per calcoli statistici

Tavola 53, p. 1185, gives the variances of $x^{\frac{1}{2}}$ and $(x+\frac{1}{2})^{\frac{1}{2}}$, where x is a Poisson variate with mean m, to 3 dec. for $m = .5(.5)1(1)4, 6(3)15$.

See also Masuyama 1951, 1951 z.

12.3 The inverse hyperbolic-sine transformation

The arguments that lead to the angular transformation of binomial data, and the square-root transformation of Poisson data, yield the transformation $y = \sinh^{-1}(\gamma x)^{\frac{1}{2}}$ for stabilizing the variance of the negative binomial distribution with fixed exponent $1/\gamma$ and expectation not too small. In practice γ is not known a priori; but the transformation may still be hoped to make variance more homogeneous and cut down skewness.

As $\gamma \to 0$, the $\sinh^{-1}$ transformation approaches the square-root transformation.

12.31 $\sinh^{-1} x^{\frac{1}{2}}$

3 dec. $x = 0(.01)1(.1)10(1)200(10)500$ No Δ Anscombe 1950 a

3 dec. $0(.01)1(.1)10(1)100$ Δ Lindley & Miller 1953(14)

12.32 $\gamma^{-\frac{1}{2}} \sinh^{-1}(\gamma x)^{\frac{1}{2}}$

2 dec. $\begin{cases} \gamma = 0(.02).1(.05).3(.1).6(.2)1 \\ x = 0(1)50(5)100(10)300 \end{cases}$ Beall 1942(250)

12.33 Special tables of the $\sinh^{-1}$ transformation

Anscombe, F. J. 1948

Biometrika 35, 246-254

If the variate x has the negative binomial distribution with mean m and exponent k, viz $(x+k) m^x k^{-k} (m+k)^{-x+k}/x! \Gamma(k)$, and $y = \sinh^{-1}[(8x+3)/(8k-6)]^{\frac{1}{2}}$, then T. 1, p. 253, gives $(k-3/4)\sinh^2 E(y) - 3/8$ to 3 dec., $4 \operatorname{var} y/\psi'(k-1)$ to 3 dec., $[\beta_1(y)]^{\frac{1}{2}}$ and $\beta_2(y) - 3$ to 2 dec., $[r(x,y)]^2$ to 3 dec.; all for $k = 2, 5$, and $m = 1(1)4, 6, 10, 20$; where ψ' is the trigamma function of section 2.5142.

Table 2, p. 254, gives $4 \operatorname{var} y_2/\psi'(k-1)$ and $4 \operatorname{var} y_3/\psi'(k-1)$ for $k = 2$ and m as in T. 1; and $\operatorname{var} y_4/\psi'(k-1)$ for k and m as in T. 1; where $y_2 = \sinh^{-1}(x/k)^{\frac{1}{2}}$, $y_3 = \sinh^{-1}[(x+\frac{1}{2})/(k-1)]^{\frac{1}{2}}$, $y_4 = \log_e(x+\frac{1}{2}k)$.

12.4 The probit transformation

See pp. 22-30. For the conversion of angles into probits, see the abstract of Finney 1952 b in section 12.15.

12.5 [see also p. 728]
12.54

12.5 The logit transformation

Basically the transformation is $x = \log_e[p/(1-p)]$, but there are minor variants. It may be noted that $x = 2\tanh^{-1}(2p-1)$; thus the tables of section 7.21 can be pressed into service as logit tables and conversely.

12.51 $x = \log_e[p/(1-p)]$

5 dec.	x	$p = .001(.001).999$	No Δ	Berkson 1953 (568)
4 dec.	$\tfrac{1}{2}x$	$.5(.001).999$	No Δ	Fisher & Yates 1957 (T.XI)
2 dec.	$5 + \tfrac{1}{2}x$	$.01(.01).97(.001).999$	No Δ	Finney 1952 b (636)

12.52 $p = (1 + e^{-x})^{-1}$

5 dec. $x = 0(.01)4.99$ No Δ Berkson 1953 (570), 1957 a (33)

12.53 Working logits and weights

The computations parallel those for working probits, pp. 22-24 above. We adopt the customary abbreviation $q = 1-p$.

Finney, D. J. 1952 b

Statistical method in biological assay

Table 9, p. 637, gives the minimum working logit $5 + \tfrac{1}{2}(x - 1/q)$ to 4 dec. for $5 + \tfrac{1}{2}x = 1.1(.1)6.5$; the range $1/2pq$ to 5 fig. and the weight $4pq$ to 5 dec. for $5 + \tfrac{1}{2}x = 1.1(.1)8.9$; the maximum working logit $5 + \tfrac{1}{2}(x + 1/p)$ to 4 dec. for $5 + \tfrac{1}{2}x = 3.5(.1)8.9$.

Table 10, pp. 338-639, gives the working logit $5 + \tfrac{1}{2}(x - 1/q + s/pq)$ to 2 dec. for $5 + \tfrac{1}{2}x = 2(.1)3(.2)7(.1)8$ and $s = 0(.05)1$, omitting values outside the range 0 to 10.

Berkson, J. 1953

Journal of the American Statistical Association 48, 565-599

Table 3, pp. 571-572, gives pq and pqx to 4 dec. for $p = 0(.001)1$.

Fisher, R. A. and Yates, F. 1957

Statistical tables

Table XI 1 gives the minimum working logit $\tfrac{1}{2}(x - 1/q)$, the range $1/2pq$, the maximum working logit $\tfrac{1}{2}(x + 1/p)$ and the weight $4pq$; all to 5 fig. for $\tfrac{1}{2}x = 0(.05)3.95$.

Berkson, J. 1957 a

Biometrics 13, 28-34

Table 2, pp. 33-34, gives p and pq to 5 dec. for $x = 0(.01)4.99$.
Reprinted in Vianelli 1959, Prontuario 148, pp. 396-397.

12.54 Special tables of the logit transformation

Finney, D. J. 1952 b

Statistical method in biological assay

Table 15, p. 645, gives $5 + \log_e \tan y$ to 2 dec. for $y = 1°(1°)89°$. The table is intended to convert angles into logits.

12.6
12.7

12.6 The loglog transformation

12.61 Tables of loglogs

For a table of $-\log_e(-\log_e x)$, and a note on Mather 1949, see section 8.3121, p. 364. No differences are given with the tables below.

4 dec. $\log_e[-\log_e(1-p)]$ $p = .002(.002).998$ Fisher & Yates 1957(T.XII)

2 dec. $\log_e[-\log_e p]$ $.001(.001).01(.01)$ Finney 1952 b (646)
                            $.97(.001).999$

12.62 Working loglogs and weights

Fisher & Yates' table leads to computations strictly parallel to those for working probits, pp. 22-24 above. Finney's table relates to a slight variant of the working-loglog method: see Finney 1952 b, pp. 577-580.

Finney, D. J. 1952 b
Statistical method in biological assay

Table 17, pp. 647-649, gives the minimum working loglog deviate $-e^{-Y}$ to 3; 4 dec. for $Y = -4.6(.1)-2.4$; $-2.3(.1)+2.5$: the range $e^{(e^Y-Y)}$ to 5 fig. and the weight $e^{2Y}/(e^{e^Y}-1)$ to 5 dec. for $Y = -1(.1)+2.5$: the maximum working loglog deviate $e^{(e^Y-Y)} - e^{-Y}$ to 4; 3 dec. for $Y = -9(.1)+1.2$; $1.3(.1)1.8$.

Fisher, R. A. and Yates, F. 1957
Statistical tables

Table XII 1 gives the minimum working loglog, $Y + e^{-Y} - e^{(e^Y-Y)}$; the range, $e^{(e^Y-Y)}$; the maximum working loglog, $Y + e^{-Y}$; all to 5 fig.: and the weight, $e^{2Y}/(e^{e^Y}-1)$, to 5 dec. or 4 fig.; for $Y = -7.5(.1)+2.4$.

12.7 The Legit* transformation

Fisher, R. A. 1950 b

<u>Biometrics</u> 6, 353-361

"Gene frequencies in a cline determined by selection and diffusion"

Let q be the solution of the differential equation $d^2q/dx^2 = 4xpq$, $p = 1-q$, subject to the two-point boundary condition $q(0) = \frac{1}{2}$, $q(+\infty) = 0$; let q' denote dq/dx. Then x, regarded as a function of q or p, is called a Legit.

Table 1, pp. 354-355, gives q to 7 dec. for $x = 0(.02)5.7$. Table 2, p. 357, gives x to 5 dec. for $q = .001(.001).01(.002).05(.005).2(.01).5$.

Table 3, p. 358, gives the maximum working Legit $x - q/q'$ to 4 dec.; the minimum working Legit $x + p/q'$ to 5 fig.; the weight $(pq)^{-1}(q')^{-2}$ to 4 dec.; all for $x = 0(.1)3$.

12.9 Other transformations

Cavalli, L. L. 1950

<u>Biometrics</u> 6, 208-220

Table 1, p. 211, gives $Y = p_E^{-1} \log_e p + q_E^{-1} \log_e q - (p_E q_E)^{-1} \log_e |p_E - p|$, where $q = 1-p$ and $q_E = 1-p_E$, to 2 dec. for
$\begin{cases} p_E = .05(.05).95 & .99 \\ p = .01, .05(.05)p_E & .01, .05(.05).95 \end{cases}$. This transformation arises from the differential equation $dp/dt = pq/(Cq - Ap)$ for the gene frequencies p, q of two alleles in competition.

---

*We adopt a capital initial in the hope of reducing confusion between Legits and logits.

13.0

SECTION 13
TABLES OF RANDOM SAMPLES

13.0 <u>Introduction</u>

The theory of statistics and probability, being based on the concept of random variable, has from its beginnings put its formulas to empirical tests. The decision to record the results of large-scale empirical random processes in tabular form is more recent.

The tabulation most frequently printed is of random decimal* digits: i.e. a random sequence of the integers 0 1 2 3 4 5 6 7 8 9 each represented with equal probability. A number of such tables are described in section 13.1, together with some account of the methods by which they were produced: section 13.6 contains more extended notes on the generation of random digits.

The result of grouping several (say 8) consecutive random digits, say 6 9 0 7 4 9 4 1, and prefixing a decimal point (i.e. 0·6907 4941) may be regarded as an 8-decimal random number between 0 and 1, i.e. a random sample, rounded down to 8 dec., from the rectangular distribution on the range [0,1]. Such random rectangular numbers (say P_i) form a basis for drawing random samples from <u>any</u> continuous distribution: denoting the cumulative distribution of a variate x by $P(x)$, we obtain random samples from this distribution by finding (by inverse interpolation, or from a table of the inverse function) solutions x_i of $P(x) = P_i$. For some important distributions, notably the normal distribution, random samples have been tabled: tables of random normal deviates are indexed in section 13.3 and random samples from a few other distributions in section 13.4. For the technique of drawing random samples from a discrete distribution see, for example, Kendall & Stuart 1958, pp. 215-216.

---

*For internal use in many computing machines, random <u>binary</u> digits (0 or 1) are demanded.

Frequently random samples from a distribution are used to obtain random samples of a derived statistic. Thus, if we wish to generate random values of the sum of squares

$$\sum_{i=1}^{n}(x_i-\bar{x})^2,$$

where the x_i are $N(0,1)$, n random normal deviates afford only one random sum of squares. In this problem we could sample from the χ^2-distribution; but if we wish to study the distribution of an analytically intractable statistic by random sampling, we need an abundant supply of random samples from the parent distribution.

The application of electronic computing machinery facilitates both the transformation of random rectangular numbers into random samples from any parent distribution and the computation of derived statistics. Thus it may be argued that such machinery obviates tables of random samples x_i and demands only a source of random digits for Monte Carlo calculations. Electronic devices have been invented that, installed on computing machines, should automatically generate random digits. Moreover, several good methods of generating, purely arithmetically, pseudo-random numbers (<u>not</u> digits) are in use; see section 13.62. These pseudo-random numbers are practically satisfactory for many statistical problems that in theory demand random rectangular numbers. For description of some of these methods, see the abstracts by D.G. Kendall on pp. 323-337 of H.A. Meyer 1956.

Random- and pseudo-random number generators obviate tables of random numbers; thus such tables will not be used in large-scale Monte Carlo work. The tables retain their utility for small-scale work; in particular, for pilot computations.

The tables abstracted in sections 13.1 - 13.5, and the methods abstracted in section 13.6, are random sampling <u>tools</u> for the statistician that wishes to randomize an experimental design or perform a sampling experiment. We have in general ignored the many published tables of random sampling <u>results</u>, intended to elicit the approximate distribution of some statistic. Occasionally we have noticed such results if they parallel theoretical results; see abstract of Hojo 1931 on pp. 92-93.

13.1 Random digits

Methods of producing tables of random digits, defined in section 13.0, vary greatly; we offer a tentative classification of such methods in sections 13.12 - 13.13. Kendall & Stuart 1958, pp. 213-218, give a useful introduction to such methods, with particular attention to pitfalls in their use; for a list of papers on the generation of random numbers, see section 13.61.

13.11 Finding list of tables of random digits

The symbol ? affixed to a number of digits indicates a table of whose randomness we have seen no evidence.

1 000 000 digits	RAND Corp. 1955
105 000	Interstate Commerce Commission 1949
100 000	M.G. Kendall & Babington Smith 1939
41 600	Tippett 1927
15 000	Fisher & Yates 1938, 1943, 1948, 1953, 1957(T.XXXIII)
15 000	Hald 1952(T.19)
10 000?	Snedecor* 1946(10)
10 000?	Snedecor* 1950(9), Ostle 1954(429), Huntsberger 1958(179), 1961
9 000	RAND Corp. 1954 a b c d
8 000	Arkin & Colton 1950(142)
8 000	Zia-ud-Din & Moin-ud-Din Siddiqi 1953
6 875	RAND Corp. 1953 a b c d
6 250	Dixon & Massey 1951(290), 1957(366)
6 000	M.G. Kendall & Babington Smith 1938
6 000	RAND Corp. 1952
4 000	Lindley & Miller 1953(12)
2 000?	Quenouille 1950(230)
2 000?	Mahalanobis 1933
1 600	Peatman & Schafer 1942
1 530	Deming 1950

---

*These two tables have the same arrangement but different digits.

13.12 Original tables of random digits

13.121 Original tables of random digits from physical sources

13.1211 Tables published as obtained

Kendall, M. G. and Babington Smith, B. 1939
Tables of random sampling numbers

100 000 digits constructed by a mechanical device: a disk with numbers 0 to 9, rotated in a darkened room at 250 r.p.m. and illuminated at irregular intervals of 3 to 4 seconds by a flash from a neon lamp.

Peatman, J. G. and Schafer, R. 1942
Journal of Psychology 14, 295-305

1600 digits compiled from the U.S. selective service drawing of 18 July 1941.

Hald, A. 1952
Statistical tables and formulas

Table 19, p. 92, gives 15 000 digits compiled from the drawings of the Danish State Interest Lottery of 1948.

Zia-ud-din, M. and Moin-ud-din Siddiqi, M. 1953
Proceedings of the Second Pakistan Statistical Conference held in January 1952 at Dacca, 91-98

8000 digits obtained by shuffling a pack of cards containing 8 each of the digits 0 to 9.

13.1212 Tables subjected to 'improvement' before publication

Interstate Commerce Commission 1949
Table of 105,000 random decimal digits

105 000 digits obtained as moving sums modulo 10 of ten digits from a waybill survey.

The RAND Corporation 1955
A million random digits with 100,000 normal deviates

A random-frequency pulse source was gated about one per second to produce 'random' numbers in a 5-place binary counter. The number 00000 was discarded; the numbers 00001 - 01001 translated into 1 - 9; the numbers

13.1212
13.13

01010 - 10011 translated into 0 - 9; the number 10100 into 0; the numbers 10101 - 11111 discarded. The resulting digits were added in consecutive overlapping pairs, and the pair totals (reduced mod 10) are the random digits published.

Because of the importance of this table, we quote a very minor caveat from Tukey 1955, pp. 569-570:

"The table explains (page xii) how the original digits were improved by adding together pairs of digits modulo 10. The author(s) inform[s] me that this was done in pairs down the present columns, starting with a line since removed. This means that users must, in principle, beware of the very occasional problem where the computations rather accurately approximate differences of cumulative sums of digits taken down columns. (I can't specify a realistic problem where this would happen.) Outside of this, very, very unlikely possibility, I know of no warnings needed beyond those indicated in the introduction to the tables."

RAND Corp. 1955, p. xj, footnote, states that ". . . the digit . . . table [is] available on IBM cards and may be purchased from The RAND Corporation, Santa Monica, California."

<u>13.122</u> Original tables of random digits produced by rearrangement

<u>13.1221</u> Rearrangement of non-random digits

Tippett, L. H. C. 1927

Random sampling numbers

41 600 digits from census reports. Tukey 1955, p. 570, reports on the authority of a private communication from Tippett that the digits are taken from the middle digits of areas of parishes in England, and improved by shuffling by organized rearrangement.

Fisher, R. A. and Yates, F. 1938, 1943, 1948, 1953, 1957

Statistical tables

Table XXXIII contains 15 000 digits obtained from the 15th-19th digits in A.J. Thompson 1924 and adjusted before publication, it having been found that there were too many sixes. For details of the construction see Fisher & Yates 1938, pp. 18-19; these details are omitted in later editions.

13.1222 Rearrangement of random digits

Mahalanobis, P. C. 1933

Indian Journal of Agricultural Science 3, 1108-1115

    2000 digits obtained by taking random samples from Tippett 1927, and rearranging them in a random manner.

Snedecor, G. W. 1946

Statistical methods

    10 000 digits obtained by punching the contents of Tippett 1927 on cards, and subjecting the cards to a process whose details are now unknown: the cards, but not the instructions, are at the Statistical Laboratory, Iowa State Univ.

Snedecor, G. W. 1950

Everyday statistics

    Another set of 10 000 digits, with a history similar to those in Snedecor 1946.

13.13 Secondary tables of random digits

Digits	Source	Published in
24 000 ?	Vielrose 1951	Czechowski et al. 1957(93)
12 000	Royo López & Ferrer Martín 1954	Royo López & Ferrer Martín 1955
10 000	Kendall & Babington Smith 1939	Vianelli 1959(483)
8 000	Kendall & Babington Smith 1939	Arkin & Colton 1950(142)
6 000	Kendall & Babington Smith 1939	Kendall & Babington Smith 1938
4 000	Kendall & Babington Smith 1939	Lindley & Miller 1953(12)
10 000 ?	Snedecor 1950	Ostle 1954(429), Huntsberger 1958(179), 1961
9 000	RAND Corp. 1955	RAND Corp. 1954 a b c d
6 875	RAND Corp. 1955	RAND Corp. 1953 a b c d
6 000	RAND Corp. 1955	RAND Corp. 1952
6 250	RAND Corp. [1955?]	Dixon & Massey 1951(290), 1957(366)
8 000	Tippett 1927	Vianelli 1959(478)
1 530	Interstate Commerce Comm. 1949	Deming 1950

13.2 Random permutations

Form the set of n! possible permutations of the integers from 1 to n. Random permutations are obtained by sampling with replacement from the distribution that assigns probability $1/n!$ to each permutation in the set. A sequence of random permutations can be generated from a sequence of random digits by filling n places with the integers from 1 to n, not permitting any integer to occur more than once; other generating methods, which are more efficient in that they yield more permutations per page of random digits, are given by Fisher & Yates 1957, pp. 34-35.

Tables of random permutations are of use in arranging n treatments at random in an experimental design, and in other problems demanding sampling without replacement.

Steinhaus, H. 1954

Instytut Matematyczny Polskiej Akademii Nauk, Rozprawy Matematyczne, No. 6

One permutation for n = 10,000 : the figures tabled are the 10,000 numbers from 0000 to 9999 in pseudo-random order.

The numbers were obtained by subjecting the numbers $4567\,n \pmod{10^4}$ to three successive permuting operations, described in the introduction.

From his introduction, it appears that Steinhaus wishes to claim an advantage for pseudo-random numbers over random numbers:

"The present table is the first to put into effect the postulate of shuffling successive natural numbers in a way that can be determined mathematically. This is what distinguishes it from the so called tables of random numbers, obtained from the observation of physical instruments, from the registers of population statistics, or from other sources not determined mathematically." (p. 15)

He further discusses this alleged advantage in Steinhaus 1954 a :

"This table differs from the tables of random numbers containing the same number of four digit arrangements in the following respect : in the tables of random numbers certain arrangements appear more than once while other arrangements do not appear at all. The removal of this defect facilitates the practical application of the new table. Moreover, reading the new table vertically we obtain a table of random numbers."
(1954 a, p. 45)

D.H. Lehmer 1954, 1955 warns that there is no evidence that tests of randomness have been applied to the permutation.

Cochran, W. G. and Cox, Gertrude M. 1950, 1957

Experimental Designs

Chapter 15: Random permutations of 9 and 16 numbers.

1000 permutations of 9 numbers were produced by running through a table of random digits, throwing out duplicates, until a whole permutation emerged. 800 permutations of 16 numbers were obtained by taking the permutation necessary to rank 16 2-digit random numbers in natural order. 200 permutations of 16 numbers were obtained by drawing numbered marbles from an urn.

Randomness of these permutations was tested by the frequency test and by Kendall's tau.

Reprinted in Vianelli 1959, Prontuari 168-169, pp. 498-512.

Fisher, R. A. and Yates, F. 1953, 1957

Statistical Tables

T. XXXIII 1 : 750 random permutations of the numbers 0-9.

T. XXXIII 2 : 200 random permutations of the numbers 0-19.

Computed by B. M. Church from Fisher & Yates' random digits. Details not given.

We have seen no evidence of the randomness of the following tables of permutations:

    Quenouille 1953(340). 480 permutations of 9 numbers.
    Quenouille 1953(344). 240 permutations of 16 numbers.
    Quenouille 1953(348). 60 permutations of 25 numbers.
    Quenouille 1953(350). 60 permutations of 36 numbers.

13.3   [see also p. 729]

13.3 Random normal numbers

The random variates drawn from normal distributions, contained in the tables described below, were all computed from random rectangular numbers, i.e. strings of random digits, by the method sketched in section 13.0. The customary form, and probably the most useful form, of a table of random normal numbers is a table of random normal deviates, i.e. random normal numbers with zero mean and unit variance; Dixon & Massey 1951, 1957 have tabled certain other forms.

The RAND Corporation   1955

A million random digits with 100,000 normal deviates

500,000 of RAND's million random digits were grouped into 100,000 5-decimal rectangular numbers; these were transformed into 100,000 3-decimal normal deviates.

RAND Corp. 1955, p. xj, footnote, states that '. . . the deviate . . . table [is] available on IBM cards and may be purchased from The RAND Corporation, Santa Monica, California.'

Wold, H.   1948

Random normal deviates

The 100,000 random digits of Kendall & Babington Smith 1939 were grouped into 25,000 4-decimal rectangular numbers; these were transformed into 25,000 2-decimal normal deviates.

Mahalanobis, P. C., Bose, Subendu Sekhar, Ray, P. R. and Banerji, S. K.
Sankhyā 1, 289-328   1934

The 41,600 random digits of Tippett 1927 were grouped into 10,400 4-decimal rectangular numbers; these were transformed into 10,400 3-decimal normal deviates.

Dixon, W. J. and Massey, F. J.   1951, 1957

Introduction to statistical analysis

Table 2 contains 5000 3-decimal random normal deviates. Table 23 contains 2500 3-decimal random normal numbers with mean 2 and variance 1; T. 24, 2500 numbers with mean 0 and variance 4.

These tables are reproduced from tables of The RAND Corporation.

[462]

Fieller, E. C., Lewis, T. and Pearson, E. S. 1955

Correlated random normal deviates

The table contains 3000 lines, each line containing the 10 2-decimal normal numbers x_{ti} ($t = 0, \ldots, 9$). For $t > 0$, the x_{ti} were computed by the formula $x_{ti} = (t/10)x_{0i} + (1 - t^2/100)z_{ti}$, where x_{0i} and z_{ti} are different normal deviates from Wold 1948. Thus the pairs x_{0i}, x_{ti} constitute a sample of 3000 pairs from a bivariate normal distribution with correlation $t/10$; and the pairs x_{ti}, x_{ui} constitute a sample of 3000 pairs from a bivariate normal distribution with correlation $tu/100$.

For each block of 50 lines the column sums, sums of squares and cross-products, and sample correlation coefficients are given.

According to the preface, p. [v], these 3000 sets "are part of a total of 25,000 which have been computed, corresponding to the 25,000 items in Wold [1948]. Its bulk has precluded the printing . . . of the whole set, but it is hoped that the portion presented . . . will suffice for illustrative purposes . . . and for the preliminary empirical investigation of unknown sampling distributions deriving from the bivariate normal population. For more extensive investigations the whole 25,000 lines, together with their summary values in sets of 50, 500, 5000 and 25,000, are available both punched on Hollerith cards and printed in the form of a Hollerith tabulator listing . . . Inquiries should be addressed to the Director, National Physical Laboratory, Teddington, Middlesex, England."

For corrigenda see Biometrika 43, 496-497.

On March 16 1956, the U.S. Office of Technical Services announced (U.S. Govt. Technical Reports 25, 120):

Random normal deviates, by Herbert A. Meyer, Ernest J. Lytle, Jr. and Landis S. Gephart. Florida. University. Statistical Laboratory, Gainesville, Fla. Mar 1955. 122 p tables. Order from OTS. $3.25.
PB 111875

Tables of a random arrangement of 25,000 six-place normal deviates, mean zero, variance one, are included with a description of the method used in their generation and with results of tests of normality and of randomness. Project no. 7060. AF WADC TR 55-125. Contract AF 33(616)-285.

Czechowski, T. et al. 1957

Tablice statystyczne

Tablica 29, pp. 105-111, contains 5000 2-decimal normal deviates from Wold 1948.

13.4
13.5

13.4 Random samples from populations other than rectangular and normal

We have excluded from this section many published investigations of the multinomial distribution by means of dice-casting etc. For the work of W.F.R. Weldon see F.N. David 1949a, p. 34. For extensive records of roulette at Monte Carlo (the gambling establishment, not the physico-mathematical method) see Marbe 1934 and works cited therein: but see also Coolidge 1925, p. 50.

Pearson, E. S. 1935 b

Biometrika 27, 333-352

Short tables of random samples, with replacement, from seven finite populations of 10,000 each, constructed to simulate a two-tailed exponential distribution and six representative Pearson curves.

K. Pearson prefaces the tables with the following apology (p. 345):

"The object of publishing the following Appendix containing 72 random samples of various sizes from diverse forms of curves is to put into the hands of those desiring to test other, old or new, criteria ready-made samples and save them from the heavy labour of obtaining afresh their own samples."

Kullback, S. 1939

Annals of Mathematical Statistics 10, 77-80

457 pairs of sequences of 300 digits (0 to 9) were drawn and the number of matchings counted. Table 1, p. 78, gives the frequency distribution of the number of matchings in the 457 pairs. Tables 3 and 4, p. 79, give the fit of the binomial $457(.9 + .1)^{300}$ to this series.

13.5 Random samples from artificial time series

Kendall, M. G. 1944

Biometrika 33, 105-122

"On autoregressive time series"

Table 3 (p. 111) gives 65 terms generated by the process

$$u_{t+2} = 1.1 u_{t+1} - 0.5 u_t + \epsilon_{t+2}$$

where ϵ_{t+2} is a random rectangular variate with range $-9.5(1)9.5$. Table 4 (p. 111) gives 2 decimal values of serial correlations

$$r_s = \frac{\sum_{j=1}^{n-s} x_j x_{j+s}}{\left[\sum_{j=1}^{n-s} x_j^2 \sum_{j=1}^{n-s} x_{j+s}^2\right]^{\frac{1}{2}}} \quad \text{for} \quad s = 1(1)30 \quad \text{for above series.}$$

A 65 term series from 2nd order scheme $u_{t+2} = 1.5 u_{t+1} - 0.9 u_t + \epsilon_{t+2}$ was formed where ϵ_{t-2} is a random rectangular variate with range $-49.5(1)49.5$. Upon this series was superposed a further random rectangular variate with range $-49.5(1)49.5$. Table 5 (p. 114) gives resultant as series (a). To this series an additional random rectangular variate with range $-199.5(1)199.5$ was added, and the resultant, divided by 10, given as series (b) in Table 5. Both (a), (b) given as integers. Table 6 (p. 114) gives 2 decimal values of serial correlations r_s for $s = 1(1)30$ for both series of Table 5.

Kendall, M. G. 1946

Contributions to the study of oscillatory time-series

The tables on pp. 10-12 give four series generated by the second-order autoregressive process $u_{t+2} + a u_{t+1} + b u_t = z_{t+2}$.

Series	Length	a	b	Distribution of z
1	480 terms	-1.1	+0.5	Rectangular
2	240	-1.2	+0.4	Rectangular
3	240	-1.1	+0.8	Rectangular
4	240	+1.0	+0.5	Rectangular

13.5

The range of the rectangular random input z_{t+2} was the integers from -49 to +49, both inclusive: sample values were obtained from two-digit random numbers from Kendall & Babington Smith 1939. u_t is given to the nearest integer.

Kendall, M. G. 1949a
Biometrika 36, 267-289
"Tables of autoregressive series"

Tables of series generated by the second-order autoregressive process above, or by the corresponding first-order process (b = 0). For convenience in reference, Kendall numbered the series consecutively with those in Kendall 1946, abstracted above.

Series	Length	a	b	Distribution of z
5 a	400 terms	-1.1	+0.5	Rectangular
5 b	400	-1.1	+0.5	Rectangular
5 c	400	-1.1	+0.5	Rectangular
5 d	400	-1.1	+0.5	Rectangular
6 a	400	-1.1	+0.8	Rectangular
6 b	400	-1.1	+0.8	Rectangular
6 c	400	-1.1	+0.8	Rectangular
6 d	400	-1.1	+0.8	Rectangular
7	502	-0.9	0	Normal
8	500	-1.1	+0.5	Series 7
9	502	-0.7	0	Normal
10	500	-1.1	+0.5	Series 9
11	502	-0.5	0	Normal
12	500	-1.1	+0.5	Series 11
13	502	-0.3	0	Normal
14	500	-1.1	+0.5	Series 13
15	502	-0.1	0	Normal
16	500	-1.1	+0.5	Series 15

Thus series 8, 10, 12, 14 and 16 are effectively samples from third-order autoregressive processes.

The input and output of series 5 and 6 are as in Kendall 1946, abstracted above. The input to series 7, 9, 11, 13 and 15 consists of 3-decimal random normal deviates from Mahalanobis et al. 1934. The output of series 7 to 16 is given to 3 dec.

Serial product moments and serial correlations are given for series 1, 2, 3, 4 (of Kendall 1946); 5 a b c d; 6 a b c d; 5 *in toto*, 6 *in toto*; 8, 10, 12, 14, 16.

Bartlett, M. S. 1950
<u>Biometrika</u> 37, 1-16

Bartlett proposes to compute smoothed periodograms by splitting a series of mn terms into m consecutive sections of n terms each, computing the periodogram intensities for each section and averaging. Alternatively, the smoothed intensity corresponding to period p may be computed from

$$I_p = 2 \sum_{s=-n+1}^{n-1} 1 - \frac{|s|}{n} C_s \cos \frac{2\pi ps}{n}$$

where C_s is the autocovariance at lag s computed from the complete series.

The tables give smoothed periodograms I_p, computed from the artificial series 1 and 3 of Kendall 1946, for q = 60 p/n = 1(1)30 ; n = 15, 30 and sometimes 45.

See also Mozart 1793.

13.6
14.1

13.6   Generation and testing of random and pseudo-random numbers

The lists of references in this section are not exhaustive: they are compiled mostly from the abstracts edited by D.G. Kendall on pp. 323-337 of H.A. Meyer 1956. The columns headed <u>Meyer</u> in these lists give the page reference to such abstracts. We have added a few later papers that have come to our notice, but have not searched systematically for such papers.

13.61   Generation of random numbers

	Meyer		Meyer
Hamaker 1948, 1949	328	G.W. Brown 1951	324
Horton 1948	328	von Neumann & Forsythe 1951	332
Horton & Smith 1949	329	Royo López & Ferrer M. 1954 a	
Walsh 1949	335	Lytle 1956	234

13.62   Generation of pseudo-random numbers

	Meyer		Meyer
Forsythe 1951	326	Taussky & Todd 1956	15
Hammer 1951	328	Bofinger & Bofinger 1958	
D.H. Lehmer 1951	331	Certaine 1958	
von Neumann & Forsythe 1951	332	Zierler 1959	
Moshman 1954		Coveyou 1960	
D.L. Johnson 1956		Rotenberg 1960	

13.63   Testing of random and pseudo-random numbers

	Meyer		Meyer
Yule 1938	337	Hammer 1951	328
Dodd 1942	325	von Neumann & Forsythe 1951	332
Hamaker 1948, 1949	328	Good 1953	326
Azorin & Wold 1950	323	Geiringer 1954	
Gruenberger 1950	327	Royo López & Ferrer M. 1954 a	
Metropolis et al. 1950	331	R.E. Greenwood 1955	327
G.W. Brown 1951	324	D.L. Johnson 1956	
Forsythe 1951	326	Lytle 1956	234
Gruenberger & Mark 1951	327	Green, Smith & Klem 1959	

SECTION 14
ACCEPTANCE SAMPLING; CONTROL CHARTS; TOLERANCE LIMITS

14.1 Acceptance sampling

A useful list of definitions is found in A.S.Q.C. 1957. The following abbreviations and symbols are used throughout the reviews below:

α Producer's risk: the probability that a sampling procedure will reject a lot, or interrupt a process, of acceptable quality. Conventionally, $\alpha = .05$.

β Consumer's risk: the probability that a sampling procedure will accept a lot, or pass a process, of rejectable quality. Conventionally, $\beta = .10$.

AOQ   Average outgoing quality: the mean per cent defective, or mean number of defects, in the output of a rectifying inspection. Usually plotted as ordinate against the incoming quality as abscissa.

AOQL[*]   Average outgoing quality limit: the supremum of the AOQ for variations in incoming quality.

AQL[*]   Acceptable quality level: the incoming (lot or process) quality that a sampling procedure is designed to accept[‡] with probability $1-\alpha$.

[The distinction between acceptable lots and acceptable process is blurred, because the binomial distribution, appropriate to process control by number of <u>defectives</u>, and the Poisson distribution, appropriate to process control by number of <u>defects</u>, are frequently used as approximations to the hypergeometric distribution, appropriate to <u>lot</u> control by number of defectives. The abbreviation AQL is made to do duty for both lots and process. Cf. LTPD and PTPD, below]

defect   Any distinguishable failure to meet specifications, discovered upon inspection. Conventionally, the number of defects per article

---

[*] Note that the letters A and L have different meanings in AOQL and AQL.
[‡] The AQL of the plans of the Dept. of Defense reflects no specific α, but is offered as a rough guide to the selection of a plan.

14.1
14.111 [see also p. 730]

produced is assumed to follow a Poisson distribution (and thus, in theory, to be unbounded).

defective   In quality-control parlance, **defective** is a substantive, denoting an article containing one or more defects. Per cent defective, accordingly, is the fraction of defectives that a sample, lot or process is found or estimated to contain. A good article, i.e. one containing no defects, is sometimes (not here) called an **effective**.

LTPD   Lot tolerance per cent defective: the incoming lot quality that a sampling procedure is designed to reject with probability $1-\beta$.

[When the probability of rejection is computed from the binomial distribution, we have preferred the designation PTPD]

OC   Operating characteristic: a graph of the probability of accepting a lot as ordinate against the incoming process quality as abscissa.

PTPD*   Process tolerance per cent defective: the incoming process quality that a sampling procedure is designed to reject with probability $1-\beta$.

14.11   Lot inspection by attributes

The published basic tools for constructing an attributes sampling procedure include the Poisson tables of section 2.12, the binomial tables of section 3.11, and the hypergeometric charts of Chung & De Lury 1950. Dodge & Romig 1959, pp. 18-19, 33-34, give useful information on the Poisson and binomial approximations to the hypergeometric distribution.

14.111   Comprehensive tables

Dodge, H.F. and Romig, H.G.   1959
Sampling inspection tables

Appendix 4, pp. 182-185, gives 531 single-sampling plans, classified by lot size, LTPD ($\beta = .10$ nominal) and process average. It is assumed that rejected lots are 100% inspected; the plans minimize the expected total amount of inspection (sampling + 100% inspection) when the incoming quality equals the process average. Lot sizes range from 21 to 100 000; LTPDs from $\frac{1}{2}$% to 10%; process averages from 0 to $\frac{1}{2}$(LTPD).

---

*PTPD is not a standard quality-control term.

Appendix 5, pp. 188-195, gives 565 double-sampling plans, classified as in App. 4; lot sizes range from 51 to 100 000 ; other parameters as in App. 4. For each plan in App. 4 and 5, the AOQL is given to 2 fig.

Appendix 6, pp. 198-204, gives 640 single-sampling plans, classified by lot size, AOQL and process average. It is assumed that rejected lots are 100% inspected; the plans minimize the expected total amount of inspection at the process average. Lot sizes range from 4 to 100 000 ; AOQLs from .1% to 10% ; process averages from 0 to (AOQL) . Appendix 7, pp. 206-218, gives 809 double-sampling plans, classified as in App. 6; lot sizes range from 16 to 100 000 ; other parameters as in App. 6. For each plan in App. 6 and 7, the LTPD (β = .10) is given to 3 dec. or 2 fig.

Appendix 1, pp. 62-109, gives OC curves for the plans of App. 6; App. 2, pp. 112-170, for the plans of App. 7. Appendix 3, pp. 172-179, gives OC curves for the following single-sampling plans:

```
c = 0   n = 2(1)16(2)34, 35(5)50(10)100(20)200(50)500
    1       3(1)20(2)50(5)100(10)200(20)500
    2       5(1)20(2)36, 35(5)100(10)160(20)280, 250(50)500
    3       8(1)20(2)36, 35(5)80(10)200(20)300(50)500
```

The tables of App. 4-7 were published in Dodge & Romig 1941, TT. 1-4, pp. 30-61; 1944, pp. 70-101.

Dept. of Defense   1959 a
Military Standard 105 B

Tables of 411 single-, 105 double-, and 104 multiple-sampling plans, classified by lot size and AQL. General arrangement and many plans in common with Freeman et al. 1948.

The AQLs in these tables are not tied to a specific producer's risk. We read on p. 4 (emphasis supplied):

"The Acceptable Quality Level (AQL) is a _nominal_ value expressed in terms of percent defective or defects per hundred units, whichever is applicable, specified for a given group of defects of a product."

The meaning of AQL in previous editions of these tables is discussed by Freeman et al. 1948, p. 144; Duncan 1959, p. 192; Cowden 1957, pp. 603-606.

14.111
14.112

Lot sizes range from 2 to 550 001 ; AQLs from .015% to 10% ; single-sample sizes from 1 to 1500 . Double- and multiple-sampling plans are chosen to match the OC curves of the single-sampling plans. Additional single-sampling plans cover AQLs, expressed in <u>defects per 100 units</u>, from 15 to 1000 ; for lower numbers of defects, use the corresponding per cent defective. Precepts for normal, reduced and tightened inspection are given.

Under certain conditions, reduced inspection may be instituted when the process average falls 3 standard deviations below the AQL : this point is given in T. 2 A, pp. 16-17, and the point 2 standard deviations below the AQL in T. 2 B, p. 18. Tightened inspection shall be instituted when the process average falls 3 standard deviations above the AQL , as shown in T. 2 C, pp. 19-21.

Tables 3 A, 3 B, pp. 22-23, index the plans by lot size and inspection level. Tables 4 A, 4 B, 4 C, pp. 24-29, give the sample sizes and acceptance and rejection numbers for normal and tightened inspection; T. 5, p. 30, for reduced inspection.

Tables 6 A - 6 Q, pp. 31-64, rehearse the sample sizes and acceptance numbers of the plans of TT. 4 A - 4 C, and plot an OC curve for each single-sampling plan: the OC curves for double- and multiple-sampling plans are stated to be essentially equivalent.

An earlier edition of these tables, MIL-STD 105 A, was widely reprinted in whole or part: we have seen reprints in Burr 1953, pp. 423-435; Cowden 1957, pp. 585-607; Duncan 1952, pp. 181-193; 1959, pp. 170-194; E.L. Grant 1952, pp. 530-541.

Freeman, H.A., Friedman, M., Mosteller, F. and Wallis, W.A.   1948
<u>Sampling inspection</u>

Tables of 103 single-, 108 double-, and 121 multiple-sampling plans, classified by lot size and approximate AQL (α = .05 nominal). Lot sizes range from 25 to 550 000 ; AQLs from .024% to 11% ; single-sample sizes from 5 to 1500 . Double- and multiple-sampling plans are chosen to match the OC curves of the single-sampling plans. Precepts for normal, reduced and tightened inspection are given.

The multiple-sampling plans satisfy $\alpha = .05$, $\beta = .10$ only approximately; T. 15.1, p. 162, gives the $1-\alpha$ and β actually achieved to 3 dec.

Tables 15.2-15.4, pp. 163-165, give the AQL ($\alpha = .05$ nominal), the PTPD (called LTPD; $\beta = .10$ nominal), and the AOQL of the plans to 3 dec. or 2 fig. Tables 15.5-15.6, pp. 166-169, gives the parameters of Wald sequential plans for which the multiple-sampling plans tabled are substitutes. Table 15.7, p. 170, gives the average sample number of the multiple-sampling plans.

Table 1, p. 217, indexes the plans by lot size and inspection level. Tables 2a-2c, facing p. 218, give the sample sizes and acceptance numbers. Table 3, pp. 219-220, indexes the plans by approximate AQL and AOQL.

Table 4, pp. 221-379, rehearses the sample sizes and acceptance numbers, and plots an OC curve for each plan. Freeman et al. state (p. x) that because of mathematical approximations, graphical interpolation, and the difficulties of four-color printing, the accuracy of these curves is limited.

Columbia Univ. 1945

Sequential analysis

Table 2.23, pp. 2.39-2.42, gives the characteristic quantities of Wald sequential tests for acceptance sampling by attributes. $\alpha = .05$; $\beta = .10$ and .50; AQL varies from .02% to 20%; PTPD from .2% to 35%.

14.112 Other tables

Horsnell, G. 1956

Journal of the Royal Statistical Society, Series A 120, 148-201

Tables of 529 single-sampling plans.

Let the cost to manufacture an item be 1, and the cost to inspect a sample of n be nc_s. Then if the probability of accepting a lot of size N of the process average quality is A, the average cost to produce an accepted item is $c = \frac{N + nc_s}{NA}$ for non-destructive testing and $c' = \frac{N + nc_s}{(N-n)A}$ for destructive testing. Horsnell seeks, for fixed PTPD, β, and process average, to minimize c or c'. He tables n, the rejection number k, NA or (N-n)A, and c or c', as a function of $1/c_s$ as follows:

14.112 [see also p. 730]

Non-destructive testing: N = 10 000

Process Average	PTPD	β	Table	Page
1%	3%	.05	1	176
		.01	8	181
	4%	.05	2	176
		.01	9	181
	5%	.05	3	177
		.01	10	182
2%	5%	.05	4	177
		.01	11	182
3%	6%	.05	5	178
	7%	.05	6	179
4%	9%	.05	7	180

Destructive testing

Process Average	PTPD	β	N = 10 000		N = 20 000	
			Table	Page	Table	Page
1%	3%	.05	12	183	22	187
		.01	17	185	27	189
	4%	.05	13	183	23	187
		.01	18	185	28	190
	5%	.05	14	183	24	188
		.01	19	185	29	190
2%	5%	.05	15	184	25	188
		.01	20	186	30	191
	6%	.05	16	184	26	189
		.01	21	186	31	191

Anscombe, F. J. 1949

<u>Journal</u> <u>of</u> the <u>Royal</u> <u>Statistical</u> <u>Society</u>, Series A 112, 180-206

Tables of 223 untruncated-sequential- and 19 truncated-sequential-sampling plans.

Tables 1-3, pp. 199-201, based on the Poisson approximation, give the characteristics, including 5 non-trivial points on the OC curve, of 27 untruncated and 19 truncated plans, which Anscombe calls <u>open</u> and <u>closed</u>.

Tables 4-5 are devoted to rectifying schemes, which Anscombe defines by the following precept (p. 193):

"From a batch of N articles, a first sample of αN articles[*] is inspected, and then further samples of βN articles[*] each. Defective articles found are removed or replaced by good ones. Inspection ceases after the first sample if no defectives have been found, or after the second sample if altogether one defective has been found, or, generally, after the $(r+1)^{th}$ sample if altogether r defectives have been found. Inspection is continued until either this stopping rule operates or the whole batch is inspected."

Table 4, pp. 202-204, gives average sample number and AOQL, as a function of the number of defectives in a batch, for consumer's risk .10 and lot tolerance defect number 5(5)30(10)60(20)100. Table 5, pp. 204-206, does the same for consumer's risk .01.

Greb, D. J. and Berrettoni, J. N. 1949

<u>Journal of the American Statistical Association</u> 44, 62-76

Table of 183 single-sampling plans.

Let m^* be the value of m that maximizes, for fixed c,

$$a = m \sum_{x=0}^{c} e^{-m} m^x / x!\ ;$$

let a^* be the corresponding maximum value of a.

Table 1, p. 65, gives m^* to 3 dec., a^*/m^* to 6 dec. and a^* to 4 dec. for c = 0(1)12. Table 2, p. 65, gives a^*/A to 0 dec. for

$$\begin{cases} A = .001 & .0025 & .005 & .0075 & .01 & .015,\ .02 & .025(.005).035 \\ c = 0(1)4 & 0(1)5 & 0(1)6 & 0(1)7 & 0(1)8 & 0(1)9 & 0(1)10 \end{cases}$$

$$\begin{cases} .04(.005).05(.01).1 \\ 0(1)10,\ 12 \end{cases}.$$

Cameron, J. M. 1952

<u>Industrial Quality Control</u> 9, no. 1, 37-39

Tables of 150 single-sampling plans.

Table 2, p. 39, based on the Poisson approximation, gives PTPD/AQL to 3 dec. for α = .05 ; β = .01, .05, .1 ; acceptance number c = 0(1)49.

---

[*] The α and β of this article have no immediate connexion with producer's and consumer's risks.

14.112
14.121
[see also pp. 730-731]

Kitagawa, T. 1952

Tables of Poisson distribution

Tables of 57 single- and 58 double-sampling plans.

Table 1, pp. 10-11, based on the Poisson approximation, gives PTPD/AQL to 3 dec. for single-sampling plans with $\begin{cases} \alpha = .1 & .1 & .01 \\ \beta = .1 & .01 & .01 \end{cases}$. Table 2, pp. 11-12, does the same for double-sampling plans.

Enrick, N.L. 1954 e

Modern Textiles Magazine 35, no. 5, 42-43, 62-63

Tables of 48 single- and 24 multiple-sampling plans, for $\alpha = .05$ nominal, AQL 1% to 10%, and lot size 200 to 3000 pieces and 300 to 3000 yards of cloth.

Golub, A. 1953

Journal of the American Statistical Association 48, 278-288

Tables of 57 single-sampling plans.

For sample size n = 5(5)40, AQL = 1%(1%)20%, PTPD = 2%(1%)40%, TT. 1-8 (so called in Golub's summary, p. 278; no numbers appear on the tables), pp. 286-288, give the acceptance numbers that minimize $\alpha+\beta$.

Peach, P. 1947

Industrial statistics and quality control

Tables of 19 single-, 13 double- and 16 truncated-sequential-sampling plans.

Tables 1-3, pp. 228-230, based on the Poisson approximation, give the characteristics of these plans for $\alpha = \beta = .05$ as a function of $R_0 = \text{PTPD/AQL}$, which takes values from 1.3 to 15.

Table 1 is reprinted in Vianelli 1959, Tavola 218, p. 1390.

Harding, H.G. and Price, S. 1957

IRE National Convention Record 1957, part 10, 54-58

Table of 1 double- and 21 triple-sampling plans. The first stage of sampling uses narrow gauge limits. Lot size from 101 to 100 000; AQL 1%, 1½%, 2½%.

Jacobs, R.M. 1957

American Machinist 101, no. 3, 113-118

Table 4, p. 117, gives 9 multiple-sampling plans for AQL = $2\frac{1}{2}\%$. Lot sizes from 65 to 8001.

Grubbs, F. E. 1949

Annals of Mathematical Statistics 20, 242-256

See abstracts on pp. 153 and 191.

14.12 Lot inspection by variables

All the variables plans that we have seen assume the normal distribution. For constructing plans with known process standard deviation, tables of the normal distribution, both direct and inverse, are basic; for unknown standard deviation, tables of non-central t (see section 4.141).

14.121 Comprehensive tables

Bowker, A. H. and Goode, H. P. 1952

Sampling inspection by variables

Tables of 163 single- and 163 double-sampling plans for use when the process standard deviation must be estimated from the sample, and 163 single- and 145 double-sampling plans for use when the process standard deviation is known; classified by lot size and AQL (α = .05 nominal). Lot sizes range from 25 to 550 000; AQLs from .024% to 11%. The plans are chosen to match the OC curves of the attributes plans in Freeman et al. 1948.

Table A, p. 164, indexes the plans by lot size and inspection level. Table B, p. 165, gives the sample sizes and acceptance criteria for single-sampling plans with unknown σ; T. C, pp. 166-167, does the same for double sampling. Table D, p. 168, gives the AQL (α = .05) and LTPD (β = .10) for these plans. Table I, p. 196, indexes the plans by approximate AQL and AOQL.

Table E, pp. 170-189, rehearses the sample sizes and acceptance numbers, and plots an alleged OC curve, for each combination of sample-size letter and AQL; the curve is to do duty for both single and double sampling; but we are not told whether it was computed for single or double

14.121
14.122   [see also pp. 730-731]

sampling, or what compromise was adopted. Table F, pp. 190-191, is an algebraic transformation of T. B that obviates the computation of square roots. Table H, pp. 194-195, gives precepts for normal, reduced and tightened inspection. Table P, pp. 206-209, gives acceptance criteria for single-sampling plans with unknown σ and two-sided specification limits.

Tables K, L, N, pp. 199-203, correspond to TT. B, C, D, but relate to plans with known σ. Table M, pp. 202-203, corresponds to T. D; specifically, it gives AQL and LTPD for each single-sampling plan for unknown σ, and for the corresponding plan for known σ.

An abridgement of these tables, containing 57 plans, is given in E.L. Grant 1952, T. S, pp. 542-543.

Dept. of Defense 1957

Military Standard 414

Tables of 180 single-sampling plans for use when the process standard deviation has been estimated from the sample variance; 180 plans for use when the process standard deviation has been estimated from the sample range; 180 plans for use when the process standard deviation is known; classified by lot size and AQL. The AQL is not tied to a fixed producer's risk; the αs used are tabled on p. 4 of Dept. of Defense 1958. Lot sizes range from 3 to 550 001 ; AQLs from .04% to 15%. Precepts for normal, reduced and tightened inspection are given.

Table A2, p. 4, indexes the plans by lot size and inspection level. Tables B1 and B2, pp. 39-40, give the sample sizes and acceptance criteria for plans with σ estimated by s. Tables B3 and B4, pp. 45-46, are a mathematically equivalent alternative to B1 and B2, requiring the estimation (by means of T. B5, pp. 47-51) of the lot % defective. Table B6, pp. 54-55, enjoins tightened inspection when too many of the last 5, 10 or 15 lots have failed to meet the AQL. Under certain conditions, when the estimated process average falls below the $\frac{1}{2}$% point of the appropriate sampling distribution centered at the AQL, reduced inspection may be instituted; this $\frac{1}{2}$% point is given in T. B7, pp. 56-57. T. B8, p. 58, permits the rejection, without the full procedure given in annotated worksheets on pp. 43-44, of lots yielding samples whose standard deviation exceeds the tabled fraction of the range between upper and lower specification limits.

Tables C1-C8, pp. 65-66, 71-77, 80-84, correspond to TT. B1-B8, but relate to plans with σ estimated by the sample range. Tables D1-D7, pp. 91-94, 99-103, 106-109, correspond to TT. B1-B7, but relate to plans with known σ.

Dept. of Defense   1958
Technical Report (Resnikoff)

A mathematical exposition of the formulas underlying Dept. of Defense 1957.

14.122   Other tables: controlling percent defective

Storer, R. L. and Davison, W. R.   1955
Industrial Quality Control 12, no. 1, 15-18

Tables of 264 single-sampling plans with process standard deviation estimated from the sample.

Table S, p. 17, gives 132 plans where the process standard deviation is estimated by the sample standard deviation. Provision is made for one- and two-sided specification limits; three inspection levels; lot size from 40 to 110 001; sample size from 5 to 200; AQL (α = .05 nominal) from .065% to 6½%. Table R, p. 16, does the same where the process standard deviation is estimated by the sample range.

Eisenhart, C., Hastay, M. W. and Wallis, W. A.   1947
Techniques of statistical analysis

Table of 150 single-sampling plans with one-sided specification and estimated process standard deviation.

Table 1.2, pp. 22-25, gives the characteristics of plans with α = .05, β = .10, AQL from .1% to 5%, and PTPD from .15% to 40%.

Chernoff, H. and Lieberman, G. J.   1957
Industrial Quality Control 13, no. 7, 5-7

Table of 137 single-sampling plans. Sample size from 3 to 200, AQL from 4% to 15%. The table implements a graphical method based on normal probability paper.

14.122
14.13

Chernoff & Lieberman state (p. 5):

"The purpose of this article is to present a graphical procedure for sampling inspection by variables which involves no computations and which also gives a check on the assumptions of normality. The results are only approximate and should be considered to be a 'quick and dirty' technique and used where such procedures are tolerable. Although the approximations are good, 'No Calc.' is not a replacement for the usual variables procedures when a contract between two parties exists and calls for inspection by variables."

Grant, E. L. 1952

Statistical quality control

Table of 110 single-sampling plans with estimated process standard deviation and one-sided specification.

Table T, pp. 544-545, due to W.G. Ireson, gives the 95%, 50% and 10% points on the OC curve (i.e. the AQL, the indifference point and the PTPD) for plans based on the rule: accept if $\bar{x} \geq L + ks$. The points are given to 4 dec. for $k = .6(.2)2(.25)2.5, 3$, and sample size $n = 5, 7, 10(5)40, 50$.

14.124 Other tables: controlling variance

McElrath, G. W. and Bearman, J. E. 1957

American Society for Quality Control, National Convention Transactions, 11, 447-462

Tables of 297 single-sample plans for controlling variability using the sample variance, and 44 plans using the sample range.

Table A, pp. 458-460, gives the characteristics of plans using the sample variance for $\alpha = .01, .05, .10$; $\beta = .01, .05, .10$; sample size $n = 3(1)31$. Table B, p. 461, gives the characteristics of plans using the sample range for $\alpha = .01, .05$; $\beta = .01, .05$; sample size $n = 2(1)12$.

Burr, I. W. 1953

Engineering statistics and quality control

Table of 148 single-sampling plans for rejecting an excessively variable product.

Table I, p. 422, gives the characteristics of plans that reject when the sample variance exceeds k, for $\alpha = \beta = .1, .05, .02, .01$; $n = 2(1)31, 40(10)100$.

14.129 Auxiliary tables for the design of variables plans

Columbia Univ. 1945

Sequential Analysis

Section 4 gives precepts for constructing Wald sequential tests for acceptance sampling by variables, for one-sided specification limits and known process standard deviation; section 5 does the same for two-sided specification limits; section 6 gives precepts for constructing Wald sequential tests for rejecting excessively variable product. The user must construct the test: no ready-made plans are given. The methods are implemented by the following tables:

Table 4.2, p. 4.24; T. 5.1, p. 5.14; and T. 61, p. 6.29 (they are identical) give $\log_e(1-\alpha) - \log_e \beta$ to 3 dec. for α, β = .001, .01(.01).05(.05).2(.1).4 . Table 5.2, pp. 5.15-5.18, gives $\log_e \cosh u$ to 3 dec. for $u = 0(.01)4$.

14.13 Continuous inspection by attributes

Dodge, H. F. 1943

Annals of Mathematical Statistics 24, 264-279

This paper contains no numerical tables. We cite it for the definition (p. 265) of a CSP-1 plan, quoted below:

"(a) At the outset, inspect 100% of the units consecutively as produced and continue such inspection until i units in succession are found clear of defects.
"(b) When i units in succession are found clear of defects, discontinue 100% inspection, and inspect only a fraction f of the units, selecting individual sample units one at a time from the flow of product, in such a manner as to assure an unbiased sample.
"(c) If a sample unit is found defective, revert immediately to a 100% inspection of succeeding units and continue until again i units in succession are found clear of defects, as in paragraph (a).
"(d) Correct or replace with good units, all defective units found."

Abacs on pages 269 and 272 illustrate the AOQL of such plans.

14.13
14.151

[see also p. 732]

Page, E. S.   1954 a

Biometrika 41, 100-115

The table on p. 115, based on the Poisson approximation, gives the average run length, i.e. the average number of pieces sampled before reaching a decision to reject, for 509 sequential plans for continuous inspection.

Storer, R. L.   1956

Industrial Quality Control 12, no. 11, 48-54

Tables of 70 CSP-1 and 70 CSP-2 plans. [See the following review for these terms] The sampling fraction varies from $\frac{1}{2}\%$ to 8%; the AOQL from .12% to 10.7%; the number of good pieces needed to restore sampling after a defective from 12 to 3200. AQL values from .015% to 10% are shown; they are purely nominal, and are not tied to any value of producer's risk.

Dept. of Defense   1959 b

Handbook 107

Tables of 308 continuous-inspection plans, classified by production rate and AQL. As in Dept. of Defense 1959 a, the AQL is not tied to a specific producer's risk. Production rates run from 2 to 110 001 units per day; AQLs from .015% to 10%.

Three classes of plans are provided, viz:

CSP-1 is the plan of Dodge 1943.

CSP-2 differs from CSP-1 in requiring 100% inspection only after two defectives are found within a run of i inspected units; it was introduced by Dodge & Torrey 1951.

CSP-A shuts down production after $a+1$ defectives are found in a day's work.

Table II, p. 5, indexes the plans by lot size and inspection level (called 'inspection interval' in the caption). Table IIIA, pp. [16-18], gives the characteristics a, f, i and the AOQL of CSP-A plans. Tables III 1 and III 2, pp. 19-20, give the characteristics f and i and the AOQL of CSP-1 and -2 plans.

A feature has been grafted on the CSP-1 and -2 plans that shuts down production after L consecutive units undergo 100% inspection without a run of *i* consecutive good units. Tables IVA and IVB, pp. 21-23, give appropriate values of L.

OC curves for CSP-2 plans are given on 33 plates, and for CSP-A plans on 40 plates; the characteristics of CSP-1 plans are stated to be essentially equivalent to those of CSP-2.

14.15 Life testing

Our principal source of references is Mendenhall 1958; we have taken a narrower view of the subject than he. Sections 1.23, 1.42, 8.2, 8.3, 9.3, 12.5, 12.6 may be consulted for related topics.

14.151 Tables of life-testing plans

Epstein, B. and Sobel, M. 1953
Journal of the American Statistical Association 48, 486-502

Table of 63 destructive sampling plans.

Let the time to failure in a destructive test be exponentially distributed with mean θ. Then if n items are exposed and the time to failure, x_i, of the first r items to fail is recorded, the statistic $r^{-1}[x_1 + \ldots + x_r + (n-r)x_r]$ is an unbiased estimate of θ, and has the same distribution as the mean of a sample of r. The rth failure out of n occurs sooner, on the average, than the rth failure out of r: T. 1, p. 490, gives the ratio of mean times to 2 fig. for r = 1(1)5, 10; n = 1(1)5(5)20; n > r.

Table 2, p. 500, gives characteristics of sampling plans with acceptable mean life θ_1 and rejectable mean life θ_2, for $\theta_2/\theta_1 = 1\frac{1}{2}(\frac{1}{2})3(1)5, 10$; $\alpha = .01, .05, .10$; $\beta = .01, .05, .10$. An abridged version is given as T. 1, p. 561, of Epstein 1954c.

Epstein, B. and Sobel, M. 1955
Annals of Mathematical Statistics 26, 82-93

Table 1, p. 85, gives the average number of pieces destroyed in Wald sequential tests with $\theta_2/\theta_1 = 1\frac{1}{2}(\frac{1}{2})3$; $\alpha = .01, .05$; $\beta = .01, .05$.

14.151
14.159
[see also p. 732]

Sobel, M. and Huyett, Marilyn J. 1957

Bell System Technical Journal 36, 537-576

"Selecting the best one of several binomial populations"

Authors' summary:

"Tables have been prepared for use in any experiment designed to select that particular one of k binomial processes or populations with the highest (long time) yield or the highest probability of success. Before experimentation the experimenter chooses two constants d^* and P^* ($0 < d^* \leq 1$; $0 \leq P^* \leq 1$) and specifies that he would like to guarantee a probability of at least P^* of a correct selection whenever the true difference between the long-time yields associated with the best and the second best processes is at least d^*. The tables show the smallest number of units required per process to be put on test to satisfy this specification. Separate tables are given for k = 2, 3, 4 and 10. Each table gives the result for $d^* = 0.05(0.05)0.50$ and for $P^* = 0.50, 0.60, 0.70, 0.75, 0.80, 0.85, 0.90, 0.95$, and 0.99. For values of d^* and P^* not considered in the tables, graphs are given on which interpolation can be carried out. Graphs have also been constructed to make possible an interpolation or extrapolation for other values of k."

14.152 Auxiliary tables of exponentials

Steffensen, J. F. 1938

Skandinavisk Aktuarietidskrift 21, 49-71

The table on pp. 52-69 gives $x/(1-e^{-x})$ to 5 dec. for $x = 0(.01)10$ with Δ. The table on pp. 70-71 gives $\log_e(1+x)$ to 7 dec. for $x = 0(.0005).1$ with Δ.

Sherman, J. and Ewell, R. B. 1942

Journal of Physical Chemistry 46, 641-662

Table 2a, pp. 645-652, gives $x/(e^x-1)$, $x^2 e^x/(e^x-1)^2$ and $-\log_e(1-e^{-x})$ to 6 dec. for $x = 0(.005)3$. Table 2b, pp. 653-659, gives the same for $x = 3(.01)8$. Table 2c, pp. 660-662, gives the same for $x = 8(.05)15$.

Peto, S. 1953

Biometrics 9, 320-335

Table A, p. [324], abridged from Steffensen 1938, gives $x/(1-e^{-x})$ to 3 dec. for $x = 0(.01)4.99$. Table B, p. [325], abridged from Sherman & Ewell 1942, gives $x^2 e^x/(e^x-1)^2$ to 3 dec. for $x = 0(.01)4.99$.

Bartlett, M. S. 1953

Philosophical Magazine 44, 249-262

Table 1, pp. 252-253, gives $1 - e^{-x}$, $1 - xe^{-x}/(1-e^{-x})$, $1 - x^2 e^{-x}/(1-e^{-x})^2$, to 4 dec.; and $2 - x^3 e^{-x}(1+e^{-x})/(1-e^{-x})^3$, $2(1-e^{-x}) - 3xe^{-x}$, to 3 dec.; all for $x = 0(.1)8$.

14.159 Other auxiliary tables

Littell, A. S. 1952

Human Biology 24, 87-116

Littell investigates, both algebraically and by empirical sampling, the bias and efficiency of standard actuarial formulas for decrement rates, compared with the estimation of such rates by maximum likelihood, when the true decrement rate is constant (i.e. the distribution of lifetimes is exponential). Tables 2-5, pp. 96-105, give his algebraic results; TT. 8-10, pp. 110-112, the result of his sampling experiment.

Freudenthal, A. M. and Gumbel, E. J. 1953

Proceedings of the Royal Society of London, Series A 216, 309-332

Table 1, p. 320, gives $x \log_{10} e$, $6^{-\frac{1}{2}} \pi x \log_{10} e$ and $[(2x)!-(x!)^2]^{\frac{1}{2}}$; all to 5 dec. for $x = .05(.05)1$.

Freudenthal, A. M. and Gumbel, E. J. 1954

Journal of the American Statistical Association 49, 575-597

Table 1, pp. 584-586, computed by Gladys R. Garabedian, gives $[(2x)!-(x!)^2]^{-\frac{1}{2}}$, $[1-x!][(2x)!-(x!)^2]^{-\frac{1}{2}}$ and $[(3x)! - 3(2x)!x! + 2(x!)^3][(2x)! - (x!)^2]^{-3/2}$; all to 4 dec. for $x = .01(.01)1(.2)2(1)5$.

Sobel, M. 1956

Bell System Technical Journal 35, 179-206

Given two processes, the units from which fail in accordance with an exponential law, Sobel seeks to select the process with the smaller failure rate. He considers three techniques for reducing the average experiment time required to reach a decision:

1. Increase the initial number of units put on test.

2. Replace each failure by a new unit from the same process.

14.159
14.21   [see also p. 732]

    3. Use a sequential test (cf. Epstein & Sobel 1955).

Seven short tables on pp. 182-187 illustrate the efficiency of these techniques.

Epstein, B.   1956 b

<u>Annals</u> <u>of</u> <u>the</u> <u>Institute</u> <u>of</u> <u>Statistical</u> <u>Mathematics</u>, Tokyo 8, 15-26

Epstein shows that in the notation of Epstein & Sobel 1953, abstracted on p. 483 above, $x_r/[\Psi(n) - \Psi(n-r)]$ is an unbiased estimate of . where the Ψ function is defined on p. 163 above. Table 1, p. 23, gives $\Psi(n) - \Psi(n-r)$ and $[\Psi(n) - \Psi(n-r)]^{-1}$ to 4 dec. for $n = 2(1)10$, $r = 1(1)n$. Table 2, p. 23, gives the efficiency of this estimate.

14.2 Control charts

14.21 Control charts for number or per cent defective

Factors for constructing these charts can easily be obtained from tables of percentage points of the Poisson and binomial distributions: see pp. 151, 195. We have in general ignored graphs and nomographs for computing binomial percentage points and three-sigma binomial limits. Such graphs are common in quality control textbooks; see also Koller 1943, pp. 20-25.

Stevens, W. L. 1948
Journal of the Royal Statistical Society, Series B 10, 54-108
"Control by gauging"

This paper is concerned with the efficiency of control by two gauges, set above and below the process mean, as compared to control by mean- and s-charts.

Set the gauges at x_1 and x_2 ($x_1 < x_2$). Let p, q, r be the expected proportions of the sample below x_1, between x_1 and x_2, above x_2; let a, b, c be the numbers observed ($a+b+c = n$).

In large samples, estimates of the process mean and standard deviation from a, b, c are normally distributed with variances $v_m \sigma^2/n$ and $v_s \sigma^2/n$. Table 3, p. 68, gives v_m and v_s, the efficiencies $Ef(m)$ and $Ef(s)$, and $[Ef(m).Ef(s)]^{\frac{1}{2}}$, for $p = r = .01(.01).5$.

Table 4, p. 69, gives all these and v_{ms} (the covariance of the estimates of the mean and standard deviation is $v_{ms} \sigma^2/n$) for $q = .1(.1).9$ and $p/(1-q) = .1(.1).5$.

Grant, E. L. 1952
Statistical quality control

Table F, p. 516, gives upper and lower 3-sigma limits for per cent defective to .01% for process fraction defective .001, .002(.002).02(.005).04(.01).1(.02).2(.05).4; and sample size 100(50)200(100)600(200)1000(500)2000(1000)5000, 10000, 20000, 50000, 100000. Reprinted in Arkin & Colton 1950, T. 18, pp. 134-135.

14.22
14.221   [see also p. 733]

14.22 Control charts for variables

14.221 Factors for constructing charts for mean and standard deviation, or for mean and range: A, B, C, D

The factors indexed below follow American practice and are essentially 3-sigma limits. British practice is to set upper (optionally lower) .1% limits (action limits) and $2\frac{1}{2}$% limits (warning limits). The latest edition of <u>British Standard</u> 600, which we have not seen, contains a standard table of factors for computing such limits.

The letters A, B, etc. were introduced in A.S.T.M. 1940 (originally issued 1935); Cowden 1957, pp. 709-711, contains a good set of definitions. The use of n in the denominator of the sample variance has persisted longer in quality control than elsewhere in statistics: it is advisable to check the definition of standard deviation before entering a table.

n =			
2(1)25	$A, A_1, A_2, B_2, B_4, d_2, d_3,$ D_2, D_4 $c_2, 1/c_2, 1/d_2$	3 dec. 4 dec.	A.S.T.M. 1951(115), Duncan 1952(628),
6(1)25	B_1, B_3	3 dec.	1959(886)
7(1)25	D_1, D_3		
2(1)25	$A, A_1, A_2, B_2, B_4, d_2$ c_2	3 dec. 4 dec.	A.S.T.M. 1940(50), Amer. Standards Assoc.
6(1)25	B_1, B_3		1941(50), 1942(39)
7(1)15	D_1, D_3	3 dec.	
2(1)15	D_2, D_4		
2(1)15	$A, A_1, A_2, B_2, B_4, d_2,$ D_2, D_4 c_2	3 dec. 4 dec.	Burr 1953(409)
6(1)15	B_1, B_3	3 dec.	
7(1)15	D_1, D_3		
2(1)25	A_2, A_3, B_4		
7(1)15	D_3	3 dec.	Juran 1951(760)
2(1)15	D_4		
6(1)25	B_3		

14.22
14.221

3(1)16(2)24, 25, 28(2)32, 35, 36, 40(5)60, 70(15)100, 125, 200	$\begin{cases} c_2, B_3 \\ A_1, B_4 \end{cases}$	$\begin{matrix} 4 \text{ dec.} \\ 3 \text{ dec.} \end{matrix}$	Bowker & Goode 1952(197)
2(1)25(5)100	A, B_3		
7(1)20	D_1		
2(1)20	D_3	2 dec.	Grant 1946(539), 1952(515)
6(1)25(5)100	B_1		
2(1)25	A, B_2		
7(1)15	D_1		
2(1)15	D_2	3 dec.	Juran 1951(762)
6(1)25	B_1		
2(1)25	$\begin{cases} A, A_1 \\ 1/(c_2 n^{\frac{1}{2}}) \end{cases}$	$\begin{matrix} 4 \text{ dec.} \\ 5 \text{ dec.} \end{matrix}$	Cowden 1957(693)
2(1)12	$\begin{cases} 1/(d_2 n^{\frac{1}{2}}) \\ A_2 \end{cases}$	$\begin{matrix} 5 \text{ dec.} \\ 4 \text{ dec.} \end{matrix}$	
2(1)25	B, B_2, B_4	4 dec.	Cowden 1957(696)
6(1)25	B_1, B_3		
2(1)12	A_0, A_2, D_4, c_2	3 dec.	Peach 1947(235)
7(1)12	D_3		
2(1)12	D, D_2, D_4	3 dec.	Cowden 1957(697)
7(1)12	D_1, D_3		
2(1)10	$\begin{cases} d_2, A_2, D_4 \\ A_2 n^{\frac{1}{2}} \end{cases}$	$\begin{matrix} 3 \text{ dec.} \\ 2 \text{ dec.} \end{matrix}$	Rice 1947(47)
7(1)10	D_3	3 dec.	
2(1)25	$\begin{cases} 1/c_2 \\ c_2 \end{cases}$	$\begin{matrix} 3 \text{ dec.} \\ 4 \text{ dec.} \end{matrix}$	Cowden 1957(691)
2(1)12	$\begin{cases} 1/d_2 \\ d_2 \end{cases}$	$\begin{matrix} 4 \text{ dec.} \\ 3 \text{ dec.} \end{matrix}$	
2(1)25(5)100	A_1, B_4	2 dec.	E.L. Grant 1946(538), 1952(514)
6(1)25(5)100	B_3		
2(1)20	A_2, D_4	2 dec.	E.L. Grant 1952(513), 1946(537)
7(1)20	D_3		

14.221 [see also pp. 733-734]
14.229

2(1)15	A_2, D_3, D_4		3 dec.	Radford & Waldvogel 1953(524)
2(1)25(5)100	d_2, c_2		4 fig.	E.L. Grant 1952(512), 1946(536); Juran 1951(761)
1(1)30(10)290	A		4 dec.	Cowden 1957(694)

14.222 Factors for constructing charts: other forms

Tables of .1% and $2\frac{1}{2}$% limits (see introduction to section 14.221) are given in O.L. Davies 1949, T. G, pp. 280-283; Dudding & Jennett 1944, TT. 2, 3, p. 59; Mothes 1952, TT. 14-17, pp. 506-509 (text in French). Tables of $\frac{1}{2}$%, 1%, 5% and 10% limits are given in E.S. Pearson & Haines 1935, p. 86.

Simon 1941, T. C, pp. 205-206, gives both 3-sigma and percentage-point limits for means and ranges for n = 2(1)15.

Burington & May 1953, T. 17.43.2, p. 236, give 3-sigma control limits, but do not use the symbols A, B, C, D.

14.229 Miscellaneous tables of control charts for variables

Eisenhart, C. 1949
<u>Industrial Quality Control</u> 6, No. 1, 24-26
    See abstract on p. 149.

Howell, J.M. 1949
<u>Annals of Mathematical Statistics</u> 20, 305-309
"Control chart for largest and smallest values"
    See abstract on p. 60.

Masuyama, M. 1955
<u>Sankhyā</u> 15, 291-294
    The table on pp. 292-293 gives roots x of $I_x(r, n-r+1) = \alpha$ to 5 fig., and roots y of P(y) = x to 4 fig., for α = .005, .025, .975, .995; n = 3(1)10; r = 1(1)n. Reprinted in Vianelli 1959, <u>Prontuario</u> 123, p. 338.

Žaludová, A. H. 1958 a
<u>Revue de Statistique Appliquée</u>, Paris, 6, no. 4, 21-31
    Three short tables setting action and warning limits for a control chart plotting individual values, and comparing the OC of such a chart with mean-range charts and with the chart of Howell 1949.

Table 1, p. 30, giving the limits for sample size 3(1)10, is reprinted in Žaludová 1958 b, T. 1, p. 577.

Page, E. S. 1955

Biometrika 42, 243-257

Table 1, pp. 255-256, gives the average run length, i.e. the average number of items inspected before taking action, under control chart rule IV (p. 246):

"Choose n, N. Take samples of size N. Take action if n consecutive points fall between the warning and action lines or if any point falls outside the action lines."

Page assumes a normal distribution whose mean, originally μ, has shifted to $\mu+\lambda\sigma$; and sets action lines at $\mu \pm B_1 N^{-\frac{1}{2}}$ and warning lines at $\mu \pm B_2 N^{-\frac{1}{2}}$. He tables the average run length for $n = 3, 4$; $B_1 = 3(.125)3.25$; $B_2 = 1(.25)2$; and $\begin{cases} N = 5 & 10(5)20 \\ \lambda = 0(.2)1.8 & 0(.2)1 \end{cases}$.

Table 2 a, p. 257, gives the average run length for $N = 5$; $n = 3$; $B_1 = 2.875(.125)3.25$; $B_2 = 1.5(.25)2$; $\lambda = 0(.2)1$; under control chart rule V (p. 251):

"Choose n, N. Plot the means of samples of size N on a chart on which are drawn two action and two warning lines. Take action if:
(i) any point falls outside the action lines,
or (ii) n consecutive points fall outside the warning lines,
or (iii) two out of any set of n consecutive points fall outside opposite warning lines."

Table 2 b, p. 257, gives the average run length under rule V, where the mean has not slipped but the standard deviation has increased to $K\sigma$, for $K = 1(.25)2.25$; N, n, B_1, B_2 as in T. 2 a.

Enrick 1957, 1958 c gives what purports to be a table of $[xV/E]^2$, where $x = x(1 - \frac{1}{2}p)$, to 0 dec. for $V = .005, .006(.002).01(.005).05, .06(.02).12$; $E = .005(.005).03(.01).06(.02).12(.03).18, .2(.05).3$; $p = .003, .01, .05$. The entries seem to have been rounded **before** squaring, and are perhaps good to 2 fig.

14.3
14.32

[see also pp. 734-735]

14.3 Tolerance limits

14.31 Non-parametric tolerance limits

Wilks, S. S. 1942

Annals of Mathematical Statistics 13, 400-409

Author's summary:

Suppose "an initial sample of n product-pieces, manufactured under a given state of statistical control, are measured with respect to a given quality characteristic. Let X be a variable which measures the given characteristic . . . . Let X_1 be the smallest and X_n the largest value of X which occurs in the initial sample. Now consider a further sample of size N. . . : (1) What is the probability that at least N_0 values of in the second sample will exceed the tolerance limit X_1 set by the first sample? (2) What is the probability that at least N_0 values of X in the second sample will lie between the two tolerance limits X_1 and X_n set by the first sample? (3) For given values of n and N and α (e.g., .99 or .95), what is the largest integer N_α such that the probability is at least α that $N_0 \geq N_\alpha$? (4) What is the limiting value of $N_\alpha/N = R_\alpha$ as N increases indefinitely?"

Table 1, p. 403, gives N_α and R_α for problem (1) for n = 10, 50, 100, 500; n/N = 0(1)2; α = .99, .95. Table 2, p. 406, does the same for problem (2).

For the use of the beta distribution in constructing tolerance limits see Murphy 1948; appropriate tables are indexed in section 3.3.

14.32 Normal tolerance limits

Eisenhart, C., Hastay, M. W. and Wallis, W. A. 1947

Techniques of statistical analysis

Chapter 2, pp. 95-110, by A.H. Bowker, treats tolerance limits for normal distributions. Table 2.1, pp. 102-107, gives factors K such that the probability is γ that at least a proportion P of the distribution will be included between $\bar{X} \pm Ks$, where $\bar{X}$ and s are estimates of the mean and standard deviation computed from a sample of N, to 3 dec. for γ = .75, .9, .95, .99; P = .75, .9, .95, .99, .999; N = 2(1)102(2)180(5)300(10)400(25)750(50)1000, ∞.

Reprinted in Dixon & Massey 1951, T. 16, pp. 334-339; Vianelli 1959, Pro. 100, pp. 269-276.

Proschan, F. 1953

<u>Journal</u> <u>of</u> <u>the</u> <u>American</u> <u>Statistical</u> <u>Association</u> 48, 550-564

Let $\bar{x}$, s be estimates of μ, σ from a sample of n. Table 2, pp. 558-559, gives factors k_i for tolerance intervals $\bar{x} \pm k_5 s$, $\bar{x} \pm k_6 \sigma$, $\mu \pm k_7 s$, that will include, on the average, 50% of the population; and for tolerance intervals $\mu \pm k_8 s$, $\bar{x} \pm k_9 \sigma$, that will include at least 50% of the population 50% of the time; all to 3 dec. for n = 2(1)30 ; 120/n = 0(1)4 . Table 3, p. 561, gives factors for tolerance intervals $\bar{x} \pm ks$ that will include, on the average, 100 a % of the population, to 3 dec. for a = .5, .75, .9, .95, .98, .99, .99 ; and n as in T. 1.

Fraser, D. A. S. and Guttman, I. 1956

<u>Annals</u> <u>of</u> <u>Mathematical</u> <u>Statistics</u> 27, 162-179

The tolerance regions tabled answer to the following problem (cf. Behrens 1929): Given a sample of n from a j-variate normal distribution, to fix an ellipsoid that will contain the next single observation with confidence β. The ellipsoid is given in the form $(\xi-\bar{w})'A^{-1}(\xi-\bar{w}) \leq c_\beta$, where ξ is the new observation, $\bar{w}$ is the sample mean, and A is the sample covariance matrix. Set $\nu = n - j$.

Table 1, p. 174, gives $c_\beta^{\frac{1}{2}}$ to 4 fig. for j = 1 ; β = .995, .99, .975, .95, .9, .75 ; ν = 1(1)30 ; 120/ν = 0(1)4 . Tables 3-5, pp. 176-178, give c_β to 4 fig. for j = 2(1)4 ; β and ν as in T. 1. Table 2, p. 175, gives b_β, used for setting one-dimensional tolerance limits in the form $|\xi - \bar{w}| \leq b_\beta$, to 4 fig. for β and ν as in T. 1.

Tables 3 and 5 are reprinted in Vianelli 1959, Prontuario 209, p. 653; T. 4 is reprinted in Vianelli 1959, T. 188, p. 1353.

Dixon, W. J. and Massey, F. J. 1957

<u>Introduction</u> <u>to</u> <u>statistical</u> <u>analysis</u>

Table 16, pp. 436-437, abridged from Eisenhart et al. 1947 (see abstract on p. 492), gives K to 3 dec. for γ = .75, .9, .95, .99 ; P = .75, .9, .95, .99, .999 ; N = 2(1)27, 30(5)100(10)200(50)300(100)1000, and N = ∞ .

14.32
15.0

Mitra, S. K. 1957

<u>Journal of the American Statistical Association</u> 52, 88-94

Table 1, p. 92, gives factors k_1 such that the probability is β that at least a proportion p of a normal distribution will be included between $\bar{x} \pm k_1 r$ where $\bar{x}$ is the mean and r is the range in a sample of size n: to 3 dec. for β = .75, .9, .95, .99; p = .75, .9, .95, .99, .999; n = 2(1)20

Table 2, p. 93, gives factors k_2 such that the probability is β that at least a proportion p will be included between $\bar{\bar{x}} \pm k_2 \bar{r}$ where $\bar{\bar{x}}$ is the grand mean and $\bar{r}$ is the mean range in N samples of size 4: to 3 dec. for N = 4(1)20(5)30(10)50(25)125, ∞; β and p as in T. 1. Table 3, p. 94, does the same for N samples of size 5, omitting N = 125.

Reprinted in Vianelli 1959, <u>Prontuari</u> 101-102, pp. 277-279.

Lieberman, G. J. 1958

<u>Industrial Quality Control</u> 14, no. 10, 7-9

Table 1, p. 8, gives factors K such that the probability is γ that at least a proportion P of a normal distribution will be less than $\bar{x} + Ks$, where $\bar{x}$ and s are estimates of the mean and standard deviation computed from a sample of size n: for P = .75, .9, .95, .99, .999 and
$\begin{cases} \gamma = .75, .9, .95 \quad .99 \\ n = 3(1)25(5)50 \quad 6(1)25(5)50 \end{cases}$

SECTION 15

TABLES FOR THE DESIGN OF EXPERIMENTS

15.0 Introduction

In this section we catalogue tables of combinatorial patterns, usable as experimental designs, and two small groups of tables concerned with the analysis of experimental data and the determination of sample size. The concepts involved in the construction of combinatorial patterns are numerous and highly specialized. Accordingly, we attempt here neither to develop these patterns from first principles nor to discuss the choice of patterns, their applicability in various experimental situations, or their comparative efficiency. Whenever we introduce a concept, we are citing the discussion of such concept in three text books, viz: Cochran & Cox 1950, 1957; Davies 1954; Kempthorne 1952.

The combinatorial patterns will be listed under two main subsections: factorial patterns (15.1) and incomplete block patterns (15.2). In 15.1 we shall list combinatorial patterns that have arisen from--or are closely associated with--experimental situations involving several factors: Cochran & Cox 1950, 122-207; 1957, 148-292; Davies 1954, 247-494; Kempthorne 1952, 234-364, 390-429. In 15.2 we shall list combinatorial patterns that have arisen where the number of experimental units (i.e. plots) available in a homogeneous group called a block is smaller than the number of experimental treatments (or varieties) to be compared: Cochran & Cox 1950, 259-390; 1957, 376-544; Davies 1954, 199-246; Kempthorne 1952, 444, 526-528. It will be immediately recognized that there is considerable overlap between these two categories: Kempthorne 1952, 430-506. This overlap will be covered by cross references: see section 15.12.

A minor section (15.3) deals with tables facilitating the choice of sample size.

15.1
15.10

15.1 Factorial patterns: Latin squares, etc.

15.10 Introduction: blocking and unblocking; effects and estimates

A factorial experiment is concerned with measuring the response of experimental units* to the simultaneous application of n factors†. If there are v_i different levels‡ at which the ith factor is to be applied, then the total number of treatment combinations is $T = v_1 v_2 \ldots v_n$.

The classification of factorial patterns adopted below reflects their use rather than their construction: for each pattern, it displays the number of factors, and the number of levels of each; the total number of experimental units; and the facilities, if any, for control of heterogeneity.

Consider first factorial patterns without control of heterogeneity. We shall call the specification of levels to be applied to an experimental unit a <u>trial</u>; normally the number of experimental units equals the number of trials, but in the execution of an experiment each trial of the factorial pattern may be replicated into several experimental units. The notation 2^7 in 32 indicates that 7 factors, each at 2 levels, appear in 32 trials; while the notation $4.3^2.2^4$ in 120 indicates that 1 factor at 4 levels, 2 at 3 levels and 4 at 2 levels, appear in 120 trials. Such designations do not completely specify the combinatorial pattern; there are, for example, several different patterns that we would denote 2^4 in 8; three of these, in the notation of N.B.S. 1957, are obtained from the identities $I = AB$, $I = ABC$, $I = ABCD$; others can be constructed <u>ad hoc</u>. In classical factorial patterns, the number of trials equals T, or is an integer multiple or submultiple of T; the catalogue below takes no special notice of exceptions to this rule.

A factorial pattern may afford single, double, or multiple control of

---

*Cochran & Cox 1950, 15; 1957, 15; Davies 1954, 583, s.v. <u>Experiment</u>.

†Cochran & Cox 1950, 122; 1957, 148; Davies 1954, 584; Kempthorne 1952, 284.

‡Cochran & Cox 1950, 122; 1957, 148; Davies 1954, 584, s.v. <u>Factor</u>; Kempthorne 1952, 234.

heterogeneity. This occurs when the trials are divided into sets in one way (when the sets are habitually called blocks*), in two ways✗ (one family of sets called rows, the other columns) or in more ways.

Patterns affording control of heterogeneity can be converted into patterns without control of heterogeneity by assigning to each block one level of a new factor; to each row one level of a first new factor, and to each column one level of a second new factor; etc. A 5 × 5 Latin square, for example, in which one factor at 5 levels appears in 25 trials with 2-way control of heterogeneity, is thus transformed into a 5^3 in 25 pattern in which three factors at 5 levels each (formerly 'letters', 'rows', 'columns') appear in the same 25 trials. It is convenient to regard the transformation that replaces each heterogeneity control by a factor as bringing the combinatorial pattern into canonical form, and the finding lists in sections 15.11 and 15.12 are arranged in order of such canonical forms; for the transformation of the example above, see section 15.115, line 342.

Similarly any factor in a factorial pattern can be transformed into a heterogeneity control.

We shall call the transformation that replaces heterogeneity controls by additional factors <u>unblocking</u>, and the inverse transformation <u>blocking</u>. Blocking must be distinguished from confounding in blocks, the operation of introducing a new factor and then blocking on it: Cochran & Cox 1950, 154; 1957, 183; Davies 1954, 581; Kempthorne 1952, 252.

The finding list below (section 15.11) displays, in five columns:

(1) a line number, for convenient reference;

(2) the canonical, i.e. fully unblocked, form of the pattern;

(3) a syncopated statement of the pattern with the heterogeneity control indicated by the author;

(4) the symmetry properties of the pattern, when these admit a brief description;

---

*Cochran & Cox 1950, 32, 154; 1957, 37, 183; Davies 1954, 580; Kempthorne 1952, 252.

✗Cochran & Cox 1950, 103; 1957, 117; Davies 1954, 581, s.v. <u>Double confounding</u>; 587, s.v. <u>Latin square</u>; Kempthorne 1952, 184, 286.

(5) references to the author index.

Canonical form. In describing the lexicographical order of canonical forms, we introduce temporary definitions to save periphrasis: The height of a factor is the number of levels at which it appears. Like factors are factors of equal height; unlike factors are factors of unequal height, and are distinguished as highest, second highest, etc. The form A precedes the form B if and only if, for some positive integer j, the numbers assigned to A and B by the first (j-1) differentiae below are equal, and the jth differentia assigns a smaller number to A than to B:

1. Height of the highest factor (cf. lines 505/506);

2. Number of unlike factors (cf. lines 229/230);

3, 4, 5. Height of the second (third, fourth) highest factor (cf. lines 322/323);

6. Number of like factors that are like the highest factor, i.e. exponent assigned to the highest factor in the notation adopted (cf. lines 1/2);

7, 8, 9. Number of like factors that are like the second (third, fourth) highest factor (cf. lines 141/142, 322/333);

10. Total number of trials (cf. lines 3/4).

A pattern may appear in the column of canonical forms in as many as three versions, viz:

(a) the immediate result of unblocking;

(b) the result of decomposing each factor in (a) into prime factors;

(c) the result of decomposing each factor in (a) into prime factors, except that factors at 4 levels are not decomposed, and factors at 16 and 64 levels are decomposed into 4^2 and 4^3 respectively.

For example, lines 516, 35, 287 present versions (a), (b), (c) of a pattern for 7 factors at 2 levels each, in 16 blocks of 4 trials each, which unblocks into 16.2^7 in 64.

Heterogeneity controls. The commonest groupings (no control, 1-way control, 2-way control) will be illustrated by reference to section 15.112, line 15. The notation 2^8 in 1.32 indicates that 8 factors, each at 2 levels, appear in a monolith of 32 trials without control of heterogeneity: the distinction between '2^8 in 32' and '2^8 in 1.32' is that the former merely recites the total number of trials, whereas the latter states that no blocking has been applied. The notation 2^5 in 8.4 indicates that 5 factors, each at 2 levels, appear in 32 trials arrayed in 8 blocks of 4 trials each. The notation 2^3 in 1.(8.4) indicates that 3 factors, each at 2 levels, appear in 32 trials arrayed in 8 rows × 4 columns; similarly, the notation 2^3 in 2.(4.4) indicates 3 such factors in 2 arrays, each of 4 rows × 4 columns.*

Patterns for q in 1.(q.q) have been listed simply as Latin squares; for q^2 in 1.(q.q), as Graeco-Latin squares; the order of the square is not specifically indicated, but is the square root of the total number of trials.† Patterns listed as Latin cubes and hypercubes can be used to afford 3-way and 4-way control of heterogeneity; for details see Kishen 1949. Patterns listed as orthogonal squares and cubes are, in fact, complete sets of orthogonal squares and cubes. While they can be used to afford 2-way and 3-way control, we consider that their chief use is as saturated patterns; we discuss them in section 15.1312 below.

Estimability properties. In order to discuss the estimability properties of factorial patterns, we introduce the concepts effect and estimate. A k-factor effect is a real-valued function whose arguments are the levels of k factors. Note that, under this definition, any (k-m) factor effect, and in particular the grand mean, is a fortiori a k-factor effect. For example, given three factors with q, r, s levels, there are qrs linearly independent effects; and these can be chosen to be: the grand mean; $(q-1)+(r-1)+(s-1)$ 1-factor effects, none of which is the grand mean; $(q-1)(r-1)+(r-1)(s-1)+(q-1)(s-1)$ 2-factor effects, none of which is a 1-factor effect; $(q-1)(r-1)(s-1)$ 3-factor effects, none of which is a 2-factor effect. The customary partition into main effects, 2-factor[‡] and

---

*The notation 4.2^2 in (4.4).2 indicates a split-plot pattern, discussed in section 15.1621 below.

†By a slight licence, 20^3 in 1.(20.20) has been listed as a Graeco-Latin square: see section 15.11, lines 365, 372, 548.

‡Some authors, following Yates, speak of a main effect as a zero-order interaction, a 2-factor interaction as a first-order interaction, and so on.

15.10

3-factor interactions is a convenient but not unique choice: Cochran & Cox 1950, 127-146; 1957, 153-175; Davies 1954, 250-257; Kempthorne 1952, 235-241, 291-298, 318-321.

An <u>estimate</u> of an effect is a homogeneous linear function of the results of the trials that, evaluated with the values of that effect for the levels of the k factors at each trial as arguments, yields unity. Given two effects, and a proposed estimate of the first effect; if such estimate, evaluated with the values of the second effect for the levels of its factors at each trial as arguments, vanishes, then such estimate is said to be <u>clear</u> of the second effect. If, for all blocks, an estimate of an effect, evaluated with ones as arguments for all trials in a block and zeroes as all other arguments, vanishes, then such estimate is said to be clear of blocks. An estimate not clear of a (factorial or block) effect is said to be <u>confounded</u> with such effect.

Given a set of n factors at $q_1, \ldots, q_n$ levels, a linearly independent set of $q_1 q_2 \ldots q_n$ effects partitioned into grand mean, 1-factor, 2-factor, ..., n-factor effects, and proposed estimates of the grand mean, 1-factor, 2-factor, ..., j-factor effects in such set; then estimates of <u>all</u> j-factor effects follow as linear combinations of such proposed estimates. If each proposed estimate of a 1-factor, ..., j-factor effect in such set is clear of the grand mean; of all 1-factor, ..., k-factor effects in such set <u>except</u> the effect being estimated; and of blocks, if the pattern discussed affords control of heterogeneity; then we say that j-factor estimates are clear of k-factor effects (and of blocks), and we write 'j c k' in the fourth column. This abbreviation applies to the pattern as originally published; we have not studied the estimability of interactions of factors (originally blocks) with other factors.

The abbreviation 'sat.' in the fourth column denotes a <u>saturated</u> pattern; loosely stated, a pattern in which only 1-factor effects are estimable; see section 15.131 below. The absence of a symbol indicates that the estimability structure can not be briefly stated. For patterns given by Brownlee et al. 1948, N.B.S. 1957, and Connor & Zelen 1959, and not marked by a symbol, the usual (not invariable) situation is that estimates of some 2-factor interactions are clear of other 2-factor interactions and of blocks; but that estimates of other 2-factor interactions are confounded with blocks: for details, see the tables cited.

The abbreviation 'part.' denotes partial confounding: a pattern divisible into replicates, with a different set of estimable effects in each replicate. Thus, in section 15.1132, line 149, the notation 4^2 in 3.4.4 indicates a pattern for 2 factors, each at 4 levels, in 48 trials; viz 3 arrays, each of 4 blocks of 4 trials. See section 15.1422; and for the connexion between partially confounded factorials and lattices, see Kempthorne 1952, 445-484.

We have excluded from section 15.11 twelve patterns of the 3^a system and 33 patterns of the 5^a system. In these patterns, due chiefly to Box, and intended for the investigation of response surfaces, the estimated effects are all terms of a second-degree polynomial, amounting to $\frac{1}{2}(a^2+3a+2)$ out of the $(2a^2+1)$ 2-factor effects of a 3^a pattern or the $(8a^2-4a+1)$ 2-factor effects of a 5^a pattern. See sections 15.138 and 15.148.

Section 15.12 is a selective list of factorial patterns that can be obtained by unblocking the incomplete block patterns in section 15.21. These patterns involve, in general, less symmetrical estimates than patterns constructed as factorials; we have not investigated what 2-factor and higher effects may be estimable. The lexicographical order is that used in section 15.11 and defined above. We do not recommend these patterns in preference to the patterns of section 15.11; accordingly, we have excluded certain patterns from section 15.12, as follows:

1. All patterns with only 2 factors are omitted.

2. All simple fractions of the $3^a.2^b$ system, viz 2^B in $[2^b]$, 3^A in $[3^a]$, $3^A.2^B$ in $[3^a 2^b]$, $A > a$, $B > b$, are omitted.

3. If $A < B < C$, and $p^a.q^b.r^c$ in A, $p^a.q^b.r^c$ in C are canonical forms in section 15.11, then $p^a.q^b.r^c$ in B is omitted.

4. If, after deleting a factor from the left-hand side of a pattern, and dividing the right-hand side by such factor, the quotient is a canonical form in section 15.11, the original (undeleted) pattern is omitted.

15.11 [see also p. 735]
15.112

15.11 Finding list of factorial patterns

The following abbreviations are used in the list below:

Latin sq. = Latin square

Graeco-L. sq. = Graeco-Latin square

ortho. sq. = complete set of orthogonal Latin squares

ortho. cubes = complete set of orthogonal Latin cubes

ortho. part'n = orthogonal partition of a Latin square

ortho. mtx. = orthogonal matrix: see Paley 1933, Plackett & Burman 1946

i c j = i-factor interactions clear of j-factor interactions

sat. = saturated: the number of blocks, plus the number of estimable main effects, equals the total number of trials

part. = partial confounding

() round brackets indicate blocking for two-way control of heterogeneity

___ underscoring indicates factors applied to split plots

For further information, consult section 15.10 above.

15.112 Factorial patterns of the 2^a system

1. 2^3 in 4
   - 2^2 in 2.2 — Kitagawa & Mitome 1953(II.1)
   - 2^3 in 1.4   1 c 1 — Kempthorne 1947(263), Davies 1954(484)
   - Latin sq. — Cayley 1890(136), 1897(56), Kitagawa & Mitome 1953(I.1)

2. 2^4 in 8
   - 2^3 in 2.4 — Kitagawa & Mitome 1953(II.2), Davies 1954(427), Cochran & Cox 1950(186, 208), 1957(215, 234)
   - 2^4 in 1.8   1 c 2 — Kempthorne 1947(263), Davies 1954(484), Cochran & Cox 1957(261, 276), Taguchi 1959(135/3)
   - 4.2 in 2.4 — Binet et al. 1955(35)

3. 2^5 in 8
   - 2^3 in 4.2 — Kitagawa & Mitome 1953(II.2)
   - 2^5 in 1.8   1 c 1 — Kempthorne 1947(263), Davies 1954(484), Cochran & Cox 1957(261, 277)
   - 4.2^3 in 1.8 — Taguchi 1959(136/4)

4. 2^5 in 16 2^4 in 2.8   1 c 2 — Cochran & Cox 1950(186, 208), 1957(215, 234), Kitagawa & Mitome 1953(II.4, II.8), Davies 1954(429), Vianelli 1959(403)

5. 2^5 in 16
- 2^5 in 1.16   2 c 2   Kempthorne 1947(263), Kitagawa & Mitome 1953(III.1), Davies 1954 (485), Cochran & Cox 1957(259, 261, 277), N.B.S. 1957(5)
- 4^2 in 2.8   Kitagawa & Mitome 1953(II.56), Binet et al. 1955(35)
- 4.2^2 in 2.8   Li 1944(461), Cochran & Cox 1950 (186, 216), 1957(215, 242), Kitagawa & Mitome 1953(II.101), Vianelli 1959(406)

6. 2^6 in 8 2^6 in 1.8   1 c 1   Davies 1954(485), Cochran & Cox 1957(261, 278)

7. 2^6 in 16
- 2^4 in 4.4   Cochran & Cox 1950(186, 209), 1957 (215, 235), Kitagawa & Mitome 1953(II.4, II.7), Davies 1954 (428), Vianelli 1959(403)
- 2^5 in 2.8   Cochran & Cox 1957(261, 277), N.B.S. 1957(5)
- 2^6 in 1.16   1 c 2   Kempthorne 1947(263), Davies 1954 (486), Cochran & Cox 1957(261, 278), N.B.S. 1957(18)
- $2^2.2$ in 2.(4.2)   Cochran & Cox 1950(256), 1957(332), Kitagawa & Mitome 1953(V.8)
- 4^2 in 4.4   Li 1944(462), Cochran & Cox 1950 (186, 215), 1957(215, 241), Kitagawa & Mitome 1953(II.58), Binet et al. 1955(35), Vianelli 1959(405)
- 4.2^2 in 4.4   Li 1944(459)
- Latin square   Cayley 1890(136), 1897(56), Euler 1923(299, 301), Fisher & Yates 1938-57(T.XV), Cochran & Cox 1950(119), 1957(145), Kitagawa & Mitome 1953(I.1), Czechowski et al. 1957(100), Vianelli 1959(398)

8. 2^6 in 32
- 2^5 in 2.16   Davies 1954(430)
- 2^6 in 1.32   2 c 3   Brownlee et al. 1948(T.2), Kitagawa & Mitome 1953(III.1), Cochran & Cox 1957(259, 261, 277), N.B.S. 1957(6)

15.112   [see also p. 735]

9. 2^6 in 32
$\begin{cases} 4.2^3 \text{ in } 2.16 & \text{Kitagawa \& Mitome 1953(II.106)} \\ 4^2.2 \text{ in } 2.16 & \text{Li 1944(466),} \\ & \quad \text{Kitagawa \& Mitome 1953(II.105)} \end{cases}$

10. 2^7 in 8
$\begin{cases} 2^7 \text{ in } 1.8 \quad \text{sat.} & \text{Davies 1954(485),} \\ & \quad \text{Cochran \& Cox 1957(261, 280)} \\ \text{ortho. mtx.} \quad \text{sat.} & \text{Plackett \& Burman 1946(323),} \\ & \quad \text{Taguchi 1959(133/1), 1960(177/2)} \end{cases}$

11. 2^7 in 16
$\begin{cases} 2^3 \text{ in } 1.(4.4) & \text{Kitagawa \& Mitome 1953(V.1)} \\ 2^4 \text{ in } 8.2 & \text{Kitagawa \& Mitome 1953(II.4, II.6)} \\ 2^5 \text{ in } 4.4 & \text{Cochran \& Cox 1957(261, 277),} \\ & \quad \text{N.B.S. 1957(5)} \\ 2^6 \text{ in } 2.8 & \text{Cochran \& Cox 1957(261, 278),} \\ & \quad \text{N.B.S. 1957(18)} \\ 2^7 \text{ in } 1.16 \quad 1 \text{ c } 2 & \text{Davies 1954(486), Cochran \& Cox} \\ & \quad \text{1957(261, 281), N.B.S. 1957(30)} \\ \underline{2^4}.\underline{2^2}.2 \text{ in } 4.2.2 & \text{Taguchi 1959(163/31)} \\ & \quad \text{(split-split-plot)} \\ \text{ortho. part'n} & \text{Kitagawa \& Mitome 1953(I.22)} \end{cases}$

12. 2^7 in 32
$\begin{cases} 2^5 \text{ in } 4.8 \quad 1 \text{ c } 2 & \text{Kempthorne 1947(265), Cochran \&} \\ & \quad \text{Cox 1950(186, 209), 1957(215,} \\ & \quad \text{235), Kitagawa \& Mitome 1953} \\ & \quad \text{(II.9, II.14), Davies 1954(429),} \\ & \quad \text{Vianelli 1959(403)} \\ 2^6 \text{ in } 2.16 \quad 2 \text{ c } 3 & \text{Cochran \& Cox 1957(259, 261, 279),} \\ & \quad \text{N.B.S. 1957(7)} \\ 2^7 \text{ in } 1.32 \quad 1 \text{ c } 2 & \text{Kempthorne 1947(263), Brownlee et} \\ & \quad \text{al. 1948(T.2), Cochran \& Cox} \\ & \quad \text{1957(261, 282), N.B.S. 1957(19)} \\ 4.2^3 \text{ in } 4.8 & \text{Li 1944(468),} \\ & \quad \text{Kitagawa \& Mitome 1953(II.104)} \\ 4.2^4 \text{ in } 2.16 \quad 2 \text{ c } 2 & \text{Cochran \& Cox 1957(270, 292)} \\ 4^2.2 \text{ in } 4.8 & \text{Li 1944(464),} \\ & \quad \text{Kitagawa \& Mitome 1953(II.125)} \end{cases}$

13. 2^7 in 64 2^7 in 1.64   3 c 3   Brownlee et al. 1948(T.3),
                                          Kitagawa & Mitome 1953(III.1),
                                          Cochran & Cox 1957(259, 261,
                                          284), N.B.S. 1957(7)

15.112

14. 2^8 in 16	2^6 in 4.4	1 c 2	Cochran & Cox 1957(261, 278), N.B.S. 1957(18)
	2^7 in 2.8		Cochran & Cox 1957(261, 281), N.B.S. 1957(30)
	2^8 in 1.16	1 c 2	Brownlee et al. 1948(270), Davies 1954(486), Cochran & Cox 1957 (261, 285), N.B.S. 1957(41)
	Graeco-Latin sq.		Euler 1923(347, 348, 448), Cochran & Cox 1950(120), 1957 (146), Vianelli 1959(402)
	ortho. partition		Kitagawa & Mitome 1953(I.22)
15. 2^8 in 32	2^3 in 1.(8.4)		Yates 1937(32)
	2^3 in 2.(4.4)		Yates 1937(32), Cochran & Cox 1950(252), 1957(328), Vianelli 1959(407)
	2^5 in 8.4		Kitagawa & Mitome 1953(II.9, II.17), Davies 1954(428)
	2^6 in 4.8		Kempthorne 1947(266), Brownlee et al. 1948(T.2), Kitagawa & Mitome 1953(III.3, III.4), Cochran & Cox 1957(261, 279), N.B.S. 1957(6)
	2^7 in 2.16		Cochran & Cox 1957(261, 282), N.B.S. 1957(20)
	2^8 in 1.32	1 c 2	Kempthorne 1947(266), Brownlee et al. 1948(T.2), Cochran & Cox 1957(261, 286), N.B.S. 1957(30)
	4.2^2 in (4.4).2		Kitagawa & Mitome 1953(IV.2)
	4.2^4 in 4.8		Brownlee et al. 1948(275), Cochran & Cox 1957(270, 292)
16. 2^8 in 64	2^6 in 4.16	1 c 2	Cochran & Cox 1950(186, 208), 1957(215, 234), Kitagawa & Mitome 1953(II.21, II.23), Davies 1954(430), Vianelli 1959(403)
	2^7 in 2.32	2 c 4	Cochran & Cox 1957(259, 261, 284), N.B.S. 1957(8)

15.112 [see also p. 735]

17. 2^8 in 64	2^8 in 1.64		2 c 2	Brownlee et al. 1948(T.3), Kitagawa & Mitome 1953(III.10), Cochran & Cox 1957(259, 261, 288), N.B.S. 1957(20)
	4^3 in 4.16			Kitagawa & Mitome 1953(II.68), Binet et al. 1955(42)
18. 2^8 in 128	2^8 in 1.128		3 c 4	Cochran & Cox 1957(259, 261, 289), N.B.S. 1957(9)
19. 2^9 in 16	2^5 in 1.(4.4)			N.B.S. 1957(5), Taguchi 1959(141/9)
	2^6 in 8.2			N.B.S. 1957(18)
	2^7 in 4.4			Cochran & Cox 1957(261, 280), N.B.S. 1957(30)
	2^8 in 2.8			Cochran & Cox 1957(261, 285), N.B.S. 1957(42)
20. 2^9 in 32	2^6 in 8.4			Brownlee et al. 1948(T.2), Cochran & Cox 1957(261, 279), N.B.S. 1957(6)
	2^7 in 4.8			Brownlee et al. 1948(T.2), Cochran & Cox 1957(261, 282), N.B.S. 1957(19)
	2^8 in 2.16			Cochran & Cox 1957(261, 286), N.B.S. 1957(31)
	2^9 in 1.32		1 c 2	Brownlee et al. 1948(T.2), N.B.S. 1957(42)
	4.2^3 in (4.4).2			Kitagawa & Mitome 1953(IV.3)
21. 2^9 in 64	2^6 in 8.8			Cochran & Cox 1950(186, 210), Kitagawa & Mitome 1953(II.22), Davies 1954(429), Vianelli 1959(404)
	2^7 in 4.16		2 c 4	Brownlee et al. 1948(T.3), Cochran & Cox 1957(259, 261, 284), N.B.S. 1957(8)
	2^8 in 2.32		2 c 2	Cochran & Cox 1957(259, 261, 288), N.B.S. 1957(22)
	2^9 in 1.64		1 c 2	Brownlee et al. 1948(T.3), N.B.S. 1957(32)

15.112

22. 2^9 in 64	$\underline{2^3}.2$ in $2.(8.4)$			Cochran & Cox 1950(257), 1957(333), Kitagawa & Mitome 1953(V.9), Vianelli 1959(409)
	$\underline{2^3}.2^2$ in $[2.2].(4.4)$			Cochran & Cox 1950(258), 1957(334), Kitagawa & Mitome 1953(V.13), Vianelli 1959(409)
	4^3 in 8.8			Kitagawa & Mitome 1953(II.62)
	$4.\underline{2^3}$ in $4.(8.2)$			Yates 1937(79), Kitagawa & Mitome 1953(V.10)
	$4.\underline{2^3}$ in $(4.4).4$			Kitagawa & Mitome 1953(IV.3)
	$4^2.\underline{2^3}$ in 4.16	2 c 2		Brownlee et al. 1948(275)
	$\underline{8}.4$ in $4.(8.2)$			Yates 1937(79)
	Latin square			Euler 1923(343), Fisher & Yates 1938-57(T.XV), Cochran & Cox 1950(119), 1957(145), Kitagawa & Mitome 1953(I.16), Czechowski et al. 1957(100), Vianelli 1959(400)
23. 2^9 in 128	2^7 in 4.32			Kitagawa & Mitome 1953(II.24)
	2^8 in 2.64	3 c 4		Cochran & Cox 1957(259, 261, 289), N.B.S. 1957(12)
	2^9 in 1.128	2 c 3		Brownlee et al. 1948(274), N.B.S. 1957(22)
24. 2^9 in 256	2^9 in 1.256	4 c 4		N.B.S. 1957(12)
25. 2^{10} in 16	2^6 in $1.(4.4)$			N.B.S. 1957(18)
	2^6 in $1.(8.2)$			N.B.S. 1957(18)
	2^8 in 4.4			Brownlee et al. 1948(290), Cochran & Cox 1957(261, 285), N.B.S. 1957(41)
	orthogonal squares			Fisher & Yates 1938-57 (T.XVI), Bose & Nair 1941 (381)*, Kitagawa & Mitome 1953(I.18), Czechowski et al. 1957(100), Vianelli 1959(400)

*In symbolic form

15.112

26. 2^{10} in 16   orthogonal matrix   [see also p. 735]   Taguchi 1960(221/46)

27. 2^{10} in 32
- 2^7 in 8.4   Brownlee et al. 1948(T.2), Cochran & Cox 1957(261, 281), N.B.S. 1957(19)
- 2^8 in 4.8   Brownlee et al. 1948(T.2), Cochran & Cox 1957(261, 286), N.B.S. 1957(31)
- 2^9 in 2.16   N.B.S. 1957(43)
- 2^{10} in 1.32   1 c 2   Brownlee et al. 1948(T.2), N.B.S. 1957(52)

28. 2^{10} in 64
- 2^4 in 1.(8.8)   Cochran & Cox 1950(252), 1957(328), Kitagawa & Mitome 1953(V.2), Vianelli 1959(407)
- 2^6 in 16.4   Davies 1954(428)
- 2^7 in 8.8   2 c 4   Brownlee et al. 1948(T.3), Kitagawa & Mitome 1953 (III.3, III.7), Cochran & Cox 1957(259, 261, 284), N.B.S. 1957(8)
- 2^8 in 4.16   2 c 2   Brownlee et al. 1948(T.3), Kitagawa & Mitome 1953 (III.12), Cochran & Cox 1957(259, 261, 287), N.B.S. 1957(21)
- 2^9 in 2.32   N.B.S. 1957(33)
- 2^{10} in 1.64   1 c 2   Brownlee et al. 1948(T.3), N.B.S. 1957(43)
- $2^{\underline{4}}.2$ in 2.(8.4)   Cochran & Cox 1950(258), 1957(334), Kitagawa & Mitome 1953(V.9), Vianelli 1959(409)
- $2^{\underline{4}}.2^2$ in [2.2].(4.4)   Cochran & Cox 1950(258), 1957(334), Kitagawa & Mitome 1953(V.14), Vianelli 1959(409)
- 4^3 in 16.4   Kitagawa & Mitome 1953 (II.65)
- 4.2^2 in 1.(8.8)   Cochran & Cox 1950(256), 1957(332)

15.112

29.	2^{10} in 64	Latin cube		Kishen 1949(41)
30.	2^{10} in 128	2^7 in 8.16		Kitagawa & Mitome 1953(II.24), Davies 1954(430)
		2^8 in 4.32	3 c 4	Cochran & Cox 1957(259, 261, 289), N.B.S. 1957(11)
		2^9 in 2.64	2 c 3	N.B.S. 1957(24)
		2^{10} in 1.128	2 c 2	Kempthorne 1947(264), Brownlee et al. 1948(274), N.B.S. 1957(34)
31.	2^{10} in 256	2^8 in 4.64		Kitagawa & Mitome 1953(II.16)
		2^9 in 2.128	4 c 4	N.B.S. 1957(17)
		2^{10} in 1.256	2 c 3	N.B.S. 1957(25)
32.	2^{11} in 12	ortho. matrix	sat.	Plackett & Burman 1946(323), Taguchi 1960(177/2)
33.	2^{11} in 16	2^7 in 1.(4.4)		N.B.S. 1957(30)
34.	2^{11} in 32	2^6 in 1.(8.4)		Brownlee et al. 1948(T.2), N.B.S. 1957(6)
		2^8 in 8.4		Brownlee et al. 1948(T.2), Cochran & Cox 1957(261, 286), N.B.S. 1957(30)
		2^9 in 4.8		Brownlee et al. 1948(T.2), N.B.S. 1957(42)
		2^{10} in 2.16		N.B.S. 1957(53)
		2^{11} in 1.32		N.B.S. 1957(58)
35.	2^{11} in 64	2^5 in 1.(8.8)		Yates 1937(35), Cochran & Cox 1950(253), 1957(329), Kitagawa & Mitome 1953(V.2), Vianelli 1959(407)
		2^7 in 16.4		Brownlee et al. 1948(T.3), Cochran & Cox 1957(261, 283), N.B.S. 1957(7)
		2^8 in 8.8		Brownlee et al. 1948(T.3), Cochran & Cox 1957(261, 287), N.B.S. 1957(21)
		2^9 in 4.16		Brownlee et al. 1948(T.3), N.B.S. 1957(33)
		2^{10} in 2.32		N.B.S. 1957(44)

15.112 [see also p. 735]

36. 2^{11} in 64	2^{11} in 1.64	1 c 2	Brownlee et al. 1948(T.3), N.B.S. 1957(53)	
	4.2^3 in 1.(8.8)		Kitagawa & Mitome 1953(V.5)	
37. 2^{11} in 128	2^7 in 16.8		Kitagawa & Mitome 1953(II.24), Davies 1954(429)	
	2^8 in 8.16	3 c 4	Kitagawa & Mitome 1953(III.3, III.8), Cochran & Cox 1957(259, 261, 288), N.B.S. 1957(10)	
	2^9 in 4.32	2 c 3	N.B.S. 1957(24)	
	2^{10} in 2.64	2 c 2	N.B.S. 1957(36)	
	2^{11} in 1.128	2 c 2	Brownlee et al. 1948(274), N.B.S. 1957(44)	
38. 2^{11} in 256	2^8 in 8.32		Kitagawa & Mitome 1953(II.27)	
	2^9 in 4.64	2 c 6	N.B.S. 1957(16)	
	2^{10} in 2.128	2 c 3	N.B.S. 1957(29)	
	2^{11} in 1.256	2 c 3	N.B.S. 1957(36)	
39. 2^{11} in 512	2^9 in 4.128		Kitagawa & Mitome 1953(II.29)	
40. 2^{12} in 32	2^7 in 1.(8.4)		Brownlee et al. 1948(T.2), N.B.S. 1957(19)	
	2^9 in 8.4		Brownlee et al. 1948(T.2), N.B.S. 1957(42)	
	2^{10} in 4.8		Brownlee et al. 1948(T.2), N.B.S. 1957(52)	
	2^{11} in 2.16		N.B.S. 1957(58)	
	2^{12} in 1.32	1 c 2	N.B.S. 1957(65)	
41. 2^{12} in 64	2^6 in 1.(8.8)		Yates 1937(35), Cochran & Cox 1950(253), 1957(329), Kitagawa & Mitome 1953(V.3), Vianelli 1959(407)	
	2^8 in 16.4		Brownlee et al. 1948(T.3), Cochran & Cox 1957(261, 287), N.B.S. 1957(20)	
	2^9 in 8.8		Brownlee et al. 1948(T.3)	
	2^{10} in 4.16		Brownlee et al. 1948(T.3), N.B.S. 1957(44)	
	2^{11} in 2.32		N.B.S. 1957(54)	

15.112

42. 2^{12} in 64 $\begin{cases} 2^{12} \text{ in } 1.64 & 1 \text{ c } 2 \quad \text{N.B.S. 1957(59)} \\ 8.2^3 \text{ in } 1.(8.8) & \quad \text{Kitagawa \& Mitome 1953(V.6)} \\ \text{Graeco-L. sq.} & \quad \text{Euler 1923(345, 346), Cochran \&} \\ & \qquad \text{Cox 1950(120), 1957(146),} \\ & \qquad \text{Vianelli 1959(402)} \end{cases}$

43. 2^{12} in 128 $\begin{cases} 2^8 \text{ in } 16.8 & \qquad \text{Cochran \& Cox 1957(261, 289),} \\ & \qquad \text{N.B.S. 1957(10)} \\ 2^9 \text{ in } 8.16 & 2 \text{ c } 3 \quad \text{Brownlee et al. 1948(274),} \\ & \qquad \text{N.B.S. 1957(23)} \\ 2^{10} \text{ in } 4.32 & 2 \text{ c } 2 \quad \text{N.B.S. 1957(35)} \\ 2^{11} \text{ in } 2.64 & 2 \text{ c } 2 \quad \text{N.B.S. 1957(46)} \\ 2^{12} \text{ in } 1.128 & 1 \text{ c } 2 \quad \text{N.B.S. 1957(55)} \end{cases}$

44. 2^{12} in 256 $\begin{cases} 2^9 \text{ in } 8.32 & 3 \text{ c } 5 \quad \text{N.B.S. 1957(15)} \\ 2^{10} \text{ in } 4.64 & 2 \text{ c } 3 \quad \text{N.B.S. 1957(28)} \\ 2^{11} \text{ in } 2.128 & 2 \text{ c } 3 \quad \text{N.B.S. 1957(40)} \\ 2^{12} \text{ in } 1.256 & 2 \text{ c } 3 \quad \text{Kempthorne 1947(264), Brownlee} \\ & \qquad \text{et al. 1948(274),} \\ & \qquad \text{N.B.S. 1957(47)} \\ \text{Latin cubes} & \qquad \text{Kishen 1949(43)} \end{cases}$

45. 2^{12} in 512 2^9 in 8.64   Kitagawa & Mitome 1953(II.30)

46. 2^{13} in 32 $\begin{cases} 2^8 \text{ in } 1.(8.4) & \qquad \text{N.B.S. 1957(30, 31)} \\ 2^{10} \text{ in } 8.4 & \qquad \text{Brownlee et al. 1948(T.2),} \\ & \qquad \text{N.B.S. 1957(52)} \\ 2^{11} \text{ in } 4.8 & \qquad \text{N.B.S. 1957(58)} \\ 2^{12} \text{ in } 2.16 & \qquad \text{N.B.S. 1957(65, 66)} \\ 2^{13} \text{ in } 1.32 & 1 \text{ c } 2 \quad \text{N.B.S. 1957(74)} \end{cases}$

47. 2^{13} in 64 $\begin{cases} 2^7 \text{ in } 1.(8.8) & 2 \text{ c } 4 \quad \text{Brownlee et al. 1948(T.3),} \\ & \qquad \text{N.B.S. 1957(8)} \\ 2^7 \text{ in } 1.(16.4) & \qquad \text{Brownlee et al. 1948(T.3),} \\ & \qquad \text{N.B.S. 1957(7)} \\ 2^9 \text{ in } 16.4 & \qquad \text{Brownlee et al. 1948(T.3),} \\ & \qquad \text{N.B.S. 1957(32)} \\ 2^{10} \text{ in } 8.8 & \qquad \text{Brownlee et al. 1948(T.3),} \\ & \qquad \text{N.B.S. 1957(43)} \\ 2^{11} \text{ in } 4.16 & \qquad \text{Brownlee et al. 1948(T.3),} \\ & \qquad \text{N.B.S. 1957(54)} \end{cases}$

15.112 [see also p. 735]

48. 2^{13} in 64	2^{12} in 2.32		N.B.S. 1957(59, 60)	
	2^{13} in 1.64	1 c 2	N.B.S. 1957(66)	
49. 2^{13} in 128	2^{6} in 2.(8.8)	part.	Yates 1937(35)	
	2^{8} in 32.4		N.B.S. 1957(9)	
	2^{9} in 16.8		N.B.S. 1957(22)	
	2^{10} in 8.16	2 c 2	Kitagawa & Mitome 1953(III.14), N.B.S. 1957(34)	
	2^{11} in 4.32	2 c 2	N.B.S. 1957(46)	
	2^{12} in 2.64		N.B.S. 1957(57)	
	2^{13} in 1.128	1 c 2	N.B.S. 1957(60)	
50. 2^{13} in 256	2^{9} in 16.16	2 c 6	N.B.S. 1957(14)	
	2^{10} in 8.32	2 c 3	N.B.S. 1957(27)	
	2^{11} in 4.64	2 c 3	N.B.S. 1957(40)	
	2^{12} in 2.128	2 c 3	N.B.S. 1957(51)	
	2^{13} in 1.256	2 c 2	Brownlee et al. 1948(274)	
51. 2^{13} in 512	2^{9} in 16.32		Kitagawa & Mitome 1953(II.31)	
52. 2^{14} in 32	2^{12} in 4.8		N.B.S. 1957(65)	
	2^{13} in 2.16		N.B.S. 1957(74, 75)	
53. 2^{14} in 64	2^{8} in 1.(8.8)		Brownlee et al. 1948(T.3), N.B.S. 1957(21)	
	2^{8} in 1.(16.4)		Brownlee et al. 1948(T.3), N.B.S. 1957(20, 21)	
	2^{10} in 16.4		Brownlee et al. 1948(T.3)	
	2^{11} in 8.8		Brownlee et al. 1948(T.3), N.B.S. 1957(53)	
	2^{12} in 4.16		N.B.S. 1957(59)	
	2^{13} in 2.32		N.B.S. 1957(66, 67)	
	2^{14} in 1.64	1 c 2	N.B.S. 1957(76)	
54. 2^{14} in 128	2^{10} in 16.8		N.B.S. 1957(34)	
	2^{11} in 8.16	2 c 2	Kitagawa & Mitome 1953(III.16), N.B.S. 1957(45)	
	2^{12} in 4.32		N.B.S. 1957(56)	
	2^{13} in 2.64		N.B.S. 1957(60, 62)	
	2^{14} in 1.128	1 c 2	N.B.S. 1957(68)	

15.112

55.	2^{14} in 256	2^9 in 32.8			N.B.S. 1957(12)
		2^{10} in 16.16	2 c 3		N.B.S. 1957(26)
		2^{11} in 8.32	2 c 3		N.B.S. 1957(39)
		2^{12} in 4.64	2 c 3		N.B.S. 1957(50)
		2^{14} in 1.256	2 c 2		N.B.S. 1957(62)
		Latin hypercube			Kishen 1949(45)
56.	2^{15} in 16	ortho. matrix	sat.		Plackett & Burman 1946(323), Taguchi 1959(137/5), 1960(178/3)
57.	2^{15} in 32	2^{13} in 4.8			N.B.S. 1957(74)
58.	2^{15} in 64	2^9 in 1.(8.8)			Brownlee et al. 1948(T.3), N.B.S. 1957(32)
		2^9 in 1.(16.4)			N.B.S. 1957(32, 33)
		2^{11} in 16.4			Brownlee et al. 1948(T.3)
		2^{12} in 8.8			N.B.S. 1957(59)
		2^{13} in 4.16			N.B.S. 1957(66, 67)
		2^{14} in 2.32			N.B.S. 1957(76, 77)
59.	2^{15} in 128	2^8 in 1.(32.4)			N.B.S. 1957(9)
		2^{11} in 16.8			N.B.S. 1957(44)
		2^{12} in 8.16			N.B.S. 1957(56)
		2^{13} in 4.32			N.B.S. 1957(60, 61)
		2^{14} in 2.64			N.B.S. 1957(68, 69)
		2^{15} in 1.128	1 c 2		N.B.S. 1957(78)
60.	2^{15} in 256	2^{10} in 32.8			N.B.S. 1957(25)
		2^{11} in 16.16	2 c 3		N.B.S. 1957(38)
		2^{12} in 8.32	2 c 3		Brownlee et al. 1948(274), N.B.S. 1957(49)
		2^{14} in 2.128			N.B.S. 1957(62, 64)
		2^{15} in 1.256	2 c 2		N.B.S. 1957(70)
61.	2^{16} in 64	2^{10} in 1.(8.8)			Brownlee et al. 1948(T.3)
		2^{13} in 8.8			N.B.S. 1957(66)
		2^{14} in 4.16			N.B.S. 1957(76, 77)
62.	2^{16} in 128	2^{12} in 16.8			N.B.S. 1957(55)
		2^{13} in 8.16			N.B.S. 1957(60, 61)
		2^{14} in 4.32			N.B.S. 1957(68, 69)

15.112 [see also p. 735]

63.	2^{16} in 128	2^{15} in 2.64		N.B.S. 1957(78, 81)
64.	2^{16} in 256	2^{11} in 32.8		N.B.S. 1957(36)
		2^{12} in 16.16	2 c 3	Kitagawa & Mitome 1953(III.19), N.B.S. 1957(48)
		2^{14} in 4.64	2 c 2	N.B.S. 1957(62, 64)
		2^{15} in 2.128	2 c 2	N.B.S. 1957(70, 73)
65.	2^{16} in 512	2^{16} in 1.512	2 c 3	Brownlee et al. 1948(274)
66.	2^{17} in 64	2^{11} in 1.(8.8)		Brownlee et al. 1948(T.3)
		2^{11} in 1.(16.4)		Brownlee et al. 1948(T.3)
		2^{14} in 8.8		N.B.S. 1957(76)
67.	2^{17} in 128	2^{10} in 1.(16.8)		N.B.S. 1957(34)
		2^{13} in 16.8		N.B.S. 1957(60)
		2^{14} in 8.16		N.B.S. 1957(68)
		2^{15} in 4.32		N.B.S. 1957(78, 80)
68.	2^{17} in 256	2^{9} in 1.(16.16)	2 c 6	N.B.S. 1957(14)
		2^{9} in 1.(32.8)		N.B.S. 1957(12, 15)
		2^{12} in 32.8		N.B.S. 1957(47)
		2^{14} in 8.32	2 c 2	N.B.S. 1957(62, 64)
		2^{15} in 4.64	2 c 2	N.B.S. 1957(70, 72)
		2^{16} in 2.128		N.B.S. 1957(82, 84)
69.	2^{18} in 128	2^{14} in 16.8		N.B.S. 1957(68)
		2^{15} in 8.16		N.B.S. 1957(78, 80)
70.	2^{18} in 256	2^{10} in 1.(16.16)	2 c 3	N.B.S. 1957(26)
		2^{10} in 1.(32.8)		N.B.S. 1957(25, 27)
		2^{14} in 16.16		N.B.S. 1957(62, 63)
		2^{15} in 8.32	2 c 2	N.B.S. 1957(70, 72)
		2^{16} in 4.64		N.B.S. 1957(82, 84)
71.	2^{19} in 20	ortho. matrix	sat.	Plackett & Burman 1946(323)
72.	2^{19} in 32	ortho. matrix		Taguchi 1960(221/46)
73.	2^{19} in 128	2^{15} in 16.8		N.B.S. 1957(78)
74.	2^{19} in 256	2^{11} in 1.(16.16)	2 c 3	N.B.S. 1957(38)
		2^{11} in 1.(32.8)		N.B.S. 1957(36, 39)
		2^{14} in 32.8		N.B.S. 1957(62)

75.	2^{19} in 256	$\begin{cases} 2^{15} \text{ in } 16.16 \\ 2^{16} \text{ in } 8.32 \end{cases}$		N.B.S. 1957(70, 71) N.B.S. 1957(82, 84)
76.	2^{19} in 512	2^{16} in 8.64	2 c 3	Brownlee et al. 1948(274)
77.	2^{20} in 256	$\begin{cases} 2^{15} \text{ in } 32.8 \\ 2^{16} \text{ in } 16.16 \end{cases}$		N.B.S. 1957(70) N.B.S. 1957(82, 83)
78.	2^{21} in 32	$4.2^7.[\underline{2^6} + \underline{4.2^2}]$ in 2.8.2		Taguchi 1959(149/17)
79.	2^{21} in 256	2^{16} in 32.8		N.B.S. 1957(32)
80.	2^{23} in 24	orthogonal matrix	sat.	Plackett & Burman
81.	2^{27} in 28			1946(323)
82.	2^{27} in 64	orthogonal squares		Fisher & Yates 1938-57 (T.XVI), Bose & Nair 1941(381)*, Kitagawa & Mitome 1953(I.8), Czechowski et al. 1957 (101), Vianelli 1959(401)
83.	2^{31} in 32	orthogonal matrix	sat.	Plackett & Burman 1946 (323), Taguchi 1960(181/6)
84.	2^{35} in 36	orthogonal matrix	sat.	Plackett & Burman
85.	2^{39} in 40			1946(323)
86.	2^{42} in 64	orthogonal cubes orthogonal matrix		Kishen 1949(32), Kitagawa & Mitome 1953(I.28) Taguchi 1960(222)
87.	2^{43} in 44			
88.	2^{47} in 48	orthogonal matrix	sat.	Plackett & Burman
89.	2^{51} in 52			1946(323)
90.	2^{55} in 56	orthogonal matrix	sat.	Plackett & Burman
91.	2^{59} in 60			1946(324)
92.	2^{63} in 64	orthogonal matrix	sat.	Plackett & Burman 1946 (324), Taguchi 1960(191)

---

*In symbolic form

15.112
15.113   [see also p. 735]

93. 2^{67} in 68   ortho. mtx. sat.   Plackett & Burman 1946(324)

94. 2^{68} in 256   ortho. sq.   Stevens 1939(90)*, Bose & Nair 1941
                                            (381)*, Kitagawa & Mitome 1953
                                                                          (I.19)*

95. 2^{71} in 72 ⎫
96. 2^{75} in 76 ⎪
97. 2^{79} in 80 ⎪
98. 2^{83} in 84 ⎬ ortho. mtx. sat.   Plackett & Burman 1946(324)
99. 2^{87} in 88 ⎪
100. 2^{95} in 96 ⎪
101. 2^{99} in 100 ⎭

102. 2^{165} in 1024 ⎫
                       ⎬ ortho. sq.   Stevens 1939(92)*
103. 2^{390} in 4096 ⎭

15.113 Factorial patterns of the 3^a system

104. 3^3 in 9
- 3^2 in 3.3   Kitagawa & Mitome 1953(II.33), Davies 1954(435), Binet et al. 1955(23)
- 3^3 in 1.9   Davies 1954(479)
- Latin sq.   Cayley 1890(136), 1897(56), Czechowski et al. 1957(100), Kitagawa & Mitome 1953(I.1), Cochran & Cox 1950(119), 1957(145)

105. 3^4 in 9
- 3^4 in 1.9   sat. Davies 1954(480)
- Graeco-L sq.   sat. Euler 1923(445), Stevens 1939(82), Cochran & Cox 1950(120), 1957(146), Vianelli 1959(402)
- ortho. sq.   sat. Fisher & Yates 1938-57(T.XVI), Bose & Nair 1941(381)*, Kitagawa & Mitome 1953(I.18), Czechowski et al. 1957(100), Vianelli 1959(400)

---

*One square printed in full; 14 additional squares obtained by cyclically permuting all rows except the first.

*In symbolic form.

106.	3^4 in 9	orthogonal matrix	sat.	Plackett & Burman 1946(325), Taguchi 1960(204/29)
107.	3^4 in 27	3^3 in 3.9	1 c 2	Yates 1937(42), K.R. Nair 1938(127), Kempthorne 1947 (268), Cochran & Cox 1950 (186, 212), 1957(215, 238), Kitagawa & Mitome 1953 (II.34, II.36), Davies 1954 (438), Vianelli 1959(404)
		3^4 in 1.27	1 c 2	Davies 1954(481), Cochran & Cox 1957(270, 290), Connor & Zelen 1959(11)
		Latin cube		Kishen 1949(22)
108.	3^5 in 27	3^3 in 9.3		Kitagawa & Mitome 1953(II.34, II.38), Davies 1954(435)
		3^4 in 3.9	1 c 2	Kitagawa & Mitome 1953 (III.23), Cochran & Cox 1957(270, 290), Connor & Zelen 1959(11)
		3^5 in 1.27		Davies 1954(482)
		Latin cube		Kishen 1949(24, 41)
109.	3^5 in 81	3^4 in 3.27		Kitagawa & Mitome 1953(II.53)
		3^5 in 1.81	2 c 2	Cochran & Cox 1957(270, 291), Connor & Zelen 1959(11)
110.	3^6 in 27	3^4 in 9.3	1 c 2	Connor & Zelen 1959(11)
111.	3^6 in 81	3^4 in 9.9	1 c 1	Yates 1937(48), K.R. Nair 1938(129), Kempthorne 1947 (261), Cochran & Cox 1950 (186, 212), 1957(215, 238), Kitagawa & Mitome 1953 (II.43), Davies 1954(439), Vianelli 1959(404)
		3^5 in 3.27	2 c 2	Cochran & Cox 1957(270, 291), Connor & Zelen 1959(12)
		3^6 in 1.81	1 c 2	Connor & Zelen 1959(5, 18)
		$\underline{3^2}.3^2$ in [3.3].(3.3)		Yates 1937(81), Kitagawa & Mitome 1953(V.15)
		$\underline{3^2}.3^2$ in 9.9		Kitagawa & Mitome 1953(V.12)

15.113

112. 3^6 in 81	$\underline{9}.3^2$ in [3.3].(3.3)			Yates 1937(81)
	Latin sq.			Euler 1923(313, 320), Fisher & Yates 1938-57(T.XV), Kitagawa & Mitome 1953 (I.16), Cochran & Cox 1950 (119), 1957(145), Czechowski et al. 1957(100), Vianelli 1959(400)
	Latin cubes			Kishen 1949(43)
113. 3^6 in 243	3^6 in 1.243		2 c 3	Connor & Zelen 1959(13)
114. 3^6 in 729	3^6 in 1.729		5 c 6	Connor & Zelen 1959(5)
115. 3^7 in 81	3^3 in 1.(9.9)			Yates 1937(49), Cochran & Cox 1950(254), 1957(330), Kitagawa & Mitome 1953(V.3), Vianelli 1959(408)
	3^5 in 9.9		1 c 3	Kitagawa & Mitome 1953 (III.24), Cochran & Cox 1957(270, 291), Connor & Zelen 1959(12)
	3^6 in 3.27		1 c 2	Connor & Zelen 1959(18)
	3^7 in 1.81		1 c 2	Connor & Zelen 1959(23)
	$3^{\underline{2}}.3^2$ in 3.(9.3)			Kitagawa & Mitome 1953(V.12)
	$3^{\underline{3}}.3$ in 3.(9.3)			Yates 1937(80)
	$9.\underline{3}.3$ in 3.(9.3)			Yates 1937(80)
	Latin hypercube			Kishen 1949(45)
116. 3^7 in 243	3^6 in 3.81		2 c 3	Connor & Zelen 1959(14)
	3^7 in 1.243		2 c 2	Connor & Zelen 1959(19)
117. 3^7 in 729	3^7 in 1.729		3 c 3	Connor & Zelen 1959(14)
118. 3^8 in 81	3^4 in 1.(9.9)			Yates 1937(49), Cochran & Cox 1950(255), 1957(331), Kitagawa & Mitome 1953(V.4, V.15), Vianelli 1959(408)
	3^5 in 27.3		1 c 3	Connor & Zelen 1959(11)
	3^6 in 9.9		1 c 2	Connor & Zelen 1959(18)
	3^7 in 3.27		1 c 2	Connor & Zelen 1959(25)
	3^8 in 1.81		1 c 2	Connor & Zelen 1959(30)

15.113

119.	3^8 in 81	9.3^2 in 1.(9.9)			Kitagawa & Mitome 1953(V.7)
		Graeco-Latin sq.			Euler 1923(314, 366, 367), Cochran & Cox 1950(120), 1957 (146), Vianelli 1959(402)
120.	3^8 in 243	3^5 in 27.9	1 c 2		Kempthorne 1947(261)
		3^6 in 9.27	2 c 3		Connor & Zelen 1959(14)
		3^7 in 3.81	2 c 2		Connor & Zelen 1959(20)
		3^8 in 1.243	2 c 2		Connor & Zelen 1959(25)
121.	3^8 in 729	3^7 in 3.243	3 c 3		Connor & Zelen 1959(17)
		3^8 in 1.729	2 c 2		Connor & Zelen 1959(20)
122.	3^9 in 81	3^6 in 27.3	1 c 2		Connor & Zelen 1959(18)
		3^7 in 9.9	1 c 2		Connor & Zelen 1959(24)
		3^8 in 3.27	1 c 2		Connor & Zelen 1959(31)
		3^9 in 1.81	1 c 2		Connor & Zelen 1959(34)
123.	3^9 in 243	3^6 in 27.9	1 c 4		Connor & Zelen 1959(13)
		3^7 in 9.27	2 c 2		Connor & Zelen 1959(20)
		3^8 in 3.81	2 c 2		Connor & Zelen 1959(26)
		3^9 in 1.243	2 c 2		Connor & Zelen 1959(31)
124.	3^9 in 729	3^6 in 27.27	1 c 2		Kempthorne 1947(261)
		3^7 in 9.81	2 c 4		Connor & Zelen 1959(17)
		3^8 in 3.243	2 c 2		Connor & Zelen 1959(23)
		3^9 in 1.729	2 c 3		Connor & Zelen 1959(26)
125.	3^{10} in 81	3^7 in 27.3	1 c 2		Connor & Zelen 1959(23)
		3^8 in 9.9	1 c 2		Connor & Zelen 1959(30)
		3^9 in 3.27	1 c 2		Connor & Zelen 1959(35)
126.	3^{10} in 243	3^7 in 27.9	1 c 3		Connor & Zelen 1959(19)
		3^8 in 9.27	2 c 2		Connor & Zelen 1959(26)
		3^9 in 3.81	2 c 2		Connor & Zelen 1959(34)
		3^{10} in 1.243	2 c 2		Connor & Zelen 1959(36)
127.	3^{10} in 729	3^7 in 27.27	2 c 4		Connor & Zelen 1959(14)
		3^8 in 9.81	2 c 2		Connor & Zelen 1959(23)
		3^9 in 3.243	2 c 3		Connor & Zelen 1959(29)
128.	3^{11} in 81	3^8 in 27.3	1 c 2		Connor & Zelen 1959(30)

15.113 [see also p. 735]
15.1132

129.	3^{11} in 81	3^9 in 9.9	1 c 2	Connor & Zelen 1959(35)
130.	3^{11} in 243	$\begin{cases} 3^8 \text{ in } 27.9 \\ 3^9 \text{ in } 9.27 \\ 3^{10} \text{ in } 3.81 \end{cases}$	1 c 3 2 c 2 1 c 3	Connor & Zelen 1959(25) Connor & Zelen 1959(32) Connor & Zelen 1959(37)
131.	3^{11} in 729	$\begin{cases} 3^8 \text{ in } 27.27 \\ 3^9 \text{ in } 9.81 \end{cases}$	2 c 2 2 c 3	Connor & Zelen 1959(20) Connor & Zelen 1959(29)
132.	3^{12} in 81	3^9 in 27.3	1 c 2	Connor & Zelen 1959(34)
133.	3^{12} in 243	$\begin{cases} 3^9 \text{ in } 27.9 \\ 3^{10} \text{ in } 9.27 \end{cases}$	1 c 3 1 c 3	Connor & Zelen 1959(31) Connor & Zelen 1959(37)
134.	3^{12} in 729	3^9 in 27.27	2 c 3	Connor & Zelen 1959(26)
135.	3^{13} in 27	$\begin{cases} \text{ortho. cubes} \\ \text{ortho. mtx.} \end{cases}$	sat. sat.	Kishen 1949(31) Plackett & Burman 1946(325), Taguchi 1960(205/30)
136.	3^{13} in 243	3^{10} in 27.9	1 c 3	Connor & Zelen 1959(36)
137.	3^{20} in 81	ortho. sq.		Fisher & Yates 1938-57(T/XVI), Stevens 1939(88), Bose & Nair 1941(381)*, Kitagawa & Mitome 1953(I.19), Czechowski et al. 1957(101)
138.	3^{40} in 81	ortho. mtx.	sat.	Plackett & Burman 1946(325), Taguchi 1959(208/33)
139.	3^{84} in 729	ortho. sq.		Stevens 1939(91)⁺, Bose & Nair Bose & Nair 1941(381)*, Kitagawa & Mitome 1953(I.20)⁺
140.	3^{328} in 6561	ortho.ssq.		Stevens 1939(92)⁺

15.1132 Factorial patterns of the $3^a.2^b$ system

141.	3.2^2 in 6	3.2 in 2.3	Binet et al. 1955(23)
142.	3.2^3 in 12	3.2^2 in 2.6	Yates 1937(58), Cochran & Cox 1950 (186, 214), 1957(215, 240), Kitagawa & Mitome 1953(II.88), Vianelli 1959(405)

---

*In symbolic form.

⁺One square printed in full; 25 additional squares obtained by cyclically permuting all rows except the first.

143.	3.2^3 in 12	$\{$ 4.3 in 2.6		Binet et al. 1955(35)
		6.2 in 2.6		Binet et al. 1955(51)
144.	3.2^4 in 24	$\{$ 3.2^3 in 2.12		Kitagawa & Mitome 1953(II.91)
		4.3.2 in 2.12		Li 1944(483), Cochran & Cox 1950 (186, 216), 1957(215, 242), Kitagawa & Mitome 1953(II.135), Vianelli 1959(406)
		6.4 in 2.12		Binet et al. 1955(52)
145.	3.2^4 in 36	3.2^4 in 1.36	2 c 2	Connor & Young 1961($T.2^43^1$)
146.	3.2^5 in 24	3.2^3 in 4.6		Yates 1937(61), Cochran & Cox 1950(186, 214), 1957(215, 240), Kitagawa & Mitome 1953(II.90), Vianelli 1959(405)
147.	3.2^5 in 48	$\{$ 3.2^4 in 2.24		Kitagawa & Mitome 1953(II.98)
		3.2^5 in 1.48	2 c 2	Connor & Young 1961($T.2^53^1$)
		4.2^2 in 3.2.8 part.		Cochran & Cox 1950(216), 1957 (215, 242), Vianelli 1959(406)
148.	3.2^6 in 16	ortho. part'n		Kitagawa & Mitome 1953(I.22)
149.	3.2^6 in 48	$\{$ 3.2^4 in 4.12		Kitagawa & Mitome 1953(II.94)
		3.2^6 in 1.48	2 c 2	Connor & Young 1961($T.2^63^1$)
		4^2 in 3.4.4 part.		Cochran & Cox 1950(215), 1957 (215, 241), Vianelli 1959(405)
		$4^2.3$ in 4.12		Li 1944(463, 487), Kitagawa & Mitome 1953(II.129)
150.	3.2^7 in 96	$\{$ 2^4 in 6.4.4 part.		Cochran & Cox 1950(209), 1957 (215, 235), Vianelli 1959(403)
		3.2^7 in 1.96	2 c 2	Connor & Young 1961($T.2^73^1$)
		4.2^3 in 3.4.8 part.		Li 1944
151.	3.2^8 in 96	3.2^8 in 1.96	2 c 2	Connor & Young 1961($T.2^83^1$)
152.	3.2^9 in 128	3.2^9 in 1.128	2 c 2	Connor & Young 1961($T.2^93^1$)
153.	$3^2.2$ in 6	3.2 in 3.2		Binet et al. 1955(23)
154.	$3^2.2$ in 9	3^2 in 4 + 5		Binet et al. 1955(59)
154a	$3^2.2^2$ in 12	4.3 in 3.4		Binet et al. 1955(35)
154b	$3^2.2^2$ in 18	$3^2.2$ in 2.9		Binet et al. 1955(24)

15.1132   [see also p. 735]

155.	$3^2.2^2$ in 18	$3^2.2$ in 2.9		Binet et al. 1955(24)
		6.3 in 2.9		Binet et al. 1955(51)
156.	$3^2.2^3$ in 36	3.2^2 in 3.2.6	part.	Yates 1937(58), Cochran & Cox 1950(214), 1957(215, 240), Vianelli 1959(405)
		$3^2.2^2$ in 2.18		Binet et al. 1955(27)
		$3^2.2^3$ in 1.36	2 c 2	Connor & Young 1961($T.2^3 3^2$)
		6^2 in 2.18		Binet et al. 1955(53)
		$6.3.2$ in 2.18		Binet et al. 1955(54)
157.	$3^2.2^4$ in 36	3.2^2 in 3.(6.2)		Kitagawa & Mitome 1953(V.10)
158.	$3^2.2^4$ in 72	$3^2.2^4$ in 1.72	2 c 2	Connor & Young 1961($T.2^4 3^2$)
159.	$3^2.2^5$ in 72	3.2^3 in 3.4.6	part.	Yates 1937(61), Cochran & Cox 1950(214), 1957(215, 240), Vianelli 1959(405)
		$3^2.2^5$ in 1.72	2 c 2	Connor & Young 1961($T.2^5 3^2$)
160.	$3^2.2^6$ in 48	6.2^3 in 12.4		Kitagawa & Mitome 1953(IV.1)
161.	$3^2.2^6$ in 96	$3^2.2^6$ in 1.96	2 c 2	Connor & Young 1961($T.2^6 3^2$)
162.	$3^2.2^6$ in 144	$4^2.3$ in 3.4.12		Li 1944
163.	$3^2.2^7$ in 144	$3^2.2^7$ in 1.144	2 c 2	Connor & Young 1961($T.2^7 3^2$)
164.	$3^2.2^8$ in 27	$3^2.2^8$ in 1.27		Taguchi 1959(145/13)
165.	$3^2.2^8$ in 144	$3^2.2^8$ in 1.144	2 c 2	Connor & Young 1961($T.2^8 3^2$)
166.	$3^3.2$ in 18	$3^2.2$ in 3.6		Yates 1937(62), Cochran & Cox 1950(186, 215), 1957(215, 241), Kitagawa & Mitome 1953 (II.114), Vianelli 1959(405)
167.	$3^3.2$ in 27	3^3 in 13 + 14		Binet et al. 1955(59)
168.	$3^3.2^2$ in 36	$3^2.2$ in 2.3.6	part.	Yates 1937(62)
		$3^2.2^2$ in 3.12		Li 1944(472), Kitagawa & Mitome 1953(II.132), Binet et al. 1955(29)
		4.3^2 in 3.12		Li 1944(476), Kitagawa & Mitome 1953(II.122)
		$6.3.2$ in 3.12		Binet et al. 1955(54)
169.	$3^3.2^2$ in 54	$3^3.2^2$ in 1.54	2 c 2	Connor & Young 1961($T.2^2 3^3$)

170.	$3^3.2^2$ in 54	$3^3.2^2$ in 1.54	2 c 2	Connor & Young 1961(T.2^23^3)
171.	$3^3.2^3$ in 36	$\underline{3}.2.2$ in 2.(6.3)		Cochran & Cox 1950(256), 1957(332), Kitagawa & Mitome 1953(V.10), Vianelli 1959(409)
		$3.\underline{3}.2$ in 3.(6.2)		Cochran & Cox 1950(257), 1957(333), Kitagawa & Mitome 1953(V.11), Vianelli 1959(409)
		$3^2.2^2$ in 6.6		Kitagawa & Mitome 1953 (II.131), Binet et al. 1955(32)
		Latin square		Fisher & Yates 1934, 1938-57(T.XV), Kitagawa & Mitome 1953(I.10), Cochran & Cox 1950(119), 1957(145), Vianelli 1959(399)
172.	$3^3.2^3$ in 72	3.2^3 in 3.4.6	part.	Yates 1937(63), Vianelli 1959(405)
		$3^2.2$ in 4.3.6	part.	Cochran & Cox 1950(215), 1957(215, 241)
		$3^2.2^2$ in 2.3.12	part.	Li 1944
		$3^3.2^3$ in 1.72	2 c 2	Connor & Young 1961(T.2^33^3)
		4.3^2 in 2.3.12	part.	Li 1944
173.	$3^3.2^4$ in 36	ortho. partition		Kitagawa & Mitome 1953(I.25)
174.	$3^3.2^4$ in 72	$3.\underline{3}.2$ in 2.3.(6.2)		Cochran & Cox 1950(257), 1957(333)
175.	$3^3.2^4$ in 108	$3^3.2^4$ in 1.108	2 c 2	Connor & Young 1961(T.2^43^3)
176.	$3^3.2^4$ in 144	$6.\underline{2}^3$ in (6.6).4		Yates 1937(78)
177.	$3^3.2^4$ in 216	$4.3.2$ in 2.12		Cochran & Cox 1950(216), 1957(215, 242), Vianelli 1959(406)
178.	$3^3.2^5$ in 72	$6.\underline{2}^2$ in (6.6).2		Kitagawa & Mitome 1953(IV.4)
179.	$3^3.2^5$ in 144	$3^3.2^5$ in 1.144	2 c 2	Connor & Young 1961(T.2^53^3)

15.1132   [see also p. 735]

180. $3^3.2^6$ in 144	$6.\underline{2^3}$ in $(6.6).4$			Yates 1937(78), Kitagawa & Mitome 1953(IV.6)
	Latin square			Fisher & Yates 1938-57(T.XV), Cochran & Cox 1950(119), 1957(146), Kitagawa & Mitome 1953(I.16), Czechowski et al. 1957(100), Vianelli 1959(400)
181. $3^3.2^6$ in 288	$3^3.2^6$ in 1.288	2 c 2		Connor & Young 1961(T.$2^6 3^3$)
182. $3^3.2^7$ in 432	$3^3.2^7$ in 1.432	2 c 2		Connor & Young 1961(T.$2^7 3^3$)
183. $3^3.2^{11}$ in 24	$\underline{3^2.2^6}.2^5$ in 8.3			Taguchi 1959(157/25)
184. $3^4.2$ in 54	$3^3.2$ in 3.18			Li 1944(480), Kitagawa & Mitome 1953(II.117)
185. $3^4.2$ in 81	$3^4.2$ in 1.81	2 c 2		Connor & Young 1961(T.$2^1 3^4$)
186. $3^4.2^2$ in 108	3^3 in 4.3.9	part.		Cochran & Cox 1950(212), 1957(215, 238), Vianelli 1959(404)
	$3^3.2$ in 2.3.18	part.		Li 1944
187. $3^4.2^2$ in 162	$3^4.2^2$ in 1.162	2 c 2		Connor & Young 1961(T.$2^2 3^4$)
188. $3^4.2^3$ in 36	$3^2.2$ in 1.(6.6)			Kitagawa & Mitome 1953(V.6)
	$3^5.2$ in 1.27 + $+ 3^4.2$ in 1.9			Taguchi 1959(154/22)
	ortho. partition			Kitagawa & Mitome 1953(I.26)
189. $3^4.2^3$ in 162	$3^4.2^3$ in 1.162	2 c 2		Connor & Young 1961(T.$2^3 3^4$)
190. $3^4.2^4$ in 162	$3^4.2^4$ in 1.162	2 c 2		Connor & Young 1961(T.$2^4 3^4$)
191. $3^4.2^5$ in 216	$3^4.2^5$ in 1.216	2 c 2		Connor & Young 1961(T.$2^5 3^4$)
192. $3^4.2^6$ in 324	$3^4.2^6$ in 1.324	2 c 2		Connor & Young 1961(T.$2^6 3^4$)
193. $3^4.2^8$ in 144	Graeco-L. sq.			Kitagawa & Mitome 1953(I.16), Cochran & Cox 1950(121), 1957(147), Vianell 1959(402)
194. $3^5.2$ in 54	$3^3.2$ in 9.6			Yates 1937, Kitagawa & Mitome 1953(II.118)
195. $3^5.2$ in 162	$3^5.2$ in 1.162	.2 c 2		Connor & Young 1961(T.$2^1 3^5$)
196. $3^5.2^2$ in 108	$3^3.2$ in 2.9.6	part.		Yates 1937
197. $3^5.2^2$ in 162	$3^5.2^2$ in 1.162	2 c 2		Connor & Young 1961(T.$2^2 3^5$)

198.	$3^5.2^3$ in 108	$6.\underline{3}^2$ in $(6.6).3$		Kitagawa & Mitome 1953(IV.7)
199.	$3^5.2^3$ in 216	$3^5.2^3$ in 1.216	2 c 2	Connor & Young 1961($T.2^3 3^5$)
200.	$3^5.2^4$ in 324	$3^5.2^4$ in 1.324	2 c 2	Connor & Young 1961($T.2^4 3^5$)
201.	$3^5.2^5$ in 432	$3^5.2^5$ in 1.432	2 c 2	Connor & Young 1961($T.2^5 3^5$)
202.	$3^6.2$ in 243	$3^6.2$ in 1.243	2 c 2	Connor & Young 1961($T.2^1 3^6$)
203.	$3^6.2^2$ in 324	3^4 in $4.9.9$	part.	Yates 1937(48), Cochran & Cox 1950(212), 1957(215, 238), Vianelli 1959(404)
204.	$3^6.2^2$ in 486	$3^6.2^2$ in 1.486	2 c 2	Connor & Young 1961($T.2^2 3^6$)
205.	$3^6.2^3$ in 27	$3^6.2^3$ in 1.27		Taguchi 1959(144/12)
206.	$3^6.2^3$ in 486	$3^6.2^3$ in 1.486	2 c 2	Connor & Young 1961($T.2^3 3^6$)
207.	$3^6.2^4$ in 486	$3^6.2^4$ in 1.486	2 c 2	Connor & Young 1961($T.2^4 3^6$)
208.	$3^7.2$ in 18	orthogonal matrix		Taguchi 1960(204/29)
209.	$3^7.2$ in 162	3^3 in $2.(9.9)$	part.	Cochran & Cox 1950(254), 1957(330), Vianelli 1959(408)
210.	$3^7.2$ in 243	$3^7.2$ in 1.243	2 c 2	Connor & Young 1961($T.2^1 3^7$)
211.	$3^7.2^2$ in 324	$\{\underline{3^3}.3$ in $4.3.(9.3)$ / $\underline{9.3}.3$ in $4.3.(9.3)\}$	part. part.	Yates 1937(80) Yates 1937(80)
212.	$3^7.2^2$ in 486	$3^7.2^2$ in 1.486	2 c 2	Connor & Young 1961($T.2^2 3^7$)
213.	$3^7.2^3$ in 486	$3^7.2^3$ in 1.486	2 c 2	Connor & Young 1961($T.2^3 3^7$)
214.	$3^8.2$ in 162	3^4 in $2.(9.9)$	part.	Cochran & Cox 1950(255), 1957(331), Vianelli 1959(408)
215.	$3^8.2$ in 243	$3^8.2$ in 1.243	2 c 2	Connor & Young 1961($T.2^1 3^8$)
216.	$3^8.2^2$ in 486	$3^8.2^2$ in 1.486	2 c 2	Connor & Young 1961($T.2^2 3^8$)
217.	$3^9.2$ in 243	$3^9.2$ in 1.243	2 c 2	Connor & Young 1961($T.2^1 3^9$)
218.	$3^{12}.2^{11}$ in 36	orthogonal matrix		Taguchi 1960(225/50)
219.	$3^{25}.2$ in 54	orthogonal matrix		Taguchi 1960(207)

15.114   [see also p. 735]
15.114 Factorial patterns with one or more factors at 4 levels

220.	4^3 in 16	4^2 in 4.4	Li 1944(462), Cochran & Cox 1950 (186, 215), 1957(215, 241), Kitagawa & Mitome 1953(II.58), Binet et al. 1955(35), Vianelli 1959(405)
		Latin sq.	Cayley 1890(136), 1897(56), Euler 1923(299, 301), Fisher & Yates 1938-57(T.XV), Cochran & Cox 1950(119), 1957(143), Kitagawa & Mitome 1953(I.1), Czechowski et al. 1957(100), Vianelli 1959(398)
221.	4^4 in 16	Graeco-L sq.	Euler 1923(347, 348, 448), Cochran & Cox 1950(120), 1957(146), Vianelli 1959(402)
		ortho. part'n	Kitagawa & Mitome 1953(I.22)
222.	4^4 in 64	4^3 in 4.16	Kitagawa & Mitome 1953(II.68), Binet et al. 1955(42)
223.	4^5 in 16	ortho. sq. sat.	Fisher & Yates 1938-57(T.XVI), Bose & Nair 1941(381)*, Kitagawa & Mitome 1953(I.18), Czechowski et al. 1957(100), Vianelli 1959(400)
		ortho. mtx. sat.	Taguchi 1960(221/46)
224.	4^5 in 64	4^3 in 16.4	Kitagawa & Mitome 1953(II.65)
		Latin cube	Kishen 1949(41)
225.	4^6 in 256	Latin cubes	Kishen 1949(43)
226.	4^7 in 256	Latin hypercube	Kishen 1949(32)
227.	4^{21} in 64	ortho. cubes sat.	Kishen 1949(32), Kitagawa & Mitome 1949(32)
		ortho. mtx. sat.	Taguchi 1960(222)
228.	4^{34} in 256	ortho. sq.	Stevens 1939(90)✗, Bose & Nair 1941(381)*, Kitagawa & Mitome 1953(I.19)✗

*In symbolic form.

✗One square printed in full; 14 additional squares obtained by cyclically permuting all rows except the first.

15.114

229.	4^{195} in 4096	ortho. sq.		Stevens 1939(92)*
230.	4.2^2 in 8	4.2 in 2.4		Binet et al. 1955(35)
231.	4.2^3 in 8	$\begin{cases} 2^3 \text{ in } 4.2 \\ \text{ortho. mtx.} \end{cases}$		Kitagawa & Mitome 1953(II.2) Taguchi 1959(136/4)
232.	4.2^3 in 16	4.2^2 in 2.8		Li 1944(461), Cochran & Cox 1950(186, 216), 1957(215, 242), Kitagawa & Mitome 1953 (II.101), Vianelli 1959(406)
233.	4.2^4 in 16	$\begin{cases} 2^4 \text{ in } 4.4 \\ \\ \underline{2^2}.2 \text{ in } 2.(4.2) \end{cases}$		Cochran & Cox 1950(186, 209), 1957(215, 235), Kitagawa & Mitome 1953(II.4, II.7), Davies 1954(428), Vianelli 1959(403) Cochran & Cox 1950(256), 1957 (**332**), Kitagawa & Mitome 1953 (V.8), Vianelli 1959(409)
234.	4.2^4 in 32	$\begin{cases} 4.2^3 \text{ in } 2.16 \\ 4.2^4 \text{ in } 1.32 \end{cases}$	 2 c 2	Kitagawa & Mitome 1953(II.106) Brownlee et al. 1948(275), Morrison 1956(5), Cochran & Cox 1957(270, 292)
235.	4.2^5 in 16	2^5 in 4.4		Cochran & Cox 1957(261, 277), N.B.S. 1957(5)
236.	4.2^5 in 32	$\begin{cases} 2^5 \text{ in } 4.8 \\ \\ 4.2^4 \text{ in } 2.16 \end{cases}$	1 c 2 2 c 2	Kempthorne 1947(265), Cochran & Cox 1950(186, 209), 1957(215, 235), Kitagawa & Mitome 1953 (II.9, II.14), Davies 1954 (429), Vianelli 1959(403) Cochran & Cox 1957(270, 292)
237.	4.2^6 in 16	2^6 in 4.4		Cochran & Cox 1957(261, 278), N.B.S. 1957(18)
238.	4.2^6 in 32	2^6 in 4.8		Kempthorne 1947(266), Brownlee et al. 1948(T.2), Kitagawa & Mitome 1953(III.3, III.4), Cochran & Cox 1957(261, 279), N.B.S. 1957(6)

---

*In symbolic form.

15.114 [see also p. 735]

239.	4.2^6 in 64	2^6 in 4.16	1 c 2	Cochran & Cox 1950(186, 208), 1957 (215, 234), Kitagawa & Mitome 1953(II.21, II.23), Davies 1954 (430), Vianelli 1959(403)
240.	4.2^7 in 16	2^7 in 4.4		Cochran & Cox 1957(261, 280), N.B.S. 1957(30)
241.	4.2^7 in 32	2^7 in 4.8		Brownlee et al. 1948(T.2), Cochran & Cox 1957(261, 282), N.B.S. 1957(19)
242.	4.2^7 in 64	2^7 in 4.16	2 c 4	Brownlee et al. 1948(T.3), Cochran & Cox 1957(259, 261, 284), N.B.S. 1957(8)
243.	4.2^7 in 128	2^7 in 4.32		Kitagawa & Mitome 1953(II.24)
244.	4.2^8 in 16	2^8 in 4.4		Brownlee et al. 1948(270), Cochran & Cox 1957(261, 285), N.B.S. 1957(41)
245.	4.2^8 in 32	2^8 in 4.8		Brownlee et al. 1948(T.2), Cochran & Cox 1957(261, 286), N.B.S. 1957(31)
246.	4.2^8 in 64	2^8 in 4.16	2 c 2	Brownlee et al. 1948(T.3), Kitagawa & Mitome 1953(III.12), Cochran & Cox 1957(259, 261, 287), N.B.S. 1957(21)
247.	4.2^8 in 128	2^8 in 4.32	3 c 4	Cochran & Cox 1957(259, 261, 289), N.B.S. 1957(11)
248.	4.2^8 in 256	2^8 in 4.64		Kitagawa & Mitome 1953(II.26)
249.	4.2^9 in 32	2^9 in 4.8		Brownlee et al. 1948(T.2), N.B.S. 1957(42)
250.	4.2^9 in 64	2^9 in 4.16		Brownlee et al. 1948(T.3), N.B.S. 1957(33)
251.	4.2^9 in 128	2^9 in 4.32	2 c 3	N.B.S. 1957(24)
252.	4.2^9 in 256	2^9 in 4.64	2 c 6	N.B.S. 1957(16)
253.	4.2^9 in 512	2^9 in 4.128		Kitagawa & Mitome 1953(II.29)
254.	4.2^{10} in 32	2^{10} in 4.8		Brownlee et al. 1948(T.2), N.B.S. 1957(52)
255.	4.2^{10} in 64	2^{10} in 4.16		Brownlee et al. 1948(T.3), N.B.S. 1957(44)

256. 4.2^{10} in 128 2^{10} in 4.32   2 c 2   N.B.S. 1957(35)

257. 4.2^{10} in 256 2^{10} in 4.64   2 c 3   N.B.S. 1957(28)

258. 4.2^{11} in 32 2^{11} in 4.8   N.B.S. 1957(58)

259. 4.2^{11} in 64 2^{11} in 4.16   Brownlee et al. 1948(T.3),
                                                      N.B.S. 1957(54)

260. 4.2^{11} in 128 2^{11} in 4.32   2 c 2   N.B.S. 1957(46)

261. 4.2^{11} in 256 2^{11} in 4.64   2 c 3   N.B.S. 1957(40)

262. 4.2^{12} in 32 2^{12} in 4.8   N.B.S. 1957(65)

263. 4.2^{12} in 64 2^{12} in 4.16   N.B.S. 1957(59)

264. 4.2^{12} in 128 2^{12} in 4.32   N.B.S. 1957(56)

265. 4.2^{12} in 256 2^{12} in 4.64   2 c 3   N.B.S. 1957(50)

266. 4.2^{13} in 32 2^{13} in 4.8   N.B.S. 1957(74)

267. 4.2^{13} in 64 2^{13} in 4.16   N.B.S. 1957(66, 67)

268. 4.2^{13} in 128 2^{13} in 4.32   N.B.S. 1957(60, 61)

269. 4.2^{14} in 64 2^{14} in 4.16   N.B.S. 1957(76, 77)

270. 4.2^{14} in 128 2^{14} in 4.32   N.B.S. 1957(68, 69)

271. 4.2^{14} in 256 2^{14} in 4.64   2 c 2   N.B.S. 1957(62, 64)

272. 4.2^{15} in 128 2^{15} in 4.32   N.B.S. 1957(78, 80)

273. 4.2^{15} in 256 2^{15} in 4.64   2 c 2   N.B.S. 1957(70, 72)

274. 4.2^{16} in 256 2^{16} in 4.64   N.B.S. 1957(82, 84)

275. $4^2.2$ in 16 4^2 in 2.8   Kitagawa & Mitome 1953(II.56),
                                                      Binet et al. 1955(35)

276. $4^2.2^2$ in 16 $\begin{cases} 4.2^2 \text{ in } 4.4 \\ 4.\underline{2}^2 \text{ in } 4.4 \end{cases}$ Li 1944(459)
                                                      Kitagawa & Mitome 1953(V.8)

277. $4^2.2^2$ in 32 $4^2.2$ in 2.16   Li 1944(466),
                                                      Kitagawa & Mitome 1953(II.125)

278. $4^2.2^3$ in 16 2^3 in 1.(4.4)   Kitagawa & Mitome 1953(V.1)

279. $4^2.2^3$ in 32 4.2^3 in 4.8   Li 1944(468),
                                                      Kitagawa & Mitome 1953(II.104)

15.114   [see also p. 735]

No.	Design	Confounding	Type	Reference
280.	$4^2.2^3$ in 64	$4^2.2^3$ in 1.64	2 c 2	Brownlee et al. 1948(275)
281.	$4^2.2^4$ in 32	2^3 in 2.(4.4)		Yates 1937(32), Cochran & Cox 1950(252), 1957(328), Vianelli 1959(407)
		4.2^4 in 4.8		Brownlee et al. 1948(275), Cochran & Cox 1957 (270, 292)
282.	$4^2.2^5$ in 16	2^5 in 1.(4.4)		N.B.S. 1957(5), Taguchi 1959(141/9)
283.	$4^2.2^5$ in 64	$\underline{2^3}.2^2$ in [2.2].(4.4)		Cochran & Cox 1950(258), 1957(334), Kitagawa & Mitome 1953(V.13), Vianelli 1959(409)
284.	$4^2.2^6$ in 16	2^6 in 1.(4.4)		N.B.S. 1957(18)
285.	$4^2.2^6$ in 64	2^6 in 16.4		Davies 1954(428)
		$\underline{2^4}.2^2$ in [2.2].(4.4)		Cochran & Cox 1950(258), 1957(334), Kitagawa & Mitome 1953(V.14), Vianelli 1959(409)
286.	$4^2.2^7$ in 16	2^7 in 1.(4.4)		N.B.S. 1957(30)
287.	$4^2.2^7$ in 64	2^7 in 16.4		Brownlee et al. 1948(T.3), Cochran & Cox 1957(261, 283), N.B.S. 1957(7)
288.	$4^2.2^7$ in 128	2^7 in 16.8		Kitagawa & Mitome 1953 (II.24), Davies 1954(429)
289.	$4^2.2^8$ in 64	2^8 in 16.4		Brownlee et al. 1948(T.3), Cochran & Cox 1957(261, 287), N.B.S. 1957(20)
290.	$4^2.2^8$ in 128	2^8 in 16.8		Cochran & Cox 1957(261, 289), N.B.S. 1957(10)
291.	$4^2.2^8$ in 256	2^8 in 16.16		Kitagawa & Mitome 1953(II.28)
292.	$4^2.2^9$ in 64	2^9 in 16.4		Brownlee et al. 1948(T.3), N.B.S. 1957(32)
293.	$4^2.2^9$ in 128	2^9 in 16.8		N.B.S. 1957(22)
294.	$4^2.2^9$ in 256	2^9 in 16.16	2 c 6	N.B.S. 1957(14)

295.	$4^2.2^9$ in 512	2^9 in 16.32		Kitagawa & Mitome 1953(II.31)
296.	$4^2.2^{10}$ in 64	2^{10} in 16.4		Brownlee et al. 1948(T.3)
297.	$4^2.2^{10}$ in 128	2^{10} in 16.8		N.B.S. 1957(34)
298.	$4^2.2^{10}$ in 256	2^{10} in 16.16	2 c 3	N.B.S. 1957(26)
299.	$4^2.2^{11}$ in 64	2^{11} in 16.4		Brownlee et al. 1948(T.3)
300.	$4^2.2^{11}$ in 128	2^{11} in 16.8		N.B.S. 1957(44)
301.	$4^2.2^{11}$ in 256	2^{11} in 16.16	2 c 3	N.B.S. 1957(38)
302.	$4^2.2^{12}$ in 128	2^{12} in 16.8		N.B.S. 1957(55)
303.	$4^2.2^{12}$ in 256	2^{12} in 16.16	2 c 3	Kitagawa & Mitome 1953(III.19), N.B.S. 1957(48)
304.	$4^2.2^{13}$ in 128	2^{13} in 16.8		N.B.S. 1957(60)
305.	$4^2.2^{14}$ in 128	2^{14} in 16.8		N.B.S. 1957(68)
306.	$4^2.2^{14}$ in 256	2^{14} in 16.16		N.B.S. 1957(62, 63)
307.	$4^2.2^{15}$ in 128	2^{15} in 16.8		N.B.S. 1957(78)
308.	$4^2.2^{15}$ in 256	2^{15} in 16.16		N.B.S. 1957(70, 71)
309.	$4^2.2^{16}$ in 256	2^{16} in 16.16		N.B.S. 1957(82, 83)
310.	$4^2.2^{17}$ in 32	$4.2^7.[\underline{2^6} + \underline{4.2^2}]$ in 2.8.2		Taguchi 1959(149/17)
311.	$4^3.2$ in 16	ortho. partition		Kitagawa & Mitome 1953(I.22)
312.	$4^3.2$ in 32	$4^2.2$ in 4.8		Li 1944(464), Kitagawa & Mitome 1953(II.125)
313.	$4^3.2^2$ in 32	$4.\underline{2^2}$ in (4.4).2		Kitagawa & Mitome 1953(IV.2)
314.	$4^3.2^3$ in 32	$4.\underline{2^3}$ in (4.4).2		Kitagawa & Mitome 1953(IV.3)
315.	$4^3.2^3$ in 64	$\{4.\underline{2^3}$ in (4.4).4 $\;\; 4^2.2^3$ in 4.16	2 c 2	Kitagawa & Mitome 1953(IV.3) Brownlee et al. 1948(275)
316.	$4^3.2^7$ in 64	2^7 in 1.(16.4)		Brownlee et al. 1948(T.3), N.B.S. 1957(7)
317.	$4^3.2^8$ in 64	2^8 in 1.(16.4)		Brownlee et al. 1948(T.3), N.B.S. 1957(20, 21)
318.	$4^3.2^9$ in 64	2^9 in 1.(16.4)		N.B.S. 1957(32, 33)

15.114
15.115   [see also pp. 735-736]

319. $4^3.2^{11}$ in 64 2^{11} in 1.(16.4)   Brownlee et al. 1948(T.3)

320. $4^4.2^9$ in 256 2^9 in 1.(16.16)   2 c 6   N.B.S. 1957(14)

321. $4^4.2^{10}$ in 256 2^{10} in 1.(16.16) 2 c 3   N.B.S. 1957(26)

322. $4^4.2^{11}$ in 256 2^{11} in 1.(16.16) 2 c 3   N.B.S. 1957(38)

323. $4^9.2$ in 32   ortho. matrix   Taguchi 1960(221/46)

324. 4.3^2 in 12   4.3 in 3.4   Binet et al. 1955(35)

325. 4.3^3 in 36 4.3^2 in 3.12   Li 1944(476), Kitagawa &
                                                             Mitome 1953(II.122)

326. 4.3^4 in 108 3^3 in 4.3.9   part.   Yates 1937(42), Cochran & Cox
                                                             1950(212), 1957(215, 238),
                                                             Vianelli 1959(404)

327. $4^3.3$ in 16   ortho. part'n   Kitagawa & Mitome 1953(I.22)

328. $4^3.3$ in 48 $\begin{cases} 4^2 \text{ in } 3.4.4 \quad\text{part.} \\ \\ 4^2.3 \text{ in } 4.12 \end{cases}$ Cochran & Cox 1950(215),
                                                                     1957(215, 241),
                                                             Vianelli 1959(405)

                                                             Li 1944(463, 487)

329. $4^3.3^2$ in 144 $4^2.3$ in 3.4.12   part.   Li 1944

330. $4^3.3^3$ in 144   Latin square   Fisher & Yates 1938-57(T.XV),
                                                             Kitagawa & Mitome 1953
                                                             (I.16), Cochran & Cox 1950
                                                             (120), 1957(146),
                                                             Czechowski et al. 1957(100),
                                                             Vianelli 1959(400)

331. $4^4.3^4$ in 144   Graeco-Latin sq.   Cochran & Cox 1950(121), 1957
                                                             (147), Kitagawa & Mitome
                                                             1953(I.16),
                                                             Vianelli 1959(402)

332. 4.3.2 in 12   4.3 in 2.6   Binet et al. 1955(35)

333. $4.3.2^2$ in 24 $\begin{cases} 6.4 \text{ in } 2.12 \\ 4.3.2 \text{ in } 2.12 \end{cases}$ Binet et al. 1955(52)

                                                             Li 1944(483), Cochran & Cox
                                                             1950(186, 216), 1957(215,
                                                             242), Kitagawa & Mitome
                                                             1953(II.135),
                                                             Vianelli 1959(406)

334. $4.3.2^3$ in 24 3.2^3 in 4.6   Cochran & Cox 1950(186, 214), 1957(215, 240), Kitagawa & Mitome 1953(II.90), Vianelli 1959(405)

335. $4.3.2^3$ in 48 4.2^2 in 3.2.8   part.   Cochran & Cox 1950(216), 1957 (215, 242), Vianelli 1959(406)

336. $4.3.2^4$ in 48 3.2^4 in 4.12   Kitagawa & Mitome 1953(II.94)

337. $4.3^2.2^3$ in 72 3.2^3 in 3.4.6   part.   Cochran & Cox 1950(214), 1957 (215, 240), Vianelli 1959(405)

338. $4.3^2.2^4$ in 48 6.2^3 in 12.4   Kitagawa & Mitome 1953(IV.1)

339. $4.3^3.2$ in 72
$\begin{cases} 3^2.2 \text{ in } 4.3.6 \quad \text{part.} \\ 4.3^2 \text{ in } 2.3.12 \quad \text{part.} \end{cases}$
      Yates 1937(63), Cochran & Cox 1950(215), 1957(215, 241), Vianelli 1959(405)
      Li 1944

340. $4.3^4.2$ in 36 $3^5.2$ in 1.27 +   Taguchi 1959(154/22)
                  + $3^4.2$ in 1.9

341. $4^2.3.2^3$ in 96 4.2^3 in 3.4.8   part.   Li 1944

<u>15.115</u> Factorial patterns with one or more factors at 5 levels

342. 5^3 in 25
$\begin{cases} 5^2 \text{ in } 5.5 \\ \text{Latin sq.} \end{cases}$
      Kitagawa & Mitome 1953(II.70)
      Cayley 1890(137), 1897(57), Euler 1923 (303, 304, 387), Fisher & Yates 1938-57(T.XV), Kitagawa & Mitome 1953(I.8), Cochran & Cox 1950(119), 1957(145), Czechowski et al. 1957(100), Vianelli 1959(398)

343. 5^4 in 25   Graeco-L sq.   Euler 1923(304, 387, 452), Cochran & Cox 1950(120), 1957(146), Vianelli 1959(402)

344. 5^4 in 125 5^3 in 5.25   K.R. Nair 1940(60), Kitagawa & Mitome 1953(II.75)

345. 5^5 in 125
$\begin{cases} 5^3 \text{ in } 25.5 \\ \text{Latin cube} \end{cases}$
      Kitagawa & Mitome 1953(II.73)
      Kishen 1949(42)

346. 5^6 in 25   ortho. sq. sat.   Euler 1923(305, 324), Fisher & Yates 1938-57(T.XVI), Stevens 1939(85), Bose & Nair 1941(381)*, Kitagawa & Mitome 1953(I.18), Czechowski et al. 1957(100), Vianelli 1959(400)

---
*In symbolic form.

15.115
15.116   [see also p. 736]

347. 5^6 in 25   ortho. matrix   sat.   Plackett & Burman 1946(325),
                                                  Taguchi 1960(224/49)

348. 5^6 in 625 $\begin{cases} 5^4 \text{ in } 25.25 \\ \text{Latin cubes} \end{cases}$ K. R. Nair 1940(63, 64)
                                                                                                    Kishen 1949(44)

349. 5^7 in 625   Latin hypercube   Kishen 1949(46)

350. 5^{31} in 125 $\begin{cases} \text{ortho. cubes} \quad \text{sat.} \\ \\ \text{ortho. matrix} \quad \text{sat.} \end{cases}$ Kishen 1949(34),
                                                                                                                                          Kitagawa & Mitome 1953(I.29)
                                                                                                                                          Plackett & Burman 1946(325)

351. 5^{52} in 625   ortho. sq.   Stevens 1939(91)*,
                                                  Bose & Nair 1941(381)⨯,
                                                  Kitagawa & Mitome 1953(I.20)*

352. 5^{378} in 15625 ortho. sq.   Stevens 1939(92)⨯

353. 5.2^3 in 20 $\begin{cases} 5.2^2 \text{ in } 2.10 \\ \\ 5.4 \text{ in } 2.10 \end{cases}$ Li 1944(469),
                                                                                                          Kitagawa & Mitome 1953(II.109)
                                                                                                          Binet et al. 1955(45)

354. 5.2^4 in 40 5.2^3 in 2.20   Kitagawa & Mitome 1953(II.111)

355. 5.2^7 in 160 2^5 in 5.4.8   part. Cochran & Cox 1950(209), 1957
                                                  (215, 235), Vianelli 1959(403)

356. 5.2^9 in 320 2^6 in 5.8.8   part. Vianelli 1959(404)

357. 5.2^{10} in 640 2^6 in 10.8.8   part. Cochran & Cox 1950(210),
                                                                1957(215, 236)

358. 5.2^{12} in 320 2^6 in 5.(8.8) part. Cochran & Cox 1950(253),
                                                  1957(329), Vianelli 1959(407)

359. $5^2.2$ in 25 5^2 in 12 + 13   Binet et al. 1955(60)

360. $5^2.2^3$ in 100 5.2^2 in 5.2.10 part. Li 1944

361. $5^3.2$ in 25   ortho. partition   Kitagawa & Mitome
                                                            1953(I.22, I.23)

362. $5^3.2^2$ in 100 5^2 in 4.5.5   part. K. R. Nair 1940

363. $5^3.2^3$ in 100   Latin square   Fisher & Yates 1938-57(T.XV),
                                                  Cochran & Cox 1950(119),
     1957(145), Kitagawa & Mitome 1953(I.16), Czechowski et al.
     1957(100), Vianelli 1959(400)

*One square printed in full; 23 additional squares obtained by cyclically permuting all rows except the first.

⨯In symbolic form.

364.	$5^4.2^8$ in 400	Graeco-L. sq.	Kitagawa & Mitome 1953(I.17)
365.	$5^5.2^{10}$ in 400	Graeco-L. sq.	Kitagawa & Mitome 1953(I.17)
366.	$5^{11}.2$ in 50	ortho. matrix	Taguchi 1960(224/49)
367.	5.3^2 in 15	5.3 in 3.5	Binet et al. 1955(45)
368.	5.3^3 in 45	5.3^2 in 3.15	Binet et al. 1955(47)
369.	$5^3.3$ in 25	ortho. part'n	Kitagawa & Mitome 1953 (I.23, I.24)
370.	$5^3.4$ in 25	ortho. part'n	Kitagawa & Mitome 1953(I.24)
371.	$5^3.4$ in 100	5^2 in 4.5.5 part.	K. R. Nair 1940
372.	$5^5.4^5$ in 400	Graeco-L. sq.	Kitagawa & Mitome 1953(I.17)
373.	5.3.2 in 15	5.3 in 7+8	Binet et al. 1955(60)
374.	$5.3.2^2$ in 30	{6.5 in 2.15 / 5.3.2 in 2.15}	Binet et al. 1955(52) / Binet et al. 1955(48)
375.	$5.3.2^3$ in 60	$5.3.2^3$ in 1.60	Morrison 1956(6)
376.	$5.3^2.2$ in 30	5.3.2 in 3.10	Binet et al. 1955(49)
377.	5.4.2 in 20	5.4 in 2.10	Binet et al. 1955(45)
378.	$5.4.2^5$ in 160	2^5 in 5.4.8 part.	Cochran & Cox 1950(209), 1957 (215, 235), Vianelli 1959(403)

15.116 Factorial patterns with one or more factors at 6 levels

379.	6^3 in 36	Latin square	Tarry 1900(197), Fisher & Yates 1934, 1938-57(T.XV), Kitagawa & Mitome 1953(I.10), Cochran & Cox 1950 (119), 1957(145), Czechowski et al. 1957(100), Vianelli 1959(399)
380.	6.2^2 in 12	6.2 in 2.6	Binet et al. 1955(51)
381.	$6^2.2$ in 36	6^2 in 2.18	Binet et al. 1955(53)
382.	$6^3.2$ in 36	ortho. partition	Kitagawa & Mitome 1953(I.25)
383.	$6^3.2^2$ in 72	$6.\underline{2}^2$ in (6.6).2	Kitagawa & Mitome 1953(IV.4)
384.	$6^3.2^3$ in 144	$6.\underline{2}^3$ in (6.6).4	Yates 1937(78), Kitagawa & Mitome 1953(IV.5, IV.6)

15.116
15.118

[see also p. 536]

385.	$6^3.3$ in 36	ortho. partition	Kitagawa & Mitome 1953(I.26)
386.	$6^3.3^2$ in 108	$6.\underline{3^2}$ in (6.6).3	Kitagawa & Mitome 1953(IV.7)
387.	$6.3.2$ in 18	6.3 in 2.9	Binet et al. 1955(51)
388.	$6.3.2^2$ in 36	$6.3.2$ in 2.18	Binet et al. 1955(54)
389.	$6.3.2^3$ in 36	$3.\underline{2^2}$ in $3.(6.2)$	Kitagawa & Mitome 1953(V.10)
390.	$6.3^2.2$ in 36	$6.3.2$ in 3.12	Binet et al. 1955(54)

391. $6.3^2.2^2$ in 36
$$\begin{cases} 3.\underline{2}.2 \text{ in } 2.(6.3) \\ 3.\underline{3}.2 \text{ in } 3.(6.2) \\ 3^2.2^2 \text{ in } 6.6 \end{cases}$$

- $3.\underline{2}.2$ in $2.(6.3)$ — Cochran & Cox 1950(256), 1957(332), Kitagawa & Mitome 1953(V.10), Vianelli 1959(409)
- $3.\underline{3}.2$ in $3.(6.2)$ — Cochran & Cox 1950(257), 1957(333), Kitagawa & Mitome 1953(V.11), Vianelli 1959(409)
- $3^2.2^2$ in 6.6 — Kitagawa & Mitome 1953(II.131), Binet et al. 1955(32)

392.	$6.3^2.2^3$ in 72	$3.\underline{3}.2$ in $2.3.(6.2)$	Cochran & Cox 1950(257), 1957(333)
393.	$6^2.3.2^3$ in 144	$6.\underline{2^3}$ in (6.6).4	Yates 1937(78)
394.	$6^2.3^2.2$ in 36	$3^2.2$ in $1.(6.6)$	Yates 1937(65), Kitagawa & Mitome 1953(V.6)
395.	$6.4.2$ in 24	6.4 in 2.12	Binet et al. 1955(52)
396.	$6.4.2^4$ in 96	2^4 in $6.4.4$ part.	Cochran & Cox 1950(209), 1957(215, 235), Vianelli 1959(403)
397.	$6.5.2$ in 30	6.5 in 2.15	Binet et al. 1955(52)

15.117 Factorial patterns with one or more factors at 7 levels

398. 7^3 in 49
$$\begin{cases} 7^2 \text{ in } 7.7 \\ \text{Latin sq.} \end{cases}$$

- 7^2 in 7.7 — Kitagawa & Mitome 1953(II.80)
- Latin sq. — Euler 1923(294, 305), Fisher & Yates 1938-57 (T.XV), Cochran & Cox 1950(119), 1957(145), Kitagawa & Mitome 1953(I.10), Czechowski et al. 1957(100), Vianelli 1959(399)

399.	7^4 in 49		Graeco-Latin sq.	Euler 1923(294, 312, 325), Norton 1939(303), Cochran & Cox 1950(120), 1957(146), Vianelli 1959(402)
400.	7^5 in 343	7^3 in 49.7		Kitagawa & Mitome 1953(II.83)
401.	7^8 in 49		ortho. squares sat.	Fisher & Yates 1938-57(T.XVI), Norton 1939(306), Bose & Nair 1941(381)*, Kitagawa & Mitome 1953(I.18), Czechowski et al. 1957(101), Vianelli 1959(400)
			ortho. matrix sat.	Plackett & Burman 1946(325)
402.	7^{57} in 343		ortho. cubes sat.	Kitagawa & Mitome 1953(I.32)
403.	7^{100} in 2401		ortho. squares	Stevens 1939(92)*
404.	7.2^2 in 14	7.2 in 2.7		Binet et al. 1955(56)
405.	7.2^3 in 28	7.2^2 in 2.14		Binet et al. 1955(57)
		7.4 in 2.14		Binet et al. 1955(57)
406.	$7^2.3^5$ in 27	$7^2.3^2$ in 1.(9.3)		Taguchi 1959(152/20)
407.	7.3.2 in 21	7.3 in 10 + 11		Binet et al. 1955(61)
408.	7.4.2 in 28	7.4 in 2.14		Binet et al. 1955(57)
409.	7.5.2 in 35	7.5 in 17 + 18		Binet et al. 1955(61)

<u>15.118</u> Factorial patterns with one or more factors at 8 levels

410.	8^3 in 64		Latin square	Euler 1923(343), Fisher & Yates 1938-57(T.XV), Cochran & Cox 1950(119), 1957(145), Kitagawa & Mitome 1953(I.16), Czechowski et al. 1957(100), Vianelli 1959(399)
411.	8^4 in 64		Graeco-L. sq.	Euler 1923(345, 346), Cochran & Cox 1950(120), 1957(146), Vianelli 1959(402)
412.	8^9 in 64		ortho. sq. sat.	Fisher & Yates 1938-57(T.XVI), Bose & Nair 1941(381)*, Kitagawa & Mitome 1953(I.18), Czechowski et al. 1957(101), Vianelli 1959(401)
413.	8.2^4 in 16	2^4 in 8.2		Kitagawa & Mitome 1953(II.4, II.6)

*In symbolic form.

15.118

414. 8.2^5 in 32 2^5 in 8.4   Kitagawa & Mitome 1953(II.9, II.17), Davies 1954(428)

415. 8.2^6 in 16 2^6 in 8.2   N.B.S. 1957(18)

416. 8.2^6 in 32 2^6 in 8.4   Brownlee et al. 1948(T.2), Cochran & Cox 1957(261, 279), N.B.S. 1957(6)

417. 8.2^6 in 64 2^6 in 8.8   Cochran & Cox 1950(186, 210), 1957(215, 236), Kitagawa & Mitome 1953(II.22), Davies 1954(429), Vianelli 1959(404)

418. 8.2^7 in 16 2^6 in 1.(8.2)   N.B.S. 1957(18)

419. 8.2^7 in 32 2^7 in 8.4   Brownlee et al. 1948(T.2), Cochran & Cox 1957(261, 281), N.B.S. 1957(19)

420. 8.2^7 in 64 2^7 in 8.8   2 c 4   Brownlee et al. 1948(T.3), Kitagawa & Mitome 1953(III.3, III.7), Cochran & Cox 1957(259, 261, 284), N.B.S. 1957(8)

421. 8.2^7 in 128 2^7 in 8.16   Kitagawa & Mitome 1953(II.24), Davies 1954(430)

422. 8.2^8 in 32 2^8 in 8.4   Brownlee et al. 1948(T.2), Cochran & Cox 1957(261, 286), N.B.S. 1957(30)

423. 8.2^8 in 64 2^8 in 8.8   Brownlee et al. 1948(T.3), Cochran & Cox 1957(261, 287), N.B.S. 1957(21)

424. 8.2^8 in 128 2^8 in 8.16   3 c 4   Kitagawa & Mitome 1953(III.3, III.8), Cochran & Cox 1957(259, 261, 288), N.B.S. 1957(10)

425. 8.2^8 in 256 2^8 in 8.32   Kitagawa & Mitome 1953(II.27)

426. 8.2^9 in 32 2^9 in 8.4   Brownlee et al. 1948(T.2), N.B.S. 1957(42)

427. 8.2^9 in 64 2^9 in 8.8   Brownlee et al. 1948(T.3), N.B.S. 1957(32)

428. 8.2^9 in 128 2^9 in 8.16   2 c 3   Bronwlee et al. 1948(274), N.B.S. 1957(23)

429. 8.2^9 in 256 2^9 in 8.32   3 c 5   N.B.S. 1957(15)

430. 8.2^9 in 512 2^9 in 8.64   Kitagawa & Mitome 1953(II.30)
431. 8.2^{10} in 32 2^{10} in 8.4   Brownlee et al. 1948(T.2),
                                                    N.B.S. 1957(52)
432. 8.2^{10} in 64 2^{10} in 8.8   N.B.S. 1957(43)
433. 8.2^{10} in 128 2^{10} in 8.16   2 c 2 Kitagawa & Mitome 1953(III.14),
                                                    N.B.S. 1957(34)
434. 8.2^{10} in 256 2^{10} in 8.32   2 c 3 N.B.S. 1957(27)
435. 8.2^{11} in 64 2^{11} in 8.8   Brownlee et al. 1948(T.3),
                                                    N.B.S. 1957(53)
436. 8.2^{11} in 128 2^{11} in 8.16   2 c 2 Kitagawa & Mitome 1953(III.16),
                                                    N.B.S. 1957(45)
437. 8.2^{11} in 256 2^{11} in 8.32   2 c 3 N.B.S. 1957(39)
438. 8.2^{12} in 64 2^{12} in 8.8   N.B.S. 1957(59)
439. 8.2^{12} in 128 2^{12} in 8.16   N.B.S. 1957(56)
440. 8.2^{12} in 256 2^{12} in 8.32   2 c 3 Brownlee et al. 1948(274),
                                                    N.B.S. 1957(49)
441. 8.2^{13} in 64 2^{13} in 8.8   N.B.S. 1957(66)
442. 8.2^{13} in 128 2^{13} in 8.16   N.B.S. 1957(60, 61)
443. 8.2^{14} in 64 2^{14} in 8.8   N.B.S. 1957(76)
444. 8.2^{14} in 128 2^{14} in 8.16   N.B.S. 1957(68)
445. 8.2^{14} in 256 2^{14} in 8.32   2 c 2 N.B.S. 1957(62, 64)
446. 8.2^{15} in 128 2^{15} in 8.16   N.B.S. 1957(78, 80)
447. 8.2^{15} in 256 2^{15} in 8.32   2 c 2 N.B.S. 1957(70, 72)
448. 8.2^{16} in 256 2^{16} in 8.32   N.B.S. 1957(82, 84)
449. 8.2^{16} in 512 2^{16} in 8.64   2 c 3 Brownlee et al. 1948(274)
450. $8^2.2^4$ in 64 2^4 in 1.(8.8)   Cochran & Cox 1950(252), 1957
                                                    (328), Kitagawa & Mitome 1953
                                                    (V.2), Vianelli 1959(407)
451. $8^2.2^5$ in 64 2^5 in 1.(8.8)   Yates 1937(35), Cochran & Cox
                                                    1950(253), 1957(329), Kitagawa
                                                    & Mitome 1953(V.2), Vianelli 1959(407)

15.118
15.119

452.	$8^2.2^7$ in 64	2^7 in 1.(8.8)	2 c 4	Brownlee et al. 1948(T.3), N.B.S. 1957(8)
453.	$8^2.2^7$ in 128	2^6 in 2.(8.8)		Yates 1937(35)
454.	$8^2.2^8$ in 64	2^8 in 1.(8.8)		Brownlee et al. 1948(T.3), N.B.S. 1957(21)
455.	$8^2.2^9$ in 64	2^9 in 1.(8.8)		Brownlee et al. 1948(T.3), N.B.S. 1957(32)
456.	$8^2.2^{10}$ in 64	2^{10} in 1.(8.8)		Brownlee et al. 1948(T.3)
457.	$8^2.2^{11}$ in 64	2^{11} in 1.(8.8)		Brownlee et al. 1948(T.3)
458.	$8^3.2^3$ in 64	8.2^3 in 1.(8.8)		Kitagawa & Mitome 1953(V.6)
459.	8.4^3 in 64	4^3 in 8.8		Kitagawa & Mitome 1953(II.62)
460.	$8.4.2^3$ in 32	2^3 in 1.(8.4)		Yates 1937(32)
461.	$8.4.2^4$ in 64	$\underline{2^3}.2$ in 2.(8.4)		Cochran & Cox 1950(257), 1957(333), Kitagawa & Mitome 1953(V.9), Vianelli 1959(409)
		$4.\underline{2^3}$ in 4.(8.2)		Yates 1937(79)
462.	$8.4.2^5$ in 64	$\underline{2^4}.2$ in 2.(8.4)		Cochran & Cox 1950(258), 1957(334), Kitagawa & Mitome 1953(V.9), Vianelli 1959(409)
463.	$8.4.2^6$ in 32	2^6 in 1.(8.4)		Brownlee et al. 1948(T.2), N.B.S. 1957(6)
464.	$8.4.2^7$ in 32	2^7 in 1.(8.4)		Brownlee et al. 1948(T.2), N.B.S. 1957(19)
465.	$8.4.2^8$ in 32	2^8 in 1.(8.4)		N.B.S. 1957(30, 31)
466.	$8.4^2.2^9$ in 128	2^9 in 1.(16.8)		N.B.S. 1957(22, 23)
467.	$8.4^2.2^{10}$ in 128	2^{10} in 1.(16.8)		N.B.S. 1957(34)
468.	$8^2.4.2$ in 64	$\underline{8}.4$ in 4.(8.2)		Yates 1937(79)
469.	$8^2.4.2^2$ in 64	4.2^2 in 1.(8.8)		Cochran & Cox 1950(256), 1957(332)
470.	$8^2.4.2^3$ in 64	4.2^3 in 1.(8.8)		Kitagawa & Mitome 1953(V.5)

471.	$8.5.2^6$ in 320	2^6 in 5.8.8	part.	Vianelli 1959(404)
472.	$8^2.5.2^6$ in 320	2^6 in 5.(8.8)	part.	Cochran & Cox 1950(253), 1957 (329), Vianelli 1959(407)

15.119 Factorial patterns with one or more factors at 9 levels

473.	9^3 in 81	Latin square		Euler 1923(313, 320), Fisher & Yates 1938-57(T.XV), Cochran & Cox 1950(119), 1957(145), Kitagawa & Mitome 1953(I.16), Czechowski et al. 1957 (100), Vianelli 1959(400)
474.	9^4 in 81	Graeco-Latin square		Euler 1923(314, 366, 367), Cochran & Cox 1950(120), 1957(146), Vianelli 1959(402)
475.	9^{10} in 81	orthogonal squares	sat.	Fisher & Yates 1938-57 (T.XVI), Stevens 1939(88), Bose & Nair 41(381)*, Kitagawa & Mitome 1953 (I.19), Czechowski et al. 1957(101), Vianelli 1959(401)
476.	9.3^3 in 27	3^3 in 9.3		Kitagawa & Mitome 1953(II.34, II.38), Davies 1954(435)
		Latin cube		Kishen 1949(24)
477.	9.3^4 in 27	3^4 in 9.3	1 c 2	Connor & Zelen 1959(11)
478.	9.3^4 in 81	3^4 in 9.9	1 c 1	Yates 1937(48), K.R.Nair 1938(129), Kempthorne 1947 (261), Cochran & Cox 1950 (186, 212), 1957(215, 238), Kitagawa & Mitome 1953 (II.43), Davies 1954(439), Vianelli 1959(404)
		$3^2.\underline{3^2}$ in 9.9		Kitagawa & Mitome 1953(V.12)
		$\underline{9}.3^2$ in [3.3].(3.3)		Yates 1937(81)
		Latin cubes		Kishen 1949(43)

---

*In symbolic form.

15.119
15.1199
[see also p. 736]

479. 9.3^5 in 81	3^5 in 9.9	1 c 3	Kitagawa & Mitome 1953 (III.24), Cochran & Cox 1957(270, 291), Connor & Zelen 1959(12)
	$3^3.3$ in 3.(9.3)		Yates 1937(80)
480. 9.3^6 in 81	3^6 in 9.9	1 c 2	Connor & Zelen 1959(18)
481. 9.3^6 in 243	3^6 in 9.27	2 c 3	Connor & Zelen 1959(14)
482. 9.3^7 in 81	3^7 in 9.9	1 c 2	Connor & Zelen 1959(24)
483. 9.3^7 in 243	3^7 in 9.27	2 c 2	Connor & Zelen 1959(20)
484. 9.3^7 in 729	3^7 in 9.81	2 c 4	Connor & Zelen 1959(17)
485. 9.3^8 in 81	3^8 in 9.9	1 c 2	Connor & Zelen 1959(30)
486. 9.3^8 in 243	3^8 in 9.27	2 c 2	Connor & Zelen 1959(26)
487. 9.3^8 in 729	3^8 in 9.81	2 c 2	Connor & Zelen 1959(23)
488. 9.3^9 in 81	3^9 in 9.9	1 c 2	Connor & Zelen 1959(35)
489. 9.3^9 in 243	3^9 in 9.27	2 c 2	Connor & Zelen 1959(32)
490. 9.3^9 in 729	3^9 in 9.81	2 c 3	Connor & Zelen 1959(29)
491. 9.3^{10} in 243	3^{10} in 9.27	1 c 3	Connor & Zelen 1959(37)
492. $9^2.3^3$ in 81	3^3 in 1.(9.9)		Yates 1937(49), Cochran & Cox 1950(254), 1957(330), Kitagawa & Mitome 1953 (V.3), Vianelli 1959(408)
	$9.3.3$ in 3.(9.3)		Yates 1937(80)
493. $9^2.3^4$ in 81	3^4 in 1.(9.9)		Yates 1937(49), Cochran & Cox 1950(254), 1957(331), Kitagawa & Mitome 1953 (V.4), Vianelli 1959(408)
494. $9^3.3^2$ in 81	9.3^2 in 1.(9.9)		Kitagawa & Mitome 1953(V.7)
495. $9.3^3.2$ in 54	$3^3.2$ in 9.6		Yates 1937, Kitagawa & Mitome 1953(II.118)
496. $9.3^3.2^2$ in 108	$3^3.2$ in 2.9.6	part.	Yates 1937
497. $9^2.3^3.2$ in 162	3^3 in 2.(9.9)	part.	Cochran & Cox 1950(254), 1957(330), Vianelli 1959(408)

498.	$9^2.3^4.2$ in 162	3^4 in 2.(9.9)	part.	Cochran & Cox 1950(255), 1957(331), Vianelli 1959(408)
499.	$9.4.3^4$ in 324	3^4 in 4.9.9	part.	Yates 1937(48), Cochran & Cox 1950(212), 1957(215, 238), Vianelli 1959(404)
500.	$9.4.3^5$ in 324	$\underline{3^3}.3$ in 4.3.(9.3)	part.	Yates 1937(80)
501.	$9^2.4.3^3$ in 324	$\underline{9}.3.3$ in 4.3.(9.3)	part.	Yates 1937(80)
502.	$9.7^2.3^3$ in 27	$7^2.3^2$ in 1.(9.3)		Taguchi 1959(152/20)
503.	$9.4.3.2^2$ in 216	4.3.2 in 9.2.12	part.	Cochran & Cox 1950(216), 1957(215, 242), Vianelli 1959(406)

15.1199 Factorial patterns with one or more factors at 10 or more levels

504.	10^3 in 100	Latin square		Fisher & Yates 1938-57(T.XV), Kitagawa & Mitome 1953(I.16), Cochran & Cox 1950(119), 1957(145), Czechowski et al. 1957(100), Vianelli 1959(400)
505.	$10.8.2^6$ in 640	2^6 in 10.8.8	part.	Cochran & Cox 1950(210), 1957(215, 236)
506.	11^3 in 121	Latin square		Fisher & Yates 1938-57(T.XV), Kitagawa & Mitome 1953(I.16), Cochran & Cox 1950(119), 1957(145), Czechowski et al. 1957(100), Vianelli 1959(400)
507.	11^4 in 121	Graeco-L. sq.		Cochran & Cox 1950(121), 1957(147), Vianelli 1959(402)
508.	11^{12} in 121	ortho. sq.	sat.	Bose & Nair 1941(381)*
509.	12^3 in 144	Latin square		Fisher & Yates 1938-57(T.XV), Kitagawa & Mitome 1953(I.16), Cochran & Cox 1950(120), 1957(146), Czechowski et al. 1957(100), Vianelli 1959(400)
510.	12^4 in 144	Graeco-L. sq.		Kitagawa & Mitome 1953(I.16), Cochran & Cox 1950(121), 1957(147), Vianelli 1959(402)

*In symbolic form.

15.1199 [see also p. 736]

511.	$12.6.2^3$ in 48	$6.\underline{2^3}$ in 12.4		Kitagawa & Mitome 1953(IV.1)
512.	16^{17} in 256	ortho. squares	sat.	Stevens 1939(90)*, Bose & Nair 1941(381)✗, Kitagawa & Mitome 1953(I.19)*
513.	16.2^6 in 64	2^6 in 16.4		Davies 1954(428)
514.	16.2^7 in 64	2^7 in 16.4		Brownlee et al. 1948(T.3), Cochran & Cox 1957(261, 283), N.B.S. 1957(7)
515.	16.2^7 in 128	2^7 in 16.8		Kitagawa & Mitome 1953 (II.24), Davies 1954(429)
516.	16.2^8 in 64	2^8 in 16.4		Brownlee et al. 1948(T.3), Cochran & Cox 1957(261, 287), N.B.S. 1957(20)
517.	16.2^8 in 128	2^8 in 16.8		Cochran & Cox 1957(261, 289), N.B.S. 1957(10)
518.	16.2^8 in 256	2^8 in 16.16		Kitagawa & Mitome 1953(II.28)
519.	16.2^9 in 64	2^9 in 16.4		Brownlee et al. 1948(T.3), N.B.S. 1957(32)
520.	16.2^9 in 128	2^9 in 16.8		N.B.S. 1957(22)
521.	16.2^9 in 256	2^9 in 16.16	2 c 6	N.B.S. 1957(14)
522.	16.2^9 in 512	2^9 in 16.32		Kitagawa & Mitome 1953(II.31)
523.	16.2^{10} in 64	2^{10} in 16.4		Brownlee et al. 1948(T.3)
524.	16.2^{10} in 128	2^{10} in 16.8		N.B.S. 1957(34)
525.	16.2^{10} in 256	2^{10} in 16.16	2 c 3	N.B.S. 1957(26)
526.	16.2^{11} in 64	2^{11} in 16.4		Brownlee et al. 1948(T.3)
527.	16.2^{11} in 128	2^{11} in 16.8		N.B.S. 1957(44)
528.	16.2^{11} in 256	2^{11} in 16.16	2 c 3	N.B.S. 1957(38)
529.	16.2^{12} in 128	2^{12} in 16.8		N.B.S. 1957(55)

---

*One square printed in full; 14 additional squares obtained by cyclically permuting all rows except the first.

✗In symbolic form.

530.	16.2^{12} in 256	2^{12} in 16.16	2 c 3	Kitagawa & Mitome 1953 (III.19), N.B.S. 1957(48)
531.	16.2^{13} in 128	2^{13} in 16.8		N.B.S. 1957(60)
532.	16.2^{14} in 128	2^{14} in 16.8		N.B.S. 1957(68)
533.	16.2^{14} in 256	2^{14} in 16.16		N.B.S. 1957(62, 63)
534.	16.2^{15} in 128	2^{15} in 16.8		N.B.S. 1957(78)
535.	16.2^{15} in 256	2^{15} in 16.16		N.B.S. 1957(70, 71)
536.	16.2^{16} in 256	2^{16} in 16.16		N.B.S. 1957(82, 83)
537.	$16^2.2^9$ in 256	2^9 in 1.(16.16)	2 c 6	N.B.S. 1957(14)
538.	$16^2.2^{10}$ in 256	2^{10} in 1.(16.16)	2 c 3	N.B.S. 1957(26)
539.	$16^2.2^{11}$ in 256	2^{11} in 1.(16.16)	2 c 3	N.B.S. 1957(38)
540.	16.4^3 in 64	$\begin{cases} 4^3 \text{ in } 16.4 \\ \text{Latin cube} \end{cases}$		Kitagawa & Mitome 1953(II.65) Kishen 1949(41)
541.	16.4^4 in 256	Latin cubes		Kishen 1949(43)
542.	$16.4.2^7$ in 64	2^7 in 1.(16.4)		Brownlee et al. 1948(T.3), N.B.S. 1957(7)
543.	$16.4.2^8$ in 64	2^8 in 1.(16.4)		Brownlee et al. 1948(T.3), N.B.S. 1957(20, 21)
544.	$16.4.2^9$ in 64	2^9 in 1.(16.4)		N.B.S. 1957(32, 33)
545.	$16.4.2^{11}$ in 64	2^{11} in 1.(16.4)		Brownlee et al. 1948(T.3)
546.	$16.8.2^9$ in 128	2^9 in 1.(16.8)		N.B.S. 1957(22, 23)
547.	$16.8.2^{10}$ in 128	2^{10} in 1.(16.8)		N.B.S. 1957(34)
548.	20^5 in 400	Graeco-Latin sq.		Kitagawa & Mitome 1953(I.17)
549.	25^{26} in 625	ortho. squares	sat.	Stevens 1939(91)*, Bose & Nair 1941(381),† Kitagawa & Mitome 1953(I.20)*
550.	25.5^3 in 125	$\begin{cases} 5^3 \text{ in } 25.5 \\ \text{Latin cube} \end{cases}$		Kitagawa & Mitome 1953(II.73) Kishen 1949(42)

*One square printed in full; 23 additional squares obtained by cyclically permuting all rows except the first.

† In symbolic form.

15.1199
15.12

[see also p. 736]

551. 25.5^4 in 625	$\begin{cases} 5^4 \text{ in } 25.25 \\ \text{Latin cubes} \end{cases}$		K. R. Nair 1940(63, 64) Kishen 1949(44)
552. 27^{28} in 729	ortho. squares	sat.	Stevens 1939(91)*, Bose & Nair 1941(381)✶, Kitagawa & Mitome 1953(I.20)*
553. 27.3^4 in 81	Latin hypercube		Kishen 1949(45)
554. 27.3^5 in 81	3^5 in 27.3	1 c 3	Connor & Zelen 1959(11)
555. 27.3^5 in 243	3^5 in 27.9	1 c 2	Kempthorne 1947(261)
556. 27.3^6 in 81	3^6 in 27.3	1 c 2	Connor & Zelen 1959(18)
557. 27.3^6 in 243	3^6 in 27.9	1 c 4	Connor & Zelen 1959(13)
558. 27.3^6 in 729	3^6 in 27.27	1 c 2	Kempthorne 1947(261)
559. 27.3^7 in 81	3^7 in 27.3	1 c 3	Connor & Zelen 1959(23)
560. 27.3^7 in 243	3^7 in 27.9	1 c 3	Connor & Zelen 1959(19)
561. 27.3^7 in 729	3^7 in 27.27	2 c 4	Connor & Zelen 1959(14)
562. 27.3^8 in 81	3^8 in 27.3	1 c 2	Connor & Zelen 1959(30)
563. 27.3^8 in 243	3^8 in 27.9	1 c 3	Connor & Zelen 1959(25)
564. 27.3^8 in 729	3^8 in 27.27	2 c 2	Connor & Zelen 1959(20)
565. 27.3^9 in 81	3^9 in 27.3	1 c 2	Connor & Zelen 1959(34)
566. 27.3^9 in 243	3^9 in 27.9	1 c 3	Connor & Zelen 1959(31)
567. 27.3^9 in 729	3^9 in 27.27	2 c 3	Connor & Zelen 1959(26)
568. 27.3^{10} in 243	3^{10} in 27.9	1 c 3	Connor & Zelen 1959(36)
569. 32^{33} in 1024	ortho. squares	sat.	Stevens 1939(92)✶
570. 32.2^8 in 128	2^8 in 32.4		N.B.S. 1957(9)
571. 32.2^9 in 256	2^9 in 32.8		N.B.S. 1957(12)
572. 32.2^{10} in 256	2^{10} in 32.8		N.B.S. 1957(25)

_____

*One square printed in full; 25 additional squares obtained by cyclically permuting all rows except the first.

✶ In symbolic form.

573.	32.2^{11} in 256	2^{11} in 32.8	N.B.S. 1957(36)
574.	32.2^{12} in 256	2^{12} in 32.8	N.B.S. 1957(47)
575.	32.2^{14} in 256	2^{14} in 32.8	N.B.S. 1957(62)
576.	32.2^{15} in 256	2^{15} in 32.8	N.B.S. 1957(70)
577.	32.2^{16} in 256	2^{16} in 32.8	N.B.S. 1957(82)
578.	$32.4.2^{8}$ in 128	2^{8} in 1.(32.4)	N.B.S. 1957(9)
579.	$32.8.2^{9}$ in 256	2^{9} in 1.(32.8)	N.B.S. 1957(12, 15)
580.	$32.8.2^{10}$ in 256	2^{10} in 1.(32.8)	N.B.S. 1957(25, 27)
581.	$32.8.2^{11}$ in 256	2^{11} in 1.(32.8)	N.B.S. 1957(36, 39)
582.	49^{50} in 2401	ortho. squares sat.	Stevens 1939(92)*
583.	49.7^{3} in 343	7^{3} in 49.7	Kitagawa & Mitome 1953(II.83)
584.	64^{65} in 4096	ortho. squares sat.	Stevens 1939(92)*
585.	64.4^{4} in 256	Latin hypercube	Kishen 1949(45)
586.	81^{82} in 6561	ortho. squares sat.	Stevens 1939(92)*
587.	125^{126} in 15625	ortho. squares sat.	Stevens 1939(92)*
588.	125.5^{4} in 625	Latin hypercube	Kishen 1949(46)

15.12 Supplementary finding list: factorial patterns derivable from incomplete block patterns

The 4-figure number following the factorial description is the line number in the finding list of incomplete block patterns, Section 15.21 below.

3^{4} in 63	1388	$3^{2}.2^{2}$ in 21	1390	$3^{2}.2^{5}$ in 24	1200
3^{4} in 72	1428	$3^{2}.2^{2}$ in 30	1253	$3^{2}.2^{5}$ in 48	{1051 / 1179}
3.2^{4} in 16	1169	$3^{2}.2^{3}$ in 24	{1017 / 1077 / 1432}	$3^{2}.2^{5}$ in 96	1435
3.2^{5} in 72	1485			$3^{2}.2^{5}$ in 120	1272
3.2^{6} in 72	1036	$3^{2}.2^{3}$ in 48	1433	$3^{2}.2^{6}$ in 216	1503
3.2^{7} in 144	1346	$3^{2}.2^{4}$ in 24	1177	$3^{2}.2^{7}$ in 48	1358
		$3^{2}.2^{4}$ in 84	1391		

*In symbolic form.

15.12

$3^2.2^7$ in 96	1205	$5^2.2^4$ in 40	1195	$5.3^2.2^3$ in 120	1453
$3^2.2^7$ in 240	1284	$5^2.2^4$ in 80	{1043, 1228}	$5.3^2.2^4$ in 80	1429
$3^3.2^2$ in 72	{1321, 1322}	$5^2.2^5$ in 40	1226	$5.3^2.2^4$ in 120	{1057, 1207}
$3^4.2$ in 36	1034	$5^2.2^5$ in 80	{1197, 1431}	$5.3^2.2^5$ in 120	1535
$3^4.2^2$ in 54	1091	$5^2.2^7$ in 320	1246	$5.3^2.2^5$ in 240	1455
$3^6.2^2$ in 162	1138	$5^2.2^8$ in 160	1245	$5.3^2.2^6$ in 120	1283
4^3 in 32	1008	$5^2.2^8$ in 720	1524	$5.3^3.2^2$ in 90	1293
4^4 in 80	1278	$5^3.2$ in 100	1198	$5.3^4.2$ in 81	1481
4^4 in 112	1401	$5^4.2^2$ in 200	1072	$5.3^4.2^4$ in 180	1296
4^4 in 144	1490	$5^4.2^2$ in 300	1155	$5^2.3.2$ in 25	1252
4^5 in 128	1065	5.3^3 in 90	{1323, 1540}	$5^2.3.2$ in 30	{1038, 1081, 1250, 1324}
$4^4.2$ in 96	1133	5.3^4 in 90	1037		
$4^4.2$ in 160	1279	5.3^6 in 270	1073	$5^2.3.2$ in 60	1196
4.3^3 in 72	1321	$5^2.3$ in 30	1012	$5^2.3.2$ in 90	1326
$4^2.3$ in 16	1169	$5^2.3^2$ in 45	1251	$5^2.3.2^2$ in 30	1106
$4.3^2.2^2$ in 24	1177	$5^2.3^2$ in 60	1215	$5^2.3.2^2$ in 60	{1041, 1124, 1227, 1342}
$4.3^2.2^2$ in 36	{1110, 1484}	$5^2.3^2$ in 90	1343		
$4.3^2.2^3$ in 24	1200	$5^2.3^2$ in 105	1400	$5^2.3.2^2$ in 90	1109
$4^2.3.2^3$ in 144	1346	$5^2.3^2$ in 135	1489	$5^2.3.2^2$ in 120	{1199, 1344}
$4^2.3.2^5$ in 96	{1132, 1220}	$5^2.3^3$ in 45	1123	$5^2.3.2^2$ in 180	1327
5^2 in 20	1174	$5^2.3^3$ in 90	1060	$5^2.3.2^3$ in 60	{1203, 1325}
5^3 in 50	1014	$5^2.3^4$ in 135	1129	$5^2.3.2^3$ in 120	{1128, 1449}
5.2^4 in 16	1173	$5^3.3^2$ in 150	1062		
5.2^4 in 20	1170	5.4^2 in 16	1173	$5^2.3^2.2$ in 90	{1044, 1488, 1514}
5.2^8 in 240	1135	5.4^2 in 20	1170		
$5^2.2$ in 30	1082	$5^2.4$ in 25	1249	$5^2.3^2.2^2$ in 60	1214
$5^2.2^2$ in 20	{1011, 1194}	$5.3.2^2$ in 15	1080	$5^2.3^2.2^2$ in 90	1108
$5^2.2^2$ in 25	1249	$5.3.2^3$ in 24	1175	$5^2.3^2.2^2$ in 120	1061
$5^2.2^2$ in 40	1013	$5.3.2^4$ in 60	1112	$5^2.3^2.2^2$ in 180	1230
$5^2.2^2$ in 50	1262	$5.3^2.2^2$ in 36	1310	$5^2.3^2.2^2$ in 270	1509
$5^2.2^2$ in 90	1482	$5.3^2.2^2$ in 120	1541	$5^2.3^2.2^3$ in 120	{1216, 1439}
$5^2.2^3$ in 40	1039	$5.3^2.2^3$ in 60	{1052, 1292, 1366, 1534}	$5^2.3^2.2^3$ in 180	1355
$5^2.2^3$ in 60	1107			$5^2.3^2.2^3$ in 360	1466
				$5^2.3^2.2^4$ in 180	1143

15.12

$5^2.3^3.2$ in 90	1126	$7^2.2^2$ in 98	1395	$7^2.3.2^3$ in 84	{1122, 1341}
$5^2.3^3.2$ in 270	1369	$7^2.2^3$ in 49	1385	$7^2.3.2^4$ in 168	1458
$5^2.3^3.2^3$ in 360	1165	$7^2.2^3$ in 56	{1384, 1437}	$7^2.3^2.2$ in 126	{1357, 1376}
$5^2.3^4.2$ in 450	1304	$7^2.2^3$ in 112	1213	$7^2.3^2.2^2$ in 252	{1240, 1519}
$5^2.3^4.2^2$ in 270	{1508, 1530}	$7^2.2^4$ in 56	1211	$7^2.3^2.2^3$ in 504	{1471, 1520}
$5^3.3.2$ in 150	{1130, 1367}	$7^2.2^5$ in 112	1438	$7^2.3^2.2^4$ in 756	1160
$5^3.3^2.2^3$ in 300	1560	$7^2.2^5$ in 168	1159	$7^2.3^4.2^3$ in 504	1473
$5.4^2.2^4$ in 240	1135	$7^2.2^5$ in 224	1239	$7^2.4.2^2$ in 56	1211
$5^2.4.2^2$ in 40	1195	$7^4.2$ in 294	1166	$7^2.4.2^3$ in 224	1239
$5^2.4.2^3$ in 40	1226	7.3^2 in 21	1076	$7.5^2.2$ in 70	1042
$5^2.4.2^3$ in 80	1197	$7^2.3$ in 21	1093	$7.5^2.2^2$ in 70	1389
$5^2.4.2^6$ in 160	1245	$7^2.3^2$ in 63	1499	$7^2.5.2^4$ in 280	1241
$5.4.3$ in 15	1080	$7^2.3^2$ in 168	1451	$7^2.5^2.2$ in 350	1295
$5^2.4.3^2$ in 60	1214	$7^2.3^3$ in 189	1501	$7.5^2.3$ in 105	{1127, 1412}
$5.4.3.2$ in 24	1175	$7^2.3^4$ in 189	1146	$7.5^2.3^2$ in 105	1399
$5^2.4.3^2.2$ in 120	1216	7.4^2 in 28	1171	$7.5^2.3^2$ in 315	{1414, 1515}
$6^2.3$ in 18	1085	$7^2.4$ in 28	1181	$7^2.5.3$ in 105	1144
$6^2.3^2$ in 54	1091	7.6^2 in 36	1312	$7^2.5.3^2$ in 105	1282
$6^2.4$ in 24	1177	7.6^2 in 42	1311	$7^2.5.3^2$ in 210	1071
$6^2.5$ in 36	1310	$7^2.6$ in 49	1382	$7.6.5$ in 35	1254
6.5^2 in 25	1252	$7.3.2^4$ in 48	1314	$7.4.3.2^2$ in 84	1114
6.5^2 in 30	1250	$7.3.2^4$ in 84	1114	$7^2.4.3.2^2$ in 168	1458
$6.3.2$ in 12	1176	$7.3^2.2$ in 54	1591	$7.5.3.2$ in 35	1254
$6^2.3.2$ in 36	1088	$7.3^2.2^2$ in 36	1312	$7.5^2.3.2$ in 210	{1368, 1413}
$6^2.4.2$ in 48	1179	$7.3^2.2^2$ in 42	1311	$7^2.5.3.2$ in 210	{1147, 1148, 1562, 1572}
$6.5.3$ in 30	1018	$7.3^2.2^3$ in 63	1387		
$6.5.3$ in 60	1180	$7.3^2.2^3$ in 84	1054	$7^2.5.3.2^2$ in 420	1378
$6.5.4$ in 24	1175	$7.3^2.2^4$ in 168	1233	$7^2.5.3.2^3$ in 420	1573
7^2 in 28	1182	$7.3^2.2^8$ in 336	1415	$7^2.5.3^2.2$ in 210	1564
7.2^4 in 28	1171	$7^2.3.2$ in 42	{1025, 1356}	$7^2.5.3^2.2$ in 630	{1521, 1574}
7.2^6 in 64	1425	$7^2.3.2$ in 49	1382	$7^2.5.3^2.2^3$ in 420	1565
$7^2.2$ in 28	1024	$7^2.3.2$ in 126	1500	8^2 in 40	1255
$7^2.2^2$ in 28	1181	$7^2.3.2^2$ in 84	1212	8^2 in 56	1386
$7^2.2^2$ in 42	1340				

[549]

15.12

$8^2.2$ in 80	1256	$10^2.2$ in 100	1266	$11^2.5.2$ in 110	{1049 1552 1569}
8.3^2 in 24	1077	10.3^2 in 30	1078	$11^2.5.2^2$ in 220	1232
$8^2.3$ in 24	1094	$10^2.3$ in 30	1106	$11^2.5.2^3$ in 220	1554
8.4^2 in 16	1183	$10^2.3^2$ in 90	1108	$11^2.5.2^3$ in 440	1467
$8^2.6$ in 48	1316	$10^2.4$ in 40	1195	$11^2.5^3.2$ in 550	1570
8.7^2 in 49	1385	$10^2.6$ in 60	1325	$11^2.5^2.3$ in 495	1167
8.7^2 in 56	1384	$10^2.7$ in 70	1389	$11^2.7.2^2$ in 462	1149
$8^2.7$ in 64	1425	$10^2.8$ in 80	1431	$11^2.7.3^2$ in 231	1411
8.3.2 in 24	1184	10.9^2 in 81	1481	11.10.9 in 99	1483
$8^2.3.2$ in 48	1095	10.4.2 in 40	1261	$11.5.3^2.2$ in 99	1483
8.4.2 in 24	1315	10.5.2 in 20	1194	$11^2.5.3.2$ in 330	{1372 1579}
8.6.3 in 72	1189	10.5.3 in 30	1324	$11^2.5.3.2^3$ in 660	1168
8.7.2 in 56	1187	10.5.4 in 40	1430	$12^2.2$ in 120	1272
8.7.4 in 56	1030	10.6.2 in 60	1092	12.4^2 in 48	1111
8.7.6 in 48	1314	10.8.2 in 80	1264	$12^2.4$ in 48	1202
9^2 in 63	1388	10.9.5 in 90	1044	$12^2.6$ in 72	1333
$9^2.3$ in 27	1098	10.9.8 in 80	1429	$12^2.8$ in 96	1435
9.4^2 in 36	1172	11.2^3 in 30	1452	$12^2.9$ in 108	1486
$9^2.4$ in 36	1191	11.2^3 in 34	1502	$12^2.10$ in 120	1535
9.5^2 in 45	1251	11.2^3 in 38	1551	12.11^2 in 121	1580
$9^2.5$ in 45	1257	$11^2.2$ in 44	1046	12.3.2 in 24	1432
$9^2.6$ in 54	1320	$11^2.2$ in 110	{1269 1553}	$12^2.4.2$ in 96	1205
9.8^2 in 64	1427	$11^2.2^2$ in 88	1048	12.4.3 in 36	{1110 1484}
$9^2.8$ in 81	1480	$11^2.3$ in 66	1047	12.6.3 in 36	1330
9.3.2 in 18	1318	$11^2.4$ in 88	1048	12.6.3 in 72	1204
$9.3^2.2$ in 18	1097	$11^2.5$ in 55	1268	12.6.4 in 24	1200
$9^2.4.2$ in 72	{1036 1192}	$11^2.5^2$ in 550	1571	12.6.5 in 60	1534
9.4.3 in 72	1321	$11^2.6$ in 66	1329	12.7.4 in 84	1114
$9.4.3^2$ in 36	1099	11.10^2 in 100	1532	12.8.2 in 48	{1186 1331}
$9^2.5.2$ in 90	1258	$11.3.2^4$ in 132	1118	12.9.2 in 72	1434
9.5.3 in 45	1100	$11^2.3.2$ in 66	1329	12.9.3 in 108	1206
9.6.2 in 36	{1178 1319}	$11^2.3.2^2$ in 121	1580	$13^2.2$ in 78	1362
9.7.3 in 63	1102	11.4.2 in 30	1452	$13^2.2$ in 156	1339
9.8.3 in 72	1103	11.4.2 in 34	1502	$13^2.2^2$ in 52	1209
9.8.7 in 63	1387	11.4.2 in 38	1551		
9.5.3.2 in 90	1105	$11.5^2.2^2$ in 100	1532		

15.12

$13^2.2^3$ in 104	{1210, 1456}	$15^2.8$ in 120	1439	$18^2.8$ in 144	1448		
$13^2.2^5$ in 208	{1236, 1458}	$15^2.10$ in 150	1542	$18^2.10$ in 180	1544		
$13^2.3$ in 39	1119	15.3.2 in 30	1537	18.4.3 in 72	1349		
$13^2.3$ in 78	1058	15.6.2 in 60	1538	18.5.3 in 90	1350		
$13^2.3^2$ in 117	{1121, 1487}	15.6.3 in 90	1275	18.9.3 in 54	1101		
		15.7.5 in 105	1127	18.12.3 in 108	1495		
$13^2.4$ in 52	1209	15.9.3 in 135	1276	$19^2.2$ in 114	1141		
$13^2.6$ in 78	1338	15.10.2 in 60	1342	$19^2.3$ in 57	1140		
$13^2.7$ in 91	1392	15.10.3 in 90	1488				
$13^2.9$ in 117	1487	$16^2.2$ in 96	1133	$19^2.3^2$ in 171	{1142, 1498, 1516}		
$13^2.10$ in 130	1536	$16^2.2$ in 160	1279	$19^2.3^2$ in 513	1518		
$13^2.3.2$ in 78	{1120, 1338, 1361}	$16^2.3$ in 48	1131	$19^2.9$ in 171	1478		
		$16^2.4$ in 64	1218	$19^2.10$ in 190	1545		
$13^2.3.2$ in 234	1364	$16^2.6$ in 96	1345	$19^2.3.2$ in 342	1517		
$13^2.3.2^3$ in 156	1363	$16^2.10$ in 160	1543	$19^2.3^2.2$ in 342	1375		
$13^2.3^2.2$ in 234	1513	16.3.2 in 48	1440	$19^2.3^2.2^3$ in 456	1472		
$13^2.3^2.2^2$ in 468	1586	$16.5.4^2$ in 80	1219	$19^2.5.2$ in 190	1545		
$13^2.4.2$ in 104	{1210, 1456}	16.7.2 in 112	1442	20.5^2 in 100	1229		
		16.7.4 in 112	1221	$20^2.8$ in 160	1450		
$13^2.4.2^3$ in 208	1236	16.8.3 in 144	1346	$20^2.10$ in 200	1550		
$13.5^2.2^4$ in 520	1247	16.8.4 in 32	1217	20.3.2 in 60	1546		
$13^2.5.2$ in 130	1536	16.9.4 in 144	1223	20.5.3 in 60	1227		
$13^2.5.2^3$ in 1040	1307	16.12.2 in 96	{1334, 1441}	20.8.4 in 160	1281		
$13^2.5.3^2.2$ in 390	1297	$17^2.2^2$ in 136	1068	20.9.5 in 180	1230		
$13.7^2.3.2^2$ in 546	1590	$17^2.2^3$ in 136	{1224, 1447}	20.10.4 in 40	1226		
$13^2.7^2.5.2$ in 910	1577	$17^2.3^2$ in 153	1491	20.10.4 in 80	1197		
$14^2.4$ in 56	1211	$17^2.4$ in 136	1068	20.12.2 in 120	{1117, 1547}		
$14^2.6$ in 84	1341	$17^2.5^2$ in 425	1308	20.15.2 in 120	1449		
$14^2.8$ in 112	1438	$17^2.8$ in 136	1447	20.16.2 in 160	{1445, 1548}		
14.7.2 in 42	1340	$17^2.9$ in 153	1491	$21^2.2$ in 126	1357		
14.7.4 in 56	1437	$17.3^3.2^2$ in 306	1139	$21^2.3$ in 189	1501		
14.8.2 in 112	1396	$17^2.4.2$ in 136	1224	$21^2.5$ in 105	1282		
$15^2.3$ in 45	1123	$17^2.5^2.2$ in 850	1309	21.7^2 in 147	1145		
$15^2.4$ in 60	1214	18.4^2 in 144	1496	21.7.2 in 42	1356		
15.5^2 in 75	1125	18.6^2 in 36	1348	21.7.3 in 63	1499		
$15^2.7$ in 105	1399	$18^2.6$ in 108	1351	21.9.3 in 189	1405		
				21.10.7 in 210	1147		

15.12
15.1311

21.14.2 in 84	1212	$29.5^2.3.2$ in 870	1162	$37^2.3^2$ in 333	1164	
					1512	
$22^2.10$ in 220	1554	$30^2.9$ in 270	1508	$37^2.9$ in 333	1512	
22.11.5 in 110	1552	$30^2.10$ in 300	1560	39.26.3 in 234	1513	
$23.3.2^6$ in 552	1153	30.6.2 in 60	1292	$40^2.4$ in 160	1245	
$24^2.3$ in 216	1151	30.6.3 in 90	1293	40.8^2 in 320	1298	
$24^2.5$ in 120	1283	30.7.5 in 210	1368	40.7.4 in 280	1563	
24.8^2 in 192	1150	30.9.3 in 270	1559	40.9.5 in 360	1466	
24.5.3 in 120	1453	30.9.5 in 270	1369	$41^2.5.2$ in 410	1300	
24.7.4 in 168	1359	30.10.6 in 60	1366		1575	
24.8.3 in 72	1096	30.12.2 in 120	1207	$41^2.5.2^3$ in 820	1576	
24.8.6 in 48	1358	30.15.3 in 90	1126	42.7^2 in 294	1397	
24.9.3 in 216	1454	30.15.4 in 120	1216	$42^2.10$ in 420	1565	
24.12.3 in 72	1113	30.15.5 in 150	1277	42.7.3 in 126	1376	
$24.16.12^2$ in 96	1220		1558	42.10.7 in 420	1378	
$25^2.5$ in 125	1286	30.18.2 in 180	1354	42.21.5 in 210	1564	
$25^2.9$ in 225	1504	30.20.3 in 180	1355	45.9^2 in 405	1303	
$25^2.3.2$ in 150	1154	30.24.2 in 240	1455	45.7.5 in 315	1515	
$25^2.4.2$ in 200	1235	$31^2.6$ in 186	1371	45.13.3 in 135	1129	
$25.5^2.3$ in 75	1285	$31^2.10$ in 310	1561	45.36.2 in 360	1566	
25.6.5 in 150	1288	$31^2.3.2$ in 186	1371	$48^2.7$ in 336	1415	
25.7.5 in 175	1289	$31^2.5.2$ in 310	1561	48.16.3 in 144	1134	
25.8.5 in 200	1290	$32^2.8$ in 256	1462	$49^2.6$ in 294	1380	
25.10.2 in 100	1555	32.5.4 in 160	1460	$49^2.7$ in 343	1417	
$26^2.6$ in 156	1363	32.7.4 in 224	1461	$49.7^2.4$ in 196	1416	
$26^2.8$ in 208	1457	32.16.3 in 96	1132	49.8.7 in 392	1418	
26.13.3 in 78	1361	32.16.4 in 128	1222	49.9.7 in 441	1419	
26.13.4 in 104	1456	32.24.2 in 192	1360	49.10.7 in 490	1420	
$27^2.9$ in 243	1507	$33^2.7$ in 231	1411	$50^2.8$ in 400	1469	
27.4.3 in 108	1506	35.7^2 in 245	1294	50.6.5 in 300	1567	
27.18.3 in 162	1353	35.21.2 in 210	1562	50.9.5 in 450	1568	
28.7^2 in 196	1238	$36^2.2$ in 288	1465	50.20.2 in 200	1231	
28.8.4 in 224	1408	$36^2.5$ in 180	1296	50.25.4 in 200	1468	
28.9.7 in 252	1240	36.9^2 in 324	1244	50.30.2 in 300	1370	
28.14.3 in 84	1122	36.6.3 in 108	1373	52.26.4 in 208	1236	
28.21.4 in 168	1458	36.6.4 in 144	1374	54.9^2 in 486	1381	
$29^2.2^3$ in 232	1459	36.9.2 in 72	1464	54.18.3 in 162	1138	
$29^2.7$ in 203	1410	36.12.3 in 108	1116	54.27.4 in 216	1237	
$29^2.8$ in 232	1459	$36.18.2^2$ in 144	1225	$55^2.10$ in 550	1570	
		36.18.3 in 108	1136	55.11.2 in 110	1569	
		36.27.2 in 216	1365	55.11.5 in 550	1571	

15.12
15.1311

55.22.2 in 220	1232	64^2.7 in 448	1424	81.9.2 in 162	1525
55.33.2 in 330	1372	64^2.8 in 512	1475	81.9.3 in 243	1526
55.44.2 in 440	1467	64.8.4 in 256	1474	81.9^2.5 in 405	1527
56.8^2 in 448	1422	64.9.8 in 576	1476	$81.10.9^2$ in 810	1529
56.8.3 in 168	1421	64.10.8 in 640	1477	81.27.3 in 243	1157
56.9.7 in 504	1471	64.32.4 in 256	1243	82^2.10 in 820	1576
56.28.8 in 224	1239	64.48.2 in 384	1379	82.41.5 in 410	1575
57^2.8 in 456	1472	70.28.2 in 280	1241	91^2.10 in 910	1577
57.19.3 in 171	1516	70.35.5 in 350	1295	100^2.9 in 900	1531
57.19.3 in 513	1518	70.42.2 in 420	1573	100.10.3 in 300	1578
57.38.3 in 342	1375	72.9^2 in 648	1479	110.11.3 in 330	1579
60.20.3 in 180	1143	72.9.3 in 216	1478	111.37.3 in 333	1164
60.24.2 in 240	1234	$73^2.3^2$ in 657	1523	121.11^2.6 in 726	1583
61^2.3 in 915	1306	73^2.9 in 657	1523	132.12.3 in 396	1584
63^2.8 in 504	1473	73^2.3.2 in 1752	1248	144.12.4 in 576	1585
63.9^2 in 567	1423	75.30.2 in 300	1242	156.13.3 in 468	1586
63.10.7 in 630	1521	78.39.5 in 390	1297	169.13^2.7 in 1183	1589
63.21.3 in 189	1146	81^2.9 in 729	1528	182.14.3 in 546	1590
63.28.2 in 504	1520				

15.13 Patterns without control of heterogeneity

15.131 Saturated patterns: weighing designs

The pattern $p^a.q^b.r^c$ in j.k is called <u>saturated</u> when:

(1) The grand mean, $a(p-1) + b(q-1) + c(r-1)$ treatment effects, and j-1 block effects are estimable, and each estimate is clear of the other effects.

(2) $a(p-1) + b(q-1) + c(r-1) + j = jk$.

15.1311 Saturated fractions of 2^a factorials

By identifying the 2^k-k-1 2-factor, 3-factor, ..., k-factor interactions in a 2^k in $[2^k]$ pattern with new factors, we obtain a (saturated) pattern for $2^{[2^k-1]}$ in $[2^k]$. Every interaction is confounded with a simple main effect. See section 15.112,

$\begin{cases} k = 2 & 3 & 4 & 5 & 6 \\ \text{line } 1 & 10 & 56 & 83 & 92 \end{cases}$. Taguchi 1959 gives diagrams of exactly what is confounded with what.

[553]

15.1312
15.141

15.1312 Complete sets of orthogonal squares and cubes

Whenever p is a prime or a power of a prime, the p-1 alphabets of a complete set of orthogonal squares, plus rows and columns, yield a saturated pattern for p^{p+1} in $[p^2]$. See section 15.11:

$$\begin{cases} p = & 3 & 4 \\ \text{line} & 105 & 25, 223 \end{cases}$$

$$\begin{cases} 5 & 7 & 8 & 9 & 11 & 16 & 25 & 27 \\ 346 & 401 & 82, 412 & 137, 475 & 508 & 94, 228, 512 & 351, 549 & 139, 552 \end{cases}$$

$$\begin{cases} 32 & 49 & 64 & 81 & 125 \\ 102, 569 & 403, 582 & 103, 229, 584 & 140, 586 & 352, 587 \end{cases}$$

Whenever p is a prime or a power of a prime, the p^2+p-2 alphabets of a complete set of orthogonal cubes, plus rows, columns and flats, yield a saturated pattern for $p^{[p^2+p+1]}$ in $[p^2]$. See section 15.11:

$$\begin{cases} p = & 3 & 4 & 5 & 7 \\ \text{line} & 136 & 86, 227 & 350 & 402 \end{cases}.$$ Analogous patterns for $p^{[p^k-1]/[p-1]}$ in $[p^k]$

can be constructed, but we have seen no tables.

15.1313 Orthogonal matrices

The construction of a square 4k 4k matrix, with elements +1 and -1, such that the inner product of any two columns vanishes, is discussed by Paley 1933, who gives methods for $4k = 2^j(p^n+1)$, p a prime; this in particular covers $4k = 4(4)88, 96(4)112$. Plackett & Burman 1946 apply these matrices (called orthogonal or Hadamard matrices) to yield saturated patterns for 2^{4k-1} in 4k. When $4k = 2^j$ these patterns reduce to those noticed in section 15.1311. See section 15.112:

$$\begin{cases} 4k = & 12 & 16 & 20 & 24 & 28 & 32\text{-}40 & 44\text{-}68 & 72\text{-}100 \\ \text{line} & 32 & 56 & 71 & 80 & 81 & 83\text{-}85 & 87\text{-}93 & 95\text{-}101 \end{cases}.$$

Taguchi 1959, 1960 gives several notable patterns of the form $3^a.2^b$ in j.k, where $jk - [2a+b+j]$ is small; see section 15.1132, lines 183, 208, 218, 219.

15.138 Patterns for investigating response surfaces

We confine our notes to central composite second-order patterns: Cochran & Cox 1957, 346-347. These patterns are fractions of 3^a or 5^a patterns, with the levels of each (quantitative) factor so chosen as to yield estimates, clear of cubic effects, of the $\frac{1}{2}(a+1)(a+2)$ grand mean, linear, and quadratic effects.

(6 effects)	5^2 in 13	Box & Hunter 1957(227), Cochran & Cox 1957(348)
	5^2 in 16	Box & Hunter 1957(227)
(10 effects)	3^3 in 15	Box & Behnken 1961(462)
	5^3 in 20	Box & Hunter 1957(227), Cochran & Cox 1957(350)
	5^3 in 23	Box & Hunter 1957(227)
(15 effects)	5^4 in 17	Hartley 1959(614)
	5^4 in 31	Box & Hunter 1957(227), Cochran & Cox 1957(370)
	5^4 in 36	Box & Hunter 1957(227)
(21 effects)	5^5 in 32	Box & Hunter 1957(227), Cochran & Cox 1957(370)
	5^5 in 36, 52, 59	Box & Hunter 1957(227)
(28 effects)	5^6 in 29	Hartley 1959(614)
	5^6 in 53	Box & Hunter 1957(227), Cochran & Cox 1957(371)
	5^6 in 59, 91, 100	Box & Hunter 1957(227)
(36 effects)	5^7 in 92, 100, 163, 177	Box & Hunter 1957(227)
(45 effects)	5^8 in 93, 100, 164, 177, 300, 324	Box & Hunter 1957(227)
(78 effects)	3^{11} in 188	Box & Behnken 1961(462)

15.14 Patterns with 1-way control of heterogeneity

15.141 Fractional factorials confounded in blocks

Extensive tables have been confined to the 2^a and 3^a systems, and are devoted to patterns in which as many 2-factor interactions as possible are estimable, clear of blocks and of other 2-factor interactions. For theory see Cochran & Cox 1957, 256-258; Davies 1954, 465-466; Kempthorne 1952, 397-406. For tables see references to Brownlee et al. 1948 and N.B.S. 1957 in section 15.112, and to Connor & Zelen 1959 in section 15.113.

15.142
15.148

15.142 Full factorials confounded in blocks

For theory see Cochran & Cox 1950, 154-207; 1957, 183-232; Davies 1954, 367-409; Kempthorne 1952, 252-270, 298-300, 321-325.

15.1421 Patterns with some interactions completely confounded

Lists of patterns adequate for many practical purposes are given by Cochran & Cox 1950, 186, 188; 1957, 215, 217. See section 15.11, lines 2, 3, 7-9, 11, 12, 15, 16, 21, 23, 28, 30, 31, 37-39, 104, 107-109, 111, 120, 124, 141-144, 146, 147, 149, 153-156, 160, 167, 168, 171, 177, 184, 194, 220, 222, 224, 230-234, 236, 239-243, 248, 253, 275-277, 279, 285, 288, 291, 295, 312, 324, 325, 328, 332-334, 336, 344, 345, 348, 353, 359, 367, 368, 373, 374, 376, 377, 380, 381, 387, 388, 390, 395, 397, 398, 400, 404, 405, 407-409, 413, 414, 417, 421, 425, 430, 459, 476, 478, 495, 513, 515, 518, 522, 540, 550, 551, 555, 558, 583.

15.1422 Balanced patterns: all interactions partly confounded

A typical pattern with partial confounding leaves main effects clear of blocks in all replications, and confounds a different set of interactions with blocks in each replication: Cochran & Cox 1950, 156-157; 1957, 185-186; Davies 1954, 380-381; Kempthorne 1952, 271-288, 300, 306-307. Lists of patterns adequate for many practical purposes are given by Cochran & Cox 1950, 186, 188; 1957, 215, 217. See section 15.11, lines 49, 147, 149, 150, 172, 186, 196, 203, 209, 211, 214, 326, 328, 329, 335, 337, 339, 341, 355-358, 360, 362, 371, 378, 396, 471, 472, 496-501, 503, 505.

15.143 Lattices

For an extended discussion of lattices, regarded as factorial patterns, see Kempthorne 1952, 445-484.

15.1431 Balanced lattices

A lattice pattern for $(q+1)$ replications of q^2 varieties (or treatments) may be regarded as a factorial pattern for q^2 in $[q+1].q.q$, with partial confounding; or, unblocked, as $[q+1].q^3$ in $[q^3+q^2]$. See section 15.21:

$\begin{cases} q = & 3 & 4 & 5 & 7 & 8 & 9 \\ \text{line} & 1099 & 1219 & 1288 & 1418 & 1476 & 1529 \end{cases}$.

15.1432 Unbalanced lattices

A lattice pattern for r replications of q^2 varieties may be regarded as a factorial pattern for q^2 in r.q.q , with some interactions estimable; or, unblocked, as q^3.r in $[rq^2]$. See section 15.21:

$$\begin{cases} q = & 6 & 6 & 10 & 12 \\ r = & 3 & 4 & 3 & 4 \\ \text{line} & 1373 & 1374 & 1578 & 1585 \end{cases}$$

15.148 Patterns for investigating response surfaces

The linear and quadratic estimates from these patterns, which are closely related to those of section 15.138, are clear of cubic effects and of blocks: note that blocks of unequal size are often used.

(6 effects)	5^2 in 2.7	Box & Hunter 1957(233)
(10 effects)	5^3 in 2.6 + 1.8	Box & Hunter 1957(233), Cochran & Cox 1957(373)
(15 effects)	$\begin{cases} 3^4 \text{ in } 3.9 \\ 5^4 \text{ in } 3.10 \end{cases}$	Box & Behnken 1960(460) Box & Hunter 1957(233), Cochran & Cox 1957(373)
(21 effects)	$\begin{cases} 5^5 \text{ in } 1.22 + 2.11 \\ 3^5 \text{ in } 2.23 \\ 5^5 \text{ in } 4.10 + 1.14 \end{cases}$	Box & Hunter 1957(233), Cochran & Cox 1957(374) Box & Behnken 1960(460) Box & Hunter 1957(233)
(28 effects)	$\begin{cases} 5^6 \text{ in } 2.20 + 1.14 \\ 3^6 \text{ in } 2.27 \\ 5^6 \text{ in } 8.9 + 1.18 \end{cases}$	Box & Hunter 1957(233), Cochran & Cox 1957(375) Box & Behnken 1960(461) Box & Hunter 1957(233)
(36 effects)	3^7 in 2.31	Box & Behnken 1960(461)
(55 effects)	$\begin{cases} 3^9 \text{ in } 10.13 \\ 3^9 \text{ in } 5.26 \end{cases}$	Box & Behnken 1960(461)
(66 effects)	3^{10} in 2.85	Box & Behnken 1960(462)
(91 effects)	3^{12} in 2.102	Box & Behnken 1960(462)
(153 effects)	$\begin{cases} 3^{16} \text{ in } 12.33 \\ 3^{16} \text{ in } 6.66 \end{cases}$	Box & Behnken 1960(463)

15.15
15.1512

15.15 Patterns with 2-way control of heterogeneity

15.151 Squares

For complete (saturated) sets of orthogonal squares and cubes see section 15.1312.

15.1511 Latin squares and cubes

A Latin square of side s is a pattern of s letters, arrayed in s rows and s columns, each letter occurring once in each row and each column: Cochran & Cox 1950, 103-112; 1957, 117-127; Davies 1954, 585-586, 159-169; Kempthorne 1952, 184-203. Such a pattern, i.e. s in 1.(s.s), can be constructed for any s, and yields s^3 in $[s^2]$ on unblocking. See section 15.11:

$\begin{cases} s = & 3 & 4 & 5 & 6 & 7 & 8 & 9 \\ \text{line} & 104 & 7,\ 220 & 342 & 171,\ 379 & 398 & 22,\ 410 & 112,\ 473 \end{cases}$

$\begin{cases} & 10 & 11 & 12 \\ & 363,\ 504 & 506 & 180,\ 330,\ 509 \end{cases}$.

For definitions and tables of Latin cubes and hypercubes see Kishen 1949.

15.1512 Enumeration and randomization of Latin squares

Latin squares of side 2, 3, 4, 5, 6, 7 have been enumerated. Euler 1782, paragraph 48 (1923, p. 385) gives the correct count of squares of side 2, 3, 4, 5, "d'après un dénombrement exact"; but omits the details of such enumeration. The earliest published enumeration of these squares that we have seen is Cayley 1890 (1897, pp. 56-57).

The squares of side 6 are enumerated by Tarry 1900 and by Fisher & Yates 1934; of side 7, by Norton 1939, with a gap filled by Sade 1951.

In the following table, arranged from Fisher & Yates 1953, pp. 18-19 (1957, pp. 21-22), the term <u>standard</u> <u>square</u> denotes a Latin square whose first row and first column are in alphabetical order; each standard square yields s!(s-1)! squares on permutation of rows and columns.

Side	Standard Squares	Total Squares
2	1	2
3	1	12
4	4	576
5	56	1 61280
6	9408	8128 51200
7	169 42080	

To select a Latin square of side s at random (Kempthorne 1952, p. 187), apply to each table the randomizing operations indicated by the letters opposite.

s = 3   The standard square is
$$\begin{array}{ccc} A & B & C \\ B & C & A \\ C & A & B \end{array}$$
apply operations E, F. Or take squares tabled in Kitagawa & Mitome 1953(I.1) and apply operation A.

s = 4 $\begin{cases} \text{Fisher \& Yates 1938-57(T.XV); A, E, G.} \\ \text{Kitagawa \& Mitome 1953(I.1-I.8); A.} \end{cases}$

s = 5 $\begin{cases} \text{Fisher \& Yates 1938-57(T.XV); B, C, E, G.} \\ \text{Kitagawa \& Mitome 1953(I.8-I.9); A, E, G.} \end{cases}$

s = 6 $\begin{cases} \text{Fisher \& Yates 1938-57(T.XV); B, C, F, G, H.} \\ \text{Kitagawa \& Mitome 1953(I.10); B, F, G, H.} \end{cases}$

s = 7   Kitagawa & Mitome 1953(I.10-I.15); B, D, F, G, H.

A   Select any printed square with equal probability.

B   Select any printed square with probability proportional to the numbers printed below the table.

C   With probability $\frac{1}{2}$, interchange rows and columns.

D   Number the rows, columns, and letters. Let $\delta_{ijk} = 1$ if letter k appears at the intersection of row i and column j, $\delta_{ijk} = 0$ otherwise. With probability 1/6, choose one of the 6 permutations of the letters i, j, k. Apply this permutation to the subscripts of the δ_{ijk}. Replace the square with the old δ_{ijk} by the square with the new δ_{ijk}.

E   Apply a random permutation to all rows except the first.

F   Apply a random permutation to all rows.

G   Apply a random permutation to all columns.

H   Apply a random permutation to all letters.

For tables of random numbers, see section 13.1; for tables of random permutations, section 13.2. It is probably most convenient to use Fisher

15.1512
15.153

& Yates 1953, 1957; this book contains under one cover enumerations of Latin squares (T.XV), random numbers (T.XXXIII), random permutations (TT.XXXIII 1, XXXIII 2).

15.1513 Graeco-Latin squares; unsaturated sets of orthogonal squares

A Graeco-Latin square is an array formed by superposing two Latin squares so that each letter of the first square occurs once with each letter of the second square: Cochran & Cox 1950, 117-118; 1957, 132-133; Davies 1954, 169-175; Kempthorne 1952, 187. Such a pattern, i.e. s^2 in 1.(s.s), yields s^4 in $[s^2]$ on unblocking. See section 15.11:

$\{$ s = 3   4   5   7   8   9   11   12
  line 105   14, 221 343   399   42, 411 119, 474 507   193, 331, 510

$\{$ 20
  364, 372, 548.

If q is a prime or a power of a prime, then the pattern q^{q+1} in $[q^2]$ has been constructed: see section 15.1312. If $q = p_1^a p_2^b \ldots p_n^z$, where the p_i are distinct primes, and $\delta = \min(p_1^a, \ldots, p_n^z)$, then the pattern $q^{\delta+1}$ in $[q^2]$ can always be constructed: Mac Neish 1922. For q = 6 this result is best possible, i.e. there is no Graeco-Latin square of side 6: Fisher & Yates 1934. For many q > 6 this result has been improved by Bose and his colleagues: in particular, if $q = 4k + 2 \geq 10$, then the pattern q^4 in $[q^2]$, i.e. a Graeco-Latin square, can be constructed.

The following table, based on Bose, Chakravarti & Knuth 1960, Bose & Shrikhande 1960, and Bose, Shrikhande & Parker 1960, asserts that the pattern q^a in $[q^2]$, i.e. a set of (a-2) orthogonal Latin squares of side q, can be constructed; either by combinatorial methods given in the papers cited, or (12^7 in 144) by ingenious computational devices. The values of a given are lower bounds, and subject to improvement.

$\{$ q = 10 12 14 18 21 22 24 26 30 33 34 38 39 42 46 50 54 57
  a =   4   7   4   4   6   4   5   4   4   5   4   4   5   4   4   7   6   9

$\{$ 58 60 62 65 66 68 69 70 74 75 76 78 80 82 84 85 86 90 92
   4   5   4   9   7   7   7   8   7   7   7   8   9   6   7   7   7   4   7

$\{$ 94 95 96 98 100 102 105 106 110 111 114 115 118 120 122 123
   4   8   8   7   7   7   8   4   4   8   4   8   4   9   4   7

$$\begin{cases} q = & 124 & 126 & 129 & 130 & 132 & 134 & 135 & 138 & 140 & 141 & 142 & 145 & 146 & 147 \\ a = & 6 & 7 & 8 & 7 & 7 & 7 & 8 & 6 & 7 & 8 & 7 & 9 & 7 & 8 \end{cases}$$

$$\begin{cases} 150 & 154 \\ 7 & 4 \end{cases}.$$

15.1514 Orthogonal partitions of Latin squares

Given the 3 factors (rows, columns, letters) of an s × s Latin square, Finney 1946 showed how to introduce a fourth factor (at less than s levels) so that the customary estimates of row, column and letter effects are clear of the fourth factor. This technique is most profitable when no Graeco-Latin square exists, i.e. for s = 6; when the added factor can be at either 2 or 3 levels. For tables see Kitagawa & Mitome 1953, I.21-I.27.

15.152 Double confounding

We confine the term 'double confounding' to factorial patterns, not published as Latin squares, where rows and columns are used as heterogeneity controls: Davies 1954, 582, s.v. confounding; Kempthorne 1952, 286. See section 15.11, lines 7, 11, 15, 19, 25, 28, 33, 34, 35, 36, 40, 41, 42, 46, 47, 49, 53, 58, 59, 61, 66, 67, 68, 70, 74, 115, 118, 119, 188, 209, 214, 278, 281, 282, 284, 286, 316-319, 320-322, 358, 394, 406, 418, 450-458, 460, 463-467, 469, 470, 472, 492-494, 497, 498, 502, 537-539, 542-547, 578-581. The line numbers not underscored above are examples of quasi-Latin squares; Cochran & Cox 1950, 241-248; 1957, 317-324.

15.153 Lattice squares

For an extended discussion of lattice squares, regarded as factorial patterns, see Kempthorne 1952, 485-506.

A lattice square pattern for (q+1) replications of q^2 varieties (or treatments) may be regarded as a factorial pattern for q^2 in [q+1].(q.q), with partial confounding; or, unblocked, as $[q+1].q^4$ in $[q^3+q^2]$. See section 15.21: $\begin{cases} q = & 3 & 4 & 5 & 7 & 8 & 9 \\ \text{line} & 1099 & 1219 & 1288 & 1418 & 1476 & 1529 \end{cases}.$

When q is odd, a lattice square pattern for $\frac{1}{2}(q+1)$ replications of q^2 varieties may be regarded as a factorial pattern for q^2 in $\frac{1}{2}[q+1].(q.q)$,

15.153
15.20

with partial confounding; or, unblocked, as $q^4 \cdot \frac{1}{2}[q+1]$ in $\frac{1}{2}[q^3+q^2]$. See

section 15.21: $\begin{cases} q = & 5 & 7 & 9 & 11 & 13 \\ \text{line} & 1285 & 1416 & 1527 & 1583 & 1589 \end{cases}$.

15.16 Factorial patterns with factors applied to split plots

For the principles of split-plot experiments see Cochran & Cox 1950, 218-240; 1957, 293-315; Kempthorne 1952, 370-389. We indicate the factors applied to split plots by underscoring: in general, the estimates of these factors have smaller errors than the estimates of whole-plot factors. The tables we have seen afford either 1-way or 2-way control of heterogeneity; the patterns are due chiefly to Yates.

15.161 Split-plot patterns with 1-way control of heterogeneity

See section 15.11, lines 11, 78, 159, 183, 276, 310, 478.

15.162 Split-plot patterns with 2-way control of heterogeneity

15.1621 Latin squares with split plots

The notation $4.\underline{2^2}$ in (4.4).2 indicates a factor at 4 levels, applied to a 4 4 Latin square, with each plot split in two and $\frac{1}{2}$ of a 2^2 applied to the splits.

See section 15.11, lines 15, 20, 22, 28, 176, 178, 180, 198, 313-315, 383, 384, 386, 393.

15.1622 Half-plaid squares

In these arrays, row effects are divided; some provide estimates of a whole-plot factor, others afford control of heterogeneity; columns afford control of heterogeneity. Cochran & Cox 1950, 248-251; 1957, 324-327. See section 15.11, lines 22, 115, 157, 171, 174, 211, 233, 389, 391, 392, 461, 462, 468, 479, 492, 500, 501.

15.1623 Plaid squares

In these arrays, row effects are divided between estimates of a whole-plot factor and control of heterogeneity; column effects are divided between estimates of a second whole-plot factor and a second control of heterogeneity. Cochran & Cox 1950, 251; 1957, 327. See section 15.11, lines 22, 28, 111, 112, 283, 285, 478.

15.2   Incomplete block patterns

15.20   Introduction

The incomplete block patterns treated in section 15.2 may be described as factorial patterns with one principal factor and one or two heterogeneity controls; where the number of trials in a block is less than the number of levels of the principal factor: Cochran & Cox 1950, 259-390; 1957, 376-544; Davies 1954, 579 s.v. Balanced, 199-246; Kempthorne 1952, 430-568. The nomenclature and notation of such patterns reflects their origin in agricultural experimentation. The levels of the principal factor are called varieties*; the trials are called plots. The customary notations are:

b = number of blocks

k = number of plots (trials) in a block

r = number of replications: i.e. number of times that each variety appears in the pattern

v = number of varieties*.

Then the total number of trials is bk = rv.

The finding list below (section 15.21) displays, in five columns:

(1) a line number, for convenient reference.

(2) the number of varieties, replications, and total trials.

(3) the blocking structure.

(4) for six popular species of symmetrical patterns, an abbreviation of the specific name.

(5) references to the author index.

The lexicographical order adopted is based on Clatworthy 1956 and is an index by number of plots in a block (Fisher & Yates 1938-57, T.XIX); we believe that block size is often the most restrictive consideration in

---

*The name varieties, or the abbreviation v, is used in the extensive tables of Fisher & Yates 1938-57(TT.XVII-XIX), Kitagawa & Mitome 1953 (T.VII), and Bose, Clatworthy & Shrikhande 1954. Many text books use the name treatments, and the somewhat unfortunate letter t: Cochran & Cox 1950, 271; 1957, 390; Davies 1954, 229; Kempthorne 1952, 529.

15.20
choosing an incomplete block pattern. The pattern A precedes the pattern B if and only if $k_A < k_B$; or $k_A = k_B$ and $v_A < v_B$; or $k_A = k_B$, $v_A = v_B$, and $r_A < r_B$.

The canonical form used in the second and third columns is 'v(r) = vr in b.k'. Thus section 15.2102, line 1001, 3(4) = 12 in 6.2, indicates a pattern for 3 varieties, each appearing 4 times, for a total of 12 trials, arrayed in 6 blocks of 2 plots each. The notation b.k undergoes modifications for patterns divisible into replicates and patterns with 2-way control of heterogeneity, discussed below.

Incomplete block patterns divisible into replicates are called resolvable. The ordinary division is into replicates containing each variety once: Cochran & Cox 1950, 318-321; 1957, 443-446; Kempthorne 1952, 537-539. Thus section 15.2102, line 1003, 4(3) = 12 in 3.2.2, indicates a pattern for 4 varieties, each appearing 3 times, for a total of 12 trials, arrayed in 6 blocks of 2 plots each; the 6 blocks are divisible into 3 pairs of blocks, each pair containing each variety once. Some patterns are divisible into replicates containing each variety twice or more: Cochran & Cox 1950, 321; 1957, 446. Thus (section 15.2102, line 1011) 5(4) = 20 in 2.5.2 : the 10 blocks fall into 2 sets of 5 blocks, and each set contains each variety twice. D. R. Cox 1958, p. 225, recommends that "a resolvable design should be used in preference to a nonresolvable one, whenever a choice exists."

The notation for most incomplete block patterns with 2-way control of heterogeneity follows that for factorial patterns with 2-way control. Thus section 15.2103, line 1081, 5(3) = 15 in 1.(5.3), indicates that 5 varieties appear 3 times in 15 trials, in an array of 5 rows × 3 columns; line 1082, 5(6) = 30 in 2.(5.3), indicates that 5 varieties appear 6 times in 30 trials, in 2 such arrays. The notation for heterogeneity control by sets of columns, $v(r) = vr$ in $(b.k_1).k_2$, is best explained by means of an example. At the top of p. 565, we quote from Bose, Clatworthy & Shrikhande 1954, p. 94 (their pattern S 12). This pattern accommodates 9 varieties, each replicated twice, in 3 blocks (i.e. rows) of 6 plots each. Rows afford the first heterogeneity control; and the two groups of 3 columns, each

15.20

```
1 4 7 | 2 5 8
2 5 8 | 3 6 9
3 6 9 | 1 4 7
  I   |  II
     reps
```

of which is a replication, afford the second control. Our notation for this pattern (section 15.2106, line 1318) is $9(2) = 18$ in $(3.2).3$, the round brackets indicating 2-way control of heterogeneity and the figure 3 after the bracket indicating the division of each half-row into 3 plots. It should be noted that $k = k_1.k_2$: i.e. that the number of plots in a block does not explicitly appear in our notation, but may be obtained by multiplying the number of levels of the second heterogeneity control by the number of splits. These patterns are discussed by Bose et al. 1954, pp. 7-8, 10-11, under the title 'replication groups orthogonal to blocks'.

The abbreviations in the fourth column are listed at the head of section 15.21; they are explained in sections 15.24, 15.25 below.

15.21
15.2102

15.21 Finding list of incomplete block patterns

The canonical form of an entry in this list is $v(r) = vr$ in b.k. Round brackets after the word 'in' denote two-way control of heterogeneity. The following abbreviations are used; absence of an abbreviation denotes a partially balanced design.

Y sq = Youden square
L sq = lattice square
2bal = doubly balanced incomplete blocks
latt = square lattice
bal. = balanced incomplete blocks
rect = rectangular lattice

15.2102 Patterns in blocks of 2 plots

1001.	$3(4) = 12$	in 6.2	bal.	Clatworthy 1956(10)
1002.	$4(2) = 8$	in 4.2		Clatworthy 1955(180, 184)
1003.	$4(3) = 12$	in 3.2.2	bal.	Cochran & Cox 1950(329), 1957(471), Kitagawa & Mitome 1953(VII.1), Vianelli 1959(414)
1004.	$4(4) = 16$	in 8.2		Clatworthy 1955(180, 184)
1005.	$4(5) = 20$	in 10.2		Clatworthy 1955(180)
1006.	$4(6) = 24$	in 12.2		Clatworthy 1955(180, 184)
1007.	$4(7) = 28$	in 14.2		Clatworthy 1955(180)
1008.	$4(8) = 32$	in 16.2		Clatworthy 1955(184)
1009.	$4(9) = 36$	in 18.2		Clatworthy 1955(180)
1010.	$4(10) = 40$	in 20.2		Clatworthy 1955(180, 184)
1011.	$5(2) = 10$	in 5.2		Clatworthy 1955(185)
1011.	$5(4) = 20$	in 2.(5.2)	Y sq	Cochran & Cox 1957(538)
		in 2.5.2	bal.	Cochran & Cox 1950(329), 1957(471), Vianelli 1959(414)
		in 10.2	bal.	Kitagawa & Mitome 1953(VII.1)
		in 10.2		Clatworthy 1955(185)
1012.	$5(6) = 30$	in 15.2		Clatworthy 1955(185)
1013.	$5(8) = 40$	in 20.2		Clatworthy 1955(185)
1014.	$5(10) = 50$	in 25.2		Clatworthy 1955(185)
1015.	$6(2) = 12$	in 6.2		Nair & Rao 1948(123)

1016.	6(3) = 18	in 9.2		Nair & Rao 1948(123), Clatworthy 1955(180)
1017.	6(4) = 24	in 12.2		Zoellner & Kempthorne 1954(178), Clatworthy 1955(180, 182), D.R. Cox 1958(241)
1018.	6(5) = 30	in 5.3.2	bal.	Cochran & Cox 1950(329), 1957(471), Kitagawa & Mitome 1953(VII.1), Vianelli 1959(414)
1019.	6(6) = 36	in 18.2		Clatworthy 1955(180)
1020.	6(7) = 42	in 21.2		Clatworthy 1955(180)
1021.	6(8) = 48	in 24.2		Clatworthy 1955(180, 182)
1022.	6(9) = 54	in 27.2		Clatworthy 1955(180)
1023.	6(10) = 60	in 30.2		Clatworthy 1955(180)
1024.	7(4) = 28	in 14.2		Zoelnner & Kempthorne 1954(178), D.R. Cox 1958(241)
1025.	7(6) = 42	in 3.(7.2)	Y sq	Cochran & Cox 1957(539)
		in 21.2	bal.	Kitagawa & Mitome 1953(VII.2), Cochran & Cox 1957(472)
1026.	8(3) = 24	in 12.2		Nair & Rao 1948(123), Zoellner & Kempthorne 1954(178), D. R. Cox 1958(241)
1027.	8(4) = 32	in 16.2		Nair & Rao 1948(123), Zoellner & Kempthorne 1954(178), Clatworthy 1955(180), D. R. Cox 1958(241)
1028.	8(5) = 40	in 20.2		Zoellner & Kempthorne 1954(178), D. R. Cox 1958(241)
1029.	8(6) = 48	in 24.2		Zoellner & Kempthorne 1954(178), Clatworthy 1955(180), D. R. Cox 1958(241)
1030.	8(7) = 56	in 7.4.2	bal.	Cochran & Cox 1950(330), 1957(473), Kitagawa & Mitome 1953(VII.2), Vianelli 1959(415)
1031.	8(8) = 64	in 32.2		Clatworthy 1955(180)
1032.	8(9) = 72	in 36.2		Clatworthy 1955(180)
1033.	8(10) = 80	in 40.2		Clatworthy 1955(180)
1034.	9(4) = 36	in 18.2		Zoellner & Kempthorne 1954(178), Clatworthy 1955(184), D. R. Cox 1958(241)

15.2102

1035.	9(6) = 54	in 27.2		Zoellner & Kempthorne 1954(178), Clatworthy 1955(180), D. R. Cox 1958(241)
1036.	9(8) = 72	⎰ in 4.(9.2)	Y sq	Cochran & Cox 1957(539)
		⎱ in 4.9.2	bal.	Cochran & Cox 1957(473)
		⎰ in 36.2	bal.	Kitagawa & Mitome 1953(VII.2)
		⎱ in 36.2		Clatworthy 1955(184)
1037.	9(10) = 90	in 45.2		Clatworthy 1955(180)
1038.	10(3) = 30	in 15.2		Zoellner & Kempthorne 1954(178), Clatworthy 1955(182)
1039.	10(4) = 40	in 20.2		Nair & Rao 1948(123), Zoellner & Kempthorne 1954(178), D. R. Cox 1958(241)
1040.	10(5) = 50	in 25.2		Nair & Rao 1948(123), Zoellner & Kempthorne 1954(178), Clatworthy 1955(180), D. R. Cox 1958(241)
1041.	10(6) = 60	in 30.2		Zoellner & Kempthorne 1954(178), Clatworthy 1955(182), D. R. Cox 1958(241)
1042.	10(7) = 70	in 35.2		Zoellner & Kempthorne 1954(178), D. R. Cox 1958(241)
1043.	10(8) = 80	in 40.2		Zoellner & Kempthorne 1954(178), Clatworthy 1955(180), D. R. Cox 1958(241)
1044.	10(9) = 90	in 9.5.2	bal.	Cochran & Cox 1950(331), 1957(475), Kitagawa & Mitome 1953(VII.3), Vianelli 1959(415)
1045.	10(10) = 100	in 50.2		Clatworthy 1955(180)
1046.	11(4) = 44	in 22.2		Zoellner & Kempthorne 1954(178), D. R. Cox 1958(241)
1047.	11(6) = 66	in 33.2		Zoellner & Kempthorne 1954(178), D. R. Cox 1958(241)
1048.	11(8) = 88	in 44.2		Zoellner & Kempthorne 1954(178), D. R. Cox 1958(241)
1049.	11(10) = 110	⎰ in 5.(11.2)	Y sq	Cochran & Cox 1957(541)
		⎨ in 5.11.2	bal.	Cochran & Cox 1957(476)
		⎱ in 55.2	bal.	Kitagawa & Mitome 1953(VII.4)

15.2102

1050. 12(3) = 36 in 18.2   Zoellner & Kempthorne 1954(178), D. R. Cox 1958(241)
1051. 12(4) = 48 in 24.2   Zoellner & Kempthorne 1954(178), D. R. Cox 1958(241)
1052. 12(5) = 60 in 30.2   Nair & Rao 1948(123), Zoellner & Kempthorne 1954(178), D. R. Cox 1958(241)
1053. 12(6) = 72 in 36.2   Nair & Rao 1948(123), Zoellner & Kempthorne 1954(178), Clatworthy 1955(180), D. R. Cox 1958(241)
1054. 12(7) = 84 in 42.2   Zoellner & Kempthorne 1954(178), D. R. Cox 1958(241)
1055. 12(8) = 96 in 48.2   Zoellner & Kempthorne 1954(178), Clatworthy 1955(180), D. R. Cox 1958(241)
1056. 12(9) = 108 in 54.2 Zoellner & Kempthorne 1954(178), Clatworthy 1955(180), D. R. Cox 1958(241)
1057. 12(10) = 120 in 60.2 Zoellner & Kempthorne 1954(178), Clatworthy 1955(180), D. R. Cox 1958(241)
1058. 13(6) = 78 in 39.2   Clatworthy 1955(185)
1059. 14(7) = 98 in 49.2   Clatworthy 1955(180)
1060. 15(6) = 90 in 45.2   Clatworthy 1955(182)
1061. 15(8) = 120 in 60.2 Clatworthy 1955(182)
1062. 15(10) = 150 in 75.2 Clatworthy 1955(180)
1063. 16(5) = 80 in 40.2   Clatworthy 1955(186)
1064. 16(6) = 96 in 48.2   Clatworthy 1955(184)
1065. 16(8) = 128 in 64.2 Clatworthy 1955(180)
1066. 16(9) = 144 in 72.2 Clatworthy 1955(184)
1067. 16(10) = 160 in 80.2 Clatworthy 1955(186, 187, 188)
1068. 17(8) = 136 in 68.2 Clatworthy 1955(185)
1069. 18(9) = 162 in 81.2 Clatworthy 1955(180)
1070. 20(10) = 200 in 100.2 Clatworthy 1955(180)
1071. 21(10) = 210 in 105.2 Clatworthy 1955(182)
1072. 25(8) = 200 in 100.2 Clatworthy 1955(184)
1073. 27(10) = 270 in 135.2 Clatworthy 1955(185, 188)
1074. 36(10) = 360 in 180.2 Clatworthy 1955(184)

15.2103

15.2103 Patterns in blocks of 3 plots

1075.	3(5) = 15	in 1.(5.3)	Y sq	Cochran & Cox 1950(388), 1957(536), Vianelli 1959(430)
1076.	3(7) = 21	in 1.(7.3)	Y sq	Cochran & Cox 1950(388), 1957(536), Vianelli 1959(430)
1077.	3(8) = 24	in 1.(8.3)	Y sq	Cochran & Cox 1950(388), 1957(536), Vianelli 1959(431)
1078.	3(10) = 30	in 1.(10.3)	Y sq	Cochran & Cox 1950(388), 1957(536), Vianelli 1959(431)
1079.	4(3) = 12	in 4.3	bal.	Fisher & Yates 1938-57(T.XIX), Czechowski et al. 1957(104), Kitagawa & Mitome 1953(VII.1)
1080.	3(4) + 1(3) = 15	in 1.(5.3)		Cochran & Cox 1957(391)
1081.	5(3) = 15	in 1.(5.3)		Clatworthy 1956(66)
		in 5.3		Clatworthy 1956(66)
1082.	5(6) = 30	in 2.(5.3)	Y sq	Cochran & Cox 1957(538)
		in 2.(5.3)		Clatworthy 1956(66)
		in 2.5.3	bal.	Cochran & Cox 1957(471)
		in 10.3	bal.	Fisher & Yates 1938-57(T.XIX), Czechowski et al. 1957(104), Kitagawa & Mitome 1953(VII.1)
		in 2.5.3		Clatworthy 1956(66)
1083.	5(9) = 45	in 3.(5.3)		Clatworthy 1956(67)
		in 3.5.3		Clatworthy 1956(67)
1084.	6(2) = 12	in 4.3		Nair & Rao 1948(125), Bose et al. 1954(139, 140)
1085.	6(3) = 18	in 1.(6.3)		Bose et al. 1954(183, 184)
		in 6.3		Bose et al. 1954(183, 184)
1086.	6(4) = 24	in 4.2.3		Bose et al. 1954(139, 140)
		in 8.3		Bose & Nair 1939(354), Clatworthy 1956(15)
1087.	6(5) = 30	in 10.3	bal.	Fisher & Yates 1938-57(T.XIX), Yates 1937(122), Bose 1939(378, 395, 397), Cochran & Cox 1950 (329), 1957(471), Kitagawa & Mitome 1953(VII.1), Czechowski et al. 1957(104), D. R. Cox 1958(226), Vianelli 1959(414)

1088.	6(6) = 36	in 2.(6.3)		Clatworthy 1956(33)	
		in 1.(12.3)		Bose et al. 1954(139, 141, 183, 185)	
		in 3.4.3		Bose et al. 1954(139, 141)	
		in 2.6.3		Clatworthy 1956(33)	
		in 12.3		Bose et al. 1954(183, 185)	
1089.	6(7) = 42	in 14.3		Clatworthy 1956(33)	
1090.	6(8) = 48	in 16.3		Bose et al. 1954(139, 141), Clatworthy 1956(33, 34)	
1091.	6(9) = 54	in 3.(6.3)		Clatworthy 1956(34)	
		in 3.6.3		Clatworthy 1956(34)	
1092.	6(10) = 60	in 10.2.3	bal.	Cochran & Cox 1950(329), 1957(472), Vianelli 1959(414)	
		in 20.3		Bose et al. 1954(139, 141)	
1093.	7(3) = 21	in 1.(7.3)	Y sq	Youden 1940(220), Cochran & Cox 1950 (379), 1957(522), Kitagawa & Mitome 1953(VIII.2), D. R. Cox 1958(234), Vianelli 1959(428)	
		in 7.3	bal.	Fisher & Yates 1938-57(T.XIX), Bose 1939(374), Cochran & Cox 1950(330), 1957(472), Kitagawa & Mitome 1953 (VII.2), Czechowski et al. 1957 (104), D. R. Cox 1958(234), Vianelli 1959(414)	
1094.	8(3) = 24	in 1.(8.3)		Bose et al. 1954(183, 185)	
		in 8.3		Bose & Nair 1939(353), Bose et al. 1954(183, 185)	
1095.	8(6) = 48	in 2.(8.3)		Clatworthy 1956(34)	
		in 16.3		Clatworthy 1956(34)	
1096.	8(9) = 72	in 1.(24.3)		Bose et al. 1954(183, 186)	
		in 3.8.3		Bose et al. 1954(183, 186)	
		in 24.3		Bose et al. 1954(183, 185)	
1097.	9(2) = 18	in 2.(3.3)	L sq	Cochran & Cox 1950(360), 1957(497), Kitagawa & Mitome 1953(VI.7), D. R. Cox 1958(236), Vianelli 1959(425)	
1098.	9(3) = 27	in 1.(9.3)		Bose et al. 1954(139, 143)	
		in 9.3		Bose & Nair 1939(352), Bose et al. 1954(139, 143)	

15.2103

1099.	$9(4) = 36$	in 4.(3.3)	L sq	Kitagawa & Mitome 1953(VI.7)
		in 4.3.3	latt	Cochran & Cox 1950(304), 1957(428), Kitagawa & Mitome 1953(VI.7), Czechowski et al. 1957(33), D.R.Cox 1958(226), Vianelli 1959(410)
		in 12.3	bal.	Fisher & Yates 1938-57(T.XIX), Yates 1936(133), Bose 1939(373), Czechowski et al. 1957(104)
		in 4.3.3		Bose et al. 1954(240, 243)
1100.	$9(5) = 45$	in 5.3.3		Bose et al. 1954(183, 186)
1101.	$9(6) = 54$	in 1.(18.3)		Bose et al. 1954(183, 187, 240, 241)
		in 6.3.3		Bose et al. 1954(183, 187, 240, 241)
		in 18.3		Bose et al. 1954(139, 143)
1102.	$9(7) = 63$	in 7.3.3		Bose et al. 1954(183, 187)
		in 21.3		Clatworthy 1956(35)
1103.	$9(8) = 72$	in 8.3.3		Bose et al. 1954(240, 242)
		in 24.3		Bose et al. 1954(240, 243), Clatworthy 1956(35)
1104.	$9(9) = 81$	in 9.3.3		Bose et al. 1954(183, 187)
		in 27.3		Bose et al. 1954(139, 144)
1105.	$9(10) = 90$	in 2.5.3.3		Clatworthy 1956(36)
		in 10.3.3		Bose et al. 1954(183, 188, 240, 242)
1106.	$10(3) = 30$	in 1.(10.3)		Bose & Shimamoto 1952(161, 162), Bose et al. 1954(229, 231), Vianelli 1959(433)
		in 10.3		Bose & Nair 1939(353), Bose & Shimamoto 1952(161, 162), Bose et al. 1954(229, 231), Clatworthy 1956(22)
1107.	$10(6) = 60$	in 20.3		Bose et al. 1954(229, 231)
1108.	$10(9) = 90$	in 3.(10.3)	Y sq	Cochran & Cox 1957(540)
		in 3.10.3	bal.	Cochran & Cox 1957(475)
		in 30.3	bal.	Fisher & Yates 1938-57(T.XIX), Yates 1936(133), Bose 1939(370, 380, 397), Cochran & Cox 1950(332), Kitagawa & Mitome 1953(VII.3), Czechowski et al. 1957(104), D. R. Cox 1958(226), Vianelli 1959(416)

1109. 10(9) = 90   in 30.3   Bose et al. 1954(229, 231)

1110. 12(3) = 36
- in 3.4.3   rect.   Robinson & Watson 1949(47), Cochran & Cox 1950(311), 1957(435), Vianelli 1959(413)
- in 12.3   Nair & Rao 1948(124)

1111. 12(4) = 48
- in 4.4.3   Bose & Shimamoto 1952(157, 158), Bose et al. 1954(139, 146), Vianelli 1959(432)
- in 16.3   Bose & Nair 1939(362), Nair & Rao 1948(124)

1112. 12(5) = 60   in 20.3   Bose & Nair 1939(355), Bose et al. 1954(183, 189), Clatworthy 1956(47)

1113. 12(6) = 72
- in 1.(24.3)   Bose et al. 1954(183, 189)
- in 24.3   Bose et al. 1954(183, 189)

1114. 12(7) = 84
- in 7.4.3   Bose et al. 1954(183, 189)
- in 28.3   Clatworthy 1956(36)

1115. 12(8) = 96   in 32.3   Bose et al. 1954(139, 148, 183, 190)

1116. 12(9) = 108
- in 1.(36.3)   Bose et al. 1954(183, 190)
- in 3.12.3   Bose et al. 1954(183, 190)

1117. 12(10) = 120
- in 2.20.3   Clatworthy 1956(57)
- in 40.3   Bose et al. 1954(183, 190, 191)

1118. 12(11) = 132   in 44.3   bal.   Bose 1939(378), Kitagawa & Mitome 1953(VII.4)

1119. 13(3) = 39
- in 1.(13.3)   Bose & Shimamoto 1952(166, 167), Bose et al. 1954(250)
- in 13.3   Bose & Shimamoto 1952(166, 167), Bose et al. 1954(250)

1120. 13(6) = 78
- in 2.(13.3)   Y sq   Cochran & Cox 1957(542)
- in 2.(13.3)   Clatworthy 1956(68)
- in 2.13.3   bal.   Cochran & Cox 1957(477)
- in 26.3   bal.   Fisher & Yates 1938-57(T.XIX), Yates 1936(133), Bose 1939(375, 397), Cochran & Cox 1950(333), Kitagawa & Mitome 1953(VII.5), Czechowski et al. 1957(104), D. R. Cox 1958 (226), Vianelli 1959(416)
- in 2.13.3   Clatworthy 1956(68)

15.2103

1121. 13(9) = 117
- in 3.(13.3) — Bose & Shimamoto 1952(166, 167), Clatworthy 1956(68)
- in 1.(39.3) — Bose et al. 1954(250, 251)
- in 3.13.3 — Bose & Shimamoto 1952(166, 167), Bose et al. 1954(250, 251), Clatworthy 1956(68)

1122. 14(6) = 84
- in 1.(28.3) — Bose et al. 1954(183, 192)
- in 28.3 — Bose et al. 1954(183, 192)

1123. 15(3) = 45
- in 1.(15.3) — Bose & Shimamoto 1952(161, 162), Vianelli 1959(433)
- in 15.3 — Bose & Shimamoto 1952(161, 162)

1124. 15(4) = 60 in 20.3 — Nair & Rao 1948(124), Bose et al. 1954(229, 236)

1125. 15(5) = 75
- in 5.5.3 — Bose et al. 1954(139, 152)
- in 25.3 — Nair & Rao 1948(124)

1126. 15(6) = 90
- in 1.(30.3) — Bose et al. 1954(183, 193)
- in 2.15.3 — Bose et al. 1954(183, 193)

1127. 15(7) = 105
- in 7.5.3 bal. Fisher & Yates 1938-57(T.XVII), Cochran & Cox 1950(334), 1957(478), Kitagawa & Mitome 1953(VII.5), Czechowski et al. 1957(102), D. R. Cox 1958(226)
- in 35.3 bal. Fisher & Yates 1938-57(T.XIX), Bose 1939(365, 373, 397, 398), Vianelli 1959(417)

1128. 15(8) = 120 in 40.3 — Bose et al. 1954(183, 194, 229, 236)

1129. 15(9) = 135
- in 1.(45.3) — Bose et al. 1954(183, 195)
- in [2.15 + 3.5].3 — Bose et al. 1954(183, 195)

1130. 15(10) = 150 in 50.3 — Bose et al. 1954(139, 153, 229, 238)

1131. 16(3) = 48
- in 1.(16.3) — Bose & Shimamoto 1952(164, 165), Bose et al. 1954(240, 245), Vianelli 1959(435)
- in 16.3 — Bose & Nair 1939(363), Bose & Shimamoto 1952(164, 165), Bose et al. 1954(240, 245)

1132. 16(6) = 96 in 1.(32.3) — Bose et al. 1954(183, 196)

1133.	$16(6) = 96$	in 2.16.3		Bose et al. 1954(183, 196), Clatworthy 1956(61)
1134.	$16(9) = 144$	$\begin{cases} \text{in } 1.(48.3) \\ \text{in } 3.16.3 \end{cases}$		Bose et al. 1954(183, 197) Bose et al. 1954(183, 197), Clatworthy 1956(61)
1135.	$16(15) = 240$	in 80.3	bal.	Bose 1939(381), Kitagawa & Mitome 1953(VII.6)
1136.	$18(6) = 108$	$\begin{cases} \text{in } 1.(36.3) \\ \text{in } 36.3 \end{cases}$		Bose et al. 1954(139, 155) Bose et al. 1954(139, 155)
1137.	$18(8) = 144$	in 48.3		Bose et al. 1954(183, 198)
1138.	$18(9) = 162$	$\begin{cases} \text{in } 1.(54.3) \\ \text{in } 54.3 \end{cases}$		Bose et al. 1954(183, 199) Bose et al. 1954(183, 199)
1139.	$18(17) = 306$	in 102.3	bal.	Bose 1939(378)
1140.	$19(3) = 57$	$\begin{cases} \text{in } 1.(19.3) \\ \text{in } 19.3 \end{cases}$		Bose et al. 1954(218) Bose et al. 1954(218)
1141.	$19(6) = 114$	in 38.3		Bose et al. 1954(218)
1142.	$19(9) = 171$	in 3.(19.3)	Y sq	Cochran & Cox 1957(542)
		in 57.3	bal.	Fisher & Yates 1938-57(T.XIX), Bose 1939(375, 376, 397), Cochran & Cox 1950(336), 1957 (479, 542), Kitagawa & Mitome 1953(VII.7), Czechowski et al. 1957(104), Vianelli 1959(418)
		in 57.3		Bose et al. 1954(218, 219)
1143.	$20(9) = 180$	$\begin{cases} \text{in } 1.(60.3) \\ \text{in } 60.3 \end{cases}$		Bose et al. 1954(183, 199) Bose et al. 1954(183, 199)
1144.	$21(5) = 105$	in 35.3		Clatworthy 1956(51)
1145.	$21(7) = 147$	in 7.7.3		Bose et al. 1954(139, 161)
1146.	$21(9) = 189$	$\begin{cases} \text{in } 1.(63.3) \\ \text{in } 63.3 \end{cases}$		Bose et al. 1954(183, 200) Bose et al. 1954(183, 200)
1147.	$21(10) = 210$	in 10.7.3	bal.	Fisher & Yates 1938-57(T.XVII), Cochran & Cox 1950(337), 1957 (479), Kitagawa & Mitome 1953 (VII.7), Vianelli 1959(419)

15.2103
15.2104

1148. 21(10) = 210	in 70.3	bal.	Fisher & Yates 1938-57(T.XIX), Bose 1939(373, 397), Czechowski et al. 1957(104)
	in 2.35.3		Clatworthy 1956(51)
1149. 22(21) = 462	in 154.3	bal.	Bose 1939(380)
1150. 24(8) = 192	in 8.8.3		Bose et al. 1954(140, 162)
1151. 24(9) = 216	in 3.24.3		Bose et al. 1954(184, 202)
1152. 24(10) = 240	in 80.3		Bose et al. 1954(184, 202)
1153. 24(23) = 552	in 184.3	bal.	Bose 1939(379)
1154. 25(6) = 150	in 2.(25.3)		Bose & Shimamoto 1952(164, 165), Vianelli 1959(435)
	in 1.(50.3)		Bose et al. 1954(240, 246)
	in 2.25.3		Bose & Shimamoto 1952(164, 165), Bose et al. 1954(240, 246)
1155. 25(12) = 300	in 100.3	bal.	Bose 1939(395), Kitagawa & Mitome 1953(VII.8)
1156. 27(5) = 135	in 45.3		Bose et al. 1954(218, 220), Bose & Clatworthy 1955(229)
1157. 27(9) = 243	in 1.(81.3)		Bose et al. 1954(140, 164)
	in 9.9.3		Bose et al. 1954(140, 164)
1158. 27(10) = 270	in 90.3		Bose et al. 1954(218, 220)
1159. 28(6) = 168	in 56.3		Clatworthy 1956(51)
1160. 28(27) = 756	in 252.3	bal.	Bose 1939(381)
1161. 30(10) = 300	in 100.3		Bose et al. 1954(140, 168)
1162. 30(29) = 870	in 290.3	bal.	Bose 1939(379)
1163. 36(7) = 252	in 84.3		Clatworthy 1956(52)
1164. 37(9) = 333	in 1.(111.3)		Bose et al. 1954(250, 255)
	in 3.37.3		Bose et al. 1954(250, 255)
1165. 45(8) = 360	in 120.3		Clatworthy 1956(53)
1166. 49(6) = 294	in 98.3		Clatworthy 1956(62)
1167. 55(9) = 495	in 165.3		Clatworthy 1956(54)
1168. 66(10) = 660	in 220.3		Clatworthy 1956(55)

15.2104 Patterns in blocks of 4 plots

1169.	2(5) + 1(6) = 16	in 1.(4.4)		Cochran & Cox 1957(391)
1170.	4(5) = 20	in 1.(5.4)	Y sq	Cochran & Cox 1950(388), 1957(536), Vianelli 1959(431)
1171.	4(7) = 28	in 1.(7.4)	Y sq	Cochran & Cox 1950(389), 1957(537), Vianelli 1959(431)
1172.	4(9) = 36	in 1.(9.4)	Y sq	Cochran & Cox 1950(389), 1957(537), Vianelli 1959(431)
1173.	4(3) + 1(4) = 16	in 1.(4.4)		Cochran & Cox 1957(391)
1174.	5(4) = 20	in 5.4	bal.	Fisher & Yates 1938-57(T.XIX), Kitagawa & Mitome 1953(VII.1), Czechowski et al. 1957(104)
1175.	4(5) + 1(4) = 24	in 1.(6.4)		Cochran & Cox 1957(391)
1176.	6(2) = 12	in (3.2).2		Bose et al. 1954(90, 92)
		in 3.4		Bose et al. 1954(90, 92), Clatworthy 1956(12)
1177.	6(4) = 24	in 1.(6.4)		Bose et al. 1954(90, 92, 183, 184)
		in 6.4		Bose et al. 1954(90, 92, 183, 184)
1178.	6(6) = 36	in (9.2).2		Bose et al. 1954(139, 141)
		in 3.3.4		Clatworthy 1956(12)
		in 9.4		Bose et al. 1954(90, 92, 139, 141)
1179.	6(8) = 48	in 2.(6.4)		Clatworthy 1956(37)
		in 1.(12.4)		Bose et al. 1954(183, 185)
		in 2.6.4		Clatworthy 1956(37)
		in 12.4		Bose et al. 1954(90, 92, 183, 185)
1180.	6(10) = 60	in 5.3.4	bal.	Cochran & Cox 1950(330), 1957(472), Kitagawa & Mitome 1953(VII.2), Vianelli 1959(414)
		in 15.4	bal.	Fisher & Yates 1938-57(T.XIX), Czechowski et al. 1957(104)
		in 15.4		Bose et al. 1954(90, 92)
1181.	7(4) = 28	in 1.(7.4)	Y sq	Youden 1940(220), Cochran & Cox 1950(379), 1957(522), D.R. Cox 1958(234), Vianelli 1959(428)

15.2104

1182. 7(4) = 28 in 7.4   bal.   Fisher & Yates 1938-57(T.XIX), Cochran & Cox 1950(330), 1957(473), Kitagawa & Mitome 1953(VII.2), Czechowski et al. 1957(104), D. R. Cox 1958(234), Vianelli 1959(414)

1183. 8(2) = 16 in 1.(4.4)   Mandel 1954(256)

1184. 8(3) = 24
- in 3.2.4   Bose et al. 1954(90, 92)
- in 6.4   Bose & Nair 1939(353), Clatworthy 1956(14)

1185. 8(4) = 32
- in 4.2.4   Bose et al. 1954(139, 142)
- in 8.4   Bose & Nair 1939(355)

1186. 8(6) = 48
- in (12.2).2   Bose et al. 1954(90, 93, 139, 142)
- in 6.2.4   Bose et al. 1954(90, 93, 139, 142)
- in 12.4   Bose et al. 1954(139, 142)

1187. 8(7) = 56
- in 14.4   2bal   Calvin 1954(83), Vianelli 1959(436)
- in 7.2.4   bal.   Fisher & Yates 1938-57(T.XVII), Cochran & Cox 1950(330), 1957(473), Kitagawa & Mitome 1953(VII.2), Czechowski et al. 1957(102), D.R.Cox 1958(226), Vianelli 1959(415)
- in 14.4   bal.   Fisher & Yates 1938-57(T.XIX), Yates 1936(133), Bose 1939(365, 395, 397, 398), Czechowski et al. 1957(104)

1188. 8(8) = 64 in 16.4   Bose et al. 1954(139, 143)

1189. 8(9) = 72
- in 3.6.4   Clatworthy 1956(15)
- in 18.4   Bose et al. 1954(90, 93)

1190. 8(10) = 80 in 20.4   Bose et al. 1954(139, 143)

1191. 9(4) = 36
- in 1.(9.4)   Bose & Shimamoto 1952(164, 165), Bose et al. 1954(183, 186, 240, 241), Vianelli 1959(435)
- in 9.4   Bose & Shimamoto(164, 165), Bose et al. 1954(183, 186, 240, 241)

1192. 9(8) = 72
- in 2.(9.4)   Y sq   Cochran & Cox 1957(539)
- in 2.(9.4)   Clatworthy 1956(37)
- in 2.9.4   bal.   Cochran & Cox 1957(474)

15.2104

1193.	$9(8) = 72$	in 18.4	bal.	Fisher & Yates 1938-57(T.XIX), Yates 1936(133), Bose 1939(370, 396, 397), Cochran & Cox 1950(331), Kitagawa & Mitome 1953(VII.2), Czechowski et al. 1957(104), D. R. Cox 1958(226), Vianelli 1959(415)
		in 18.4		Bose et al. 1954(240, 241)
1194.	$10(2) = 20$	in (5.2).2		Bose & Shimamoto 1952(161, 162), Bose et al. 1954(229), Vianelli 1959(433)
		in 5.4		Bose & Shimamoto 1952(161, 162), Bose et al. 1954(229)
1195.	$10(4) = 40$	in 1.(10.4)		Bose & Shimamoto 1952(161, 162), Bose et al. 1954(90, 94, 229, 230, 232), Vianelli 1959(433)
		in 2.5.4		Bose et al. 1954(90, 94, 229, 230)
		in 10.4		Bose & Shimamoto 1952(161, 162), Bose et al. 1954(229, 232)
1196.	$10(6) = 60$	in 15.4	bal.	Fisher & Yates 1938-57(T.XIX), Yates 1936(133), Bose 1939(394, 397), Cochran & Cox 1950(332), 1957(475), Kitagawa & Mitome 1953(VII.3), Czechowski et al. 1957(104), D. R. Cox 1958(227), Vianelli 1959(416)
		in 15.4		Bose et al. 1954(229, 230)
1197.	$10(8) = 80$	in 1.(20.4)		Bose et al. 1954(183, 188)
		in 20.4		Bose et al. 1954(183, 188, 229, 230, 232)
1198.	$10(10) = 100$	in 25.4		Bose et al. 1954(229, 230)
1199.	$10(12) = 120$	in 30.4	2bal	Calvin 1954(84), Vianelli 1959(436)
1200.	$12(2) = 24$	in 1.(6.4)		Mandel 1954(256)
1201.	$12(3) = 36$	in 9.4		Nair & Rao 1948(126), Bose & Shimamoto 1952(157, 158), Bose et al. 1954 (139, 146), Vianelli 1959(432)
1202.	$12(4) = 48$	in 1.(12.4)		Bose & Shimamoto 1952(157, 158), Bose et al. 1954(183, 188), Vianelli 1959(432)
		in 12.4		Bose & Shimamoto 1952(157, 158), Bose et al. 1954(183, 188)

15.2104

1203.	12(5) = 60	in 15.4		Bose et al. 1954(90, 96)
1204.	12(6) = 72	in 6.3.4		Bose et al. 1954(139, 147)
1205.	12(8) = 96	in 2.(12.4)		Clatworthy 1956(37)
		in 8.3.4		Nair & Rao 1948(128)
		in 24.4		Clatworthy 1956(37)
1206.	12(9) = 108	in 9.3.4		Bose et al. 1954(139, 140)
1207.	12(10) = 120	in (30.2).2		Bose et al. 1954(183, 191)
		in 30.4		Bose et al. 1954(90, 98, 183, 191)
1208.	12(11) = 132	in 33.4	bal.	Bose 1939(371), Kitagawa & Mitome 1953(VII.4)
1209.	13(4) = 52	in 1.(13.4)	Y sq	Youden 1937(47), 1940(225), Cochran & Cox 1950(380), 1957(523), Kitagawa & Mitome 1953(VIII.2), D. R. Cox 1958(234), Vianelli 1959(428)
		in 13.4	bal.	Fisher & Yates 1938-57(T.XIX), Yates 1936(132), Bose 1939(369, 383), Cochran & Cox 1950(333), 1957(477), Kitagawa & Mitome 1953 (VII.5), Czechowski et al. 1957 (33, 104), D. R. Cox 1958(234), Vianelli 1959(417)
1210.	13(8) = 104	in 2.(13.4)		Bose & Shimamoto 1952(166, 167)
		in 1.(26.4)		Bose et al. 1954(250, 251)
		in 2.13.4		Bose & Shimamoto 1952(166, 167), Bose et al. 1954(250, 251)
1211.	14(4) = 56	in 1.(14.4)		Bose & Shimamoto 1952(157, 158), Bose et al. 1954(183, 191), Vianelli 1959(432)
		in 14.4		Bose & Shimamoto 1952(157, 158), Bose et al. 1954(183, 191)
1212.	14(6) = 84	in (21.2).2		Bose et al. 1954(90, 100)
		in 3.7.4		Bose et al. 1954(90, 100)
1213.	14(8) = 112	in 28.4		Bose et al. 1954(183, 192)
1214.	15(4) = 60	in 1.(15.4)		Bose & Shimamoto 1952(157, 158), Bose et al. 1954(183, 193), Cochran & Cox 1957(518), Vianelli 1959(432)

1215.	$15(4) = 60$	in 15.4		Bose & Nair 1939(358, 366), Bose & Shimamoto 1952(157, 158), Bose et al. 1954(183, 193)
1216.	$15(8) = 120$	in 1.(30.4)		Bose et al. 1954(183, 194)
		in 2.15.4		Bose et al. 1954(183, 194)
		in 30.4		Bose et al. 1954(183, 194), Vartak 1955(434)
1217.	$16(2) = 32$	in 1.(8.4)		Mandel 1954(256)
1218.	$16(4) = 64$	in 1.(16.4)		Bose et al. 1954(139, 153)
		in 4.4.4		Bose et al. 1954(139, 153)
1219.	$16(5) = 80$	in 5.(4.4)	L sq	Cochran & Cox 1950(360), 1957(497), Kitagawa & Mitome 1953(VI.7), D.R. Cox 1958(236), Vianelli 1959(425)
		in 5.4.4	latt	Cochran & Cox 1950(304), 1957(428), Kitagawa & Mitome 1953(VI.7), D.R. Cox 1958(226), Vianelli 1959(410)
		in 20.4	bal.	Fisher & Yates 1938-57(T.XIX), Yates 1936(133), Bose 1939(385), Kitagawa & Mitome 1953(VII.6), Czechowski et al. 1957(104)
1220.	$16(6) = 96$	in (24.2).2		Bose et al. 1954(183, 196)
		in 6.4.4		Bose et al. 1954(183, 196)
1221.	$16(7) = 112$	in 7.4.4		Bose et al. 1954(90, 103, 183, 197)
		in 28.4		Bose et al. 1954(240, 244)
1222.	$16(8) = 128$	in 1.(32.4)		Bose et al. 1954(240, 246)
		in 32.4		Bose et al. 1954(139, 153, 240, 246)
1223.	$16(9) = 144$	in 9.4.4		Bose et al. 1954(183, 198), Clatworthy 1956(38)
		in 36.4		Bose et al. 1954(240, 245)
1224.	$17(8) = 136$	in 2.(17.4)		Bose & Shimamoto 1952(166, 167)
		in 1.(34.4)		Bose et al. 1954(250, 252)
		in 2.17.4		Bose & Shimamoto 1952(166, 167), Bose et al. 1954(250, 252)
1225.	$18(8) = 144$	in (36.2).2		Bose et al. 1954(91, 105)
		in 4.9.4		Bose et al. 1954(91, 105)
1226.	$20(2) = 40$	in 1.(10.4)		Mandel 1954(256)

15.2104
15.2105

1227.	$20(3) = 60$	in 3.5.4	rect.	Robinson & Watson 1949(47), Cochran & Cox 1950(311), 1957(435), Vianelli 1959(413)
1228.	$20(4) = 80$	in 20.4		Nair & Rao 1948(124)
1229.	$20(5) = 100$	in 5.5.4		Bose et al. 1954(139, 158)
		in 25.4		Nair & Rao 1948(124)
1230.	$20(9) = 180$	in 9.5.4		Bose et al. 1954(91, 108)
		in 45.4		Bose et al. 1954(184, 200)
1231.	$20(10) = 200$	in (50.2).2		Bose et al. 1954(159, 160)
		in 10.5.4		Bose et al. 1954(159, 160)
1232.	$22(10) = 220$	in (55.2).2		Bose et al. 1954(91, 111)
		in 5.11.4		Bose et al. 1954(91, 111)
1233.	$24(7) = 168$	in 42.4		Bose et al. 1954(184, 201)
1234.	$24(10) = 240$	in (60.2).2		Bose et al. 1954(184, 203)
		in 60.4		Bose et al. 1954(184, 203)
1235.	$25(8) = 200$	in 2.(25.4)	Y sq	Cochran & Cox 1957(543)
		in 50.4	bal.	Fisher & Yates 1938-57(T.XIX), Bose 1939(383, 397), Cochran & Cox 1950(338), 1957(480, 543), Kitagawa & Mitome 1953(VII.9), Czechowski et al. 1957(104), Vianelli 1959(420)
1236.	$26(8) = 208$	in 1.(52.4)		Bose et al. 1954(184, 204)
		in 52.4		Bose et al. 1954(184, 204)
1237.	$27(8) = 216$	in 1.(54.4)		Bose et al. 1954(184, 205)
		in 54.4		Bose et al. 1954(184, 205)
1238.	$28(7) = 196$	in 7.7.4		Bose et al. 1954(140, 165)
1239.	$28(8) = 224$	in 1.(56.4)		Bose et al. 1954(184, 205)
		in 8.7.4		Bose et al. 1954(184, 205)
1240.	$28(9) = 252$	in 9.7.4	bal.	Fisher & Yates 1938-57(T.XVII), Cochran & Cox 1950(339), 1957 (481), Kitagawa & Mitome 1953 (VII.9), Czechowski et al. 1957 (103), Vianelli 1959(420)
		in 63.4	bal.	Fisher & Yates 1938-57(T.XIX), Bose 1939(385, 397), Czechowski et al. 1957(104)

1241. $28(10) = 280$ $\begin{cases} \text{in } (70.2).2 \\ \text{in } 10.7.4 \end{cases}$ Bose et al. 1954(184, 206)
Bose et al. 1954(184, 206)

1242. $30(10) = 300$ $\begin{cases} \text{in } (75.2).2 \\ \text{in } [2.30 + 1.15].4 \end{cases}$ Bose et al. 1954(184, 207)
Bose et al. 1954(184, 207)

1243. $32(8) = 256$ $\begin{cases} \text{in } 1.(64.4) \\ \text{in } 8.8.4 \end{cases}$ Bose et al. 1954(140, 168)
Bose et al. 1954(140, 168)

1244. $36(9) = 324$ $\begin{cases} \text{in } 9.9.4 \\ \text{in } 81.4 \end{cases}$ Bose et al. 1954(140, 170)
Bose & Nair 1939(363)

1245. $40(4) = 160$ $\begin{cases} \text{in } 1.(40.4) \\ \text{in } 40.4 \end{cases}$ Bose et al. 1954(218, 221)
Bose et al. 1954(218, 221), Clatworthy 1954(53)

1246. $40(8) = 320$ in 80.4 Bose et al. 1954(218, 222)

1247. $40(13) = 520$ in 130.4 bal. Bose 1939(387)

1248. $73(24) = 1752$ in 438.4 bal. Bose 1939(383)

15.2105 Patterns in blocks of 5 plots

1249. $3(6) + 1(7) = 25$ in 1.(5.5) Cochran & Cox 1957(391)

1250. $5(6) = 30$ in 1.(6.5) Y sq Cochran & Cox 1950(389), 1957(537), Vianelli 1959(431)

1251. $5(9) = 45$ in 1.(9.5) Y sq Cochran & Cox 1950(389), 1957(537), Vianelli 1959(431)

1252. $5(4) + 1(5) = 25$ in 1.(5.5) Cochran & Cox 1957(391)

1253. $6(5) = 30$ in 6.5 bal. Fisher & Yates 1938-57(T.XIX), Kitagawa & Mitome 1953(VII.2), Czechowski et al. 1957(104)

1254. $5(6) + 1(5) = 35$ in 1.(7.5) Cochran & Cox 1957(391)

1255. $8(5) = 40$ in 8.5 Clatworthy 1956(38, 39)

1256. $8(10) = 80$ in 2.8.5 Clatworthy 1956(39)

1257. $9(5) = 45$ $\begin{cases} \text{in } 1.(9.5) \\ \text{in } 9.5 \end{cases}$ Bose et al. 1954(240, 244)
Bose et al. 1954(240, 244), Clatworthy 1956(39)

1258. $9(10) = 90$ $\begin{cases} \text{in } 2.(9.5) \quad \text{Y sq} \\ \text{in } 2.9.5 \quad \text{bal.} \end{cases}$ Cochran & Cox 1957(540)
Cochran & Cox 1957(474)

15.2105

1259.	9(10) = 90	in 18.5	bal.	Fisher & Yates 1938-57(T.XIX), Cochran & Cox 1950(331), Kitagawa & Mitome 1953(VII.2), Czechowski et al. 1957(104), D. R. Cox 1958(227), Vianelli 1959(415)
		in 2.9.5		Clatworthy 1956(39)
		in 18.5		Bose et al. 1954(240, 244)
1260.	10(3) = 30	in 6.5		Bose & Nair 1939(371), Bose & Shimamoto 1952(161, 162), Bose et al. 1954(229, 231), Vianelli 1959(433)
1261.	10(4) = 40	in 2.4.5		Bose & Shimamoto 1952(157, 158), Bose et al. 1954(139, 144), Vianelli 1959(432)
1262.	10(5) = 50	in 10.5		Clatworthy 1956(40)
1263.	10(6) = 60	in 12.5		Bose et al. 1954(139, 144, 229, 232)
1264.	10(8) = 80	in 8.2.5		Bose et al. 1954(139, 145)
1265.	10(9) = 90	in 18.5	bal.	Fisher & Yates 1938-57(T.XIX), Yates 1936(133), Bose 1939(395, 397), Cochran & Cox 1950(332), 1957(475), Kitagawa & Mitome 1953(VII.3), Czechowski et al. 1957(104), D. R. Cox 1958(227), Vianelli 1959(416)
		in 18.5		Bose et al. 1954(229, 232)
1266.	10(10) = 100	in 2.10.5		Clatworthy 1956(40)
		in 20.5		Bose et al. 1954(139, 145)
1267.	10(18) = 180	in 36.5	2bal	Calvin 1954(85), Vianelli 1959(436)
1268.	11(5) = 55	in 1.(11.5)	Y sq	Youden 1940(225), Cochran & Cox 1950(379), 1957(522), Kitagawa & Mitome 1953(VIII.2), D. R. Cox 1958(234), Vianelli 1959(428)
		in 11.5	bal.	Fisher & Yates 1938-57(T.XIX), Yates 1936(131), Bose 1939(390, 397), Cochran & Cox 1950(332), 1957(476), Kitagawa & Mitome 1953(VII.4), Czechowski et al. 1957(104), D. R. Cox 1958(234), Vianelli 1959(416)
1269.	11(10) = 110	in 22.5	bal.	Kitagawa & Mitome 1953(VII.4)
1270.	12(5) = 60	in 12.5		Clatworthy 1956(40, 41)

1271.	12(5) = 60	in 12.5	Clatworthy 1956(40, 41)
1272.	12(10) = 120	in 2.12.5	Clatworthy 1956(41)
1273.	15(2) = 30	in 6.5	Bose & Shimamoto 1952(161, 162), Bose et al. 1954(229, 235), Vianelli 1959(433)
1274.	15(4) = 60	in 12.5	Bose et al. 1954(229, 235)
1275.	15(6) = 90	in 6.3.5	Bose et al. 1954(139, 152)
1276.	15(9) = 135	in 9.3.5	Bose et al. 1954(139, 152)
1277.	15(10) = 150	in 1.(30.5)	Bose et al. 1954(183, 196)
		in 2.15.5	Bose et al. 1954(183, 196)
1278.	16(5) = 80	in 16.5	Clatworthy 1956(57)
1279.	16(10) = 160	in 2.16.5	Clatworthy 1956(58)
1280.	20(4) = 80	in 16.5	Bose & Shimamoto 1952(157, 158), Bose et al. 1954(139, 158), Vianelli 1959(432)
1281.	20(8) = 160	in 8.4.5	Bose et al. 1954(139, 159)
1282.	21(5) = 105	in 1.(21.5) Y sq	Youden 1937(47), 1940(227), Cochran & Cox 1950(385), 1957(528), D. R. Cox 1958(234), Vianelli 1959(429)
		in 21.5 bal.	Fisher & Yates 1938-57(T.XIX), Yates 1936(132), Cochran & Cox 1950(337), 1957(479, 528), Kitagawa & Mitome 1953(VII.8), Czechowski et al. 1957(104), D. R. Cox 1958(234), Vianelli 1959(419)
1283.	24(5) = 120	in 1.(24.5)	Bose & Shimamoto 1952(157, 158), Bose et al. 1954(184, 201), Vianelli 1959(432)
		in 24.5	Bose & Nair 1939(358), Bose & Shimamoto 1952(157, 158), Bose et al. 1954(184, 201)
1284.	24(10) = 240	in 48.5	Bose et al. 1954(184, 203)
1285.	25(3) = 75	in 3.(5.5) L sq	Cochran & Cox 1950(361), 1957(498), Kempthorne 1952(485), Kitagawa & Mitome 1953(VI.7), D. R. Cox 1958 236), Vianelli 1959(425)
1286.	25(5) = 125	in 1.(25.5)	Bose et al. 1954(140, 164)

15.2105
15.2106

1287.	$25(5) = 125$	in 5.5.5		Bose et al. 1954(140, 164)
1288.	$25(6) = 150$	in 6.(5.5)	L sq	Kitagawa & Mitome 1953(VI.7)
		in 6.5.5	latt	Cochran & Cox 1950(304), 1957(428), Kitagawa & Mitome 1953(VI.2, VII.9), D. R. Cox 1958(227), Vianelli 1959(410)
		in 30.5	bal.	Fisher & Yates 1938-57(T.XIX), Yates 1936(133), Bose 1939(389) Czechowski et al. 1957(104)
1289.	$25(7) = 175$	in 7.5.5		Bose et al. 1954(184, 203)
1290.	$25(8) = 200$	in 8.5.5		Bose et al. 1954(184, 204)
1291.	$25(10) = 250$	in 50.5		Bose et al. 1954(140, 164)
1292.	$30(2) = 60$	in 2.6.5	rect	Kitagawa & Mitome 1953(VI.3)
1293.	$30(3) = 90$	in 3.6.5	rect	Robinson & Watson 1949(47), Cochran & Cox 1950(311), 1957(435), Vianelli 1959(413)
1294.	$35(7) = 245$	in 7.7.5		Bose et al. 1954(140, 169)
1295.	$35(10) = 350$	in 1.(70.5)		Bose et al. 1954(184, 208)
		in 2.35.5		Bose et al. 1954(184, 208)
1296.	$36(5) = 180$	in 1.(36.5)		Bose et al. 1954(240, 247)
		in 36.5		Bose et al. 1954(240, 247)
1297.	$39(10) = 390$	in 1.(78.5)		Bose et al. 1954(184, 209)
		in 2.39.5		Bose et al. 1954(184, 209)
1298.	$40(8) = 320$	in 8.8.5		Bose et al. 1954(140, 171)
1299.	$40(9) = 360$	in 72.5		Bose et al. 1954(184, 210)
1300.	$41(10) = 410$	in 2.(41.5)	Y sq	Cochran & Cox 1957(544)
		in 82.5	bal.	Fisher & Yates 1938-57(T.XIX), Bose 1939(387, 397), Cochran & Cox 1950(342), 1957(482, 544), Kitagawa & Mitome 1953(VII.11), Czechowski et al. 1957(104), Vianelli 1959(422)
1301.	$45(3) = 135$	in 27.5		Bose et al. 1954(218, 222)
1302.	$45(6) = 270$	in 54.5		Bose et al. 1954(218, 222)
1303.	$45(9) = 405$	in 9.9.5		Bose et al. 1954(140, 172)
		in 81.5		Bose et al. 1954(218, 222)

1304. 45(10) = 450   in 90.5   Bose et al. 1954(184, 211)

1305. 45(11) = 495   in 99.5   bal.   Bose 1939(389),
                                          Kitagawa & Mitome 1953(389)

1306. 61(15) = 915   in 183.5   bal.   Bose 1939(387)

1307. 65(16) = 1040 in 208.5   bal.   Bose 1939(389)

1308. 85(5) = 425   in 85.5   Bose et al. 1954(218, 228)

1309. 85(10) = 850   in 170.5   Bose et al. 1954(218, 228)

15.2106 Patterns in blocks of 6 plots

1310. 4(7) + 1(8) = 36 in 1.(6.6)   Cochran & Cox 1957(391)

1311. 6(7) = 42   in 1.(7.6) Y sq   Cochran & Cox 1950(390), 1957
                                          (538), Vianelli 1959(431)

1312. 6(5) + 1(6) = 36 in 1.(6.6)   Cochran & Cox 1957(391)

1313. 7(6) = 42   in 7.6   bal.   Fisher & Yates 1938-57(T.XIX),
                                          Kitagawa & Mitome 1953(VII.2),
                                          Czechowski et al. 1957(104)

1314. 6(7) + 1(6) = 48 in 1.(8.6)   Cochran & Cox 1957(391)

1315. 8(3) = 24   { in (4.2).3   Bose et al. 1954(90, 93),
                                           Cochran & Cox 1957(518)
                     { in 4.6   Bose et al. 1954(90, 93)

1316. 8(6) = 48   { in 1.(8.6)   Bose et al. 1954(90, 93)
                     { in 2.4.6   Bose et al. 1954(90, 93)

1317. 8(9) = 72   in 12.6   Bose et al. 1954(90, 93),
                                           Clatworthy 1956(42)

1318. 9(2) = 18   { in (3.2).3   Bose et al. 1954(90, 94)
                     { in 3.6   Bose et al. 1954(90, 94)

1319. 9(4) = 36   { in (6.2).3   Bose et al. 1954(240, 243)
                     { in 6.6   Bose et al. 1954(90, 94, 240,
                                           243), Clatworthy 1956(13)

1320. 9(6) = 54   { in 1.(9.6)   Bose et al. 1954(90, 94, 139, 144)
                     { in 3.3.6   Bose et al. 1954(90, 94)
                     { in 9.6   Bose et al. 1954(139, 144)

1321. 9(8) = 72   in 4.3.6   bal.   Cochran & Cox 1950(331), 1957
                                          (474), Kitagawa & Mitome 1953
                                          (VII.3), Vianelli 1959(415)

15.2106

1322.	9(8) = 72	in 12.6	bal.	Fisher & Yates 1938-57(T.XIX), Czechowski et al. 1957(104), D. R. Cox 1958(227)
		in 12.6		Bose et al. 1954(90, 94, 240, 243)
1323.	9(10) = 90	in 15.6		Bose et al. 1954(90, 94)
1324.	10(3) = 30	in (5.3).2		Bose et al. 1954(229, 233)
		in 5.6		Bose et al. 1954(229, 233)
1325.	10(6) = 60	in 1.(10.6)		Bose et al. 1954(90, 95, 229, 233, 234)
		in 2.5.6		Bose et al. 1954(90, 95, 229, 233)
		in 10.6		Bose et al. 1954(229, 233, 234), Clatworthy 1956(10, 42)
1326.	10(9) = 90	in 15.6	bal.	Fisher & Yates 1938-57(T.XIX), Cochran & Cox 1950(332), 1957(476), Kitagawa & Mitome 1953 (VII.3), Czechowski et al. 1957 (104), D. R. Cox 1958(227), Vianelli 1959(416)
		in 15.6		Bose et al. 1954(229, 234)
1327.	10(18) = 180	in 30.6	2bal	Calvin 1954(86), Vianelli 1959(436)
1328.	6(2) + 5(1) = 17	in 2.6 + 1.5		Youden & Connor 1953(128)
1329.	11(6) = 66	in 1.(11.6)	Y sq	Youden 1940(225), Cochran & Cox 1950(380), 1957(523), Kitagawa & Mitome 1953(VIII.2), D.R. Cox 1958(234), Vianelli 1959(428)
		in 11.6	bal.	Fisher & Yates 1938-57(T.XIX), Cochran & Cox 1950(333), 1957 (476), Kitagawa & Mitome 1953 (VII.4), Czechowski et al. 1957 (104), D. R. Cox 1958(234)
1330.	12(3) = 36	in (6.3).2		Bose et al. 1954(90, 96), Cochran & Cox 1957(519)
		in 3.2.6		Bose et al. 1954(90, 96), Cochran & Cox 1957(519)
1331.	12(4) = 48	in (8.2).3		Bose et al. 1954(139, 146)
		in 2.4.6		Bose et al. 1954(139, 146)
1332.	12(5) = 60	in 10.6		Bose et al. 1954(90, 97)

15.2106

1333.	$12(6) = 72$	in 1.(12.6)		Bose et al. 1954(90, 97, 139, 147)
		in 3.4.6		Bose et al. 1954(139, 147)
		in 12.6		Bose & Nair 1939(362), Bose et al. 1954(90, 97)
1334.	$12(8) = 96$	in (16.2).3		Bose et al. 1954(139, 148)
		in 8.2.6		Bose et al. 1954(139, 148)
1335.	$12(9) = 108$	in 18.6		Bose et al. 1954(90, 98)
1336.	$12(10) = 120$	in (20.2).3		Bose et al. 1954(139, 150)
		in 20.6		Bose et al. 1954(90, 98, 139, 149, 150)
1337.	$12(11) = 132$	in 22.6	2bal	Calvin 1954(87), Vianelli 1959(436)
		in 22.6	bal.	Kitagawa & Mitome 1953(VII.4)
1338.	$13(6) = 78$	in 1.(13.6)		Clatworthy 1956(69)
		in 13.6		Clatworthy 1956(69)
1339.	$13(12) = 156$	in 26.6	bal.	Kitagawa & Mitome 1953(VII.5)
1340.	$14(3) = 42$	in (7.2).3		Bose et al. 1954(90, 99)
		in 7.6		Bose & Connor 1952(372), Bose et al. 1954(90, 99)
1341.	$14(6) = 84$	in 1.(14.6)		Bose et al. 1954(90, 100)
		in 2.7.6		Bose et al. 1954(90, 100)
1342.	$15(4) = 60$	in (10.2).3		Bose et al. 1954(90, 101, 229, 235)
		in 2.5.6		Bose et al. 1954(90, 101)
		in 10.6		Bose & Nair 1939(371), Bose & Shimamoto 1952(161, 162), Bose et al. 1954(90, 101, 229, 235), Vianelli 1959(433)
1343.	$15(6) = 90$	in 15.6		Clatworthy 1956(42)
1344.	$15(8) = 120$	in 20.6		Bose et al. 1954(90, 102)
1345.	$16(6) = 96$	in 1.(16.6)	Y sq	Youden 1940(221, 226), Cochran & Cox 1950(382), 1957(525), Kitagawa & Mitome 1953(VIII.3), Vianelli 1959(429)
		in 16.6	bal.	Fisher & Yates 1938-57(T.XIX), Yates 1936(132), Bose 1939(391, 397), Cochran & Cox 1950(335), 1957(478, 525), Kitagawa & Mitome 1953(VII.6), Czechowski et al. 1957(104), Vianelli 1959(417)

15.2106

1346.	16(9) = 144	in 3.8.6	bal.	Cochran & Cox 1957(478)
		in 24.6	bal.	Fisher & Yates 1938-57(T.XIX), Cochran & Cox 1950(335), Kitagawa & Mitome 1953(VII.6), Czechowski et al. 1957(104), D. R. Cox 1958(227), Vianelli 1959(418)
1347.	10(2) + 7(1) = 27	in 2.6 + 3.5		Cochran & Cox 1957(464)
1348.	18(2) = 36	in 1.(6.6)		Mandel 1954(256)
1349.	18(4) = 72	in 4.3.6		Bose et al. 1954(90, 104)
1350.	18(5) = 90	in 5.3.6		Bose et al. 1954(90, 104), Nair & Rao 1948(128)
1351.	18(6) = 108	in 1.(18.6)		Bose et al. 1954(139, 155)
		in 6.3.6		Bose et al. 1954(139, 155)
1352.	18(8) = 144	in 24.6		Bose et al. 1954(91, 105)
1353.	18(9) = 162	in (27.3).2		Bose et al. 1954(139, 151)
		in 9.3.6		Bose et al. 1954(139, 151)
1354.	18(10) = 180	in (30.2).3		Bose et al. 1954(191, 206)
		in 10.3.6		Bose et al. 1954(191, 206)
1355.	20(9) = 180	in (30.3).2		Bose et al. 1954(91, 109)
		in 30.6		Bose et al. 1954(91, 109)
1356.	21(2) = 42	in (7.2).3		Bose & Shimamoto 1952(161, 162), Bose et al. 1954(229, 238), Vianelli 1959(433)
		in 7.6		Bose & Shimamoto 1952(161, 162), Bose et al. 1954(229, 238)
1357.	21(6) = 126	in (21.2).3		Bose et al. 1954(91, 110)
		in 3.7.6		Bose et al. 1954(91, 110)
1358.	24(2) = 48	in 1.(8.6)		Mandel 1954(256)
1359.	24(7) = 168	in 7.4.6		Bose et al. 1954(91, 113)
1360.	24(8) = 192	in (32.2).3		Bose et al. 1954(140, 163)
		in 8.4.6		Bose et al. 1954(140, 163)
1361.	26(3) = 78	in (13.3).2		Bose & Shimamoto 1952(161, 163), Bose et al. 1954(218, 219), Vianelli 1959(434)

15.2106

1362.	$26(3) = 78$	in 13.6		Bose & Nair 1939(371), Bose & Shimamoto 1952(161, 163), Bose et al. 1954(218, 219)
1363.	$26(6) = 156$	$\begin{cases} \text{in } 1.(26.6) \\ \text{in } 2.13.6 \\ \text{in } 26.6 \end{cases}$		Bose et al. 1954(91, 114, 218, 219) Bose et al. 1954(91, 114) Bose et al. 1954(218, 219)
1364.	$26(9) = 234$	in 39.6		Bose et al. 1954(218, 220)
1365.	$27(8) = 216$	$\begin{cases} \text{in } (36.2).3 \\ \text{in } 36.6 \end{cases}$		Bose et al. 1954(90, 116) Bose et al. 1954(90, 116)
1366.	$39(2) = 60$	in 1.(10.6)		Mandel 1954(256)
1367.	$30(5) = 150$	in 25.6		Bose et al. 1954(140, 166)
1368.	$30(7) = 210$	in 7.5.6		Bose et al. 1954(91, 117)
1369.	$30(9) = 270$	in 9.5.6		Bose et al. 1954(91, 117)
1370.	$30(10) = 300$	$\begin{cases} \text{in } (50.2).3 \\ \text{in } 2.25.6 \end{cases}$		Bose et al. 1954(140, 167) Bose et al. 1954(140, 167)
1371.	$31(6) = 186$	$\begin{cases} \text{in } 1.(31.6) \\ \\ \text{in } 31.6 \end{cases}$	Y sq bal.	Youden 1937(47), 1940(227), Cochran & Cox 1950(386), 1957(530), Vianelli 1959(430) Fisher & Yates 1938-57(T.XIX), Cochran & Cox 1950(340), 1957(482, 530), Kitagawa & Mitome 1953 (VII.10), Vianelli 1959(421), Czechowski et al. 1957(104)
1372.	$33(10) = 330$	$\begin{cases} \text{in } (55.2).3 \\ \text{in } 5.11.6 \end{cases}$		Bose et al. 1954(91, 121) Bose et al. 1954(91, 121)
1373.	$36(3) = 108$	in 3.6.6	latt	Cochran & Cox 1950(309), 1957(433), Vianelli 1959(412)
1374.	$36(4) = 144$	in 4.6.6	latt	Kitagawa & Mitome 1953(VI.4)
1375.	$38(9) = 342$	$\begin{cases} \text{in } (57.3).2 \\ \text{in } 3.19.6 \end{cases}$		Bose et al. 1954(91, 122) Bose et al. 1954(91, 122)
1376.	$42(3) = 126$	in 3.7.6	rect	Robinson & Watson 1949(48), Cochran & Cox 1950(312), 1957(436), Vianelli 1959(413)
1377.	$42(7) = 294$	in 7.7.6		Bose et al. 1954(140, 171)
1378.	$42(10) = 420$	in 10.7.6		Bose et al. 1954(91, 125)

1379. 48(8) = 384 $\begin{cases} \text{in } (64.2).3 & \text{Bose et al. } 1954(140, 173) \\ \text{in } 8.8.6 & \text{Bose et al. } 1954(140, 173) \end{cases}$

1380. 49(6) = 294 $\begin{cases} \text{in } 1.(49.6) & \text{Bose et al. } 1954(240, 247) \\ \text{in } 49.6 & \text{Bose \& Nair } 1939(364), \text{ Bose et al. } 1954 \\ & \hspace{6em} (240, 247) \end{cases}$

1381. 54(9) = 486   in 9.9.6   Bose et al. 1954(140, 175)

15.2107 Patterns in blocks of 7 plots

1382. 5(8) + 1(9) = 49   in 1.(7.7)   Cochran & Cox 1957(391)

1383. 7(3) = 21   in 3.7   bal. Yates 1936(131)

1384. 7(8) = 56   in 1.(8.7)   Y sq Cochran & Cox 1950(390), 1957
                                             (538), Vianelli 1959(431)

1385. 7(6) + 1(7) = 49   in 1.(7.7)   Cochran & Cox 1957(391)

1386. 8(7) = 56   in 8.7   bal. Fisher & Yates 1938-57(T.XIX),
                                             Czechowski et al. 1957(104)

1387. 7(8) + 1(7) = 63   in 1.(9.7)   Cochran & Cox 1957(391)

1388. 9(7) = 63   in 9.7   Clatworthy 1956(10, 43)

1389. 10(7) = 70 $\begin{cases} \text{in } 1.(10.7) & \text{Bose et al. } 1954(229, 234) \\ \text{in } 10.7 & \text{Bose et al. } 1954(229, 234) \end{cases}$

1390. 9(2) + 3(1) = 21 in 3.7   Cochran & Cox 1957(465)

1391. 12(7) = 84   in 12.7   Clatworthy 1956(43, 44)

1392. 13(7) = 91 $\begin{cases} \text{in } 1.(13.7) & \text{Clatworthy } 1956(69) \\ \text{in } 13.7 & \text{Clatworthy } 1956(69) \end{cases}$

1393. 14(4) = 56   in 8.7   Bose & Nair 1939(361), Nair & Rao
                                             1948(126), Bose & Shimamoto
                                             1952(157, 158), Bose et al.
                                             1954(139, 150), Vianelli 1959
                                             (432)

1394. 14(6) = 84   in 12.7   Bose et al. 1954(139, 150)

1395. 14(7) = 98   in 14.7   Clatworthy 1956(**44**)

1396. 14(8) = 112   in 8.2.7   Bose et al. 1954(139, 151)

1397. 14(10) = 140   in 20.7   Bose et al. 1954(139, 151)

1398. 14(13) = 182   in 26.7   bal. Kitagawa & Mitome 1953(VII.5)

1399. 15(7) = 105   in 1.(15.7) Y sq Youden 1940(226), Cochran & Cox
                                             1950(381), 1957(524), Kitagawa
                                             & Mitome 1953(VIII.3), D.R. Cox
                                             1958(234), Vianelli 1959(428)

15.2106
15.2107

1400. 15(7) = 105 in 15.7   bal.   Fisher & Yates 1938-57(T.XIX), Bose 1939(364, 397, 398), Cochran & Cox 1950(334), 1957(478, 524), Kitagawa & Mitome 1953(VII.5), Czechowski et al. 1957(104), D. R. Cox 1958(234), Vianelli 1959(417)

1401. 16(7) = 112 in 16.7   Clatworthy 1956(63)

1402. 18(7) = 126 in 18.7   Clatworthy 1956(45)

1403. 19(7) = 133 in 19.7   bal.   Bose 1939(391)

1404. 21(6) = 126 in 18.7   Bose et al. 1954(139, 161)

1405. 21(9) = 189 in 9.3.7   Bose et al. 1954(140, 162)

1406. 21(10) = 210 in 30.7 bal.   Fisher & Yates 1938-57(T.XIX), Cochran & Cox 1950(358), 1957(480), Kitagawa & Mitome 1953(VII.8), Czechowski et al. 1957(104), Vianelli 1959(419)

1407. 28(2) = 56 in 8.7   Bose & Shimamoto 1952(161, 162), Bose et al. 1954(229, 238)

1408. 28(8) = 224 in 8.4.7   Bose et al. 1954(140, 166)

1409. 28(9) = 252 in 36.7   bal.   Fisher & Yates 1938-57(T.XIX), Bose 1939(397), Cochran & Cox 1950(340), 1957(482), Kitagawa & Mitome 1953 (VII.10), Czechowski et al. 1957 (104), Vianelli 1959(421)

1410. 29(7) = 203 { in 1.(29.7)   Bose & Shimamoto 1952(166, 167), Bose et al. 1954(250, 254)
                  { in 29.7   Bose & Shimamoto 1952(166, 167), Bose et al. 1954(250, 254)

1411. 33(7) = 231 { in 1.(33.7)   Bose et al. 1954(184, 208)
                  { in 33.7   Bose et al. 1954(184, 208)

1412. 35(3) = 105 in 15.7   Bose & Shimamoto 1952(161, 163), Bose et al. 1954(218, 220), Vianelli 1959(434)

1413. 35(6) = 210 in 30.7   Bose et al. 1954(218, 220)

1414. 35(9) = 315 in 45.7   Bose et al. 1954(218, 221)

1415. 48(7) = 336 { in 1.(48.7)   Bose et al. 1954(184, 212)
                  { in 48.7   Bose & Nair 1939(359), Bose et al. 1954(184, 212)

1416. $49(4) = 196$ in 4.(7.7)   L sq   Cochran & Cox 1950(361), 1957(498), Kitagawa & Mitome 1953(VI.7), Vianelli 1959(425)

1417. $49(7) = 343$
- in 1.(49.7)   Bose et al. 1954(140, 174)
- in 7.7.7   Bose et al. 1954(140, 174)

1418. $49(8) = 392$
- in 8.(7.7)   L sq   Kitagawa & Mitome 1953(VI.7)
- in 8.7.7   latt   Cochran & Cox 1950(305), 1957(429), Kitagawa & Mitome 1953(VII.12), Vianelli 1959(410)
- in 56.7   bal.   Fisher & Yates 1938-57(T.XIX), Czechowski et al. 1957(104)

1419. $49(9) = 441$ in 9.7.7   Bose et al. 1954(184, 213)

1420. $49(10) = 490$ in 10.7.7   Bose et al. 1954(184, 214)

1421. $56(3) = 168$ in 3.8.7   rect   Robinson & Watson 1949(48), Cochran & Cox 1950(312), 1957(436), Vianelli 1959(413)

1422. $56(8) = 448$ in 8.8.7   Bose et al. 1954(140, 177)

1423. $63(9) = 567$ in 9.9.7   Bose et al. 1954(140, 178)

1424. $64(7) = 448$
- in 1.(64.7)   Bose et al. 1954(240, 248)
- in 64.7   Bose & Nair 1939(364), Bose et al. 1954(240, 248)

15.2108 Patterns in blocks of 8 plots

1425. $6(9) + 1(10) = 64$ in 1.(8.8)   Cochran & Cox 1957(391)

1426. $8(7) = 56$ in 7.8   bal.   Kitagawa & Mitome 1953(VII.2)

1427. $8(7) + 1(8) = 64$ in 1.(8.8)   Cochran & Cox 1957(391)

1428. $9(8) = 72$ in 9.8   bal.   Fisher & Yates 1938-57(T.XIX), Kitagawa & Mitome 1953(VII.3), Czechowski et al. 1957(104)

1429. $8(9) + 1(8) = 80$ in 1.(10.8)   Cochran & Cox 1957(391)

1430. $10(4) = 40$
- in (5.4).2   Bose et al. 1954(90, 95)
- in 5.8   Bose et al. 1954(90, 95)

1431. $10(8) = 80$
- in 1.(10.8)   Bose et al. 1954(90, 95)
- in 2.5.8   Bose et al. 1954(90, 95)

1432. $12(2) = 24$
- in (3.2).4   Bose et al. 1954(90, 95)
- in 3.8   Bose et al. 1954(90, 95)

1433.	12(4) = 48	in 6.8		Bose et al. 1954(90, 96)
1434.	12(6) = 72	in (9.2).4		Bose et al. 1954(139, 148)
		in 9.8		Bose et al. 1954(90, 97, 139, 148)
1435.	12(8) = 96	in 1.(12.8)		Bose et al. 1954(90, 98)
		in 12.8		Bose et al. 1954(90, 98)
1436.	12(10) = 120	in 15.8		Bose et al. 1954(90, 99)
1437.	14(4) = 56	in (7.4).2		Bose et al. 1954(90, 100)
		in 7.8		Bose et al. 1954(90, 100)
1438.	14(8) = 112	in 1.(14.8)		Bose et al. 1954(90, 101)
		in 2.7.8		Bose et al. 1954(90, 101)
		in 14.8		Clatworthy 1956(45)
1439.	15(8) = 120	in 1.(15.8)	Y sq	Youden 1940(226), Cochran & Cox 1950(382), 1957(525), Kitagawa & Mitome 1953(VIII.3), D.R.Cox 1958(234), Vianelli 1959(429)
		in 15.8	bal.	Fisher & Yates 1938-57(T.XIX), Cochran & Cox 1950(334), 1957 (478, 525), Kitagawa & Mitome 1953(VII.5), Czechowski et al. 1957(104), D. R. Cox 1958(234), Vianelli 1959(417)
1440.	16(3) = 48	in 3.2.8		Bose et al. 1954(90, 103)
1441.	16(6) = 96	in (12.2).4		Bose et al. 1954(90, 103, 139, 153)
		in 6.2.8		Bose et al. 1954(90, 103)
		in 12.8		Bose et al. 1954(139, 153)
1442.	16(7) = 112	in 7.2.8		Bose et al. 1954(90, 104)
1443.	16(8) = 128	in 8.2.8		Bose et al. 1954(139, 154)
1444.	16(9) = 144	in 18.8		Bose et al. 1954(90, 104)
1445.	16(10) = 160	in (20.2).4		Bose et al. 1954(139, 154)
		in 20.8		Bose et al. 1954(139, 154)
1446.	16(15) = 240	in 30.8	2bal	Calvin 1954(88), Vianelli 1959 (436)
1447.	17(8) = 136	in 1.(17.8)		Bose & Shimamoto 1952(166, 167), Bose et al. 1954(250, 253)
		in 17.8		Bose et al. 1954(250, 253)

15.2108

1448.	$18(8) = 144$	in 1.(18.8)	Bose et al. 1954(91, 105)
		in 2.9.8	Bose et al. 1954(91, 105)
1449.	$20(6) = 120$	in (15.2).4	Bose et al. 1954(91, 107)
		in 15.8	Bose et al. 1954(91, 107)
1450.	$20(8) = 160$	in 1.(20.8)	Bose et al. 1954(91, 108)
		in 4.5.8	Bose et al. 1954(91, 108)
1451.	$21(8) = 168$	in 21.8	Clatworthy 1956(45)
1452.	$8(2) + 14(1) = 30$	in 2.8 + 2.7	Youden & Connor 1953(129)
1453.	$24(5) = 120$	in 5.3.8	Bose et al. 1954(91, 112)
1454.	$24(9) = 216$	in 9.3.8	Bose et al. 1954(140, 163)
1455.	$24(10) = 240$	in (30.2).4	Bose et al. 1954(91, 113)
		in 10.3.8	Bose et al. 1954(91, 113)
1456.	$26(4) = 104$	in (13.4).2	Bose et al. 1954(91, 114)
		in 13.8	Bose et al. 1954(91, 114)
1457.	$26(8) = 208$	in 1.(26.8)	Bose et al. 1954(91, 115)
		in 2.13.8	Bose et al. 1954(91, 115)
1458.	$28(6) = 168$	in (21.4).2	Bose et al. 1954(91, 116)
		in 3.7.8	Bose et al. 1954(91, 116)
1459.	$29(8) = 232$	in 1.(29.8)	Bose & Shimamoto 1952(166, 167), Bose et al. 1954(250, 254)
		in 29.8	Bose & Nair 1939(367), Bose & Shimamoto 1952(166, 167), Bose et al. 1954(250, 254)
1460.	$32(5) = 160$	in 5.4.8	Bose et al. 1954(91, 120)
1461.	$32(7) = 224$	in 7.4.8	Bose et al. 1954(91, 120)
1462.	$32(8) = 256$	in 1.(32.8)	Bose et al. 1954(140, 169)
		in 8.4.8	Bose et al. 1954(140, 169)
1463.	$32(10) = 320$	in 40.8	Bose et al. 1954(91, 120)
1464.	$36(2) = 72$	in (9.2).4	Bose & Shimamoto 1952(161, 162), Bose et al. 1954(229, 239)
		in 9.8	Bose & Shimamoto 1952(161, 162), Bose et al. 1954(229, 239)
1465.	$36(8) = 288$	in (36.2).4	Bose et al. 1954(91, 122)
		in 4.9.8	Bose et al. 1954(91, 122)

1466.	40(9) = 360	in 9.5.8		Bose et al. 1954(91, 124)
1467.	44(10) = 440	in (55.2).4		Bose et al. 1954(91, 127)
		in 5.11.8		
1468.	50(4) = 200	in (25.4).2		Bose & Shimamoto 1952(161, 163), Bose et al. 1954(218, 223), Vianelli 1959(434)
		in 25.8		Bose & Shimamoto 1952(161, 163), Bose et al. 1954(218, 223)
1469.	50(8) = 400	in 1.(50.8)		Bose et al. 1954(91, 130)
		in 2.25.8		Clatworthy 1956(58)
		in 50.8		Bose et al. 1954(91, 130)
1470.	56(7) = 392	in 49.8		Bose et al. 1954(140, 176)
1471.	56(9) = 504	in 9.7.8		Bose et al. 1954(91, 133)
1472.	57(8) = 456	in 1.(57.8)	Y sq	Cochran & Cox 1957(533)
		in 57.8	bal.	Fisher & Yates 1938-57(T.XIX), Cochran & Cox 1950(343), 1957 (482, 533), Kitagawa & Mitome 1953(VII.12), Czechowski et al. 1957(104), Vianelli 1959(423)
1473.	63(8) = 504	in 1.(63.8)		Bose et al. 1954(184, 215)
		in 63.8		Bose & Nair 1939(359), Bose et al. 1954(184, 215)
1474.	64(4) = 256	in 4.8.8	latt	Cochran 1943(746)
1475.	64(8) = 512	in 1.(64.8)		Bose et al. 1954(140, 179)
		in 8.8.8		Bose et al. 1954(140, 179)
1476.	64(9) = 576	in 9.(8.8)	L sq	Cochran & Cox 1950(362), 1957(499), Kitagawa & Mitome 1953(VI.8), Vianelli 1959(425)
		in 9.8.8	latt	Cochran & Cox 1950(306), 1957(430), Kitagawa & Mitome 1953(VII.13), Vianelli 1959(411)
		in 72.8	bal.	Fisher & Yates 1938-57(T.XIX), Czechowski et al. 1957(104)

15.2108
15.2109

1477. 64(10) = 640   in 10.8.8   Bose et al. 1954(184, 216)

1478. 72(3) = 216   in 3.9.8   rect   Robinson & Watson 1949(49), Cochran & Cox 1950(313), 1957(437), Vianelli 1959(413)

1479. 72(9) = 648   in 9.9.8   Bose et al. 1954(140, 181)

<u>15.2109</u>   Patterns in blocks of 9 plots

1480. 7(10) + 1(11) = 81   in 1.(9.9)   Cochran & Cox 1957(391)

1481. 9(8) + 1(9) = 81   in 1.(9.9)   Cochran & Cox 1957(391)

1482. 10(9) = 90   in 10.9   bal.   Fisher & Yates 1938-57(T.XIX), Kitagawa & Mitome 1953(VII.4), Czechowski et al. 1957(104)

1483. 9(10) + 1(9) = 99   in 1.(11.9)   Cochran & Cox 1957(391)

1484. 12(3) = 36
- in (4.3).3   Bose et al. 1954(90, 96)
- in 4.9   Bose et al. 1954(90, 96)

1485. 12(6) = 72   in 8.9   Bose et al. 1954(90, 98)

1486. 12(9) = 108
- in 1.(12.9)   Bose et al. 1954(90, 98)
- in 12.9   Bose et al. 1954(90, 98), Clatworthy 1956(46)

1487. 13(9) = 117
- in 1.(13.9)   Y sq   Youden 1940(225), Cochran & Cox 1950(381), 1957(524), Kitagawa & Mitome 1953 (VIII.3), Vianelli 1959(428), D. R. Cox 1958(234)
- in 13.9   bal.   Fisher & Yates 1938-57(T.XIX), Cochran & Cox 1950(333), 1957(477), Kitagawa & Mitome 1953 (VII.5), Czechowski et al. 1957(104), D. R. Cox 1958(234), Vianelli 1959(417)

1488. 15(6) = 90
- in (10.3).3   Bose et al. 1954(90, 102, 229, 237)
- in 2.5.9   Bose et al. 1954(90, 102)
- in 10.9   Bose et al. 1954(229, 237)

1489. 15(9) = 135   in 15.9   Clatworthy 1956(46)

1490. 16(9) = 144   in 16.9   Vartak 1955(434), Clatworthy 1956(47, 64)

1491. 17(9) = 153   in 1.(17.9)   Bose et al. 1954(250, 253)

1492.	17(9) = 153	in 17.9		Bose et al. 1954(250, 253)
1493.	18(4) = 72	in 8.9		Clatworthy 1956(16)
1494.	18(5) = 90	in 10.9		Bose et al. 1954(90, 105)
1495.	18(6) = 108	in (12.3).3		Bose et al. 1954(139, 156)
		in 12.9		Bose et al. 1954(139, 156)
1496.	18(8) = 144	in 4.4.9		Bose et al. 1954(139, 156)
1497.	18(10) = 180	in 20.9		Bose et al. 1954(91, 106, 139, 157)
1498.	19(9) = 171	in 1.(19.9)	Y sq	Youden 1940(227), Cochran & Cox 1950(383), 1957(526), Vianelli 1959(429)
		in 19.9	bal.	Fisher & Yates 1938-57(T.XIX), Bose 1939(397), Cochran & Cox 1950(336), 1957(479, 526), Kitagawa & Mitome 1953(VII.7), Czechowski et al. 1957(104), Vianelli 1959(418)
1499.	21(3) = 63	in (7.3).3		Bose et al. 1954(91, 109)
		in 7.9		Bose et al. 1954(91, 109)
1500.	21(6) = 126	in 14.9		Bose et al. 1954(91, 110)
1501.	21(9) = 189	in (21.3).3		Bose et al. 1954(91, 110)
		in 3.7.9		Bose et al. 1954(91, 110)
1502.	12(2) + 10(1) = 34	in 2.9 + 2.8		Youden & Connor 1953(130)
1503.	24(9) = 216	in 24.9		Clatworthy 1956(47)
1504.	25(9) = 225	in 1.(25.9)	Y sq	Fisher & Yates 1943-57(T.XVII), Cochran & Cox 1957(529)
		in 25.9	bal.	Fisher & Yates 1938-57(T.XIX), Cochran & Cox 1950(339), 1957(480, 529), Czechowski et al. 1957(104), Kitagawa & Mitome 1953(VII.9), Vianelli 1959(420)
1505.	26(9) = 234	in 26.9		Bose & Nair 1939(359)
1506.	27(4) = 108	in 4.3.9		Bose et al. 1954(91, 115)
1507.	27(9) = 243	in 1.(27.9)		Bose et al. 1954(140, 165)
		in 9.3.9		Bose et al. 1954(140, 165)
1508.	30(9) = 270	in 1.(30.9)		Bose et al. 1954(91, 118)

15.2109
15.2110

1509. 36(9) = 270   in 30.9   Bose et al. 1954(91, 118)

1510. 36(7) = 252   in 28.9   Bose et al. 1954(229, 236)

1511. 36(8) = 288   in 32.9   Bose et al. 1954(140, 170)

1512. 37(9) = 333 $\begin{cases} \text{in } 1.(37.9) \quad \text{Y sq} \\ \text{in } 37.9 \qquad \text{bal.} \end{cases}$ Cochran & Cox 1950(387), 1957(532), Vianelli 1959(430)

Fisher & Yates 1938-57(T.XIX), Bose 1939(393, 397), Cochran & Cox 1950(341), 1957(482, 532), Kitagawa & Mitome 1953(VII.10), Czechowski et al. 1957(104), Vianelli 1959(422)

1513. 39(6) = 234 $\begin{cases} \text{in } (26.3).3 \\ \text{in } 2.13.9 \end{cases}$ Bose et al. 1954(91, 123)

Bose et al. 1954(91, 123)

1514. 45(2) = 90   in 10.9   Bose & Shimamoto 1952(161, 162), Bose et al. 1954(229, 240)

1515. 45(7) = 315   in 7.5.9   Bose et al. 1954(91, 128)

1516. 57(3) = 171 $\begin{cases} \text{in } (19.3).3 \\ \text{in } 19.9 \end{cases}$ Bose & Shimamoto 1952(161, 163), Bose et al. 1954(218, 223), Vianelli 1959(434)

Bose & Nair 1939(371), Bose & Shimamoto 1952(161, 163), Bose et al. 1954(218, 223)

1517. 57(6) = 342   in 38.9   Bose et al. 1954(218, 224)

1518. 57(9) = 513 $\begin{cases} \text{in } 1.(57.9) \\ \text{in } 3.19.9 \\ \text{in } 57.9 \end{cases}$ Bose et al. 1954(91, 134, 218, 224)

Bose et al. 1954(91, 134)

Bose et al. 1954(218, 224)

1519. 63(4) = 252   in 28.9   Bose & Shimamoto 1952(161, 163), Bose et al. 1954(218, 225), Vianelli 1959(434)

1520. 63(8) = 504   in 2.28.9   Clatworthy 1956(58)

1521. 63(10) = 630 in 10.7.9   Bose et al. 1954(91, 135)

1522. 72(8) = 576   in 64.9   Bose et al. 1954(140, 180)

1523. 73(9) = 657 $\begin{cases} \text{in } 1.(73.9) \quad \text{Y sq} \\ \text{in } 73.9 \qquad \text{bal.} \end{cases}$ Cochran & Cox 1957(534)

Fisher & Yates 1938-57(T.XIX), Cochran & Cox 1950(344), 1957(482, 534), Kitagawa & Mitome 1953 (VII.13), Czechowski et al. 1957 (104), Vianelli 1959(423)

1524.	$80(9) = 720$	in 80.9		Bose & Nair 1939(359), Bose et al. 1954(184, 217)
1525.	$81(2) = 162$	in 2.9.9	latt	Cox & Eckhardt 1940(7)
1526.	$81(3) = 243$	in 3.9.9	latt	Cox & Eckhardt 1940(23)
1527.	$81(5) = 405$	in 5.(9.9)	L sq	Cochran & Cox 1950(363), 1957(500), Kitagawa & Mitome 1953(VI.9), Vianelli 1959(426)
1528.	$81(9) = 729$	in 1.(81.9)		Bose et al. 1954(140, 182)
		in 9.9.9		Bose et al. 1954(140, 182)
1529.	$81(10) = 810$	in 10.(9.9)	L sq	Kitagawa & Mitome 1953(VI.9)
		in 10.9.9	latt	Cochran & Cox 1950(307), 1957(431), Kitagawa & Mitome 1953(VII.14), Vianelli 1959(411)
		in 90.9	bal.	Fisher & Yates 1938-57(T.XIX), Czechowski et al. 1957(104)
1530.	$90(3) = 270$	in 3.10.9	**rect**	Robinson & Watson 1949(50), Cochran & Cox 1950(314), 1957(437)
1531.	$100(9) = 900$	in 1.(100.9)		Bose et al. 1954(240, 249)
		in 100.9		Bose & Nair 1939(364), Bose et al. 1954(240, 249)

15.2110 Patterns in blocks of 10 plots

1532.	$10(9) + 1(10) = 100$	in 1.(10.10)		Cochran & Cox 1957(391)
1533.	$11(10) = 110$	in 11.10	bal.	Fisher & Yates 1938-57(T.XIX), Kitagawa & Mitome 1953 (VII.4), Czechowski et al. 1957(104)
1534.	$12(5) = 60$	in (6.5).2		Bose et al. 1954(90, 97)
		in 6.10		Bose et al. 1954(90, 97)
1535.	$12(10) = 120$	in 1.(12.10)		Bose et al. 1954(90, 99)
		in 2.6.10		Bose et al. 1954(90, 99)
1536.	$13(10) = 130$	in 1.(13.10)		Bose et al. 1954(250, 252)
		in 13.10		Bose et al. 1954(250, 252)
1537.	$15(2) = 30$	in (3.2).5		Bose et al. 1954(90, 101)
		in 3.10		Bose et al. 1954(90, 101)
1538.	$15(4) = 60$	in (6.2).5		Bose et al. 1954(229, 236)

15.2110

1539.	15(4) = 60	in (6.2).5		Bose et al. 1954(229, 236)
		in 6.10		Bose et al. 1954(229, 236)
1540.	15(6) = 90	in 9.10		Bose et al. 1954(90, 102)
1541.	15(8) = 120	in 12.10		Bose et al. 1954(90, 102, 229, 236)
1542.	15(10) = 150	in 1.(15.10)		Bose et al. 1954(90, 102)
		in 3.5.10		Bose et al. 1954(90, 102)
1543.	16(10) = 160	in 1.(16.10)	Y sq	Youden 1940(226), Cochran & Cox 1950(383), 1957(526), Kitagawa & Mitome 1953(VIII.3), Vianelli 1959(429)
		in 16.10	bal.	Fisher & Yates 1938-57(T.XIX), Cochran & Cox 1950(335), 1957 (479, 526), Kitagawa & Mitome 1953(VII.6), Czechowski et al. 1957(104), Vianelli 1959(418)
1544.	18(10) = 180	in 1.(18.10)		Bose et al. 1954(91, 106)
		in 2.9.10		Bose et al. 1954(91, 106)
		in 18.10		Clatworthy 1956(48)
1545.	19(10) = 190	in 1.(19.10)	Y sq	Youden 1940(227), Cochran & Cox 1950(384), 1957(527), Vianelli 1959(429)
		in 19.10	bal.	Fisher & Yates 1938-57(T.XIX), Cochran & Cox 1950(337), 1957 (479, 527), Kitagawa & Mitome 1953(VII.7), Czechowski et al. 1957(104), Vianelli 1959(419)
1546.	20(3) = 60	in 3.2.10		Bose et al. 1954(91, 107)
1547.	20(6) = 120	in (12.2).5		Bose et al. 1954(91, 108, 139, 158)
		in 12.10		Bose et al. 1954(91, 108, 139, 158)
1548.	20(8) = 160	in (16.2).5		Bose et al. 1954(139, 159)
		in 4.4.10		Bose et al. 1954(139, 159)
1549.	20(9) = 180	in 18.10		Bose et al. 1954(91, 109)
1550.	20(10) = 200	in 1.(20.10)		Bose et al. 1954(139, 160)
		in 20.10		Bose et al. 1954(139, 160)
1551.	16(2) + 6(1) = 38	in 2.10 + 2.9		Youden & Connor 1953(130)
1552.	22(5) = 110	in (11.5).2		Bose et al. 1954(91, 111)

15.2110

1553.	22(5) = 110	in 11.10		Bose et al. 1954(91, 111)
1554.	22(10) = 220	in 1.(22.10)		Bose et al. 1954(91, 112)
		in 2.11.10		Bose et al. 1954(91, 112)
1555.	25(4) = 100	in (10.2).5		Bose et al. 1954(91, 114)
		in 2.5.10		Bose et al. 1954(91, 114)
1556.	25(8) = 200	in 20.10		Bose et al. 1954(91, 114)
1557.	27(10) = 270	in 27.10		Clatworthy 1956(48)
1558.	30(5) = 150	in (15.5).2		Bose et al. 1954(91, 117)
		in 5.3.10		Bose et al. 1954(91, 117)
1559.	30(9) = 270	in 3.9.10		Bose et al. 1954(140, 167)
1560.	30(10) = 300	in 1.(30.10)		Bose et al. 1954(91, 119)
		in 10.3.10		Bose et al. 1954(91, 119)
1561.	31(10) = 310	in 1.(31.10)	Y sq	Cochran & Cox 1957(531)
		in 31.10	bal.	Fisher & Yates 1938-57(T.XIX), Cochran & Cox 1950(341), 1957 (482, 531), Kitagawa & Mitome 1953(VII.10), Czechowski et al. 1957(104), Vianelli 1959(421)
		in 31.10		Bose & Nair 1939(367)
1562.	35(6) = 210	in (21.2).5		Bose et al. 1954(91, 121)
		in 3.7.10		Bose et al. 1954(91, 121)
1563.	40(7) = 280	in 7.4.10		Bose et al. 1954(91, 123)
1564.	42(5) = 210	in (21.5).2		Bose et al. 1954(91, 124)
		in 21.10		Bose et al. 1954(91, 124)
1565.	42(10) = 420	in 1.(42.10)		Bose et al. 1954(91, 126)
		in 2.21.10		Bose et al. 1954(91, 126)
1566.	45(8) = 360	in (36.2).5		Bose et al. 1954(91, 129)
		in 4.9.10		Bose et al. 1954(91, 129)
1567.	50(6) = 300	in 6.5.10		Bose et al. 1954(91, 130)
1568.	50(9) = 450	in 9.5.10		Bose et al. 1954(91, 131)
1569.	55(2) = 110	in (11.2).5		Bose & Shimamoto 1952(161, 162), Bose et al. 1954(229, 240)
		in 11.10		Bose & Shimamoto 1952(161, 162), Bose et al. 1954(229, 240)
1570.	55(10) = 550	in 1.(55.10)		Bose et al. 1954(91, 132)

15.2110
15.2411

1571.	55(10) = 550	in 5.11.10		Bose et al. 1954(91, 132)
1572.	70(3) = 210	in 21.10		Bose & Nair 1939(371), Bose & Shimamoto 1952(161, 163), Bose et al. 1954(218, 225), Vianelli 1959(434)
1573.	70(6) = 420	in (42.2).5		Bose et al. 1954(218, 226)
		in 2.21.10		Bose et al. 1954(218, 226)
1574.	70(9) = 630	in 63.10		Bose et al. 1954(218, 227)
1575.	82(5) = 410	in (41.5).2		Bose et al. 1954(218, 227)
		in 41.10		Bose & Shimamoto 1952(161, 163), Bose et al. 1954(218, 227), Vianelli 1959(434)
1576.	82(10) = 820	in 1.(82.10)		Bose et al. 1954(91, 137)
		in 2.41.10		Clatworthy 1956(58)
		in 82.10		Bose et al. 1954(91, 137)
1577.	91(10) = 910	in 1.(91.10)	Y sq	Cochran & Cox 1957(535)
		in 91.10	bal.	Fisher & Yates 1938-57(T.XIX), Cochran & Cox 1950(345), 1957 (482, 535), Kitagawa & Mitome 1953(VII.15), Czechowski et al. 1957(104), Vianelli 1959(424)
1578.	100(3) = 300	in 3.10.10	latt	Cochran & Cox 1950(809), 1957(433), Vianelli 1959(412)
1579.	110(3) = 330	in 3.11.10	rect	Robinson & Watson 1949(51)

<u>15.212</u> Patterns in blocks of 11 or more plots

1580.	11(10) + 1(11) = 121	in 1.(11.11)		Cochran & Cox 1957(391)
1581.	22(6) = 132	in 12.11		Nair & Rao 1948(127)
1582.	23(11) = 253	in 23.11	bal.	Bose 1939(391), Kitagawa & Mitome 1953(VII.8)
1583.	121(6) = 726	in 6.(11.11)	L sq	Cochran & Cox 1950(364), 1957(501), Kitagawa & Mitome 1953(VI.10), Vianelli 1959(426)
1584.	132(3) = 396	in 3.12.11	rect	Robinson & Watson 1949(52)
1585.	144(4) = 576	in 4.12.12	latt	Cochran & Cox 1950(310), 1957(434), Vianelli 1959(412)

1586.	156(3) = 468	in 3.13.12	rect	Robinson & Watson 1949(54)
1587.	27(13) = 351	in 27.13	bal.	Bose 1939(391), Kitagawa & Mitome 1953(VII.9)
1588.	39(9) = 351	in 27.13		Nair & Rao 1948(127)
1589.	169(7) = 1183	in 7.(13.13)	L sq	Cochran & Cox 1950(366), 1957(503), Kitagawa & Mitome 1953(VI.11), Vianelli 1959(427)
1590.	182(3) = 546	in 3.14.13	**rect**	Robinson & Watson 1949(55)
1591.	12(2) + 30(1) = 54	in 3.18		Youden & Connor 1953(137)

15.24 Incomplete block patterns with 1-way control of heterogeneity

15.241 Balanced incomplete block patterns

15.2411 Balance on pairs

In a classical balanced incomplete block (or incomplete randomized block) pattern, each pair of varieties occurs together in an equal number of blocks: Cochran & Cox 1950, 259, 315; 1957, 376, 439; Davies 1954, 579; Kempthorne 1952, 444, 527. A simple example is the pattern 5(4) = 20 in 10.2 (section 15.2102, line 1011) where the 10 blocks are exactly the $\binom{5}{2}$ pairs of varieties. The best known table is Fisher & Yates 1938-57(TT.XVII, XVIII, XIX); the most extensive table that we have seen is Kitagawa & Mitome 1953, T. VII.

Balanced incomplete block patterns are indicated by the abbreviation **bal.** in the finding list above. See section 15.21, lines 1001, 1003, 1011, 1018, 1023, 1030, 1036, 1044, 1049, 1079, 1082, 1087, 1092, 1093, 1099, 1108, 1118, 1120, 1127, 1135, 1139, 1142, 1147-1149, 1153, 1155, 1160, 1162, 1174, 1180, 1182, 1187, 1192, 1193, 1196, 1208, 1209, 1219, 1235, 1239, 1240, 1247, 1248, 1253, 1258, 1259, 1265, 1267-1269, 1282, 1288, 1300, 1305-1307, 1313, 1321, 1322, 1326, 1329, 1337, 1339, 1345, 1346, 1371, 1383, 1386, 1398, 1400, 1403, 1406, 1409, 1418, 1426, 1428, 1439, 1472, 1476, 1482, 1487, 1498, 1504, 1512, 1523, 1529, 1533, 1543, 1545, 1561, 1577, 1582, 1587.

15.2412 Balance on pairs and triples

Calvin 1954 has introduced a class of patterns, which he calls doubly balanced incomplete blocks; in these, not only each pair but each triple of varieties occurs together in an equal number of blocks: Cochran & Cox 1957, 440-441. These patterns are indicated by the abbreviation <u>2bal</u> in the finding list above. See section 15.21, lines 1187, 1199, 1267, 1327, 1337, 1446.

15.243 Square lattices

Some patterns for k^2 varieties in r replications, each of k blocks of k plots, are square lattices: Cochran & Cox 1950, 274-292; 1957, 396-415; Davies 1954, 220-226; Kempthorne 1952, 445-484. When $r = k+1$, the lattice is a balanced incomplete block pattern. For a list of lattice patterns, see section 15.143 above; they are indicated by the abbreviation <u>latt</u> in section 15.21.

15.244 Rectangular lattices

Some patterns for k(k+m) varieties in r replications, each of k+m blocks of k plots, are rectangular lattices: Cochran & Cox 1950, 292-299; 1957, 415-422; Kempthorne 1952, 507-525. In the tables we have seen, m = 1 and $r \geq 3$. Rectangular lattices are indicated by the abbreviation <u>rect</u> in the finding list above. See section 15.21:

k =	3	4	5	6	7	8	9	10	11
line	1110	1227	1292, 1293	1376	1421	1478	1530	1579	1584

12	13
1586	1590

15.245 Cubic lattices

Some patterns for k^3 varieties in r replications, each of k^2 blocks of k plots, are cubic lattices: Cochran & Cox 1950, 299-303; 1957, 422-426; Kempthorne 1952, 446, 456. We have seen no extensive tables of patterns; Cochran & Cox give a short illustrative table.

Kempthorne 1952, 447, 446-459, gives the analysis of a 5-dimensional lattice in some detail: we call this account to the attention of anyone wishing to test 243 varieties in blocks of 3 plots, or 1024 varieties in blocks of 4 plots.

15.246 Chain blocks

In a chain block pattern (Cochran & Cox 1957, 463-468) some varieties occur twice, others once. The patterns possess very little symmetry; their construction is ordinarily left to the experimenter. We have seen several ready-made chain block patterns: section 15.21, lines 1328, 1347, 1390, 1452, 1502, 1551.

15.248 Partially balanced incomplete blocks

For the definition of this extensive class of patterns see Cochran & Cox 1957, 453-454; Kempthorne 1952, 549. For $3 \leq k \leq 10$, the standard table is Bose, Clatworthy & Shrikhande 1954*; for $k = 2$, Clatworthy 1955. Clatworthy 1956 gives a number of additional patterns.

The patterns listed without a name in section 15.21, and not listed in section 15.246, 15.2513 or 15.256, are partially balanced incomplete block patterns.

15.25 Patterns with 2-way control of heterogeneity

15.251 Youden squares and generalizations

N.B. In section 15.21, the label Y sq is applied, without distinction, to the patterns of sections 15.2511 and 15.2512.

15.2511 Incomplete Latin squares

A Youden square, or incomplete Latin square, is an array formed from k columns of a b × b Latin square, so that the rows are the blocks of a balanced incomplete block pattern: Cochran & Cox 1950, 370-390; 1957, 507-544; Davies 1954, 592, 214-215; Kempthorne 1952, 539-541. See section 15.21, lines 1093, 1181, 1209, 1268, 1282, 1345, 1371, 1399, 1439, 1472, 1487, 1498, 1504, 1512, 1523, 1543, 1545, 1561, 1577.

C.A.B. Smith & Hartley 1949 give a method for converting any balanced incomplete block pattern for which $b = v$ and $k = r$ into a Youden square:

---

*This table is cited in section 15.21 as Bose et al. 1954.

15.2512
15.2582

this method can be applied to section 15.21, lines 1313, 1386, 1403, 1428, 1533, 1582, 1587.

15.2512 Redundant Latin squares

Cochran & Cox 1950, 375-378; 1957, 513-515; give several patterns, analysable in the same manner as Youden squares, but for which $k > v$. In the finding list above, these patterns are entered with k and b interchanged. For example, Cochran & Cox 1950, p. 388, T. 13.17; 1957, p. 536, T. 13.17; for which $b = v = 3$ and $k = r = 7$, appears in section 15.2103, line 1076, as $3(7) = 21$ in $1.(7.3)$, although the description $7(3) = 21$ in $1.(3.7)$, i.e. 3 blocks of 7 plots with 2-way control of heterogeneity, is logically defensible. See section 15.21, lines 1075-1078, 1170-1172, 1250, 1251, 1311, 1384.

Cochran & Cox 1957, 538-541, following an unpublished report by M. K. Rupp, give several resolvable patterns divided into b/v replicates; each replicate comprises k columns of a $v \times v$ Latin square, and Youden-square analysis applies. For example, Cochran & Cox 1957, p. 538, T.13.6a, for which $b = 10$, $k = 2$, $r = 4$, $v = 5$, appears in section 15.2102, line 1011, as $5(4) = 20$ in $2.(5.2)$. See section 15.21, lines 1011, 1025, 1036, 1049, 1082, 1108, 1120, 1142, 1192, 1235, 1258.

15.2513 Mutilated Latin squares

Cochran & Cox 1957, 391, 516-517, give patterns formed by adding one row and one column to a Latin square, by deleting one row and one column, and by adding one row and deleting one column: see section 15.21, lines 1080, 1169, 1173, 1175, 1249, 1252, 1254, 1310, 1312, 1314, 1382, 1385, 1387, 1425, 1427, 1429, 1480, 1481, 1483, 1532, 1580.

15.253 Lattice squares

Some patterns for k^2 varieties in r replications, each of k rows $\times$ k columns, are lattice squares*: Cochran & Cox 1950, 346-369; 1957, 483-506; Kempthorne 1952, 485-506. For a list of lattice squares, see section

---

*The patterns described as lattice squares by Davies 1954, 586, 220-226, afford only 1-way control of heterogeneity and are in fact square lattices.

15.153 above; lattice squares are indicated by the abbreviation L sq in section 15.21.

15.256 Generalized chain blocks

Cochran & Cox 1957, 517, describe a class of patterns due to Mandel 1954, in which $r = 2$, and chain-block analysis applies both by rows and by columns: see section 15.21, lines 1183, 1200, 1217, 1226, 1348, 1358, 1366.

15.258 Partially balanced incomplete block patterns with 2-way control of heterogeneity

For notation see section 15.20 above. For discussion see Bose, Clatworthy & Shrikhande 1954, 7-8, 10-11, under the title 'replication groups orthogonal to blocks'; and Cochran & Cox 1957, 518-519.

15.2581 Single columns are replications

See section 15.21, lines 1082, 1083, 1085, 1088, 1091, 1094-1096, 1098, 1101, 1106, 1113, 1116, 1119-1123, 1126, 1129, 1131, 1132, 1134, 1136, 1138, 1140, 1143, 1146, 1154, 1157, 1164, 1177, 1179, 1191, 1192, 1195, 1197, 1202, 1205, 1210, 1211, 1214, 1216-1218, 1222, 1224, 1226, 1236, 1237, 1239, 1243, 1245, 1257, 1277, 1283, 1286, 1295-1297, 1300, 1316, 1320, 1325, 1333, 1338, 1341, 1345, 1351, 1358, 1366, 1389, 1392, 1410, 1411, 1415, 1417, 1424, 1431, 1435, 1438, 1447, 1448, 1450, 1457, 1459, 1462, 1469, 1473, 1475, 1486, 1491, 1507, 1508, 1518, 1531, 1535, 1536, 1542, 1544, 1550, 1576.

15.2582 Groups of columns are replications

See section 15.21, lines 1176, 1178, 1186, 1194, 1207, 1212, 1220, 1225, 1231, 1232, 1234, 1241, 1242, 1315, 1318, 1319, 1324, 1330, 1331, 1336, 1340, 1342, 1353-1357, 1360, 1361, 1363, 1365, 1370, 1372, 1375, 1379, 1380, 1430, 1432, 1434, 1436, 1441, 1445, 1449, 1455, 1456, 1458, 1464, 1465, 1467, 1468, 1484, 1495, 1499, 1501, 1513, 1516, 1534, 1537-1539, 1547, 1548, 1552, 1573, 1575.

15.28 Tables for the analysis of incomplete block experiments

A judge effects $\frac{1}{2}nv(v-1)$ paired comparisons, arranged as n repetitions of the incomplete block pattern $v(v-1) = v^2-v$ in $\binom{v}{2}.2$. Assume that the judge assigns preference Π_i to the ith variety, and that in a paired comparison between the ith and j^{th} varieties he assigns rank 1 to the ith (and rank 2 to the jth) with probability $\Pi_i/(\Pi_i+\Pi_j)$. Bradley & Terry 1952, p. 326, obtain maximum-likelihood estimates of the Π_i, and a likelihood-ratio test of the hypothesis that all the Π_i are equal. These estimates and this test are functions of the r_i, the sum of the ranks (1 or 2) assigned to the ith variety in its n(v-1) presentations.

The estimates and test have been tabled, for all attainable r_i, for v and n as follows:

v	n	
v = 3	n = 1(1)10	Bradley & Terry 1952(336), Vianelli 1959(437)
4	1(1)6	Bradley & Terry 1952(340), Vianelli 1959(440)
4	7, 8	Bradley 1954(509), Vianelli 1959(445)
5	1(1)5	Bradley 1954(518), Vianelli 1959(454)

When v = 3, n = 2(1)5, Bradley & Terry 1952, p. 345, give a table for assessing a joint likelihood-ratio test based on the ratings of 2, 3, 4 or 5 judges. Reprinted in Vianelli 1959, Pro. 160, p. 473.

15.3 Tables facilitating the choice of sample size

Dorfman, R. 1943

Annals of Mathematical Statistics 14, 436-440

"The detection of defective members of large populations"

Blood samples are taken from N individuals for the purpose of detecting the presence of an antigen. Samples from n individuals are pooled in a group test; if the group test is positive, the individuals in that group must be retested. If p is the probability of the presence of the antigen in an individual, then the chance that a group test is positive is $1-(1-p)^n$; the expected total number of tests is $N/n + n(1-[1-p]^n)$; and the number of tests per individual tested is $1 + n^{-1} - (1-p)^n = C$, say.

Figure 1, p. 439, gives graphs of C as a function of n for $p = .01, .02, .05, .1, .15$; these have minimum C at the optimum group size n^*, which is the integer nearest the root of $dC/dn = 0$. Table 1, p. 438, gives these n^*, and C_{min} and $1 - C_{min}$ to 2 dec., for $p = .01(.01).1, .12, .13, .15(.05).3$.

Brooks, S. H. 1955

Journal of the American Statistical Association 50, 398-415

"The estimation of an optimum subsampling number"

A population consists of N primary units, each containing an equal number M of secondary units. If y_{ij} is a characteristic attached to the jth secondary of the ith primary, define:
the population mean $\bar{\bar{Y}} = (NM)^{-1} \Sigma\Sigma y_{ij}$, the primary mean $\bar{Y}_i = M^{-1} \Sigma y_{ij}$,
the within-primaries mean square $S_w^2 = \sum (y_{ij} - \bar{Y}_i)^2 / N(M-1)$,
the between-primaries mean square $S_w^2 + MS_h^2 = M \sum (\bar{Y}_i - \bar{\bar{Y}})^2 / (N-1)$.

Draw a sample of n primaries at random and draw, from each sampled primary, a sample of m secondaries at random. The grand sample mean $\bar{\bar{y}}$ is then an unbiased estimate of $\bar{\bar{Y}}$. If the total cost of sampling is given by the linear cost function

$$C = C_u n + C_c nm, \qquad (1)$$

then the variance of $\bar{\bar{y}}$ is minimized, for prescribed cost C, by taking

15.3 [see also pp. 736-737]

$m_{op} = (C_u/C_c)^{\frac{1}{2}}(S_w/S_h)$ and then computing n_{op} from (1).

In order to estimate m_{op}, draw a pilot sample of h primaries and k secondaries per primary; denote the between-primaries and within-primaries mean squares of the hk sampled y_{ij} by s_b^2 and s_w^2 respectively. Then a consistent estimate of m_{op} is

$$m_{op} = (kC_u/C_c)^{\frac{1}{2}}(s_b^2/s_w^2 - 1)^{-\frac{1}{2}}. \qquad (2)$$

Define the relative precision of the pilot sample to be the ratio of the variance of $\bar{\bar{y}}$ obtained from the true m_{op} to the variance of $\bar{\bar{y}}$ obtained from m_{op} from (2). Table 1, p. 407, gives those integer values of h, k and hk that achieve 90% expected relative precision and for which the pilot sample cost $C' = C_u h + C_c hk$ is a minimum: for $C_u/C_c = 1, 2, 4, 8, 16, 32, 64, 100$; $s_w^2/s_h^2 = \frac{1}{2}, 1(1)4(2)8(4)16(8)32(16)64$.

For any selected m_0, and n_0 then determined from (1), Table 3, p. 412, gives the range of variance ratios s_w^2/s_h^2 for which the relative precision resulting from the use of m_0 is at least 90%: for $m_0 = 1(1)10, 12, 16, 25, 50, 100$; $C_u/C_c = 1, 2, 4, 8, 16, 32, 64, 100$.

Grundy, P. M., Healy, M. J. R. and Rees, D. H. 1956
Journal of the Royal Statistical Society, Series B 18, 32-55
"Economic choice of the amount of experimentation"

The financial gain from a new process with process average θ is assumed to be $k'\theta$; the cost of performing n experiments to estimate θ is assumed to be kn. An initial set of n_1 experiments has yielded an estimate x_1 of θ, with variance σ^2/n_1; n_2 further experiments will yield an estimate x_2, with variance σ^2/n_2: to fix an optimum n_2.

If the new process will be adopted for $n_1 x_1 + n_2 x_2 > 0$, then the monetary risk attached to n_2 is $kn_2 - k'\theta P[(n_1 x_1 + n_2 \theta)/\sigma n_2^{\frac{1}{2}}]$, where P(X) is the standardized normal integral. The fiducial expectation of the risk is $kn_1 Q(X,N)$, where $Q(X,N) = N - \lambda[XP(XB^{\frac{1}{2}}) + B^{-\frac{1}{2}}Z(XB^{\frac{1}{2}})]$, $N = n_2/n_1$, $B = (N+1)/N$, $\lambda = k'\sigma/kn_1^{3/2}$, $X = x_1 n_1^{\frac{1}{2}}/\sigma$ and Z is the standardized normal ordinate. Table 1, p. 36, gives Q(X,N) to 6 dec. for $X = -2.5, -2(.25)0$; $N = .5(.5)4(1)6(2)10(5)20, \infty$. The function Q(X,N) can be reduced to the univariate function $\omega(\eta) = \eta P(\eta) + Z(\eta)$, where $\eta = XB^{\frac{1}{2}}$. Table 2 gives $\omega(\eta)$ to 6 dec. for $\eta = 0(.1)4.4$, with first and second differentials. Based on this table, the nomogram on p. 37 yields recommended values of n_2: that is, it gives N as a function of X and λ. Tables 3, 4, 5, p. 39, give respectively the probability of a wrong decision, the expected amount of additional experimentation, and the ratio of expected second-stage cost to cost of initial experimentation: all for $\log_{10}\lambda = 2(.5)3.5$; $|\Theta| = 0(.1)1(.5)2.5$.* Tables 6, 7, 8, p. 40, give certain ratios of the quantities in TT. 3-5; T. 9, p. 40, gives the risk function for the minimax fixed N; all for λ and $|\Theta|$ as in TT. 3-5.

The tables of Brooks 1955, reviewed on pp. 611-612 above, are reprinted in Vianelli 1959, Pro. 176, pp. 526-527.

See also Thionet 1955; Vianelli 1959, Pri. 173-175, 178-180, pp. 511-525, 530-534.

* $\Theta = \theta n_1^{\frac{1}{2}}/\sigma$.

16.00
16.023

SECTION 16

SUNDRY MATHEMATICAL TABLES

16.00 Introduction

This section catalogues the tables of mathematical functions that we have found included in the collections of statistical tables listed in Appendix 2, and a selection of larger tables of these functions. The user that seeks more details regarding some function can find the relevant article in Fletcher et al. 1946 by canceling the '16.' from our subsection number and inserting a decimal point to the right of the second digit remaining: for example, our 16.0618 refers to Fletcher's 6•18.

The works recommended below as 'serviceable modern tables' are in print (as of 1960) and have good reputations. In assessing reputations we have leaned heavily on the bold entries in Fletcher et al. 1946; for their use of bold type, see their p. 11.

16.02 Powers: positive, negative and fractional

Barlow 1941 and B.A. 9, 1940 are serviceable modern tables. Barlow 1930 differs from Barlow 1941 only in omitting arguments from 10001 to 12500.

16.0211 Squares: exact values

1(1)12500	Barlow 1941
1(1)2000	Graf & Henning 1953(76)
1(.01)10(1)100	Lindley & Miller 1953(14)
1(1)1000	Glover 1923(436), Siegel 1956(289)
1(1)100; 101(1)1000	Vianelli 1959(3; 7)
1(1)999	Fisher & Yates 1938 etc.(T.XXVII), Hald 1952(T.15), Burington & May 1953(293), E.S. Pearson & Hartley 1954(226)
100(1)999	Arkin & Colton 1950(27), Czechowski et al. 1957(131)

[614]

1(1)100   K. Pearson 1930a(38)

1(1)100   K. Pearson 1931a(258)

16.0213 Squares: abridged values

3; 2 dec. 1(.01)3.16; 3.17(.01)9.99 No Δ Dixon & Massey 1951(346), 1957(472)

16.0214 Quadratic polynomials

For $1-x^2$ see section 7.31. No differences are given with the following tables.

$x - x^2$	exact	.001(.001).999	Vianelli 1959(155)
$x - x^2$	exact	.01(.01).99	Dixon & Massey 1957(465)
$x - x^2$	exact	0(.01).5	E.S. Pearson & Hartley 1954(234)
$1 - 2x + x^2$	exact	0(.01).5	E.S. Pearson & Hartley 1954(234), Vianelli 1959(562)
$(n^2-1)/12$	3 dec.	2(1)1000	Vianelli 1959(107)

16.022 Cubes

The tables below give exact values.

1(1)12500   Barlow 1941

1(1)1000   Glover 1923(436), Arkin & Colton 1950(68)

1(1)100   Vianelli 1959(3)

101(1)1000   Vianelli 1959(7)

100(1)999   Czechowski et al. 1957(133)

1(1)100   K. Pearson 1930a(38), 1931a(258), Pearson & Hartley 1954(222)

16.023 Higher integral powers: n^p

$p =$	$n =$		
2(1)12	1(1)1099	exact	
13(1)20	1(1)299	exact	
21(1)27, 30(10)50	1(1)120	exact	B.A. 9, 1940
28, 29, 31(1)39, 41(1)49	1(1)120	exact or 21+ fig.	
4	1(1)1000	exact	Barlow 1941(2)
4(1)10	1(1)100	exact	Barlow 1941(252)
11(1)20	1(1)10		

16.023
16.0262   [see also pp. 737-738]

p =	n =		
5(1)10	1(1)100	exact	K. Pearson 1931a(258)
4(1)8	1(1)100	exact	Glover 1923(432), Vianelli 1959(3)
4	1(1)100	exact	⎫
5, 6, 7	1(1)100	exact or 7-8 fig.	⎬ Arkin & Colton 1950(78)
8	1(1)100	exact or 7-9 fig.	⎭
4(1)7	1(1)100	exact	K. Pearson 1930a(38), E.S. Pearson & Hartley 1954(222)

16.0233 kn^4

K. Pearson 1930a(102) gives a table of kn^4 for

$\begin{cases} n = 2(1)7 & 8 & 9 & 10, 11 & 12 & 13, 14 & 15(1)19 \\ k = 1(1)400 & 1(1)350 & 1(1)300 & 1(1)250 & 1(1)200 & 1(1)150 & 1(1)100 \end{cases}$.

16.024 Reciprocals

20 dec.	1(1)100	No Δ	K. Pearson 1931a(244)
7 fig.	1(1)12500	Δ	Barlow 1941
9 dec.	1(1)1000	No Δ	Glover 1923(436)
9 dec.	1(1)100; 101(1)1000	No Δ	Vianelli 1959(20; 22)
7 fig.	50(1)101(2)199, 200	No Δ	K. Pearson 1931a(244)
6 dec.	1(.01)2.5	Δ	⎫ Fisher & Yates 1938 etc.(T.XXIX)
6 dec.	2.5(.01)9.99	IΔ	⎭
8 dec.	100(1)999	No Δ	Arkin & Colton 1950(104)
6 dec.	1(.01)9.99	No Δ	Czechowski et al. 1957(139)
7 dec.	1(1)100	No Δ	E.S. Pearson & Hartley 1954(228)
5 dec.	1(.001)1.1(.01)9.99	No Δ	Dixon & Massey 1957(470)
5 fig.	1(1)999	No Δ	Burington & May 1953(293)
5 dec.	1(.01)9.99	No Δ	Hald 1952(T.17)
4 dec.	1(.01)5	Δ	⎫ Lindley & Miller 1953(T.9)
5 dec.	5(.01)10; 1(1)100	Δ; No Δ	⎭

16.0244 Miscellaneous rational functions

$(1+x)/2x$	5 dec.	.002(.002)1	No Δ	Vianelli 1959(112)
$(1+x)/(1-x)$	5 dec.	.002(.002).998	No Δ	Vianelli 1959(114)

Bush & Mosteller 1955, T. C, p. 346, computed by D.G. Hays and T.R. Wilson, gives $nx^n(1-x^n)^{-1}$ to 3 dec. for $x = .5(.01).99$ and $n = 1(1)10$. Reprinted in Vianelli 1959, Pro. 142, p. 376.

16.025 n^{-p}

9 dec.; 7 fig.	$p = 2, 3$	$n = 1(1)10; 11(1)100$	Vianelli 1959(20)
7 fig.		$1(1)3$ $50(1)101(2)199, 200$	K. Pearson 1931a (244)

16.0261 Square roots

7 dec.	1(1)1000	No Δ	Barlow 1941
6 dec.	1000(1)12500(10)125000	Δ	
7 dec.	1(1)1000	No Δ	Glover 1923(436)
7 dec.	1(.01)10(.1)100	No Δ	Kelley 1948(204)
8 fig.	1(1)1000	No Δ	Kelley 1938(132)
6 dec.	1(1)1000	No Δ	Vianelli 1959(28)
7 dec.	1(1)100	No Δ	E.S. Pearson & Hartley 1954(228)
4 dec.	1000(1)10000(10)99990	No Δ	Arkin & Colton 1950(28)
4 dec.	1(1)2000	No Δ	Graf & Henning 1953(76)
4 dec.	1(1)1000	No Δ	Siegel 1956(289)
5 fig.	100(1)1000(10)9990	PPMD	Fisher & Yates 1938 etc.(T.XXVIII)
5 fig.	1(1)1000(10)9990	No Δ	Burington & May 1953(293)
3 dec.	100(1)1000(10)9990	No Δ	Czechowski et al. 1957(135)
4 dec.	1(.01)10(.1)99.9	No Δ	Hald 1952(T.16)
4 dec.	1(.01)10(.1)99.9	No Δ	Dixon & Massey 1957(474)
4; 3 d.	1(.01)4; 10(.1)100	Δ	Lindley & Miller 1953(T.9)
3; 4 d.	4(.01)10; 0(1)100	No Δ	

16.0262 Reciprocal square roots: $x^{-\frac{1}{2}}$

7 dec.	1(1)1000	No Δ	Barlow 1941
8 dec.	1(1)100	No Δ	Glover 1923(488)
8 dec.	50(1)101(2)199, 200	No Δ	K. Pearson 1931a (244)
7 dec.	1(1)100	No Δ	E.S. Pearson & Hartley 1954(228)
4 dec.	1(.01)10(.1)25	Δ	Lindley & Miller 1953(T.9)
5 dec.	25(.1)100; 1(1)100	Δ; No Δ	

16.0263
16.044

[see also pp. 737-738]

16.0263 $n^{\frac{1}{2}}p$

7 fig. $-p = 1(1)6$ $n = 50(1)101(2)199, 200$ K. Pearson 1931 a (244)

16.0264   Square roots of rational functions

For $(npq)^{\frac{1}{2}}$, see section 3.53; for $(pq/n)^{\frac{1}{2}}$, see section 3.54; for $(1-x^2)^{\frac{1}{2}}$, see section 7.32; for $(1-x^2)^{-\frac{1}{2}}$, see section 7.33.

The order in the following list is chronological. Tables of $.6745\,n^{-\frac{1}{2}}$ and similar functions are common in books on the combination of observations; we have not searched for them.

$.67450[n(n-1)]^{-\frac{1}{2}}$
$.67450(n-1)^{-\frac{1}{2}}$
$.84535[n(n-1)]^{-\frac{1}{2}}$ 5 dec.
$.84535\,n^{-1}(n-1)^{-\frac{1}{2}}$ $1(1)100$ Glover 1923(488)
$[n(n-1)]^{-\frac{1}{2}}$ 8 dec.

$.67449\,n^{-\frac{1}{2}}$, $.47694\,n^{-\frac{1}{2}}$ 5 dec. $1(1)1000$ K. Pearson 1930 a (12)

$x(1+2x^2)^{\frac{1}{2}}$ [with Δ^3]   7 dec. $0(.01).5$ K. Pearson 1930 a (18)

$[x^3/24(x-1)]^{\frac{1}{2}}$ 4 dec. $1.05(.001)1.5$ K. Pearson 1930 a (232)

$[x(1-x)]^{\frac{1}{2}}$ 8 dec. $0(.0001)1$ Kelley 1938(14), 1948(38)

$\psi^2/2n \pm [(\psi^2/2n)(2x + \psi^2/2n)]^{\frac{1}{2}}$ 6 dec.
$\psi = 1.959964, 2.575829$
$x = .001, .005, .01, .02(.02).1, .2(.2)1(1)10$ Kitagawa 1942(149),
$n = 50(10)100(100)500, 1000, 2000(2000)10000,$ Czechowski et al. 1957(69)
     $20000, 50000, 100\,000$

$[x(1-x)]^{\frac{1}{2}}$ 4 dec. $0(.005).5$ Burington & May 1953(258)

$[x(1-x)]^{\frac{1}{2}}$ 5 dec. $0(.01).5$ Pearson & Hartley 1954(234)

$[x(1-x)]^{\frac{1}{2}}$ 4 dec. $0(.01)1$ Dixon & Massey 1957(465)

$n^{\frac{1}{2}}(n-1)^{-\frac{1}{2}}$ 7 dec. $2(1)1000$ Vianelli 1959(38)

$[x(1-x)]^{\frac{1}{2}}$ 5 dec. $0(.01).5,$ Vianelli 1959(562)
                              $.91(.002).99(.001)1$

$[x(1-x)/y(1-y)]^{\frac{1}{2}}$ 2 dec.   Vianelli 1959(538)
$x, y = .001(.001).006(.002).01(.001).016(.002)$
     $.03(.005).1(.01).14(.02).3(.05).5$

16.027 Cube roots

7 dec.	1(1)1000	No Δ	Barlow 1941
6 dec.	1000(1)12500	Δ	
7 dec.	1(.01)10(.1)100(1)1000	No Δ	Kelley 1948(204)
7 dec.	1(1)1000	No Δ	Glover 1923(436)
7 fig.	1(1)1000(10)10^4(100)10^5	No Δ	Arkin & Colton 1950(68)
6 dec.	1(1)1000	No Δ	Vianelli 1959(28)

16.03 Factorials; binomial coefficients; partitions

For factorials, see sections 2.5112, 2.5113, 2.5122, 2.5134, 2.5135; for binomial coefficients, see sections 3.21, 5.31.

16.036 Partitions

Fix, Evelyn and Hodges, J. L. Jr. 1955

Annals of Mathematical Statistics 26, 301-312

Table 1, pp. 304-307, gives $A_0(u,m)$, the number of ways in which it is possible to choose exactly m non-negative integers whose sum shall not exceed u, as an exact integer for m = 2(1)12 and u = 0(1)100. Table 2, pp. 308-312, gives

$$A_2(u,m) = \sum_{v=0}^{u} [A_0(v,2) - A_0(v-1, 2)]A_0(u-v, m)$$

as an exact integer for m = 1(1)11 and u = 0(1)75.

For application to the distribution of the Wilcoxon two-sample test see the paper.

Reprinted in Vianelli 1959, *Prontuario* 126, pp. 341-344.

16.04 Sums of powers; Bernoulli numbers; differences of zero

16.041 Bernoulli numbers

$\log_{10} B_n$ 7 dec.   n = 1(1)200   Glover 1923(431), Vianelli 1959(1025)

16.044 Sums of powers of integers

For description of Bailey 1931 see Fletcher et al. 1946, p. 54.

The tables listed in sections 16.0444 and 16.0445 give exact values.

16.0444 $S_n(x) = \sum_{r=1}^{x} r^n$

$n = 1(1)8$	$x = 1(1)100$	Vianelli 1959(16)
$n = 1(1)7$	$x = 1(1)100$	K. Pearson 1930a(40), Pearson & Hartley 1954(224)
$n = 1(1)6$	$x = 1(1)40$	Czechowski et al. 1957(14)

16.0445 Sums of powers of odd integers: $S_n(2x+1) - 2^n S_n(x)$

$n = 1(1)6$	$2x+1 = 1(2)199$	F.A. Ross 1925, J.G. Smith 1934(498)
$n = 1(1)6$	$2x+1 = 1(2)69$	Czechowski et al. 1957(142)

16.046 Sums of reciprocals: $\phi(x) = \sum_{r=1}^{x} r^{-1}$

10 dec.	$x = 1(1)450$	Glover 1923(456)
10 dec.	$x = 1(1)100$	Vianelli 1959(20)

16.0492 Differences and derivatives of zero

$\Delta^p 0^{n+p}/(n+p)!$ 12 dec. $n = 0(1)20$ $p = 0(1)20$ K. Pearson 1931a(246)

$\left. \begin{array}{l} D^p 0^{(n)}/p! \\ \Delta^p 0^n/p! \end{array} \right\}$ exact $n = 1(1)12$ $p = 1(1)n$ Vianelli 1959(1483)

See also section 5.9.

16.05 Mathematical constants; roots of equations

We have not attempted to describe the contents of the tables of constants appended to many of the volumes listed in Appendix 2.

16.0564 Solution of cubic equations

Vianelli 1959 reduces cubic equations to the form $y^2 - y + R - B/y = 0$. **Prontuario** 300, pp. 1080-1085, gives $B/y + y - y^2$ to 4 dec. for $y = 0(.01)1$ and $B = 0(.0001).001(.001).01(.01).05, .1$.

16.06 Common logarithms and related functions

A.J. Thompson 1924 (2 volumes!) and J.T. Peters 1922 are serviceable modern tables. The radix table (cf. Fletcher et al. 1946, pp. 114-115) of Spenceley et al. 1952 combines 22-place common **and** natural logs in one handy octavo volume.

16.061 $\log_{10} x$

23 dec.	$1(1)9999$; $1+m.10^{-n}$, $m = 1(1)9999$, $n = 7, 11$	No Δ	Spenceley et al. 1952
20 dec.	$10^4(1)10^5$	δ^{2n}	A. J. Thompson 1924
10 dec.	$1(1)10^4$	Δ	J. T. Peters 1922
7 dec.	10 000 (1) 100 009	fPPd	Glover 1923(496)
5 dec.	$1(1)10009$	fPPd	Arkin & Colton 1950(80)
5 dec.	$100(1)999$	PPMD	Fisher & Yates 1938 etc. (T.XXV)
5 dec.	$1(1)1000$	No Δ	Vianelli 1959(43)
5 dec.	$100(1)999$	No Δ	Czechowski et al. 1957(145)
4 dec.	$110(1)1100$	PPMD	Hald 1952(T.18)
4 dec.	$100(1)999$	PPMD	Burington & May 1953(308)
4 dec.	$1(.01)10$	Δ	Lindley & Miller 1953(16)

16.0618 $\log_{10} \log_{10} x$

For $\log_e \log_e x$, see section 8.3121.

5 dec. $1.001(.001)1.15(.002)1.3(.005)1.5(.01)$ No Δ Vianelli 1959(64)
$2(.02)3(.05)5(.1)10(.5)20(1)50(2)100(5)$
$200(10)300(20)500(50)1000(100)2000(200)$
$5000(500)10000(1000)20000(5000)40000(10000)100000$

16.064 10^x

4; 3 dec. $0(.001).3$; $.3(.001)1$ Δ Lindley & Miller 1953(14)

16.068 Addition and subtraction logarithms

The arrangement of Wittstein 1866, while not conforming to the standard recommended by Fletcher et al. 1946, p. 117, is perhaps the easiest for the occasional computer: Wittstein effectively tables $B = \log(1+x)$ with argument $A = \log x$.

$10 + x =$

$\log(1 + 10^x)$	7 dec.	$3(.1)4(.01)6(.001)8(.0001)$ $14(.01)16(.1)17$	PPd	Wittstein 1866
$\log(1 + 10^x)$	5 dec.	$5.1(.1)6.5(.05)8(.005)10$	No Δ	Vianelli 1959(53)

16.068
16.101

$-\log(1-10^x)$ 5 dec. $10 + x = 5(.2)6(.02)8.2(.01)$ No Δ   Vianelli 1959(56)
                         $9.6(.005)9.95(.001)$
                         $9.996(.0005)9.999(.0001)9.9994(.00001)9.99999$

16.07 Natural trigonometric functions

B.A. 1, 1931 and N.B.S. 1943 (argument in radians) and N.B.S. 1949 z (argument in degrees) are serviceable modern tables.

16.0711 $\sin x$ and $\cos x$ (radians)

15 dec.	0(.1)50	No Δ	B.A. 1, 1931
11 dec.	0(.001)1.6		
9 dec.	0(.0001)2(.1)10	No Δ	N.B.S. 1953 y
10 dec.	0(.1)10(1)100	No Δ	Becker & Van Orstrand 1924
5 dec.	0(.0001).1(.001)1.6	v	

16.0712 $\tan x$ and $\cot x$ (radians)

For $\tan^{-1} x$ see section 10.0432.

10 dec.	0(.1)10	No Δ	N.B.S. 1943
8 fig.	0(.0001)2	δ^2	

16.0721 $\sin x$ (degrees)

30 dec.	$0(1°)90°$	No Δ	N.B.S. 1949 z
15 dec.	$0(0°.01)90°$	δ^2	
5 dec.	$0(0°.1)90°$	$PP_6 MD$	Fisher & Yates 1938 etc.(T.XXXI)
5 dec.	$0(1°)90°$	No Δ	Vianelli 1959(71)
4 dec.	$0(10')90°$	No Δ	Arkin & Colton 1950(146)
4 dec.	$0(1°)90°$	No Δ	Burington & May 1953(303)

16.0722 $\tan x$ (degrees)

$\tan x$	5 dec.	$0(0°.1)60°$	$PP_6 MD$	Fisher & Yates 1938 etc.
	3 fig.	$60°(1°)89°$	No Δ	(T.XXXII)
$(90°-x)\tan x$	3 dec.	$60°(0°.1)90°$	$PP_6 MD$	
$\tan x$	5 dec.	$0(1°)90°$	No Δ	Vianelli 1959(71)
$\tan x$	4 dec.	$0(10')90°$	No Δ	Arkin & Colton 1950(146)
$\tan x$	4 dec.	$0(1°)90°$	No Δ	Burington & May 1953(303)

16.0723   sec x (degrees)

6 dec. $0(1°)90°$ No Δ Vianelli 1959(71)

4 dec. $0(1°)90°$ No Δ Burington & May 1953(303)

16.073   Tables with centesimal argument

$\sin 2\pi x$, $\tan 2\pi x$, $\sec 2\pi x$ 5 dec. $0(.001).25$ No Δ Vianelli 1959(72)

16.0735 $\sin[\pi(2x-\tfrac{1}{2})]$

3 dec. $.25(.001).499$ No Δ Vianelli 1959(1366)

16.079   Miscellaneous circular functions

$\sin^2 x$, $\cos^2 x$ — 6 dec.

$\cos^3 x$, $\cos^4 x$, $(\cos x)^{\frac{1}{2}}$, $(\cos x)^{3/2}$

$(1-\tan x)/(1+\tan x)$ — 4 dec.

$\sin x \cos^m x$, $m = -\tfrac{1}{2}, +1(1)3$

$0(\tfrac{1}{2}°)90°$ Vianelli 1959(1039)

16.08   <u>Logarithms of trigonometric functions</u>

$\log \sin x$, $\log \tan x$, $\log \sec x$ 5 dec. $0(1')90°$ No Δ Vianelli 1959(77)

S and T   5 decimals critical   to 240"   Vianelli 1959(1134)

16.10   <u>Exponential and hyperbolic functions</u>

   N.B.S. 1951 z, 1955 z are serviceable modern tables of exponentials; N.B.S. 1943, 1953 y, 1955 y of hyperbolic functions.

16.101 e^x

18 dec. $0(10^{-6})10^{-4}(10^{-4})1$
15 dec. $1(10^{-4})2.5(.001)5$
12 dec. $5(.01)10$
19 fig. $1(1)100$

No Δ N.B.S. 1951 z

6 dec. $0(.001)3(.01)6.9$
9 fig. $6.91(.01)15, 1(1)50$

No Δ Becker & Van Orstrand 1924

5 fig. $0(.01)5.5(.05)9.95$ No Δ Vianelli 1959(1026)

   For $x/(e^x-1)$ etc., see section 14.152.

16.102
16.112
[see also p. 738]

16.102 e^{-x}

18 dec.	$0(10^{-6})10^{-4}(10^{-4})2.4999$	No Δ	N.B.S. 1951 z
19 fig.	1(1)100	No Δ	N.B.S. 1951 z
20 dec.	2.5(.001)10	No Δ	N.B.S. 1955 z
7 dec.	0(.001)3(.01)15	No Δ	Becker & Van Orstrand 1924
9 fig.	0(1)50	No Δ	Becker & Van Orstrand 1924
6; 7 dec.	0(.01)5.49; 5.5(.05)9.95	No Δ	Vianelli 1959(1026)

For $(1-e^{-x})/x$ etc., see section 14.152.

16.103 $\log_{10} e^x$

7 dec.	0(.001)3(.01)15	Becker & Van Orstrand 1924
5 dec.	0(.01)5.5(.05)9.95	Vianelli 1959(1026)

16.1041 sinh x and cosh x

15 dec.	0(.1)10	No Δ	B.A. 1, 1931
9 dec.	0(.0001)2(.1)10	No Δ	N.B.S. 1953 y
9 fig.	2(.001)10	No Δ	N.B.S. 1955 y
5 dec.	0(.0001).1(.001)3	v	Becker & Van Orstrand 1924
4 dec.	3(.01)6	v	Becker & Van Orstrand 1924
7 dec.	0(.02)2	No Δ	Fisher & Yates 1957(T.XIV)
5 fig.	0(.01)3(.1)10	No Δ	Vianelli 1959(1032)

16.1043 coth x

For tanh x see section 7.211.

10 dec.	0(.1)10	No Δ	N.B.S. 1943
8 d.; 8 f.	0(.0001).1; 1(.0022)2	δ^2	N.B.S. 1943
5 fig.	0(.0001).1(.001)3(.01)5.3	v	Becker & Van Orstrand 1924
4 dec.	0(.01)3(.1)5.3	No Δ	Vianelli 1959(1032)

16.1051 log sinh x and log cosh x

5 dec.	0(.0001).1(.001)3(.01)6	v	Becker & Van Orstrand 1924
5 dec.	0(.01)3(.1)10	No Δ	Vianelli 1959(1032)

For $\log_e \cosh x$ see section 14.129, p. 481.

16.1052 log tanh x and log coth x

5 dec.   0(.0001).1(.001)3(.01)6   v   Becker & Van Orstrand 1924

5 dec.   0(.01)3(.1)6   No △   Vianelli 1959(1032)

16.11 Natural logarithms

N.B.S. 1953 z, 1958 are the two volumes of a serviceable modern table. The radix table of Spenceley et al. 1952 combines 22-place natural *and* common logs in one handy octavo volume. The volumes cited by Fletcher et al. 1946 as New York W.P.A. 1941 a, b are out of print.

16.111 Natural logarithms of integers

48 dec. 1(1)146(primes)10000   No △   Peters & Stein 1922

23 dec. 1(1)9999   No △   Spenceley et al. 1952

5 dec.   1(1)1000   v   ⎫
                                                  ⎬ Becker & Van Orstrand 1924
5 dec.   1000(primes)10000   No △   ⎭

5 dec.   11(1)1109   No △   Arkin & Colton 1950(100),
                                                  Burington & May 1953(304)

5 dec.   1(1)1000   No △   Vianelli 1959(48)

16.112 Natural logarithms of decimal numbers

23 dec. $1 + m \cdot 10^{-n}$,   No △   Spenceley et al. 1952
         $m = 1(1)9999$, $n = 7, 11$

16 dec. 0(.0001)5   No △   N.B.S. 1953 z

16 dec. 5(.0001)10   No △   N.B.S. 1958

8 dec.   1(.01)10   No △   Kelley 1948(204)

8 dec.   1.0025(.0025)1.15   No △   Vianelli 1959(1433)

5 dec.   0(.0001).1(.001)1(.01)10(.1)99.9   PPMD Fisher & Yates 1957
                                                  (T.XXVI)

5 dec.   .1(.001)1(.01)10(.1)99.9   PPMD   Fisher & Yates 1948, 1953
                                                  (T.XXVI)

5 dec.   1(.01)10(.1)99.9   PPMD   Fisher & Yates 1938, 1943
                                                  (T.XXVI)

5 dec.   .1(.001)1(.01)10(.1)99.9   No △   Czechowski 1957(147)

5 dec.   1(.005)2(.01)10   PPMD   Pearson & H 1954(236)

3; 5 d. .01(.01).99; 1(.01)10.09   No △   Arkin & Colton 1950(100),
                                                  Burington & M 1953(304)

16.112 16.2311					
3 dec.	.1(.01)1(.1)9.9, 19, 19.8, 95, 99		No Δ	Dixon & Massey	1951(332)
3 dec.	.1(.01)1(.1)9.9		No Δ	Dixon & Massey	1957(434)

16.1149 Miscellaneous natural logarithms

Bush & Mosteller 1955, T. B, pp. 344-345, computed by Cleo Youtz and Lotte Bailyn, gives $[\log_e(1-x) - \log_e(1-y)]/(y-x)$ to 3 dec. for $y = .7(.01).99$ and $x = y(.01).99$. Reprinted in Vianelli 1959, Pro. 141, p. 375.

For $\log_e \log_e x$ and $x.\log_e x$, see section 8.312; for $\log_e \frac{1-x}{y}$, see section 14.129.

16.12 Combinations of circular and hyperbolic functions

16.1231 Segmental functions

Fisher & Yates 1953, T. XIV2, table the four functions
$$F_j(x) = \sum_{t=0}^{\infty} x^{4t+j}/(4t+j)!$$
to 7 dec. for $x = 0(.02)1$. These functions may be written in terms of circular and hyperbolic functions:
$2F_0(x) = \cosh x + \cos x$, $2F_1(x) = \sinh x + \sin x$, $2F_2(x) = \cosh x - \cos x$,
$2F_3(x) = \sinh x - \sin x$. No differences are given: the relation
$d^t F_j(x)/dx^t = F_{3t+j}(x)$, where the subscript must be reduced modulo 4, can be used for interpolation.

Fisher & Yates 1957, T. XIV, table inter al. the same functions to 7 dec. for $x = 0(.02)2$.

16.1232 $e^{-x} \sin y$ and $e^{-x} \cos y$

5-6 fig. $x = .1(.1)1(1)7$ $y = 0(1°)90°$ Vianelli 1959(1043)

16.17 Bessel functions

The several editions of Jahnke & Emde contain good collections of 4-figure tables of Bessel functions; the later editions are preferable. Jahnke & Emde 1960, pp. 309-312, give a serviceable hand-list of tables and formularies for Bessel functions.

We call attention to <u>A guide to tables of Bessel functions</u> (Bateman & Archibald 1944).

16.21 Elliptic functions

For some tables of the theta function ϑ_4, see section 9.411.

16.219 Functions related to theta functions

Bush & Mosteller 1955, T. A, pp. 340-343, gives $\sum_{\gamma=0}^{\infty} a^{\frac{1}{2}\gamma(\gamma+1)} \beta^{\gamma}$ and $\sum_{\gamma=0}^{\infty} \gamma a^{\frac{1}{2}\gamma(\gamma+1)} \beta^{\gamma}$ to 4 dec. for $a = \cdot 5(\cdot 02) \cdot 98$ and $\beta = \cdot 5(\cdot 02)1$. Reprinted in Vianelli 1959, Pri. 139-140, pp. 371-374.

Bush & Mosteller 1955, T. D, pp. 348-355, computed by D.G. Hays, gives $\sum_{\gamma=0}^{\Gamma} a^{\gamma} \beta [1 - a^{\gamma}\beta]^{-1}$ and $\sum_{\gamma=0}^{\Gamma} \gamma a^{\gamma} \beta [1 - a^{\gamma}\beta]^{-1}$ to 2 dec. for $\Gamma = 4(4)16$ and $a, \beta = \cdot 5(\cdot 05) \cdot 7(\cdot 02) \cdot 98$. Reprinted in Vianelli 1959, Pri. 143-144, pp. 377-384.

16.23 Interpolation; numerical integration

For the fitting of frequency curves see sections 10.00-10.29; for regression and other curves, section 11.

N.B.S. 1944 and A.J. Thompson 1943 are serviceable modern tables of Lagrange and Everett interpolation coefficients respectively.

16.2311 Lagrange interpolation coefficients

11-point	10 dec.	$-5(.1)+5$	
10-point	10 dec.	$-4(.1)+5$	N.B.S. 1944
9-point	10 dec.	$-4(.1)+4$	
8-point	11 dec.	$.1(.1).9$	Kelley 1938(130), 1948(200)
8-point	10 dec.	$-3(.1)0(.001)1(.1)4$	
7-point	10 dec.	$-3(.1)-1(.001)+1(.1)3$	N.B.S. 1944
6-point	10 dec.	$-2(.01)0(.001)1(.01)3$	
6-point	10 dec.	$0(.01)1$	Kelley 1938(128), 1948(198)
5-point	10 dec.	$-2(.001)+2$	
4-point	10 dec.	$-1(.001)0(.0001)1(.001)2$	N.B.S. 1944
4-point	10 dec.	$0(.001)1$	Kelley 1938(118)
4-point	7 dec.	$0(.001)1$	Kelley 1948(188)

16.2311
16.2369

3-point   9 dec. $-1(.0001)+1$ N.B.S. 1944

3-point   5 dec. $-.5(.0001)+.5$ Kelley 1948(138)

See also K. Pearson 1920 a, b ; for Lagrange coefficients for special unequally spaced pivotal points, see section 16.2319 below.

16.2313   Gregory-Newton interpolation coefficients:

i.e. $\binom{x}{m}$ for fractional x

5 dec. $m = 2(1)6$ $x = 0(.01)1$ No Δ Vianelli 1959(1079)

5 dec. $2(1)5$ $0(.01)1$ Δ Glover 1923(412)

5 dec. $2(1)5$ $0(.01)1$ No Δ Czechowski et al. 1957(130), Vianelli 1959(998)

16.2315   Stirling interpolation coefficients

Glover 1923, pp. 414-415, tables the 2^d-5th Stirling coefficients for even and mean odd differences to 5 dec. for $x = 0(.01)1$ with Δ. Reprinted (without Δ) in Vianelli 1959, Pro. 280, p. 1000.

16.2316   Bessel interpolation coefficients

Glover 1923, pp. 416-417, tables the 2^d-5th Bessel coefficients for odd and mean even differences to 5 dec. for $x = 0(.01)1$ with Δ. Reprinted (without Δ) in Vianelli 1959, Pro. 281, p. 1002.

16.2317   Everett interpolation coefficients

Exact	2^d-8th	$x = 0(.01)1$	δ^{2n}	
Exact	2^d-12th	$0(.1)1$	δ^{2n}	
9-13 dec.	2^d-16th	$0(.2)1$	δ^{2n}	A. J. Thompson 1943
10 dec.	2^d-8th	$0(.001)1$	δ^2	
5 dec.	2^d-6th	$0(.01)1$	Δ	Glover 1923(418)
5 dec.	2^d-6th	$0(.01)1$	No Δ	Vianelli 1959(1002)

16.2319   Lagrange interpolation coefficients
for special unequally spaced pivotal points

Comrie & Hartley 1941 give 7-point Lagrange coefficients, based on the argument $120/\nu$, to 7 dec. They permit interpolation of a function, tabled for $120/\nu = 0(1)6(2)12$, to intermediate arguments $\nu = 16(1)119$.

Richardson 1946, following a suggestion by Simaika 1942, gives 3-point Lagrange coefficients, based on the argument $\log q$, to 3 dec. They permit interpolation of a set of percentage points, tabled for $q = .005, .01, .025, .05, .1$, to intermediate significance levels $q = .007(.0001).011(.0002).019, .0175(.0005).038(.01).075$.

16.234 Numerical differentiation

Glover 1923, pp. 420-427, gives coefficients for computing the first derivative: from Gregory-Newton, Stirling and Bessel differences, through the 5th difference; from Everett differences, through the 6th difference: all to 5 dec. for $x = 0(.01)1$ with Δ. Reprinted (without Δ) in Vianelli 1959, Pri. 279-282, pp. 999, 1001, 1003, 1005.

16.236 Numerical integration

Fisher & Yates 1938 etc., T. XXIV, gives integration coefficients as follows: Newton-Cotes coefficients for $3, 5, 7$ ordinates as exact vulgar fractions. Lagrange coefficients for integrating over p strips using p+3 ordinates, $p = 2(2)6$: as exact vulgar fractions. Coefficients of $\delta^{2m}f(0)$ in $(2nw)^{-1} \int_{-nw}^{+nw} f(x) \, dx$, $m = 0(1)5$, $n = 1(1)5$: exact or 7+ dec. The first 16 coefficients in Gregory's integration formula: exact or 5+ dec.

According to Sheppard 1900 z, p. 270, eq. (43),
$$\int_0^p y(x) \, dx \cong \tfrac{1}{2}y(0) + y(1) + \ldots + y(p-1) + \tfrac{1}{2}y(p) + C_1[y(1) - y(0) - y(p) + y(p-1)]$$
$$+ C_2[y(2) - y(1) - y(p-1) + y(p-2)] + C_3[y(3) - y(2) - y(p-2) + y(p-3)]$$
where the C s are rational functions of p.

Tables for S & B II, T. 43, p. 234, gives C_1, C_2, C_3 to 7 dec. for $p = 6(1)100$.

See also Irwin 1923.

16.2369 Lubbock coefficients

8 dec. $1^{st}-6^{th}$ $s = 2(1)12$ Glover 1923(430)

APPENDICES

1. Supplement to the Descriptive Catalogue 630
2. Contents of Books of Tables 739
3. Material Treated both in this *Guide* and in Fletcher et al. 1946. 786

1.1121
1.2121

APPENDIX I
SUPPLEMENT TO THE DESCRIPTIVE CATALOGUE

__1.1121__ $P(x) - \tfrac{1}{2}$
[see also pp. 4-5]

6 dec.	0(.01)3.99	D^n, $n = 1, 3, 4, 5$	Nemčinov 1946(118)
5 dec.	0(.01)3	No Δ	Ferber 1949(486)
4 dec.	0(.01)4(.05)5	D^7	Burington & May 1953(267)

__1.1131__ $2Q(x)$
[see also pp. 5-6]

4 dec.   0(.01).3(.02)1.5(.05)   No Δ Royo López & Ferrer Martín
         3.5(.1)3.9   1954(201), 1955(28)

__1.1141__ $Q(x)$
[see also pp. 6-7]

4; 5 dec.   0(.01)3.2(.1)4; 3.5   No Δ Siegel 1956(247)

__1.1211__ $2P(2^{\frac{1}{2}}x) - 1$
[see also p. 11]

5; 6, 7 dec.	0(.01)3; 2, 2.6	Δ	Jahnke & Emde 1960(31)
4; 5, 6, 7 dec.	0(.01)3.49; 1.6, 3, 3.1	No Δ	Vianelli 1959(133)

__1.1217__ $P(2^{-\frac{1}{2}}x) - \tfrac{1}{2}$

5 dec.   0(.01)3   No Δ Noether 1956(445)

__1.131__ $Q(x)/Z(x)$
[see also p. 13]

5 dec.   -3(.01)+4(.05)5   No Δ Birnbaum et al. 1955

__1.132__ $Z(x)/Q(x)$
[see also p. 13]

5; 6 fig.   -5(.05)-4(.01)+4(.05)5; +0.31   No Δ Birnbaum et al. 1955

1.139 Miscellaneous ratios
[see also p. 13]

Clark, F. E. 1957

Journal of the American Statistical Association 52, 527-536

When the normal distribution is truncated at a and b, the mean and variance of the truncated distribution are $\mu_{ab} = [Z(b)-Z(a)]/[Q(b)-Q(a)]$, $\sigma^2_{ab} = 1 - \mu^2_{ab} + [bZ(b)-aZ(a)]/[Q(b)-Q(a)]$.

Table 1, p. 529, gives μ_{ab} and σ_{ab} to 4 dec. for b = 0(.25)3, a = -b(.25)+.5.

Jahnke & Emde 1960, p. 32, table $2\pi^{-\frac{1}{2}}e^{x^2}\int_x^\infty e^{-t^2}dt$ to 5 dec. for x = 3(.02)5 with $\frac{1}{2}\Delta$.

1.2111 x(P) for general P
[see also p. 19]

5 dec.   P = .001(.001).999   No Δ Berkson 1955(540),
                                            Vianelli 1959(1261)

4 dec.   Q = .001(.001).03(.002).05(.05).5 No Δ Lindley & Miller 1953(5)

1.2112 Tail area: x(P) for small Q
[see also p. 20]

4 dec.   Q = 10^{-i}, $\frac{1}{2} \cdot 10^{-i}$, i = 2(1)5   Lindley & Miller 1953(5)

1.2113 x(P) for P = m/n

Fisher & Yates 1938, 1943, 1948, 1953 (T. X), 1957 (T. IX 1), give x(1-Q) to 4 dec.: where Q is a vulgar fraction, in lowest terms, not exceeding $\frac{1}{2}$, and with denominator not exceeding 30.

1.2121 x(1-$\frac{1}{2}$p) for general p
[see also p. 20]

6 dec.   p = .01(.01).99   No Δ Czechowski et al. 1957(44),
                                         Vianelli 1959(137)

5 dec.   .01(.01)1   No Δ Royo López & Ferrer Martín
                                         1954(202), 1955(29)

3 dec.   .05, .1(.1)1 No Δ Dixon & Massey 1957(381)

1.2122
1.233

1.2122 Double tail area: $x(1-\tfrac{1}{2}p)$ for small p
[see also p. 20]

5 dec. $p = 10^{-i}$, $i = 3(1)9$ Czechowski et al. 1957(44)

3 dec. $p = 10^{-i}$, $i = 2(1)4$ Dixon & Massey 1957(381)

1.222 Ratio of area to bounding ordinate as a function of area
[see also p. 22]

See also Leverett 1947.

1.2312 Full single-entry tables of working probits
[see also pp. 24-26]

$1/Z$, Z^2/PQ	4 dec.	$Y = 1.5(.01)8.5$
P, Q	5 dec.	
$1/Q - 1/P - (Y-5)/Z$	5; 4, 3 dec.	$Y = 5(.01)8.5$; 7.26, 8.17

Vianelli 1959(138)

$Y - P/Z$	4 dec.	$Y = 1(.01)6.42$
$Y + Q/Z$	4 dec.	$Y = 3.58(.01)9$
$1/Z$, Z^2/PQ	4 dec.	$Y = 1(.01)9$

Vianelli 1959(147)

$Y - P/Z$	3 dec.	$Y = 1.6(.1)6.4$
$Y + Q/Z$	3 dec.	$Y = 3.6(.1)8.4$
$1/Z$; Z^2/PQ	4 fig.; 4 dec.	$Y = 1.6(.1)8.4$

Bliss 1938(197)

1.2313 Other tables of working probits
[see also pp. 26-27]

Berkson 1955, T. 4, pp. [542-543], gives Z^2/PQ and $Z^2(Y-5)/PQ$ to 4 dec. for $P = .001(.001).999$. Berkson 1957b, T. 2, pp. [418-419]; Vianelli 1959, Pro. 146, pp. 389-390; give the same to 5 dec.

Berkson 1957b, T. 1, pp. 414-417, gives Z^2/PQ and $Z^2(Y-5)/PQ$ to 5 dec. for $P = r/n$, $r = 0(1)[\tfrac{1}{2}n]$, $n = 2(1)50$. The figures shown for $r = 0$ were calculated for $r = \tfrac{1}{2}$. Reprinted in Vianelli 1959, Pro. 145, pp. 385-388.

1.232 Tables of probits: $Y = x(P) + 5$
[see also p. 27]

4 dec. $P = .001(.001).999$ No Δ Vianelli 1959(145)

1.233 Tables for special problems of probit analysis

[see also pp. 28-30]

Sampford, M. R. 1952b

Biometrics 8, 307-369

Review by L. A. Aroian (MTAC 8, 21; abridged):

Tables of Z/Q, $(Z^2-QZx)/Q^2$ and $(ZQ-Z^2x+QZx^2)/Q^2$ are given to 5 dec. for $x = -5(.1)-3(.01)+3(.1)5$.

Aitchison, J. and Brown, J. A. C. 1954

Biometrika 41, 338-343

Review by R. L. Anderson (MTAC 10, 48):

A popular model for the quantitative response u to a stimulus of concentration x is: $u = HP(\alpha + \beta x) + \epsilon$, where H is the maximum expected response; $P(\alpha + \beta x)$ is the normal cumulative based on a linear function of x; and ϵ is $N(0, \sigma^2)$, independent of x. The authors consider a model of the form:
$$\ln u = v = \ln H + \ln [P(\alpha + \beta x)] + \epsilon = K + \ln P + \epsilon,$$
i.e., Var (u) is proportional to $[E(u)]^2$. Iterative equations, familiar to probit users, are set up to estimate α, β, and K.

A table of working probits, to facilitate the solution of these equations, is given for initial guessed probits, $Y = a_0 + b_0 x = 1.0(0.1)9.0$. This table contains values of the minimum working probit to 4D; of the auxiliary variable, P/Z to 4D or 5S; and of the weighting factor, Z^2/P^2 to 4D. The working probit for this model is $y = \left(Y - \frac{P}{Z} \ln P\right) + (v - K)\frac{P}{Z}$.

Horn, Henry, J. 1956

Biometrics 12, 311-322

Review by Paul Meier (MTAC 12, 72):

Let $F(x)$ be a cumulative distribution function and $x_1 \leq \ldots \leq x_n$, and let $p_i = r_i/n_i$ be a binomial estimate of $F(x_i)$ based on n_i independent trials. W.R. Thompson 1947 has given a moving-average method for estimating the value of m such that $F(m) = \frac{1}{2}$, i.e. the LD_{50} or ED_{50} in dose-response assays: in this method only the k values of p_i most nearly bracketing $\frac{1}{2}$ are used. Weil 1952 gives tables to facilitate the calculation of m and approximate confidence limits as a function of $r_{h+1}, \ldots, r_{h+k}$ in the special case $n_i = n$ and $\log x_{k+1} = \log x_k + d$. If $k = 4$, $n = 4$ or 5, and the values of x are in geometric progression with ratio $10^{\frac{1}{2}}$ or $10^{1/3}$, then

1.233
1.332

the present tables give directly the estimate of m and approximate confidence limits, both to 3 fig., for the more probable of the n^k possible outcomes $(r_{n+1}, \ldots, r_{n+k})$.

See also Worcester & Wilson 1943.

1.3211 $\quad Z^{(n)}(x) = \dfrac{d^n Z(x)}{dx^n}$

[see also pp. 33-34]

5 dec.	n = 2(1)8	x = 0(.01)4.99	Glover 1923(392)
5 dec.	2(1)8	0(.01)4.49	Vianelli 1959(989)
5 dec.	2(1)4	0(.01)3.99	Nemčinov 1946(118)
4 dec.	1(1)6	0(.01)4(.05)5	Burington & May 1953(267)

1.3249 Miscellaneous tetrachorics

[see also pp. 35-36]

Vianelli 1959, T. 198, p. 1384, gives p_n and q_n, defined on p. 35 of this Guide, to 5 dec. for n = 7(1)24.

1.325 Gram-Charlier series of Type A

[see also pp. 36-37]

Lefèvre, J. 1952

Skandinavisk Aktuarietidskrift 35, 161-187

Review by C. C. Craig (MTAC 8, 83):

This paper contains a table of values of

$$B_n(u) = \int_0^\infty t e^{-tu} d\phi^{(n)}(t)$$

in which $\phi(t)$ is the cumulative normal frequency function in standard units. Values are given for $n = 0, 3, 4, 6$ for $u = 0(.1)3$ to 5D.

1.332 Truncated normal distribution

[see also pp. 39-45]

Cohen, A. C., Jr. and Woodward, John 1953

Biometrics 9, 489-497

Review by Leo Katz (MTAC 9, 29):

A normal variable, x, with frequency function,

$$\phi(x) = (2\pi)^{-\frac{1}{2}}\sigma^{-1} \exp{(x - m)^2/2\sigma^2},$$

for all real x, truncated (below) at x_0', gives the truncated variable, x', with frequency function, $f(x') = \phi(x')/I_0(x')$, $x_0' \leq x'$, where $I_0(x') = \int_{x_0'}^{\infty} \phi(x)dx$. It is required to estimate m and σ on the basis of a sample of n observations on $x = x' - x_0'$.

Pearson and Lee[1] gave estimates based on the first two observed moments. FISHER[2] showed their estimates to be maximum likelihood estimates. The authors estimate $\xi = \dfrac{x_0' - m}{\sigma}$ and σ by the relations (equivalent to the earlier ones)

(1) $$\frac{n\sum x^2}{2(\sum x)^2} = \frac{1}{2}\left[\frac{1}{z - \xi}\right]\left[\frac{1}{z - \xi} - \xi\right] = g(\xi),$$

and

(2) $$\sigma \equiv \frac{\sum x}{n}\left[\frac{1}{z - \xi}\right] = \frac{\sum x}{n} h(\xi),$$

where $z = \phi(\xi)/I_0(\xi)$. To facilitate computations they give tables of $h(\xi)$ and $g(\xi)$ to 8D except for the largest values of ξ (where 7D and 6D are given) for $\xi = -4.(.1)-2.5(.01).5(.1)3$. The authors suggest using (1) to estimate $\hat{\xi}$ and (2) with $\hat{\xi}$ from (1) to estimate $\hat{\sigma}$.

The variances of the estimates and the correlation coefficient between the estimates are given by

$$\mathrm{var}(\hat{\xi}) = \frac{\sigma^2}{n} \frac{1 - z(z - \xi)}{[1 - z(z - \xi)][2 - \xi(z - \xi)] - [z - \xi]^2} = \frac{\sigma^2}{n} W'(\xi),$$

$$\mathrm{var}(\hat{\sigma}) = \frac{1}{n} \frac{2 - \xi(z - \xi)}{[1 - z(z - \xi)][2 - \xi(z - \xi)] - [z - \xi]^2} = \frac{1}{n} w'(\xi),$$

and

$$\rho(\hat{\xi}, \hat{\sigma}) = \frac{z - \xi}{\sqrt{[1 - z(z - \xi)][2 - \xi(z - \xi)]}}.$$

Tables of $W'(\xi)$ and $w'(\xi)$ are given to 6D and of $\rho(\xi, \sigma)$ to 4D for $\xi = -3.(.1)2$.

[1] K. Pearson & Lee 1908   [2] R.A. Fisher 1931

1.332
1.42

Clark, F. E. 1957

See abstract on p. 631

See also Francis 1946.

1.41 Median and other quantiles
[see also pp. 50-53]

Ogawa, J. 1951

Osaka Mathematical Journal 3, 175-213

Table 6.3, p. 197, gives the large-sample best linear unbiased estimates of the mean of a normal distribution based on n quantiles, n = 2(1)10, with quantiles and weights to 3 dec.; the weights have not been forced to sum to 1. Table 6.2, p. 197, gives the efficiency of these estimates to 4 dec.

Table 6.6, p. 200, gives the large-sample best linear unbiased estimates of the standard deviation of a normal distribution based on n quantiles, n = 2, 4, 6; and based on n quantiles and the known population mean, n = 3, 5; with quantiles and weights to 3 dec. Table 6.5, p. 200, gives the efficiency of these estimates to 3 dec.

Cadwell, J. H. 1952 z

Biometrika 39, 207-211

Normal approximation to distribution of quantiles in large normal samples. Tables 1 and 2, p. 209, illustrate approximate moment fit for distribution of median in odd and even samples respectively.

David, F. N. and Johnson, N. L. 1956

Journal of the Royal Statistical Society, Series B 18, 1-31.

Review by I. R. Savage (MTAC 12,68):

A survey of test procedures based on medians, quartiles, interquartile ranges, etc., is made. Many small tables of approximate percentage points, computed by the method of moments, are presented. In particular, to test the hypothesis of equality of the means of two normal universes with equal but unknown variances, David & Johnson propose the test criterion $T = (X_{M1}-X_{M2})(N_1+N_2)/[N_1(X_{U1}-X_{L1}) + N_2(X_{U2}-X_{L2})]$ where N_j, X_{Mj}, X_{Uj}, X_{Lj} are the sample size, the median, and the upper and lower quartiles in samples from the jth universe. 1%, 2½% and 5% points for T are given to 3 dec. for N_1, N_2 = 11(4)51.

Dixon, W. J. 1957

Annals of Mathematical Statistics 28, 806-809

Table 1, p. 807, gives, inter al., the variance and efficiency of the median to 3 fig. in samples of size 2(1)20.

1.42 Distribution of straggler and straggler deviate; powers of normal tail area

[see also pp. 54-62]

K.R. Nair 1948 b, T. 2, described on p. 59 above, is reprinted in Vianelli 1959, Prontuario 81, p. 228.

Ruben, Harold 1954

Biometrika 41, 200-227

Ruben's method rests on the evaluation of the integral

$$\phi(s; \alpha; \beta, \gamma) = \int_{-\infty}^{+\infty} x^s Z^\alpha P^\beta Q^\gamma dx.$$

The integral for integer s can be reduced to the integral for s = 0 by a somewhat complicated recursion; the necessary coefficients are tabled on pp. 205-208 for s = 1(1)10.

Table 1, pp. 222-223, gives the relative surface contents of regular hyperspherical simplices with dimensionality n and primary bounding angles $\cos^{-1} 1/x$, for x = 2(1)12 and n = 0(1)[51-x]. At least 5 fig., but not more than 10 dec., are tabled.

Table 2, pp. 224-225, gives the first 10 moments of the largest member of a normal sample of size 1(1)50 to 9 fig. Table 3, p. 226, gives μ_2 of such largest member to 8 fig. and μ_3, μ_4, β_1, $\beta_2 - 3$, $\mu_2^{\frac{1}{2}}$ to 7 fig.

David, H. A. 1956

Biometrika 43, 85-91

Review by L. A. Aroian (MTAC 12, 75):

The author is concerned with statistical applications of the following theorem in probability. Let $P_{ij...k}$ denote the joint probability of $r (\leq n)$ events A_i, A_j, $\cdots$, A_k; and S_r the sum of the $\binom{n}{r}$ P's with r different subscripts; then the probability $p_{m,n}$ of the realization of at least m events out of n is given by

$$p_{m,n} = \sum_{l=0}^{n-m} (-1)^l \binom{m+l-1}{m-1} S_{m+l}.$$

This result is applied to the distribution of the extreme deviate from the sample mean, and the maximum F-ratio s^2_{max}/s^2. The author's table 1 gives the value of $P(\lambda)$, the probability to 3D of rejecting the largest of n observations by the use of $x_{max} - \bar{x}$ at the 5 per cent level of significance, when all the observations are normal with unit variance, $n - 1$ have mean μ and one has mean $\mu + \lambda$, for $\lambda = 1(1)4$, $n = 3(1)10, 12, 15, 20, 25$.

1.42
1.44

In table 2 are given the probabilities to 4D for various designs of size $l \times l$, (randomized blocks, Latin squares, and Graeco-Latin squares) with which the ordered F-ratio $F_{(t)}$ exceeds the upper percentage point F_α of the corresponding random F-ratio where $l = 3, t = 1, 2, 3; l = 5, 7, 9, \infty; t = 1(1)4, \alpha = .05$, and 0.01.

In table 3 are given the upper 100α percentage points, to 3S usually, of the maximum F-ratio followed by the corresponding $100\alpha/n$ percentage points of F for the n-factor designs of table 2 where $l = 3, 5, 6(1)9, n = 2, 3, 4, \alpha = 0.05$ and 0.01.

Teichroew, D. 1956

Annals of Mathematical Statistics 27, 410-426

Table 3, p. 422, gives $\psi(a) = \int_{-\infty}^{+\infty} [P(x)Q(x)]^a dx$ to 25 dec. for $a = 1(1)10$. Table 4, pp. 423-426, gives $\psi(a,b) = \int_{-\infty}^{+\infty} \int_{-\infty}^{x} [P(x)]^a [Q(\xi)]^b d\xi\, dx$ to 25 dec. for $a, b = 1(1)9$; missing values may be supplied from $\psi(a,b) = \psi(b,a)$.

1.43 Ordered normal deviates; normal ranking scores; moments of order statistics; order statistic tests
[see also pp. 62-68]

Anis, A. A. 1955

Biometrika 42, 96-101

Review by J. L. Hodges, Jr. (MTAC 10, 233):

Let $x_1, x_2, \ldots, x_n$ be independent standard normal variates, let $S_r = x_1 + x_2 + \ldots + x_r$ and let $U_n = \max[S_1, S_2, \ldots, S_n]$. It was shown by Anis & Lloyd 1953 that $E(U_n) = (2\pi)^{-\frac{1}{2}} \sum_{r=1}^{n-1} r^{-\frac{1}{2}}$ which is known to be $(2\pi)^{-\frac{1}{2}}[2(n-1)^{\frac{1}{2}} + \zeta(\frac{1}{2})] + O(n^{-\frac{1}{2}})$. In the present paper it is shown that $E(U_n^2) = \frac{1}{2}(n+1) + (2\pi)^{-\frac{1}{2}} \sum_{r=1}^{n-2} \sum_{s=1}^{r} [s(r-s+1)]^{-\frac{1}{2}}$. There is a table of $E(U_n)$, $E(U_n^2)$ and $\sigma(U_n)$ to 4 dec. for $n = 3(1)25$.

Anis approximates the double sum by a double integral in order to write $E(U_n^2)$ $n - \pi^{-\frac{1}{2}} n^{\frac{1}{2}} (2+2^{\frac{1}{2}})$ to terms of order $n^{\frac{1}{2}}$. This formula, together with that for $E(U_n)$ mentioned above, implies

$$\sigma(U_n) \sim [(\pi-2)n/\pi]^{\frac{1}{2}} - [\pi(\pi-2)]^{-\frac{1}{2}}[1 + 2^{-\frac{1}{2}} + \zeta(\frac{1}{2})]$$

which does not agree well with the tabled values. Using another method,

Hodges finds $\sum_{i=1}^{n-1}[(n-i)/i]^{\frac{1}{2}} \sim \frac{1}{2}\pi n + n^{\frac{1}{2}}\zeta(\frac{1}{2})$ with the aid of which the double sum may be evaluated. We get $E(U_n^2) \sim n + 2n^{\frac{1}{2}}\zeta(\frac{1}{2})/\pi$ and hence find constant term zero in the expression for $\sigma(U_n)$. This is in agreement with the tabled values, which exceed $.60281\,n^{\frac{1}{2}}$ by about $.478\,n^{-\frac{1}{2}}$.

Teichroew, D. 1956

Annals of Mathematical Statistics 27, 410-426

Let x_i be the ith order statistic, counted from the right end, of a sample of N from a unit normal distribution. Table 1, p. 416, gives $E(x_i)$ to 10 dec. for $N = 1(1)20$; $i = 1(1)N$. Table 2, pp. 417-422, gives $E(x_i x_j)$ to 10 dec. for $N = 1(1)20$; $i, j = 1(1)N$. Missing values may be supplied from $E(x_i) = -E(x_{N-i+1})$; $E(x_i x_j) = E(x_j x_i) = E(x_{N-i+1} x_{N-j+1})$.

Reprinted in Vianelli 1959, <u>Prontuari</u> 103-104, pp. 280-286.

Sarhan, A. E. and Greenberg, B. G. 1956

Annals of Mathematical Statistics 27, 427-451

Table 1, pp. 428-433, gives the variances and covariances of order statistics in samples of sizes up to 20 from a standard normal distribution to 10 dec. They are derived from the moments and product-moments of Teichroew 1956; missing values are to be supplied by symmetry, as outlined in the previous abstract.

Reprinted in Vianelli 1959, <u>Prontuario</u> 105, pp. 287-290.

See also M.G. Kendall 1954.

1.44 Studentized distributions involving order statistics
[see also pp. 68-69]

Dixon, W. J. 1951

Annals of Mathematical Statistics 22, 68-76

Review by J. E. Walsh (MTAC 6, 22):

Let $x_1, \ldots, x_n$ represent the values of a sample of size n from a normal population, arranged in increasing order of magnitude. Set $r_{ij} = (x_n - x_{n-i})/(x_n - x_{j+1})$. The six tables of this paper present upper $100\alpha\%$ points of r_{ij} for $i = 1, 2$; $j = 0(1)2$; $n = [2+i+j](1)30$; $\alpha = .005, .01, .02, .05, .1(.1).9, .95$.

1.44
1.51

Nair, K. R.   1952

Biometrika 39, 189-191

"Tables of percentage points of the 'Studentized' extreme deviate from the sample mean"

For definitions see the abstract of Nair 1948 b on p. 69 above.

Table 1 A, p. 190, gives lower percentage points of t' to 2 dec. for $p = .001, .005, .01, .025, .1$; $120/\nu = 0(4)12$.

Table 1 B, pp. 190-191, gives upper percentage points of t' to 2 dec. for $1-p = .001, .005, .01, .025, .05, .1$; $\begin{cases} \nu = 10(1)20 \\ 120/\nu = 0(1)6 \end{cases}$.

Pillai, K. C. S. and Ramachandran, K. V.   1954

Annals of Mathematical Statistics 25, 565-572

Review by W. J. Dixon (MTAC 10, 43):

Tables to 2D are given for (1), the 95th percentile of distribution of $q_n = x_n/x$; (2), the 5th, and (3), the 95th percentiles of the distribution of $u_n = |x_n/s|$ where x_n is the largest of a sample of n observations from a normal population with zero mean and unit variance and s is independently distributed with ν degrees of freedom, for (1): $n = 1(1)8$: $\nu = 3(1)10(2)20, 24, 30, 40, 60, 120, \infty$; (2): $n = 1(1)10$: $\nu = 1(1)5(5)20, 24, 30, 40, 60, 120, \infty$; (3): $n = 1(1)8$: $\nu = 5(5)20, 24, 30, 40, 60, 120, \infty$, respectively.

A table used in the derivation of the above tables is given to 8S for coefficients $a_i^{(k)}$ which are defined by

$$\left(\int_{-\infty}^{x} \frac{e^{-t^2/2}}{\sqrt{2\pi}} dt\right)^k = e^{-kx^2/6} \left(\sum_{i=0}^{\infty} a_i^{(k)} x^i\right)$$

for $i = 0(1)30$; $k = 1(1)7$.

Halperin, Max, Greenhouse, Samuel W., Cornfield, Jerome and Zalokar, Julia

Journal of the American Statistical Association 50, 185-195   1955

Let $x_1, \ldots, x_k$ be independent unit normal deviates. Let s^2 be distributed as χ^2/ν and be independent of the x_i. Define $d = s^{-1} \max_i |x_i - \bar{x}|$.

Tables 1-2, pp. 187-188, give (fairly close) upper and lower bounds to the upper 5% and 1% points of d for $k = 3(1)10(5)20(10)40, 60$; $\nu = 3(1)10(5)20$, $120/\nu = 0(1)4(2)8$; $\nu \geq k$.

Reprinted in Vianelli 1959, Tavola LIV, p. 1197.

David, H. A. 1956*b*

Biometrika 43, 449-451

Review by C. A. Bennett (MTAC 12, 68):

The upper $100\alpha\%$ points of the extreme studentized deviate in a random sample of n observations, based on an independent estimate of variance with ν d.f., are given to 2; 1 dec. for α = .1, .05, .025, .01, .005; .001 : all for n = 3(1)12 and ν = 10(1)20, $120/\nu$ = 0(1)6. Accuracy to one unit in the last place recorded is claimed. This is a revision and slight extension of the original tables by K.R. Nair 1952.

Vianelli, S. 1959

Prontuari per calcoli statistici

Prontuario 82, pp. 229-231, combines T.6A of Nair 1948b and T.1B of Nair 1952, and thus gives lower percentage points of t' (defined on p. 69 above) to 2 dec. for p = .01, .05; $120/\nu$ = 0(4)12: and upper percentage points of t' to 2 dec. for 1-p = .001, .005, .01, .025, .05, .1; ν = 10(1)20, $120/\nu$ = 0(1)6.

See also Smirnov 1941.

1.51 Measures of location [based on order statistics]
[see also pp. 70-71]

Gupta, A. K. 1952

Biometrika 39, 260-273

Let $x_1 \leq \ldots \leq x_k$ be the k smallest order statistics in a sample of n from a normal population. Write the minimum-variance unbiased linear estimate of the mean as $\sum \beta_i x_i$. Table 3, p. 267, gives the β_i to 5 dec. for n = 3(1)10 and k = 2(1)[n-1]. Reprinted in Vianelli 1959, Prontuario 73, p. 219. [N.B. Vianelli omits the minus signs affecting some values of β_1 and β_2; these can be restored by noting that $\sum \beta_i = 1$.]

Table 5, p. 269, gives $\sigma^{-2}\text{Var}(\sum \beta_i x_i)$ to 5 dec. for n and k as in T. 3.

Sarhan, A. E. and Greenberg, B. G. 1956

Annals of Mathematical Statistics 27, 427-451

Table 2, pp. 434-440, gives inter al. the coefficients of the minimum-variance unbiased linear estimate of the mean of a normal distribution, based on any m consecutive order statistics from a sample of n,

1.51
1.531

$2 < \underline{m} < \underline{n} < 10$, to 8 dec. Table 3, pp. 442-444, gives inter al. σ^{-2} times the variance of such estimates to 8 dec. Table 4, pp. 446-447, gives inter al. the efficiencies of such estimates to 4 dec.

A.K. Gupta 1952 pointed out that the mean and standard deviation of a normal distribution could be estimated from a censored sample by regressing the available observations on the expected values of corresponding normal order statistics. Table 5, p. 448, gives inter al. the variances of such estimates of the mean to 8 dec. and their efficiencies, relative to the estimates of T. 2, to 4 dec. for n = 10 ; the lowest efficiency tabled is 89.83%.

Tables 2-4 are reprinted in Vianelli 1959, <u>Prontuari</u> 106-108, pp. 291-297.

Dixon, W. J. 1957

<u>Annals</u> <u>of</u> <u>Mathematical</u> <u>Statistics</u> 28, 806-809

Table 1, p. 807, gives inter al. the variance and efficiency of the mean of two symmetrically placed order statistics, approximating the 27th and 73^{d} percentiles; the variance and efficiency of the mean of all order statistics except the first and last; the ratio of this efficiency to the efficiency of the Best Linear Systematic Statistic of A.K. Gupta 1952; all to 3 fig. in samples of size 2(1)20.

Sarhan, A. E. and Greenberg, B. G. 1958

<u>Annals</u> <u>of</u> <u>Mathematical</u> <u>Statistics</u> 29, 79-105.

Table 1, pp. 80-87, continues T. 2 of Sarhan & Greenberg 1956 for $2 < \underline{m} < \underline{n}$, $11 < \underline{n} < 15$, to 4 dec. Table 2, pp. 88-90, and T. 3, pp. 92-96, continue TT. 3-4 of Sarhan & Greenberg 1956 for $11 < \underline{n} < 15$ to 4 dec. Table 4, pp. 98-99, continues T. 5 of Sarhan & Greenberg 1956, giving variances to 8 dec. and efficiencies to 4 dec. for n = 12, 15 ; the lowest efficiencies tabled are 84.66% for the mean and 86.75% for the standard deviation.

1.52 The mean deviation from mean and median and related measures
[see also pp. 71-75]

Cadwell, J. H. 1954 y

Biometrika 41, 12-18

Review by E. J. Gumbel (MTAC 9, 31):

Cadwell wants to obtain properties of the mean deviation m that are analogous to the well-known properties of the standard deviation σ of a normal distribution. To this end, he approximates the distribution of the quotient m/σ by the χ^2 distribution: he matches the first two moments of the two distributions, with a small discrepancy for the third moment.

Let $\bar{m}(k,n)$ be the average of k mean deviations, each of a sample of size n from a normal population with standard deviation σ. Then Cadwell shows that $c[\bar{m}(k,n)/\sigma]^{1.8}$ has approximately the χ^2 distribution with ν d.f. Table 1 gives c to 4 fig., ν to 1 dec., $E[\bar{m}(k,n)/\sigma]$ to 4 dec., $Var[\bar{m}(k,n)/\sigma]$ to 5 dec.; all for $k = 1(1)10$ and $n = 4(1)10$. Table 2 gives the same for $k = 1(1)5$ and $n = 10(5)50$. Table 3 gives the lower and upper $2\frac{1}{2}\%$ and 5% points of $\bar{m}(k,n)$ to 3 dec. for $k = 1(1)10$ and $n = 4(1)10$: the values are exact for $k = 1$, and for other k the error will not exceed .003. For $k > 10$ a normal approximation can be used.

See also Zitek 1954.

1.531 Variate differences
[see also pp. 75-80]

Ferber, R. 1949

Statistical techniques in market research

Table 18, p. 525, abridged from Hart 1942, gives upper and lower 1% and 5% points of δ^2/s'^2 to 4 dec. for $n = 4(1)60$. Reprinted in Vianelli 1959, Prontuario 117, p. 319.

David, H. A. 1955

Biometrika 42, 512-515

Review by F. E. Grubbs (MTAC 12, 79):

This publication deals with some of the statistical properties of moving ranges, moving maxima, and moving minima which are of course based on overlapping data. In particular, an expression is obtained for the correlation between two ranges in overlapping samples in terms of means, variances, and covariances of order statistics. With the derived result and the hypothesis of a normal population, the efficiency of mean moving ranges is studied.

1.531
1.541

Letting $x_{(1)}, x_{(2)}, \cdots, x_{(N)}$ represent the observations ordered in time, we may say that $x_{(i)}, x_{(i+1)}, \cdots, x_{(i+n-1)}$ is the i-th moving sample of size n and designate $m_i(n), m_i'(n)$, and $w_i(n)$ as the corresponding maximum, minimum, and sample or group moving range respectively. Writing the order statistics of a sample of n as $x_1 \geq x_2 \geq x_3 \cdots \geq x_n$, values are given to 4D by David for the correlation between $m_1(n)$ and $m_{1+d}(n)$ for $n = 2(1)5$ and $d \leq (n-1)$ and for the correlation between two ranges w_1 and $w_1 + d$ in overlapping samples in terms of known moments and cross-product expected values of the x_i. Now the first mean moving range is defined as

$$\bar{\omega}_i = \sum_{i=1}^{N-n+1} \frac{\omega_i}{N-n+1}$$

and a mean moving range $\bar{\omega}_r$ can be constructed on every r-th sample. David then studies the efficiency of $\bar{\omega}_r$, i.e.,

$$\text{Eff} = \text{var } S'/\text{var } \bar{\omega}_r$$

where S' is the usual (total sample) estimate of population standard deviation

$$S' = \frac{\Gamma(N-1/2)}{\sqrt{2}\,\Gamma(N/2)} \sqrt{\sum_{i=1}^{N} (x_i - \bar{x})^2},$$

which is tabulated to 3D for $N = 10(10)60, 50, 100, 200, \infty$, and for $n = 2(1)5$ for $\bar{\omega}_1$; for $n = 4$ for $\bar{\omega}_2$; for $n = 5$ for $\bar{\omega}$, the ordinary mean range estimation; and ω^* the most efficient (non-overlapping) weighted mean range estimator [1].

In the absence of trends, $\bar{\omega}_1(5)$ is found to be considerably more efficient than $\bar{\omega}_1(2)$, the ordinary mean successive difference. The statistic $\bar{\omega}_1(2)$ is less influenced by trends in observed data, however, than the statistics $\bar{\omega}_1(n)$ based on larger n. Compared with $\bar{\omega}$ for groups of 5 observations each and ω^*, the moving range estimators, $\bar{\omega}$, are seen to gain in efficiency as the total sample size N increases. The relatively low efficiency of the mean moving range for small total sample size N and small group size n is due no doubt to the lack of weighting.

Moore, P. G.   1955

<u>Journal of the American Statistical Association</u> 50, 434-456

Review by R. L. Anderson (MTAC 10, 237):

Three estimators of the parent population variance (σ^2), based on samples of n, are compared: the usual sample variance (s^2) and two definitions based on successive differences,

$$\delta^2 = \sum_{i=1}^{n-1} (x_{i+1} - x_i)^2/(n-1),$$

$$\eta^2 = \sum_{i=1}^{m} (x_{2i} - x_{2i-1})^2/2m,$$

where η^2 is only used for $n = 2m$.

---

[1] Grubbs & Weaver 1947

The first four moments of each estimator are derived. The relative efficiencies as estimators of σ^2 (for fixed mean) are tabulated to 2D for $n = 5(5)25$, ∞ and various values of the kurtosis parameter, $\beta_2 (= \mu_4/\sigma^4)$.

The bias of the estimators is studied when the population mean shifts; only s^2 is seriously biased. Also the variance of s^2 increases materially, the amount being inversely proportional to β_2.

Since δ^2 has good properties for shifting populations, its distribution is studied. This distribution is best approximated by a Type VI Pearson curve but is close enough to a Type III (χ^2) for most purposes. Four types of parent populations are considered to represent different values of β_2 and the skewness parameter $\beta_1 (= \mu_3/\sigma^3)$: Normal ($\beta_1 = 0, \beta_2 = 3$), Rectangular ($\beta_1 = 0, \beta_2 = 1.8$), Double Exponential ($\beta_1 = 0, \beta_2 = 6$), and Type III ($\beta_1 = 1, \beta_2 = 6; \beta_1 = \frac{2}{3}, \beta_2 = 4$). Exact and Type III and Type VI approximate distributions of δ^2/σ^2 are considered for $n = 10(10)50, 75, 100$. The following tabulations are given:

(1) Values of β_1 and β_2 to 4D
(2) Prob $[(\delta^2/\sigma^2) < \xi]$ for selected ξ to 5D for normal parent only
(3) Upper 5% points for δ^2/σ^2 to 2D for normal parent only.

Values of β_1 and β_2 of s^2 and log s^2 are also tabulated for normal (to 2D) and Type III parent populations (to 4D). The log transformation is also advocated for δ^2.

Four examples are given to illustrate the usefulness of δ^2 as an estimator of σ^2.

1.541 The range
[see also pp. 82-88]

Burington, R. S. and May, D. C. 1953

Probability and statistics

Table 13.86.1, p. 167, based on E.S. Pearson 1932, 1941, gives d_n to 3; 2; 1 dec. for $n = 2(1)20$; 30, 50(25)100; 150, 200 : $\sigma_{R|n}$ to 3 dec. for $n = 2(1)20$; the upper and lower .1%, $\frac{1}{2}$%, 1%, $2\frac{1}{2}$%, 5% and 10% points of R to 2 dec. for $n = 2(1)20$.

Cadwell, J. H. 1954 z

Annals of Mathematical Statistics 25, 803-806

Review by L.A. Aroian (MTAC 9, 219):

Cadwell gives an asymptotic expression for the probability integral of range for samples from a symmetrical unimodal population. He investigates its accuracy for a normal parent population for sample sizes of 20 to 100. Over this range errors are small and by use of a given table of corrections the probability integral can be found with a maximum error of 10^{-4}. The

1.541
1.544

upper and lower .1%, ½%, 1%, 5%, 10%, 25%, 50% points are given to 3 fig. for sample sizes of 20(20)100 of the range w from a normal population with unit standard deviation and also the mean value of w for the same sample sizes.

Noether, G. E. 1955

Journal of the American Statistical Association 50, 1040-1055

Review by C. A. Bennett (MTAC 11, 27):

This expository article summarizes methods which use the range instead of the standard deviation: (1) to obtain confidence intervals for the mean μ of a normal population, the difference of two means, and (2) to estimate the standard deviation σ of a normal population. Table 1 gives for $N = 2(1)100$ the appropriate subsample size, the necessary factors to 3 fig. to obtain an unbiased estimate of σ and to 2 dec. to find 90% and 98% upper and lower confidence limits. This table is based on the optimum procedure given by Grubbs & Weaver 1947. It is pointed out that the loss in efficiency due to the use of equal subsamples is slight compared to the gain in computational ease. For use in obtaining confidence intervals for the mean and difference of means, TT. 1 and 2 of Jackson & Ross 1955, which were derived from earlier tables by Lord, are recommended.

Dixon, W. J. 1957

Annals of Mathematical Statistics 28, 806-809

Table 2, p. 808, gives inter al. d_n^{-1}, $d_n^{-2}\sigma_{R|n}^2$ and the efficiency of the range, all to 3 fig. in samples of size $2(1)20$.

Moore, P. G. 1957

Biometrika 44, 482-489

"The two-sample t-test based on range"

H.A. David 1951 has shown that the best linear estimate of σ, based on the ranges w_1, w_2 of two samples of sizes n_1, n_2, is $\left[\sigma_{R|n_2}^2 d_{n_1} w_1 + \sigma_{R|n_1}^2 d_{n_2} w_2\right] \Big/ \left[\sigma_{R|n_2}^2 d_{n_1}^2 + \sigma_{R|n_1}^2 d_{n_2}^2\right]$. Table 2, p. 489, gives $\sigma_{R|n_1}^2 d_{n_2} \Big/ \sigma_{R|n_2}^2 d_{n_1}$ and $\left[\sigma_{R|n_2}^2 d_{n_1}^2 + \sigma_{R|n_1}^2 d_{n_2}^2\right] \Big/ \sigma_{R|n_2}^2 d_{n_1}^2$ to 3 dec. for $n_1 = 2(1)19$, $n_2 = [n_1+1](1)20$.

Reprinted in Vianelli 1959, **Prontuario** 88, p. 251.

Vianelli, S. 1959

Prontuari per calcoli statistici

Prontuario 75, p. 221, based on Hald 1952, gives $R(p,n)$ to 2 dec. for $p = .0005, .001, .005, .01, .025, .05, .1(.1).5$; $1-p$ = ditto; $n = 2(1)20$.

Prontuario 76, p. 221, gives inter al. d_n to 5 dec.; $\sigma_{R|n}$ to 4 dec.; $\sigma^2_{R|n}$ to 5 dec.; β_1 to 4 dec.; β_2 to 3 dec.; and $d_n^{-1}\sigma_{R|n}$ to 3 dec.; all for $n = 2(1)20$. Vianelli credits this table to Hald 1952: apparently, however, $d_n^{-1}\sigma_{R|n}$ is from Hald, and the other functions from Hartley & Pearson 1951.

See also Moriguti 1954; Sibuya & Toda 1957.

1.542 The midrange
[see also pp. 88-90]

Dixon, W. J. 1957

Annals of Mathematical Statistics 28, 806-809

Table 1, p. 807, gives inter al. the variance and efficiency of the midrange to 3 dec. in samples of size $2(1)20$.

1.544 Measures of dispersion based on selected order statistics
[see also pp. 92-95]

Gupta, A. K. 1952

Biometrika 39, 260-273

Let $x_1 \leq \ldots \leq x_k$ be the k smallest order statistics in a sample of n from a normal population. Write the minimum-variance unbiased linear estimate of the standard deviation as $\sum \gamma_i x_i$. Table 4, p. 268, gives the γ_i to 5 dec. for $n = 3(1)10$ and $k = 2(1)[n-1]$. Reprinted in Vianelli 1959, Prontuario 74, p. 220.

Table 6, p. 269, gives $\sigma^{-2} \text{Var}(\sum \gamma_i x_i)$ to 5 dec. for n and k as in T. 4.

Sarhan, A. E. 1955

Annals of Mathematical Statistics 26, 576-592

Review by J. E. Walsh (MTAC 12, 80):

The data considered consist of singly and doubly censored samples from specified probability density functions. The best linear estimates of the population mean and standard deviation using sample order statistics, their variances and their efficiencies (compared to best [linear] estimates using uncensored samples) are derived for rectangular and exponential distributions and arbitrary sample size n. The same results are given for sample

1.544
1.5511

sizes up to 5 for other types of distributions. The symmetric distributions considered are U-shaped, rectangular, parabolic, triangular, normal, and double [two-tailed] exponential. Let the smallest r_1 and the largest r_2 sample elements be missing. Table 1 contains coefficients for the best linear estimates of the mean and standard deviation for these symmetric distributions in the singly censored case. Table 2 contains the variances and efficiencies for the estimates of T. 1. For TT. 1, 2, $r_1 = 0$, $r_2 = 1(1)[n-2]$, $n = 3(1)5$. Tables 3 and 4 contain results of the same nature as TT. 1, 2, for the case of doubly censored samples. In TT. 3, 4, $r_1 = 1$, $r_2 = 1(1)[n-1]$, $n = 3, 4$. Results are also computed for a third-degree-polynomial skew distribution. Table 5 contains coefficients for the best linear estimates of the mean and standard deviation of this skew distribution for singly and doubly censored samples. Table 6 contains the variances and efficiencies of the estimates of T. 5. Table 7 contains a comparison of the variances of the estimates for the skew distribution with the variances of the corresponding estimates for the exponential distribution. For TT. 5-7, $n = 3(1)5$, and various combinations of r_1, r_2 are considered. With the exception of efficiencies, the results in all tables are expressed to 7 dec. Efficiencies are expressed to .01%.

Sarhan, A. E. and Greenberg, B. G. 1956

Annals of Mathematical Statistics 27, 427-451

Table 2, pp. 434-440, gives inter al. the coefficients of the minimum-variance unbiased linear estimate of the standard deviation of a normal distribution, based on any m consecutive order statistics from a sample of n, $2 \leq m \leq n \leq 10$, to 8 dec. Table 3, pp. 442-444, gives inter al. σ^{-2} times the variances of such estimates, and σ^{-2} times the covariances of such estimates with corresponding estimates of the mean, to 8 dec. Table 4, pp. 446-447, gives inter al. the efficiencies of such estimates.

A.K. Gupta 1952 pointed out that the mean and standard deviation of a normal distribution could be estimated from a censored sample by regressing the available observations on the expected values of the corresponding normal order statistics. Table 5, p. 448, gives inter al. the variances of such estimates of the standard deviation to 8 dec. and their efficiencies, relative to the estimates of T. 2, to 4 dec. for $n = 10$; the lowest efficiency tabled is 90.17%.

Tables 2-4 are reprinted in Vianelli 1959, Prontuari 106-108, pp. 291-297.

Dixon, W. J. 1957

<u>Annals of Mathematical Statistics</u> 28, 806-809

Table 2, p. 808, gives inter al. the coefficients used in certain simple linear estimates of the standard deviation based on 4, 6 or 8 order statistics to 4 fig.; the variance and efficiency of these estimates to 3 fig.; the ratio of this efficiency to the efficiency of the Best Linear Systematic Statistic of Sarhan & Greenberg 1956 to 3 dec.; all for samples of size 6(1)20.

Sarhan & Greenberg 1958 continue the tables of Sarhan & Greenberg 1956 up to n = 15: see abstract in section 1.51, p. △△△.

See also Cadwell 1953 z.

1.5511 The studentized range
[see also pp. 95-99]

David, H. A., Hartley, H. O. and Pearson, E. S. 1954

<u>Biometrika</u> 41, 482-493

The statistic tabled in this paper is not <u>the</u> studentized range; see review by C. A. Bennett on p. △△△ below.

Duncan, D. B. 1955 a

<u>Biometrics</u> 11, 1-42

Review by Frank Massey (MTAC 10, 45; abridged):

Let $q(p, n_2) = w/s$ where w is the range of p independent normal variables having the same mean and unit standard deviation, and $n_2 s^2$ is distributed independently of w as chi square with n_2 d.f.

Table II labeled "Significant studentized ranged for a 5% level new multiple range test" lists the $(.95)^{p-1}$ percentiles of the sampling distributions of $q(p, n_2)$ for p = 2(1)10(2)20, 50, 100 and n_2 = 1(1)20(2)30, 40, 60, 100, ∞. Table III lists the $(.99)^{p-1}$ percentiles for the same distributions. These tables are used in performing comparisons among the means of equal sized samples from p populations. Tables of the 95th and 99th (same percentiles for all p) percentiles of $q(p, n_2)$ are given in Biometrika Tables [E.S. Pearson & Hartley 1954, T. 29].

1.5512
1.552

1.5512 Studentized statistics with range in denominator
[see also pp. 99-104]

Jackson, J. E. and Ross, Eleanor L. 1955
Journal of the American Statistical Association 50, 416-433
'Extended tables for use with the "G" test for means'

As a variant, mathematically trivial but allegedly easier to compute, of the u-tests of Lord 1947, 1950, Jackson & Ross propose the one-sample statistic $G_1 = |\bar{X}-\mu|/\bar{R}$, where $\bar{X}$ and $\bar{R}$ are the mean and mean range computed from m subsamples each of size n; and the two-sample statistic $G_2 = |\bar{X}_1-\bar{X}_2|/\bar{R}$, where the means $\bar{X}_1$, $\bar{X}_2$ are computed from m_1, m_2 subsamples, and the mean range $\bar{R}$ from m_1+m_2 subsamples, each of size n.

Table 1, p. 418, gives the upper 1%, 5% and 10% points of G_1 to 2 dec. for m = 1(1)15, n = 2(1)15. Table 2, pp. 419-432, gives the upper 1%, 5% and 10% points of G_2 to 2 dec. for m_1 = 1(1)m_2, m_2 = 1(1)15, n = 2(1)15. The tables are commendably well reproduced from typescript. Unfortunately, in many parts of the tables 2 dec. yield only 1 fig.

Reprinted in Vianelli 1949, Prontuari 85-86, pp. 234-248.

Bliss, C. I., Cochran, W. G. and Tukey, J. W. 1956

Biometrika 43, 418-422

Review by J. E. Walsh (MTAC 12, 68):

The data consist of k independent samples of n measurements each. Under the null hypothesis, the measurements are assumed to be normally distributed and have the same variance. The alternative hypothesis of interest is that one of the measurements is too large or too small to be consistent with the null hypothesis: i.e. the problem is that of rejecting an outlying observation. To perform the proposed test first compute the range for each sample. Let T equal the largest of the k ranges divided by the sum of all the ranges. If T is significantly large, the null hypothesis is rejected and the aberrant observation is determined by inspection and rejected. Table 1 contains upper 5% points of T to 3 dec. for $k = 2(1)10$ and $n = 2(1)10$. Table 2 contains upper 5% points of $(k+2)T$ to 2 dec. for $k = 10, 12, 15, 20, 50$; $n = 2(1)10$.

Moore, P. G. 1957

Biometrika 44, 482-489

Let $\bar{x}_i$ and w_i be the means and ranges of two samples of size n_i ($i = 1, 2$). Table 1, pp. 487-488, gives upper 1%, 2%, 5% and 10% points of $|\bar{x}_1 - \bar{x}_2|/(w_1 + w_2)$ to 3 dec. for $n_1 = 2(1)20$, $n_2 = n_1(1)20$.

Reprinted in Vianelli 1959, Prontuario 87, pp. 249-250.

1.552 Analysis of variance based on range

[see also pp. 104-106]

Cox, D. R. 1949

Journal of the Royal Statistical Society, Series B 11, 101-114

Given two normal populations I and II with variances σ_1^2 and σ_2^2. To test the hypothesis $H_0 [\sigma_1^2 = \sigma_2^2]$ against $H_1 [\sigma_2^2 = \psi\sigma_1^2]$ the following sequential procedure is suggested: take successive pairs of samples of size size m from the two populations. Let R_{j1}, R_{j2} be the ranges of the jth such samples. Define $Q_n = (R_{12} + R_{22} + \ldots + R_{n2})/(R_{11} + R_{21} + \ldots + R_{n1})$. Using the χ^2 approximation to range discussed on p. 107 of this Guide, Cox obtains critical limits for Q_n. Table 5, p. 111, gives these to 3 dec.:

1.552
1.78

for m = 4, 8 ; α = .01, .05 ; β = .05 ; $\begin{cases} \psi = 1.5 & 2 \\ n = 1(1)10(2)30 & 1(1)10(2)20 \end{cases}$.

1.5531 Approximations [to distribution of range] for moderate n

Duncan, A. J. 1955

<u>Industrial Quality Control</u> 11, no. 5, 18-22

Review by C.A. Bennett (MTAC 9, 217):

Duncan tabulated the quantities d_2^* and ν such that $E(\bar{R}/d_2^*)^2 = \sigma^2$ and the mean and variance of $\nu^{\frac{1}{2}}\bar{R}/d_2^*\sigma$ are those of a χ^2-distribution with d.f., where $\bar{R}$ is the mean of the ranges of k samples of n from a normal distribution with variance σ^2. He gives d_2^* to 2 dec. and ν to 1 dec. for k = 1(1)20(5)30, 50 and n = 2(1)7 . This is an extension of Table I of H.A. David 1951, which in turn is an extension, with minor corrections, of the original table given by Patnaik 1950. The methods used for the extension are not indicated.

See also Cadwell 1953 y .

1.69 Miscellaneous tests of normality
[see also pp. 117-118]

David, H. A., Hartley, H. O. and Pearson, E. S. 1954

<u>Biometrika</u> 41, 482-493

Review by C.A. Bennett (MTAC 9, 216):

Percentage points of the distribution of w/s , where w is the range and s the sample standard deviation of a sample of n from a normal population, are obtained partly by the use of an exact distribution applicable in the case of small samples, and partly by fitting a curve of the Pearson system to the first four moments for selected values of n and interpolating between these results. The final table gives upper 100 α% points for n = 3(1)20(10)60(20)100(50)200, 500, 1000 and α = .1, .05, .025, .01, .005 to 3 dec. for n ≤ 9 and 2 dec. for n > 9 ; and lower percentage points for the same values of α and n ≥ 10 . The use of this ratio as a test of homogeneity in both large and small samples is considered, and it is compared to the ratio of the mean deviation to the sample standard deviation and the sample fourth standard moment as a test of normality.

1.78 Miscellaneous tables connected with the bivariate normal distribution
[see also pp. 123-125]

K. Pearson 1925, T. 28, p. 189, gives the correlation between r and s_1 to 4 dec. for ρ = 0(.1)1 and n = 2(1)25, 50, 100, 400, ∞ . Reprinted in Tables for S & B II, T. 28, p. 180.

For some numerical work on the distribution of the ratio of two correlated normal variables, see Merrill 1928.

N.B.S. 1954

Table of salvo kill probabilities for square targets

Review by J.L. Hodges, Jr. (MTAC 10, 43):

Let $[x, y, \sigma]$ denote the circular normal distribution centered at (x, y) with standard deviation σ in each direction. A salvo of N bombs is "centered" at (α, β), which has the distribution $[0, y_0, \sigma_A]$. The bombs of the salvo are independently distributed according to $[\alpha, \beta, \sigma_R]$. If a bomb hits the square target ($x^2 \leq 1$, $y^2 \leq 1$), there is chance P_K of a kill. Assuming the bombs act independently, the chance P_{SK} that the salvo kills the target is computed for $P_K = .1(.3)1$; $y_0 = 0, 1, 2, 4, 7, 11, 16, 22$; $\sigma_A, \sigma_R = 1, 2, 4, 7, 11, 16, 22$; $N = 1, 5, 10, 25, 50(50)200$. The entries are 4 dec. with possible error of 2 in the last place. In an introduction, A.D. Hestenes warns the user against interpolating in this quintuple-entry table, except in the P_K direction for small N.

Fan, C. T. 1954

Psychometrika 10, 231-237

Review by John E. Walsh (MTAC 10, 50; abridged):

The table whose construction Fan describes provides a means of translating the observed proportions of success, p_H and p_L, in the high and low 27% groups, into measures of item difficulty (p and Δ) and of item discrimination (r). Values are tabled for p and r as functions of p_H and p_L. These results are based on K. Pearson's tables of the normal bivariate surface. A second item difficulty index, Δ, can be determined from p and is expressed in terms of a normal deviate with mean 13 and standard deviation 4. The final complete table is not given in this paper but has been published by the Educational Testing Service, Princeton, N.J. Preliminary computations in the table construction are presented in T. 3, which furnishes p_H and p_L to 4 dec. for $.5 \leq p \leq .9713$ and $r = .05(.05)1$.

Welsh, G. S. 1955

Psychometrika 20, 83-85

Review by J. R. Vatnsdal (MTAC 10, 234):

A table is given for obtaining tetrachoric r when it is possible to make cuts at the medians. The computing chart of Chesire, Saffir & Thurstone 1933 is used to set up a table which can be used if the proportion in the plus-plus cell is known. The table gives r_{tet} to 3 dec., for proportions to 3 dec.

1.79
1.9

See also Hoyt 1947; Davidoff & Goheen 1953.

1.79 The trivariate normal distribution
[see p. 125]

See also Nabeya 1952.

1.795 The normal distribution in more than three variates

Bechhofer, R. E. 1954
Annals of Mathematical Statistics 25, 16-39

The following account leans heavily on page 25 of Bechhofer:

Tables have been prepared to assist the experimenter in designing and interpreting experiments for ranking means.

Table 1, pp. 30-36, is to be used for designing experiments involving k normal populations to decide which t populations have the largest (or smallest) population means. The table gives to 4 dec. the value of $d = N^{\frac{1}{2}}$ associated with the probabilities .05(.05).8(.02).9(.01).99, .995, .999, .9995 for $k = 2(1)10$, $11(1)15$, $t = 1(1)[\frac{1}{2}k]$, $[k-9](1)5$. The table is based on the least favorable configuration of the population means which, for picking the t largest, is given by $\mu_{[k]} - \mu_{[k-t+1]} = 0$, $\mu_{[k-t+1]} - \mu_{[k-t]} = \lambda\sigma$, $\mu_{[k-t]} - \mu_{[1]} = 0$; and for picking the t smallest, is given by the same expressions with t replaced by k-t. The values of d were obtained by inverse linear interpolation in unpublished N.B.S. tables.

Table 2, p. 37, is a special table to be used for designing experiments involving 3 normal populations to decide which one has the largest, which the second largest, and which the smallest population mean. The table gives to 4 dec. the value of $d = N^{\frac{1}{2}}\lambda$ associated with the probabilities 1/6, .2(.05).8(.02).9(.01).99. The least favorable configuration $\mu_{[3]} - \mu_{[2]} = \mu_{[2]} - \mu_{[1]} = \lambda\sigma$ is assumed throughout. The values of d were obtained by inverse interpolation in bivariate normal tables now incorporated in N.B.S. 1959.

Somerville, P. N. 1954

Biometrika 41, 420-429

Review by Ingram Olkin (MTAC 10, 47):

Let
$$f(y) \equiv f(y_1, \cdots, y_k) = (2\pi)^{-k/2}|\Sigma|^{-k/2} \exp -\tfrac{1}{2} \operatorname{tr} y\Sigma^{-1}y',$$
where $y = (y_1, \cdots, y_k)$, $\Sigma = (\sigma_{ij})$, $\sigma_{ii} = 1$, $\sigma_{ij} = \tfrac{1}{2}$ and let $F_k(x) = \int_R f(y)\pi dy_i$, where $R: -\infty < y_i < x$, $i = 1, \cdots, k$. In Table 3.1 the author computes $F_k(x)$, $x = 0(.1)2(.5)3$ to 5D for $k = 1, 2$; 4D for $k = 3, 4$; 3D for $k = 5$. $F_1(x)$ is available in a number of tables. $F_2(x)$ was computed using Table VIII of part II of the Pearson Tables [1], and $F_1(x)$. For $F_k(x)$, $k = 3, 4, 5$, an expansion of $f(y)$ in terms of Hermite polynomials was used [2].

1.9 Miscellanea [miscellaneous tables of the normal distribution]

[see also p. 129]

Kolmogorov, A. N., Švesikov, A. A. and Gubler, I. A. 1945

Review by R. C. Archibald (MTAC 2, 69):

This issue of the *Trudy* is devoted to a collection of articles on the theory of gunnery, edited by Kolmogorov. There are four tables (p. 96–106). **Table 1** gives values of

$$H(m, a) = 1 - e^{-a}\left(1 + \frac{a}{1!} + \frac{a^2}{2!} + \cdots + \frac{a^{m-1}}{(m-1)!}\right),$$

for $a = [.1(.1)15; 5D]$, $m = 1(1)11$.

For the same range T. 2 gives the values of

$$\nabla^2 H(m, a) = H(m, a) - 2H(m-1, a) + H(m-2, a).$$

In **T. 3**, values are given for

$$\tau(u) = \frac{1}{2u}\int_{-\infty}^{+\infty} \ln\{1 - \tfrac{1}{2}[\Phi(z+u) - \Phi(z-u)]\}dz$$

for $u = [0(.05)1(1)5; 3D]$, with Δ; $\Phi(z) = \dfrac{2\rho}{\sqrt{\pi}}\int_0^z e^{\rho^2 z^2} du$, $\rho = .476936\ldots$

T. 4 gives the values of $P(N) = \Phi(s_0) - 2s_0\phi(s_0)$, $\phi(z) = e^{\rho^2 z^2}/\pi^{\frac{1}{2}}$, $s_0(N) = \rho^{-1}(\tfrac{3}{4}\pi^{\frac{1}{2}})N^{\frac{1}{2}}$, and $d^2(N) = \tfrac{1}{5}s_0^2$, for $N = [0(.1)1(.2)5(.5)10; 3S]$ with Δ.

On p. 104–106 are graphs of (a) $H(m, a)$, $m = 1(1)10$, $a = 0(1)15$; (b) $R(k, c, n)$, $k = \tfrac{1}{2}$, $c = 4/15$, $n = 2(2)6, 12(6)24$, $0 < a < 4.8$, the values of R being given in a table on p. 33 for given values of n and a; (c) R, $k = \tfrac{1}{2}$, $c = 4/15$, $n = 3, 6(6)24$, $0 < a < 2.4$, the values of R being given on p. 34 for values of n and a.

Springer, C. H. 1951

See abstract in section 10.039, p. 717.

See also Finney 1956.

---

[1] K. Pearson 1931a   [2] Kibble 1945, p. 15

2.112
2.21
2.112 Areas under the Pearson Type III curve
[see also p. 132]

Vianelli 1959, Pro. CIC, pp. 597-598, abridged from Salvosa 1930, gives Pr[t ≤ x] to 4 dec. for α_3 = .1(.2)1.1 and x = -3.65(.05)+2(.1)4(.2)6.6.

2.1131 $p(\chi^2, \nu)$
[see also pp. 132-133]

4 dec. ν = 1(1)29 χ^2 = 1(1)30   Gnedenko 1950(378), Czechowski 1957(75)

2.1134 Ratio of integral to bounding ordinate [of the χ^2 distribution]
[see also p. 133]

A revised version of the tables of Kotani et al. 1938 appears in Kotani et al. 1955.

2.122 [the cumulative Poisson distribution:] 1 - P(m,i)
[see also p. 135]

Kolmogorov, A. N., Švesikov, A. A. and Gubler, I. A.   1945

See review by R. C. Archibald on p. 655 above.

2.129 Miscellaneous tables of the cumulative Poisson distribution
[see also p. 136]

Palm, C.   1947

Table of the Erlang loss formula

Review by W. Feller (MTAC 3, 98):

The simplest situation in the theory of automatic telephone exchanges can be described as follows. Incoming calls are "randomly distributed" (that is, at least during the "busy hour"). The probability that exactly k calls will originate during time t is given by the POISSON expression $e^{-at}(at)^k/k!$ where $a > 0$ is a constant. Moreover, it is assumed that the lengths of the individual conversations are statistically independent and that the probability of any conversation lasting for time t or more is $e^{-(t/b)}$. Suppose there are n circuits available so that an incoming call is served whenever a circuit is free. If no circuit is free, the call is "lost" and has no influence on the future traffic. Erlang's formula[1] for the proportion of lost calls is,

$$E_{1,n}(A) = A_n(A_0 + A_1 + \cdots + A_n)^{-1}, A = ab,$$

where we put for abbreviation $A_k = (ab)^k/k!$ (the notation is not standard and the general conditions of applicability of Erlang's formula are still discussed). The tables are double-entry 6D tables with n = 1(1)150, and A = .05(.05)1(.1)20(.5)30(1)50, 52(4)100. Only entries which are significantly different from zero are tabulated, and thus the effective range

---

[1] Erlang 1917, 1918; Fry 1928

of A decreases with increasing n. For $n = 100$ entries occur only for $A \geqslant 56$, and for $n = 150$ we have the single entry $E_{1,150}(100) = .000\,001$. The tabular intervals are selected so as to permit linear interpolation in A with an accuracy to four or five decimals. The sixth decimals are said to be accurate throughout the tables. These have been computed from the recurrence formula $E_{1,n}(A)\{n + AE_{1,n-1}(A)\} = AE_{1,n-1}(A)$.

Brockmeyer, E., Halstrøm, H. L. and Jensen, A. 1948

Transactions of the Danish Academy of Technical Sciences, 1948, no. 2

Review by D.H. Lehmer (MTAC 6, 84):

There are small tables related to the Poisson distribution included in this work. Among those of some general use are

(a) 6D values of $m^x e^{-m}/x!$ for $x = 0(1)14$ and $m = -2(.1)0$. This table is on p. 137.
(b) 6D values of $(-m)^x e^m/x!$ for $x = 0(1)20$, $m = 2.1(.1)4$ on p. 195–196.
(c) 6D values of the real and imaginary parts of the solutions of
$$z \exp(z) = \alpha \exp(-\alpha) \exp(2\pi i k/40)$$
for $\alpha = 0(.1)1$, $k = 0(1)20$ (p. 197–198).
(d) BROCKMEYER's table, described in RMT 965.

Katz, L. 1955

Annals of Mathematical Statistics 26, 512-517

Review by W. J. Dixon (MTAC 10, 237):

Tables are given to 5D for the expression,

$$\frac{(N-1)!}{(N-i)!} \sum_{M=0}^{N-1-i} N^M/M!,$$

for $i = 0, 1$; $N = 2(1)20(2)40(5)100$. A formula is given for large N. These quantities are the probabilities referred to in the title for $i = 0$ corresponding to a general case and $i = 1$ to the "hollow" case defined in the article. They are computed from values of P given in Molina's tables [1] by the formula,

$$(N-1)! e^N P(N; N-i-1)/(N-i)^N.$$

See also Makabe & Morimura 1955.

2.21 Ordinates of the Pearson Type III frequency curve
[see also p. 137]

Vianelli 1959, Pro. 200, pp. 599-600, abridged from Salvosa 1930, gives $y(t)$ to 4 dec. for $\alpha_3 = .1(.2)1.1$ and $x = -3.9(.1)-3(.05)+3(.1)4(.2)6.8$.

---
[1] Molina 1942

2.3114
2.322

2.3114 Short tables [of percentage points of χ^2]
[see also p. 143]

2 dec. $\nu = 1(1)30$ $p = .01(.02).05, .1;$ Royo López & Ferrer Martin
                        1-p = ditto   1955(30)

2.313 Percentage points of χ
[see also p. 144]

Vianelli 1959, T. 193, p. 1360, credited to Pearson & Lee, gives $[\chi^2(\frac{1}{2},\nu)]^{\frac{1}{2}}$ to 7; 3 dec. for $\nu = 1(1)11; 12(1)15$.

2.315 Confidence limits for σ
[see also pp. 148-150]

Croxton, F. E. and Cowden, D. J.   1946
<u>Industrial Quality Control</u> 3, July 1946, 18-21

Table 1 gives upper and lower .1%, $\frac{1}{2}$%, 1%, $2\frac{1}{2}$%, 5%, 10% points of s'/σ for $N = 2(1)30$ to 3 dec.; T. 2 does the same for σ/s'.

Pearson & Hartley 1954, T. 35, p. 184, gives inter al. upper and lower confidence limits for σ/s to 3 fig. for confidence levels 95%, 99% and $\nu = 1(1)20(5)50(10)100$. An identical table appears in Vianelli 1959, T. IL, p. 1182.

2.319 Miscellaneous percentage points of χ^2
[see also pp. 150-151]

Eisenhart, Hastay & Wallis 1947, TT. 8.1, 8.2, pp. 272-273, 276-277, give $\rho(\alpha,\beta,\nu) = \chi^2(\alpha,\nu)/\chi^2(1-\beta,\nu)$ to 4-5 fig. for $\alpha = •01, •05;$
$\beta = •005, •01, •025, •05, •1, •25, •5; 1-\beta = $ ditto;
$\nu = 1(1)30(10)100, 120, \infty$.

Reprinted in Vianelli 1959, TT. 221-222, pp. 1399-1400.

Cameron, J. M.   1952
<u>Industrial Quality Control</u> 9, no. 1, 37-39

Table 1, p. 38, gives $\frac{1}{2}\chi^2(p,\nu)$ to 3 dec. for $\nu = 2(2)100$; $p = .005, .01, .025, .05, .1, .25, .5; 1-p = $ ditto.

Burington, R. S. and May, D. C. 1953

Probability and statistics

Table 11.10.1, p. 101, gives the ordinate and abscissa of the χ^2 distribution with 2 d.f. Specifically it tables $\chi = (-2 \log_e p)^{\frac{1}{2}}$ and p_χ to 4 dec. for $1-p = 0(.05)1$.

See also Bhatt 1953.

2.322 Confidence limits for m [Poisson expectation]
[see also pp. 151-153]

E. S. Pearson & Hartley 1954, T. 40, p. 203, gives Poisson confidence limits, defined in the abstract of Garwood 1936 on p. 152 of this Guide, to 3 fig. or 2 dec. for $c = 0(1)30(5)50$ and $p = .001, .005, .01, .025, .05$. Reprinted in Vianelli 1959, Pro. 193, p. 578.

Girshick, M. A., Rubin, H. and Sitgreaves, R. 1955

Annals of Mathematical Statistics 26, 276-285

Review by Frank Massey (MTAC 10, 236):

This paper gives a method of estimating the parameter λ in a Poisson distribution by gradually expanding the area (or time) observed until a fixed number M of events occur. If a_M denotes the necessary area, and γ a constant, then confidence limits for λ are $b/[(1 + \gamma)a_M]$ and $b/[(1 - \gamma)a_M]$ with confidence level

$$\alpha = \int_{b/(1+\gamma)}^{b/(1-\gamma)} \frac{x^{M-1}e^{-x}}{(M-1)!} dx.$$

A preassigned bound on the percent error in estimating λ is given by 100γ. For fixed M and γ, the value $b = b^* = \frac{M(1-\gamma^2)}{2\gamma} \log \frac{1+\gamma}{1-\gamma}$ maximizes α, and, for fixed γ, M may be chosen as the least integer so that by using b^*, α is just larger than a preassigned level of confidence.

Table I, using $b = b^*$, gives values of α to 4D for $M = 2(2)40$, $\gamma = .01, .05, .10, .20$. Most of these tabled α turn out to be small. The largest, for $M = 40$, are .0503, .2479, .4733, .7988 corresponding to the four values of γ. An approximation formula is $\sqrt{4M}\left(\frac{\gamma}{2} + \frac{7}{48}\gamma^3\right) = z_\alpha$ where z_α is the $100(1 + \alpha)/2$ percentile of the unit normal distribution. A listing for $\alpha = .90, .95, .99$ and $\gamma = .05, .10$ shows this approximation as good for these cases as a more complicated one which in turn checks with the tabled values for $M = 40$.

2.41
2.5112

2.41 Laguerre series

[see also pp. 154-155]

Jahnke & Emde 1948(32), 1960(105), gives $e^{\frac{1}{2}x}D^n(x^n e^{-x})/n!$ to 4 fig. for $n = 0(1)10$ and $x = 0(.1)1(.25)6(1)14(2)34$.

Grad, A. and Solomon, H. 1955

Annals of Mathematical Statistics 26, 464-477

Review by W. J. Dixon (MTAC 10, 234):

Tables I and II of this paper are probabilities $P(Q_k < t)$ for $k = 2, 3$ and $t = .1(.1)1(.5)2(1)5$ to 4S where $Q_k = \sum_{i=1}^{k} a_i x_i^2$ and x_i are normally and independently distributed with zero mean and unit variance, $\sum a_i = 1$, $a_i > 0$. The probabilities for $k = 2$ are given for $a_1 = .5(.1).9, .95, .99, 1$ and for $k = 3$ for $(a_1, a_2) = (1/3, 1/3), (.4, .3), (.4, .4), (.5, .3), (.6, .2), (.5, .4), (.6, .3), (.7, .2), (.8, .1)$. Various approximations are investigated for accuracy. Use of the tables is illustrated with several applications.

Huitson, A. 1955

Biometrika 42, 417-479

Review by J. E. Walsh (MTAC 12, 71):

Consider estimation of a linear function of r unknown population variances $\sigma_1, \cdots, \sigma_r^2$. The estimation data consists of the corresponding statistics, $s_1^2, \cdots, s_r^2$. The s_i^2 are mutually independent and s_i^2/σ_i^2 has a χ^2 distribution with f_i degrees of freedom $(i = 1, \cdots, r)$. The problem is to obtain approximate confidence limits for $\sum_1^r \lambda_i \sigma_i^2$, where the constants λ_i are arbitrary but specified. These confidence limits are obtained by determining an approximate expression for the statistic $y = y(s_1^2, \cdots, s_r^2, P)$ with the property that $Pr[\sum \lambda_i s_i^2 / \sum \lambda_i \sigma_i^2 < y] = P$. Let ξ denote the $100P$ percentage point of the normal distribution with zero mean and unit variance and $V_{mn} = (\sum \lambda_i^m s_i^{2m} f_i^{-n})/(\sum \lambda_i s_i^2)^m$. Then, to terms of order $1/f_i^2$, $y = h_0 + h_1 + h_2 + h_3$. Here $h_0 = 1$, $h_1 = \xi\sqrt{2V_{21}}$,

$$h_2 = \xi^2\sqrt{2V_{21}} - 4V_{23}/3V_{21}] - 2V_{32}/3V_{21},$$

$$h_3 = \sqrt{2V_{21}}\{\xi[-4V_{32}(3V_{21})^{-1} + 9V_{43}\sqrt{2}\,V_{21})^{-2} - V_{22}(V_{21})^{-1}$$

$$- 23(V_{32}/3)^2 V_{21})^{-3}] + \xi^3[-8V_{32}(3V_{21})^{-1}$$

$$+ 2V_{21} - 16(V_{32}/3)^2(V_{21})^{-3} + 5V_{43}(\sqrt{2}\,V_{21})^{-2}.$$

For $r = 2$, the λ_i of the same sign, and $f_i \geq 16$, two decimal accuracy appears to be obtainable from the relation $y = h_0 + h_1, + h_2$. Tables 1-4 contain 2D values of y for $r = 2$; $(\lambda_1 s_1^2 + \lambda_2 s_2^2) = 0(.1)1.; f_1, f_2 = 16, 36, 144, \infty; P = .01, .05, .95, .99$; and $\lambda_2 \lambda_2 > 0$.

2.42 Non-central χ^2

[see also pp. 155-157]

Burington, R. S. and May, D. C. 1953

See abstract of T. 11.11.1 on p. 125 above.

Abdel-Aty, S. H. 1954 z

Biometrika 41, 538-540

"Approximate formulae for the percentage points of the non-central χ^2 distribution"

Review by Julius Lieblein (MTAC 9, 218):

The main point of this note is to derive the two approximation formulas of the title and compare them numerically with the exact values. The cube root transformation of Wilson & Hilferty [1931] turns out to be the appropriate one for transforming the non-central χ^2 variate, χ'^2, to an approximately normal variate y. The expansion of Cornish & Fisher 1937 is then used to obtain a formula for the 100 α% point of Y (the variate y after standardization) in ascending powers of $r^{-\frac{1}{2}}$, where $r = f + \lambda$ and f is the number of d.f. and λ is the [non-centrality] parameter. This formula is written out to the term in $r^{-3/2}$, with the coefficients expressed in terms of quantities whose values are given for α = .001, .005, .01, .05, .1(.1).5. The numerical comparison with a few exact values indicates that even for small degrees of freedom and moderate values of λ the expansion agrees very well for both the upper and lower 5% points. The probability integral approximation for Y is obtained by straightforward application of the Edgeworth expansion. Numerical comparison with exact values for a few values of f and λ show agreement to 3 dec. or better.

2.5111 x!

[see also p. 160]

7 dec.	0(.001)1	No Δ	Vianelli 1959(1007)
7 dec.	0(.01)1	No Δ	E.S. Pearson & Hartley 1954(234)
4-5 fig.	0(.01)3.99	No Δ	Burington & May 1953(291)
4-5 fig.	0(.02)4	$\frac{1}{2}\Delta$	Jahnke & Emde 1960(15)

2.5112 n!

[see also p. 161]

exact	1(1)20	Siegel 1956(287)
6 fig.	1(1)300	Czechowski et al. 1957(143)

2.5121
2.613

2.5121 $1/x!$
[see also p. 161]

5 fig. $-3(.02)+3$ $\frac{1}{2}\Delta$ Jahnke & Emde 1960(16)

2.5131 $\log_{10} x!$
[see also p. 162]

10 dec. $1(.1)4(.2)69$ No Δ Vianelli 1959(1011)

7 dec. $0(.001)1$ No Δ Glover 1923(464)

7 dec. $0(.01)1$ No Δ Pearson & H. 1954(234), Vianelli 1959(562)

4 dec. $0(.01)1$ No Δ Burington & May 1953(292)

2.5134 $\log_{10} n!$
[see also p. 163]

10 dec. $1(1)1113$ Vianelli 1959(1011)

7 dec. $1(1)1000$ Vianelli 1959(59)

5 dec. $1(1)300$ Czechowski et al. 1957(143)

4 dec. $0(1)999$ Graf & Henning 1953(70)

2.514 The polygamma functions
[see also p. 163]

A number of somewhat special combinations of digamma and trigamma functions are described in section 10.035, p. 395.

2.529 Moments of other functions of χ^2
[see also pp. 172-173]

Chandrasekhar, S. 1949

<u>Proceedings of the Cambridge Philosophical Society</u> 45, 219-224

The following unsigned review is taken from MTAC 4, 19:

The function β_p under tabulation is the p-th moment of $W(\beta)$, where

$$W(\beta) = 2\beta\pi^{-1} \int_0^\infty e^{-u} y \sin \beta y\, dy, \quad u = y^{3/n}.$$

The function β_p is given explicitly by

$$\beta_p = 2\pi^{-1}(p+1)\Gamma(p)\Gamma(1 - np/3) \sin \tfrac{1}{2}\pi p$$

and is tabulated to 5S for

$n = 1.6, p = .25(.25)1.75, 1.80, 1.85$
$n = 2, p = .25(.25)1.25(.05)1.45, 1.475$
$n = 3, p = .2(.2).8, .9, .95, .975$
$n = 4, p = .1(.1).5, .55, .575$
$n = 6, p = .1(.1).4, .45, .475$
$n = 8, p = .1(.1).3, .325, .35, .36$
$n = 10, p = .1, .2, .25, .275, .280$

On p. 222, there is a 5S table of

$$[\tfrac{2}{3}\pi n(n+3)^{-1}\Gamma(3/n) \sin (\tfrac{3}{2}\pi n^{-1})]^{n/3}$$

for $n = 1.51, 1.52(.02)1.6(.1)2(.5)4(2)10(5)25$.

<u>2.611</u> Exact tests of significance in 2 × 2 tables
[see also pp. 174-176]

Siegel 1956, T. I, pp. 256-270, abridged from Finney 1948, gives the integer b* for m = 3(1)15 and n = 3(1)m, but omits the probability P^*_{b*}.

<u>2.612</u> Approximate tests of significance in 2 × 2 tables
[see also p. 176]

See also Jurgensen 1947.

<u>2.613</u> Power of the χ^2 test in 2 × 2 tables
[see also pp. 177-179]

Chandra Sekar, C., Agarwala, S. P. and Chakraborty, P. N. 1955
<u>Sankhyā</u> 15, 381-390
"On the power function of a test of significance for the difference between two proportions"

In the notation of p. 174 of this <u>Guide</u>, let $E(a) = mp_1$, $E(b) = np_2$, $m = n = \tfrac{1}{2}N$.

Table 4, pp. 387-388, gives the power of the test of the hypothesis $p_1 = p_2$, at nominal significance level .05 (the actual significance levels range from .00055 to .04023) for $p_1 = .1(.1).9$, $p_2 = .1(.1).5$, $N = 10(10)40(20)100, 200, 400$. Table 5, pp. 389-390, gives the power of

2.613
3.319

the test of the hypothesis $p_1 = p_2$ against the one-sided alternative $p_1 > p_2$ at nominal significance level .025 (the actual significance levels range from .00027 to .02012) for $p_2 = .1(.1)p_1$; p_1 and N as in T. 4.

[N.B. The powers in TT. 4-5 are consistent with the null distribution of Swaroop 1938. The tables of Dixon & Mood 1946 are germane only in the limit as the frequencies a and b become Poisson variables: see review of Przyborowski & Wilenski 1940, p. 315, above]

2.63 Fully specified tables

Williams, C. A., Jr. 1950
Journal of the American Statistical Association 45, 77-86
Review by Z. W. Birnbaum (MTAC 5, 150):

This paper contains a presentation and discussion, intended mainly for the non-mathematical user of tests of goodness of fit, of the results of Mann & Wald 1942 that make it possible to optimize the number of classes and the choice of class-boundaries for the chi-square test. Tables are given showing, for the 1% and 5% significance level and sample sizes N = 200(50)1000(100)1500, 2000, the optimum number k of classes, as well as some information related to the power of the test. Williams suggests that the tabled values of k may be halved with little loss of power.

2.69 Miscellaneous goodness-of-fit tests involving χ^2
[see also pp. 179-180]

See also Akaike 1954.

3.11 The cumulative binomial sum
[see also p. 183]

3.111 Upper tail

7 dec.   p = .1(.1).5   n = 2(1)49   Czechowski et al. 1957(50)

3.112 Lower tail

5 dec.   p = .01, .02(.02).1(.1).5   n = 5(5)30   Vianelli 1959(551)

3.12 Tables of the incomplete-beta-function ratio
[see also p. 185]

Dixon, W. J. 1953

Annals of Mathematical Statistics 24, 467-473

Table 1, p. 468, gives $I_x(N-i, i+1) + I_{1-x}(N-i, i+1)$ to 5 dec. for
$x = .5(.05).95$ and $\begin{cases} N = 5(1)8 & 9(1)11 & 10(1)14 & 15, 16 & 17(1)20 & 20 & 25 \\ i = 0 & 1 & 2 & 3 & 4 & 5, 6 & 7 \end{cases}$

$\begin{cases} 30 & 35 & 40 & 45 & 50 & 60 & 70 & 80 & 90 & 100 \\ 9 & 11 & 13 & 15 & 17 & 20 & 26 & 30 & 35 & 39 \end{cases}$. Table 2, p. 469, does the same

for $\begin{cases} N = 8(1)11 & 12(1)14 & 15(1)17 & 18(1)20 & 25 & 30 & 35 & 40 & 45 & 50 & 60 & 70 \\ i = 0 & 1 & 2 & 3 & 5 & 7 & 9 & 11 & 13 & 15 & 19 & 23 \end{cases}$

$\begin{cases} 80 & 90 & 100 \\ 28 & 32 & 36 \end{cases}$.

Dept. of Defense 1958

Technical Report (Resnikoff)

Appendix 1, pp. 20-22, gives $I_x(a,b)$ to 7 dec. for $x = 0(.01)5$ and $a = b = 2.250, 3.037, 5.053, 8.678, 10.493, 11.5, 12.306, 14.118, 16.5, 17.743, 21.368, 30.428, 36.5, 41.301, 63.046, 74, 82.979, 99$.

3.21 Binomial coefficients

In addition to the tables listed on p. 186 above, see Sakoda & Cohen 1957.

3.319 Miscellaneous percentage points of the beta distribution
[see also pp. 192-193]

R. E. Clark 1953 gives roots x of $I_x(c, n-c+1) = \alpha$ to 4 fig. for $c = 1(1)n$; $n = 10(10)50$; $\alpha = .005, .01, .025, .05$. Reprinted in Vianelli 1959, Pro. 212, pp. 675-686.

Hald 1952, T. XI, is abridged in Czechowski et al. 1957, T. 6, pp. 55-58, viz: $\begin{cases} c = 0(1)10(2)30, 40, 60, 100, 200, 500, \infty \\ n-c = 1(1)10(2)30, 40, 60, 100, 200, 500, \infty \end{cases}$.

Royo López & Ferrer Martín 1954(88), 1955(18) give inter al. p_L and p_U to 4 dec. for $\alpha = .025$, $n = 10(1)50$, $c = 0(1)n$.

3.319
3.42

Sterne, T. E. 1954

<u>Biometrika</u> 41, 275-278

By abandoning Clopper & Pearson's requirement that the two tails excluded from a 100(1-2α)% confidence region have each size not exceeding α, Sterne obtains binomial confidence intervals that are shorter than those of Clopper & Pearson 1934. The table on p. 277 gives p_L and p_U to 2 fig. for α = .05, .25; n = 1(1)10; c = 0(1)n.

Crow, E. L. 1956

<u>Biometrika</u> 43, 423-435

Table 1, pp. 428-431, based on a modification of the method of Sterne 1954, gives p_L to 3 dec. for α = .005, .025, .05; n = 1(1)30; c = 0(1)n: p_U is to be obtained by symmetry. Reprinted in Vianelli 1959, <u>Prontuario</u> 122, pp. 334-337.

Bargmann, Rolf 1955

<u>Mitteilungsblatt für mathematische Statistik</u> 7, 1-24

Review by Ingram Olkin (MTAC 12, 73):

Let $P_k = I_{.10}((k-1)/2, \frac{1}{2})$,

$$P(k, r, n) = \sum_{i=r}^{n} \binom{n-k+1}{i-k+1} P_k^{i-k+1}(1-P_k)^{n-i} = I_{P_k}(r-k+1, n-r+1),$$

where $I_x(a, b)$ is the incomplete Beta function. Table I gives the values of P_k to 4D for k = 2(1)16. (P_5 = .1495 and not the .1459 entered in the table, P_{12} = .2549 instead of .2548). The principal table gives integral values of r approximately satisfying $P(k, r, n) = \alpha$ for α = .001, .01, .05, .10, .25, .50; k = 2(1)12. The tabled r is usually the largest integer satisfying $P(k, r, n) \leq \alpha$. In some cases the author uses the next lower value of r when the resulting $P(k, r, n) - \alpha$ is small; in these cases the value of r is printed in italics. The rules followed by the author in deciding whether $P(k, r, n) - \alpha$ is small are not clear.

Masuyama, M. 1955

<u>Sankhyā</u> 15, 291-294

See abstract on p. △△△.

3.42 Approximations to the cumulative binomial summation
[see also pp. 198-199]

Freudenberg, K. 1951

Metron 16, 285-310

Review by S. B. Littauer (MTAC 7, 89):

The principal objective of this paper is to determine the smallest sample size required in order to obtain a given degree of approximation to the binomial distribution by means of the Poisson distribution and by means of the following, so to speak, second order approximating function

(1) $$B_{\lambda, m} = \frac{e^{-\lambda}\lambda^m}{m!}\left[1 - \frac{(m-\lambda)^2 - m}{2s}\right],$$

where s is sample size, $\lambda = sq$ is the mean number of occurrences of an event of probability q in random samples of size s; λ is also the mean of the approximating Poisson distribution and $m \equiv$ an integer. The mean of $B_{\lambda, m}$ is also independent of s, and its variance is $\lambda(1 - \lambda s^{-1})$. In spite of the fact that the zeroth moment of $B_{\lambda, m}$ is unity independent of s, it is not a distribution, since some of its values can be negative. Since these negative probabilities are, however, exceedingly small, they do not vitiate the practical usefulness of $B_{\lambda, m}$ as an approximation to the Poisson and binomial distributions. In fact, the $B_{\lambda, m}$ require much smaller sample sizes for equivalent approximation to the binomial distribution than does the Poisson distribution. The Poisson and binomial are in this notation denoted as follows:

(2) $$A_{\lambda, m} = e^{-\lambda}\lambda^m/m!$$

(3) $$C_{\lambda, m} = \binom{s}{m}\left(\frac{\lambda}{s}\right)^m\left(1 - \frac{\lambda}{s}\right)^{s-m}.$$

The following tables are of principal interest: on p. 307 values of $s_{\rho, A}$ and $s_{\rho, B}$ as functions of $\lambda = \frac{1}{2}(\frac{1}{2})10$, which are respectively lower bounds for s such that the deviations of $A_{\lambda, m}$ and $B_{\lambda, m}$ from $C_{\lambda, m} \leq \rho = .001$. $s_{\rho, A}$ varies from 141.5 to 640.8, where correspondingly $s_{\rho, B}$ goes from 7.8 to 78.8. On p. 308 $s_{\rho, A}$ and $s_{\rho, B}$ are given as functions of $q = .01(.01).10$, again for $\rho = .001$. Here $s_{\rho, A}$ varies from 406.0 to 4912.2 while $s_{\rho, B}$ goes from 0 to 34.1.

Auxiliary tables of some interest are found on p. 292–294. $A_{\lambda, m}$, $B_{\lambda, m}$, $C_{\lambda, m}$, $(A_{\lambda, m} - C_{\lambda, m})$, and $(B_{\lambda, m} - C_{\lambda, m})$ are given for $s = 50$ and 100 and for $\lambda = \frac{1}{2}, m = 0(1)7, > 7; \lambda = 1, \frac{3}{2}, m = 0(1)10, > 10; \lambda = 2, m = 0(1)11, > 11; \lambda = 3, m = 0(1)12, > 12; \lambda = 4, m = 0(1)13, > 13$.

For the approximation $B_{\lambda, m}$ to be of practical use in sampling, extensive tables would be required; the present table has limited usefulness. The reader's attention may be called to errors in formula (11) p. 289 and the

3.42
3.6

subsequent argument. The right side of (11) should be λ and the subsequent argument can be corrected so as to yield this result. The findings of the paper are not affected by these errors.

3.43   Approximations to the incomplete beta-function

In addition to the tables described on pp. 199-200 above, see Cadwell 1952 y.

3.52   Moments of the beta distribution
[see also p. 204]

Woo, T. L.   1929

<u>Biometrika</u> 21, 1-66

See abstract on p. 192 above.

3.53   Moments of the binomial distribution
[see also pp. 204-205]

Grab, E. L. and Savage, I. R.   1954

<u>Journal</u> <u>of</u> <u>the</u> <u>American</u> <u>Statistical</u> <u>Association</u> 49, 169-177

The random variable X is said, following Stephan 1945, to have a positive Bernoulli distribution if $\Pr[X = x] = (1-q^n)^{-1} \binom{n}{x} p^x q^{n-x}$, $x = 1(1)n$, $q = 1-p$.

Table 1, pp. 174-176, gives $E(1/X \mid n,p) = (1-q^n)^{-1} \sum_{x=1}^{n} \binom{n}{x} x^{-1} p^x q^{n-x}$

to 5 dec. for $\begin{cases} n = 2(1)20 & 21(1)30 \\ p = .01, .05(.05).95, .99 & .01, .05(.05).5 \end{cases}$.

Fieller, E. C. and Hartley, H. O.   1954

<u>Biometrika</u> 41, 494-501

Review by Ingram Olkin (MTAC 10, 46):

Let x and y be jointly distributed, and let $f(x, \theta)$, $g(y, \phi)$ be the respective marginal distributions. Information about the control variable x is available, e.g., $f(x, \theta)$ may be completely known or θ may be known. The authors are concerned with using this information in the estimation of ϕ for the case where $X_1, \cdots, X_5$ are observations from an $N(\mu, 1)$ population, $x = \sum (X_i - \bar{X})^2$, $y = X_{\max} - X_{\min}$; the distribution of y is assumed unknown and that of x known. In the determination of the variance of the proposed estimate, the expectation of the reciprocal of a binomial variate, excluding the zero class, is needed.

Values of (1) $E'(np, n) = \sum_{r=1}^{n} \frac{1}{r} \binom{n}{r} p^r q^{n-r}$ for $m \equiv np = 1(1)10$, $n = 25$, 50, ∞ are computed directly. For $m = 12(2)20$, $n = 25, 50, 100$,

(2) $E'(m, n) \simeq \dfrac{1}{m+p} + \dfrac{1}{(m+p)(m+2p)} + \dfrac{2!}{(m+p)(m+2p)(m+3p)} + \cdots$

is used; for $m = 25(5)45$, $n = 50, 100$ a control is used with (2). $E'(m, \infty)$ may be calculated from $E'(m, \infty) = e^{-m}\{\text{Ei}(m) - \log_e m - \gamma\}$ using available tables [1, 2]. For large np, the approximation $E' \simeq \dfrac{1}{m-q} + \dfrac{q}{(m-q)^3} + \dfrac{q(q+1)}{(m-q)^4}$ is given. Values of (1) for $m = 1(1)10$ are given to 4D, all others to 5D.

See also Bosse 1948; Raff 1956.

3.6 Tables for the binomial sign test
[see also p. 206]

Nair, K. R. 1940 a

Sankhyā 4, 551-558

One may assert that a population median lies between the kth and $(n-k+1)^{\text{th}}$ order statistics in a sample of n with confidence $1 - 2I_{\frac{1}{2}}(n-k+1, k)$. Table 1, pp. 556-557, gives $I_{\frac{1}{2}}(n-k+1, k)$ to 4 dec. for $n = 6(1)81$, $n-k+1 < 50$, and such k as make $1 - 2I_{\frac{1}{2}}(n-k+1, k)$ just greater than .95 and .99.

An abridgement to 3 dec. for $n = 6(1)65$ is given in Dixon & Massey 1951, T. 25, p. 360; 1957, T. 35, p. 462.

Cole 1945 tables $2P(\frac{1}{2}, n, c+\frac{1}{2})$ to 5 dec. for $n = 0(1)35$ and $c = 0(1)12$.

Walter, E. 1954

Mitteilungsblatt für mathematische Statistik 6, 170-179

Review by Cyrus Derman (MTAC 10, 49; abridged):

Tests of significance based on statistics which can assume only a finite number of different values usually do not attain their asserted level of significance. It is always possible to modify the test (by randomizing) to attain the asserted level of significance. Walter is interested, however, in a modification based on the observations. He

---

[1] B.A. 1, 1931   [2] Jahnke & Emde 1948

3.6
4.1313

considers the sign test for testing the null hypothesis that the distribution function is symmetric about zero. The modified sign test is defined as follows. Let y equal the number of negative observations in the sample of n; let y' be the rank of the negative observation with largest absolute value among the absolute values of all the observations; let y" be the rank of the negative observation with second largest absolute value; etc. The null hypothesis is to be rejected if $y < y_0$, or if $y = y_0$ and $y' < y_0'$, or if $y = y_0$, $y' = y_0'$ and $y" < y_0"$, etc. In T. 1 values of y_0, y_0', etc. are given for n = 5(1)25, α = .01, .05, and for one- and two-sided tests: the two-sided test considers either the number of positive observations or the number of negative observations, whichever is smaller.

In T. 3 the effective significance levels of the sign test and modified sign test are given for n = 5(1)30.

Dixon & Massey 1951, T. 10, p. 324, is reprinted in Czechowski et al. 1957, T. 22, p. 92.

Walsh 1951 gives illustrative values of the size of the binomial sign test when one or both of the following conditions are relaxed: that all populations sampled have a common median; that no population sampled has a finite lump of probability at the median.

Aoyama 1953, p. 84, gives the distribution of the sign test to 4 dec. or 2 fig. in samples of 6, 12, 20, 30, 42, 56, 72.

Dixon, W. J. 1953

Annals of Mathematical Statistics 24, 467-473

Table 3, p. 471, gives illustrative values of the power-efficiency of the sign test to 3 dec. for normal alternatives in samples of 5, 10, 20.

See also Mac Stewart 1941.

4.1111 $P(t,\nu)$

[see also p. 208]

3; 5 dec. ν = 1(1)20; ∞ t = 0(.1)2(.2)6   Gnedenko 1950(382),
                                                                                                           Czechowski et al. 1957(78)

4.1113 $2Q(t,\nu)$

[see also p. 209]

Vianelli 1959(170) gives $2Q(t,\nu)$ to 4 dec. for substantially the same arguments as given for $P(t,\nu)$ by Hartley & Pearson 1950, catalogued on p. 208 above.

3.6
4.1313

<u>4.112</u> Tables with argument $t\nu^{-\frac{1}{2}}$
[see also p. 209]

5 dec. $\nu+1 = 2(1)10$ $z = 0(.1)3$ K. Pearson 1930 a (36)

<u>4.115</u> Approximations to the t integral

In addition to the works cited on p. 210 above, see Chu 1956.

<u>4.13</u> Percentage points of the t-distribution

<u>4.1311</u> Double tail percentage points: $t(Q,\nu)$ as a function of 2Q
[see also pp. 211-212]

$\nu = $; [120/ν =]   2Q =

3 dec.	1(1)30; [0(1)4]	.001, .01, .02, .05, .1(.1).9	Burington & May 1953(283), Czechowski et al. 1957(77), Vianelli 1959(173)
3; 5 dec.	1(1)30; ∞	.01, .02, .05, .1(.1).9	Ferber 1949(487)
2 dec.	1(1)10; [0(1)6(2)12]	.001, .002, .01, .02, .05, .1, .25	Lindley & Miller 1953(6)
3 dec.	1(1)30; [0(1)4]	.001, .01, .02, .05, .1, .2	Siegel 1956(248)

See also Olekiewicz 1951 b.

<u>4.1312</u> Double tail percentage points: $t(Q,\nu)$ as a function of 1 - 2Q
[see also p. 213]

2 dec.	1(1)30, ∞	.1, .2, .5, .9, .95, .98, .99	Royo López & Ferrer Martín 1955(31)
2 dec. 3 dec.	1(1)20 21(1)30(5)50(10) 100(20)200(100)500, 1000, ∞	.95, .99, .999	Graf & Henning 1953(61)

<u>4.1313</u> Single tail percentage points: $t(Q,\nu)$ as a function of Q
[see also p. 213]

3 dec.   1(1)30; [0(1)4]   .0005, .005, .01, .025, .05, .1   Siegel 1956(248)

4.132 Approximations to percentage points of t
[see also pp. 213-216]

Burrau, Ø. 1943

<u>Matematisk Tidskrift</u> 1943 B, 9-16

Review by S. S. Wilks (MTAC 2, 74):

Suppose $x_1, x_2, \cdots, x_n$ are values of x in a sample of n items from a normal population with mean a and standard deviation σ. W. S. GOSSET, under the pseudonym "Student,"[1] in 1908 conjectured that the sampling distribution of

$$z = \frac{\bar{x} - a}{\sqrt{\frac{1}{n}\sum_{i=1}^{n}(x_i - \bar{x})^2}}$$

was given by[2]

(1) $$f(z)dz = C(1 + z^2)^{-\frac{1}{2}n}dz$$

where C is a constant, such that $\int_{-\infty}^{\infty} f(z)dz = 1$. This conjecture was verified in 1925 by R. A. FISHER.[3] Gosset originally tabulated values of $\int_{-\infty}^{z} f(v)dv$ for which $z = .1(.1)3$, and $n = 4(1)10$. Fisher[4] has tabulated values of $t = z\sqrt{(n-1)}$ for which

$$\int_{-z\sqrt{(n-1)}}^{+z\sqrt{(n-1)}} f(v)dv = .1(.1).9, .95, .98, .99,$$

and for $n = 1(1)30$.

Mr. Burrau tabulates values of the integral

(2) $$\int_{-z\sqrt{(n-1)}}^{z\sqrt{(n-1)}} f(v)dv$$

for which $z\sqrt{(n-1)} = 0(.2)5$, and for $n = 4(2)12$ and ∞. The case for $n = \infty$ is, of course, equivalent to finding the values of the Gaussian integral $\frac{1}{\sqrt{2\pi}}\int_{-t}^{t} e^{-\frac{1}{2}z^2}dz$ for $t = 0(.2)5$.

He then observes that since the mean value of z^2, i.e. $E(z^2)$, is $1/(n-3)$, thus making $E[(n-1)z^2] = 1$, one might consider the tabulation of the integral

(3) $$\int_{-z\sqrt{(n-3)}}^{z\sqrt{(n-3)}} f(v)dv$$

for $z\sqrt{(n-3)} = 0(.2)5$, and for $n = 4(2)12$ and ∞, rather than the analogous tabulation of (2), with the hope of obtaining values which change relatively little as n changes. The tabulation of (3) is made and the resulting values are considerably more constant with respect to changes in n than the analogous tabulations of (2). For example, the values of

---

[1] Gosset 1908; Gosset 1942, pp. 11-34

[2] The z being used here is that originally used by "Student," and is being used throughout this review. Mr. Burrau, although actually writing about "Student's" problem, unfortunately denoted $z/n^{\frac{1}{2}}$ (in "Student's" sense) by the letter z.

[3] R. A. Fisher 1925 y   [4] Fisher & Yates 1938 etc. (T. III)

(2) and (3) for $z\sqrt{(n-1)} = 2$ and $z\sqrt{(n-3)} = 2$ are:

n	Value of (2) for $z\sqrt{(n-1)} = 2$	Value of (3) for $z\sqrt{(n-3)} = 2$
4	0.861	0.959
6	0.898	0.950
8	0.914	0.950
10	0.923	0.950
12	0.929	0.954
∞	0.954	0.954.

This indicates that the "significance level" corresponding to $\pm z\sqrt{(n-3)}$ is approximately 95 per cent for all values of $n \geqq 4$. The author does not discuss the problem for $n < 4$.

4.139 Miscellaneous significance tests related to t
[see also pp. 217-218]

Sengupta 1953 tables $(\nu+1)^{-1} + \nu/(\nu+1)[t(Q,\nu)]^2$ to 4 dec. for $2Q = .01, .05$; $N = \nu+1 = 2(1)30(10)60, \infty$.

Banerjee, D. P. 1954

Journal of the Indian Society of Agricultural Statistics 6, 93-100

Review by C. C. Craig (MTAC 11, 28):

For samples of N from a normal bivariate universe, Banerjee has tabled to 3 fig. the upper 80%, 90%, 95% and 99% points of the distribution of the ratio of the two sample standard deviations for N = 3(1)30 and ρ = 0(.1).9, where ρ is the universe coefficient of correlation. For ρ = 0 the values given in a high proportion of cases are one less in the third figure than the 3 fig. square root of the corresponding variance ratio, F, given in the standard tables.

Proschan, F. 1953

See review on p. 493 above.

See also Burrau 1954.

4.1421 The confluent hypergeometric function $M(\alpha, \gamma, x)$ for general α and γ
[see also p. 227]

8 fig.	$\gamma = .1(.1)1$	$\alpha = -1(.1)+1$	$x = .1(.1)10$	Slater 1960(132)
8 fig.	.1(.1)1	-1(.1)+1	1(1)10	Slater 1953
7 dec.	-4(.2)+1	-11(.2)+2	1	Slater 1960(234)

4.152
4.19

4.152 Welch's test
[see also pp. 237-239]

Trickett, W. H., Welch, B. L. and James, G. S.   1956

Biometrika 43, 203-205

(For definitions, see abstract of Aspin 1949, p. 237, above)

The table on pp. 204-205 gives upper $2\frac{1}{2}\%$ points of v for ν_1, $\nu_2 = 8$, 10, 12, 15, 20, ∞; and upper $\frac{1}{2}\%$ points for $120/\nu_1$, $120/\nu_2 = 0, 4(2)12$; all to 2 dec. for $\lambda_1\sigma_1^2/(\lambda_1\sigma_1^2 + \lambda_2\sigma_2^2) = 0(.1)1$.

Reprinted in Vianelli 1959, Prontuario 96, pp. 261-262.

4.17 Multivariate t-distributions
[see also pp. 241-242]

Dunnett, C. W.   1955

Journal of the American Statistical Association 50, 1096-1121

Suppose there are available N_0 observations on a control and N_i observations on the ith of p treatments. Let X_{ij} denote the jth observation under treatment i (i = 0 indicates the control), and set $X_{i.} = N_i^{-1} \sum_j X_{ij}$. Assume that the X_{ij} are normally and independently distributed with means m_i and variance σ^2; and let s^2 be an estimate of σ^2, independent of the X_i, and based on ν d.f. Set $sz_i = [X_{i.} - X_{0.} - (m_i - m_0)][N_i^{-1} + N_0^{-1}]^{-\frac{1}{2}}$. Dunnett contemplates confidence statements of the forms max z_i < d', max$|z_i|$ < d".

Table 1, pp. 117-118, gives d' to 2 dec. for p = 1(1)9; $\nu = 5(1)20$, $120/\nu = 0(1)6$; confidence coefficient 95% and 99%. Table 2, pp. 119-120, does the same for d". The tables embody certain approximations, and are strictly valid only when all N_i are equal.

Reprinted in Vianelli 1959, Prontuario 170, pp. 513-514.

See also Nomachi 1955.

4.18 Confidence intervals for a weighted mean

James, G. S.   1956

Biometrika 43, 304-321

Let $x = (w_1x_1 + w_2x_2)/(w_1 + w_2)$ be the mean of two independent normally distributed variates x_1 and x_2, which have the same expected value μ, and variances $\lambda_1 \sigma_1^2$ and $\lambda_2 \sigma_2^2$; the weight $w_i = 1/(\lambda_i s_i^2)$, where the s_i^2 are distributed as $\sigma_i^2 \chi^2 / \nu_i$, independently of each other and of the x_i. Set $r = w_1/(w_1+w_2)$, $u = (x-\mu)(w_1+w_2)^{\frac{1}{2}}$.

Table 1, p. 318, gives upper 5% points of u for ν_1, $\nu_2 = 6(2)10(5)20$, ∞; T. 2, p. 319, gives upper $2\frac{1}{2}$% points for ν_1, $\nu_2 = 8(2)12, 15, 20, \infty$; TT. 3, 4, pp. 320, 321, give upper 1% and $\frac{1}{2}$% points for $120/\nu_1$, $120/\nu_2 = 0$, $4(2)12$: all to 2 dec. for $r = 0(.1)1$.

4.19 Miscellaneous tabulations connected with t
[see also pp. 242-244]

Walsh, J. E. 1949 z

Journal of the American Statistical Association 44, 122-125

Review by Carl F. Kossack (MTAC 4, 151):

Table 1 gives the approximate number of sample values 'wasted' if, when the population variance is known, one uses a t-test (estimating variance from sample) in place of the appropriate normal deviate test, when testing whether the population mean differs from a given constant. 5%, $2\frac{1}{2}$%, 1% and $\frac{1}{2}$% significance points are tabled to 2 fig. for both the one-sided and the symmetrical test. 'Waste' is defined in terms of equal power functions.

See also Olekiewicz 1950.

4.231 / 4.33 Tables of percentage points of F and z
[see also pp. 248-251]

4.2311 Tables with material from both families A and B

Q =		$\nu_1 = [\ 120/\nu_1 =]$	$\nu_2 = [\ 120/\nu_2 =]$	
2 dec.	.001, .01, .05, .1, .2	1(1)6 [0(5)20]	1(1)30 [0(1)4]	
		1(1)10(5)20, 30, 50, 100, 200, 500, ∞	1(1)20(2)30(10)60(20)100, 200, 500, ∞	⎫ ⎬ Vianelli 1959(179) ⎭
	.0005			
3 fig.	.005, .025	1(1)20(2)30(5)50, 60(20)100, 200, 500, ∞	1(1)30(2)50(5)70(10)100(25)150, 200, 300, 500, 1000, ∞	

4.2312 Tables of family A

2 dec.	.001, .01, .025, .05	1(1)8, 10 [0(5)10]	1(1)30(2)40 [0(1)3]	Lindley & Miller 1953(8)
		1(1)10(5)20, 30, 50, 100, 200, 500, ∞	1(1)20(2)30(10)60(20)100, 200, 500, ∞	⎫ ⎬ Czechowski et al. 1957(83) ⎭
	.1			
3 fig.	.01, .05	1(1)20(2)30(5)50, 60(20)100, 200, 500, ∞	1(1)30(2)50, 60(20)100(25)150, 200, 300, 500, 1000, ∞	

4.2313 Tables of family B

| 1 dec. | .001, .01, .05 | 1(1)6, 8, ∞ | 1(1)30, ∞ | Royo López & Ferrer Martín 1955(32) |

4.331 Tables of percentage points of z

| 4 dec. | .001, .01, .05, .1, .2 | 1(1)6 [0(5)20] | 1(1)30 [0(1)4] | Vianelli 1959(176) |
| 4 dec. | .01, .05, .1 | 1(1)6 [0(5)20] | 1(1)30 [0(1)4] | Czechowski et al. 1957(80) |

                                                    4.2311   [4.331]
                                                    4.241

4.239 Miscellaneous combinations of percentage points of F
       [see also pp. 253-254]

Snedecor 1946, T. 13.6, is reprinted in Ferber 1949, T. 24, pp. 520-521.

For Eisenhart, Hastay & Wallis 1947, TT. 8.3-8.4, see abstract on p. 258 above.

Linder, Arthur 1951

Statistische Methoden

Review by C. C. Craig (MTAC 6, 155; abridged):

Besides standard statistical tables, this book includes a table for the direct evaluation of the significance levels of regression coefficients: 5%, 1%, .1% points are given to 4 fig. for 1(1)6 indepdndent variables and for $\nu = 1(1)30$, $120/\nu = 1(1)4$.

4.241 Power of the F-test, in terms of the central F-distribution
       [see also pp. 256-258]

Moriguti, Sigeiti 1954

Reports of Statistical Application Research,
Union of Japanese Scientists and Engineers, 3, 29-41

Review by C. C. Craig (MTAC 11, 32):

If, as in the case of an analysis of variance with random effects, one has a mean square V_1 with ν_1 d.f., whose expected value is $\sigma^2 + n\nu$, such that $\nu_1 V_1 / \sigma^2$ obeys a χ^2 distribution; and one also has an independent mean square V with ν d.f., whose expected value is σ^2, such that $\nu V/\sigma^2$ obeys a χ^2 distribution; then it is of interest to determine confidence intervals for the variance component v, n being a known constant and σ^2 a nuisance parameter. Moriguti derives his approximation formulas for $100(1-\alpha)\%$ confidence limits in which for each limit two parameters enter linearly. These parameters are tabled for $\alpha = .1$ to 4 fig. or 3 dec. for $\nu = 6(2)12$, $120/\nu = 2(2)10$; and $\nu_1 = 1(1)6(2)12$, $120/\nu_1 = 0(2)10$. Comparisons are made with previously obtained approximations which favor the present one.

Ray, W. D. 1956

Biometrika 43, 388-403

Review by R. L. Anderson (MTAC 12, 77):

Given the one-way classification model

$$x_{ti} = a + b_t + z_{ti} (t = 1, \cdots, k; i = 1, \cdots, n),$$

where the z_{ti} are uncorrelated with zero means and equal variances, σ^2, the b_t are

4.241
4.242

treatment parameters ($\sum b_t = 0$). We consider the non-centrality parameter $\delta = \sum b_t/k\sigma^2$ and the testing statistic

$$G = \frac{\text{Treatment Sum of Squares}}{\text{Error Sum of Squares}}.$$

Two hypotheses are considered: $H_0(\delta = 0)$ and H_1 (δ some specified value $\neq 0$). A sequential sampling procedure is set up (consider successive values of n) so that after n samples per treatment one of these 3 decisions is made:

Result of n samples per treatment		Decision
$Me^{-\lambda/2} \geq$	$(1-\beta)/\alpha$	accept H_1
$Me^{-\lambda/2} \leq$	$\beta/(1-\alpha)$	accept H_0
$\beta/(1-\alpha) < Me^{-\lambda/2} <$	$(1-\beta)/\alpha$	take more samples

M is a confluent hypergeometric function,

$$M\left[\frac{kn-1}{2}, \frac{k-1}{2}; G/2(1+G)\right];$$

$\lambda = kn\delta$; and α and β are the usual Type I and Type II error rates.

Let $M(\bar{G})e^{-\lambda/2} = (1-\beta)/\alpha$ and $M(\underline{G})e^{-\lambda/2} = \beta/(1-\alpha)$. Values of $\underline{G}$ and $\bar{G}$ to 3D have been computed for $\alpha = \beta = .05$; selected values of n; for $\delta = 0.5$, $k = 2(1)7$; for $\delta = 1.0$, $k = 2(1)10$; for $\delta = 2.0$, $k = 2(1)6$. Certain of the values of $\underline{G}$ and $\bar{G}$ were computed by approximation methods. Tables 7 and 8 present some comparisons of exact and approximate results.

The results are extended to a two-way classification model with k blocks and n treatments (note the unfortunate interchange of n and k from the one-way model). In this case $kn - 1$ is changed to $k(n-1)$ and $k-1$ to $n-1$ in M. Tables of $\underline{G}$ and $\bar{G}$ generally to 3D are presented as before, except that for $\delta = 0.5$ and 1.0, $k = 2(1)8$. Conjectural expected sample sizes are presented to go with these tables, based on a generalization of Wald's formula.

See also N. L. Johnson 1953.

4.242 Power of the F-test, in terms of the non-central F-distribution
[see also pp. 258-260]

Ura, Shoji 1954

Reports of Statistical Application Research,
Union of Japanese Scientists and Engineers, 3, 23-28

Review by C. C. Craig (MTAC 11, 32):

Ura extends the inverse tables of E. Lehmer 1944 for probabilities of errors of the second kind in the variance ratio test ordinarily used in the analysis of variance. He develops a formula for the necessary power function which he credits to J. Yamauti (apparently hitherto unpublished) and employs it to tabulate values of the quantity $\psi = [(\nu_1+1)/\nu_1]^{\frac{1}{2}}\varphi$, where φ is the quantity tabled by Lehmer and introduced by P.C. Tang 1938, for which the significance level is .05 and the probability of an error of the second kind is .1. Values are given to 2 dec. for $\nu_1 = 1(1)10$, $120/\nu_1 = 0(1)6(2)12$; $\nu_2 = 2(2)20$, $120/\nu_2 = 0(1)6$.

Hodges, J. L., Jr. 1955

Annals of Mathematical Statistics 26, 648–653

Review by L. A. Aroian (MTAC 11, 31):

Nicholson [1] has derived a closed expression for B, the noncentral beta-distribution in case b is an integer:

$$B(x; a, b, \lambda) = 1 - e^{-\lambda x}\{I_{1-x}(a, b) + (1-x)^a \sum_{j=1}^{b-1} [x(1-x)\lambda]^j (P_j/j!)\}$$

where

$$P_j = \sum_{k=0}^{b-j-1} \left[(-1)^k \binom{b-j-1}{k} \frac{(a+b-1)(a+b-2)\cdots(a+j)}{(b-j-1)!(a+j+k)}\right](1-x)^k,$$

and $I_x(a, b)$ is the beta-distribution. The author proves that $P_j = \sum_{t=0}^{b-j-1} \binom{A+t}{t}x^t$, $A = a + j - 1$. A table of $\binom{A+t}{t}$ to 7S is provided for $A = .5(1)19.5$, $t = 1(1)18$. The author compares these direct methods for computing B with Tang's recursion formula [2].

Marakthavalli, N.

Sankhyā 15, 321–330.

Review by R. L. Anderson (MTAC 12, 74):

Given a sample of n observations $\{x_i\}$ from normal populations with common variance σ^2 but unequal means $\{a_i\}$. Then $\sum_{i=1}^{n} x_i^2/\sigma^2$ is distributed as a non-central χ'^2 with n degrees of freedom and non-centrality parameter, $\lambda = \sum_{i=1}^{n} a_i^2/\sigma^2$. The author considers the non-central F'-distribution, which is the ratio of a non-central $\chi_1'^2$ and a central χ^2 with respective degrees of freedom, ν_1 and ν_2, under the null hypothesis, $H_0: \lambda = \lambda_0$. A table is presented to determine the rejection region for an unbiased test of H_0 against the two-sided alternative $\lambda \neq \lambda_0$. The region is: $F' \leq a_1$ or $\geq a_2$, where Prob $(F' \leq a_1 | H_0) = \alpha_1$ and Prob $(F' \geq a_2 | H_0) = \alpha_2$, $\alpha_1 + \alpha_2 = \alpha = .05$. This table presents values of α_1 and α_2 to 3D, a_1 generally to 2D and a_2 to 1D, for $\nu_1 = 1(1)8$, $\nu_2 = 4(4)20, 40, 60, \infty$ and

$$\varphi = \sqrt{\frac{\lambda_0}{\nu_1+1}} = 1, 2.$$

The author also indicates how this table can be used for analysis of variance, correlation ratio, multiple correlation, and discriminant analysis problems.

---

[1] W. L. Nicholson 1954   [2] Tang 1938

4.242
4.28

Fox, M. 1956

Annals of Mathematical Statistics 27, 484-497

Review by H. A. Freeman (MTAC 12, 75):

Write f_1 and f_2 for the degrees of freedom of numerator and denominator, respectively, of F. Let α be the size and β the power of F. Write

$$\phi = [S_e^*/(f_1 + 1)\sigma^2]^{\frac{1}{2}},$$

where S_e^* is the value of S_b^2 when the observable random variables are replaced by their expectations under the alternative hypothesis, and S_b^2 is the sum of squares in the numerator of F. Eight charts show constant ϕ for all combinations of $\beta = .5, .7, .8, .9$, and $\alpha = .01, .05$. The charts are designed to show readily the combination of f_1 and f_2 required to obtain power β against a given alternative. A detailed numerical example, two nomograms facilitating interpolation in β for both values of α, and references to earlier tables are included.

4.251 The largest variance ratio

[see also p. 261]

Ramachandran, K. V. 1956

Annals of Mathematical Statistics 27, 521-528

Review by R. L. Anderson (MTAC 12, 75):

The author first considers the general problem of determining the a_i in the equation

$$P\left[\frac{S_i}{S} \leq a_i t_i/m; i = 1, 2, \cdots, k\right] = 1 - \alpha,$$

where the S_i and S are mutually independent; S_i/σ^2 is distributed as χ^2 with t_i d.f. and S^2/σ^2 as χ^2 with m d.f. A method of evaluating P is presented and then a special method is developed for the case $t_i = t$, for all i. In the latter case P becomes

$$P\left[u = \frac{S_{\max}}{S} \leq b\right],$$

where $S_{\max}$ is the maximum of the S_i and $b = at/m$ ($a_i = a$, for all i). Upper 5% points of u are given to 2D for $k = 2$; $t = 1(1)4(2)12, 16, 20$; and

$$m = 5(1)8(2)12(4)24, \infty.$$

4.254 The ranking of mean squares

Bechhofer, R. E. and Sobel, M. 1954

Annals of Mathematical Statistics 25, 273-289

Review by J. L. Hodges, Jr. (MTAC 9, 34):

Let U, V, W, X be independent χ^2 random variables, each with ν d.f. The paper provides 5 dec. tables of $\Pr[U > \epsilon V]$, $\Pr[U > \epsilon V, \epsilon W]$, $\Pr[U, V > \epsilon W]$, $\Pr[U > \epsilon V > \epsilon^2 W]$ and $\Pr[U > \epsilon V, \epsilon W, \epsilon X]$ for $\nu = 1(1)20$ and $\epsilon = 1.2(.2)2.2$. The tables provide the confidence coefficients of certain statements about the order of the variances of normal populations.

See also H. A. David 1956.

4.28 Effect of non-normality, heteroscedasticity and dependence on the distribution of F

[see also pp. 264-269]

Gayen, A. K. 1950 y

Biometrika 37, 236-255

Review by C. F. Kossack (MTAC 5, 144):

The author is interested in the non-normal sampling distribution of the two statistics: w = the ratio of two estimates of the variance obtained in a one-way classification for the analysis of variance (homogeneity), v = the variance ratio of two independent samples (compatibility). To introduce non-normal populations the EDGEWORTH series,

$$f(x) = \varphi(x) - \frac{\lambda_3}{6}\varphi^{(3)}(x) + \frac{\lambda_4}{24}\varphi^{(4)}(x) + \frac{\lambda_3^2}{72}\varphi^{(6)}(x),$$

is used. With an Edgeworth series population the author evolves the frequency densities of the two statistics. These distributions are then related to their corresponding "normal" sampling distributions and correction terms for the skewness and excess are determined. The coefficients of $\lambda_3 (= \sqrt{\beta_1})$ and $\lambda_4 (= \beta_2 - 3)$ are tabulated in Table 4 (p. 252-255) to 4D for the 5 percent points of Fisher's Z for the degrees of freedom $\nu_1 = 1(1)6, 8, 12, 24, \infty$; $\nu_2 = 1(1)6, 8, 12, 20, 24, 30, 40, 60, 120, \infty$. These tables enable one to compute the error one would make in using the ordinary tables associated with the normal population distribution of Z for the 5 percent level if he were aware of the λ_3 and λ_4 of his population and was willing to assume an Edgeworth type distribution. In this connection the author notes that the Edgeworth restriction would not be serious if the size of sample was sufficient to enable one to neglect terms of order N^{-3}. If λ_3 and λ_4 are unknown one might use their sample estimate in the formulas to get an indication of the size of error involved.

4.28
5.4
Weibull, M. 1953

Skandinavisk Aktuarietidskrift 36, no. 1-2, supplement, 106 pp.

Review by C.A. Bennett (MTAC 9, 217):

This long paper is principally concerned with the application of the standard statistics used in exact significance tests in samples from normal [universes] to samples from stratified universes in which the distributions within each stratum are normal with variances constant in all strata but with means different in different strata. It begins with a discussion of non-central χ^2, t- and F-statistics and includes an illustrative table of the non-central F distribution in which both the χ^2s appearing in the numerator and the denominator are non-central: 1% and 5% points of this F are given to 2 dec. for d.f. for the numerator b_1 = 1, 2, 4, 8; d.f. for the denominator b_2 = 1, 2, 4, 8, 16, 32, ∞; but only for $b_1 < b_2$; and the non-centrality parameters β_1 and β_2 = 0, $\frac{1}{2}$, 1.

4.331 Tables of percentage points of z

See pp. 250, **676**.

Section 5. Various discrete and geometrical probabilities

We call attention to the highly compressed information about many discrete distributions in Haight 1955, pp. 22-27.

5.21 Contagious distributions
[see also pp. 278-282]

Douglas, J. B. 1955

Biometrics 11, 145-173

For description of tables of Poisson moments, see section 2.53, p. △△△, above; for their application to Neyman's Type A contagious distribution, see Douglas 1955, pp. 150-153; Vianelli 1959, p. 1333.

See also Bailey 1953, Tanner 1953.

5.31 Tables of $^{N}H_{r}$

In addition to tables listed on p. 284 above, see Hodges 1953.

5.32 The negative binomial distribution
[see also p. 285]

For the truncated negative binomial distribution, see Sampford 1955.

Vianelli, S. 1959

Prontuari per calcoli statistici

Setting $r = 0$ in equation (5.3.1), p. 284, above, we obtain $\Pr[x = 0] = (1+d)^{-h/d}$. Prontuario 195, pp. 580-587, tables this to 7 dec. for $h = .01(.01).8$ and $d = .01(.01).5$; Travola 173, pp. 1331-1332, gives $\frac{h}{d} \log_e (1+d)$ to 5 dec. for △△△. $h = .02(.02).8$ and $d = .02(.02).5$

See also Taguti 1952.

5.4 Multinomials
[see also p. 287]

Wishart, J. 1949

Biometrika 36, 47-58

Review by Joe J. Livers (MTAC 4, 210):

The univariate Bernoulli and Pascal multinomial distributions are first considered. Using cumulant generating functions recurrence relations are obtained from which cumulants to order four are recorded.

Bivariate cumulants to order four are found by recurrence formulae paralleling the univariate case and are also recorded.

Extension to the multivariate case follows from the simpler univariate and bivariate cases. Of importance is the use of a notation that makes the corresponding cumulants of the Bernoulli and Pascal distributions greatly resemble each other. A complete list of auxiliary patterns and the cumulants to the fourth order is given for a particular case (5 × 4 × 3 × 2) of the 4-variate multinomial Bernoulli distribution.

There are misprints on pp. 52-53.

5.6 Probabilities in card games

5.61 Packs of 52 cards

Pearson, K. 1924

Biometrika 16, 172-188

"On a certain double hypergeometrical series and its representation by continuous frequency surfaces"

If all 52! shufflings of a pack are equally likely, and the pack is dealt into 4 hands, then the probability that North (say) have s spades and South have t spades is

$$39!\,26!\,(13!)^3 / 52!\,s!\,t!\,(13-s)!\,(13-t)!\,(13-s-t)!\,(13+s+t)!$$

Table 3, p. 186, compares this with the results of actual play at whist; T. 4, p. 187, with hands shuffled and dealt but not played.

Table 1, p. 184, gives the standard deviation of s given t; T. 2, p. 185, the (hypergeometric) marginal distribution of s; T. 5, p. 188, the conditional distribution of s given t.

Carver, H. C. 1931

Annals of Mathematical Statistics 2, 82-98

Shuffle and deal a pack as in the previous abstract. Table 1, p. 83, gives the marginal distribution of s as exact vulgar fractions and to 5 dec.

Assign a count c to the North hand by scoring 4 for each ace, 3 for each king, 2 for each queen, 1 for each knave, 0 for each other card; and adding the scores. Table 2, p. 84, gives the distribution of c as exact vulgar fractions.

Greenwood, R. E. 1953

Journal of the American Statistical Association 48, 88-93

"Probabilities of certain solitaire card games"

Two identical packs of 52 cards, each pack containing 4 cards each numbered 1(1)13, are randomly and independently shuffled. One card is simultaneously dealt from the top of each pack, and the numbers on the two cards so dealt are compared; if they are equal a match is scored. This process is repeated through the two packs, yielding a total matching score between 0 and 52 ; a score of 51 is impossible.

Table 2, p. 91, gives the probability of a matching score of 0(1)7 to 3-6 dec., and some Gram-Charlier approximations thereto.

Borel, É. et Chéron, A. 1940

<u>Théorie mathématique du Bridge</u>

Review by Mark Eudey (MTAC 2, 305):

The title-page description shows the impracticability of attempting to list all the tables in this exhaustive work. These tables fall into 3 main groups: 1, a priori; 2, bidding; 3, play. In the first group the a priori probabilities of all distributions of the 4 suits among the 4 players, of the suits between the 2 partnerships, of the distributions of aces, aces and kings, etc., and the probabilities of voids, singletons, etc., are given. In the second group similar tables are given except now we know the 13 cards in the bidder's hand. The third group of tables mainly covers the probabilities after the dummy has been exposed and we know 26 cards, plus derived tables to cover the cases where part of a suit has been played and we are interested in the distribution of the remainder.

The basic probabilities were calculated as exact vulgar fractions, with the aid of a Pascal triangle complete up to $\binom{52}{26}$ which Borel & Chéron possess, and are given generally to about 6 or 7 dec., but up to 15 dec. in some cases. Formulae for making the calculations are given in the text, and the use of the Tables in evaluating a hand or selecting the best method of play is illustrated.

Borel & Chéron point out in the first chapter that ordinary shuffling is quite apt to give results that are far from the random ordering which is assumed in their later calculations.

5.62 Other packs of cards

Huntington, E. V. 1937

<u>Journal of Parapsychology</u> 1, 292-294

"A rating table for card-matching experiments"

Two identical packs of 25 cards, each pack containing 5 cards in each of 5 suits, are randomly and independently shuffled. One card is simultaneously dealt from the top of each pack, and the cards so dealt are compared; if they are of the same suit, a match is scored. This process is repeated through the two packs, yielding a total matching score, x, between 0 and 25 ; x cannot equal 24.

5.62

Sterne 1937 showed that $E(x) = 5$, $\mu_2(x) = 25/6$, $\mu_3(x) = 125/46$, $\mu_4(x) = 3625/69$. Hence Huntington obtains the moments of the average score, $\bar{x}$, in n games; and approximates the distribution of $\bar{x}$ by a Gram-Charlier series.

The table on p. 293 gives $\Pr[\bar{x} \geq S]$ to 3-4 dec. for n = 4, 8, 16, 100 and selected values of S 4; the table on p. 294 gives upper 100 Q% points of $\bar{x}$ to 2 dec. for $Q = 1/75, 1/500, 10^{-4}$; n = 1, 4(4)20, 50, 100.

Huntington, E. V. 1937 a

Science 86, 499-500

"Exact probabilities in certain card-matching problems"

The problem of Huntington 1937 is solved exactly for certain packs of less than 25 cards. Table A, p. 499, gives the probabilities of all attainable matching scores for a pack of 3 suits of 3 cards each, as exact vulgar fractions and to 5 dec. Table B, p. 500, gives the probabilities of all attainable matching scores for a pack of 4 suits of 4 cards each, as exact vulgar fractions and to 8 dec.

Sterne, T. E. 1937

Science 86, 500-501

"The solution of a problem in probability"

Table 1, p. 500, gives binomial and Pearson Type I approximations to the distribution of matching scores for a pack of 5 suits of 5 cards each to 4 dec. Table 2, p. 500, gives the exact distribution of matching scores for a pack of 4 suits of 4 cards each, and a Pearson Type I approximation, both to 4 dec.

Greville, T. N. E. 1938

Journal of Parapsychology 2, 55-59

"Exact probabilities for the matching hypothesis"

For definition of score x see the abstract of Huntington 1937, above. $\Pr[x = X]$ is tabled on p. 59 as exact vulgar fractions and to 7 dec.; the denominator of the fractions, 623 360 743 125 120, is given on p. 57.

Joseph Albert Greenwood 1938 gives the variance of certain card-matching distributions, but no numerical tables.

Spitz, J. C. 1953

Statistica Neerlandica 7, 23-40

Review by J. H. B. Kemperman (MTAC 9, 35):

Let p_r be the probability of exactly r matches in a random matching of two similar decks of 3 distinct cards. Then, $p_0 = \frac{1}{3}$, $p_1 = \frac{1}{2}$, $p_2 = 0$, $p_3 = \frac{1}{6}$. In a series of n such random matchings, let $R = r_1 + \cdots + r_n$ be the total number of matches ($0 \leq R \leq 3n$) and let $P_{n,a} = \Pr(R \geq a)$. Obviously,

(1) $$P_{n,a} = \tfrac{1}{3} P_{n-1,a} + \tfrac{1}{2} P_{n-1,a-1} + \tfrac{1}{6} P_{n-1,a-3}$$

from which the value $P_{n,a}$ is tabulated to 3D for $n = 1(1)30$ and $a = 0(1)3n$. If in an actual experiment the number $R = a$ of matches is such that $P_{n,a} \leq .05$ (say), one rejects the hypothesis that the matchings were random.

For large n, by the central limit theorem, R is approximately normal. Now, R has a mean $\mu = n$, a variance $\sigma^2 = n$ and a skewness $\gamma_1 = 1/\sqrt{n}$. Thus, for large n, $t = (R - \tfrac{1}{2} - n)/\sqrt{n}$ is about $N(0, 1)$. For $n = 30$, the resulting approximation to $P_{n,a}$ appears to be fairly good. An even better approximation is obtained by replacing t by a type III variable with $\mu = \sigma = 1$, $\gamma_1 = 1/\sqrt{n}$; then $P_{n,a}$ can be readily computed from SALVOSA's tables.[1]

The reviewer would expect a good approximation to $P_{n,a}$ by replacing R by a Poisson variable with parameter n which has $\mu = \sigma^2 = n$, $\gamma_1 = 1/\sqrt{n}$.

Gilbert, E. J. 1956

Psychometrika 21, 253-266

Review by T. N. E. Greville (MTAC 12, 70):

Each of two decks of cards consists of s suits of c cards per suit. One deck is thoroughly shuffled, and each card is then compared with the card in the same ordinal position in the other deck. If the two cards compared are of the suit, a 'matching' is recorded. The probability of h or more matchings is tabled to 5 dec. for

$\begin{cases} c = 1, 2 & 3 & 4, 5 & 6, 7 & 8(1)12 \\ s = 2(1)11 & 2(1)8 & 2(1)5 & 2, 3 & 2 \end{cases}$; in each case for all possible

values of h. The fitting of various standard statistical distributions as an approximation to other cases is discussed. It is concluded that the normal distribution is frequently wide of the mark, and that the Gram-Charlier type B is the most generally accurate.

---

[1] Salvosa 1930

5.9 Miscellanea [miscellaneous discrete distributions]
[see also pp. 292-294]

Chapman, D. G. 1955

Biometrika 42, 279-290

Review by R. L. Anderson (MTAC 12, 73):

A population consists of two classes with sizes X_i and Y_i ($X_i + Y_i = N_i$) at times t_i ($i = 0, 1$). Samples of n_0 and n_1 are taken from this population at t_0 and t_1, respectively. Between these two periods, $R_x = X_1 - X_0$ are removed from the X-class and $R_y = Y_1 - Y_0$ from the Y-class. Formulas are given to estimate N_0 and X_0 and the variances of the estimates. Table 1 presents values to 2D of the optimum sample ratio n_1/n_0 to estimate N_0 when $R_y = 0$ for selected values of (R_x/X_0): .01, .02, .05, and .1(.1).5. Table 2 presents simliar values to 3D to estimate X_0 for selected values of (X_0/N_0): .1(.1).9; and (R_x/N_0): .01, .02, and .05(.05).25.

The author also compares the cost of the above method relative to the familiar tagging or capture-recapture procedure. Table 3 presents minimum values to 2D of the relative cost of tagging to the cost of classification such that the variance of N_0 is greater for the tagging procedure; this table is given for the selected values (X_0/N_0) and (R_x/N_0) used in Table 2.

6.113 Bartlett's modification of the L_1 test

[see also pp. 300-302]

C. M. Thompson & Merrington 1946, TT. 1-2, described on p. 302 above, are reprinted in Vianelli 1959, Pro. 95, p. 260; T. 3 is reprinted in Vianelli 1959, T. 71, p. 1206.

6.114 Other tests of homogeneity of variance

In addition to works reviewed on pp. 302-303 above, see Moran 1953.

6.119 Miscellanea
[miscellaneous tests applied to univariate normal populations]
[see also p. 304]

Walsh, J. E. 1952

Journal of the American Statistical Association 47, 191-201

"Operating characteristics for tests of the stability of a normal population"

A sample of n observations x_i has been drawn from $N(\mu, \sigma^2)$: it is desired to test whether the sample conforms to the control standards $\mu = \mu_0$, $\sigma = \sigma_0$. Walsh constructs simple critical regions based on: $\bar{x}$ and s; s and $t = n^{\frac{1}{2}}(\bar{x}-\mu_0)/s$; $\bar{x}$ and $s' = [n^{-1}\sum(x_i-\mu_0)^2]^{\frac{1}{2}}$.

Table 1, p. 193, gives limits for t and s/σ to 4 fig. for n = 3, 5; α = .005, .01(.01).05 : T. 2, p. 194, gives limits for $\bar{x}-\mu_0$ and s'/σ for the same n and α. Tables 3-4, pp. 196-199, give power to 3-4 dec. for the three tests against the 34 alternatives $n^{\frac{1}{2}}(\mu-\mu_0)/\sigma$ = 0(1)4 ; σ/σ_0 = 1/8, 1/4, ½, 1, 2, 4, 8 : for α = .01, n = 3, 5 : and, for the test based on $\bar{x}$ and s, for α = .005, .01(.01).05; n = 3, 5, 7.

Bhattacharya, P. K. 1955

Calcutta Statistical Association, <u>Bulletin</u> 6, 73-90

Review by J. L. Hodges, Jr. (MTAC 12, 67):

Let $x_1, \ldots, x_n$ be a sample from $N(m, \sigma^2)$ and let $n\bar{x} = \sum x_i$, $ns_0^2 = \sum(x_i-m_0)^2$, $(n-1)s^2 = \sum(x_i - \bar{x})^2$, $st = n^{\frac{1}{2}}(\bar{x} - m_0)$. The problem of testing simultaneously H_1 [m = m_0] against m > m_0, and H_2 [$\sigma = \sigma_0$] against $\sigma > \sigma_0$, is regarded as a four-decision problem: we reject H_1 when $\bar{x} > c$, and H_2 when $s^2 > k$, where c and k are fixed by requiring that Pr[reject H_1] = Pr[reject H_2] and that Pr[accept both] = 0.95, when m = m_0 and $\sigma = \sigma_0$. A second procedure is obtained by replacing $\bar{x}$ by t and s^2 by s_0^2. For both procedures we are given the probabilities of the four decisions to 3 dec. for n = 5(5)20 ; σ^2/σ_0^2 = 1(.5)3.5 ; $n^{\frac{1}{2}}(m-m_0)$ = 0, .1, .4, 1, 1.5.

Reiter, S. 1956

<u>Journal of the American Statistical Association</u> 51, 481-488

Review by H. A. Freeman (MTAC 12, 70):

Let n_1, n_2 be sizes of independent random samples: let $\theta = n_1/n_2$. Enter Table 1 with θ = 1, 1.5, 2, 5, 10, 100, and for a required bounded relative error (in the sense of Blackwell & Girshick 1954, pp. 316-323) K = 1.01, 1.02, 1.05, 1.1, 1.25, 1.3, 1.5, find β^* to 4 dec. to be used in the 'best' estimate $\beta^*\hat{\sigma}_1^2/\hat{\sigma}_2^2$ of σ_1^2/σ_2^2, where $n_1\hat{\sigma}_1^2/\sigma_1^2$ and $n_2\hat{\sigma}_2^2/\sigma_2^2$ are required to be independently distributed as χ^2 with n_1 and n_2 d.f. The confidence coefficient associated with the estimate is then read from one of seven charts that show it as a function of n_1, n_2 and K (K as above, excluding K = 1.01).

6.12 Tests applied to bivariate normal populations

[see also pp. 305-306]

Marriott, F. H. C. 1952

Biometrika 39, 58-64

Review by I. Olkin (MTAC 7, 26).

Given two sets of variates $x_1, \cdots, x_p$ and $y_1, \cdots, y_q$ ($p \leq q$) and $n + 1$ observations on each set, the canonical correlations are defined as the p roots ($l_1^2 \geq l_2^2 \geq \cdots \geq l_p^2$) of the determinantal equation $|Q - l^2 T| = 0$, where T is the dispersion matrix of the y's; W is the dispersion matrix of the y's with the x's eliminated; $Q = T - W$ is the dispersion matrix due to regression.

The exact null distribution of the largest root l_1^2 is given by Roy[1] in a recursive form involving multiple integrals. The author gives the exact distribution of l_1^2 for $p = 2$ and $p = 3$, $q = 4$. A significance test which is exact for practical purposes is given for $p = 3$ and $p = 4$, $q = 5$. 5% and 1% significance levels to 2D (3 or 4S) for $\frac{1}{2} n l_1^2$ (n large) are given for $p = 2$, $q = 2(1)12, 21$; $p = 3$, $q = 3(1)12, 21$; $p = 4$, $q = 5$.

An approximate test,

$$\chi^2_{[D]} = - \{n - \tfrac{1}{2}(p + q + 1)\} \log (1 - l_1^2)$$

where $D = p + q - 1 + \tfrac{1}{2}\{(p-1)(q-1)\}^{\frac{1}{2}}$, based on Wilks' criterion[2] is proposed. 5% and 1% significance levels to 2D (3 or 4S) for $\frac{1}{2} n l_1^2$ (n large) are given for $p = 2$, $q = 2, 6, 12, 21$; $p = 3$, $q = 3, 6, 12, 21$; $p = 4$, $q = 5$ which compare favorably with the exact values. The derivation of the test is omitted, with reference to the author's unpublished thesis.[3] The 5% points of l_1^2 for the exact and approximate tests for $p = 2$, $q = 5$ are given to 2 or 3D for $n = 10, 20, 50, 100, \infty$, which indicate good results for n down to 20.

Chowdhury, S. B. 1954

Sankhyā 14, 71-80

Review by Julius Lieblein (MTAC 9, 220):

This continuation of an earlier paper by P. K. Bose [1] includes two short tables for determining the shortest unbiased upper and lower 95% confidence intervals for a parameter Δ^2 used in testing the equality of means of two p-variate normal populations. Let the (known) variance matrices of the two populations be $A = (\alpha_{ij})$, $A' = (\alpha'_{ij})$, and their unknown vector means be $\{\alpha_i\}$, $\{\alpha_i'\}$. In terms of these quantities,

---

[1] S. N. Roy 1945   [2] Wilks 1932

[3] F. H. C. Marriott, *The Analysis and Interpretation of Multiple Measurements.* 1951, University of Aberdeen.

[1] P. K. Bose 1947

$$\Delta^2 = p^{-1}\sum\sum \beta^{ij}(\alpha_i - \alpha_i')(\alpha_j - \alpha_j'),$$

where β^{ij} are the elements of the inverse matrix $(\beta_{ij})^{-1}$, $\beta_{ij} = (n'\alpha_{ij} + n\alpha'_{ij})/(n+n')$. To test the equality of means a random sample is drawn from each population, of, say, respective sizes n, n', and sample means $\{a_i\}$, $\{a_i'\}$. One then computes a sample statistic D_i^2 (to which the "classical D^2" statistic introduced by MAHALANOBIS is related by $D^2 = D_1^2 - 2/\bar{n}$), corresponding to Δ^2, by means of $D_1^2 = p^{-1}\sum\sum \beta^{ij}(a_i - a_i')(a_j - a_j')$, and from this, the quantities $l^2 = kD_1^2/2$, $L = (\bar{n})^{\frac{1}{2}}l$, where $\bar{n}$ is the harmonic mean of n, n'.

Table 1 gives, for $p = 1(1)10$ and for 14 values of L ranging from $L = 2$ to $L = 400$, the lower 95% confidence limit λ for the parameter λ related to Δ^2 in the same way as l is related to D_1^2, for all alternatives $\lambda > \lambda_0$. Here λ_0 represents the hypothesis to be tested, that is, λ_0 represents the bound for λ which would assure that Δ^2 is sufficiently small to make the means equal within desirable tolerances. The values of $\lambda = \lambda(L)$ are the solutions, for the given values of L, of

$$\int_0^L \psi(x, \lambda(L))dx = 1 - \alpha,$$

where $\alpha = .95$, and

$$\psi(L, \lambda) = L^{p/2}\lambda^{-(p-2)/2}e^{-(L^2+\lambda^2)/2}I_{(p-1)/2}(L\lambda)$$

(where I is the imaginary Bessel function) is the probability density function of L.

Table 2 gives, for the same values of p and L, the upper 95% confidence limit $\bar{\lambda}$, for all alternatives $\lambda < \lambda_0$, in precisely analogous manner to Table 1, with the limits of integration 0 to L in the equation for $\bar{\lambda}(L)$ being replaced by L to ∞.

Chowdhury, S. B. 1956

Calcutta Statistical Association, Bulletin 6, 181-188

Review by C. C. Craig (MTAC 12, 74):

In samples of n_1 and n_2 from normal p-variate universes, in order to test the hypothesis that the two universes have the same set of variances and covariances, the fundamental test statistics are the p roots, [here] called p-statistics, of the determinantal equation $|A'_{ij} - K^2 A''_{ij}| = 0$, where A' and A'' are the variance-covariance matrices of the two samples. With L defined as the largest root K^2, and $\vartheta = (n_1-1)L/[(n_2-1)+(n_1-1)L]$, Chowdhury gives the 5% and 1% points of ϑ to 2 dec. for $p = 2$ and $n_1 = n_2 = 9(1)14(2)26$. This extends the table of Nanda 1951. In addition, for $p = 3$, Chowdhury gives the 5% and 1% points of ϑ to 3 dec. for n_1, $n_2 = 5(1)9$; $n_1 = n_2 = 10(1)15(2)21$.

6.132
6.31

6.132 Other tests of homogeneity [of multivariate normal populations]
[see also pp. 308-311]

Foster, F. G. and Rees, D. H.   1957

Biometrika 44, 237-247

Let x_{im}, y_{in}, $i = 1(1)k$, $m = 1(1)\nu_1$, $n = 1(1)\nu_2$, be independent unit normal deviates. Define $a_{ij} = \sum x_{im} x_{jm}$, $b_{ij} = \sum y_{in} y_{jn}$. Let x be the largest root of the determinantal equation $|\theta a_{ij} + (\theta-1) b_{ij}| = 0$. Then the distribution of x is given by

$$I_x(k; p, q) = K \int_0^x d\theta_k \int_0^{\theta_k} d\theta_{k-1} \cdots \int_0^{\theta_2} d\theta_1 \prod_{i=1}^{k} \theta_i^{p-1} (1-\theta_i)^{q-1} \prod_{i>j} (\theta_i - \theta_j)$$

where $p = \frac{1}{2}(\nu_2 - k + 1)$, $q = \frac{1}{2}(|\nu_1 - k| + 1)$.

The table on pp. 245-247 gives the upper 100 α % points of x to 4 dec. for $k = 2$; α = .01, .05(.05).2; $p = \frac{1}{2}$, 1(1)10; $q = 2(1)20(5)50$, 60, 80.

Reprinted in Vianelli 1959, Pro. 213, pp. 687-689.

Foster, F. G.   1957

Biometrika 44, 441-453

The table on pp. 444-453 gives the upper 100 α % points of x to 4 dec. for $k = 3$; α = .01, .05(.05)2; $p = \frac{1}{2}(\frac{1}{2})4$; $q = 1(1)96$.

Reprinted in Vianelli 1959, Pro. 213, pp. 690-698.

Foster, F. G.   1958

Biometrika 45, 492-503

The table on pp. 494-503 gives the upper 100 α % points of x to 4 dec. for $k = 4$; α, p, q as in Foster 1957.

6.21 Tests applied to binomial populations
[see also p. 314]

Swineford, Frances   1949

Psychometrika 14, 183-187

Review by Leo Katz (MTAC 4, 150):

The two tables are designed to determine the least common size, N, for each of two samples in testing the hypothesis that the difference in the two population proportions, $p_1 - p_2$, is at least d_t. It is assumed that the sample proportions, p_1' and p_2', are distributed normally; so that the appropriate test is a one-tailed test of the hypothesis $p_1 - p_2 = d_t$. Then $N = 5.4119(p_1 q_1 + p_2 q_2)(d_0 - d_t)^{-2}$ at the 1% point and approximately half as much (1/2.0003) at the 5% point, where $d_0 = p_1' - p_2'$. The tables give N for the 1% points only.

Table 1 gives $N' = 10.8238 \, pq \, (d_0 - d_t)^{-2}$, where $2p = p_1' + p_2'$, to 0 dec. for $p = .1(.05).9$ and $|d_0 - d_t| = .05(.002).08(.005).135$. Table 2 gives the correction [factor] $(p_1 q_1 + p_2 q_2)/2pq$ to 3 dec. or 3 fig. for $p = .1(.05).9$ and $d_t = .1(.05).5$.

Patnaik, P. B. 1954

Sankhyā 14, 187-202

Review by K.J. Arnold (MTAC 10, 44; abridged):

Patnaik is concerned with the following problem: "if there is a random sample of N_1 individuals from a population, of which x_1 have the characteristic A, and a random sample of N_2 from a second population, of which x_2 have A, then it is desired to test whether the chance of possessing A is the same in the two populations". Consideration is restricted to cases in which Poisson distributions are acceptable approximations to the binomial distributions which give the probabilities for various values of X_1 and of X_2. The problem thus becomes one of testing, using one observation from each Poisson distribution, the hypothesis that the ratio of the parameters of the two Poisson distributions is a given constant.

The idea of a "test with minimal bias" is introduced. The power curve must be at or below the nominal size at the parameter value for the hypothesis under test, derivatives at this point must satisfy certain conditions, and among tests satisfying these conditions, that one is chosen which minimizes the length of the interval of parameter values for which the power curve is below the nominal size. The requirement of minimal bias is imposed on the conditional regions, making the test that of a binomial variate. Critical values for a binomial variate are given (in T. 6) for α (nominal size) = .05, .1; ρ (parameter of the binomial) = $\frac{1}{2}$, 3/5, 2/3, 3/4; n (index of the binomial) = 6(1)25.

See also De Finetti 1954; Chandra Sekar, Agarwala & Chakraborty 1955.

6.31 Tests applied to Poisson populations
[see also pp. 315-316]

Vianelli, S. 1959

Prontuari per calcoli statistici

Distributions of the number of plants per quadrat often differ from the Poisson by exhibiting too many vacant quadrats and too few quadrats with exactly one plant. To test such discrepancies P. G. Moore has proposed to compute the statistic $\varphi_3 = 2 n_0 n_2 n_1^{-2}$, where n_i is the observed number of quadrats with i plants, and to reject the Poisson hypothesis if

6.31
7.32

φ_3 exceed its mean by more than twice its standard error. Moore exhibits the approximate formulas: $M_\varphi = 1 + 3/np_1$, $\sigma_\varphi^2 = n^{-1}(p_0^{-1} + 4p_1^{-1} + p_2^{-1})$, where n is the sample size and $p_i = e^{-m} m^i/i!$

Based on these approximations, Pro. 192, pp. 576-577, gives σ_φ to 6 dec. and $M_\varphi + 2\sigma_\varphi$ to 5 dec. for m = .5(.5)6 and n = 25(25)1000.

See also Klerk-Grobben & Prins 1954.

6.4 Tests applied to rectangular populations
[see also p. 318]

Broadbent, S. R. 1955

Biometrika 42, 45-57

Review by L. A. Aroian (MTAC 10, 235):

Broadbent considers populations consisting of normally distributed components, particularly if the means of the components differ by a constant amount or quantum, 2δ. The hypothesis that the means of the components are equally spaced is called the quantum hypothesis. The problems treated are the estimation of the quantum which determines the spacing of the modes, the estimate of the scatter within each subdivision, and a test of the quantum hypothesis. Actually Broadbent tests the hypothesis of a rectangular distribution in a subdivision, against the alternative of the quantum hypothesis. This test makes use of the statistic s^2/δ^2 where s^2 is the lumped variance. A table of the critical values of s^2/δ^2 is given for sample sizes n = 20(5)100(50)1000 to 4 fig. for the 5%, 1%, and .1% probability levels.

Bartholomew, D. J. 1956

Biometrika 43, 64-78

Review by K. J. Arnold (MTAC 12, 78):

The paper is concerned with a sequential test of $a = 0$ against $a = a_0 \neq 0$ for the class of processes for which Pr {event in $(T, T + dT)$} $= \mu(\mu T)^a dT + 0(dT)$, $(a > -1)$. At the cost of one observation, a transformation eliminates the nuisance parameter μ. The Wald formulas for the operating characteristic and expected sample numbers are then applied. Table 3 gives $E_a(n \mid a_0)$ for various values of a_0 and a, and for $(\alpha, \beta) = 0.01, 0.05$ where

(1) $$E_a(n \mid a_0) = \frac{L(a) \log B + [1 - L(a)] \log A}{\log(1 + a_0) - a_0/(1 + a)},$$

$A = (1 - \beta)/\alpha$, $B = \beta/(1 - \alpha)$, $L(a) = (A^{h(a)} - 1)/(A^{h(a)} - B^{h(a)})$, $h(a)$ is the non-zero solution of $1 + ha_0/(1 + a) = (1 + a_0)^h$. The value a' of a for which (1) is indeterminate is near that which maximizes $E_a(n \mid a_0) \cdot E_{a'}(n \mid a_0) = -\log A \log B / [\log(1 + a_0)]^2$. Table 4 gives $E_a(n \mid a_0)$ and $E_{a_0}(n \mid a_0)$ for various values of a_0 and for $(\alpha, \beta) = 0.01, 0.025, 0.05$. Integral values are given in each case. Table 5 lists percentage savings in sample size in using the sequential test rather than a corresponding fixed sample size test.

7.131 Moments of r

In addition to works reviewed on p. 325 above, see K. Pearson 1925, reviewed on p. 652 above.

7.211 tanh x
[see also pp. 330-331]

5 dec.	0(.0001)1(.001)3(.01)6	v	Becker & Van Orstrand 1924
5 dec.	0(.01)3(.1)6.5	No Δ	Vianelli 1959(1032)
4; 5 dec.	0(.01)1.79; 1.8(.01)3(.1)4.9	No Δ	Czechowski et al. 1957 (88), Vianelli 1959(198)
4; 5 dec.	0(.01)1.79; 1.8(.01)3	No Δ	Ferber 1949(522)

7.212 $\tanh^{-1} x$
[see also p. 331]

5 dec.	0(.01).99	No Δ	Dixon & Massey 1957(468)
4 dec.	0(.01).9(.001).999	No Δ	Arkin & Colton 1950(122)
4; 3 dec.	0(.001).749; .75(.01).999	No Δ	Vianelli 1959(196)

7.3 Tables of $1-r^2$ and related functions
[see also p. 334]

7.31 $1-r^2$

6 fig.	0(.0001).9999	Miner 1922(30)
4 dec.	0(.01)1	Bingham 1937(258), Arkin & Colton 1950(138), Dixon & Massey 1957(465), Vianelli 1959(562)

7.32 $(1-r^2)^{\frac{1}{2}}$

6 fig.	0(.0001).9999	Miner 1922(7)
6 dec.	0(.001).999	Vianelli 1959(717)
5 dec.	0(.01).91(.002).99(.001)1	Vianelli 1959(562)

| 7.33 |
| 7.91 |

7.33 $(1-r^2)^{-\frac{1}{2}}$

5 dec.   0(.01).91(.002).99(.001).999   Vianelli 1959(562)

7.51   Circular serial correlation

[see also pp. 336-338]

Ferber, R.   1949

Statistical techniques in market research

     Table 17, p. 524, based on R. L. Anderson 1942, gives upper and lower 1% and 5% points of $_1R_n$ to 3 dec. for n = 5(1)15(5)100. Reprinted in Vianelli 1959, Prontuario 116, p. 319.

7.54   Correlation between time series

Hannan, E. J.   1955

Biometrika 42, 316-326

Review by W. J. Dixon (MTAC 11, 27):

     This article is concerned with tests for correlation between two time series x_t and y_t with serially correlated normal residuals. The estimates compared are: (1) the partial correlation between x_{2t} and y_{2t} when the effects of $(y_{2t-1} + y_{2t+1})$, x_{2t-1} and x_{2t+1} have been removed; (2) the ordinary correlation coefficient r between the two series, and (3) the partial correlation between x_t and y_t given x_{t-1} and y_{t-1}. The asymptotic efficiencies of these statistics are compared under the conditions: (a) the residual process from the regression of y_t and x_t is independent of the x_t process and comes from a Gaussian Markov process; (b) the two series are Markovian and are correlated through correlated errors; (c) same as (b) but with second order autoregression. The paper shows statistic (3) to lead to the asymptotically most efficient test for conditions (a), (b), and (c), except for some cases under (c) where the first partial correlation of the x_t process is high and positive. The criterion used for comparison requires the evaluation of the quantities:

$$\frac{(1 + p_1 p_2)(1 - p_2^2)}{(1 - p_1 p_2)(1 - p_1^2)}$$

and

$$\frac{1}{2} \frac{(1 - p_1^2 p_2^2)(1 + p_1 p_2)^2}{(1 - p_1^4)(1 - p_2^4)}$$

which are tabulated to 2D for $p_1, p_2 = -.8(.2).8$ and

$$\frac{(1 + p_1 p_2 - b)^2 [1 + p_1 p_2 - b(p_1^2 + p_1 p_2)][1 - p_1 p_2 - b(p_1 p_2 - p_1^2)]}{2(1 - p_1^4)(1 - p_2^2)(1 - b)[1 + p_2^2 - b(1 - p_2^2)]}$$

which is tabulated to 2D for $p_1 = p_2 = .4$ and all combinations of $p_1, p_2 = .6, .8$ for $b = -.6(.2).6$.

7.6 Covariance; covariance ratio; regression coefficient

[see also pp. 340-342]

Ronge, F. 1954

<u>Mitteilungsblatt für Mathematische Statistik</u> 6, 221-232

Review by C. F. Kossack (MTAC 10, 239):

In this paper Ronge introduces an additive adjustment method for estimating the total of some population from a sample when information is available from some previous period. Thus $Y_2^* = XN(\bar{y}-\bar{x})$, where Y_2^* is the estimate for the second period, X is the total for the population at the first period, N is the number in the population, $\bar{y}$ is the mean of the sample for the second period, and $\bar{x}$ is the mean of the same sample for the first period.

A comparison of the efficiency of this estimate with the standard ratio estimate is made. This relative efficiency is found to depend upon the correlation, r, between observations made on two different occasions, the ratio, s, of the standard deviations of individual observations, and the ratio, f, of the two means. For small s and small f or for large s and large f, the additive adjustment method is found to be more efficient than the ratio estimate. Tables of the relative efficiency to 4 dec. for r = .6(.1).9, .95; s = .4(.2)2; and f = .6(.2)2.4 are given.

Ihm, P. 1955

<u>Mitteilungsblatt für mathematische Statistik</u> 7, 46-52

Review by E. J. Gumbel (MTAC 11, 29):

A significance test is constructed for the composite hypothesis $H_0: \sigma_1 = \sigma_2$; $\rho = 0$ for a bivariate normal distribution p and a sample size N. The alternative H_1 is $\sigma_1 \neq \sigma_2$ or $\rho \neq 0$ or both. Let $L = p_0/p_1$ be the likelihood ratio. The test function used is $Z = L^{2/N}$. In order to test H_0 against H_1, the author calculates the probability $P(Z \leq Z_0 | H_0) = \alpha$. If $Z \leq Z_0$, then H_0 is rejected and H_1 is accepted. If $Z > Z_0$, then H_1^* is rejected and H_0^* accepted, where H_1^* and H_0^* stand respectively for $\rho^2 \gtreqless \rho^{*2}$. A table gives Z_0 to 3D as a function of $N = 3(1)30$, 40, 60, 120 for $\alpha = .01$ and $\alpha = .05$.

7.9 Other measures of correlation

[see also pp. 343-344]

For non-parametric tests of association see section 9.33 (in appendix).

7.91 Biserial correlation

Pearson, K. 1931a

<u>Tables for S & B</u> II

For abstract of TT. 16, 16 bis, see p. 16.

7.91
8.21

Dingman, H. F. 1954

Psychometrika 19, 257-259

> Review by J. R. Vatnsdal (MTAC 10, 240):
>
> A computing chart is given for quick estimation of a point biserial correlation coefficient when a normally distributed continuous variable is artificially dichotomized at the median. Use is made of Chesire, Saffir & Thurstone 1933 and Michael, Perry & Guilford 1952 to make a chart which is claimed to have an accuracy to 1 dec.

Tate, R. F. 1955

Biometrika 42, 205-216

> Review by Leo Katz (MTAC 10, 240):
>
> Karl Pearson's biserial correlation coefficient r^* is shown to be consistent and asymptotically normal as an estimate of ρ, the bivariate normal correlation. If ω is the (standardized) point of dichotomy, and $p(\omega) = \int_\omega^\infty (2\pi)^{-\frac{1}{2}} e^{-t^2/2} dt$, the distribution of r^* depends upon ρ and p. Soper [1] obtained the asymptotic variance,
>
> $$AV(r^*) = \frac{1}{n}\left\{\rho^4 + \rho^2\left[\frac{pq\omega^2}{\lambda^2} + \frac{(2p-1)\omega}{\lambda} - 5/2\right] + \frac{pq}{\lambda^2}\right\},$$
>
> which is a function of ρ and p only, λ being the normal density with argument ω.
>
> Table 2 gives $\{nAV(r^*)\}^{\frac{1}{2}}$, the square root of the expression in curly brackets above, to 3D (3 and 4S) for $\pm \rho = 0(.10)1$ and p or $(1-p) = .05(.05).5$.

7.94 Tetrachoric correlation

Pearson, K. 1913 a

Biometrika 9, 22-28

> Pearson's approximate formula (v), p. 24, for the probable error of tetrachoric r, contains a factor $(1-r^2)^{\frac{1}{2}}[1-(2\pi^{-1}\sin^{-1} r)^2]^{\frac{1}{2}}$. Table 1, p. 26, computed by Julia Bell, gives this factor to 4 dec. for r = 0(.01)1.
>
> Reprinted in Tables for S & B I, T. 23, p. 35.

Woo, T. L. 1929

> See abstract on p. 192.

Welch, G. S. 1955

> See review by J. R. Vatnsdal on p. ∆∆∆/

---

[1] Soper 1915

7.91
8.21

Vianelli 1959, T. 202, p. 1366, gives $\sin[\pi(2x-\frac{1}{2})]$ to 3 dec. for $x = .25(.001).499$.

See also Davidoff & Goheen 1953; Hayes 1943.

7.99 Miscellanea

Fan, C. T. 1954

See review by J. E. Walsh on p. △△△.

8.14 Paired comparisons; inconsistent triads
[see also pp. 355-356]

Aoyama, Hirojiro 1953

Annals of the Institute of Statistical Mathematics, Tokyo 4, 83-87

"On a test in paired comparisons"

Each of n persons is asked to rate his own occupation with respect to the occupations of the other n-1 persons. To each of the $k = \binom{n}{2}$ paired comparisons is attached a score of +1 if both persons rated their own occupation higher; of -1 if both rated their own occupation lower; of 0 if they agreed. On the null hypothesis, that over and under rating are equally likely, S+k has a binomial distribution with $p = \frac{1}{2}$. $\Pr[S \geq S_0]$ is tabled to 4 dec. or 2 fig. for $n = 3(1)9$ and all attainable non-negative S_0.

8.2 Order statistics from non-normal parents
[see also pp. 360-363]

8.21 Continuous rectangular parent

Hyrenius, H. 1953

Journal of the American Statistical Association 48, 534-545

"On the use of ranges, cross-ranges and extremes in comparing small samples"

Given two samples from a continuous rectangular population, call the sample with the smaller minimum observation the first sample. Let N_i, u_i, v_i be the number of observations, the minimum observation and the maximum observation in the ith sample. Define $T = (u_2-u_1)/(v_1-u_1)$, $U = (v_2-u_2)/(v_1-u_1)$, $V = T+U$.

Table 1, p. 542, gives the upper 1%, 5% and 10% points of T to 2 dec.; Table 2, p. 543, gives the lower $\frac{1}{2}$%, $2\frac{1}{2}$% and 5% points of U to 2 dec. and the upper $\frac{1}{2}$%, $2\frac{1}{2}$% and 5% points of U to 3 fig.; Table 3, p. 544, does the

8.21
8.27

same for V : all for N_1 , N_2 = 2(1)6(2)10 . [N.B. In caption of T. 2, for 0.5 read 0.005]

Reprinted, including the error in the caption of T. 2, in Vianelli 1959, <u>Prontuari</u> 83-84, pp. 232-233.

Murty, V. N. 1955

<u>Journal of the American Statistical Association</u> 50, 1136-1141

Review by J. R. Vatnsdal (MTAC 12, 81):

These tables are designed to test the hypothesis that two samples have been drawn from the same rectangular population with a known lower bound. Let L_1 be the maximum in a sample of size m , and L_2 the maximum in a sample of size n . The distribution of $u = L_1/L_2$ is obtained, and is independent of the parameter. The power of the test is calculated and 5% points of u for samples of size m , n = 2(1)10 are computed to 3 dec.

See also Lal & Mishra 1955.

8.23 K. Pearson's frequency curves as parents

Sarhan, A. E. and Greenberg, B. G. 1957

<u>Journal of the American Statistical Association</u> 52, 58-87

The tables herein permit the estimation of the mean (equivalently of the standard deviation) of a one-tailed exponential distribution with known endpoint; and the joint estimation of the mean and standard deviation (equivalently, of the mean and endpoint, or of the standard deviation and endpoint) of a one-tailed exponential distribution with unknown endpoint; all by means of any m consecutive order statistics from a sample of n . For both problems, $2 \leq m \leq n \leq 10$; for unknown endpoint, it is tacitly assumed that we know whether the distribution is bounded on the left or the right.

For known endpoint, T. 1, pp. 68-69, gives the coefficients for the best linear estimate of the standard deviation as exact vulgar fractions; T. 2, pp. [70-71], gives σ^{-2} times the variance of such estimates as exact vulgar fractions; T. 3, pp. 72-73, gives the efficiency of such estimates, relative to the use of all n order statistics, to 4 dec.

For unknown endpoint, TT. 5-7, pp. 74-80, give the coefficients for the best linear estimates of the endpoint, the standard deviation, and the mean as exact vulgar fractions; TT. 8, 12, pp. 81-85, give σ^{-2} times the variance of such estimates of the endpoint and the mean to 7 dec.; T. 10,

p. 83, gives σ^{-2} times the variance of such estimates of the standard deviation as exact vulgar fractions; TT. 9, 11, 13, pp. 82, 84, 86, give the efficiency of such estimates, relative to the use of all n order statistics, to 4 dec.

Reprinted in Vianelli 1959, Pri. 109-111, 113-115, pp. 298-318.

8.24 Iterated exponential parent

See Lieblein & Salzer 1957.

8.27 Comparison of ranges from various parents

David, H. A. 1954

Biometrika 41, 463-468

Review by C. A. Bennett (MTAC 9, 222; abridged):

David obtains explicit expressions for the probability integral $P(w|n)$ and the expectation $E(w|n)$ of the range w of a sample of n for several non-normal distributions with varying degrees of skewness and/or kurtosis. Values of $1 - P(w|n)$ for $w = k(n, \alpha)\sigma_x$, where $k(n, \alpha)$ is the 100 $\alpha\%$ point of w/σ for the normal distribution, are given to 3 dec. in T. 1 for $\alpha = .01, .05$; $n = 2(2)12$; and various degrees of skewness in the populations considered. Table 2 gives the ratio $E(w|n)/d_n\sigma_x$ to 3 dec., where d_n is the expected range, in standard units, of a sample of n from the normal distribution, for $n = 2(2)12$ and those distributions for which $E(w|n)$ was obtained explicitly.

Cox, D. R. 1954a

Biometrika 41, 469-481

Review by L.A. Aroian (MTAC 9, 221; abridged):

Cox is interested in the mean range and the coefficient of variation of the range in small samples of 2(1)5 from non-normal populations. He considers different types of populations covering a wide range of values of β_1 and β_2: symmetrical and unsymmetrical mixed normal distributions, the normal distribution, the rectangular distribution, exponential type distributions, the Pearson system, and the numerical results of Shone 1949 for 5 populations of discrete values. Based on these results he provides a table for the normalized mean range and for the coefficient of variation of the range to 3 dec. for sample sizes of 2(1)5, for $\beta_2 = 1(.2)2(.5)5(1)9$; β_1 is not a determining factor.

8.28
8.321

8.28 Arbitrary parent

Plackett, R. L. 1947

Biometrika 34, 120-122

"Limits of the ratio of mean range to standard deviation"

Plackett proves that, for any continuous parent,
$$0 \leq d_n \leq D_n = 2^{\frac{1}{2}}n[(2n-1)^{-1} - B(n,n)]^{\frac{1}{2}}.$$
The table on p. 122 gives D_n, $(n+\frac{1}{2})^{\frac{1}{2}}$ and $12^{\frac{1}{2}}(n-1)/(n+1)$ to 5 dec.; and normal-theory d_n to 3 dec.; all for $n = 2(1)12$.

Hartley, H. O. and David, H. A. 1954

Annals of Mathematical Statistics 25, 85-99

Review by C. A. Bennett (MTAC 9, 33; abridged):

Hartley & David extend the theory of universal upper and lower bounds for $E(w_n)$, and universal upper bounds for $E(x_n)$, where x_n is the standardized extreme variate and w_n the standardized range of a sample of n. Table 1 gives the upper bound of $E(x_n)$ to 4 dec. for any population for $n = 2(1)20$. Table 2 gives the universal lower bound for $E(w_n)$ over distributions with finite range $|x| \leq X$ to 3 dec. for
$$\begin{cases} n = & 2(1)12 \quad 12(2)20 \\ X = & 1(1)5 \\ p = x^2/(1+x^2) = & .95(.01).99 \end{cases}$$
It is shown that universal upper bounds previously computed for symmetric populations with infinite range are also applicable to non-symmetric populations; and to finite range unless $X < X_n$. X_n is tabled to 3 dec. for $n = 2(1)20$, and the algebraic form for the bound when $X < X_n$ is given.

Thompson, G. W. 1955

Biometrika 42, 268-269.

Review by J. R. Vatnsdal (MTAC 10, 243):

This table gives upper and lower bounds (distribution free) for the ratio of the range w to standard deviation estimate s, both from the same sample of size n. It can be shown that the upper bound of w/s is $2^{\frac{1}{2}}(n-1)^{\frac{1}{2}}$ and the lower bound of w/s is $2(n-1)^{\frac{1}{2}}n^{-\frac{1}{2}}$ for n even and $2n^{\frac{1}{2}}(n+1)^{-\frac{1}{2}}$ for n odd.

A table is given for $n = 3(1)20(10)100(50)200, 500, 1000$ to 3 dec. For samples of size 3 from a normal population, a table of upper and lower percentage points of w/s is given for percentages 0, 0.5, 1, 2.5, 5, and 10, to 5 dec.

These can be used for routine checks of computation of s.

8.3122 $-x \cdot \log_e x$

5 dec.   0(.0001).01(.001)1   No Δ   N.B.S. 1953(26), Vianelli 1959(1058)

4 dec.   0(.01)1   No Δ   Bartlett 1952(230)

8.3129 Miscellaneous combinations of $x \log x$

Cavalli 1950, T. 1, p. 211, tables
$[(1-y)\log_e x + y \log_e(1-x) - \log_e(y-x)]/y(1-y)$ to 2 dec. for
$x = .01, .05(.05).95$; $y = .05(.05).95, .99$; $x \neq y$.

Bartlett 1952, T. 1, p. 230, tables inter al. $-x \cdot \log_e x + (x-1)\log_e(1-x)$, $x(\log_e x)^2$ and $x(1-x)[\log_e x - \log_e(1-x)]^2$, all to 4 dec. for $x = 0(.01)1$.

Sampford 1955 tables $x(x-1)^{-1}\log_e x$ to 4 dec. for $x = 0(.01)1$.

8.319 Miscellaneous tables of the distribution of the extreme value

Kimball, B. F. 1956

<u>Annals of Mathematical Statistics</u> 27, 758-767

The table on p. 763 gives $b_n = \left[\sum_{r=2}^{n}(r^2-r)^{-1}\sum_{i=1}^{r-1}(-)^{i+1}\Delta^i \log 1\right]^{-1}$

to 4 dec. for $n = 2(1)112$.

8.321 [the mth extreme value:] direct tables

[see also p. 365]

Gumbel 1937y, T. 4, p. 318, gives a short table of the asymptotic distributions of the second and third extremes.

9.
9.118

Section 9. Non-parametric tests
[see also p. 366]

The ready-reckoner reviewed below may be of use in non-parametric statistics.

Chambers, E. G. 1952

Statistical calculation for beginners

Review by C. C. Craig (MTAC 6, 154; abridged):

Besides standard statistical tables, usually in shortened form, there are some auxiliary tables, viz:

$[mn(m+n-2)/(m+n)]^{\frac{1}{2}}$ 2 dec. $m, n = 10(1)50$

$(n^3-n)/6$ $n = 10(1)69$

$n(n-1)(2n+5)$ $n = 2(1)60$

$n(n-1)(n-2)$ $n = 3(1)50$

$mn(m-1)(n-1)/4$ $m = 2(1)6 ;\; n = 3(1)15$

$\binom{n}{2} m(m-1) / (m-2)^2$ $m = 3(1)6 ;\; n = 3(1)15$

$\binom{n}{2}\binom{m}{2}(m-3) / 2(m-2)$ $m = 4(1)6 ;\; n = 2(1)15$

$m^2(n^3-n)/12$ $m = 3(1)6 ;\; n = 3(1)15$

9.1 Tests of location

For test of equality of two Poisson means see Przyborowski & Wileński 1940, abstracted on p. 315 above. See also Romaní 1956.

9.111 The Walsh test
[see also pp. 366-367]

Walsh 1949c, T. 1, is reprinted (without column showing efficiencies) in Vianelli 1959, Pro. 125, p. 340.

9.112 The signed-rank test
[see also pp. 367-368]

Dixon & Massey 1957, T. 19, pp. 443-444, gives the distribution of T to 3 dec. for $n = 1(1)20$ and a varying range of T approximately extending from the two-tailed 1% point to the two-tailed 25% point. Thus,

for $\begin{cases} N = 1(1)3 & 8 & 9 & 20 \\ T = 0 & 0(1)9 & 1(1)12 & 37(1)74 \end{cases}$.

9.118 Tests of regression

Daniels, H. E. 1954

Annals of Mathematical Statistics 25, 499-513

Review by J. L. Hodges, Jr. (MTAC 9, 213):

Consider the usual regression model with $y_i = \alpha_0 + \beta_0 x_i + \delta_i$, where we assume only that the x_i are distinct and that the errors δ_i are independent and satisfy $\Pr[\delta_i > 0] = \Pr[\delta_i < 0] = \frac{1}{2}$. The lines $\alpha = -x_i \beta + y_i$ divide the (α, β) plane into polygonal regions, some of them open. Let m denote the minimum number of lines that must be crossed to escape from (α_0, β_0) into an open polygon. The distribution of m is given to 3 dec. for $n = 3(1)30$, providing confidence regions for (α_0, β_0) and tests for an hypothesis specifying both parameters. The power of the m-test is compared with certain competitors, and the complications arising when not all x_i are distinct are examined.

Cox, D. R. and Stuart, A. 1955

Biometrika 42, 80-95

Review by Leo Katz (MTAC 10, 241):

In a linear regression model with $y_i = \alpha + \Delta i + \epsilon_i$, $(i = 1, 2, \cdots, N)$, the ϵ_i are independent standardized normal variates. The null hypothesis, $\Delta = 0$, is to be tested against alternatives $\Delta > 0$, using a (quick) test based on comparisons of *independent* pairs of the observations. It is shown that the best weighted sign test is given by $S_1 = \sum_{K=1}^{N/2} (N - 2K + 1) h_{K, N-K+1}$, and the best unweighted sign test is given by $S_3 = \sum_{K=1}^{N/3} h_{K, \frac{2N}{3} + K}$ where, for $i < j$, $h_{ij} = 1$ if $y_i > y_j$ and $h_{ij} = 0$ otherwise. Since the asymptotic efficiency of S_3 relative to S_1 is 96%, the authors recommend use of S_3. It is also established that the asymptotic relative efficiency of S_3 compared to the best parametric test based on b, the sample regression coefficient, is 83%.

Table 3 gives the exact power of S_3 to 3D for sample sizes $N = 15(15)135$, significance level the largest value $\leq .05$, and alternatives given by $p = .50, .49, .45(.05).05$, where $p = \Phi\left(\frac{-\sqrt{2N}\Delta}{3}\right)$, the unit normal c.d.f. The values of p are identified as the true significance level α, and the rows $a, b, \cdots, j$, give powers for the various Δ's used. Table 4 gives the exact power of the b test to the same accuracy for the same alternatives. Table 3 also shows, in parentheses, the values of the corresponding standardized regression coefficient, Δ, to 4D. The smallest ratio of powers tabulated is 50.3%, for $N = 30$, $p = .25$ and $\Delta = .0477$.

9.121
9.14

9.121 The Wilcoxon test
[see also pp. 368-369]

Fix, Evelyn and Hodges, J. L., Jr. 1955

See abstract on p. 619 above.

Dixon & Massey 1957, T. 20, pp. 445-449, gives the lower half of the (symmetrical) distribution of T for $1 \leq m \leq n \leq 10$.

Vianelli, S. 1959

Prontuari per calcoli statistici

Prontuario 120, p. 327, is a table corresponding to Siegel's abridgement of Kruskal & Wallis, reviewed on p. 371 above: but for two samples of sizes $n_1 = 1(1)n_2$, $n_2 = 3(1)8$. Vianelli credits this table to Kruskal & Wallis 1952, which paper contains no such table. Apparently, the table under review was obtained from T. 1 of Mann & Whitney 1947, reviewed on p. 368 above, by selecting the attainable values of U that yielded tail sums near .1, .05 and .01, and converting the Us into Hs by the formula $H = 3[2U - n_1 n_2]^2 / n_1 n_2 (n_1 + n_2 + 1)$, where Vianelli's n_1, n_2 are Mann & Whitney's m, n. This formula follows from (1.6) of p. 370 above after expressing the $\bar{n}_i$ in terms of U. Since Mann & Whitney tabled single-tail sums, it is necessary to multiply the tabled P by 2 to obtain $\Pr[H \geq x]$.

9.129 Other two-sample tests
[see also p. 370]

Hyrenius, H. 1953

Journal of the American Statistical Association 48, 534-545

See abstract on p. 699 above.

Rosenbaum, S. 1954

Annals of Mathematical Statistics 25, 146-150

Draw a sample of m and a sample of n from a continuous population. The probability that exactly r values out of n are greater than the greatest value of the sample of n is obtained from the formulas of Wilks 1942 and is $n\binom{m}{r} B(n+m-r, r+1)$.

Rosenbaum tables x such that $\Pr[r \geq x] \geq \alpha > \Pr[r > x]$ for $\alpha = .01, .05$; $m, n = 2(1)50$.

Reprinted in Vianelli 1959, Prontuario 130, pp. 349-352.

Epstein, B. 1954 *a*

Annals of Mathematical Statistics 25, 762-768

Draw two samples each of size n from a continuous population. Let U_r^n be the number of values in the second sample that exceed the rth smallest value in the first sample. Then $\Pr[U_r^n = x] = \binom{n-x+r-1}{r-1}\binom{n-r+x}{x} / \binom{2n}{n}$.

Table 1, pp. 763-765, gives $\Pr[U_r^n \geq x]$ to 3 fig. or 4 dec. for $n = 2(1)20$ and all attainable r and x. Several symmetries are exploited to severely abridge the table: T. 2, p. 766, gives an unabridged table for n = 5 and thus facilitates restoration of missing values.

Table 1 is reprinted in Vianelli 1959, Prontuario 132, pp. 357-358.

9.13 Three-sample tests
[see also pp. 370-371]

Vianelli, S. 1959

Prontuari per calcoli statistici

Prontuario 120, pp. 328-329, credited to Kruskal & Wallis 1952, is substantially identical with Siegel 1956, T. Q, pp. 282-283, described on p. 371 above.

See also Hiraga, Morimura & Watanabe 1954.

9.14 k-sample tests
[see also pp. 371-372]

Jonckheere, A. R. 1954

Biometrika 41, 133-145

Review by Z. W. Birnbaum (MTAC 10, 48):

Let $X_i, i = 1, 2, \cdots, k$, be independent random variables with the continuous cumulative probability functions $F_i(x)$, and let $X_{ij}, j = 1, 2, \cdots, m_i$, be samples of size m_i of the X_i. To test the hypothesis that $X_1, \cdots, X_k$ have the same distribution $(H_0): F_1(x) = F_2(x) = \cdots = F_k(x)$ against the alternative that these random variables are stochastically increasing $(A): F_1(x) > F_2(x) > \cdots > F_k(x)$ (on p. 134 the sense of these inequalities is inverted, as it is in the footnote on p. 135), the author considers the following statistic: Let p_{ij} = number of pairs X_{ir}, X_{js} such that $X_{ir} < X_{js}$, among all possible $m_i m_j$ pairs, for given $i = 1, \cdots, k-1$;

9.14
9.311

$j = i+1, \cdots, k$. The test statistic is $S = 2 \sum_{i=j}^{k-1} \sum_{j=1+1}^{k} p_{ij} - \sum_{i=1}^{k-1} \sum_{j=i+1}^{k} m_i m_j$, identical with a statistic considered by M. G. KENDALL [1] and closely related to the rank correlation coefficient.

In his study of the probability distribution of S under (H_0) the author obtains the first four cumulants, studies their extreme values, and obtains asymptotic distributions under several assumptions. He finds in particular that if at least two of the sample sizes m_i tend to infinity, then the limit distribution of S is normal, and he proposes a somewhat better approximation by STUDENT's distribution. Finally, the exact distribution for small samples is given.

Table 3 gives $\Pr\{S \geq S_0\}$ to 3 or 4S for k samples, each of size m, covering the range: $k = 3$ and $m = 2, 3, 4, 5$; $k = 4$ and $m = 2, 3, 4$; $k = 5$ and $m = 2, 3$; $k = 6$ and $m = 2$; $S_0 = 0(2) \cdots$ until $\Pr\{S \geq S_0\}$ becomes negligible.

9.2   Tests of spread
[see also p. 373]

Hyrenius, H.   1953

Journal of the American Statistical Association 48, 534-545

See abstract on p. 699 above.

Rosenbaum, S.   1953

Annals of Mathematical Statistics 24, 663-668

"Tables for a nonparametric test of dispersion"

Draw a sample of m and a sample of n from a continuous population. The probability that exactly r values out of m lie outside the extreme values of the sample of n is obtained from the formulas of Wilks 1942 and is $n(n-1)\binom{m}{r} B(n+m-r-1, r+2)$.

Rosenbaum tables x such that $\Pr[r \geq x] \geq \alpha > \Pr[r > x]$ for $\alpha = .01, .05$; m, n = 2(1)50.

Reprinted in Vianelli 1959, Prontuario 131, pp. 353-356.

Kamat, A. R.   1956

Biometrika 43, 377-387

Review by J. E. Walsh (MTAC 12, 82):

The data consist of two independent samples of sizes m and n with $m \geq n$. These sample values are unequal. The null hypothesis asserts that these samples are from the same population. The alternative hypothesis of interest is that the samples are from populations with different dispersions. The samples are pooled and their values ranked. Let R_j be the

range of the ranks for the sample of [size] j. The test statistic is
$D_{n,m} = R_n - R_m - m$. Table 1 contains the largest integer values D such
that $\Pr[D_{n,m} \le D_\alpha] \le \alpha$, for $100\alpha = .5, 1, 2.5, 5$ and $m+n = 7(1)20$.
Table 2 contains the mean of $D_{n,m}$ to 4 dec., and $\mu_2^{\frac{1}{2}}, \mu_2, \beta_1^{\frac{1}{2}}, \beta_2$ of $D_{n,m}$
to 3 dec., for $m+n = 16$ and $n = 2(1)8$. Table 3 contains 3 dec. values of
$\mu_2^{\frac{1}{2}}, \mu_2, \beta_2$ of $D_{n,m}$ for $n = m = 2(1)5, 7, 10, 15, \infty$. Table 4 contains
moments and standardized deviates for the limiting distribution of
$d = D_{n,m} - m$ for $m/(m+n) = p = .5(.1).9$. The contents of T. 4 are useful
in approximating percentage points of $D_{n,m}$ when $m+n > 20$. Table 5 contains
a comparison of the approximate and true $2\frac{1}{2}\%$ points when $m+n = 20$,
$p = .5(.1).9$, $m = 10(2)18$. Table 6 contains corrected and standardized
upper and lower $2\frac{1}{2}\%$ points for d to 2 dec. when $\mu = .5(.05).75$.

See also Sadowski 1955.

9.311 Runs

[see also pp. 374-376]

The second table on p. 232 of Mosteller 1941 is reprinted in Vianelli 1959, T. 97, p. 1248; Vianelli credits it to P. S. Olmstead.

Krishna Iyer, P. V. and Rao, A. S. P. 1953

Journal of the Indian Society of Agricultural Statistics 5, 29-77

Review by Carl F. Kossack (MTAC 9, 30; abridged):

The purpose of this paper is to investigate the distribution of the number of ascending, descending and stationary runs in a sequence of n observations: in both the infinite case, where a particular value has a given probability of occurrence; and the finite case, where one knows the number of times a given value has occurred. For each of the three types of runs, the various related configurations are tabulated along with their probability of occurrence in the sequence and the number possible. Variances and covariances for k kinds of elements with equal probabilities of occurrence are tabled to 4 dec. for $k = 2(1)5, 10$; $n = 30(10)50(25)100$: for ascending, stationary, and descending runs as well as for the total number of runs. The paper also lists algebraic expressions for the covariances of runs of lengths p and q for $\begin{cases} p = 1(1)3 & 4 \\ q = p(1)5 & 4 \end{cases}$; of runs of length p [exactly] and q or more for $p, q = 1(1)4$; of runs of length p or more and q or more for $p = 1(1)3$ and $q = [p+1](1)4$. For junctions the actual distributions are given for $n = 4(1)7$.

9.311
9.319

Foster, F. G. and Stuart, A. 1954

Journal of the Royal Statistical Society, Series B 16, 1-22

Review by C.F. Kossack (MTAC 9, 212):

Two distribution-free tests making use of the statistics involving the upper records and the lower records for the randomness of a series are introduced in this paper. An observation in a time series is called a lower (upper) record if it is smaller (greater) than all previous observations in the series. The statistics used are s, the sum of the number of upper and lower records in a series and d, the difference between the number of upper and lower records. The hypothesis of randomness of a series can be tested by the critical region $s > s_0$ or $|d| > d_0$. The first test provides a test against trend in dispersion while the second tests against a trend in location. Comparable round-trip tests (S and D) are also introduced involving counting the records in the series in both directions.

The joint distribution of s and d is obtained under the null hypothesis and the two variables are shown to be independently distributed as well as being asymptotically normal. In the round-trip case the corresponding statistics are established to be asymptotically normally distributed and the variances have been tabulated.

The powers of the tests against trend in the mean have been computed using empirical sampling techniques on a high-speed computer.

The following tables are given:

T. 1. $\Pr[s < s_0]$ to 3-4 dec. for $n = 3(1)15$ and the normal approximation for $n = 15$.

T. 2. $\Pr[d < d_0]$ to 3-4 dec. for $n = 3(1)6$ with normal approximation for $n = 6$.

T. 3. Mean and standard error of s and standard error of d to 3 dec. for $n = 10(5)100$ with approximation for $n = 100$.

T. 4. The standard error of D to 2 dec. based on 1000 samples of size n for $n = 10, 25(25)125$.

T. 5. The correlation between d and d' to 2 dec. for $n = 10, 25(25)125$.

T. 6. The power of the d test against normal regression at the 5% level for trends $\Delta = .01(.01).07$ and sample sizes $n = 25(25)125$ based on 1000 samples in each case.

T. 7. The power of the D test against normal regression at the 5% level for trends $\Delta = .01(.01).06, .08$, and sample size $n = 25(25)125$, based on 1000 samples in each case.

Barton, D. E. and David, F. N. 1957

Biometrika 44, 168-178

Place r_1 ones, r_2 twos, r_3 threes at random on a line: for example, with $r_1 = r_2 = r_3 = 2$, a possible arrangement is 1 3 3 2 2 1. Each block of like elements counts as a run: in the example there are 4 runs.

Table 1, p. 177, gives the individual terms of the distribution of the number of runs as exact vulgar fractions for all r_i such that $r_1 \geq r_2 \geq r_3 \geq 1$ and $6 \leq r_1+r_2+r_3 \leq 12$. Table 2, p. 173, does the same for the distribution of runs of four kinds of elements.

Reprinted in Vianelli 1959, Prontuario 138, pp. 369-370.

Olmstead, P. S. 1958

Bell System Technical Journal 37, 55-82

"Runs determined in a sample by an arbitrary cut"

Author's summary:

This paper, after making a critical review of the literature pertaining to runs above and below in a fixed sample, provides the following extensions:

1. Sample arrangement distributions for runs of length at least s on one, each and either side of <u>any selected</u> cut for samples of 10 and 20.

2. Sample arrangement distributions for runs of length at least s on one, each and either side of the <u>median</u> for samples of 10, 20, 40, 60, 100, and 200.

3. Sample arrangement distributions for runs of length at least s on each side of <u>all possible</u> cuts for samples of 10, 20, 40 and 100.

4. Asymptotic values of the probabilities of such arrangements when the sample size and length of run are large.

5. Convenient charts and tables for probabilities of .01, .1, .5, .9 and .99.

See also Olekiewicz 1951a.

9.319 Other tests of randomness

Barton, D. E. and David, F. N. 1956

Biometrika 43, 104-112

Review by K. J. Arnold (MTAC 12, 77):

The present paper is one of a set of papers by the same authors dealing with ordered intervals [1]. On a line of unit length, $(n - 1)$ points are dropped randomly

1. Barton & David 1956 z

9.319
9.33

forming n intervals. Let the distances from one end of the line be x_i ($i = 1, 2, \cdots, n-1$) with $x_1 \leq x_2 \leq \cdots \leq x_{n-1}$. Let $d_1 = x_1$,

$$d_i = x_i - x_{i-1} (1 < i < n-1), \qquad d_n = 1 - x_{n-1}.$$

Let $\{g_i\}$ be $\{d_i\}$ with g_1 the smallest d_i, g_2 the next larger, etc. Let r_i be the solution of $d_{r_i} = g_i$. Let r_i' be the solution of $d_i = g_{r_i'}$. The paper deals with distributions of (i)

(i) $R = \sum_{i=1}^{n} i r_i'$,   (ii) $Y = \sum_{i=1}^{n} i d_i$,   (iii) $G = \sum_{i=s}^{n} r_i g_i$,

(iii') $G_{st} = \sum_{i=s}^{t} r_i g_i$,   and   (iv) $G^* = n \sum_{i=1}^{n} d_i g_i$.

R and Y are disposed of quickly by references to papers by other authors and by references to the use of the normal approximations to R and to $n - Y$ for moderate or large n. Moments of G_{st} are discussed. Four moments of G are displayed. Table 3 gives μ_1'/n to 4D, μ_2/n to 5D, $n\beta_1$ to 3D and $n(\beta_2 - 3)$ to 3D for $n = 5(5)25$, ∞; $p = s/n = 1/5((1/5)4/5$. μ_1' is the 1st moment, $\beta_1 = \mu_3^2/\mu_2^3$, $\beta_2 = \mu_4/\mu_2^2$, μ_2, μ_3, μ_4 are central moments. For G^*, μ_1^1, μ_2, β_1 and β_2 are displayed as functions of n and values of β_1 and β_2 are given to 2D in Table 5 for $n = 2(1)10, 20, 40, 100, 400, 1000, \infty$. Percentage points of G^* to 3D and of standardized G^* to 2D are given in Table 6 for $n = 40, 100, 400, 1000$, and for percentiles 0.5, 1, 2.5, 5, 95, 97.5, 99, 99.5.

Barton, D. E. and David, F. N. 1956 z.

Journal of the Royal Statistical Society, Series B 18, 79-94

Review by C. C. Craig (MTAC 12, 80):

In this paper, which is a further investigation of the distribution of the ordered intervals obtained by dropping n-1 points at random on a line segment of unit length, some short tables are included. First the values of $\beta_1^{\frac{1}{2}}$ and β_2 are given to 4 dec. for the distribution of the rth greatest interval for n = 5(5)25 and r = 1(1)n. Next the coefficient of correlation is given between the two shortest intervals and between the first and third shortest intervals to 3 dec. for n = 2(1)10, 15, 25, 50, 100, ∞. Then the limiting values of $(n\beta_1)^{\frac{1}{2}}$, of $n\beta_2$, and of n^2 times the variance divided by the expected value are given to 4-5 fig. for the deciles. Turning to the distribution of ratios of two intervals, the upper and lower 1%, 5%, 10% points are given for the ratio of the smallest to the largest interval to 3 fig. for n = 3(1)10(5)20. Finally upper 5%, 2½%, 1% and ½% points for the interquartile differences of interval length are given to 4-5 dec. for n = 7(4)23.

9.32 Two-sample tests [of randomness]
[see also pp. 377-379]

Takashima, Michio 1955

<u>Bulletin of Mathematical Statistics</u> (Fukuoka) 6, no. 1-2, pp. 17-23

Review by Joan R. Rosenblatt (MTAC 11, 33):

In testing an ordered arrangement of m+n objects of two kinds (say m A s and n B s) for randomness, one may use as test criterion the length of the longest run, or of the longest A-run. Takashima has tabulated the critical run-lengths for tests based on these criteria.

Let $Q(t)$ be the probability that there appears at least one run (of A s or B s) of length t or longer. Let $Q_1(u)$ be the probability that there appears at least one A-run of length u or longer.

The tables give the smallest integers t, u such that $Q(t) \leq \alpha$, $Q_1(u) \leq \alpha$, for α = .01, .05; m, n = 1(1)25. Calculations were based on the investigation of Mood 1940 z.

See also Krisha Iyer & Singh 1955.

9.33 Tests of association

Olmstead, P. S. and Tukey, J. W. 1947

<u>Annals of Mathematical Statistics</u> 28, 495-513

Divide a scatter diagram containing 2N points (x,y) into quadrants by a vertical line through the median of x and an horizontal line through the median of y. Attach the sign + to the 1st and 3^d quadrants; − to the 2^d and 4th. Beginning at the right-hand side of the diagram, count in (toward the left) along the observations until forced to cross the horizontal median. Write down the number of observations met before this crossing, attaching the sign + if they lay in a + quadrant and − if in a − quadrant. Turn the diagram through $90°$ and repeat (three times). The test statistic is the algebraic sum of the four terms thus written down.

Table 2, p. 502, gives the distribution of the absolute value, S, of the test statistic to 4 dec. for 2N = 2(2)10, 14, ∞; S = 0(1)30. Dixon & Massey 1957, T. 30 d, p. 469, give an abridged table to 3 dec. for 2N = 6, 8, 10, 14, ∞; S = 8(1)23.

9.33
10.002
Hodges, J. L., Jr. 1955y

<u>Annals of Mathematical Statistics</u> 26, 523-527

"A bivariate sign test"

Imagine a circle on whose circumference 2n equally spaced points are given. Distribute n + signs and n - signs among these, subject to the condition that each diameter have one + and one - sign; call each such arrangement a <u>cycle</u>. A cycle is rotatable into 2n positions, each of which counts as a cycle. For each position count the number t of - signs in the upper semicircle; let k be the minimum of the 2n values of t.

Hodges finds that, for $x < n/3$, $\Pr[k \leq x] = (n-2x)2^{1-n}\binom{n}{x}$. He tables this on p. 526 to 5 dec. for $n = 1(1)30$.

Reprinted in Vianelli 1959, <u>Prontuario</u> 124, p. 339.

9.4111 Tables of $\vartheta_4(0, e^{-2z^2})$

Smirnov's table, listed on p. 380 above, is reprinted in Gnedenko 1950, pp. 384-385; Czechowski et al. 1957, T. 20, p. 90.

9.4113 Inverse tables: z as a function of $\vartheta_4(0, e^{-2z^2})$
[see also p. 381]

Siegel 1956, T. M, p. 279, based on Smirnov 1948, gives z to 2 dec. for $p = \vartheta_4(0, e^{-2z^2}) = .9, .95, .975, .99, .995, .999$.

9.4121 [exact] Distribution of D_n
[see also pp. 381-382]

Massey 1950, T. 1, p. 118, is reprinted in Czechowski et al. 1957, T. 18, p. 89.

Z. W. Birnbaum 1952, T. 1, pp. 428-430, is reprinted in Vianelli 1959, Pro. 133, pp. 359-361.

L. H. Miller 1956, T. 1, pp. 113-115, is reprinted in Vianelli 1959, Pro. 134, pp. 362-363, with captions that allege (not altogether accurately: cf. pp. 382, 384, above) that it is a table of lower (100-2p)% points of D_n.

9.4122 [exact] Distribution of $D_{m,n}$
[see also pp. 382-384]

Massey 1951 a, T. 1, pp. 126-127, is reprinted in Czechowski et al. 1957, T. 19, p. 89.

9.414 Three-sample Kolmogorov-Smirnov tests

For extension of the $D_{n,m}$ test to three samples see H. T. David 1958 and works there cited. David gives the exact and asymptotic distributions of $\max[D_{n_1,n_2}, D_{n_1,n_3}, D_{n_2,n_3}]$ without numerical tables.

9.4151 One-sided tests of one sample against a population
[see also p. 384]

L. H. Miller 1956, T. 1, pp. 113-115, is reprinted in Vianelli 1959, Pro. 134, pp. 362-363, with captions that allege (not altogether accurately: cf. pp. 382, 384 above) that it is a table of lower $(100-2p)\%$ points of D_n.

9.4152 One-sided tests of slippage between two samples
[see also p. 385]

Siegel 1956, T. L, p. 378, is reprinted in Vianelli 1959, Pro. 136, p. 367.

Section 10. Systems of frequency curves;
moments and other symmetric functions

We call attention to the highly compressed information about many frequency curves in Haight 1955.

10.002 Percentage points of Pearson curves
[see also p. 389]

Vianelli, S. 1959

Prontuari per calcoli statistici

Prontuario 198, pp. 595-596, is abridged from Pearson & Hartley 1954, T. 42, by omitting the upper and lower $2\frac{1}{2}\%$ points.

10.023
10.095
10.009 Miscellaneous tables of Pearson curves

Raj, Des 1955

Sankhyā 15, 191-196

Review by Benjamin Epstein (MTAC 10, 243):

Suppose that we want to estimate the proportion, P, of a population falling in the interval (a, b). Raj compares the large sample relative efficiency of two methods for estimating this proportion: one method consisting of measuring each item in the sample, the other method consisting of using go & not-go gauges set at the values a and b. Tables for the proportion P and the efficiency are given for the Type III, Cauchy, and normal distributions, for the case where a and b are symmetrically located about the mean of the distribution, and for the one-sided case where $b = \infty$. In the normal case efficiencies are tabulated to 3 dec. for those values $a = \kappa\sigma+\mu$ for which $\kappa = 0(.05)3$.

10.023 Auxiliary tables for computing ordinates . . .
of the Type II distribution
[see also p. 394]

Vianelli, S. 1959

Prontuari per calcoli statistici

Prontuario 197, p. 594, gives inter al. $(5\beta_2 - 9)/(6 - 2\beta_2)$ and $2\beta_2/(3 - \beta_2)$ to 5 dec. for $\beta_2 = 1.8(.025)2.975$.

10.033 Auxiliary tables for computing ordinates
of the Type III distribution

Elderton 1903a, T. 1, p. 270, described on p. 395 above, is reprinted in Vianelli 1959, T. 176, p. 1336.

10.035 The fitting of Type III curves by the method of maximum likelihood
[see also p. 395]

Cohen, A. C. Jr. 1955

Journal of the American Statistical Association 50, 1122-1135

Review by Leo Katz (MTAC 12, 71):

The radial error (from the true mean) in a p-variate normal distribution with covariance matrix $\sigma^2 I$, is distributed, essentially, as χ^2 with p degrees of freedom. The maximum likelihood estimate of σ is to be computed on the basis of a truncated sample of n observations in the range $0 \leq r \leq r_0$, e.g., as in a circular target. Setting $\xi_0 = r_0/\sigma$, the maximum likelihood estimate is obtained by solving

$$\sum_1^n r_0{}^2/nr_0{}^2 = \frac{1}{\xi_0}\left[\frac{p}{\xi_0} - \frac{\vartheta_p(\xi_0)}{1 - I_p(\xi_0)}\right] = g_p(\xi_0),$$

for ξ_0 and converting to the equivalent σ where $\vartheta_p(\xi_0)$ is the standard χ^2 density at the point ξ_0 for p d.f. and $I_p(\xi_0)$ is the upper tail area of the same distribution. Table I gives $G_p(\xi)$ for $p = 2$ and $p = 3$ to 5D for $\xi = .0(.1)4(.5)5$.

<u>10.039</u> Miscellaneous tables of the Type III distribution
[see also pp. 396-397]

Springer, C. H. 1951

<u>Industrial Quality Control</u> 8, no. 1, 36-39

Tables 1-2, pp. 36-37, give $W = 2/k - B/(Y-1)$, where $Y = C^{k^2/(4-k^2)} e^{2kB/(4-k^2)}$, to 2 dec. for $B = 2.5(.5)5.5$ and

$$\begin{cases} \pm k = 0 & .3, .7, 1.1 \\ C = .005, .01, .02, .05, .1, .2, .5, 1 & .01, .05, .2, .5, 1. \end{cases}$$

<u>10.0431</u> The $G(r,\nu)$ integrals
[see also pp. 397-398]

Vianelli, S. 1959

<u>Prontuari per calcoli statistici</u>

Prontuario 201, pp. 601-612, gives $\log_{10} F(r,\nu)$ for $r = 1(1)48$; and $\log_{10} H(r,\nu)$ for $r = 2(1)48$; all to 7 dec. without differences for $\phi = 0(1°)45°$.

<u>10.073</u> Auxiliary tables for computing ordinates . . .
of the Type VII distribution
[see also p. 400]

Vianelli, S. 1959

<u>Prontuari per calcoli statistici</u>

Prontuario 197, p. 594, gives inter al. $(5\beta_2 - 9)/(6 - 2\beta_2)$ and $2\beta_2/(3 - \beta_2)$ to 5 dec. for $\beta_2 = 3.025(.025)7$.

<u>10.095</u> The fitting of Type IX curves by methods other than moments
Downton, F. 1954

<u>Annals of Mathematical Statistics</u> 25, 303-316

Table 3, p. 311, gives coefficients for estimating the mean of a right triangular distribution from order statistics, to 4 dec. in samples of $n = 2(1)10$; T. 4, p. 311, does the same for the standard deviation. Table 5, p. 311, gives the variance and covariance of these estimates to 4 fig.

10.097
10.222

10.097 Distribution of statistics in samples from Type IX populations
[see also p. 402]

Downton, F. 1954

Annals of Mathematical Statistics 25, 303-316

Table 1, p. 309, gives the mean values of order statistics from a right triangular distribution to 5 dec. in samples of $n = 2(1)10$. Table 2, p. 310, gives the variances and covariances of these order statistics to 5 dec.

10.109 Generalizations of the Type X distribution

Johnson, N. L. 1954

Trabajos de Estadística 5, 283-291

See abstract in section 10.222 below.

10.18 Generalizations of K. Pearson's frequency curves

Karmazina, Lena N. 1954

For further information about Karmazina's tables, see the review by David A. Pope (MTAC 13, 58). The account below is largely adapted from Pope.

The Jacobi polynomials $G_n(p,q,x)$, as tabled by Karmazina, may be characterized by the properties that they are orthogonal on [0,1] with weight function $x^{q-1}(1-x)^{p-q}$, and are so normalized that the coefficient of x^n in G_n is unity.

The table on pp. 15-214 gives $G_n(p,q,x)$ to 7 dec. for $n = 1(1)5$, $p = 1.1(.1)3$, $q = .1(.1)1$, $x = 0(.01)1$. The table on p. 217 gives $G_n(1,1,x)$ to 7 dec. for $n = 1(1)5$, $x = 0(.01)1$; these polynomials are Legendre polynomials with an unusual normalization.

The table on pp. 221-233 gives the coefficients in $G_n(p,q,x)$ to 7 fig. for $n = 1(1)5$, $p = 1.1(.1)3$, $q = .1(.1)1$; the table on pp. 237-249 gives the zeroes of $G_n(p,q,x)$ to 7 dec. for the same arguments. The tables on p. 218 give coefficients in $G_n(1,1,x)$ to 7 fig., and zeroes to 7 dec., both for $n = 1(1)5$.

When the values of G_n are simple terminating fractions, the table prints a terminal string of zeroes if the function is negative, and of nines if the function is positive. Pope conjectures that the table was computed to 8 or 9 dec. on a tabulator with nines-complement arithmetic, and truncated to 7 dec. without rounding. Spot checks confirm the conjecture, and support Karmazina's claim (p. 6) of accuracy to 2.10^{-7} or better.

Typographical error: in formula on p. 13, <u>for</u> (-1) <u>read</u> $(-1)^j$. N.B.: in all formulas, C_n^j represents the binomial coefficient $n!/j!(n-j)!$

10.22 Translation systems

See also Masuyama 1952.

10.221 Polynomial translation
[see also p. 411]

For a detailed numerical example of translation by a quartic polynomial, see Baker 1934, p. 118.

10.222 Logarithmic and hyperbolic translation
[see also pp. 412-414]

Johnson, N. L. 1954

<u>Trabajos de Estadística</u> 5, 283-291

Johnson applies the logarithmic and hyperbolic translations of Johnson 1949, reviewed on p. 412 above, to the distribution $p(z) = \frac{1}{2}e^{-|z|}$, yielding the families S'_L, S'_B, S'_U.

Table 1, p. 285, gives selected points on the S'_L line in the β_1-β_2 plane to 4 dec. Table 2, p. 286, gives selected points on the boundary between unimodal and W-shaped S'_B curves to 3 dec. Table 3a, p. 289, gives illustrative values of the difference between a fitted and a true distribution, caused by an error in the parameters; T. 3b, p. 289, does the same for the curves of Johnson 1949.

10.2221
10.261
10.2221 S_L curves: i.e. the log-normal distribution

Broadbent, S. P. 1956 a

Biometrika 43, 404-417

Broadbent standardizes the log-normal distribution to have lower limit 0, mean 1, and standard deviation s. Table 1, pp. 408-409, gives upper and lower 1% and 5% points to 4 dec. for s = 0(.001).15 with IΔ.

Reprinted in Vianelli 1959, Prontuario 206, pp. 632-633.

10.24   Cumulative distribution functions
[see also pp. 416-417]

Topp, C. W. and Leone, F. C. 1955

Journal of the American Statistical Association 50, 209-219

Topp & Leone consider the cumulative distributions
$F(x) = a(2x/b - x^2/b^2)^r + (1-a)x/b$, $0 \leq x \leq b$, $0 < r < 1$, $0 < a < 1$,
whose frequency curves are J-shaped. Figure 1, p. 213, plots these distributions on the chart of Craig 1936 z (cf. Hatke 1949, fig. 1). Table 1, pp. 214-215, gives $\alpha_3^2 = \beta_1$ and $\delta = (2\beta_2 - 3\beta_1 + 6)/(\beta_2 + 3)$ to 3 dec. for a = .05(.05)1 and r = .01, .02, .05, .08(.01).1(.05).95.

Table 1 is reprinted in Vianelli 1959, Prontuario 203, pp. 620-621.

10.261   The continuous rectangular distribution [and its convolutions]
[see also p. 418]

Grimsey, A. H. R. 1945

Philosophical Magazine 36, 294-295

Let $P_n = 2\pi^{-1}\int_0^\infty (\sin v)^n v^{-n} dv$, $U_n = 2\pi^{-1}\int_0^\infty (\sin v)^{n+2} v^{-n} dv$, $Q_n = (6/\pi n)^{\frac{1}{2}}$

The table on p. 295 gives P_n, Q_n and U_n to 8 dec.; and P_n as exact vulgar fractions; all for n = 1(1)12.

Packer, L. R. 1951

Journal of the Institute of Actuaries Students' Society 10, 52-61

Review by C. C. Craig (MTAC 6, 88):

Packer rederives, using difference methods, the well-known distribution function for the sum of n randomly drawn values from the continuous rectangular universe on the interval (0,1). The cumulative distribution function F(x,n) is tabled to 6 dec. for n = 1(1)12 and

$x = .5(.5)6$. As an application, Packer gives a table of the exact probabilities that the error in the sum of 12 numbers, each rounded off to the nearest unit, will lie in the interval $(-x, +x)$ for $x = .5(.5)6$.

David, F. N. and Johnson, N. L. 1954

Biometrika 41, 228-240

Review by A.C. Cohen, Jr. (MTAC, 9, 211):

This paper is concerned with censored samples in which the smallest k of n observations are fully measured, while of the (n-k) largest observations, it is known only that $x > x_k$. David & Johnson consider a population described by a continuous random variable with p.d.f. $f(t)$, and with distribution function $F(x) = \int_{-\infty}^{x} f(t)\,dt$. Arranged in ascending order of magnitude, observations of a (complete) random sample from a population of this type correspond to $x_1, \ldots, x_n$. Variables $X_1, \ldots, X_n$, defined by the equations $F(X_r) = r/(n+1)$, are introduced and approximations to the moments of the x_r are found by expanding x_r about X_r in an inverse Taylor series. Central moments and cumulants are then obtained from these non-central moments. With the first four moments or cumulants known, functions of the Pearson system are available for approximating distributions of the x_r and functions of them.

The first four orders of cumulants of $F(x_r)$ are listed in Table 1. The first four orders of cumulants of the ordered variables x_r are listed in T. 2 with an accuracy of $O[(n+2)^{-3}]$. Sums of powers and products of the first k natural numbers which enter into calculation of the cumulants of TT. 1-2, are given in T. 3.

The application of results obtained in the paper is illustrated by determining upper and lower 5% points of $\bar{x}_k$, which is the mean of the first k observations (in ascending order of magnitude) of a censored sample from a normal population. Calculations are given to 3 dec. for $n = 10$, $k = 2(2)8$ and for $n = 20$, $k = 4(4)16$.

Broadbent, S. R. 1954

Biometrika 41, 330-337

"The quotient of a rectangular or triangular and a general variate"

Let y_1, y_2 be independent rectangular variates with range $-\alpha$ to $+\alpha$; let x be a normal variate, independent of y_1, y_2, with mean 1 and variance β^2.

10.261
10.36

Table 1, p. 336, gives upper and lower 1% and 5% points of $(1+y_1)/x$ to 3 dec. for $\alpha = 0(.02).1$, $\beta = 0(.01).05$. Table 2, p. 336, gives upper and lower 1% and 5% points of $(1+y_1+y_2)/x$ to 3 dec. for $\alpha = 0(.01).05$, $\beta = 0(.01).05$.

Reprinted in Vianelli 1959, Tavole 184-185, p. 1348.

10.262 The discrete rectangular distribution [and its convolutions]
[see also p. 418]

Tsao, C. K. 1956
Annals of Mathematical Statistics 27, 703-712

Tables 1-4, pp. 705-711, give the lower half of the (symmetrical) cumulative distribution of the sum of m k-point uniform random variables to 7 dec. or 5 fig. for $k = 3(1)6$ and $m = 1(1)20$. For $m = 20$ a normal approximation is also given.

Reprinted (some of the small probabilities curtailed to 1-3 fig.) in Vianelli 1959, Prontuario 298, pp. 1073-1078.

See also Mitropol'skii 1955.

10.28 The circular normal distribution
[see also pp. 418-419]

Gumbel et al. 1953, T. 2, p. 140, is reprinted, omitting the δ^2 columns, in Vianelli 1959, T. 205, p. 1369. Table 3, p. 146, is reprinted as Pro. 219, pp. 721-722 [!] Table 4, pp. 147-150, is reprinted, omitting the δ^2 columns, as Pro. 220, pp. 723-724.

Gumbel 1954a, T. 1, p. 271, is reprinted in Vianelli 1959, T. 206, p. 1370.

Greenwood & Durand 1955, T. 1, p. 235, is reprinted in Vianelli 1959, Pro. 293, p. 1051.

10.311 Comprehensive tables of symmetric functions

For errors in F. N. David & Kendall 1951, 1955, described on pp. pp. 422-423 above, see Biometrika 45, 292.

10.331 Sheppard's corrections

In addition to works cited on pp. 426-427 above, see Daniels 1947; Taguti 1951.

10.343 Sampling cumulants of k-statistics

[see also p. 430]

David, F. N. 1949

Biometrika 36, 383-393

Review by C. C. Craig (MTAC 5, 72; abridged):

This paper contains a table of the terms that must be added to the product cumulant of order hj of the rth and s^{th} k-statistics in order to give the corresponding product moment. The table gives all terms for $h+j = 4, 5$; for $h+j = 6$, only the terms involving n^{-3}.

10.36 Multivariate symmetric functions

[see also pp. 431-432]

Bennett, J. H. 1956

Journal of the Royal Statistical Society, Series B 18, 104-112

Review by C. C. Craig (MTAC 12, 83):

Though there is some discussion of partitions in more than two dimensions, the greater part of this paper is concerned with bipartitions, extending and clarifying the original work by R.A. Fisher 1950 c. The first table gives values of certain auxiliary functions and the second gives the numbers of bipartitions of $n = 1(1)8$ with given marginal partitions.

Hooke, R. 1956

Annals of Mathematical Statistics 27, 55-79

Review by Leo Katz (MTAC 12, 84):

Generalized symmetric functions, (g.s.m.'s) of the elements in an $r \times c$ array are defined as averages of restricted monomial symmetric functions; the restrictions apply to location of the elements in rows and columns. Thus, the g.s.m. of degree $(a + b + c)$ with two elements in the same row and one in a distinct row and a distinct column is

$$\begin{bmatrix} a & b & - \\ - & - & c \end{bmatrix} = \frac{1}{r^{(2)}c^{(3)}} \sum{'} x_{ij}^a x_{ik}^b x_{mn}^c$$

where $\sum{'}$ indicates summation over distinct subscripts.

Bipolykays are defined by a non-commutative multiplication of the polykays, a family of polynomials in symmetric functions defined by Tukey [1] in generalization of Fisher's cumulants. Therefore, the bipolykays are linear combinations, with integer coefficients, of the g.s.m.'s.

---

1. Tukey 1950

10.36
11.19

Conversion formulas for the representation of bipolykays in terms of g.s.m.'s and the reverse are given in the text for functions of degree 1 and 2; Table 1 gives the coefficients for degree 3 and Table 2 for degree 4. Multiplication formulas for bipolykays of degree 1 by bipolykays of degree up to 3 are given in the text and, in Table 3, for products of two bipolykays both of degree 2. Table 4 gives the variances and covariances of the bipolykays

$$\begin{pmatrix} 1 & 1 \\ - & - \end{pmatrix}, \quad \begin{pmatrix} 1 & - \\ 1 & - \end{pmatrix} \text{ and } \begin{pmatrix} 2 & - \\ - & - \end{pmatrix}$$

for matrix bisamples selected by taking from an $R \times C$ population matrix, independently, samples of r from the R rows and c from the C columns and forming the indicated submatrices.

10.4 Čebyšev's inequality and related inequalities
[see also pp. 433-434]

Vianelli, S. 1959

Prontuari per calcoli statistici

Prontuario 39, p. 167, taken from Bignardi 1947, gives $1-k^{-2}+kK^{-3}$ to 4 dec. for $\begin{cases} K = 2(1)7 & 4.472136 & 8(1)10, \infty \\ k = 1.5(.1)K & 1.5(.1)4.4 & 1.5(.1)7.5 \end{cases}$.

Prontuario 40, pp. 168-169, gives $1 - 2h^h(h+1)/(h+2)^{h+1}k^2$ to 6 dec. for $h = 0(1)5, \infty$; $k = 1(.1)4.5$.

11.119 Miscellaneous tables of orthogonal polynomials
[see also pp. 439-440]

Nemčinov, V. S. 1946

Tables 1-5, pp. 112-116, give orthogonal polynomials for $r = 1(1)5$ and $n = 2(1)20$; each table is devoted to one value of r. Table 6, p. 117, gives the sums of squares of these polynomials.

Nemčinov, apparently following Čebyšev, normalizes the rth polynomial with unit coefficient of x^r: this results in half- and quarter-integer values of both the polynomials and the sums of squares.

11.12 Power polynomials
[see also p. 440]

n = 5(1)100	r = 1(1)4	Nielsen & Goldstein 1947
2(1)100	1, 2	Berjman 1941 c (214)
4(1)100	3, 4	Berjman 1942 a (208)
6(2)100	5	Berjman 1942 b (442)
7(2)99	5	Berjman 1942 a (214)
4(1)50	2	Quartey 1945

11.125 The fitting of power polynomials without moments

The coefficient a_s in a polynomial of degree r, defined on p. 440 above, can be expressed (by a simple intermediate calculation involving orthogonal polynomials) as a linear combination of the ordinates: for example, if $a_0 + a_1 x + a_2 x^2$ is to be fitted to 7 ordinates, then

$$a_0 = [7y_0 + 6(y_{-1} + y_1) + 3(y_{-2} + y_2) - 2(y_{-3} + y_3)]/21,$$
$$a_1 = [(y_1 - y_{-1}) + 2(y_2 - y_{-2}) + 3(y_3 - y_{-3})]/28,$$
$$a_2 = [-4y_0 - 3(y_{-1} + y_1) + 5(y_{-3} + y_3)]/84.$$

Vianelli 1959, Prontuari 22-24, pp. 116-120, gives the necessary coefficients as exact vulgar fractions for $r = 1(1)3$; $s = 0(1)4$; $n = [r+1](1)20$.

11.19 Miscellaneous polynomial fitting
[see also p. 441]

Guest, P. G. 1951 a

<u>Annals of Mathematical Statistics</u> 22, 537-548

Review by K. J. Arnold (MTAC 6, 228):

Given n equally spaced values of x and the corresponding n observed values of y, it is desired to fit a polynomial $u_p(x) = \sum_{j=0}^{p} b_{pj} x^j$ ($p < n$) by a method of weighted grouping. To obtain an unbiased estimate of $b_{pp} = a_p$, the n points are divided into not more than $2p + 1$ groups of successive points, the number of points in the ith and in the $(n - i + 1)$th groups being equal. The sum of the y's in each of $p + 1$ of these groups is to be assigned a non-zero weight. Because of symmetry, not more than $(p + 2)/2$ different weights are involved if p is even. If p is odd, $p + 1$ different weights are involved but half of these are the negatives of the other half. The groupings are determined for each n and p in such a way that the variance

11.19
11.28

of a_p, assuming equal variances of the y's, is a minimum. To estimate b_{pj}, $j = 0, 1, \cdots, p - 1$, the a_j, $j = 0, 1, \cdots, p - 1$ are calculated and the estimates of b_{pj} are obtained from the relation, estimate of $b_{pj} = a_j + \beta_{j+1,j} a_{j+1} + \cdots + \beta_{pj} a_p$. (Alternate β's are zero.)

A table (p. 541–545) gives the groupings, the weights (exact), the β's (β_{20} and β_{31}, exact; β_{42} and β_{53}, 10S; β_{40} and β_{51}, 9S) for $n = 7(1)55$, $j = 0(1)5$.

The relative (compared with least square procedure) efficiencies of the estimates of b_{pj} are discussed and a table of some limiting relative efficiencies is given. The lowest limiting relative efficiency appearing in the table is .889 for estimates of a_1. Computational schedules and an example are given.

Scott, J. F. and Small, V. J. 1955

Journal of the Royal Statistical Society, Series B 17, 105-144

Review by Frank Massey (MTAC 11, 33):

Tables 1 and 2 give for a number of values of two parameters, ρ_1 and ρ_2 (representing serial correlations of independent and residual variables) factors that facilitate the computations in the formula for the asymptotic variance of an estimate of a regression slope in a trend-reduced Markov time series. Two smoothing formulas, each extending over 2k+1 terms, were considered: (i) a moving average of 3 separated terms, and (ii) an equally weighted moving average. Tables 1 and 2 give results to 2 dec. for k = 1(1)3, 5, 10, ∞ ; ρ_1, ρ_2 = 0(.1).9 ; thence one can make an optimum choice of k if the values of the other parameters are known.

Tables 5 and 6 give, for the same arguments, correction factors to 2 dec. for converting a classical estimate of this same regression slope into an estimate that is adjusted for autocorrelations in the series.

Tables 3 and 4 give, for the above smoothing formulas, the first serial correlation of the reduced series to 2 dec. for the same values of k in terms of the Markov parameter ρ = 0(.1).9 . Used inversely, these tables may be used to obtain estimates of ρ from the observed serial correlations.

11.24 Tests of significance in harmonic analysis
[see also p. 443]

Eisenhart, C., Hastay, M. W. and Wallis, W. A. 1947

Techniques of statistical analysis

Table 15.1, pp. 390-391, by Churchill Eisenhart and Herbert Solomon, gives 95% and 99% points of Cochran's statistic for r = 1(1)10, 16, 36, 144, ∞ ; n = 2(1)10, 120/n = 0(1)6(2)12 . Reprinted in Dixon & Massey 1951, T. 17, pp. 340-341; 1957, T. 17, pp. 438-439; Vianelli 1959, Pri. 93-94, p. 259.

11.28 Miscellaneous tables for harmonic analysis

Labrouste, H. et Labrouste, Y., Mme   1943 a , b

Analyse des graphiques résultant de la superposition des sinusoïdes

Review by A. Zygmund (MTAC 2, 305):

The problem discussed by the authors is that of harmonic analysis. Suppose that a function $y = y(x)$ is a sum of a finite, though unknown, number of simple oscillations. How may one find the periods, the phases, and the amplitudes of the terms? The authors' approach to the problems is as follows: Suppose that

(1) $\qquad y = a \sin(\theta x + \phi) + a' \sin(\theta' x + \phi') + a'' \sin(\theta'' x + \phi'') + \cdots;$

if such a representation is possible, it is certainly unique. Let us consider $2m + 1$ equidistant points $x_0 - m, x_0 - (m-1), \cdots, x_0, \cdots, x_0 + m$ symmetric with respect to the point x_0, and let $y_{-m}, y_{-(m-1)}, \cdots, y_0, \cdots, y_m$ be the corresponding ordinates of the curve. We write

$$\theta = 2\pi/n, \qquad \theta' = 2\pi/n', \cdots, \qquad \alpha_m = 2\cos(2\pi m/n), \qquad \alpha'_m = 2\cos(2\pi m/n'), \cdots$$

If we set $Y_\mu = y_\mu + y_{-\mu}$, we get

$$Y_\mu = \alpha_\mu a \sin(\theta x_0 + \phi) + \alpha'_\mu a' \sin(\theta' x_0 + \phi') + \cdots,$$

so that any linear combination

$$R_m = K_0 y_0 + K_1 Y_1 + \cdots + K_m Y_m$$

of the quantities $y_0, Y_1, \cdots, Y_m$ can be written in the form

$$R_m = \rho_m a \sin(\theta x_0 + \phi) + \rho'_m a' \sin(\theta' x_0 + \phi') + \cdots$$

where

$$\rho_m = K_0 + K_1 \alpha_1 + \cdots + K_m \alpha_m, \qquad \rho'_m = K_0 + K_1 \alpha'_1 + \cdots + K_m \alpha'_m, \cdots.$$

It follows that (with x_0 replaced by x) the phases and the periods of the terms composing R_m are the same as the phases and the periods of the terms of y. If the K's are so chosen that all the numbers ρ except one—say except ρ_m—are very small, the graph of R_m/ρ_m gives the first term on the right of (1). To each period $\theta_0 = 2\pi/n_0$ corresponds a "selective" combination R_m. The numbers K_m are the Fourier coefficients of a function large in the neighborhood of the point θ_0 and small elsewhere. The simplest combinations R_m are $s_m = y_{-m} + \cdots + y_0 + \cdots + y_{m-1} + y_m$. The corresponding multipliers of the amplitudes are then

$$\sigma_m = 1 + \alpha_1 + \alpha_2 + \cdots + \alpha_m = \sin(2m+1)\frac{\pi}{n} \Big/ \sin\frac{\pi}{n}.$$

Tables I and II, p. 91–142, of **a** give the values, to 3D, of α_m and σ_m respectively for $m = 0(1)20$, $n = .5(.01)3(.02)5(.05)10(.1)15(.2)25(.5)50(1)100(5)200(10)500$; $m = 21(1)40$, $n = 4(.02)5[\text{then as above}]500$; $m = 41(1)50$, $n = 50(1)100(5)200(10)500$. Other tables give values of multipliers corresponding to combinations whose basic element is not $y_\mu + y_{-\mu}$ but $y_\mu - y_{-\mu}$.

In **b** the authors investigate in great detail products (superpositions) of simple combinations and give graphs of the ratio ρ/ρ max as functions of period n. Here ρ is the amplitude multiplier and ρ max is its maximum.

The title-page states that H. Labrouste is a professor in the Faculty of Sciences of the University of Paris and that Mme. Y. Labrouste is an associate physicist at the Institut de Physique du Globe, of the University of Paris.

12.11
13.3

12.11 $\sin^{-1} x^{\frac{1}{2}}$: degrees
[see also p. 445]

$0°.01$ $0(.001).999$ No Δ   Vianelli 1959(208)

12.13 $\sin^{-1} x^{\frac{1}{2}}$: radians
[see also p. 445]

4 dec. $2\sin^{-1} x^{\frac{1}{2}}$ $0(.001).999$ No Δ   Vianelli 1959(206)
4 dec. $2\sin^{-1} x^{\frac{1}{2}}$ $0(.01)1$ No Δ   Dixon & Massey 1957(465)

12.54 Special tables of the logit transformation
[see also p. 451]

Wilson, E. B. and Worcester, Jane   1943 a
Proceedings of the National Academy of Sciences, U.S.A. 29, 79-85
  Assume that the observed proportion, P, reacting at log dose x, follows the logistic curve $P = \frac{1}{2} + \frac{1}{2}\tanh \alpha(x-x_0-\gamma)$, where $x_0+\gamma$ is the ED_{50}. Expose 10 subjects each at log dose $x_0-\frac{1}{2}c$ and $x_0+\frac{1}{2}c$; let P_1, P_2 be the observed proportions. Table 1, p. 83, gives $2\gamma/c$ and $2c\alpha$ together with their standard errors for $P_1 = 0(.1)1$, $P_2 = .4(.1)1$.

Wilson, E. B. and Worcester, Jane   1943 b
Proceedings of the National Academy of Sciences, U.S.A. 29, 207-212
  Assume that the expected proportion, P, reacting at log dose x, follows the logistic curve $P = \frac{1}{2} + \frac{1}{2}\tanh \alpha(x-x_0-\gamma)$, where $x_0+\gamma$ is the ED_{50}. Expose n subjects each at log dose x_0-c, x_0, x_0+c; let P_1, P_2, P_3 be the observed proportions; set $A = 2(P_1+P_2+P_3) - 3$, $B = P_3 - P_1$. Then maximum likelihood estimates of α and γ are given by $\hat{\alpha}c = \tanh^{-1}[2B/(X^2-AX+2)]$, $\hat{\gamma}/c = (-\tanh^{-1}X)/\hat{\alpha}c$; where X is the root of the cubic $3X^3 - 4AX^2 + (6+A^2-4B^2)X - 2A = 0$. The table on pp. 208-210 gives $\hat{\gamma}/c$ and $\hat{\alpha}c$ to 2 dec. for $B = .3(.05).95$ and $A = 0(.01)[2.9-2B]$.

13.11 Finding list of tables of random digits
[see also p. 456]

The symbol ? affixed to a number of digits indicates a table of whose randomness we have seen no evidence.

250 000 digits	Royo López & Ferrer Martín 1954(3)
88 000 ??	Vielrose 1951
24 000 ?	Czechowski et al. 1957(93)
18 000	Vianelli 1959(478, 483)
12 000	Royo López & Ferrer Martín 1955(3)

[N.B. Czechowski et al. 1957, T. 31, pp. 114-123, is not a table of random numbers]

13.1211 Tables [of random digits] published as obtained
[see also p. 457]

Royo López, J. y Ferrer Martín, S. 1954

The table on pp. 3-127 contains 250 000 digits compiled from the Spanish National Lottery by methods set forth in Royo Lopez & Ferrer Martin 1954a.

13.3 Random normal numbers
[see also pp. 462-463]

Royo López & Ferrer Martín 1954, T. 13, pp. 203-207, gives 2500 random normal deviates from Wold 1948.

Vianelli, S. 1959

Prontuari per calcoli statistici

Prontuario 166, pp. 488-492, contains 2500 random normal deviates from Wold 1948. Prontuario 167, pp. 493-497, contains 250 decuples of correlated normal deviates from Fieller et al. 1955.

[??] We have not seen this 24-page book. The page size is the same as that of Czechowski et al. 1957; presumably the typography is similar. If so, Vielrose contains the 24 000 digits reprinted by Czechowski, and between 34 000 and 68 000 additional digits.

14.11
14.129

14.11 Lot inspection by variables
[see also pp. 470-477]

14.111 Comprehensive tables

Dodge & Romig 1959, Appendices 4-7, are reprinted in Vianelli 1959, Pri. 225-228, pp. 785-815.

Freeman et al. 1948, TT. 1-3, pp. 417-420, are reprinted in Vianelli 1959, Pri. 229-234, pp. 816-820.

14.112 Other tables

Golub, A. 1953

Journal of the American Statistical Association 48, 278-288

Tables 2-6, 8 (i.e. sample sizes 10(5)30, 40) are reprinted in Vianelli 1959, Pro. 237, pp. 827-829.

Horsnell, G. 1954

Applied Statistics 3, 150-158

Review by F. E. Grubbs (MTAC 10, 235; abridged):

The present work by Horsnell covers the case of sampling Poisson populations, i.e., percentage defective less than about 10, and gives some very useful tables for several acceptance and rejection probability levels. In Horsnell's notation, $P(\alpha)$ is the percentage defective for which the chance of acceptance of the lot is α and $P(\beta)$ is the percentage defective for which the chance of acceptance of the lot is β. Horsnell gives tables for the Poisson case for $\alpha = .99$ and $\alpha = .95$ and $\beta = .1, .05, .01$ which are tabulated usually to 4 fig. in the form $K = 1(1)20$ against $NP(.99)$, $NP(.95)$, $P(.1)/P(.99)$, $P(.05)/P(.99)$, $P(.01)/P(.99)$, $P(.1)/P(.95)$, $P(.05)/P(.95)$, $P(.01)/P(.95)$ and $NP(.50)$ [; N being the sample size and K the acceptance number].

14.12 Lot inspection by variables
[see also pp. 477-481]

14.121 Comprehensive tables

Lieberman, G. J. and Resnikoff, G. J. 1955

Journal of the American Statistical Association 50, 457-516

A preliminary issue, including 17 pages of mathematical introduction, of the tables of MIL-STD 414 (reviewed on p. 478 above). Reprinted, in substance, in Vianelli 1959, Prontuari 238-243, pp. 830-846. A concordance follows:

Lieberman & Resnikoff	MIL-STD 414	Vianelli
T. 1, p. [474]	T. D 3, pp. 99-100	Pro. 238, p. 830
T. 2, p. [475]	T. D 5, p. 103	Pro. 239, p. 831
T. 3, p. 476	T. B 3, p. 45	Pro. 240, p. 832
T. 4, pp. 477-481	T. B 5, pp. 47-51	Pro. 242, pp. 833-839 [!]
T. 5, p. 482	T. C 3, p. 71*	Pro. 241, p. 832
T. 6, pp. 483-487	T. C 5, pp. 73-77	Pro. 243, pp. 840-846 [!]

Figures 1-16, pp. [488-515], give OC curves for plans with process standard deviation estimated by sample standard deviation; OC curves for plans with process standard deviation known or estimated from sample range are stated to be essentially equivalent. Reprinted in MIL-STD 414, T. A 3, pp. 6-33.

14.123 Other tables: controlling the mean

Barraclough, Elizabeth D. and Page, E. S. 1959
<u>Biometrika</u> 46, 169-177

Table 1, p. 174, gives the characteristics of a Wald sequential test of $\mu = \mu_0$ against $\mu = \mu_1$ for $(\mu_1-\mu_0)/\sigma = \cdot 5(\cdot 5)2$; $\alpha = \cdot 001, \cdot 005, \cdot 01, \cdot 05$; $\beta = \cdot 05, \cdot 1(\cdot 1)\cdot 7$. Table 2, p. 175, gives the average sampling number of such tests to 2 dec. for the same arguments.

14.129 Auxiliary tables for the design of variables plans

N.B.S. 1951

See abstract on p. 303 above.

Dixon, W. J. and Massey., F. J. 1957
<u>Introduction to statistical analysis</u>

Table 14, p. 434, gives inter al. $\log_e[(1-\beta)/\alpha]$ to 3 dec. for $\alpha, \beta = .001, .005, .01(.01).05(.05).25$.

Vianelli, S. 1959
<u>Prontuari per calcoli statistici</u>

<u>Prontuario</u> 250, p. 852, gives $\log_e[(1-\beta)/\alpha]$ to 5 dec. for $\alpha, \beta = .005(.005).1(.01).12, .15, .2$. <u>Tavola</u> 229, p. 1411, gives $\log_e[(1-\alpha)/\beta]$ to 5 dec. for the same arguments.

---

*Minor variations

14.151
14.229

14.151 Tables of life-testing plans
[see also pp. 483-484]

Epstein, B. 1954c

Annals of Mathematical Statistics 25, 555-564

Review by E.J. Gumbel (MTAC 10, 44; abridged):

With n items on a life test, it is decided in advance that the experiment is terminated either at a fixed truncation time T_U or at the time $X(r_0, n)$ necessary for the r_0^{th} failure, whichever is smaller (stopping rule), and it is assumed that the distribution of lives X ($X \geq U$) is exponential with unknown mean θ. If $X(r_0, n) < T_0$ the null hypothesis $\theta = \theta_0$ is rejected; otherwise accepted. Two cases are considered, non-replacement and replacement from the same population.

Table 1 gives the values of r and $x^2_{1-\alpha}(2r)\big/(2r)$ such that the test based on using $\vartheta_{r,n} > C = \theta_0 x^2_{1-\alpha}(2r)\big/2r$ as acceptance region for $\theta = \theta_0$ will have the probabilities $L(\theta_0) = 1-\alpha$ and $L(\theta_1) \geq \beta$ with $\alpha = .01, .05, .1$ and $\beta = .01, .05, .1$ for $\theta_0/\theta_1 = 1.5(.5)3(1)5, 10$. (The heading of the last column should be $\beta = .1$ instead of $\beta = .01$.) Table 2 gives some integer values of $n = [\theta_0 x^2_{1-\alpha}(2r_0)\big/2T_0]$ to be used in truncated non-replacement procedures for $\alpha = .01, .05$; $\beta = .01, .05$; $\theta_0/\theta_1 = 2, 3, 5$; $\theta_0/T_0 = 3, 5, 10, 20$.

Sarhan, A. E. and Greenberg, B. G. 1957

Journal of the American Statistical Association 52, 58-87

Table 4, p. 73, gives the ratio of the mean times $\dfrac{r^{th} \text{ failure out of n}}{n^{th} \text{ failure out of n}}$ under exponential life testing as exact vulgar franctions for $r = 1(1)n$ and $n = 2(1)10$. Reprinted in Vianelli 1959, **Prontuario** 112, p. 303.

14.21 Control charts for number or per cent defective
[see also p. 487]

Schrock, E. M. 1957

Quality control and industrial statistics

[For Table 11.7a, see p. 205 above]

Table 11.7b, pp. 91-137, gives upper and lower three-sigma control limits for number of defects for $n = 100(100)4000(200)5000(500)10000$ and $p = .001, .0025, .005(.005).02(.01).1(.02).2(.05).5$.

14.221 Factors for constructing charts for mean and standard deviation, or for mean and range: A, B, C, D
[see also pp. 488-490]

n =

2(1)25, 30(10)50(25)100	A_1, B_4	2 dec.	Graf & Henning 1953(67)
7(1)25, 30(10)50(25)100	B_3		
2(1)16	d_2	3 dec.	Graf & Henning 1953(66)
2(1)15	A_2, D_4	2 dec.	Graf & Henning 1953(69)
7(1)15	D_3		

14.222 Factors for constructing charts: other forms
[see also p. 490]

American Standards Association 1941, p. 40, gives factors for 0.1% and $\frac{1}{2}$% control limits.

Graf & Henning 1953, T. 9, p. 68, gives factors for upper and lower $\frac{1}{2}$% and $2\frac{1}{2}$% limits for mean and range charts to 2 dec. for n = 2(1)10, with text in German.

14.229 Miscellaneous tables of control charts for variables
[see also pp. 490-491]

Page, E. S. 1954
<u>Journal of the Royal Statistical Society</u>, Series B 16, 131-135
"Control charts for the mean of a normal population"

If an upper control limit is placed at $\mu + B\sigma n^{-\frac{1}{2}}$, then, when the process is operating at mean μ, the average number of articles L inspected between rectifying actions is $L_0 = n/Q(B)$, where Q is the normal tail area. If the process mean shifts to $\mu + k\sigma$, L will drop to $L_1 = n/Q(B - kn^{\frac{1}{2}})$.

Table 1, p. 133, gives values of n and B that minimize L_1 for fixed L_0 and k, and consequent values of L_1: $10^{-3} L_0$ = 2, 5(5)20(20)60; k = .2(.1)1.8. Table 2, p. 134, gives values of n and B that maximize L_0 for fixed L_1 and k, and consequent values of L_0: L_1 = 10(5)25(25)100; k = .1(.1)1.

Reprinted in Vianelli 1959, <u>Prontuario</u> 248, pp. 849-850.

14.229
15.114

Shimada, Shozo 1954

Reports of Statistical Application Research,
Union of Japanese Scientists and Engineers, 3, 70-74

Review by Seymour Geisser (MTAC 10, 50):

Let x_1, x_2, x_3, x_4 be independent and normally distributed random variables with means m_1, m_2, m_3, m_4 and common variance σ^2. The range R is defined as the difference between the largest and smallest value among the four successive observations.

If $m_1 = m_2 = m_3 = m_4$, then the range provides an estimate of σ. On the other hand, if not all the m_i are equal then the range may include the effect of the variability of the m_i in its estimation of σ.

Shimada finds the distribution of R when the m_i are not all equal. He then plots the power curves of the range chart for three special cases, approximating a twofold integral by using "Circular Probability Paper" (Leone & Topp 1952).

He considers the following three cases:
Case 1. $m_i = \alpha + (i-1)\eta$ where α and η are constants.
Case 2. $m_a = m_b = m_c + \delta = m_d + \delta$ (abcd represents some permutation of the integers 1234).
Case 3. $m_a = m_b = m_c = m_d + \delta$.

Table 1 gives the probability that a point plotted on a range chart falls outside a specified control limit for the three different cases. Table 2 gives the probability that x_i and x_j take on the smallest and largest value for Case 1 where $\eta = .05$.

Some typographical errors:
1. Fig. 2 Case 1 should be Fig. 3 Case 2.
2. Fig. 3 Case 2 should be Fig. 2 Case 1.
3. in Table 2, the headings Max. and Min. should be interchanged.

14.31 Non-parametric tolerance limits
[see also p. 492]

Burington, R. S. and May, D. C. 1953

Probability and statistics

Table 13.88.1, p. 168, gives $1 - nh^{n-1} + (n-1)h^n$ to 2 fig. for $n = 5(5)30(10)50(25)100$ and $h = .8, .9, .95, .99$.

Rosenbaum, S. 1953

Annals of Mathematical Statistics 24, 663-668

See abstract on p. ΔΔΔ above.

Rosenbaum, S. 1954

<u>Annals</u> <u>of</u> <u>Mathematical</u> <u>Statistics</u> 25, 146-150

See abstract on p. 706 above.

Somerville, P. N. 1958

Annals of Mathematical Statistics 29, 599-601

Table 1, p. 600, gives integer values of $m = r+s$ such that we may assert with confidence at least γ that $100\,P\%$ of a population (with continuous cumulative distribution function) lie between the rth smallest and the sth largest of a sample of n: for $\gamma = .5, .75, .9, .95, .99$; $P = .5, .75, .9, .95, .99$; $n = 50(5)100(10)150, 170, 200(100)1000$.

Table 2, p. 600, gives the confidence γ with which we may assert that $100\,P\%$ of the population lie between <u>the</u> smallest and largest of a sample of n: for $\begin{cases} P = .5 & .75 & .9 & .95, .99 \\ n = 3(1)12 & 3(1)20(5)30 & 3(1)20(5)30(10)80 & 3(1)20(5)30(10)100 \end{cases}$.

15.11 Finding list of factorial patterns

For notation see page 502, above.

15.112 Factorial patterns of the 2^a system

8[a]	2^6 in 32	4.2^4 in 1.32	2 c 2	Vianelli 1959(476)
10.	2^7 in 8			
32.	2^{11} in 12			
56.	2^{15} in 16	ortho. matrix sat.		Vianelli 1959(475)
71.	2^{19} in 20			

15.1132 Factorial patterns of the $3^a.2^b$ system

144.	3.2^4 in 24	3.2^4 in 1.24	2 c 2	Vianelli 1959(477)
		$\underline{4}.3$ in 3.4.2		Vianelli 1959(474)

15.114 Factorial patterns with one or more factors at 4 levels

234.	4.2^4 in 32	4.2^4 in 1.32	2 c 2	Vianelli 1959(476)
326[a]	$4^2.3$ in 24	$\underline{4}.3$ in 3.4.2		Vianelli 1959(474)

15.115
16.0244

15.115 Factorial patterns with one or more factors at 5 levels

375[a]	$5.3^2.2^3$ in 120	10.$\underline{6}$ in 10.6.2	Vianelli 1959(474)
376[a]	$5^2.3.2$ in 60	6.$\underline{5}$ in 6.5.2	Vianelli 1959(474)

15.116 Factorial patterns with one or more factors at 6 levels

396[a]	6.5^2 in 60	6.$\underline{5}$ in 6.5.2	Vianelli 1959(474)

15.117 Factorial patterns with one or more factors at 7 levels

407[a]	$7.3^4.2^2$ in 504	28.$\underline{9}$ in 28.9.2	Vianelli 1959(474)
408[a]	$7.4.3^4$ in 504		

15.1199 Factorial patterns with one or more factors at 10 or more levels

504[a]	10.6^2 in 120	10.$\underline{6}$ in 10.6.2	Vianelli 1959(474)
568[a]	28.9^2 in 504	28.$\underline{9}$ in 28.9.2	Vianelli 1959(474)

15.2413 Balanced incomplete block patterns in which a second factor is confounded with replications

Vianelli 1959, Pro. 161, p. 474, gives patterns (due to Bose 1956) for 4(6) = 24 in 3.4.2 , 5(12) = 60 in 6.5.2 , 6(20) = 120 in 10.6.2 and 9(56) = 504 in 28.9.2 ; each group of 2 replications is to be assigned to a different operator, in order to assess each operator's ability in making paired comparisons.

15.3 Tables facilitating the choice of sample size

Johnson, N. L. 1957

<u>Biometrika</u> 44, 518-523

It is desired to obtain a sample from a stratified population in such a way that there are exactly m_i individuals from the ith stratum (i = 1, 2, ..., k) . It is more convenient to take a random sample from the whole population, and to ascertain subsequently the strata to which the chosen individuals belong: therefore, a first sample of N individuals is chosen without regard to stratification, and any shortfall is made up by a further set of samples, one sample from each deficient stratum. Let c be the cost per individual in the first sample; c_i , the cost per individual in the subsequent sample, if any, from the ith stratum; c'_i, the worth of an individual in excess of the quota for its stratum. Assume that the composition of the first sample is multinomially distributed with expectation Np_i in the ith stratum.

Johnson's table relates to the symmetrical problem: all $p_i = 1/k$, all $m_i = m$, all $c_i = dc$, all $c'_i = d'c$. Table 1, pp. 521-522, gives the optimum N for $d = 1.25, 1.5(.5)3$; $d' = 0, .25, .7, .9$;

$$\begin{cases} k = & 2 & 3 & 4 & 5 & 6 & 7 & 8 & 9 & 10 \\ m = & \begin{cases} 25 & 17 & 13 & 10 & 8 & 7 & 6 & 6 & 5 \\ 50 & 33 & 25 & 20 & 17 & 14 & 13 & 11 & 10 \\ 100 & 67 & 50 & 40 & 33 & 29 & 25 & 22 & 20 \\ 250 & 167 & 125 & 100 & 83 & 71 & 63 & 56 & 50 \end{cases} \end{cases}$$

Table 2, p. 523, computed by Miss E. J. Smith, gives the ratio of the cost for optimum N to the cost for $N = 0$: for $d = 1.5, 2.5, 3$;

$d' = 0, .1, .5$;

	$k =$	2	3	4	5	10
		25	17	13	10	5
	$m =$	50	33	25	20	10
		250	167	125	100	50

Reprinted in Vianelli 1959, Pro. 172, pp. 519-520.

Seelbinder, B. M. 1953
Annals of Mathematical Statistics 24, 640-649

Seelbinder gives a method for determining the size of the first sample for use in the two-stage procedure of Stein 1945 for estimating the mean of a normal population with unknown variance. Let n_0 = size of first sample, n = combined size of first and second samples; fix d and α so that the desired accuracy is specified by $Pr[|\bar{x}-\mu| > d] < \alpha$. Tables IA-IV B, pp. 648-649, give $E(n)$ to 3 fig. for $n_0 = 10(10)60, 80, 120, 240$; $d/\sigma = .01(.01).1(.1)1$; $\alpha = .1, .05, .02, .01$.

Reprinted in Vianelli 1959, Pro. 177, pp. 528-529.

<u>16.0244</u> Miscellaneous rational functions

$q^3(3q+2\alpha)/2\alpha^2(1+\alpha)$ 2-3 fig.
$3 + 3q/\alpha$ 0 dec.
$q = .01(.01).3$;
$= .01(.01).12$

Thionet 1955(177),
Vianelli 1959(521)

16.0244
16.0568

$4(1-p)/S^2p$ 2-3 fig. $p = .01(.01).5$;   Vianelli 1959(530)
                                           $S = .01(.01).1$

$4p(1-p)/s^2$ 0 dec. $p = .01(.01).5$;   Vianelli 1959(534)
                                             $s = .01(.005).035$

16.0264   Square roots of rational functions

$[4(1-p)/Np]^{\frac{1}{2}}$ 3 dec. $p = .01(.01).5$;   Vianelli 1959(532)
                                    $10^{-3}N = .5(.1)1(1)10$

$kh^{\frac{1}{2}}(1 - k/N)^{-\frac{1}{2}}$ 0 dec. $\begin{cases} h = 1\frac{1}{2}, 2, 4, 9, 16, 25 \\ k = 1\frac{1}{2}, 2, 4, 8, 16, 32 \\ 10^{-3}N = .1(.1)1(\frac{1}{2})10(1)20(5) \\ 50(10)100(50)200, 300, 500 \end{cases}$ Vianelli 1959(523)

16.0318   Iron numbers

     Let $f(n)$ equal the fractional part of $\frac{1}{2}(5^{\frac{1}{2}}-1)n$. Write the 10^4 pairs $[n, f(n)]$, $n = 1(1)10^4$, on cards; sort the cards in ascending order of $f(n)$. The resulting sequence of integers n is named <u>iron numbers</u> (liczby żelazne) by Steinhaus 1956, who offers it as a tool in constructing representative samples. The sequence is printed in Czechowski et al. 1957, T. 31, pp. 114-123.

16.0568   Golden numbers

$\frac{1}{2}(5^{\frac{1}{2}}-1)n$ 4 dec. $n = 1(1)1000$ Czechowski et al. 1957(112)

16.102 e^{-x}

5 dec. $0(.01)3(.05)4(.1)6(.25)7(.5)10$ No Δ Burington & May 1953(288)

APPENDIX II

CONTENTS OF BOOKS OF TABLES

Arkin, H. and Colton, R. R. 1950
Tables for Statisticians

    Table 1, pp. 26-27. Squares.   16.0211/614

    T. 2, pp. 28-67. Square roots.   16.0261/617

    T. 3, pp. 68-77. Cubes and cube roots.   16.022/615   16.027/619

    T. 4, pp. 78-79. Powers.   16.023/616

    T. 5, pp. 80-99. Common logarithms.   16.061/621

    T. 6, pp. 100-103. Natural logarithms.   16.111/625   16.112/625

    T. 7, pp. 104-107. Reciprocals.   16.024/616

    T. 8, p. 108. Factorials.   2.5112/161

    T. 9, pp. 109-113. Logarithms of factorials.   2.5134/163

    T. 10, p. 114. Areas of the normal curve.   1.1121/5

    T. 11, p. 115. Ordinates of the normal curve.   1.1162/10

    T. 12, p. 116. [Percentage points of t-distribution]   4.1311/212

    T. 13, pp. 117-120. [Percentage points of F-distribution]   4.2314/249

    T. 14, p. 121. [Percentage points of χ^2-distribution]   2.3113/142

    T. 15, p. 122. [Fisher's r to z transformation]   7.212/695

    T. 16, p. 123. [Fisher's z to r transformation]   7.211/331

    T. 17, pp. 124-133. Poisson distribution.   2.221/138

    T. 18, pp. 134-135. Confidence limits for binomial distribution. 14.21/487(Grant)

    T. 19, pp. 136-137. Sample size required for finite populations, for confidence limits and specified reliability limits in per cent.

    T. 20, pp. 138-139. Functions of r.   7.31/695

    T. 21, p. 140. [Percentage] points for r and R.   4.239/254(Snedecor)   7.1221/323

    T. 22, p. 141. Probability of occurrence of statistical deviations of different magnitudes relative to the standard error.   1.1131/6   1.1132/6

    T. 23, pp. 142-145. Random numbers.   13.11/456   13.13/459

    T. 24, pp. 146-151. Natural trigonometric functions. 16.0721/622   16.0722/622

Burington, R. S. and May, D. C. 1953

Probability and Statistics

 Table 10.2.1, p. 82. Cumulative normal distribution.   1.115/8

 T. 10.8.1, p. 88. Probability of occurrence of deviations.
1.1131/5   1.2121/20

 T. 10.8.3, p. 89. Normal distribution.   1.111/3   1.115/8   1.1161/9

 T. 11.10.1, p. 101. Normal probability integral over a circular disk.
2.319/659

 T. 11,11.1, pp. 102-105. Normal probability integral over a circular disk.
1.78/125   2.42/661

 T. 13.58.1, p. 155. Relation of sample standard deviation and population standard deviation.   2.315/148(Pearson)

 T. 13.76.1, p. 162. [Percentage points of the correlation coefficient]
7.1223/324

 T. 13.86.1, p. 167. Distribution of sample ranges.
1.541/88   1.541/645

 T. 13.88.1, p. 168. Probability p that fraction h of population values lies within the range of a sample of size n.   14.31/734

 T. 14.57.1, pp. 192-193; T. 14.57.2, pp. 194-195. Confidence interval for proportion of successes.   3.321/194

 T. 17.43.2, p. 236. Numerical coefficients for estimating central lines and control limits.   14.222/490

 T. I, pp. 247-250. Binomial distribution function.   3.221/187

 T. II, pp. 251-254. Summed binomial distribution function.   3.111/183

 T. III, pp. 255-257. Incomplete Beta-function ratio.   3.311/191

 T. IV, p. 257. Binomial coefficients.   3.211/186

 TT. V, VI, pp. 259-262. [Standard deviation of binomial distribution]
3.53/205   16.0264/618

 T. VII, pp. 259-262. Poisson distribution function.   2.221/138

 T. VIII, pp. 263-266. Summed Poisson distribution function.   2.122/135

 T. IX, pp. 267-275. Normal distribution.
1.1161/10   1.1121/630   1.3211/634

 T. X, pp. 276-279. [Percentage points of F]   4.2314/249

 T. XI, pp. 280-282. [Percentage points of z]   4.331/251

 T. XII, p. 283. t-distribution: Interval -t to +t containing all but "fraction" ϵ of distribution.   4.1311/671

 T. XIII, p. 284. t-distribution: cumulative distribution function.
4.1111/208

T. XIV, pp. 286-287. [percentage points of χ^2]   2.3113/142

T. XV, pp. 288-289. Values of e^{-x}.   16.102/738

T. XVI, p. 290. Factorials and logarithms of factorials.
2.5112/161   2.5134/163

T. XVII, p. 291. Gamma function.   2.5111/661

T. XVIII, p. 292. Common logarithms of gamma function.   2.5131/662

T. XIX, p. 292. Factorials and their reciprocals.
2.5112/161   2.5122/162

T. XX, pp. 293-302. Squares, square roots, and reciprocals.
16.0211/614   16.024/616   16.026/617

T. XXI, p. 303. Natural trigonometric functions.
16.0721/622   16.0722/622   16.0723/623

T. XXII, pp. 304-307. Natural logarithms of numbers.
16.111/625   16.112/625

T. XXIII, pp. 308-309. Common logarithms of numbers.   16.061/621

Czechowski, T. et al. 1957

Tablice statystyczne

Tablica 1, p. 41. Ordinates of the normal distribution.   1.1161/9

T. 2, pp. 42-43. Areas of the normal distribution.   1.1141/6   1.115/8

T. 3, p. 44. Normal deviates.   1.2121/631   1.2122/632

T. 4, pp. 44-49. Individual binomial frequencies.   3.221/187

T. 5, pp. 50-54. Cumulative binomial sum.   3.111/664

T. 6, pp. 55-58. Binomial confidence limits.   3.321/665(Hald)

T. 7, pp. 59-63. Individual Poisson frequencies.   2.221/138

T. 8, pp. 64-68. Cumulative Poisson distribution.   2.122/135

T. 9, pp. 69-74. Poisson confidence limits.   16.0264/618

T. 10, p. 75. Percentage points of χ^2.   2.3113/142

T. 11, pp. 75-76. The χ^2 distribution.   2.1131/656

T. 12, p. 77. Percentage points of t.   4.1311/671

T. 13, pp. 78-79. Probability integral of t.

T. 14, pp. 80-82. Percentage points of z.   4.331/676

T. 15, pp. 83-87. Percentage points of F.   4.2312/676

T. 16, p. 88. Percentage points of r.   7.1221/323

T. 17, p. 88. tanh x.   7.211/695

T. 18, p. 89. Kolmogorov distribution of D_n.   9.4121/714(Massey)

T. 19, p. 89. Kolmogorov distribution of $D_{m,n}$.   9.4122/715

Czechowski et al.
Dixon & Massey

    T. 20, p. 90. Asymptotic Kolmogorov distribution.   9.4111/714

    T. 21, pp. 91-92. Significance points of the distribution of runs.
9.32/379

    T. 22, p. 92. Significance points of the sign test.
3.6/670(Dixon & Massey)

    T. 23, pp. 93-99. Random numbers.   13.13/459   13.11/729

    T. 24, p. 100. Latin squares.
15.112/503   15.112/507   15.113/518   15.1132/523   15.114/526   15.115/533
15.116/535   15.117/536   15.118/537   15.119/541

    T. 25, pp. 100-101. Complete sets of orthogonal Latin squares.
15.112/507   15.112/515   15.112/516   15.113/520   15.114/516   15.115/533
15.117/537   15.118/537   15.119/541

    T. 26, pp. 102-103. Balanced incomplete blocks: combinatorial solutions.
15.2103/574   15.2103/575   15.2104/578   15.2104/582   15.2109/599

    T. 27, p. 104. Balanced incomplete blocks: index by number of replications.

    T. 28, p. 104. Balanced incomplete blocks: index by number of units in a block.
15.2103/570(3)   15.2104/578(2)   15.2105/586(2)   15.2107/594   15.2110/601
       /571   /579(2)   .2106/587   .2108/594   /602(2)
       /572(2)   /580   /588(3)   /595   /603
       /573   /581   /589   /597(2)   /604
       /574   /582(2)   /590   15.2109/598(2)
       /575   15.2105/583   /591   /599(2)
       /576   /584(3)   15.2107/592   /600(2)
15.2104/577(2)   /585   /593(3)   /601

    T. 29, pp. 105-111. Random normal numbers.   13.3/463

    T. 30, pp. 112-113. Golden numbers.   16.0568/738

    T. 31, pp. 114-123. Iron numbers.   16.0318/738

    T. 32, p. 124. Binomial coefficients.   3.211/186

    T. 33, pp. 124-129. Logarithms of binomial coefficients.   3.212/186

    T. 34, p. 130. Binomial, i.e. Newton, interpolation coefficients.
16.2313/628

    T. 35, pp. 131-132. Squares.   16.0211/614

    T. 36, pp. 133-134. Cubes.   16.022/615

    T. 37, pp. 135-138. Square roots.   16.0261/617

    T. 38, pp. 139-140. Reciprocals.   16.024/616

    T. 39, p. 141. Sums of powers of integers.   16.0444/620

    T. 40, p. 142. Sums of powers of odd integers.   16.0445/620

    T. 41, pp. 143-144. Factorials.   2.5112/661   2.5134/662

    T. 42, pp. 145-146. Logarithms.   16.061/621

    T. 43, pp. 147-152. Natural logarithms.   16.112/625

    T, 44, p. 153. Some mathematical constants.

Dixon, W. J. and Massey, F. J. 1951

Introduction to Statistical Analysis (first edition)

Table 17.6, p. 261. Critical values [of the] rank-correlation coefficient. 8.112/350

T. 1, pp. 290-294. Random numbers. 13.11/456 13.13/459

T. 2, pp. 295-304. Random normal numbers, $\mu = 0$, $\sigma = 1$. 13.3/462

T. 3, p. 305. Ordinates, and areas between $-z$ and $+z$, of the normal curve. 1.111/3

T. 4, p. 306. Areas of the normal curve. 1.115/8

T. 5, p. 307. Values of t. 4.1313/213

T. 6 a, p. 308. Percentiles of the χ^2 distribution. 2.3112/141

T. 6 b, p. 309. Percentiles of the χ^2/d.f. distribution. 2.312/143

T. 7, pp. 310-313. F distribution. 4.2312/248

T. 8 a 1, p. 314. Simplified statistics: percentile estimates: mean. 1.51/71

T. 8 a 2, p. 314. Simplified statistics: percentile estimates: standard deviation. 1.544/94

T. 8 a 3, p. 314. Simplified statistics: percentile estimates: mean and standard deviation. 1.544/94

T. 8 b 1, p. 315. Simplified statistics: estimates of dispersion in small samples: efficiency of range, w. 1.541/87

T. 8 b 2, p. 315. Simplified statistics: estimates of dispersion in small samples: mean deviation estimate of σ. 1.52/75

T. 8 b 3, p. 315. Simplified statistics: estimates of dispersion in small samples: modified linear estimate of σ. 1.544/94

T. 8 c 1, p. 316. Simplified statistics: substitute t ratios: upper percentiles for $\tau_1 = (\bar{X}-\mu)/w$. 1.5512/102

T. 8 c 2, p. 316. Simplified statistics: substitute t ratios: upper percentiles for $\tau_d = (\bar{X}_1-\bar{X}_2)/(w_1+w_2)$. 1.5512/102

T. 8 c 3, p. 316. Simplified statistics: substitute t ratios: upper percentiles for $\tau_2 = (X_1+X_N-2\mu)/2w$. 1.5512/102

T. 8 d, pp. 317-318. Simplified statistics: substitute F ratio, upper percentile values. 1.5512/102

T. 8 e, p. 319. Simplified statistics: criteria for testing for extreme mean. 1.5512/103

T. 9, pp. 320-323. Confidence belts for proportions. 3.321/194

T. 10, p. 324. Critical values of r for the sign test. 3.6/206(Dixon & Mood) 3.6/670

Dixon & Massey 1951
Dixon & Massey 1957

T. 11, pp. 325-326. [2½% and 97½% points for runs among elements in two samples]   9.32/378

T. 12, p. 327. Confidence belts for the correlation coefficient.   7.121/323(David)

T. 13, pp. 328-331. [Power function of F tests]   4.242/260

T. 14, p. 332. Natural logarithms.   16.112/626

T. 15, p. 333. Cumulative Poisson distribution.   2.121/135

T. 16, pp. 334-339. Tolerance factors for normal distributions.   1.19/18   14.32/492(Eisenhart et al.)

T. 17, pp. 340-341. Critical values for Cochran's test.   11.24/726(Eisenhart et al.)

T. 18, pp. 342-343. [Percentage points of the studentized range]   1.5511/98

T. 19, pp. 344-345. Determination of second sample size.   4.1413/222(Harris et al.)

T. 20, pp. 346-347. Four-place [sic] squares of numbers.   16.0213/615

T. 21, p. 348. Percentiles of the distribution of d .   9.4121/382

T. 22, p. 349. Possible populations for sampling experiments.

T. 23, pp. 350-354. Random normal numbers, $\mu = 2$, $\sigma = 1$.   13.3/462

T. 24, pp. 355-359. Random normal numbers, $\mu = 0$, $\sigma = 2$.   13.3/462

T. 25, p. 360. Confidence intervals for the median.   3.6/669(Nair)

T. 26, p. 361. Some one-sided and symmetrical tests for $N < 10$.   9.111/367

Dixon, W. J. and Massey, F. J. 1957

<u>Introduction</u> <u>to</u> <u>Statistical</u> <u>Analysis</u> (second edition)

Table 1, pp. 366-370. Random numbers.   13.11/456   13.13/459

T. 2, pp. 371-380. Random normal numbers, $\mu = 0$, $\sigma = 1$.   13.3/462

T. 3, p. 381. Normal distribution.   1.111/3   1.1131/5   1.1161/10   1.221/21   1.2121/631   1.2122/632

T. 4, pp. 382-383. Cumulative normal distribution.   1.1141/6   1.2111/19

T. 5, p. 384. Percentiles of the t distribution.   4.1313/213

T. 6 a, p. 385. Percentiles of the χ^2 distribution.   2.3112/141

T. 6 b, pp. 386-387. Percentiles of the $\chi^2/d.f.$ distribution.   2.312/143

T. 7, pp. 388-403. Percentiles of the F distribution.   4.2314/249   4.2311/248

Dixon & Massey 1951
Dixon & Massey 1957

T. 8 a 1, p. 404. Simplified statistics: percentile estimates: mean. 1.51/71

T. 8 a 2, p. 404. Simplified statistics: percentile estimates: standard deviation.   1.544/94

T. 8 a 3, p. 404. Simplified statistics: percentile estimates: mean and standard deviation.   1.544/94

T. 8 b 1, pp. 404-405. Simplified statistics: estimates of mean and dispersion in small samples: unbiased estimate of using w (variance to be multiplied by σ^2); percentiles of the distribution of w/σ; unbiased estimate of σ based on s.   1.541/88

T. 8 b 2, p. 405. Simplified statistics: estimates of mean and dispersion in small samples: mean deviation estimate of σ.   1.52/75

T. 8 b 3, p. 406. Simplified statistics: estimates of mean and dispersion in small samples: modified linear estimate of σ (variance to be multiplied by σ^2).   1.544/94

T. 8 b 4, p. 406. Simplified statistics: estimates of mean and dispersion in small samples: several estimates of the mean (variance to be multiplied by σ^2).   1.542/90

T. 8 b 5, p. 407. Simplified statistics: estimates of mean and dispersion in small samples: mean values of the order statistics.

T. 8 b 6, p. 407. Simplified statistics: estimates of mean and dispersion in small samples: best linear estimate of σ.   1.544/95

TT. 8 c 1- 8 c 3, pp. 408-409. Simplified statistics: substitute t ratios: [as on p. 743 above]   1.5512/104

T. 8 d, pp. 410-411. Simplified statistics: substitute F ratio, ratio of ranges.   1.5512/104

T. 8 e, p. 412. Simplified statistics: criteria for testing for extreme mean.   1.5512/104

T. 9, pp. 413-416. Confidence belts for proportions.   3.321/194

T. 10 a, p. 417. Critical values of r for the sign test.   3.6/206(Dixon & Mood)

T. 10 b, pp. 418-420. Distribution of the sign test.   3.119/183   3.6/206

T. 11, pp. 421-422. Distribution of the total number of runs.   9.32/379

T. 12, pp. 423-425. Power of one- and two-sample tests of the mean [power of the t-test)   4.1213/223

T. 13, pp. 426-433. Power of the analysis-of-variance test.   4.242/260(Pearson & Hartley)

T. 14, p. 434. Natural logarithms.   16.112/626   14.129/731

T. 15, p. 435. Cumulative Poisson distribution.   2.121/135

Dixon & Massey
Fisher & Yates

    T. 16, pp. 436-437. Tolerance factors for normal distributions.
1.19/18   14.32/493

    T. 17, pp. 438-439. Critical values for Cochran's test.
11.24/726(Eisenhart et al.)

    T. 18, pp. 440-442. Percentiles of the distribution of $q = w/s$.
1.5511/99

    T. 19, pp. 443-444. Distribution of the signed rank statistic T.
9.112/704

    T. 20, pp. 445-449. Distribution of the rank sum T'.   9.121/706

    T. 21, p. 450. Percentiles of the distribution of d.   9.4121/382

    T. 22, p. 451. Possible populations for sampling experiments.

    T. 23, pp. 452-456. Random normal numbers, $\mu = 2$, $\sigma = 1$.   13.3/462

    T. 24, pp. 457-461. Random normal numbers, $\mu = 0$, $\sigma = 2$.   13.3/462

    T. 25, p. 462. Confidence intervals for the median.
9.111/367   3.6/669

    T. 26, p. 463. Some one-sided and symmetrical significance tests for $N \leq 10$.   9.111/367

    T. 27, p. 464. Confidence belts for the correlation coefficient.
7.12/323(David)

    T. 28, p. 465. Various functions of a proportion p.
16.0214/615   16.0264/618   7.31/695   12.13/718

    T. 29 a, p. 466. Binomial probabilities.   3.221/187

    T. 29 b, p. 467. Binomial coefficients.   3.211/186

    T. 29 c, p. 467. Factorials.   2.5112/161

    T. 30 a, p. 468. Percentiles of the distribution of r when $o = 0$.

    T. 30 b, p. 468. Values of $[z = \tanh^{-1} r]$.   7.212/695

    T. 30 c, p. 469. Distribution of $\sum d_i^2$ used in the [Spearman] rank-correlation coefficient.   8.1112/349

    T. 30 d, p. 469. Distribution of quadrant sum for corner test for association for samples of size 2 N.   9.33/713(Olmstead & Tukey)

    T. 31, pp. 470-471. Table of reciprocals.   16.024/616

    T. 32, pp. 472-473. Four-place [sic] squares of numbers.   16.0213/615

    T. 33, pp. 474-479. Table of square roots.   16.026/617

Fisher, R. A. and Yates, F.   1938, 1943, 1948, 1953

Statistical Tables

    Unless indicated by a footnote, the following tables appeared in the first (1938) edition. We give the page numbers of the fourth (1953) edition.

Dixon & Massey
Fisher & Yates

References in the text of this <u>Guide</u> cite <u>table</u> numbers, which remained unchanged from 1938 to 1953.

    T. I, p. 39. The normal distribution.   1.2121/20   1.2122/20

    T. II, p. 39. Ordinates of the normal distribution.   1.1161/10

    T. III, p. 40. Distribution of t.   4.1311/211   4.132/672(ftn.)

    T. IV, p. 41. Distribution of χ^2.   2.3113/142(2)

    T. V, pp. 42, 44, 46, 48, 50. Distribution of z.
4.331/250   4.333/270   4.35/274(Cochran)   4.35/274(Carter)

    T. V, pp. 43, 45, 47, 49, 51. Variance ratio: Per Cent. points of e^{2z}.
4.231/247   4.2311/248   4.2313/249   4.233/252

[2]T. V1, p. 52. Significance of difference between two means.
4.151/234   4.151/234(Sukhatme)   4.151/235(Sukhatme et al.)   4.152/238(Aspin)

[2]T. V2, p. 53. Significance of difference between two means: one component of error distributed normally, the other in Student's distribution.
4.151/233(Fisher)   4.152/238(Aspin)

    T. VI, p. 54. Values of the correlation coefficient for different levels of significance.   7.1221/323

    T. VII, p. 54. Transformation of r to z.   7.211/331

    T. VIII, p. 55. Tests of significance for 2 × 2 contingency tables.
2.612/176

[2]T. VIII1, p. 56. Binomial and Poisson distributions: limits of the expectation.   3.322/195

[2]T. VIII2, p. 57. Densities of organisms estimated by the dilution method.
6.32/316

[4]T. VIII3, p. 58. Significance of leading periodic components.
11.24/443

[4]T. VIII4, p. 59. The normal probability integral.   1.1141/6

    T. IX, pp. 60-62. Probits.   1.2111/19   1.232/27

    T. X, p. 62. Simple quantiles of the normal distribution.
1.41/50   1.2113/631

    T. XI, p. 63. Probits: weighting coefficients and probit values to be used in adjustments of special accuracy.   1.2312/25

[3]T. XI1, pp. 64-65. Probits: weighting coefficients for use when there is (a) natural mortality.   1.233/30

    T. XII, p. 66. Transformation of percentages to degrees.   12.11/445

---

[2]First included in 1943 edition.

[3]First included in 1948 edition.

[4]First included in 1953 edition.

Fisher & Yates 1938-53
Fisher & Yates 1957

    T. XIII, p. 66. Transformation of proper fractions to degrees.
12.12/445

    T. XIV, p. 67. Angular transformation: angular values for final adjustments.   12.14/446

    [3]T. XIV1, p. 68. Scores for linkage data from intercrosses.

    [4]T. XIV2, p. 69. Segmental functions.   16.1231/626

    T. XV, pp. 70-72. Latin squares.
15.112/503   15.112/507   15.113/518   15.1132/523   15.114/526   15.115/533
15.116/535   15.117/536   15.118/537   15.119/541   15.1199/543(3)
15.1512/559(3)   15.1512/560

    T. XVI, pp. 72-73. Complete sets of orthogonal Latin squares.
15.112/507   15.112/515   15.112/516   15.113/520   15.114/526   15.115/533
15.117/537   15.118/537   15.119/541

    T. XVII, p. 74. Balanced incomplete blocks: combinatorial solutions.
15.2103/574   15.2103/575   15.2104/578   15.2104/582   15.2109/599
15.2411/605

    T. XVIII, p. 75. Balanced incomplete blocks: index by number of replications.   15.20/563(ftn.)   15.2411/605

    T. XIX, p. 75. Balanced incomplete blocks: index by number of units in a block.
15.2103/570(3)   15.2104/578(2)   15.2105/586(2)   15.2107/594   15.2110/601
       /571   /579(2)   .2106/587   .2108/594   /602(2)
       /572(2)   /580   /588(3)   /595   /603
       /573   /581   /589   /597(2)   /604
       /574   /582(2)   /590   15.2109/598(2)   15.2411/605
       /575   15.2105/583   /591   /599(2)
       /576   /584(3)   15.2107/592   /600(2)
15.2104/577(2)   /585   /593(3)   /601

    T. XX, pp. 76-77. Scores for ordinal (or ranked) data.
1.43/64   9.122/369(Terry)

    T. XXI, p. 76. Sums of squares of deviations tabulated.   1.43/64

    T. XXII, pp. 78-79. Initial differences of powers of natural numbers.
5.9/293

    T. XXIII, pp. 80-90. Orthogonal polynomials.   11.111/437   11.112/438

    T. XXIV, p. 91. Calculation of integrals from equally spaced ordinates.
16.236/629

    T. XXV, pp. 92-93. Logarithms.   16.061/621

    T. XXVI, pp. 94-99. Natural logarithms.   10.102/402   16.112/625(2)

    T. XXVII, pp. 100-101. Squares.   16.0211/614

    T. XXVIII, pp. 102-105. Square roots.   16.0261/617

---

[3]First included in 1948 edition.

[4]First included in 1953 edition.

Fisher & Yates 1938-53
Fisher & Yates 1957

    T. XXIX, pp. 106-107. Reciprocals.   16.024/616

    T. XXX, pp. 108-109. Factorials.   2.5112/161   2.5134/163

    T. XXXI, pp. 110-111. Natural sines.   16.0721/622

    T. XXXII, pp. 112-113. Natural tangents.   16.0722/622

    T. XXXIII, pp. 114-119. Random numbers.
13.11/456   13.1221/458   15.1512/560

[4]T. XXXIII 1, pp. 120-121. Random permutations of 10 numbers.
13.2/461   15.1512/560

[4]T. XXXIII 2, pp. 122-123. Random permutations of 20 numbers.
13.2/461   15.1512/560

    T. XXXIV, pp. 124-126. Constants, weights and measures, etc.

Several tables in the fourth (1953) edition were renumbered in the fifth (1957) edition. A concordance follows:

1953: T. V 1	V 2	VI	VII	VIII 4	X	XI	XI 1	XII	XIII
1957: T. VI	VI 2	VII	VIII 1	II 1	IX 1	IX 2	IX 3	X	X 1

1953: T. XIV	XIV 1	XIV 2		1957: T. II 1	VI	VI 2	VII	VII 1
1957: T. X 2	XIII	XIV	.	1953: T. VIII 4	V 1	V 2	VI	VII

1957: IX 1	IX 2	IX 3	X	X 1	X 2	XIII	XIV
1953: X	XI	XI 1	XII	XIII	XIV	XIV 1	XIV 2 .

Tables VI 1, XI, XI 1, XII, XII 1, XIII 1 of the 1957 edition were first included in that edition.

Fisher, R. A. and Yates, F. 1957

Statistical Tables

    T. I, p. 42, The normal distribution.   1.2121/20   1.2122/20

    T. II, p. 42. Ordinates of the normal distribution.   1.1161/10

    T. II 1, p. 43. The normal probability integral.   1.1141/6

    T. III, p. 44. Distribution of t.   4.1311/211   4.132/672(ftn.)

    T. IV, p. 45. Distribution of χ^2.   2.313/141

    T. V, pp. 46, 48, 50, 52, 54. Distribution of z.
4.331/250   4.333/271   4.35/274(Cochran)   4.35/274(Carter)

    T. V, pp. 47, 49, 51, 53, 55. Variance ratio: Per Cent. points of e^{2z}.
4.231/247   4.2311/248   4.233/252

    T. VI, p. 56. Significance of difference between two means.   4.151/234

[5]T. VI 1, p. 57. Significance of difference between two means: Behrens' test —odd degrees of freedom.   4.151/236

---

[4]First included in 1953 edition.

[5]First included in 1957 edition.

Fisher & Yates 1957

T. VI2, p. 58. Significance of difference between two means: one component of error distributed normally, the other in Student's distribution. 4.151/233(Fisher)   4.151/236(headnote)

T. VII, p. 59. Values of the correlation coefficient for different levels of significance.

T. VII1, p. 59. Transformation of   to .   7.211/331

T. VIII, p. 60. Tests of significance for 2 2 contingency tables. 2.612/176

T. VIII 1, p. 61. Binomial and Poisson distributions: limits of the expectation.   3.322/195

T. VIII 2, p. 62. Densities of organisms estimated by the dilution method. 6.32/316

T. VIII 3, p. 63. Significance of leading periodic components. 11.24/443

T. IX, pp. 64-66. Probits.   1.2111/19   1.232/27

T. IX 1, p. 66. Simple quantiles of the normal distribution. 1.41/50   1.2113/631

T. IX 2, p. 67. Probits: weighting coefficients and probit values to be used for final adjustments.   1.2312/25

T. IX 3, pp. 68-69. Probits: weighting coefficients for use when there is (a) natural mortality.   1.233/30

T. X, p. 70. Transformation of percentages to degrees.   12.11/445

T. X 1, p. 70. Transformation of proper fractions to degrees. 12.12/445

T. X 2, p. 71. Angular values for final adjustments.   12.14/446

[5]T. XI, p. 72. Logits: the logit or , transformation.   12.51/450

[5]T. XI 1, p. 73. Logits: weighting coefficients and logit values to be used for final adjustments.   12.53/450

[5]T. XII, p. 74. Complementary loglog transformation. 8.3121/364   12.61/452

[5]T. XII 1, p. 75. Complementary loglog transformation: working values. 12.62/452/intro.)   12.62/452

T. XIII, p. 76. Scores for linkage data from intercrosses.

[5]T. XIII 1, p. 77. Product ratios for different recombination fractions.

T. XIV, pp. 78-79. Segmental functions.   16.1041/624   16.1231/626

---

[5]First included in 1957 edition.

T. XV, pp. 80-82. Latin squares.
15.112/503   15.112/507   15.113/518   15.1132/523   15.114/526   15.115/533
15.116/535   15.117/536   15.118/537   15.119/541   15.1199/543(3)
15.1512/559(3)   15.1512/560

T. XVI, pp. 82-83. Complete sets of orthogonal Latin squares.
15.112/507   15.112/515   15.112/516   15.113/520   15.114/526   15.115/533
15.117/537   15.118/537   15.119/541

T. XVII, p. 84. Balanced incomplete blocks: combinatorial solutions.
15.2103/574   15.2103/575   15.2104/578   15.2104/582   15.2109/599
15.2411/605

T. XVIII, p. 85. Balanced incomplete blocks: index by number of replications.   15.20/563(ftn.)   15.2411/605

T. XIX, p. 85. Balanced incomplete blocks: index by number of units in a block.
15.2103/570(3)   15.2104/578(2)   15.2105/586(2)   15.2107/594   15.2110/601
    /571   /579(2)   .2106/587   .2108/594   /602(2)
    /572(2)   /580   /588(3)   /595   /603
    /573   /581   /589   /597(2)   /604
    /574   /582(2)   /590   15.2109/598(2)   15.2411/605
    /575   15.2105/583   /591   /599(2)
    /576   /584(3)   15.2107/592   /600(2)
15.2104/577(2)   /585   /593(3)   /601

T. XX, pp. 86-87. Scores for ordinal (or ranked) data.   1.43/64

T. XXI, p. 86. Sums of squares of mean deviations tabulated.   1.43/64

T. XXII, pp. 88-89. Initial differences of powers of natural numbers.
5.9/293

T. XXIII, pp. 90-100. Orthogonal polynomials.   11.111/437

T. XXIV, p. 101. Calculation of integrals from equally spaced ordinates.
16.236/629

T. XXV, pp. 102-103. Logarithms.   16.061/621

T. XXVI, pp. 104-111. Natural logarithms.   10.102/402   16.112/625

T. XXVII, pp. 112-113. Squares.   16.0211/624

T. XXVIII, pp. 114-117. Square roots.   16.0261/617

T. XXIX, pp. 118-119. Reciprocals.   16.024/616

T. XXX, pp. 120-121. Factorials.   2.5112/161   2.5134/163

T. XXXI, pp. 122-123. Natural sines.   16.0721/622

T. XXXII, pp. 124-125. Natural tangents.   16.0722/622

T. XXXIII, pp. 126-131. Random numbers.
13.11/456   13.1221/458   15.1512/560

T. XXXIII 1, pp. 132-133. Random permutations of 10 numbers.
13.2/461   15.1512/560

Fisher & Yates
Graf & Henning

T. XXXIII 2, pp. 134-135. Random permutations of 20 numbers.
13.2/461   15.1512/560

T. XXXIV, pp. 136-138. Constants, weights and measures, etc.

Glover, J. W. 1923

Tables of applied mathematics

The tables are not numbered: we cite the page numbers of the 1923 edition. Fletcher et al. 1946 cite the 1930 edition, which we have not seen: statistical tables begin on p. 392 of the 1923 edition, which corresponds to p. 394 of the 1930 edition (and so on up).

pages 392-411. Areas, ordinates and second to eighth derivatives of the normal curve of error.   1.1121/4   1.3211/634

pp. 412-413. Newton's binomial interpolation coefficients.
16.2313/628

pp. 414-415. Stirling's central difference interpolation coefficients.
16.2315/628

pp. 416-417. Bessel's interpolation coefficients.   16.2316/628

pp. 418-419. Everett's central difference interpolation coefficients.
16.2317/628

pp. 420-421. Newton's binomial interpolation coefficients for computing the first derivative.   16.234/629

pp. 422-423. Stirling's central difference interpolation coefficients for computing the first derivative.   16.234/629

pp. 424-425. Bessel's interpolation coefficients for computing the first derivative.   16.234/629

pp. 426-427. Everett's central difference interpolation coefficients for computing the first derivative.   16.234/629

pp. 428-429. Coefficients of leading major differences to compute leading minor differences when six (four) ordinates are given and fifth (third) difference interpolation is used.

p. 430. Coefficients in Lubbock's summation formula.   16.2369/629

p. 431. Logarithms of Bernoulli's numbers from $n = 1$ to $n = 200$.
16.041/619

pp. 432-435. First eight powers of integers from 1 to 100.   16.023/616

pp. 436-455. Squares, cubes, square roots, cube roots, and reciprocals of integers from 1 to 1000.
16.021/614   16.022/615   16.024/616   16.0261/617   16.027/619

pp. 456-458. Sum of reciprocals of first n integers from $n = 1$ to $n = 450$.
16.046/620

pp. 459-463. Logarithm of the number of combinations of n things r at a time from $n = 5$ to $n = 64$.   3.212/186 pp. 464-467. Logarithm of the gamma function from n = 1 to n = 2 by intervals of .001.   2.5131/663

pp. 469-481. Mean value table to fit exponential growth curve $y = ar^x$ to n observed statistical ordinates.   11.3/444

pp. 482-486. Logarithm of factorial n from n = 1 to n = 1000 by unit intervals.   2.5134/163

p. 487. Leading differences of 0^n; coefficients of x^{n+r} in expansion of $(e^x-1)^n$; coefficients of x^r in expansion of e^{e^x}/e; coefficients of x^r in expansion of $e^x(1-x)^{-n}$.

pp. 488-489. Factors for computing probable errors.   16.0262/617

p. 490. Table of constants.

pp. 492-675. Common logarithms of numbers from 1 to 100,000 to seven places of decimals.   16.061/621

Graf, U. and Henning, H. J. 1954

<u>Formeln</u> <u>und</u> <u>Tabellen</u>

Tabelle 1, pp. 58-59. Ordinates of the Gaussian normal distribution.   1.1161/10

T. 2, p. 60. Integral of the Gaussian normal distribution.   1.111/3

T. 3, p. 61. Limits of integration of the t-distribution as a function of the degrees of freedom n and the statistical confidence $S\%$.   4.1312/671

T. 4, pp. 62-64. Limits of integration of the F-distribution for the statistical confidence $\bar{S} = 95\%$ (99%, 99.9%) as a function of the degrees of freedom n_1 and n_2.   4.2313/249

T. 5, p. 65. Limits of integration of the χ^2-distribution as a function of the degrees of freedom n and the statistical confidence $S\%$.   2.3114/143

T. 6, p. 66. Connexion between the root-mean-square deviation of the population and the average value, $\bar{s}$, of the root-mean-square deviations s of a series of samples [of size] N.   2.522/170

T. 7, p. 67. Factors for determining 3-sigma control limits for $\bar{x}$- and s-charts from the average value, $\bar{s}$, of the root-mean-square deviations s of a series of samples [of size] N.   14.221/733

T. 8, p. 66 [sic]. Connexion between the root-mean-square deviation of the population and the average value $\bar{R}$ of the ranges R of a series of samples [of size] N.   14.221/733

T. 9, p. 68. Factors for arranging $\bar{x}$- and R-charts to determine control limits and warning limits.   14.222/733

T. 10, p. 69. Factors for determining 3-sigma control limits for $\bar{x}$- and R-charts from the average value $\bar{R}$ of the ranges R of a series of samples [of size] N.   14.221/733

Hald
Jahnke & Emde

    T. 11, pp. 70-75. Logarithms of factorials from 0! to 999!   2.5134/662

    T. 12, pp. 76-90. Squares and square roots from 1 to 2000.
16.0211/614   16.026/617

Hald, A. 1952*

Statistical tables and formulas

    Table 1, p. 33. The normal distribution function.   1.1161/9

    T. 2, pp. 34-35. The cumulative normal distribution function.
1.1141/6   1.115/8

    T. 3, pp. 36-38. Fractiles +·5 (probits)$^{\curlyvee}$ of the normal distribution.
1.2111/19

    T. 4, p. 39. Fractiles of the distribution.   4.1311/211   4.1314/213

    T. 5, pp. 40-43. Fractiles of the χ^2 distribution.   2.3111/140

    T. 6, pp. 44-46. Fractiles of the χ^2/f distribution.   2.312/143

    T. 7, pp. 47-59. Fractiles of the ^{2} [F] distribution.   4.2311/248

    T. 8, pp. 60-61. Distribution of the range.   1.541/87

    T. 9, pp. 62-63. The one-sided truncated normal distribution.
1.332/40(2)

    T. 10, pp. 64-65. The one-sided censored‡ normal distribution.
1.332/41(2)   1.332/42

    T. 11, pp. 66-69. Two-sided confidence limits for the probability in the binomial distribution.   3.321/194   3.319/665

    T. 12, pp. 70-71. [Angular transformation]   12.13/445

    T. 13, pp. 72-75. Logarithms of n!   2.5134/163

    T. 14, pp. 76-81. Logarithms of binomial coefficients.   3.212/186

    T. 15, pp. 82-83. Squares.   16.0211/614

    T. 16, pp. 84-87. Square roots.   16.0261/617

    T. 17, pp. 88-89. Reciprocals.   16.024/616

    T. 18, pp. 90-91. Logarithms.   16.061/621

$\parallel$T. 19, pp. 92-97. Random sampling numbers.   13.11/456   13.1211/457

---

*Pages 38-91 of Hald 1952 correspond to pp. 19-77 of Hald 1948 b.

$^{\curlyvee}$Kendall & Buckland 1957 (p. 113) define fractile:

"A term introduced by Hald 1948[b] to denote the variate value below which lies a given fraction of the cumulative frequency. This term is synonymous with the more generally used term quantile . . . and the necessity for its coining is not clear."

‡But cf. Kendall & Buckland 1957 (p. 37) s.v. Censoring:
"It also seems better to avoid speaking of censored distributions."

$\parallel$Not in Hald 1948 b.

Hald
Jahnke & Emde

Jahnke, P. R. E. and Emde, F.

<u>Funktionentafeln</u> / <u>Tables of functions</u>

We do not attempt a complete listing of the contents of these volumes, but confine our notes to tables of statistical interest. It must be remembered, when referring to the number of places listed for these tables, that Emde neither strove for nor achieved end-figure accuracy: "As in the old [1909] edition no attempt has been made to obtain more than a reasonable degree of accuracy. The last decimal place in the values given must be regarded as uncertain. Any one who requires greater accuracy must consult tables giving a further decimal place." (preface to 1933 edition)

R. C. Archibald and L. J. Comrie made a serious attempt to purge the 1945 reprint of end-figure errors.

In addition to the numerical tables indexed in this <u>Guide</u>, we would call attention to the exposition, formulas and graphs given by Jahnke & Emde.

First edition (1909)

Chapter 8, pp. 26-31. The Γ- function.
2.5111/160   2.5131/162   2.5141/164

Cap. 9, pp. 31-42. The Gaussian error function (x).
1.1211/11   1.3212/34

Cap. 10, pp. 43-45. The Pearsonian function $F(r,)$.   10.0431/398

Cap. 13, pp. 90-174. Bessel functions <u>alias</u> cylinder functions.

Second edition (1933)

Cap. 12, pp. 86-96. Factorial function.
2.5111/160   2.5141/164   2.5142/164

¶ A graph of the incomplete Γ-function appears on p. 96.

Cap. 13, pp. 97-113. Error integral and related functions.
1.1211/11   1.3212/34   1.811/126   5.21/278(Wagner

Cap. 18, pp. 192-318. Bessel functions.

Third edition (1938)

Cap. 2, pp. 9-23. Factorial function.
2.5111/160   2.5121/161   2.5141/164   2.5142/164

¶ A graph of the incomplete Γ-function appears on p. 23.

Cap. 3, pp. 23-40. Error integral and related functions.
1.1211/11   1.3212/34   1.811/126   5.21/278(Wagner)

Cap. 8, pp. 126-268. Bessel functions.

Cap. 10, pp. 275-282. Confluent hypergeometric functions. Formulas and graphs; no numerical tables.

Jahnke & Emde
Kitagawa & Mitome

Fourth edition (1948); fifth edition (1952)

 Cap. 2, pp. 9-23. Factorial function.
2.5111/160 2.5121/161 2.5141/164 2.5142/164
 ¶ A graph of the incomplete Γ-function appears on p. 23.

 Cap. 3, pp. 23-41. Error integral and related functions.
1.1211/11 1.322/34 1.811/126 5.21/278(Wagner) 2.41/660

 Cap. 8, pp. 125-264. Bessel functions.

 Cap. 10, pp. 271-278. Confluent hypergeometric functions. Formulas and graphs; no numerical tables.

Sixth edition (1960)

 Cap. 1, pp. 4-16. The gamma functions.
2.5141/164 2.5142/164 2.5111/661 2.5121/662
 ¶ A graph of the incomplete Γ-function appears on p. 14.

 Cap. 3, pp. 26-36. The error function and related functions.
1.115/8 1.1161/9 1.811/126 1.1211/630 1.139/631

 Cap. 7, pp. 96-109. Orthogonal polynomials: Tschebyscheff, Laguerre, Hermite. 1.322/34 2.41/660

 Cap. 9, pp. 131-262. The Bessel functions.

 Cap. 10, pp. 276-285. The confluent hypergeometric functions. Formulas and graphs; no numerical tables.

 "Literatur": pp. 300-314.

Kelley, T. L. 1938

The Kelley Statistical Tables

 Table 1, pp. 14-114. Normal distribution, simple correlation, and probability functions.
1.2111/19 1.2112/20 1.221/21(2) 7.32/334 16.0264/618

 T. 2, pp. 116-117. Giving P,--the probability that, for a given n, a divergence as great as χ^2 will arise as a matter of chance.
1.1131/6 2.1133/133

 T. 3, pp. 118-127. Cubic interpolation coefficients. 16.2311/627

 T. 4, pp. 128-129. Quintic interpolation coefficients. 16.2311/627

 T. 5, p. 130. Septic [i.e. 8-point Lagrange] interpolation coefficients. 16.2311/627

 T. 6, pp. 132-125. Square roots. 16.0261/617

Kelley, T. L. 1948

The Kelley Statistical Tables (second edition)

T. F, p. 32. P values for various F's and various combinations of degrees of freedom. 4.27/264

T. 1, p. 36. Supplementary values. 1.2112/20 1.221/21

T. 1, pp. 38-137. Normal distribution, simple correlation, and probability functions. 1.2111/19 1.221/21 7.32/334 16.0264/618

T. 2, pp. 138-187. Three-point interpolation coefficients. 16.2311/628

T. 3, pp. 188-197. Four-point interpolation coefficients. 16.2311/627

T. 4, pp. 198-199. Six-point interpolation coefficients. 16.2311/627

T. 5, p. 200. Eight-point interpolation coefficients. 16.2311/627

T. 6, pp. 201-203. Giving P,—the probability that, for a given n, a divergence as great as χ^2 will arise as a matter of chance. 1.1131/6 2.1133/133

T. 7, pp. 204-221. Square and cube roots and natural logarithms. 16.0261/617 16.027/619 16.112/625

T. 8, p. 222. Table of values. 4.27/264

T. 9, p. 223. Constants frequently needed.

Kitagawa, T. and Mitome, M. 1953

Tables for the design of factorial experiments

T. I, 38 pages. Latin squares and cubes.

15.112/502	15.113/516(2)	15.114/532(3)	15.117/536	15.1199/545
/503	/520(2)	.115/533(2)	/537(2)	/546
/504	15.1132/523(2)	/534(4)	15.118/537(2)	15.1512/559(5)
/507	/524(3)	/535(5)	.119/541(2)	.1514/561
/515(2)	15.114/526(4)	15.116/535(2)	.1199/543(3)	
/516	/531	/536	/544	

T. II, 136 pages. Factorial designs.

15.112/502(4)	15.112/510(2)	15.1132/523	15.114/532(2)	15.118/538(4)
/503(4)	/511	/524(2)	/533(2)	/539
/504(6)	/512	15.114/526(3)	15.115/533(2)	/540
/505(2)	15.113/516	/527(5)	/534(2)	15.119/541(2)
/506(2)	/517(4)	/528(4)	15.116/536	/542
/507(2)	15.1132/520	/529(3)	.117/536	15.1199/544(3)
/508	/521(6)	/530(2)	/537	/545(2)
/509(2)	/522(3)	/531(2)	15.118/537	

T. III, 25 pages. Fractional replication in factorial designs.

15.112/503	15.112/510	15.114/527	15.119/542
/504	/512(2)	/528	.1199/545
/505	/514	/531	
/506	15.113/517	15.118/538(2)	
/508(2)	/518	/539(2)	

Kitagawa & Mitome
Pearson & Hartley

    T. IV, 7 pages. Factorial designs with split-plot confounding.
15.112/505   15.112/506(2)   15.1132/522   15.1132/523   15.1132/524
15.1132/525   15.114/531(3)   15.114/533   15.116/535(2)   15.116/536
15.1199/544

    T. V, 15 pages. Factorial designs confounded in quasi-Latin squares.
15.112/503   15.112/510(2)   15.1132/523(2)   15.118/539(2)
/504   /511   /524   /540(4)
/506(2)   15.113/517(3)   15.114/529(2)   15.119/541
/508   /518(3)   /530(2)   /542(2)
/509   /519   15.116/536(4)

    T. VI, 13 pages. Lattice designs.
15.2103/571   15.2103/572(2)   15.2104/581(2)   15.2105/585   15.2105/586(3)
15.2107/594   15.2108/597   15.2109/601   15.212/604   15.212/605

    T. VII, 15 pages. Balanced incomplete block designs.
15.20/563(ftn.)   15.2103/574   15.2105/583   15.2106/591   15.2109/600(2)
.2102/566(2)   /575(3)   /584(4)   .2107/592   /601
/567(3)   /576   /585   /593(3)   15.2110/601
/568(3)   15.2104/577(2)   /586(2)   /594   /602(2)
15.2103/570(3)   /578(2)   /587   15.2108/594(2) /603
/571   /579(2) 15.2106/587(2) /595   /604
/572   /580(2)   /588(2)   /597(2)   15.212/604
/573(2)   /581   /589(3)   15.2109/598(2) /605
                  /582(2)   /590   /599(2)   15.2411/605

    T. VIII, 4 pages. Youden's squares.
15.2103/571   15.2104/580   15.2105/584   15.2106/588   15.2106/589
15.2107/592   15.2108/595   15.2109/598   15.2110/602

Lindley, D. V. and Miller, J. C. P. 1953

Cambridge Elementary Statistical Tables

    T. 1, pp. 4-5. Normal distribution function.   1.115/8   1.1161/10

    T. 2, p. 5. Percentage points of the normal distribution.
1.2111/631   1.2112/631

    T. 3, p. 6. Percentage points of the -distribution.   4.1311/671

    T. 4, p. 6. Transformation of the correlation coefficient.   7.212/331

    T. 5, p. 7. Percentage points of the χ^2-distribution.   2.3112/141

    T. 6, p. 7. Conversion of range to standard deviation.   1.541/87

    T. 7, pp. 8-11. 5% ($2\frac{1}{2}$%, 1%, 0.1%) points of the -distribution.
4.2312/676

    T. 8, pp. 12-13. Random sampling numbers.   13.11/456   13.13/459

    T. 9, pp. 14-33. Square roots and other functions.
12.13/445   12.31/449   16.021/614   16.024/616   16.0261/617   16.0262/617
16.061/621   16.064/621

    T. 10, p. 34. Logarithms of factorials.   2.5134/163

Pearson, E. S. and Hartley, H. O. 1954

<u>Biometrika Tables for Statisticians</u>

T. 1, pp. 104-110. The normal probability function: the integral P(X) and ordinate Z(X) in terms of the standardized deviate X.
1.115/7   1.1161/9   10.13/404

T. 2, p. 111. The normal probability function: values of $-\log Q(X) = -\log[1 - P(X)]$ for large values of X.   1.1144/7   10.13/404

T. 3, p. 111. The normal probability function: values of X for extreme values of Q and P.   1.2111/19   10.13/404

T. 4, p. 112. The normal probability function: values of X in terms of Q and P.   1.2111/19   10.13/404

T. 5, p. 113. The normal probability function: value of Z in terms of Q and P.   1.221/21   10.13/404

T. 6, pp. 114-121. The normal probability function: table for probit analysis.   1.2312/24   10.13/404

T. 7, pp. 122-129. Probability integral of the χ^2-distribution and the cumulative sum of the Poisson distribution.
2.1131/132   2.121/135   2.311/140   10.13/404

T. 8, pp. 130-131. Percentage points of the χ^2-distribution.
2.3112/141   10.032/395(Thompson)   10.13/404

T. 9, pp. 132-134. Probability integral, $P(t|\nu)$, of the t-distribution.
4.1111/208   4.1313/213   10.071/400(Hartley & Pearson)   10.072/400   10.13/404

T. 10, p. 135. Chart for determining the power function of the t-test.
4.1412/221   4.242/260(Pearson & Hartley 1951)   10.13/404

T. 11, pp. 136-137. Test for comparisons involving two variances which must be separately estimated.
4.151/236(Fisher & Yates)   4.152/238(Aspin)   10.13/404

T. 12, p. 138. Percentage points of the t-distribution.
4.1311/211   10.072/400   10.13/404

T. 13, p. 138. Percentage points for the distribution of the correlation coefficient, r, when $\rho = 0$.   7.1221/323   10.022/394

T. 14, p. 139. The z-transformation of the correlation coefficient: $z = \tanh^{-1} r$.   7.212/331

T. 15, pp. 140-141. Chart giving confidence limits for the population correlation coefficient, ρ, given the sample coefficient, r.
7.121/323(David)

T. 16, pp. 142-155. Percentage points of the B-distribution.
3.0/181(ftn.)   3.311/191(Thompson)   10.012/392(Thompson)   10.13/404

T. 17, p. 156. Chart for determining the probability levels of the incomplete B-function, $I_x(a,b)$.   3.12/185(Hartley & Fitch)   10.13/404

Pearson & Hartley
K. Pearson (Tables for S & B I)

 T. 18, pp. 157-163. Percentage points of the F-distribution (variance ratio).   4.2311/248   10.062/399   10.13/404

 T. 19, p. 164. Percentage points of the largest variance ratio, s^2_{max}/s^2_0. 4.251/261(Nair)   10.13/404

 T. 20, p. 164. Moment constants of the mean deviation and of the range. 1.52/73(Vianelli)   1.52/75   1.541/88   10.13/404

 T. 21, p. 165. Percentage points of the distribution of the mean deviation. 1.52/74(Pearson, Godwin & Hartley)   10.13/404

 T. 22, p. 165. Percentage points of the distribution of the range. 1.541/88   10.13/404

 T. 23, pp. 166-171. Probability integral of the range, W, in normal samples of size n.   1.541/85(Pearson & Hartley 1942)   10.13/404

 T. 24, p. 172. Percentage points of the extreme standardized deviate from population mean.   1.42/62   10.13/404

 T. 25, p. 172. Percentage points of the extreme standardized deviate from sample mean.   1.42/59(Nair)   10.13/404

 T. 26, p. 173. Percentage points of the extreme studentized deviate from sample mean.   1.44/69   10.13/404

 T. 27, p. 174. Mean range in normal samples of size n. 1.541/83(Tippett)   10.13/404

 T. 28, p. 175. Mean positions of ranked normal deviates (normal order statistics).   1.43/67   10.13/404

 T. 29, pp. 176-177. Percentage points of the studentized range. 1.5511/99   10.13/404   1.5511/649(Duncan)

 T. 30, p. 178. Tables for analysis of variance based on range. 1.552/106(David)   10.13/404

 T. 31, p. 179. Percentage points of the ratio, s^2_{max}/s^2_{min}. 4.253/262(David)   10.13/404

 TT. 32-33, pp. 180-182. Test for homogeneity of variance. 6.113/300(intro.)   6.113/302(Thompson & Merrington)   10.13/404

 T. 34, pp. 183-184. Tests for departure from normality. 1.611/112(Pearson)   1.613/113(Pearson)   1.623/114(Geary & Pearson) 1.631/115(Geary)   1.633/116(Geary)   10.13/404

 T. 35, p. 184. Moments of $s/\sigma = \chi/\sqrt{\nu}$ and factors for determining confidence limits for σ.   2.315/658   2.522/171   10.13/404

 T. 36, p. 183. Test for the significance of the difference between two Poisson variables.   6.31/316

 T. 37, pp. 186-187. Individual terms of certain binomial distributions. 3.221/187

T. 38, pp. 188-193. Significance tests in a 2 × 2 contingency table. 2.611/175(Finney)

T. 39, pp. 194-202. Individual terms ... of the Poisson distribution. 2.221/138

T. 40, p. 203. Confidence limits for the expectation of a Poisson variable. 2.322/153

T. 41, pp. 204-205. Chart providing confidence limits for p in binomial sampling, given a sample fraction c/n.   3.321/193(Clopper & Pearson)

T. 42, pp. 206-209. Percentage points of Pearson curves for given β_1, β_2, expressed in standard measure.   10.00/389   10.002/389

T. 43, p. 210. Chart relating the type of Pearson frequency curve to the values of β_1, β_2.   10.00/389   10.004/390(Rhind)

T. 44, p. 211. Distribution of Spearman's rank correlation coefficient, r_s, in random rankings.   8.1112/349

T. 45, p. 211. Distribution of Kendall's rank correlation coefficient, t_k, in random rankings.   8.1212/352

T. 46, p. 211. Distribution of the concordance coefficient, W, in random rankings.   8.1312/354

T. 47, pp. 212-221. Orthogonal polynomials.   11.110/437   11.111/437

T. 48, pp. 222-223. Powers of integers.   16.022/615   16.023/616

T. 49, pp. 224-225. Sums of powers of integers.

T. 50, pp. 226-227. Squares of integers.   16.0211/614

T. 51, pp. 228-233. Factorials of integers, their logarithms; square roots; and their reciprocals.
2.5112/161   2.5134/163   16.024/616   16.0261/617   16.0262/617

T. 52, pp. 234-235. Miscellaneous functions of p and q = 1-p over the unit range.
7.31/334   7.32/334   7.33/334   16.0214/615   16.0264/618   2.5112/161
2.5134/163

T. 53, pp. 236-237. Natural logarithms.   16.112/625

T. 54, p. 238. Useful constants.

Pearson, K. 1930 a

Tables for S & B, I

T. 1, p. 1. Table of deviates of the normal curve for each permille of frequency.   1.2111/19   10.13/404

T. 2, pp. 2-8. [The normal curve:] area and ordinate in terms of abscissa.
1.115/7   1.1161/9   10.13/404

T. 3, pp. 9-10. [The normal curve:] abscissa and ordinate in terms of difference of areas.   1.2121/20   1.221/21   10.13/404

K. Pearson (Tables for S&B I)

T. 4, p. 11. Extension of table of the probability integral $F = \frac{1}{2}(1-\alpha)$.
1.1144/7   10.13/404

T. 5, pp. 12-18. Probable errors of means and standard deviations.
10.13/404   16.0264/618

T. 6, p. 18. Probable errors of coefficient of variation.
10.13/404   16.0264/618

T. 7, p. 19. Abac for determining the probable errors of correlation coefficients.

T. 8, pp. 20-21. Values of $1-r^2$ from $r = \cdot 001$ to $\cdot 999$.   7.31/334

T. 9, pp. 22-23. Values of the incomplete normal moment function.
1.31/31   10.13/404

T. 10, p. 24. Diagram of generalised 'probable error'; table of generalised 'probable errors'.   2.313/144

T. 11, p. 25. Constants of normal curve from moments of tail about stump.
1.332/40(2)   1.332/43   10.13/404

T. 12, pp. 26-27. Test for goodness of fit. Values of P.
2.1132/132   10.13/404

TT. 13-16, pp. 29-30. [Test for goodness of fit:] auxiliary tables A, B, C and D.   2.119/133   2.5135/163   10.13/404

T. 17, p. 31. Values of $(-\log P)$ corresponding to given values of χ^2 in a fourfold table.   2.119/134

T. 18, p. 31. Values of $(-\log P)$, entering with r and $_0\rho_r$.   1.31/32

T. 19, p. 32. Values of χ^2 corresponding to the values of $(-\log P)$ in Table 18.   1.31/32

T. 20, p. 32. Values of $\log \chi^2$ corresponding to the values of r and $_0\rho_r$ in Tables 17 and 18.   1.31/32

T. 21, p. 33. Abac to determine $_0\rho_r$.   1.139/13

T. 22, p. 34. Abac to determine r_p.   1.31/32

T. 23, p. 35. Approximate values of probable error of r from a four-fold correlation table: values of χ_r^2 for values of r.   7.94/698(Pearson 1913 a)

T. 24, p. 35. Approximate values of probable error of r from a four-fold correlation table: values of χ for values of $\frac{1}{2}(1+\alpha)$.   1.2313/26

T. 25, p. 36. ... the probability that the mean of a sample of n, drawn at random from a normal population, will not exceed (in algebraic sense) the mean of the population by more than z times the standard deviation of the sample.
4.112/671   10.13/404

T. 26, p. 37. Table for use in plotting Type III curves.
10.033/395(Elderton)

T. 27, pp. 38-39. Powers of natural numbers.
16.0211/615   16.022/615   16.023/616

T. 28, pp. 40-41. Sums of powers of natural numbers.   16.0444/620

T. 29, pp. 42-51. Tetrachoric functions for four-fold correlation tables.
1.2111/19   1.221/21   1.3241/35   7.94/343(intro.)   10.13/404

T. 30, pp. 52-57. Supplementary tables for determining high correlations from tetrachoric groupings.   1.721/120

T. 31, pp. 58-61. The Γ-function.   2.5131/162

TT. 32, 33 A, 33 B, pp. 62-64. Subtense from arc and chord in the case of the common catenary.*

T. 34, p. 65. Diagram to find correlation r from mean contingency on the hypothesis of a normal distribution.

T. 35, p. 66. Diagram to determine the type of a frequency distribution from a knowledge of the constants β_1 and β_2. Customary values of β_1 and β_2.
10.004/390(Rhind)

T. 36, p. 67. Diagram showing distribution of frequency types for high values for β_1 and β_2.   10.004/390(Rhind)

T. 37, pp. 68-69. To find the probable error of β_1.
10.007/391(intro.)   10.007/391(Rhind 1909)

T. 38, pp. 70-71. To find the probable error of β_2.
10.007/391(intro.)   10.007/391(Rhind 1909)

T. 39, pp. 72-73. To find the correlation in errors of β_1 and β_2.
10.007/391(intro.)   10.007/391(Rhind 1909)

T. 40, pp. 74-75. To find the probable error of the distance from mean to mode.   10.007/391(intro.)   10.007/391(Rhind 1909)

T. 41, pp. 76-77. To find the probable error of the skewness sk.
10.007/391(intro.)   10.007/391(Rhind 1909)

T. 42, pp. 78-79. To give values of β_3, β_4, β_5 and β_6 in terms of β_1 and β_2 on the assumption that the frequency falls into one or other of Pearson's types.   10.006/391(Rhind)

T. 43, pp. 80-81. Probable error of criterion x_2.
10.007/391(intro.)   10.007/391(Rhind 1910)

T. 44, pp. 82-83. To find probable frequency type: values of $1 \cdot 177 N^{\frac{1}{2}} \Sigma_1$ for given values of β_1, β_2 ([semi-axis minor] of probability ellipse)
10.007/391(intro.)   10.007/391(Rhind 1910)

T. 45, pp. 84-85. To find probable frequency type: values of $1 \cdot 177 N^{\frac{1}{2}} \Sigma_2$ for given values of β_1, β_2 ([semi-axis major] of probability ellipse)
10.007/391(intro.)   10.007/391(Rhind 1910)

T. 46, pp. 86-87. To find probable frequency type: angle between major axis and axis of β_2 (probability ellipse) measured in degrees.
10.007/391(intro.)   10.007/391(Rhind 1910)

---

*See Fletcher et al. 1946, article 10.47, p. 172.

K. Pearson (Tables for S&B I)
K. Pearson (Tables for S&B II)

 T. 47, p. 88. Diagram determining the probability of a given type of frequency.   10.007/391(intro.)   10.007/391(Rhind 1910)

 T. 48, pp. 90-97. Percentage frequency of successes in a second sample after drawing successes in a first sample.   5.1/277(Greenwood)

 T. 49, pp. 98-101. Logarithms of factorials.   2.5134/163

 T. 50, pp. 102-111. Fourth moments of subgroup frequencies. 16.0233/616

 T. 51, pp. 113-121. Tables of $e^{-m}m^x/x!$: general term of Poisson's exponential expansion ("law of small numbers").   2.221/138

 T. 52, pp. 122-124. Table of Poisson-exponential for cell frequencies 1 to 30.   2.121/135   2.122/135

 T. 53, p. 125. Angles, arcs and decimals of degrees.

 T. 54, pp. 126-142. The $G(r,\nu)$ integrals.   10.0431/397(B.A.)

 T. 55, p. 143. Miscellaneous constants.

Pearson, K. 1931 a

<u>Tables</u> <u>for</u> <u>S</u> & <u>B</u>, II

 Frontispice. Types of skew frequency for values of β_1 and β_2. 10.004/390(Pearson 1916)

 T. 1, p. 1. Table of ordinates of the normal curve for each permille of frequency.   1.221/21   10.13/404

 T. 2, pp. 2-10. Tables of normal curve functions to each permille of frequency.   1.2111/19   1.221/21   1.222/22   10.13/404

 T. 3, pp. 11-15. Table of ratio: area to bounding ordinate of normal curve. 1.131/13   10.13/404

 T. 4, pp. 16-72. Table for ascertaining the significance of the correlation ratio.   3.19/193(Woo)   10.13/404

 TT. 5-6, p. 73: for assisting the computing of tetrachoric functions. 1.3249/35   1.3249/36   10.13/404

 T. 7, pp. 74-77. Table of the first twenty tetrachoric functions. 1.1141/6   1.1161/9   1.3241/35   10.13/404

 TT. 8-9, pp. 78-137. Table for finding the volumes of the normal bivariate surface: positive (negative) correlation.   1.721/210   1.722/120

 T. 10, pp. 138-144: for finding the frequency of the first product moment: table of the auxiliary function T_m.   7.6/340(Pearson et al. 1929)

 T. 11a, p. 145. Means and standard deviations of r and of $\sigma_1\sigma_2 r/\Sigma_1\Sigma_2$. 7.6/342(Pearson et al. 1929)

 T. 11b, p. 145. β_1 and β_2 for r and v.   7.6/342(Pearson et al. 1929)

 T. 12, p. 146. Table of Gaussian "tail" functions, "tail" larger than "body".   1.332/40(2)   1.332/42(Lee)   10.13/404

K. Pearson (Tables for S&B I)
K. Pearson (Tables for S&B II)

T. 13, p. 147. Values of the 11th and 12th normal moment functions.
1.31/32   10.13/404

T. 14, pp. 148-155. [Values of β_3, β_4, β_5 and β_6 in terms of β_1 and β_2 for Pearson-type curves]   10.006/391(Yasukawa)   10.007/391

T. 15, pp. 156-157. Ratio of the standard error of the mode to the standard error of the mean [for Pearson-type curves]
10.007/391   10.007/392(Yasukawa)

TT. 16, 16 bis, p. 158. Standard error of a correlation coefficient found from a bi-serial table.   1.19/16   7.91/697

T. 17, p. 159. Values of the constants of the frequency distribution of the standard deviations of samples drawn at random from a normal population.
2.522/168(Pearson 1915 b)   10.13/404

T. 18, p. 159. Values of the constants of the frequency distributions of the intervals between the first three individuals in samples from a normal population.   1.543/91   10.13/404

TT. 19-20, pp. 160-161. Values of . . . the probability that the first and second (second and third) individuals in a random sample of n differ by more than   times the standard-deviation of the original population.
1.543/91   10.13/404

T. 21, pp. 162-163. Probability integral of distribution of largest individual in samples of size n taken from normal population.
1.42/56(Tippett)   10.13/404

T. 21 bis, p. 164. Magnitude in terms of the parent-population standard deviation of the individual in samples of various sizes, which will only be exceeded in the following percentage of cases.   1.42/58

T. 22, pp. 165-166. Mean range of samples of size n taken from normal population (given in terms of standard deviation)
1.541/83(Tippett)   10.13/404

T. 23, p. 167. Distribution constants of range in samples from symmetrical curves; from a Type III curve.

T. 24, p. 168. Permille of samples from symmetrical curves with range greater than multiples of population standard deviation; permille of samples from a Type III curve with ranges in excess of certain values.

TT. 25, 25 bis, pp. 169-178. [Probability integral of the Pearson symmetrical curves]
3.12/185(Pearson & Stoessiger)   10.021/393(Pearson & Stoessiger)
10.071/400(Pearson & Stoessiger)

T. 26, p. 179. To assist in computing the modal ordinate of Pearson Types II and VII curves, and also the constants of curves of frequency in the case of the correlation coefficient.   3.51/204

K. Pearson (Tables for S&B II)
Siegel

T. 27, p. 180. Values of the correlation of standard deviations in samples for different values of variates correlation in a sampled normal population.   1.78/123   2.522/169

T. 28, p. 180. Values of the correlation of the standard deviation of one variate with the correlation of both variates in samples of different sizes taken from a normal population.   1.78/652

T. 29, p. 181. Ratio of mean S.D. of arrays in samples to array S.D. in sampled population.   2.522/168(Pearson 1925 a)

T. 30, p. 181. Values of the ratio [of the standard deviation of standard deviations in samples to its asymptotic value] for various sized samples.   2.522/168(Pearson 1925 a)

T. 31, pp. 182-184. [Constants for r-curves in samples of 2, 3, 4, 25, 50, 100, 200 and 400 from a normal population]   10.29/420(Soper et al.)

T. 32, pp. 185-211. Distribution of correlation coefficient in small samples: ordinates and constants of frequency curves.   7.11/321(Soper et al.)   10.29/420(Soper et al.)

T. 33, pp. 212-218. To assist the calculation of the ordinates of the correlation frequency curves from expansion formulae.   7.11/321(Soper et al.)

T. 34, pp. 219-220. To assist the use of the . . . curves [of T. 32] in the case of other frequencies with like β_1 and β_2.   10.29/420(Soper et al.)

T. 35, pp. 221-223. To determine the probability that a small sample has been drawn from a normal population with a specified mean and standard deviation.   6.111/296(Neyman & Pearson)   10.13/404

T. 36, p. 223. To find P for samples from the ratio P_λ/λ for n = 10 and beyond.   6.111/296(Neyman & Pearson)   10.13/404

TT. 37 a, 37 b, pp. 223-224. Single test of the probability that two samples have been drawn from the same population.   6.111/296(Pearson & Neyman)   10.13/404

T. 37 bis, p. 224. 5% and 1% points for $\beta_1^{\frac{1}{2}}$, β_1 and β_2.   1.613/113(E.S. Pearson 1930 a)   1.623/114(E.S. Pearson 1930 a)

TT. 38-41, pp. 225-231. Tables to assist the computation of abruptness coefficients for the correction of moment-coefficients in the case of asymptotic frequency.   10.332/428(Pearse)

T. 42, pp. 232-223. Table for calculating occipital index from occipital arc and chord.   16.0264/618

T. 43, p. 234. Coefficients for Sheppard's quadrature formula (c).   16.236/629

T. 44, p. 235: for testing correlations in variate difference method: table of functions for geometrical decadence.   7.53/339

K. Pearson (Tables for S&B II)
Siegel

T. 45, pp. 236-238. Evaluation of the integral $\int_0^\theta \cos^{n+1}\theta \, d\theta$ for high values of n.
3.44/202   10.023/394(Wishart 1925a)   10.073/400(Wishart 1925a)

T. 46, p. 239. For use in computing the integral . . . for values of n less than those provided for in Table 45.
3.44/202   10.023/394(Wishart 1925a)   10.073/400(Wishart 1925a)

T. 47, pp. 240-243. [Table to obtain the coefficients in the expansion formula for the incomplete B-function ratio, . . . i.e. the probability integral of Pearson's curve Type I]   3.44/202   10.013/392(Wishart)

T. 48, pp. 244-245. Table of reciprocals of powers of n
10.013/392(Wishart)   16.024/616   16.025/617   16.0262/617

T. 49, pp. 246-247. Values of differences of the powers of zero.
5.9/293(Fisher & Yates)   16.0492/620

T. 50, pp. 248-251. Table of deviates of Type I curves measured from the mean in terms of the standard deviation.   3.319/192   10.012/392(Pearson)

T. 51a, p. 252. Table for determining the mode $\check{r}$ of a frequency distribution of considerable size n when the correlation in the sampled population is known to be ρ.   7.133/327(Soper et al.)

T. 51b, p. 253. Table for determining the "most likely" value $\hat{\rho}$ of the correlation in a sampled population from the knowledge of the correlation in a sample of size n, when n is considerable and it is legitimate to distribute ignorance equally.   7.19/329(Soper et al.)

TT. 52, 52 bis, pp. 254-255. [Tables to determine the mean and variance of squared multiple correlation in samples]   7.4/335(Pearson 1931b)

T. 53, p. 256. Inverse factorials (Glaisher's table)   2.5122/162

T. 54, p. 257. Reciprocals of the first hundred integers, to 12, 16 and 20 decimal places.   16.024/616

T. 55, pp. 258-261. The first ten powers of the first hundred natural numbers.   16.0211/615   16.022/615   16.023/616

T. 56, p. 262. Constants occasionally useful.

Siegel, S. 1956

Nonparametric Statistics

T. A, p. 247. Table of probabilities associated with values as extreme as observed values of z in the normal distribution.   1.1141/630

T. B, p. 248. Table of critical values of t.   4.1311/671   4.1313/671

T. C, p. 249. Table of critical values of chi square.   2.3113/142

T. D, p. 250. Table of probabilities associated with values as small as observed values of x in the binomial test.   3.6/206(Walker & Lev)

Siegel
Vianelli

T. E, p. 251. Table of critical values of D in the Kolmogorov-Smirnov one-sample test. 9.4121/381(Massey 1951)

T. F, pp. 252-253. Table of critical values of r in the runs test. 9.32/379

T. G, p. 254. Table of critical values of T in the Wilcoxon matched-pairs signed-ranks test. 9.112/368(Wilcoxon 1949)

T. H, p. 255. Table of critical values for the Walsh test. 9.111/367(Walsh 1949 c)

T. I, pp. 256-270. Table of critical values of D (or C) in the Fisher test. 2.611/663 9.123/370

T. J, pp. 271-273. Table of probabilities associated with values as small as observed values of U in the Mann-Whitney test. 9.121/368(Mann & Whitney)

T. K, pp. 274-277. Table of critical values of U in the Mann-Whitney test. 9.121/369

T. L, p. 278. Table of critical values of K_D in the Kolmogorov-Smirnov two-sample test. 9.4122/378(Massey 1951 a) 9.4152/385(Goodman)

T. M, p. 279. Table of critical values of D in the Kolmogorov-Smirnov two-sample test. 9.4113/714

T. N, pp. 280-281. Table of probabilities associated with values as large as observed values of χ_r^2 in the Friedman two-way analysis of variance by ranks. 8.1313/354(Friedman)

T. O, pp. 282-283. Table of probabilities associated with values as large as observed values of in the Kruskal-Wallis one-way analysis of variance by ranks. 9.13/371(Kruskal & Wallis) 9.13/707(Vianelli)

T. P, p. 284. Table of critical values of . . . the Spearman rank correlation coefficient. 8.112/350

T. Q, p. 285. Table of probabilities associated with values as large as observed values of S in the Kendall rank correlation coefficient. 8.1212/352

T. R, p. 286. Table of critical values of s in the Kendall coefficient of concordance. 8.132/355(Friedman)

T. S, p. 287. Table of factorials. 2.5112/661

T. T, p. 288. Table of binomial coefficients.

T. U, pp. 289-301. Table of squares and square roots. 16.021/614 16.026/617

Vianelli, S. 1959

Prontuari per calcoli statistici

Prontuario 1, pp. 3-6. First 8 powers of the natural numbers from 1 to 100. 16.0211/614 16.022/615 16.023/616

Pro. 2, pp. 7-15. First 3 powers of the natural numbers from 101 to 1000.
16.0211/614   16.022/615

Pro. 3, pp. 16-19. Sums of the first 8 powers of the natural numbers from 1 to 100.   16.0444/620

Pro. 4, pp. 20-21. First 3 powers of reciprocals and sums of reciprocals of the natural numbers from 1 to 100.   16.024/616   16.025/617   16.046/620

Pro. 5, pp. 22-27. Reciprocals of the natural numbers from 101 to 1000. 16.024/616

Pro. 6, pp. 28-37. Square and cube roots of the natural numbers from 1 to 1000.   16.0261/617   16.027/619

Pro. 7, pp. 38-42. Square root of the ratio n/(n-1) for n from 1 to 1000. 16.0264/618

Pro. 8, pp. 43-47. Decimal logarithms of the natural numbers from 1 to 1000.   16.061/621

Pro. 9, pp. 48-52. Neperian logarithms of the natural numbers from 1 to 1000.   16.111/625

Pro. 10, pp. 53-55. Addition logarithms.   16.068/621

Pro. 11, pp. 56-58. Subtraction logarithms.   16.068/622

Pro. 12, pp. 59-63. Decimal logarithms of factorials of the natural numbers from 1 to 1000.   2.5134/662

Pro. 13, pp. 64-66. Values of log log x for x from 1.001 to 100 000. 16.0618/621

Pro. 14, pp. 67-68. Proportional parts.

Pro. 15, pp. 69-70. Lengths of circular arcs.

Pro. 16, p. 71. Values of trigonometric functions for each degree from 1° to 90°.   16.0721/622   16.0722/622

Pro. 17, pp. 72-76. Values of trigonometric functions of arcs in thousandths of a circumference.   16.073/623

Pro. 18, pp. 77-106. Decimal logarithms of trigonometric functions for each minute of arc from 0 to 90°.   16.08/623

Pro. 19, pp. 107-111. Variances of the first n natural numbers for n from 1 to 1000.   16.0214/615

Pro. 20, pp. 112-113. Values of Pareto's index α corresponding to values of the concentration ratio of statistical distributions of hyperbolic form. 16.0244/616

Pro. 21, pp. 114-115. Values of Gini's index δ corresponding to values of the concentration ratio of statistical distributions of hyperbolic form. 16.0246/616

Pro. 22, p. 116. Weights to apply to observed data in order to estimate, by the method of least squares, the parameters of a straight line. 11.125/725

Pro. 23, pp. 117-118. Weights to apply to observed data in order to estimate, by the method of least squares, the parameters of a parabola of the second degree.   11.125/725

Pro. 24, pp. 118-120. Weights to apply to observed data in order to estimate, by the method of least squares, the parameters of a parabola of the third degree.   11.125/725

Pro. 25, pp. 121-132. Ordinates of the normal curve.   1.1161/9

Pro. 26, p. 133. Probability of a deviation between $\pm 2^{\frac{1}{2}}u$, according to the normal curve.   1.1211/630

Pro. 27, p. 134. Probability of a deviation between $\pm k$, according to the normal curve.   1.111/3

Pro. 28, p. 135. Probability of a deviation not exceeding $2^{\frac{1}{2}}u$, according to the normal curve.

Pro. 29, p. 136. Probability of a deviation not exceeding k, according to the normal curve.   1.115/8

Pro. 30, p. 137. Values of k as a function of the area under the normal curve included between $\pm k$.   1.2121/631

Pro. 31, p. 137. Values of k as a function of the area under the normal curve not included between $\pm k$.   1.2121/631

Pro. 32, pp. 138-144. Values of certain constants for the estimation of a straight line by means of empirical probits.   1.2312/632

Pro. 33, pp. 145-146. Values of empirical probits corresponding to observed relative frequencies p.   1.232/632

Pro. 34, pp. 147-154. Values of minimum and maximum working probits, of the range 1/Z, and of the weight Z^2/PQ, for the estimation of the parameters of a straight line.   1.2312/632

Pro. XXV [i.e. 35], pp. 155-157. Products, P, of the complementary probabilities p and q = 1-p, and values of the real positive roots of certain equations of the second degree.   16.0214/615

Pro. 36, pp. 158-163. Statistical constants of some binomial probability distributions.   3.53/205

Pri. 37-38, pp. 164-166. [Tables for Vercelli's method of moving averages]

Pro. 39, p. 167. Minimum relative frequencies of a deviation between $\pm k_{m,2}$.   10.4/724

Pro. 40, pp. 168-169. Minimum relative frequencies according to Dantzig's inequality.   10.4/724

Pro. 41, pp. 170-172. Probability that the absolute value of the t statistic exceed t_r.   4.1113/670

Pro. 42, p. 173. Theoretical values of the t statistic that have a probability P of being exceeded by chance.   4.1311/671

Pri. 43-47, pp. 174-178. Theoretical values of the z statistic that have a probability .001 (.01, .05, .1, .2) of being exceeded by chance.   4.331/676

Pri. 48-55, pp. 179-191. Theoretical values of the F statistic that have a probability .001 (.01, .05, .1, .2; .025, .005, .0005) of being exceeded by chance.   4.2311/676

Pro. 56, p. 192. Theoretical values of the χ^2 statistic that have a probability P of being exceeded by chance.   2.3113/142

Pro. 57, pp. 193-195. Probability that a value of the χ^2 statistic exceed χ_r^2.   2.1132/132

Pro. 58, pp. 196-197. Values of the statistic $[z = \tanh^{-1} r]$ corresponding to certain values of the correlation coefficient r.   7.212/695

Pro. 59, p. 198. Theoretical values of the correlation coefficient that have a probability P of being exceeded by chance.   7.1221/323

Pro. 60, p. 198. Values of the correlation coefficient corresponding to certain values of the statistic $[z = \tanh^{-1} r]$.   7.211/695

Pro. 61, p. 199. Probability that a value of $s = \frac{1}{2}n(n-1)t_k$ exceed or equal s_r for certain values of n.   8.1212/352

Pro. 62, pp. 199-200. Probability that a value of $S = m^2(n^3-n)w/12$ exceed or equal S_r for certain values of m and n.   8.1312/353(Kendall & 8.1312/353(Kendall & Babington Smith)

Pro. 63, p. 201. Values of the terms . . . for correcting the probability that a value of the t statistic exceed t_r when the statistical universe is asymmetrical.   4.143/231(Gayen)

Pro. 64, p. 202. Values of Sukhatme's function d for testing the significance of the difference between two arithmetic means.   4.151/234(Fisher & Yates)

Pro. 65, pp. 203-205. Theoretical values of the ratio χ^2/ν that have a probability P of being exceeded by chance.   2.312/143

Pro. 66, pp. 206-207. Values of the function $\varphi = 2\sin^{-1} p^{\frac{1}{2}}$ expressed in radians.   12.13/728

Pro. 67, pp. 208-209. Values of the function $\theta = \sin^{-1} p^{\frac{1}{2}}$ expressed in degrees.   12.11/728

Pro. 68, pp. 210-214. Probability that a value of the mean absolute deviation m, calculated on a sample of n observations drawn from a normal universe with unit variance, be not exceeded by chance. 1.52/74(Pearson et al.)

Pro. 69, p. 215. Estimates of the standardized points of truncation, z, as functions of the parameter q, for truncated normal distributions. 1.332/40

Vianelli

Pro. 70, p. 216. Values of certain functions of z for estimating and for calculating the variances, covariance and correlation coefficient of the estimates of M and in truncated normal distributions.   1.332/40

Pro. 71, p. 217. Estimates of the standardized point of truncation, z, as functions of the parameters q and p, in censured normal distributions. 1.332/41

Pro. 72, p. 218. Values of certain functions of z for estimating σ and for calculating the variances, covariance and correlation coefficient of the estimates of M and σ in censured normal distributions.   1.332/41   1.332/42

Pro. 73, p. 219. Coefficients for the best linear estimate of the arithmetic mean of a normal distribution, based on censured samples. 1.51/641(Gupta)

Pro. 74, p. 220. Coefficients for the best linear estimate of the root-mean-square deviation of a normal distribution, based on censured samples. 1.544/647(Gupta)

Pro. 75, p. 221. Values of the standardized range from samples drawn from a normal distribution, corresponding to certain probability integrals P. 1.541/647

Pro. 76, p. 221. Statistical constants of the sampling distributions of the range, W, and of the mean absolute deviation, s, in samples drawn from a standard normal distribution.   1.52/73   1.541/647

Pro. 77, p. 222. Mean values, M_W, of the sampling distributions of the standardized range in samples drawn from a normal distribution.   1.541/83

Pro. 78, pp. 223-225. Values of the functions P_n, a_n, b_n, for determining the probability that the studentized range not exceed Q.   1.5511/97

Pro. 79, pp. 226-227. Values of the studentized range that have a probability P of being not exceeded or of being exceeded by chance. 1.5511/99(Pearson & Hartley)

Pro. 80, p. 228. Values of the parameters and c for analysis of variance based on the range.   1.552/106(David)

Pro. 81, p. 228. Theoretical values of the standardized extreme deviate that have a probability P of being not exceeded or of being exceeded by chance. 1.42/637

Pro. 82, pp. 229-231. Theoretical values of the standardized [i.e. studentized] extreme deviate that have a probability P of being not exceeded or of being exceeded by chance.   1.44/641

Pro. 83, p. 232. Values of the ratio T that have a probability P of being exceeded by chance.   8.21/700(Hyrenius)

Pro. 84, pp. 232-233. Theoretical values of Hyrenius' ratios U and V that have a probability P of being exceeded by chance.   8.21/700(Hyrenius)

Pro. 85, p. 234. Absolute values of the statistic G_1 that have a probability P of being exceeded by chance.   1.5512/650(Jackson & Ross)

Pro. 86, pp. 235-248. Absolute values of the statistic G_2 that have a probability P of being exceeded by chance.   1.5512/651(Jackson & Ross)

Pro. 87, pp. 249-250. Absolute values of the statistic U, for comparing the means of two samples, that have a probability P of being exceeded by chance. 1.5512/651(Moore)

Pro. 88, p. 251. Values of the functions $f(n_1, n_2)$ and (n_1, n_2) for estimating the root-mean-square deviation of a universe by means of the weighted sum of the ranges of two samples.   1.541/646(Moore)

Pro. 89, pp. 252-254. Values of the power of certain tests of significance of the difference (p_1-p_2) between two proportions.
2.613/663(Chandra Sekar et al.)

Pro. 90, p. 255. Values of Patnaik's function $k(p_1, p_2)$.   2.613/177

Pro. 91, pp. 256-257. Values of the power, as a function of h and x, of a test of the hypothesis $p_1 = p_2$ against the two-sided alternative $p_1 < p_2$ and $p_1 > p_2$, and of the hypothesis $p_1 \leq p_2$ against the alternative $p_1 > p_2$. 2.613/177

Pro. 92, p. 258. Values of the statistic L_1 corresponding to the mean size, n, of m samples, for testing the hypothesis that these samples came from normal distributions with equal variances.   6.112/299(Nayer)

Pri. 93, 94, p. 259. Values of the statistic g that have a probability $0.048750 < P \leq 0.05$ ($0.00995 < P \leq 0.01$) of being exceeded, by chance, by the largest ratio $g_r = \sigma_r^2 / \sum \sigma_r^2$.   11.24/726(Eisenhart et al.)

Pro. 95, p. 260. Values of the statistic L_2, as a function of m and c_1, for testing the hypothesis that m samples come from normal distributions with equal variances.   6.113/688(Thompson & Merrington)

Pro. 96, pp. 261-262. Theoretical values of the statistic v, for comparing statistical constants depending on two estimated variances, that have a probability P of being exceeded in absolute value by chance.
4.152/674(Trickett et al.)

Pro. 97, p. 263. Comparison of some values of the non-central t statistic, obtained by three different methods.   4.1419/225(Harley)

Pro. 98, pp. 263-267. Values of the function $\lambda(\nu, t_0, P)$ for calculating values of the non-central t-statistic that have a probability P of being exceeded by chance.   4.1414/225(Johnson & Welch)

Pro. IC [i.e. 99], p. 268. Values of $k_1 = t[1+n^{-1}]^{\frac{1}{2}}$ for determining the tolerance limits $M \pm k_1 \sigma$ that include the fraction β of the values of a normally distributed universe.   4.139/217

Pro. 100, pp. 269-276. Values of k corresponding to a probability that at least a fraction β of the values of a normally distributed universe be comprised between the tolerance limits $M \pm k\sigma$.   14.32/492(Eisenhart et al.)

Vianelli

Pro. 101, p. 277. Values of k_2 corresponding to a probability α that at least a fraction β of the values of a normally distributed universe be comprised between $M \pm k_2 w$.   14.32/494(Mitra)

Pro. 102, pp. 278-279. Values of k_3 corresponding to a probability α that at least a fraction β of the values of a normally distributed universe be comprised between $M \pm k_3 \bar{w}$.   14.32/494(Mitra)

Pro. 103, p. 280. Theoretical mean values of the ith order statistics in samples of n drawn from a normal distribution.   1.43/639(Teichroew)

Pro. 104, pp. 281-286. Theoretical mean values of the squares of the ith order statistics and of the products of the ith and j^{th} order statistics in samples of n drawn from a normal distribution.   1.43/639(Teichroew)

Pro. 105, pp. 287-290. Theoretical variances and covariances of the order statistics in samples of n drawn from a normal distribution.
1.43/639(Sarhan & Greenberg)

Pro. 106, pp. 291-294. Coefficients for the best linear estimate, from a censured sample, of the mean and of the root-mean-square deviation of a normal distribution.   1.51/642(Sarhan & Greenberg)   1.544/648(Sarhan & Greenberg)

Pro. 107, pp. 295-296. Variances and covariances of the estimates, from censured samples, of the arithmetic mean and of the root-mean-square deviation of a normal distribution.
1.51/642(Sarhan & Greenberg)   1.544/648(Sarhan & Greenberg)

Pro. 108, p. 297. Percentage efficiency of the estimate of the arithmetic mean and of the root-mean-square deviation, obtained from censured samples.
1.51/642(Sarhan & Greenberg)   1.544/648(Sarhan & Greenberg)

Pro. 109, pp. 298-300. Coefficients for the best linear estimate, from censured samples, of the root-mean-square deviation of a one-tailed exponential distribution with one sole parameter.   8.23/701(Sarhan & Greenberg)

Pro. 110, pp. 301-302. Variances of the estimates, from censured samples, of the root-mean-square deviation of a one-tailed exponential distribution with one sole parameter.   8.23/701(Sarhan & Greenberg)

Pro. 111, p. 303. Per cent efficiency of the estimates, obtained from censured samples, of the root-mean-square deviation of a one-tailed exponential distribution with one sole parameter.   8.23/701(Sarhan & Greenberg)

Pro. 112, p. 303. Proportional reduction in the theoretical waiting time to observe the first n-r failures in samples, censured on one side, drawn from a one-tailed exponential distribution with one sole parameter.
14.151/732(Sarhan & Greenberg)

Pro. 113, pp. 304-312. Coefficients for the best linear estimates, from censured samples, of the origin m, of the arithmetic mean and of the root-mean-square deviation of an exponential distribution with two parameters.
8.23/701(Sarhan & Greenberg)

Pro. 114, pp. 313-315. Variances of the linear estimates, from censured samples, of the origin m, of the arithmetic mean, and of the root-mean-square deviation of an exponential distribution with two parameters.
8.23/701(Sarhan & Greenberg)

Pro. 115, pp. 316-318. Per cent efficiency of the estimates, obtained from censured samples, of the origin m, of the arithmetic mean, and of the root-mean-square deviation of a one-tailed exponential distribution with two parameters.
8.23/701(Sarhan & Greenberg)

Pro. 116, p. 319. Values of the circular autocorrelation coefficient that have a probability P of being not exceeded or of being exceeded by chance.
7.51/696(Ferber)

Pro. 117, p. 319. Values of the statistic $D^2_{\Delta,2}/\sigma^2$ that have a probability P of being not exceeded or of being exceeded by chance. 1.531/643(Ferber)

Pro. 118, p. 320. Values of the ratio between two ranges that have a probability P of being not exceeded by chance. 1.5512/101(Link)

Pro. 119, pp. 321-326. Probability integral of the standardized range for samples drawn from a normal distribution. 1.541/85(Pearson & Hartley)

Pro. 120, pp. 327-329. Probabilities close to 0.10, to 0.05 and to 0.01 that, in analysis of variance based on ranks, the statistic H exceed H_r.
9.121/706   9.13/707

Pro. 121, pp. 330-333. Absolute values of a statistic u, derived from two estimates of means or ratios, that have a probability P of being exceeded by chance. 4.18/675(James)

Pro. 122, pp. 334-337. Confidence limits, with coefficient P, for the estimate of a proportion by the ratios, r/n, obtained from samples of size n.
3.321/666(Crow)

Pro. 123, p. 338. Control limits for single values drawn from an arbitrary continuous distribution and from a normal distribution. 14.229/490(Masuyama)

Pro. 124, p. 339. Probability that the statistic k, in the sign test for differences between correlated variates, not exceed the value k_r.
9.33/714(Hodges)

Pro. 125, p. 340. Theoretical values of Walsh's statistic for testing the difference between paired observations. 9.111/704(Walsh)

Pro. 126, pp. 341-344. Values of the parameters $A_0(u,m)$ and $A_2(u,m)$ for calculating the exact probabilities corresponding to the sampling distribution of Wilcoxon's statistic. 16.036/619(Fix & Hodges)

Pro. 127, p. 344. Theoretical values of Wilcoxon's statistic for testing the significance of differences between paired variates.
9.12/368(Wilcoxon 1949)

Pro. 128, pp. 345-346. Probability that the Mann-Whitney statistic for comparing two samples not exceed the value U_r. 9.121/368(Mann & Whitney)

Vianelli

Pro. 129, pp. 347-348. Theoretical values of the Mann-Whitney statistic that have a probability P of being not exceeded by chance.   9.121/369(Siegel)

Pro. 130, pp. 349-352. Minimum number of values of one sample of size m that are greater than the largest value of a second sample of size n. 9.129/707(Rosenbaum)

Pro. 131, pp. 353-356. Minimum number of values of one sample of size m that have a probability less than P of lying outside the extreme values of a second sample of size n.   9.2/708(Rosenbaum)

Pro. 132, pp. 357-358. Probability that x values of one sample not exceed the r smallest values of a second sample, of the same size n, drawn from the same distribution.   9.129/707(Epstein)

Pro. 133, pp. 359-361. Probability that Kolmogorov's statistic not exceed the ratio c/n.   9.4121/714(Birnbaum)

Pro. 134, pp. 362-363. Theoretical values of Kolmogorov's statistic that have a probability P of being not exceeded by chance. 9.4121/714(Miller)   9.4151/715(Miller)

Pro. 135, pp. 364-366. Probability that the statistics d_r and d'_r, for comparing the cumulative distributions of two ordered samples, not exceed the ratio c/m.   9.4122/715(Tsao)

Pro. 136, p. 367. Theoretical values of the Kolmogorov-Smirnov statistic, for comparing the cumulative distributions of two ordered samples, that have a probability P of being exceeded by chance. 9.4122/715(Siegel)   9.4152/715(Siegel)

Pro. 137, p. 368. Theoretical values of the Wald-Wolfowitz statistic that have a probability P of being exceeded or of being not exceeded by chance. 9.32/379(Siegel)

Pro. 138, pp. 369-370. Values for calculating the probability that a sequence of T groups occurs in r elements of three or four species. 9.311/711(Barton & David)

Pro. 139, pp. 371-372. Values of the function $f_1(\alpha,\beta)$, corresponding to certain values of α and β, for estimating the parameter from the mean length of sequences of like responses.   16.219/627(Bush & Mosteller)

Pro. 140, pp. 373-374. Values of the function $f_2(\alpha,\beta)$, corresponding to certain values of α and β, for estimating the mean square error of the mean length of a sequence of like responses.   16.219/627(Bush & Mosteller)

Pro. 141, p. 375. Values of the function $\varphi(\alpha,\beta)$, corresponding to certain values of α and β, for estimating the parameter from the total number of like responses.   16.1149/626(Bush & Mosteller)

Pro. 142, p. 376. Values of the function $\psi(\alpha,v)$, corresponding to certain values of v and α, for facilitating the calculation of the estimate of the parameter by the method of maximum likelihood. 16.0247/617(Bush & Mosteller)

Pro. 143, pp. 377-380. Values of the function $_1(\ ,\ ,\ )$, corresponding to certain values of , , , for simultaneously estimating, by the method of maximum likelihood, the parameters and q_0 in the expression $q_v = {}^v q_0$. 16.219/627(Bush & Mosteller)

Pro. 144, pp. 381-384. Values of the function $_2(\ ,\ ,\ )$, corresponding to certain values of , , , for estimating, by the method of maximum likelihood, the parameter in the expression $q_v = {}^v q_0$. 16.219/627(Bush & Mosteller)

Pro. 145, pp. 385-388. Values of the weights w and [the weighted abscissas] wX, as functions of r, for estimating the parameters of the straight line by the method of normits. 1.2313/632(Berkson 1957b)

Pro. 146, pp. 389-390. Values of the weights w and [the weighted abscissas] wX, as functions of , for estimating the parameters of the straight line by the method of normits. 1.2313/632 (Berkson 1955)

Pro. 147, pp. 391-395. Values of antinormits, or values of p corresponding to in normit analysis. 1.115/8

Pro. 148, pp. 396-397. Values of antilogits, p, and of weights, w, for estimating the parameters of the logistic function by the method of maximum likelihood. 12.53/451(Berkson)

Pro. CIL [i.e. 149], pp. 398-401. Latin squares for experimental designs.

15.112/503	15.1132/523	15.115/534	15.119/541(2)
/507(2)	/524	.116/535	.1199/543(3)
/515	15.114/526(2)	.117/536	
15.113/516	/532	/537	
/518	15.115/533(2)	15.118/537(2)	

Pro. 150, p. 401. Graeco-Latin squares for experimental designs.

15.112/505 15.112/511 15.113/516 15.113/519 15.1132/524 15.114/526
15.114/532 15.115/533 15.117/537 15.118/537 15.119/541 15.1199/543(2)

Pro. 151, pp. 403-406. Factorial designs with confounding of effects.

15.112/502	15.113/517(2)	15.1132/525	15.115/534(2)	15.119/541
/503(3)	.1132/520	.114/526	/535	/543(2)
/504	/521(5)	/527(3)	15.116/536	
/505(2)	/522(3)	/528	.118/538	
/506	/523(2)	/532(3)	/539(2)	
/507(2)	/524	/533(3)	/541(2)	

Pro. 152, pp. 407-409. Factorial designs with confounding of effects by means of quasi-Latin squares.

15.112/505 15.112/508(3) 15.112/509 15.112/510 15.113/518(2)
15.1132/523(2) 15.1132/525(2) 15.114/527 15.114/530(3) 15.116/536(2)
15.118/540(2) 15.119/542(3) 15.119/543

Pro. 153, pp. 410-413. Lattice designs.

15.2103/572 15.2104/581 15.2104/582 15.2105/586(2) 15.2106/591(2)
15.2107/594(2) 15.2108/597 15.2108/598 15.2109/601 15.2110/604
15.212/604

Vianelli

Pro. 154, pp. 414-424. Balanced incomplete block designs.
15.2102/567(2)   15.2103/575(2)   15.2105/584(2)   15.2106/590   15.2109/599(2)
       /568   .2104/575   /585   /591   /600(2)
15.2103/570   /578(2)   /586   15.2107/593(3) 15.2110/602(2)
       /572   /579(2)   15.2106/587   .2108/595   /603
       /573(3)   /580   /588   /597(2)   /604
       /574   /582(2)   /589   15.2109/598

Pro. 155, pp. 425-427. Experimental designs with lattice squares.
15.2104/581   15.2105/585   15.2107/594   15.2108/597   15.2109/601   15.212/604

Pro. 156, pp. 428-431. Incomplete Latin squares.
15.2103/570(4)   15.2104/576(4)   15.2104/580   15.2105/583(2)   15.2105/584
15.2105/585   15.2106/587   15.2106/588   15.2106/589   15.2106/591
15.2106/592(3)   15.2108/595   15.2109/598   15.2109/599   15.2109/600
15.2110/602(2)

Pro. 157, pp. 432-435. Partially balanced incomplete block designs with two associate classes.
15.2103/574(2)   15.2103/576   15.2104/578   15.2104/579(4)   15.2104/580(2)
15.2105/584   15.2105/585(2)   15.2106/589   15.2106/590(2)   15.2106/593
15.2109/600   15.2110/604

Pro. 158, p. 436. Doubly balanced incomplete block designs.
15.2104/578   15.2104/579   15.2108/595

Pro. 159, pp. 437-473. Values and estimates of the parameters, for incomplete block designs, by the method of paired comparisons.   15.28/610(4)

Pro. 160, p. 473. Values of the parameter $_cB_1$ and of the corresponding probability P, in the method of paired observations with three treatments and several groups of equal size.   15.28/610

Pro. 161, p. 474. Experimental designs with paired observations for testing agreement between the judgement of several operators.
15.1132/735(2)   15.115/736(2)   15.116/736(2)   15.117/736   15.1199/736(2)
15.2413/736

Pro. 162, p. 475. Best factorial designs for certain values of the index, λ, of the orthogonal array.   15.112/735

Pro. 163, pp. 476-477. Coefficients for estimating the effects in experimental designs with fractional replication.
15.112/735   15.1132/735   15.114/735

Pro. 164, pp. 478-482. L.H. Tippett's random numbers.
13.11/729   13.13/459

Pro. 165, pp. 483-487. M.G. Kendall & B. Babington Smith's random numbers.
13.11/729   13.13/459

Pro. 166, pp. 488-492. Random normal deviates.   13.3/729

Pro. 167, pp. 493-497. Correlated random normal deviates.   13.3/729

Pro. 168, pp. 498-502. Random permutations of the natural numbers from 1 to 9.   13.2/461(Cochran & Cox)

Pro. 169, pp. 503-512. Random permutations of the natural numbers from 1 to 16. 13.2/461(Cochran & Cox)

Pro. 170, pp. 513-514. Values of the statistic for comparing the mean of the results of p treatments with the mean of the results of a control experiment. 4.17/694(Dunnett)

Pro. 171, pp. 515-518. Values of the lower and upper limits, $\underline{G}$ and $\overline{G}$, for a sequential procedure for analysis of variance, with $\alpha = \beta = .05$, applied to experimental designs.

Pro. 172, pp. 519-520. Best size of a first random sample, in order to obtain a final stratified sample of size m. 15.3/737(Johnson)

Pro. 173, p. 521. Values of the parameters mC_0^2 and $1/f$ for sampling a markedly skew universe in two strata. 16.0244/737

Pro. 174, p. 522. Values of the parameters mC_0^2 and $1/f$ for sampling a universe following Pareto's hyperbolic distribution, in two strata.

Pro. 175, pp. 523-525. Number of secondary units of a random sample to draw from each primary unit in an optimal two-stage sampling plan. 16.0264/738

Pro. 176, pp. 526-527. Design of a pilot sample for estimating the best number of subsampling units, with theoretical relative precision equal to 90%. 15.3/613(Brooks)

Pro. 177, pp. 528-529. Parameter for selecting, by the minimax principle, the best size of the first part of a two-stage sampling plan, for certain values of P. 15.3/737(Seelbinder)

Pro. 178, pp. 530-531. Size of a random sample corresponding to certain values of the proportion p and of the degree of relative precision S with confidence coefficient $P = 0.95$. 16.0244/738

Pro. 179, pp. 532-533. Per cent relative precision ($\pm$) of the estimate of p, with confidence coefficient $P = 0.95$, for certain values of the size n of the random sample. 16.0264/738

Pro. 180, p. 534. Size of a random sample corresponding to certain values of the proportion p and of the degree of absolute precision s with confidence coefficient $P = 0.95$. 16.0244/738

Pro. 181, pp. 535-537. Values of $3[p(1-p)/n]^{\frac{1}{2}}$ corresponding to certain values of p and n. 3.54/205

Pro. 182, pp. 538-541. Values of $z = [p(1-p)/q(1-q)]^{\frac{1}{2}}$ for certain values of p and q between 0.001 and 0.500. 16.0264/618

Pro. 183, pp. 542-543. Uniform series of repeated trials.

Pro. 184, pp. 544-548. Values of certain functions concerning a series of n repeated trials of an unfair game.

Pro. 185, pp. 549-550. Binomial probability distribution for certain values of p and n. 3.221/187

Pro. 186, pp. 551-552. Cumulative binomial probability distribution for certain values of p and n.   3.112/664

Pro. 187, pp. 553-556. Binomial coefficients $\binom{n}{x}$ for n from 1 to 100 and x from 1 to 10.   5.31/284

Pro. 188, pp. 557-561. Decimal logarithms of binomial coefficients for n from 1 to 100 and x from 1 to $\frac{1}{2}[n+1]$.   3.212/186

Pro. 189, pp. 562-563. Values of certain functions of p and of $q = 1-p$, for p in the interval from 0 to 1.
16.0214/615   16.0264/618   2.5131/662   7.31/695   7.32/695   7.33/696

Pro. 190, pp. 564-569. Probability distributions given by Poisson's exponential for certain mean values, M, of the variable.   2.221/138

Pro. 191, pp. 570-575. Cumulative probability distributions given by Poisson's exponential for certain mean values, M, of the variable.   2.122/135

Pro. 192, pp. 576-577. Values . . . for testing the hypothesis of a distribution of plants or animals, according to Poisson's exponential, in a specified area.   6.31/674

Pro. 193, p. 578. Confidence limits for the mean value of a Poisson exponential distribution.   2.322/659

Pro. 194, p. 579. Values for testing the significance of the difference between the frequencies of two variables assumed to follow Poisson's exponential law.   6.31/316(Pearson & Hartley)

Pro. 195, pp. 580-587. Polya's [i.e. Pólya's] probability distributions for rare events with weak interdependence, according to the scheme of contagion.   5.32/683

Pro. 196, pp. 588-593. Coefficients p_i and q_i for calculating the probabilities of Neyman's Type A contagious distribution by the recurrence formula.   2.53/173

Pro. 197, p. 594. Values of the parameters m and $A^2 = a^2/\mu_2$ of a Pearson Type II [or VII] distribution for certain values of β_2.
10.023/716   10.073/717

Pro. 198, pp. 595-596. Values of the standardized deviates of certain Pearson distributions—for fixed values of β_1 and β_2—that have a probability of being not exceeded or of being exceeded by chance.   10.002/715

Pro. CIC [i.e. 199], pp. 597-598. Probability of a deviation not greater than k for asymmetrical Pearson Type III curves.   2.112/656

Pro. 200, pp. 599-600. Values of ordinates of an asymmetrical Pearson Type III curve.   2.21/657

Pro. 201, pp. 601-612. Logarithms of two functions for calculating the integral $G(r,\nu)$ and for fitting a Pearson Type IV curve to a statistical distribution.   10.0431/717

Pro. 202, pp. 613-619. Values of $tg^{-1}z$ and of $Log(1+z^2)$ for calculating ordinates of a Pearson Type IV curve.   10.0432/398(Comrie)

Pro. 203, pp. 620-621. Values of certain parameters of a family of cumulative frequency curves.   10.24/720(Topp & Leone)

Pro. 204, pp. 622-623. Values of the constants β_3, β_4, β_5 and β_6 corresponding to certain values of β_1 and β_2 for distributions of the several Pearson types.   10.006/391(Rhind)

Pro. 205, pp. 624-631. Orthogonal polynomials.   11.111/437

Pro. 206, pp. 632-633. Values of a logarithmico-normal distribution that have a probability P of being not exceeded or of being exceeded by chance.   10.2221/720(Broadbent)

Pri. 207-208, pp. 634-652. [Tables for periodogram analysis]

Pro. 209, p. 653. Tolerance factors for samples, of size n, drawn from multivariate normal distributions with unknown means and variance matrices.   14.32/493(Fraser & Guttman)

Pro. 210, pp. 654-660. Values of the functions T(h,a) for calculating the double probability integral of a bivariate normal function.   1.723/121(Owen)

Pro. 211, pp. 661-674. Values of the variable x, in the incomplete Beta function, that have a probability P of being not exceeded by chance.   3.311/191(Thompson)

Pro. 212, pp. 675-686. Values, multiplied by 10 000, of the variable x in the incomplete Beta function $I_x(a, n-a+1)$, corresponding to certain values of n and a.   3.319/665(R.E.Clark 1953)

Pro. 213, pp. 687-698. Values of the variable x, in the generalized Beta function, that have, in multivariate analysis, a probability P of being not exceeded by chance.   6.132/692(Foster & Rees)   6.132/692(Foster 1957)

Pro. 214, pp. 699-700. Values of the first ten incomplete moments of the normal probability curve.   1.31/31(Pearson & Lee)

Pro. 215, p. 701. Values of certain indices of divergence from normality that have a probability P of being not exceeded or of being exceeded by chance.   1.613/113(Pearson)   1.623/114(Geary & Pearson)   1.633/116(Geary)

Pro. 216, pp. 702-711. Values of tetrachoric functions and of the parameter h for calculating the correlation coefficient of a dichotomic table.   1.3241/35

Pro. 217, pp. 712-716. Values of the parameter d/N for determining tetrachoric correlation coefficients, r_t, greater than 0.80.   1.721/120

Pro. 218, pp. 717-720. Values of the function $(1-r^2)^{\frac{1}{2}}$ for r from 0.001 to 0.999.   7.32/695

Pro. 219, pp. 721-722. Values of the function $\psi(\alpha)$ corresponding to certain values of the parameter k and of the variable of the circular normal distribution.   10.28/722(Gumbel et al.)

Pro. 220, pp. 723-724. Values of the probability integral of the circular normal function.   10.28/722(Gumbel et al.)

Pro. 221, pp. 725-732. Probability that, in a succession of m+n elements, the number of homogeneous sequences be less than or equal to u' by chance. 9.32/379

Pro. 222, pp. 733-735. Values of the greatest integer u' for which $\Pr[u \leq u'] \leq P$ for $P < 0.50$ and values of the smallest integer u' for which $\Pr[u \leq u'] \geq P$ for $P > 0.50$.   9.32/379

Pro. 223, pp. 736-746. Rounded values (not, however, less than zero) of the quantities $S = p + 3[p(1-p)/n]^{\frac{1}{2}}$, $I = p - 3[p(1-p)/n]^{\frac{1}{2}}$, nS, np and nI for certain values of n and p.   14.21/732(Schrock)

Pro. 224, pp. 747-784. Minimum sizes, n_1 and n_2, of two samples for testing—with probability $P = 0.05$—the significance of the difference between the corresponding relative frequencies of a fixed attribute.

Pri. 225-228, pp. 785-815. Values of the parameters of certain simple (double) acceptance sampling plans with consumer's risk and several values of LTPD (AOQL).   14.111/730(Dodge & Romig)

Pro. 229, p. 816. Sample size letter for inspection level and lot size. 14.111/730(Freeman et al.)

Pri. 230-232, pp. 816-819. Simple (double, sequential) sampling plans for certain values of the acceptable quality level and for each sample size letter. 14.111/730(Freeman et al.)

Pro. 233, p. 820. Sample size letters for certain AOQL and AQL classes. 14.111/730(Freeman et al.)

Pro. 234, p. 820. Lot sizes that certain simple sampling plans for small lots control with several AOQLs.   14.111/730(Freeman et al.)

Pri. 235-236, pp. 821-826. Values of p_1 (p_2) for which $P = 0.95$ ($P = 0.10$) is the probability of obtaining not more than c defective elements in a random sample of size n, drawn from a binomial universe containing a fraction p_1 (p_2) of defective elements.   3.311/191(Grubbs)

Pro. 237, pp. 827-829. Acceptance number corresponding to the maximum sum of the probabilities p_1 and p_2 (with $p_1 < p_2$) of assigning a lot, respectively, to one or the other of two categories, by a sample of size n. 14.112/730(Golub)

Pro. 238, p. 830. Values of the parameters of certain simple sampling plans by variables with known mean square deviation. 14.121/731(Lieberman & Resnikoff)

Pro. 239, p. 831. Values for estimating the percentage of defective elements in a lot by means of simple sampling plans by variables with known mean square deviation.   14.121/731(Lieberman & Resnikoff)

Pro. 240, p. 832. Percentage values of p* for certain simple sampling plans by variables with known mean square deviation. 14.121/731(Lieberman & Resnikoff)

Pro. 241, p. 832. Percentage values of p* for certain simple sampling plans based on the mean range.   14.121/731(Lieberman & Resnikoff)

Pro. 242, pp. 833-839. Values for estimating the percentage of defective elements in a lot by means of simple sampling plans by variables with unknown mean square deviation.   14.121/731(Lieberman & Resnikoff)

Pro. 243, pp. 840-846. Values for estimating the percentage of defective elements in a lot by means of simple sampling plans based on the mean range. 14.121/731(Lieberman & Resnikoff)

Pro. 244, p. 847. Limits of the mean percentage of defective elements contained in lots: to be used in determining at what moment to reduce or reinforce the amount of inspection.

Pro. 245, p. 847. Values of the coefficients C_p and C_n for control of the coefficient of variation.

Pro. 246, p. 848. Values of the coefficients A_p and B_p for control of means and mean square deviations.

Pro. 247, p. 848. Values of the coefficients A_p', D_p' and D_p for control of means and ranges.

Pro. 248, pp. 849-850. Sample size and control limits for the mean of a normal universe that minimize the function L_1 or maximize the function L_0. 14.229/733(Page)

Pro. CCIL [i.e. 249], p. 851. Probability P of eliminating a tool after a period of time greater than h intervals, in industrial processes that tend to cause variable quality of product as a result of tool wear.   1.9/129(Tippett)

Pro. 250, p. 852. Values of $\log[(1-\beta)/\alpha]$ for sequential analyses of the results of a production process.   14.129/731

Pro. 251, pp. 853-854. Elements for evaluating the optimum size of each experiment and the risk function of the decisions, according to the minimax principle, in tests of a new production process.

Pri. 252-255, pp. 855-862. [Tables from queueing theory]

Pri. 256-257, pp. 863-872. Coefficients for conversion of the value of the lira, based on wholesale-price indices (on cost-of-living indices).

Pri. 258-273, pp. 873-976. [Tables of compound interest and annuities certain]

Pri. 274-277, pp. 977-988. [Mortality tables and commutation columns]

Pro. 278, pp. 989-997. Values of the derivates—from the second to the eighth—of the normal curve $f(t)$.   1.1161/9   1.3211/634

Pri. 279-282, pp. 998-1005. Coefficients in Newton's (Stirling's, Bessel's, Everett's) interpolation formula for calculating the values of the function $f(x)$ and of its derivate $f'(x)$.
16.2313/628   16.2315/628   16.2316/628   16.2317/628   16.234/629

Pro. 283, pp. 1006-1010. Values of the function $\Gamma(x)$ for x between 1 and 2. 2.5111/661

Vianelli

Pro. 284, pp. 1011-1017. Decimal logarithms of the function $\Gamma(x)$ for certain values of x greater than 2. 2.5131/662   2.5134/662

Pro. 285, pp. 1018-1024. Values of the digamma and trigamma functions for x between 0 and 20. 2.5141/164   2.5142/164

Pro. 286, p. 1025. Decimal logarithms of Bernoulli's numbers. 16.041/619

Pro. 287, pp. 1026-1031. Values of exponential functions and of their decimal logarithms.   16.101/623   16.102/624   16.103/624

Pro. 288, pp. 1032-1038. Values of hyperbolic functions and of their decimal logarithms.
16.1041/624   16.1043/624   16.1051/624   16.1052/625   7.211/695

Pro. 289, pp. 1039-1042. Values of certain functions of goniometric functions.   16.079/623

Pro. 290, pp. 1043-1046. Values of the damped sinusoid $e^{-x}\sin\alpha = e^{-x}\cos\beta$. 16.1232/626

Pro. 291, pp. 1047-1048. Legendre polynomials.

Pro. 292, pp. 1049-1050. Bessel functions.

Pro. 293, p. 1051. Values of the probability $P(r,n) = \int_0^\infty [J_0(x)]^n J_1(rx)\, dx$ corresponding to certain values of r and n.   10.28/722(Greenwood & Durand)

Pro. 294, pp. 1052-1057. Values of the inverse of the cumulative probability function of extremes $y = -\log(-\log_y)$.   8.3121/364

Pro. 295, pp. 1058-1059. Values of the probability density of extremes as a function of the cumulative probability.   8.3122/703

Pro. 296, pp. 1060-1062. Values of the cumulative probability function and of the density function of the reduced range.   8.331/365

Pro. 297, pp. 1063-1072. Values of the function S(h,a,b) for calculating the probability integral of a trivariate normal function.   1.79/125

Pro. 298, pp. 1073-1078. Values of the function G(m+r ; 1/k ; m) corresponding to certain values of r, m and k.   10.262/722(Tsao)

Pro. CCIC [i.e. 299], p. 1079. Values of binomial coefficients $\binom{m}{x}$ for m between 0 and 1, and integer x from 1 to 6.   16.2313/628

Pro. 300, pp. 1080-1085. Values for the resolution of equations of the third degree with positive real roots of sum equal to unity.   16.0564/620

¶ Besides the 300 Prontuari listed above, Vianelli gives 263 Tavole, i.e. short tables, usually illustrating a method or a calculation. We list below those Tavole that are of more general application: the titles affixed are our own descriptions.

T. 8, p. 1134. S and T functions.   16.08/623

T. 43, p. 1175. 5% and 1% points of Spearman's Rho.   8.112/350(Siegel)

T. IL [i.e. 49], p. 1182. Moments of $s/\sigma = \chi \nu^{-\frac{1}{2}}$ and factors for determining confidence limits for σ.
2.522/171   2.315/658(Pearson & Hartley)

T. 53, p. 1185. Variance of $x^{\frac{1}{2}}$ and of $(x+\frac{1}{2})^{\frac{1}{2}}$.   12.2/448

T. LIV [i.e. 64], p. 1197. Percentage points of the studentized maximum extreme deviate.   1.44/640(Halperin et al.)

T. 71, p. 1206. Test for heterogeneity of variance: table to facilitate interpolation in Pro. 95.   6.113/688(Thompson & Merrington)

T. 77, p. 1216. Expected values of normal order statistics.   1.43/67

T. 86, p. 1229. Binomial sign test.   3.6/206(Walker & Lev)

T. 97, p. 1248. Distribution of runs.   9.311/709(Mosteller)

T. 107, p. 1261. Table of normits.   1.2111/631

T. 173, p. 1331-1332. $hd^{-1} \log(1+d)$.   5.32/683

T. 176, p. 1336. Table for use in plotting Pearson Type III curves.
10.033/716(Elderton)

T. 184, p. 1348. Percentage points of the quotient of a rectangular by a normal variate.   10.261/722(Broadbent)

T. 185, p. 1348. Percentage points of the quotient of a triangular by a normal variate.   10.261/722(Broadbent)

T. 188, p. 1353. Normal tolerance regions in 3 dimensions.
14.32/433(Fraser & Guttman)

T. 193, p. 1360. Generalized probable errors.   2.313/658

T. 198, p. 1364. Coefficients in the recurrence formula for tetrachoric functions.   1.3249/634

T. 202, p. 1366. Tetrachoric correlation for median dichotomies.
16.0735/623   7.94/699

T. 205, p. 1369. To estimate the parameter k in a circular normal distribution.   10.28/722(Gumbel et al.)

T. 206, p. 1370. Integral of circular normal distribution over $30°$ sectors.
10.28/722(Gumbel 1954)

TT. 218-219, pp. 1390-1392. Tables for the design of single-sampling plans.
2.322/153(Grubbs)   14.112/476(Peach)

TT. 221-222, pp. 1399-1400. $\rho(\alpha,\beta,\nu) = \chi^2(\alpha,\nu)/\chi^2(1-\beta,\nu)$.
2.319/658(Eisenhart et al.)

T. 228, p. 1408. Percentage points of Tippett's distribution of trends in quality.   1.9/129(Tippett)

T. 229, p. 1411. $\log[\beta/(1-\alpha)]$ 14.129/731

T. 240, p. 1433. $\log_e(1+i)$ 16.112/625

T. 252, pp. 1466-1467. Hermite polynomials.   1.323/34

T. 260, p. 1483. Differences and derivatives of zero.   16.0492/620

APPENDIX III

MATERIAL TREATED BOTH IN THIS GUIDE AND IN FLETCHER et al. 1946

There is considerable overlap between the contents of this Guide to Tables in Mathematical Statistics and the contents of Fletcher, Miller & Rosenhead's An Index of Mathematical Tables (cited, here and throughout, as Fletcher et al. 1946). The table below lists the locations of comparable material, arranged in order of our section numbers.

We would emphasize that this comparison has been made with the first (1946) edition of Fletcher. We have been privileged to see a ms. copy of the 'Index of named functions' of the second edition, i.e. a much enlarged version of the 'Index to part I' on pp. 445-450 of Fletcher et al. 1946; apparently the coverage of statistical tables will be more extensive in the second edition than in the first.

For citations to Fletcher et al. 1946 in the text and Introduction of this Guide, see the Author Index below.

This Guide		Fletcher et al. 1946		subject
section	pages	article	pages	
1.111	3	⁺15.241	218	normal distribution: area
1.1121	4-5, 630	⁺15.231	217	
*1.1122	5	⁺15.243	218	
1.1131	5-6, 630	15.242	218	
1.1141	6-7, 630	⁺15.233	218	
*1.1142	7	⁺15.244	218	
*1.1143	7	⁺15.251	219	
*1.1144	7	⁺15.252	219	
1.115	7	15.232	217-218	
1.1161	9	15.14	214-215	normal distribution: ordinate
1.1162	10	15.16	215	
1.1211	11	15.211	215-216	normal distribution: area
*1.1212	11	⁺15.221	217	
1.1213	11	⁺15.212	216	
*1.1214	12	15.222	217	
*1.1215	12	⁺15.223	217	
*1.1216	12	⁺15.226	217	
*1.1221	12	⁺15.261	219	
*1.1222	12	15.262	219	

---

*All tables catalogued in the marked section of this Guide are also catalogued in Fletcher et al. 1946.

⁺All tables catalogued in the marked article of Fletcher et al. 1946 are also catalogued in this Guide.

1.131	13, 630	⊁15.33	220⎰	normal distribution:
1.132	13	⊁15.37	221⎱	ratio of area to ordinate
1.2111	19, 631⎞			
1.2121	20, 631⎬	15.411	221-222	
*1.213	21 ⎠			
1.2112	20, 631	15.413	222	normal distribution:
1.2122	20, 632			inverse tables
1.221	21	15.42	223	
1.222	22, 632	15.43	223	
1.31	31-32	⎧ 15.7	228-229⎰	normal distribution: moments
		⎩⊁15.71	229 ⎱	
1.3211	33-34, 634	⊁15.523	227	
*1.3212	34	⊁15.522	226	
*1.3213	34	⊁15.524	227	normal distribution:
1.322	34	⊁15.64	228	derivatives
*1.3241	35	15.54	227	
1.331	37-38	⊁15.511	224	normal distribution: integrals
1.332	39-45, 635-636	15.513	224	
1.721	120	⊁15.517	226	normal distribution: volumes
1.722	120			
1.811	126	15.281	219	error integral:
1.812	126	15.312	220	imaginary argument
*1.813	126			
1.82	127	15.8	230	Fresnel integrals
2.111	131-132	⊁14.81	206	
2.112	132, 656	23.87	366	
2.1132	132-133	⊁15.731	229-230	incomplete gamma function
2.1133	133	⊁15.72	229	
2.1134	133, 656	14.83	207	
2.121	134-135	⊁23.862	365	Poisson distribution:
2.122	135, 656			cumulative sum
2.21	137, 657	23.87	366	Pearson curve, Type III
2.221	138	⊁23.861	365	Poisson distribution:
				individual terms
2.3112	141			
2.3113	141-142	15.74	230	χ^2 distribution: inverse tables
2.313	144, 658			
2.5111	160, 661	14.1	199	factorials
2.5112	161, 661	3.11	34	
*2.5113	161	3.51	37	double factorials
2.5121	161, 662	⊁14.15	199	reciprocals of factorials
2.5122	161-162	3.12	34-35	
2.5131	162, 662	14.2	199-200	
2.5132	162			
*2.5133	162	⊁14.22	200	logarithms of factorials
2.5134	163, 662	3.14	35	
*2.5135	163	⊁3.52	37	logarithms of double factorials
2.5141	163	14.4	202-203	digamma function

This *Guide* section	pages	Fletcher et al. 1946 article	pages	subject
2.5142	164	⨯14.5	204	polygamma functions
3.12	185, 665	14.91	208	incomplete beta function
3.211	186	3.31	36	
5.31	284, 683			binomial coefficients
3.212	186	3.32	36	
3.311	191	⨯14.92	208	incomplete beta function: inverse tables
3.44	200–202	14.98	211	incomplete beta function: trigonometric form
*3.51	204	14.7	205–206	complete beta function
3.52	204, 668	14.91	208	incomplete beta function: constants of curves
4.1111	208, 670	14.96	210	Student's t
4.1114	209			
4.112	209, 671	14.95	210	Student's z
4.1311	211–212, 671	14.97	210	Student's t: inverse tables
4.1422	227	22.56	336	confluent hypergeometric function
4.2312	248, 676			
4.2313	249, 676	14.93	209	Snedecor's F
4.2314	249–250			
4.331	250–251, 676	14.94	209	Fisher's z
7.211	330–331, 695	10.42	171	$\tanh x$
7.212	331, 695	11.63	181	$\tanh^{-1} x$
7.32	334, 695	2.6411	30	$(1-x^2)^{\frac{1}{2}}$
7.33	334, 696	2.6416	30	$(1-x^2)^{-\frac{1}{2}}$
10.0431	397–398, 717	22.72	341	Pearson's G(r,) integrals
11.11	436–440	23.82	363–364	orthogonal polynomials
11.12	440	23.81	362–363	power polynomials
11.21	442	24.11	368	angles: $2\pi r/s$
11.22	442	24.21	368	$\cos 2\pi r/s$ and $\sin 2\pi r/s$
*11.23	443	24.23	369	$a \cos 2\pi r/s$ and $a \sin 2\pi r/s$
11.3	444	23.83	364	fitting of exponentials
14.152	484–485	10.38	170	Einstein functions
16.0211	614–615	2.11	24–25	squares
16.022	615	2.2	25	cubes
16.023	615–616	2.31	25–26	higher powers
*16.0233	616	2.32	26–27	
16.024	616	2.42	27–28	reciprocals
16.0261	617	2.61	28–29	square roots
16.0262	617	2.62	29	reciprocal square roots
16.0263	618	2.63	30	$n^{\frac{1}{2}p}$

*All tables catalogued in the marked section of this *Guide* are also catalogued in Fletcher et al. 1946.

⨯All tables catalogued in the marked article of Fletcher et al. 1946 are also catalogued in this *Guide*.

16.041	619	4.12	46	Bernoulli numbers
16.0444	620	4.44	54	sums of powers of integers
16.061	621	6.1	111	common logarithms
16.068	621–622	6.85	118–119	addition and subtraction logarithms
16.0711	622	7.11	122–123	$\sin x$ and $\cos x$ (radians)
16.0712	622	7.12	123	$\tan x$ (radians)
		7.13	123	$\cot x$ (radians)
16.101	623	10.12	166	e^x
16.102	624, 738	10.22	167–168	e^{-x}
16.103	624	10.31	168	$\log_{10} e^x$
16.1041	624	10.41	170–171	$\sinh x$ and $\cosh x$
16.1043	624	10.43	171	$\coth x$
16.1051	624	10.51	173	$\log \sinh x$ and $\log \cosh x$
16.1052	625	10.52	173–174	$\log \tanh x$ and $\log \coth x$
16.111	625	$\begin{cases}11.11\\11.12\end{cases}$	$\begin{matrix}176\text{–}177\\177\end{matrix}$	natural logarithms of integers
16.112	625–626	11.2	177–178	natural logarithms of decimal numbers
16.2311	627–628	$\begin{cases}23.111\\23.112\end{cases}$	$\begin{matrix}348\text{–}349\\349\end{matrix}$	Lagrange interpolation coefficients
16.2313	628	23.13	350	Gregory-Newton interpolation coefficients
16.2315	628	23.151	351	Stirling interpolation coefficients
16.2316	628	23.161	352–353	Bessel interpolation coefficients
16.2317	628	23.17	353–354	Everett interpolation coefficients
16.234	629	$\begin{cases}23.41\\23.42\\23.43\\23.44\end{cases}$	357	numerical differentiation
16.2369	629	23.69	362	Lubbock coefficients

INDICES

Author Index 790
Subject Index 953

AUTHOR INDEX

A.S.Q.C.   See American Society for Quality Control

A.S.T.M.   See American Society for Testing Materials

ABDEL-ATY, S. H.

   1954   Tables of generalized k-statistics. Biometrika 41, 253-260.
       10.342/430

   1954 z Approximate formulae for the percentage points and the probability integral of the non-central χ^2 distribution. Biometrika 41, 538-540.
       $2.42/661_M$

ABELSON, R. M.   See Bradley 1954

ABRAMOV, A. A.   АБРАМОВ А. А.

   1954   Таблицы ln Γ[z] в комплексной области.
       Москва: Издат. Акад. Наук СССР.   2.5152/164

AGARWALA, S. P.   See Chandra Sekar, Agarwala & Chakraborty

AIKEN, HOWARD HATHAWAY   See Harvard Univ. 1949, 1952, 1955

AITCHISON, J. and BROWN, J. A. C.

   1954   An estimation problem in quantitative assay. Biometrika 41, 338-343.
       $1.233/633_M$

   1957   The lognormal distribution:
       with special reference to its uses in economics.
       University of Cambridge, Department of Applied Economics, Monographs, 5.
       Cambridge Univ. Press, etc.
       10.2221/412   10.2221/413   10.2221/414   10.2222/414

AITKEN, ALEXANDER C.

   1933   On the graduation of data by the orthogonal polynomials of least squares. Proc. Roy. Soc. Edinburgh 53, 54-78.
       11.110/436   11.110/437   11.113/438   11.113/439

   1939   Statistical mathematics. First edition.
       (Second edition 1942, 3d 1944, 4th 1945, 5th 1947, 6th 1949, 7th 1952)
       Edinburgh, etc.: Oliver and Boyd; New York: Interscience.
       $1.1121/5_F$ 5.1/275   5.1/276(intro.)   5.22/282(Maritz)   11.113/439

AKAIKE, HIROTUGU

   1954   An approximation to the density function.
       Ann. Inst. Statist. Math. (Tokyo) 6, 127-132.

¶ Not abstracted; apparent classification 2.69.
See Mathematical Reviews 16, 726.

ALBERT, G. E. and JOHNSON, R. B.

    1951   On the estimation of central intervals which contain assigned proportions of a normal distribution. Ann. Math. Statist. 22, 596-599.
    6.119/304

ALLAN, FRANCES E.

    1930 a   The general form of the orthogonal polynomials for simple series, with proofs of their simple properties.
    Proc. Roy. Soc. Edinburgh 50, 310-320.   11.110/437

    1930 b   A percentile table of the relation between the true and the observed correlation coefficient from a sample of 4.
    Proc. Cambridge Philos. Soc. 26, 536-537.   7.22/331

AMEMIYA, A.

    See Kotani & Amemiya; Kotani, Amemiya, Ishiguro & Kimura; Kotani, Amemiya & Simose

AMERICAN SOCIETY for QUALITY CONTROL

    1957   ASQC Standard A2-1957:
    Definitions and symbols for acceptance sampling by attributes.
    Indust. Quality Control 14, no. 5, 5-6.   14.1/469

AMERICAN SOCIETY FOR TESTING MATERIALS

    1940   ASTM manual on presentation of data.
    14.221/488(intro.)   14.221/488

    1951   ASTM manual on quality control of materials, prepared by ASTM Committee E-11 . . . Special technical publication 15-C. (Reprint 1956, incorporating "Tentative recommended practice for choice of sample size to estimate the average quality of a lot or process")
    Philadelphia: the Society.   14.221/488

AMERICAN STANDARDS ASSOCIATION, New York

    1941   Guide for quality control and control chart method of analyzing data. (War Standard Z 1.1, Z 1.2-1941)   14.221/488   14.222/733

    1942   Control chart method of controlling quality during production.
    (War Standard Z 1.3-1942)   14.221/488

ANDERSON, RICHARD L.

    1942   Distribution of the serial correlation coefficient.
    Ann. Math. Statist. 13, 1-13.   7.51/336   7.51/696(Ferber)

    See Binet, Leslie, Weiner & Anderson

ANDERSON, R. L. and ANDERSON, T. W.

    1950   Distribution of the circular correlation coefficient for residuals from a fitted Fourier series. Ann. Math. Statist. 21, 59-81.
        7.51/338

ANDERSON, R. L. and BANCROFT, T. A.

    1952   Statistical theory in research.
        New York, London, etc.: McGraw-Hill Book Co.
        2.3112/141   4.2314/249

ANDERSON, R. L. and HOUSEMAN, E. E.

    1942   Tables of orthogonal polynomial values extended to n = 104.
        Iowa State College, Agricultural Experiment Station, Bulletin 297.
        $11.111/437_F$ 11.112/438(Fisher & Yates)   11.112/438

ANDERSON, THEODORE WILBUR

    See Anderson, R. L. & Anderson, T. W.; Villars & Anderson

ANDERSON, T. W. and DARLING, D. A.

    1952   Asymptotic theory of certain "goodness of fit" criteria based on stochastic processes. Ann. Math. Statist. 23, 193-212.   9.4212/386

ANIS, A. A.

    *1955   The variance of the maximum of partial sums of a finite number of independent normal variates. Biometrika 42, 96-101. $1.43/638_M$

ANSCOMBE, F. J.

    1948   The transformation of Poisson, binomial and negative-binomial data. Biometrika 35, 246-254.   12.15/446   12.2/448   12.33/449

    1949   Tables of sequential inspection schemes to control fraction defective. J. Roy. Statist. Soc. Ser. A 112, 180-206.   14.112/474

    1950   Sampling theory of the negative binomial and logarithmic series distributions. Biometrika 37, 358-382.   5.21/278(intro.)

    1950a   Table of the hyperbolic transformation $\sinh^{-1}\sqrt{x}$.
        J. Roy. Statist. Soc. Ser. A. 113, 228-229.   12.31/449

AOYAMA, H.

    1953   On a test in paired comparisons.
        Ann. Inst. Statist. Math. (Tokyo) 4, 83-87.   3.6/670   8.14/699

ARCHIBALD, R. C.

    See Bateman & Archibald; Harvard Univ. 1949; Jahnke & Emde 1938 (1945 reprint)

A[RCHIBALD], R. C. and J[OFFE], S. A.

    1947   RMT 428 [Review of Boll 1947] Math. Tables Aids Comput. 2, 336-338.

ARKIN, H. and COLTON, R. R.

    1950   Tables for statisticians. New York: Barnes & Noble.
¶ For contents of this book, see p. 739 above.   1j

ARLEY, N.

    1940   On the distribution of relative errors from a normal population of errors: a discussion of some problems in the theory of errors. Mat.-Fys. Medd. Danske Vid. Selsk. 18, No. 3.
4.132/215   4.132/216(Arley & Buch)   7.1223/323

ARLEY, N. and BUCH, K. R.

    1950   Introduction to the theory of probability and statistics. New York [and London]: John Wiley & Sons.
1.115/8   1.1161/9   1.2122/20   4.1311/212   4.132/216   4.2314/250
7.1223/324

ARMSEN, P.

    1955   Tables for significance tests of 2×2 contingency tables. Biometrika 42, 494-511. Reprinted as New Statistical Tables, No. 21. A new form of table for significance tests in a 2×2 contingency table.
2.611/176   5.1/275

[U. S.] ARMY. ORDNANCE CORPS

    1952   Tables of the cumulative binomial probabilities. Ordnance Corps Pamphlet ORDP 20-1. Available from Library of Congress.
3.111/183

AROIAN, L. A.

    1943   A new approximation to the levels of significance of the chi-square distribution. Ann. Math. Statist. 14, 93-95.   2.313/146

    1947   The probability function of the product of two normally distributed variables. Ann. Math. Statist. 18, 265-271.   1.78/124

    1948   The fourth degree exponential distribution function. Ann. Math. Statist. 19, 589-592.
10.23/415   10.23/416(intro.)   10.23/416(O'Toole)

See N.B.S. 1959

ASPIN, ALICE A.

    1948   An examination and further development of a formula arising in the problem of comparing two mean values. Biometrika 35, 88-96.
4.152/237

    1949   Tables for use in comparisons whose accuracy involves two variances, separately estimated. Biometrika 36, 290-293. See also Welch 1949.
4.152/238   4.152/674(Trickett et al.)

ASTIN, A. V.

    See Connor & Young 1961; Connor & Zelen 1959;
N.B.S. 1953 y,z, 1954, 1955 y,z, 1957, 1958, 1959

ASTRACHAN, M.   See Sampling by Variables I.

AUBLE, D.

   1953   Extended tables for the Mann-Whitney statistic. Indiana Univ., Institute of Educational Research, Bulletin, 1, No. 2.
      9.121/369(Siegel)

AYLING, JOAN   See Kendall, M. G. 1949

AZORIN, F. and WOLD, H.

   1950   Product sums and modulus sums of H. Wold's normal deviates. Sumas de productos y sumas de módulos de las desviantes normales de H. Wold. Trabajos Estadística 1, 5-28.   13.63/468

B. A. = British Association for the Advancement of Science

   In references, B. A. 1916 means British Association Report (q.v.) for 1916, and similarly for other years. B.A. 1, 1931 means British Association Mathematical Tables (q.v.), Vol. 1, 1931, and similarly for other volumes.

BABINGTON SMITH, B.

   See Kendall, M. G. & Babington Smith;
   Kendall, M. G., Kendall, S. F. H. & Babington Smith

BAILEY, J. L., Jr.

   1931   A table to facilitate the fitting of certain logistic curves. Ann. Math. Statist. 2, 355-359.   16.044/619

BAILEY, NORMAN T. J.

   1953   The total size of a general stochastic epidemic. Biometrika 40, 177-185.
      ¶ Not abstracted; apparent classification 5.2.
   See Mathematical Reviews 14, 1101.

BAKER, G. A.

   1934   Transformation of non-normal frequency distributions into normal frequency distributions. Ann. Math. Statist. 5, 113-123.
      10.221/719

BALDWIN, ELIZABETH M.

   1946   Tables of percentage points of the t-distribution. Biometrika 33, 362 only.   4.1311/212

[U. S.] BALLISTIC RESEARCH LABORATORIES

   See Army Ordnance Corps

BANCROFT, T. A.

   1944   On biases in estimation due to the use of preliminary tests of significance. Ann. Math. Statist. 15, 190-204.
      2.521/166   4.19/243

   See Anderson, R. L. & Bancroft

BANERJEE, D. P.

*1954 A note on the distribution of the ratio of sample standard deviations in random samples of any size from a bi-variate correlated normal population. J. Indian Soc. Agric. Statist. 6, 93-100.
4.139/673$_M$

BANERJEE, S. K.

1936 The one-tenth per cent level of the ratio of variances. Sankhyā 2, 425-428.   4.2313/249   4.233/252

See Mahalanobis, Bose, Ray & Banerji

BARGMANN, ROLF

*1955 Signifikanzuntersuchungen der einfachen Struktur in der Faktorenanalyse. Mitteilungsbl. Math. Statist. 7, 1-24.   3.319/666$_M$

BARK, L. S.   See Karmazina 1954; Karpov 1954, 1958

BARLOW, P.

1930 Barlow's tables of squares, cubes, square roots, cube roots and reciprocals of all integer numbers up to 10000.
Third edition, by L. J. Comrie. London: E. & F.N. Spon.
2.5112/161$_F$ 3.211/186$_F$

1941 Barlow's tables of squares, cubes, square roots, cube roots and reciprocals of all integer numbers up to 12500.
Fourth edition, by L. J. Comrie. London: E. & F.N. Spon.
2.5112/161$_F$ 3.211/186$_F$ 16.0211/614$_F$ 16.022/615$_F$ 16.023/615(2)$_F$
16.024/626$_F$ 16.0261/617$_F$ 16.0262/617$_F$ 16.027/619$_F$

¶ We are aware of five piratical editions—three American, one Chinese, one Russian—of Barlow 1930 or 1941; there are probably others.

BARNARD, G. A.

1947 z 2 × 2 tables. A note on E.S. Pearson's paper [Pearson, E.S. 1947]. Biometrika 34, 168-169.

BARRACLOUGH, ELIZABETH D. and PAGE, E. S.

1959 Tables for Wald tests for the mean of a normal distribution. Biometrika 46, 169-177.   14.123/731

BARTHOLOMEW, D. J.

1956 A sequential test of randomness for events occurring in time or space. Biometrika 43, 64-78.   6.4/694$_M$

BARTLETT, JAMES H., Jr.

1931 Orbital valency. Phys. Rev. 37, 507-531.   2.1134/133$_F$

BARTLETT, MAURICE S.

    1936   The square root transformation in the analysis of variance. J. Roy. Statist. Soc. Suppl. 3, 68-78.   12.2/448

    1937   Properties of sufficiency and statistical tests. Proc. Roy. Soc. London Ser. A 160, 268-282.   6.113/300(intro)

    1937 a   Subsampling for attributes. J. Roy. Statist. Soc. Suppl. 4, 131-135.   12.13/445

    1938   Further aspects of the theory of multiple regression. Proc. Cambridge Philos. Soc. 34, 33-40.   6.1/295   6.133/311

    1941   The statistical significance of canonical correlations. Biometrika 32, 29-37.   6.133/312   6.4/318

    1950   Periodogram analysis and continuous spectra. Biometrika 37, 1-12.   11.29/443   13.5/467

    1952   The statistical significance of odd bits of information. Biometrika 34, 228-237.   8.3129/703

    1953   On the statistical estimation of mean life-times. Phil. Mag. 44, 249-262.   14.152/485

    See Wishart & Bartlett

BARTLETT, M. S. and KENDALL, D. G.

    1946   The statistical analysis of variance-heterogeneity and the logarithmic transformation. J. Roy. Statist. Soc. Suppl. 8, 128-138.   4.34/273

BARTON, D. E. and DAVID, F. N.

    *1956   Tests for randomness of points on a line. Biometrika 43, 104-112. $9.319/711_M$

    *1956 z   Some notes on ordered random intervals. J. Roy. Statist. Soc. Ser. B 18, 79-94.   9.319/711(ftn.) $9.319/712_M$

    1957   Multiple runs. Biometrika 44, 168-178.   9.311/711

    1958   A test for birth-order effect. Ann. Human Genetics 22, 250-257.   9.2/373

BARTON, D. E. and DENNIS, K. E.

    1952   The conditions under which Gram-Charlier and Edgeworth curves are positive definite and unimodal. Biometrika 39, 425-427.   10.211/409

BATEMAN, G. I.

    1948   On the power of the longest run as a test for randomness in a sequence of alternatives. Biometrika 35, 97-112.   9.32/378

    1950   The power of the χ^2 index of dispersion test when Neyman's contagious distribution is the alternate hypothesis. Biometrika 37, 59-63.   5.21/280   5.21/281(Thomas)

BATEMAN, H. and ARCHIBALD, R. C.

   1944   A guide to tables of Bessel functions.
       Math. Tables Aids Comput. 1, 205-308.   16.17/626   xxxvij

BATEN, W. D.

   1938   Elementary mathematical statistics. New York: John Wiley & Sons;
       London: Chapman & Hall.   1.115/8

BAYES, THOMAS   See Wishart 1927

BEALL, G.

   1942   The transformation of data from entomological field experiments so
       that the analysis of variance becomes applicable.
       Biometrika 32, 243-262.   12.32/449

BEARMAN, J. E.   See Sampling by Variables III.

BECHHOFER, R. E.

   1954   A single-sample multiple decision procedure for ranking means of
       normal populations with known variances. Ann. Math. Statist. 25, 16-39.
       1.795/654

BECHHOFER, R. E. and SOBEL, MILTON

   1954   A single-sample multiple decision procedure for ranking variances of
       normal populations. Ann. Math. Statist. 25, 273-289.   4.254/681

BECKER, G. F. and Van ORSTRAND, C. E.

   1924   Smithsonian mathematical tables: Hyperbolic functions.
       Third reprint. (Original edition 1909, first reprint 1911, 2d 1920,
       4th 1931, 5th 1942) Washington: Smithsonian Institution.
       $16.0711/622_F$ $16.101/623_F$ $16.102/624_F$ $16.103/624_F$ $16.1041/624_F$
       $16.1043/624_F$ $16.1051/624_F$ $16.1052/625_F$ $16.111/625_F$ $7.211/695$

BEER, MARGARET T.   See Pearson, K. 1934

BEHNKEN, D. W.   See Box & Behnken

BEHRENS, W. U.

   1929   Ein Beitrag zur Fehlerberechnung bei wenigen Beobachtungen.
       Landwirtschaftliche Jahrbücher 68, 807-837.
       4.11/208   4.1112/209   4.113/210   4.1312/213   4.151/234   14.32/493

BELL, JULIA   See K. Pearson 1913 a, 1930 a

BELLAVITIS, G.

   1874   Tavole numeriche del logaritmo-integrale ossia dell' esponenziale-
       integrale, e di altri integrali Euleriani.
       Ist. Veneto Sci. Lett. Arti, Memorie 18, 125-162. $2.5133/162_F$

BELLINSON, H. R.   See von Neumann, Kent, Bellinson & Hart

BENNETT, J. H.

*1956   Partitions in more than one dimension.
J. Roy. Statist. Soc. Ser. B 18, 104-112.   10.36/723$_M$

BERGE, P. O.

1932   Über das Theorem von Tchebycheff und andere Grenzen einer Wahrscheinlichkeitsfunktion. Skand. Aktuarietidskr. 15, 65-77.
10.4/433

BERJMAN, ELENA

1941 a   Una solución de ajustamiento per minimos cuadrados según los polinomios de Gauss, mediante la determinación y tabulación de los coefficientes parametricos de funciones parabolicas de 1°. a 5°. grado, para series hasta de 100 bases. Capítulo I.
An. Soc. Ci. Argentina 132, 34-48.

1941 b   Una solución de ajustamiento, etc. Capítulo II.
An. Soc. Ci. Argentina 132, 104-117.

1941 c   Una solución de ajustamiento, etc. Capítulo III.
An. Soc. Ci. Argentina 132, 212-217.   11.12/725

1942 a   Una solución de ajustamiento, etc.
An. Soc. Ci. Argentina 133, 208-215.   11.12/725(2)

1942 b   Una solución de ajustamiento, etc.
An. Soc. Ci. Argentina 133, 442-445.   11.12/725

1942 c   (Una solución de ajustamiento) Capítulo IV.
An. Soc. Ci. Argentina 133, 446-456.

BERKSON, J.

1953   A statistically precise and relatively simple method of estimating the bio-assay with quantal response, based on the logistic function.
J. Amer. Statist. Assoc. 48, 565-599.
12.51/450   12.52/450   12.53/450

1955   Estimate of the integrated normal curve by minimum normit chi-square, with particular reference to bio-assay.
J. Amer. Statist. Assoc. 50, 529-547.
1.115/8   1.2111/631   1.2313/632

1957 a   Tables for the maximum likelihood estimates of the logistic function.
Biometrics 13, 28-34.   12.52/450   12.53/451

1957 b   Tables for use in estimating the normal distribution function by normit analysis. Biometrika 44, 411-435.   1.2313/632(2)

BERRETTONI, J. N.   See Greb & Berrettoni

BHATE, D. H.

1951   A note on the significance levels of the distribution of the mean of a rectangular population. Calcutta Statist. Assoc. Bull. 3, 172-173.
10.261/418

BHATT, N. M.

*1953   Sextiles and octiles with the ordinates at these quantiles of the
standardized Pearson's type III distribution.
J. Maharaja Sayajirao Univ. Baroda 2, no. 2, 117-124.
¶ Not abstracted; apparent classification 2.319.
See Mathematical Reviews 15, 971.

BHATTACHARYA, P. K.

1955   Joint test for the mean and variance of a normal population.
Calcutta Statist. Assoc. Bull. 6, 73-90. $6.119/689_M$

BHUYIA, K. L.   See Mahalanobis, Bose, Ray & Banerji 1934

BICKLEY, W. G.   See B.A. 9, 1940

BIENAYMÉ, J.

1853   Considerations à l'appui de la découverte de Laplace sur la loi de
probabilité dans la méthode des moindres carrés.
C. R. Acad. Sci. Paris 37, 309-324.
Reprinted, not literatim, as Bienaymé 1867.   10.4/433

1867   Considerations à l'appui de la découverte de Laplace sur la loi de
probabilité dans la méthode des moindres carrés.
J. Math. Pures Appl. 12, 158-176.   10.4/433

See Čebyšev 1858

BIGNARDI, F.

*1947   Sulle generalizzazioni della disuguaglianza di Bienaymé-Tchebycheff
nello studio delle distribuzioni di frequenza. Annali della Facoltà di
Economia e Commercio dell' Università di Palermo, No. 1.
10.4/724(Vianelli)

BINET, F. E., LESLIE, S. T., WEINER, S. and ANDERSON, R. L.

1955   Analysis of confounded factorial experiments in single replications.
North Carolina Agric. Expt. Sta., Tech. Bull. 113.
15.112/502   15.112/503(2)   15.112/506   15.113/506   15.1132/520
15.1132/521(7)   15.1132/522(7)   15.1132/523   15.114/526(2)
15.114/527   15.114/529   15.114/532(3)   15.115/534(2)   15.115/535(7)
15.116/535(2)   15.116/536(6)   15.117/537(6)

BINGHAM, W. V.

1937   Aptitudes and aptitude testing.
New York and London: Harper & Bros., etc.
(Reprint 1942, including a list of 'recent publications')
¶ The notations '2d edition', '4th edition', etc., on back of t.p.
of some copies, do not denote revisions.
7.31/695

BIRGE, R. T. and SHEA, J. D.

   1927   A rapid method for calculating the least squares solution of a polynomial of any degree. Univ. California Publ. Math. 2, 67-118. Issued in wrappers with Condon 1927.   11.111/437

BIRGE, R. T. and WEINBERG, J. W.

   1947   Least-squares' fitting of data by means of polynomials. Rev. Mod. Phys. 19, 298-360.   11.110/437   11.111/437

BIRNBAUM, Z. W.

   1952   Numerical tabulation of the distribution of Kolmogorov's statistic for finite sample size. J. Amer. Statist. Assoc. 47, 425-441.
    9.4121/382   9.3121/714

BIRNBAUM, Z. W. and TINGEY, F. H.

   1951   One-sided confidence contours for probability distribution functions. Ann. Math. Statist. 22, 592-596.   9.4151/384

[BIRNBAUM, Z. W. et al.]

   *1955   The normal probability function: Tables of certain area-ordinate ratios and of their reciprocals. Biometrika 42, 217-222.   1.131/630   1.132/630   1ij

BISHOP, D. J.

   1939   On a comprehensive test for the homogeneity of variances and covariances in multivariate problems. Biometrika 31, 31-55.
    6.1/295   6.12/306(Box)   6.132/309   6.132/311(Box)

BISHOP, D. J. and NAIR, U. S.

   1939   A note on certain methods of testing for the homogeneity of a set of estimated variances. J. Roy. Statist. Soc. Suppl. 6, 89-99.
    6.113/300(intro.)   6.113/300

BLACK, A. N.

   1950   Weighted probits and their use. Biometrika 37, 158-167.
    1.2312/26

BLACKWELL, D. and GIRSHICK, M. A.

   1954   Theory of games and statistical decisions. New York [and London]: John Wiley & Sons.   6.119/689(Reiter)

BLANCH, GERTRUDE   See N.B.S. 1959; Teichroew 1956

BLISS, C. I.   БЛИСС Ч. И.

   1935   The calculation of the dosage-mortality curve. Annals of Applied Biology 22, 134-167.
    1.23/22   1.2312/25   1.232/27

1937   Анализ данных полевого опыта, выраженных в процентах.
Перевод А А Любищева.
The analysis of field experimental data expressed in percentages.
Защита Растений (Plant Protection) № 12, 67-77.   12.11/445

1937 a   The calculation of the time-mortality curve.
Annals of Applied Biology 24, 815-852.   See Stevens 1937 a.
2.522/170

1938   The determination of the dosage-mortality curve from small numbers.
Quart. J. Pharmacy Pharmacol. 11, 192-216.   1.2312/632

1948   Estimation of the mean and its error from incomplete Poisson distribution. Connecticut Agricultural Experiment Station, Bulletin 513.
2.229/138

See Fisher, R.A. & Yates 1938, etc., TT. IX, XI

BLISS, C. I., COCHRAN, W. G. and TUKEY, J. W.

*1956   A rejection criterion based upon the range. Biometrika 43, 418-422.
$1.5512/651_M$

BLOMQVIST, N.

1950   On a measure of dependence between two random variables.
Ann. Math. Statist. 21, 593-600.   8.19/357

1951   Some tests based on dichotomization.
Ann. Math. Statist. 22, 362-371.   8.19/358

BOFINGER, EVE and BOFINGER, V. I.

*1958   A periodic property of pseudo-random sequences.
J. Assoc. Comput. Mach. 5, 261-265.   13.62/468

BOFINGER, V. I.   See Bofinger, Eve & Bofinger, V. I.

BOLL, MARCEL

*1947   Tables numeriques universelles des laboratoires et bureaux d'etudes.
Paris: Dunod.

¶ The following quotation from Archibald & Joffe 1947, p. 337, may serve both as apology for not indexing the relevant tables of Boll and as warning to the statistician that may be tempted to use that "well-printed, well-arranged, and excellently indexed volume";

"Wholly random checking [by Joffe] of a few of the tables . . . shows that they are highly unreliable, displaying not only defective proof reading, but also carelessness and inadequate checking of basic calculations. Hence the reliability of no table in the volume should be assumed without careful checking. It looks as if Hayashi's throne has been lost to a Frenchman."

BONYNGE, W.   See Fisher, A. 1922

BOREL, É. et CHERON, A.

    1940   Théorie mathématique du Bridge a la portée de tous:
134 tableaux de probabilités avec leurs modes d'emploi. Formules simples.
Applications. Environ 4000 probabilités.
Monographies des probabilités, fasc. 5.   Paris: Gauthier-Villars.
$5.61/685_M$

BORTKIEWICZ, L. v.

    1898   Das Gesetz der kleinen Zahlen, von Dr. L. von Bortkewitsch.
Leipzig: B. G. Teubner.   2.221/138

    See Charlier 1947

BOSE, P. K.

    1947   Parametric relations (recursions) in multivariate distributions.
Sankhyā 8, 167-171.   6.131/307(Bose 1949)

    1947a   On recursion formulae, tables and Bessel function populations associated with the distribution of (for) classical D^2- statistic.
Sankhyā 8, 235-248.   6.131/306   6.12/690(ftn.)

    1949   Incomplete probability integral tables connected with studentised D^2-statistic.   Calcutta Statist. Assoc. Bull. 2, 131-137.
6.131/307   6.131/307(Rao)

    1951   Corrigenda: On the construction of incomplete probability integral tables of the classical D^2-statistic. Sankhyā 11, 96 only.
6.131/307(Bose 1947)

BOSE, RAJ CHANDRA

    1936   On the exact distribution and moment-coefficients of the D^2-statistic.   Sankhyā 2, 143-154.   6.131/306(P.K. Bose)

    1939   On the construction of balanced incomplete block designs.
Ann. Eugenics 9, 353-400.

    15.2103/570   15.2103/575(3)   15.2105/586   15.212/604
            /571   /576(6)   /587(3)   /605
            /572(2)   15.2104/580(2)   15.2106/589
            /573(3)   /582   .2107/593(2)
            /574   /583(2)   .2109/600

    1956   Paired comparison designs for testing concordance between judges.
Biometrika 43, 113-121.   15.2413/736(Vianelli)

    See Connor & Young 1961; Connor & Zelen 1959

BOSE, R. C., CHAKRAVARTI, I. M. and KNUTH, D. E.

    1960   On methods of constructing sets of mutually orthogonal Latin squares using a computer. I. Technometrics 2, 507-516.   15.1513/560

    1961   On methods of constructing sets of mutually orthogonal Latin squares using a computer. II. Technometrics 3, 111-117.

BOSE, R. C., CLATWORTHY, W. H. and SHRIKHANDE, S. S.

    1954   Tables of partially balanced designs with two associate classes.
North Carolina Agric. Expt. Sta., Tech. Bull. 107.

```
   15.20/563(ftn.)      15.2104/579(12)      15.2106/589(16)      15.2109/598(9)
        /564                  /580(13)              /590(19)              /599(15)
        /565                  /581(17)              /591(18)              /600(16)
   15.2103/570(4)             /583(9)               /592(5)               /601(5)
         /571(11)       15.2105/583(2)       15.2107/592(6)        15.2110/601(9)
         /572(14)             /584(6)               /593(13)              /602(17)
         /573(12)             /585(12)              /594(8)               /603(25)
         /574(15)             /586(17)       15.2108/594(6)                /604(7)
         /575(17)             /587(3)               /595(21)       15.248/607
         /576(12)       15.2106/587(12)             /596(27)        .258/609
   15.2104/577(9)             /588(13)              /597(11)
         /578(9)                                    /598(2)
```

BOSE, R. C. and CONNOR, W. S.

    1952   Combinatorial properties of group divisible incomplete block designs.
Ann. Math. Statist. 23, 367-383.   15.2106/589

BOSE, R. C. and NAIR, K. R.

    1939   Partially balanced incomplete block designs. Sankhyā 4, 337-372.

```
   15.2103/570         15.2104/583         15.2107/592        15.2109/600
         /571(2)            .2105/584            /593               /601
         /572                    /585            /594        15.2110/603
         /573(2)       15.2106/589(2)     15.2108/596
   15.2104/578(2)            /591                /597
         /581                 /592                /599
```

    1941   On complete sets of Latin squares. Sankhyā 5, 361-382.

    15.112/507   15.112/515   15.112/516   15.113/520(2)   15.114/526(2)

    15.115/533   15.115/534   15.117/537   15.118/537   15.119/541

    15.1199/543   15.1199/544   15.1199/545   15.1199/546

BOSE, R. C. and SHIMAMOTO, T.

    1952   Classification and analysis of partially balanced incomplete block
designs with two associate classes.
J. Amer. Statist. Assoc. 47, 151-184.

```
   15.2103/572(2)      15.2104/580(5)      15.2106/591        15.2109/600(3)
         /573(3)             /581(3)            .2107/592           .2110/603(2)
         /574(6)       15.2105/584(2)            /593(4)             /604(2)
         /576(2)             /585(3)       15.2108/595
   15.2104/578(2)      15.2106/589                /596(4)
         /579(7)             /590(3)              /597(2)
```

BOSE, R. C. and SHRIKHANDE, S. S.

    1960   On the construction of sets of mutually orthogonal Latin squares and
the falsity of a conjecture of Euler.
Trans. Amer. Math. Soc. 95, 191-209.

    ¶ Presumably, this is the paper cited by Bose, Shrikhande & Parker 1960, p. 203, as Bose, Shrikhande & Bhattacharya: 'On the construction of pairwise orthogonal squares and the falsity of a conjecture of Euler.
15.1513/560

BOSE, R. C., SHRIKHANDE, S. S. and PARKER, E. T.

    1960   Further results on the construction of mutually orthogonal Latin squares and the falsity of Euler's conjecture.
Canadian J. Math. 12, 189-203.   15.1513/560

BOSE, SUBHENDU SEKHAR

    1934   Tables for testing the significance (of coefficient) of linear regression in the case of time-series and other single-valued samples. Sankhyā 1, 277-288.   7.1229/324

    See Mahalanobis 1933 a; Mahalanobis, Bose, Ray & Banerji.

BOSSE, LOTHAR

    *1948   Tafel der Funktion $n = f(i,p) = \dfrac{90\,000 \cdot (1-p)}{i^2 \cdot p}$ zur Bestimmung der Beobachtungszahl n, wenn die Zufallfehlergrenzen $(3 \cdot \sigma_v)$ einen bestimmten Prozentsatz i der vorgegebenen Warscheinlichkeit p nicht überschreiten sollen. Statist. Vierteljschr. 1, 38-39.

    Not abstracted; apparent classification 3.53.
See Mathematical Reviews 12, 13.

BOURGET, H.   See Hermite 1908

BOWKER, A. H.

    1946   Computation of factors for tolerance limits on a normal distribution when the sample is large. Ann. Math. Statist. 17, 238-240.   1.19/17

    See Eisenhart et al. 1947

BOWKER, A. H. and GOODE, H. P.

    1952   Sampling inspection by variables.
New York, London, etc.: McGraw-Hill Book Co.
14.121/477   14.221/489

BOWLEY, A. L.

    1928   F. Y. Edgeworth's contributions to mathematical statistics.
London: Royal Statistical Society.   10.221/411

    1937   Elements of statistics. Sixth edition.
(First edition 1901, 3d 1916, 4th 1920)
London: P.S. King & Son; New York: Charles Scribner's Sons.
$1.1121/5_F$

    See Edgeworth & Bowley

BOX, G. E. P.

    1949   A general distribution theory for a class of likelihood criteria.
Biometrika 36, 317-346.
6.1/295   6.113/302   6.12/306   6.132/311   6.133/312   10.00/389

    1953   Non-normality and tests of variance. Biometrika 40, 318-335.
4.28/264

1954   Some theorems on quadratic forms applied in the study of analysis of variance problems, I.
Effect of inequality of variance in the one-way classification.
Ann. Math. Statist. 25, 290-302.   4.28/268   4.28/269(Box 1954 a)

1954 a   Some theorems on quadratic forms applied in the study of analysis of variance problems, II. Effects of inequality of variance and of correlation between errors in the two-way classification.
Ann. Math. Statist. 25, 484-498.   4.28/269

See Davies 1954

BOX, G. E. P. and BEHNKEN, D. W.

1960   Some new three level designs for the study of quantitative variables.
Technometrics 2, 455-475.   15.138/555(2)   15.148/558(8)

BOX, G. E. P. and HUNTER, J. S.

1957   Multi-factor experimental designs for exploring response surfaces.
Ann. Math. Statist. 28, 195-241.   15.138/555(12)   15.148/557(7)

BOYD, MARIANNE, Mrs   See Bradley 1954

BRADLEY, R. A.

1954   Rank analysis of incomplete block designs. II.
Additional tables for the method of paired comparisons.
Biometrika 41, 502-537.
¶ For I see Bradley & Terry 1952.
15.28/610(2)

1955   Rank analysis of incomplete block designs. III. Some large-sample results on estimation and power for a method of paired comparisons.
Biometrika 42, 450-470.

BRADLEY, R. A. and TERRY, M. E.

1952   Rank analysis of incomplete block designs. I.
The method of paired comparisons. Biometrika 39, 324-345.
¶ For II, III see Bradley 1954, 1955.
15.28/610(4)

BRAMLEY-MOORE, E., Miss   See B.A. 1899

BRAMLEY-MOORE, L., Miss   See B.A. 1899

BRANDNER, F. A.

1933   A test of the significance of the difference of (in) the correlation coefficients in normal bivariate samples. Biometrika 25, 102-109.
7.19/329   7.29/332

BRIGGS, HENRY and GELLIBRAND, H.

*1633   Trigonometria Britannica. Gouda.   xxxxj

BRIGGS, LYMAN J.   See N.B.S. 1944

British Association
Brownlee

BRITISH ASSOCIATION MATHEMATICAL TABLES

The various volumes are cited as B.A. 1, 1931, etc. All are now published for the Royal Society (which has assumed the duties of the British Association's former Committee on Mathematical Tables) by the Cambridge University Press.

1, 1931 Circular & hyperbolic functions;
exponential & sine & cosine integrals;
factorial function & allied functions;
Hermitian probability functions. (Second edition 1946, 3d 1951)
$1.1142/7_F$ $1.1162/10_F$ $1.3213/34_F$ $1.331/37$ $1.331/38$ $1.332/40$
$1.332/43$ $2.5111/160(2)$ $2.5141/163_F$ $2.5142/164(3)_F$ $16.07/622$
$16.0711/622_F$ $16.1041/624_F$ $3.42/669(\text{ftn.})$

7, 1939 The probability integral, by W.F. Sheppard. See Sheppard 1939

9, 1940 Table of powers giving integral powers of integers. Initiated by J.W.L. Glaisher; extended by W.G. Bickley, C.B. Gwyther, J.C.P. Miller, E.J. Ternouth. $5.31/284_F$ $16.023/615_F$

10, 1952 Bessel functions. Part 2, functions of positive integer order.
$5.23/282(\text{Jørgensen})$ $5.23/282(\text{Theil})$

BRITISH ASSOCIATION REPORTS

These are cited by the year of the meeting to which they relate; the year of publication is sometimes different. Reports are cited simply in the form B.A. 1916, and similarly for other years. We give below, for the volumes cited, the place of meeting, the pages of the report of the Tables Committee, and the usual cross-references.

1896   Liverpool   70-82 $10.0431/398(\text{ftn.})$

1899   Dover   65-120* $10.0431/397_F$

1916   Newcastle-on-Tyne   59-126 $2.5133/162_F$

1926   Oxford   273-297 $4.1422/227(4)_F$

1927   Leeds   220-254 $4.1422/227(7)_F$ $4.1422/228_F$

1928   Glasgow   305-340 $1.331/38$

BROADBENT, S. R.

1954   The quotient of a rectangular or triangular and a general variate. Biometrika 41, 330-337. $10.261/721$

1955   Quantum hypotheses. Biometrika 42, 45-57. $6.4/694_M$

1956 a   Lognormal approximation to products and quotients. Biometrika 43, 404-417. $10.2221/710$

---

*The table seems to have been reprinted as a pamphlet. Jahrbuch über die Fortschritte der Mathematik, 31/1900, p. 309, and Morant 1939, p. 6, do not agree about the details.

BROCKMEYER, E., HALSTRØM, H. L. and JENSEN, ARNE

    1948   The life and works of A. K. Erlang.
             Trans. Danish Acad. Tech. Sci. 1948, no. 2.
             (added t.p.: Copenhagen Telephone Co.) $2.129/657_M$

BROOKES, B. C. and DICK, W. F. L.

    1951   Introduction to statistical method. London: Heinemann.
             2.3112/141   4.1311/212

BROOKNER, R. J.   See Wald & Brookner

BROOKS, C. E. P.

    1924   The difference-periodogram--
           a method for the rapid determination of short periodicities.
           Proc. Roy. Soc. London Ser. A 105, 346-359.   11.29/443

BROOKS, S. H.

    1955   The estimation of an optimum subsampling number.
           J. Amer. Statist. Assoc. 50, 398-415.   15.3/611   15.3/613

BROWN, FREDERICK   See Bowley 1928; Edgeworth 1924

BROWN, GEORGE H.   See Ferber 1949

BROWN, GEORGE W.

    1939   On the power of the L_1 test for equality of several variances.
           Ann. Math. Statist. 10, 119-128.   4.241/256

    1951   History of RAND's random digits--summary.
           In N.B.S. 1951 a, pp. 31-32.   13.61/468   13.63/468

BROWN, JAMES ALEXANDER CAMPBELL   See Aitchison & Brown

BROWN, SAMUEL

    1873   On the application of the binomial law to statistical enquiries,
           illustrated by the law of growth of man at different ages.
           J. Inst. Actuar. 17, 340-351.   3.221/187

BROWNLEE, JOHN

    1923   Log $\Gamma(x)$ from $x = 1$ to $50\cdot 9$ by intervals of $\cdot 01$.
           Tracts for Computers, 9. Cambridge Univ. Press, etc. $2.5131/162_F$

BROWNLEE, KENNETH ALEXANDER

    1946   Industrial experimentation. (Second edition 1947, 4th 1949)
           London: H.M.S.O. Reprint of 1947 edition, Brooklyn (New York):
           Chemical Pub. Co.: this is what we have seen.
           2.3113/142   4.1311/212   4.2313/249   7.1221/323

BROWNLEE, K. A., KELLY, B. K. and LORAINE, P. K.

    1948   Fractional replication arrangements for factorial experiments with factors at two levels. Biometrika 35, 268-282.

        15.10/500   15.112/508(6)   15.112/514(3)   15.114/531(5)   15.1199/545(3)
          .112/503   /509(7)   /515   /532   .141/555
            /504(2)   /510(8)   15.114/527(2)   15.118/538(7)
            /505(2)   /511(8)   /528(9)   /539(4)
            /506(6)   /512(5)   /529   /540(7)
            /507(3)   /513(4)   /530(4)   15.1199/544(5)

BRUNO, F. FAA' di   See Faà di Bruno, F.

BRUNS, E. H.

    1906   Wahrscheinlichkeitsrechnung und Kollektivmasslehre. B. G. Teubners Sammlung von Lehrbüchern auf dem Gebiete der mathematischen Wissenschaften mit Einschluss ihrer Anwendungen, 17. Leipzig, etc.: B. G. Teubner. $1.3212/34_F$

BRUNT, D.

    1931   The combination of observations. Second edition. (First edition 1917) Cambridge Univ. Press, etc. $1.1211/11_F$

BUCH, K. R.   See Arley & Buch

BUCKLAND, W. R.   See Kendall, M.G. & Buckland

BÜLBRING, EDITH   See Burn 1937 b

BURGESS, D.   See Charlier 1947

BURGESS, J.

    1898   On the definite integral $\frac{2}{\sqrt{\pi}} \int_{\epsilon}^{t} e^{-2} dt$ $\left[\text{i.e. } \frac{2}{\sqrt{\pi}} \int_{0}^{t} e^{-t^2} dt \right]$, with extended tables of values. Trans. Roy. Soc. Edinburgh 39, 257-321.

        $1.1211/11(4)_F$ $1.1214/12_F$

BURGESS, R. W.

    1927   Introduction to the mathematics of statistics. Boston, etc.: Houghton Mifflin. $1.1121/4_F$ 2.21/137

BURINGTON, R. S. and MAY, D. C.

    1953   Handbook of probability and statistics with tables. Sandusky (Ohio): Handbook Publishers.

        ¶ For contents of this book, see pp. 740-741 above.

    1j(2)

BURMAN, J. P.   See Plackett & Burman

BURN, J. H.

    1937 a   Biological standardization. Oxford Univ. Press.   3.112/183

    1937 b   Biologische Auswertungsmethoden, trans. Edith Bülbring. Berlin: Julius Springer.   3.112/183

BURN, J. H., FINNEY, D. J. and GOODWIN, L. G.

   1950   Biological standardization. Second edition.
      (For first edition, see Burn 1937 a,b) Oxford Univ. Press.
      3.112/183

BURNSIDE, W.

   1928   Theory of probability. Cambridge Univ. Press, etc.   1.1211/11

BURR, I. W.

   1942   Cumulative frequency functions. Ann. Math. Statist. 13, 215-232.
      10.24/416

   1953   Engineering statistics and quality control.
      New York, London, etc.: Mc Graw-Hill Book Co.
      14.111/472(Dept. of Defense)   14.124/480   14.221/488

BURRAU, ØYVIND

  *1943   Middelfejlen som Usikkerhedsmaal. Mat. Tidskrift 1943 B, 9-16.
      $4.132/672_M$

  *1954   On the weight of a physically determined quantity.
      Geodætisk Inst., København, Medd. no. 28.
        ¶ Not abstracted: apparent classification 4.139.
      See Mathematical Reviews 17, 91.

BURTON, R. C.   See Connor & Young 1961; Connor & Zelen 1959; N.B.S. 1957

BURUNOVA, NINA MIHAĬLOVNA   БУРУНОВА НИНА МИХАЙЛОВНА

   1959   Справочник по математическим таблицам. Дполнение № 1.
      Москва: Издат. Акад. Наук СССР.   xxxxij(3)

   1960   A guide to mathematical tables: Supplement No. 1, trans. D. G. Fry.
      Oxford, New York, etc.: Pergamon.   xxxxij

BUSH, R. R. and MOSTELLER, F.

   1955   Stochastic models for learning.
      New York [and London]: John Wiley & Sons.
      16.0244/617   16.1149/626   16.219/627(2)

CADWELL, J. H.

   1951   The bivariate normal integral. Biometrika 38, 59-72.   1.72/120

   1952 y   An approximation to the symmetrical incomplete beta function.
      Biometrika 39, 204-207.
        ¶ Not abstracted; apparent classification 3.43.
      See Mathematical Reviews 13, 961.

   1952 z   The distribution of quantiles of small samples.
      Biometrika 39, 207-211.   1.41/636

1953 y Approximating to the distributions of measures of dispersion by a power of χ^2. Biometrika 40, 336-346.
  ⁋ Not abstracted; apparent classification 1.5531.
  See Mathematical Reviews 15, 452.

1953 z The distribution of quasi-ranges in samples from a normal population. Ann. Math. Statist. 24, 603-613.
  ⁋ Not abstracted; apparent classification 1.544.
  See Mathematical Reviews 15, 452.

1954 y The statistical treatment of mean deviation. Biometrika 41, 12-18.
  $1.52/643_M$

1954 z The probability integral of range for samples from a symmetrical unimodal population. Ann. Math. Statist. 25, 803-806. $1.541/645_M$

CALVIN, L. D.

1954 Doubly balanced incomplete designs for experiments in which the treatment effects are correlated. Biometrics 10, 61-88.
  15.2104/578   15.2104/579   15.2104/584   15.2106/588   15.2106/589
  15.2108/595   15.2412/606

CAMERON, J. M.

1952 Tables for constructing and for computing the operating characteristics of single-sampling plans.
  Indust. Quality Control 9, No. 1, 37-39.   14.112/475

  See Connor & Zelen 1959

CAMP, B. H.

1931 The mathematical part of elementary statistics.
  New York: D.C. Heath.
  1.115/8   1.1161/10   1.133/13   1.3211/33   1.43/63

1948 Generalization to N dimensions of inequalities of the Tchebycheff type. Ann. Math. Statist. 19, 568-574.   10.4/433

CAMPBELL, G. A.

1923 Probability curves showing Poisson's exponential summation.
  Bell System Tech. J. 2, No. 1, 95-113.   2.322/151

CAMPBELL, NORMAN

1930 Fitting observations to a curve. Phil. Mag. 10, 745-758.
  11.12/440(5)

CAPUANO, RUTH   See N.B.S. 1954

CAQUOT, A.   See Mothes 1952

CARLSON, J. L.

1932 A study of the distribution of means estimated from small samples by the method of maximum likelihood for Pearson's Type II curve. Ann. Math. Statist. 3, 86-107.   10.021/393

CARTER, A. H.

    1947   Approximation to percentage points of the z-distribution.
           Biometrika 34, 352-358.   4.35/274

CARVER, H. C.

    1924   Frequency curves. In Rietz, Carver et al. 1924, pp. 92-119.
           10.2/407   10.331/426

    1931   The interdependence of sampling and frequency distribution theory.
           Ann. Math. Statist. 2, 82-98.
           ¶ Signed editorial.
           5.61/684

    1933   Editorial: Note on the computation and modification of moments.
           Ann. Math. Statist. 4, 229-239.   10.333/428

    See Rietz, Carver et al.

CAVALLI, L. L.

    1950   Recent applications of biometrical methods in genetics. (2)
           The analysis of selection curves. Biometrics 6, 208-220.
           12.9/453   8.3129/703

CAVE, B. M.   See Soper, Young, Cave, Lee & Pearson

CAYLEY, A.

    1858   A memoir on the symmetric functions of the roots of an equation.
           Philos. Trans. Roy. Soc. London 147(1857), 489-496 + 5 folding plates.
           Reprinted in Cayley 1889, pp. 417-439; addenda, pp. 602-603.
           10.3123/424   10.3124/425

    1885   Tables of the symmetric functions of the roots, to the degree 10, for the form $1 + bx + \frac{cx^2}{1.2} + \ldots = (1-\alpha x)(1-\beta x)(1-\gamma x)\ldots$.
           Amer. J. Math. 7, 47-56. Reprinted in Cayley 1896 pp. 263-272.
           10.3123/424

    1889   Collected mathematical papers, vol. 2. Cambridge Univ. Press, etc.
           10.3123/424   10.3124/425

    1890   On Latin squares. Messenger of Math. 19, 135-137.
           Reprinted in Cayley 1897, pp. 55-57.
           15.112/502   15.112/503   15.113/516   15.114/526   15.115/533
           15.1512/558

    1896   Collected mathematical papers, vol. 12. Cambridge Univ. Press, etc.
           10.3123/424

    1897   Collected mathematical papers, vol. 13. Cambridge Univ. Press, etc.
           15.112/502   15.112/503   15.113/516   15.114/526   15.115/533
           15.1512/558

Čebyšev
Charlier

ČEBYŠEV, P. L.   ЧЕБЫШЕВЪ П. Л.

1855   О непрерывныхъ дробяхъ.
       Ученыя записки И. Акад. Наукъ, 1-му и 3-му одт.,[*] 3, 636-664.
       Reprinted in Čebyšev 1947, pp. 103-126.   11.110/436(ftn.)

1858   Sur les fractions continues, trans. I.J. Bienaymé.
       J. Math. Pures. Appl. 3, 289-323. Reprinted in Čebyšev 1899, pp. 201-230.
       11.110/436(ftn.)

1859   Sur l'interpolation par la méthode des moindres carrés.
       Académie I. des Sciences, St.-Pétersbourg, Mémoires, 1, No. 15.
       Reprinted in Čebyšev 1899, pp. 471-498;
       Russian translation in Čebyšev 1947, pp. 314-341.   11.110/437

1860   Sur le développement des fonctions à une seule variable.
       Acad. I. des Sciences, St.-Pétersbourg, Bull. 1, 193-200.
       Reprinted in Čebyšev 1899, pp. 501-508;
       Russian translation in Čebyšev 1947, pp. 335-341.   11.110/436(ftn.)

1867 a О среднихъ величинахъ. Мат. Сб. 2, одт. 2, 1-9. Reprinted, with
       modernized spelling and trivial editorial changes, in Čebyšev 1947,
       pp. 431-437.

1867 b Des valeurs moyennes, tr. N. Khanikof.
       J. Math. Pures Appl. 12, 177-184.
       Reprinted, not literatim, in Čebyšev 1899, pp. 685-694.   10.4/433

1873   О функціяхъ, наименѣе уклоняющихся отъ нуля.
       Записки И. Акад. Наук 22, Приложеніе № 1, 1-32.
       Reprinted in Čebyšev 1948, pp. 24-48.

1874 a Sur les quadratures. J. Math. Pures Appl. 19, 19-34.
       Reprinted in Čebyšev 1907, pp. 165-180;
       Russian translation in Čebyšev 1948, pp. 49-62.   11.110/436(ftn.)

1874 b Sur les fonctions qui diffèrent le moins possible de zéro,
       tr. N. Khanikov. J. Math. Pures Appl. 19, 319-346.
       Reprinted in Čebyšev 1907, pp. 189-215.   11.110/436(ftn.)

1899   Oeuvres, ed. A. Markoff et N. Sonin, Tome 1.
       St.-Pétersbourg, etc.: Académie I. des Sciences.
       10.4/433   11.110/437

1907   Oeuvres, ed. A. Markoff et N. Sonin, Tome 2. St. Pétersbourg, etc.

1947   Полное собрание сочинений, Том 2. Москва и Ленинград:
       Издат. Акад. Наук СССР.   10.4/431   11.110/437

1948   Полное собрание сочинений, Том 3. Москва и Ленинград:
       Издат. Акад. Наук СССР.

_____

[*] See Gregory 1943, page 68, column 1, lines 10-22.

CERTAINE, J. E.

 *1958 On sequences of pseudo-random numbers of maximal length.
  J. Assoc. Comput. Mach. 5, 353. 13.62/468

CHAKRABORTY, P. N. See Chandra Sekar, Agarwala & Chakraborty

CHAKRAVARTI, I. M. See Bose, Chakravarti & Knuth; Rao & Chakravarti

CHAMBERS, E. G.

 *1952 Statistical calculation for beginners. Second edition.
  (First edition 1940) Cambridge Univ. Press, etc. $9.1/704_M$

CHANDLER, K. N.

 1952 The duration and frequency of record values.
  J. Roy. Statist. Soc. Ser. B 14, 220-228. 1.43/67

CHANDRA SEKAR, C. See Pearson, E.S. & Chandra Sekar

CHANDRA SEKAR, C., AGARWALA, S. P. and CHAKRABORTY, P. N.

 1955 On the power function of a test of significance for the difference between two proportions. Sankhyā 15, 381-390. 2.613/663

CHANDRASEKHAR, S.

 *1949 On a class of probability distributions.
  Proc. Cambridge Philos. Soc. 45, 219-224. $2.529/662_M$

CHAPMAN, D. G.

 1950 Some two sample tests. Ann. Math. Statist. 21, 601-606.
  4.151/235

 1955 Population estimation based on change of composition caused by a selective removal. Biometrika 42, 279-290. $5.9/688_M$

 1956 Estimating the parameters of a truncated gamma distribution.
  Ann. Math. Statist. 27, 498-506. 10.035/395

CHAPMAN, S. See Mian & Chapman

CHARLIER, C. V. L.

 1905 a Über das Fehlergesetz. Ark. Mat. Astr. Fys. 2, No. 8.
  10.211/408

 1906 Researches into the theory of probability.
  Lunds Univ. Årsskr. N. F. Avd. 2 (Acta Univ. Lund. N.S.) 1, No. 5 = K. Fysiogr. Sällsk. i Lund Handl. N. F. (Acta R. Soc. Physiogr. Lund.) 16, No. 5 = Medd. Lunds Astr. Obs. (2) No. 4.
  $1.111/3_F$ $1.3211/34_F$ 2.221/138

 1908 Weiteres über das Fehlergesetz. Ark. Mat. Astr. Fys. 4, No. 13 = Medd. Lunds Astr. Obs., No. 34. 5.23/282(Theil)

1920 Vorlesungen über die Grundzüge der mathematischen Statistik.
Second edition. Lund: Verlag Scientia. (Reprint 1931.
Lund: C. W. K. Gleerup. First edition-no tables-1910.
Grunddragen af den matematiska statistiken.
Lund: Statsvetenskaplig Tidskrifts Expedition)
1.2111/19 1.3211/34$_F$

1928 A new form of the frequency function.
Lunds Univ. Årsskr. N.F. Avd. 2 (Acta Univ. Lund. N.S.) 24, No. 8
= Kungl. Fysiogr. Sällsk. i Lund Handl. N.F.
(Acta Reg. Soc. Physiogr. Lund) 39, No. 8
= Medd. Lunds Astr. Obs. Ser. 2, 51.   10.213/410

1931 Application de la théorie des probabilités a l'astronomie. Tome 2,
Fasc. 4 of Traité du calcul des probabilités et de ses applications:
Les applications de la théorie des probabilités aux sciences
mathématiques et aux sciences physiques, ed. É. Borel. Paris:
Gauthier-Villars.   10.213/410

1947 Elements of mathematical statistics . . . . [New York: Hafner]
1.115/7 1.1161/9 1.2111/19 1.3211/33 2.221/138
10.211/409(Shenton)

CHAUDHURI, S. B.   See Chowdhury

CHAUDHURY, S. B.   See Chowdhury

CHAUVENET, W.

1874 A manual of spherical and practical astronomy, etc., Vol. II. Theory
and use of astronomical instruments; method of least squares. Fifth
edition. (First edition 1863) Philadelphia: J. B. Lippincott;
London: Trübner.   1.1222/12$_F$

CHEBYSHEV, P. L.   See Čebyšëv

CHERNOFF, H. and LIEBERMAN, G. J.

1957 Sampling inspection by variables with no calculations.
Indust. Quality Control 13, no. 7, 5-7.   14.122/479

CHERON, A.   See Borel & Cheron

CHESIRE, L., SAFFIR, M. and THURSTONE, L. L.

*1933 Computing diagrams for the tetrachoric correlation coefficient.
Chicago: Univ. of Chicago Bookstore.   7.91/698(Dingman)

CHIFOS, P.   See Spenceley et al. 1952

CHOWDHURY, S. B., alias S. B. CHAUDHURI

1954 The most powerful unbiased critical regions and the shortest
unbiased confidence intervals associated with the distribution of
classical D^2-statistic. Sankhyā 14, 71-80.   6.12/690$_M$

1956 Statistical tables and certain (some) recurrence relations connected
with (for) p-statistics. Calcutta Statist. Assoc. Bull. 6, 181-188.
6.12/691$_M$

CHU, J. T.

    1956   Errors in normal approximations to the t, τ , and similar types of distribution. Ann. Math. Statist. 27, 780-789.
         ¶ Not abstracted; apparent classification 4.115.
    See Mathematical Reviews 18, 423.

CHUNG, J. H. and De LURY, D. B.

    1950   Confidence limits for the hypergeometric distribution.
    Univ. of Toronto Press; [London: Oxford Univ. Press]
    5.1/276(intro.)   14.11/470

CHURCH, A. E. R.   See Pearson, K. 1930 b, 1934

CHURCH, B. M.   See Fisher, R.A. & Yates 1953, 1957, TT. XXXIII 1, XXXIII 2.

CLARK, CHARLES E.

    1953   An introduction to statistics.
    New York [and London]: John Wiley & Sons.
    1.1121/4   2.3113/142   4.1311/211   4.2314/249

CLARK, FRANK EUGENE

    1957   Truncation to meet requirements on means.
    J. Amer. Statist. Assoc. 52, 527-536.   1.139/631   1.332/636

CLARK, R. E.

    1953   Percentage points of the incomplete beta function.
    J. Amer. Statist. Assoc. 48, 831-843.   3.319/665

CLATWORTHY, W. H.

    1954   A symmetrical configuration which is a partially balanced incomplete block design. Proc. Amer. Math. Soc. 5, 47-55.   15.2104/583

    1955   Partially balanced incomplete block designs with two associate classes and two treatments per block.
    J. Res. Nat. Bur. Standards 54, 177-190.
    15.2102/566(13)   15.2102/567(12)   15.2102/568(7)   15.2102/569(21)
    15.248/607

    1956   Contributions on partially balanced incomplete block designs with two associate classes.
    National Bureau of Standards, Applied Mathematics Series 47.
    Washington: G.P.O.

```
15.20/563         15.2103/576(7)    15.2106/588       15.2109/599(3)
  .2102/566         .2104/577(4)        /589(3)            /600
  .2103/570(7)      /578(3)         15.2107/592(4)    15.2110/602
       /571(8)      /581                /593(2)            /603
       /572(3)    15.2105/583(3)    15.2108/595            /604
       /573(5)      /584(4)            /596           15.248/647
       /574(2)      /585(4)            /597
       /575(3)    15.2106/587(2)    15.2109/598(3)
```

    See Bose & Clatworthy; Bose, Clatworthy & Shrikhande; N.B.S. 1957

CLEAVER, F. H.   See Sheppard 1939

Clopper
Columbia

CLOPPER, C. J. and PEARSON, E. S.

    1934   The use of confidence or fiducial limits illustrated in the case of the binomial. Biometrika 26, 404-413.
        3.321/193   3.321/194(Eisenhart et al.)   3.321/194(Dixon & Massey)
        3.319/666(Sterne)

COCHRAN, WILLIAM G.

    1937   The efficiencies of the binomial series tests of significance of a mean and of a correlation coefficient. J. Roy. Statist. Soc. 100, 69-73.
        3.33/196

    1941   The distribution of the largest of a set of estimated variances as a fraction of their total. Ann. Eugenics 11, 47-52.   11.24/443

    1943   Some additional lattice square designs.
    Iowa Agric. Expt. Sta., Res. Bull. 318.   15.2108/597

    1952   The χ^2 test of goodness of fit. Ann. Math. Statist. 23, 315-345.
        9.4/380

    See Bliss, Cochran & Tukey

COCHRAN, W. G. and COX, G. M.

    1950   Experimental designs.
    New York: John Wiley & Sons; London: Chapman & Hall.

4.1413/222	15.114/526(3)	15.152/561	15.2106/589(2)
13.2/461	/527(4)	.16/562	/590
15.0/495(3)	/528	.1622/562	/591(4)
.10/496(3 ftn.)	/530(3)	.1623/563	15.2107/592(2)
/497	/532(5)	.20/563	/593(3)
/497(2 ftn.)	/533(4)	/563(ftn.)	/594(2)
/500	15.115/533(2)	/564(2)	15.2108/595(2)
15.112/502(4)	/534(4)	15.2102/566(2)	/597(3)
/503(5)	/535	/567(2)	/598
/504	15.116/535	/568	15.2109/598(2)
/505(3)	/536(4)	15.2103/570(5)	/599(3)
/506	15.117/536	/571(4)	/600(3)
/507(3)	/537	/572(2)	/601(3)
/508(4)	15.118/537(2)	/573(2)	15.2110/602(4)
/509	/538	/575	/603
/510	/539(2)	15.2104/577(5)	/604
/511	/540(3)	/578(2)	15.212/604(2)
15.113/516(2)	/541	/579(2)	/605
/517(2)	15.119/541(3)	/580(2)	15.2411/605
/518(3)	/542(3)	/581(2)	.243/606
/519	/543(3)	/582(2)	.244/606
15.1132/520	15.1199/543(4)	/583(2)	.245/606
/521(5)	.142/556	15.2105/584(4)	.2511/607
/522(3)	.1421/556	/585(3)	.2512/608(2)
/523(6)	.1422/556(2)	/586(3)	.253/608
/524(3)	.1511/558	15.2106/587(2)	
/525(2)	.1513/560	/588(3)	

    1957   Experimental designs. Second Edition.
    New York [and London]: John Wiley & Sons.

13.2/461	15.112/503(10)	15.112/510(3)	15.1132/521(5)
15.0/495(3)	/504(9)	/511(3)	/522(3)
.10/496(3 ftn.)	/505(10)	15.113/516(2)	/523(6)
/497	/506(9)	/517(6)	/524(3)
/497(2 ftn.)	/507(5)	/518(4)	/525(2)
/500	/508(8)	/519	15.114/526(3)
15.112/502(4)	/509(5)	15.1132/520	[more]

[816]

15.114/527(8)	15.141/555	15.2104/579	15.2110/601
/528(8)	.142/556	/580(3)	/602(4)
/530(7)	.1421/556	/581(2)	/603(2)
/532(5)	.1422/556(2)	/582(3)	/604(2)
/533(4)	.148/557(4)	/583(5)	15.212/604(3)
15.115/533(2)	.1511/558	/584(3)	/605
/534(4)	.1513/560	15.2105/585(3)	15.2411/605
/535	.152/561	/586(4)	.2412/606
15.116/535	.161/562	15.2106/587(5)	.243/606
/536(4)	.1622/562	/588(5)	.244/606
15.117/536	.1623/563	/589(2)	.245/606
/537	.20/563	/590(2)	.246/607
15.118/537(2)	/563(ftn.)	/591(4)	.248/607
/538(7)	/564(2)	15.2107/592(6)	.2511/607
/539(2)	15.2102/566(2)	/593(3)	.2512/608(4)
/540(2)	/567(4)	/594(2)	.2513/608
/541	/568(5)	15.2108/594(5)	.253/608
15.119/541(3)	15.2103/570(8)	/595(2)	.256/608
/542(4)	/571(4)	/597(4)	.258/608
/543(3)	/572(4)	/598	xxxvij
15.1199/543(4)	/573(4)	15.2109/598(4)	xxxxiv
/544(3)	/575(3)	/599(4)	
15.138/554	15.2104/577(8)	/600(4)	
/555(5)	/578(2)	/601(3)	

COCHRANE, DONALD   See Orcutt & Cochrane

COHEN, A. C., Jr.

   1955   Maximum likelihood estimation of the dispersion parameter of a chi-distributed radial error from truncated and censored samples with applications to target analysis. J. Amer. Statist. Assoc. 50, 1122-1135. $10.035/716_M$

COHEN, A. C., Jr. and WOODWARD, JOHN

   *1953   Tables of Pearson-Lee-Fisher functions of singly truncated normal distributions. Biometrics 9, 489-497. $1.332/635_M$

COHEN, BURTON H.   See Sakoda & Cohen

COHN, FRIEDA S.   See Swed, F. S.

COLCORD, C. G. and DEMING, LOLA S.

   1936   The one-tenth percent level of 'Z'. Sankhyā 2, 423-424. $4.331/251_F$ 4.233/252(Banerjee)   4.333/270

   See Fisher, R. A. & Yates 1938 etc., T. V

COLE, La MONT C.

   *1945   A simple test of the hypothesis that alternative events are equally probable. Ecology 26, 204.   3.6/669

COLTON, R. R.   See Arkin & Colton

COLUMBIA UNIVERSITY. STATISTICAL RESEARCH GROUP

   1945   Sequential analysis of statistical data: Applications. New York: Columbia Univ. Press.   14.111/473   14.129/481

   See Eisenhart, Hastay & Wallis; Freeman, Friedman, Mosteller & Wallis

COMRIE, LESLIE JOHN

    1938   Tables of $\tan^{-1} x$ and $\log(1+x^2)$ to assist in the calculation of the ordinates of a Pearson Type IV curve. Tracts for Computers, 23. Cambridge Univ. Press.   10.0432/398   10.073/400

    See B. A. 1, 1931; Barlow; Jahnke & Emde 1938 (1945 reprint); Milne-Thomson & Comrie; Pearson, E.S. 1932; Thompson, C.M. 1941a

COMRIE, L. J. and HARTLEY, H. O.

    1941   Table of Lagrangian coefficients for harmonic interpolation in certain tables of percentage points. Biometrika 32, 183-186.
    16.2319/628

COMRIE, PHYLLIS BETTY   See Pearson, E.S. 1931

CONDON, E.

    1927   The rapid fitting of a certain class of empirical formulae by the method of least squares. Univ. California Publ. Math. 2, 55-66. Issued in wrappers with Birge & Shea 1927.

CONDON, E. U.   See N.B.S. 1951z

CONDORCET, M. J. A. N. C., marquis de

    1785   Essai sur l'application de l'analyse à la probabilité des décisions rendues à la pluralité des voix. Paris: Imprimerie Royale.
    5.1/276(Greenwood)

CONNOR, L. R.   See Davies 1954

CONNOR, W. S.   See Bose & Connor; N.B.S. 1957; Youden & Connor

CONNOR, W. S. and YOUNG, SHIRLEY

    1961   Fractional factorial designs for experiments with factors at two and three levels.
    National Bureau of Standards, Applied Mathematics Series 58.
    Washington: G.P.O.
    15.1132/521(6)   15.1132/522(7)   15.1132/523(4)   15.1132/524(10)
    15.1132/525(13)

CONNOR, W. S. and ZELEN, M.

    1959   Fractional factorial experiment designs for factors at three levels.
    National Bureau of Standards, Applied Mathematics Series 54.
    Washington: G.P.O.
    15.10/500   15.113/517(6)   15.113/518(12)   15.113/519(27)
    15.113/520(11)   15.119/541   15.119/542(13)   15.1199/546(13)
    15.141/555

CONRAD, S. H. and KRAUSE, RUTH H.

    1937a   An extension of the Kelley-Wood and Kondo-Elderton tables of abscissae of the unit normal curve, for areas ($\frac{1}{2}\alpha$) between .4500 and .49999 99999. Journal of Experimental Education 5, 278-285.
    $1.2112/20_F$

1937 b Krause & Conrad, A seven-decimal table of the area under the unit normal curve, for abscissae expressed in terms of *P.E.*
Psychômetrika 2, 55-66. $1.1221/12_F$ $1.1222/12_F$

1938 a Krause & Conrad, New and extended tables of the unit normal curve:
I. Areas ($\frac{1}{2}\alpha$) corresponding to abscissae from .00 to 10.0 *PE*;
II. Abscissae (*x*/*PE*) corresponding to areas ($\frac{1}{2}\alpha$) from .000 to .49999 99999 9. Journal of Psychology 5, 397-424.
$1.1221/12_F$ $1.213/21_F$

1938 b Students' tables of the unit normal curve for abscissae expressed in terms of the probable error or *PE*: I. Areas corresponding to abscissae. II. Abscissae corresponding to areas.
Journal of Educational Psychology 29, 491-500. $1.1221/12_F$ $1.213/21_F$

COOK, M. B.

1951 Bi-variate *k*-statistics and cumulants of their joint sampling distribution. Biometrika 38, 179-195.
7.13/324   7.14/328   7.6/342   10.36/431(2)

COOLIDGE, J. L.

1925 An introduction to mathematical probability. Oxford Univ. Press.
13.4/464(intro.)

COOPERATIVE STUDY.   See Soper, Young, Cave, Lee & Pearson 1917

CORBET, A. S.   See Fisher, R.A., Corbet & Williams

CORNELL, R. G.   See Bradley 1954

CORNFIELD, JEROME   See Halperin, Greenhouse, Cornfield & Zalokar

CORNFIELD, J. and MANTEL, N.

1950 Some new aspects of the application of maximum likelihood to the calculation of the dosage response curve.
J. Amer. Statist. Assoc. 45, 181-210.   1.2313/27

CORNISH, E. A. and FISHER, R. A.

1937 Moments and cumulants in the specification of distributions.
Rev. Inst. Internat. Statist. 5, 307-320. (Issued Jan. 1938) Reprinted, with extensive emendations by Fisher (p. 312 reset), as paper 30 in Fisher, R.A. 1950.
2.314/147(Goldberg & Levine)   4.132/216(Goldberg & Levine)
4.151/233(Fisher)   10.221/411

COUSINS, W. R.   See Davies 1954

COVEYOU, R. R.

1960 Serial correlation in the generation of pseudo-random numbers.
J. Assoc. Comput. Mach. 7, 72-74.   13.62/468

COWDEN, D. J.

    1957   Statistical methods in quality control.
Englewood Cliffs (New Jersey): Prentice-Hall.
14.111/471(Dept. of Defense)   14.111/472(Dept. of Defense)
14.221/488(intro.)   14.221/489(4)

    See Croxton & Cowden

COX, DAVID R.

    1948   A note on the asymptotic distribution of the range.
Biometrika 35, 310-315.   1.5532/110

    1949   The use of the range in sequential analysis.
J. Roy. Statist. Soc. Ser. B 11, 101-114.
1.5531/107   1.5531/108(Pearson)   1.552/651

    1954 a   The mean and coefficient of variation of range in small samples from non-normal populations. Biometrika 41, 469-481. $8.27/701_M$

    1958   Planning of experiments. New York [and London]: John Wiley & Sons.
15.20/564   15.2103/571(3)   15.2104/580(2)   15.2106/590
.2102/567(7)   /573   /581(2)   .2107/592
/568(9)   /574   15.2105/584(4)   /593
/569(8)   15.2104/577   /585(3)   15.2108/595
15.2103/570   /578(2)   /588(4)   .2109/598(2)

    See Benson & Cox

COX, D. R. and STUART, A.

    *1955   Some quick sign tests for trend in location and dispersion.
Biometrika 42, 80-95. $9.118/705_M$

COX, GERTRUDE MARY   See Cochran & Cox

COX, G. M. and ECKHARDT, R. C.

    1940   The analysis of lattice and triple lattice experiments in corn varietal tests. I. Construction and numerical analysis.
Iowa Agric. Expt. Sta., Res. Bull. 281, 5-44.   15.2109/601(2)

CRAIG, CECIL C.

    1929   Sampling when the parent population is of Pearson's Type III.
Biometrika 21, 287-293.   10.037/396(Pearson)

    1936   On the frequency function of xy. Ann. Math. Statist. 7, 1-15.
1.78/124

    1936 z   A new exposition and chart for the Pearson system of frequency curves. Ann. Math. Statist. 7, 16-28.
1.78/124(Aroian)   10.004/390   10.24/416(Hatke)
10.24/720(Topp & Leone)

    1941   Note on the distribution of non-central t with an application.
Ann. Math. Statist. 12, 224-228.   4.1412/220

CRAMÉR, HARALD

    1946   Mathematical methods of statistics. Princeton Univ. Press;
Uppsala (Sweden): Almqvist & Wiksell-Hugo Geber.
(The Uppsala edition appeared in 1945)
1.1131/5   1.115/8   1.1161/9   1.2121/20   1.3211/33   2.3113/142
4.1311/211

CRATHORNE, A. R.   See Rietz, Carver et al.

CROW, EDWIN L.

    1956   Confidence limits for a' proportion. Biometrika 43, 423-435.
3.319/666

CROW, ELEANOR G.   See Crow, Edwin L. 1956

CROXTON, F. E. and COWDEN, D. J.

    1931   Applied general statistics. New York: Prentice-Hall.   1.1121/4

   *1946   Tables to facilitate computation of sampling limits of s, and
fiducial limits of sigma. Indust. Quality Control, July 1946, 18-21.
2.315/658

CRUICKSHANK, D. W. J.   See James 1956; Trickett, Welch & James 1956

CRUM, W. L.   See Rietz, Carver et al.

CULLIS, C. E.   See B.A. 1896

CURETON, EDWARD E.   See Gordon, Loveland & Cureton

CURTISS, J. H.

    1946   A note on some single sampling plans requiring the inspection of a
small number of items. Ann. Math. Statist. 17, 62-70.   3.119/184

CZECHOWSKI, T., FISZ, M., IWIŃSKI, T., LANGE, O., SADOWSKI, W. i ZASEPA, R.

    1957   Tablice statystyczne. Warszawa: Panstwowe Wydawnictwo Naukowe.
¶ For contents of this book, see pp. 741-742 above.
1j(2)

CZUBER, E.

    1914   Wahrscheinlichkeitsrechnung und ihre Anwendung auf
Fehlerausgleichung, Statistik und Lebensversicherung.
Third edition, Vol. 1. (First edition 1903)   1.3212/34

DALE, H. H.   See Durham, Gaddum & Marchal 1929

DALE, J. B.

   *1903   Five-figure tables of mathematical functions. London: Arnold.
[fifty]

DALY, J. F.

    1946   On the use of the sample range in an analogue of Student's t-test.
Ann. Math. Statist. 17, 71-74.   1.5512/99

Daniels
David

DANIELS, H. E.

    1941   A property of the distribution of extremes. Biometrika 32, 194-195.
        1.42/59

    1947   Grouping corrections for high autocorrelations.
        J. Roy. Statist. Soc. Suppl. 9, 245-249.
          ¶ Not abstracted; apparent classification 10.331.
        See Mathematical Reviews 9, 294.

    1952   The covering circle of a sample from a circular normal distribution.
        Biometrika 39, 137-143.   5.5/291

    1954   A distribution-free test for regression parameters.
        Ann. Math. Statist. 25, 499-513. $9.118/705_M$

DANNEMILLER, CARROLL   See Connor & Young 1961

DARLING, D. A.

    1957   The Kolmogorov-Smirnov, Cramér-von Mises tests.
        Ann. Math. Statist. 28, 823-838.   9.4/380   xxxxiij   xxxxiv

    See Anderson, T.W. & Darling

DARMOIS, G.

    1928   Statistique mathématique. Paris: Gaston (Octave) Doin.
        $2.221/138_F$

DAVENPORT, C. B. and EKAS, M. P.

    1936   Statistical methods in biology, medicine and psychology.
        Fourth edition. (First edition, by Davenport, 1899, 2d 1904, 3d 1914.
        The second edition was entitled:
        Statistical methods with special reference to biological variation)
        New York: John Wiley & Sons. $1.1121/4_F$ $1.1162/10_F$ $7.32/334_F$

DAVID, FLORENCE N.

    1934   On the P_{λ_n} test for randomness: Remarks, further illustration, and table of P_{λ_n} (for given values of $-\log_{10} \lambda_n$). Biometrika 26, 1-11.
        2.119/134

    1938   Tables of the ordinates and probability integral of the correlation coefficient in small samples. Cambridge Univ. Press, etc.
        7.1/320   7.11/321(Soper et al.)   7.11/322   7.121/322   10.29/420(ftn.)

    1947   A χ^2 'smooth' test for goodness of fit. Biometrika 34, 299-310.
        2.69/180

    1949   Note on the application of Fisher's k-statistics.
        Biometrika 36, 383-393.   10.137/405 $10.343/723_M$

    1949 a   Probability theory for statistical methods.
        Cambridge Univ. Press, etc.   13.4/464(intro)

1950   Two combinatorial tests of whether a sample has come from a given population. Biometrika 37, 97-110.
5.9/293(Fisher & Yates)   5.9/293   9.9/387

See Barton & David; Pearson, K., Stouffer & David

DAVID, F. N. and JOHNSON, N. L.

1951a   The effect of non-normality on the power function of the F-test in the analysis of variance. Biometrika 38, 43-57.
4.28/267   4.28/267(David & Johnson 1951b)

1951b   The sensitivity of analysis of variance tests with respect to random between groups variation. Trabajos Estadíst. 2, 179-188.   4.28/267

*1954   Statistical treatment of censored data. Part I. Fundamental formulae. Biometrika 41, 228-240. $10.261/721_M$

*1956   Some tests of significance with ordered variables.
J. Roy. Statist. Soc. Ser. B 18, 1-31.   1.41/636

DAVID, F. N. and KENDALL, M. G.

1949   Tables of symmetric functions. I. Biometrika 36, 431-449.
10.31/421   10.311/422   10.32/425   10.34/429

1951   Tables of symmetric functions. II and III. Biometrika 38, 435-462.
10.31/421(ftn.)   10.311/422   10.312/424   10.3121/424(ftn.)
10.3123/424(2 ftn.)   10.311/722

1953   Tables of symmetric functions. IV. Biometrika 40, 427-446.
10.311/422

1955   Tables of symmetric functions, V. Biometrika 42, 223-242.
10.311/422   10.311/722

DAVID, HERBERT ARON

1951   Further applications of range to the analysis of variance. Biometrika 38, 393-409.   1.552/105   1.541/646(Moore)

1952   Upper 5 and 1% points of the maximum F-ratio. Biometrika 39, 422-424.   4.253/261(Hartley)   4.253/261

1954   The distribution of the range in certain non-normal populations. Biometrika 41, 463-468. $8.27/701_M$

*1955   A note on moving ranges. Biometrika 42, 512-515. $1.531/643_M$

*1956   On the application to statistics of an elementary theorem in probability. Biometrika 43, 85-91. $1.42/637_M$

1956a   The ranking of variances in normal populations.
J. Amer. Statist. Assoc. 51, 621-626.
     ⁋ Not abstracted; apparent classification 4.254.
See Mathematical Reviews 18, 521.

David, H. A.
Deming

 *1956 b Revised upper percentage points of the extreme studentized deviate from the sample mean. Biometrika 43, 449-451. 1.44/641

 See Hartley & David

DAVID, H. A., HARTLEY, H. O. and PEARSON, E. S.

 1954 The distribution of the ratio, in a single normal sample, of range to standard deviation. Biometrika 41, 482-493.
 1.5511/649 1.69/652$_M$

DAVID, HERBERT THEODORE

 1958 A three-sample Kolmogorov-Smirnov test.
 Ann. Math. Statist. 29, 842-851. 9.414/715

DAVID, STEPHEN T., KENDALL, M. G. and STUART, A.

 1951 Some questions of distribution in the theory of rank correlation. Biometrika 38, 131-140. 8.1111/349 8.1112/349

DAVIDOFF, M. D. and GOHEEN, H. W.

 1953 A table for the rapid determination of the tetrachoric correlation coefficient. Psychometrika 18, 115-121.
  ¶ Not abstracted; apparent classifications 1.78, 7.94.

DAVIES, OWEN L.

 1934 On asymptotic formulae for the hypergeometric series: II. Hypergeometric series in which the fourth element, x, is not necessarily unity. Biometrika 26, 59-107. 10.18/406

 1949 Statistical methods in research and production: with special reference to the chemical industry. Second edition. (First edition 1947) London, etc.: Oliver and Boyd; [New York: Hafner]
 1.1131/6 1.1141/7 2.3112/141 3.322/195(Fisher & Yates 1943)
 4.1311/212 4.2312/248 7.1221/323 14.222/490

 1954 The design and analysis of industrial experiments.
 London, etc.: Oliver & Boyd; [New York: Hafner]

15.0/495(3)	15.112/505(3)	15.114/530(2)	15.1513/560
.10/496(3 ftn.)	/506	.118/538(3)	.152/561
/497	/509	.119/541(2)	.20/563
/497(2 ftn.)	/510	.1199/544(2)	/563(ftn.)
/500	15.113/516(3)	.141/555	15.2411/605
15.112/502(5)	/517(3)	.142/556	.243/606
/503(2)	15.114/527(2)	.1422/556	.248/607
/504(2)	/528	.1511/558	.253/608(ftn.)

DAVIES, O. L. and PEARSON, E. S.

 1934 Methods of estimating from samples the population standard deviation. J. Roy. Statist. Soc., Supplement 1, 76-93.
 1.52/73 1.541/85 2.522/169

DAVIS, GEORGE See Davis, H.T. 1933

DAVIS, HAROLD T.

    1933   Tables of the higher mathematical function, Vol. 1.
            Bloomington (Indiana): Principia Press.
            $2.5111/160(2)_F$ $2.5121/161_F$ $2.5131/162(4)_F$ $2.5141/163(5)_F$
            $2.5141/164_F$ xxxxj   xxxxvij

    1933 a   Polynomial approximation by the method of least squares.
            Ann. Math. Statist. 4, 155-195.   11.110/437   11.12/440(3)

    1935   Tables of the higher mathematical functions, vol. 2.
            Bloomington (Indiana) [i.e. Evanston (Illinois)]: Principia Press.
            $2.5142/164_F$ 11.110/437   11.12/440(7)   xxxxj

            See N.B.S. 1943; see also p. 951

DAVIS, H. T. and NELSON, W. F. C.

    1935   Elements of statistics with applications to economic data.
            Bloomington (Indiana): Principia Press. $1.1121/4_F$ $2.1132/132_F$

DAVISON, W. R.   See Storer & Davison

DAWSON, H. G.

    1898   On the numerical value of $\int_0^h e^{x^2} dx$.
            Proc. London Math. Soc. 29, 519-522.   1.811/26

DE, S. P.   See Mahalanobis, Bose, Ray & Banerji 1934

DECKER, F. F.

    1910   The symmetric function tables of the fifteenthic: including an historical summary of symmetric functions as relating to symmetric function tables.
            Carnegie Institution of Washington, Publication no. 120.
            Washington: the Inst.   10.3123/424

De FINETTI, B.

    1934   Sulla legge di probabilità degli estremi.
            Metron 9, No. 3-4, 127-138.   1.42/58

    1954   Concetti sul "comportamento induttivo" illustrati su di un esempio.
            Statistica (Bologna) 14, 350-378.
            ¶ Not abstracted; apparent classification 6.21.
            See Mathematical Reviews 16, 840.

De LURY, D. B.

    1950   Values and integrals of the orthogonal polynomials up to n = 26.
            Univ. of Toronto Press, etc.   11.111/437   11.119/439

            See Chung & De Lury

DEMING, LOLA S.

            See Colcord & Deming; Connor & Young 1961; Connor & Zelen 1959; Grab & Savage 1954; N.B.S. 1952, 1957

DEMING, WILLIAM EDWARDS

    1950   Some theory of sampling. New York [and London]: John Wiley & Sons.
           3.321/193(Clopper & Pearson)   4.331/251   13.11/456   13.13/459

DE MORGAN, AUGUSTUS

    1838   An essay on probabilities, and on their application to life contingencies and insurance offices.
           London: Longman, Brown, Green & Longmans. $1.1211/11_F$

    1845   Theory of probabilities. In Encyclopaedia Metropolitana 2, 393-490.
           (Original edition 1837: 2, 359-468) $1.1211/11_F$ $1.1216/12_F$

DENNIS, K. E.   See Barton & Dennis

[U.S.] DEPARTMENT of DEFENSE

    1957   MIL-STD-414, 11 June 1957, Superseding: ORD-M 608-10, June 1954; NAVORD OSTD 80, 8 May 1952. Military Standard: Sampling procedures and tables for inspection by variables for percent defective. Washington: G.P.O.
           14.121/478   14.121/479(Dept. of Defense 1958)
           14.121/730(Lieberman & Resnikoff)

    1958   Mathematical and statistical principles underlying Military Standard 414: Technical report. Sampling procedures and tables for inspection by variables for percent defective, 15 October 1958. Washington: [G.P.O.]
           14.121/478(Dept. of Defense 1957)   14.121/479   3.12/665

    1959 a   MIL-STD-105 B, 31 December 1958, Superseding: MIL-STD-105 A, 11 September 1950; Appendix, 27 April 1955. Military Standard: Sampling procedures and tables for inspection by attributes. Washington: G.P.O.   14.111/471   14.13/482(Dept. of Defense 1959 b)

    1959 b   Inspection and quality control handbook (Interim) H 107. Single-level continuous sampling procedures and tables for inspection by attributes, 30 April 1959. Washington: G.P.O.   14.13/482

DES RAJ   See Raj, Des

di BRUNO, F. FAA'   See Faà di Bruno, F.

DICK, W. F. L.   See Brookes, B.C. & Dick

DICKERSON, WILLIE S., Mrs   See Bradley 1954

DICKSON, CHARLOTTE   See Fisher, A. 1922

DINGMAN, H. F.

    *1954   A computing chart for the point biserial correlation coefficient. Psychometrika 19, 257-259. $7.91/698_M$

DIXON, WILFRID J.

    1940   A criterion for testing the hypothesis that two samples are from the same population (Problem of two samples).
           Ann. Math. Statist. 11, 199-204.   9.129/370   9.32/377

1951   Ratios involving extreme values. Ann. Math. Statist. 22, 68-78.
1.44/639$_M$

1954   Power under normality of several nonparametric tests.
Ann. Math. Statist. 25, 610-614.   9.413/384

1957   Estimates of the mean and standard deviation of a binomial population. Ann. Math. Statist. 28, 806-809.
1.51/642   1.541/646   1.542/647   1.542/649

DIXON, W. J. and MASSEY, F. J.

1951   Introduction to statistical analysis. (Second edition 1957) New York, London, etc.: McGraw-Hill Book Co.
¶ For contents of the 1951 edition, see pp. 743-744 above; for the 1957 edition, pp. 744-746.
xxxvij   xxxxiv   1j

DIXON, W. J. and MOOD, A. M.

1946   The statistical sign test. J. Amer. Statist. Assoc. 41, 557-566.
3.6/206

DODD, E. L.

1942   Certain tests for randomness applied to data grouped into small sets. Econometrica 10, 249-257.   5.9/293   13.63/468

DODGE, H. F.

1943   A sampling inspection plan for continuous production.
Ann. Math. Statist. 14, 264-279.
14.13/481   14.13/482(Dept. of Defense)

DODGE, H. F. and ROMIG, H. G.

1941   Single sampling and double sampling inspection tables.
Bell System Tech. J. 20, 1-61.   14.111/471(Dodge & Romig 1959)

1944   Sampling inspection tables: Single and double sampling.
New York: John Wiley & Sons; London: Chapman & Hall.
14.111/471(Dodge & Romig 1959)

1959   Sampling inspection tables: single and double sampling.
Second edition. New York [and London]: John Wiley & Sons.
14.11/470   14.111/470   14.111/730

DODGE, H. F. and TORREY, M. N.

1951   Additional continuous inspection plans.
Indust. Quality Control 7, no. 5, 7-12.   14.13/482(Dept. of Defense)

DONATO, V.   See Miller, L. H. 1956

DONOVAN, J. F. P.   See James 1956; Trickett, Welch & James 1956

DORFMAN, R.

1943   The detection of defective members of large populations.
Ann. Math. Statist. 14, 436-440.   15.3/611

DOUGLAS, J. B.

    1955   Fitting the Neyman Type A (two parameter) contagious distribution. Biometrics 11, 149-173.   2.53/173   5.21/683

DOWNTON, F.

    1954   Least-squares estimates using ordered observations. Ann. Math. Statist. 25, 303-316.   10.095/717   10.097/717

DRESSEL, P. L.

    1940   Statistical seminvariants and their estimates with particular emphasis on their relation to algebraic invariants. Ann. Math. Statist. 11, 33-57.   10.39/432

DREYER, K.   See Hald 1948

DRION, E. F.

    1952   Some distribution-free tests for the difference between two empirical cumulative distribution functions. Ann. Math. Statist. 23, 563-574.   9.4122/383

DUARTE, F. J.

    1927   Nouvelles tables de log n! à 33 décimales depuis n = 1 jusqu'à n = 3000. Genève: Kundig, etc. $2.5134/163_F$

    1933   Nouvelles tables logarithmiques à 36 décimales. Paris: Gauthier-Villars. $2.5134/163_F$

DUBROFF, RUTH   See Fox 1956

DUDDING, B. P. and JENNETT, W. J.

    1944   Quality control chart technique when manufacturing to a specification. London: General Electric Co. 1.1141/7   2.321/151   14.222/490

DUFFELL, J. H.

    1909   Tables of the Γ-function. Biometrika 7, 43-47. $2.5131/162_F$

DUNCAN, ACHESON J.

    1952   Quality control and industrial statistics. (Second edition 1959) Homewood (Illinois): Richard D. Irwin.
    1.115/8   1.1161/10   1.541/87   1.613/113   1.623/114   1.633/117
    2.3111/140   4.1311/211   4.2311/248   14.111/471(Dept. of Defense)
    14.111/472(Dept. of Defense)   14.221/488

  *1955   The use of ranges in comparing variabilities. Indust. Quality Control 11, no. 5, 18-22. $1.5531/652_M$

    See Smith, J. G. & Duncan

DUNCAN, DAVID B.

  *1955 a   Multiple range and multiple F tests. Biometrics 11, 1-42. $1.5511/649_M$

DUNLAP, J. W. and KURTZ, A. K.

   1932   Handbook of statistical nomographs, tables, and formulas.
Yonkers-on-Hudson [i.e. Yonkers] (New York): World.   xxxxix

DUNNETT, CHARLES W.

   1955   A multiple comparison procedure for comparing several treatments with a control. J. Amer. Statist. Assoc. 50, 1096-1121.   4.17/674

DUNNETT, C. W. and SOBEL, M.

   1954   A bivariate generalization of Student's t-distribution, with tables for certain special cases. Biometrika 41, 153-169.
4.17/241   4.17/241(Dunnett & Sobel 1955)

   1955   Approximation to the probability integral and certain percentage points of a multivariate analogue of Student's t-distribution. Biometrika 42, 258-260.   4.17/241

du PASQUIER, L. G.   See Euler 1923

DURAND, DAVID   See Greenwood & Durand; Gumbel, Greenwood & Durand

DURAND, D. and GREENWOOD, J. A.

   1957   Random unit vectors II: Usefulness of Gram-Charlier and related series in approximating distributions. Ann. Math. Statist. 28, 978-986.
2.41/154

DURBIN, J.   See Watson & Durbin.

DURBIN, J. and WATSON, G. S.

   1950   Testing for serial correlation in least squares regression. I. Biometrika 37, 409-428.   1.531/78(Durbin & Watson 1951)

   1951   Testing for serial correlation in least squares regression. II. Biometrika 38, 159-178.   1.531/78   7.52/338

DURFEE, W. P.

   1882a   Tables of the symmetric functions of the twelfthic.
Amer. J. Math. 5, 45-61.   10.3123/424   10.3124/425

   1882b   The tabulation of symmetric functions.
Amer. J. Math. 5, 348-349 + folding plate.
10.3123/424   10.3124/425

   1887   Symmetric functions of the 14^{ic}. Amer. J. Math. 9, [278-296].
10.3123/424

DURHAM, FLORENCE M., GADDUM, J. H. and MARCHAL, J. E.

   1929   Reports on biological standards II.
Toxicity tests for novarsenobenzene (neosalvarsan).
Medical Research Council (Gt. Brit.) Special Report Series, 128.
London: H.M.S.O.   3.112/183

DWIGHT, H. B.

    1941   Mathematical tables of elementary and some higher mathematical functions including trigonometric functions of decimals of degrees and logarithms. New York and London: Mc Graw-Hill Book Co.
            $1.1211/11_F$ $3.211/186_F$

    1947   Tables of integrals and other mathematical data. (First edition 1934) New York: Macmillan. $1.111/3$

    See N.B.S. 1953 z

DWYER, P. S.

    1937   Moments of any rational integral isobaric sample moment function(s). Ann. Math. Statist. 8, 21-65. $10.3121/424$ $10.349/430$

    1938 a   Combined expansions of products of symmetric power sums and of sums of symmetric power products with application to sampling. I. Ann. Math. Statist. 9, 1-47. $10.349/430$

    1938 b   On combined expansions of products of symmetric power sums and of sums of symmetric power products with applications to sampling. II. Ann. Math. Statist. 9, 97-132. $10.349/430$

DYŠKO, A. L.   See Karpov 1958

DYSON, R. M.   See Bartlett, M. S. 1950

EANDI, A. M. V.-   See Vassalli-Eandi, A. M.

EAST, D. A.   See Daniels 1952

ECKHARDT, R. C.   See Cox & Eckhardt

EDGEWORTH, F. Y.

    1898   On the representation of statistics by mathematical formulae. Part I. J. Roy. Statist. Soc. 61, 670-700.

    1899 a   On the representation of statistics by mathematical formulae. Part II. J. Roy. Statist. Soc. 62, 125-140.

    1899 b   On the representation of statistics by mathematical formulae. Part III. J. Roy. Statist. Soc. 62, 373-385.

    1899 c   On the representation of statistics by mathematical formulae. Part IV. J. Roy. Statist. Soc. 62, 534-555.
            $10.221/411$(Edgeworth 1924)

    1900   On the representation of statistics by mathematical formulae. Supplement. J. Roy. Statist. Soc. 63, 72-81.

    1905   The law of error. Trans. Cambridge Philos. Soc. 20, 36-65.
            $10.211/408$

    1905 a   The law of error. Part II. Trans. Cambridge Philos. Soc. 20, 113-141.

1913   A method of representing statistics by analytical geometry. In Proceedings of the fifth international congress of mathematicians (Cambridge, 22-28 August 1912), ed. E.W. Hobson & A.E.H. Love, vol. 2, pp. 427-440. Cambridge Univ. Press, etc.   10.221/411(Edgeworth 1924)

1914 a   On the use of analytical geometry to represent certain kinds of statistics. J. Roy. Statist. Soc. 72, 300-312.
10.221/411(Edgeworth 1924)

1914 b   On the use of analytical geometry to represent certain kinds of statistics. Subsection 2. Moderately abnormal curves.
J. Roy. Statist. Soc. 72, 415-432.

1914 c   On the use of analytical geometry to represent certain kinds of statistics. Subsection III. Very abnormal curves.
J. Roy. Statist. Soc. 72, 653-671.

1914 d   On the use of analytical geometry to represent certain kinds of statistics. Section II. The method of percentiles.
J. Roy. Statist. Soc. 72, 724-749.

1914 e   On the use of analytical geometry to represent certain kinds of statistics. Section III. Frequency-Surfaces.
J. Roy. Statist. Soc. 72, 838-852.

1924   Untried methods of representing frequency.
J. Roy. Statist. Soc. 87, 571-594.   10.221/411

See Bowley 1928

EDGEWORTH, F. Y. and BOWLEY, A. L.

1902   Methods of representing statistics of wages and other groups not fulfilling the normal law of error. J. Roy. Statist. Soc. 65, 325-354.
1.325/36

EDMONDSON, MARGARET   See Box 1954, 1954 a

EGAN, U. N.   See Pollak & Egan

EISENHART, C.

1944   RMT 129 [Review of Anderson & Houseman 1942]
Math. Tables Aids Comput. 1, 148-150.   11.110/437

1949   Probability center lines for standard deviation and range charts.
Indust. Quality Control 6, no. 1, 24-26.   2.315/149   14.229/490

See Swed & Eisenhart

EISENHART, C., HASTAY, M. W. and WALLIS, W. A.

1947   Selected techniques of statistical analysis for scientific and industrial research and production and management engineering.
New York and London: McGraw-Hill Book Co.
1.1141/6   1.19/18   1.2111/19   3.321/194   4.241/258   14.122/479
14.32/492   14.32/493(Dixon & Massey)   4.239/677   11.24/726   1ij

EKAS, M. P.   See Davenport & Ekas

ELDERTON, ETHEL M.

    See Kondo & Elderton; Pearson, K. 1925 a, 1930 b; Pearson, K. & Elderton; Pearson, K., Jeffery & Elderton

ELDERTON, WILLIAM PALIN

    1902   Tables for testing the goodness of fit of theory to observation (Tables for testing curve fitting) Biometrika 1, 155-163.
        2.1132/132$_F$ 2.119/133   2.5135/163$_F$

    1903 a Graduation and analysis of a sickness table. Biometrika 2, 260-272.
        10.033/395

    1906   Frequency curves and correlation. First edition. (Second edition 1927) London: Charles and Edwin Layton.
        10.00/388(ftn.)

    1938   Frequency curves and correlation. Third edition. Cambridge Univ. Press, etc. (Fourth edition 1954, [New York: Dover?])

10.00/389	10.04/397	10.09/402	10.211/408
.004/390	.05/398	.10/402	/409(Barton & Dennis)
.01/392	.06/399(2)	.11/403	10.212/409
.013/392	.07/399	.12/403	.23/415(Wolfenden)
.02/393	.08/401	.13/404	.332/427(Pairman & Pearson)
.03/394			

ELFVING, G.

    1947   The asymptotical distribution of range in samples from a normal population. Biometrika 34, 111-119.
        1.5532/108   1.5532/109(ftn.)   1.5532/110(Cox 1948)

EL SHANAWANY, M. R.

    1936   An illustration of the accuracy of the χ^2 approximation. Biometrika 28, 179-187.   5.4/287(Neyman & Pearson)   5.4/287

    See Pearson, E.S. 1932

EMDE, F.   See Jahnke & Emde

ENO FOUNDATION for HIGHWAY TRAFFIC CONTROL

    See Greenshields, Schapiro & Ericksen 1947; Greenshields & Weida 1952

ENRICK, N. R.

    1953 a Quality control through statistical methods [I.] Modern Textiles Mag. 34, no. 10, 32-33, 49-51.

    1953 b Quality control through statistical methods II. Measurement of reliability and its uses. Modern Textiles Mag. 34, no. 11, 43-46.

    1953 c Quality control through statistical methods III. What is wrong with "maximum variation"? Modern Textiles Mag. 34, no. 12, 79-82.

    1954 a Quality control through statistical methods IV. Process analysis. Modern Textiles Mag. 35, no. 1, 49, 52-54.

1954 b   Quality control through statistical methods V. Control charts.
    Modern Textiles Mag. 35, no. 2, 35, 72, 74-75.

1954 c   Quality control through statistical methods VI.
    Additional control chart applications.
    Modern Textiles Mag. 35, no. 3, 44, 74-77.

1954 d   Quality control through statistical methods VII.
    Sampling economically and effectively.
    Modern Textiles Mag. 35, no. 4, 44, 46, 50, 52, 56.

1954 e   Quality control through statistical methods VIII.
    Ready-made sampling plans. Modern Textiles Mag. 35, no. 5, 42-43, 62-63.
    14.112/476

1957   Tables showing how much quality control testing to do.
    Quality Control and Applied Statistics   2, 347-350 (code 221:Ai)
    14.229/491

1957 a   Qualitätsüberwachung durch statistische Methoden. [I-a]
    Reyon Zellwolle 7, 107-110.

1957 b   Qualitätsüberwachung durch statistische Methoden. [I-b]
    Reyon Zellwolle 7, 183-186.

1958 a   Qualitätsüberwachung durch statistische Methoden. II.
    Reyon Zellwolle 8, 50-54, 56.

1958 b   Qualitätsüberwachung durch statistische Methoden. III.
    Reyon Zellwolle 8, 284-286, 288.

1958 c   Qualitätsüberwachung durch statistische Methoden. IV.
    Reyon Zellwolle 8, 371-376.   14.229/491

ENTWISLE, P. C. V., Mrs   See David, F.N. 1938

EPPERSON, E. R.   See Spenceley, Spenceley & Epperson

EPSTEIN, B.

1954 a   Tables for the distribution of the number of exceedances.
    Ann. Math. Statist. 25, 762-768.   9.129/707

1954 c   Truncated life tests in the exponential case.
    Ann. Math. Statist. 25, 555-564.
    14.151/483(Epstein & Sobel 1953) $14.151/732_M$

1956 b   Simple estimators of the parameters of exponential distributions
    when samples are censored. Ann. Inst. Statist. Math. 8, 15-26.
    14.159/486

EPSTEIN, B. and SOBEL, M.

1953   Life testing. J. Amer. Statist. Assoc. 48, 486-502.
    14.141/483   14.159/486(Epstein)

1955   Sequential life tests in the exponential case.
    Ann. Math. Statist. 26, 82-93.   14.151/483   14.159/486(Sobel)

ERDÉLYI, A., MAGNUS, W., OBERHETTINGER, F. and TRICOMI, F. G.

    1953   Higher transcendental functions, Vol. 1.
          New York, London, etc.: Mc Graw-Hill Book Co.   5.1/275(2)

ERICKSEN, E. L.   See Greenshields, Schapiro & Ericksen

ERLANG, A. K.

    *1917   Løsning af nogle Problemer fra Sandsynlighedsregningen af Betydning for de automatiske Telefoncentraler.
          Elektroteknikeren (Copenhagen) 13, 5-13.   2.129/656(ftn.)

    *1918   Lösung einiger Probleme der Wahrscheinlichkeitsrechnung von Bedeutung für die selbstättigen Fernsprechämter.
          Elektrotechn. Z. 39, 504-508.   2.129/656(ftn.)

    See Brockmeyer, Halstrøm & Jensen 1948

EUDY, M.   See Lehmer, E. 1944

EULER, L.

    1782   Recherches sur une nouvelle espèce de quarrés magiques. Verhandelingen uitgegeven door het zeeuwsch Genootschap der Wetenschappen te Vlissingen (Middelburg)* 9, 85-239.
          What we have seen is the reprint in Euler 1923, pp. 291-392.
          15.1512/558

    1923   Commentationes algebraicae ad theoriam combinationum et probabilitatum pertinentes, ed. L. G. du Pasquier.
          (L. Euleri Opera Omnia, Ser. 1, Vol. 7)
          Lipsiae et Berolini [i.e. Leipzig and Berlin]: B. G. Teubner.
          15.112/503   15.112/505   15.112/507   15.112/511   15.113/516
          15.113/518   15.113/519   15.114/526(2)   15.115/533(2)   15.117/536
          15.118/537(2)   15.119/541(2)   15.1512/558

EVERITT, P. F.

    1910   Tables of the tetrachoric functions for fourfold correlation tables.
          Biometrika 7, 437-451. $1.2111/19_F$ $1.221/21_F$ $1.3241/35_F$

    1912   Supplementary tables for finding the correlation coefficient from tetrachoric groupings. Biometrika 8, 385-395. $1.721/120_F$

EWELL, R. B.   See Sherman & Ewell

FAA' di BRUNO, F.

    1876   Théorie des formes binaires.
          Turin: Brero; Paris: Gauthier-Villars; etc., etc.   10.3123/424

    1881   Einleitung in die Theorie der binären Formen, tr. Th. Walter.
          Leipzig: B.G. Teubner.   10.3123/424

---

*See Gregory 1943, p. 3012, column 1, lines 13-18.

FADDEEVA, V. N. and TERENT'EV, N. M.   ФАДДЕЕВА В. Н. и ТЕРЕНТЬЕВ Н. М.

1954   Таблицы значений $w(z) = e^{-z^2}\left(1 + \frac{2i}{\sqrt{\pi}}\int_0^z e^{t^2} dt\right)$ (интеграла вероятностей) от комплексного аргумента. Москва: Гос. Издат. Техн.-Теорет. Лит.
1.83/127

¶ According to <u>Mathematical Reviews</u> 18, 155, Morris D. Friedman published an English translation of [the text of] the above work in 1956.

FAN, C. T.

*1954   Note on the construction of an item analysis table for the high-low-27-per-cent method. Psychometrika 19, 231-237.
1.78/653$_M$ 7.99/699

FAWCETT, C. D., Miss   See B.A. 1899

FËDOROVA, RIMMA MAKSIMOVNA   See Lebedev & Fëdorova

FELDMAN, H. M.

1932   The distributions of the precision constant and its square in samples of n from a normal population. Ann. Math. Statist. 3, 20-31.
2.529/172

FELLER, W.

1950   An introduction to probability theory and its applications, Vol. I. First edition. (Second edition 1957)
New York [and London]: John Wiley & Sons.   1.115/8   1.1161/9

FERBER, ROBERT

1949   Statistical techniques in market research.
New York, London, etc.: Mc Graw-Hill Book Co.
1.541/86   2.3113/142   4.2314/249   8.112/350(Snedecor)   12.11/445
1.1121/630   1.531/643   4.1311/671   4.239/677(Snedecor)   7.211/695
7.51/696

FERRER, S.   See Royo Lopez & Ferrer Martin

FERRER MARTIN, SEBASTIAN   See Royo Lopez & Ferrer Martin

FERRIS, C. D., GRUBBS, F. E. and WEAVER, C. L.

1946   Operating characteristics for the common statistical tests of significance. Ann. Math. Statist. 17, 178-197.   1.19/14   2.43/158

FERTIG, J. W. and LEARY, MARGARET V.

1936   On a method of testing the hypothesis that an observed sample of variables and of size has been drawn from a specified population of the same number of variables.
(Tests of significance for multivariate samples).
Ann. Math. Statist. 7, 113-121.   6.132/308

FERTIG, J. W. and PROEHL, ELIZABETH A.

    1937   A test of a sample variance based on both tail ends of the distribution. Ann. Math. Statist. 8, 193-205.   2.119/134

FESTINGER, L.

    1946   The significance of difference between means without reference to the frequency distribution function. Psychometrika 11, 97-105.
    9.121/368

FIEDLER, W.   See Salmon 1877

FIELLER, E. C.

    1931   The duration of play. Biometrika 22, 377-404.   5.9/292

    1932   The distribution of the index in a normal bivariate population. Biometrika 24, 428-440.   4.27/263(Paulson)

    See Pearson, K. 1930 b, 1934

FIELLER, E. C. and HARTLEY, H. O.

    1954   Sampling with control variables. Biometrika 41, 494-501.
    $3.53/668_M$

FIELLER, E. C., LEWIS, T. and PEARSON, E. S.

    1955   Correlated random normal deviates: 3000 sets of deviates, each giving 9 random pairs with correlations 0·1(0·1)0·9, compiled from Herman Wold's table of random normal deviates (Tract No. 25). Tracts for computers, no. 26. Cambridge Univ. Press, etc.
    13.3/463   13.3/729(Vianelli)

FILON, L. N. G.   See Pearson, K. & Filon

FINANCIAL PUBLISHING CO., Boston   See Gushee 1942

FINETTI, B. de   See De Finetti, B.

FINNEY, D. J.

    1941   The joint distribution of variance ratios based on a common error mean square. Ann. Eugenics 11, 136-140.   4.251/261

    1941 a   On the distribution of a variate whose logarithm is normally distributed. J. Roy. Statist. Soc. Suppl. 7, 155-161.
    10.2221/413(Aitchison & Brown)

    1946   Orthogonal partitions of the 6 × 6 Latin squares. Ann. Eugenics 13, 184-196.   15.1514/561

    1947   Probit analysis. First edition. Cambridge Univ. Press, etc.
    1.2311/24   1.2312/25   1.232/27(2)   1.233/28

    1948   The Fisher-Yates test of significance in 2 × 2 contingency tables. Biometrika 35, 145-156.
    2.611/175   2.611/175(Latscha)   2.611/176(Armsen)   5.1/275
    2.611/663(Siegel)

1949   The truncated binomial distribution. Ann. Eugenics 14, 319-328.
6.22/314

1949 a The estimation of the parameters of tolerance distributions.
Biometrika 36, 239-256.   1.233/28

1952 a Probit analysis. Second edition. Cambridge Univ. Press, etc.
1.23/22   1.2311/24   1.2312/25   1.232/27   1.233/29   2.3113/142
4.1311/211

1952 b Statistical methods in biological assay.
London: Charles Griffin; New York: Hafner.
1.2311/24   1.2312/25   1.232/27   2.3114/143   4.1311/212   4.151/236
4.2314/250   12.11/445   12.14/446   12.15/447   12.4/449   12.51/450
12.53/450   12.54/451   12.61/452   12.62/452(intro.)   12.62/452

*1956   The consequences of selection for a variate subject to errors of
measurement. Rev. Inst. Internat. Statist. 24, 1-10.
¶ Not abstracted; apparent classification 1.9.
See Mathematical Reviews 19, 781.

See Burn, Finney & Goodwin; Fisher, R.A. & Yates 1948, 1953, T. XI 1,
1957, T. IX 3; Latscha 1953

FINNEY, D. J. and STEVENS, W. L.

1948   A table for the calculation of working probits and weights in probit
analysis. Biometrika 35, 191-201.   1.2312/24

FISHER, ARNE

1922   The mathematical theory of probabilities and its application to
frequency curves and statistical methods, Vol. 1. Second edition.
(First edition—no tables—1915) New York, etc.: Macmillan.
$1.115/8_F$ $1.3211/34_F$

1922 a An elementary treatise on frequency curves and their application in
the analysis of death curves and life tables, trans. E. A. Vigfusson.
New York: Macmillan.   2.41/154

FISHER, RONALD AYLMER

1915   Frequency distribution of the values of the correlation
coefficient(s) in (of) samples from an indefinitely large population.
Biometrika 10, 507-521.
7.1/319   7.11/320   7.13/324   7.131/325   7.131/325(ftn.)   7.134/327

1921   On the "probable error" of a coefficient of correlation deduced from
a small sample. Metron 1, No. 4, 1-32.
7.2/330 $7.212/331_F$ 7.23/331

1922   On the mathematical foundations of theoretical statistics.
Philos. Trans. Roy. Soc. London A 222, 309-368.
Reprinted, with author's emendations, as paper 10 in Fisher, R.A. 1950.
6.32/316(Fisher & Yates)   10.024/394   10.035/395(intro.)
10.23/415(intro.)(2)   10.331/426

Fisher, R. A.
Fitch

1925   Statistical methods for research workers. (Second edition 1928, 3d 1930, 4th 1932, 5th 1934, 6th 1936, 7th 1938, 8th 1941, 9th 1944, 10th 1946, 11th 1950, 12th 1954)
Edinburth, etc.: Oliver and Boyd; New York: Hafner.
4.1311/211$_F$ 4.35/273

¶ We have not, in general, indexed the tables in this book that are reproduced (usually with additions) in Fisher & Yates 1938.

1925 y   Applications of "Student's" distribution.
Metron 5, no. 3, 90-104.   4.132/672(ftn.)

1925 z   Expansion of "Student's" integral in powers of n^{-1}.
Metron 5, No. 3, 109-112.   4.115/210(Gosset)

1928   The general sampling distribution of the multiple correlation coefficient. Proc. Roy. Soc. London Ser. A 121, 654-673.
Reprinted as paper 14 in Fisher, R.A. 1950.   2.42/155   7.4/335

1929 a   Tests of significance in harmonic analysis.
Proc. Roy. Soc. London Ser. A 125, 54-59.
Reprinted, with author's emendations, as paper 16 in Fisher, R.A. 1950.
11.24/443

1929 b   Moments and product moments of sampling distributions.
Proc. London Math. Soc. 30, 199-238. Reprinted, with author's emendations affecting, inter al., the tables on pp. 210, 211, 212, 213, as paper 20 in Fisher, R.A. 1950.
1.52/72   2.69/180(Hoel)   10.34/429   10.34/430   10.341/430
10.343/430

1930   The moments of the distribution for normal samples of measures of departures from normality. Proc. Roy. Soc. London Ser. A 130, 16-28.
Reprinted as paper 21 in Fisher, R.A. 1950.   1.611/112(intro.)

1931   B.A. 1, 1931, pp. xxvj-xxxv. Reprinted under the title 'The sampling error of estimated deviates, together with other illustrations of the properties and applications of the integrals and derivatives of the normal error function' as paper 23 in Fisher, R.A. 1950.
1.332/40

1938   The statistical utilization of multiple measurements.
Ann. Eugenics 8, 376-386. Reprinted, with author's emendations amounting to a rewriting of pp. 383-386, as paper 33 in Fisher, R.A. 1950.
2.42/156(Fisher 1928)

1940 a   On the similarity of the distributions found for the test of significance in harmonic analysis, and in Stevens's problem in geometrical probability. Ann. Eugenics 10, 14-17.
Reprinted as paper 37 in Fisher, R.A. 1950.   11.24/443

1941   The asymptotic approach to Behrens' integral, with further tables for the d test of significance. Ann. Eugenics 11, 141-172.
4.151/232   4.151/234(Sukhatme)   4.151/235(Sukhatme et al.)

[See also p. 951]   Fisher, R. A.
                                 Fitch

   1941 a   The negative binomial distribution. Ann. Eugenics 11, 182-187.
            Reprinted as paper 38 in Fisher, R.A. 1950.   5.32/285(Sichel)

   1950   Contributions to mathematical statistics.
           New York [and London]: John Wiley & Sons.
           5.33/286(Fisher et al.)   10.221/411(Cornish & Fisher)   10.331/426(ftn.)
           10.343/430   11.24/443

   1950 a   The significance of deviations from expectation in a Poisson series.
           Biometrics 6, 17-24.   6.31/315

   1950 b   Gene frequencies in a cline determined by selection and diffusion.
           Biometrics 6, 353-361.   12.7/453

 *1950 c   A class of enumerations of importance in genetics.
           Proc. Roy. Soc. London, Ser. B 136, 509-520.   10.36/723(Bennett)

           See Cornish & Fisher; Gosset 1925

FISHER, R. A., CORBET, A. S. and WILLIAMS, C. B.

   1943   The relation between the number of species and the number of
           individuals in a random sample of an animal population.
           J. Animal Ecology 12, 42-58.
           Pages 54-57 are reprinted as paper 43 in Fisher, R.A. 1950.
           5.33/285

FISHER, R. A. and HEALY, M. J. R.

   1956   New tables of Behrens' test of significance.
           J. Roy. Statist. Soc. Ser. B 18, 212-216.
           4.151/236   4.151/236(Fisher & Yates)

FISHER, R. A. and TIPPETT, L. H. C.

   1928   Limiting forms of the frequency distribution of the largest or
           smallest member of a sample. Proc. Cambridge Philos. Soc. 24, 180-190.
           Reprinted as paper 15 in Fisher, R.A. 1950.   8.3/364

FISHER, R. A. and YATES, F.

   1934   The 6 x 6 Latin squares. Proc. Cambridge Philos. Soc. 30, 492-507.
           15.116/535   15.1512/558

   1938   Statistical tables for biological, agricultural and medical research.
           First edition. (Second edition 1943, 3d 1948, 4th 1953, 5th 1957)
           London, etc.: Oliver and Boyd; New York: Hafner.
           ¶ For contents of the 1938-1953 editions, see pp. 746-749 above; for
           the 1957 edition, pp. 749-752.
           xxxxiv   1j

FISZ, M.   See Czechowski et al.

FITCH, E. R.   See Hartley & Fitch

Fix
Freeman

FIX, EVELYN

 1949 Tables of noncentral χ^2. Univ. California Publ. Statist. 1, 15-19.
  2.42/156

 See N.B.S. 1959

FIX, E. and HODGES, J. L., jr.

 1955 Significance probabilities of the Wilcoxon test.
  Ann. Math. Statist. 26, 301-312. 16.036/619 9.121/706

FLETCHER, A.

 1948 Guide to tables of elliptic functions.
  Math. Tables Aids Comput. 3, 229-281.
  9.411/380 9.4112/381 xxxvij

FLETCHER, A., MILLER, J. C. P. and ROSENHEAD, L.

 1946 An Index of mathematical tables. First edition.
  (A second edition is promised for 1961 or 1962)
  London: Scientific Computing Service; New York: McGraw-Hill Book Co.
  ¶ For a table of parallel passages in this <u>Guide</u> and Fletcher et al. 1946, see pp. 786-789 above.

1.0/1	5.9/293(Fisher & Yates)	11.22/442
.121/10	7.21/330	.23/443
.122/12	.3/334	16.00/614(2)
.13/13	10.013/392	.044/619
.331/38(2)	.0431/398(ftn.)	.068/621(intro.)
.812/126(ftn.)	.101/402	.11/625
.813/126(ftn.)	.102/402	xxxviij
.82/127	.103/403	xxxxj
2.113/132	.123/403	xxxxij
.51/160(3)	.331/427	xxxxvij
.515/164	11.110/436(ftn.)	xxxxix
3.21/186	.2/442	[fifty]
5.1/275	.21/442	1iij(2)

FLORIDA. UNIVERSITY, GAINESVILLE. STATISTICAL LABORATORY See Mayer 1956

FOK, V. A. See Faddeeva & Terent'ev 1954

FORSYTH, A. R. See B.A. 1896, 1899

FORSYTHE, GEORGE ELMER

 1951 Generation and testing of random digits at the National Bureau of Standards, Los Angeles. <u>In</u> N.B.S. 1951 a, pp. 34-35.
  13.62/468 13.63/468

 See von Neumann & Forsythe

FORSYTHE, WILLIAM ELMER

 1954 Smithsonian physical tables. Ninth edition.
  Smithsonian Miscellaneous Collections, 120.
  Washington: Smithsonian Institution. (for 8th ed. see Fowle 1933)

FORTHAL, LILLIAN See N.B.S. 1959

FOSTER, BILLY L. See Kruskal & Wallis 1952

FOSTER, F. G.

    1957   Upper percentage points of the generalized beta distribution. II. Biometrika 44, 441-453.   6.132/692   6.132/692(Foster 1958)

    1958   Upper percentage points of the generalized beta distribution. III. Biometrika 45, 492-503.   6.132/692

FOSTER, F. G.; and REES, D. H.

    1957   Upper percentage points of the generalized beta distribution. I. Biometrika 44, 237-247.   6.132/692

FOSTER, F. G. and STUART, A.

    *1954   Distribution-free tests in time-series based on the breaking of records. J. Roy. Statist. Soc. Ser. B 16, 1-22. $9.311/710_M$

FOSTER, R. M.   See Peirce 1929

FOWLE, F. E.

    1933   Smithsonian physical tables. Eighth edition.
(First edition 1896, 5th 1910, 6th 1914, 7th 1919; for 9th see Forsythe, W.E. 1954)
Smithsonian Miscellaneous Collections, 88.
Washington: Smithsonian Institution. $1.1222/12_F$

FOX, M.

    1956   Charts of the power of the F-test. Ann. Math. Statist. 27, 484-497. $4.242/680_M$

FRANCIS, M. G., Miss   See David, F.N. 1938

FRANCIS, V. J.

    *1946   On the distribution of the sum of n sample values drawn from a truncated normal population. J. Roy. Statist. Soc. Suppl. 8, 223-232.
¶ Not abstracted; apparent classification 1.332.
See Mathematical Reviews 9, 47.

FRANKEL, L. R.   See Hotelling & Frankel

FRANKLAND, F. W.   See Fisher, A. 1922

FRASER, D. A. S. and GUTTMAN, I.

    1956   Tolerance regions. Ann. Math. Statist. 27, 162-179.   14.32/493

FREEMAN, H. A.

    1942   Industrial statistics: Statistical technique applied to problems in industrial research and quality control.
New York [and London]: John Wiley & Sons.
1.1121/4   1.541/85   1.613/113(Pearson)   1.623/114(Geary & Pearson)
1.633/116   2.3113/142   4.1311/211   4.2314/249   6.112/298(Mahalanobis)

    See Columbia Univ. 1945

FREEMAN, H. A., FRIEDMAN, M., MOSTELLER, F. and WALLIS, W. A.

    1948   Sampling inspection: Principles, procedures and tables for single, double and sequential sampling in acceptance inspection and quality control based on percent defective.
New York and London: McGraw-Hill Book Co.
14.111/471(Dept. of Defense)   14.111/472   14.121/477   14.111/730 1ij

FREEMAN, LEONA   See N.B.S. 1954

FREUDENBERG, K.

    *1951   Die Grenzen für die Anwendbarkeit des Gesetzes der kleinen Zahlen. Metron 16, 285-310. $3.42/667_M$

FREUDENTHAL, A. M. and GUMBEL, E. J.

    1953   On the statistical interpretation of fatigue tests.
Proc. Roy. Soc. London Ser. A 216, 309-332.   14.159/485

    1954   Minimum life in fatigue. J. Amer. Statist. Assoc. 49, 575-597.
14.159/485

FRIEDMAN, MILTON

    1937   The use of ranks to avoid the assumption of normality implicit in the analysis of variance. J. Amer. Statist. Assoc. 32, 675-701.
8.1313/354   8.132/354(Friedman 1940)   9.14/371

    1940   A comparison of alternative tests of significance for the problem of m rankings. Ann. Math. Statist. 11, 86-92.   8.10/348   8.132/354

    See Freeman, Friedman, Mosteller & Wallis

FRIEDMAN, MORRIS D.   See Faddeeva & Terent'ev 1954

FRY, D. G.   See Burunova 1960; Lebedev & Fëdorova 1960

FRY, M., Miss   See B.A. 1899

FRY, THORNTON C.

    1928   Probability and its engineering uses.
New York [i.e. Princeton (New Jersey), London, etc.]: D. Van Nostrand.
$1.1131/5_F$ $1.115/8_F$ $1.1161/9_F$ $1.2122/20_F$ $1.3211/33_F$ $2.122/135_F$
$2.221/138(2)_F$ $2.3113/142_F$ $2.5112/161_F$ $3.211/186_F$ 2.129/656(ftn.)

FÜRST, DARIO

    1955   Nota alla tabella per il calcolo dei coefficienti di regressione quadratica. Statistica (Bologna) 15, 492-495. See also Quartey 1955.

FYoDOROVA, RIMMA MAXIMOVNA   See Lebedev & Fedorova

GADD, J. O., Jr.   See Harvard Univ. 1949, 1955

GADDUM, J. H.   See Durham, Gaddum & Marchal

GALLOWAY, T.

 1839 A treatise on probability. In Encyclopaedia Britannica.
  Seventh edition. Edinburgh: A. & C. Black (Reissued in book form.
  What we have seen is Galloway's article in the eighth edition, 1859,
  vol. 18, pp. 588-635. The table appears on p. 636) 1.1211/11

GALTON, F.

 1902 The most suitable proportion between the values of first and second
  prizes. Biometrika 1, 385-390.

GALVANI, L.

 1931 Contributi alla determinazione degli indici di variabilità per
  alcuni tipi di distribuzione. Metron 9, No. 1, 3-45. 10.29/420

GARABEDIAN, GLADYS R. See Freudenthal & Gumbel 1954

GARDNER, A.

 1952 Greenwood's "Problem of Intervals": an exact solution for $N = 3$.
  J. Roy. Statist. Soc. Ser. B 14, 135-139. 5.5/291

GARDNER, R. S. See Crow, Edwin L. 1956

GARRETT, H. E.

 1937 Statistics in psychology and education. Second edition.
  (First edition 1926) New York, London, etc.: Longmans, Green.
  $1.1121/4_F$ $1.1162/10_F$ $1.1221/12_F$ $1.221/21_F$ $7.32/334_F$

GARWOOD, F.

 1936 Fiducial limits for the Poisson distribution.
  Biometrika 28, 437-442.
  2.322/152 2.322/153(Pearson & Hartley) 2.322/659(Pearson & Hartley)

 1940 An application of the theory of probability to the operation of
  vehicular-controlled traffic signals.
  J. Roy. Statist. Soc. Suppl. 7, 65-77. 5.21/278

 1941 The application of maximum likelihood to (and) dosage-mortality
  curves. Biometrika 32, 46-58. 1.2313/27(Cornfield & Mantel)

 1947 The variance of the overlap of geometrical figures, with reference
  to a bombing problem. Biometrika 34, 1-17. 5.5/289

GAUSS, C. F., alias K. F. GAUSS

 1813 Disqvitiones generales circa seriem infinitam
$$1 + \frac{\alpha\beta}{1.\gamma}x + \frac{\alpha(\alpha+1)\beta(\beta+1)}{1.2.\gamma(\gamma+1)}xx + \frac{\alpha(\alpha+1)(\alpha+2)\beta(\beta+1)(\beta+2)}{1.2.3.\gamma(\gamma+1)(\gamma+2)}x^3 + \text{etc.}$$
  Commentationes Societatis R. Scientiarvm Gottingensis Recentiores, 2,
  classis mathematicae.
   Reprinted in Gauss 1866, pp. 125-162. The typography of the reprint
  is better than that of the original, and far better than that of Davis
  1933. $2.5131/162_F$ $2.5141/163_F$

1866   Carl Friedrich Gauss Werke, Vol. 3, ed. [E.C.J.] Schering. Göttingen: K. Gesellschaft der Wissenschaften.

GAYEN, A. K.

1949   The distribution of 'Student's' t in random samples of any size drawn from non-normal universes. Biometrika 36, 353-369.   4.143/230

*1950 y   The distribution of the variance ratio in random samples of any size drawn from non-normal universes. Biometrika 37, 236-255.   4.28/681$_M$

*1950 z   Significance of difference between the means of two non-normal samples. Biometrika 37, 399-408.   4.143/231

1951   The frequency distribution of the product-moment correlation coefficient in samples of any size drawn from non-normal universes. Biometrika 38, 219-247.   7.14/328   7.23/331

GEARY, R. C.

1930   The frequency distribution of the quotient of two normal variates. J. Roy. Statist. Soc. 93, 442-446.   4.27/263(Paulson)

1935   The ratio of the mean deviation to the standard deviation as a (new) test of normality. Biometrika 27, 310-332. See also E.S. Pearson 1935 b; Geary 1935 a.   1.621/114   1.69/117

1935 a   Note on the correlation between β_2 and w'. Biometrika 27, 353-355. See also Geary 1935; E.S. Pearson 1935 b.

1936   The distribution of "Student's" ratio for non-normal samples. J. Roy. Statist. Soc. Suppl. 3, 178-184.   4.143/230

1936 a   Moments of the ratio of the mean deviation to the standard deviation for normal samples. Biometrika 28, 295-305. 1.52/72   1.631/115   1.633/116

1947 b   Testing for normality. Biometrika 34, 209-242. 1.6/111   1.631/116   4.143/230   4.28/266

GEARY, R. C. and PEARSON, E. S.

1938   Tests of normality. New Statistical Tables, 1. London: Biometrika Office, University College. Contains tables from E.S. Pearson 1930 a(248), Williams 1935(271), Geary 1935(330), Geary 1936(303, 304, 305), and E.S. Pearson 1936(306, 307); with new matter. 1.611/112(Pearson)   1.613/113(Pearson)   1.623/114   1.623/114(Duncan) 1.631/115(Geary)   1.633/116(Geary)

GEBELEIN, H.   See Schuler & Gebelein

GEIRINGER, HILDA

1954   On the statistical investigation of transcendental numbers. <u>In</u>

Studies in mathematics and mechanics presented to Richard von Mises, etc., pp. 310-322. New York: Academic Press.   13.63/468

GELLIBRAND, H.   See Briggs & Gellibrand

GEPHART, LANDIS S.   See Meyer, Lytle & Gephart

GERMOND, H. H.   See N.B.S. 1959

GEROULD, WINIFRED   See Gregory, Winifred

GEVORKIANTZ, S. R.

   1928   Determination of forest seed quality. Journal of Forestry 26, 1043-1046.   3.54/205

GHOSH, H.   See Mahalanobis, Bose, Ray & Banerji 1934

GILBERT, EDGAR J.

   *1956   The matching problem. Psychometrika 21, 253-266. $5.62/687_M$

GILBERT, J. P.   See Kruskal & Wallis 1952

GINI, C.

   1938   Di una formula comprensiva delle medie. Metron 13, No. 2, 3-22.   10.37/432(2)

   1949   Le medie di campioni. Metron 15, 13-28.   10.37/432

GINTZLER, LEONE   See Lehmer, E. 1944

GIRSHICK, M. A.   See Blackwell & Girshick; Columbia Univ. 1945

GIRSHICK, M. A., RUBIN, H. and SITGREAVES, R.

   1955   Estimates of bounded relative error in particle counting. Ann. Math. Statist. 26, 276-285. $2.322/659_M$

GJEDDEBÆK, N. F.

   1949   Contribution to the study of grouped observations. Application of the method of maximum likelihood in case of normally distributed observations. Skand. Aktuarietidskr. 32, 135-159.   1.332/45

GLAISHER, J. W. L.

   1871   On a class of definite integrals—Part II. Phil. Mag. 42, 421-436. $1.1214/12_F$

   1883   Tables of the exponential function. Trans. Cambridge Philos. Soc. 13, 243-272. $2.5122/162_F$

   1917   Table of binomial-theorem coefficients. Messenger Math. 47, 97-107. $3.211/186_F$

   See B.A. 1896, 1899; B.A. 9, 1940

GLANZ, M.   See Miller, L.H. 1956

GLOVER, J. W.

    1923   Tables of applied mathematics in finance, insurance, statistics.
              (Second edition 1930)
              Ann Arbor (Michigan): George Wahr [i.e. Wahr's bookstore]
                  ¶ For partial contents of this book, see pp. 752-753 above.
            1j(2)

    See Rietz, Carver et al.

GNEDENKO, B. V.   ГНЕДЕНКО Б. В.

    1950   Курс теорий вероятностей, и др.
              Москва и Ленинград: Гос. Издат. Техн.-Теорет. Лит.
              2.1131/656   4.1111/670   9.4111/714

    1954   Проверка неизменности распределения вероятностей в двух неазависимых выборках. Kriterien für die Unveränderlichkeit der Wahrscheinlichkeitsverteilung von zwei unabhängigen Stichprobenreihen. [So translated by editors of Math.Nachr.; a perhaps better translation is Nachprüfung der Unveränderlichkeit der Warscheinlichkeitsverteilung bei zwei unabhängigen Stichproben.] Math. Nachr. 12, 29-66.   9.4152/385

GNEDENKO, B. V. and KOROLYuK, V. S.   ГНЕДЕНКО Б. В. и КОРОЛЮК В. С.

    1951   О максимальном расхождении двух эмпирических распределений.
              Докл. Акад. Наук СССР 80, 525-528.   9.4152/385(Goodman)

GNEDENKO, B. V. and STUDNEV, Yu. P.   ГНЕДЕНКО Б. В. і СТУДНЕВ Ю. П.
                                              ГНЕДЕНКО Б. В. и СТУДНЕВ Ю. П.

    1952   Порівняння ефективності деяких методів перевіки однорідності статистичного матеріалу. Сровение эффективности некоторых методов проверки однородности статистического материала.
              Доповіді Акад. Наук Україн. РСР 1952, 359-363.
              9.4152/385(Gnedenko 1954)

GODWIN, H. J.

    1948   A further note on the mean deviation. Biometrika 35, 304-309.
              1.52/72

    1949 a   On the estimation of dispersion by linear systematic statistics.
              Biometrika 36, 92-100.   1.42/60   1.43/66   1.544/93

    1949 b   Some low moments of order statistics.
              Ann. Math. Statist. 20, 279-285.   1.43/66(Hastings et al.)   1.43/67

    1955   On generalizations of Tchebychef's inequality.
              J. Amer. Statist. Assoc. 50, 923-945.   10.4/434(ftn.)

    See Pearson, E.S., Godwin & Hartley

GOEDICKE, V. A.

    1953   Introduction to the theory of statistics. New York: Harper & Bros.
              1.1121/4   1.1161/9   2.3113/142

GOHEEN, H. W.   See Davidoff & Goheen

GOLDBERG, H. and LEVINE, HARRIET

    1946   Approximate formulas for the percentage points and normalization of t and χ^2. Ann. Math. Statist. 17, 216-225.   2.314/147   4.132/216

GOLUB, A.

    1953   Designing single sampling inspection plans when the sample size is fixed. J. Amer. Statist. Assoc. 48, 278-288.
    14.112/476   14.112/730

GOOD, I. J.

    1953   The serial test for sampling numbers and other tests of randomness. Proc. Cambridge Philos. Soc. 49, 276-284.   13.63/468

GOODMAN, L. A.

    1954   Kolmogorov-Smirnov tests for psychological research. Psych. Bull. 51, 160-168.   9.4152/385

GOODWIN, E. T. and STATON, J.

    1948   Table of $\int_0^\infty \frac{e^{-u^2}}{u+x} du$ (a certain integral)

    Quart. J. Mech. Appl. Math. 1, 319-326.   1.9/129

GOODWIN, L. G.   See Burn, Finney & Goodwin

GORDON, A. R.   See Miller, W.L. & Gordon

GORDON, M. H., LOVELAND, E. H., and CURETON, E. E.

    1952   An extended table of chi-square for two degrees of freedom, for use in combining probabilities from independent samples. Psychometrika 17, 311-316.   2.3115/143

GOSSET, W. S., called STUDENT

    1908a   The probable error of a mean. Biometrika 6, 1-28. Reprinted in Gosset 1942, pp. 11-34.   4.112/209   4.132/672(ftn.)

    1917   Tables for estimating the probability that the mean of a unique sample of observations lies between $-\infty$ and any given distance of the mean of the population from which the sample is drawn. Biometrika 11, 414-417. Reprinted in Gosset 1942, pp. 61-64.
    4.112/209

    1925   New tables for testing the significance of observations. Metron 5, No. 3, 105-108 and 113/120. See also Fisher, R.A. 1925 z.
    $1.115/7_F$ 4.1111/208   4.115/210

    1942   "Student's" collected papers. London: Biometrika Office, University College.
    4.1111/208   4.1114/209   4.112/209(2)   4.115/210(Gosset 1925)
    4.132/672(ftn.)

Goulden
Gregory

GOULDEN, C. H.

    1939   Methods of statistical analysis.
       New York: John Wiley & Sons; London: Chapman & Hall.
       $2.3113/142_F$ $4.1311/212_F$ 4.2314/249   11.113/438

    1952   Methods of statistical analysis. Second edition.
       New York [and London]: John Wiley & Sons.   2.3113/142

GRAB, EDWIN L., and SAVAGE, I. RICHARD

    1954   Tables of the expected value of 1/X for positive Bernoulli and Poisson variables. J. Amer. Statist. Assoc. 49, 169-177.
       2.53/173   3.53/668

GRAD, ARTHUR and SOLOMON, HERBERT

    1955   Distribution of quadratic forms and some applications.
       Ann. Math. Statist. 26, 464-477. $2.41/660_M$

GRAF, U. und HENNING, H. J.

    1953   Formeln und Tabellen der mathematischen Statistik.
       Berlin, etc.: Springer.
          ¶ For contents of this book, see pp. 753-754 above.   lj

GRAM, J. P.

    1883   Ueber die Entwickelung reeller Functionen in Reihen mittelst (Reihenentwickelungen nach) der Methode der kleinsten Quadrate.
       J. Reine Angew. Math. 94, 41-73.   10.211/408   11.110/436

GRANT, ALISON M.

    1952   Some properties of runs in smoothed random series.
       Biometrika 39, 198-204.   9.311/375

GRANT, EUGENE L.

    1946   Statistical quality control. (Second edition 1952)
       New York and London: Mc Graw-Hill Book Co.
       1.115/8   1.541/86   2.121/135   14.111/472(Dept. of Defense)
       14.121/478(Bowker & Goode)   14.122/480   14.21/487   14.221/489(3)
       14.221/490

GREB, D. J. and BERRETTONI, J. N.

    1949   AOQL single sampling plans from a single chart and table.
       J. Amer. Statist. Assoc. 44, 62-76.   14.112/475

GREEN, B. L., Jr., SMITH, J. E. KEITH and KLEM, LAURA

    1959   Empirical tests of an additive random number generator.
       J. Assoc. Comput. Mach. 6, 527-537.   13.63/468

GREENBERG, B. G.   See Sarhan & Greenberg

GREENHOUSE, SAMUEL W.   See Halperin, Greenhouse, Cornfield & Zalokar

GREENLEAF, H. E. H.

    1932   Curve approximation by means of functions analogous to the Hermite polynomials. Ann. Math. Statist. 3, 204-255.   11.19/441

GREENSHIELDS, B. D., SCHAPIRO, D. and ERICKSEN, E. L.

    1947   Traffic performance at urban street intersections. Yale Bureau of Highway Traffic, Technical Report No. 1. New Haven (Connecticut): the Bureau.   2.122/135

GREENSHIELDS, B. D. and WEIDA, F. M.

    1952   Statistics with applications to highway traffic analyses. Saugatuck (Connecticut): Eno Foundation for Highway Traffic Control. 2.122/135

GREENWOOD, JOSEPH ALBERT

    1938   Variance of a general matching problem. Ann. Math. Statist. 9, 56-59.   5.62/686

GREENWOOD, JOSEPH ARTHUR

    See Charlier 1947; Durand & Greenwood; Gumbel & Greenwood; Gumbel, Greenwood & Durand; N.B.S. 1953

GREENWOOD, JOSEPH ARTHUR and DURAND, DAVID

    1955   The distribution of length and components of the sum of n random unit vectors. Ann. Math. Statist. 26, 233-246. 10.28/419   10.28/722

    1960   Aids for fitting the gamma distribution by maximum likelihood. Technometrics 2, 55-65.   10.035/395(intro.)   10.035/395   10.055/398

GREENWOOD, MAJOR (Junior)

    1913   On errors of random sampling in certain cases not suitable for the application of a "normal" curve of frequency. Biometrika 9, 69-90. 5.1/276

    1946   The statistical study of infectious diseases. J. Roy. Statist. Soc. 109, 85-109.   5.5/291(Gardner)

    See Pearson, K. 1930 a, T. 49

GREENWOOD, ROBERT E.

    1953   Probabilities of certain solitaire card games. J. Amer. Statist. Assoc. 48, 88-93.   5.61/684

    1955   Coupon collector's test for random digits. Math. Tables Aids Comput. 9, 1-5.   13.63/468

GREGORY, WINIFRED

    1932   List of the serial publications of foreign governments. New York: H.W. Wilson.

    1943   Union list of serials in the libraries of United States and Canada. Second edition. New York: H.W. Wilson.

GREINER, R.   See Ranke & Greiner

GREVILLE, T. N. E.

    1938   Exact probabilities for the matching hypothesis.
J. Parapsychol. 2, 55-59.   5.62/686

GRIMSEY, A. H. R.

    1945   On the accumulation of chance effects and the Gaussian frequency distribution. Phil. Mag. 36, 294-295.   10.261/720

GROBBEN, GERDA KLERK-   See Klerk-Grobben, Gerda

GRONOW, D. C. G.

    1951   Test for the significance of the difference between means in two normal populations having unequal variances. Biometrika 38, 252-256.   4.152/238

GROSSMAN, A.   See N.B.S. 1954

GRUBBS, F. E.

    1944   On the distribution of the radial standard deviation.
Ann. Math. Statist. 15, 75-81.   2.313/144

    1949   On designing single sampling inspection plans.
Ann. Math. Statist. 20, 242-256.   2.322/153   3.311/191   14.112/477

    1950   Sample criteria for testing outlying observations.
Ann. Math. Statist. 21, 27-58.   1.42/61

    See Army Ordnance Corps 1952; Ferris, Grubbs & Weaver; Morse & Grubbs

GRUBBS, F. E. and WEAVER, C. L.

    *1947   The best unbiased estimate of population standard deviation based on group ranges. J. Amer. Statist. Assoc. 42, 224-241.
1.531/644(ftn.)   1.541/646(Noether)

GRUENBERGER, F.

    1950   Tests of random digits. Math. Tables Aids Comput. 4, 244-245.
13.63/468

GRUENBERGER, F. and MARK, A. M.

    1951   The d^2 test of random digits. Math. Tables Aids Comput. 5, 109-110.
13.63/468

GRUND, F. J.   See Hirsch 1831

GRUNDY, P. M.

    1951   The expected frequencies in a sample of an animal population in which the abundances of species are log-normally distributed. Part I. Biometrika 38, 427-434.   5.21/278(Wagner)   5.21/281

GRUNDY, P. M., HEALY, M. J. R. and REES, D. H.

    1956   Economic choice of the amount of experimentation.
J. Roy. Statist. Soc. Ser. B 18, 32-55.   15.3/612

GRYLLS, V. G., Mrs   See Thompson, C. M.

GUBLER, I. A.   See Kolmogorov, Svešikov & Gubler

GUEST, P. G.

    1951   The estimation of standard error from successive finite differences. J. Roy. Statist. Soc. Ser. B 13, 233-237.   1.531/76(intro.)

    1951 a   The fitting of polynomials by the method of weighted grouping. Ann. Math. Statist. 22, 537-548. $11.19/725_M$

GUILFORD, J. P.   See Michael, Perry & Guilford

GUILFORD, J. P. and LYONS, T. C.

    1942   On determining the reliability and significance of a tetrachoric coefficient of correlation. Psychometrika 7, 243-246.   7.94/343

GUMBEL, E. J.

    1935   Les valeurs extrêmes des distributions statistiques. Ann. Inst. H. Poincaré 5, 115-158.   8.3111/364

    1937 y   Les intervalles extrêmes entre les émissions radio-actives. I. J. Phys. Radium 8, 321-329.   8.321/703

    1937 z   Les intervalles extrêmes entre les émissions radio-actives. II. J. Phys. Radium 8, 446-452.

    1941   The return period of flood flows. Ann. Math. Statist. 12, 163-190. 8.3111/364   8.3112/364

    1943   On serial numbers. Ann. Math. Statist. 14, 163-178.   1.41/53

    1947   The distribution of the range. Ann. Math. Statist. 18, 384-412. 1.5532/108   1.5532/109(ftn.)   1.5532/110(Cox)   8.331/365   8.332/365

    1949   Probability tables for the range. Biometrika 36, 142-148. 8.331/365   8.332/365

    1954   Statistical theory of extreme variables and some practical applications. [U.S.] National Bureau of Standards, Applied Mathematics Series, 33. Washington: G.P.O.   8.3/364

    1954 z   Applications of the circular normal distribution. J. Amer. Statist. Assoc. 49, 267-279.   10.28/419   10.28/722

    1958   Statistics of extremes. New York: Columbia Univ. Press; London, etc.: Oxford Univ. Press.   8.3/364

    See Freudenthal & Gumbel; N.B.S. 1953

GUMBEL, E. J. and GREENWOOD, J. A.

    1951   Table of the asymptotic distribution of the second extreme. Ann. Math. Statist. 22, 121-124.   8.3211/365

GUMBEL, E. J., GREENWOOD, J. A. and DURAND, D.

    1953   The circular normal distribution: theory and tables.
J. Amer. Statist. Assoc. 48, 131-152.   10.28/419   10.28/722

GUPTA, A. K.

    1952   Estimation of the mean and standard deviation of a normal population from a censored sample. Biometrika 39, 260-273.
1.332/43   1.51/641   1.51/642(Sarhan & Greenberg 1956)
1.51/642(Dixon 1957)   1.544/647   1.544/648(Sarhan & Greenberg)

GUPTA, JITENDRAMOHAN SEN   See Mahalanobis 1933 a; Sengupta, J. M.

GUPTA, SHANTI S. and SOBEL, M.

    1957   On a statistic which arises in selection and ranking problems.
Ann. Math. Statist. 28, 957-967.   4.17/242

GUSHEE, C. H.

    1942   Financial compound interest and annuity tables.
Boston: Financial Publishing Co.; London: Routledge [& Kegan Paul]
xxxxiij

GUTTMAN, I.   See Fraser & Guttman

GWYTHER, C. E.   See B.A. 9, 1940

HAGOOD, MARGARET J.

    1941   Statistics for Sociologists. New York: Reynal and Hitchcock.
1.1121/4

HAIGHT, F. A.

    1955   Index to the distributions of mathematical statistics.[No tables]
Auckland (New Zealand): Auckland University College.
5./682   10./715

HAINES, JOAN   See Pearson, E.S. & Haines

HALD, A. H.

    1946   Den afstumpede normale Fordeling.
Matematisk Tidskrift, B 1946, 83-91.   1.332/41(ftn.)

    1948   The decomposition of a series of observations composed of a trend, a periodic movement and a stochastic variable. Copenhagen: G.E.C. Gad.
11.119/439

    1948 b   Statistiske Metoder: Tabel- og Formelsamling.
København: Privat Ingeniørfond.
❡ For contents of this book, see p. 754 above.   1j

    1949   Maximum likelihood estimation of the parameters of a normal distribution which is truncated at a known point.
Skand. Aktuarietidskr. 32, 119-134.
1.332/40(3)   1.332/41(3)   1.332/42

1952   Statistical tables and formulas.
New York [and London]: John Wiley & Sons.
¶ For contents of this book, see p. 754 above.   1j

HALD, A. and SINKBÆK, S. A.

1950   A table of percentage points of the χ^2-distribution.
Skand. Aktuarietidskr. 33, 168-175.   2.3111/140

HALDANE, J. B. S.

1938   The approximate normalization of a class of frequency distributions.
Biometrika 29, 392-404.   2.314/145   3.42/199

1940 a   The mean and variance of χ^2, when used as a test of homogeneity, when expectations are small. Biometrika 31, 346-355.
2.61/174   2.62/179(intro.)

1940 b   The cumulants and moments of the binomial distribution, and the cumulants of χ^2 for a $(n \times 2)$-fold table. Biometrika 31, 392-396.
2.61/174   2.62/179(intro.)

1945 a   Moments of r and χ^2 for a fourfold table in the absence of association. Biometrika 33, 231-233.   2.61/174

1945 b   The use of χ^2 as a test of homogeneity in a $(n \times 2)$-fold table when expectations are small. Biometrika 33, 234-238.   2.62/179(intro.)

HALE, A. W.

1882   Tables for facilitating the determination of empirical formulae.
Amer. J. Math. 5, 342-347.   11.12/440

HALL, P.

1927 a   Multiple and partial correlation coefficients in the case of an n-fold variate system. Biometrika 19, 100-109.   7.4/335(Pearson)

1927 b   The distribution of means for samples of size N drawn from a population in which the variate takes values between 0 and 1, all such values being equally probable (values). Biometrika 19, 240-244.
10.017/393(Irwin)

HALPERIN, MAX, GREENHOUSE, SAMUEL W., CORNFIELD, JEROME and ZALOKAR, JULIA

1955   Tables of percentage points for the studentized maximum absolute deviate in normal samples. J. Amer. Statist. Assoc. 50, 185-195.
1.44/640

HALSTRØM, H. L.   See Brockmeyer, Halstrøm & Jensen

HALTON THOMSON, D.   See Thomson, D.H.

HAMAKER, H. C.

1948   Toevals-cijfers. Statistica [Neerlandica] 2, 97-106.
13.61/468   13.63/468

1949   A simple technique for producing random sampling numbers.
Nederl. Akad. Wetensch. Proc. 52, 145-150.   13.61/468   13.63/468

HAMMER, P. C.

    1951   The mid-square method of generating [pseudo-random] digits.
        In N.B.S. 1951, p. 33.   13.62/468   13.63/468

HAMMERSLEY, J. M.

    1953   On counters with random dead time. I.
        Proc. Cambridge Philos. Soc. 49, 623-637.
        ¶ Not abstracted; apparent classification 5.9.
        See Mathematical Reviews 15, 139.

HANAFI, A. R.   See Kerawala & Hanafi

HANDBOOK of CHEMISTRY and PHYSICS   See Hodgman

HANNAN, E. J.

   *1955   An exact test for correlation between time series.
        Biometrika 42, 316-326.   7.54/696$_M$

HANSEN, JOAN M.   See Mainland & Murray 1952

HANSMANN, G. H.

    1934   On certain non-normal symmetrical frequency distributions.
        Biometrika 26, 129-195.   10.18/405

HARDING, H. G. and PRICE, S.

    1957   Narrow limit gage sampling procedure.
        IRE National Convention Record [5], 1957, part 10, 54-58.
        14.112/476

HARLEY, BETTY I.

   *1954   A note on the probability integral of the correlation coefficient.
        Biometrika 41, 278-280.
        ¶ Not abstracted; apparent classification 7.29.
        See Mathematical Reviews 15, 971.

    1957   Relations between the distributions of non-central t and of a transformed correlation coefficient. Biometrika 44, 219-224.
        7.134/327

HARLEY, ROBERT   See B.A. 1896, 1899

HARRIS, H. F.   See B.A. 1896

HARRIS, MARILYN, HORVITZ, D. G. and MOOD, A. M.

    1948   On the determination of sample sizes in designing experiments (sampling size in experiment design).
        J. Amer. Statist. Assoc. 43, 391-402.   4.1413/222

HARSHBARGER, B.

   *1947   Rectangular lattices. Virginia Agr. Expt. Sta. Mem. 1.
        Not abstracted: apparent classification 15.244.

HART, B. I.

    1942   Significance levels for the ratio of the mean square sucsessive difference to the variance. Ann. Math. Statist. 13, 445-447.
1.531/76(intro.)   1.531/77   1.531/80(Kamat)   1.531/643(Ferber)

    See von Neumann, Kent, Bellinson & Hart

HART, B. I. and von NEUMANN, J.

    1942   Tabulation of the probabilities for the ratio of the mean square successive difference to the variance. Ann. Math. Statist. 13, 207-214.
1.531/76(intro.)   1.531/76(Young)   1.531/76   1.531/80(Kamat)

HARTLEY, H. O.

    1938   Studentization and large-sample theory. J. Roy. Statist. Soc. Suppl. 5, 80-88.
4.251/261(Finney)   4.252/261

    1940   Testing the homogeneity of a set of variances. Biometrika 31, 249-255.
6.1/295   6.113/300   6.113/301(Thompson & Merrington)

    1942   The range in random samples. Biometrika 32, 334-348.   1.549/95

    1944   Studentization, or the elimination of the standard deviation of the parent population from the random sample-distribution of statistics. Biometrika 33, 173-180.   3.43/200

    1950   The use of range in analysis of variance. Biometrika 37, 271-280.
1.5511/98   1.552/104   1.552/105(David)   1.552/106(Moshman)   1.5531/107(Patnaik)

    1950 a   The maximum F-ratio as a short-cut test for heterogeneity of variance. Biometrika 37, 308-312.   4.253/262

    1953   Corrigenda: Tables of percentage points of the 'studentized' range [May 1952]. Biometrika 40, 236.
1.5511/98(May)   1.5511/98   1.5511/99(Pearson & Hartley)

    1959   Smallest composite designs for quadratic response surfaces. Biometrics 15, 611-624.   15.138/555(2)

    See Comrie & Hartley; David, H.A., Hartley & Pearson; Fieller & Hartley; Hirschfeld; Pearson, E.S., Godwin & Hartley; N.B.S. 1952; Pearson, E.S. & Hartley; Sheppard 1939; Smith, C.A.B. & Hartley; Thompson, C.M. 1941 a; Thompson & Merrington 1946

HARTLEY, H. O. and DAVID, H. A.

    1954   Universal bounds for mean range and extreme observation. Ann. Math. Statist. 25, 85-99. $8.28/702_M$

HARTLEY, H. O. and FITCH, E. R.

    1951   A chart for the incomplete Beta-function and the cumulative binomial distribution. Biometrika 38, 423-426.
3.119/184   3.12/185   3.311/191

HARTLEY, H. O. and PEARSON, E. S.

    1950   Tables of the χ^2-integral and of the cumulative Poisson distribution.
Biometrika 37, 313-325.   2.1/131   2.1131/132   2.121/135

    1950 a   Table of the probability integral of the t-distribution.
Biometrika 37, 168-172.
4.1111/208   10.071/400   4.1113/670(Vianelli)

    1951   Moment constants for the distribution of range in normal samples.
Biometrika 38, 463-464.   1.541/87   1.541/647(Vianelli)

HARVARD UNIVERSITY. COMPUTATION LABORATORY

    1949   Tables of inverse hyperbolic functions.
The Annals of the Computation Laboratory of Harvard University, 20.
Cambridge (Massachusetts): Harvard Univ. Press;
London: Oxford Univ. Press.   7.212/331

    1952   Tables of the error function and of its first twenty derivatives.
The Annals of the Computation Laboratory of Harvard University, 23.
Cambridge (Massachusetts): Harvard Univ. Press;
London: Oxford Univ. Press.
1.1121/4   1.1161/9   1.321/32   1.3211/33(6)   10.211/408

    1955   Tables of the cumulative binomial probability distribution.
The Annals of the Computation Laboratory of Harvard University, 35.
Cambridge (Massachusetts): Harvard Univ. Press;
London: Oxford Univ. Press.   3.111/183

HASTAY, M.W.   See Eisenhart, Hastay & Wallis

HASTINGS, C. Jr.   See N.B.S. 1959

HASTINGS, C., Jr., HAYWARD, JEANNE T. and WONG, J. P., Jr.

    1955   Approximations for digital computers.   Princeton Univ. Press;
London: Oxford Univ. Press.   1.19/15   1.2119/20

HASTINGS, C., Jr., MOSTELLER, F., TUKEY, J. W. and WINSOR, C. P.

    1947   Low moments for small samples:
a comparative study of order statistics.
Ann. Math. Statist. 18, 413-426.   1.43/65   8.21/360

HATKE, MARY AGNES, Sister

    1949   A certain cumulative probability function.
Ann. Math. Statist. 20, 461-463.   10.24/416   10.24/720(Topp & Leone)

HAYASHI, K.[*]

    1926   Sieben- und mehrstellige Tafeln der Kreis- und Hyperbelfunktionen
und deren Produkte sowie der Gammafunktion: nebst einem Anhang

---

[*]According to Fletcher et al. 1946, p. 12: "The inaccuracy of certain tables is well known to computers. We may mention particularly the tables of Hayashi, to the errors in which several writers have alluded."

Interpolations- und sonstige Formeln. Berlin: Julius Springer.
7.211/330$_F$ 7.212/331$_F$ xxxxviij(2)

1932   Berichtigungen in Hayashis sieben [sic] u. mehrstellige Tafeln (1926). Fukuoka [Japan].

HAYES, S. P., Jr.

1943   Tables of the standard error of tetrachoric correlation coefficient. Psychometrika 8, 193-203.
¶ Not abstracted; apparent classification 7.94.
See Mathematical Reviews 5, 42.

HAYWARD, JEANNE T.   See Hastings, Hayward & Wong

HAZELWOOD, R. L.   See Peck & Hazelwood

HEALY, M. J. R.

1949   Routine computation of biological assay(s) involving a routine response. Biometrics 5, 330-334.   1.233/29

See Fisher, R. A. & Healy; Grundy, Healy & Rees

HEILFRON, C.   See Pollak & Heilfron

HENDERSON, JAMES

1922   On expansions in tetrachoric functions. Biometrika 14, 157-185. 1.325/37

See B.A. 1, 1931; Pearson, K. & Elderton 1923

HENDRICKS, W. A.

1936   An approximation to "Student's" distribution. Ann. Math. Statist. 7, 210-221.   4.132/213

HERMITE, C.

1864 a   Sur un nouveau développement en série des fonctions. C. R. Acad. Sci. Paris 58, 93-100.
Reprinted in Hermite 1908, pp. 293-300.   11.110/436(ftn.)

1864 b   Sur un nouveau développement en série des fonctions. C. R. Acad. Sci. Paris 58, 266-273.
Reprinted in Hermite 1908, pp. 301-308.   11.110/436(ftn.)

1908   Œuvres, Tome 2. Paris: Gauthier-Villars.

HERON, D.

1910   Biometric notes: I.
An abac for determining the probable errors of correlation coefficients.
II. On the probable error of a correlation coefficient.
Biometrika 7, 410-412.   7.132/326

HESTENES, A. D.   See Teichroew 1956

HIJIKATA, K.   See Kotani, Amemiya, Ishiguro & Kimura 1955

HILFERTY, MARGARET M.   See Wilson & Hilferty

HIMSWORTH, F. R.   See Davies 1954

HIRAGA, Y., MORIMURA, H. and WATANABE, H.

    1954   Tables for three-sample test.
        Ann. Inst. Statist. Math. (Tokyo) 5, 97-102.
            ☞ Not abstracted; apparent classification 9.13.
        See Mathematical Reviews 15. 970.

HIRSCH, MEYER (or MEIER)

    1804   Sammlung von Beispielen, Formeln und Aufgaben aus der Buchstabenrechnung und Algebra. First edition.
        (Eighth edition 1853; this is what we have seen. 20th ed. 1890)
        Berlin: Duncker und Humblot.
            ☞ This work contains no tables of symmetric functions.
            According to Grund (Hirsch 1831, p. [v]), a "translation of [Hirsch 1804] has been published in England, by Rev. S.[!]A. Ross;" we have not seen this translation, or any catalogue entry therefor.

    1809   Sammlung von Aufgaben aus der Theorie der algebraischen Gleichungen. Erster Theil. (added t.p.: Fortsetzung der Sammlung von Beyspielen, Formeln und Aufgaben aus der Buchstabenrechnung und Algebra. Erster Theil) Berlin: Duncker und Humblot.
            ☞ Cayley 1889, p. 417, confuses this work with Hirsch 1804; Decker 1910, p. 5, cites it to the year 1808, and gives it the title of Hirsch 1827.
        10.3123/424

    1827   Hirsch's collection of examples, formulæ, & calculations, on the literal calculus and algebra. Translated from the German, by the Rev. J.A. Ross, A.M., translator of Hirsch's integral tables.
        London: Black, Young, and Young.
            ☞ Notwithstanding the title page, this is a translation of Hirsch 1809 and not of Hirsch 1804.
        10.3123/424

    1831   A collection of arithmetical and algebraic problems and formulæ; by Meier Hirsch: translated from the original German, and adapted to the use of the American Student.
        By Francis J. Grund, teacher of mathematics.
        Boston: Carter, Hendee and Babcock.
            ☞ A translation of Hirsch 1804, augmented by 10 pages of examples in arithmetic and 2 pages of promiscuous examples in fractions. We do not know what edition of Hirsch 1804 Grund worked from. This work contains no tables of symmetric functions.

HIRSCHFELD, H. O.

    1937   The distribution of the ratio of covariance estimates in two samples drawn from normal bivariate populations. Biometrika 29, 65-79.
        7.6/340

HOADLEY, MARIAN F.   See Pearson, K. 1934

HODGES, J. L., Jr.

    1955   On the non central beta-distribution.
        Ann. Math. Statist. 26, 648-653.   4.242/679   5.31/683

  *1955y   A bivariate sign test. Ann. Math. Statist. 26, 523-527.
        9.33/714

    See Fix & Hodges

HODGMAN, C. D.

    1957   C.R.C. standard mathematical tables. Eleventh edition.
        Cleveland (Ohio): Chemical Rubber Pub. Co.
        ¶ Many of the mathematical tables (but not the actuarial tables) in this book are also in the Handbook of Chemistry and Physics, editied by Hodgman: same publisher.
        1.1121/4   1.1131/6   1.1132/6   1.1161/9   1.3211/34   2.1132/133
        2.5111/160   2.5112/161(2)   2.5122/162   2.5134/163   3.211/186

HOEFFDING, W.

    1948   A non-parametric test of independence.
        Ann. Math. Statist. 19, 546-557.   8.19/356

    1951   "Optimum" nonparametric tests. In Neyman, J., ed., Proceedings of the second Berkeley symposium on mathematical statistics and probability, pp. 83-92. Berkeley, etc.: Univ. of California Press; London: Cambridge Univ. Press.   9.122/369(Terry)

HOEL, P. G.

    1943   On indices of dispersion. Ann. Math. Statist. 14, 155-162.
        2.69/180

    1947   Introduction to mathematical statistics. (Second edition 1954) New York [and London]: John Wiley & Sons.
        1.1121/4   1.1161/9   2.3113/142   4.1311/211   4.2314/249

HOJO, T.

    1931   Distribution of the mean, quartiles and interquartile distance in samples from a normal population. Biometrika 23, 315-360.
        1.40/48(ftn.)   1.41/51   1.41/52(K. & M.V. Pearson)   1.42/57
        1.544/92   13.0/455

    1933   A further note on the relation between the median and the quartiles in small samples from a normal population. Biometrika 23, 79-90.
        1.41/53

    See Pearson, K. 1931 z

HOLBROOK, LYSBETH   See Kendall, M.G. 1946

HOLZINGER, K. J.

    1925   Statistical tables for students in education and psychology.
Univ. of Chicago Press.   1.1121/4   1.2111/19

    1925 a   Tables of the probable error of the coefficient of correlation as found by the product moment method. Tracts for Computers, 12. Cambridge Univ. Press, etc.   7.132/326

    1928   Statistical methods for students in education. Boston; etc.: Ginn.
$1.2111/19_F$ $1.221/21_F$

HOOKE, ROBERT

    1956   Symmetric functions of a two-way array.
Ann. Math. Statist. 27, 55-79. $10.36/723_M$

HORENSTEIN, W.   See N.B.S. 1949

HORN, HENRY J.

    *1956   Simplified LD_{50} (or ED_{50}) calculations. Biometrics 12, 311-322.
$1.233/633_M$

HORSNELL, GARETH

    *1954   The determination of single-sample schemes for percentage defectives.
Appl. Statist. 3, 150-158. $14.112/730_M$

    1956   Economic acceptance sampling schemes.
J. Roy. Statist. Soc. Ser. A 120, 148-191; and discussion, 192-201.
14.112/473

HORTON, H. B.

    1948   A method for obtaining random numbers.
Ann. Math. Statist. 19, 81-85.   13.61/468

    See Interstate Commerce Commission 1949

HORTON, H. B. and SMITH, R. TYNES, iij

    1949   A direct method for producing random digits in any number system.
Ann. Math. Statist. 20, 82-90.   13.61/468

HORVITZ, D.G.   See Harris, Horvitz & Mood

HOTELLING, H.

    1939   Tubes and spheres in n-spaces, and a class of statistical problems.
Amer. J. Math. 61, 440-460.   11.3/444(Keeping)

    1953   New light on the correlation coefficient and its transforms.
J. Roy. Statist. Soc. Ser. B 15, 193-232.   7.2/330

HOTELLING, H. and FRANKEL, L. R.

    1938   The transformation (distribution) of statistics to simplify their distribution. Ann. Math. Statist. 9, 87-96.
4.132/214   4.132/216(Goldberg & Levine)

HOUSEHOLDER, A.S.   See N.B.S. 1951 a

HOUSEMAN, E. E.   See Anderson, R.L. & Houseman

HOWARD, M.   See Teichroew 1956

HOWELL, J. M.

    1949   Control chart for largest and smallest values.
        Ann. Math. Statist. 20, 305-309.
        1.42/60   1.542/88(intro.)   14.229/490   14.229/490(Žaludová)

    1950   Errata to "Control chart for largest and smallest values"
        [Howell 1949] Ann. Math. Statist. 21, 615-616.   1.42/61(Howell 1949)

HOYT, RAY S.

    1947   Probability functions for the modulus and angle of the normal
        complex variate. Bell System Tech. J. 26, 318-359.
        ¶ Not abstracted; apparent classification 1.78.
        See Mathematical Reviews 8, 522.

HSU, C. T.

    1940   On samples from a normal bivariate population.
        Ann. Math. Statist. 11, 410-426.   6.12/305

HUBER, M.   See Darmois 1928

HUITSON, A.

   *1955   A method of assigning confidence limits to linear combinations of
        variances. Biometrika 42, 417-479. $2.41/660_M$

HUNTINGTON, E. V.

    1937   A rating table for card-matching experiments.
        J. Parapsychol. 1, 292-294.
        5.62/685   5.62/686(Huntington 1937 a)   5.62/686(Greville)

    1937 a   Exact probabilities in certain card-matching problems.
        Science 86, 499-500.   5.62/686(Huntington 1937)   5.62/686
        See Rietz, Carver et al.

HUNTSBERGER, D. V.

    1958   Elementary principles of statistics, part 1.
        Dubuque (Iowa): Wm. C. Brown.   13.11/456   13.13/459

   *1961   Elements of statistical inference.
        Englewood Cliffs (New Jersey): Allyn & Bacon.   13.11/456   13.13/459

HUYETT, MARILYN J.   See Gupta & Sobel 1957; Sobel & Huyett

HYRENIUS, H.

    1953   On the use of ranges, cross-ranges and extremes in comparing small
        samples. J. Amer. Statist. Assoc. 48, 534-545.
        8.21/699   9.129/706   9.2/708

IHM, P.

    1955   Ein Kriterium für zwei Typen zweidimensionaler Normalverteilungen. Mitteilungsbl. Math. Statist. 7, 46-52. $7.6/697_M$

[U.S.] INTERSTATE COMMERCE COMMISSION,
BUREAU OF TRANSPORT ECONOMICS AND STATISTICS

    1949   Table of 105,000 random decimal digits. Statement No. 4914, File No. 261-A-1. Washington, D.C.

        Unpublished: a copy exists in the Library of the Association of American Railroads, Bureau of Railway Economics, Washington. 13.11/456   13.1212/457   13.13/459

IRESON, W. G.   See E.L. Grant 1946 (1952 ed.)

IRWIN, J. O.

    1923   On quadrature and cubature: or, on methods of determining approximately single and double integrals. Tracts for Computers, 10. Cambridge Univ. Press, etc.   1.71/119   16.236/629

    1925 a   The further theory of Francis Galton's individual difference problem. Biometrika 17, 100-128. See also Pearson, K. 1925 z.   1.543/90

    1925 b   On a criterion for the rejection of outlying observations. Biometrika 17, 238-250.   1.543/91

    1927   On the frequency distribution of the means of (in) samples from a population having any law of frequency with finite moments, with special reference to Pearson's Type II. Biometrika 19, 225-239. 10.017/393(Irwin 1930)

    1930   On the frequency distribution of the means of samples from populations of certain of Pearson's types. Metron 8, No. 4, 51-105. 10.017/393

    1943   A table of the variance of $\sqrt{x}$ when x has a Poisson distribution. J. Roy. Statist. Soc. 106, 143-144.   12.2/448

    See B.A. 1, 1931; Greenwood, M. 1946; Hall 1927 a; Pearson, E.S. 1927; Sheppard 1939

IRWIN, J. O. and KENDALL, M. G.

    1944   Sampling moments of moments for a finite population. Ann. Eugenics 12, 138-142.   10.35/431

ISHIGURO, E.   See Kotani, Amemiya, Ishiguro & Kimura

ISSERLIS, L.   See Greenwood, M. 1956

IWASZKIEWICZ, KAROLINA, Miss   See Neyman, Iwaszkiewicz & Kołodziejczyk

IWAZURO, MIEKO, Miss   See Hiraga, Morimura & Watanabe 1954

IWIŃSKI, T.   See Czechowski et al.

IYER, P. V. K.   See Krishna Iyer, P.V.

JACKMAN, MARIAN C   See Bliss 1948

JACKSON, J. E. and ROSS, ELEANOR L.

   1955   Extended tables for use with the "G" test for means.
J. Amer. Statist. Assoc. 50, 416-433.
1.541/646(Noether)   1.5512/650

JACOBS, R. M.

   1957   Lower administrative costs ... closer supervision ...
faster scheduling ... past-performance record ...
recurring-defects record, with ...
New multiple-sampling plan for quality control.
Amer. Machinist 101, no. 3, 113-118.   14.112/477

JAHNKE, P. R. E. and EMDE, F.

   1909   Funktionentafeln mit Formeln und Kurven. Mathematisch-physikalische Schrifte für Ingenieure und Studierende, herausgegeben von E. Jahnke, 5. Leipzig, etc.: B. G. Teubner.
      ☞ For partial contents of this book, see p. 755 above.
[fifty]   1j(2)

   1933   Funktionentafeln mit Formeln und Kurven, zweite/neubearbeitete Auflage.
Tables of functions with formulae and curves, second (revised) edition.
Leipzig, etc.: B. G. Teubner.
      ☞ For partial contents of this book, see p. 755 above.
      ☞ For reprint of pp. 1-76 of this book, see the following entry.

   1938   Funktionentafeln mit Formeln und Kurven, dritte/neubearbeitete Auflage.
Tables of functions with formulae and curves, third (revised) edition.
Leipzig, etc.: B. G. Teubner.
(Reprint 1941. New York: G. E. Stechert, etc. Reprint 1943, incorporating pp. 1-76 of Jahnke & Emde 1933. New York: Dover. Reprint 1945, incorporating pp. 1-76 of Jahnke & Emde 1933, with corrections and new pages 302-304 (called fourth edition on t.p.) New York: Dover; [London: Constable])
      ☞ For partial contents of this book, see p. 755 above.

   1948   Jahnke-Emde Tafeln höherer Funktionen bearbeitet von . . . Fritz Emde, vierte/neubearbeitete Auflage. Jahnke-Emde tables of higher functions treated by . . . Fritz Emde, fourth (revised) edition.
Leipzig: B. G. Teubner. (Fifth edition, under assistance of . . . Friedrich Lösch, 1952. Leipzig [i.e. Stuttgart]: B. G. Teubner. This is what we have seen)
      ☞ For partial contents of this book, see p. 756 above.

   1960   See the following entry.

JAHNKE, P. R. E., EMDE, F., and LÖSCH, F.

    1960   Tafeln höherer Funktionen. Tables of higher functions. Sixth edition.
        Stuttgart: B. G. Teubner; New York, etc.: Mc Graw-Hill Book Co.
        * For partial contents of this book, see p. 756 above.
        xxxxj

JAKOWLEWA-ZUBRZYCKE, LUDMIŁA   See Steinhaus 1954

JAMES, G. S.

    1956   On the accuracy of weighted means and ratios.
        Biometrika 43, 304-321.   4.18/674

    See Trickett, Welch & James

JAMES, S. F.   See Orcutt & James

JEFFERY, G. B.   See Pearson, K., Jeffery & Elderton

JEFFREYS, HAROLD, Sir

    1939   The law of error and the combination of observations.
        Philos. Trans. Roy. Soc. London Ser. A 237, 231-271.   10.23/415

    1948   Theory of probability. Second edition. (First edition 1939)
        Oxford Univ. Press.   1.19/14   2.319/150   4.139/218   4.339/272

JENKINS, T. N.

    1932   A short method and tables for the calculation of the average and standard deviation of logarithmic distributions.
        Ann. Math. Statist. 3, 45-55.   10.2221/413

JENNETT, W. J.   See Dudding & Jennett

JENNETT, W. J. and WELCH, B. L.

    1939   The control of proportion defective as judged by a single quality characteristic varying on a continuous scale.
        J. Roy. Statist. Soc. Suppl. 6, 80-88.   2.522/171

JENSEN, ARNE   See Brockmeyer, Halstrøm & Jensen

JOFFE, S. A.   See Archibald & Joffe

JOHNSEN, MADELINE   See N.B.S. 1959

JOHNSON, D. L.

    1956   Generating and testing pseudo random numbers on the IBM Type 701.
        Math. Tables Aids. Comput. 10, 8-13.   13.62/468   13.63/468

JOHNSON, NORMAN L.

    1949 a   Systems of frequency curves generated by methods of translation.
        Biometrika 36, 149-176.
        10.222/412   10.2222/414   10.2223/414   10.222/719(Johnson 1954)

    1949 b   Bivariate distributions based on simple translation systems.
        Biometrika 36, 297-304.   10.229/414

1953   Some notes on the application of sequential methods in the analysis of variance. Ann. Math. Statist. 24, 614-623.
¶ Not abstracted; apparent classification 4.241.
See Mathematical Reviews 15, 638.

1954   Systems of frequency curves derived from the first law of Laplace. Trabajos Estadíst. 4, 283-291.   10.109/718   10.222/719

1954a [Review:] Tables for the design of factorial experiments. Biometrika 41, 567 only.

1957   Optimum sampling for quota fulfilment. Biometrika 44, 518-523.
15.3/736

See David, F.N. & Johnson

JOHNSON, N. L. and TETLEY, H.

1949   Statistics, an intermediate textbook. Vol. 1.
Cambridge Univ. Press, etc.   1.115/8   1.1161/9   3.229/189(Seal)

1950   Statistics, an intermediate textbook. Vol. II.
Cambridge Univ. Press, etc.
1.52/75   1.541/86   2.3112/141   4.1311/211   4.2314/249

JOHNSON, N. L. and WELCH, B. L.

1939   On the calculation of the cumulants of the χ-distribution. Biometrika 31, 216-218.   2.522/171

1940   Applications of the non-central t-distribution. Biometrika 31, 362-389.   4.141/219   4.1414/224

JOHNSON, PALMER O.

1949   Statistical methods in research. New York: Prentice-Hall.
1.115/8   2.3113/142   4.1311/211   4.2314/249   6.112/299(Nayer)
7.131/325   7.23/332

JOHNSON, R. B.   See Albert & Johnson

JOHNSON, WILLIAM WALTER   See Wittstein 1866

JONCKHEERE, A. R.

1954   A distribution-free k-sample test against ordered alternatives. Biometrika 41, 133-145.   9.14/707$_M$

JONES, A. E.

1946   A useful method for the routine estimation of dispersion from large samples. Biometrika 33, 274-282.   1.533/81

JONES, C. W.   See Miller, J.C.P. 1954

JONES, D. CARADOC

1924   A first course in statistics. Second edition. (First edition 1921)
London: G. Bell & Sons. $1.111/3_F$ $1.115/8_F$ $2.1132/133_F$

JONES, H. GERTRUDE   See Heron 1910

JONES, HOWARD L.

    1953   Approximating the mode from weighted sample values.
J. Amer. Statist. Assoc. 48, 113-130.   8.23/363   10.074/401

JORDAN, C.

    1921   Sur une série de polynomes dont chaque somme partielle représente la meilleure approximation d'un degré donné suivant la méthode des moindres carrés. Proc. London Math. Soc. 20, 297-325.   11.111/437

    1932   Approximation and graduation according to the principle of least squares by orthogonal polynomials. Ann. Math. Statist. 3, 257-357.
$5.31/284_F$ 11.113/439   11.13/441

JØRGENSEN, N. R.

    1915   Note sur la fonction de répartition de Type B de M. Charlier.
Ark. Mat. Astr. Fys. 10, No. 15.   5.23/282(Theil)   10.212/409

    1916   Undersøgelser over Frequensflader og Korrelation. København: Arnold Busck.
$1.115/7_F$ $1.1161/9_F$ 1.3211/33   1.323/34   2.5132/162   5.23/282   5.23/282(Theil)   10.212/409

    See Charlier 1947

JURAN, J. M.

    1951   Quality control handbook.
New York, London, etc.: McGraw-Hill Book Co.
2.3113/142   4.1311/211   4.2314/249   14.221/488   14.221/489   14.221/490

JURGENSEN, C. E.

    1947   Table for determining phi coefficients. Psychometrika 12, 17-29.
    * Not abstracted; apparent classification 2.612.
See Mathematical Reviews 8, 477.

KÄÄR, J.   See Wold 1948

KAARSEMAKER, L., Miss, and van WIJNGAARDEN, A.

    1953   Tables for use in rank correlation. Statistica Neerlandica 7, 41-54.
8.1212/352   8.122/352   8.125/353

KAC, M., KIEFER, J. and WOLFOWITZ, J.

    1955   On tests of normality and other tests of goodness of fit based on distance methods. Ann. Math. Statist. 26, 189-211.
9.4/380   9.429/386

KAMAT, A. R.

 1953a On the mean successive difference and its ratio to the root mean square. Biometrika 40, 116-127. 1.531/76 (intro.) 1.531/79

 1953b Some properties of estimates for the standard deviation based on deviations from the mean and variate differences.
J. Roy. Statist. Soc. Ser. B 15, 233-240. 1.52/72 1.531/76(intro.)

 *1956 A two-sample distribution-free test. Biometrika 43, 377-387.
$9.2/708_M$

KAPLAN, E. L.

 1952 Tensor notation and the sampling cumulants of k-statistics. Biometrika 39, 319-323. 10.36/432

KARLIN, M. See N.B.S. 1949

KARMAZINA, LENA NIKOLAEVNA КАРМАЗИНА ЛЕНА НИКОЛАЕВНА

 1954 Таблицы полиномов Якоби. Москва: Издат. Акад. Наук СССР.
10.18/405 10.18/718

KARPOV, K. A. КАРПОВ К. А.

 1954 Таблицы функции $w(z) = e^{-z^2} \int_0^z e^{x^2} dx$ в комплексной области.
Москва: Издат. Акад. Наук СССР. 1.83/128

 1958 Таблицы функции $F(z) = \int_0^z e^{x^2} dx$ в комплексной области.
Москва: Издат. Акад. Наук СССР. 1.83/128

KATZ, L.

 1952 The distribution of the number of isolates in a social group.
Ann. Math. Statist. 23, 271-276. 5.9/294

 1952a RMT 946[K] (review of Sichel 1951)
Math. Tables Aids Comput. 6, 24 only. 5.32/285(Sichel)

 1955 Probability of indecomposability of a random mapping function.
Ann. Math. Statist. 26, 512-517. $2.129/657_M$

KEEPING, E. S.

 1951 A significance test for exponential regression.
Ann. Math. Statist. 22, 180-198. 11.3/444

 See Kenney & Keeping

KEITH SMITH, J. E. See Green, Smith & Klem

KELLEY, TRUMAN L.

 1923 Statistical method. New York: Macmillan.
1.2111/19 1.221/21 10.004/390(Pearson) 10.06/399

Kelley
Kendall

    1938   The Kelley statistical tables. New York: Macmillan.
              ☛ For contents of this book, see p. 756 above.   1j

    1947   Fundamentals of statistics.
        Cambridge (Massachusetts): Harvard Univ. Press;
        London: Oxford Univ. Press.
        1.2111/19   1.221/21   1.2313/26   4.27/264   12.13/445

    1948   The Kelley statistical tables, second edition.
        Cambridge (Massachusetts): Harvard Univ. Press.
            ☛ For contents of this book, see p. 757 above.   1j

    See Rietz, Carver et al.

KELLY, B. K.   See Brownlee, Kelly & Loraine

KELVIN, WILLIAM THOMSON, Lord

    1890   Mathematical and physical papers, Vol. 3.
        Cambridge Univ. Press, etc.   1.1211/11$_F$

KEMP, C. D. and KEMP, A. W.

    1956   Generalized hypergeometric distributions.
        J. Roy. Statist. Soc. Ser. B 18, 202-211.   10.2/407

KEMPTHORNE, O.

    1947   A simple approach to confounding and fractional replication
        in factorial experiments. Biometrika 34, 255-272.
        15.112/502(3)   15.112/503(2)   15.112/504(2)   15.112/511
        15.113/517(2)   15.113/519(2)   15.114/527(2)   15.1199/546(2)
        15.141/555

    1952   The design and analysis of experiments.
    New York [and London]: John Wiley & Sons.

4.2312/248	15.10/501	15.152/561	15.244/606
.242/258(Tang)	.142/556	.153/561	.245/606(2)
15.0/495(4)	.1422/556	.16/562	.248/607
.10/496(3 ftn.)	.143/556	.20/563	.2511/607
/497	.1511/558	/563(ftn.)	.253/608
/497(2 ftn.)	.1512/559	/564	
/500	.1513/560	15.243/606	

    See Zoellner & Kempthorne

KENDALL, DAVID G.

    See Bartlett, M.S. & Kendall; Meyer 1956

KENDALL, MAURICE G.

    1938   A new measure of rank correlation. Biometrika 30, 81-93.
        8.1211/352

    1938 a   The conditions under which Sheppard's corrections are valid.
        J. Roy. Statist. Soc. 101, 592-605.   10.331/427

1940 b   Proof of Fisher's rules for ascertaining the sampling semi-invariants of k-statistics. Ann. Eugenics 10, 215-222.   10.34/430

1940 c   The derivation of multivariate sampling formulae from univariate formulae by symbolic operation. Ann. Eugenics 10, 392-402.
10.32/426

1942   Partial rank correlation. Biometrika 32, 277-283.   8.15/356

1943   The advanced theory of statistics, vol. 1. First edition. (Second edition 1945, 3d 1947, 4th 1948, 5th 1952. For subsequent editions see Kendall & Stuart 1958) London: Charles Griffin; New York: Hafner.
¶ For a short time the U.S. agent was J.B. Lippincott of Philadelphia. In the Lippincott issue the t.p. is a cancel and the table of contents is missing.
1.115/8   1.1161/9   4.1111/208   4.331/251$_F$ 7.2/330   8.1111/349
8.1112/349   8.1211/352   8.1212/352
8.1312/353(Kendall & Babington Smith)
8.14/356(Kendall & Babington Smith)   10.004/390(Pearson)   10.32/426(3)
10.341/430   10.343/430(2)   xxxviij

1944   On autoregressive time series. Biometrika 33, 105-122.
13.5/465

1946   Contributions to the study of oscillatory time-series. National Institute of Economic and Social Research, Occasional Papers, 9. Cambridge Univ. Press, etc.
13.5/465   13.5/466(Kendall 1949 a)   13.5/467(Kendall 1949 a)
13.5/467(Bartlett)

1946 a   The advanced theory of statistics, vol. 2. First edition.
(Second edition 1947, 3d 1951)
London: Charles Griffin; New York: Hafner.
10.332/428(ftn.)   11.19/441

1948   Rank correlation methods. London: Charles Griffin; New York: Hafner.
8.1/345   8.10/345   8.1112/349   8.1212/352
8.1312/353(Kendall & Babington Smith)   8.132/355(Friedman)
8.14/356(Kendall & Babington Smith)

1949   Rank and product-moment correlation. Biometrika 36, 177-193.

1949 a   Tables of autoregressive series. Biometrika 36, 267-289.
13.5/466

1952   Moment-statistics in samples from a finite population.
Biometrika 39, 14-16.   10.35/431

*1954   Two problems in sets of measurements. Biometrika 41, 560-564.
❡ Not abstracted; apparent classification 1.43.
See Mathematical Reviews 16, 498.

Kendall, M. G.
Kiefer

1955 Rank correlation methods. Second edition. London: Charles Griffin; New York: Hafner.
8.1/345 8.10/345 8.1112/349 8.1212/352
8.1312/353(Kendall & Babington Smith) 8.132/355(Friedman)
8.14/356(Kendall & Babington Smith) 8.15/356 xxxviij

See David, F.N. & Kendall; David, S.T., Kendall & Stuart; Irwin & Kendall; Yule & Kendall

KENDALL, M. G. and BABINGTON SMITH, B.

1938 Randomness and random sampling numbers.
J. Roy. Statist. Soc. 101, 147-166.
13.11/456 13.13/459

1939 Tables of random sampling numbers. Tracts for computers, 24. Cambridge Univ. Press, etc.
13.11/456 13.1211/457 13.13/459 13.3/462(Wold)
13.5/466(Kendall 1946)

1939a The problem of m rankings. Ann. Math. Statist. 10, 275-287.
8.10/348 8.1312/353 8.132/354(Friedman)

1940 On the method of paired comparisons. Biometrika 31, 324-345.
8.14/355

KENDALL, M. G. and BUCKLAND, W. R.

1957 A dictionary of statistical terms.
Edinburgh, etc.: Oliver and Boyd; [New York: Hafner]
1.40/49(ftn.) 1.533/81(ftn.) 10.28/418 754(ftn.)

KENDALL, M. G., KENDALL, SHEILA F. H. and BABINGTON SMITH, B.

1939 The distribution of Spearman's coefficient of rank correlation in a universe in which all rankings occur an equal number of times.
Biometrika 30, 251-273. 8.1111/349

KENDALL, M. G. and STUART, A.

1958 The advanced theory of statistics, vol. 1.
London: Charles Griffin; New York: Hafner.
1.1161/9 8.3/364 10.32/426(2) 10.343/430(2)
.5532/109(ftn.) 10.06/399 .34/429 11.113/438(intro.)
4.331/251 .221/412 /430 13.0/454
5.1/275 .24/416 10.341/430 .1/456
7.132/326

KENDALL, SHEILA F. H.

See Kendall, M.G., Kendall, Sheila F.H. & Babington Smith, B.

KENNEY, J. F.

1939a Mathematics of statistics, Vol. I. New York: D. Van Nostrand.
1.1121/4 1.1161/9

1939b Mathematics of statistics, Vol. II. New York: D. Van Nostrand.
1.1121/4 1.1161/9 2.3113/142$_F$ 4.2314/249$_F$

KENNEY, J. F. and KEEPING, E. S.

    1951   Mathematics of statistics, Part II. Second edition.
              New York [i.e. Princeton (New Jersey)], London, etc.: D. Van Nostrand.
              1.1121/4   1.1161/9   2.3113/142   4.1311/211   4.2314/249

KENT, FREDERICK C. and KENT, MAUDE E.

    1926   Compound interest and annuity tables: Values of all functions to ten
              decimal places for 1-100, 1-200, 1-300 years;
              rates of interest 1/4 of 1 per cent to $10\frac{1}{2}$ per cent;
              conversion factors and logarithms.
              New York and London: Mc Graw-Hill Book Co.   xxxxiij

KENT, MAUDE E.   See Kent, Frederick C. & Kent, Maude E.

KENT, R. H.   See von Neumann, Kent, Bellinson & Hart

KERAWALA, S. M.

    1941   Table of monomial symmetric functions of weight 9.
              Proc. Nat. Acad. Sci. India, Sect. A 11, 51-55.
              10.3121/424   10.3121/424(ftn.)

    1941 a   A rapid method for calculating the least squares solution of a
               polynomial of degree not exceeding the fifth.
               Indian J. Phys. 15, 241-276
               = Proc. Indian Assoc. Cultiv. Sci. 24, 241-276.   11.12/440

KERAWALA, S. M. and HANAFI, A. R.

    1941   The table of symmetric functions of weight 10.
              Proc. Nat. Acad. Sci. India, Sect. A 11, 56-63.   10.3121/424

    1942   Table of monomial symmetric functions of weight 11.
              Proc. Nat. Acad. Sci. India, Sect. A 12, 81-96.
              10.3121/424   10.3121/424(ftn.)

    1948   Table of monomial symmetric functions of weight 12 in terms of
              power-sums. Sankhyā 8, 345-359.   10.3121/424

KERRIDGE, S.   See Jahnke & Emde 1948, pp. 28-31

KEYFITZ, N.

    1938   Graduation by a truncated normal. Ann. Math. Statist. 9, 66-67.
              1.332/43

KHANIKOF, N.   See Čebyšev 1867 b, 1874 b

KIBBLE, W. F.

    1945   An extension of a theorem of Mehler's on Hermite polynomials.
              Proc. Cambridge Philos. Soc. 41, 12-15.   1.79/655(ftn.)

KIEFER, J.   See Kac, Kiefer & Wolfowitz

KIMBALL, B. F.

    1944   Note on asymptotic value of probability distribution of sum of random variables which are greater than a set of arbitrarily chosen numbers. Ann. Math. Statist. 15, 423-427.   1.332/45

    1956   The bias in certain estimates of the parameters of the extreme-value distribution. Ann. Math. Statist. 27, 758-767.   8.319/703

    See Lieblein & Salzer 1957

KIMURA, T.   See Kotani, Amemiya, Ishiguro & Kimura

KING, EDGAR P.

    1952   The operating characteristic of the control chart for sample means. Ann. Math. Statist. 23, 384-395.

    ¶ Not abstracted; apparent classification 14.229.

    See Mathematical Reviews 14, 297.

KING, TH. O.   See Kruskal & Wallis 1952

KINGSFORD, MARY   See Pearson, K. 1934

KIRK, R. E.   See Bradley 1954

KIRKPATRICK, M. H., Mrs   See Bradley & Terry 1952

KISHEN, K.

    1949   On the construction of Latin and hyper-Graeco-Latin cubes and hypercubes. J. Indian Soc. Agric. Statist. 2, 20-48.
    15.112/509   15.112/511   15.112/513   15.112/515   15.113/517(2)
    15.113/518(2)   15.113/520   15.114/526(4)   15.115/533   15.115/534(3)
    15.119/541(2)   15.1199/545(3)   15.1199/546(2)   15.1199/547(2)
    15.1511/558

KITAGAWA, T.

    1952   Tables of Poisson distribution. Tokyo: Baifukan.
    2.221/138   2.322/153   14.112/476   16.0264/618

KITAGAWA, T., KITAHARA, T., NOMACHI, Y. and WATANABE, N.

    1953   On the determination of sample size from the two sample theoretical formulation. Bull. Math. Statist. 5, No. 3~4, 35-45.   6.115/303

KITAGAWA, T. and MITOME, M.

    1953   Tables for the design of factorial experiments. Tokyo: Baifukan; Rutland (Vermont): Charles E. Tuttle.

    ¶ According to N. L. Johnson 1954 a, this book contains x + 292 pages of text (in Japanese) and 253 pages of tables. What we have seen is the 1955 reprint (New York: Dover; printed, however, in Japan) which contains 7 pages of front matter in English and 38 + 136 + 25 + 7 + 15 + 13 + 15 + 4 [= 253] pages of tables, apparently identical with the tables reviewed by Johnson.

    ¶ For contents of this book, see pp. 757-758 above.   1j

KITAHARA, T.   See Kitagawa, Kitahara, Nomachi & Watanabe

KLEM, LAURA   See Green, Smith & Klem

KLERK-GROBBEN, GERDA en PRINS, H. J.

    1954   Toets voor de gelijkheid von twee kleine kansen met behulp van even grote steekproeven en het onderscheidingsvermogen van deze toets. Statistica Neerlandica 8, 7-20.
¶ Not abstracted; apparent classification 6.32.
See Mathematical Reviews 16, 499.

KNUTH, D. E.   See Bose, Chakravarti & Knuth

KOGEKNIKOV, N.   See Kolmogorov 1933

KOLLER, S.

    1943   Graphische Tafeln zur Beurteilung statistischer Zahlen. Second edition. (First edition 1940) Dresden, etc.: Theodor Steinkopff. (Third edition 1953. Darmstadt: Dietrich Steinkopff. What we have seen is the 1945 reprint of the 2d ed. Ann Arbor (Michigan): J. W. Edwards)
¶ For serious use of the nomographs in this book, the German prints, which are bound to lie flat, are preferable.
1.1121/5   1.1161/10   14.21/487   xxxxiv

KOLMOGOROV, A.

    1933   Sulla determinazione empirica di una legge di distribuzione.
Giorn. Ist. Ital. Attuari 4, 83-91.   9.411/380   9.4111/380

    1941   Confidence limits for an unknown distribution function.
Ann. Math. Statist. 12, 461-463.   9.4113/381

KOLMOGOROV, A. N., SVEŠIKOV, A. A. and GUBLER, I. A.
КОЛМОГОРОВ А. Н., СВЕШНИКОВ А. А. и ГУБЛЕР И. А.

    1945   Сборник статей по теории стрельбы. I. Труды Мат. Инст. Стеклов. 12.
1.9/655   2.122/656

KOŁODZIEJCZYK, ST.   See Neyman, Iwaszkiewicz & Kołodziejczyk

KONDO, T.

    1930   A theory of the sampling distribution of standard deviations.
Biometrika 22, 36-64.   2.522/168

    *1954   Evaluation of some ω_n^2 distribution.
J. Gakugei Coll. Tokushima Univ. (Nat. Sci.) 4, 45-47.
¶ Not abstracted; apparent classification 9.429.
See Mathematical Reviews 15, 885.

KONDO, T. and ELDERTON, ETHEL M.

    1931   Table of normal curve functions to each permille [sic] of frequency (table of functions of the normal curve to ten decimal places).
Biometrika 22, 368-376. $1.2111/19_F$ $1.221/21_F$ $1.222/22_F$

KOOPMANS, TJALLING

    1942   Serial correlation and quadratic forms in normal variables.
Ann. Math. Statist. 13, 14-33.   7.5/336

KOROLYuK, V. S. and YaROŠEVS'KIĬ, B. I. {КОРОЛЮК В. С. і ЯРОШЕВСЬКИЙ Б. І.
КОРОЛЮК В. С. и ЯРОШЕВСКИЙ Б. И.

    1951   Вивчення максимального розходження двох емпіричних розподілів. Изучение максимального расхождения двух эмпирических распределений. Доповіді Акад. Наук Україн. РСР 1951, 243-247.   9.4122/382

KOTANI, M. and AMEMIYA, A.

    1940   Tables of integrals useful for the calculations of molecular energies. II. Proc. Phys.-Math. Soc. Japan 22, extra no.
    2.1134/133

KOTANI, M., AMEMIYA, A., ISHIGURO, E. and KIMURA, T.

    1955   Table of molecular integrals. Tokyo: Maruzen.   2.1134/656

KOTANI, M., AMEMIYA, A. and SIMOSE, T.

    1938   Tables of integrals useful for the calculations of molecular energies. Proc. Phys.-Math. Soc. Japan 20, extra no.
    $2.1134/133_F$ 2.1134/656

    ¶ Neither Kotani, Amemiya & Simose 1938 nor Kotani & Amemiya 1940 will necessarily be found in bound sets of Proc. Phys.-Math. Soc. Japan. According to Fletcher et al. 1946, p. 12: "A record degree of inaccuracy appears to be reached in the two papers by Kotani [et al.], where the 1940 paper gives errata occupying about 20 per cent of the space taken by the 1938 paper, and introduces fresh errors."

KRAMER, C. Y.   See Bradley 1954

KRAMP, C.

    An VII [i.e. 1799] Analyse des réfractions astronomiques et terrestres. Leipsic: E.B. Schwikkert; Paris: Armand Koenig.
    $1.1216/12_F$ xxxxj

KRAUSE, RUTH H.   See Conrad & Krause

KRISHNA IYER, P. V.

    1950   The theory of probability distributions of points on a lattice. Ann. Math. Statist. 21, 198-217.   5.5/290

KRISHNA IYER, P. V. and RAO, A. S. P.

    *1953   Theory of the probability distribution of runs in a sequence of observations. J. Indian Soc. Agric. Statist. 5, 29-77. $9.311/709_M$

KRISHNA IYER, P. V.; and SINGH, B. N.

    *1955   On certain probability distributions arising from a sequence of observations and their applications.

J. Indian Soc. Agric. Statist. 7, 127-168.
¶ Not abstracted; apparent classification 9.32.
See Mathematical Reviews 19, 188.

KRISHNAN NAIR, A. N.   See Nair, A.N.K.

KRUSKAL, W. H.

1957 z Historical notes on the Wilcoxon unpaired two-sample test.
J. Amer. Statist. Assoc. 52, 356-360.   9.121/368(intro.)

KRUSKAL, W. H. and WALLIS, W. A.

1952   Use of ranks in one-criterion variance analysis.
J. Amer. Statist. Assoc. 47, 583-621.
9.121/368(intro.)   9.13/370   9.14/372   9.121/706(Vianelli)
9.13/707(Vianelli)

1953   Errata: Kruskal & Wallis 1952, etc., etc.
J. Amer. Statist. Assoc. 48, 907-911.
9.121/369(Van der Reyden)   9.13/371(Kruskal & Wallis 1952)
9.14/372(Rijkoort)

KULLBACK, S.

1937   On certain distributions derived from the multinomial distribution.
Ann. Math. Statist. 8, 127-144.   5.4/287

1939   Note on a matching problem. Ann. Math. Statist. 10, 77-80.
13.4/464

KUNISAWA, K., MAKABE, H. and MORIMURA, H.

1951   Tables of confidence bands for the population distribution function
1. Rep. Statist. Appl. Res. Un. Jap. Sci. Engrs. 1, No. 2. 23-44.
9.4111/380

KUROIWA, YOKO, Miss   See Masuyama & Kuroiwa

KUSIAKOWA, MARIA   See Steinhaus 1954

LAADI, H.   See Wold 1948

LABROUSTE, H. et LABROUSTE, Y., Mme

*1943 a Analyse des graphiques résultant de la superposition de sinusoïdes:
Tables numériques précédées d'un exposé de la méthode d'analyse par
combinaisons linéaires d'ordonnées. Paris: Presses Univ. de France.
11.28/727$_M$

*1943 b Analyse des graphiques résultant de la superposition de sinusoïdes:
Atlas de courbes de selectivité. Supplément aux tables numériques.
Paris: Presses Univ. de France.   11.28/727$_M$

LABROUSTE, Y., Mme   See Labrouste, H. & Labrouste, Y., Mme

LABY, BETTY   See Douglas 1955

LAGRANGE, J. L.

[1774*] Mémoire sur l'utilité de la méthode de prendre le milieu entre les résultats de plusieurs observations; dans lequel on examine les avantages de cette méthode par le calcul des probabilités; & où l'on résoud differens problèmes relatifs a cette matière
Mem. Accad. Sci. Torino [5] (1770/73) (Misc. Taurin. V), classe Mathématique, 167-232.
Reprinted, not literatim, in Lagrange 1868, pp. 173-234.
10.261/418   10.262/418   iij

1868   Œuvres, vol. 2.   Paris: Gauthier-Villars.   10.261/418
10.262/418

LAL, D. N.; and MISHRA, D.

*1955   Distribution of the ratio of the logarithm of any one of the ranges of samples from a rectangular population to the sum of the logarithms of each of them. J. Indian Soc. Agric. Statist. 7, 179-186.
¶ Not abstracted; apparent classification 8.21.
See Mathematical Reviews 19, 73.

LANG, E. D.   See Rushton & Lang

LANGDON, W. H. and ORE, Ø.

1930   Semi-invariants and Sheppard's correction.
Annals of Mathematics 31, 230-232.   10.331/427

LANGE, O.   See Czechowski et al.

LARGUIER, E. H.

1936   On a method for evaluating the moments of a Bernoulli distribution.
Ann. Math. Statist. 7, 191-195.   3.53/204

LARMOR, IDA   See David, F.N. 1938

LARSEN, L. H.   See Schuler & Gebelein 1955a, 1955b

LATSCHA, R.

1953   Tests of significance in a 2 × 2 table: extension of Finney's table.
Biometrika 40, 74-86.   2.611/175   2.611/176(Armsen)   5.1/275

LATSHAW, V. V.   See Davis, H.T. 1933

LAWLEY, O. N.   See Hsu & Lawley

LEARY, MARGARET V.   See Fertig & Leary

LEBEDEV, A. V. and FËDOROVA, RIMMA MAKSIMOVNA
ЛЕБЕДЕВ А. В. и ФЕДОРОВА РИММА МАКСИМОВНА

1956   Справочник по математическим таблицам.
Москва: Издат. Акад. Наук СССР.   1.82/127   xxxxij

1960   A guide to mathematical tables, trans. D. G. Fry.
Oxford, New York, etc.: Pergamon.   xxxxij

---

*The date 1774 is given by Vassalli-Eandi 1816, p. IX.

LEDWIRTH, JOAN, Mrs   See Bradley 1954

LEE, ALICE

   1914   Table of the Gaussian "tail" functions; when the "tail" is larger than the body. Biometrika 10, 208-214. $1.332/42_F$

   1917   Further supplementary tables for determining high (tetrachoric) correlations from tetrachoric groupings. Biometrika 11, 284-291. $1.722/120_F$

   1925   Table of the first twenty tetrachoric functions to seven decimal places. Biometrika 17, 343-354.
$1.1141/6_F$ $1.1161/9_F$ $1.3241/35_F$ $1.3249/35$

   1927   Supplementary tables for determining correlation from tetrachoric groupings (tetrachoric correlations). Biometrika 19, 354-404. $1.721/120_F$

See B.A. 1896, 1899; Jahnke & Emde 1909, pp. 44-45; Pearson, K. 1902 a; Pearson, K. 1930 a, T. 54; Pearson, K. & Lee; Soper, Young, Cave, Lee & Pearson

LEFEVRE, J.

  *1952   Application de la théorie collective du risque a la réassurance 'Excess-Loss'. Skand. Aktuarietidskr. 35, 161-187. $1.325/634_M$

LEGENDRE, A. M.

   1826   Traité des fonctions elliptiques et des intégrales Eulériennes, avec des tables pour en faciliter le calcul numérique. Vol. 2. Paris: Huzard-Courcier. $1.1215/12_F$ $2.5131/162_F$

     See Duffell 1909

LEGENDRE, A. M. and PEARSON, K.

   1921   Tables of the logarithms of the complete Γ-function to twelve figures. Tracts for Computers, 4. Cambridge Univ. Press, etc. $2.5131/162_F$

LEHMER, DERRICK HENRY

   1941   Guide to tables in the theory of numbers.
Bull. Nat. Res. Council, 105.   xxxvij

   1951   Mathematical methods in large-scale computing units.
In Proceedings of a second symposium on large-scale digital calculating machinery jointly sponsored by the Navy Department Bureau of Ordnance and Harvard University at the Computation Laboratory, 13-16 September 1949, pp. 141-146.
The Annals of the Computation Laboratory of Harvard University, 26.
Cambridge (Massachusetts): Harvard Univ. Press;
London: Oxford Univ. Press.   13.62/468

   1954   [review of Steinhaus 1954] Math. Reviews 15, 636 only.
13.2/461(Steinhaus 1954)

Lehmer, D. H.
Littell

 1955 [review of Steinhaus 1954 a] Math. Reviews 16, 628 only.
  13.2/461(Steinhaus 1954)

LEHMER, EMMA

 1944 Inverse tables of probabilities of errors of the second kind. Ann. Math. Statist. 15, 388-398.
  4.24/256(ftn.) 4.242/259 4.242/260(Dixon & Massey) 5.1/275
  6.131/307(Bose 1949)

LENGYEL, B. A.

 1939 On testing the hypothesis that two samples have been drawn from a common normal population. Ann. Math. Statist. 10, 365-375.
  6.132/308

LEONE, FRED C. See Sampling by Variables II.; Topp & Leone

LEONE, F. C. and TOPP, C. W.

 1952 Circular probability paper. Indust. Quality Control 9, no. 3, 10-16.
  14.229/734

LeROUX, J. M.

 1931 A study of the distribution of the variance in small samples. Biometrika 23, 134-190. 2.521/166

LESCISIN, ANNA See Davis, H.T. 1933

LESLIE, S. T. See Binet, Leslie, Weiner & Anderson

LEV, J. See Walker & Lev

LEVENE, HOWARD

 1952 On the power function of tests of randomness based on runs up and down. Ann. Math. Statist. 23, 34-56. 9.311/375

LEVERETT, H. M.

 1947 Table of mean deviates for various portions of the unit normal distribution. Psychometrika 12, 141-152.
  ⁕ Not abstracted; apparent classification 1.222.
  See Mathematical Reviews 9, 47.

LEVINE, HARRIET See Goldberg & Levine; Eisenhart, Hastay & Wallis

LEVY, H. and ROTH, L.

 1936 Elements of probability. Oxford Univ. Press. $1.1211/11_F$

LEWIS, EDWARD ERWIN

 1953 Methods of statistical analysis in economics and business. Boston, etc.: Houghton Mifflin.
  1.1121/4 2.3113/142 4.1311/212 4.2314/249

LEWIS, T.

    1953   99·9 and 0·1% points of the χ^2 distribution. Biometrika 40, 421-426.
        2.3114/143   2.3115/143

    See Fieller, Lewis & Pearson

LI, J. C. R.

    1944   Design and analysis of some confounded factorial experiments.
        Iowa Agric. Expt. Sta., Res. Bull. 333.
        15.112/503(2)   15.112/504(3)   15.1132/521(3)   15.1132/522(3)
        15.1132/523(2)   15.1132/524   15.114/526   15.114/527   15.114/529(3)
        15.114/531   15.114/532(4)   15.114/533(2)   15.115/534(2)

LIEBERMAN, G. J.

    1958   Tables for one-sided statistical tolerance limits.
        Indust. Quality Control 14, no. 10, 7-9.   14.32/494

    See Chernoff & Lieberman; Resnikoff & Lieberman

LIEBERMAN, G. J. and RESNIKOFF, G. J.

    1955   Sampling plans for inspection by variables.
        J. Amer. Statist. Assoc. 50, 457-516.   14.121/730

LIEBLEIN, J.   See Gumbel 1954; N. B. S. 1952

LIEBLEIN, J. and SALZER, H. E.

    1957   Table of the first moment of ranked extremes.
        J. Res. Nat. Bur. Standards 59, 203-206.
        ⁋ Not abstracted; apparent classification 8.24.

LINDER, ARTHUR

    *1951   Statistische Methoden für Naturwissenschafter, Mediziner und
        Ingenieure. Second edition. (First edition 1945) Basel: Birkhäuser.
        4.239/677$_M$

LINDLEY, D. V. and MILLER, J. C. P.

    1953   Cambridge elementary statistical tables. Cambridge Univ. Press, etc.
        ⁋ For contents of this book, see p. 758 above.   1j

LINK, R. F.

    1950   The sampling distribution of the ratio of two ranges from
        independent samples. Ann. Math. Statist. 21, 112-116.
        1.5512/101   1.5512/103(Dixon & Massey)   1.5512/103(Pillai 1951a)
        1.5512/104(Dixon & Massey)

LINTON, JEAN   See N.B.S. 1959

LITTAUER, S. B.   See Peach & Littauer

LITTELL, A. S.

    1952   Estimation of the T-year survival rate from follow-up studies over a
        limited period of time. Human Biol. 24, 87-116.   14.159/485

LIVERMORE, J. R.

    1934   The interrelations of various probability tables and a modification of Student's probability table for the argument "t"
J. Amer. Soc. Agronomy 26, 665-673.   4.114/210

LIVERMORE, J. R. and NEELY, W.

    1933   The determination of the number of samples necessary to measure differences with varying degrees of precision.
J. Amer. Soc. of Agronomy 25, 573-578.   1.19/14

LODGE, A.   See B.A. 1896, 1899; B.A. 1, 1931

LOE, ALICE G.   See Romig 1953

LÖSCH, F.   See Jahnke & Emde 1948 (1952 edition); Jahnke, Emde & Lösch

LONDON. UNIVERSITY COLLEGE. DEPARTMENT OF STATISTICS

    See Birnbaum et al. 1955

LÓPEZ, J. R.   See Royo López & Ferrer Martin

LORAINE, P. K.   See Brownlee, Kelly & Loraine

LORD, E.

    1947   The use of range in place of standard deviation in the t-test.
Biometrika 34, 41-67.
1.233/30   1.5512/99   1.5512/101(Dixon & Massey)
1.5512/102(Dixon & Massey)   1.5512/103(Pillai 1951 b)
1.5512/104(Dixon & Massey)

    1950   Power of the modified t-test (u-test) based on range.
Biometrika 37, 64-77.   1.5512/101   4.1412/221

LOVASICH, JEAN L.   See N.B.S. 1959

LOVE, H. H.

    1924   A modification of Student's table (method) for use in interpreting experimental results. J. Amer. Soc. Agronomy 16, 68-73.   4.114/210

LOVELAND, EDWARD H.   See Gordon, Loveland & Cureton

LOWAN, A. N.   See N.B.S. 1943, 1953 y, z, 1958

LUNDWALL, E.   See Wold 1948

LURY, D. B. de   See Chung & De Lury; De Lury

LYERLY, S. B.

    1952   The average Spearman rank correlation coefficient.
Psychometrika 17, 421-428.   8.119/351

LYONS, T. C.   See Guilford & Lyons

LYTLE, E. J., Jr.

    1956   A description of the generation and testing of a set of random normal deviates. In Meyer, H.A., 1956, pp. 234-248.
    13.61/468   13.63/468

    See Meyer, Lytle & Gephart

McCALLAN, S. E. A.   See Wilcoxon & McCallan

McCALLAN, S. E. A. and WILCOXON, F.

    1932   The precision of spore germination tests.
    Contributions from Boyce Thompson Institute 4, 233-243.   3.42/199

McCRADY, M. H.

    1915   The numerical interpretation of fermentation-tube results (Treatment of fermentation by bacteria).
    J. Infectious Diseases 17, 183-212.   6.32/317

McELRATH, G. W. and BEARMAN, J. E.

    1957   See Sampling by Variables III.

McINTYRE, G. A.

    1952   A method for unbiased selective sampling, using ranked sets.
    Austral. J. Agric. Res. 3, 385-390.   8.23/362

McKAY, A. T.

    1932   A Bessel function distribution. Biometrika 24, 39-44.
    10.25/417

    1933 a   The distribution of $\sqrt{\beta_1}$ in samples of four from a normal universe. Biometrika 25, 204-210. See also Pearson, K. 1933 z.   1.612/112

    1933 b   The distribution of β_2 in samples of 4 from a normal universe. Biometrika 25, 411-415.   1.622/114

    1935   The distribution of the difference between the extreme observation and the sample mean in samples of n from a normal universe. Biometrika 27, 466-471.   1.42/61(Grubbs)(2)   1.43/63

McKAY, A. T. and PEARSON, E. S.

    1933   A note on the distribution of range in samples of n. Biometrika 25, 415-420.   1.541/84

MACKENZIE, W. A., Miss   See Gosset 1925

McLEARN, IDA   See Woo 1929

MacMAHON, P. A.

    1884   Symmetric functions of the 13^{ic}. Amer. J. Math. 6, 289-300.
    10.3123/424

Mac Mahon
Marbe

    1915   Combinatory analysis, vol. 1. Cambridge Univ. Press, etc.
           10.311/423   10.3123/424   10.32

McMULLEN, L.   See Gosset 1942

MacNEISH, H. F.

    1922   Euler squares. Ann. Math. 23, 221-227.   15.1513/560

McNEMAR, Q.

    1949   Psychological statistics. First edition. (Second edition 1955)
           New York [and London]: John Wiley & Sons.   1.1141/6

Mac STEWART, W.

    1941   A note on the power of the sign test.
           Ann. Math. Statist. 12, 236-239.
               Not abstracted; apparent classification 3.6.
           See Mathematical Reviews 3, 8.

MADOW, W. G.

    1945   Note on the distribution of the serial correlation coefficient.
           Ann. Math. Statist. 16, 308-310.   7.29/333(Quenouille)

MAGNUS, W.   See Erdélyi, Magnus, Oberhettinger & Tricomi

MAHALANOBIS, P. C.

    1932   Statistical notes for agricultural workers. No. 3.
           Auxiliary tables for Fisher's z-test in analysis of variance.
           Indian J. Agricultural Sci. 2, 679-693.
           4.2314/250   4.232/250   4.332/251

    1933   Statistical notes for agricultural workers, 14.
           The use of random sampling numbers in agricultural experiments.
           Indian J. Agric. Sci. 3, 1108-1115.   13.11/456   13.1222/459

    1933 a   Tables for L-tests. Sankhyā 1, 109-122.
           6.112/298(intro.)   6.112/298

MAHALANOBIS, P. C., BOSE, S. S., RAY, P. R. and BANERJI, S. K.

    1934   Tables of random samples from a normal population.
           Sankhyā 1, 289-328.   13.3/462   13.5/466(Kendall 1949 a)

MAINLAND, D. and MURRAY, I. M.

    1952   Tables for use in fourfold contingency tests. Science 116, 591-594.
           2.611/175

MAKABE, H.   See Kunisawa, Makabe & Morimura

MAKABE, HAJIME; and MORIMURA, HIDENORI

    1955   A normal approximation to Poisson distribution.
           Rep. Statist. Appl. Res. Un. Jap. Sci. Engrs. 4, 37-46.
               Not abstracted; apparent classification 2.129.
           See Mathematical Reviews 17, 756.

MALLOWS, C. L.

   1956   Generalizations of Tchebycheff's inequalities.
      J. Roy. Statist. Soc. Ser. B 18, 139-168; and discussion, 168-176.
      10.4/434(ftn.)

MANDEL, J.

   1954   Chain block designs with two-way control of heterogeneity.
      Biometrics 10, 251-272.
      15.2104/578   15.2104/579   15.2104/580(2)   15.2106/590(2)
      15.2106/591   15.256/608

MANIYa, G. M.   МАНИЯ Г. М.

   1953   Практическое применение оценки максимума двусторонних уклонений
      эмпирической кривой распределения интервале роста теоретического закона.
      Сообщ. Акад. Наук Грузин. ССР 14, 521-524.   9.4121/382

MANN, H. B.

   1949   Analysis and design of experiments:
      Analysis of variance and analysis of variance designs. New York: Dover.
      4.2314/249   4.242/258(Tang)

MANN, H. B. and WALD, A.

   1942   On the choice of the number of class intervals in the application of
      the Chi Square test. Ann. Math. Statist. 13, 306-317.
      9.4/380   2.63/664(Williams)

MANN, H. B. and WHITNEY, D. R.

   1947   On a test of whether one of two random variables is stochastically
      larger than the other. Ann. Math. Statist. 18, 50-60.
      9.121/368   9.121/706(Vianelli)

MANTEL, N.   See Cornfield & Mantel

MARAKATHAVALLI, N.

  *1954   The distribution of t_1 and its applications.
      J. Madras Univ. Sect. B 24, 251-272.
      * Not abstracted; apparent classification 4.1419.
      See Mathematical Reviews 16, 602.

  *1955   Unbiased test for a specified value of the parameter in the non-
      central F-distribution. Sankhyā 15, 321-330.   4.242/679

MARBE, K.

   1934   Grundfragen der angewandten Warscheinlichkeitsrechnung und
      theoretischen Statistik. München, etc.: C.H. [Oscar] Beck.
      13.4/464(intro.)

MARCHAL, J. E.   See Durham, Gaddum & Marchal

MARCINKIEWICZ, J.

    1938   Sur une propriété de la loi de Gauss. Math. Z. 44, 612-618.
            10.211/408

MARITZ, J. S.

    1950   On the validity of inferences drawn from the fitting of Poisson and negative binomial distributions to observed accident data. Psychological Bull. 47, 434-443.   5.22/282

MARK, A. M.   See Gruenberger & Mark

MARKOV, A. A.

    1888   Table des valeurs de l'intégrale $\int_{x}^{\infty} e^{-t^2} dt$.
            St.-Pétersbourg: Académie I. des Sciences; etc. $1.1214/12_F$
            See Čebyšev 1899

MARKS, S.   See Teichroew 1956

MARRIOTT, F. H. C.

    1952   Tests of significance in canonical analysis. Biometrika 39, 58-64.
            $6.133/312_M$ 6.12/690

MARTIN, CELIA   See N.B.S. 1952

MARTIN, E. S.

    1934   On corrections for the moment coefficients of frequency distributions when the start of the frequency is one of the characteristics to be determined (Moment corrections with unknown curve-start). Biometrika 26, 12-58
            10.332/428

MARTÍN, S. F.   See Royo López & Ferrer Martín

MASSEY, F. J.

    1950   A note on the estimation of a distribution function by confidence limits. Ann. Math. Statist. 21, 116-119.
            9.4121/381   9.4121/381(Massey 1951)   9.4121/714

    1951   The Kolmogorov-Smirnov test for goodness of fit.
            J. Amer. Statist. Assoc. 46, 68-78.   9.4121/381

    1951a   The distribution of the maximum deviation between two sample cumulative step functions. Ann. Math. Statist. 22, 125-128.
            9.4122/383   9.4122/715

    1952   Distribution table for the deviation between two sample cumulatives. Ann. Math. Statist. 23, 435-441.   9.4122/383
            See Dixon & Massey

MASUYAMA, M.

 1951 An improved binomial probability paper and its use with tables.
  Rep. Statist. Appl. Res. Un. Jap. Sci. Engrs. 1, no. 2, 15-22 (1 plate)
   ¶ Not abstracted; apparent classification 12.2.
  See Mathematical Reviews 13, 961.

 1951 z Revision of the tables in "An improved binomial probability paper and its use with tables" [Masuyama 1951].
  Rep. Statist. Appl. Res. Un. Jap. Sci. Engrs. 1, no. 3, 32-33.
   ¶ Not abstracted; apparent classification 12.2.
  See Mathematical Reviews 14, 487.

 1952 A graphical method of estimating parameters in Kapteyn distributions.
  Rep. Statist. Appl. Res. Un. Jap. Sci. Engrs. 1, no. 4, 32-34.
   ¶ Not abstracted; apparent classification 10.22.

 1955 Tables of two-sided 5% and 1% control limits for individual observations of the r-th order. Sankhyā 15, 291-294.
  14.229/490 3.319/666

MASUYAMA, M. and KUROIWA, YOKO, Miss

 1951 Table for the likelihood solutions of gamma distribution and its medical applications.
  Rep. Statist. Appl. Res. Un. Jap. Sci. Engrs. 1, No. 1, 18-23.
  10.035/395 10.055/398

MATHER, K.

 1943 Statistical analysis in biology. London: Methuen; New York: Interscience. 2.3113/142 4.1311/211 4.2313/249

 1949 The analysis of extinction time data in bioassay. Biometrics 5, 127-143. 8.3121/364 12.61/452

MATHISEN, H. C.

 1943 A method of testing the hypothesis that two samples are from the same population. Ann. Math. Statist. 14, 188-194. 9.123/370

MAUCHLY, J. W.

 1940 Significance test for sphericity of a normal n-variate distribution. Ann. Math. Statist. 11, 204-209. 6.139/313

MAY, DONALD CURTIS See Burington & May

MAY, JOYCE M.

 1952 Extended and corrected tables of the upper percentage points of the "Studentized" range. Biometrika 39, 192-193. See also Hartley 1953.
  1.5511/98 1.5511/98(Pillai) 1.5511/98(Hartley)
  1.5511/99(Pearson & Hartley)

 See Pearson & Hartley 1954

MAYER, JOHANN TOBIAS, 1723-1762

 1750 Abhandlung über die Umwälzung des Monds um seine Axe, und die scheinbare Bewegung der Mondsflecken. Worinnen der Grund einer verbesserten Mondsbeschreibung aus neuen Beobachtungen geleget wird. Kosmographische Nachrichten und Sammlungen*, 1748, 52-183.
   11.12/440(ftn.)

MELLER, N. A. See Karpov 1954

MELLOR, J. W.

 1931 Higher mathematics for students of chemistry and physics. (First edition 1902, 4th 1913) London: Longmans. Reprinted 1946. New York: Dover; [London: Constable] $1.1222/13_F$

MENDENHALL, W.

 1958 A bibliography [sic] on life testing and related topics. Biometrika 45, 521-543. 14.15/483 xxxxiij xxxxiv

MERRILL, A. S.

 1928 Frequency distribution of an index when both the components follow the normal law. Biometrika 20 A, 53-63. 1.78/653

MERRIMAN, M.

 1910 A text-book on the method of least squares. Eighth edition. (First edition 1877, 4th 1884) New York: John Wiley & Sons. $1.1222/13_F$

MERRINGTON, MAXINE

 1941 Numerical approximations to the percentage points of the χ^2 distribution. Biometrika 32, 200-202. 2.314/145

 1942 Table of percentage points of the t-distribution. Biometrika 32, 300 only. $4.1311/212_F$

   See Pearson, E.S. 1947; Pearson, E.S. & Merrington

MERRINGTON, MAXINE and THOMPSON, CATHERINE M.

 1943 Tables of percentage points of the inverted beta (F) distribution. Biometrika 33, 73-88. 4.231/247 $4.2312/248_F$ 4.233/252

MESCAL, CATHERINE See Mainland & Murray 1952

METROPOLIS, N., REITWIESNER, G. and von NEUMANN, J.

 1950 Statistical treatment of first 2000 decimal digits of e and π calculated on the ENIAC. Math. Tables Aids Comput. 4, 109-111.
   13.63/468

---

*See Gregory 1943, page 1515, column 1, last 3 lines.

MEYER, HERBERT A.

    1956   Symposium on Monte Carlo methods held at the University of Florida; conducted by the Statistical Laboratory, sponsored by Wright Air Development Center of the Air Research and Development Command; March 16 and 17, 1954. New York [and London]: John Wiley & Sons. 13.0/455   13.6/468   1ij

MEYER, HERBERT A., LYTLE, ERNEST J., Jr. and GEPHART, LANDIS S.

    *1955   Random normal deviates.   10.3/463

MICHAEL, W. P., PERRY, N. C. and GUILFORD, J. P.

    *1952   The estimation of a point biserial coefficient of correlation from a phi coefficient. British J. Statist. Psychol. 5, 139-150. 7.91/698 (Dingman)

MIDDLETON, DAVID   See Harvard Univ. 1952

MIHAĬLOVA, V. I.   See Faddeeva & Terent'ev 1954

MILES, S. R.

    1934   A new table of odds based on Student's table of probability. J. Amer. Soc. Agronomy 26, 341-346 + folding plate.   4.114/210

MILLS, FREDERICK C.

    1938   Statistical methods. Second edition (First edition 1924, 3d 1955) New York: Holt, [Rinehart & Winston].   1.1121/4

MILLER, F. L., Jr.   See Connor & Zelen 1959; N.B.S. 1957

MILLER, JEFFERY CHARLES PERCY

    1954   Table of binomial coefficients. Royal Society Mathematical Tables, 3. Cambridge Univ. Press, etc.   3.211/186   5.31/284

    See B.A. 9, 1940; Fletcher, Miller & Rosenhead; Lindley & Miller; N.B.S. 1952

MILLER, LESLIE H.

    1956   Table of percentage points of Kolmogorov statistics. J. Amer. Statist. Assoc. 51, 111-121. 9.4121/382   9.4151/384   9.4121/714   9.4151/715

MILLER, W. LASH and GORDON, A. R.

    1931   Numerical evaluation of infinite series and integrals which arise in certain problems of linear heat flow, electrochemical diffusion, etc. J. Phys. Chem. 35, 2785-2884.   1.812/126

MILLS, JOHN P.

    1926   Table of the ratio: Area to bounding ordinate for any portion of normal curve. Biometrika 18, 395-400. $1.131/13_F$

MILNE, WILLIAM EDMUND

    1949   Numerical calculus: approximations, interpolation, finite differences, numerical integration, and curve fitting. Princeton Univ. Press; London: Oxford Univ. Press.   1.321/32

MILNE-THOMSON, L. M.

    1932   The zeta function of Jacobi. A seven-decimal table of $Z(u|m)$ at interval $0\cdot01$ for u and for values of m $(= k^2)$ from $0\cdot1$ to $1\cdot0$. Proc. Roy. Soc. Edinburgh 52, 236-250.   7.211/331

    1950   Jacobian elliptic function tables: a guide to practical computation with elliptic functions and integrals together with tables of sn u, cn u, dn u, $Z(u)$. New York: Dover.   7.211/331

    See B.A. 1, 1931

MILNE-THOMSON, L. M. and COMRIE, L. J.

    *1931   Standard four-figure mathematical tables. London: Macmillan. [fifty]

MINER, J. R.

    1922   Tables of $\sqrt{1-r^2}$ and $1-r^2$ for use in partial correlation and in trigonometry. Baltimore: The Johns Hopkins Press. 7.31/695   7.32/695$_F$

MISES, H. v.   See Geiringer, H.

MISES, R. v.   See von Mises, R.

MISHRA, D.   See Lal & Mishra

MITCHELL, A. C. G. and ZEMANSKY, M. W.

    1934   Resonance radiation and excited atoms. Cambridge Univ. Press, etc. 1.812/126$_F$

MITOME, M.   See Kitagawa & Mitome

MITRA, S. K.

    1957   Table for tolerance limits for a normal population based on sample mean and range or mean range. J. Amer. Statist. Assoc. 52, 88-94. 14.32/494

MITROPOL'SKIĬ, A. K.   МИТРОПОЛЬСКИЙ А. К.

    1955   Об определителях распределения ряда натуральных чисел. Успехи Мат. Наук 10, вып. 4, 143-144.
    ¶ Not abstracted; apparent classification 10.262.
    See Mathematical Reviews 17, 702.

MIYAKE, KIMIKO   See Kitagawa 1952

MOCK, O.   See Teichroew 1956

MODE, E. B.

   1951   Elements of statistics. Second edition. (First edition 1941)
      New York: Prentice-Hall.
      1.1121/4   1.1161/10   2.3113/142   3.211/186   4.1313/213   4.2314/249
      7.1221/323

MOLINA, E. C.

   1942   Poisson's exponential binomial limit.
      New York [i.e. Princeton (New Jersey), London, etc.]: D. Van Nostrand.
      $2.122/135_F$ $2.221/138_F$ 2.129/657(ftn.)

MOOD, A. M.

   1939   Note on the L_1 test for many samples.
      Ann. Math. Statist. 10, 187-190.   6.112/299

   1940 z The distribution theory of runs. Ann. Math. Statist. 11, 367-392.
      9.311/374(intro.)

   1950   Introduction to the theory of statistics.
      New York, London, etc.: McGraw-Hill Book Co.
      1.115/8   1.1161/10   1.2112/20   2.3112/141   4.1311/211   4.2312/248

      See Dixon & Mood; Harris, Horvitz & Mood

MOORE, E. BRAMLEY-, Miss   See B.A. 1899

MOORE, G. H.   See Wallis & Moore

MOORE, G. H. and WALLIS, W. A.

   1943   Time series significance tests based on signs of differences.
      J. Amer. Statist. Assoc. 38, 153-164.   9.312/376

MOORE, L. BRAMLEY-, Miss   See B.A. 1899

MOORE, P. G.

   1955   The properties of the mean square successive difference in samples
      from various populations. J. Amer. Statist. Assoc. 50, 434-456.
      $1.531/644_M$

   1957   The two-sample t-test based on range. Biometrika 44, 482-489.
      1.541/646   1.5512/651

MORAN, P. A. P.

   1948   Rank correlation and product-moment correlation.
      Biometrika 35, 203-206.   8.114/351

   1950   Recent developments in ranking theory.
      J. Roy. Statist. Soc. Ser. B 12, 153-162; and discussion, pp. 182-191.
      8.10/347

   1950 a A curvilinear ranking test.
      J. Roy. Statist. Soc. Ser. B 12, 292-295.   8.19/358

   1951   Partial and multiple rank correlation. Biometrika 38, 26-32.
      8.15/356

1953 The random division of an interval. III.
J. Roy. Statist. Soc. Ser. B 15, 77-80.
• Not abstracted; apparent classification 6.114.
See Mathematical Reviews 15, 237.

MORANT, G. M.

1939 A bibliography [sic] of the statistical and other writings of Karl Pearson. London: Biometrika office, University College.

MORGAN, A. de   See De Morgan, A.

MORGAN, W. A.

1939 A test for the significance of the difference between the two variances in a sample from a normal bivariate population.
Biometrika 31, 13-19.   4.28/265

MORIGUTI, S.

1951 Extremal properties of extreme value distribution.
Ann. Math. Statist. 22, 523-536.   10.9/435

1952 A lower bound for a probability moment of any absolutely continuous distribution with finite variance. Ann. Math. Statist. 23, 286-289.
10.4/434

*1954 Confidence limits for a variance component.
Rep. Statist. Appl. Res. Un. Jap. Sci. Engrs. 3, 29-41. $4.241/677_M$

1954 z Bounds for second moments of the sample range.
Rep. Statist. Appl. Res. Un. Jap; Sci. Engrs. 3, 57-64.
• Not abstracted; apparent classification 1.541.
See Mathematical Reviews 16, 840.

MORIMURA, HIDENORI

See Hiraga, Morimura & Watanabe; Kunisawa, Makabe & Morimura; Makabe & Morimura

MORRISON, D. R.   See Noether 1956

MORRISON, M.

1956 Fractional replication for mixed series. Biometrics 12, 1-19.
15.114/527   15.115/535

MORSE, A. P. and GRUBBS, F. E.

1947 The estimation of dispersion from differences.
Ann. Math. Statist. 18, 194-214.   1.531/76(intro.)   1.531/77

MORTON, ROBERT LEE

1928 Statistical tables. New York, etc.: Silver Burdett.
Also issued as the appendix to Morton's Laboratory exercises in educational statistics with tables. Same date and publisher.
1.1121/4   1.1162/10

MOSES, L. E.

 1952 A two-sample test. Psychometrika 17, 239-247. 9.2/373

MOSHMAN, J.

 1952 Testing a straggler mean in a two way classification using the range.
  Ann. Math. Statist. 23, 126-132. 1.552/106

 1953 Critical values of the log-normal distribution.
  J. Amer. Statist. Assoc. 48, 600-609. 10.2221/413

 *1954 The generation of pseudo random numbers on a decimal calculator.
  J. Assoc. Comput. Mach. 1, 88-91. 13.62/468

MOSKOWITZ, D. See Rosenbach, Whitman & Moskowitz

MOSTELLER, F.

 1941 Note an application of runs to quality control charts.
  Ann. Math. Statist. 12, 228-232.
  9.311/374(intro.) 9.311/374 9.311/709

 1946 On some useful "inefficient" statistics.
  Ann. Math. Statist. 17, 377-408.
  1.51/71 1.541/87 8.19/358(Blomqvist)

 1948 A k-sample slippage test for an extreme population.
  Ann. Math. Statist. 19, 58-65. 9.14/371

 See Bush & Mosteller; Freeman, Friedman, Mosteller & Wallis;
 Harvard Univ. 1955; Hastings, Mosteller, Tukey & Winsor

MOSTELLER, F. and TUKEY, J. W.

 1950 Significance levels for a k-sample slippage test.
  Ann. Math. Statist. 21, 120-123. 9.14/371(Mosteller 1948)

MOTHES, J.

 1952 Techniques modernes de contrôle des fabrications, Vol. 1.
  Paris: Dunod. 2.121/134 2.221/138 14.222/490

MOUL, MARGARET See Pearson, K. 1930 b

MOZART, W. A.

 *1793 Anleitung, Walzer oder Schleifer mit zwei Würfeln zu componiren, so
  viele man will, ohne etwas von der Musik oder Composition zu verstehen.
  Instruction pour composer, etc. Instruction to compose, etc.
  Instruzione per comporre, ecc. Berlin, etc.: J. J. Hummel.
  ¶ Not abstracted; apparent classification 13.5.

MULHOLLAND, H. P.

 1951 On distributions for which the Hartley-Khamis solution of the
  moment-problem is exact. Biometrika 38, 74-89. 10.261/418

MURPHY, R. B.

 1948 Non-parametric tolerance limits. Ann. Math. Statist. 19, 581-589.
  14.31/492

Murray
National Bureau of Standards

MURRAY, I. M.   See Mainland & Murray

MURRAY, J. A., Miss   See Douglas 1955

MURTY, V. N.

    1955   The distribution of the quotient of maximum values in samples from a rectangular distribution. J. Amer. Statist. Assoc. 50, 1136-1141.
          $8.21/700_M$

N. B. S.   See National Bureau of Standards

NABEYA, S.

    1951   Absolute moments in 2-dimensional normal distribution. Ann. Inst. Statist. Math., Tokyo 3, 2-6.   1.73/123

    1952   Absolute moments in 3-dimensional normal distribution. Ann. Inst. Statist. Math. (Tokyo) 4, 15-30.
        ¶ Not abstracted; apparent classification 1.79.
    See Mathematical Reviews 14, 569.

NAG, A. C.   See Mahalanobis, Bose, Ray & Banerji 1934

NAIR, K. RAGHAVAN

    1938   On a method of getting confounded arrangements in the general symmetrical type of experiment. Sankhyā 4, 121-138.
        15.113/517(2)   15.119/541

    1940   Balanced confounded arrangement(s) for the 5^m type of (in a factorial) experiment. Sankhyā 5, 57-70.
        15.115/533   15.115/534(2)   15.115/535   15.1199/546

    1940 a   Tables of confidence intervals for the median in samples from any continuous population. Sankhyā 4, 551-558.   3.6/669

    1948 a   The studentized form of the extreme mean square test in the analysis of variance. Biometrika 35, 16-31.   4.251/261   4.252/261

    1948 b   The distribution of the extreme deviate from the sample mean and its studentized form. Biometrika 35, 118-144.
        1.42/59   1.42/61(Grubbs)   1.44/69   1.42/637   1.44/640(Nair 1952)

    1949   A further note on the mean deviation from the median. Biometrika 36, 234-235.   1.52/74

    1950   Efficiencies of certain linear systematic statistics for estimating dispersion from normal samples. Biometrika 37, 182-183.
        1.52/72   1.52/74   1.533/81

    *1952   Table of percentage points of the 'Studentized' extreme deviate from the sample mean. Biometrika 39, 189-191.   1.44/640

    See Bose & Nair

NAIR, K. R. and RAO, C. R.

    1948   Confounding in asymmetrical factorial designs.
        J. Roy. Statist. Soc. Ser. B 10, 109-131.
        15.2102/566   15.2102/567(3)   15.2102/568(2)   15.2102/569(2)
        15.2103/570   15.2103/573(2)   15.2103/574(2)   15.2104/579
        15.2104/580   15.2104/582(2)   15.2106/590   15.2107/592   15.2110/604
        15.212/605

NAIR, U. S.

    1936   The standard error of Gini's mean difference.
        Biometrika 28, 428-436.   1.532/80

    1939   The application of the moment function in the study of distribution
        laws in statistics. Biometrika 30, 274-294.
        6.113/300(Bishop & Nair)

    See Bishop & Nair

NAKAMURA, T.   See Kotani, Amemiya, Ishiguro & Kimura 1955

NANDA, D. N.

    1951   Probability distribution tables of the lerger [!] root of a
        determinantal equation with two roots.
        J. Indian Soc. Agric. Statist. 3, 175-177.
        6.12/306   6.12/691(Chowdhury 1956)

NATH, PRAN

    1951   Confluent hypergeometric function. Sankhyā 11, 153-166.
        4.1422/227   4.1422/228

[U.S.] NATIONAL BUREAU of STANDARDS

    1941   Tables of probability functions, Vol. I.
        (Reissued 1954: Tables of the error function and its derivative.
        Applied Mathematics Series 41. Washington: G.P.O.)
        1.1211/11   1.1213/11

    1941a   Miscellaneous physical tables:
        Planck's radiation functions and electronic functions.
        New York: Federal Works Agency, Work Projects Administration.
        $7.33/334_F$

    1942   Tables of probability functions, Vol. II.
        (Reissued 1953: Tables of normal probability functions.
        Applied Mathematics Series 23. Washington: G.P.O.)
        1.111/3   1.1131/5   1.1161/9(2)   1.1211/11(3 ftn.)

    1943   Table of circular and hyperbolic tangents and cotangents for radian
        arguments. New York: Columbia Univ. Press.
        $7.211/330_F$ 16.07/622 $16.0712/622_F$ 16.10/623 $16.1043/624_F$

National Bureau of Standards
Newman

1944   Tables of Lagrangian interpolation coefficients.
New York: Columbia Univ. Press.
16.23/627   16.2311/627(3)$_F$ 16.2311/628$_F$

1949   Tables of the confluent hypergeometric function F(n/2, $\frac{1}{2}$, x) and related functions. Applied Mathematics Series 3. Washington: G.P.O.
4.1422/227   4.1423/228   5.1/275

1949 z   Table of sines and cosines to fifteen places at hundredths of a degree. Applied Mathematics Series 5. Washington: G.P.O.
7.32/334(intro.)   16.07/622   16.0721/622   xxxxj

1950   Tables of the binomial probability distribution.
Applied Mathematics Series 6. Washington: G.P.O.
3.111/183   3.221/187

1951   Tables to facilitate sequential t-tests.
Applied Mathematics Series 7. Washington: G.P.O.
5.1/275   6.115/303   14.129/731

1951 a   Monte Carlo method: Proceedings of a symposium held June 29, 30, and July 1, 1949, in Los Angeles, California, under the sponsorship of the RAND Corporation, and the National Bureau of Standards, with the cooperation of the Oak Ridge National Laboratory.
Applied Mathematics Series 12. Washington: G.P.O.

1951 z   Tables of the exponential function e^x.
Applied Mathematics Series 14. Washington: G.P.O.
10.101/402   16.10/623   16.101/623   16.102/624

1952   A guide to tables of the normal probability integral.
Applied Mathematics Series 21. Washington: G.P.O.
xxxviij   xxxxij   xxxxvij

1953   Probability tables for the analysis of extreme-value data.
Applied Mathematics Series 22. Washington: G.P.O.
8.3/364   8.3111/364   8.3121/364   8.322/365   8.331/365   8.332/365
8.3122/703

1953 y   Tables of circular and hyperbolic sines and cosines for radian arguments. Applied Mathematics Series 36. Washington: G.P.O.
16.0711/622   16.10/623   16.1041/624

1953 z   Table of natural logarithms for arguments between zero and five to sixteen decimal places. Applied Mathematics Series 31. Washington: G.P.O.   10.102/402   16.11/625   16.112/625

*1954   Tables of salvo kill probabilities for square targets.
Applied Mathematics Series 44. Washington: G.P.O.   1.78/653$_M$

1954a   Table of the gamma function for complex arguments.
Applied Mathematics Series 34. Washington: G.P.O.   2.5152/164

1955 y   Table of hyperbolic sines and cosines: x = 2 to x = 10.
Applied Mathematics Series 45. Washington: G.P.O.
16.10/623   16.1041/624

1955 z Table of the descending exponential: $x = 2.5$ to $x = 10$.
Applied Mathematics Series 46. Washington: G.P.O.
10.101/402   16.10/623   16.102/624

1957   Fractional factorial experiment designs for factors at two levels.
Applied Mathematics Series 48. Washington: G.P.O.

15.10/500	15.112/509(15)	15.114/528(13)	15.1199/544(11)
.112/503(4)	/510(15)	/529(19)	/545(15)
/504(4)	/511(20)	/530(6)	/546(3)
/505(7)	/512(25)	/531(15)	/547(9)
/506(13)	/513(26)	15.118/538(12)	15.141/555
/507(6)	/514(27)	/539(17)	
/508(8)	/515(5)	/540(8)	

1958   Table of natural logarithms for arguments between five and ten to sixteen decimal places. Applied Mathematics Series 53.
Washington: G.P.O.   16.11/625   16.112/625

1959   Tables of the bivariate normal distribution function and related functions. Applied Mathematics Series 50. Washington: G.P.O.
1.72/120(4)   1.721/120   1.721/120(ftn.)   1.722/120   1.722/120(ftn.)
1.723/121

See Connor & Young 1961; Connor & Zelen 1959; Gumbel 1954; Salzer 1951

NATU, N. P.   See Sukhatme, Thawani, Pendharkar & Natu

NAYER, P. P. N.

1936   An investigation into the application of Neyman and Pearson's L_1 test, with tables of percentage limits.
Statist. Research Memoirs 1, 38-51.   6.112/299

NEELY, W.   See Livermore & Neely

NELSON, W. F. C.   See Davis, H.T. & Nelson

NEMČINOV, V. S.   НЕМЧИНОВ В. С.

1946   Полиномы Чебышева и математическая статистика.
Москва: Издание Московской сельскохозяйственной академии.
1.1161/9   1.1121/630   1.3211/634   11.119/724

NEUMANN, J. v.   See von Neumann, J.

NEVILLE, E. H.   See B.A. 1, 1931

NEW YORK W. P. A.

Tables so indexed by Fletcher et al. 1946, pp. 420-421, if germane to the present work, will be found under National Bureau of Standards.

NEWMAN, D.

1939   The distribution of range in samples from a normal population, expressed in terms of an independent estimate of standard deviation.
Biometrika 31, 20-30.   1.5511/96   1.5511/98(Dixon & Massey)

*1951   Computational methods useful in analyzing series of binary data.
Amer. J. Psychol. 54, 252-262.   8.3123/365

NEYMAN, J.

    1926   On the correlation of the mean and the variance in samples drawn from an "infinite" population. Biometrika 18, 401-413.   10.9/434

    1941   On a statistical problem arising in routine analyses and in sampling inspections of mass production. Ann. Math. Statist. 12, 46-76.
            6.114/302   10.059/399

    1950   First course in probability and statistics, vol. 1.
            New York: Holt [Rinehart & Winston].   1.1121/5   1.2111/19

NEYMAN, J., IWASZKIEWICZ, K. and KOŁODZIEJCZYK, ST.

    1935   Statistical problems in agricultural experimentation.
            J. Roy. Statist. Soc. Suppl. 2, 107-180.   4.1412/220

NEYMAN, J. and PEARSON, E. S.

    1928 a   On the use and interpretation of certain test criteria for purposes of statistical inference, Part I. Biometrika 20 A, 175-240.
            6.111/296   8.21/360

    1928 b   On the use and interpretation of certain test criteria for purposes of statistical inference, Part II. Biometrika 20 A, 263-294.

    1931 a   Further notes on the χ^2 distribution. Biometrika 22, 298-305.
            2.69/179   5.4/287

    1931 b   O zagadnienu k prob. On the problem of k samples.
            Bull. Acad. Polon. Sci. Ser. A 1931, 460-481.
            6.112/297   6.112/298(Mahalanobis)

NEYMAN, J. and TOKARSKA, B.

    1936   Errors of the second kind in testing "Student's" hypothesis.
            J. Amer. Statist. Assoc. 31, 318-326.   4.1413/222

NICHOLAS, F. J., Miss   See Bowley 1928

NICHOLSON, C.

    1943   The probability integral for two variables. Biometrika 33, 59-72.
            1.72/120   1.723/121

NICHOLSON, W. L.

    1954   A computing formula for the power of the analysis of variance test.
            Ann. Math. Statist. 25, 607-610.   4.242/679(ftn.)

NIELSEN, K. L. and GOLDSTEIN, L.

    1947   An algorithm for least squares.
            J. Math. Phys. (Cambridge) 26, 120-132.   11.12/725

NOETHER, GOTTFRIED E.

    *1955   Use of the range instead of the standard deviation.
            J. Amer. Statist. Assoc. 50, 1040-1055. $1.541/646_M$

    1956   Two sequential tests against trend.
            J. Amer. Statist. Assoc. 51, 440-450.   6.21/314   1.1217/630

NOETHER, MAX   See Faa di Bruno 1881

NOMACHI, Y.

   *1955   Auxiliary tables for the applications of -dimensional -distributions to certain class of empirical functions. Bull. Math. Statist. (Fukuoka) 6, no. 1-2, 25-47.
       ¶ Not abstracted; apparent classification 4.17.

   See Kitagawa, Kitahara, Nomachi & Watanabe

NORTON, H. W.

   1939   The 7 7 squares. Ann. Eugenics 9, 269-307.
      15.117/537(2)   15.1512/558

   1952   MTA 199. (Errata in Fisher & Yates 1948) Math. Tables Aids Comput. 6, 35-38.
      2.3113/142(ftn.)   4.1311/211(ftn.)   4.233/252(Fisher & Yates)
      4.233/252(Merrington & Thompson)   4.333/270

   See Fisher, R.A. & Yates 1938 etc., T. V

NOTTINGHAM, W. B.

   1936   Thermionic emission from tungsten and thoriated tungsten filaments. Phys. Rev. 49, 78-97. $1.813/126_F$

OBERG, E. N.

   1947   Approximate formulas for the radii of circles which include a specified fraction of a normal bivariate distribution. Ann. Math. Statist. 18, 442-447.   5.5/290

OBERHETTINGER, F.   See Erdelyi, Magnus, Oberhettinger & Tricomi

OGAWA, JUNJIRO

   1951   Contributions to the theory of systematic statistics, I. Osaka Math. J. 3, 175-213.   1.41/636

   1952   Contributions to the theory of systematic statistics, II. Large sample theoretical treatment of some problems arising from dosage and time mortality curve. Osaka Math. J. 4, 41-[69].

OKAMOTO, M.

   1955   Fit of a Poisson distribution by the index of dispersion. Osaka Math. J. 7, 7-13.
      ¶ Not abstracted; apparent classification 2.53.
   See Mathematical Reviews 17, 53.

OLDS, E. G.

   1938   Distributions of sums of squares of rank differences for small numbers of individuals. Ann. Math. Statist. 9, 133-148.
      8.1111/349   8.1113/349   8.112/349   8.112/350(Olds 1949)
      8.112/350(Siegel)

1949   The 5% significance levels for sums of squares of rank differences and a correction [to Olds 1938]. Ann. Math. Statist. 20, 117-118.
8.112/350   8.112/350(Siegel)

OLEKIEWICZ, M.

1950   Tables of significance limits for the largest critical ratio out of k ratios. Tablice wartości granicznych dla największego z pośród k ilorazów sprawdzenioych.
Ann. Univ. Mariae Curie-Skłodowska, Sect. A 4, 115-121.
* Not abstracted; apparent classification 4.19.
See Mathematical Reviews 13, 360.

1951a   Tables of expected values and variances of numbers of runs in random sequences with probabilities of exceeding expected values. Tablice wartości oczekiwanych i dyspersji liczby ogniw w sekwencjach losowych z podaniem prawdopodobienstw przekroczenia wartości oczekiwanych.
Таблицы математических ожиданий и дисперсий числа звеньей в случайных последовательностях альтернатив с приведением вероятностей превышения математических ожиданий.
Ann. Univ. Mariae Curie-Skłodowska, Sect. A 5, 147-159.
Not abstracted; apparent classification 9.311.
See Mathematical Reviews 15, 140.

1951b   An extended table of Student's t-distribution for one-sided and two-sided tests of significance at 5% and 1% probability levels. Rozszerzona tablica rozkładu t Student'a dla sprawdzianów jedno- i obustronnych przy 5%-owym i 1%-owym ryzykach błędów.
Расширенная таблица распределения функций Студента к 5%-му и 1%-му уровню вероятности при двухсторонних и односторонних статистических критериях. Ann. Univ. Mariae Curie-Skłodowska, Sect. A 5, 161-163.
Not abstracted; apparent classification 4.131.
See Mathematical Reviews 15, 140.

OLMSTEAD, P. S.

1946   Distribution of sample arrangements for runs up and down.
Ann. Math. Statist. 17, 24-33.   9.311/374   9.312/376(intro.)

1958   Runs determined in a sample by an orbitrary cut.
Bell System Tech. J. 37, 55-82.   9.311/711
See Mosteller 1941

OLMSTEAD, P. S. and TUKEY, J. W.

1947   A corner test for association. Ann. Math. Statist. 18, 495-513.
8.19/359

OLSHEN, A. C.

1938   Transformations of the Pearson Type III distribution.
Ann. Math. Statist. 9, 176-200.   10.031/395   10.039/396

OPPOLZER, T. v.

 1880 Lehrbuch zur Bahnbestimmung der Kometen und Planeten, Vol. II.
  Leipzig: Wilhelm Engelmann. $1.1212/11_F$

ORE, Ø. See Langdon & Ore

ORSTRAND, C. E. van

 See Becker & Van Orstrand; Van Orstrand, C.E.

OSTLE, B.

 1954 Statistics in research:
  basic concepts and techniques for research workers.
  Ames (Iowa): Iowa State [Univ.] Press. 13.11/456 13.13/459

O'TOOLE, A. L.

 1931 On symmetric functions and symmetric functions of symmetric
  functions. Ann. Math. Statist. 2, 102-149. 10.3121/424

 1932 On symmetric functions of more than one variable and of frequency
  functions. Ann. Math. Statist. 3, 56-63.

 1933 On the system of curves for which the method of moments is the best
  method of fitting. Ann. Math. Statist. 4, 1-29. 10.23/416

 1933a A method of determining the constants in the bimodal fourth degree
  exponential function. Ann. Math. Statist. 4, 79-93.
  10.23/416(O'Toole 1933)

OSGOOD, W. F. See Peirce 1929

OWEN, D. B.

 1956 Tables for computing bivariate normal probabilities.
  Ann. Math. Statist. 27, 1075-1090. 1.723/121 1.79/125(Steck)
  See N.B.S. 1959

OZAKI, EIKO, Miss See Hirago, Morimura & Watanabe 1954

PACKER, L. R.

 *1951 The distribution of the sum of *n* rectangular variates I.
  J. Inst. Actuar. Students' Soc. 10, 52-61. $10.261/720_M$

PAGE, E. S.

 *1954 Control charts for the mean of a normal population.
  J. Roy. Statist. Soc. Ser. B 16, 131-135. $14.229/733_M$

 1954a Continuous inspection schemes. Biometrika 41, 100-115.
  14.13/482

 1955 Control charts with warning lines. Biometrika 42, 243-257.
  14.229/491

1955a   A test for a change in a parameter occurring at an unknown point. Biometrika 42, 523-527.   9.9/387$_M$

See Barraclough & Page

PAIRMAN, ELEANOR

1919   Tables of the digamma and trigamma functions. Tracts for Computers, I. Cambridge Univ. Press, etc. 2.5141/164$_F$ 2.5142/164$_F$

PAIRMAN, ELEANOR and PEARSON, K.

1919   On corrections for the moment-coefficients of limited range frequency distributions when there are finite or infinite ordinates and any slopes at the terminals of the range. Biometrika 12, 231-258. 10.332/427   10.332/428(Pearse)

PALEY, R. E. A. C.

1933   On orthogonal matrices. J. Math. Phys. 12, 311-320. 15.1313/554

PALM, CONRAD RUDOLF AGATON, called CONNY PALM

*1947   Table of the Erlang loss formula. Tables of telephone traffic formulae, no. I. Stockholm: C.E. Fritz. 2.129/656$_M$

PARKER, E. T.   See Bose, Shrikhande & Parker

PATNAIK, P. B.

1948   The power function of the test for the difference between two proportions in a 2 × 2 table. Biometrika 35, 157-175.   2.613/177

1949   The non-central χ^2- and F-distributions and their applications. Biometrika 36, 202-232.   2.42/157

1950   The use of mean range as an estimator of variance in statistical tests. Biometrika 37, 78-87. 1.5512/104(Terpstra)   1.552/106(Moshman)   1.5531/107 1.5531/108(Pearson)

1954   A test of significance of a difference between two sample proportions when the proportions are very small. Sankhyā 14, 187-202. 6.21/693$_M$ **[For 1955 see p. 951]**

PAULSON, E.

1942   An approximate normalization of the Analysis Of Variance distribution. Ann. Math. Statist. 13, 233-235.   4.27/263

PEACH, P.

1947   An introduction to industrial statistics and quality control. Second edition. (First edition 1945) Raleigh (North Carolina): Edwards & Broughton. 4.1311/211   4.2314/250   14.112/476   14.221/489

PEACH, P. and LITTAUER, S. B.

    1946   A note on sampling inspection. Ann. Math. Statist. 17, 81-84.
            2.315/150

PEARL, R.

    1940   Introduction to medical biometry and statistics. Third edition. (First edition 1923, 2d 1930) Philadelphia and London: W. B. Saunders.
            1.1131/6   1.1132/6

    See Fisher, Arne, 1922 a.

PEARSE, GERTRUDE E.

    1928   On corrections for the moment-coefficients of (asymptotic) frequency distributions when there are infinite ordinates at one or both of the terminals of the range. Biometrika 20 A, 314-355.
            10.332/427(Pairman & Pearson)   10.332/427   10.332/428(Martin)

PEARSON, EGON S.

    1922   Table of the logarithms of the complete Γ-function (for arguments 2 to 1200, i.e. beyond Legendre's range). Tracts for Computers, 8. Cambridge Univ. Press, etc. $2.5131/162_F$ $2.5134/163_F$

    1925   Bayes' theorem examined in the light of experimental sampling. Biometrika 17, 388-442.   3.319/192   10.012/392

    1926   A further note on the distribution of range in samples taken from a normal population. Biometrika 18, 173-194.
            1.42/57   1.541/83(Tippett)   1.541/83   1.541/84(Pearson 1932)
            1.542/88

    1927   Editorial note (on two preceding papers) [i.e. Irwin 1927, Hall 1927 a] Biometrika 19, 244-245.

    1929   Some notes on sampling tests with two variables. Biometrika 21, 337-360.   4.21/246

    1929 b   Note on Dr Craig's paper [Craig 1929]. Biometrika 21, 294-302.
            10.037/396

    1930 a   A further development of tests for normality. Biometrika 22, 239-249.
            1.613/112   1.621/113   1.623/114   1.623/114(Geary & Pearson)
            1.623/114(Smith & Duncan)

    1931 z   Note on tests for normality. Biometrika 22, 423-424.
            1.621/114(Geary)

    1932   The percentage limits for the distribution of range in samples from a normal population. Biometrika 24, 404-417.
            1.541/84   1.541/85(Mc Kay & Pearson)   1.541/86(Ferber)
            1.5511/96(Newman)   10.00/388

    1935 b   A comparison of β_2 and Mr. Geary's w_n criteria [Geary 1935, 1935 a]. Biometrika 27, 333-352.   13.4/464

Pearson, E. S.
Pearson, K.

    1935 c   The application of statistical methods to industrial standardisation and quality control. London: British Standards Institution.
          ¶ At head of pages: 'No. 600-1935'.
          1.1141/7   1.52/73   2.315/148   4.133/217   4.1419/226

    1936   Note on probability levels for $\sqrt{b_1}$. Biometrika 28, 306-307.
          1.611/112   1.613/113

    1947   The choice of statistical tests illustrated on the interpretation of data classes in a 2 × 2 table. Biometrika 34, 139-167.
          See also Barnard 1947 z.
          2.611/175(Finney)   2.613/177(Patnaik)   2.613/178(Pearson & Merrington)

    1952   Comparison of two approximation to the distribution of the range in small samples from normal populations. Biometrika 39, 130-136.
          1.5531/108

    See Birnbaum et al. 1955; Clopper & Pearson; Comrie 1938; David, F.N. 1938; David, H.A., Hartley & Pearson; Davies & Pearson; Fieller, Lewis & Pearson; Geary & Pearson; Gosset 1942; Hartley & Pearson; Kendall, M.G. & Babington Smith 1939; McKay & Pearson; Merrington & Thompson 1943; Neyman & Pearson; Pearson, K. 1948; Pearson, K. & Pearson, E.S.; Pitman 1939; Thompson, A.J. 1943; Thompson, C.M. & Merrington 1946; Wold 1948

PEARSON, E. S. and ADYANTHĀYA, N. K.

    1929   The distribution of frequency constants in small samples from non-normal symmetrical populations. Biometrika 21, 259-286.
          1.5512/101(Walsh)

PEARSON, E. S. and CHANDRA SEKHAR, C.

    1936   The efficiency of statistical tools and a criterion for the rejection of outlying observations. Biometrika 28, 308-319.
          See also Scott 1936.   1.44/68

PEARSON, E. S., GODWIN, H. J. and HARTLEY, H. O.

    1945   The probability integral of the mean deviation. Biometrika 33, 252-265. (E.S. Pearson, Editorial note, pp. 252-253. H.J. Godwin, On the distribution of the estimate of mean deviation obtained from samples from a normal population, pp. 254-256. H.O. Hartley, Appendix: Note on the calculation of the distribution of the mean deviation in normal samples, pp. 257-258. Tables of the probability integral of the mean deviation in normal samples, pp. 260-265)   1.52/74

PEARSON, E. S. and HAINES, JOAN

    1935   The use of range in place of standard deviation in small samples. J. Roy. Statist. Soc. Suppl. 2, 83-98.   2.315/149   14.222/490

PEARSON, E. S. and HARTLEY, H. O.

    1942   The probability integral of the range in samples of n observations from a normal population. Biometrika 32, 301-310.
        1.541/85   1.541/86(Ferber)   1.5512/99

    1943   Tables of the probability integral of the studentized range. Biometrika 33, 89-99.
        1.5511/97   1.5511/98(Dixon & Massey)   1.5511/99(Pearson & Hartley 1954)

    1951   Charts of the power function for analysis of variance tests, derived from the non-central F-distribution. Biometrika 38, 112-130.
        4.1412/221(Pearson & Hartley 1954)   4.242/260   xxxxiv

    1954   Biometrika tables for statisticians, vol. 1. Cambridge Univ. Press, etc.
        ¶ For contents of this book, see pp. 759-761 above.
        xxxxj   xxxxix   1j

    1954a   Biometrika tables for statisticians, vol. I: Position with regard to tables omitted from present volume. One loose leaf, circulated with Pearson & Hartley 1954.
        10.007/391(intro.)

PEARSON, E. S. and MERRINGTON, MAXINE

    1948 2×2 tables; the power function of the test on a randomized experiment. Biometrika 35, 331-345.   2.613/178

    1951   Tables of the 5% and 0·5% points of Pearson curves . . . expressed in standard measure. Biometrika 38, 4-10.   10.002/389

PEARSON, E. S. and NEYMAN, J.

    1930   O zagadnienu dwóch prób. On the problem of two samples. Bull. Acad. Polon. Sci. Sér. A 1930, 73-96.
        6.111/297   6.112/298(Mahalanobis)

PEARSON, E. S. and SUKHATME, A. V.

    1935   An illustration of the use of fiducial limits in determining the characteristics of a sampled batch. Sankhyā 2, 13-32.
        4.133/216   4.239/253

PEARSON, KARL

    1895   Contributions to the mathematical theory of evolution. II. Skew variation in homogeneous material. Philos. Trans. Roy. Soc. London Ser. A 186, 343-414. (Reprinted in Pearson, K. 1948, pp. 41-112)
        10.00/388   10.00/388(ftn.)   10.01/392   10.02/393   10.03/394
        10.04/397   10.13/404   10.19/406

Pearson, K.

1900   On the criterion that a given system of deviations from the probable in the case of a correlated system of variables is such that it can be reasonably supposed to have arisen from random sampling.
Phil. Mag. 50, 157-175. Reprinted in Pearson, K. 1948, pp. 339-357.
2.6/174   9.4/380

1901   Mathematical contributions to the theory of evolution. X.
Supplement to a memoir [Pearson 1895] on skew variation.
Philos. Trans. Roy. Soc. London Ser. A 197, 443-459.
(Reprinted in Pearson, K. 1948, pp. 359-375)
10.05/398   10.06/399   10.19/406

1902   Note on Francis Galton's problem [Galton 1902].
Biometrika 1, 390-399.   1.543/90

1902 z   On the systematic fitting of curves to observations and measurements.
Biometrika 1, 265-303.   10.23/415

1905   "Das Fehlergesetz und seine Verallgemeinerungen durch Fechner und Pearson." [Ranke & Greiner 1904] (Skew variation) A rejoinder.
Biometrika 4, 169-212.   10.00/388(ftn.)

1906   Mathematical contributions to the theory of evolution, XV.
A mathematical theory of random migration.
Drapers' Company Research Memoirs, Biometric Series, 3.
Cambridge Univ. Press, etc. (original imprint London: Dulau)
2.41/154

1913 a   On the probable error of a coefficient of correlation as found from a fourfold table. Biometrika 9, 22-27.
1.2313/26   7.94/343   7.94/698

1915 b   On the distributions of the standard deviations of small samples: Appendix I to papers by 'Student' [Gosset 1908 a] and R.A. Fisher [1915].
Biometrika 10, 522-529.
(Unsigned editorial: attributed to Pearson by Morant 1939, p. 42)
2.522/167

1916   Mathematical contributions to the theory of evolution. XIX.
Second supplement to a memoir [Pearson 1895; 1901] on skew variation.
Philos. Trans. Roy. Soc. London Ser. A 216, 429-457.
(Reprinted in Pearson, K. 1948, pp. 529-557)
10.00/388(ftn.)   10.004/390   10.07/399   10.08/401   10.09/402
10.10/402   10.11/403   10.12/403   10.19/406

1919   On generalized Tchebycheff theorems in the mathematical theory of statistics. Biometrika 12, 284-296.   10.4/433

1920   On the probable errors of frequency constants. Part 3.
Biometrika 13, 113-132.
(Unsigned editorial: attributed to Pearson by Morant 1939, p. 47)
1.40/49   1.41/51   1.51/70   1.544/92

Pearson, K.

1920 a On the construction of tables and on interpolation. Part I.
Uni-variate tables. Tracts for Computers, 2.
Cambridge Univ. Press, etc.   16.2311/628

1920 b On the construction of tables and on interpolation. Part II.
Bi-variate tables. Tracts for Computers, 3.
Cambridge Univ. Press, etc.   16.2311/628

1921 Table of ordinates of the normal curve for each permille [sic] of
frequency. Biometrika 13, 426-428.
(Unsigned editorial: attributed to Pearson by Morant 1939, p. 49.
The table was computed by H. E. Soper) $1.221/21_F$

1921 a On a general method of determining the successive terms in a skew
regression line. Biometrika 13, 296-300.   10.9/434(Neyman)

1922 Tables of the incomplete Γ-function.
London: H. M. Stationery Office.
(Reissue 1934, London: Biometrika Office, University College.
Presently published by Cambridge Univ. Press)   2.111/131

1923 y Notes on skew frequency surfaces. Biometrika 15, 222-230.
10.19/406

1923 z On non-skew frequency surfaces. Biometrika 15, 231-244.
10.19/406

1924 On a certain double hypergeometric series and its representation by
continuous frequency surfaces. Biometrika 16, 172-188.
5.1/276(intro.)

1924 z Note on Professor Romanovsky's generalisation of My frequency curves.
Biometrika 16, 116-117.   10.18/405(Romanovsky)

1925 a Further contributions to the theory of small samples.
Biometrika 17, 176-200.
1.78/124(Pearson 1931 a)   2.522/168   1.78/652

1925 z On the multiple correlation of brothers, being a note on Mr J.O.
Irwin's memoir [Irwin 1925 a], and on my statement of the application
of Galton's difference problem to the determination of the degree of
relationship of brothers, made in August 1902 [K. Pearson 1902]
Biometrika 17, 129-141.

1928 On a method of ascertaining limits to the actual number of marked
members in a population of given size from a sample.
Biometrika 20 A, 149-174 + folding plate.
5.1/276(intro.)   5.1/277

1930 a Tables for statisticians and biometricians, part 1, third edition.
(First edition 1914, second 1924) [Cambridge Univ. Press, etc.]
(Original imprint London: Biometric Laboratory, University College)
¶ For contents of this book, see pp. 761-764 above.
xxxxj [fifty] 1j

Pearson, K.
Pearson, M. V.

1930 b On the remaining tables for determining the volumes of a (the) bivariate normal surface. Biometrika 22, 1-34.
(Unsigned editorial: attributed to Pearson by Morant 1939, p. 62)
1.722/120

1931 a Tables for statisticians and biometricians, part 2.
[Cambridge Univ. Press, etc.]
(Original imprint London: Biometric Laboratory, University College)
¶ For contents of this book, see pp. 764-767 above.
xxxxj [fifty] 1j

1931 b Appendix to a paper by Dr. Wishart [i.e. Wishart 1931]. Tables of the mean and squared standard deviation of the square of a multiple correlation coefficient. Biometrika 22, 362-367.
(Unsigned editorial: attributed to Pearson by Morant 1939, p. 64)
5.1/276(intro.) 7.4/335

1931 c On the nature of the relationship between two of "Student's" variates (z_1 and z_2) when samples are taken from a bivariate normal population. Biometrika 22, 405-422. 4.19/242

1931 d Some properties of "Student's" z: correlation, regression and scedasticity of z with the mean and standard deviation of the sample. Biometrika 23, 1-9. 4.16/240

1931 z Appendix to a paper by Professor Tokishige Hojo: On the standard error of the median to a third approximation when the median is found from a sample of size n = 2p + 1, and the parent population is normal and of standard deviation σ. Biometrika 23, 361-363. 1.41/53(Hojo)

1933 On a method of determining whether a sample of size n supposed to have been drawn from a parent population having a known probability integral has been drawn at random (General criterion for random sampling) Biometrika 25, 379-410. 7.11/322

1933 z Note on Mr McKay's paper [McKay 1933 a]. Biometrika 25, 210-213.

1934 Tables of the incomplete Beta-function:
prepared under the direction of and edited by K. Pearson, F.R.S.
[Cambridge Univ. Press]
(Original imprint London: "Biometrika" Office, University College)
3.12/185$_F$ 3.51/204$_F$ 3.52/204

1948 Karl Pearson's early statistical papers. Cambridge Univ. Press, etc.
10.00/388 10.00/388(ftn.) 10.004/390(Pearson 1916) 10.01/392
10.02/393 10.03/394 10.04/397 10.05/398 10.06/399 10.07/399
10.08/401 10.09/402 10.10/402 10.11/403 10.12/403 10.13/404

See Arkin & Colton 1950, T. 9; B.A. 1896, 1899; Brownlee, J. 1923; Holzinger 1925 a; Irwin 1923; Lee 1925; Legendre & Pearson; Morant 1939; N.B.S. 1959; Pairman & Pearson; Pearson, E.S. 1935 b; Soper, Young, Cave, Lee & Pearson; Tippett 1927; Woo 1929

PEARSON, K. and ELDERTON, ETHEL M.

    1923   On the variate difference method. Biometrika 14, 281-310.
            5.1/276(intro.)   7.53/339

    1908   On the generalised probable error in multiple correlation.
            Biometrika 6, 59-68. $1.31/31(intro.)_F$ $1.31/31_F$ 1.332/43

PEARSON, K. and PEARSON, E. S.

    1922   On polychoric coefficients of correlation. Biometrika 14, 127-156.
            1.3249/36

PEARSON, K. and PEARSON, MARGARET V.

    1931   On the mean character and variance of a ranked individual, and on the mean and variance of the intervals between ranked individuals. Biometrika 23, 364-397.
            1.41/52   1.41/53(Hojo)   1.43/63   1.541/84

    1932   On the mean character and variance of a ranked individual, and on the mean and variance of the intervals between ranked individuals. II. Case of certain skew curves (Rank-variates and rank-intervals) Biometrika 24, 203-279.
            8.23/362   10.087/401   10.097/402   10.107/403   10.117/403

    1935   On the numerical evaluation of high order incomplete Eulerian integrals. Biometrika 27, 409-423.   2.41/154   2.41/155   3.43/199

PEARSON, K. and SOPER, H. E.

    1921   See Pearson, K. 1921

PEARSON, K. and STEOSSIGER, BRENDA

    1931   Tables of the probability integrals of symmetrical frequency curves in the case of low powers such as arise in the theory of small samples. Biometrika 22, 253-283.   3.12/185   10.021/393   10.071/400

PEARSON, K., JEFFERY, G. B. and ELDERTON, ETHEL M.

    1929   On the distribution of the first product moment-coefficient, in samples drawn from an indefinitely large normal population. Biometrika 21, 164-201.
            7.6/340   7.6/342   10.25/417(intro.)
            10.25/417(Pearson, Stouffer & David)

PEARSON, K., STOUFFER, S. A. and DAVID, F. N.

    1932   Further applications in statistics of the $T_m(x)$ Bessel function. Biometrika 24, 293-350.
            7.6/340(Pearson, Jeffery & Elderton)   10.25/417

PEARSON, K. and YOUNG, A. W.

    1918   On the (higher order normal) product-moments of various orders of the normal correlation surface of two variates. Biometrika 12, 86-92.
            1.73/122

PEARSON, MARGARET V.   See Pearson, K. & Pearson, Margaret V.

PEATMAN, J. G. and SHAFER, R.

    1942   A table of random numbers from selective service numbers.
J. Psychology 14, 295-305.   13.11/456   13.1211/457

PECK, L. G. and HAZELWOOD, R. L.

    1958   Finite queuing tables. Operations Research Society of America, Publications in Operations Research, 2.
New York [and London]: John Wiley & Sons.
    ¶ Not abstracted; apparent classification 5.1.
See Mathematical Reviews 19, 1146.

PEIRCE, B. O.

    1929   A short table of integrals. Third edition [by W.F. Osgood] (First edition 1899, 2d 1910, 4th, by R.M. Foster, 1956)
Boston, London, etc.: Ginn.   1.1211/11

PEISER, A. M.

    1943   Asymptotic formulas for significance levels of certain distributions. Ann. Math. Statist. 14, 56-62.
2.314/146   2.314/147(Goldberg & Levine)   4.132/215

PENDHARKAR, V. G.   See Sukhatma, Thawani, Pendharkar & Natu

PEPPER, J.

    1929   Studies in the theory of sampling. Biometrika 21, 231-258.
10.36/432

    1932   The sampling distribution of the third moment coefficient—an experiment. Biometrika 24, 55-64.   1.611/112(intro.)

    See Pearson, K. 1934

PERLO, V.

    1933   On the distribution of Student's ratio for samples of three drawn from a rectangular distribution. Biometrika 25, 203-204.   4.143/229

PERRY, N. C.   See Michael, Perry & Guilford

PERSONS, W. M.   See Rietz, Carver et al.

PETERS, CHARLES CLINTON and Van VOORHIS, W. R.

    1940   Statistical procedures and their mathematical bases.
New York and London: Mc Graw-Hill Book Co.
$1.1121/4_F$ $1.1161/10_F$ $1.2111/19_F$ $1.221/21_F$ $1.323/34_F$ 1.3249/35
1.631/116   1.633/117   2.1132/132   4.1111/208   4.115/210(Gosset)
4.1311/211   4.239/253   7.93/343

PETERS, JOHANN THEODOR, called JEAN PETERS

    1919   Hilfstafeln zur Zehnstelligen [sic] Logarithmentafel.
Berlin: Preussische Landesaufnahme. (Reissue 1957 with added t.p.: Auxiliary tables to the ten-place logarithm table.
New York: Frederick Ungar)

1922 Zehnstellige Logarithmentafel. Erster Band. Zehnstellige Logarithmen der Zahlen von 1 bis 100 000 nebst einem Anhang mathematischer Tafeln. Berlin: Reichsamt für Landesaufnahme. (Reissue 1957 with added t.p.: Ten-place logarithm table. Volume one. Ten-place logarithms of the numbers from 1 to 100 000 together with an appendix of mathematical tables. New York: Frederick Ungar)

Pages III-XXVIII and 1-162 of the appendix, entitled Mathematische Tafeln. Bearbeitet und berechnet von Prof. Dr. J. Peters und Dr. J. Stein (in 1957 issue also: Mathematical tables edited and computed by Prof. Dr. J. Peters and Dr. J. Stein) have been cited as Peters & Stein 1922. For C.J. Hyman's English translation of Peters & Stein's text, see Peters 1919 (1957 issue), appendix, pp. 16-36; for a list of errors in their tables, <u>ibid</u>., pp. 50-53.

2.5112/161$_F$ 2.5122/162$_F$ 2.5134/163$_F$ 3.211/186$_F$ 16.06/620
16.061/621$_F$ 16.111/625$_F$

PETERS, J. and STEIN, J.

1922   See Peters, J. 1922

PETO, S.

1953   A dose-response equation for the invasion of micro-organisms. Biometrics 9, 320-335.   14.152/484

PETTIGREW, H. N.   See Connor & Zelen 1959; N.B.S. 1957

PHILLIPPI, NANCY, Miss   See Bradley 1954

PICARD, E.   See Hermite 1908

PIEDEN, MARGARET   See N.B.S. 1959

PILLAI, K. C. S.

1950   On the distribution(s) of midrange and semi-range in samples from a normal population. Ann. Math. Statist. 21, 100-105.   1.542/89

1951 a   Some notes on ordered samples from a normal population. Sankhyā 11, 23-28.   1.5512/103

1951 b   On the distribution of an analogue of Student's t. Ann. Math. Statist. 22, 469-472.   1.5512/103

1952   On the distribution of 'Studentized' range. Biometrika 39, 194-195.   1.5511/98

1956   On the distribution of the largest or the smallest root of a matrix in multivariate analysis. Biometrika 43, 122-127.   6.12/306

PILLAI, K. C. S. and RAMACHANDRAN, K. V.

1954   On the distribution of the ratio of the iTH observation in an ordered sample from a normal population to an independent estimate of the standard deviation (studentized order statistic). Ann. Math. Statist. 25, 565-572.   1.44/640$_M$

PILLAI, P. S. B.   See Sukhatme 1938

PILLING, D. E., Miss   See James 1956; Trickett, Welch & James 1956

PITMAN, E. J. G.

   1937 a Significance tests which may be applied to samples from any populations. J. Roy. Statist. Soc. Suppl. 4, 119-130.

   1937 b Significance tests which may be applied to samples from any populations. II. The correlation coefficient test.
J. Roy. Statist. Soc. Suppl. 4, 225-232.   7.99/344

   1939 Tests of hypotheses concerning location and scale parameters. Biometrika 31, 200-215.   6.113/300(intro.)   6.113/301

   See Robbins, H.E. & Pitman

PIZANIS, S.   See Spenceley et al. 1952

PLACKETT, R. L.

   1946 Literature on testing the equality of variances and covariances in normal populations. J. Roy. Statist. Soc. 109, 457-468.   6.1/295

   1947 Limits of the ratio of mean range to standard deviation. Biometrika 34, 120-122.   8.28/702

PLACKETT, R. L. and BURMAN, J. P.

   1946 The design of optimum multifactorial experiments. Biometrika 33, 305-325.
15.112/504   15.112/509   15.112/513   15.112/514   15.112/515(6)
15.112/516(2)   15.113/517   15.113/520(2)   15.115/534(2)   15.117/537
15.1313/554

POGGENDORFF, J. C.

   1926 J.C. Poggendorff's biographisch-literarisches Handwörterbuch, usw. Vol. 5 (1904-1922), ed. P. Weinmeister.
Leipzig, etc.: Verlag Chemie GmbH.

POLLAK, L. W.

   1926 Rechentafeln zur harmonischen Analyse.
Leipzig: Johann Ambrosius Barth.   11.23/443$_F$

   1929 Études géophysiques faites à Prague. II.
Manuel de l'analyse harmonique. Czechoslovak Republic, Státní úřad statistický, Statistique tchécoslovaque[*], 54.
= Pražké studie geofysikální. II. Rukověť harmonické analysy. Czechoslovak Republic, Státní úřad statistický, Československá statistiká[‡], 54.

     ¶ Fletcher et al. 1946, p. 426, report a German edition entitled Handweiser zur harmonischen Analyse.   11.22/442$_F$

---

[*] See Gregory 1932, page 168, column 1, lines 41-66.

[‡] See Gregory 1932, page 167, column 3, lines 45-61.

POLLAK, L. W. and EGAN, U. N.

    1949 a   All term guide for harmonic analysis and synthesis using 3 to 24; 26, 28, 30, 34, 36, 38, 42, 44, 46, 52, 60, 68, 76, 84 and 92 equidistant values. An Roinn Tionnscail agus Tráchtála, An tSeirbhís Mheteoraíochta, Foillseacháin Géofisice, Iml. II. [Ireland,] Department of Industry and Commerce, Meteorological Service, Geophysical Publications, Vol. II. Dublin: Stationery Office.
        11.2/442   11.22/442

    1949 b   Eight-place supplement to Harmonic analysis and synthesis schedules for three to one hundred equidistant values of empiric function. Dublin Institute for Advanced Studies, School of Cosmic Physics (Institiúid Árd-leinn Bhailé Átha Cliath, Scoil na Fisice Cosmai), Geophysical Memoirs, 1.   11.21/442   11.22/442

POLLAK, L. W. and HEILFRON, C.

    1947   Harmonic analysis and synthesis schedules for three to one hundred equidistant values of empiric functions. An Roinn Tionnscail agus Trachtala, An tSeirbhis Mheteoraiochta, Foillseachain Geofisice, Iml. I. [Ireland,] Department of Industry and Commerce, Meteorological Service, Geophysical Publications, Vol. I. Dublin: Stationery Office.
        11.21/442(2)   11.22/442

POTIN, L.

    1925   Formules et tables numériques relatives aux fonctions circulaires, hyperboliques, elliptiques. Paris: Gauthier-Villars, etc.
        $2.5113/161_F$ $3.211/186_F$

POTTER, MURIEL   See Kendall, M.G. 1946

POZNAHIRKO, N. A.   See Faddeeva & Terent'ev 1954

PRENTISS, N.   See Miller, L.H. 1956

PRETORIUS, S. J.

    1930   Skew bivariate frequency surfaces, examined in the light of numerical illustrations. Biometrika 22, 109-223.
        10.19/406   10.219/410   10.22/410

    See K. Pearson 1931 a, 1934

PRICE, IRENE   See Davis, H.T. 1933

PRICE, S.   See Harding & Price

PRINS, H. J.   See Klerk-Grobben & Prins

PROEHL, ELIZABETH A.   See Fertig & Proehl

PROSCHAN, F.

    1953   Confidence and tolerance intervals for the normal distribution. J. Amer. Statist. Assoc. 48, 550-564.   14.32/493   4.139/673

PRZYBOROWSKI, J. and WILEŃSKI, H.

    1935   Statistical principles of routine work in testing clover seed for dodder. Biometrika 27, 273-292.   2.121/135   2.122/135   2.322/152

    1940   Homogeneity of results in testing samples from Poisson series: with an application to testing clover seed for dodder. Biometrika 31, 313-323.
            6.31/315   6.31/316(Pearson & Hartley)   2.613/664   9.1/704

PUTZ, ELOISE   See N.B.S. 1959

QUARTEY, JAMES

    1955   Table of inverted matrices for the solution of quadratic regression coefficients. Statistica (Bologna) 15, 491 only. See also Fűrst 1955. 11.12/725

QUENOUILLE, M. H.

    1948   Some results in the testing of serial correlation coefficients. Biometrika 35, 261-267.   7.29/333   7.52/338

    1949   Problems in plane sampling. Ann. Math. Statist. 20, 355-375. 7.6/341

    1950   Introductory statistics. London and New York: [Pergamon]
            1.1141/7   2.3112/141   4.1311/212   4.2314/249   13.11/456

    1951   The variate-difference method in theory and practice. Rev. Inst. Internat. Statist. 19, 121-129.   7.53/339

    1952   Associated measurements.
        London: Butterworth; New York: Academic Press.
        2.3112/141   4.1311/212   4.2314/249   7.1222/323   7.211/331
        11.111/437

    1953   The design and analysis of experiment. London: Charles Griffin; New York: Hafner.   2.3112/141   4.1311/212   4.2314/249   13.2/461(4)

RAND CORPORATION.

    Works of this author are listed as if spelled <u>Rand Corporation</u>.

RADFORD, J. F. and WALDVOGEL, R. K.

    1953   Quality control of resistance welding by statistical methods. Welding Journal 32, 521-526.   14.221/490

RAFF, M. S.

    1951   The distribution of blocks in an uncongested stream of automobile traffic. J. Amer. Statist. Assoc. 46, 114-123.   5.21/280

    1956   On approximating the point binomial.
J. Amer. Statist. Assoc. 51, 293-303.
¶ Not abstracted; apparent classification 3.53.
See Mathematical Reviews 18, 160.

RAJ, DES

    1955   Relative efficiency of gauging and exact measurement in estimating the proportion of a population between limits. Sankhyā 15, 191-196. $10.009/716_M$

RAMACHANDRAN, K. V.

    1956   On the simultaneous analysis of variance test.
Ann. Math. Statist. 27, 521-528. $4.251/680_M$
See Pillai & Ramachandran

RAND CORPORATION, the

    1952   Random digits (1-6000). J. Amer. Statist. Assoc. 47, 710-714.
13.11/456   13.13/459

    1953 a   Random digits (6001-6100) [i.e. 6001-6500].
J. Amer. Statist. Assoc. 48, 167 only.   13.11/456   13.13/459

    1953 b   Random digits (6501-6875). J. Amer. Statist. Assoc. 48, 383 only.
13.11/456   13.13/459

    1953 c   Random digits (6876-8125). J. Amer. Statist. Assoc. 48, 672 only.
13.11/456   13.13/459

    1953 d   Random digits (9001-13750) [i.e. 8126-12875].
J. Amer. Statist. Assoc. 48, 931-934.   13.11/456   13.13/459

    1954 a   Random digits (12876-15125). J. Amer. Statist. Assoc. 49, 206-207.
13.11/456   13.13/459

    1954 b   Random digits (15126-17375). J. Amer. Statist. Assoc. 49, 410-411.
13.11/456   13.13/459

    1954 c   Random digits (17376-20875). J. Amer. Statist. Assoc. 49, 682-684.
13.11/456   13.13/459

    1954 d   Random digits (20876-21875). J. Amer. Statist. Assoc. 49, 928 only.
13.11/456   13.13/459

    1955   A million random digits with 100,000 normal deviates.
Glencoe (Illinois): The Free Press.
13.11/456   13.1212/457   13.13/459(2)   13.3/462

See Dixon & Massey 1951, 1957, TT. 1, 2, 23, 24

RANKE, K. E. und GREINER, R.

    1904   Das Fehlergesetz und seine Verallgemeinerungen durch Fechner und Pearson in ihrer Tragweite für die Anthropologie.
Arch. Anthropol. 2, 295-331.

RAO, C. RADHAKRISHNA

    1948   Tests of significance in multivariate analysis.
Biometrika 35, 58-79.   6.1/295

    1949   On some problems arising out of discrimination with multiple characters. Sankhyā 9, 343-366.   6.131/307

RAO, C. RADHAKRISHNA and CHAKRAVARTI, I. M.

    1956   Some small sample tests of significance for a Poisson distribution.
Biometrics 12, 264-282.   6.31/316

RAY, P. R.   See Mahalanobis, Bose, Ray & Banerji

RAY, W. D.

    *1956   Sequential analysis applied to certain experimental designs in the analysis of variance. Biometrika 43, 388-403. $4.241/677_M$

REES, D. H.   See Foster & Rees; Grundy, Healy & Rees

REES, MINA   See Erdélyi, Magnus, Oberhettinger & Tricomi

REITER, S.

    1956   Estimates of bounded relative error for the ratio of variance(s) of normal distributions. J. Amer. Statist. Assoc. 51, 481-488.
$6.119/689_M$

REITWIESNER, G.   See Metropolis, Reitwiesner & von Neumann

RENYI, A.

    1953   On the theory of order statistics. К теории вариационных рядоб.
Acta Math. Acad. Sci. Hungar. 4, 191-231.   9.416/385

RESNIKOFF, GEORGE J.   See Dept. of Defense 1958; Lieberman & Resnikoff

RESNIKOFF, G. J. and LIEBERMAN, G. J.

    1957   Tables of the non-central t-Distribution:
Density function, cumulative distribution function and percentage points.
Stanford (California) Univ. Press; London: Oxford Univ. Press.
4.1411/219   4.1414/225

RHIND, A.

    1909   Tables to facilitate the computation of the probable errors of the chief (frequency) constants of skew frequency distributions.
Biometrika 7, 127-147.
10.00/388(ftn.)   10.004/389   10.006/391   10.007/391

1910   Additional tables and diagram for the determination of the errors of type of frequency distributions. Second paper.
(Probable errors of frequency types) Biometrika 7, 386-397.
10.007/391

RHODES, E. C.

1923   On a certain skew correlation surface. Biometrika 14, 355-377.
10.19/406

RHODES, IDA   See N.B.S. 1951

RICE, W. B.

1947   Control charts in factory management.
New York [and London]: John Wiley & Sons.   14.221/489

RICHARDSON, CLARENCE HUDSON

1934   An introduction to statistical analysis.
(Revised editions 1935, 1944)   New York: Harcourt, Brace [& World]
1.1121/4

RICHARDSON, HAROLD W. P.   See Pearson, K. 1934

RICHARDSON, J. T.

1946   A table of Lagrangian coefficients for logarithmic interpolation of standard statistical tables to obtain other probability levels.
J. Roy. Statist. Soc. Suppl. 8, 212-215.   16.2319/629

RIDER, P. R.

1929   On the distribution of the ratio of mean to standard deviation in small samples from non-normal universes. Biometrika 21, 124-143.
4.143/228   8.22/361   10.262/418

1931   On small samples from certain non-normal universes.
Ann. Math. Statist. 2, 48-65.   4.143/229

1939   An introduction to modern statistical methods.
New York: John Wiley & Sons; London: Chapman & Hall.
$1.1141/6(2)_F$ $1.1161/10_F$ $1.2111/19_F$ $1.2112/20_F$ $2.3113/142_F$
$4.1311/211$ $4.331/251_F$

1951   The distribution of the range in samples from a discrete rectangular population. J. Amer. Statist. Assoc. 46, 375-378.   8.22/361

1951a   The distribution of the quotient of ranges in samples from a rectangular population. J. Amer. Statist. Assoc. 46, 502-507.
8.21/360

RIETZ, H. L.

1924   See Rietz, Carver et al.

{RIETZ, H. L., CARVER, H. C., CRATHORNE, A. R., CRUM, W. L., GLOVER, J. W.,
{HUNTINGTON, E. V., KELLEY, T. L., PERSONS, W. M. and YOUNG, A. A.

    1924   Handbook of mathematical statistics. Boston, etc.: Houghton Mifflin.
        $1.1121/4_F$ $1.1161/9_F$ $1.3211/33_F$

RIJKOORT, P.J.

    1952   A generalisation of Wilcoxon's test.
        Nederl. Akad. Wetensch. Proc. Ser. A 55, 394-404.   9.14/372

    1953   A generalisation of Wilcoxon's test: Errata.
        Nederl. Akad. Wetensch. Proc. Serl. A 56, 407 only.
        9.14/372(Rijkoort 1952)

RIOS, SIXTO   See Royo López & Ferrer Martín 1954

ROA, EMETERIO, born 1896

    [1923*]   A number of new generating functions with applications to statistics, etc. Lancaster (Pennsylvania): [printed by] Lancaster Press.
        10.218/410

ROBBINS, FRANK

    1917   Factorials and allied products with their logarithms.
        Trans. Roy. Soc. Edinburgh 52, 167-174.
        $2.5112/161_F$ $2.5122/162$ $2.5134/163_F$ $2.5135/163_F$

ROBBINS, HERBERT E.

    1944   On the measure of a random set. Ann. Math. Statist. 15, 70-74.
        5.5/289(Garwood)

    1945   On the measure of a random set. II. Ann. Math. Statist. 16, 342-347.
        5.5/289(Garwood)

ROBBINS, HERBERT E. and PITMAN, E. J. G.

    1949   Application of the method of mixtures to quadratic forms in normal variates. Ann. Math. Statist. 20, 552-560.
        2.41/154   4.28/268

ROBINSON, GEORGE   See Whittaker & Robinson

ROBINSON, H. F. and WATSON, G. S.

    1949   An analysis of simple and triple rectangular lattice designs.
        North Carolina Agric. Expt. Sta., Tech. Bull. 88.
        15.2104/582   15.2105/586   15.2106/591   15.2107/594   15.2109/601
        15.2110/604   15.212/604   15.212/605(2)

ROBINSON, JULIA   See Lehmer, E. 1944

ROE, EDWARD DRAKE, Junior

    1904a   On complete symmetric functions. Parts I and II.
        Amer. Math. Monthly 11, 156-163.

---

*Date assigned by Carver 1924, p. 116.

1904 b   On complete symmetric functions. Part III.
   Amer. Math. Monthly 11, 179-184.
   10.3125/425   10.3126/425   10.3127/425   10.3128/425   10.3129/425

ROE, JOSEPHINE ROBINSON

   1931*   Interfunctional expressibility problems of symmetric functions.
   [Ph.D. thesis,] Syracuse Univ., 1918.
   Text, [7] + 46 pp., printed at Cambridge (Massachusetts);
   atlas, 23 plates, printed at New York.   10.311/423

ROESER, H. M.

   1920   Adjustment of parabolic and linear curves to observations taken at equal intervals of the independent variable.
   Sci. Papers Bur. Standards 16, 363-375. $11.12/440_F$

ROMANÍ, J.

   1956   Tests no parametricos en forma secuencial.
   Trabajos Estadíst. 7, 43-96.
   ¶ Not abstracted; apparent classification 9.1.
   See Mathematical Reviews 18, 427.

ROMANOVSKY, V.

   1924   Generalisation of some types of the frequency curves of Professor [Karl] Pearson. Biometrika 16, 106-116.
   2.41/154   10.18/405   10.218/410

   1928   On the criteria that two given samples belong to the same normal population (On the different coefficients of racial likeness)
   Metron 7, No. 3, 3-46.   4.16/240   4.26/263

ROMIG, H. G.

   1953   50-100 binomial tables: New York [and London]: John Wiley & Sons.
   3.112/183   3.221/187

   See Dodge & Romig

RONGE, F.

   1954   Die Verhältnisschätzung (ratio estimate) nach der Methode des "Veränderungsfaktors" und der "additiven Veränderungsgrösse".
   Mitteilungsbl. Math. Statist. 6, 221-232. $7.6/697_M$

RORTY, M. C.   See Fisher, A. 1922

ROSEBRUGH, T. R.   See Miller, W.L. & Rosebrugh

---

*See [U.S.] Library of Congress, Catalog of printed cards, vol. 127, p. 333, cards 31-6253 and 31-34635.

ROSEN, N.

 1931a Calculation of interaction between atoms with s-electrons.
  Phys. Rev. 38, 255-276. 2.1134/133

 1931b The normal state of the hydrogen molecule. Phys. Rev. 38, 2099-2114.
  $2.1134/133_F$

ROSENBACH, J. B., WHITMAN, E. A. and MOSKOWITZ, D.

 1943 Mathematical tables. [Second edition] (First edition 1937)
  Boston, London, etc.: Ginn. 1.1121/4

ROSENBAUM, S.

 1953 Tables for a nonparametric test of dispersion.
  Ann. Math. Statist. 24, 663-668. 9.2/708 14.31/734

 1954 Tables for a nonparametric test of location.
  Ann. Math. Statist. 25, 146-150. 9.129/706 14.31/735

ROSENHEAD, L. See Fletcher, Miller & Rosenhead

ROSENTHAL, A. See Teichroew 1956

ROSS, ELEANOR L. See Jackson & Ross

ROSS, F. A.

 1925 Formulae for facilitating computations in time series analysis.
  J. Amer. Statist. Assoc. 20, 75-79. 16.0445/620

ROSS, J. A. See Hirsch 1827

ROSSER, J. B.

 1948 Theory and application of $\int_0^z e^{-x^2}\,dx$ and $\int_0^z e^{-p^2 y^2}\,dy \int_0^y e^{-x^2}\,dx$.
  Brooklyn (New York): Mapleton House. 1.812/126 1.82/127

 See Teichroew 1956

ROTENBERG, A.

 1960 A new pseudo-random number generator.
  J. Assoc. Comput. Mach. 7, 75-77. 13.62/468

ROTH, L. See Levy & Roth

ROY, S. N.

 1945 The individual sampling distribution(s) of the maximum, the minimum and any intermediate of the p-statistics on the null-hypothesis.
  Sankhyā 7, 133-158. 6.12/690(ftn.)

ROYAL SOCIETY See Miller, J.C.P. 1954

ROYNANE, J. H., Miss   See Douglas 1955

ROYO, J.   See Royo López & Ferrer Martín

ROYO LÓPEZ, J. y FERRER MARTÍN, S.

    1954   Tablas estadísticas: números aleatorios, errores de muestreo y distribución normal. Madrid: Consejo Superior de Investigaciones Cientificas, Instituto de Investigaciones Estadísticas.
        1.1161/10   3.211/186   13.13/459   1.1131/630   1.2121/631   3.319/665
        13.11/729   13.1211/729   13.3/729

    1954a   Tabla de números aleatorios obtenida de los números de la Lotería Nacional Española. Trabajos Estadíst. 4, 247-256.
        13.61/468   13.63/468   13.1211/729

    1955   Tablas auxiliares de estadística. Madrid: Consejo Superior de Investigaciones Cientificas, Instituto de Investigaciones Estadísticas.
        1.1161/10   3.211/186   13.13/459   1.1131/630   1.2121/631   2.3114/658
        3.319/665   4.1312/671   4.2313/676   13.11/729

RUBEN, HAROLD

    1954   On the moments of order statistics in samples from normal populations. Biometrika 41, 200-227.   1.42/637

RUBIN, HERMAN

    1945   On the distribution of the serial correlation coefficient. Ann. Math. Statist. 16, 211-215.   7.51/337

    See Girshick, Rubin & Sitgreaves

RUIZ, A. L., Mrs   See Bradley & Terry 1952

RUSHTON, S.

    1952   On sequential tests of the equality of variances of two normal populations with known means. Sankhyā 12, 63-78.   6.114/302

RUSHTON, S. and LANG, E. D.

    1954   Tables of the confluent hypergeometric function. Sankhyā 13, 377-411.   4.1422/227(6)   4.1422/228(2)

RUSSELL, J. B.

    1933   A table of Hermite functions. J. Math. Phys. 12, 291-297.
        1.332/34   1.332/34(Jahnke & Emde)

RUSSELL, T. S., Mrs   See Bradley & Terry 1952

SADE, A.

    1951   An omission in Norton's list of 7 × 7 squares. Ann. Math. Statist. 22, 306-307.   15.1512/558

SADLER, D. H.   See B.A. 1, 1931

SADOWSKI, W.

   1955   O nieparametrycznym teście na porównywanie rozsiewów.
О непараметрическом критерии для сравнивания разбросов.
Zastos. Mat. 2, 161-171.
      Not abstracted; apparent classification 9.2.
    See Mathematical Reviews 16, 103.

     See Czechowski et al.

SAFFIR, M.   See Chesire, Saffir & Thurstone

SAKODA, J. M. and COHEN, B. H.

   1957   Exact probabilities for contingency tables using binomial coefficients. Psychometrika 22, 83-86.
    ¶ Not abstracted; apparent classification 3.211.
    See Mathematical Reviews 19, 73.

SALMON, G.

   1876   Lessons introductory to the modern higher algebra. Third edition. (First edition—no tables—1859) Dublin: Hodges, Foster.
10.3123/424

   1877   Vorlesungen über die Algebra der linearen Transformationen, tr. W. Fiedler. Second edition. Leipzig: B.G. Teubner.
10.3123/424

SALTON, G.   See Harvard Univ. 1955

SALVOSA, L. R.

   1930   Tables of Pearson's Type III function.
Ann. Math. Statist. 1, 191-198 + 125 pages of tables (May 1930); pp. 128-187 of tables (August 1930).
1.115/7   2.0/130   2.112/132   2.21/137   2.41/154   2.41/155
10.031/394   10.039/396   10.218/410   2.112/656(Vianelli)
2.21/657(Vianelli)   5.62/687(ftn.)

  [1930*z] Generalizations of the normal curve of error, etc. Ann Arbor (Michigan): Edwards Bros.   10.218/410

SALZER, H. E.

   1951   Tables of $n!$ and $\Gamma(n+\frac{1}{2})$ for the first thousand values of n. National Bureau of Standards, Applied Mathematical Series 16. Washington: G.P.O.   2.5111/160   2.5112/161

    See Lieblein & Salzer; N.B.S. 1951, 1952, 1953, 1954

SAMPFORD, M. R.

   1952 b   The estimation of response-time distributions. II. Multi-stimulus distributions. Biometrics 8, 307-369.   1.233/633$_M$

---

*Date conjectured.

*1955   The truncated negative binomial distribution. Biometrika 42, 58-69.
5.32/683   8.3129/703

SAMPLING by VARIABLES

1957   Four papers read in the Detroit convention of the American Society for Quality Control.
      I. One-way protection on $\overline{X}$, [by] Max Astrachan.
      II. Two-way protection on $\overline{X}$, [by] Fred C. Leone.
      III. Protection on variability, [by] Gayle W. Mc Elrath [and] Jacob E. Bearman.
      IV. Available sampling plans, [by] R. L. Storer.
ASQC, National Convention Trans. 11 (1957), 429-470.   14.124/480

SANDELIUS, M.   See Wold 1948

SANDON, F.

1946   Scores for ranked data in school examination practice.
Ann. Eugenics 13, 118-121.   1.43/64

SARHAN, A. E.

1955   Estimation of the mean and standard deviation by order statistics. Part III. Ann. Math. Statist. 26, 576-592.   1.544/647$_M$

SARHAN, A. E. and GREENBERG, B. G.

1956   Estimation of location and scale parameters by order statistics from singly and doubly censored samples. Part I.
The normal distribution up to samples of size 10.
Ann. Math. Statist. 27, 427-451.
1.43/639   1.51/641   1.51/642(Sarhan & Greenberg 1958)   1.544/648
1.544/649(Dixon)   1.544/649(Sarhan & Greenberg 1958)

1957   Tables for best linear estimates by order statistics of the parameters of single exponential distributions from singly and doubly censored samples. J. Amer. Statist. Assoc. 52, 58-88.
8.23/700   14.151/732

1958   Estimation of location and scale parameters by order statistics from singly and doubly censored samples. Part II.
Tables for the normal distribution for samples of size $11 \leq n \leq 15$.
Ann. Math. Statist. 29, 79-105.   1.51/642   1.544/649

SASULY, M.

1934   Trend analysis of statistics: theory and technique.
Washington: Brookings Institution.
5.31/284$_F$ 11.110/437   11.111/437   11.113/438   11.13/441
11.19/441(ftn.)

SATAKOPAN, V.   See Fisher, R.A. & Yates 1948, 1953, 1957, T. XXIII

SAVAGE, I. RICHARD

    1953   Bibliography [sic] of nonparametric statistics and related topics. J. Amer. Statist. Assoc. 48, 844-906.
        8./345   9./366   10.4/434(ftn.)   xxxxiij

    See Grab & Savage

SCARBOROUGH, J. B. and WAGNER, ROBERT WANNER

    1948   Fundamentals of statistics. Boston, etc.: Ginn.
        1.1161/10   1.3211/34

SCHAPIRO, D.   See Greenshields, Schapiro & Ericksen

SCHEFFÉ, H.

    1942   On the ratio of the variances of two normal populations. Ann. Math. Statist. 13, 371-388.   4.241/257

    1943   On solutions of the Behrens-Fisher problem, based on the t-distribution. Ann. Math. Statist. 14, 35-44.
        4.153/239   4.153/240(Walsh)

SCHERING, E. C. J.   See Gauss 1866

SCHNEIDER, B.   See Schuler & Gebelein 1955a, 1955b

SCHROCK, E. M.

    1957   Quality control and statistical methods. Second edition. (First edition 1950) New York: Reinhold; London: Chapman & Hall.
        3.54/205   14.21/732

SCHULER, M. and GEBELEIN, H.

    1955a   Fünfstellige Tabellen zu den elliptischen Funktionen dargestellt mittels des Jacobischen parameters q. Five place tables of elliptical functions based on Jacobi's parameter q. Berlin, etc.: Springer.
        9.4112/381

    1955b   Acht- und neunstellige Tabellen zu den elliptischen Funktionen dargestellt mittels des Jacobischen parameters q. Eight and nine place tables of elliptical functions based on Jacobi's parameter q. Berlin, etc.: Springer.   9.4112/381

SCOTT, ELIZABETH   See Neyman 1941

SCOTT, J. F. and SMALL, V. J.

    *1955   A numerical investigation of least squares regression involving trend-reduced Markoff series. J. Roy. Statist. Soc. Ser. B 17, 105-114.
        $11.19/726_M$

SCOTT, J. M. C.

    1936   Appendix [to Pearson, E.S. & Chandra Sekhar 1936] Biometrika 28, 319-320.   1.44/68(Pearson & Chandra Sekhar)

SEAL, HILARY LATHAM

    1941   Tests of a mortality table graduation. J. Inst. Actuar. 71, 5-67.
        1.1133/6   2.3115/143   3.229/188

    See Comrie 1938

SEELBINDER, B. M.

    1953   On Stein's two-stage sampling scheme.
    Ann. Math. Statist. 24, 640-649.   15.3/737

SEKAR, C. C.   See Pearson, E.S. & Chandra Sekar, C.

SEMON, W. L.

    1951a   MTE 194 (review of Glover 1923)   Math. Tables Aids Comput. 5, 228.
    1.3211/33(Glover)

    1951b   MTE 195 (review of Jørgensen 1916)
    Math. Tables Aids Comput. 5, 228-230.   1.3211/33(Jørgensen)

    See Harvard Univ. 1952

SEN, S.   See Mahalanobis, Bose, Ray & Banerji 1934

SENDERS, VIRGINIA R.

    1958   Measurement and statistics. New York: Oxford Univ. Press.
    8.3123/365   9.112/368(Wilcoxon 1949)   9.121/369(White)   9.32/379
    9.4121/382(Dixon & Massey)

SENGUPTA, J. M.

    1953   Significance level of $\Sigma x^2/(\Sigma x)^2$ based on Student's distribution.
    Sankhyā 12, 363 only.   4.139/673

SHANAWANY, M. R. El   See El Shanawany, M.R.

SHAW, J. B.

    1907   Synopsis of linear associative algebra: a report on its natural
    development and results reached up to the present time.
    Carnegie Institution of Washington, Publication no. 78.
    Washington: the Inst.   10.3123/424(ftn.)

SHEA, J. D.   See Birge & Shea

SHELLEY, B.   See Davis, H.T. 1933

SHENTON, L. R.

    1949   On the efficiency of the method of moments and Neyman's Type A
    distribution. Biometrika 36, 450-454.   5.21/279   5.21/280(Bateman)

    1951   Efficiency of the method of moments and the Gram-Charlier Type A
    distribution. Biometrika 38, 58-73.   10.211/409   10.23/415

SHEPPARD, W. F.

    1898   On the application of the theory of error to cases of normal distribution and normal correlation.
Philos. Trans. Roy. Soc. London Ser. A 192, 101-167.
1.1162/10   1.19/15   1.2121/20(2)

    1898 a   On the calculation of the most probable values of frequency-constants, for data arranged according to equidistant divisions of a scale. Proc. London Math. Soc. 29, 353-380.   10.331/426

    1899   On the statistical rejection of extreme variations, single or correlated. (Normal variation and normal correlation)
Proc. London Math. Soc. 31, 70-99.   1.42/54

    1900   On the calculation of the double-integral expressing normal correlation. Trans. Cambridge Philos. Soc. 19, 23-66.
1.72/120   1.729/122   7.92/343

    1900 z   Some quadrature-formulæ. Proc. London Math. Soc. 32, 258-277.
16.236/629

    1903   New tables of the probability integral. Biometrika 2, 174-190.
$1.115/7_F$ $1.1161/9_F$ $1.2121/20_F$ 1.221/21

    1907   Table of deviates of the normal curve (for each permille [sic] of frequency). Biometrika 5, 404-406. $1.2111/19_F$

    1907 a   The calculation of moments of a frequency-distribution.
Biometrika 5, 450-459.   10.331/426

    1939   The probability integral. British Association for the Advancement of Science, Mathematical Tables, 7. Cambridge Univ. Press, etc.
$1.1143/7(2)_F$ $1.1144/7(2)_F$ $1.131/13(2)_F$ $1.132/13_F$ 1.331/38

SHERMAN, J.

    1951   MTE 181 [List of errata in Anderson & Houseman 1942]
Math. Tables Aids Comput. 5, 81.   11.112/438(Anderson & Houseman)

SHERMAN, J. and EWELL, R. B.

    1942   A six-place table of the Einstein functions.
J. Phys. Chem. 46, 641-662.
14.152/484   14.152/484(Peto)   xxxxiij

SHEWHART, W. A.

    1928 b   Significance of an observed range. Journal of Forestry 26, 899-905.
1.541/83

    1931   Economic control of quality of manufactured product.
New York [i.e. Princeton (New Jersey), London, etc.]: D. Van Nostrand.
1.1121/5

SHIMADA, SHOZO

    *1954   Power of R-charts.
Rep. Statist. Appl. Res. Un. Jap. Sci. Engrs. 3, 70-74. $14.229/734_M$

SHIMAMOTO, T.   See Bose & Shimamoto

SHOHAT, J.   See N.B.S. 1944

SHOOK, B. L.

    1930   Synopsis of elementary mathematical statistics.
        Ann. Math. Statist. 1, 1-41 + 224-259.   3.229/188

SHRIKHANDE, S. S.

    See Bose, Clatworthy & Shrikhande; Bose & Shrikhande;
    Bose, Shrikhande & Parker

SIBUYA, M. and TODA, H.

    1957   Tables of the probability density function of range in normal samples. Ann. Inst. Statist. Math., Tokyo 8, 155-165.
        ¶ Not abstracted; apparent classification 1.541.
    See Mathematical Reviews 19, 583.

SICHEL, H. S.

    1949   The method of frequency-moments and its application to Type VII populations. Biometrika 36, 404-425.   10.074/401

    1951   The estimation of the parameters of a negative binomial distribution with special reference to psychological data. Psychometrika 16, 107-127.
    5.32/285

SIEGEL, S.

    1956   Nonparametric statistics for the behavioral sciences.
    New York, London, etc.: McGraw-Hill Book Co.
        ⋆ For contents of this book, see pp. 767-768 above.
    xxxxiv   1j

SILLITTO, G. P.

    1947   The distribution of Kendall's τ coefficient of rank correlation in rankings containing ties. Biometrika 34, 36-40 + folding plate.
    8.123/353

    See Davies 1954

SILVER, R.   See Kruskal & Wallis 1952

SILVERSTONE, H.

    1950   A note on the cumulation of Kendall's S-distribution.
    Biometrika 37, 231-235.   8.10/347

SIMAIKA, J. B.

    1942   Interpolation for fresh probability levels between the standard table levels of a function. Biometrika 32, 263-276.
    16.2319/629(Richardson)

SIMM, JOYCE   See Miller, J.C.P. 1954

SIMON, L. E.

    1941   An engineers' manual of statistical methods.
        New York: John Wiley & Sons; London: Chapman & Hall.   14.222/490

        See Army Ordnance Corps 1952

SIMOSE, T.

        See Kotani, Amemiya, Ishiguro & Kimura 1955; Kotani, Amemiya & Simose

SINGER, T.   See Harvard Univ. 1949, 1955

SINGH, B. N.   See Krishna Iyer & Singh

SINKBÆK, S. A.   See Hald & Sinkbæk

SIRAŽDINOV, S. H.   СИРАЖДИНОВ С. Х.

    1956   Об оценках с наименьшим смещением при биномиальном распределении.
        Concerning estimations with minimum bias for a binomial distribution.
        Теор. Вероятност. и Применен. 1, 168-174.
        Not abstracted; apparent classification 6.22.
        See Mathematical Reviews 17, 757.

SISTER M. AGNES   See Hatke, M. A.

SITGREAVES, ROSEDITH   See Girshick, Rubin & Sitgreaves

SKELLAM, J. G.

    1946   The frequency distribution of the difference between two Poisson variates belonging to different populations.
        J. Roy. Statist. Soc. 109, 296 only.   5.23/282(Theil)

SLATER, LUCY J.

    1953   On the evaluation of the confluent hypergeometric function.
        Proc. Cambridge Philos. Soc. 49, 612-622.
        4.142/226   4.1421/227   4.1421/673

    1960   Confluent hypergeometric functions. Cambridge Univ. Press.
        4.1421/673(2)

SLUCKIĬ, E. E.   СЛУЦКИЙ Е. Е.

    1950   Таблицы для вычисления неполной Г-функции и функции вероятности χ^2.
        Москва и Ленинград: Издат. Акад. Наук СССР.   2.111/131

SMALL, V. J.   See Scott, J.F. & Small

SMIRNOV, N.

    1939   On the estimation of the discrepancy between empirical curves of distribution for two independent samples. Bulletin Mathématique de l'Université de Moscou, série internationale, 2, No. 2.
        9.411/380   9.4111/380

1939 z   Об уклонениях эмпирической кривой распределения.
Sur les écarts de la courbe de distribution empirique.
Rec. Math. (Мат. Сб.) 6, 3-26.
   Contains no tables; not to be confused with the preceding entry.

1941   On the estimation of the maximum term in a series of observations.
Dokl. Akad. Nauk SSSR 33, 346-350.
   ¶ Not abstracted; apparent classification 1.44.
See Mathematical Reviews 5, 127.

1948   Table for estimating the goodness of fit of empirical distributions.
Ann. Math. Statist. 19, 279-281.   9.4111/380   9.4113/714(Siegel)

SMITH, B. B.

See Kendall, M.G. & Babington Smith, B.;
Kendall, M.G., Kendall, S.F.H. & Babington Smith, B.

SMITH, CEDRIC A. B. and HARTLEY, H. O.

1948   The construction of Youden squares.
J. Roy. Statist. Soc. Ser. B 10, 262-263.   15.2511/607

SMITH, E. J., Miss   See Johnson, N.L. 1957

SMITH, ED SINCLAIR

1953   Binomial, normal and Poisson probabilities.
Bel Air (Maryland): published by the author.
2.122/135   2.129/136   3.111/183   3.211/186   3.212/186   3.42/199

SMITH, J. E. KEITH   See Green, Smith & Klem

SMITH, JAMES G.

1934   Elementary statistics. New York: Henry Holt.
1.1121/4   1.1162/10   16.0645/620

SMITH, JAMES G. and DUNCAN, ACHESON J.

1944   Elementary statistics and applications.
New York and London: McGraw-Hill Book Co.   1.1121/4

1945   Sampling statistics and applications.
New York and London: McGraw-Hill Book Co.
1.1121/4   1.1161/10   1.3211/34   1.541/86   1.613/113(Williams)
1.623/114   1.633/117   2.3113/142   4.1311/211   4.2314/249   7.211/331

SMITH, R. TYNES, iij

See Horton & Smith; Interstate Commerce Commission 1949

SNEDECOR, G. W.

1934   Calculation and interpretation of analysis of variance and
covariance. Ames (Iowa): Collegiate Press. $4.1311/212_F$ 4.2314/250

Snedecor
Steck

    1946   Statistical methods, fourth edition.
       (First edition 1937, 2d 1938, 3d 1940, 5th 1956)
       Ames (Iowa): Iowa State [Univ.] Press.
       1.1121/4   1.1161/10   1.541/86   2.3113/142(2)   3.321/193   4.1311/212
       4.2314/249   4.239/254   7.1221/323   7.4/335   8.112/350   11.111/437
       12.11/445   13.11/456   13.1222/459

    1950   Everyday statistics: facts and fallacies.
       Dubuque (Iowa): Wm. C. Brown.   13.11/456   13.1222/459   13.13/459

SOBEL, M.

    1956   Statistical techniques for reducing the experiment time in
       reliability studies. Bell System Tech. J. 35, 179-202.   14.159/485

       See Bechhofer & Sobel; Dunnett & Sobel; Epstein & Sobel; Gupta & Sobel

SOBEL, M. and HUYETT, MARILYN J.

    1957   Selecting the best one of several binomial populations.
       Bell System Tech. J. 36, 537-576.   14.151/484

SOMERVILLE, P. N.

    1954   Some problems of optimum sampling. Biometrika 41, 420-429.
       $1.795/655_M$

    1958   Tables for obtaining non-parametric tolerance limits.
       Ann. Math. Statist. 29, 599-601.   14.229/735

SONIN, N.   See Čebyšëv 1899

SOPER, H. E.

    1913   On the probable error of the correlation coefficient to a second
       approximation. Biometrika 9, 91-115.
       7.11/320(Fisher)   7.13/324   7.131/325   7.133/327(Soper et al. 1917)

    1914   Tables of Poisson's exponential binomial limit.
       Biometrika 10, 25-35. $2.221/138_F$

    1915   On the probable error of the biserial expression for the correlation
       coefficient. Biometrika 10, 384-390.
       1.19/16   7.91/343   7.91/698(Tate)

    1921   The numerical evaluation of the incomplete B-function, or of the
       integral $\int_0^x x^{p-1}(1-x)^{q-1}\,dx$ for ranges of x between 0 and 1.
       Tracts for Computers, 7. Cambridge Univ. Press, etc.
       3.44/200
       See Pearson, K. 1921

SOPER, H. E., YOUNG, A. W., CAVE, B. M., LEE, A. and PEARSON, K.

   1917   On the distribution of the correlation coefficient in small samples. Appendix II to the papers of "Student" and R.A. Fisher. A cooperative study. Biometrika 11, 328-413.
       7.1/320   7.11/321   7.13/324   7.133/326   7.19/328   10.29/420

SPAIN. CONSEJO SUPERIOR de INVESTIGACIONES CIENTIFICAS. INSTITUTO de INVESTIGACIONES ESTADISTICAS.

   See Royo López & Ferrer Martín 1954, 1955

SPŁAWA-NEYMANN, J.   See Neyman, J.

SPEARMAN, C.

   1906   'Footrule' for measuring correlation. British J. Psychology 2, 89-108.   8.10/345

SPENCELEY, GEORGE WELLINGTON; SPENCELEY, RHEBA MURRAY; and EPPERSON, EUGENE RHODES

   1952   Smithsonian logarithmic tables to base e and base 10. Smithsonian Miscellaneous Collections, 118. Washington: Smithsonian Institution.
      16.06/620   16.061/621   16.11/625   16.111/625   16.112/625

SPENCELEY, RHEBA MURRAY   See Spenceley, G.W., Spenceley, R.M. & Epperson

SPIERS, BETTY O.   See Pearson, K. 1934

SPITZ, J. C.

   1953   Het matchen in de psychologie. Statistica Neerlandica 7, 23-40.
      5.62/687$_M$

SPRACHER, F. A., Mrs   See Bradley & Terry 1952

SPRINGER, C. H.

   1951   A method for determining the most economic position of a process mean. Indust. Quality Control 8, no. 1, 36-39.
      1.9/655   10.039/717

STANLEY, J. P. and WILKES, M. V.

   1950   Table of the reciprocal of the gamma function for complex argument. [Toronto]: Computation Centre, Univ. of Toronto.   2.5151/164

STATON, J.   See Goodwin, E.T. & Staton

STECK, G. P.

   1958   A table for computing trivariate normal probabilities. Ann. Math. Statist. 29, 780-800.   1.79/125

Steffensen
Sukhatme

STEFFENSEN, J. F.

*1916   Note sur la fonction de type B de M. Charlier. Svenska Aktuarieförenigens Tidskrift 3, 226-228.   10.212/409

1922   A correlation-formula. Skand. Aktuarietidskr. 5, 73-91.   10.19/406

1923   Matematisk Iagttagelseslaere. København: G.E.C. Gad.   10.19/406

1938   A table of the function $G(x) = \dfrac{x}{1-e^{-x}}$ and its applications to problems in compound interest. Skand. Aktuarietidskr. 21, 47-71.   14.152/484   14.152/484(Peto)

STEIN, C.

1945   A two-sample test for a linear hypothesis whose power is independent of the variance. Ann. Math. Statist. 16, 243-258.   15.3/737(Seelbinder)

STEIN, J.   See Peters, J. 1922

STEIN, MILTON   See N.B.S. 1954

STEINHAUS, H.

1954   Tablica liczb przetasowanych czterocyfrowych. Таблица перетасованных четырехзначных чисел. Table of shuffled four-digit numbers. Rozprawy Mat., 6.   13.2/460

1954 a   Liczby przetasowane. Перетасованные числа. Shuffled numbers. Zastos. Mat. 2, 34-45.   13.2/460(Steinhaus 1954)

See Czechowski et al. 1957; see also p. 951

STEPHAN, F. F.

1945   The expected value and variance of the reciprocal and other negative powers of a positive Bernoullian variate. Ann. Math. Statist. 16, 50-61.   3.53/668(Grab & Savage)

STERNE, T. E.

1937   The solution of a problem in probability. Science 86, 500-501.   5.62/686

1954   Some remarks on confidence or fiducial limits. Biometrika 41, 275-278.   3.319/666   3.319/666(Crow)

STEVENS, WILFRED L.

1937 b   Significance of grouping. Ann. Eugenics 8, 57-73.   5.21/281(Thomas)   5.9/293(Fisher & Yates)

1937 z   A test for uniovular twins in mice. Ann. Eugenics 8, 70-73.   5.9/293(Fisher & Yates)   [For 1937 a, see p. 951]

1939   The completely orthogonalized Latin square. Ann. Eugenics 9, 82-93.
    15.112/516(2)   15.113/515   15.113/520(3)   15.114/526   15.114/527
    15.115/533   15.115/534(2)   15.117/537   15.119/541   15.1199/544
    15.1199/545   15.1199/547(4)

1948   Control by gauging. J. Roy. Statist. Soc. Ser. B 10, 54-108.
    14.21/487

1951   Asymptotic regression. Biometrics 7, 247-267.   11.3/444

1953   Tables of the angular transformation. Biometrika 40, 70-73.
    12.15/447

    See Finney & Stevens; Fisher, R.A. 1940 a;
Fisher, R.A. & Yates 1943, 1948, 1953, 1957, TT. VIII 1, VIII 2

STEVENS, WILLIAM H. S.   See Interstate Commerce Commission 1949

STOESSIGER, BRENDA

    See Pearson, K. 1934; Pearson, K. & Stoessiger

STORER, R. L.

1956   The use of continuous sampling in ammunition procurement.
    Indust. Quality Control 12, no. 11, 48-54.   14.13/482

    See Sampling by Variables IV.

STORER, R. L. and DAVISON, W. R.

1955   Simplified procedures for sampling inspection by variables.
    Indust. Quality Control 12, no. 1, 15-18.   14.122/479

STOUFFER, S. A.   See Pearson K., Stouffer & David

STRONG, P. F.   See Harvard Univ. 1949, 1955

STUART, A.

1952   The power of two difference-sign tests.
    J. Amer. Statist. Assoc. 47, 416-424.   9.312/377(Moore & Wallis)

    See Cox, D.R. & Stuart; David, S.T., Kendall & Stuart;
Kendall, M.G. & Stuart

STUDENT   See Gosset, W.S.

STUDNEV, Yu. P.   See Gnedenko & Studnev

STUMPFF, K.

1937   Grundlagen und Methoden der Periodenforschung.
    Berlin: Julius Springer.
    (Reprinted 1945. Ann Arbor (Michigan): J.W. Edwards)   11.2/442

SUKHATME, INDUMATI   See Sukhatme, P.V. 1938

SUKHATME, P. V.

1935   A contribution to the problem of two samples.
    Proc. Indian Acad. Sci. Sect. A 2, 584-604.   6.111/297

Sukhatme, P. V.
Tanner

    1938   On Fisher and Behrens' test of significance for the difference in means of two normal samples. Sankhyā 4, 39-48.
        4.12/210   4.151/234

    1938 a   On bipartitional functions.
        Philos. Trans. Roy. Soc. London, Ser. A 237, 375-409.   10.311/424

        See also Fisher, R.A. & Yates 1943, 1948, 1953, 1957, T. V 1;
Pearson, E. S. & Sukhatme

SUKHATME, P. V., THAWANI, V. D., PENDHARKAR, V. G. and NATU, N. P.

    1951   Revised tables for the d-Test of significance.
        J. Indian Soc. Agric. Statist. 3, 9-23.
        4.151/234(Sukhatme 1938)   4.151/235

SUNDRUM, R. M.

    1953   Moments of the rank correlation coefficient τ in the general case. Biometrika 40, 409-420.   8.10/347

SVEŠIKOV, A. A.   See Kolmogorov. Svešikov & Gubler

SWAROOP, S.

    1938   Tables of the exact values of probabilities for testing the significance of differences between proportions based on pairs of small samples. Sankhyā 4, 73-84.   2.611/174   2.613/664(ftn.)

SWARTZ, GANELLE, Mrs   See Bradley 1954

SWED, FRIEDA S. and EISENHART, C.

    1943   Tables for testing randomness of grouping in a sequence of alternatives. Ann. Math. Statist. 14, 66-67.
        9.311/376   9.32/377   9.32/378(Bateman)   9.32/378(Wilks)
        9.32/378(Dixon & Massey)   9.32/379(Siegel)   9.32/379(Czechowski et al.)
        9.32/379(Dixon & Massey)   9.32/379(Senders)

SWEENY, LUCILE   See Terrill & Sweeny

SWINEFORD, FRANCES

    *1949   Further notes on differences between percentages. Psychometrika 14, 183-187.   6.21/692$_M$

TABLES FOR S & B I   See Pearson, K. 1930 a

TABLES FOR S & B II   See Pearson, K. 1931 a

TABLICE STATYSTYCZNE   See Czechowski et al. 1957

TAGUCHI, GEN-ICHI, alias Gen-Iti Taguti

    1951   On bias of sample mean and sample variance due to rounding or grouping. Rep. Statist. Appl. Res. Un. Jap. Sci. Engrs. 1, no. 2, 9-14.
        ¶ Not abstracted; apparent classification 10.331.
        See Mathematical Reviews 13, 569.

1952   Tables of 5% and 1% points for the Pólya-Eggenberger's distribution function.
Rep. Statist. Appl. Res. Un. Jap. Sci. Engrs. 2, 27-32.
¶ Not abstracted; apparent classification 5.32.
See Mathematical Reviews 14, 775.

1959   Linear graphs for orthogonal arrays and their applications to experimental designs with the aid of various techniques.
Rep. Statist. Appl. Res. Un. Jap. Sci. Engrs. 6, 133-175.
15.112/502(2)   15.112/504(2)   15.112/506   15.112/513   15.112/515
15.113/520   15.1132/522   15.1132/524(2)   15.1132/525   15.114/527
15.114/531   15.114/533   15.117/537   15.119/543   15.1311/553
15.1313/554

1960   Tables of orthogonal arrays and linear graphs.
Rep. Statist. Appl. Res. Un. Jap. Sci. Engrs. 6, 176-227 = 7, 1-52.
15.112/504   15.112/508   15.112/509   15.112/513   15.112/514
15.112/515(3)   15.113/517   15.113/520   15.1132/525(3)   15.114/526
15.114/532   15.115/534   15.115/535   15.1313/554

TAKASHIMA, MICHIO

*1955   Tables for testing randomness by means of lengths of runs.
Bull. Math. Statist. (Fukuoka) 6, no. 1-2, 17-23.   9.32/713$_M$

TALLQVIST, HJ.

1908*   Tafeln der abgeleiteten und zugeordneten Kugelfunctionen erster Art.
Soc. Sci. Fenn. Acta 33, No. 9.   7.32/334   1iij

TALLQVIST, TH.   See Tallqvist, Hj. 1908

TANG, P. C.

1938   The power function of the analysis of variance tests with tables and illustrations of their use.
Statist. Research Memoirs 2, 126-149 + 8 pages of tables.
4.24/256(ftn.)   4.242/258   5.1/275   6.131/307(Bose 1949)
4.242/679(ftn.)

TANNER, J. C.

1953   A problem of interference between two queues. Biometrika 40, 58-69.
¶ Not abstracted; apparent classification 5.2.
See Mathematical Reviews 14, 1102.

---

*We have not seen a copy of this memoir in original wrappers. The title page of the memoir bears no date; the title page of Tomus 33 is dated MCMVIII. Poggendorff 1926, p. 1239, and the New York Public Library ascribe this memoir to 1908; Fletcher et al. 1946, p. 436, to 1906. Jahrbuch über die Fortschritte der Mathematik 35, p. [III], states that Tomus 33 issued in 1904; but this appears to be the date of issue of the first number, and not of the whole.

TARRY, G.

    1900   Le problème des 36 officiers. Association Française pour l'avancement des sciences, Compte rendu, 29, 2de partie, 170-203.
        15.116/535   15.1512/558

TATE, R. F.

    *1955   The theory of correlation between two continuous variates when one is dichotomized. Biometrika 42, 205-216.   7.91/698$_M$

    See Birnbaum et al. 1955

TATTERSFIELD, F.   See Finney 1947, 1952 a

TAUSSKY, OLGA and TODD, J.

    1956   Generation of pseudo-random numbers.
        In Meyer, H.A. 1956, pp. 15-28.   13.62/468

TAYLOR, O. E., Mrs   See Miller, J.C.P. 1954

TCHEBYCHEF, P. L.   See Čebyšev

TECHNIQUES MODERNES DE CONTRÔLE DES FABRICATIONS.   See Mothes 1952

TEICHROEW, D.

    1955 y   Numerical Analysis Research unpublished statistical tables. J. Amer. Statist. Assoc. 30, 550-556.
        ¶ Not abstracted: see MTAC 10, 232.

    1956   Tables of expected values of order statistics and products of order statistics for samples of size twenty and less from the normal distribution. Ann. Math. Statist. 27, 410-426.
        1.42/638   1.43/639   1.43/639(Sarhan & Greenberg)

TERAZAWA, K.

    1917   On the oscillations of the deep-sea surface caused by a local disturbance. Sci. Rep. Tôhoku Univ. Ser. 1, 6, 169-181.
        1.812/126$_F$ 1.813/126$_F$

TERNOUTH, E. J.   See B.A. 9, 1940

TERPSTRA, T. J.

    1952   A confidence interval for the probability that a normally distributed variable exceeds a given value, based on the mean and the mean range of a number of samples. Appl. Sci. Res. A 3, 297-307.
        1.5512/104

TERRILL, H. M. and SWEENY, LUCILE

    1944 a   An extension of Dawson's table of the integral of e^{x^2}. J. Franklin Inst. 237, 495-497.   1.811/126$_F$

    1944 b   Table of the integral of e^{x^2}. J. Franklin Inst. 238, 220-222.
        1.811/126$_F$

TERRY, M. E.

    1952   Some rank order tests which are most powerful against specific parametric alternatives. Ann. Math. Statist. 23, 346-366.
    9.122/369

    See Bradley & Terry

TESKE, A. F.   See Bradley & Terry 1952

TETLEY, H.   See Johnson, N.L. & Tetley

THAWANI, V. D.   See Sukhatme, Thawani, Pendharkar & Natu

THEIL, H.

    1952   On the time shape of economic microvariables and the Munich business test. Rev. Inst. Internat. Statist. 20, 105-120.   5.23/282

THIELE, T. N.

    1903   The theory of observations. London: C. and E. Layton. What we have seen is the reprint in Ann. Math. Statist. 2, [3 unnumbered leaves following 164] + 165-308. Page 18 of the original is p. 182 of the reprint. $1.1122/5_F$ $1.3213/34_F$

THIONET, P.

    1955   Un problème de sondage parmi les éléments dont la distribution est tres dissymètrique. J. Soc. Statist. Paris 96, 192-206.
    16.0244/737

THOMAS, MARJORIE

    1949   A generalization of Poisson's binomial limit for use in ecology. Biometrika 36, 18-25.   5.21/279   5.21/281(Thomas 1951)

    1951   Some tests for randomness in plant populations. Biometrika 38, 102-111.   5.21/281

THOMAS, RICHARD   See Miller, L.H. 1956

THOMPSON, ALEXANDER JOHN

    1924   Logarithmetica Britannica: being a standard table of logarithms to twenty decimal places of the numbers 10,000 to 100,000. Cambridge Univ. Press, etc. Issued in parts as follows:

    Part 1. Numbers 10,000 to 20,000. Tracts for Computers, 19. 1934
    Part 2. Numbers 20,000 to 30,000. Tracts for Computers, 22. 1952
    Part 3. Numbers 30,000 to 40,000. Tracts for Computers, 21. 1937
    Part 4. Numbers 40,000 to 50,000. Tracts for Computers, 16. 1928
    Part 5. Numbers 50,000 to 60,000. Tracts for Computers, 17. 1931
    Part 6. Numbers 60,000 to 70,000. Tracts for Computers, 18. 1933
    Part 7. Numbers 70,000 to 80,000. Tracts for Computers, 20. 1935
    Part 8. Numbers 80,000 to 90,000. Tracts for Computers, 14. 1927
    Part 9. Numbers 90,000 to 100,000. Tracts for Computers, 11. 1924

    With the possible exception of Part 2, copies exist with the Tract No. omitted from the wrapper and the words Subscription Issue added. Reissued in 2 volumes, 1954. Vol. 1 comprises parts 1-4; vol. 2 comprises parts 5-9.
    13.1221/458(Fisher & Yates)   16.06/620 $16.061/621_F$

Thompson, A. J.
Tricomi

    1943   Tables of the coefficients of Everett's central-difference interpolation formula. Second edition. (First edition 1921) Tracts for Computers, 5. Cambridge Univ. Press, etc.
16.23/627   16.2317/628$_F$

    See B.A. 1, 1931

THOMPSON, CATHERINE M.

    1941a   Tables of percentage points of the incomplete Beta-function. Biometrika 32, 151-181.   3.311/191$_F$ 10.012/392

    1941b   Table of percentage points of the χ^2 distribution. Biometrika 32, 187-191.   2.3112/141$_F$ 8.322/365   10.032/395

    See David, F.N. 1938; Merrington & Thompson

THOMPSON, CATHERINE M. and MERRINGTON, MAXINE

    1946   Tables for testing the homogeneity of a set of estimated variances. Prefatory note by H.O. Hartley and E.S. Pearson. Biometrika 33, 296-304.
6.113/300(intro.)   6.113/301   6.113/688

THOMPSON, G. W.

    *1955   Bounds for the ratio of range to standard deviation. Biometrika 42, 268-269.   8.28/702$_M$

THOMPSON, H. R.

    1951   Truncated lognormal distributions I. Solution by moments. Biometrika 38, 414-422.   1.332/44   10.2221/413

THOMPSON, JEAN H.

    See Pearson, E.S. & Hartley 1954; Pearson, E.S. & Merrington 1951.

THOMPSON, W. R.

    1935   On a criterion (criteria) for the rejection of (outlying) observations and the distribution of the ratio of deviation to sample standard deviation. Ann. Math. Statist. 6, 214-219.
1.44/68   1.44/68(Pearson & Chandra Sekhar)

THOMSON, D. HALTON

    1947   Approximate formulae for the percentage points of the incomplete beta function and of the χ^2 distribution. Biometrika 34, 368-372.
2.314/147   3.41/197

THOMSON, LOUIS MELVILLE MILNE-   See Milne-Thomson, L.M.

THORNTON, G. R.

    1943   The significance of rank difference coefficients of correlation. Psychometrika 8, 211-222.   8.112/350

THURSTONE, L. L.   See Chesire, Saffir & Thurstone

TINGEY, F. H.   See Birnbaum & Tingey

TIPPETT, L. H. C.

    1925   On the extreme individuals and the range of samples taken from a normal population. Biometrika 17, 364-387.
            1.42/56   1.42/59(David)   1.42/61(Grubbs)   1.541/83   1.541/83(Pearson)
            1.541/86(Lindley & Miller)

    1927   Random sampling numbers. Tracts for computers, 15. Cambridge Univ. Press, etc.
            13.11/456   13.1221/458   13.1222/459(Mahalanobis)
            13.1222/459(Snedecor 1946)   13.3/462(Mahalanobis et al.)

    1932   A modified method of counting particles. Proc. Roy. Soc. London Ser. A 137, 434-446.   2.229/139(Bliss)

    1944   The control of industrial processes subject to trends in quality. Biometrika 33, 163-172.   1.9/129

    1952   The methods of statistics. Fourth edition. (First edition 1931, 2d 1937) London: Williams & Norgate; New York: John Wiley & Sons.
            1.2121/20   1.541/88   2.3114/143   2.312/143   4.1311/212

    See Fisher, R.A. & Tippett; Pearson, K. & Pearson, M.V. 1931

TODA, H.   See Sibuya & Toda

TODD, JOHN   See Taussky & Todd

TODD, OLGA   See Taussky, O.

TOKARSKA, B.   See Neyman & Tokarska

TOPP, C. W.   See Leone & Topp

TOPP, C. W. and LEONE, F. C.

    1955   A family of J-shaped frequency functions. J. Amer. Statist. Assoc. 50, 209-219.   10.24/720

TORREY, M. N.   See Dodge & Torrey

TRELOAR, A. E.

    1942   Random sampling distribution. Minneapolis (Minnesota): Burgess.
            4.1113/209   4.2314/249   7.212/331

TRELOAR, A. E. and WILDER, MARIAN A.

    1934   The adequacy of "Student's" criterion of deviations in small sample means. Ann. Math. Statist. 5, 324-341.   4.139/217

TREON, R.   See Spenceley et al. 1952

TRICKETT, W. H., WELCH, B. L. and JAMES, G. S.

    1956   Further critical values for the means problem. Biometrika 43, 203-205.   4.152/674

TRICOMI, F. G.   See Erdélyi, Magnus, Oberhettinger & Tricomi

TSAO, C. K.

    1954   An extension of Massey's distribution of the maximum deviation between two sample cumulative step functions.
Ann. Math. Statist. 25, 587-592.   9.4122/383

    1956   Distribution of the sum in random samples from a discrete population. Ann. Math. Statist. 27, 703-712.   10.262/722

TSCHEBYSCHEW, P. L.   See Čebyšev

TUKEY, J. W.

    1950   Some sampling simplified. J. Amer. Statist. Assoc. 45, 501-519.
10.34/429   10.35/431   10.36/723(ftn.)

    1955   [Review of RAND Corp. 1955]
J. Operations Res. Soc. Amer. 3, 568-571.
13.1212/458(RAND Corp.)   13.1221/458(Tippett)

    See Bliss, Cochran & Tukey; Hastings, Mosteller, Tukey & Winsor; Mosteller & Tukey; Olmstead & Tukey

U. S.

    For publications of organs of the U. S. Government see the name of the specific agency.

UEDA, Y.   See Watanabe, Y. 1952

UHLER, H. S.

    1937   A new table of reciprocals of factorials and some derived numbers.
Trans. Connecticut Acad. Arts Sci. 32, 381-434. $2.5122/161_F$

    1944   Exact values of the first 200 factorials.
New Haven (Connecticut): [Published by the author] $2.5112/161_F$

UNITED STATES

    For publications of organs of the U.S. Government see the name of the specific agency.

URA, SHOJI

    *1954   A table of the power function of the analysis of variance tests.
Rep. Statist. Appl. Res. Un. Jap. Sci. Engrs. 3, 23-28. $4.242/678_M$

URANISI, H.

    1950   The distribution of statistics drawn from the Gram-Charlier type A population. Bull. Math. Statist. (Fukuoka) 4, 1-14.   4.143/231

USPENSKY, J. V.

    1937   Introduction to mathematical probability.
New York and London: McGraw-Hill Book Co. $1.1121/5_F$

Van Der REYDEN, D.

    1943   Curve fitting by the orthogonal polynomials of least squares. Onderstepoort J. Vet. Sci. Animal Industry 18, 355-404.
           11.110/437   11.111/437

    1952   A simple statistical significance test. Rhodesia Agric. J. 49, 96-104.   9.121/369

Van ORSTRAND, C.E.

    1921   Tables of the exponential function and of the circular sine & cosine to radian argument. Mem. Nat. Acad. Sci. U.S.A. 14, No. 5.
           $2.5122/161_F$

    See Becker & Van Orstrand; N.B.S. 1953 y

van UVEN, M. J.

    1947 a   Extension of *Pearson*'s probability distributions to two variables. I. Nederl. Akad. Wetensch. Proc. 50, 1063-1070.   10.19/406

    1947 b   Extension of *Pearson*'s probability distributions to two variables. II. Nederl. Akad. Wetensch. Proc. 50, 1252-1264.   10.19/406

    1948 a   Extension of *Pearson*'s probability distributions to two variables. III. Nederl. Akad. Wetensch. Proc. 51, 41-52.   10.19/406

    1948 b   Extension of *Pearson*'s probability distributions to two variables. IV. Nederl. Akad. Wetensch. Proc. 51, 191-196.   10.19/406

Van VOORHIS, W. R.   See Peters, C.C. & Van Voorhis

Van WIJNGAARDEN, A.

    1950   Table of the cumulative symmetric binomial distribution. Nederl. Akad. Wetensch. Proc. 53, 857-868
          = Indagationes Math. 12, 301-312.   3.119/184

    See Kaarsemaker & van Wijngaarden

VARTAK, M. N.

    1955   On an application of Kronecker product of matrices to statistical designs. Ann. Math. Statist. 26, 420-438.
           15.2104/581   15.2109/598

VASSALLI-EANDI, A. M.

    1816   Mémoire historique. Mem. Accad. Sci. Torino 22(1813/14), V-XI.

VIANELLI, S.

    1959   Prontuari per calcoli statistici: Tavole numeriche e complementi. Palermo, etc.: Abbaco.
         ¶ For partial contents of this book, see pp. 768-785 above.
         xxxviij   xxxxj   xxxxiv   [fifty]   1j(2)

VICKERS, T.   See Miller, J.C.P. 1954

VIELROSE, EGON

   *1951   Tablice liczb losowych. Warszawa: Główny Urzad Statystyozny.
           13.13/459   13.11/729

VIGFUSSON, E. A.

    See Fisher, Arne, 1922 a. [Vigfusson seems to have been an allonym of Fisher's.]

von MISES, R.

   1918   Über die „Ganzzahligkeit" der Atomgewichte und verwandte Fragen.
         Phys. Zs. 19, 490-500.   10.28/419

   1931   Wahrscheinlichkeitsrechnung und ihre Anwendung in der Statistik und theoretische Physik.
         Vorlesungen aus dem Gebiete der angewandte Mathematik: I. Band. Leipzig, etc.: Franz Deuticke. (Reprinted 1945, New York: Mary S. Rosenberg)   10.004/390(Rhind)

von NEUMANN, J.

   1941   Distribution of the (a) ratio of the mean square successive difference to the variance. Ann. Math. Statist. 12, 367-395.
         1.531/76(intro.)   1.531/76(Hart & von Neumann)

   1942   A further remark concerning the distribution of the ratio of the mean square successive difference to the variance.
         Ann. Math. Statist. 13, 86-88.   1.531/76(intro.)

    See Hart & von Neumann; Metropolis, Reitwiesner & von Neumann

von NEUMANN, J. [and FORSYTHE, G.E.]

   1951   Various techniques used in connection with random digits.
         In N.B.S. 1951 a, pp. 36-38.   13.61/468   13.62/468   13.63/468

von NEUMANN, J., KENT, R. H., BELLINSON, H. R. and HART, B. I.

   1941   The mean square successive difference.
         Ann. Math. Statist. 12, 153-162.   1.531/76(intro.)

VOORHIS, W. R. Van   See Peters, C.C. & Van Voorhis

WAGNER, KARL WILLY

   1913   Theorie der dielektrischen Nachwirkung.
         Elektrotechn. Z. 34, 1279-1281.   5.21/278

WAGNER, ROBERT WARNER   See Scarborough & Wagner

WAITE, H.

   1910   Mosquitoes and malaria. A study of the relation between the number of mosquitoes in a locality and the malaria rate. Biometrika 7, 421-436.
         5.9/292

WALD, A.   See Mann & Wald; N.B.S. 1949

WALD, A. and BROOKNER, R. J.

    1941   On the distribution of Wilks' statistic for testing the independence of several groups of variates. Ann. Math. Statist. 12, 137-152.
        6.1/295   6.133/311

WALD, A. and WOLFOWITZ, J.

    1945   Sampling inspection plans for continuous production which insure a prescribed limit on the outgoing quality. Ann. Math. Statist. 16, 30-49.
        2.53/173   5.21/279

    1946   Tolerance limits for a normal distribution. Ann. Math. Statist. 17, 208-215.   1.19/17

WALDVOGEL, R. K.   See Radford & Waldvogel

WALKER, GLADYS   See N.B.S. 1959

WALKER, HELEN M.

    1943   Elementary statistical methods. New York: Henry Holt.
        1.1121/5   1.1161/10

WALKER, HELEN M. and LEV, JOSEPH

    1953   Statistical inference. New York: Holt, [Rinehart & Winston]
        3.6/206

WALLIS, W. ALLEN

    See Columbia Univ. 1945; Eisenhart, Hastay & Wallis; Freeman, Friedman, Mosteller & Wallis; Kruskal & Wallis; Moore & Wallis

WALLIS, W. A. and MOORE, G. H.

    1941   A significance test for time series and other ordered observations. Technical Paper 1. New York: National Bureau of Economic Research.
        9.312/376   9.312/376(Wallis & Moore 1941 a)

    1941 a   A significance test for time series analysis.
        J. Amer. Statist. Assoc. 36, 401-409.   9.312/376

WALSH, J. E.

    1946   Some significance tests based on order statistics.
        Ann. Math. Statist. 17, 44-52.   1.43/64

    1947 a   Concerning the effect of intraclass correlation on certain significance tests. Ann. Math. Statist. 18, 88-96.
        2.43/158   4.19/243   4.28/266

    1949   Concerning compound randomization in the binary system.
        Ann. Math. Statist. 20, 580-589.   13.61/468

    1949 a   On the range-midrange test and some tests with bounded significance levels. Ann. Math. Statist. 20, 257-267.   1.5512/101   1.5512/102

    1949 b   On the power function of the "best" t-test solution of the Behrens-Fisher problem. Ann. Math. Statist. 20, 616-618.   4.153/240

1949 c Some significance tests for the median which are valid under very general conditions. Ann. Math. Statist. 20, 64-81.
9.111/366   9.112/367   9.111/704

1949 d Applications of some significance tests for the median which are valid under very general conditions.
J. Amer. Statist. Assoc. 44, 342-355.   9.111/367(Walsh 1949 c)

*1949 z On the 'information' lost by using a t-test when the population variance is known. J. Amer. Statist. Assoc. 44, 122-125. $4.19/675_M$

1950 Large sample tests and confidence intervals for mortality rates.
J. Amer. Statist. Assoc. 45, 225-237.   9.111/367

1951 Some bounded significance level properties of the equal-tail sign test. Ann. Math. Statist. 22, 408-417.   3.6/670

1952 Operating characteristics for tests of the stability of a normal population. J. Amer. Statist. Assoc. 47, 191-201.
4.1413/223   6.119/688

1959 Comments on "The simplest signed rank tests".
J. Amer. Statist. Assoc. 54, 213-224.   9.112/367(ftn.)

WALTER, EDWARD

*1954 Über die Ausnutzung der Irrtumswahrscheinlichkeit.
Mitteilungsbl. Math. Statist. 6, 170-179. $3.6/669_M$

WALTER, TH.   See Faà di Bruno 1881

WARWICK, B. L.

1932 Probability tables for Mendelian ratios with small numbers. Texas Agricultural Experiment Station, Bulletin 463.
3.221/187   3.33/195

WATANABE, H.   See Hiraga, Morimura & Watanabe

WATANABE, N.   See Kitagawa, Nomachi & Watanabe

WATANABE, Y.

1952 On the w^2 distribution. J. Gakugei Coll. Tokushima Univ. 2, 21-30.
9.4211/386

WATSON, EARNEST CHARLES   See Erdélyi, Magnus, Oberhettinger & Tricomi

WATSON, GEOFFREY STUART   See Durbin & Watson; Robinson & Watson

WATSON, G. S. and DURBIN, J.

1951 Exact tests of serial correlation using noncircular statistics.
Ann. Math. Statist. 22, 446-451.
1.531/79(Durbin & Watson)   1.531/79   7.52/338

WEATHERBURN, C. E.

    1946   A first course in mathematical statistics.
        Cambridge Univ. Press, etc.
        1.1121/5   1.1161/10   2.3113/142   4.1311/212   4.2314/250   7.1221/323
        7.211/331

WEAVER, C. L.   See Ferris, Grubbs & Weaver

WEGENER, G. S.   See Hald 1948

WEIBULL, MARTIN

    *1953   The distribution of t- and F-statistics and of correlation and regression coefficients in stratified samples from normal populations with different means.
        Skand. Aktuarietidskr. 36, no. 1-2, supplement, 106 pp.   4.28/682$_M$

WEICH, S.   See von Mises 1918

WEIDA, F. M.   See Greenshields & Weida

WEILER, H.

    *1954   A new type of control chart limits for means, ranges, and sequential runs. J. Amer. Statist. Assoc. 49, 298-314.
        ¶ Not abstracted; apparent classification 14.229. See MTAC 9, 31.

WEINBERG, J. W.   See Birge & Weinberg

WEINER, S.   See Binet, Leslie, Weiner & Anderson

WEINMEISTER, P.   See Poggendorff 1926.

WELCH, B. L.

    1938 a   The significance of the difference between two means when the population variances are unequal. Biometrika 29, 350-361.
        4.152/239(Gronow)

    1938 b   On tests for homogeneity. Biometrika 30, 149-158.   2.62/179

    1947   The (a) generalization of 'Student's' problem when several different population variances are involved. Biometrika 34, 28-35.
        4.152/237(Aspin)

    1949   Appendix (to Mrs Aspin's paper): Further note on Mrs Aspin's tables and on certain approximations to the tabled function.
        Biometrika 36, 293-296. See also Aspin 1949

    See Jennett & Welch; Johnson, N.L. & Welch; Morant, G.M. 1939; Trickett, Welch & James

WELSH, G. S.

    *1955   A tabular method of obtaining tetrachoric r with median-cut variables. Psychometrika 20, 83-85.   1.78/653$_M$ 7.94/698

WELDON, W. F. R.   See David, F.N. 1949

WESTENBERG, J.

    1947 a   Mathematics of pollen diagrams. I.
        Nederl. Akad. Wetensch. Proc. 50, 509-520.   9.123/370

    1947 b   Mathematics of pollen diagrams. II.
        Nederl. Akad. Wetensch. Proc. 50, 640-648.   9.123/370

    1948   Significance test for median and interquartile range in samples from continuous populations of any form.
        Nederl. Akad. Wetensch. Proc. 51, 252-261.   9.123/370

    1950   A tabulation of the median test for unequal samples.
        Nederl. Akad. Wetensch. Proc. 53, 77-82 = Indagationes Math. 12, 8-13.
        9.123/370

    1952   A tabulation of the median test with comments and corrections to previous papers. Nederl. Akad. Wetensch. Proc. Ser. A 55, 10-15 = Indagationes Math. 14, 10-15.   9.123/370

WHITAKER, LUCY

    1914   On the Poisson law of small numbers. Biometrika 10, 36-71.
        2.121/135   2.122/135   3.229/188

WHITE, COLIN

    1952   The use of ranks in a test of significance for comparing two treatments. Biometrics 8, 33-41.   9.121/369

WHITE, ROBERT F.   See Zoellner & Kempthorne 1954

WHITFIELD, J. W.

    1949   Intra-class rank correlation. Biometrika 36, 463-467.   8.19/357

    1953   The distribution of total rank value for one particular object in m rankings of n objects. British J. Statist. Psychol. 6, 35-40.
        10.262/418

WHITMAN, E. A.   See Rosenbach, Whitman & Moskowitz

WHITNEY, D. R.

    1951   A bivariate extension of the U statistic.
        Ann. Math. Statist. 22, 274-282.   9.13/370

    See Mann & Whitney

WHITTAKER, EDMUND TAYLOR, Sir, and ROBINSON, GEORGE

    1924   The calculus of observations: a treatise on numerical mathematics. London, etc.: Blackie and Son.
        1.1211/11   11.110/436(ftn.)   11.19/441

WICKSELL, S. D.

    1925   The corpuscle problem: a mathematical study of a biometric problem. Biometrika 17, 84-99.   5.5/288

    1926   The corpuscle problem: second memoir. Case of ellipsoidal corpuscles. Biometrika 18, 151-172.   5.5/288

WIJNGARDEN, A. van   See Kaarsemaker & Van Wijngarden

WILCOXON, F.
- 1945   Individual comparisons by ranking methods. Biometrics Bull. 1, 80-83.   9.112/367   9.112/368   9.121/368
- 1947   Probability tables for individual comparisons by ranking methods. Biometrics 3, 119-122.   9.112/368   9.121/369
- 1949   Some rapid approximate statistical procedures. New York: American Cyanamid Co.   9.112/368

    See McCallan & Wilcoxon

WILCOXON, F. and McCALLAN, S. E. A.
- 1939   Theoretical principles underlying laboratory toxicity tests of fungicides. Contributions from Boyce Thompson Institute 10, 329-338.   1.233/30

WILDER, MARIAN A.   See Treloar & Wilder

WILEŃSKI, H.   See Przyborowski & Wileński

WILKES, MAURICE VINCENT   See Stanley & Wilkes

WILKS, S. S.
- 1932   Certain generalizations in the analysis of variance. Biometrika 24, 471-494.   6.12/690(ftn.)
- 1935   On the independence of k sets of normally distributed statistical variables. Econometrica 3, 309-326.   6.1/295   6.133/312(Wald & Brookner)   6.133/312(Box)
- 1938   Shortest average confidence intervals from large samples. Ann. Math. Statist. 9, 166-175.   2.322/153(Kitagawa)
- 1941   Determination of sample sizes for setting tolerance limits. Ann. Math. Statist. 12, 91-96.   6.119/304(Albert & Johnson)
- 1942   Statistical prediction with special reference to the problem of tolerance limits. Ann. Math. Statist. 13, 400-409.   14.31/492
- 1946   Sample criteria for testing equality of means, equality of variances, and equality of covariances in a normal multivariate distribution. Ann. Math. Statist. 17, 257-281.   6.132/310   6.139/313(Mauchly)
- 1948   Elementary statistical methods. Princeton Univ. Press; London: Oxford Univ. Press.   1.115/8   3.321/193(Clopper & Pearson)   4.1311/212   9.32/378

    See Friedman, Milton 1937

WILLIAMS, C. A., Jr.
- *1950   On the choice of the number and width of classes for the chi-square test of goodness of fit. J. Amer. Statist. Assoc. 45, 77-86.   9.4/380   2.63/664$_M$

WILLIAMS, C. B.   See Fisher, R.A., Corbet & Williams

WILLIAMS, J. D.

    1946   An approximation to the probability integral. Ann. Math. Statist. 17, 363-365.   1.19/18

WILLIAMS, P.

    1935   Note on the sampling distribution of $\sqrt{\beta_1}$, where the population is normal. Biometrika 27, 269-271.   1.613/113

WILSON, EDWIN BIDWELL and HILFERTY, MARGARET M.

    1931   The distribution of chi-square. Proc. Nat. Acad. Sci. U.S.A. 17, 684-688.   4.27/263(Paulson)

WILSON, E. B. and WORCESTER, JANE

    1943 a   The determination of L.D.50 and its sampling error in bio-assay. Proc. Nat. Acad. Sci. U.S.A. 29, 79-85.   12.54/728

    1943 b   A table determining L.D.50 or the fifty per cent end-point. Proc. Nat. Acad. Sci. U.S.A. 29, 207-212.   12.54/728

WILSON, T. R. C.   See Shewhart 1928 a

WINSOR, C. P.   See Hastings, Mosteller, Tukey & Winsor

WINSTEN, C. B.

    1946   Inequalities in terms of mean range. Biometrika 33, 283-295. 10.9/435

WISE, M. E.

    1950   The incomplete beta function as a contour integral and a rapidly converging series for its inverse. Biometrika 37, 208-218. 3.41/198

WISHART, J.

    1925 a   Determination of $\int_0^\theta \cos^{n+1}\theta \, d\theta$ for large values of n, and its application to the probability integral of symmetrical frequency curves. Biometrika 17, 68-78.   3.44/200   10.023/394

    1925 b   Further consideration of the integral $\int_0^\theta \cos^{n+1}\theta \, d\theta$ for large values of n. Biometrika 17, 469-472.   3.44/201   10.023/394

    1926   On Romanovsky's generalized frequency curves. Biometrika 18, 221-228.   10.18/405(Romanovsky)

    1927   On the approximate quadrature of certain skew curves, with an account of the researches of Thomas Bayes. Biometrika 19, 1-38. 3.44/202   3.44/202(Pearson)   10.013/392

1930   The derivation of certain high order sampling product moments from a normal population. Biometrika 22, 224-238.   10.343/430

1931   The mean and second moment coefficient of the multiple correlation coefficient, in samples from a normal population.
Biometrika 22, 353-362.   7.4/335(Pearson)

1947   The cumulants of the z and of the logarithmic χ^2 and t distributions. Biometrika 34, 170-178.   4.34/272

*1949   Cumulants of multivariate multinomial distributions.
Biometrika 36, 47-58. $5.4/683_M$

1952   Moment coefficients of the k-statistics in samples from a finite population. Biometrika 39, 1-16.   10.342/430(3)   10.35/431

See B.A. 1, 1931; Gosset 1942; Pearson, K. 1931 b, 1934

WITTSTEIN, T. L.

1866   Logarithmes de Gauss à sept décimales pour servir à trouver le logarithme de la somme ou de la différence de deux nombres, leurs logarithmes étant donnés, arrangés d'après une nouvelle méthode: Ouvrage servant de supplément a toute table ordinaire de logarithmes à sept décimales. Siebenstellige Gaussische Logarithmen zur Auffindung des Logarithmus der Summe oder Differenz zweier Zahlen, deren Logarithmen gegeben sind. In neuer Anordnung: Ein Supplement zu jeder gewöhnlichen Tafel siebenstelliger Logarithmen. Hannover: Hahn.
¶ Reproduced and in print as: Johnson, William W., ed. [!] Addition and subtraction logarithms to seven decimal places. Chicago: Charles T. Powner: 1943. Carried in Powner's catalogue as "Johnson's Addition & Subtraction Logarithms (Gaussian Tables)".
10.013/392(Elderton)   16.068/621(intro.) $16.068/621_F$

WOLD, H.

1934   Sheppard's correction formulae in several variables.
Skand. Aktaurietidskr. 17, 248-255.   10.36/431(2)

1948   Random normal deviates: 25,000 items compiled from Tract No. 24 (M.G. Kendall & B. Babington Smith's tables of random sampling numbers). Tracts for computers, no. 25. Cambridge Univ. Press, etc.
13.3/462   13.3/463(Fieller et al.)   13.3/463(Czechowski et al.)
13.3/729(Royo López & Ferrer Martín)   13.3/729(Vianelli)

See Azorín & Wold

WOLFENDEN, H. H.

1942   The fundamental principles of mathematical statistics; with special reference to the requirements of actuaries and vital statisticians; and an outline of a course in graduation. [Chicago: Society of Actuaries]; Toronto: Macmillan.   10.23/415   11.19/441(ftn.)

WOLFOWITZ, J.   See Kac, Kiefer & Wolfowitz; Wald & Wolfowitz

WOLLETT, M. F. C.   See Miller, J.C.P. 1954

WOLMAN, W.   See Arkin & Colton 1950, T. 19 B

WONG, J. P., Jr.   See Hastings, Hayward & Wong

WOO, MARY   See N.B.S. 1959

WOO, T. L.

   1929 Tables for ascertaining the significance of non-significance of association measured by the correlation ratio. Biometrika 21, 1-66.*
     3.319/192 7.4/335 3.52/668 7.94/668

WOODBURY, M. A.

   1940 Rank correlation when there are equal variates. Ann. Math. Statist. 11, 358-362. 8.113/351

WOODWARD, JOHN   See Cohen & Woodward

WOOLHOUSE, W. S. B.

   1864 On interpolation, summation, and the adjustment of numerical tables. Assurance Mag. & J. Inst. Actuar. 11, 301-332. 3.211/186

WORCESTER, JANE   See Wilson & Worcester

YANG, SIMON   See Craig 1936

YaROŠEVS'KIĬ, B. I.   See Korolyuk & Yaroševs'kiĭ

YASUKAWA, K.

   1925 On the means, standard deviations, correlations, and frequency distributions of functions of variates. Biometrika 17, 211-237. 10.229/414

   1926 * On the probable error of the mode of skew frequency distributions. Biometrika 18, 263-292 + 2 plates. 10.006/391 10.007/392

YATES, F.

   *1934 Contingency tables involving small numbers and the χ^2 test. J. Roy. Statist. Soc. Suppl. 1, 217-235. 2.6/174

   1936 Incomplete randomized blocks. Ann. Eugenics 7, 121-140. 15.2105/586 15.2107/592

   1937 The design and analysis of factorial experiments. Imperial Bureau of Soil Science, Technical Communication 35. Harpenden (Herts): the Bureau.
     15.112/505(2) 15.112/507(2) 15.112/509 15.112/510 15.112/512
     15.113/517(3) 15.113/518(5) 15.1132/520 15.1132/522(4)

---

*Morant 1939, p. 61, attributes pp. 1-8 of this memoir to K. Pearson.

15.1132/523(2) 15.1132/524(3) 15.1132/525(3) 15.114/530
15.114/532 15.116/535 15.116/536(2) 15.118/539 15.118/540(4)
15.119/541 15.119/542(6) 15.119/543(3) 15.2103/570

*1940   Lattice squares. J. Agric. Sci. 30, 672-687.
Not abstracted: apparent classification 15.253.

See Fisher, R.A. & Yates

YORK, E. J.   See Miller, J.C.P. 1954

YOST, E. K.   See Dixon & Massey 1951, 1957, TT. 8a2, 8a3

YOUDEN, W. J.

1937   Use of incomplete block replications in estimating tobacco mosaic virus. Contrib. Boyce Thompson Inst. 9, 41-48.   15.2106/591

1940   Experimental designs to increase accuracy of greenhouse studies. Contrib. Boyce Thompson Inst. 11, 219-228.
15.2103/571 14.2104/577 15.2106/588 15.2106/589 15.2106/591
15.2107/592 15.2108/595 15.2109/598 15.2109/599 15.2110/602(2)

1951   Statistical methods for chemists.
New York [and London]: John Wiley & Sons.   4.1311/212   4.2314/249

YOUDEN, W. J. and CONNOR, W. S.

1953   The chain block design. Biometrics 9, 127-140.
15.2106/588 15.2108/596 15.2109/602 15.212/605

YOUNG, A. W.

See Pearson, K. & Young; Soper, Young, Cave, Lee & Pearson

YOUNG, ALLYN A.   See Rietz, Carver et al.

YOUNG, JOHN WESLEY   See Burgess, R.W. 1927

YOUNG, L. C.

1941   On randomness in ordered sequences. Ann. Math. Statist. 12, 293-300.
1.531/76

YOWELL, E. C.   See Teichroew 1956

YUAN, P. T.

1933   On the logarithmic frequency distribution and the semi-logarithmic correlation surface. Ann. Math. Statist. 4, 30-74.
10.2221/413   10.2221/414   10.229/414

YULE, G. U.

1938   A test of Tippett's random sampling numbers.
J. Roy. Statist. Soc. 101, 167-172.   13.63/468

See B.A. 1896; Greenwood, M. & Yule

YULE, G. U. and KENDALL, MAURICE G.

    1937   An introduction to the theory of statistics. Eleventh edition. (First edition, by Yule, 1911) London: Griffin.
             1.1131/5_F 1.115/8   1.1161/9   2.1132/133   4.1111/208

    1950   An introduction to the theory of statistics. Fourteenth edition. London: Charles Griffin; New York: Hafner.
             1.115/8   1.1161/9   4.2313/249   4.331/251

ZALOKAR, JULIA   See Halperin, Greenhouse, Cornfield & Zalokar

ŽALUDOVÁ, A. H., Mme

    1958 a   Statistical quality control of production processes using individual sample values. Bull. Inst. Internat. Statist. 36, no. 3, 573-578.
             14.229/490

    1958 b   Méthode de contrôle statistique de qualité basée sur les valeurs individuelles de l'échantillon. Revue Statist. Appl. 6, no. 4, 21-31.
             14.229/491(Žaludová 1958 a)

ZAPPA, G.

    1940   Oservazioni sopra le medie combinatorie. Metron 14, 31-53.
             10.37/432

ZASĘPA, R.   See Czechowski et al.

ZELEN, M.   See Connor & Zelen; N.B.S. 1957

ZEMANSKY, M.W.   See Mitchell & Zemansky

ZIA-ud-DIN, M.

    1940   Tables of symmetric functions for statistical purposes. Proc. Nat. Acad. Sci. India, Sect. A 10, 53-60.   10.3121/424

    1954   Expression of the k-statistics k_9 and k_{10} in terms of power sums and sample moments. Ann. Math. Statist. 25, 800-803.   10.341/430

ZIA-ud-DIN, M. and MOIN-ud-DIN SIDDIQI, M.

    1953   Random sampling numbers. In Proceedings of the second Pakistan Statistical Conference held in the University of Dacca, pp. 91-98. [Lahore: Punjab Educational Press]   13.11/456   13.1211/457

ZIERLER, N.

    1959   Linear recurring sequences. J. Soc. Indust. Appl. Math. 7, 31-48.
             13.62/468

ZITEK, F.

    1954   O pewnych estymatorach odchylenia standardowego.
           О некоторых оценках стандартного отклонения.
           On certain estimators of standard deviation. Zastos. Mat. 1, 342-353.
             Not abstracted; apparent classification 1.52.
           See Mathematical Reviews 16, 103.

ZOELLNER, J. A. and KEMPTHORNE, O.

    1954   Incomplete block designs with blocks of two plots.
        Iowa Agric. Expt. Sta., Res. Bull. 418.
        15.2102/567(7)   15.2102/568(10)   15.2102/569(8)

ZUCKER, RUTH   See N.B.S. 1954

ADDENDA

DAVIS, H. T. and FISHER, VERA

   *1949   A bibliography and index[†] of mathematical tables.
       Evanston (Illinois): copyright[‡] by Harold T. Davis.   xxxxij

PATNAIK, P. B.

   1955   Hypotheses concerning the means of observations in normal samples.
      Sankhya 15, 343-372. $4.1414/225_M$

SCHÜTTE, KARL

   1955   Index mathematischer Tafelwerke und Tabellen aus allen Gebieten der
      Naturwissenschaften. Index of mathematical tables from all branches of
      sciences. München: R. Oldenbourg.   xxxxij

STEINHAUS, H.   ШТАЙНХАУЗ Г.

   1956   Liczby złote i żelazne. Золотне и железные числа.
      On golden and iron numbers. Zastos. Mat. 3, 51-65.   16.0568/738

STEVENS, W. L.

   1937a Appendix [to Bliss 1937a]: The truncated normal distribution.
      Annals of Applied Biology 24, 847-850.
      1.132/13   1.332/41   1.332/42

[†] Our secondary sources are unusually discrepant:
MTAC 4, 77: Harold T. Davis & Vera Fisher, A Bibliography of Mathematical Tables.
Lebedev & Fëdorova 1956, p. [III]: H. T. Davis, V. Fisher "A bibliography an [!]
     index of mathematical tables."
Lebedev & Fëdorova 1960, p. [III]: A Bibliography and Index of Mathematical
     Tables by H. T. Davies [!] and V. Fisher.

[‡] Sic the three secondary sources cited supra. A search of the Catalog of
[U.S.] Copyright Entries for 1949 and 1950 does not confirm this copyright.

SUBJECT INDEX

A, B, C, D: factors for constructing control charts 488-490, 733 (§14.221)

AOQ   184, 469

AOQL   469

AQL   469

   nominal   471

ASN   475

abruptness, corrections to moments for   427 (§10.332)

accelerated life testing   485

acceptable quality level   469

acceptance numbers   472, 473

acceptance sampling   469-486, 730-732 (§14.1*)

actuarial formulas, for decrement rates, efficiency of   485

addition and subtraction logarithms   621-622 (§16.063)

additive adjustment, to population estimates   697

after effect function   278

agreement, coefficient of   356

algebraic semi-invariants, yielding statistical semi-invariants   432

alienation, coefficient of   334, 695 (§7.32)

α, i.e. producer's risk   469

α-index, Pareto's   616

alphabets, in orthogonal squares and cubes   554

American practice for control charts   488

amount of experimentation   612

amplitude, hyperbolic   410

*This section is subdivided; see Table of Contents.

analysis of incomplete block experiments   610 (§15.28)

analysis of variance   243, 246, 298, 677

  based on range   104-106, 651-652 (§1.552)

  by ranks   370

  component-of-variance model

    effect of heteroscedasticity on power of F-test in   267

    effect of non-normality on power of F-test in   267

    power of F-test under   255

  F-test

    effect of non-normality on power function of   267

    for comparison of group means   264

  in terms of intra-class correlation coefficient   330

  linear-hypothesis model

    effect of non-normality on power of F-test in   267

    power of F-test under   255

  non-centrality yielding specified power, approximations to   259

  one-way classification   298

    effect of inequality of variance on distribution of F   268

    power of F-test in   255

    sensitivity of, to random between groups variation   267

  two-way classification

    effect of correlation between errors on distribution of F   269

    effect of inequality of variance on distribution of F   269

angles

  converted into logits   447, 451

  converted into probits   447, 449

  vulgar fractions of 2π 442 (§11.21)

  working   446 (§12.14)

[953]

angular transformation
approximation

angular transformation 445-447, 728 (§12.1*)
  of binomial data 449
  (SEE ALSO inverse hyperbolic-sine transformation)

antilogarithms 621 (§16.064)

antiloglogs 364

antilogits 450 (§12.52)

approximation
  Bartlett's
    to distribution of Bartlett's modified L_1 test 300, 301
    to percentage points of Bartlett's modified L_1 test 300
  beta, to distribution:
    of concordance coefficient 354
    of likelihood test 297
  beta, to distribution, under permutations:
    of product-moment correlation coefficient 344
    of Spearman's rank correlation coefficient 344
  beta, to percentage points of k-sample generalization of Wilcoxon's test 372
  χ^2, to distribution:
    of mean deviation 643
    of mean range 652
    of mean square successive difference 645
  χ^2, to multinomial distribution:
    not used 287
    not improved by Yates' correction 287
  χ^2, to percentage points:
    of Bartlett's test 273
    of concordance coefficient 354
    of determinantal ratio 311
    of Friedman's concordance score 354
    of goodness-of-fit χ^2 for runs up and down 376
    of k-sample generalization of Wilcoxon test 372
    of Neyman and Pearson's L_1 299
    of test for equality of means of correlated variates 311
    of test for equality of means, variances and covariances of correlated variates 311
    of test for equality of variances and covariance of correlated variates 311
    of z^2 273
  Cornish-Fisher
    to percentage points of non-central χ^2 661
    to percentage points of z, yielding approximate percentage points of beta distribution 198
    to weighted sum of χ^2 variates 660
    to z-distribution 272, 274
  Gram-Charlier
    to distribution of matching scores 685, 686
    in multivariate t-tables 674
    to binomial distribution 188
    to confidence limits for Poisson parameter 659
  normal, to distribution:
    of average mean deviation 643
    of correlation coefficient 321
    of log χ^2 261
    of matching scores, fits poorly 687
    of mean matching score 687
    of ordered intervals 712
    of sum of 20 rectangular variates 722
  normal, to percentage points:
    of Spearman's rank correlation coefficient 349, 350
    of two-sample test of randomness 378
  parabolic, to regression of standard deviation on mean 434
  Pearson Type I
    to distribution of matching scores 686
    to test of equality of covariances 310
  Pearson Type III
    to binomial distribution 189
    to distribution of mean matching score 687
    to distribution of mean square successive difference 645
  Pearson Type VI, to distribution of mean square successive difference 645

---

*This section is subdivided; see Table of Contents.

[approximation]

Pearson Type VII, to distribution of k-sample paired comparison slippage test 708

Poisson
  to binomial distribution 188, 189, 482, 667, 693
  to distribution of mean matching score, not used 687
  to hypergeometric distribution 469, 470
  to percentage points of runs up and down 375

Poisson-Charlier, to binomial distribution 667

to beta distribution and binomial distribution, interchangeable 198

to binomial distribution 198-199, 667-668 (§3.42)

to binomial distribution and incomplete beta function, interchangeable 198

to bounds for percentage points of variate differences 79

to confidence intervals for Poisson parameter 153

to cumulants of $\log \chi^2$ 273

to distribution:
  of concordance coefficient 348
  of correlation coefficient 226
  of Galton differences 91
  of k-sample slippage test 371
  of largest canonical correlation 312
  of likelihood tests 295
  of mean square successive difference 76-77
  of non-central χ^2 157
  of non-central t 226
  of product of correlated normal variables 125
  of range, in normal samples 107-110, 652 (§1.553*)
  of ratios involving ordered mean squares 261
  of ratio of two independent quadratic forms 268
  of Spearman's rank correlation 349 (§8.113); 346
  of straggler deviate 60
  of t 210
  of 3-sample ranking test 371
  of Welch's two-sample test 239

to duration of play 293

to efficiency of Scheffe's two-sample test 240

to estimated variance of sample standard deviation 170

to F-distribution 263-264 (§4.27); 246, 273

to incomplete beta-function 197-200, 667-668 (§3.4*)

to incomplete beta-function and binomial distribution, interchangeable 198

to Jeffreys' significance criterion K for t 218

to mean of straggler 59

to moments:
  of mean square successive difference 77
  of sample standard deviation 168

to non-centrality for specified power in analysis of variance 259

to normal:
  central area 18
  ordinate 15
  tail area 15

to percentage points:
  of $b_1^{\frac{1}{2}}$ 112
  of b_2 114
  of beta distribution 197-198 (§3.41); 150
  of bivariate normal distribution 290
  of χ^2 144-147 (§2.314); 263
  of concordance coefficient 354, 355
  of largest determinantal root 690
  of mean square successive difference 76
  of non-central χ^2 157
  of normal distribution 20 (§1.2119)
  of range 84
  of ratio of largest to smallest mean square, by percentage points of range in normal samples, inadequate 262
  of smallest variance ratio 261
  of straggler deviate 59
  of studentized mean square successive difference 80
  of t 213-216, 672-673 (§4.132)
  of test of equality of covariance matrices 309
  of 3-dimensional sphericity test 313

approximation
beta-variate

[approximation
 to distribution]
   of z   273-274 (§4.35)
   to power of F-test, under
     linear-hypothesis model, for
     non-normal variates   267
   to probable error of tetrachoric
     correlation   698
   to t-distribution
     210, 671 (§4.115); 273
   to test for multiple rank
     correlation   356
   Yates', to percentage points of χ^2
     in 2 × 2 tables   176
arc-sine transformation (SEE angular
   transformation; inverse
   hyperbolic-sine transformation)
area
   of normal distribution
     3-8, 10-13, 630 (§1.1*); 40
   of Pearson curves   389 (§10.001)
     Type I   392 (§10.011)
     Type II   393 (§10.021)
     Type III 132, 656 (§2.112);
               394-395 (§10.031)
     Type VI   399 (§10.061)
     Type VII 400 (§10.071)
     Type X   402 (§10.101)
   ratio of area to bounding ordinate
     as a function of
       22, 632 (§1.222)
arithmetic mean, as a biplane
   combinatorial power mean   432
artificial time series, random
   samples from   465-467 (§ 13.5)
ascending, descending and stationary
   runs   709
Aspin, Alice A., tables not related
   to Fisher-Behrens test   238
assay, 6-point, probit analysis   29
associates, in social group   294
associated spherical harmonics   334
association
   non-parametric tests for
     713-714 (§ 9.33)
   quadrant test for   357

asymptotic distribution
   normal, of k-sample paired-
     comparison slippage test   708
   of Cramer-Mises-Smirnov test
     386 (§9.421*)
   of extreme values
     364-365, 703 (§8.3*); 366
   of range   365 (§8.33)
   of runs   711
asymptotic uncorrelated normal
   distribution, of two tests
   of randomness   710
asymptotic variance, of estimated
   regression   726
attributes
   continuous inspection by
     481-483 (§14.13)
   lot inspection by
     470-477, 730 (§14.11*)
augmented monomial symmetric
   functions   421, 430, 431
autocovariance, of samples from
   autoregressive time series   467
autoregressive time series
     465-466, 696
auxiliary functions: S and T   623
averages, unweighted   71
average deviation
   (mean deviation)
average outgoing quality   184, 469
average outgoing quality limit   469
average number of articles inspected
     733
average quality protection
   (average outgoing quality)
average run length, in quality
   control   491
average sample number   475

b = number of blocks, in incomplete
   block patterns
     563, 564, 566, 607, 608
B, C, D, A: factors for constructing
   control charts
     488-490, 733 (§14.221)
$b_1^{\frac{1}{2}}$ 111-113 (§1.61*)
b_2 113-114 (§1.62*)

approximation
beta-variate

balanced confounding, in factorial patterns   556 (§15.1422)

balanced incomplete block patterns   605-606, 736 (§15.241*); 566

balanced lattices   556 (§15.1431)

balanced lattice square
(SEE square lattice)

Bartlett's test of homogeneity of variance   300-302, 608 (§6.113)

   power of   256

Behrens' integral   232

Behrens-Fisher test   232-236 (§4.151)

   Aspin's tables not related to   238

Bernoulli distribution   204, 668

   multinomial   683

   positive, negative moments of   668
(SEE ALSO binomial distribution)

Bernoulli numbers   619 (§16.041); 427

Bessel functions   626 (§16.17); 416

   I   157, 282, 306, 307, 417, 418, 419, 691

   J   419

   K   365 (§8.33*); 108, 109, 110, 124, 340

   Pearson's T_m 417

Bessel-function approximation, to distribution of range   110

Bessel-function distributions   417 (§10.25)

Bessel interpolation coefficients   628 (§16.2316)

Bessel interpolation formula, first derivative of   629

best linear systematic statistic   642, 649

β, i.e. consumer's risk   469

β_1 and β_2

   correlation ellipse of   391

   estimated, of Pearson curves, standard errors of   391

   of correlation coefficient   420

beta coefficients
(SEE $b_1^{\frac{1}{2}}$; b_2; β_1; β_2)

beta distribution   297, 308, 309, 310, 393, 492

   approximations to, as approximations to binomial distribution   198

   as approximation to distribution:

     of concordance coefficient   348, 354

     of likelihood test   297

     of rank-sum test   369

     of 3-dimensional sphericity test   313

   as approximation to randomization distribution:

     of product-moment correlation coefficient   344

     of Spearman's rank correlation coefficient   344

   generalized   692

   in constructing tolerance limits   492

   inverted   245-269, 676-682 (§4.2*)

   moments of   204, 668 (§3.52); 192, 203

   non-central   679

   of second kind   245

   percentage points of   246, 690

     approximate   150

   yields exact distribution:

     of test for equality of means, variances and covariances of correlated variates   310

     of test for equality of variances and covariances of correlated variates   310

(SEE ALSO beta function, incomplete)

beta-function

   complete   204 (§3.51)

   incomplete   181-206, 664-670 (§3*); 37, 50, 62, 246, 275, 389

     approximations to   273

     distribution of t from non-normal parents expanded in   230

     integrals involving   297

     trigonometric form of   200, 201

     yielding power of F-test   255

   noncentral   679

beta-function distribution, of circular serial correlation coefficient   337

beta-variate   354

bias, of estimated variance, after preliminary significance test 166

Bienayme-Cebysev inequality 433

bimodal distribution, produced by fourth-degree exponential 415

binary digits, random 454

binomial and negative binomial distribution, factitious resemblance between cumulants of 684

binomial coefficients 186, 665 (§3.21*); 284, 683 (§5.31); 436, 619
  as weights in polynomial fitting 441
  expansion of orthogonal polynomials in 439
  interpolation 628 (§16.2313)
  notation for 719

binomial dispersion 180

binomial distribution 181-206, 664-670 (§3*); 469, 470
  angular transformation of 449
  moments of 446
  approximations to, as approximations to incomplete beta function 198
  as approximation:
    to distribution of matching scores 686
    to hypergeometric distribution 469, 470
  confidence intervals for 193-195 (§3.32); 190
  control charts for 487
  estimators of parameters in 314 (§6.22)
  factorial moments of 203
  in process control by numbers of defectives 469
  moments of 204-205, 668-669 (§§3.53-3.54); 203
  multiple comparison of parameters in 484
  negative 285, 683 (§5.32); 278, 286, 449
    inverse hyperbolic-sine transformation of, moments of 449
    truncated 683

---
*This section is subdivided; see Table of Contents.

  percentage points of 195-196 (§3.33)
  Poisson approximation to 482
  Poisson distribution as limit of 182
  posterior 276
  tests applied to 314, 692-693 (§6.21)
  truncated 314
  Wald sequential test of parameter in 473-474

binomial estimate 633

binomial frequencies, individual 187

binomial index of dispersion (SEE index of dispersion)

binomial interpolation coefficients 628 (§16.2313)

binomial moments 439

binomial populations
  estimators of parameters in 314 (§6.22)
  multiple choice among 484
  tests applied to 314, 692-693 (§6.21)

binomial process, test of, one-sample, from two-sample test of randomness of occurrence 376

binomial sign test 206, 669-670 (§3.6); 183, 184, 315, 366
  of significance of mean and of correlation coefficient, efficiencies of 196

binomial success rate, estimated, moments of 205

binomial theorem 405

binomial variate 693

bipartitions 723

biplane combinatorial power mean 432 (§10.37)

bipolykays 723
  in terms of g.s.m.'s 724
  variances and covariances of 724

birth-order effect, test for 373

bisamples, matrix 724

biserial correlation 343, 397-698 (§7.91); 16
  rank 359

bivariate cumulants 683

bivariate distribution   319

bivariate frequency surfaces

   Pearson   406 (§10.19)

   translation of   414 (§10.229)

bivariate Gram-Charlier series
                    410 (§10.219)

bivariate hypergeometric distribution
                    276

bivariate multinomial distribution
   cumulants of   683

bivariate normal distribution
   119-125, 652-654 (§1.7*); 241,
   291, 302, 324, 340, 343, 346,
      347, 370, 433, 697

   circularly symmetric   653

     covering circle of sample from
                    291

     volume over elliptical regions
                    660

   moment inequalities applied to   433

   percentage points of, approximate
                    290

   random samples from   463

   tests applied to
     305-306, 690-691 (§6.12); 319-
     329, 695 (§7.1*); 351 (§8.114)

   test for independence and
     homoscedasticity   697

bivariate Pearson frequency
   surfaces   406 (§10.19)

bivariate Poisson distribution
            282 (§5.22)

bivariate Sheppard corrections   431

bivariate sign test   714

bivariate t-distribution   241

bivariate translation systems   414

blocks

   confounding in, blocking
     distinguished from   497

   factorial patterns confounded in
     555-556 (§§15.141-15.142*)

   in factorial patterns   497, 499

     estimate clear of   500

     estimate confounded with   500

     estimates of linear and quadratic
       effects clear of cubic effects
       and   557

     interaction with factors   500

     main effects estimable clear of
                    556

   of unequal size   557

   in incomplete block patterns   495,
                    563

     replication groups orthogonal to
                  565, 609

   blocks and other 2-factor interactions,
     2-factor interactions estimable
     clear of   555

block effects   553

blocking   496-497

   distinguished from confounding in
     blocks   497

   in factorial patterns   499

     for 2-way control of heterogeneity
                    502

   (SEE ALSO heterogeneity, control of)

blood tests of pooled samples   611

bounds

   for cumulative distribution   433

   for expectation of straggler   435

   upper, for straggler   702

   upper and lower

     for range   702

     for ratio of range to standard
       deviation   702

bounded relative error   689

Box, G. E. P.

   suggests omission of preliminary
     tests of homoscedasticity   264

   survey of tests applied to normal
     distributions   295

Bravais correlation coefficient
   (SEE product-moment correlation
   coefficient)

bridge, probabilities in   685

British practice for control charts,
   contrasted with American   488

Burr's cumulative distribution
   functions   416-417, 720 (§10.24)

C, D, A, B
chi-square test

C, D, A, B: factors for constructing
    control charts   488-490, 733
                            (§14.221)

CSP-1   481, 482, 483

CSP-2   482, 483

CSP-A   482, 483

canonical correlations   311-313
                 (§6.133); 318, 690
canonical form
  of factorial patterns   497
    lexicographical order of   498
    multiple appearances in list of
                            498
  of incomplete block patterns   564

capture-recapture method   688

card games, probabilities in
        684-687 (§5.6*)

Cauchy distribution, estimated by
    gauging   716

Cauchy-Riemann differential equations
                              127
Cavalli's transformation   453

Cebysev, P. L., inventor of
    polynomials   436

Cebysev's inequality
    433-434, 724 (§10.4)

Cebysev polynomials   436

Cebysev's quadrature formula   436

Cebysev-Hermite polynomials
                     34 (§1.223)
censored samples
  order statistics from   647
    probability integral
      transformation of   721
  regression estimates of mean and
    standard deviation from   642-648

censoring (SEE truncation)

center (SEE midrange)

centile   49

central area, under normal
    distribution   3 (§1.111)

central composite second-order
    patterns, for investigating
    response surfaces   554

---

*This section is subdivided;
see Table of Contents.

central differences, integration
    formula using   629

central F-distribution, power of
    F-test in terms of   256

central moments   426

central tendency
    (SEE Location, measures of)

chain blocks   607 (§15.246), 609
                              (§15.256)
  generalized   609 (§15.256)

characteristic function
  of $\log \chi^2$ 272
  of Pearson curves   226

characteristic root (SEE
    determinantal equation, root of)

Charlier, C. V. L., theta-function
                             282
  moments of are divergent   409

Charlier series   407-419 (§10.21*)

charts for mean and range, factors for
    constructing   733

Chebyshev, P. L. (SEE Cebysev)

χ, i.e. $(\chi^2)^{\frac{1}{2}}$
  moments of   167-171 (§2.522)
  non-central, moments of   226
  percentage points   144, 658

χ-distribution, as approximation to
    distribution of mean range
                            107, 108
χ^2 257
  approximate formulas for percentage
    points and normalization of   147
  approximations to percentage points
    of   144
  as square contingency
                   343 (§7.92); 356
  central   256
  distribution of, from stratified
    universe   682
  effect of sample correlation on   158
  goodness-of-fit, computed from runs
    up and down
    χ^2 approximation to   376
    not a gamma variable   376
  goodness-of-fit, Pearson   174-180,
         663-664 (§2.6*); 315, 376
  index-of-dispersion

[960]

$[\chi^2]$

Jeffreys' significance criterion K for 150

logarithm of

  approximately normally distributed 261

  characteristic function of 272

  cumulants of 272-273 (§4.34)

moments of 166-167, 662-663 (§2.521)

moments of functions of 165-173 (§2.52*)

non-central 155-157, 660 (§2.42); 256, 661, 679, 682

percentage points of 140-151, 658-659 (§2.31); 246, 365, 491

percentage points of, as approximation to percentage points:

  of Bartlett's test 273

  of beta distribution 150, 197-198

  of concordance coefficient 354

  of Neyman and Pearson's L_1 299

  of z 273

percentage points of Wilson-Hilferty approximation to 263

power of 158-159 (§2.43)

χ^2 approximation

  to multinomial distribution

    not used 287

    not improved by Yates' correction 287

  to multivariate tests of homogeneity of variance 295

χ^2 distribution 130, 132-134, 140-147, 150-151, 165-173, 656, 658-659 (§2*); 37, 207, 264, 677

  approximation to percentage points of 146, 197

  as approximation, inadequate, to two-sample test of randomness 377

  as approximation to distribution:

    of Bartlett's modified L_1 test 300

    of concordance coefficient 348

    of determinantal ratio 311

    of Friedman's concordance score 354

    of goodness-of-fit χ^2 for runs up and down 376

    of largest canonical correlation 312

    of likelihood tests 295

    of non-central χ^2 157

    of range 107-108

    of mean deviation 643

    of mean range 652

  as approximation to multinomial distribution, percentage error of 287

  as approximation to two-sample test of randomness, inadequate 377

  as limit of F-distribution 182, 245

  in 2 × k contingency tables 179

  non-central 155-157, 660 (§2.42); 288

  with 2 d.f. 125

  order statistics from 260-262, 681 (§4.25*)

  ordinates of 137 (§2.20)

  random samples from 455

  ratio of integral to bounding ordinate of 133, 656 (§2.1134); 717

  scale parameter of, maximum likelihood estimate of 716

  square-root transformation of, variance of 448

  yields asymptotic distribution of test:

    for equality of means of correlated variates 310

    for equality of means, variances and covariances of correlated variates 311

    for equality of variances and covariances of correlated variates 311

χ^2 integral, incomplete normal moments in terms of 31

χ^2 smooth test, F. N. David's, of goodness of fit 180

χ^2 test

  index-of-dispersion 180, 315, 316

    power of, against Neyman's contagious distribution, Type A 280

  of goodness of fit 174-180, 663-664 (§26*); 315, 366, 380

    inefficiency of 380

  power of 158

chi-square variates
confounding

χ^2 variates

    independent   443

    weighted sum of, approximate normalization of   660

χ_r^2, Friedman's concordance score   354 (§8.1313)

    percentage points of   355

    choice of sample size   611-613, 736-737 (§15.3); 495

circular and hyperbolic functions combinations of   626 (§16.12*)

circular distribution   418-419, 722 (§10.28)

circular functions   622 (§16.071*); 334, 699

    inverse   398 (§10.0432), 445 (§12.11-12.13); 343

    (SEE ALSO trigonometric functions)

circular normal distribution (Mises)   418-419, 722 (§10.28)

circular serial correlation   336-338, 696 (§7.51)

    of residuals from fitted Fourier series   338

circular triads
    (SEE inconsistent triads)

circularly symmetric bivariate normal distribution   653

    volume over elliptical regions   660

classes, number of, in χ^2 test of goodness of fit   664

clear   500

clear of blocks

    estimate   500

    main effects estimable   556

clear of blocks and cubic effects, estimates of linear and quadratic effects   557

clear of blocks and other 2-factor interactions, 2-factor interactions estimable   555

clear of cubic effects, estimates of linear and quadratic effects   555

---

*This section is subdivided; see Table of Contents.

clear of cubic effects and blocks, estimates of linear and quadratic effects   557

clear of factorial effect, estimate   500

Cochran, W. G., test of homogeneity of variance   303, 443, 726

Cochran, W. G. and Fisher, R. A., approximation to percentage points of z, yielding approximate percentage points of beta distribution   198

co-cumulants   430

    of k-statistics   723

        multivariate   432

    Sheppard corrections to, vanish   431

coefficients

    binomial (SEE binomial coefficients)

    confidence   681, 689

    Fourier   727

    in fitted polynomials, as linear combinations of ordinates   725

    interpolation
        (SEE interpolation coefficients)

    Lubbock   629 (§16.2369)

coefficient of concordance, Kendall's W   353-355 (§8.13); 348

coefficient of correlation
    (SEE correlation coefficient)

coefficient of variation   74

    confidence interval for   225

    of j-statistics   81

    of mean deviation from mean, of normal distribution   74

    of mean deviation from median, of normal distribution   74

    of range   87, 701

        effect of β_2 on   701

    of standard deviation   87

collections of constants, not catalogued   620

cologarithms   109, 311, 403

columns

    in factorial patterns   497, 499, 561, 562

    in incomplete block patterns   564

combinatorial patterns, for
  experimental designs
  495-609, 735-736 (§§15.1-15.25*)

combinations
  (SEE binomial coefficients)

combinatorial power mean
  biplane 432
  monoplane 432

common logarithms 620-622 (§16.06*)

comparisons, paired 355-356, 699
                                  (§8.14)
  signed-rank test for 367

compatible events, probability of 637

complementary loglog transformation
                                  24
  (SEE ALSO loglog transformation)

complete beta-function 204 (§3.51)

complete gamma-function 160-164
                         660-662 (§2.51)

complete (saturated) sets:
  of orthogonal cubes
            554 (§15.1312); 499, 502
  of orthogonal hypercubes, not tabled
                                  554
  of orthogonal squares
            554 (§15.1312); 499, 502

completely balanced lattice square
  (SEE square lattice)

complex argument
  error integral of 126-128 (§1.8*)
  Γ-function of 164 (§2.515*)

composite second-order patterns,
  central, for investigating
  response surfaces 554

component-of-variance model, power of
  F-test under 255

compound Poisson distribution
  278-282, 683 (§5.21), 285, 683
                                  (§5.32)

concordance coefficient W, Kendall's
            353-355 (§8.13*); 347
  moments of 348

concordance score, Kendall's 347

concurrent deviation
  (SEE Kendall's Tau)

confidence intervals
  approximate, for median effective
    (lethal) dose 633
  based on mean range 104
  based on t-distribution 216-217
                                  (§4.133)
    for batch mean 216
  for binomial distribution 193-195
                         (§3.32); 190, 191
  for coefficient of variation 225
  for correlation coefficient
                         322-323 (§7.121)
  for Λ^2 690
  for difference of means 239, 646
  for hypergeometric distribution 276
  for maximum studentized treatment
    effect 674
  for mean of normal distribution
                         103, 217, 303, 646
    effect of intraclass correlation
      on 243
  for median lethal dose 30
    approximate 633
  for median of symmetrical
    distribution 367
  for Poisson parameter 151-153,
                         659 (§2.322)
  for range 646
  for ratio of variances
                         257, 258, 266, 689
  for standard deviation
                         148-150, 658 (§2.315)
  for variance 158
    of finite population 253
  for weighted mean 674-675 (§4.18)
  logarithmically shortest, for ratio
    of variances 258

confidence statements, about order of
  normal variances 681

confluent hypergeometric function
  200, 673 (§4.142*); 124, 219,
                         275, 678
  inverse tables 303

confounded factorials and lattices
                                  501
confounding 500
  double 561 (§15.152); 497
  in saturated fractions of 2^a
    factorial patterns 553
  partial 556 (§15.1422); 502, 561
    connexion with lattices 501
  total 556 (§15.1421)

confounding in blocks
covariance

confounding in blocks
  blocking distinguished from   497

constants, tables of, contents of,
    not described   620

consumer's risk   469

contagious distributions   278-282,
    683 (§5.21), 285, 683 (§5.32)

contingency, mean square
    343 (§7.92); 356

contingency tables, χ^2 test in
    174-179, 663-664 (§§2.61-2.62*)
  2 × 2, test of independence in   275

continuity correction, to distribution
    of concordance coefficient
    354, 355

continuous inspection by attributes
    481-483 (§14.13)

continuous rectangular distribution
    418, 720-722 (§10.261)
  as limit of discrete rectangular
    distribution   418
  order statistics from
    360, 699-700 (§8.21)

contour integral, incomplete beta
    function as a   198

contour moments   433

control by gauging   487

control charts
    487-491, 732-734 (§14.2*)
  for largest and smallest values
    60, 490
  for range   83
  for standard deviation   148-149

control chart limits for number
    defective   151

control of heterogeneity
  in factorial patterns   496
    absent   533-555 (§15.13*)
    1-way   555-557 (§15.14*)
    2-way   558-562 (§15.15*); 502

---
*This section is subdivided;
see Table of Contents.

  in incomplete block patterns   563
    1-way   605-607, 736 (§15.24*)
    2-way
      607-609 (§15.25*); 564, 565, 566

corner test, for association   713
  (SEE ALSO median test)

Cornish, E. A., and Fisher, R. A.,
    polynomial translation
    146, 147, 233, 242, 411-412

Cornish-Fisher approximation
  to percentage points:
    of non-central χ^2 661
    of t   215
  to weighted sum of χ^2 variates   660
  to z-distribution   272, 274

corpuscle problem, Wicksell's   288

corrected probit (SEE working probit)

corrections
  Euler-MacLaurin   427
  for continuity   344, 354
    Yates', to χ^2, not useful in
      multinomial approximation   287
  for grouping
    of discrete distributions
      428-429 (§10.333)
    Sheppard's 426-427, 722 (§10.331)

correlated normal variables
  ratio of   653
  ratio of standard deviations of   673
  two; three   55

correlated random normal deviates   463

correlated variances, test for
    difference between   265

correlated variates
  equality of means, variances and
    covariances of, test for   310
  equality of means of, F-test for
    310
  equality of variances and
    covariances of, test for   310

correlation   319-344, 695-699 (§7*)
  between β_1 and β_2,
    in Pearson curves   391
  between correlation coefficient and
    standard deviation   652
  between mean and variance   434
  between order statistics   67
  between quantiles   67
  between ranges in overlapping
    samples   643
  between ranked intervals   712
  between standard deviations   123
  between time series   696 (§7.54)
  between t-values in correlated
    samples   242
  biserial
    343, 697-698 (§7.91); 15, 16
  canonical
    311-313 (§6.133); 318, 690
  curvilinear   253
  effect of, on χ^2 158
  intra-class, effect of
    on confidence interval for mean
      243
    on percentage points of t   243
  multiple   335 (§7.4)
  non-parametric measures of
    345-359, 699 (§8.1*)
  normal, in population, distribution
    of r_s under   351
  partial   696
  point biserial   698
  product-moment   336, 345, 346
    distribution of, as approximation
      to distribution of Spearman's
      rank correlation   346
  rank   345-359, 699 (§8.1*); 343, 360
  serial   336-339 (§7.5*)
    from first order Markov process,
      hyperbolic-tangent
      transformation of   333
    non-circular   338 (§7.52); 79
    of independent and residual
      variables   726
    of residuals from regression   78
    of samples from autoregressive
      time series   467
  Spearman's rank   348-352 (§8.11)
  tetrachoric   343-344, 698-699
    (§7.94)
    for median dichotomies   653
  triserial   653
  within sample, effect of, on χ^2 158
correlation coefficient
    319-329, 695 (§7.1*); 119
  β_1 and β_2 of   420
  between ranges   105
  correlation between standard
    deviation and   652
  distribution of
    as approximation to distribution
      of non-central t   224
    under permutations   344
  equality of, test for   332
  intra-class, hyperbolic-tangent
    transformation of   330
  maximum likelihood estimate of   328
  mean of   420
  mode of   326 (§7.133); 420
  moments of
    325, 695 (§7.131); 226, 342
  multiple   335 (§7.4); 155, 254
  standard deviation of   420
  test of   305
correlation ellipse of β_1 and β_2 391
correlation ratio   343 (§7.93); 253
  of variance on mean   434
  unbiased estimate of   254
correlation surface
    (SEE bivariate frequency surface)
correlogram   341
cost function, linear   611
Cotes integration coefficients   629
count, in bridge hands   684
covariance   340-342, 697 (§7.6)
  in bivariate Poisson distribution
    282
  of estimates of mean and standard
    deviation
    in control by gauging   487
  of truncated normal distribution
    40
  of order statistics   65, 67, 639
  of runs   709
  test for constancy of   295
  vanishing of, and variances,
    equality of, test for   313

covariance estimates
deviate

covariance estimates   340

covariance matrix (matrices)
    308, 312, 690, 691, 696, 716

  equality of, test of   309

  residual, after regression   311

covariance ratio   340

covering circle, of sample from circularly symmetric bivariate normal distribution   291

Cramer-Mises-Smirnov test   386 (§9.42*)

Criterion x, for Pearson curves   391

critical run lengths   713

cubature, numerical (SEE Irwin 1923 in Author Index)

cubes

  Latin   499, 554, 558

  orthogonal   499, 554

cubes of numbers   615 (§16.022)

cube roots   619 (§16.027)

cubic effects, estimates of linear and quadratic effects clear of   555

cubic effects and blocks, estimates of linear and quadratic effects clear of   557

cubic equations, solution of   620 (§16.0564)

cubic lattices   606-607 (§15.245)

cubic polynomials   704

cumulants   425-426 (§10.32); 432

  bivariate   683

  multivariate   429

  of binomial and negative binomial distributions, factitious resemblance between   684

  of bivariate multinomial distribution   683

  of χ 171

  of contagious distribution   281

  of F   263

  of k-sample paired-comparison slippage test   708

  of k-statistics   430-723 (§10.343)

  of linear systematic statistics   63

---

*This section is subdivided; see Table of Contents.

  of $\log \chi^2$ 272-273 (§4.34)

  of $\log L$ 299

  of multinomial distribution   683

  of order statistics, series for   721

  of t^2 240

  of z 272-273 (§4.34)

  Sheppard's corrections to   427

cumulant-generating function   407

  of multinomial distribution   683

cumulative binomial sum   183-184, 664 (§3.11)

  approximations to   198-199, 667-668 (§3.42)

cumulative distribution   454

  bounds for   433

  confidence interval for, under normality   104

cumulative distribution functions (Burr)   416-417, 720 (§10.24); 633

cumulative moments   416

cumulative Poisson distribution   134-136, 656-657 (§2.12)

cumulative sums:

  of normal deviates   638

  of RAND Corporation's digits, differences of, not necessarily random   458

curtailed inspection (SEE sequential sampling)

curtosis (SEE kurtosis)

curve fitting

  4 parameters to 6 moments   416

  4 parameters to 7 moments   416

  frequency curves   388-420, 715-722 (§§10.0-10.2*)

  other than frequency curves   436-444, 724-727 (§11*)

curvilinear correlation   253

curvilinear ranking test   358

curvilinear regression   436-444, 724-727 (§11*)

cycle, in bivariate sign test   714

D, A, B, C: factors for constructing control charts 488-490, 733 (§14.221)

D: normal deviate 14

$D_{m,n}$ 382-384, 715 (§9.4122)

D_n 381-382, 714 (§9.4121)

D^2-statistic 306-308 (§6.131); 691

damped sinusoidal functions 626 (§16.1232)

David, F. N., χ^2 smooth test of goodness of fit 180

Dawson's integral 126-128 (§1.81*)

decile (SEE quantile)

decimal digits, random 454

decimal logarithms 620-622 (§16.06*)

decimal numbers, natural logarithms of 625-626 (§16.112)

decrement rates, actuarial formulas for, efficiency of 485

defects 469

  number of

    lot inspection by 472

    process control of, by Poisson distribution 469

defectives 470

  number of

    control charts for 487, 732 (§14.21)

    lot inspection by 470-477, 730 (§14.11*)

    process control of, binomial distribution in 469

defective members of large populations, detection of 611

deficit, curvilinear 358

DeForest, E. L., method of moving averages 441

degenerate cases of Pearson curves

  Type I 393 (§10.018)

  Type II 394 (§10.028)

  Type III 396 (§10.038)

  Type IV 398 (§10.048)

  Type V 398 (§10.058)

  Type VI 399 (§10.068)

  Type VII 401 (§10.078)

  Type VIII 401 (§10.088)

  Type IX 402 (§10.098)

  Type XI 403 (§10.118)

  Type XII 404 (§10.128)

degrees of freedom 105, 107

  of χ^2 31

  of numerator and denominator of F 680

Δ^2, confidence intervals for 690

density function, of order statistics 62

dependence between mean squares, effect of, on F distribution 264-269, 681-682 (§4.28)

derivatives, of normal distribution 32-34, 634 (§1.321*)

derivatives of zero · 620 (§16.0492)

design of experiments 475-613, 735-737 (§15*)

destructive tests, Wald sequential 483

destructive testing 483-486, 732 (§14.15*); 474

determinantal equation 690, 692

  largest root of 312

  percentage points of 691, 692

  percentage points of, approximate 690

  root of 690

  percentage points of 690

determinantal ratio, percentage points of, approximate 311

determination, coefficient of 614-615 (§§16.0211-16.0213)

deviate

  extreme 61, 69

  normal, ordered 62-68, 638-639 (§1.43); 49

  studentized 49

  normal equivalent 22, 24

  (SEE ALSO probit)

  straggler 54-62, 637-638 (§1.42); 49

  studentized 49

dice-casting
Einstein functions

dice-casting, Weldon's experiments 464

dichotomization tests 358, 370

difference between proportions, test for 692
  power of 663

differences of cumulative sums of RAND Corporation's digits, not necessarily random 458

differences of log 1 703

difference of means
  confidence interval for 646
  tests for, in face of unequal variances 231-239, 674 (§4.15*)

differences of zero 620 (§16.0492); 293, 294

differential coefficients of zero 620 (§16.0492)

differential equation of Pearson curves 388

differentiation, numerical 629 (§16.234)

digamma function 163-164 (§2.5141); 395, 486

digits, random 456-459, 729 (§13.1*), 468 (§13.61)

dilution method 316, 317

disarray, coefficient of (SEE Kendall's Tau)

discrepancies from Poisson distribution, test for 693
  approximate 694

discrete distributions 275-287, 292-294, 683-688 (§5*); 407
  Charlier's Type B valid only as 409
  corrections to, for grouping 428-429 (§10.333)
  technique of drawing random samples from 454

discrete rectangular distribution 417-418, 722 (§10.262); 407
  order statistics from 361 (§8.22)
  range in samples from 293

---
*This section is subdivided; see Table of Contents.

dispersion
  binomial 180
  estimated from:
    linear systematic statistics 93
    selected order statistics 92-95, 647-649 (§1.544)
  index of 180, 281
  χ^2 test of 180, 315, 316
  χ^2 test of, power of 280
  measures of 71-88, 92-95, 107-110, 643-649, 652 (§1.5*)
    based on order statistics 92-95, 647-649 (§1.544)
    based on selected quantiles 636
  non-parametric tests of 373, 708-709 (§9.2)
  trend in, in time series, test against 710

dispersion matrix (SEE covariance matrix)

distance (SEE D^2-statistic)

distribution-free (SEE non-parametric)

distribution-free recurrence formula, for moments of order statistics 62

division, random, of line segment 291

Dodge, H. F., CSP, i.e. continuous inspection by attributes 481, 482

dosage mortality curve 23

dose
  critical 22
  minimum 22
  tolerance 22

double confounding 561 (§15.152); 496, 497 ftn.

double control of heterogeneity, in factorial experiments 496

double-entry tables, of working probits 24 (§1.2311)

double exponential distribution
  Gumbel ($\beta_1 = 1.139$, $\beta_2 = 5.4$) 364, 703 (§8.311*); 701 (§8.24)
  Laplace ($\beta_1 = 0$, $\beta_2 = 6$) 402, 464, 645, 647, 648

double factorials 161 (§2.5113), 163 (§2.5135)

double hypergeometric series 684

double integrals
   (SEE Irwin 1923 in Author Index)

double sampling
   by attributes   471, 472, 476
   by variables   477

double tail area, of normal
   distribution
      5-6, 630 (§1.113); 20

double tail percentage points
   of normal distribution
      20, 632 (§1.2122); 2
   of t   211-213, 671 (§§4.1311-4.1312)

double-tailed test (SEE two-sided test)

doubly balanced incomplete block
   patterns   606 (§15.2412); 566

doubly censored samples   647, 648

duration of play   292

Edgeworth series
   37, 147, 154, 216, 230, 408, 409
   distribution of F in samples from
                                681
   distribution of t in samples from
                                230

effect, in factorial patterns
                             496, 499
   block   503
   estimable   501
   estimate clear of   500
   estimate confounded with   500
   estimate of   500
   linear combination of   500
   partition of   499
      into main effects and
      interactions, not unique   500
   k-factor   499
   treatment   553

effective (a substantive)   470

efficiency   470
   asymptotic, of tests for correlation
      between time series   696
   of accelerated life testing   486
   of actuarial formulas for decrement
      rates   485
   of best linear estimate of
      dispersion, for normal
      distribution   94, 95

of confidence intervals for mean of
   normal distribution with range
   in denominator   103
of estimate of mean, in control by
   gauging   487
of estimate of standard deviation,
   in control by gauging   487
of gauging, applied to normal
   distribution   716
of j-statistics, for normal
   distribution   81, 82
of mean deviation   74, 94
of mean moving ranges   643, 644
of mean square successive difference
                                645
of median   636
of midrange   90, 647
of non-parametric tests of
   regression   705
of order statistics   71, 642
   as estimates of moments
                          94, 95, 362
   from censored samples   647
   from exponential distribution
                             700, 701
of quartile range, for normal
   distribution   93
of random sampling   341
of range, for normal distribution
                              87, 88, 94
of range-midrange test, relative to
   t-test   101
of ratio estimate   697
of Scheffe's two-sample test   240
of stratified sampling   341
of substitute t-test based on order
   statistics   64
of systematic sampling   341
of systematic statistics   87
of t-test, relative to normal
   deviate test   675
of variate differences   77
of Wald sequential test of
   rectangular distribution   695
of Walsh test   367
of weighted grouping   726

eigenvalue (SEE determinantal
   equation, root of)

Eigenwert (SEE determinantal equation,
   root of)

eighth moment, finite, of Pearson
   curves   390

Einstein functions   484-485 (§14.152)

[969]

Elderton
factor

Elderton's 'first main type', i.e.
　　Pearson's Type I
　　　　392-393 (§10.01*)

Elderton's 'normal curve of error',
　　i.e. the normal distribution
　　<u>qua</u> Pearson curve
　　　　404-405 (§10.13*)

Elderton's 'second main type', i.e.
　　Pearson's Type IV
　　　　397-398, 717 (§10.04*)

Elderton's 'third main type', i.e.
　　Pearson's Type VI　399 (§10.06*)

ellipsoidal regions, integral of
　　spherically symmetric trivariate
　　normal distribution over　660

elliptic functions　627 (§16.21*)

elliptic theta functions
　　380-381, 714 (§9.41*); 627

elliptical regions, integral of
　　circularly symmetric bivariate
　　normal distribution over　660

empirical distribution function　380

empirical sampling investigation of
　　power of tests of randomness
　　　　710

endpoint, of exponential distribution
　　estimated from order statistics
　　　　700
enumeration and randomization of
　　Latin squares　558-560 (§15.1512)

equality of variances, sequential test
　　for, based on ranges　651

equations

　　cubic　620

　　determinantal　690, 692

equivalent deviate (SEE normal
　　equivalent deviate; logit;
　　probit)

Erf, erf, ambiguous notations　1

Erlang distribution　483

Erlang loss formula　656

　　recurrence formula for computing
　　　　657
errata, lists of　252 (§4.233);
　　270-271 (§4.333); 438 (§11.112);
　　　　3, 66, 183 ftn., 187 ftn.

---

　　*This section is subdivided;
see Table of Contents.

error

　　of second kind　222
　　probable　12, 14
　　relative　215
　　standard, of estimates of variance
　　　　170
error function　1

error integral　10-12, 630 (§1.121*)

　　inverse　19-21, 631-632 (§1.21*)
　　of complex argument　126-128 (§1.8*)

estimable effects　501

estimability, in factorial
　　experiments　499

estimate　496, 499, 500

　　from dilution series　317
　　of censored normal distribution　43
　　of negative binomial distribution
　　　　285
　　minimum-variance unbiased linear
　　　　641
　　of optimum subsampling number　611
　　of linear and quadratic effects
　　　　clear of cubic effects　555
　　　　clear of cubic effects and blocks
　　　　　　557
　　of parameters in binomial
　　　　populations　205 (§3.54), 314
　　of standard deviation　169
　　　　standard error of　170
　　of variance　166

Euler-MacLaurin summation formula
　　　　427, 428

Everett interpolation coefficients
　　　　628 (§16.2317); 627

Everett interpolation formula
　　first derivative of　629

excess, coefficient of
　　(SEE b_2; kurtosis)

expectation, fiducial, of risk　612

experiments, design of
　　　　495-613, 735-737 (§15*)

experimental units　496

exponential curves, fitting of

　　frequency curves
　　　　(SEE exponential distribution)
　　regression etc.　444 (§ 11.3)

exponential distribution
    402-413, 718 (§10.10*); 396
  censored samples from,
    order statistics from   647
  double
    Gumbel ($\beta_1 = 1.139$, $\beta_2 = 5.4$)
      364, 703 (§8.311*); 701 (§8.24)
    Laplace ($\beta_1 = 0$, $\beta_2 = 6$)
      402, 464, 645, 647, 648
  fourth-degree 415-416 (§10.23); 483
  Gini's mean difference in samples
    from 80
  in life testing 732
  iterated 364-365, 703 (§8.31*)
    order statistics from 701 (§8.24)
  order statistics from 647, 648
  two-tailed 402, 645
    censored samples from,
      order statistics from 647
    logarithmic and hyperbolic
      translation of 719
    simulated random samples from 464
exponential functions
    623-624, 738 (§§16.101-16.103);
    450 (§12.52); 364, 365, 402, 404
  base ten 621 (§16.064)
  descending
    624, 738 (§16.102); 402 (§10.101)
  in life testing 484-485 (§14.152)
exponential regression 444 (§11.3)
extreme deviate 61, 69
extreme values, asymptotic theory of
    364-365, 703 (§8.3*); 366

F, i.e. variance ratio
    245-269, 676-682 (§4.2*)
  degrees of freedom of numerator and
    denominator of 680
  distribution of, from stratified
    universe 682
  kurtosis of 263
  non-central
    258-260, 678-682 (§4.242)
    both numerator and denominator
      non-central 682
  percentage points of decimal
    logarithm of 251 (§4.332)
  skewness of 263

$F^{\frac{1}{2}}$, percentage points of 250 (§4.232)
$F(r,\nu)$ 397-398, 717
F-distribution 245-269, 676-682
    (§4.2*); 181, 182, 389, 399
  approximations to
    263-264 (§4.27); 246, 273
  as approximation to distribution:
    of likelihood tests 295
    of ratio of two independent
      quadratic forms 268
    of test for homogeneity of
      variance 295
  χ^2 distribution as limit of 182
  non-central 258-260, 678-682
    (§4.242); 256, 307
  practical advantages over
    z-distribution 270
  theoretical advantages of
    z-distribution over 270
F-ratio, maximum[*] 262 (§4.253)
F-test
  as approximation to test for multiple
    rank correlation 356
  for comparison of group means 264
  power of 255-260, 677-680 (§4.24*)
    compared with power of test for
      difference between correlated
      variances 265
    effect of heteroscedasticity on
      267
    effect of non-normality on 267
  preliminary, effect on estimate of
    regression 243
  substitute, from ratio of ranges
    103
factor, in combinatorial patterns
    495, 496, 498
  applied to split plots
    562-563 (§15.16); 502
  applied to whole plots 562
  interaction of blocks with 500
  principal, in incomplete block
    patterns 563

_____

[*] to be distinguished from the largest variance ratio of §4.251.

factorial
generating function

factorial, i.e. combinatorial pattern (for experimental designs) 496-563, 735-736 (§15.1*); 495, 564
  overlap with incomplete block patterns 495, 547-553 (§15.12); 556-557 (§15.143); 561-562 (§15.153); 606 (§15.243); 608-609 (§15.253)

factorial, i.e. Γ-function 160-164, 661-662 (§2.51*); 485, 619

factorial moments 438, 439, 441
  of binomial distribution 203

factorial moment generating function, of hypergeometric distribution 275

factorial polynomials 438-439 (§11.113); 441 (§11.13)

fiducial distribution theory 232

fiducial expectation of risk 612

fiducial limits
  for batch mean 216
  for batch standard deviation 253
  for binomial distribution 193
  for Poisson parameter 152
  for relative potency in 6-point assay 30
  for variance of finite population 253
  (SEE ALSO confidence intervals*)

figurate numbers 284, 683 (§5.31)

finite population
  k-statistics 431 (§10.35)
  sampling from 275, 431
  variance of, confidence interval for 253

first absolute moment, of normal distribution, about population mean 45

first law of Laplace, logarithmic and hyperbolic translation of 719

---

*But see M. G. Kendall 1946, p. 83

*This section is subdivided; see Table of Contents.

Fisher, R. A.
  approximation to percentage points of χ^2 144, 145, 147
    applied to yield approximate fiducial limits for Poisson parameter 153
  distribution (SEE F-distribution; z-distribution)
  Hh functions 37-38 (§1.331)
  hyperbolic-tangent transformation 330-333, 695 (§7.2*)
  z, i.e. $\frac{1}{2}\log_e F$ 270-284, 676 (§4.3*); 354
  z, i.e. $\tanh^{-1} r$ 330-333, 695 (§7.2*)

Fisher-Behrens test 232-236 (§4.151)
  Aspin's tables not related to 238

Fisher-Cochran approximation to percentage points of z 198
  yielding approximate percentage points of beta distribution 198

Fisher, R. A. and Cornish, E. A. (SEE Cornish and Fisher)

fitting (SEE curve-fitting)

5-dimensional lattice 606

flats, in orthogonal cubes 554

four-decision problem, simultaneous test of mean and variance as 689

fourfold contingency tables (SEE 2×2 tables)

Fourier analysis 442-443, 726-727 (§11.2*)

Fourier Series, residuals from, circular serial correlation of 338

fourth-degree exponential distribution 415-416 (§10.23)

fourth moments 616 (§16.0233)

fourth powers, multiples of 616 (§16.0233)

fractile 49

fractions
  of 3^a and 5^2 factorial patterns 554-555 (§15.138), 557 (§15.148)
  saturated, of 2^a factorial patterns 553 (§15.1311)

fraction defective (SEE defective)

fractional powers 617-619 (§§16.026-16.027)

fractional factorials confounded in blocks 555 (§15.141)

frequency curves 388-420, 715-722 (§§10.0-10.2*); 434-435 (§10.0); 627

frequency moments 401

frequency surfaces 406 (§10.19), 414 (§10.229); 119-125, 652-654 (§1.7*); 241

Fresnel integrals 127 (§1.82)

Friedman's concordance score 354 (§8.1313)

full factorials confounded in blocks 556 (§15.142*)

$G(r,\nu)$ integrals 397-398, 717 (§10.0431)

g.s.m.'s, i.e. generalized symmetric functions 723

in terms of bipolykays 724

G-test, for means 650

Galton differences 90-91 (§1.543)

Galton's individual difference problem 90-91 (§1.543)

Galton-McAllister distribution 412-414, 720 (§10.2221); 44, 281

gambler's ruin 292

(SEE ALSO sequential sampling)

gamma distribution
 (SEE χ^2 distribution; -function, incomplete; Pearson curve, Type III)

Γ-function
 complete 160-164, 661-662 (§2.51*); 410
 incomplete 131-133 (§§2.111, 2.113*); 31, 37, 130, 155, 275, 312
  as confluent hypergeometric function 226
  incomplete beta function expanded in series of 199
 logarithmic derivative of 163-164 (§2.5141)
 of complex argument 164 (§2.515*)
 product of 295

Γ-function ratios 165, 297, 298, 299

gauging
 control by 487
 to estimate proportion of a population 716

Gauss-Laplace distribution
 (SEE normal distribution)

Gaussian error distribution
 (SEE normal distribution)

Gaussian hypergeometric function 275-277 (§5.1); 112, 114, 123, 355, 406

(SEE ALSO incomplete beta-function)

Gaussian logarithms 621-622 (§16.063); 392

Gaussian Markov process 696

Geary's ratio
 first absolute moment to standard deviation 115-117 (§1.63*)
 k^{th} absolute moment to k^{th} power of standard deviation 111

Gebelein's tables 381

gene frequencies determined by selection and diffusion 453

generalized beta distribution 692

generalized chain blocks 609 (§15.256)

generalized hypergeometric distribution 407

generalized mean, Gini's 432 (§10.37)

generalized Poisson distributions 278-283, 683 (§5.2*)

generalized probable error 144

generalized symmetric functions 723

generating function
 cumulant 407
 factorial moment, of hypergeometric distribution 275
 moment 407
 of test for birth-order effect 373
 probability, of hypergeometric distribution 275

generating identity, of homogeneous
  power sums   421

generation of pseudo-random numbers
  468 (§13.62)

generation of random numbers
  468 (§13.61)

geometric mean, as a biplane
  combinatorial power mean   432

geometrical probabilities
  288-291 (§5.5)

Gesetz der kleinen Zahlen
  (SEE Poisson distribution)

Gibrat distribution
  (SEE Log-normal distribution)

Gini's generalized means   432 (§10.37)

Gini's mean difference   80 (§1.532)

golden numbers   738 (§16.0568)

goodness of fit
  χ^2 test of   174-180, 663-664
    (§2.6*); 315, 366, 376
  non-parametric tests of
    380-386, 713-714 (§9.4*)

grade correlation   351 (§8.114)

graduation, moving-average, of time
  series   441, 726

Graeco-Latin square   560-561
    (§15.1513); 499, 502, 638
  loosely construed   499
  maximum F-ratio in   638
  of side 6, does not exist   561

Gram, J. P., orthogonal polynomials
    436

Gram-Charlier approximation, to
  distribution of matching scores
    685, 686

Gram-Charlier series
  407-410 (§10.21); 32, 146, 154,
    267, 686
  bivariate   410
  generalized   410
  Type A   36-37, 634 (§1.325),
    407-409 (§10.211); 32, 125,
    154, 199, 201, 274
    distribution of t in samples from
      231
  Type B   409 (§10.212); 199, 687
  Type C   410 (§10.213)

*This section is subdivided;
see Table of Contents.

Gram-Schmidt process, for orthogonal
  polynomials   441

grand mean   500, 553
  as a factorial effect   499

Gregory's integration formula   629

Gregory-Newton interpolation
  coefficients   628 (§16.2313)

Gregory-Newton interpolation formula,
  first derivative of   629

group, social, isolates in   294

groups of variates, Wilks' test for
  independence of   295, 311

grouped distributions, moments from
  426-429, 722 (§10.33*)

grouped moments, of log-normal
  distribution   44

grouped observations, from normal
  distribution   45

grouped range   95

grouping
  corrections to moments for   428
  weighted, polynomial fitting by   725

gudermannian   410

Gumbel, E. J., limiting distributions
  of extreme values usable in
  finite samples   364

$H(r,\nu)$ 397-398, 717 (§10.0431)

Hadamard matrices   554 (§15.1313)

half-integer powers   618 (§16.0263)

half-plaid squares   562 (§15.1622)

Hankel function   365 (§8.33*); 108,
  109, 110, 124, 340

Hardy, G. F., Sir, summation method
  438-439 (§11.113)

harmonic analysis
  442-443, 726-727 (§11.2*)
  residuals in, circular serial
    correlation between   338

harmonic interpolation   628

harmonic mean   691

Helmert distribution
  (SEE chi-square distribution;
  Pearson curve, Type III)

Hermite functions
34 (§1.322); 32, 33, 232

Hermite polynomials 34 (§1.323); 33, 230, 275, 408, 410, 436, 655

  as confluent hypergeometric functions 226

  confluent hypergeometric function not disguised as 227

  recurrence formula for 35

Hermite series 32

heterogeneity, control of

  in factorial patterns
496, 497, 498, 499

    absent 553-555 (§15.13*)

    1-way
555-557 (§15.14*), 562 (§15.161)

    2-way
558-563 (§§15.15*, 15.162*); 502, 564

  in incomplete block patterns 563

    1-way 605-607, 736 (§15.24*)

    2-way
607-609 (§15.25*); 564, 565, 566

heteroscedasticity, effect of, on distribution of F
264-269, 681-682 (§4.28)

heteroscedasticity, short-cut test for 262

hexagamma function 164 (§2.5142)

Hh functions (§1.331); 33, 38, 41, 218

  as confluent hypergeometric functions 226

high contact
(SEE abruptness, corrections for)

Hilferty-Wilson approximation to percentage points of chi-square
145, 147, 263

  applied to yield approximate fiducial limits for Poisson parameter 153

histograms 405

Hoeffding, W., test for independence 356

Hojo's integrals
51-53, 57-58, 60, 89

  replaced by discrete sums 95

homogeneity

  of variance, tests of 302, 681, 688

  tests of, applied to multivariate normal populations
306-311, 692 (§§6.131-6.132)

homogeneous product sums 421

homoscedasticity

  Bartlett's test of, power of 256

  Cochran's test of 443

  preliminary tests of, Box suggests omission of 264

  sequential test of, based on ranges 651

homoscedasticity and independence, test of, in bivariate normal distribution 697

Hotelling, H., extensions of hyperbolic-tangent transformation 330

hyper-Graeco-Latin Square 554 (§15.1312), 499, 561

hyperbolic amplitude 410

hyperbolic functions
330-331, 695 (§7.211); 624-625 (§§16.1041-16.1042)

  combinations of circular and 626 (§16.12)

  inverse 331, 695 (§7.212)

  natural logarithms of 481, 624

hyperbolic logarithms
625-626 (§16.11*)

hyperbolic translation
414 (§§10.2222-10.2223); 412

hyperbolic-sine transformation, inverse 449 (§12.3*)

hyperbolic-tangent transformation
330-333, 695 (§7.2*)

<u>alias</u> logit transformation 450

hypergeometric distribution
275-277 (§5.1); 175, 469, 470

  binomial approximation to 469, 470

  confidence limits for 276

  generalized 407

  in lot control by number of defectives 469

  of χ^2 test of goodness of fit
174, 175

  Poisson approximation to 469, 470

hypergeometric function
k-factor effect

hypergeometric function
   confluent   200, 673 (§4.142*); 124, 219, 275, 678
     inverse tables   303
   Gaussian   275-277 (§5.1); 112, 114, 123, 335, 406
     no direct tables   275, 276

hypergeometric series, double   684

hyperspherical simplices, relative surface contents of   637

hypothesis, quantum   694

imaginary Bessel function
   (SEE Bessel function I; Bessel function K)

incomplete beta-function
   181-206, 664-670 (§3*); 37, 50, 62, 246, 275, 284, 389
   approximations to   199-200, 668 (§3.43); 273
   as approximations to binomial distribution   198
   F-distribution in terms of   246
   integrals involving   297
   series of, distribution of t from non-normal parents expanded in   230
   trigonometric form of   200, 201
   yielding power of F-test   255
   yielding probability integral of t   208

incomplete-beta-function ratio
   185, 665 (§3.12); 181, 208, 230
   percentage points of   191-194, 665-666 (§3.31); 490

incomplete block experiments, tables for analysis of   610 (§15.28)

incomplete block patterns
   563-609, 736 (§15.2*); 495
   factorial patterns obtained by unblocking   547-553 (§15.12); 501
   overlap with factorial patterns   547-553 (§15.12); 556-557 (§15.143); 561-562 (§15.153); 606 (§15.243); 608-609 (§15.253); 495

incomplete Γ-function
   131-134 (§§2.111, 2.113*); 31, 37, 130, 155, 275, 312, 389

*This section is subdivided; see Table of Contents.

as confluent hypergeometric function   226
   incomplete beta-function expanded in series of   199

incomplete Latin squares   607 (§15.2511)

incomplete normal moments   31-32 (§1.31); 37, 38, 42, 130, 200, 201

incomplete randomized block patterns   605 (§15.2411)

inconsistent triads   355-356, 699 (§8.14)

independence
   Hoeffding's test of   356
   in 2×2 tables, χ^2 test of   174-179, 663-664 (§2.61*)
   of k groups of variates, Wilks' test of   295, 311
   of two tests of randomness   710

independence and homoscedasticity, test for, in bivariate normal distribution   697

independent variables, serial correlation of   726

index of dispersion   180, 281
   χ^2 test of   180, 315, 316
   power of   280

individual observations, control chart for   490

inefficiency of method of moments   428

inefficient statistics   71

inequalities, for tail area   435

infinite population, k-statistics from   429-431, 723 (§10.34)

information matrix   41

inspection
   continuous, by attributes   481-483 (§14.13)
   lot
     by attributes   470-477, 730 (§14.11*)
     by variables   477-481, 730-731 (§14.12*)
   normal   472, 478
   100%   481
   reduced   472, 478
   tightened   472, 478

inspection plan, single-sample 150

integrals

  of orthogonal polynomials 439

  repeated normal 37-38 (§1.331)

integration, numerical 629 (§16.236)

interactions

  i-factor clear of j-factor 502

  in saturated fractions of 2^a factorial patterns 553

  numeration of, confusing 499

  of blocks with factors 500

  other 2-factor, 2-factor interactions estimable clear of blocks and 555

  partition of factorial effects into main effects and, not unique 500

  2-factor, estimable, clear of blocks and other 2-factor interactions 555

interpolation

  in quintuple-entry table, warned against 653

  in tables of confluent hypergeometric functions, laborious 227

  inverse 354, 454

  Lagrange 627-628 (§16.2311); 128

    unequal intervals 628-629 (§16.2319)

interpolation coefficients 627-629 (§§16.2311-16.2319)

intra-class correlation, effect of

  on confidence interval for mean 243

  on percentage points of t 243

intra-class correlation coefficient 330

  hyperbolic-tangent transformation of 330

intra-class rank correlation 357

inverse circular functions 398 (§10.0432); 445, 728 (§§12.11-12.13); 343

inverse hyperbolic functions 331, 695 (§7.212); 449 (§12.3*)

inverse hyperbolic-sine transformation 449 (§12.3*)

  (SEE ALSO angular transformation)

inverse hyperbolic-tangent transformation (SEE hyperbolic-tangent transformation)

inverse interpolation 354, 454

inverse sine transformation (SEE angular transformation)

inverted beta distribution 245-258, 262-264, 676-680 (§4.2*); 399 (§10.06*)

  (SEE ALSO beta-distribution; F-distribution)

iron numbers 738 (§16.0318)

isolates, in social group 294

iterated exponential 364 (§8.3111)

  logarithm of 364 (§8.3112)

iterated exponential distribution, order statistics from 701 (§8.24)

j c k, i.e. j-factor estimates clear of k-factor effects 500, 502

J-shaped frequency curves 720

j-statistics, Nair's 81-82 (§1.533); 72

Jacobi polynomials 79, 405, 718

Jeffreys, H., significance criterion K

  for χ^2 150

  for normal distribution 14

  for t 218

  for z 272

Johnson, N. L., logarithmic and hyperbolic translation 412-414, 719-720 (§10.222*); 407

joint-moment (SEE product-moment)

Jones, A. E., j-statistics 81

k-factor effect 499

k = number of plots
log-normal distribution

k = number of plots in a block, in incomplete block patterns 563, 564, 565, 566, 606, 607, 608

k-sample tests

  Neyman and Pearson's 297-299 (§6.113)

  non-parametric, of location 371-372, 707-708 (§9.14)

k-statistics

  finite population 431 (§10.35)

  in terms of power sums 430

  infinite population 429-431, 723 (§10.34*)

  multivariate 429, 431

  sampling cumulants of 430, 723

Kendall, M. G.

  concordance coefficient W 353-355 (§8.13*); 348

  concordance score 347

  rank correlation Tau 352-353 (§8.12*); 346, 347

Kluyver's integral 419

Kolmogorov-Smirnov tests 380-385, 714-715 (§9.41*)

kurtosis

  effect of

    on efficiency of mean square successive difference 645

    on distribution of range 701

    of χ^2 645

    of F 263

    of $\log \chi^2$ 645

    of mean square successive difference 645

    of rth greatest interval 712

    of t^2 240

ℓ-statistics 430 (§10.342); 429

LD_{50} 30, 633

LTPD 469-470

  *This section is subdivided; see Table of Contents.

lag correlation (SEE serial correlation)

Lagrange integration coefficients 629

Lagrange interpolation 128

Lagrange interpolation coefficients 627-629 (§§16.2311, 16.2319)

Laguerre polynomials 405

  as confluent hypergeometric functions 226

Laguerre series 154-155, 660 (§2.41); 268, 301, 312

  incomplete beta function expanded in 199

Laplace, first law of, logarithmic and hyperbolic translation of 719

Laplace, P. S., two-tailed exponential distribution 402, 645

  censored samples from, order statistics from 464

  logarithmic and hyperbolic translation of 719

Laplace transformation of derivates of normal distribution 634

Laplace-Everett interpolation coefficients 627 (§16.2317)

Laplace-Everett interpolation formula, first derivative of 629

Laplace-Gauss distribution (SEE normal distribution)

Laplacean distribution (SEE normal distribution)

largest mean square, ratio of, to smallest mean square 262 (§4.253)

largest root, of determinantal equation 312

largest variance ratio 261, 681 (§4.251); 637, 638

latent root (SEE determinantal equation, root of)

Latin cubes 499, 558
  orthogonal, complete sets of 502, 554

Latin squares   558-561 (§15.151*);
                496, 497, 499, 502, 638

  enumeration and randomization of
                558-560 (§15.1512)

  factorial patterns not published as
                561
  incomplete   607 (§15.2511)
  maximum F-ratio in   638
  mutilated   608 (§15.2513)
  orthogonal
    complete (saturated) sets of
                554 (§15.1312); 502
    unsaturated sets of
                560-561 (§15.1513)
  orthogonal partitions of
                561 (§15.1514); 502
  redundant   608
  with split plots   562 (§15.1621)

lattices
  as factorial patterns
                556-557 (§15.143*); 501, 536
  as incomplete block patterns
                606-607 (§§15.243-15.245); 566
  cubic   606-607 (§15.245)
  5-dimensional   606
  rectangular   606 (§15.244)
  square   606

lattice squares
  as factorial patterns
                561-562 (§15.153)
  as incomplete block patterns
                608-609 (§15.253); 566
  square lattices miscalled   608 ftn.

law of small numbers
    (SEE Poisson distribution)

least-squares fit, of polynomials
                436-441 (§11.1*)

Legendre functions, associated   334

Legendre polynomials   718

Legits, confused with logits   453

Legit transformation   453 (§12.7)

letters, as factor in Latin square
    pattern   497, 561

levels, of a factor   496

Lexis ratio (SEE index of dispersion)

k = number of plots
log-normal distribution

liczby przetasowane   460

liczby żelazne   738 (§16.0318)

liczby złote   738 (§16.0568)

life testing   483-486, 732 (§14.15*)

likelihood-ratio tests   298
  in paired comparisons   610

likelihood criterion   308, 312, 315

likelihood tests
        295-316, 318, 688-695 (§6*)
  for equality of correlation
        coefficients   332

limits of ratio of mean range to
    standard deviation   702

line-segment, random division of   291

linear-hypothesis model, power of
    F-test under   255

linear and quadratic effects,
    estimates of
  clear of cubic effects   555
  clear of cubic effects and blocks
                557

linear cost function   611

linear systematic statistic
  best   642, 649
  efficiency of   81
  (SEE ALSO systematic statistic)

location
  measures of
    based on order statistics
                70-71, 641-642 (§1.51)
    based on selected quantiles   636
  non-parametric tests of
                366-372, 704-708 (§9.1*)
  trend in, in time series, test
        against   710

location parameter, not estimable by
    frequency moments   401

lods (SEE logits)

$\log_{10} F$, percentage points of
                251 (§4.332)

log-normal distribution
                412-414, 720 (§10.1221); 281
  truncated   44

log area
Mayer

log area, under normal distribution
    6 (§1.1133), 7 (§§1.1143-1.1144),
    12 (§1.1216)

log χ^2
  approximately normally distributed
    261
  characteristic function of 272
  cumulants of 272-273 (§4.34)
  moments of 645

logarithms
  addition 392
  addition and subtraction
    621-622 (§16.068)
  base 2 365
  common 620-622 (§16.06*)
  Gaussian 621-622 (§16.068); 392
  natural
    625-626 (§16.11*); 364, 365, 731
    of decimal fractions
      625-626 (§16.112); 402
    of hyperbolic functions 481, 684
    of trigonometric functions 451
    radix tables of 625
    square roots of 659
  of binomial coefficients
    186 (§3.212)
  of double factorials 163 (§2.5135)
  of error integral 12 (§1.1216)
  of exponential functions
    364 (§8.3112), 624 (§16.103)
  of factorials 162-163, 662 (§2.513*)
    of complex argument 164 (§2.5152)
  of hyperbolic functions 624-625
    (§§16.1051-16.1052)
  of logarithms 364 (§8.3121), 452
    (§12.61), 621 (§16.0618)
  of normal tail area
    7 (§§1.1143-1.1144)
  of rational functions
    331, 695 (§7.212); 398
    (§10.0432); 450 (§12.51)
  radix tables of 620
  Weidenbach's 403

logarithmic functions
    364-365, 703 (§8.312*)

logarithmic growth curve (SEE logit
  transformation)

logarithmic-normal distribution (SEE
  log-normal distribution)

*This section is subdivided;
see Table of Contents.

logarithmic-series distribution
    285-286 (§5.33)

logarithmic transformation
    (SEE logarithmic translation;
    hyperbolic-tangent
    transformation; inverse
    hyperbolic-sine transformation)

logarithmic translation
    412-414, 720 (§10.2221)
  bivariate 414

logarithmically shortest confidence
  intervals, for ratio of
  population variances 258

logistic tolerance curve,
  fitted from 2 and 3 doses 728

logits
  confused with Legits 453
  converted from angles 447, 451
  working 450-451 (§12.53)

logit transformation
    450-451, 728 (§12.5*); 24

loglogs 452 (§12.61)
  working 452 (§12.62)

loglog transformation 452 (§12.6*)
  complementary 24

lognormal (SEE log-normal)

lologs 621 (§16.0618)
  (SEE ALSO loglogs)

lot control, by number of defectives,
  hypergeometric distribution in
    469

lot inspection
  by attributes
    470-477, 730 (§14.11*)
  by variables
    477-481, 730-731 (§14.12*)

lot percentage defective, estimated
  from sample mean and standard
  deviation 478

lot quality protection
  (SEE average outgoing quality)

lot size 472, 478, 479

lot tolerance percent defective 470

lower bounds
  for ratio of range to standard
    deviation 702

for range 702

lower control limit (SEE control chart)

lower quartile 47, 48

   (SEE ALSO quartile)

lower records, as test for randomness 710

Lower straggler 46, 47, 89

Lubbock coefficients 629 (§16.2369)

lumped variance 694

m-test, of regression 705

m^{th} extreme 364, 703 (§8.32*)

m^{th} values (SEE extreme values; order statistics)

m rankings, problem of
   (SEE concordance coefficient)

main effects, estimable clear of blocks 556

main effects and interactions, partition of factorial effects into, not unique 500

MacLaurin-Euler summation formula 427, 428

Mahalanobis' D^2-statistic 306-308 (§6.131); 691

Mahalanobis' generalized distance 306-308 (§6.131); 691

malaria, mosquitoes and 292

manifold classification
   (SEE contingency table)

Mann-Whitney two-sample test of location 368-369, 706 (§9.121)

   (SEE ALSO Wilcoxon test)

Marcinkiewicz, J., polynomials are not cumulant generating functions 408

marginal distribution 684

marginal partitions 223

Markov process

   first-order, serial correlation from, hyperbolic-tangent transformation of 333

   Gaussian 696

   trend reduced, estimated regression 726

matching 684-687

matrices

   covariance 690, 691, 716
   Hadamard 554 (§15.1313); 502
   orthogonal 554 (§15.1313); 502

matrix bisamples 724

maxima, moving 643

maximum absolute deviate, studentized 641

maximum expected response 633

maximum F-ratio* 262 (§4.253)

maximum likelihood, method of

   fitting of normal distribution by, in censored samples 43, 44

   fitting of negative binomial distribution by 285

   fitting of Pearson curves by

      Type III 395, 716-717 (§10.035)
      Type V 398 (§10.055)

   fitting of truncated normal distribution by 40

   identical with method of moments, for fourth-degree exponential distribution 415

maximum likelihood estimates

   from dilution series 317

   in paired comparisons 610

   of population correlation coefficient 328

   of precision constant 172

maximum relative discrepancy 385 (§9.416)

maximum studentized treatment effect, confidence interval for 674

maximum working angle 446

maximum working logit 450

maximum working loglog 452

maximum working probit 23

Mayer, Tobias, method of zero sum 439

─────────
*to be distinguished from the largest variance ratio of §4.251.

mean
molecular theory

mean
  arithmetic, as a biplane combinatorial power mean   432
  combinatorial power   432 (§10.37)
  confidence interval for   303, 646
    effect of intraclass correlation on   243
  control charts for   487
  difference of, tests for, in face of unequal variances   231-239, 674 (§4.15*)
  efficiency of   642
  equality of, test of   297
    in face of unequal variances   231-239, 674 (§4.15*)
  geometric, as a biplane combinatorial power mean   432
  Gini's generalized   432 (§10.37)
  harmonic   691
  lot inspection by variables controlling   731 (§14.123)
  ranking of   654
  moments about   425, 426
  of correlated variates, F-test for equality of   310
    asymptotic χ^2 distribution of   311
  of correlation coefficient   420
  of exponential distribution, estimated from order statistics   700
  of log-normal distribution   414
  of multiple correlation coefficient   335
  of normal distribution
    confidence interval for   303
    shift in, effect on range of   734
    Wald sequential test for   481
  of order statistics   65
  of right triangular distribution, estimated from order statistics   717
  of tolerance distribution (probit analysis)   22
  regression of standard deviation on, approximate parabolic   434
  trend in   710
  variances and covariances, of correlated variables, test for equality of   310
  weighted, confidence intervals for   674-675 (§4.18)

―――――――

*This section is subdivided; see Table of Contents.

mean absolute error
  (SEE mean deviation)
mean and mode, of Pearson curves
  standard error of distance between Type VII, when coincident   399
mean and range, control chart for   488-490, 733 (§14.221)
mean and standard deviation, control chart for   488-490, 733 (§14.221)
mean and variance
  equality of, test of   305
  of normal distribution, simultaneous test of   689
  of standard deviation   171
mean deviation   71-75, 643 (§1.52); 94
  as test of normality   115-117 (§1.63*)
mean difference, Gini's   80 (§1.532)
mean orthogonal moments   439
mean moving range   644
  efficiency of   643
mean range   82, 85
  approximate distribution of   107
  effect of β_2 on   701
  inequalities for tail area in terms of   435
  ratio of to standard deviation, limits of   702
mean squares   245
  between primaries   611
  cumulative   302, 303
  dependence between, effect of, on F-distribution   264-269, 681-682 (§4.28)
  independent   302
  ordered, ratios involving   260-262, 681 (§4.25*)
  ranking of   681 (§4.254)
mean square backward difference   77
mean square contingency   343 (§7.92); 356

[982]

mean square deviation (SEE variance)

mean square error (SEE variance)

mean square successive difference 79

    studentized 80

    (SEE ALSO von Neumann's ratio)

mean successive difference 76, 644

    ratio to root mean square 79

mechanical quadrature
    (SEE numerical integration)

media biplana combinatoria potenziata 432

media monoplana combinatoria potenziata 432

median 50-53, 636 (§1.41); 46, 47, 70

    cut at, runs determined by 711

    in samples from discrete rectangular distribution 361

    mean deviation from 71

    of maximum of n normal deviates 58

    of straggler 58

    of symmetrical distribution, confidence intervals for 206, 669-670 (§3.6); 367

median dichotomy 698

    tetrachoric correlation for 653

median effective dose 30, 633

median lethal dose 30, 633

median test, two-sample, of location 370 (§9.123)

Mendelian ratios 195

mesokurtosis (SEE kurtosis)

method of moments

    efficiency of, for Gram-Charlier series, type A 409

    interpreted 415

mid-rank method (SEE tied ranks)

midpercentile 70

midrange 88-90 (§1.542); 47, 48, 70

    variance and efficiency of 647

midrange-range test 101

Military Standard:

    105 A 472

    105 B 471

    414 478

milligones 623 (§16.073)

Mills' ratio 13, 630 (§1.13*)

    as function of area 22, 632 (§1.222)

minima, moving 643

minimum-variance unbiased linear estimate 641

    of standard deviation 647

minimum working angle 446

minimum working logit 450

minimum working loglog 452

minimum working probit 23

Mises, Cramér and Smirnov, test of goodness of fit 386 (§9.42)

mode

    estimation of, from weighted sample values 363

    of correlation coefficient

    of Pearson curves, standard error of 392

    of t-distribution, estimated by order statistics 363

mode and mean, of Pearson curves

    standard error of distance between 391

    Type VII, when coincident 399

model, in analysis of variance 255

modification, non-randomized, of binomial sign test 669

modified Bessel functions (SEE Bessel functions I; Bessel functions K)

modified mean (SEE midrange; stragglers, rejection of)

molecular theory, integrals in 133, 656 (§2.1134)

moments
multivariate multinomial distribution

moments
    about the mean   425
    about the origin   425, 429
    binomial   439
    central, of median   50
    contour   433
    determining the type of a Pearson curve from   389
    eighth, finite, of Pearson curves   390
    factorial   438, 439, 441
        of binomial distribution   203
    fitting of power polynomials without   725
    first absolute, of normal distribution, about population mean   45
    fourth, auxiliary table for computing   616 (§16.0233)
    fourth standard, as test of normality   113-114 (§1.62)
    frequency   401
    from grouped distributions   426-429 (§10.33*)
    grouped, of log-normal distribution   44
    incomplete, of Poisson distribution   279
    incomplete normal   31-32 (§1.31); 37, 38, 130, 200, 201
    method of
        applied to exponential curves   444
        approximate percentage points computed by   636
        efficiency of   394, 409
        identical with method of maximum likelihood, for fourth-degree exponential distribution   415
        inefficiency of   428
    multivariate   431, 723-724 (§10.36)
    negative   399
        of positive Bernoulli distribution   668
        of standard deviation   172
    normal, incomplete   31-32 (§1.31); 37, 38, 130, 200, 201
    of angular transformation of binomial distribution   446
    of $b_1^{\frac{1}{2}}$ 111-112 (§1.611)
    of b_2 113-114 (§1.621)
    of beta distribution   204, 668 (§3.52); 192, 203

    *This section is subdivided; see Table of Contents.

of binomial distribution   204, 205, 668-669 (§3.53-3.54); 203
of bivariate normal distribution   122-123 (§1.73)
of χ 167-171 (§2.522)
of χ^2 165-173 (§2.52*); 645
of correlation coefficient   324-327, 695 (§7.13*); 226, 342
of covariance   342
of estimated binomial success ratio   205 (§3.54)
of F   262-263 (§4.26)
of Geary's ratio   115-116 (§1.631)
of inverse hyperbolic-sine transformation of negative binomial distribution   448
of Johnson's hyperbolic curves of infinite range   414
of Kendall's rank correlation coefficient   353 (§8.1125)
of likelihood tests   295
of $\log \chi^2$ 645
of log-normal distribution   412, 413
of matching distribution   686
of mean deviation   72
of mean Spearman rank correlation   352
of median   636
of moments   430-431 (§10.349); 405
of multiple correlation coefficient   335
of Neyman & Pearson's k-sample tests   298, 299
of non-central χ 266
of non-central t   226
of normal distribution   31-32 (§1.31); 122-123 (§1.73); 404 (§10.136)
of order statistics
    from normal distribution   62-68, 638-639 (§1.43); 48, 637
    from rectangular distribution   65
    series for   721
of Pearson curves   391 (§10.006); 388
    determination of type from   389-391 (§10.004)
    Type I   393 (§10.016)
    Type III   396 (§10.036)
of Poisson distribution   173 (§2.53); 683
of rth greatest interval   712
of range   87, 645, 646, 647
    misattributed   647

[moments]
  of rank correlation
    Kendall's   347, 353
    Spearman's 346
  of regression coefficient, from non-normal parents 342
  of standard deviation 165, 166, 167, 168, 169, 170, 171
  of straggler 637
  of substitute Geary's ratio:
    based on population mean 116
    based on randomization 118
  of t (Student's) 240 (§4.16)
  of $t = r(1-r^2)^{-\frac{1}{2}}$ 327 (§7.134)
  of test for birth-order effect 373
  of test of equality of covariance matrices 309
  of tests of randomness 710
  of variance 165, 166
  of weighted sum of ordered intervals 712
  of $z = \tanh^{-1} r$ 331-332 (§7.23)
  orthogonal, mean 439
  serial product, of samples from autoregressive time series 467
  third standard, as test of normality 111-113 (§1.61*)
moment-coefficients (SEE moments)
moment estimates, of parameters of negative binomial distribution 285
moment generating function 312, 407
  factorial, of hypergeometric distribution 275
moment inequalities 433-434, 724 (§10.4)
moment ratio
  (SEE $b_1^{\frac{1}{2}}$; b_2; Geary's ratio)
monolith 498
monomial symmetric functions 421, 425, 432
  augmented 421, 430
  expectation of 425
  restricted 723
monoplane combinatorial power mean 432
Monte Carlo, casino, roulette at 464
Monte Carlo, physico-mathematical method 455

mortality among controls, weighting coefficients in presence of 28
mosquitoes and malaria 292
moving averages, method of 441
  for estimating median effective (lethal) dose 633
  in smoothing of time series 726
moving maxima 643
moving ranges 643
moving minima 643
multi-factorial design (SEE factorial, i.e. combinatorial pattern)
multinomial distribution 287, 683, 684 (§5.4)
multiple classification (SEE contingency table)
multiple comparison, of binomial proportions 484
multiple control of heterogeneity, in factorial experiments 496
multiple correlation coefficient 335 (§7.4); 155, 254
multiples of fourth powers 616 (§16.0233)
multiples of trigonometric functions 443 (§11.23)
multiple rank correlation 356 (§8.15)
multiple range test 649
multiple sampling, by attributes 471, 472, 473, 476, 477
multiple-valued decision (SEE sequential sampling)
multivariate analysis 306-313, 692 (§6.13); 295
multivariate cumulants 429
multivariate k-statistics 429, 431
multivariate moments 431, 723-724 (§10.36)
  (SEE ALSO co-cumulant; product-moment)
multivariate multinomial distribution 684

multivariate normal distribution
normal distribution

multivariate normal distribution   125, 654-655 (§1.79-1.795); 266, 288

  tests applied to
    306-313, 692 (§6.13)

multivariate normal tolerance ellipsoids   493

multivariate symmetric functions
    431, 723-724 (§10.36)

multivariate t-distribution
    241, 242 (§4.17); 674

multivariate tetrachoric series   655

mutilated Latin squares   608 (§15.2513)

Nachwirkungsfunktion   278

Nair, K. R., j-statistics
    81-82 (§1.533)

natural logarithms
    625-626 (§16.11*); 364, 365, 731

  of decimal fractions
    625-626 (§16.112); 402

  of hyperbolic functions   481

  of logarithmic functions
    364 (§8.3121)

  of rational functions
    331, 695 (§7.212); 450 (§12.51)

  of trigonometric functions   451

  radix tables of   625

  square roots of   659

natural mortality, weighting coefficients in presence of   28

natural trigonometric functions
    622-623 (§16.07*); 442 (§11.22)

negative binomial distribution   285, 683 (§5.32); 278, 284, 286, 449

  and binomial distribution, factitious resemblance between moments of   684

  inverse hyperbolic-sine transformation of, moments of   449

negative exponential distribution (SEE exponential distribution)

negative frequencies, in Gram-Charlier series   409

---

*This section is subdivided; see Table of Contents.

negative moments   399

  of positive binomial distribution   668

  of standard deviation   172

negative multinomial distribution   683

negative powers   617 (§16.025)

Newton interpolation coefficients
    628 (§16.2313)

Newton interpolation formula, first derivative of   629

Newton's series, in polynomial fitting   439

Newton-Bessel interpolation coefficients   628 (§16.2316)

Newton-Bessel interpolation formula, first derivative of   629

Newton-Cotes integration coefficients   629

Newton-Stirling interpolation coefficients   628 (§16.2315)

Newton-Stirling interpolation formula, first derivative of   629

Neyman, J., contagious distribution, Type A   279, 280, 683

Neyman and Pearson's k-sample tests   297-299 (§6.112)

  Bartlett's modification of   300-302, 688 (§6.113)

Neyman and Pearson's one- and two-sample tests   296-297 (§6.111)

nines, terminal strings of, evidence of method of computation   719

nomogram

  for mean values of normal order statistics   64

  for probit analysis   29

non-central beta-distribution   679

non-central χ, moments of   226

non-central χ^2 distribution   155-157, 660 (§2.42); 256, 288, 661, 679, 682

  with 2 d.f.   125

non-central χ^2 statistics, quotient of   682

non-central F-distribution 258-260, 678-680 (§4.242); 256, 275, 308
  power of F-test in terms of 258, 678
  with both numerator and denominator non-central 682
non-central t-distribution 218-225 (§4.141*); 200, 477, 682
non-centrality 256
  for specified power in analysis of variance, approximations to 259
non-circular serial correlation 338 (§7.52)
non-destructive testing 474
non-determination, coefficient of 334, 695 (§7.31)
non-linear correlation
  (SEE correlation ratio)
non-linear regression 436-444, 724-727 (§11*)
non-monotone translation functions, difficulties arising from 410
non-negative frequency function
  represented by Edgeworth series 409
  represented by Gram-Charlier series, type A 409
non-normal distributions
  correlation coefficient in samples from 328 (§7.14)
  order statistics from 360, 699 (§8.2)
  regression coefficient in samples from 342
  t in samples from 200 (§4.143)
non-normality, effect of
  on distribution of F 264-269, 681-684 (§4.28)
  on power function of F-test 267
non-parametric measures of correlation 345-353, 356-359 (§8.1*); 343-344, 698-699 (§§7.92, 7.94)
non-parametric tests 366-387, 704-705 (§9*)
non-parametric tolerance limits 492, 734-735 (§14.31); 706, 708, 733
non-random digits, rearrangement of, yielding random digits 458

normal bivariate distribution (SEE bivariate normal distribution)
normal deviates 2, 14, 63, 144
  correlated random 463
  in polynomial translation, varying notation for 411
  independent 336, 640, 692
  ordered 62-68, 638-639 (§1.43)
  random 454
normal deviate test 367
  compared with t-test 217, 675
  power of 14
normal distribution 1-129, 630-655 (§1*); 373, 377, 396, 398, 404, 477
  as approximation to binomial distribution 188, 197, 199
  as approximation to distribution:
    of correlation coefficient 321
    of Galton differences 91
    of $\log \chi^2$ 261
    of matching scores, fits poorly 687
    of mean matching score 687
    of ordered intervals 712
    of product-moment correlation coefficient 321
    of product of correlated normal variables 125
    of range 84
    of runs 377
    of sum of 20 rectangular variates 722
    of t 210
    of 2-sample test of randomness 378
    of $z = \tfrac{1}{2}\log_e F$ 270
    of $z = \tanh^{-1} r$ 330
  as approximation to translated Pearson type III curve 397
  as asymptotic distribution:
    of k-sample paired-comparison slippage test 708
    of test of randomness 710
  as Pearson curve 388

normal distribution
orthogonal partitions

[normal distribution]
  bivariate   119-125, 652-654 (§1.7*); 241, 291, 302, 324, 340, 343, 346, 347, 370, 433, 697
    circularly symmetric, covering circle of sample from   291
    circularly symmetric, volume over elliptical regions   660
    percentage points of, approximate   290
    random samples from   463
    tests applied to 305-306, 690-691 (§6.12); 319-329, 695 (§7.1*); 351 (§8.114)
  circular (Mises)   418-419, 722 (§10.28); 653
  contour-moment inequalities applied to   434
  control charts for   488-491, 733-734 (§14.22*)
  estimated by gauging   716
  fitting of, by statistics other than the sample mean and variance   404
  grouped   56
  incomplete moments of   31-32 (§1.31); 37, 38, 42, 200, 201
  k-statistics from   430
  mean of
    confidence interval for   303
    Wald sequential test for   481
  moment inequalities applied to   433
  moments of   122-123 (§1.73); 404
    incomplete   31-32 (§1.31); 37, 38, 42, 200, 201
  multivariate   125, 654-655 (§§1.79-1.795); 288, 716
    tests applied to   306-313, 692 (§6.13*)
  order statistics from   62-69, 638-641 (§§1.43-1.44); 92-95, 647-649 (§1.544); 261, 369
  ordinates of   9-10 (§1.116*); 404
  percentage points of   19-21, 631-632 (§1.21*); 404, 491
    as approximation to percentage points of Spearman's rank correlation coefficient   349, 350
    percentage points of, in control charts   490, 733 (§14.222)
  qua Pearson curve   404-405 (§10.13*)
  random samples from   462-463, 729 (§13.3); 454

─────────
*This section is subdivided; see Table of Contents.

  tests applied to   111-113, 652 (§1.6*); 295-313, 688-692 (§6.1*)
  tolerance limits   492-494 (§14.32); 17, 18
  trivariate, spherically symmetric, integral over ellipsoidal regions   660
  yielding approximate confidence intervals for binomial distribution   195

normal equivalent deviate   22, 24

  (SEE ALSO probit)

normal inspection   472, 478

normal moments, incomplete   31-32 (§1.31); 37, 130

normal probability paper   479

normal tail area, powers of   54-62, 637-638 (§1.42)

normal tolerance limits   492-494 (§14.32); 17, 18

normal variables
  correlated, ratio of standard deviations of   673
  independent   434

normal variate   1, 12, 67, 263
  in polynomial translation, varying notation for   411
  independent standard   638, 705
  ratio of rectangular variate to   722
  ratio of triangular variate to   722

normality, tests of   111-118, 652 (§1.6*)
normalization
  awkward, of orthogonal polynomials   724
  of χ^2 147
  of frequency functions (SEE Cornish-Fisher transformation; translation; transformations)
  of t   213

number defective, control charts for   487, 732 (§14.21); 151

numerical differentiation   629 (§16.234)

numerical integration   629 (§16.236)

OC 470

OC curves 471, 472, 473, 477, 731

occupancy problems 293, 294

odd integers, sums of powers of 620 (§16.0445)

offset circle probabilities 125

ω^2 test, of goodness of fit 386 (§9.42*)

100% inspection 481

one-sample test of binomial process, derived from two-sample test of randomness of occurrence 376

one-sided specification limits, lot inspection by variables for 479, 481

one-sided tests, Kolmogorov-Smirnov 384-385, 715 (§9.415*)

one-way control of heterogeneity

  factorial patterns with 555-557 (§15.14*)

  split-plot 562 (§15.161)

  incomplete block patterns with 605-607, 736 (§15.24*)

open sequential scheme, operating characteristic of 474

operating characteristic

  of open sequential scheme 474

  of test for stability of a normal population 688

operating characteristic curves 471, 472, 473, 477

  of sampling plans by variables 731

optimum estimates, of parameters of negative binomial distribution 285

optimum sample ratio 688

optimum weighted averages 70

order, of Latin squares 497

order statistics 65, 71, 642

  from χ^2 distribution 260-262, 681 (§4.25)

  from non-normal distributions 360-363, 699-702 (§8.2*)

  from normal distribution 62-69; 638-641 (§§1.43-1.44); 261, 369

  measures of dispersion based on 92-95, 647-649 (§1.544)

  measures of location based on 70-71, 641-642 (§1.51)

  from parabolic distribution 647

  from Pearson curves 362-363, 700-701 (§8.23)

    Type IX 717, 718

  from rectangular distribution 360-361, 699-700 (§§8.21-8.22)

  moments of 65

  from right triangular distribution 717, 718

  from t-distribution 363

ordered deviate 49

ordered intervals 711, 712

ordered mean squares 260-262, 681 (§4.25*)

ordinates

  of bivariate normal distribution 119 (§1.71)

  of χ^2 distribution 137 (§2.20)

  of distribution of correlation coefficient 320-322 (§7.11)

  of normal distribution 9-10 (§1.116*); 21, 632 (§1.221); 409

    bivariate 119 (§1.71)

  of Pearson curves 389 (§10.001)

    Type III 137, 657 (§2.21); 394-395 (§10.031)

    Type VII 400 (§10.071)

    Type X 402 (§10.101)

  of t-distribution 210 (§4.12)

  of z-distribution 270

orthogonal cubes, complete (saturated) sets of 554 (§15.1312); 499, 502

orthogonal functions (SEE Hermite polynomials; Hermite functions; harmonic analysis; Jacobi polynomials; Laguerre series)

orthogonal hypercubes, complete sets of, not tabled 554

orthogonal matrices 554 (§15.1313); 502

orthogonal moments, mean 439

orthogonal partitions of Latin squares 561 (§15.1514); 502, 651

orthogonal polynomials
percentage points

orthogonal polynomials
  436-441, 724-726 (§11.11*)

  integrals of   439
  untouched by human hand   439

orthogonal squares

  complete (saturated) sets of
    499, 502, 554 (§15.1312)
  obtained by cyclic permutation   516, 520, 526, 534, 544, 545, 546
  published in symbolic form
    507, 515, 516, 520, 527, 533, 534, 537, 541, 543, 544, 545, 546, 547
  unsaturated sets of
    560, 561 (§15.1513)
  (SEE ALSO Latin squares; Graeco-Latin squares)

orthogonal to blocks, replication groups   565

outlying observations (SEE stragglers)

overlap, of geometrical figure   289

overlapping samples, correlation between ranges in   643

p-statistics, i.e. determinantal roots   691

PTPD   469, 470

paired comparisons
  355-356, 699 (§8.14)

  in incomplete block experiments
    736 (§15.2413); 610 (§15.28)
  k-sample slippage test based on   708
  signed-rank test for   367

parabolic distribution   648

  order statistics from   647

parabolic cylinder function   34

parabolic regression, of variance on mean, approximate   434

Pareto distribution (SEE Pearson curve, Type XI)

Pareto index   617

partial confounding, in factorial patterns
  556 (§15.1422); 501, 502, 561

  connexion with lattices   501

*This section is subdivided; see Table of Contents.

partial correlation, between time series   696

partial rank correlation   356 (§8.15)

partial regression coefficients   243

partially balanced incomplete blocks
  607 (§15.248); 609 (§15.258*)

partially balanced lattice square (SEE lattice square; square lattice)

partitions   619 (§16.036); 293

  in two dimensions   723
  marginal   723

Pascal distribution (SEE negative binomial distribution)

Pascal triangle   685

  (SEE ALSO binomial coefficients)

patterns, combinatorial
  495-609, 735-736 (§§15.1-15.25*)

Pearson, E. S.:

  and Hartley, notation for normal distribution 1
  and Neyman (SEE Neyman and Pearson)

Pearson, K.:

  Bessel function T_m 417
  biserial correlation
    343, 697-698 (§7.91)
  bivariate frequency surfaces   406
  χ^2 test of goodness of fit
    174-180, 663-664 (§2.6*); 315
  coefficient of correlation
    (SEE product-moment correlation coefficient)
  Criterion   391
  discountenances Romanovsky's generalized Pearson curves   405
  frequency curves
    (SEE Pearson curves)
  frequency surfaces, bivariate   406
  integrals   397-398, 717 (§10.0431)
  measure of skewness   391
  notation for incomplete Γ-function
    131, 312
  tetrachoric correlation
    343-344, 698-699 (§7.94); 119-122 (§1.72*); 653

[990]

Pearson curves   388-406, 715-719
      (§§10.0-10.1*); 80, 84, 85
  characteristic functions of   226
  Craig's criterion for   124
    applied to Burr's cumulative
      distribution functions   416
  generalizations of
      405-406, 718-719 (§10.18)
  necessarily unimodal   415
  order statistics from
      362-363, 700-701 (§8.23)
  simulated, random sample from   464
  Types I, ..., XII (Table of
      Contents at §§10.01*-10.12*)
  Type I   392-393 (§10.01*); 79, 84,
      297, 298, 299, 300, 310, 686
    percentage points of
        392 (§10.012); 192
  Type II
      393-394, 716 (§10.02*); 76, 350
  Type III   394-397, 716-717 (§10.03*);
      132, 656 (§2.112); 137, 657
      (§2.21); 79, 107, 130, 146, 155,
      157, 189, 687
    estimated by gauging   716
    translated   107
  Type IV
      397-398, 717 (§10.04*); 114, 409
  Type V   398-399 (§10.05*); 302
    percentage points of
        398 (§10.052); 658
  Type VII   399-401, 717 (§10.07*);
      112, 113, 409
    estimated by gauging   716
pentagamma function   164 (§2.5142)
per cent defective   276
  control charts for
      487, 732 (§14.21)
  lot inspection by variables
      controlling   479-480 (§14.122)
percentage points
  approximate
    computed by method of moments   636
    of largest determinantal root   690
    of non-central χ^2 661
    of weighted sum of χ^2 variates
        660
    of z   273, 274 (§4.35)

orthogonal polynomials
percentage points
  asymptotic
    of Cramer-Mises, Smirnov test
        386 (§9.4212)
    of Kolmogorov-Smirnov test   381,
        714 (§9.4113)
    of mth extreme value   365 (§8.322)
    of range   365 (§9.332)
    of the extreme value
        364-365, 703 (§8.312*)
  interpretation of   629
  of $b_1^{\frac{1}{2}}$ 112-113 (§1.613)
  of b_2 114 (§1.623)
  of Bartlett's modified
      L_1 test   300, 301, 302
    approximated by percentage points
      of chi-square   273
  of beta distribution   191-193,
      665-666 (§3.31*); 246, 490
    approximate   197-198 (§3.41)
  of binomial distribution
      195-196 (§3.33); 693
  of bivariate normal distribution,
      approximate   290
  of χ 144, 658 (§2.312)
  of χ^2 140-151, 658-659 (§2.31*);
      246, 365, 491
  of χ^2, approximate   144-147 (§2.314)
    applied to yield approximate
      fiducial limits for Poisson
      parameter   153
  of χ^2, as approximation to
      percentage points of:
    beta distribution   150, 197, 198
    Friedman's concordance score   354
    goodness-of-fit χ^2 for runs up and
      down   376
    k-sample generalization of
      Wilcoxon test   372
    Neyman and Pearson's L_1 299
    non-central χ^2 157
  of χ^2/ν 143 (§2.312)
  of circular serial correlation
      337, 338, 695
  of Cochran's test of homogeneity of
      variance   443, 726
  of concordance coefficient
      354-355 (§8.132)
  of correlation coefficient
      322-324 (§7.12*)
  of D^2 307
  of difference between proportions
      692, 693

percentage points
point biserial

[percentage points]
  of F   246-254, 676-677 (§4.23*); 299
    as approximation to percentage points of k-sample generalization of Wilcoxon test   372
  of $F^{\frac{1}{2}}$ 250 (§4.232)
  of Fisher-Behrens test   233
  of Friedman's concordance score   355
  of Geary's ratio   116-117 (§1.633)
  of goodness-of-fit χ^2 for runs up and down   376
  of incomplete beta function   191-193, 665-666 (§3.31*); 246, 490
  of k-sample generalization of Wilcoxon test, approximate   372
  of Kendall's Tau   352 (§8.122)
  of Kolmogorov-Smirnov tests   381, 382, 383, 384, 385
  of Kruskal-Wallis test, 2-sample   706
  of largest canonical correlation, asymptotic   313
  of largest determinantal root   691, 692
  of largest harmonic amplitude   443
  of largest variance ratio   261, 680
  of $\log_{10} F$ 251 (§4.332); 270
  of log-normal distribution   413
  of lumped variance   694
  of maximum F-ratio   262
  of mean deviation from mean   74
  of mean square successive difference   79, 645
    approximate   76
  of midrange   100
  of multiple correlation coefficient   156
  of Neyman and Pearson's L_0 298
  of Neyman's and Pearson's L_1 299
  of non-central χ^2 157
  of non-central t   224 (§4.1414)
  of normal distribution   19-21, 631-632 (§1.21*); 404 (§10.132); 2, 490, 491
    as approximation to percentage points of Spearman's rank correlation coefficient   349, 350
    in control charts   490, 733 (§14.222)

*This section is subdivided; see Table of Contents.

  of Pearson curves   389, 715 (§10.002)
    Type I   392 (§10.012)
    Type II   394 (§10.022)
    Type III   395 (§10.032)
    Type V   398 (§10.052)
    Type VI   399 (§10.062)
    Type VII   400 (§10.072)
  of Poisson distribution   151 (§2.321)
  of quartile range of ranked intervals   712
  of range   83, 84, 85, 87, 88, 645, 646, 647
    as approximation, inadequate, to percentage points of ratio of largest to smallest mean square   262
    reciprocals of   86
  of range-midrange test   101, 102, 103
  of rank correlation coefficient
    Kendall's   352 (§8.122)
    Spearman's   349-350 (§8.112)
  of ratio:
    of correlated normal variates   673
    of dependent ranges   103, 104
    of independent ranges   101, 102, 103, 104
    of largest to smallest mean square   262
    of range to standard deviation   702
    of ranked intervals   712
    of rectangular variate to normal variate   722
    of triangular variate to normal variate   722
  of regression coefficients   677
    alleged, from percentage points of correlation coefficient   324
  of root of determinantal equation   306
  of runs   710
    in two samples   713
    up and down   375
  of serial correlation
    circular   337, 338, 695
    non-circular   79
  of signed-rank test   368
  of slippage test   699

[percentage points]
  of smallest variance   261
    as approximation to percentage points of smallest variance ratio   261
  of Spearman's Rho   349-350 (§8.112)
  of straggler   56, 58, 62
  of straggler deviate   59, 61
  of studentized D^2 307
  of studentized extreme deviate   640
  of studentized mean square successive difference   80
  of studentized range   96, 97, 98, 99, 649
  of substitute Geary's ratio, based on randomization   118
  of substitute t-test:
    based on order statistics   64
    with range in denominator   99-102, 104
  of t   210-218, 671 (§4.13*); 246
    as approximation to percentage points of Welch's two-sample test   239
  of test of equality:
    of covariance matrices, approximate   309
    of means, variances and covariances of correlated variates   310, 311
    of means of correlated variates   310, 311
  of test of goodness of fit of exponential curves   444
  of test of ratio of Poisson parameters   693
  of test of sphericity
    three-dimensional, approximate   313
    two-dimensional   697
  of tetrachoric correlation, alleged   344
  of the extreme value   364 (§8.3121)
  of two-sample slippage test   707
  of two-sample test of randomness   378, 379
  of variate differences, bounds for   79
  of weighted sum of ordered intervals   712
  of Wilcoxon two-sample test   368, 369, 706
  of Wilks' criterion, as approximation to percentage points of largest determinantal root   609
  of $z = \frac{1}{2}\log_e F$ 250-251, 270-272, 675-676, 682 (§4.33*); 298
    approximate   273-274 (§4.35)
    as approximation to percentage points of concordance coefficient   354
    Edgeworth correction to   681
    Fisher-Cochran approximation to, yielding approximate percentage points of beta distribution   198
  of z^2, approximated by percentage points of chi-square   273
  of $z = \tanh^{-1} r$ 331 (§7.22)

percentage standard deviation (SEE coefficient of variation)

percentiles   49, 640, 649

performance characteristic (SEE operating characteristic)

periods, unknown, in harmonic analysis   727

periodic components, significance of   443, 726 (§11.24)

periodograms   443 (§11.29)
  of samples from autoregressive time series   467

permutations
  distribution of product-moment correlation coefficient under   344
  distribution of Spearman's rank correlation coefficient under   344
  of rows and columns   559
  pseudo-random   460
  random   460-461 (§13.2)

physical sources, random digits from   457

pilot sample, relative precision of   612

Plackett-Burman patterns, factorial   554

plaid squares   562-563 (§15.1623)

play, duration of   292

plots   495, 563

point binomial
  (SEE binomial distribution)

point biserial correlation   698

point count
probability-ratio test

point count, in bridge hands   684

Poisson-Charlier series   409 (§10.212)

Poisson distribution
    134-136, 656-657 (§2.12*); 138-139 (§2.22*); 151-153 (§2.32*); 173 (§2.53); 130, 280, 289, 469, 470, 476, 667, 693

  as approximation:
    in acceptance sampling
        474, 475, 476
    to binomial distribution
        188, 189, 482, 667, 693
    to distribution of mean matching score, not used   687
    to distribution of runs up and down   375
    to hypergeometric distribution
        469, 470
  as limit of binomial distribution   182
  bivariate   282 (§5.22)
  control charts for   487
  estimates of parameters in
    316-317 (§6.32)
  finite differences of, no table of   409
  generalized   278-283, 683 (§5.2*)
  in ecology   279
  in lot control   730
  in process control by number of defects   469, 470
  moments of   173 (§2.53); 683
    incomplete   279
  of exposed to risk, weighting coefficients for   28
  positive   173
  second difference of   655
  square-root transformation of   449
    variance of   448
  tests applied to
    315-316, 693-694 (§6.31)
  truncated   138, 139

Poisson's exponential limit
    (SEE Poisson distribution)

Poisson's integral   126 (§1.81*)

Poisson variates   282

  difference between
    282-283 (§5.23); 409

---

    *This section is subdivided; see Table of Contents.

Polya's distribution
    (SEE contagious distributions; negative binomial distribution)

Polya-Eggenburger distribution (SEE negative binomial distribution)

polygamma functions
    163-164, 662 (§2.514); 272, 410

polykays, i.e. l-statistics
    430, 723 (§10.342); 429

polynomials
  Čebyšëv   436
  Čebyšëv-Hermite   34 (§1.323)
    (SEE ALSO polynomials, Hermite)
  cubic   704
  factorial   441 (§11.13)
  fitting of   436-441, 724-726
        (§11.1*)
  Gram   436-440 (§11.11*)
  Hermite   34 (§1.323); 33, 408, 410, 436, 441, 655
    as confluent hypergeometric functions   226
    recurrence formula for   35
  Jacobi   79, 405, 718
  Laguerre   405
    as confluent hypergeometric functions   226
    (SEE ALSO Laguerre series)
  Legendre   718
  orthogonal   436-440 (§11.11*)
    integrals of   439
  power   440, 725 (§11.12)
  quadratic   334, 695 (§7.31); 615
        (§16.0214); 704
  Sonin   405
  trigonometric
    442-443, 726, 727 (§11.2*)

polynomial translation
    411-412, 719 (§10.221); 233, 397

  bivarate   414

polynomial trend
    (SEE polynomials, fitting of)

pooled samples, for blood tests   611

pooling, effect of, on estimate of regression   243

positive Bernoulli distribution, negative moments of   668

positive Poisson distribution   173

positive skewness (SEE skewness)

posterior binomial probability distribution   276

powers (of numbers)   614-619, 737-738 (§16.02)
   of normal tail area   54-62, 637-638 (§1.42)
   sums of   619-620 (§§16.044*, 16.046)

power (of a test)
   of Bartlett's L_1 test for equality of variances   256
   of χ^2 test
     in terms of central χ^2 distribution   158-159 (§2.43)
     in terms of non central χ^2 distribution   157
     in 2 2 tables   177-179, 663-664 (§2.613)
   of index of dispersion   280
   of F-test   255-260, 677-680 (§4.24*)
     compared with power of test for difference between correlated variances   265
     effect of heteroscedasticity on   267
     effect of non-normality on   267
   of Kolmogorov-Smirnov test, two-sided   384 (§9.413)
   of normal deviate test for mean   14
   of range chart, in samples of four, against shift in mean   734
   of range-midrange test   101
   of ratio of independent ranges, as substitute F-test   103
   of runs up and down   375
   of sign test   387
   of slippage test:
     for rectangular distribution   318
     for time series   387
   of t-test   220-223 (§4.1412-4.1413); 218
   of test for difference:
     between correlated variances, compared with power of F-test   265
     between proportions   663
   of test of equality:
     of Poisson parameters   315
     of ranges of rectangular populations   700
   of tests of randomness   710
   of test of ratio of Poisson parameters   693
   of test of specified correlation coefficient   305
   of two-sided Kolmogorov-Smirnov test   384 (§9.413)
   of Walsh test   367
   specified, in analysis of variance, non-centrality yielding approximations to   259

power-efficiency of Walsh test   367

power curve   693
   (SEE ALSO OC curve)

power function (SEE power (of a test))

power mean
   biplane combinatorial   432
   monoplane combinatorial   432

power moment (SEE moment)

power polynomials   441, 725 (§11.12*)

power products, of sample moments   432

power series   63

power sums, i.e. symmetric functions   421, 429
   k-statistics in terms of   430 (§10.341)

precision
   expected relative   612
   modulus of (SEE probable error)
   of pilot sample, relative   612

precision constant   172

preliminary tests of homoscedasticity, Box suggests omission of   264

primary sampling units   611

probable error   12, 13 (§1.122*); 21 (§1.213); 14
   approximate, of tetrachoric correlation   698
   generalized   144
   of correlation coefficient   326 (§7.132)
   (SEE ALSO standard error)

probability-moment
   (SEE frequency moment)

probability-ratio test
   (SEE likelihood-ratio test)

probability center lines
randomness, tests of

probability center lines,
    for standard deviation and range
    charts   149

probability generating function
  of hypergeometric distribution   275
  of negative binomial distribution   284

probability integral, normal
    (SEE normal distribution)

probability limits
    (confidence intervals;
    control charts; fiducial limits;
    tolerance limits)

probits   27, 632 (§1.232); 2, 23
  converted from angles   447, 449
  provisional   27
  weighted   26
  working
       24-27, 632 (§1.231*); 23, 446

probit analysis
       22-30, 632-634 (§1.23*); 13

probit transformation   449 (§12.4)

process average   478

process control
  by number of defectives, binomial
    distribution in   469
  by number of defects, Poisson
    distribution in   469

process tolerance per cent defective   470

producer's risk   469

product-moments
  of k-statistics   723
  serial, of samples from
    autoregressive time series   467
  (SEE ALSO co-cumulants)

product-moment coefficient T_m
    functions   417

product-moment correlation   336, 345

product-moment correlation
    coefficient
       319-329, 695 (§7.1*)

---

*This section is subdivided;
see Table of Contents.

distribution of, as approximation to
    distribution of Spearman's rank
    correlation coefficient   346

product of correlated normal
    variables
  cumulative distribution of   125
  frequency function of   124

product sums, homogeneous   421

proportions, difference between, test
    for   692

provisional probit   27

pseudo-random digits, pseudo-
    random-number generators do
    not necessarily yield   455

pseudo-random numbers   455, 460
  generation of   468 (§13.62)
  testing of   468 (§13.63)

pseudo-random permutation   460

ψ-function
       163-164 (§2.5141); 272, 285, 486

quadrant test, for association   357, 713

quadratic effects,
    estimates of linear and
  clear of cubic effects   555
  clear of cubic effects and blocks   557

quadratic estimator
    (SEE standard deviation)

quadratic forms, ratio of   268

quadratic mean
    (SEE standard deviation)

quadratic polynomials   334, 695
       (§7.31); 615 (§16.0214); 704

quadrature, numerical
       629 (§16.236*); 56, 321
  Cebysev's formula   436

quality control
       469-494, 730-735 (§14*)
  binomial distribution in   182

quantal response (SEE probit analysis)

quantiles   50-53, 636 (§1.41)

   estimate of dispersion of normal distribution from   94

   simultaneous estimate of mean and dispersion of normal distribution from   94

   standard error of   57

   (SEE ALSO median; percentiles; quartiles)

quantile range   50

   standard error of   51

quantum hypothesis   694

quantitative response, probit analysis for   633

quartic polynomial, translation by   719

quartiles   46, 47, 48, 51, 63, 636

quartile deviation
   (SEE quartile range)

quartile range   51, 92, 636

   efficiency of, for normal distribution   93

   of ranked intervals   712

quasi-factorial design (SEE lattice)

quasi-Latin squares   561-563 (§§15.152, 15.162*)

quintuple-entry table, interpolation in, warned against   653

quotient

   of correlated normal variables   653

   of non-central χ^2 statistics   682

   of ranges, in samples from continuous rectangular distribution   360

r = number of replications   563, 564, 566, 606, 607, 608, 609

radial standard deviation   144

radix table

   of common logarithms   620

   of natural logarithms   625

RAND Corporation's digits, differences of cumulative sums of, not necessarily random   458

random digits   454-455 (§13.1*); 460, 462

random division of line-segment   291

random matching   687

random normal numbers   462-463, 729 (§13.3); 454

random numbers   559, 560

   allegedly obtainable from shuffled numbers   460

   generation of   468 (§13.61); 456

   normal   462-463, 729 (§13.3); 454

   rectangular (SEE random digits)

   testing of   468 (§13.63)

random permutations   460-461 (§13.2); 559, 560

random rankings

   moments of concordance coefficient under   348

   moments of Kendall's Tau under   347

   moments of Spearman's Rho under   346

random samples   454-468, 729 (§13*)

   from artificial time series   465-467 (§13.5)

random sampling, efficiency of   341

random series, runs in smoothed   375

random shuffling, seldom achieved   685

random time series   465-467 (§13.5); 375

random variables, independent   707

random walk (SEE sequential sampling)

randomization of Latin squares   558-560 (§15.15*)

randomization test, for correlation coefficient

   product-moment   344

   Spearman's rank   344

randomized blocks   223

   incomplete   605

   maximum F-ratio in   638

randomness, tests of

   applied to random and pseudo-random numbers   468 (§13.63)

   non-parametric   374-379, 709-714 (§9.3*)

   power of   378

range
redundant Latin squares

range   82-88, 645-647 (§1.541);
    95-110, 649-652 (§§1.549-1.55);
        47, 48, 89, 94
  asymptotic distribution of
        365 (§8.33*)
  coefficient of variation of   701
  control chart for mean and
        488-490, 733 (§14.221)
  cumulative sums of   303
  dependent, ratio of   103, 104
  distribution of
    approximated by χ^2- distribution
        107
    in samples from discrete
        rectangular distribution   361
    in samples from various
        distributions, compared
        701 (§8.27)
  efficiency of   646
  efficiency of mean moving   643, 644
  grouped   95
  in acceptance sampling   478, 479
  in angular transformation   446
  in logit transformation   450
  in loglog transformation   452
  in overlapping samples, correlation
        between   643
  in probit analysis   23, 24
  in samples from continuous
        rectangular distribution   360
  in samples from discrete
        rectangular distribution   293
  in samples of four   734
  interquartile   51, 92, 636
  mean   82, 109, 701
    inequalities for tail area in
        terms of   435
  mean moving   644
  moving   643
  percentage points of   83
    as approximation, inadequate, to
        percentage points of ratio of
        largest to smallest mean square
        262
  quartile   50, 92, 636
    efficiency of, for normal
        distribution   93
    standard error of   51
  quotient of   101, 102, 104
    in samples from continuous
        rectangular distribution   36

―――――――――
*This section is subdivided;
see Table of Contents.

  ratio of to standard deviation
    as test of normality   652
    percentage points of   702
    upper and lower bounds for   702
  reduced   109, 110
  upper and lower bounds for   702
range-midrange test   88, 101, 102
range and mean, control charts for
        488-490, 733 (§14.221)
range charts, probability center lines
        for   149
ranks
  analysis of variance by   370
  tied, treatment of
    in Kendall's rank correlation
        353 (§8.123)
    in Spearman's rank correlation
        351 (§8.113)
    in three-sample rank-sum test   371
rank correlation
        345-359, 699 (§8.1*); 343, 366
  as k-sample slippage test   708
rank intervals   48
rank-sum tests
        368-369, 706 (§§9.121-9.122)
ranked normal deviates
        (SEE order statistics, from
        normal distribution)
rankings
  of means   645, 654
  of mean squares   681 (§4.254)
  of normal samples   46
    (SEE ALSO order statistics, from
        normal distribution)
  perfectly curvilinear   358
random
  moments of concordance
        coefficient under   348
  moments of Kendall's Tau under
        347
  moments of Spearman's Rho under
        346
ranking score   62
ranking test, curvilinear   358

ratios

  of area to bounding ordinate
    of χ^2 distribution
        133, 656 (§2.1134); 717
    of normal distribution   13, 630
        (§1.13*); 22, 652 (§1.222)
  of dependent ranges   103, 104
  of Γ-functions   165
  of independent ranges   101, 102, 104
  of mean range to standard deviation
        702
  of ordered mean squares
        260-262, 681 (§4.25*)
  of Poisson parameters, test of   693
  of quadratic forms   268
  of range to standard deviation   702
  of ranked intervals   712
  of standard deviations of
    correlated normal variables   673
  of triangular variate to normal
    variate, percentage points of
        722
  of variances, confidence intervals
    for   257, 258

ratio estimate   697

rational functions   616-617, 738
        (§16.0244); 704, 724

  square roots of
    204-205, 668-669 (§§3.53-3.54);
    334, 695-696 (§§7.32-7.33); 618,
        738 (§16.0264); 704

raw moment (SEE Sheppard's corrections)

ready-reckoner, possibly useful in
  non-parametric statistics   704

rearrangement

  of non-random digits, random digits
    produced by   458 (§13.1221)

  of random digits, random digits
    produced by   459 (§13.1222)

reciprocals   616 (§16.024)

  of factorials
        161-162, 662 (§2.512*)
  of complex argument   164 (§2.5151)

  of square roots   617 (§16.0262)

  sums of   620 (§16.046)

  (SEE ALSO digamma function)

reciprocal powers   617 (§16.025)

reciprocal square roots
        617 (§16.0262)

records, lower and upper, as test for
  randomness   710

rectangular distribution

  continuous   418, 720-722 (§10.26*);
        393, 394, 645, 647, 648

    Gini's mean difference in samples
      from   80

    mean square successive difference
      in samples from   645

    order statistics from   360,
        699-700 (§8.21); 65, 647, 648

    samples of 3 from, distribution of
      t in   229

    tests applied to
        318, 694-695 (§6.4)

  discrete   418, 722 (§10.262); 407

    order statistics from   361 (§8.22)

    random samples from   456-459, 729
        (§13.1); 454, 455, 462, 465

    range in samples from   293

    samples of 4 from, distribution of
      t in   229

rectangular lattices
        606 (§15.244); 566

rectangular variate, ratio of to
  normal variate, percentage
  points of   722

rectifying inspection (SEE average
  outgoing quality limit)

rectifying sampling   475

recurrence formula

  distribution-free, for moments of
    order statistics   62

  for binomial moments   204

  for Hermite polynomials   35

  for ordinates of distribution of
    product-moment correlation
    coefficient   319

  for tetrachoric functions   35

reduced derivatives, of Mill's ratio
        38

reduced inspection   472, 478

redundant Latin squares   608

regression
sampling plans

regression
- dispersion matrix due to   690
- estimated
  - after preliminary F-test   243
  - after preliminary t-test   243
  - in trend-reduced Markov process   726
- non-parametric tests of   705 (§9.118)
- of standard deviation on mean, approximate parabolic   434
- residual covariance matrix after   311
- serial correlation of residuals from   78
- tests of   705 (§9.118)
- variate differences of residuals from   78

regression coefficient
  340-342, 697 (§7.6)
- percentage points of   677
  - alleged, from percentage points of correlation coefficient   324
- standardized   705

regression curves, fitting of
  436-444, 724-727 (§11*)

regression estimates of mean and standard deviation, from censored sample   642, 648

regression model, linear   705

rejection of outlying observations
  (see the next entry)

rejection of stragglers   68
- bivariate normal   55
- by means of Galton differences   91
- by means of range   651
- by means of ratio of ranges   103, 104
- trivariate normal   55
- univariate normal   54

rejection, short-cut, of samples with excessive standard deviation   478

rejection number   472, 473

―――――――――
*This section is subdivided;
see Table of Contents.

relative discrepancy, maximum
  385 (§9.416)

relative precision, of pilot sample   612

relative variance
  (SEE coefficient of variation)

repeated integrals, of error integral   38

repeated normal integrals
  37-38 (§1.331)

replacement, in life testing   732

replicates
- factorial patterns divisible into   501
- incomplete block patterns divisible into   564

replication groups orthogonal to blocks   565, 609

reputations of mathematical tables   614

residuals from fitted Fourier series, circular serial correlation of   338

residual covariance matrix, after regression   311

residual variables, serial correlation of   726

response surfaces, combinatorial patterns for investigating
  554-555 (§15.138); 557 (§15.148); 501

restricted monomial symmetric functions   723

results of random sampling, tables of, ignored   455

return period (of floods), logarithm of   364 (§8.3112)

Rho (SEE Spearman's rank correlation)

risk
- consumer's   46′
- monetary   612
- producer's   469

risk function   613

right triangular distribution, order statistics from   717

[1000]

roots
  cube   619 (§16.027)
  of algebraic equations
        620, 738 (§16.05*)
  of determinantal equations
    largest   312
    percentage points of   306
  reciprocal square   617 (§16.0262)
  square   617 (§16.0261)
  of rational functions
        618, 738 (§16.0264)
root mean square, as a biplane
    combinatorial power mean   432
root-mean-square deviation
    (SEE standard deviation)
root-mean-square error (SEE standard
    deviation; variance)
roulette, at Monte Carlo casino   464
round-trip tests, of randomness in
    time series   710
rows
  in factorial patterns   497, 499, 561
  in incomplete block patterns   564
  in orthogonal squares and cubes   554
runs   374-376, 709-711 (§9.311)
  in smoothed random series   375
  in two samples, percentage points
    of   713

s-test (SEE χ^2 test)

S and T functions   623

salvo kill probabilities, for square
    targets   653

samples
  pilot, relative precision of   612
  pooled, for blood tests   611
  random   454-468, 729 (§13*)
  with excessive standard deviation,
    short-cut rejection of   478

sample cumulative distribution
    function   386

sample mean, percentage points of
    studentized extreme deviate
    from   640

sample ratio, optimum   688

sample sizes   472, 473, 479
  choice of
        611-613, 736-737 (§15.3); 495
  least   692
sample variance   245
  moments of   165
  two-tailed test of   134
sampling
  acceptance   469-486, 730-732 (§14.1*)
  double
    by attributes   471, 472, 476
    by variables   477
  from finite population   275
  multiple, by attributes
        471, 472, 473, 476, 477
  random, efficiency of   341
  rectifying   475
  sequential   678
    truncated, by attributes   474, 476
    (SEE ALSO sampling, Wald sequential)
  single   184, 191
    by attributes
        470, 471, 472, 473, 475, 476,
                477, 478, 479, 480
  stratified, efficiency of   341
  systematic, efficiency of   341
  triple, by attributes   476
  unbiased selective, using ranked
    sets   362
  Wald sequential
    by attributes   473-474
    by variables   481

sampling co-cumulants of multivariate
    k-statistics   432

sampling cumulants of k-statistics
        430, 723 (§10.343)
sampling inspection
        469-486, 730-732 (§14.1*)

sampling plans
  destructive   483
  double   471, 472, 476
  multiple   471, 472, 473, 476, 477
  single   470, 471, 472, 473, 476,
                478, 479, 480
  triple   476
  truncated sequential   474
  untruncated sequential   474

sampling units
sphericity

sampling units
  primary   611
  secondary   611

sampling variance
    (SEE standard error)

saturated factorial patterns
    553-554 (§15.131*); 499, 500, 502

scedasticity (SEE heteroscedasticity; homoscedasticity)

Scheffe's test   239 (§4.153)

Schuler's tables   381

screening inspection
    (SEE acceptance sampling)

second heterogeneity control   565

second-order patterns, central composite, for investigating response surfaces   554

second law of Laplace
    (SEE normal distribution)

secondary sampling units   611

segmental functions   626 (§16.1231)

semi-interquartile range
    (SEE quartile range)

semi-invariants
  algebraic and statistical   432
  (SEE ALSO cumulants; k-statistics)

semi-range   89
  (SEE ALSO range)

sensitivity data (SEE probit analysis)

sequential analysis   473

sequential probability-ratio test
    (SEE sequential test, Wald)

sequential sampling   678
  truncated, by attributes   474, 476
  untruncated   474
  Wald, by attributes   473-474
  Wald, by variables   481

sequential t-tests   303-304 (§6.115); 226

---

*This section is subdivided; see Table of Contents.

sequential test   694
  for equality of variances   302
    based on ranges   651
  in life testing   486
  Wald
    of binomial population   314
    of mean of normal distribution   731
    of rectangular distribution   694

serial correlation   336-339 (§7.5*)
  from first-order Markov process, hyperbolic-tangent transformation of   333
  of independent and residual variables   726
  non-circular   338 (§7.52); 79
  of residuals from regression   78
  of samples from autoregressive time series   467

serial product moments, of samples from autoregressive time series   467

Sheppard, W. F.:
  corrections   426-427, 722 (§10.331); 428
    bivariate   431
    to multivariate co-cumulants, vanish   431
    to multivariate cumulants   431
    (SEE ALSO corrections for grouping)
  notation for normal distribution   3
  quadrature formula   629

shift rule, for moments   425

short-cut rejection of samples with excessive standard deviation   478

shuffled numbers   460

shuffling, random, seldom achieved   685

sigmoid curve (SEE probit analysis; logit transformation)

sign test
  binomial   206, 669-679 (§3.6); 183, 184, 315, 699
  bivariate   714
  of regression   705
  power of   387

signed-rank test of location   367-368, 704 (§9.112)

significance, tests of, in harmonic
   analysis 443, 726 (§11.24)

significance level (SEE percentage
   points; Jeffreys' criterion K)

simple lattice design
   (SEE square lattice)

simplices, hyperspherical, relative
   surface contents of 637

sines and cosines of vulgar fractions
   of 2π 442 (§11.22)

 multiples of 443 (§11.23)

single control of heterogeneity,
   in factorial experiments 496

single-entry tables,
   of working probits 24

single sampling 150, 184, 191

 by attributes
   470, 471, 472, 473, 475, 476
 by variables 477, 478, 479, 480
 (SEE ALSO double sampling)

single-tail test (SEE one-sided test)

singly censored samples 647

six-point assay, probit analysis of
   29

skew regression
   (SEE regression curves)

skewness

 of χ^2 645
 of F 263
 of $\log \chi^2$ 645
 of mean square successive difference
   645
 of rth greatest interval 712
 of t^2 240
 Pearson measure of 391
 population
  distribution of range insensitive
   to 701
  universal upper bounds for
   straggler and range unaffected
   by 702
  (SEE ALSO $b_1^{\frac{1}{2}}$)

slippage test

 for time series 387
 k-sample, based on paired
   comparisons 708

Kolmogorov-Smirnov
   385, 715 (§9.4152)

 2-sample, percentage points of 707

small numbers, law of
   (SEE Poisson distribution)

smallest mean square, ratio of largest
   mean square to 262 (§4.253)

smallest variance, percentage points
   of 261

 as approximation to percentage
   points of smallest variance
   ratio 261

smallest variance ratio
   261-262 (§4.252)

Smirnov tests
   (see the following two entries)

Smirnov, Cramer and Mises, test of
   goodness of fit 386 (§9.42)

Smirnov-Kolmogorov tests of goodness
   of fit
   380-385, 714-715 (§9.41*)

smooth test of goodness of fit, χ^2,
   F. N. David's 180

smoothing (SEE moving averages;
   regression curves)

Snedecor, G. W., F-distribution
   245-269, 676-682 (§4.2*)

 (SEE ALSO F; F-distribution)

social group, isolates in 294

solitaire (cards) 684

Sonin polynomials 405

Spearman, C. E.

 foot rule 345

 rank correlation, Rho
   348-352 (§8.11); 345, 346

specification limits, in acceptance
   sampling by variables 479

spectrum (SEE periodogram)

spherical harmonics, associated 334

spherically symmetric trivariate
   normal distribution, integral
   over ellipsoidal regions 660

sphericity, test for 313

 2-dimensional 697

split plots
systematic statistics

split plots, factorial patterns with
   factors applied to
   562-563 (§15.16*); 499 ftn., 502

split-plot design, analysis of
   variance of, by range   106

split-split plot   504

spread, non-parametric tests of
   373, 708-709 (§9.2)

squares, i.e. combinatorial patterns
   558-561 (§15.151*)

  half-plaid   562 (§15.1622)

  incomplete Latin   607-608 (§15.2511)

  lattice   561-562 (§15.153);
         608-609 (§15.253)

  orthogonal, complete (saturated)
    sets of   554 (§15.1312)

  plaid   562 (§15.1623)

  with split plots   562 (§15.162*)

  Youden   607-608 (§15.251*)

squares, of numbers   614-615
   (§§16.0211, 16.0213)

square contingency, mean
   343 (§7.92); 356

square lattices   556-557 (§15.143*);
         606 (§15.243)

  miscalled lattice squares   608 ftn.

square roots   617-618   (§16.026*)

  not computed   478

  of natural cologarithms   659

  of rational functions
    204-205, 668-669 (§§3.53-3.54);
    334, 695-696 (§§7.32-7.33); 618,
    738 (§16.0264); 704

  reciprocal   617 (§16.0262)

square targets, salvo kill
   probabilities for   653

square-root transformation   448 (§12.2)

  as degenerate case of inverse
    hyperbolic-sine-transformation
        449

stability of normal population, tests
   of   688

standard deviation   643

  coefficient of variation of   87

  *This section is subdivided;
see Table of Contents.

  computation of, routine checks of
    702

  confidence interval for
    148-150, 658 (§2.315)

  control chart for   148, 149, 487

  control chart for mean and
    488-490, 733 (§14.221)

  correlation between   123

  correlation between correlation
    coefficient and   652

  denominator of, in quality control
    488

  estimated by variate differences   75

  excessive, short-cut rejection of
    samples with   478

  limits of ratio of mean range to
    702

  minimum-variance unbiased linear
    estimate of   646, 647

  moments of   165, 166, 167, 168,
         169, 170, 171

    negative   172

  of correlated normal variables,
    ratio of   673

  of exponential distribution,
    estimated from order statistics
      700

  of right triangular distribution,
    estimated from order statistics
      717

  of tolerance distribution
    (probit analysis)   22

  radial   144

  range estimate of, biased by shift
    in mean   734

  ratio of mean range to   702

  regression of, on mean, approximate
    parabolic   434

  (SEE ALSO variance)

standard deviation and range charts,
   probability center lines for
    149

standard errors

  of coefficient of variation   405

  of constants of Pearson curves
    391-392 (§10.007)

  of correlation coefficient   420

  of estimates of standard deviation
    93, 170

  of estimates of variance   170

  of Gini's mean difference   80

  of quartiles   51

  of quartile range   51

  of t   240

standard square, Latin 558

Stirling interpolation coefficients
    628 (§16.2315)

Stirling interpolation formula
    first derivative of 629

Stirling's numbers, of first and
    second kind 620 (§16.0492)

straggler 54-62, 637-638 (§1.42);
    46, 68, 88, 89

  asymptotic distribution of
    364-375 (§8.31*)
  bounds on expectation of 435
  control chart for 490
  in analysis of variance by range
    106
  percentage points of 56, 58
  rejection of
    by mean of Galton differences 91
    by means of range 68
    by means of ratio of ranges
      103, 104, 651
  universal upper and lower bounds for
    702

straggler deviate 49

straggler mean 106

stratified sampling, efficiency of
    341

stratified universe
  distribution of χ^2 from 682
  distribution of F from 682
  distribution of t from 682

Student's distribution
    (SEE t-distribution)

Student's hypothesis (SEE t-test)

Student's t 215

Student's z-distribution
    209, 671 (§4.112)
  (SEE ALSO t-distribution)

studentization 69, 200

studentized D^2 307

studentized deviate, extreme 641

studentized distributions, involving
    order statistics
    68-69, 639-641 (§1.44); 49

studentized mean square successive
    difference 80

studentized mean successive difference
    80

studentized range
    95-98, 649 (§1.5511); 49

studentized statistics, with range in
    denominator
    98-104, 650-651 (§1.5512)

studentized straggler deviate
    49, 640-641

subsampling number, estimation of
    optimum 611

subtraction logarithms
    621-622 (§16.063)

substitute t-test, using order
    statistics 64

sums
  of powers of integers
    619 (§16.044*); 721
  of reciprocals 620 (§16.046)
  of squares of orthogonal polynomials
    724
  power 421
    k-statistics in terms of
      430 (§10.341)

summation formula
  Euler-MacLaurin 428
  Lubbock 629

summation method, of fitting
    polynomials 438-439 (§11.113)

surfaces, response 501, 554, 557

symmetric functions
    421-432, 722-724 (§10.3*)

symmetrical balanced incomplete block
    pattern 607

symmetrical distribution 648

symmetrical frequency curves,
    Hansmann's 405

symmetrical test (SEE two-sided test)

symmetry, in factorial patterns 497

systematic sampling, efficiency of
    341

systematic statistics 81
  as measures of dispersion
    92-95, 647-649 (§1.544)
  best linear 642, 649

t (Student's) test

t (Student's)   215
  approximations to distribution of
      210, 273, 671
  approximations to percentage points
    of   672
  confidence intervals based on
    percentage points of   216
    effect of intra-class correlation
      on   243
  distribution of   207-244, 670-675
      (§4.1*); 320, 344, 389, 708
   in correlated samples, correlation
     between   242
   in samples from discrete rectangular
     distribution   418
  H. Jeffreys' significance criterion
    K for   218
  non-central
      218-225 (§4.141*); 477, 682
  normalization of   213
  percentage points of
      210-218, 671 (§4.13*); 246

$t = r(1-r^2)^{-\frac{1}{2}}$
  for testing Spearman's Rho   346
  moments of   327 (§7.134)

t: unfortunate synonym for
    v = number of varieties   563

T: sum of ranks   367, 368

t-distribution   207-244, 670-675
     (§4.1*); 320, 344, 389, 708
  approximations to   210, 273, 671
  as approximation to distribution:
    of k-sample paired-comparison
      slippage test   708
    of Spearman's rank correlation
      346
    of Welch's two-sample test   239
  as special case of F-distribution
     245
  bivariate   232
   (SEE ALSO t-distribution,
    multivariate)
  central, as degenerate case of non-
    central t-distribution   219

    *This section is subdivided;
see Table of Contents.

  expanded in Gram-Charlier series,
    type A   201
  in rejection of stragglers   68
  mode of, estimated by order
    statistics   363
  multivariate   241-242, 674 (§4.17*)
  non-central
    218-225 (§4.141*); 226, 477, 682

t-test
  compared with normal deviate test
    217
  efficiency of Walsh test, relative
    to   367
  power of   218, 220, 221
  preliminary, effect of on estimate
    of regression   243
  sequential   303-304 (§6.115); 226
  solves the 2-sample problem for
    equal variances   231
  standardized error of   221
  substitute
    based on mean range   650
    based on median and quartiles   636
    based on range   101
    two-sample, based on range
      646, 651
    using order statistics   64

t^2, skewness and kurtosis of   240

table (see under the function tabled)

tagging (of animal populations)   688

tail area
  inequalities for, in terms of mean
    range   435
  of normal distribution
    20, 630-632 (§§1.2112, 1.2122)
  powers of   54-62, 637-638 (§1.42)

tantile   49
  (SEE ALSO quantiles)

Tau (SEE Kendall's rank correlation)

Taylor series   721

Tchebycheff, P. L.   (SEE Cebysev)

t (Student's)
test

test
  Bartlett's, for equality of variances, power of   256
  Behrens-Fisher   232-236 (§4.151)
    Aspin's tables not related to   238
  binomial sign
      206, 669-670 (§3.6); 366
    of significance of correlation coefficient, efficiency of   196
    of significance of mean, efficiency of   196
  bivariate sign   714
  χ^2, of goodness of fit
      174-180, 663-664 (§2.6*); 366
    inefficiency of   380
    smooth, F. N. David's   180
  χ^2, of index of dispersion   180
    power of   280
  Cochran's, of homogeneity of variance   443 (§11.24); 303
  Cramer-Mises-Smirnov   386 (§9.42)
  curvilinear ranking   358
  dichotomization   358, 370
  exact, in 2 × 2 tables
      174-176, 663 (§2.611)
  Fisher-Behrens   232-236 (§4.151)
    Aspin's tables not related to   238
  for birth-order effect   373
  for difference between:
    correlated variances   265
    means, in face of unequal variances   231-240, 674 (§4.15*)
    proportions   663
  for heterogeneity of variance, short-cut   262
  Hoeffding's, of independence   356
  in time series, based on signs of differences   376
  in 2 × 2 tables   174
  k-sample, non-parametric, of location
      371-372, 707-709 (§9.14)
  Kolmogorov-Smirnov
      380-385, 714-715 (§9.41*)
  likelihood
      295-316, 318-319, 688-695 (§6*)
  Mann-Whitney   368-369, 706 (§9.121)
  median   370 (§9.123)
  non-parametric
      366-372, 704-708 (§9*)
  of equality:
    of correlations   332
    of covariance matrices
        295, 309, 310
    of means   297
    of means, variances and covariances of correlated variates   310
    of means and variances   297, 305
    of means of correlated variates, essentially an F-test   310
    of Poisson parameters   315, 316
    of variances   297
    of variances and covariances of correlated variates   310
  of equality of variances and a specified correlation coefficient   305
  of goodness of fit
      380-386, 714-715 (§9.4)
    χ^2 174-180, 663-664 (§2.6); 366, 380
    of exponential curves   444
  of homogeneity, applied to multivariate normal populations   306-311, 692 (§§6.13-6.132)
  of homogeneity of variance
      297-303, 688 (§§6.112-6.114)
  of largest harmonic component
      443, 726 (§11.24)
  of normality   111-118, 652 (§1.6*)
  of randomness
      374-379, 709-714 (§9.3*)
    applied to random and pseudo-random numbers   468 (§13.63)
  of regression   705 (§9.118)
  of slippage between two samples
      382-385, 715 (§§9.4122, 9.4152)
  of specified correlation coefficient   305
  of sphericity   313
  of spread   373, 708-709 (§9.2)
  of stability of a normal population   688
  ω^2 386 (§9.42)
  one-sample, non-parametric
    of binomial process, from two-sample test of randomness of occurrence   376
    of location   366-368, 704 (§9.11*)

tests
trigonometric form

[test
 non-parametric]
   of randomness of occurrence
        374-377, 709-712 (§9.31*)
   one-sided, Kolmogorov-Smirnov
        384-385, 715 (§9.415*)
   order statistic   62-68, 638-639
        (§1.43)
   preliminary, of homoscedasticity,
     Box suggests omission of   264
   quadrant, for association   357
   randomization, for correlation
     coefficient
       product-moment   344
       Spearman's rank   344
   range-midrange   101
   ranking, curvilinear   358
   Scheffe's two-sample   239 (§4.153); 232
   sequential   694
   sign
     binomial   206, 669-670 (§36); 366
     unweighted; weighted   705
   signed-rank (Wilcoxon)
        367-368, 704 (§9.112)
   slippage, k-sample   371
   three-sample
     Kolmogorov-Smirnov   715 (§9.414)
     non-parametric, of location   370
   two-sample, non-parametric
     of homogeneity
       382-385, 715 (§§9.4122, 9.4152)
     of location
       368-370, 706-707 (§9.12)
     of randomness of occurrence
       377-379, 713 (§9.32)
     of randomness of occurrence,
       yielding one-sample test of
       binomial process   376
   two-sided, Kolmogorov-Smirnov
       380-384, 714-715 (§§9.411-9.413*)
   Wald sequential
     destructive   483
     for mean of normal distribution
       481

_____
*This section is subdivided; see Table of Contents.

   of binomial proportion
        314, 473, 474
   Walsh   366-367, 704 (§9.111)
   Welch's two-sample
        237-239, 674 (§4.152); 232
     deprecated   236
   Wilcoxon
     one-sample   367-368, 704 (§9.112)
     two-sample   368-369, 706 (§9.121); 619 (§16.036)
   Wilks', for independence of k
     groups of variates   295, 311
   with minimal bias   693

testing
   destructive   474, 483
   of random and pseudo-random numbers
        468 (§13.63)

tetrachoric correlation
        343-344, 698-699 (§7.94)
   for median dichotomies   653

tetrachoric functions
        35-36, 634 (§1.324*); 343, 408

tetrachoric series, multivariate   655

tetragamma functions   164

theta functions   617 (§16.21)
   Charlier   282, 409
     moments of are divergent   409
   elliptic   380-381, 714 (§9.411*)

third standard moment, as test of
   normality   111-113 (§1.61*)

three-dimensional lattice
        606-607 (§15.245)

three-sigma limits, American practice
   for control charts   488

tied ranks, treatment of
   in Kendall's rank correlation
     coefficient   353 (§8.123)
   in Spearman's rank correlation
     coefficient   351 (§8.113)
   in three-sample rank-sum test   371

tightened inspection   472, 478

time lag (SEE serial correlation)

time series
  artificial, random samples from 465-467 (§13.5)
  correlation between 696 (§7.54)
  Markov 726
  order statistics from, moments of 67
  random 375
  slippage test for 387
  stationary 336
  test for, based on runs 376
  test for randomness in 710

tolerance distribution
  (SEE probit analysis)

tolerance ellipsoids, multivariate normal 493

tolerance limits
  non-parametric 492, 734-735 (§14.31); 706, 708, 733
    use of beta-function in constructing 492
  normal 492-494 (§14.32); 17, 18

total confounding, in factorial patterns 556 (§15.1421)

total inspection
  (SEE acceptance sampling)

transformation
  angular 445-447, 728 (§12.1*); 24
  Cavalli's 453
  hyperbolic-tangent 330-333, 695 (§7.2*)
    alias logit transformation 450
  inverse hyperbolic-sine 449 (§12.3*)
  Legit 453 (§12.7)
  logit 450-451, 728 (§12.5*); 24
  loglog 452 (§12.6*)
    complementary 24
  probit 22-30, 632-634 (§1.23*); 449 (§12.4)
  square-root 448 (§12.2)
  variance-stabilizing 330
  (SEE ALSO translation)

translation, of frequency curves 410-414, 719-720 (§10.22*)
  non-monotone 410
  polynomial 411-412, 719 (§10.221); 396, 397
    Cornish-Fisher 233
  (SEE ALSO transformation)

translation of origin, for moments 425

treatments, experimental 495
  synonym of varieties 563

treatment effects 553

trends 644

trend fitting (SEE moving averages; regression curves; variate differences)

trend in dispersion, in time series, test against 710

trend in location, in time series, test against 710

trend-reduced Markov process, estimated regression in 726

triads, inconsistent 355-356, 699 (§8.14)

trial
  in factorial patterns 496
    total number of 498
  in incomplete block patterns 563

triangle, Pascal 685

triangular distribution 648
  discrete, samples of 4 from, distribution of t in 229
  order statistics from 647
  right, order statistics from 717, 718

triangular variate, ratio of to normal variate, percentage points of 722

triserial correlation 653

trigamma function 164

trigonometric and hyperbolic functions combined 626 (§16.12)

trigonometric form of incomplete beta-function 200, 201

trigonometric functions
variance

trigonometric functions
    622-623 (§16.07*); 442 (§12.22)

  inverse 445, 728 (§§12.11-12.13)

  logarithms of 623-625 (§16.08)

    natural 451

  multiples of 443 (§11.23)

  natural logarithms of 451

trigonometric polynomials
    442-443, 726-727 (§11.2*)

triple sampling, by attributes 476

trivariate normal distribution
    125, 654 (§1.79)

  spherically symmetric, integral over ellipsoidal regions 660

truncated distribution 631

  binomial 314

  generalized hypergeometric 407

  log-normal 44

  negative binomial 683

  normal
    39-45, 635-636 (§1.332); 631

  Poisson 138

truncated life-testing plans 732

truncated sequential sampling, by attributes 474-476

truncation rules, for Gram-Charlier series 328, 407-408

Tschebyschew, P. L. (SEE Čebyšëv)

Tukey, J. W.

  differences of cumulative sums of RAND Corporation digits not necessarily random 458

  polykays 429

2 × 2 tables 174-179, 663-664 (§2.61*)

  test of independence in 275

2 × k table, distribution of χ^2 in 179

2-factor effects, estimability of, in factorial patterns derived from incomplete block patterns, not studied 501

2-factor interactions, estimable, clear of blocks and other 2-factor interactions 555

---

*This section is subdivided; see Table of Contents.

two-sample tests

  of equality of means, in face of unequal variances 231-239, 674 (§4.15*)

  of randomness of occurrence, yielding one-sample test of binomial process 376

two-sided specification limits, lot inspection by variables for 479

two-sided tests, Kolmogorov-Smirnov 380-384, 714-715 (§§9.411-9.413*)

two-stage sampling, sample size in 303

two-tailed exponential distribution

  logarithmic and hyperbolic translation of 719

  mean square successive difference in samples from 645

  order statistics from 647

two-tailed test, for variance 134

two-way control of heterogeneity

  factorial patterns with 558-562 (§15.15*)

  split-plot 562-563 (§15.162*)

  incomplete block patterns with 607-609 (§15.25*)

Type A contagious distribution, Neyman's 279, 280, 683

Type A, B, C series (SEE Gram-Charlier series, Type A, B, C)

Type I, ..., XII distribution (SEE Pearson curves, Type I, ..., XII)

type I error (SEE producer's risk)

type II error (SEE consumer's risk)

type of Hansmann's generalized Pearson curves 406

Type of Pearson curves 388

  determination from moments 389-391 (§10.004)

U-shaped distributions 648

  censored samples from, order statistics from 647

  discrete, samples of 4 from, distribution of t in 229

u-test, i.e. substitute t-test 650

unadjusted moment
    (SEE Sheppard's corrections)

unbalanced lattices   557 (§15.1432)

unbiased selective sampling, using ranked sets   362

unblocking   496, 497, 498
  applied to incomplete block patterns   501
  applied to lattice squares   561

unequal variances, two-sample tests for difference of means, in face of   231-239, 674 (§4.15*)

uniform distribution
    (SEE rectangular distribution)

unimodal frequency functions, represented by Edgeworth series   409

units, experimental   496

unit normal variates   67
  independent   692
  (SEE ALSO normal deviate)

unitary symmetric functions   421

unsaturated sets of orthogonal squares   560 (§15.1513)

untruncated sequential-sampling plans   474

unweighted averages   71

unweighted sign test   705

upper bounds
  for range   702
  for ratio of range to standard deviation   702
  for straggler   702

upper control limit
    (SEE control chart)

upper quartile   47, 48, 63
  (SEE ALSO quartile)

upper records, as test for randomness   710

upper straggler   46, 47, 88, 89

variables
  control charts for   488-491, 733-734 (§14.22*)
  lot inspection by   477-481, 730-731 (§14.12*)

variance
  asymptotic
    of biserial correlation   698
    of estimated regression   726
  component of   225
  correlated, test for difference between   265
  denominator of, in quality control   488
  equality of
    and covariances, vanishing of, test of   313
    Bartlett's L_1 test of, power of   256
    sequential test of   302
    sequential test of, based on ranges   651
    test of   305
  estimated, bias of, after preliminary significance test   166
  heterogeneity of
    effect of, on distribution of F   264-269, 681-682 (§4.28)
    short-cut test for   262
  homogeneity of, tests of
    Cochran's   443
    multivariate   306-692 (§§6.131-6.132)
    preliminary, Box suggests omission of   264
    univariate   296-303, 688 (§§6.112-6.114)
  lot inspection by variables controlling   480 (§14.124)
  lumped   694
  moments of   165, 166
  of estimates of mean   642
  of finite population, confidence interval for   253
  of log-normal distribution   414
  of maximum-likelihood estimates of parameters of Pearson curves, Type III   395
  of median   636
  of midrange   647
  of multiple correlation coefficient   335
  of order statistics   65, 67
  of range   82
  of round-trip tests of randomness   710
  of square-root transformation of χ^2 distribution   448

variance
z

[variance]
  of standard deviation 448
  ranking of 681
  two-tailed test for 134
  unequal
    effect of, on distribution of F 264-269, 681-682 (§4.28)
    two-sample tests for difference of means, in face of 231-239, 674 (§4.15*)

variance-analysis
  (SEE analysis of variance)

variance-covariance matrix
  (SEE covariance matrix)

variance-ratio distribution
  (SEE F-distribution; beta distribution; Pearson curve, Type VI)

variance-ratio test (SEE F-test)

variances and covariances
  of bipolykays 724
  of correlated variates, test for equality of 310
  of order statistics 639
    from right triangular distribution 717

variance and mean of normal distribution, simultaneous test of 689

variance component 255
  confidence intervals for 677

variance ratio 245
  confidence intervals for 689
  largest 261, 681 (§4.251)
  smallest 262 (§4.252)

variance-stabilizing transformation 330

variate differences
  as measure of dispersion 75-80, 643-645 (§1.531)
  as measure of serial correlation 339 (§7.53)
  as non-parametric tests of randomness 376, 377 (§9.312)

---

*This section is subdivided; see Table of Contents.

variate transformations 445-453, 728 (§12*)
  (SEE ALSO translation)

variation, coefficient of 74
  confidence interval for 225
  of j-statistics 81
  of range 701

varieties, in incomplete block patterns 495, 563

von Mises, Cramer and Smirnov, test of goodness of fit 386 (§9.42*)

von Neumann's ratio of mean square successive difference to variance 76, 79, 643

W, Kendall's concordance coefficient 353-355 (§8.13*); 347, 348

Wald sequential sampling
  by attributes 473, 474
  by variables 481

Wald sequential tests 473, 481
  destructive 483
  of binomial proportion 314
  of mean of normal distribution 731
  of rectangular distribution 694
  t-test 303

Walsh test, non-parametric, of location 366-367, 704 (§9.111)
  not coextensive with signed-rank test 367

waste incurred by using t-test when population variance is known 675

Weibull distribution, life testing from 485

Weidenbach's logarithm 403

weighing designs 553-554 (§15.31*)

weight of a symmetric function 421

weighted grouping, polynomial fitting by 725

weighted mean, confidence intervals for 674-675 (§4.18)

weighted mean range estimator   644

weighted probits   24, 26

weighted sign test   705

weighted sum:

   of ordered intervals, moments of   712

   of χ^2 variates, approximate normalization of   660

weighting coefficient

   binomial, in polynomial fitting   441

   in angular transformation, is constant   446

   in logit transformation   450-451 (§12.53)

   in loglog transformation   452 (§12.62)

   in probit analysis   24-27, 632 (1.231*); 23, 29, 343

      in presence of natural mortality   28

      with Poisson distribution of exposed to risk   28

Welch, B. L., two-sample test   237-239, 674 (§4.152); 232

   deprecated   236

Weldon, W. F. R., dice-casting experiments   464

whist, hand played at   684

whole-plot factors, in split-plot experiment   562

Wicksell, S. D., corpuscle problem   288

Wilcoxon, F.

   one-sample signed-rank test   367-368, 704 (§9.112)

   two-sample rank sum test   368-369, 706 (§9.121); 619

      k-sample generalization of   372

Wilks, S. S., test for independence of k groups of variates   311-313 (§6.133); 295

   percentage points of, as approximations to percentage points of largest determinantal root   690

Wilson, E. B. and Hilferty, Margaret M., approximation to percentage points of chi-square   145, 147, 263

   applied to yield fiducial limits for Poisson parameter   153

Woolhouse, W. S. B., summation formula   429

Work, Milton, point count in bridge hands   684

working angles   446 (§12.14)

working logits   450-451 (§12.53)

working loglogs   452 (§12.62)

working probits   24-27, 632 (§1.23*); 23, 486

X^2; X^2 distribution; X^2 test (SEE χ^2 distribution etc. (alphabetized under chi))

Yates, F.

   continuity correction   176

   does not improve χ^2 approximation to multinomial distribution   287

   distribution of χ^2 in 2 2 tables

      approximate   176

      exact   174

Youden squares   607-608 (§15.251*); 566

$z = \frac{1}{2} \log_e F$ 270-274, 682 (§4.3*)

   connexion with $z = \tanh^{-1} r$ 330

   Edgeworth corrections to   681

   Jeffreys' significance criterion K for   272

   percentage points of   250-251, 676 (§4.331); 270-271 (§4.333); 298

      approximate   273

      as approximation to percentage points of concordance coefficient   354

      Fisher-Cochran approximation to, yielding approximate percentage points of beta distribution   198

$z = \tanh^{-1} r$ 330-333, 695 (§7.2*)

z, Student's
zero sum

z, Student's   209, 671 (§4.112)
  (SEE ALSO t (Student's))

z-distribution

  Fisher's (SEE beta distribution;
    F-distribution; $z = \frac{1}{2} \log_e F$)

  Student's   209, 671 (§4.112)
    (SEE ALSO t-distribution)

z-test (SEE F-test)

z-transformation (SEE hyperbolic-
    tangent transformation)

Zener cards   685, 686

zero

  derivatives, of   620 (§16.0492)

  differences of
    620 (§16.0492); 293, 294

zero sum, method of   440

3/22/2010

Retain